Fire Engineering's Handbook for Firefighter I & II 2019 Update

FIRE ENGINEERING'S HANDBOOK FOR FIREFIGHTER I & II

2019 UPDATE

Fire Engineering Books & Videos
1421 South Sheridan Road
Tulsa, Oklahoma 74112-6600 USA

Phone: 918.835.3161
Fax: 918.831.9555
800.752.9764
+1.918.831.9421
info@fireengineeringbooks.com
www.fireengineeringbooks.com

Senior Vice President Eric Schlett
Operations Manager Holly Fournier
Sales Manager Josh Neal
Managing Editor Mark Haugh
Production Manager Tony Quinn
Technical Editor Glenn Corbett
Development Editor Chris Barton
Book Designer Sheila Brock
Cover Designer Beth Rose

Printed in the United States of America

1 2 3 4 5 23 22 21 20 19

CONTENTS

FIREFIGHTER I

1 THE TRADITIONS AND MISSION OF THE FIRE SERVICE

2 FIRE SERVICE HISTORY

3 FIRE DEPARTMENT ORGANIZATION

4 FIRE DEPARTMENT COMMUNICATIONS

5 FIRE BEHAVIOR

6 FIRE EXTINGUISHERS

7 BUILDING CONSTRUCTION

14 VENTILATION

15 WATER SUPPLY AND HOSE

18 VEHICLE FIRES

19 SEARCH AND RESCUE

20 BASIC FIRE ATTACK

21 SALVAGE AND OVERHAUL

22 EMERGENCY MEDICAL RESPONSE

FIREFIGHTER II

23 INCIDENT COMMAND SYSTEM

24 ADVANCED COMMUNICATIONS

25 PRE-INCIDENT PLANNING

26 FIRE PROTECTION SYSTEMS

27 ADVANCED FIRE ATTACK

28 ORIGIN AND CAUSE INVESTIGATION

29 FIRE PREVENTION AND PUBLIC EDUCATION

30 VEHICLE EXTRICATION

31 SUPPORT OF TECHNICAL RESCUE TEAMS

GLENN CORBETT, EDITOR

GLENN P. CORBETT

is a licensed fire protection engineer, former assistant chief of the Waldwick (N.J.) Fire Department, associate professor of fire science at John Jay College of Criminal Justice in New York City, and technical editor for *Fire Engineering*. He previously held the position of administrator of engineering services with the San Antonio Fire Department. Corbett has a master of engineering degree in fire protection engineering from Worcester Polytechnic Institute in Massachusetts. He authored two chapters on fire prevention/protection in *The Fire Chief's Handbook, 5th Edition* (Fire Engineering, 1995) and is the coauthor of the late Francis L. Brannigan's *Building Construction for the Fire Service, 5th Edition*. Corbett is the recipient of the 2013 Tom Brennan Lifetime Achievement Award.

CONTRIBUTING AUTHORS

ANTHONY AVILLO

retired in March 2015 after a 30-year career in the fire service. Avillo was a deputy chief in North Hudson (NJ) Regional Fire & Rescue, assigned as 1st Platoon regional tour commander. He holds a master's degree in National Security Studies and a BS degree in fire science from New Jersey City University. He is currently the director and deputy fire marshal at the Monmouth County (NJ) Fire Academy. He is an adjunct professor at New Jersey City University. He is an FDIC instructor and member of the FDIC advisory board and the editorial advisory board of *Fire Engineering*. He is the author of *Fireground Strategies*, 3rd edition, and *Fireground Strategies Workbook* volumes I, II, and III. Avillo also coauthored with Chief Ed Flood *Full Contact Leadership*. Avillo has also contributed to both volumes of the *Pass It On* books by Billy Goldfeder. He also has a regular column in *FireRescue* magazine called "Progress Reports." He was a collaborator in the *Tactical Perspectives* DVD series and has had issued the DVDs *Control of Fireground Operations* and F*orging a Culture of Safety*. Avillo also cohosts a radio show, Fireground Strategies and *Other Stuff from the Street* with Chief Jim Duffy on Fire Engineering BlogTalk Radio. Avillo was recipient of the 2012 Fire Engineering/ISFSI George D. Post Fire Instructor of the Year Award.

JOHN M. BEST

is a lecturer for John Jay College of Criminal Justice and has facilitated a public safety consultancy for more than 25 years. During his 40-year career in the fire and rescue service, Best has served as fire chief of Walt Disney World and Reedy Creek Emergency Services in Florida and as deputy fire and rescue chief of Montgomery County (Md.) Fire Rescue Service. He is a doctoral candidate with the American College of Education, has a master's degree in organizational management from the University of Phoenix, a bachelor's degree in business administration from Columbia Union College, and an associate's degree in fire science from Montgomery College. He received a presidential appointment to the *America Burning* Recommissioned: America at Risk panel and is the recipient of three fire rescue governor's citations from the State of Maryland. He served on the National Fire Protection Association (NFPA) Fire Service Training Committee and is a member of the International Association of Fire Chiefs, and the International Society of Fire Service Instructors.

STEVE BERNOCCO

is a 22-year veteran of the Seattle Fire Department, where he is the captain of Engine 38. He has served as a training officer and has been involved in developing operational skills training for his department. Captain Bernocco has written numerous articles and is coauthor of *Air Management for the Fire Service*. He is an international instructor of air management, firefighter safety and survival, and fireground strategy and tactics. He is proud to be one of the "Seattle Guys."

ALAN BRUNACINI (1937–2017)

joined the Phoenix (AZ) Fire Department in 1958. He was promoted to fire chief in 1978 and retired in 2006 after 48 years of service. He is a graduate of the fire protection technology program at Oklahoma State University. Brunacini has a bachelor's degree and an MPA from Arizona State University. He is the past Chairman of the Board of the NFPA and the NFPA Fire Service Occupational Safety Committee (standard 1500). He is chairman of the NFPA Career Fire Service Career Organization and Deployment Committee (standard 1710). He is the author of *Fire Command* (2002), *Command Safety* (2004), *Timeless Tactical Truths* (2003), *Essentials of Fire Department Customer Service* (1996), and *The Anatomy and Physiology of Leadership* (2006). He and his son John own and operate the fire service website www.bshifter.com.

JOHN BRUNACINI

is a retired fire captain from the Phoenix (AZ) Fire Department. He served 26 years as a firefighter and engineer and his last 14 years as captain. As a captain, he developed and managed the Phoenix Fire Network, worked in the field as the company officer on Ladder 11 B-shift for nine years, and managed the Command Training Center his last three years, where he developed the curriculum media packages used for command and battalion training. He is currently instructing and developing curriculum packages for the Blue Card Command Certification process and www.bshifter.com.

NICK BRUNACINI

is a retired 29-year veteran of the Phoenix (AZ) Fire Department. He worked seven years as a firefighter on engine companies, was promoted to captain, and worked nine years on a ladder company. He served as battalion chief for five years. He was promoted to shift commander and for five years ran the PFD's Command Training Center.

DON CANNON

(1937-2010) was a noted fire historian, history professor at Saint Peter's College in New Jersey, and an adjunct fire science instructor at John Jay College of Criminal Justice. Cannon joined Yonkers (N.Y.) Fire Department Auxiliary in 1962, became chief in 1975, served with the Rockaway Point Fire Department in Queens, N.Y., was a 1960s Bronx FDNY Auxiliary, and was with the Midway (S.C.) Fire Department. He was the author of the firefighting history book Heritage of Flames (1977).

MICHAEL N. CIAMPO

is a 33-year veteran of the fire service. He is a lieutenant in the Fire Department of the City of New York and served with the District of Columbia Fire Department. He has a bachelor's degree in fire science from John Jay College of Criminal Justice. He is the lead instructor for FDIC's H.O.T. Truck Essentials program and a contributing editor and editorial advisor for *Fire Engineering*. His column "On Fire" is the back-page feature.

LARRY COLLINS

has served since 2016 as chief of the Special Operations & Hazardous Materials Section for the California Governor's Office of Emergency Services (Cal OES) Fire and Rescue Division. He spent 36 years as a member of the Los Angeles County Fire Department (LACoFD), where he served as battalion chief, captain, engineer, and firefighter/paramedic. He was a fire captain assigned to a specialized Rescue/USAR fire station for 20 years, in one of the most active fire department Rescue/USAR units in the nation, responding to daily technical rescues and multi-alarm fires across Los Angeles County. He served as a task force leader of the California US&R Task Force 2 (CA-TF2), the LACoFD's US&R task force that responds to local, regional, national, and international disasters. Over the past three decades, he has authored dozens of articles and cases studies in *Fire Engineering* as well as the Pennwell textbook series *Technical Rescue Operations* and the Rescue chapter of *Fire Chiefs Handbook*.

JAY COMELLA

entered the Oakland (CA) Fire Department in 1989 and is a retired captain. He is preceded in the OFD by his father, Bob; uncle, Tom; and grandfather, Cosmo. He graduated from the University of California. Comella has conducted extensive research and testing regarding fire streams, is published in *Fire Engineering*, and is an engine company H.O.T. instructor at FDIC.

DAVID DALRYMPLE

is a career EMS provider for the RWJUH Emergency Medical Services in New Brunswick, N.J. and he is a Rescue Services Captain for Clinton EMS/Rescue in Clinton, N.J. Dave is the education chair of the Transportation Emergency Rescue Committee US (TERC). In 2007 he received the Harvey Grant award for excellence in rescue. Certified as a fire service instructor by the State of New Jersey, Dave has been actively teaching transportation rescue topics for over 14 years. He is currently the lead instructor for vehicle rescue programs at the Hunterdon County Emergency Services Training Center and is the executive educator for RoadwayRescue LLC. He also currently writes the "Extrication Tactics" column for *Fire Engineering* magazine.

PAUL T. DANSBACH

is the fire marshal in the Bureau of Fire Safety in the Borough of Rutherford, NJ. Paul has worked in code enforcement for the past 39 years, the past 32 as fire marshal and is also a fire instructor at the Bergen County Law and Public Safety Institute, Fire Academy, Bergen County, NJ. He has been a member of the Rutherford Fire Department for 40 years and is a past chief of the department, having served 10 years as a chief officer. Paul also teaches C.E.U. classes for fire inspectors, fire subcode and building subcode officials for Kean and Rutgers Universities. He has been a H.O.T. and classroom instructor at FDIC, is an editorial advisor for *Fire Engineering* magazine, and an advisory board member of FDIC. He is the coauthor of the *Study Guide for Fireground Size-Up*. He is also a member of the New Jersey Fire Safety Commission's Code Advisory Council and is a past president of the Bergen County (NJ) Fire Chief's Association.

STEVE M. DELISI

retired after 27 years in the fire service, most recently as the deputy chief for the Virginia Air National Guard Fire Rescue at the Richmond International Airport. He is a hazardous materials specialist, an adjunct instructor for the Virginia Department of Fire Programs, and a former member of the NFPA committee on hazardous materials protective clothing. He began his career in hazardous materials response in 1982 as a member of the Newport News (Va.) Fire Department Hazmat Team. He has also served as a hazardous materials officer for the Virginia Department of Emergency Management. In that capacity, he provided on-scene assistance to hazmat first responders for an area that included 20 jurisdictions. He has a master's degree in public safety leadership.

JASON EMERY

is a 29-year veteran of the fire service with the Waterbury (CT) Fire Department and serves as captain of the city's heavy rescue/hazmat company. He has lectured throughout the United States and internationally on AFVs over the past 14 years and has published numerous articles on the subject. He is the lead SME for the National Fire Protection Association's AFV safety program.

He also serves as a volunteer EMT for his hometown ambulance service and has been a certified fire service instructor since 1997 and is a member of ISFSI. He is an instructor for the Burrville Regional Fire Training School and is an FDIC lecturer. He holds a bachelor's degree in fire science from University of New Haven.

FRANK L. FIRE

is an accomplished author with more than 30 years' experience teaching the chemistry of hazardous materials to firefighters and other emergency responders. He retired from his position as executive vice president, marketing and international, at Americhem, Inc. and is currently a member of the editorial advisory board for *Fire Engineering*. He is the author of *The Common Sense Approach to Hazardous Materials* (Fire Engineering 2009) and its study guide, *Chemical Data Notebook: A User's Manual* (Fire Engineering 1992), and *Combustibility of Plastics* (1992), as well as coauthor of *SARA Title III: Intent and Implementation of Hazardous Materials Regulations* (1990). He has also published more than 120 articles on individual hazardous materials and other fire service topics.

TIMOTHY J. FLANNERY

has a fire service career of more than 30 years. Currently an adjunct professor at John Jay College of Criminal Justice, as well as Passaic County Community College in New Jersey, he spent 15 years as an active firefighter with the Oakland (NJ) Fire Department. He has served as Lieutenant, Training Officer, Rescue Squad Captain, and Fire Inspector, later becoming Senior Fire Instructor at Bergen County (NJ) Police and Fire Academy. He then became Director of the Middlesex County Fire Training Center in central New Jersey. Flannery holds a bachelor's degree in fire service administration from Empire State College/SUNY, a master's degree in protection management from John Jay College, and certifications as Fire Official, Level 2 Fire Instructor, Level 3 Incident Management, Officer Level 1, and is trained as a NIMS/ICS instructor up to ICS-400. He has been involved with the New Jersey State Fire Commission's Training and Education Advisory Council for more than 15 years.

CHRISTOPHER FLATLEY

is a veteran of the FDNY, currently a lieutenant assigned to Ladder Company 21 in the Hell's Kitchen section of Manhattan. Prior to promotion he worked in Ladder Company 2 in Midtown. He is a master exercise practitioner on the Exercise Design Team through the Center for Terrorism and Disaster Preparedness. He is a member of the incident management team and family assistance units. He twice served as chief of the Blauvelt Volunteer Fire Company and is currently the assistant chief and training coordinator. He is a nationally certified fire instructor 1 and an instructor at the Rockland County Fire Training Center, Pomona, NY, and holds a degree in fire protection technology. He has written numerous articles for *Fire Engineering*.

RICHARD A. FRITZ

has been an active member of the fire service since 1973. He has served with various career and volunteer fire departments, including the Muscatine and Davenport (Iowa) fire departments; Penfield, N.Y., and Hampton, Ill.; Union Fire Company No. 1 in Carlisle, Pa., and Williamsport (Md.) Company No. 2 volunteer fire departments. Until recently, he was a training captain and battalion chief of training for the High Point (N.C.) Fire Department. He has served on the NFPA 1001 committee. He is the author of *Tools of the Trade: Firefighting Hand Tools and Their Use* (Fire Engineering 1997) and four supporting training videos on firefighting hand tools.

MIKE GAGLIANO

has 25 years of fire/crash/rescue experience with the Seattle (WA) Fire Department and the United States Air Force. He is the captain of Ladder 5 and a member of the Seattle Fire Department's Critical Decision/Tactical Training Group. He has written numerous fire service articles and is coauthor of *Air Management for the Fire Service* and coauthor of *Challenges of the Firefighter Marriage*. He is a member of the *Fire Engineering* editorial advisory board and the FDIC educational advisory board; a director for the Cyanide Poisoning Treatment Coalition; and a national instructor on air management, fireground tactics, leadership, and company officer development.

SEAN GRAY

currently holds the rank of captain with Cobb County Fire and Emergency Services in Metro Atlanta and has been in the fire service since 1993. He has a bachelor's degree in fire safety engineering. He has been a member of multiple technical panels involving firefighter safety research and is an appointed member of the UL Firefighter Safety Research Institute Advisory Board. Sean is a NFPA committee member for Fire Hose and Fire Service Training Facilities. He recently coauthored a PennWell Publications DVD and book titled *The Evolving Fireground*. He has also been published in multiple fire service magazines, is an FDIC H.O.T. lead instructor, runs the website StopBelievingStartKnowing.com, and delivers evidence-based tactics training courses across the United States.

CHARLES JENNINGS, PH.D., MIFireE, CFO

is a faculty member at John Jay College of Criminal Justice and is director of the College's Christian Regenhard Center for Emergency Response Studies. He began in the fire service in 1979 and served in operational and administrative positions from firefighter, line officer, director, and commissioner in West Elmira, NY; Ithaca, NY; Hillandale (Montgomery County, MD) and White Plains, NY, where he served as the deputy commissioner for public safety from 2002–2008. He has conducted research on numerous topics, many related to public policy issues and the fire service. He earned his doctoral degree in city and regional planning from Cornell University.

PHILLIP JOSE

is a 21-year veteran of the Seattle (Wash.) Fire Department where he is a Battalion Chief. He has served as a lieutenant and captain in the Training Division, and is a member of the SFD Operational Skills Enhancement Development Team. Chief Jose teaches nationally on tactics, air management, and firefighter safety. He is coauthor of *Air Management for the Fire Service* (Fire Engineering 2008) and numerous articles in fire service publications.

RON KANTERMAN

is currently a career fire chief in New London County, CT. He entered the fire service in 1975 in Brooklyn, NY, and then took a position in 1989 as assistant chief and then chief of fire protection for a Fortune 100 pharmaceutical company until 2008. He holds a BA in fire service administration and an MS in fire protection management, both from John Jay College. He holds an additional master's degree in environmental science from the New Jersey Institute of Technology. He is an adjunct professor of fire protection and emergency management for various two- and four-year colleges and teaches at the undergraduate and graduate levels. He holds memberships with numerous national, state, and local organizations and is the administrator for the National Fire Academy Alumni Association. He also teaches as an adjunct instructor for the National Fire Academy. He has been published more than 60 times and contributes to Fireengineering.com regularly through their blog page under "Chief Kanterman's Journal." He also serves on the Educational Advisory Committee for FDIC. He is an avid volunteer with the National Fallen Firefighters Foundation and works as the chief of operations each year at the National Memorial Weekend in Emmitsburg, MD. He is also the lead instructor for the Local Assistance State Teams (LAST) program.

TOM KIURSKI

began his fire service career in 1981 and has served for more than 26 years with Livonia (MI) Fire and Rescue, retiring as the training coordinator. He has worked for the Brighton (MI) Area Fire Authority since 2013, starting and running a high school fire academy, leading cadets to state certification. He is the author of *Creating a Fire Safe Community: A Guide for Fire Safety Educators*. Tom has written more than 150 fire safety articles that have appeared in *Fire Engineering* and other fire service publications. He has an associate's degree in fire science, a bachelor's degree in fire and safety engineering technology, and a master's degree in public administration.

JERRY KNAPP

is the chief of the Rockland County (NY) hazmat team and a former nationally certified paramedic. His is a 39-year veteran firefighter/EMT with the West Haverstraw (NY) Fire Department and a training officer at the Rockland County Fire Training Center in Pomona, NY. He is a member of the technical panel for the Underwriters Laboratories research on interior and exterior fire attack at residential fires. He has a degree in fire protection and is an adjunct professor of fire technology. He is a frequent contributor to *Fire Engineering*.

RICK LASKY

is a 34-year veteran of the fire service and served as chief of the Lewisville (TX) Fire Department for 12 years. He began his career as a firefighter in the suburbs on the southwest side of Chicago. He received the 1996 International Society of Fire Service Instructors Innovator of the Year award for his part in developing the Saving Our Own program. He served as the co-lead instructor for the H.O.T. Firefighter Survival program at FDIC for more than 10 years, is a member of the *Fire Engineering* magazine editorial advisory board, and also serves on the FDIC advisory board. He is the author of *Pride and Ownership: A Firefighter's Love of the Job* (2006) and is the cohost of the radio show "The Command Post" heard on Fire Engineering Talk Radio.

DENNIS LEGEAR

a retired captain from Oakland, CA, holds both an AS degree in fire science and bachelor's in vocational education from CSU, Long Beach. He held the position of water supply officer in the OFD and advised the engine and truck committees in pump spec., hose and equipment layout, and compartment/hose bed spec. as it related to fire stream development. He was a hose and nozzle instructor at many in-house academies and an adjunct professor in the same capacity at the College of the Siskiyous in California. His fire service career included time at Cal Fire, Kings County, and the City of Oakland. He is owner of LeGear Engineering F.D. Consulting (L.E.F.D.C), which has provided technical advice to some of the largest fire agencies, including Portland, Seattle, Boston, and New York. He has consulted with water municipalities assisting in the development of emergency response plans for a variety of water infrastructure problems related to seismic events and other disasters. He is an author and proposal reviewer for Fire Engineering Books and Videos.

DOUG LEIHBACHER

is a 28-year veteran of the fire service and retired as an assistant chief in the Yonkers (N.Y.) Fire Department. He has an associate's degree in fire protection technology and a bachelor's degree in education. He is a certified fire instructor and municipal training officer and has served as senior instructor for the YFD Probationary Firefighter Training School and as an adjunct instructor at the Westchester County Training Academy. He has certifications in incident safety, code enforcement, and fire investigation, and he is a hazardous materials technician. He has also been a classroom and H.O.T. instructor at FDIC as well as a contributing author for *Fire Engineering* since 1994 and www.FireNuggets.com since 2000.

MIKE MCEVOY, PHD, NRP, RN, CCRN

is the EMS coordinator for Saratoga County, New York, a paramedic supervisor with Clifton Park & Halfmoon Ambulance, and the EMS editor for *Fire Engineering* magazine. He is a nurse clinician in cardiothoracic surgical intensive care at Albany Medical Center, where he also chairs the Resuscitation Committee and teaches critical care medicine. Mike is an author and popular speaker at EMS, fire, and medical conferences worldwide.

W. JACK MILLER

has been involved in the fire service and fire protection equipment for more than 40 years. He is regional sales manager, Northeast, for Amerex Corporation, a worldwide leader in the manufacture of hand portable and wheeled fire extinguishers and fire suppression systems. He was previously secretary of the Wyckoff (NJ) Fire Department and president of Protection Fire Company No. 1 Association. He is a recipient of the Dana Hannon Distinguished Service Award.

JOHN MITTENDORF

is a 30-year veteran of the Los Angeles City Fire Department who retired as a battalion chief in 1993. He is a former member of the fire science advisory boards of five colleges and the author of numerous fireground articles in U.S. and European fire service-related magazines. He is the author of *Ventilation Methods and Techniques* (1988), *Truck Company Operations* (Fire Engineering 1998), and *Facing the Promotional Interview* (Fire Engineering 2003). He lectures in the United States and United Kingdom on strategy and tactics, truck company operations, fireground operations, ventilation operations, and the complete fire officer. He is a member of the *Fire Engineering* editorial advisory board.

JACK J. MURPHY

is a retired fire marshal and a former deputy chief of the Leonia (NJ) Fire Department. He is the chairman of the New York City High-Rise Fire Safety Directors Association, a member of the National Fire Protection Association High-Rise Building Safety Advisory, and the 1620 Pre-Incident Planning Committees. He has published fire service articles and authored *RICS: Rapid Incident Command System Field Handbook* and coauthored *Bridging the Gap: Fire Safety and Green Buildings*. He contributes articles to *Fire Engineering* magazine and is a member of its editorial board. He was the recipient of the 2012 Fire Engineering Tom Brennan Lifetime Achievement Award.

MIKE NASTA

is a retired deputy chief of the Newark (NJ) Fire Department, with 31 years as the citywide tour commander. He is also a retired member of the South Hackensack (NJ) Volunteer Fire Department, having served multiple terms as department chief, and is a member of the Upper Saddle River (NJ) Fire Department. He is a New Jersey state-certified level II fire instructor and a senior fire instructor at the Bergen County (NJ) Law & Public Safety Institute. He coauthored the fireground officer development course with Anthony Avillo and has written numerous fire service articles. He is an FDIC H.O.T. coordinator and was the lead instructor for Truck Company Search at FDIC. He is a member of the *Fire Engineering* advisory board.

CASEY PHILLIPS

is a 20-year veteran of the fire service, serving the last 13 years with the Seattle (Wash.) Fire Department. He currently holds the position of captain on Engine 28, and is a member of the SFD Operational Skills Enhancement Development Team. Captain Phillips is coauthor of *Air Management for the Fire Service* (Fire Engineering 2008) and teaches across the country on air management and firefighter safety.

TIM PILLSWORTH

is a past chief of the Winona Lake Engine Company No. 2 in Orange County, N.Y., and has been an active volunteer firefighter and EMT since 1986. He is a past presenter at FDIC East and a contributor to *Fire Engineering*. He is employed as a civil engineer and project manager with the U.S. Army Corps of Engineers at West Point, N.Y.

FOREST REEDER

began his fire service career in 1979. He currently serves as battalion chief/director of training & safety with the Pleasantview Fire Protection District and is the director of training for Southwest United Fire Districts. In these capacities, he is responsible for the design, implementation, and coordination of in-service training activities as well as a full-service fire training academy program. He holds numerous Illinois fire service certifications and has an associate of applied science degree in fire science technology, a bachelor's degree in fire department administration, and a master's degree in public safety administration from Lewis University.

FRANK RICCI

is an advisory board member for *Fire Engineering* and FDIC. He is a battalion chief and drillmaster for the New Haven (CT) Fire Department and cohost of the *Politics & Tactics* radio show. He is a contributing author to the Daily Caller News Foundation. Frank has won a landmark Supreme Court case regarding fire department promotions and he has testified before Congress. He has been a lead consultant for several Yale studies and has worked on a heavy rescue unit covering Bethesda and Chevy Chase, MD. Ricci has filmed several Fire Engineering Training Minutes segments and has been a contributing author to several books and has authored several DVDs including the *Firefighter Survival Techniques* and *Tactical Perspectives* series, *Command, Ventilation, Search, Fire Attack*, and *Mayday*, for Fire Engineering.

DAN SHERIDAN

began his fire service career in 1986 with the Fire Department of the City of New York where he is a battalion chief. He is a division supervisor with the FDNY Incident Management Team in the operations section. Sheridan serves as the executive officer to the chief of training and is a national level II instructor. Previously Sheridan worked as captain of the 6th Division, on Tower Ladder 17 in the South Bronx, in Squad 41, in SOC as a lieutenant, and on Engine 58. He has been cited eight times for bravery, three individual and five unit awards.

JEFF SHUPE

has been a career firefighter for 31 years and is a member of the Cleveland (Ohio) Fire Department. He has also served as a volunteer firefighter. He is an Ohio-certified fire instructor and has served as training coordinator for volunteer and career fire departments. He has an associate's degree in fire technology and attended the University of Cincinnati fire protection program. He is an FDIC H.O.T. team member for engine company operations and a classroom presenter.

VICTOR STAGNARO

has served the National Fallen Firefighters Foundation (NFFF) as managing director of Family and Fire Service Programs, Marketing, and Business Management since 2010. He has been involved with the foundation since 1998, when he served as the incident commander for the National Fallen Firefighters Memorial Weekend for several years, and he served as a member of the initial New York Response Team on 9/11 for the Foundation. He was hired by the Prince George's County Fire/Emergency Medical Services Department in 1985 and retired in early 2010 as the deputy fire chief of emergency operations. He oversaw Emergency Operations Command and was responsible for the 44 fire/EMS stations, all technical and special operations, and the Advanced Emergency Medical Services office. Stagnaro is a contributing author to *Fire Engineering* magazine.

MARK A. SULCOV

is a 33-year veteran of the fire service with the Fort Lee (N.J.) Fire Department. He has been a lieutenant for several years and is serving as assistant training officer, the department safety officer, and a member of the technical rope rescue team. He is a New Jersey state-certified fire inspector, level II fire instructor, and a senior fire instructor at the Bergen County (N.J.) Law & Public Safety Institute at the Fire Academy. He is also a member of the Fair Lawn (N.J.) Fire Department.

ROBERT TILL

is a licensed professional engineer and associate professor teaching and researching fire science at John Jay College of Criminal Justice. In 2001 he received his Ph.D. in fire protection engineering from Worcester Polytechnic Institute, where his thesis addressed the evaluation of buildings for fire department intervention. He is special expert member of the NFPA Standard for Fixed Guideway Transit and Passenger Rail Systems (NFPA 130) Committee and is chairman of the Controls Reliability task group. He is also a member of the Society of Fire Protection Engineers Fire Department Operations task group.

CHRISTINE WAGNER

is an on-scene coordinator for the U.S. Environmental Protection Agency. She takes part in environmental cleanup projects for Region 3. Based in Richmond, Va., Region 3 serves Delaware, the District of Columbia, Maryland, Pennsylvania, Virginia, and West Virginia. Along with her work in the field, Wagner provides training in hazardous materials and emergency response to fire marshals, emergency responders, and emergency management personnel.

BECKI WHITE

is an assistant chief with the Eden Prairie Fire Department overseeing the Training and Fire Prevention Divisions. She has a master's degree in education from St. Mary's University and has completed the Executive Fire Officer program with the National Fire Academy. White is also an advisory board member for Fire Engineering and FDIC International conference and the Public Safety Educational Advisory Group for UL-FSRI.

RICHARD S. WOLFSON, CFEI

is the senior fire and explosion cause analysis expert with Atlantic Professional Services of Clifton, N.J. He is a member of the International Association of Arson Investigators and is a qualified expert witness in federal and state courts in New Jersey and New York. He has taught at the Bergen County (N.J.) Law and Public Safety Institute at the Fire Academy and is a volunteer firefighter and fire investigator with the City of Garfield (N.J. Fire Department.

JOHN WORON

a 26-year fire service veteran, is an assistant chief with the Middletown (Conn.) Fire Department. He's a certified Fire Officer IV, Fire Service Instructor II, Safety Officer, and Hazardous Materials Technician. Woron holds a bachelor's degree in fire administration, is an adjunct senior instructor for the Connecticut Fire Academy, and the director of training for the Middlesex (Conn.) Regional Fire School. He is a Task Force Leader of the CT-TF-1 USAR Team.

ALAN R. YOUNG (1947–2013)

retired in 2002 from the Sacramento (Calif.) City Fire Department as an assistant chief with 30 years of fire service experience. His last assignment was as chief of communications for the Sacramento Regional Fire and EMS Communication Center. He graduated from Sacramento State University with a bachelor's degree in business administration and is a certified fire officer with the California State Fire Marshal's Office. He has an extensive background in education concentrating in the command field. He has taught emergency operational command procedures at several community colleges.

ACKNOWLEDGMENTS

This project wouldn't be complete without acknowledging and thanking those who had a hand in its creation.

Fire Engineering **magazine: Bobby Halton, Editor in Chief; Diane Feldman, Executive Editor; and Nate DeMarse, Photo Editor, for their editorial assistance and support.**

Special thanks to these members of the fire service community who served as reviewers and provided invaluable guidance and feedback:

Jack Alderton, Chief, Morris County Fire Academy (NJ)

Brian Brush, training division sergeant, Edmond Fire Rescue (OK)

John Buckheit, Battalion Chief, FDNY (NY)

Mike Cardwell, Asst. Chief, Urbandale FD (IL)

David C. Evans, director, department of fire services, Massachusetts Firefighting Academy

Charles L. French Jr, district chief, Tulsa FD (OK)

Shane Futura, education trustee, 50th State of Fraternal Order of Leatherhead Society (FOOLS) (HI)

Al Gerber, Senior Fire Instructor, Bergen County Law & Public Safety Institute (NJ)

Bill Gustin, captain, Miami-Dade Fire Rescue Department (FL

Jim Hester, fire academy manager and rescue program coordinator, Eastern Oklahoma County Technology Center

Tyler Hice, fire captain, Platte Valley Fire Department, Kersey (CO)

Justin A Holck, fire program training specialist I, Central Oregon Region, Department of Public Safety Standards & Training

Thomas Jenkins, fire chief, Rogers FD (AR)

Alan Joos, training division chief, Nebraska State Fire Marshal Agency

Steve Kerber, Deputy Chief, College Park Volunteer FD (IL)

Ralph McNemar, program coordinator, West Virginia University State Fire Academy

Nick Morgan, Firefighter, St. Louis FD (MO)

Ted Nee, command training center manager, Sandia National Labs, Albuquerque NM; former deputy chief, Albuquerque FD (NM)

Philip Oakes, national program director/trainer, the National Association of State Fire Marshals

Jake Rhodes, fire chief, Kingman Fire Department (AZ)

Cary Roccaforte, Program Coordinator, ESTI Recruit Training Academy (TX)

Rob Schnepp, Captain/Paramedic, Alameda County FD (CA)

Kevin Shea, Glacier Rescue Solutions

Jake Sheridan, Broken Arrow Fire Department (OK)

Jack Sullivan, director of training for the Emergency Responder Safety Institute, a committee of the Cumberland Valley Volunteer Firemen's Association.

Timothy Tyner, training major, Broken Arrow Fire Department (OK)

Michael L. Walker, deputy chief of operations, Oklahoma City Fire Department (OK)

Devon Wells, Chief, Hood River FD (OR)

Mike Witten, Captain, Tulsa FD (OK)

Tim Zehnder, director of fire science, Mid-Plains Community College, McCook, Nebraska

Many thanks to these individuals, fire departments, academies, organizations, and state entities that generously gave their time and cooperation:

Additional Technical Contributors

Penny Adams, Center Manager, Sacramento Regional Fire EMS Communication Center (CA)

Brian Arnold, Captain, Oklahoma City FD (OK)

Peter Bertocchi, CFI, FDNY (NY)

David Collado, Firefighter, FDNY (NY)

Rick Dempsey, CFI, Eastchester FD (NY)

Benjamin Dockstader, Firefighter, Brook Park FD (OH)

Bill Gustin, CFI, Miami-Dade FD (FL)

Robert Halpin, Captain, Carmel FD (Ret.) (IN)

William Hicks, Asst. Professor, Fire and Safety Engineering Technology, Eastern Kentucky University (KY)

Leigh Hollins, CFI, Cedar Hammock Fire Rescue (FL)

International Society of Fire Service Instructors (ISFSI)

Jim Jadkowski, CFI, Westchester County Dept. of Emergency Services (NY)

Adam Knight, Engineer, Kootenai County Fire & Rescue (ID)

Ralph McNemar, USAF Fire Protection (Ret.)

Charles Metcalf, Captain, DeSoto FD Dist. 8, Mansfield (LA)

Richard Nagle, NYS Office of Fire Prevention and Control (NY)

Brian Ohleth, CFI, Stamford FD (CT)

Chris Pepler, Firefighter, Torrington FD (CT)

Glen Rudner, Hazardous Materials Officer, Virginia Dept. of Emergency Management (VA)

David Russell, CFI, FDNY (NY)

John Van Voorhies, CFI, White Plains FD (NY)

Photo/Video/Drill Assistance

Marion W. Anderson, Lieutenant, Mask Service Unit, FDNY (NY)

Andrew Antoni, Firefighter, Mask Service Unit, FDNY (NY)

Don Arterburn, Chief, Rockland County Technical Rescue Team (NY)

David and Lisa Barthold, Waldwick EMS (NJ)

Thomas Bierds, Instructor, Rockland County Fire Training Center (NY)

Peter Byrne, Coordinator for Safety and Compliance, Rockland County (NY)

Dannie Caldwell, Captain, Tulsa FD (OK)

Sharon Corbett, Waldwick EMS (NJ)

Dusty Danderson, Captain, Collinsville FD (OK)

George Drescher, Deputy Fire Coordinator, Rockland County (NY)

Brian Duddy, Photographer, Central Nyack FD (NY)

Jason Emery, Lieutenant, Waterbury FD (CT)

Rick Fisher, Major, Broken Arrow FD (OK)

Phil Greer, Firefighter, Dan Leghorn Engine Company #1, Orange Lake Fire District, Newburgh (NY)

Kristopher M. Grills, Firefighter, Mask Service Unit, FDNY (NY)

Peter Grosbek, Chief, Blauvelt Volunteer FD (NY)

James Hester, Captain, Tinker Air Force Base FD (OK)

Thomas F. Hoey Jr., Firefighter, Winona Lake Engine Company #2 (NY)

Chris Hohol, Live Fire Training by Fire LLC

Tom Jenkins, Chief, Sand Springs FD (OK)

Nathaniel Johnson, Firefighter, Mask Service Unit, FDNY (NY)

Mark Joslin, Chief, Sand Springs FD (OK)

John P. Kay, Detroit FD (MI)

Kyle Knapp, Firefighter, West Haverstraw FD (NY)

Ryan Lawson, Training Officer, Broken Arrow FD (OK)

Dan Madrzykowski, Fire Protection Engineer, National Institute of Standards and Technology

Chris Mancuso, Haight Fire Equipment Supply

John Marlar, Captain, Tulsa FD (OK)

Ray McCormack, Lieutenant, FDNY (NY)

Neil Moses, Firefighter, Waldwick FD (NJ)

Bryan Myrick, Training Officer, Broken Arrow FD (OK)

Paul Newton, Chief, Glenpool FD (OK)

Terrell Ogilvie, Firefighter, Glenpool FD (OK)

Frank Ogno, MES Northeast

John Peeters, Chief, Tomahawk FD (WI)

Mike Percival, Captain, Collinsville FD (OK)

Tony Piontek, Live Fire Training by Fire LLC

Rick Pisani, Asst. Chief, Orange Lake Fire District (NY)

Larry Rauch, Chief, Bergen County Law and Public Safety Institute (NJ)

Mike Reise, Captain, Broken Arrow FD (OK)

Jennie Ross, Firefighter, Collinsville FD (OK)

James Russell, Firefighter, Mask Service Unit, FDNY (NY)

Justin Shelby, Captain, Collinsville FD (OK)

Vinni Spampinato, Chief, Orange Lake Fire District (NY)

Andy Teeter, Chief, Tulsa FD (OK)

Mike Wood, Chief, Sand Springs FD (OK)

Gordon Wren, Director of Fire and Emergency Services, Rockland County (NY)

Russell Young, Chief, Collinsville FD (OK)

Bergen County Law and Public Safety Institute (NJ)

Bergenfield FD (NY)

Bergenfield Fire Training Center (NY)

Blauvelt Volunteer FD (NY)

Broken Arrow FD (OK)

Chelsea FD (MI)

Collinsville FD (OK)

Detroit FD (MI)

Eastern Oklahoma County Fire Training Center (OK)

Fort Bliss FD (TX)

Glenpool FD (OK)

Lewisville FD (TX)

New York FD (NY)

Orangeburg Volunteer FD (NY)

Pittsfield Township FD (MI)

Rockland County Fire Training Center (NY)

Rockland County Haz-Mat Response Team (NY)

Sand Springs FD (OK)

Superior Township FD (MI)

Tulsa FD (OK)

West Nyack Volunteer FD (NY)

Ypsilanti Township FD (MI)

INTRODUCTION

It's no coincidence that the creation of one of the first American training schools for firefighters centered on the use of a then-new piece of equipment—the "Pompier" scaling ladder. Introduced as part of a new "life-saving corps" school of the New York City Fire Department (FDNY) in 1887, this new ladder with a gooseneck hook at the top and a set of rungs on a center beam allowed individual firefighters to vertically ascend the entire height of a building, traveling from window to window.

Captain Chris Hoell, one of many early and important contributors to the fire service training legacy.

It is also no fluke that it was Chris Hoell, a St. Louis fire captain, who introduced the scaling ladder to the United States and the FDNY. Hoell had earlier created his own life-saving corps in the St. Louis Fire Department in 1877. He invented the ladder belt, still used by many fire departments today. The use of a scaling ladder requires upper body strength as well as a significant amount of inner strength and fortitude. The scaling ladder was used for decades by many fire academies to instill confidence in each new recruit.

Hoell's efforts are just one example of how the fire service has improved over the years. Training is at the heart of these improvements, founded upon the shared history, experience, and innovation of firefighters from across the United States. We learn from each other, improving techniques and developing new tools. Many of the training protocols that you will be taught came at the greatest price—*the supreme sacrifice*. It is up to you to study these "lessons learned" and to use them in your career. It is also your responsibility to stay current with new protocols and equipment as the fire service continues to evolve during your career.

Fire Engineering's Handbook for Firefighter I and II that you hold in your hands was created upon this solid foundation of history, experience, and innovation. It builds upon the works of such fire service luminaries as Fred Shepperd, Warren Isman, Frank Brannigan, Lloyd Layman, Ray Downey, Andy Fredericks, Tom Brennan, and many others (as you advance in your career, learn about these individuals and their contributions). It also builds upon the legacy of *Fire Engineering* magazine, the most respected fire service journal and one of the oldest trade publications in the nation, with its roots dating back to 1877.

There are nearly a thousand years of collective experience among the authors of this *Handbook*. They are among the best educators in the fire service. As editor, I asked each of them to prepare a state-of-the-art chapter, to provide you with the most effective and safest techniques available. Reading and practicing the skills contained in these pages will lead you down the path to becoming an effective firefighter. It will also help you to perform the most important of duties, to save lives: the lives of the citizens you have sworn to protect as well as you and your fellow firefighters.

How to use the Handbook

Fire Engineering's Handbook is organized into two sections: *Firefighter I* and *Firefighter II*. The chapter topics are ordered so that students learn the basic components and safety issues of firefighting first--from fire behavior all the way up to fire streams--then pulls all of that training together to apply it in Fire Attack and more complex topics. Each chapter ends with study questions for you to review. The Fireground Notes provide short tips of the trade and words of wisdom from fire training greats. Many authors have provided personal Lessons from the Fireground to underscore their topic issues.

In many chapters you will also see a special Skill Drill icon. This points you to the spiral-bound Skill Drill Book for the chapter's related drills. Each skill drill includes step-by-step photos and instructions. We hope that you will take the smaller (and lighter!) skill drill book out on the training field with you when you practice your knot tying, ladder-raising, hose-stretching, and many other skills.

Updates for 2019

The *Fire Engineering Handbook* 2019 update has been adjusted to include all changes to knowledge and skill requirements for the NFPA 1001 Standard for Firefighter Professional Qualifications, 2019 Edition. In addition, this edition of the *Handbook* has been updated to include the influence of the latest research into fire behavior, specifically how firefighting tactics influence the flow paths of a working fire.

The NFPA Knowledge Correlations following this introduction will demonstrate the coverage for both standards.

Finally, we encourage instructors and students alike to read the following position statement from the International Society of Fire Instructors (ISFSI) on Fire Dynamics Research in Tactical Operations, and to stay updated on these discussions going forward, so that you can be informed on these important new factors in firefighting safety.

I'm sure that you are very excited as you begin your fire service career, wanting to learn all you can about firefighting. I know I was when I began my firefighting journey, just as it was for my father when he became a firefighter. I encourage you to keep that excitement burning inside of you in the years to come—never stop learning and training. Learn from others and share what you have learned. Remember, it's a brotherhood and sisterhood that you have joined—we depend upon each other.

Welcome to the fire service family!

Glenn P. Corbett, editor

Waldwick, N.J.

March 2019

Editor's note: *A project of this size and scope is the result of the work of many talented hands. I would like to express my sincere gratitude to the core group who made it happen—dedicated professionals who spent days and many nights and weekends toiling away on this effort. They endured countless twice-weekly conference calls and endless email improvements, ensuring that this* Handbook *and its associated materials would set a new standard for excellence. I can't thank these people enough for their efforts on the 2008 edition: Mary McGee, Marla Patterson, Brad Epperley, Sheila Brock, Mark Haugh, Jerry Naylis, Steve Hill, Tony Quinn, Susan Ormston, Jane Green, Tim Flannery, and Jason Emery. In addition to Mark Haugh and Tony Quinn, the 2019 edition brought Chris Barton to this text; their efforts resulted in an even better cutting-edge handbook for America's new firefighters.*

I also want to send a "tip of the helmet" to our authors—thanks to each and every one of them. They know how important this book is to America's new firefighters.

International Society of Fire Service Instructors
Position Statement on Fire Dynamic Research in Tactical Operations

Given the latest research in fire dynamics and the potential impact on firefighter safety, the ISFSI board of directors unanimously releases this position statement. The ISFSI encourages all fire departments to incorporate the fire dynamics research into their tactical operations through any and all means necessary.

"Given the potential to improve firefighter safety, we believe fire departments should review their tactical operations and update their tactics. The ISFSI is proud to stand behind those conducting this research and will continue to support their efforts," said ISFSI president Doug Cline.

The International Society of Fire Service Instructors (ISFSI) states its position on the importance of recent research in fire dynamics and firefighting tactics, as conducted by the National Institute of Standards & Technology (NIST) and Underwriters Laboratories (UL). The ISFSI believes that fire departments should take action to adjust their tactical plans and training programs to incorporate this research into their emergency response operations. Additionally, the ISFSI believes that professional standards should be updated to reflect the latest information in fire dynamic research as soon as possible.

Size-Up. Size-up must occur at every fire. Consideration must be given to the resources available and situational conditions, such as weather, fire location, size of the fire and building, and the construction features. A tactical plan for that fire must be developed, communicated and implemented.

Ventilation. Fire departments should manage and control the openings to the structure to limit fire growth and spread, and to control the flow path of inlet air and fire gases during tactical operations. All ventilation must be coordinated with suppression activities. Uncontrolled ventilation allows additional oxygen into the structure which may result in a rapid increase in the fire development and increased risk to firefighters due to increased heat release rates.

Suppression. Given the fuel rich environment that the fire service operates in, water should be applied to the fire as soon as possible. In many cases, water application through an exterior opening into a fire compartment may be the best first action, prior to committing firefighting resources to the interior.

Fire departments should cool the interior spaces of a fire building with water from the safest location possible, prior to committing personnel into spaces with, or adjacent to, fully developed or smoldering (ventilation limited) fire conditions.

Rapid Intervention. Fire department rapid intervention procedures should be updated to provide water on the fire as soon as possible and ventilation openings controlled during firefighter "Mayday" incidents.

Tactical Applications. Fire departments should consider revised tactical guidelines for suppression, such as the S.L.I.C.E. - R.S. acronym. This stands for the following:

Sequential actions

Size up

Locate the fire

Identify and control the flow path

Cool the heated space from a safe location

Extinguish

Actions of opportunity that may occur at any time

Rescue

Salvage

The ISFSI acknowledges the resistance to change, but believes it is important for fire departments to act on this information in a timely manner to enhance firefighter safety. Additionally, the ISFSI believes that chief officers, instructors, company officers, firefighters and all entities that support the fire service should make incorporating the latest in fire dynamic research into regular tactical operations a high priority.

We encourage instructors and students to visit ISFSI's Online Training Courses (http://learn.isfsi.org) and *Fire Engineering's* Fire Dynamics Topic Center (www.fireengineering.com/fire-dynamics.html) for new information and instruction on fire dynamics topics and firefighter safety.

NOTE: The tactical recommendations in this statement are offered as options for educational and safety consideration. *As always, readers should follow the standard operating guidelines of their department and authority having jurisdiction (AHJ).*

NFPA 1001 KNOWLEDGE CORRELATIONS

JPR	FFI Requisite Knowledge	Chapter	Page(s)
4.1.1	Organization of FD	3	30–32
	Role of FF I in the organization	3	34–35
	Mission of fire service	1, 3	4, 30
	FD's standard operating procedures and rules and regulations	3	42–43
	Role of other agencies as they related to the FD	3	43–46
	The value of the fire department mission and to reduce firefighter line-of-duty injuries and fatalities	17	480–482
	The role of other agencies as they relate to the fire department	3, 17	43–490
	The signs and symptoms of behavioral and emotional distress	10, 17	191, 498–500, 535–537, 541
	Aspects of the FD's member assistance program	3, 17	41,96–497, 500
	Importance of physical fitness and a healthy lifestyle to the performance and duties of a firefighter	3, 17	41, 482, 500
	Critical aspects of NFPA 1500 as applied to FF I	3, 17	41–44, 183, 187, 192, 492–493
4.2.1	Procedures for reporting an emergency	4	50
	Departmental SOPs for taking and receiving alarms	4	50–51
	Radio codes or procedures	4	63
	Information needs of dispatch center	4	50–52
4.2.2	Fire department procedures for answering nonemergency telephone calls	4	52–53
4.2.3	Departmental radio procedures and etiquette for routine traffic, emergency traffic, and emergency evacuation signals	4	53, 62–63
4.2.4	Personnel accountability systems	4, 17	63–65, 532
	Emergency communication procedures	17	533–535
	Emergency evacuation methods	4, 17	65, 515, 521–522
4.3.1	Conditions that require respiratory protection	10	179–182
	Uses and limitations of SCBA	10	179–182
	Components of SCBA	10	183–188
	Donning procedures	10	188
	Breathing techniques	10	193
	Indications for and emergency procedures used with SCBA	10	196–200
	Physical requirements of the SCBA wearer	10	183
4.3.2	Mounting and dismounting procedures for riding fire apparatus	17	502
	Hazards and ways to avoid hazards associated with riding apparatus	17	501
	Prohibited practices	17	501
	Types of department personal protective equipment and the means of usage	9, 17	156–161, 502
4.3.3	Potential hazards involved in operating on emergency scenes including vehicle traffic, utilities, and environmental conditions	17	503
	Proper procedures for dismounting apparatus in traffic	17	502
	Procedures for safe operation at emergency scenes	17	502
	Protective equipment available for members' safety on emergency scenes	17	505
	Work zone designations	17	503
4.3.4	Basic construction of typical doors, windows, and walls within the department's community or service area	7	111–116

JPR	FFI Requisite Knowledge	Chapter	Page(s)
	Operation of doors, windows, and locks	12	234, 238–259
	Dangers associated with forcing entry through doors, windows, and walls	12	243–261
4.3.5	Personnel accountability systems	4	61
	Communication procedures	17, 20	491–495, 613
	Emergency evacuation methods	4, 17	60–62, 504, 492–544
	What constitutes a safe haven	17	518
	Elements that create or indicate a hazard	17	516–526
	Emergency procedures for loss of air supply	10	196–200
4.3.6	Parts of a ladder	13	269–273
	Hazards associated with setting up ladders	13	266–567
	What constitutes a stable foundation for ladder placement	13	283–284
	Different angles for various tasks	13	281–284
	Climbing techniques	13	297–298
	Safety limits to the degree of angulation	13	284
	What constitutes a reliable structural component for top placement	13	299
4.3.7	Principles of fire streams as they relate to fighting automobile fires	18	566–567
	Precautions to be followed when advancing hose lines toward an automobile	18	565–566
	Observable results that a fire stream has been properly applied	18	563–567
	Identifying alternative fuels and the hazards associated with them	18	575–576
	Dangerous conditions created during an automobile fire	18	550–562, 564–565
	Common types of accidents or injuries related to fighting automobile fires and how to avoid them	18	581–587, 589
	How to access locked passenger, trunk, and engine compartments	18	569–572
	Methods for overhauling an automobile	18	573
4.3.8	Types of attack lines and water streams appropriate for attacking stacked, piled materials and outdoor fires	20	665–668
	Dangers—such as collapse—associated with stacked and piled materials	20	664–668
	Various extinguishing agents and their effect on different material configurations	6, 20	92–96, 668
	Tools and methods to use in breaking up various types of materials	20	646
	Difficulties related to complete extinguishment of stacked and piled materials	20	667–668
	Water application methods for exposure protection and fire extinguishment	20	662, 667–668
	Dangers such as exposure to toxic or hazardous materials associated with storage building and container fires	20	663–668
	Obvious signs of origin and cause	21	686–687
	Techniques for the preservation of fire cause evidence	21	686–687
4.3.9	Use of forcible entry tools during rescue operations	12, 17	256–259, 542–543
	Ladder operations for rescue	13	297–298
	Psychological effects of operating in obscured conditions and ways to manage them	19	546
	Methods to determine if an area is tenable	19	595
	Primary and secondary search techniques	19	596–601
	Team members' roles and goals	19	598–600
	Methods to use and indicators of finding victims	19	604–607
	Victim removal methods (including various carries)	19	604–607
	Considerations related to respiratory protection	17, 19	546, 587

JPR	FFI Requisite Knowledge	Chapter	Page(s)
4.3.10	Principles of fire streams	15, 16	388–390, 443, 518–521
	Types, design, operation, nozzle pressure effects, and flow capabilities of nozzles	16	445–449
	Precautions to be followed when advancing hose lines to a fire	16	458
	Observable results that a fire stream has been properly applied	16	458–467
	Dangerous building conditions created by fire	7, 20	111–112, 614–617
	Principles of exposure protection	16, 20	471, 643–644
	Potential long-term consequences of exposure to products of combustion	10	179–182
	Physical states of matter in which fuels are found	5	77–79
	Common types of accidents or injuries and their causes	17	483–490
	Application of each size and type of attack line	15, 16	406–407, 449–452
	Role of the backup team in fire attack situations	16, 20, 30	466–467, 629–630, 612–613
	Attack and control techniques for grade level and above and below grade levels	17, 20, 27	516–517, 651–653, 797–798, 799–800
	Exposing hidden fires	20, 21	644–645, 680
4.3.11	Principles, advantages, limitations and effects of horizontal, mechanical, and hydraulic ventilation	14, 16, 20	324–325, 470–471, 641–643
	Safety considerations when venting a structure	14	327
	Fire behavior in a structure	5, 14	86–90, 327–332
	Products of combustion found in a structure fire	10, 14	184, 318–319, 327, 365
	Signs, causes, effects, and prevention of backdrafts	5	88, 90
	Relationship of oxygen concentration to life safety and fire growth	5, 14	76–90, 324–330
4.3.12	Methods of heat transfer	5	83–85
	Principles of thermal layering within a structure on fire	5, 14, 20	87, 330–332, 639–641
	Techniques and safety precautions for venting flat roofs, pitched roofs, and basements	27	797–801
	Basic indicators of potential collapse or roof failure	7	131–133
	Effects of construction type and elapsed time under fire conditions on structure integrity	7, 14	112, 135–137, 342–343
	Advantages and disadvantages of vertical and trench/strip ventilation	14	334–335
4.3.13	Types of fire attack lines and water application devices most effective for overhaul	16	472
	Water application methods for extinguishment that limit water damage	16	472
	Types of tools and methods used to expose hidden fire	21	675–676
	Dangers associated with overhaul	21	680–682
	Obvious signs of area of origin or signs of arson	21	682–685
	Reasons for protection of fire scene	21	686–687
4.3.14	Purpose of property conservation and its value to the public	21	674–675
	Methods used to protect property	21	656–657
	Types of and uses for salvage covers	21	676–677
	Operations at properties protected with automatic sprinklers	21	679–680
	How to stop the flow of water from an automatic sprinkler head	21	679–680
	Identification of the main control valve on an automatic sprinkler system	21	674–675
	Forcible entry issues related to salvage	21	680
	Procedures for protecting possible areas of origin and potential evidence	21	686–687

JPR	FFI Requisite Knowledge	Chapter	Page(s)
4.3.15	Loading and off-loading procedures for mobile water supply apparatus	15	398–403
	Fire hydrant operation	15	394–398
	Suitable static water supply sources, procedures, and protocol for connecting to various water sources	15	398–400
4.3.16	Classifications of fire	6	90
	Types of, rating systems for, and risks associated with each class of fire	6	94–98
	Operating methods of and limitations of portable extinguishers	6	98–103
4.3.17	Safety principles and practices	17	505
	Power supply capacity and limitations	17	505
	Light deployment methods	17	505–506
4.3.18	Properties, principles, and safety concerns for electricity, gas, and water systems	17	508–509
	Utility disconnect methods and associated dangers	17	508
	Use of required safety equipment	17	510
4.3.19	Types of ground cover fires	20	665–667
	Parts of ground cover fires	20	665–667
	Methods to contain or suppress	20	665–667
	Safety principles and practices	20	665–667
4.3.20	Knot types and usage	8	141–144
	Difference between life safety and utility rope	8, 14	142–148, 353
	Reasons for placing rope out of service	8	140–149
	Types of knots to use for given tools, ropes, or situations	8	145–148
	Hoisting methods for tools and equipment	8	144–147
	Using rope to support response activities	8, 13, 14	145–151, 272, 329, 347
4.5.1	Types of cleaning methods for various tools and equipment	8, 9, 10, 11, 13	148–150, 171–172, 200, 215–216, 223–224, 269
	Correct use of cleaning solvents	11	215–216
	Manufacturer's or departmental guidelines for cleaning equipment and tools	11, 20	216, 223, 646
4.5.2	Departmental procedures for noting defective hose and removing from service	15	420–421
	Cleaning methods	15	420–421
	Hose rolls and loads	15	426–431

JPR	FFII Requisite Knowledge	Chapter	Page(s)
5.1.1	Responsibilities of the FF II in assuming and transferring command within an incident management system	23, 27	717–725, 726–728, 835–836
	Performing assigned duties in conformance with applicable NFPA and other safety regulations and AHJ procedures	17, 27	512, 834
	Role of FF II within the organization	23, 27, 31	729, 837, 906
5.2.1	Content requirements for basic incident reports		
	Purpose and usefulness of accurate reports	24	736–738
	Consequences of inaccurate reports	24	736–738
	How to obtain necessary information	24	736–738
	Required coding procedures	24	736–738
5.2.2	SOPs for alarm assignments	24	740
	Fire department radio communication procedures	24	738–740
5.3.1	Methods by which foam prevents or controls a hazard	27	931–940
	Principles by which foam is generated	27	821–830
	Causes for poor foam generation and corrective measures	27	827
	Difference between hydrocarbon and polar solvent fuels and the concentrates that work on each	27	822–824
	Characteristics, uses, and limitations of fire-fighting foams	27	821–823
	Advantages and disadvantages of using fog nozzles versus foam nozzles for foam application	27	937–938
	Foam stream application techniques	27	821–830
	Hazards associated with foam usage	27	829–830
	Methods to reduce or avoid hazards	27	829–830
5.3.2	Selection of the nozzle and hose for fire attack, given different fire situations	16	472–475
	Selection of adapters and appliances to be used for specific fireground situations	16	472–475
	Dangerous building conditions created by fire and fire suppression activities	27	906–931
	Indicators of building collapse	7, 27	112, 797–798
	Effects of fire and fire suppression activities on wood, masonry (brick, block, stone), cast iron, steel, reinforced concrete, gypsum wallboard, glass, and plaster on lath	7	112
	Search and rescue and ventilation procedures	19, 27	577–591, 797–800
	Indicators of structural instability	7	131–132
	Suppression approaches and practices for various types of structural fires	16, 27	452–565, 797–813
	Association between specific tools and special forcible entry needs	12	235–238
5.3.3	Characteristics of pressurized flammable gases	27	813–821
	Elements of a gas cylinder	27	817–818
	Effects of heat and pressure on closed cylinders	27	818–821
	Boiling liquid expanding vapor explosion (BLEVE) signs and effects	27	812
	How to identify safe havens before approaching flammable gas cylinder fires	27	819
	Water stream usage and demands for pressurized cylinder fires	27	819–821
	What to do if the fire is prematurely extinguished	27	819–821
	Valve types and their operation	27	813–821
	Alternative actions related to various hazards	27	807, 811–812
	When to retreat	27	819–829
5.3.4	Methods to assess origin and cause	28	839–841

JPR	FFII Requisite Knowledge	Chapter	Page(s)
	Types of evidence	28	839
	Means to protect various types of evidence	28	840–841
	Role and relationship of FF IIs, criminal investigators, and insurance investigators in fire investigations	28	834
	Effects and problems associated with removing property or evidence from the scene	28	835, 840
5.4.1	Fire department's role at a vehicle accident	30	867–868
	Points of strength and weakness in auto body construction	30	868–874
	Dangers associated with vehicle components and systems	30	872–874
	Uses and limitations of hand and power extrication equipment	30	875–880
	Safety procedures when using various types of extrication equipment	30	879–880
5.4.2	Firefighter's role at a technical rescue operation	31	904
	Hazards associated with technical rescue operations	31	904–957
	Types and uses for rescue tools	31	904–947
	Rescue practices and goals	31	904–957
5.5.1	Organizational policy and procedures	29	846–849
	Common causes of fire and their prevention	29	960–961
	Importance of a fire safety survey and public fire education programs to fire department public relations and the community	29	850–857
	Referral procedures	29	863
5.5.2	Parts of informational materials and how to use them	29	858
	Basic presentation skills	29	863–864
	Departmental standard operating procedures for giving fire station tours	29	862
5.5.3	Sources of water supply for fire protection	25	753–756
	Fundamentals of fire suppression and detection systems	26	775–785
	Common symbols used in diagramming construction features, utilities, hazards, and fire protection systems	25, 26	753–756, 768–775, 789–790
	Departmental requirements for a preincident survey and form completion	25	749
	Importance of accurate diagrams	25	749
5.5.4	Types of cleaning methods for various tools and equipment	8, 9, 10, 11, 13	148–150, 171–172, 200, 215–216, 223, 269
	Correct use of cleaning solvents	11	216
	Manufacturer and departmental guidelines for maintaining equipment and its documentation	11	215–216
	Problem-reporting practices	30	899
5.5.5	Procedures for safely conducting hose service testing	15	421–424
	Indicators that dictate any hose be removed from service	15	421–424
	Recording procedures for hose test results	15	421–424

FIREFIGHTER I

The Traditions and Mission of the Fire Service

by Rick Lasky

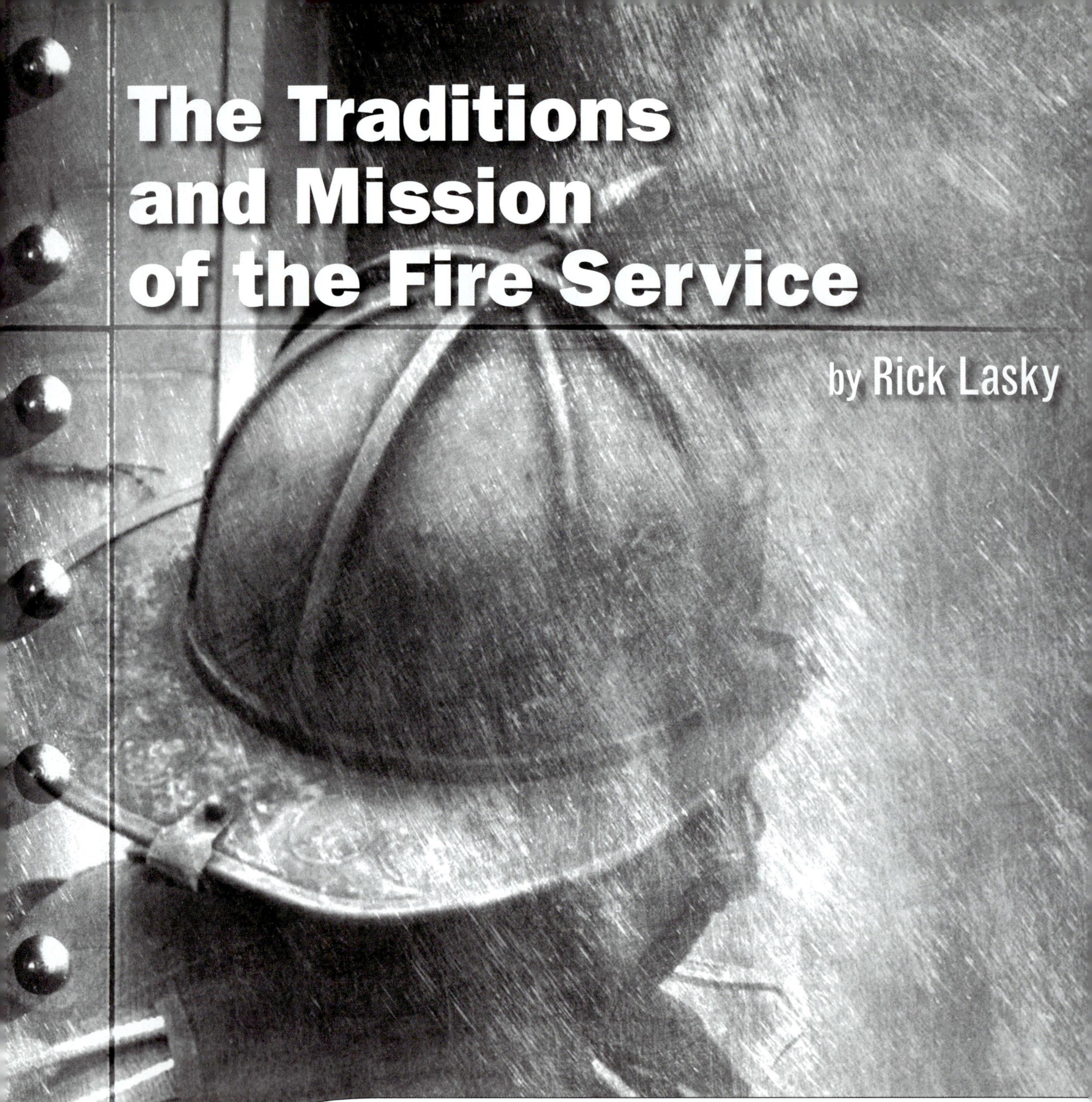

This chapter provides required knowledge items for the following NFPA Standard 1001 Job Performance Requirements:

FFI 4.1.1

OBJECTIVES

Upon completion of this chapter, you should be able to do the following:

- Identify important historical figures and their relation to the fire service
- Describe what it means to be a firefighter
- List at least three fire service-related traditions
- Understand your commitment to being a firefighter

INTRODUCTION

When you look at today's fire service—our apparatus, tools, equipment, and technology—it's hard to imagine how it was way back when. Look at pictures from the olden days or read history books and you'll be amazed at what firefighters used. Considering that they had to carry almost everything to fires, including the first fire pump, we have made amazing advances. We've moved from fire buckets and **bucket brigades** to firefighters pulling their engines, hose tenders, or ladder tenders behind them, to horses pulling steam fire engines (fig. 1–1). Back then, more than just prestige came with saying one was a volunteer or paid firefighter. Firefighters knew they were making a commitment, changing their lifestyle, and joining a family. There was nothing like it.

Fig. 1–1. Horse-drawn steam fire engines were used from the 1860s to the 1920s.

EARLY VOLUNTEERS

There was a time when most volunteer firefighters were socially established or of affluence. Famous volunteer firefighters included George Washington, Thomas Jefferson, even Benedict Arnold. Firefighters were the *who's who* of the community, anyone of importance, and especially those who wanted to climb politically or socially. Let's not forget Benjamin Franklin, often called the Father of the American Fire Service (fig. 1–2). Franklin was born on January 17, 1706. He was the first American diplomat, an ambassador, storyteller, published scientist, postmaster general, inventor, and economist. He organized the first lending library, medical center, scientific society, philosophical society, volunteer fire department, and fire insurance company. He invented the lightning rod, bifocals, the static electric generator, and the Franklin stove. He was the only American founding father who signed all four major documents that established American independence: the Declaration of Independence, the treaty with France, the treaty with England, and the Constitution. (Punishment for signing these documents if caught by the British was hanging.) Showing bravery, intelligence, wit, creativity, and a will to make the world better, Franklin exhibited the personality traits of a firefighter.

Fig. 1–2. Benjamin Franklin, one of our most revered founding fathers, created America's first fire company.

GLORY DAYS

There was a glamour associated with the fire service. How one's firehouse looked, the detail given to a fire engine, and artwork associated with one's speaking trumpet or company crest were all areas of pride (fig. 1–3). **Musters** and contests determined who had the best fire company, who could get water first, and whose fire stream could go the farthest. All created exciting competition, but sometimes tussles between competing fire companies resulted. Nevertheless, whenever the bell rang, a runner or torchbearer shouted as firefighters ran down the street pulling their equipment to a fire. Residents, business owners, and children gave in to the excitement of watching their firefighters.

Fig. 1–3. The first fire trumpet, or speaking horn, was introduced in 1749 in New York City for fire department officers to amplify their voices. The first "working" trumpets (used at fire scenes) were made of toleware and later made of brass. Presentation or "parade" trumpets were nickel- or silver-plated (as seen in this photograph) or even solid silver or glass. By the 1800s they were only used for ceremonies. The use of trumpets as the insignia of rank is one of our oldest traditions.

With time, the prestige placed upon the fire service by the public began to splinter and fragment. The gallantry of horse-drawn fire engines racing down cobblestone streets, their bells ringing, and the seriousness on firefighters' faces meant a little less to those of high society. Many still held the fire department in high regard, but growing numbers considered firefighting a profession for the poor or those in need of a job. Firefighter and police officer wages were some of the lowest.

BIRTH OF THE MODERN FIRE SERVICE

So began the filling of firefighting positions by people who were not looking for financial gains but rather community service. Those who volunteered at local fire departments soon realized their commitments. When the bell rang, a firefighter was expected to stop everything to help others. Those who joined the fire service to provide a family income realized that a regular paycheck was not necessarily a big paycheck. The fire service has historically fought for every penny to run operations and, in some cases, survive. Firefighters had to be creative in budgeting and acquiring funds to protect those they swore to serve.

> **FIREGROUND NOTE**
>
> Our mission is very clear: to be there for people when they really need us.

As a city's population grew, so did the need to properly and safely protect those who lived there. Civilian and business populations grew steadily, but fire department budgets remained the same or decreased. North American fire departments often found themselves fighting for better equipment and more staffing—a fight we still wage. One case was in New York City. Chief Edward F. Croker served as chief of the Fire Department of the City of New York from 1899 to 1911 (fig. 1–4). When faced with budget and manpower cuts, Croker wrote and read the following statement to the city powers:

> *I have no ambition in the world but one, and that is to be a fireman. The position may, in the eyes of some, appear to be a lowly one; but we who know the work which a fireman has to do believe that his is a noble calling. There is an adage which states that 'Nothing can be destroyed except by fire.' We strive to preserve from destruction the wealth of the world, which is the product of the industry of men, necessary for the comfort of both the rich and the poor. We are the defenders from fire, of the art, which has beautified the world, the product of the genius of men and the means of refinement of mankind. But, above all, our proudest endeavor is to save lives of men—the work of God himself. Under the impulse of such thoughts, the nobility of the occupation thrills us and stimulates us to deeds of daring, even at the supreme sacrifice. Such considerations may not strike the average mind, but they are sufficient to fill to the limit our ambition in life and to make us serve the general purpose of human society.*

Fig. 1–4. Chief Edward F. Croker

That statement, though dated, still serves to define the passion associated with being a firefighter. "I have no ambition in the world but one, and that is to be a fireman." Man, oh man, that is the kind of desire you need in your chest to be good at what we do. This job isn't for everybody. *It takes someone special to be a firefighter.*

WHAT IS THE FIRE SERVICE?

FFI 4.1.1 But what do the words "fire service" mean to you? What does the word "firefighter" mean? The difference in definitions between a civilian and a firefighter might surprise you. Many times civilians ask us what we do, and often we explain about fighting fires, rescuing people, etc. But what is it really all about? How did we get here? First, if you're not willing to examine our heritage, nothing else about the fire service will make sense to you. That's the only way to identify what we're all about. But the fire service has evolved into more than fighting fires. We're trained in emergency medical services, hazardous materials, specialized rescue, weapons of mass destruction, and just about everything else you can come up with if it involves helping people. We have become journeymen of public service.

The fire service is based on something bigger than any of us realized when we entered our careers as firefighters. Whether you'll serve as a volunteer, paid-on-call, or paid firefighter, you'll realize that it takes a special person with special qualities to do what we do. This person must dig deeply into our history to research where we began and how we got here. This person must be willing to protect our heritage, defend our fire service family, and bring the honor that the job deserves. (See fig. 1–5.)

Fig. 1–5. Christian Michael Otto Regenhard, 28, was a probationary firefighter on September 11th, 2001 and was one of the 343 members of the FDNY killed at the World Trade Center. Pictured here are his mother, father, and sister at his memorial service held in St. Patrick's Cathedral. As a legacy to Christian, the Regenhard family created the "Skyscraper Safety Campaign" which advocates for safer high-rises through better building codes as well as improved emergency responder safety and communications. Will you study the 9/11 disaster and other emergency responses where firefighters have made the supreme sacrifice, and will you learn from them? (Courtesy of Susan Watts/New York Daily News)

WHAT DOES IT TAKE TO BE A FIREFIGHTER?

So what does it take to be a firefighter? What kind of person? It takes a person who lives within a system of values built upon family. Everything we do in the fire service has to do with family, whether it's protecting someone's home, children, belongings, or business. Every time we answer a call, it has to do with family. Now, that can be a slight problem for those who don't live by a set of values. These are the people who constantly seem to be banging into walls. They blame others and have excuses for everything. Fortunately for the fire service, it doesn't take long for these rarities to realize that family is the foundation of our success. They support it or move on. To expand on that success, you must expand your value system to include pride associated with ownership, honor bound to loyalty, and integrity, which requires you to be honest and a person of character. Once you grab these values, good things start to happen.

As a rookie firefighter, you must understand that you create your own legacy from the moment you walk into the firehouse. It's hard to make a second first impression. Ideally, you've led a good life so far and haven't muddied your resume by doing something that "seemed like the thing to do" when you were young. Ideally, you've been hanging with the right people. Constantly ask yourself, "Why did I get into the fire service, and why did I want to be a firefighter?" If your answers are to help people, to make a difference, to be part of a special family, then you're on the right track. Remember three things to be successful:

1. Do what's right.
2. Do your best.
3. Treat others the way you want to be treated.

Doing what's right means being a person of integrity. *Doing your best* means giving the job the effort it deserves and remembering why you chose this profession. *Treating people the way you would want to be treated* means treating everyone like family. People are watching. You might think otherwise, but they are. Show your fire department that you will do what it takes to be part of their team and family. Prove that they made the right choice in bringing you on as a volunteer or paid firefighter. Never lose that vision or passion.

You should have one other passion: **safety**. Make no mistake about it—firefighting is a dangerous profession. While vast improvements in equipment and procedures have been made over the last several years, injuries and death are still a part of the business. It is up to you to practice safe habits and to look out for the safety of others. Safety is the job of *every* firefighter.

FIRE SERVICE TRADITIONS

Some fire service traditions go back to its birth in North America. The traditions we must examine created passion for the fire service. Wetting down a new apparatus, pushing out an old rig and pushing in a new one, and opening a new firehouse all have importance. The swearing in and badge pinnings are just a start. It's the first step. Company logos, insignias, and tools all require understanding of their needs and importance.

How about our uniform? Do you know what it represents? It's disappointing when a fellow firefighter at a funeral says, "It's a shame. It seems we only wear our **Class A uniforms** for funerals." My usual reply is, "You're right. It is a shame . . . shame on you. That uniform is not just a funeral uniform. It represents our profession, our heritage. It stands as a symbol for what we are all about and for those who bled, sweat, and went before us. It stands for something bigger than we are. It represents something special. It's a uniform of which few are unfamiliar. When someone sees you standing straight and proud in that uniform, he or she knows you are firefighter, one of America's bravest" (fig. 1–6).

Fig. 1–6. The Class A uniform represents our profession, heritage, and all who came before us.

WHERE DO YOU GO FROM HERE?

So now what? Where do you go from here? Now that you're a new firefighter, what's your next step? You are taking that first step in climbing the mountain we call the fire service. It's full of challenges, tasks, and experiences. But each step comes with a view so awesome that, at times, it's indescribable. Sometimes that feeling will be challenged. Now and then, it will come from someone close to you in your own firehouse.

While addressing a rookie or probationary school academy graduation, I often have the candidates sit right in the front row. I ask them to close their eyes and feel that feeling of uncontrollable excitement, the one that fills them with energy. After all that time spent on classroom work, the drill field, and the physical fitness, it's graduation day. "Where will I go next?" "Where will I go to work?" "Which firehouse will I be assigned?" "Which rig?" "The engine, the ladder truck, a rescue, or the ambulance?" "Who will be my partner, and who will be my officer?" The excitement is overwhelming. You can feel it throughout your body.

Then I ask them to flash forward about two years when they're sitting at the firehouse, around the kitchen table, or on drill night complaining about the chief or their pay. *What happened to them? What changed? Or maybe the better question is whom are they hanging around? What happened to their passion? What happened to their love for the job? Who poisoned that feeling?* There's one thing I've never understood about firefighters. There is not a firefighter alive who would let a burglar break into his or her home to steal things. There isn't a firefighter who would let somebody break into his or her car and steal from them without wanting the police to arrest, prosecute, and hang the person by his toenails. But firefighters let somebody reach into their hearts and steal their passion. When I was going through such a time, a friend told me that I was being weak and letting the other side win. He said people who are weak let other people steal from them. Never allow anyone to reach into your heart and steal your passion about this wonderful profession. Defend it with everything you have. Don't ever let go of that feeling.

WHAT NEXT?

As you work through this first phase of your firefighting career, ask yourself the following questions. Allow them to guide you into and through this journey.

1. Are you ready for the challenge?
2. If you are, are you ready for the reward the challenge brings?
3. Are you ready to protect what was accomplished by those before us to make the fire service great?
4. Are you ready to defend and protect our fire service family?
5. Will you remember that you represent all of us, in and out of the firehouse, on and off duty? Whether you wear a Maltese Cross on your shirt or not, people will know that you are one of us and base their opinions of all of us by your actions. They will expect you to be a professional. Do not tarnish our image.
6. Are you willing to commit to a lifelong pursuit of continuing education, hands-on training, and mentoring those who come after you with that knowledge and those skills? Will you show them the way?
7. Will you commit to memory that "Never forgetting means never forgetting"?
8. Can you commit to a life led by values such as pride, honor, and integrity and allow those to be served by a foundation on family values?
9. Will you remember to treat retired firefighters with honor and respect, remember what they did for us, and always make them feel welcome in your firehouse? They earned it.
10. Will you train as if your life depends on it? It does.
11. Will you be a brother or sister to our fire service family? It takes a 24-hour, seven-day commitment. It's not a part-time thing.
12. Will you remember what you owe the public, and that the public doesn't owe you jack? When you raised your right hand and took an oath, pinned on that badge, or put on that fire helmet, you became a public servant.
13. Will you develop a list of mentors, people who will steer, train, and develop you, and get you fired up about the job?
14. Will you treat your firehouse like it is your home?
15. Will you take care of your apparatus, tools, and equipment? Remember, it's not just an image thing. Think about it.
16. Will you brag about the fire service both off duty and on duty?
17. Will you be that person of integrity and of character?
18. Will you show up on time for your shift or drill night? It says a lot about you and whether we can count on you. And if you're sick, stay home. If you're not, come to work!
19. Will you wear your uniform and be proud of it? It stands for a lot more than any one person. It represents our heritage.

20. Will you celebrate and participate enthusiastically in the ceremonies and traditions that promote and protect our fire service?
21. Will you always remain physically and mentally prepared to do whatever it takes?
22. Will you become a student of our fire service history and protect our heritage?
23. Will you adhere to the following rules to combat complacency?
    a. Every time we go out the door, we're going to a fire. No matter what, never let down your guard and always be prepared for the fight.
    b. When we arrive on the scene, there is no fire only if *we* say there is no fire. Listen to civilians, but don't allow yourself to be snookered or fooled.
    c. There is nobody in the building only if *we* say no one is in the building. Even when occupants are standing outside, we still search the building for fire and occupants who might be unaccounted for or overlooked.
    d. The fire is not out unless *we* say it's out.
24. Will you build relationships within your community, city, or township?
25. If you become a company officer, will you lead with distinction, remember it takes courage to lead, and make decisions because they are right, not popular? Popular decisions hurt and kill firefighters. Be their leader first, then be their buddy.
26. If you make it to chief, will you always remember where you came from?
27. Will you learn from the past and study line-of-duty deaths so history won't repeat itself?
28. Will you leave it a little better for the next guy?
29. Will you profess that the fire service is the greatest profession in the world?
30. Will you be as prepared as possible and train, train, train?
31. Will you think safety, safety, safety?
32. Will you do everything you can so everyone goes home?

Will you agree to live up to the high standards of being a firefighter?

If you can't or don't think you can make that commitment, then don't. It's probably time for you to make a different choice. Not everyone can be a firefighter. If you want the kind of life that is dedicated to protecting men, women, children, and their possessions, then welcome, brother or sister!

Good luck, be safe, and enjoy the journey.

LESSON FROM THE FIREGROUND

My love for the job came from people such as Chief Jack MacCastland who taught my Firefighter I class when I was 18 years old. Mac said, "You want to be a good firefighter? You need to know building construction and fire behavior. You have to know how the building is going to react with the fire and how the fire is going to react with the building. Anyone can go out there and chop, but you still need to know building construction and fire behavior." How very true. The first time I saw Mac was at a fire. Here was this big guy, leather helmet all bent up in the front, pulling drywall with a pike pole in one hand and using his chin to open and close the nozzle with the other hand. I said, "Man, I want to be like this guy."

Then there was Chief Eddy Enright who taught me about caring for and respecting your guys. He'd always say, "Look for their positives, Rick. Anyone can pick out the negatives. You do that and the negatives tend to go away. Catch them doing something right. It's easy to catch them doing something wrong."

One of my best friends, Chief Tom Freeman, one of the smartest firefighters I've ever met, also shared his knowledge with me. He said things such as, "A good officer or incident commander is the one who can predict his next alarm. Any mope in a white helmet can stand outside and handle what he's got right now and burn it to the ground, but it's the guy who can predict his next alarm, knows when he needs more resources before he runs out, where the fire's going, and can think out of that box who does well."

MY POINT

Perhaps the most important thing my mentors taught me was that to be a good firefighter you must have core values such as pride, honor, and integrity. And every last bit of it starts with integrity.

Rick Lasky

QUESTIONS

1. ____________________ is often called the Father of the American Fire Service.
2. What are some of the areas that firefighters need to be trained in to be proficient at their jobs?
3. Throughout your career as a firefighter, you should remember three things to be successful. Name them.
4. When should we wear Class A uniforms?
5. Is everyone cut out to be a firefighter? What questions can we ask ourselves to be sure we are ready to be firefighters?

Fire Service History

by Don Cannon, updated by Glenn Corbett

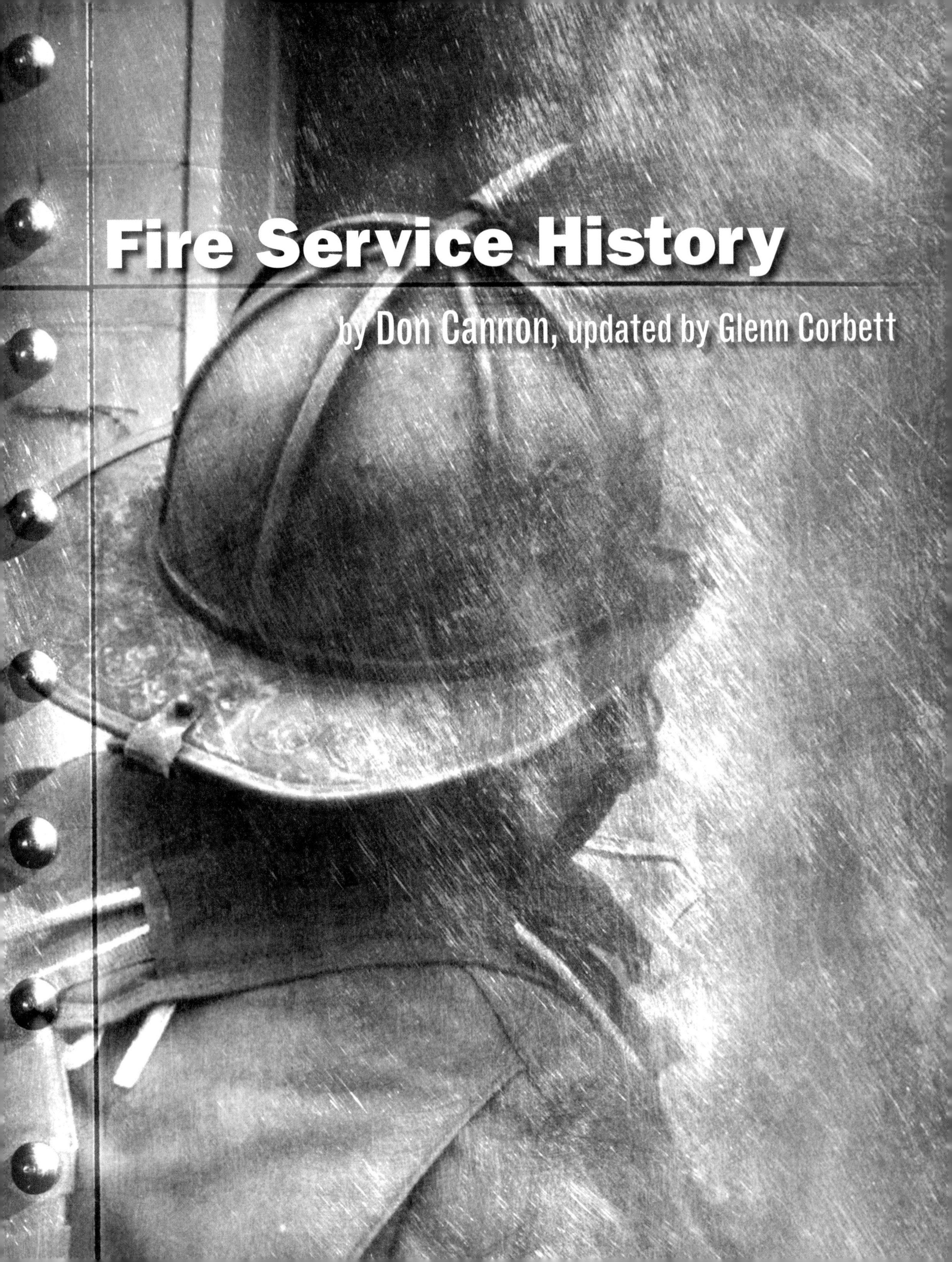

OBJECTIVES

Upon completion of this chapter, you should be able to do the following:

- Identify the origins of the fire service from the 5th century A.D. through the 14th century A.D.
- Identify major events that impacted the fire service during the 18th century
- Identify major events that impacted the fire service during the 19th century
- Identify major events that impacted the fire service during the 20th century
- Identify major events that are impacting the fire service during the 21st century

INTRODUCTION

The ancient Greeks recognized that the essential forces that act upon our physical world are fire, air, earth, and water. And thus mankind's ability to control and direct each of those factors has determined the success or failure of communities since they were first established.

5TH CENTURY A.D. TO 14TH CENTURY A.D.

While there were many cities in the ancient world of India and Mesopotamia and the classical world of the Greeks, Rome provided us with the first model of a major urban society in the West. Its population of more than a million made it the largest city in the West. The Emperor Augustus (63 B.C.–14 A.D.) is credited with transforming Rome from a city of brick into a city of marble by sponsoring a planned infrastructure of buildings, monuments, roads, aqueducts, and sewers. He is also credited with creating the **Vigiles**, the watchmen of the city and first public force of firefighters (fig. 2–1). Equipped with primitive though familiar apparatus and tools, the Vigiles coped with dozens of fires a day and fought several conflagrations, the most notable of which occurred in 68 A.D. during the reign of the Emperor Nero. With the collapse of the Empire, the cities of Rome declined as well and the Vigiles passed into history. Not until the age of Charles Dickens's 19th-century London would there be another city of similar size in the West. Inevitably for centuries thereafter, fire continued to ravage Europe's surviving small towns and settlements during the Middle Ages.

Fig. 2–1. The Vigiles of Rome were the first organized group of firefighters in recorded world history. Besides buckets of water, they actually used a primitive pump to apply water to a burning building. They also used hooks to tear down buildings, creating fire breaks—a technique that was subsequently used for more than 1,000 years.

15TH CENTURY A.D. TO 18TH CENTURY A.D.

The emergence of the modern world in the 16th century gave new impetus to the idea of commercial expansion. Great imperial cities began to grow again and connect to overseas settlements in every hemisphere. Colonial American towns were the offspring of those forces and each became subject to the same opportunities and dangers as its sponsoring community in the old world. Within a few years of the establishment of Jamestown, Plymouth, and Boston in the 17th century, the residents of each town considered fleeing back to England after they encountered midwinter fires. Adriaen Block, the Dutch explorer and trader, never intended to establish the first settlement in the area that was to become known as New Amsterdam when he and his crew were forced to spend the winter at the foot of Manhattan. Their ship the *Tyger* burned to the waterline late in 1613, a dozen years before Fort Amsterdam was constructed in the same area. Quickly, Peter Stuyvesant, governor of Nieuw Amsterdam, devoted himself to fire prevention by appointing a night patrol known as "the prowlers" or **Rattle Watch** (1648) and fire wardens to inspect chimneys, thatched roofs, and household cooking fires. Before the English took over the town in 1664, Stuyvesant required every householder to have buckets at hand and posted hooks and ladders for use by civilians in case of fire.

The **Great Fire of London**, which destroyed almost 80% of that city in 1666, exerted a direct influence on William Penn's original design for the city of Philadelphia. The wide streets, large tracts of land, and squares were intended to make the City of Brotherly Love a "*greene country towne . . . which will never be burnt, and always be wholesome.*" Philadelphia escaped a major conflagration. Other towns were not so lucky. Charleston, South Carolina became known as a "city of fires." In Boston, one blaze after another, many suspected as arson, destroyed more than 100 buildings during a three-year period. Finally, capped in 1678 by the loss of "fourscore of thy . . . dwelling houses and seventy of thy warehouses in a ruinous heap," Boston imported fire engines from England and appointed men to operate them.

Articles in Benjamin Franklin's newspaper, the ***Philadelphia Gazette***, led to the forming of America's first volunteer fire brigade, the Union Fire Company (1736). A year later in 1737, six years after two London-built **Newsham hand pumpers** arrived in New York from London, 30 "strong, able, discreet, honest, and sober men" were appointed the first firemen in New York City (fig. 2–2). In a parallel, private-sector effort to cushion the economic impact of fire losses, The Friendly Society of Charleston (1736) and the Philadelphia Contributionship (1752), America's first fire insurance companies were organized. Beginning in the 1750s, first in Philadelphia, insurance companies began to display **fire marks** on structures covered by them. Fire marks were wood or iron plaques with a unique design identifying the insurer of a building. They were initially displayed as advertisements that deterred vandals who feared being tracked down, and they encouraged fire companies to work hard in hopes of a future contribution in the annual distribution of awards by the various insurance companies (fig. 2–3).

Fig. 2–2. One of two Newsham engines New York City imported in 1731 from London, now on display in the American Museum of Firefighting in Hudson, New York. The two wooden handles on the sides of the engine—called "brakes"—were lifted up and down to pump the engine.

Fig. 2–3. Fire marks were used as forms of advertisement. This one of the mid-19th century is from an insurance company called the Fire Association of Philadelphia, and portrays a wooden hydrant and coil of hose.

One myth historians have recently laid to rest is that early American firefighters would only work on structures where fire marks were clearly displayed. As the myth goes, a purse or reward was associated with getting first water on such an insured property. It was said, falsely, that fire companies would fight each other to prevent later arriving companies from getting such a reward, even as in some cases the fire building burned to the ground. American and British historians have conclusively proved that fire brigades provided services to all who were in need, regardless of insurance. On the one hand, this was a practical matter intended to prevent extension of fire to their client properties. On the other hand, services to the uninsured were provided out of goodwill and for the positive effect it had on a company's image. In other words, fire marks were primarily posted as advertising only. Insurance companies did contribute annually to various fire companies, in some cases supplying them with needed equipment to sustain their operations, but contributions for specific fires were rare.

Arson—intentionally set fires—became a serious problem in 18th century New York. On two occasions, New York suffered a rash of arson fires attributed to discontented slaves and their sympathizers. In each case, the town's new fire companies responded. The first, in 1712, was put down quickly. In 1741, the New York Negro Plot hysteria in the city, partially triggered by a series of fires of unknown origin, culminated in the Fort George conflagration at the lower end of Manhattan. As a result, 17 blacks and 4 whites were hanged, and 13 slaves were burned at the stake. By 1790, New York City had the highest percentage of black residents other than Charleston.

Many volunteer engine companies were organized prior to the American Revolution. Shortly before the Revolution began, American inventors began to manufacture and sell their own hand pumpers. This was a clear indication that the colonial economic system was becoming less dependent upon Britain. The various trading and manufacturing restrictions imposed by London between 1763 and 1775 became increasingly viewed by merchants, mainly in the coastal cities, as an effort to strangle the emerging American economy. In the 10 years prior to the Declaration of Independence, many colonial firefighters active as Sons of Liberty seized and destroyed ships in Newport and Providence in opposition to the British Stamp Tax and trade policies. Firemen participated in the Boston Tea Party and tossed and burned cargoes of tea in Annapolis, Maryland, Greenwich, New Jersey, and Providence in protest against a British-imposed monopoly. In spring 1770, in what has become known as The Boston Massacre, a cry of "Fire!" intended to cause the town's church and fire bells to bring others to the scene caused British troops to open fire on an unarmed group of colonial protesters, killing several.

Real shooting began in April 1775. Fire was a cruel weapon. British forces withdrawing from skirmishes at Lexington and Concord lit up houses as they retreated back to Boston. In May, 30 buildings were set afire in the dock district of Boston. In mid-June 1775, British warships set fire to 380 dwellings in Charlestown, Massachusetts, as the Battle of Bunker Hill proceeded. Beginning in fall 1775, moving down the coast from Falmouth (now Portland), Maine, continuing south to Norfolk, Virginia, and ultimately to Charleston, the British leveled towns by naval bombardment, shelling from the seaside as shore parties set waterfront structures afire. The Revolutionaries were not guilt-free, for in the winter of 1775–76 as they withdrew from their failed invasion of Quebec, burning homes, barns, and mills along the way.

The American Revolution was a brutal war, all around. In his Declaration of Independence, Thomas Jefferson condemned King George III for using fire as a terror weapon, writing: "He has plundered our seas, ravaged our coasts, burnt our towns, and destroyed the lives of our people." Jefferson's words did not deter the redcoats. In fall 1777, the British set fire to every house in Kingston, New York, as punishment for the citizens' support of the Revolution. In late September 1776, less than a week after the British occupied Manhattan, fire originating in the Fighting Cocks Tavern on Whitehall Street and several other locations in the lower west side of the island ultimately destroyed almost 500 buildings, a quarter of the city. Revolutionary sympathizers, including, as many believe, Nathan Hale ("I have but one life to lose for my country . . ."), are suspected of torching the Great Fire of 1776. At this point, New York City emptied, becoming a virtual ghost town for the remainder of the war. Historians are uncertain whether Jacob Stoutenburgh, the chief of department, and the rest of his volunteer firemen in New York left the city to join Washington's forces.

In 1778, a second Great Fire in Manhattan consumed another fifty buildings. Remaining Tory sympathizers and British naval personnel fought the fire. A year later, the British captured and burned Norfolk. Wherever they went, the Brits were ruthless adversaries. Mel Gibson's film *The Patriot* gets pretty close to the reality of the Revolution, as backcountry Americans were subject to terrifying experiences throughout the war. Frontier farms and communities all along the ridgeline of the Appalachians from the Mohawk Valley in New York to the Wyoming Valley in Pennsylvania, down through the Carolinas were lit up as Tories and their American Indian allies burned more than 1,000 homes and slew 300 settlers.

Finally, it was over. The Treaty of Paris recognized American independence in 1783. The Continental Army marched back into previously occupied territory, survivors staggered out of their prison ships, citizens returned to the cities from their rural refuges, and local fire departments began to reorganize themselves under the charters of their newly independent states.

In the early years following independence, towns in the new West were commonly overcrowded with shoddily built housing and mercantile facilities. Seven-eighths of New Orleans, the export point for everything being sent down the Mississippi River, was destroyed by fire on Good Friday 1788. A second conflagration seven years later burned out an additional 200 structures in New Orleans's French Quarter. In 1805, the entire village of Detroit was virtually wiped out, save for one building. Older towns were being transformed into industrial centers and as a result experienced ever-larger fires. Paterson, New Jersey, America's first planned industrial city, saw its first mill destroyed by fire in 1807. Philadelphia was particularly hard hit in this period but soon undertook a massive water-supply project that became the model for other cities.

19TH CENTURY

During this period, efforts were made to make fire ground operations somewhat more organized as officers using **speaking trumpets** attempted to direct ladder companies, hand-pumped engines, and informal salvage operations. Hundreds of new volunteer fire companies were formed, many including hose companies (the first was introduced in Philadelphia in 1803) using American-designed **"gooseneck"-style, hand-pumped fire engines** with solid-metal play pipes much like modern-day deck guns. Some engine companies were equipped with more maneuverable leather fire hoses held together by copper rivets. In New York, the first primitive American fireboat, a hand-rowed barge on which hand-pumped Engine 42 was mounted, went into operation. During this period, Philadelphia and New York insurance companies organized the first paid fire patrol units to perform salvage work. Other cities followed. In 2006, following 203 years of service, the New York Fire Patrol became the last of all the insurance-supported salvage corps to operate in the United States.

Once again, during the War of 1812, America's War for Economic Independence, periodic raids caught border towns such as Buffalo in the crossfire. Early in the war, a dozen suspicious fires broke out in Washington, D.C., and American forces burned York, now Toronto, Canada. In retaliation, the British burned government buildings in Washington, D.C., and Baltimore in a campaign commemorated in our national anthem.

Following the war, the nation embarked on a period of explosive growth in the cities and across the continent. Augmented by the Erie Canal, which connected the interior of the country to the sea, the port of New York became the hub of the Atlantic trade, undergoing a period of frenzied building. **New York City's Great Fire of December 1835** consumed more than 650 buildings, including most of the Wall Street financial area and the lower east side wharf and dock district (fig. 2–4). The flames were so bright that mutual aid companies responded from as far as Philadelphia, 90 miles away. The 1776 and 1835 blazes wiped out most of old New York, making room for new construction. Unfortunately, completion of New York's Croton water project proved too late to contain the conflagration of 1835. Across the East River, Brooklyn had its great fire in 1848.

Fig. 2–4. This is a scene near Wall Street in New York City after the Great Fire of 1835. Although this fire wiped out much of New York's financial district, the area was quickly rebuilt.

New York Common Councilmen attempted to blame effective volunteer Chief James Gulick for the extent of the Great Fire and replaced him with a political appointee. In response, a significant number of volunteers simply quit and formed the **Resigned Firemen's Association**. This was the first time any American firefighter walked off the job. With Gulick's encouragement, however, most were convinced to return shortly thereafter. They ousted the anti-Gulick members of the Council and assured their own candidate, Cornelius Anderson, the position of chief of department.

Two years after the 1835 fire in New York, the entire nation fell into a deep financial depression. As was, and remains the case today, unemployment and idleness converted into civic unrest and a significant increase in runs and workers for firemen.

The frontier West went through its own growth pangs as settlements extended from the Mississippi Valley to the Pacific Slope. St. Louis, Missouri, lost 3 lives, 430 buildings, 23 steamers, 9 flat boats, and several barges in one of a series of major fires during the 1840s. Out on California's Gold Coast, a series of seven major fires swept San Francisco between 1849 and 1851, including one that destroyed 3,000 buildings. On April 19, 1863, only five years after its founding, much of Denver was destroyed. The day after, a **brick ordinance** was enacted requiring all new buildings be constructed of brick or stone.

The 1830s through the 1850s marked the golden years of the traditional pre-modern volunteer fire system. Coast to coast, from San Francisco's Pacific Fire Company to Chicago's Washington Volunteers, Boston's Cataract Engine, New York's Hibernia Hook and Ladder, to Protection Engine in Yonkers, New York, firehouses became social centers for working- and middle-class members. Parades, civic celebrations, balls, festivals, competitive hand-pumping contests, and a cycle of visits and conventions offered countless opportunities to showcase the enthusiasm, joy, pride, and devotion of America's volunteer fire companies before an adoring public. In most places, any person having served seven years of duty as a volunteer fireman was rewarded most tangibly by being made an **exempt fireman**, meaning he was forever excused from jury and militia duty. Such older life members of fire companies often banded together to form an exempt fire company, expected only to show up for parades and civic functions or at the scenes of general-alarm incidents. The iconic image of **Mose the Fireman,** the somewhat crude but big-hearted and brave fireman, and Bowery B'hoy, the urban hero of his day, and his girlfriend Lizzie became standard features of the American theater for the next half century.

Through this period, buildings grew taller and bigger. The introduction of iron-fronted structures by James Bogardus, heavy mill construction made possible by the increased use of iron interior beams, and the introduction of the Otis elevator changed the face of urban America. Warehouses, mills, lofts, foundries, and increasingly tall office buildings became common. Their heavy fire loads substantially increased the burdens on firemen in the number of responses, severity of fires, and threats to civilian and line-of-duty personnel. At the same time, balloon-constructed housing and shoddily built, multiple-occupancy housing turned many neighborhoods into tinderboxes. Tenement fires became frequent and killed hundreds. Complicating those threats was the uncontrolled storage of combustibles in residential-area warehouses. A New York City sperm oil storage warehouse destroyed 300 buildings in 1845. The Jennings warehouse fire and collapse claimed the lives of 11 New York City volunteers in 1854. The Pemberton Mills fire in Lawrence, Massachusetts, killed 115 occupants. Nonetheless, in the absence of effective training cycles and multi-unit coordination, volunteer operations generally remained chaotic.

As fiercely held ethnic, religious, and neighborhood loyalties pitted members of fire companies against one another, traditionally healthy competition got out of hand as some companies fought each other. In the 1830s, native-born Philadelphia firemen fought Irish newcomers. In 1834, firefighters responded to the Charlestown, Massachusetts, Convent and elite boarding school that had been set afire by an anti-immigrant mob. In the 1840s, mobs in the City of Brotherly Love destroyed the Hibernia fire station, Catholic churches, a market, and many Irish homes. Nativist and anti-immigrant riots in Baltimore (1856) and Washington (1857) divided the firefighting community.

Some departments became entangled in party politics, notably in New York City. Organized on a bottom-up, democratic basis, elections allowed popular but unskilled officers to hold rank. By the late 1840s, the controversial Tammany leader, **William M. "Boss" Tweed**, began his rise to political power as foreman of Americus Engine 6 on Henry Street in lower Manhattan (fig. 2–5). Many of the more than 4,500 volunteers in the city were among his ardent supporters.

In the larger cities, it seemed that the fire bells were ringing all the time and that the scale of fires was growing enormously. Suburban flight of older, middle-class volunteers forced a lowering of recruiting standards in the companies they left behind. Active members found it difficult to run their businesses or keep full-time jobs and respond to fires at the same time. Line-of-duty deaths and fire losses rose rapidly. Under these circumstances, it became increasingly difficult for volunteers to keep pace with the growing fire challenges of the big cities. There were no formal training cycles or multi-unit drills.

Fig. 2–5. This is a helmet frontispiece from Americus Engine Company 6 of New York, made famous by its Foreman (Captain), William "Boss" Tweed. A tiger was their symbol—they even tied a real one to their hand pumper during a parade!

On the other hand, the fire service, inventors, and civic leaders were not idle. In Great Britain, **James Braidwood**, credited with organizing the first effective paid fire department when he was appointed superintendent of the London Fire Brigade in 1833, offered a model for an effective urban fire department. In America by the early 1840s, a number of circumstances began to cast doubt on the traditional volunteer system as inventor and philanthropist Peter Cooper proposed a paid fire service for New York City.

During the 1850s, New York insurance companies began pressuring the traditional volunteer system to modernize by urging the appointment of new assistant engineers to cover newly settled uptown districts. They hired Alfred E. Baker, a former reporter for the *New York Herald*, as fire marshal in spring 1854. He was tasked "to examine into the origin of and other matters connected with fires" and to continue supporting the fire patrol. Nonetheless, in the face of legitimate appeals to change over to horse-drawn steamers made by progressive fire service leaders, insurance companies, and urban reformers, a majority of traditionalists within many departments resisted.

Real modernization began as the American industrial revolution gained traction. In 1852, Boston installed and employed a central telegraph office and street box fire-alarm telegraph system invented by **William F. Channing**. At the same time, patents were issued for the first sprinkler-perforated pipe systems, which were installed shortly thereafter. Early in 1853, advocates of theories of professional, business, and military models of organization convinced Cincinnati to introduce the horse-drawn Latta brothers' steam fire engine, **Uncle Joe Ross**, and to become the first city to employ a top-down, centralized, paid department in 1853. Six years later, New York insurance companies purchased the first two steamers put into service by the volunteers (fig. 2–6). Soon, every major city began to switch to the horse-drawn steamer. The spirit of that moment was captured most vividly by the Currier and Ives lithograph, *The Life of a Fireman. The New Era. Steam and Muscle* (fig. 2–7).

Fig. 2–6. A beautiful steam fire engine of the Passaic Engine Company Number One of Paterson, New Jersey. This heavy, hand-drawn "steamer" was built in 1874.

Fig. 2–7. Nathaniel Currier of the famed lithography firm, *Currier and Ives*, created this image called "Steam and Muscle." It portrays the transition from hand-pumped engines to steam fire engines in New York City.

On April 12, 1861, Fort Sumter in Charleston Harbor was shelled by supporters of secession, and the Civil War began. Southern fire departments were at a profound disadvantage because all their hand-drawn and steam fire apparatus and much of their equipment were manufactured in the North. Thus Southern departments, unable to replace worn-out and destroyed equipment, found themselves unable to protect their towns from fire. By Christmas 1861, much of Charleston had been destroyed by relentless fire from Union gunboats. Over the duration of the war, coastal communities of the South as well as Mississippi Valley cities, such as Vicksburg and New Orleans, were severely damaged and their fire departments disabled. In retaliation late in the war, several arson-related fires struck the lower Manhattan hotel district, part of a "Southern conspiracy" *to burn New York City*.

The involvement of firemen in the Civil War was complicated. Within days of the shelling of Fort Sumter, large numbers of New York City, Boston, and Philadelphia firefighters enlisted in the Union Army. The members of so-called **Fire Zouave** regiments wore red firemen's shirts, colorful jackets, baggy trousers, reddish-brown boots, and red caps known as "fezzes." At the same time, great numbers of firemen from Savannah, Georgia, Charleston, and other Southern towns enlisted in various Confederate units. Atlanta recruited its own Fire Zouaves. Most Zouave units took heavy casualties during the conflict and soon returned. Younger men, in many cases marginally employed, filled the ranks (fig. 2–8).

On the home front, firefighting became even more difficult than it had been prior to the war. When heavy losses forced the Union to draft young men for the army, riots broke out in many northern cities. Opposition to the draft was expressed in dozens of other cities, from Brooklyn to Staten Island, Bloomfield, New Jersey, Detroit, and as far away as Wisconsin. The most serious **draft riots** occurred in July 1863 in New York City, where pre-existing tensions between Irish and black workers over jobs and housing had precipitated conflict long before the war. Historians disagree on the level of involvement of members of Black Joke Engine 33 in carrying the riots forward once they began, but they agree that individual volunteers were on every side of the incident. Chief of Department John Decker was almost hanged by rioters when he ordered fire companies to put out fires. Many firemen and their companies were praised for their courageous actions in turning back mobs, rescuing people, and fighting fires. Martin Scorsese's film *Gangs of New York* contains many historical inaccuracies, not the least of which is its overemphasis of the role of New York City's volunteer firemen as mob participants (fig. 2–9).

Fig. 2–8. Gettysburg monument to firefighters

In the final days of the conflict, General William T. Sherman captured and burned the city of Atlanta, portrayed so dramatically in the film *Gone With the Wind*. He then commenced his March to the Sea, torching everything in his path and leaving a swath of total destruction from Savannah to Columbia, South Carolina. When Richmond, Virginia, burned in the chaotic final days shortly before General Robert E. Lee's surrender, many Confederate records were lost. In spring 1867, New York City's Volunteer Association sent a **hose reel** to their colleagues in war-torn Columbia. The first was lost in a hurricane on its way. The New Yorkers then took up a second collection, which enabled the South Carolinians to purchase their own piece of apparatus.

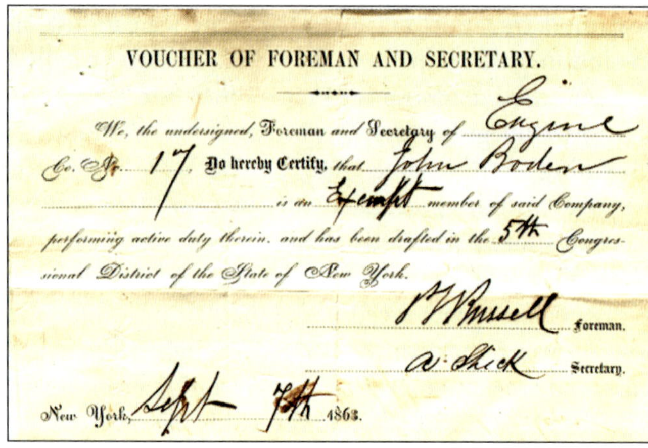

VOUCHER OF FOREMAN AND SECRETARY.

We, the undersigned, Foreman and Secretary of Engine Co. No. 17, Do hereby Certify, that John Boden is an Exempt member of said Company, performing active duty therein, and has been drafted in the 5th Congressional District of the State of New York.

P W Russell Foreman.

A. Shick Secretary.

New York, Sept 7th 1863.

Fig. 2–9. In order to keep New York City's firemen from having to serve in the military as a result of the draft (many of whom were already "exempted" from militia duty), the city paid "substitutes" to take the firefighters place. This firefighter, John Boden of Engine 17, was replaced by a man named John Wiilliams, as seen in this certificate.

With generational continuity broken by the loss of a substantial number of older and experienced members, the Civil War put an end to New York City's neighborhood-based, volunteer, locally controlled fire department. Amidst bitterness and controversy, New York's volunteers were dismissed in 1865. They were replaced with a paid **Metropolitan Fire Department** under the direction of commissioners appointed by the insurance-influenced, rural-based, state legislature. Hand-drawn apparatus was replaced by horse-drawn and steam-powered. **Alexander T. Shaler**, appointed as New York's fire commissioner in 1866, is credited with creating New York's paid department. In 1870 under the terms of the Tweed Charter, the **Fire Department of the City of New York** was established, thus restoring home rule to the City of New York. Departments across the country quickly followed the Cincinnati and New York models. Philadelphia went paid at year's end in 1870, and Los Angeles followed the same route in 1886.

Between the end of the Civil War and the First World War, the population of the United States more than doubled. In that environment, the American fire service coped with fires of increasing magnitude and deadliness. During this period, major technological and managerial innovations were introduced to cut response time, improve firefighting capacity, and strengthen fire-safety laws.

Cincinnati, Philadelphia, New York, and other cities established fire department and hospital-related ambulance services based upon the military model that emerged from the carnage of the Civil War. (During the Civil War, volunteer fire companies in Philadelphia built their own ambulances to transport wounded members coming back from the war on the railroad to city hospitals.) In that same era, the Boston Fire Department commissioned into service the **William F. Flanders** (1873), the first American steam-powered fireboat, followed in 1875 by FDNY's **William F. Havemeyer**. In 1868, **Daniel Hayes** of the San Francisco Fire Department developed the first successful aerial extension ladder truck (all ladders up to that point were individual ground ladders) that allowed firefighters to reach the upper portions of many of the newest, tallest buildings. Additional improvements saved time in hitching up fire horses. Sliding poles were installed in some New York firehouses. Cotton-covered, rubber-lined, more flexible fire hose replaced traditional, awkward, riveted-leather hoses. One of the first medals for bravery, the **James Gordon Bennett Medal**, was introduced in New York City in 1869.

Despite the efforts of inventors, fire department managers, and firefighters, a series of conflagrations swept Portland, Maine (1866), Boston (1872), Seattle (1889), and Paterson (1902) in the United States. In Canada, a fire in the city of St. John, New Brunswick laid waste to 1,600 buildings in 1877. The most serious of the late-19th-century fires occurred in the fall of 1871. The Chicago Fire raged for four days in early October 1871, destroyed almost 20,000 buildings in a four-square-mile area in the city center, and left 100,000 people homeless (fig. 2–10). Chicago's recovery was so swift that it became known as one of the most dynamic urban communities in the country. Years after the blaze, a news reporter confessed that he had made up the story blaming Mrs. O'Leary's cow for starting the blaze. Chicago's Fire Department Training Academy is now located on the former O'Leary property. Early October continues to be celebrated as Fire Prevention Week. A prairie fire in northeastern Wisconsin and upper Michigan swept more than 1.5 million acres. The town of Peshtigo was destroyed, and between 1,500 and 2,500 people were killed. The event drew little media attention, overshadowed as it was by the Chicago conflagration.

In the light of common problems facing the firefighting profession in the late 19th century, a number of organizations were established to coordinate fire prevention and operations across the nation. The **National Board of Fire Underwriters** was established by a consortium of insurance companies in 1866 and began to urge communities to upgrade their firefighting forces and building regulations. The **National Association of Fire Engineers**, later known as the International Association of Fire Chiefs, was formed in 1873. Founded by William H. Merrill in the spring of 1894, the Underwriters' Electrical Bureau, later known as the **Underwriters' Laboratories**, began laboratory testing of materials for combustibility in the upper floors of a Chicago Fire Patrol house. The **National Fire Protection Association** was established in 1896 with the original intention of standardizing codes for sprinkler and electrical systems.

Fig. 2–10. The Great Fire of Chicago was one of the largest conflagrations in American history (second only to the San Francisco Earthquake and Fire of 1906). These are the ruins at Dearborn and Madison Streets.

20TH CENTURY

As the 20th century dawned, the **Baltimore fire of 1904** raged for more than two days with mutual aid companies responding from as far away as Philadelphia, Washington, New York City, Wilmington, and Atlantic City. Tragically, most mutual aid companies found that the lack of standard fire hose couplings and threads and the absence of adaptors limited their usefulness. The Baltimore conflagration was the precipitating event for the creation of a new **model building code** (a generically written code that could be adopted by any city in the United States) known as the **National Building Code** (fig 2–11).

Fig. 2–11. The Great Baltimore Fire of 1904 led to creation of the first comprehensive standardized building code. Firefighters were not successful in standardizing hose threads; many responding mutual aid fire departments, including New York City, were unable to attach their hoses to Baltimore's hydrants. There are still a multiplicity of hose threads across the United States today.

Individual catastrophes such as the **Brooklyn Theater fire** (1876), the **Iroquois Theater fire** (fig. 2–12) in Chicago (1903), and the **Triangle Shirtwaist Factory fire** in lower Manhattan (1911) led to improved laws for high-occupancy structures. A series of marine disasters were led by one in Hoboken, New Jersey (1900), followed by another on the excursion boat *General Slocum* in the East River in New York (1904), and a third on Chicago's excursion boat *Eastland* (1915). Several "high-rise" fires in New York City, one in the Home Insurance building (1898) and the other in the Equitable Life building (1912), pointed out the need for better fire-prevention regulations in office buildings. **San Francisco's Great Earthquake and Fire of 1906** destroyed more than 28,000 buildings and caused the death of Chief of Department Dennis T. Sullivan (fig. 2–13). Originally reported to have cost fewer than 400 lives, researchers now suspect the loss of life to be in excess of 3,000 after revealing a great cover-up by the city after the event. It was the largest fire in American history.

At the same time, a generation of experienced fire service leaders introduced revolutionary changes in apparatus, fire prevention laws, and technology. In 1901, New York City's chief of department was assigned a gasoline-driven **Locomobile** (fig. 2–14). By 1922, the last horse-drawn apparatus made its final run. **Rescue companies** were established in major cities, and **high-pressure water systems** aided central business district fire companies. Following the Triangle Shirtwaist fire that cost 146 workers' lives, political leaders from the Progressive Movement, advised by fire department personnel, introduced factory legislation intended to guarantee greater safety in industrial workplaces. The **International Association of Firefighters** was established (1918), followed by the founding of fire officer associations intended to support their professional interests.

Fig. 2–12. The Iroquois Theater was supposedly "fireproof" (this is never an accurate term). 602 people died in a fire during a performance of "Mr. Bluebeard" soon after the new theater opened in 1903. It still stands today as our country's worst theater disaster. This image shows victims' bodies being removed in a hose wagon.

Fig. 2–13. The San Francisco Earthquake and Fire of 1906 was America's largest conflagration. It destroyed nearly 28,000 buildings, including City Hall as seen in this photograph.

Fig. 2–14. The transition to motorized apparatus was a major turning point in the fire service. This image shows the brand new "motor" Fire Engine Company One of San Antonio, Texas. The term "motor" is still with us today in many parts of the United States where the operators of fire engines are called motor pump operators (m.p.o.).

Along with the rest of the country, the firefighting community was swept into the vortex of each of America's 20th century wars. Before and after America's entry into the first World War, German saboteurs created numerous problems for firefighters from coast to coast as explosions lit secondary fires. In July 1916 on **Black Tom Island**, a major munitions depot situated in the Hudson River near Jersey City, two million pounds of explosives were detonated with such force that it could be felt 90 miles away. In January 1917, another explosion at the Canadian Car and Foundry in the New Jersey Meadowlands near Lyndhurst set off more than a half million, high-explosive shells. There were no deaths among the 1,400 people in the facility, who were warned building by building by company switchboard operator Tessie McNamara, to flee for their lives. In April 1917, the same month that the United States entered World War I, the Eddystone Ammunition Corporation in Pennsylvania blew up with the loss of 133 lives. In May 1918, six months before the Armistice that ended World War I, a TNT explosion in an Oakdale, Pennsylvania, chemical factory killed 200. A 1917 collision between two ships, one carrying wartime munitions, in the Canadian port of Halifax, Nova Scotia, resulted in a massive explosion which killed nearly 2,000 people and injured 9,000.

The war accelerated the migrations of hundreds of thousands of immigrants and rural black Americans seeking jobs in northern heavy industry. Auto plants in the Detroit area and steel mills in western Pennsylvania and Indiana began hiring workers without regard to race or ethnicity. Competition for jobs and housing exploded into race riots in dozens of cities including Atlanta, Omaha, Nebraska, and Tulsa, Oklahoma. Peaking in 1919, the Chicago Race Riots caused 38 deaths and left a thousand people homeless. The Tulsa Race Riots burned more than 35 city blocks with almost 1,300 buildings and, depending on the source, killed anywhere from 75 to 300. As usual, firefighters in each affected city were caught in the middle.

The period between the Great Depression and WWII was marked as an era of lean budgets, the extended use of older equipment, deferred apparatus acquisition, manpower shortages, and little or no increases in salaries. New York Mayor Fiorello LaGuardia, a fire buff himself, attempted to keep the department together by promising better pensions if members put up with the lean years of the 1930s. At the same time, the first two-way radios were making their way into many departments. Most were assigned to rescue companies, marine units, and senior command officers. In one unanticipated side effect of the Depression, many who might have gone into other lines of work or professions entered the fire service seeking security. Thus potential doctors, dentists, and teachers, and others, found themselves in the ranks.

In March 1937, a natural gas explosion leveled the **New London, Texas, Junior-Senior High School** and killed 298 students and teachers. Because natural gas is odorless, this explosion led to the requirement of using mercaptan as an odorant so leaks can be detected quickly. It was flammable hydrogen gas (used for buoyancy) and a highly combustible outer surface that spelled disaster for the airship *Hindenburg* when it caught fire at Naval Air Station Lakehurst in New Jersey, killing 36 people in May of 1937. On December 1, 1958, a fire in Chicago's Our Lady of the Angels parochial school killed 92 children and three nuns. An open stairwell allowed fire to rapidly spread from the basement of the second floor where the fatalities occurred.

When the Japanese attacked the naval base at Pearl Harbor, three civilian firefighters from the Honolulu Fire Department were among the 2,400 people killed. In New York in February 1942, a welding accident burned and sank the *USS Lafayette*, a former French luxury liner named the SS *Normandie*. In 1943, the Liberty ship *El Estero*, loaded with munitions, took fire while loading at Bayonne, New Jersey. With a cargo capable of devastating the entire lower harbor, the ship was towed to deeper water and scuttled by the heroic intervention of FDNY fireboats and Coast Guard vessels. On the West Coast, 328 ammunition loaders were killed when the Port Chicago Naval Magazine, 35 miles (56 km) north of San Francisco, blew up. On the civilian front, a fire in the **Cocoanut Grove Night Club** in Boston resulted in 492 deaths and injuries to hundreds in November 1942. Late in the war, the **Ringling Bros. and Barnum & Bailey circus tent** caught fire in Hartford, Connecticut, killing 167 and injuring nearly 500 in early summer 1944 (fig. 2–15). Just before V-J Day in 1945, FDNY responded when a B-25 bomber lost in fog crashed into the 79th floor of the **Empire State Building**. The crew and at least 11 office workers died.

Fig. 2–15. As the band played the *Stars and Stripes* to signal a fire to circus workers, attendees scrambled to get out from under the big tent and away from the quickly spreading fire. The July 1944 fire in Hartford cost many lives. Here firefighters wet down the ruins of the tent which had been soaked with paraffin and kerosene to prevent rotting of the canvas.

After the end of World War II, American manufacturers and Boeing Airplane Co. dominated the airways. The incredible growth of private aviation, the introduction of jet aircraft, and reasonable ticket pricing, enabled tourists to travel in relative comfort and ease. As the industry grew, the number of fires related to air crashes rose, commencing in 1947 with the loss of 52 when a United Airlines DC-6 went down near Bryce Canyon, Utah. The collision of two aircraft in the crowded airspace approaching LaGuardia and Idelwild International (now known as JFK) airports in December 1960 resulted in more than 130 deaths when one plane came down in Staten Island and the other in a crowded Brooklyn neighborhood. Prior to the World Trade Center attacks, the deadliest crash was an American Airlines flight at O'Hare International Airport in Chicago, which killed all 271 people on board (1979). More reliable equipment, effective inspections and maintenance, and rapid, advance firefighting tactics when aircraft hit the ground have reduced crash and death tolls for American flights.

On December 7, 1946, a fire moved quickly through Atlanta's Winecoff Hotel, killing 119 people. Starting on the third floor and moving rapidly up through the structure because of a lack of fire doors and other fire containment features, it became America's worst hotel fire. Other significant hotel fires include the MGM Grand Hotel-Casino fire in Las Vegas (1980), where 87 died, and the Hotel Dupont Plaza fire in Puerto Rico where 97 were killed in an arson fire on New Year's Eve in 1986.

In April 1947, an explosion rocked the port of Texas City. The French Liberty ship ***Grandcamp***, loaded with ammonium nitrate fertilizer, caught fire and blew up, killing 26 Texas City volunteer firefighters (fig. 2–16). It was the largest loss of firefighters prior to the attack at the World Trade Center on September 11, 2001.

Fig. 2–16. A large group of vehicles damaged in the 1947 Texas City explosion. A series of explosions over two days killed 576 people, including 26 Texas City volunteer firefighters. The disaster began with an initial explosion of ammonium nitrate in the ship Grandcamp on April 16, 1947.

Throughout the century, San Francisco continued to be battered by earthquakes. One in spring 1957 registered 5.3 on the Richter scale. In October 1989, Americans watched the developing tragedy on television as the most severe earthquake since 1906 hit the City by the Bay just before the beginning of the third game of the World Series at Candlestick Park. Quickly, San Francisco's *Fireboat Phoenix* went into operation, supplying saltwater through large-diameter hoses. Hundreds of civilian volunteers stretched lines to land companies fighting the ensuing Marina District fire.

The 1960s was an era of confusing and often violent social, racial, and economic confrontation. Beginning in the early 1960s, the fire service entered what is commonly known as **The War Years**, or the Long Hot Summers. A complicated set of factors, including frustration over poor living conditions, the emptying of some neighborhoods to the suburbs, arson, and bitterness, led to riots, burning, and looting in more than 100 American inner cities from Watts in Los Angeles to Newark, Detroit, East New York, and the South Bronx in New York City. In 1950, the number of alarms in New York had been only 62,021, rising to 261,131 annually by 1970. As portrayed by Dennis Smith in *Report From Engine Company 82*, some companies made 7,000 to 8,000 runs per year. Canopies were constructed over apparatus cabs, and members rode inside for fear of "air mail" and shooting. The orgy of destruction eventually diminished in the 1980s, and now, in many of those same neighborhoods, firefighters are viewed much more favorably.

Several reports emanated from the period. The first was from the National Advisory Commission on Civil Disorders, also known as the Kerner Commission (1968), which assessed the underlying causes for the period's civil unrest. The second, the National Commission on Fire Prevention and Control summary report on the period, ***America Burning*** (1973), noted that the mission of the American fire service had expanded rapidly beyond fire prevention, suppression, and investigation. Recognizing that firefighters are the first line of defense for citizens facing any type of disaster or emergency, the report presented an agenda for improving firefighters' ability to deal with such situations in the future.

In the words of author Terry Golway, the period after the 1960s was dominated by a "Search for Equity" in recruiting, promoting, and retaining personnel in the fire service. Since the 1970s there has been some progress toward gender, ethnic, and racial equity in recruiting, holding, and promoting firefighters from every dimension of American society. In 1973, Arlington County, Virginia, made headlines by hiring its first woman career firefighter. By the opening years of the 21st century, the number of women holding career-firefighter and officer positions exceeded 7,000. During the same time, it is estimated that there were more than 40,000 women volunteer firefighters. The 100th anniversary celebration of the San Francisco Earthquake and Fire was hosted by Chief of Department Joanne Hayes-White, a 14-year veteran and married mother to three sons. Rochelle "Rocky" Jones, whose career spans more than 20 years and whose company was part of the first alarm assignment for the Twin Towers, later earned the rank of FDNY battalion commander. Nonetheless, there were still fewer than 50 women in FDNY at the opening of the 21st century.

The situation facing black Americans is equally challenging. FDNY remains 90% white, while the Boston, Philadelphia, and Chicago departments are between 70% and 75% white. Tensions over promotions in the Saint Louis Department, led by a black chief of department, remain high, a common situation in many cities. Van Davis Jr., the first black appointment to Los Angeles County Fire Department in 1953, served 20 years, during which time no blacks were promoted to upper rank. Many fire departments in large and small communities across the nation still struggle to resolve challenges relating to gender and racial staffing, recruiting, and promotion standards.

Beginning in the 1960s, some firefighters underpaid, overworked, and convinced that they were getting little respect from politicians and residents, grew frustrated and angry. In the face of no-strike laws but encouraged by their national union, the International Association of Firefighters (IAFF), uniformed personnel went on strike in New York City, Yonkers, Chicago, New Hampshire, California, and elsewhere. Improved working conditions, binding arbitration agreements, better equipment, and some improvement in contracts diminished the number of similar job actions over time.

In the 1970s, the completion of the interstate highway system and the return of Vietnam veterans willing to serve made it possible to establish a standardized, national **Emergency Medical System** staffed by paramedics and emergency medical technicians. The onset of the high-rise and highway strip mall era in the 1950s was accompanied by cost efficiencies in architectural design and construction, as well as by major deficiencies in fire safety. In such manner, the lives of firefighters trying to effect rescues of workers and residents were placed at risk. Major blazes occurred in the **McCormick Place Exhibition Hall** in Chicago (1967) and **1 New York Plaza** (1970). Philadelphia officials described the 38-story **Meridian Plaza blaze** (1991) as the "most significant fire of the 20th century." Two significant hazardous materials incidents: propane BLEVEs (boiling liquid expanding vapor explosion, described in chapter 23, Analyzing the Hazmat/WMD Incident) occurred in the 1970s. In 1973, a rail tank car of propane BLEVEd in Kingman, Arizona, killed 12, including 11 firefighters. A few years later in 1978 at Waverly, Tennessee, another propane rail tank car exploded and killed 16. These incidents and others led to changes in rail tank cars: there have not been any rail tank car BLEVEs in the last several years. A more recent significant hazardous materials incident occurred in the small Canadian town of Lac-Mégantic, Quebec, in 2013 when a train carrying flammable "Bakken" crude oil derailed and ignited, killing at least 42 people and destroying more than 30 buildings (fig. 2–17).

Fig. 2–17. In 2013, a train carrying Bakken crude oil derailed, causing a hazmat incident in the small Canadian town of Lac-Mégantic, Quebec. (Photo courtesy of Sûreté du Québec)

Three notable fires in nightclubs ("places of public assembly") occurred around the last quarter of the 20th century. On May 28th, 1977, a fire erupted in the Zebra Room of the **Beverly Hills Supper Club** in Southgate, Kentucky. One hundred sixty-five people were killed, many of the victims caught in a crowd crush (jam of people) at exits as fire and smoke moved into the Cabaret Room where a large audience was waiting to hear 1970s singing sensation, John Davidson. On March 25th, 1990, an arson fire in the **Happyland Social Club** in the Bronx, New York, killed 87 people. Polyurethane foam lining the walls of **The Station nightclub** in West Warwick, Rhode Island, was ignited by indoor pyrotechnics used during a performance of the band, Great White, on February 20, 2003. One hundred people eventually died as a result of the Station fire; lethal conditions in the nightclub developed in about 90 seconds.

21ST CENTURY AND BEYOND

With increasingly intrusive human use of naturally forested areas, wildfires are a growing hazard in most regions of the United States, posing a threat to life and property. Thus, **wildland firefighting** is a major and most significant professional specialty, particularly in western states. The **Mann Gulch Fire** in the Helena Forest in Montana killed 16 smoke jumpers (1949); the **Rattlesnake Fire** in the Mendocino Forest took 15 firefighters (1953); and in the summer of 1994, 14 wildland firefighters died when shifting 70-mile-per-hour winds trapped them and fire swept over their position in **South Canyon Glenwood Springs, Colorado**. In fall 2007, out-of-hand wildfires destroyed thousands of homes and dislocated hundreds of thousands, resulting in major economic losses in Southern California.

Recurrent and more frequent fires involving major commitments of personnel and equipment spurred many incident command models in the late 20th century. **Firefighting Resources of Southern California Organized for Potential Emergencies (FIRESCOPE)** initiated a version of ICS for wildfire operations that was a major improvement over previous systems. Under the leadership of Alan Brunacini, FIRESCOPE evolved into the **Incident Management System**. Inspired by Capt. Ron Gore, the first **hazmat** team formally went online in Jacksonville in January 1977, inspiring the introduction of such units all over the nation. In 1989, the **National Urban Search and Rescue System** established regional task forces under the jurisdiction of the **Federal Emergency Management Agency (FEMA)** designed to coordinate emergency services during major disasters.

The attacks on the World Trade Center **Twin Towers** (1993 and 2001) were not the first terrorist events American firefighters faced. Left-wing extremists include those who destroyed the Los Angeles Times building (1910), anarchists who bombed Wall Street (1920) (fig. 2–18), and revolutionary Weathermen (1960s) whose bomb factory exploded in New York's Greenwich Village in 1970. Racists bombed homes, schools, and churches from the post-Civil War KKK years throughout the Civil Rights era of the 1960s and later. During the long, hot summer of 1964, Mt. Zion Church was burned to the ground, one of 20 black Mississippi churches to be firebombed that year. The FBI initiated the MIBURN (Mississippi Burning) investigation. Violent nationalists from Croatia and Puerto Rico (FALN) were active in the 1970s. Anti-government, right-wing radicals bombed the **Alfred P. Murrah Building** in Oklahoma City (1995). The firefighting community was on the front line in each case.

Fires in aging commercial buildings, some occupied, others abandoned and deteriorated, continue to take an increasingly heavy toll. A fire in the basement of a lower Manhattan loft killed 12 in 1966. Warehouse fires killed four firefighters in Seattle (1995) and six in Worcester, Massachusetts (1999). After nine of their members died in a collapsing furniture store in June 2007, the Charleston Fire Department requested FDNY's Emerald Society Pipes & Drums to play at the memorial service.

Fig. 2–18. Many Americans think that the terrorist attacks of September 11, 2001, were the first on American soil. There have been many attacks through the country's history, including this one on Wall Street in 1920. Damage to this building's (the J.P. Morgan bank) exterior wall, near the corner of Wall and Broad Streets, is still visible.

The destruction of the World Trade Center Towers and the Pentagon on September 11, 2001, was the most despicable act of mass murder ever perpetrated in the United States (fig. 2–19). At the Towers, it is estimated that nearly 15,000 were led to safety prior to the collapse. The retired FDNY Fireboat John Jay Harvey, which was maintained by a historical association of private citizens and retired firefighters, responded immediately and supplied the only water at the scene for more than 96 hours. Hundreds of ambulances from New Jersey were summoned to staging areas in the region, with many being sent into **Ground Zero**. Fire, police, and emergency medical service units and personnel began to respond from as far away as New Brunswick, Canada, Florida, and Oregon. Ultimately, four FEMA Urban Search and Rescue Task Forces were sent to Washington and six to New York. Among the approximately 3,000 dead were 343 FDNY members, 1 fire patrolman, 23 New York Police Department officers, and 37 Port Authority Police Department personnel. FDNY losses included First Deputy Commissioner and

Fig. 2–19. This image shows the initial response of the New York City Fire Department on 9/11. It shows the Marriott Hotel on the left along West Street. The north tower of the World Trade Center is out of view to the far left. (Courtesy of FDNY Photo Unit)

former Chief of Department William M. Feehan, Chief of Department Peter Ganci, FDNY Chaplain Mychal Judge, as well as officers and firefighters of every rank, amounting to a loss of more than 4,400 years of formal and intuitive knowledge.

The outpouring of sympathy and tangible support from around the world for the city in general and for FDNY in particular was astounding. Among dozens of fund-raising efforts around the world, the citizens of Columbia, South Carolina, presented FDNY with a tiller aerial now operating as Ladder 101 in Brooklyn, a grateful thank you for the support sent to that city in 1867. The Spirit of Louisiana, donated by citizens of that state, was assigned to Engine 283 in Brownsville. In the early stages of the Hurricane Katrina disaster, FDNY sent The Spirit of Louisiana to New Orleans, where it remains a part of that city's reviving department.

What remains and what has changed? Today's **first responders** are aware that they are primary targets of terrorists. And yet the roles of firefighters, police, and EMS personnel, both career and volunteer, remain to rescue those at risk, to contain disasters, and to comfort survivors.

QUESTIONS

1. As early as 1648, Peter Stuyvesant, the governor of New Amsterdam, implemented the use of fire wardens and strong fire prevention techniques and imposed requirements for homeowners. Nearly 360 years later and following countless large loss fires, is American society more conscious of the effects of fire or better prepared to prevent them?

2. Around 1852 the first patents were issued for the first sprinkler-perforated pipe systems. Almost 156 years after the first patents were issued, the American fire service still battles to educate politicians, model code developers, builders, businesses, and homeowners to adopt requirements to protect all structures. Discuss why the efforts to do so have been and continue to be difficult.

3. Name four professional organizations that were developed in the late 19th century. What positive effects have these organizations had on the fire service?

4. A document titled ________________ was first published in 1973 in an attempt to focus attention on the nation's fire problem.

5. Two national programs were created from this document. What are they and what are their current responsibilities?

6. George Santayana, the author of "Reason in Common Sense" is quoted as saying, "Those who cannot learn from the past are doomed to repeat it." The history of the American fire service is riddled with accounts of large fires resulting in property loss, and loss of life to civilians and firefighters. Many of the circumstances surrounding those events mirror one another. Why does it appear that the American fire service has failed to take heed of our past and take the necessary steps to prevent such losses in the future?

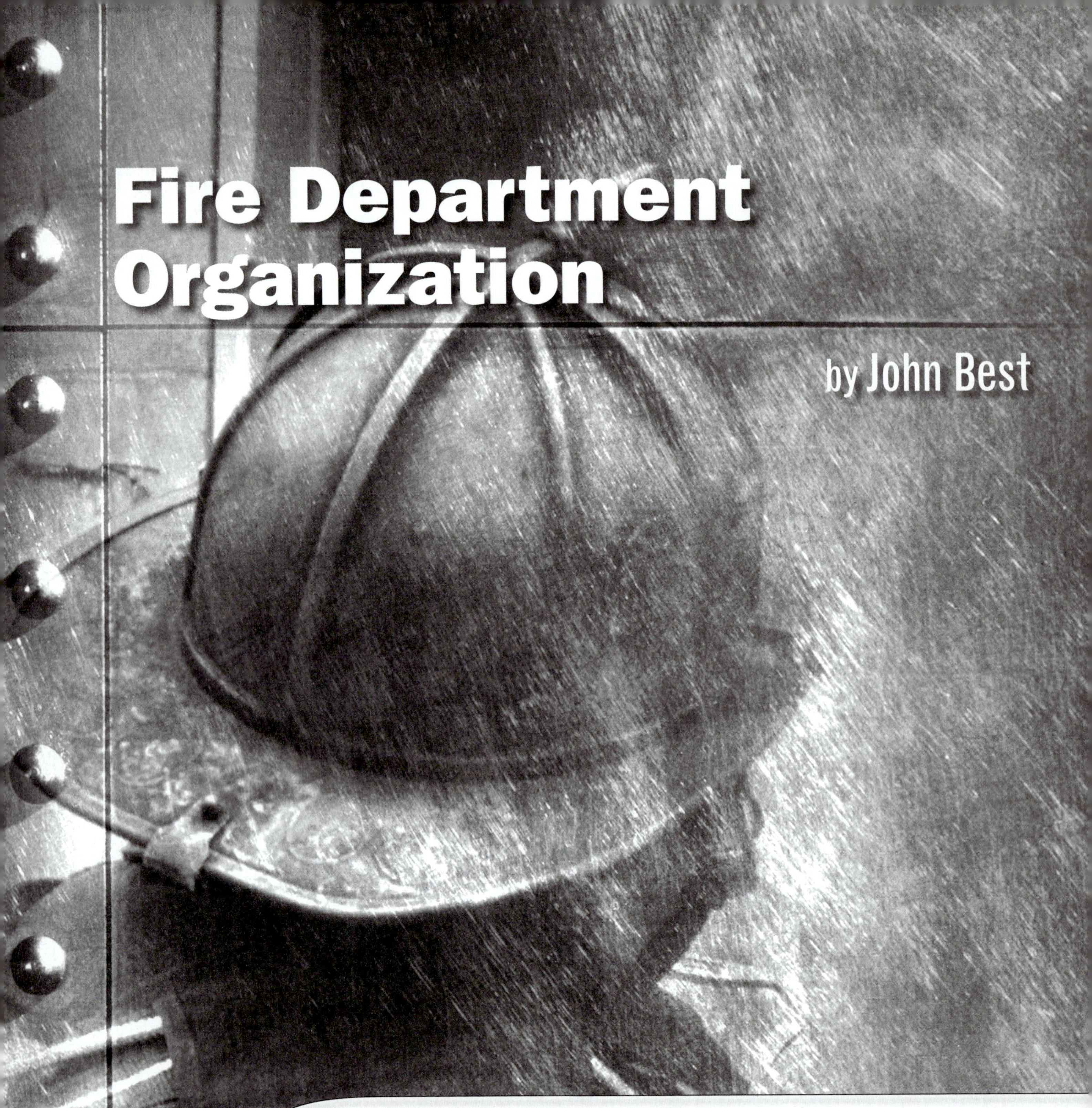

Fire Department Organization

by John Best

This chapter provides required knowledge items for the following NFPA Standard 1001 Job Performance Requirements:

FFI 4.1.1

3

OBJECTIVES

Upon completion of this chapter, you should be able to do the following:

- Describe the organizational structures of small, medium, and large departments
- Identify the six types of fire department units, and the roles they play within the department
- Identify and describe the roles and responsibilities of the firefighter and fire officer positions
- Identify additional functions the fire department performs
- Identify fire department member assistance programs
- Identify the policies, procedures, regulations, and by-laws under which fire departments operate
- List and describe the roles of other agencies that impact the operation of the local fire department

INTRODUCTION

Over the last 400 years, the fire service has evolved to the point where the fire suppression function has become but one of many important missions of the fire department organization. The modern fire department responds to numerous all-hazard responsibilities including but not limited to fire suppression; **emergency medical services (EMS)**; specialty rescue operations; hazardous materials mitigation; life- and fire-safety education; the administration and enforcement of codes and standards; and emergency communications.

This evolution continues to enhance the fire service's relationship with a number of allied organizations representing the ever-increasing spectrum of responsibilities addressed by fire departments and other emergency organizations. The following organizations represent a sampling of the major organizations closely associated with the fire service:

- *National Fire Protection Association (NFPA)*, a nonprofit organization whose purpose is to reduce the worldwide burden of fire and other hazards on the quality of life by providing and advocating consensus codes, standards, research, training, and education

- *International Association of Fire Chiefs (IAFC),* composed of members considered the leading experts in firefighting, emergency medical services, terrorism response, hazardous materials, natural disasters, search and rescue, and public safety legislation.
- *International Association of Fire Fighters (IAFF),* working to promote a safe and healthy work environment, reasonable working conditions, the research and treatment of burns and other related health problems common to career firefighters.
- *National Volunteer Fire Council (NVFC),* composed of state volunteer firefighter organizations involved in providing training programs, public education, conferences, and the compilation of pertinent statistics.

Additional allied organizations and more in-depth descriptions appear later in this chapter.

The fire service addresses operational all-hazard responsibilities through a basic fire company–firefighter organization overseen by a quasi-military leadership system. The companies are assigned to the number of stations needed to meet the community's response time or **standard of cover** requirements.

Companies are generally organized by function such as an engine, ladder truck, emergency medical, rescue, or specialty service. These companies and other fire department functions are supervised and staffed by the appropriate complement of officers and firefighters.

This chapter describes the firefighter as the basic staffing unit of the fire department and the many roles that firefighters have in the modern fire department organization.

FIRE DEPARTMENT ORGANIZATION

The fire department must organize in a manner to provide the oversight and resources required to exercise its authority to meet its responsibilities to the community.

Mission

FFI 4.1.1 A fire department should develop its mission statement commensurate with the expectations and requirements of the community it serves. Elements typically included in a fire department mission statement are summarized by the NFPA 1201, *Standard for Providing Emergency Services to the Public,* as follows:

- Save lives
- Prevent or mitigate fires, injuries, and emergencies
- Work through a system of emergency management
- Extinguish fires
- Minimize damage to property and the environment
- Protect critical infrastructure
- Perform emergency medical services
- Protect the community from hazardous situations
- Perform response to and mitigation of events of terrorism
- Perform rescue services
- Perform other community-related services

A fire department mission statement is usually part of a master plan that also includes the department's vision, values, goals, and objectives. This plan should clearly communicate these factors to the organization's members and customers.

Organizational structure

FFI 4.1.1 The organizational structure of a fire department is influenced by a multitude of factors. These factors include, but are not limited to, its mission, the communities it serves and their expectations, the size and workload of the organization, available funding, and the political landscape.

This organizational structure ranges from large and complex, as in a major municipality or county; to moderate, as in a medium-size borough or city; to small, as in a single-station, small-town, all-volunteer fire company.

Generally, fire departments are organized as full-time paid, part-time paid, volunteer, volunteer-paid-on-call, or a combination of the four. They may be operated by a federal entity, a municipal entity, a special taxing district, a not-for-profit corporation, or a for-profit corporation.

The various demands placed on the majority of fire departments have made fire-suppression activity one of the smallest percentages of their workload: less than 10% for most departments. For example, fire departments that provide EMS, including advanced life support, patient transport, or both, find their EMS workload ranging from 60% to 75%.

Most fire departments deploy their resources to effectively address their community's **fire problem**. This involves organizing a cycle of resources, including the following:

- *Public education*, so that citizens are aware of hazards, how to prevent them, and what to do should the peril occur
- *Engineering and code enforcement*, so that fire and life safety is an inherent part of the community's infrastructure, and where there is a violation, compliance is achieved
- *Fire suppression*, needed when there is a failure in the education and engineering and code enforcement parts of the cycle, so that the emergency can be resolved
- *Fire investigation*, in which the incident is documented, the cause determined accidental or intentional, and steps are taken so it will not happen again

The results of fire investigation suggest new needs and goals for public education; code modifications and changes; fire department training, resources, and deployment; and identification of the community's fire problem.

According to the NFPA, there were 1,345,500 fires reported during 2015 involving $14.3 billion in property damage. Figure 3–1 provides an overview of these fire types:

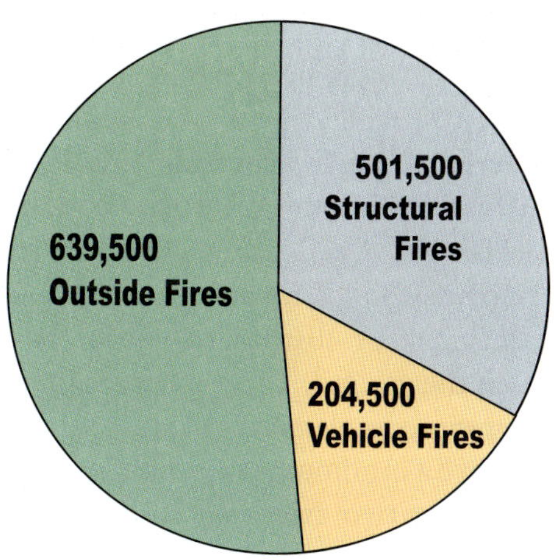

Fig. 3–1. 1,345,500 fires were reported to the National Fire Protection Association in 2015. Fires are typically less then 10% of the workload for any fire department.

Whether the fire department is career or volunteer; a large, complex entity; or a small, municipal, single-station organization, the firefighter is the primary position and resource in the fulfillment of the associated missions (figs. 3–2, 3–3, 3–4).

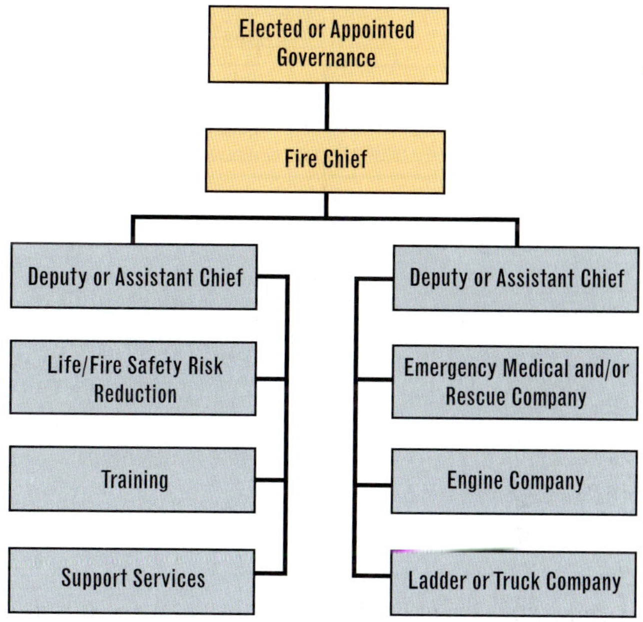

Fig. 3–2. Typical organizational structure of a small fire department

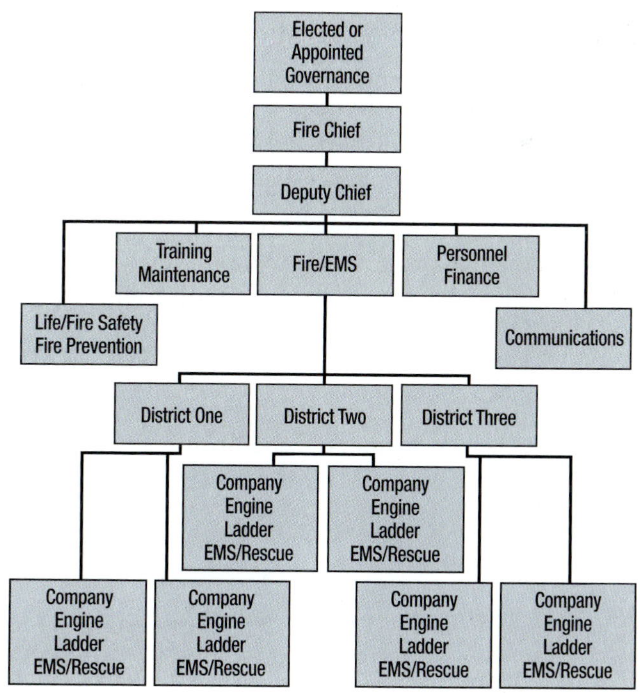

Fig. 3–3. Typical organizational structure of a medium-size fire department

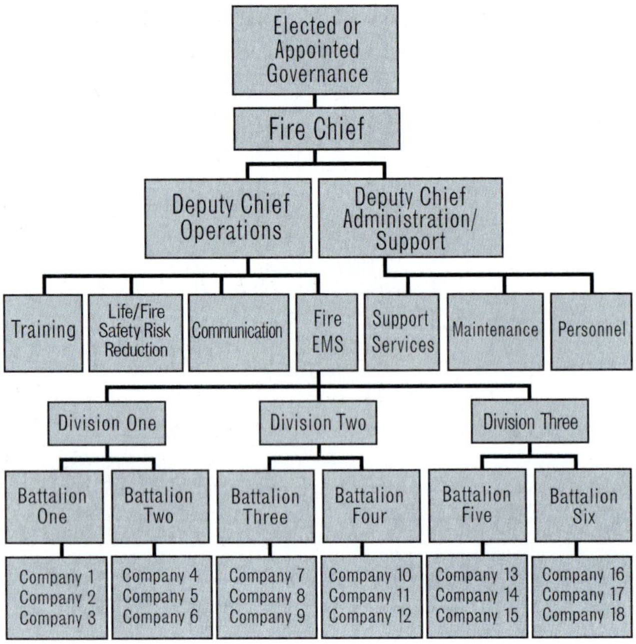

Fig. 3–4. Typical organizational structure of large fire department

THE COMPANY

As defined by NFPA 1201, *Standard for Providing Emergency Services to the Public*, the company is a group of members that is the following:

1. Under the direct supervision of an officer or leader
2. Trained and equipped to perform assigned tasks
3. Usually organized and identified as engine companies, ladder companies, rescue companies, or squad companies
4. Usually operates with one piece of fire apparatus (pumper, ladder truck, elevating platform, rescue, squad, or ambulance)
5. Arrives at the incident scene on fire apparatus or assembles at the scene prior to assignment

Engine company

An **engine company** is composed of an apparatus or vehicle, generally referred to as an engine or pumper, staffed by firefighters and a supervisor (officer) trained and assigned to deliver water, deploy hoselines, and execute other associated tactics to extinguish fires.

The engine company (pumper) vehicle, as seen in figure 3–5, is defined by NFPA 1901, *Standard for Automotive Fire Apparatus*, as fire apparatus with a permanently mounted pump of at least 750 gpm (2,839 L/min) capacity, a tank carrying at least 300 gal (1,136 L) of water, and a hose body, whose primary purpose is to combat structure and associated fires.

Fig. 3–5. The engine company is a common unit found at almost every fire station in the United States.

Many engine companies are assigned, equipped, and trained to perform rescue, extrication, and/or emergency medical services.

Ladder or truck company

A **ladder** or **truck company** is composed of an aerial apparatus or vehicle, staffed by firefighters and a supervisor (officer) trained and assigned to place ladders, accomplish search and rescue, perform ventilation and forcible-entry operations, secure utilities, and perform salvage or overhaul at fire and other emergency scenes.

The ladder or truck company aerial fire apparatus, as seen in figure 3–6, is defined by NFPA 1901, *Standard for Automotive Fire Apparatus*, as a vehicle equipped with an **aerial ladder**, **elevating platform**, **aerial ladder platform**, or **water tower** that is designed and equipped to support firefighting and rescue operations by positioning personnel, handling materials, providing continuous egress, or discharging water at positions elevated from the ground. This vehicle, as defined, must have an extensive complement of hand and power tools and should have a minimum of 115 ft (35 m) of ground ladders, including two extension ladders, two straight ladders with folding roof hooks, and an attic ladder.

Fig. 3–6. A ladder or truck company. These units can come with a wide variety of fixed aerial devices.

As with the engine company, ladder or truck companies may be assigned, equipped, and trained to perform rescue, extrication, and/or EMS.

Rescue or squad company

A **rescue** or **squad company** is composed of a vehicle carrying specialized equipment, as seen in figure 3–7, staffed by firefighters and a supervisor (officer) trained and assigned to perform forcible entry, search and rescue, and other specialized **tasks** at fire scenes. Most communities hold their rescue or squad companies responsible for confined space and **rope rescues**, vehicle accident extrications, and other technical rescue evolutions, although these tasks may be assigned to ladder companies.

Rescue or squad companies may also provide emergency medical services.

Fig. 3–7. Rescue companies can come in many shapes and sizes; however, large rescue units like this are regularly found in urban environments.

Ambulance or emergency medical services company

Many fire-service-based ambulance or EMS companies are composed of a transport vehicle, as seen in figure 3–8, carrying specialized medical equipment, and staffed by firefighters trained and assigned to provide treatment to patients, including first aid, cardiopulmonary resuscitation, basic life support (BLS) or advanced life support (ALS), and other medical procedures prior to arrival at a health care facility.

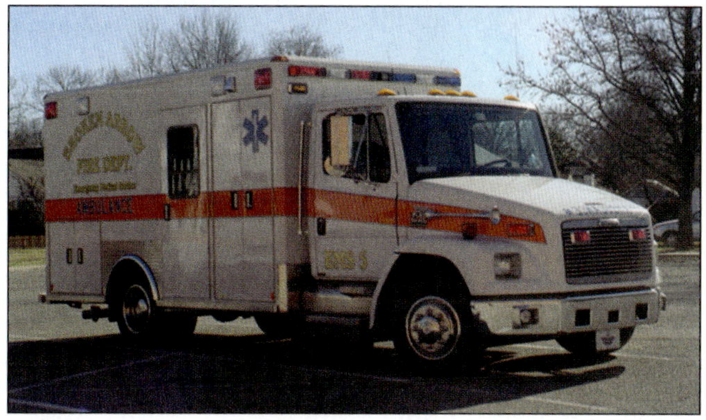

Fig. 3–8. Ambulances, or EMS units, may be staffed by firefighters or paramedics from private businesses.

Although there are numerous emergency medical staffing and response models depending on the community served, many fire departments utilize EMS-trained firefighters to staff their ambulance or EMS companies. These firefighters may be assigned to medical or firefighting duties.

Combination or specialty companies

The fire department is responsible for providing a wide spectrum of services to their respective communities. This diversity of services can mandate a variety of unique approaches, including the use of quints, **mobile water supply apparatus (water tenders)**, wildland firefighting apparatus, hazardous materials units, light and breathing air vehicles, aircraft rescue and firefighting companies, and other resources tailored to a specific community's needs.

The quint. The **quint** is defined by the NFPA 1901, *Standard for Automotive Fire Apparatus*, as fire apparatus with a permanently mounted fire pump, a water tank, a hose storage area, an aerial ladder or elevating platform with a permanently mounted waterway, and a complement of ground ladders (fig. 3–9). Basically, it combines a pumper and a ladder truck. This vehicle is staffed with firefighters and an officer trained and assigned to perform engine company and truck company evolutions.

Fig. 3–9. Quints are combination units that can do the work of both engine and ladder companies since they have a fire pump, a water tank, hose storage area, aerial ladder, and ground ladders.

Mobile water supply apparatus. NFPA 1901, *Standard for Automotive Fire Apparatus*, defines a mobile water supply apparatus (**water tender**) as a vehicle designed primarily for transporting (pickup, transporting, and delivering) water to fire emergency scenes to be applied by other vehicles or pumping equipment (fig. 3–10). This company is staffed and operated by firefighters.

Fig. 3–10. Mobile water supply apparatus, commonly referred to as tenders, help fire departments fight fires in areas without fire hydrants.

THE FIREFIGHTER

FFI 4.1.1 A firefighter, as defined by NFPA 1001, *Standard for Fire Fighter Professional Qualifications*, is an individual possessing the knowledge and skills to function as an integral member of a firefighting team, under supervision, in hazardous conditions.

In addition, firefighters should be knowledgeable of their fire department organization, its mission, standard operating procedures, and rules and regulations. They should understand the importance of physical fitness and a healthy lifestyle as it relates to the performance of their duties.

Firefighter roles may be expanded as the mission of a fire department evolves in its community. These roles may include but are not limited to emergency medical first responder, paramedic, rescue specialist, driver/operator, fire investigation or prevention officer, and fire rescue communicator/dispatcher.

Many state, county, and local entities have established minimum qualifications for firefighters within their jurisdictions. The NFPA, through its various technical committees and associated consensus processes, has established numerous training and qualification standards addressing firefighter duties and responsibilities. These standards are widely accepted by fire department and emergency services organizations worldwide and are adopted as the basis for the firefighter minimum qualifications.

NFPA 1001, *Standard for Fire Fighter Professional Qualifications*, addresses firefighter minimum knowledge, skills, and abilities. These minimum qualifications are divided into **Firefighter I** and **Firefighter II** designations, generally based on the level of supervision required and the ability to operate independently.

The following list identifies examples of typical minimum skills and abilities required of a Firefighter I (fig. 3–11):

The Firefighter I has the ability to do the following:

- Don and use personal protective clothing
- Hoist tools and equipment using ropes
- Locate information in department documents, standards, and codes
- Operate fire department communications, telephone, and intercom equipment
- Control breathing, use self-contained breathing apparatus, and replace air cylinders
- Use provided safety equipment
- Transport and operate hand and power tools to force entry and ventilate
- Operate as a team member in vision-obscured conditions
- Carry, raise, and extend ladders

- Identify automobile fuel types; assess and control fuel leaks
- Operate fire attack hoses; open, close, and adjust the flow and pattern on nozzles
- Apply water for maximum effectiveness
- Hand lay and connect water supply hose and appliances
- Recognize inherent hazards related to a material's configuration
- Search for and expose hidden fires; assess patterns for origin determination
- Evaluate for complete fire extinguishment
- Conduct property conservation, protection, and salvage evolutions
- Recognize obvious signs of arson and area of fire origin
- Stop fire sprinkler system water flow and operate main control valves
- Select and operate correct portable fire extinguishers based on fire classification
- Operate fire department power supply and lighting equipment
- Identify and operate building utility-control devices
- Suppress ground cover fires; determine exposure threat based on fire spread potential
- Clean, inspect, and maintain fire department tools and equipment

Fig. 3–11. A firefighter must be able to perform a wide variety of skills in order to be successful. The skills that a firefighter may be called to do can be found in NFPA 1001.

Firefighter I may be required to drive and operate fire department vehicles and apparatus. This may include fire department vehicles equipped with a fire pump, an aerial device, a **tiller**, and a mobile water supply, including wildland fire apparatus and aircraft firefighting and rescue apparatus.

Driving and operating fire department vehicles, whether on a nonemergency basis or in response to an emergency, requires significant knowledge, skill, and ability, including the following:

- Perform and document routine tests, inspections, and servicing function
- Drive, operate, and position fire apparatus in adverse conditions
- Deploy, energize, and monitor all systems associated with the vehicle
- Position a pumper to do the following: operate at a fire hydrant and at a static water source; power transfer from vehicle engine to pump, draft, operate pumper pressure-control systems; operate auxiliary cooling systems; make the transition between internal and external water sources; and assemble hoseline, nozzles, valves, and appliances
- Maneuver and position an aerial apparatus so that the apparatus is positioned for correct aerial device deployment
- Connect a water supply to a master stream device and control an elevated nozzle manually or remotely
- Apply the principles of tiller operation; methods of communication with the driver; and effects of general steering reactions, night driving, negotiating intersections, and manufacturer operation limitations
- Apply procedures for establishing a water shuttle dump site and principles of water transfer between multiple portable water tanks

NFPA 1002, *Standard on Fire Apparatus Driver/Operator Professional Qualifications*, provides the minimum requirements for the Firefighter I assigned to drive and operate fire department vehicles.

In addition to the typical minimum skills and abilities required of a Firefighter I, a Firefighter II must have the ability to do the following (fig. 3–12):

- Determine the need for command, and organize and coordinate incident management
- Complete basic incident reports including the operation of fire department computers
- Determine necessary applicable codes and standards
- Operate fire department communications equipment
- Extinguish an ignitable liquid fire, including through the application of firefighting foam
- Assemble a team and choose an effective fire attack technique
- Coordinate a structural interior attack hoseline as part of a team
- Execute effective fire attack advances and retreats based on condition changes
- Locate a fire's area of origin, recognize possible causes, and protect evidence
- Extricate a victim entrapped in a motor vehicle accident
- Operate hand and power tools for forcible entry and rescue
- Perform fire-safety surveys including completion of appropriate forms and reports
- Conduct life and fire-safety presentations and tours
- Identify the components of fire suppression and detection systems
- Safely conduct hose service testing
- Support technical rescue teams

Fig. 3–12. To meet the requirements of NFPA 1001 for Firefighter II, a firefighter must be willing and able to take on additional challenges and roles. Here two firefighters perform vehicle extrication.

FIRE DEPARTMENT OFFICERS

Fire departments are comprised of multiple levels of supervision depending on the size, authority, and responsibility of the organization. Although there are many variations, these supervisor levels range from company officer to chief officer.

Company officer

The **company officer** is responsible for the supervision of the personnel and resources assigned to a single or multiunit engine, ladder, or squad company. The quasi-military ranks of sergeant, lieutenant, and captain are typically associated with the company officer or the **NFPA Fire Officer I** designation.

The company officer must have an understanding of the organizational structure of the fire department, including its geographical characteristics and response districts. This station-level supervisor must be familiar with departmental operations, as well as safety and administrative procedures.

The company officer must be able do the following:

- Effectively communicate in writing, operate in an information management system, and effectively operate at all levels in the applicable incident-management system
- Assign tasks or responsibilities to unit members so that the instructions are complete, clear, and concise, with safety considerations addressed and the desired outcomes communicated
- Address community inquiries, project the role of the department, and deliver "all- hazards" education programs
- Recommend changes to existing departmental policies and/or implement new departmental policies at the unit level
- Perform a fire investigation to determine preliminary causes, secure incident scenes, and preserve evidence, as well as provide pertinent information to a fire investigator
- Supervise emergency operations, conduct pre-incident planning, and deploy assigned resources as required
- Integrate safety plans, policies, and procedures into daily and emergency activities. Don appropriate

levels of personal protective equipment to ensure a safe work environment for all assigned members

Battalion or district chief

The **battalion** or **district chief** (typically **NFPA Fire Officer II**) is responsible for the supervision of the personnel and resources associated with multiple units/stations or organizational disciplines within the fire department. This supervisor must be familiar with the organization of all applicable levels of government; their enabling and regulatory legislation; and the functions of other bureaus, divisions, agencies, organizations, and entities as they relate to the fire department.

The battalion or district chief must do the following:

- Initiate actions to enhance or correct member performance within department policies and procedures
- Develop policies and procedures identifying challenges and providing solutions
- Prepare organizational unit budgets and news releases
- Conduct inspections to identify hazards and address code violations
- Conduct fire investigations to determine origin and preliminary cause
- Supervise multiunit emergency operations, conduct pre-incident planning, and deploy assigned resources
- Produce operational plans so that required resources and their assignments are obtained and plans are carried out in compliance with approved safety procedures, resulting in the mitigation of an incident
- Review injury, accident, and health exposure reports to identify unsafe work conditions to prevent reoccurrence

Deputy or assistant chief

The **deputy** or **assistant fire chief** (typically **NFPA Fire Officer III**) is responsible for the supervision of the personnel and resources associated with a major functional discipline (training, emergency medical services, community risk reduction, etc.) within the fire department. This supervisor must keep current with national and international trends and developments related to the fire service and be familiar with public and private organizations that interact with the fire and emergency services.

The deputy or assistant fire chief must do the following:

- Have the ability to research, use evaluative methods, analyze data, communicate orally and in writing, and motivate members
- Establish procedures for hiring, assigning, promoting, and encouraging development of members
- Establish member assignments to maximize efficiency in accordance with policies and procedures
- Prepare community awareness programs
- Prepare divisional or departmental budgets, solicit bids, plan for resource allocation, and work with information management systems
- Evaluate inspection programs to determine effectiveness and develop public safety plans
- Evaluate and identify construction, alarm, detection, and suppression features that contribute to or prevent the spread of fire, heat, and smoke
- Develop pre-incident plans
- Manage multiagency planning, deployment, and operations so that the required resources are determined and the resources are assigned to mitigate an incident
- Develop, manage, and evaluate a departmental safety program
- Develop a measurable accident and injury prevention program

Fire chief (chief engineer, fire commissioner)

The **fire chief** (also known as **chief engineer** or **commissioner** in some jurisdictions) (**NFPA Fire Officer IV**) must serve as a department head and must possess advanced administrative, financial, communications, political, legal, managerial, analytical, and information management skills and abilities.

The fire commissioner or fire chief must do the following:

- Administer and evaluate job performance requirements

- Appraise the fire department's human resource demographics
- Organize and implement recruitment, selection, and placement processes consistent with law and current best practices
- Project a positive image of the fire department to the community
- Attend, participate in, and assume a leadership role in community events to enhance the image of the fire department
- Accomplish long-range planning and fiscal projections
- Develop a comprehensive disaster plan that integrates other agencies' resources to effectively mitigate the impact on a community
- Maintain, develop, and provide leadership for a risk-management program to reduce department injuries and property damage
- Interface with other department heads, a city manager (in some jurisdictions), and elected officials

ADDITIONAL FIRE DEPARTMENT FUNCTIONS

As discussed under fire department organizational structure above, modern fire and emergency service organizations provide numerous functions and services in addition to their typical fire suppression responsibilities. These "additional" functions, depending on jurisdictional needs, may include EMS, community risk reduction (fire-prevention and life-safety education, code enforcement, fire investigation), training, fleet and equipment maintenance, special operations, and administrative or logistical support (fig. 3–13).

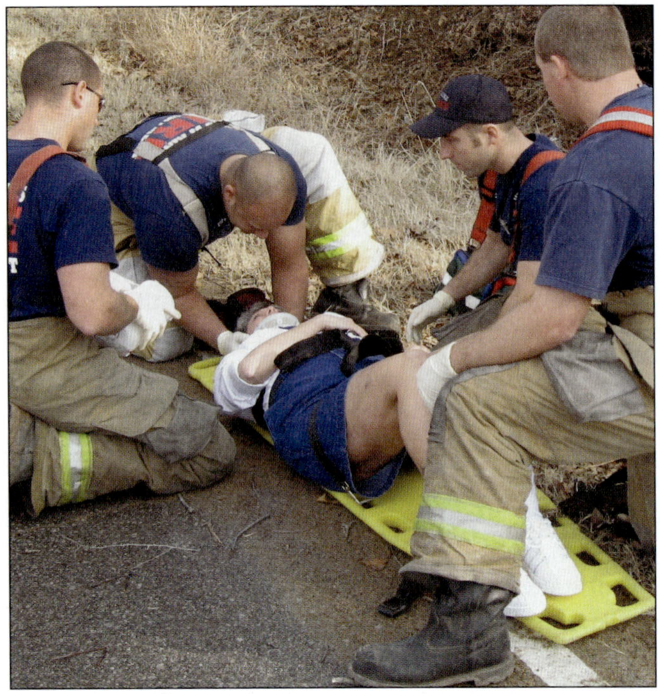

Fig. 3–13. For the average firefighter, over 50% of responses will be to emergency medical incidents. Firefighters must realize that they have responsibilities to both fire suppression and emergency medicine.

Emergency medical services

Many fire departments provide EMS to their community. This evolution to a fire-department-based EMS generally depends on the department having the required infrastructure to provide this service, an existing central dispatch resource, strategically placed stations, a centralized training resource, and a 24-hour career and/or volunteer staff.

Most fire-department-based EMS systems are staffed with firefighters with specialized EMS training, skills, knowledge, and certification, depending on community and jurisdiction needs and provider-level requirements. These provider levels are typically categorized as follows:

- **Emergency medical responder** (formerly first responder)—Provision of first aid and cardiopulmonary resuscitation (CPR)
- **Emergency medical technician** (formerly EMT Basic, EMT-B)—Considered to be the minimum level of certification required to function on an ambulance. For fire-department-based EMS systems, EMT certification is typically required during Fire Fighter I and II training.
- **Advanced medical technician** (AMT, formerly EMT-Intermediate/85, EMT-I/85)—This

provider level serves to bridge the practical skills and capabilities of EMTs and paramedics. AMTs are frequently found where the paramedic provider level is not provided.

- **Paramedic** (formerly EMT-Intermediate/99, EMT-I/99, EMT paramedic)—The care and skill capability of this certification level provides for the administration of various medications and invasive EMS procedures under medical direction.

These personnel operate under the auspices of a medical director within state and/or jurisdictional performance protocols. The emergency medical responder, EMT and AMT provider levels are usually components of a basic life support (BLS) system. The paramedic provider level is part of an advanced life support (ALS) system.

Other major considerations for communities and jurisdictions receiving fire department-based EMS is whether the patient transport function should be handled by the fire department or another service or contractor, and whether the fire department should provide BLS or ALS services. Although there are numerous variations of fire department EMS delivery systems, the major options are responding with the appropriate level of service:

- BLS staff and resources to stabilize the patient and transport to a medical facility as required
- ALS staff and resources to stabilize the patient and transport to a medical facility as required
- BLS or ALS staff and resources to stabilize the patient, and another agency transporting the patient to a medical facility. The fire department response in this variation may be accomplished by the closest company staffed with the appropriately certified personnel including an engine company, a ladder company, a squad company, or a vehicle specially designed for this purpose.

NFPA 450, *Guide for Emergency Medical Services and Systems*, provides an excellent reference for community or jurisdictional EMS programs.

Community risk reduction

The transition of fire department responsibilities to an all-hazard approach necessitates the need for progressive fire service organizations to reassign their life- and fire-safety efforts from the traditional fire-prevention program to **community risk reduction**. This reassignment more accurately addresses the contemporary educational, regulatory, and enforcement requirements of a modern fire department involved in such activities as life- and fire-safety education, hazardous materials, EMS, technical rescue, homeland security, code application and enforcement, fire investigation, and **special operations**.

Firefighters may be assigned significant roles in a fire department's community risk-reduction program commensurate with their training, experience, and certification (fig. 3–14). The following standards identify the minimum knowledge, skills, and abilities required of firefighters assigned to community risk-reduction responsibilities:

NFPA 1031, *Standard for Professional Qualifications for Fire Inspector and Plan Examiner*

NFPA 1033, *Standard for Professional Qualifications for Fire Investigator*

NFPA 1035, *Standard for Professional Qualifications for Public Fire and Life Safety Educator*

Fig. 3–14. Although many fire departments have dedicated community risk reduction personnel, it is the responsibility of each firefighter to preach fire prevention and risk reduction.

Special operations

Fire departments organize specialized operational response units to address unique challenges within their respective communities or jurisdictions (fig. 3–15). These specialized units may include hazardous materials mitigation, swift water rescue, ice rescue, confined space, technical rescue, rapid intervention teams, and explosives disposal. Firefighters assigned to these units must have specialized knowledge, skills, and abilities.

Fig. 3–15. Firefighters are responsible for a wide variety of skills involving technical rescue and hazardous materials. Here firefighters operate a boat to perform a water rescue.

Training

FFI 4.1.1 A fire department, regardless of size and mission, provides a training resource in its organizational structure. This resource ranges from an individual in a small single-station department or volunteer organization to a complete organizational division headed by a chief officer in a large metropolitan fire department.

The responsibility is the same, however: administering and documenting all training activities. The department's training officer frequently serves as the organization's safety officer. This assists the department in complying with applicable state or federal Occupational Safety and Health Administration (OSHA) requirements and addresses NFPA 1500, *Fire Department Occupational Safety and Health Program*.

The fire department training function administers and documents the training and continuing education programs for all members of the organization including recruits, probationary firefighters, company officers, and chief officers (see fig. 3–16). Ongoing training for updating firefighter skills and providing specialized training such as fire investigation, specialized rescue techniques, and code enforcement are all part of a fire academy's training responsibilities.

Fig. 3–16. In order to keep skills and knowledge accurate and up to date, firefighters must receive continuing education and training on all the different types of responses they may be responsible for. Here a firefighter trains on a circular saw under the supervision of a senior firefighter.

Communication/dispatch

A number of fire departments provide a 9-1-1 communications/dispatch resource for their community or jurisdiction (fig. 3–17). Most **public safety answering points (PSAP)** are facilitated by local law enforcement where calls for assistance are initially received and then transferred to the fire department for dispatch. There are variations to this process that include the fire department providing the initial answering point.

Fig. 3–17. 9-1-1 Centers are responsible for receiving emergency calls and dispatching the appropriate agencies and units to the emergencies.

A PSAP may be staffed with civilian personnel, law enforcement personnel, uniformed firefighters, or an appropriate combination as determined by the local community or jurisdiction. NFPA 1061, *Standard for*

Professional Qualifications for Public Safety Telecommunicator, provides the minimum qualifications required of individuals assigned to a 9-1-1 communications/dispatch center.

Apparatus and equipment acquisition and maintenance

Ensuring that apparatus and equipment are in good working order for emergency response and use is an obvious priority for a fire department. The resource and organization provided by a fire department for this program depends on the size and service demands of the community or jurisdiction and local policy.

Operational fire department personnel work closely with qualified fire apparatus- and equipment-maintenance personnel to formulate specifications for acquisition of new vehicles, tools, and appliances.

Maintenance facilities and personnel for fire department apparatus and equipment may be shared with other municipal organizations in larger communities or outsourced by smaller organizations. NFPA 1071, *Standard for Emergency Vehicle Technician Professional Qualifications*, is the generally accepted standard for the minimum qualifications for fire apparatus-maintenance personnel.

WELLNESS, RISK-MANAGEMENT, AND RECRUITMENT PROGRAMS

FFI 4.1.1 The most valuable resource of a fire department is its staff. Commensurate with local, state, and federal requirements, fire departments or their human resources offices typically provide programs for health and wellness, risk management, employee assistance and recruitment/retention, as well as other ongoing member support efforts.

Physical fitness and a healthy lifestyle

Firefighters must recognize the importance of physical fitness and a healthy lifestyle as it relates to their duties and responsibilities. Routine participation in a prescribed physical fitness program coupled with a healthy lifestyle can enhance firefighters' ability to perform required tasks and activities safely and reduce their risk of injury and premature death.

Progressive fire departments require an exercise program that includes an annual fitness assessment. The focus of this assessment is typically aerobic capacity, body composition, strength, endurance, and flexibility (fig. 3–18).

Fig. 3–18. Regular exercise, proper diet, and medical checkups are essential to all firefighters. Firefighters must have the discipline to follow a healthy lifestyle that provides for a long career and healthy retirement. (Courtesy of Frank Ricci)

Maintaining a healthy lifestyle and level of fitness is not only a benefit to firefighters but also encourages the confidence of fire department team members and the individuals the firefighters are called on to serve.

The National Fire Protection Association, through its NFPA 1582, *Standard on Comprehensive Occupational Medical Program for Fire Departments*, and 1583, *Standard on Health-Related Fitness Programs for Fire Department Members*, provides excellent job-related, references for firefighters. Additionally, the Joint Labor Management Wellness-Fitness Initiative, developed by the IAFF and the IAFC, provides a tool to be used in conjunction with NFPA 1583.

Risk management

Whether a fire department is an independent corporation, a municipal body, or a federal entity, risk management for members and employees is a high priority. Risk-management programs, for example, may encompass defining an emergency incident safety culture, establishing a departmental safety committee and safety officer, medical and workers' compensation insurance, substance abuse programs, and an **employee assistance program (EAP)**. Many of these programs are addressed in NFPA 1500, *Standard on Fire Department Occupational Safety and Health Program*.

Recruitment and retention

Fire department entities are finding it more and more difficult to attract and retain volunteer members. This phenomenon is the result of a number of factors, including increased time and training requirements, increased incident call volume, and aging communities, to name a few. The U.S. Fire Administration (USFA), the Department of Homeland Security (DHS), and the National Volunteer Fire Council (NVFC) recently published a comprehensive study addressing the decline in volunteer participation in the fire service (*Retention and Recruitment for the Volunteer Emergency Services: Challenges and Solutions*. U.S. Fire Administration, #FA-310, May 2007).

This study outlines a number of incentives designed to attract and retain volunteer firefighter candidates, many of which could be applied to career candidates. These incentives, depending on the jurisdiction, might include a **length of service award program (LOSAP)**, pay per call or per hour, tax exemptions or deductions, health insurance, tuition assistance, and low-interest housing loans.

POLICIES, PROCEDURES, REGULATIONS, AND BYLAWS

As with any organization or business, the fire department must operate within applicable policies, procedures, regulations, or bylaws. These documents may be promulgated from within the fire department or adopted by an outside entity with application to the fire department. They may be operational or administrative in nature.

Policies and procedures

FFI 4.1.1 Policies and procedures are typically used by the fire department to document the "how-to" in a consistent manner so that all involved understand what is expected of them and their coworkers and how to accomplish it. These documents may be used to communicate administrative or operational matters, are usually codified for easy filing, retrieval and access, and provide for easy updating.

Standard operating procedures (SOPs) or **standard operating guidelines (SOGs)** typically address operational matters such as hose loads, water supply, high-rise operations, and safety requirements. Policies are more often associated with administrative matters such as time and attendance and personnel matters, and are generally promulgated from management or senior staff.

Regulations

Regulations are legal requirements and generally have enforcement or compliance requirements. OSHA and the Department of Labor (DOL), for example, use regulations to communicate their legal requirements.

INCIDENT COMMAND SYSTEM

For most functions performed by the fire service, the typical organizational model does fine. Recent events have caused the fire service to adopt organizational systems that conform to many federal and state regulations when responding to emergency incidents. The **incident command system (ICS)** is an organizational system that provides command and control at these emergency incidents and meets the requirements of government regulations.

The ICS has been used in various styles by the fire service since the early 1970s. The latest method allows the fire service to operate at any type of incident, natural or man-made, and coordinate operations with many other organizations not directly connected to the fire service. ICS provides for improved communications, management of resources, effective span of control, unity of command, effective division of labor, and a coordinated action plan. How ICS works with structural firefighting and hazardous materials incidents will be covered in later chapters and a more detailed explanation of ICS will be covered in Firefighter II.

A discussion of ICS typically covers command staff, general staff, functional areas, and facilities.

The command staff consists of the incident commander, safety officer, information officer, and liaison officer.

- The **incident commander** is the individual responsible for the management of all incident operations at the incident site.
- The **safety officer** is a member of the command staff responsible for monitoring and assessing safety hazards or unsafe situations and for developing measures for ensuring personnel safety.
- The **information officer** is a member of the command staff responsible for interfacing with

the public and media or with other agencies that require information directly from the incident.

- The **liaison officer** is a member of the command staff responsible for coordinating with representatives from cooperating and assisting agencies.

The general staff is a group of incident management personnel reporting to the incident commander. The **general staff** consists of the operations section, planning section, logistics section, and finance/administration section.

- The **operations section** is responsible for all tactical operations at the incident. Includes branches, divisions and/or groups, task forces, strike teams, single resources, and staging areas (fig. 3–19).
- The **planning section** is responsible for the collection, evaluation, and dissemination of tactical information related to the incident, and for the preparation and documentation of incident action plans. This section also maintains information on the current and forecasted situation as well as the status of resources assigned to the incident. This includes the situation, resource, documentation, and demobilization units, as well as technical specialists.
- The **logistics section** is responsible for providing facilities, services, and materials for the incident.
- The **finance/administration section** chief is responsible for all incident costs and financial considerations. This includes the time unit, procurement unit, compensation/claims unit, and cost unit.

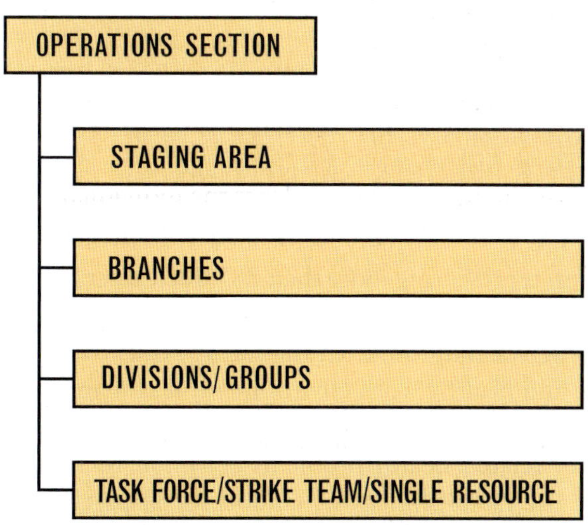

Fig. 3–19. Operations section positions

Common incident facilities include the following:

- The **incident command post** (ICP) is the location where the primary command functions are executed. The ICP may be located within the incident base or other incident facilities.
- The **staging area** is a location set up at an incident where resources can be placed while awaiting a tactical assignment. Staging areas are managed by the operations section.
- **Base** is the location where primary logistics functions for an incident are coordinated and administered. There is only one base per incident.
- **Camp** is a geographical site within the general incident area but separate from the incident base. It is equipped and staffed to provide sleeping, food, water, and sanitary services to incident personnel.

ALLIED ORGANIZATIONS AND THE FIREFIGHTER

FFI 4.1.1 The firefighter's duties and responsibilities interface with and are influenced by a number of external agencies and organizations. These entities range from neighboring fire departments to international regulatory organizations.

Mutual aid plans and automatic aid agreements

Because no single fire department can be expected to possess the resources required to resolve every possible emergency that may occur in their jurisdiction, or because the resources of a neighboring fire department may be geographically closer to an emergency, a fire department may enter into **automatic aid** or **mutual aid** agreements with neighboring organizations.

Mutual aid agreements between fire departments involve reciprocal assistance under a prearranged plan. As an example, when a target hazard or specialized hazard exists in one fire department's responsible jurisdiction that potentially exceeds that fire department's resources, a reciprocal agreement is prearranged, and the neighboring fire department responds for assistance. This type of agreement is also used in the event the closest fire department to a particular incident is committed to another response.

Automatic aid is a plan developed by two or more fire departments for the immediate, joint response to incidents. Automatic aid plans are typically developed when resources required for an initial response or fire attack are not available from the responsible or closest fire department, or neighboring fire departments are geographically closer than the responsible fire department.

These aid plans and agreements delineate requirements and expectations of each of the participating organizations, ranging from indemnity, training levels, and the number of personnel that will respond, to incident command responsibilities. Compensation arrangements are typically addressed when there is a disparity in the frequency and resources routinely responding from one fire department to another.

National Fire Protection Association (NFPA)

The NFPA was established in 1896 as a nonprofit organization to reduce the worldwide burden of fire and other hazards on the quality of life by providing and advocating consensus codes, standards, research, training, and education. The NFPA currently has more than 81,000 members, representing more than 80 national trade and professional organizations. The NFPA's 300 codes and standards influence every building, process, service, design, and installation in the United States and many other countries.

NFPA standards that have an influence on the firefighter, depending on their duties and responsibilities, include, but are not limited to, the following:

- NFPA 1001, *Standard for Fire Fighter Professional Qualifications*
- NFPA 1002, *Standard on Fire Apparatus Driver/Operator Professional Qualifications*
- NFPA 1003, *Standard for Airport Fire Fighter Professional Qualifications*
- NFPA 1005, *Standard for Professional Qualifications for Marine Fire Fighting for Land-Based Fire Fighters*
- NFPA 1006, *Standard for Rescue Technician Professional Qualifications*
- NFPA 1031, *Standard for Professional Qualifications for Fire Inspector and Plan Examiner*
- NFPA 1033, *Standard for Professional Qualifications for Fire Investigator*
- NFPA 1035, *Standard for Professional Qualifications for Public Fire and Life Safety Educator*
- NFPA 1051, *Standard for Wildland Fire Fighter Professional Qualifications*
- NFPA 1061, *Standard for Professional Qualifications for Public Safety Telecommunicator*
- NFPA 1081, *Standard for Industrial Fire Brigade Member Professional Qualifications*
- NFPA 1500, *Standard on Fire Department Occupational Safety and Health Program*
- NFPA 1581, *Standard on Fire Department Infection Control Program*
- NFPA 1582, *Standard on Comprehensive Occupational Medical Program for Fire Departments*
- NFPA 1583, *Standard on Health-Related Fitness Programs for Fire Department Members*

International Association of Fire Chiefs (IAFC)

The IAFC was established in 1873 to represent the leadership of more than 1.2 million firefighters. Members of the IAFC are considered the leading experts in firefighting, EMS, terrorism response, hazardous materials, natural disasters, search and rescue, and public safety legislation.

Some of the IAFC programs having direct interface with firefighters include the following:

- National Fire Fighter Near-Miss Reporting System
- Vehicle Safety
- Wellness Fitness
- Hazardous Materials

International Association of Fire Fighters (IAFF)

The IAFF was officially established in 1918 and is a labor union representing more than 287,000 career firefighters and emergency medical personnel. With its headquarters in Washington, DC, the IAFF works to promote a safe and healthy work environment, reasonable working conditions, the research and treatment of burns and other related health problems common to firefighters, and the establishment of schools for imparting knowledge of modern and improved methods of firefighting and prevention.

National Volunteer Fire Council (NVFC)

The NVFC pursues the interests of volunteer firefighters and volunteer fire departments. It is an organization comprising state firefighter organizations involved in providing training programs, public education, conferences, and the compilation of pertinent statistics. The NVFC represents its members' interests to the U.S. Congress and federal agencies.

Department of Homeland Security (DHS)

DHS was established after the September 11, 2001, terrorist attack, when duplications and gaps were exploited in federal, state, and local government resources. DHS integrates these resources to protect the American homeland, including the Federal Emergency Management Agency (FEMA), the USFA, and the National Fire Academy. The primary benefit to firefighters and fire departments is grant funding, incident preplanning and coordination, equipment, and training.

Occupational Safety and Health Administration (OSHA)

OSHA, part of the U.S. Department of Labor, was established to ensure safe working conditions. Many states have established a local OSHA entity to enforce the federal regulations. Whether enforced at the state or federal level, OSHA workplace requirements apply to volunteer and career fire departments.

National Institute for Occupational Safety and Health (NIOSH)

NIOSH supports the OSHA through research and educational training. The Institute recommends occupational safety and health standards and undertakes investigations of firefighter and other occupational incidents when requested.

Insurance Services Office (ISO)

ISO is a nonprofit voluntary association of insurers that gathers data for the purpose of setting fire insurance rates. Using its *Grading Schedule for Municipal Fire Protection*, ISO provides an analysis of public fire services and their ability to defend the community against a major fire. The analysis includes a review of the fire department, available water supply, the fire service communication system, and the fire-safety control capabilities.

STATISTICAL OVERVIEW

The NFPA reported that there were 29,727 fire departments in 2015. These departments were staffed as follows:

All career	2,651
Mostly career	1,893
Mostly volunteer	5,421
All volunteer	19,762

Tables 3–1, 3–2, and 3–3 provide a 5-year presentation of pertinent fire department and firefighter data, including the number of career and volunteer firefighters, the number of fire departments, the number of fire and nonfire incidents firefighters responded to, firefighter injuries, and firefighter deaths. Career firefighters include full-time, uniformed firefighters regardless of assignment, but do not include state, federal, or private fire-brigade personnel. Volunteer firefighters include active part-time and paid on-call.

Table. 3–1. Firefighter and fire department statistics

Year	Career firefighters	Volunteer firefighters	Total firefighters	Number of fire departments
2011	344,050	756,400	1,110,450	30,145
2012	345,950	783,300	1,129,250	30,100
2013	345,600	786,150	1,140,750	30,052
2014	346,150	788,250	1,134,400	29,980
2015	345,600	814,850	1,160,450	29,727

Source: NFPA

Table. 3–2. Incidents and firefighter injuries

Year	Fire incidents	Fire incident injuries	Nonfire incidents	Nonfire incident injuries
2011	1,389,500	30,505	28,708,500	14,905
2012	1,375,000	31,490	31,854,000	13,820
2013	1,240,000	29,760	30,404,505	12,535
2014	1,298,000	27,015	30,346,500	14,595
2015	1,345,500	29,130	32,257,000	14,320

Source: NFPA

Table. 3–3. Career and volunteer firefighter fatalities

Year	Total	Career firefighters	Volunteer firefighters	Nonmunicipal
2012	64	23	30	11
2013	97	25	41	31
2014	64	23	34	7
2015	68	24	32	12
2016	69	19	39	11

Source: NFPA

Most career firefighters (71%) protect communities of 25,000 or more people, while 95% of volunteer firefighters protect communities with fewer than 25,000 people. More than half of volunteer firefighters are located in rural fire departments protecting fewer than 2,500 people.

QUESTIONS

1. Discuss the differences between policies and procedures, operating guidelines, and regulations.
2. What are the differences between mutual aid agreements and automatic aid agreements? In what instances would one or both be used?
3. Explain the different levels of officers and their responsibilities (for example, fire chief, deputy chief, battalion chief, company officer).
4. A ________________________ is a type of apparatus with a permanently mounted fire pump, a water tank, a hose storage area, an aerial ladder or elevating platform with a permanently mounted waterway, and a complement of ground ladders.
5. Hazardous materials, swift water rescue, confined space, and rope rescue are all examples of ________________________ operations.
6. Name the different EMS certification levels and what general level of care each can provide.
7. The most valuable resource a fire department has is its ____________________.
8. What levels of firefighting certifications exist, and what skills are taught to each level?

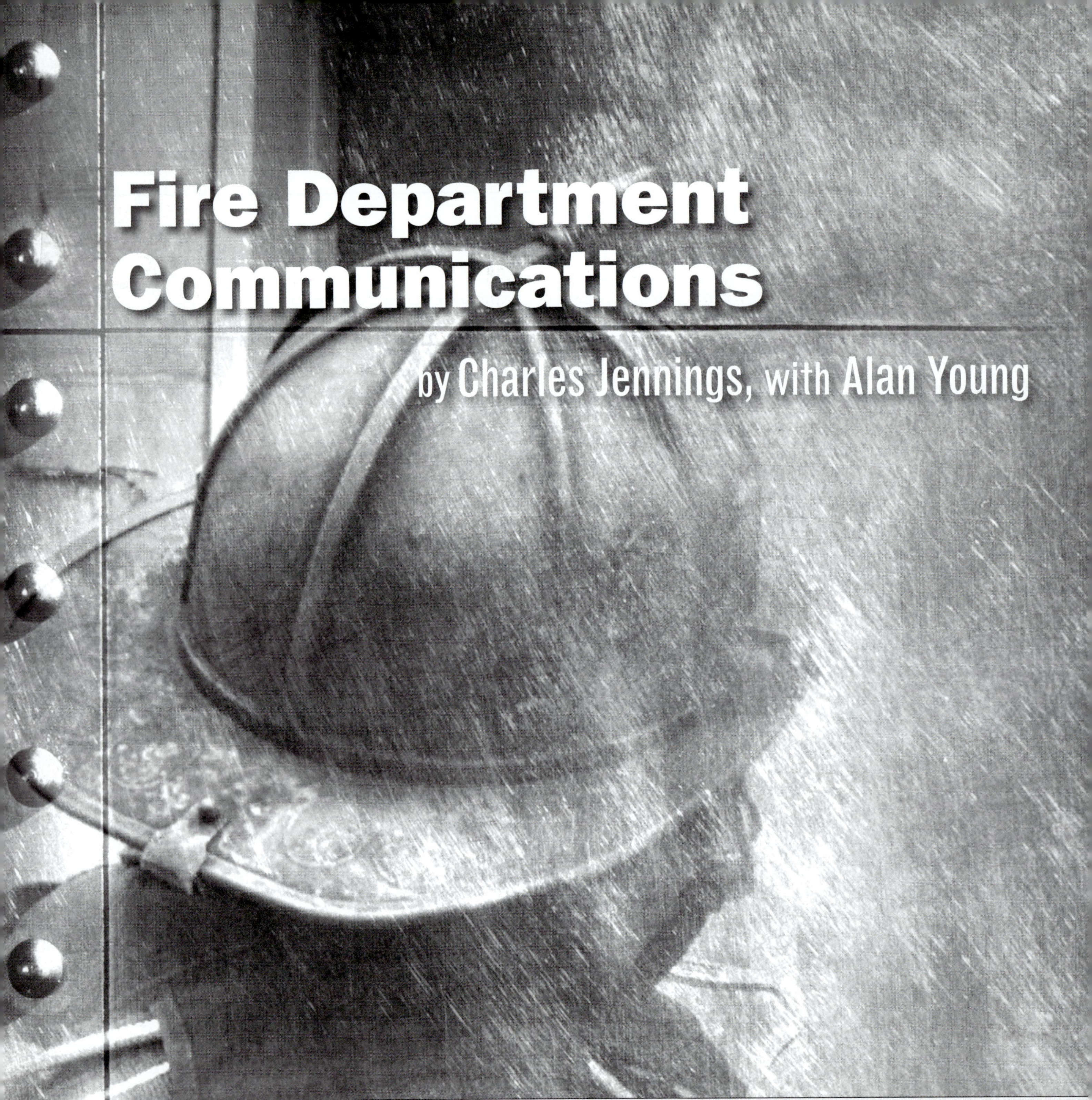

Fire Department Communications

by Charles Jennings, with Alan Young

This chapter provides required knowledge items for the following NFPA Standard 1001 Job Performance Requirements:

FFI 4.2.1

FFI 4.2.2

FFI 4.2.3

FFI 4.2.4

4

This chapter contains Skill Drills. When you see this icon, refer to your Skill Drill book for step-by-step instructions. SKILL DRILL

OBJECTIVES

Upon completion of this chapter, you will be able to do the following:

- Identify the procedures to follow when receiving an alarm or reporting an emergency
- Identify how to receive an alarm via telephone or in-person
- Identify how emergency equipment is dispatched
- Identify emergency communication procedures to be used at an emergency incident
- Describe the policy and procedures for ordering and transmitting multiple alarms
- Define the purpose and function of all alarm receiving instruments and personnel alerting equipment provided in the fire department
- Identify and describe personnel accountability systems (PAS)

INTRODUCTION

Communication within a fire department, its community, or field of service appears at a glance to be a simple matter of picking up the phone, and then all is well. In reality, it is a complex system designed to meet any and all criteria for a limitless number of circumstances and scenarios. There are several standards organizations that develop model codes, standards, and procedures for communications centers. Communications centers of all sizes should strive to conform to standards required by the National Fire Protection Association (NFPA), National Emergency Number Association, and Associated Public-Safety Communications Officials.[1]

Communication systems are a cornerstone of public fire protection, and their importance is well recognized. The system must work when all else does not. Accurate reporting, dispatching, and effective responses by fire companies and other agencies are essential to save both lives and property. With all of these elements—24-hour service, meeting industry standards, and ensuring the security of resources—the business of public safety communications is indeed an intricate one. The information in this chapter shows how the multifaceted fire communications systems work and interact to provide the public a greater degree of safety. This chapter also contains information to provide context for the Job Performance Requirements in the standard. The standard requires familiarity with your own department's policies and equipment.

INITIATING AND RECEIVING ALARMS

FFI 4.2.1 The primary purpose of the public fire service communication system is to quickly and authoritatively accept reports of emergencies, and dispatch the appropriate units in a timely fashion. There are a number of technological aids that have come into use to facilitate the job of the public safety dispatcher or telecommunicator. Before we consider the entire system, we will focus on the initiation and receipt of alarms.

From the standpoint of the firefighter assigned to a company, perhaps the most fundamental aspect of fire service communications is initiating and receiving alarms.

There are several means of reporting alarms. These methods can include the following:

- Municipal fire alarm
- Station alerting system
- Telephone
  - Land line
  - Mobile
  - Cellular
  - Voice over Internet Protocol (VoIP)
  - Telephone and Relay Services (TRS)
  - Direct Connection Lines
- Private fire alarm
- Still (walk-in) alarm

Each of these systems may be in use within a community. As technology progresses, the relative importance of each of these systems has changed over time. Each technology brings its own limitations and challenges. The earliest means for reporting of emergencies was the municipal fire alarm system. These systems, which originated in the mid-1800s, were widely used in urbanized areas, especially before the widespread utilization of telephones. They were based on telegraph technology. Many remain in service today, although almost no new systems are being installed because of the cost to maintain a dedicated network of wiring and infrastructure. In some areas, frequent false alarms led to removal of some boxes, or their replacement with voice-capable systems that could be used to potentially detect false alarms.

These municipal box systems reported a coded signal to the local receiving point (usually fire dispatch or even multiple fire stations). Personnel decipher the coded signal which translates back to an intersection or an address. These systems also permitted interconnection with fire alarm systems installed in major buildings. These systems were highly reliable and fast. Each box had the capacity to permit communication back to the dispatch center via telegraph, or later, voice. In many older areas of the country, these systems remain in service (fig. 4–1).

Fig. 4–1. Typical municipal telegraph-type fire alarm reporting box

A modern variant on these systems uses telephone systems distributed throughout a community that dial directly into the fire alarm or dispatch center (fig. 4–2).

Telephones have been a mainstay for the reporting of alarms in most of the country since the turn of the century. The community fire reporting system relies on people with telephones to report emergencies. Through the advent of 9-1-1 systems in 1968, a common emergency reporting number was established. Using this system enabled the development of technologies to improve the efficiency and effectiveness of emergency reporting. According to the National Emergency Number Association, approximately 93% of the U.S. population was covered by some level of 9-1-1 service in the year 2000. Improvements to 9-1-1 include the ability to route calls to a designated public safety answering point (PSAP) based on the location of the telephone number, and enhanced 9-1-1, which permits identification and location information to be associated with the phone number to be displayed at the PSAP.

Direct-line phones, also known as "hotlines" are dedicated phones that connect directly to the communication center or other agency. These phones are typically on a dedicated handset, and the connection can be via a dedicated phone line, or a "quick dialer" using a conventional phone line (fig. 4–3).

The widespread use of wireless (cellular) phones has created a challenge for 9-1-1 systems. While land-line phones obtain location information from a central database, the wireless caller can be anywhere. The Federal Communications Commission (FCC) requires that wireless calls be routed to the nearest PSAP based on the cell tower they connect to. Once the call is routed, the caller's number and latitude and longitude are provided to the PSAP. Wireless location is referred to as Phase I and Phase II. Phase I location only refers to the tower from which the call originated. Phase II location includes a latitude and longitude of the calling phone. Cellular carriers are working to improve the accuracy of this Phase II location information. The FCC is implementing enhanced requirements for accuracy currently within 50 to 300 meters.[2]

Until all these improvements are implemented nationwide, dispatch centers will continue to be frustrated by receiving calls from people who may not be in their service area, not know their location, or not know the location of an incident. Wireless phones also increase the workload on communication centers, as multiple reports of an incident are common, particularly for events such as traffic accidents on busy highways.

Wireless callers may also connect to a neighboring jurisdiction's PSAP, as their signal may reach a tower outside that jurisdiction. This is common along borders of densely populated areas, especially along boundaries such as rivers where a clear line of sight makes this a common occurrence.

Fig. 4–2. A municipal alarm system using dedicated 9-1-1 phones

Fig. 4–3. Direct-dial telephone in a fire station

The recent adoption of "internet-based" telephones, operating on Voice over Internet Protocol (VoIP) has complicated the task of associating a location with a particular telephone number. Some of these phone numbers can be associated with a portable computer, and have the capability of "following" the user wherever they go. Requirements are in place to require identification of the callback number and "home" location of the VoIP number to a local PSAP, but these are not entirely complete.

To accommodate the needs of deaf or hearing impaired members of the public, PSAPs should be equipped with devices to permit communication. Typically, this is accomplished through TTY, or teletypewriter services. However, several other specialized technologies may also be in use. There are also telecommunications relay services (TRS) that can perform this service is the communication center does not have this equipment.

Private fire alarm systems are used in commercial and multifamily buildings, but are gaining in popularity in private homes, as well. These alarm systems can range from simple smoke detectors to complex arrangements of smoke and heat detection, manual pull boxes, water flow alarms, and related equipment. Private fire alarms are privately maintained systems, whether required by local codes or not, that are monitored by a central station or other arrangement which in turn transmits the alarm to the fire services.

For many departments, automatic fire alarms are the most common type of fire-related incident. This increase in workload as a result of accidental, unintentional, and false alarms is offset by the early notification afforded by such systems. Earlier detection and reporting of fires leads to smaller, more easily controlled fires, this fulfilling a valuable public policy purpose. Some jurisdictions penalize owners of systems with excessive false alarms, which is a way of encouraging proper system maintenance and reducing the number of unnecessary responses.

The final method of reporting alarms is via a "still" or walk-in alarm reported directly to a fire company or personnel at a fire station. Often, people living nearby a fire station will go to the fire station rather than report an emergency by telephone. In other cases, passersby may walk in to report an incident near the fire station itself. The author vividly remembers one case in which a car on fire was driven up to the front of a firehouse—much to the surprise of personnel in the station.

Direct notification to a fire station of some type of incident still occurs every day in every community and will continue to be a reporting method. Firefighters need to understand that this vital information must be obtained from the person relaying the event and transferred to the communications center as rapidly and efficiently as possible. Firefighters must follow standard operating guidelines for information gathering for an event and use the proper procedure to transfer that information. Whatever the means of transferring that information—by radio, direct-dial hard-wired telephone, or a computer application—the process must be always thorough and precise (fig. 4–4).

Fig. 4–4. Firefighters must be able to gather the information reported directly to them.

RECEIVING CALLS

FFI 4.2.2 While emergency calls are intended to be received at the designated communications center accessible via the 9-1-1 system, the fire station telephone is still an important component of the emergency communication system. As a firefighter, you must be familiar with procedures for receiving emergency information form the public, other agencies, or your own communications center. While we refer to telephones in this section, these communications may be received via a dedicated phone line, intercom, or a publicly-accessible phone number (fire department or fire station "business" line). It is important to remember that any time the phone rings in a fire station, it could be an emergency.

Calls for assistance are generally received at a communications center. They can also be received while fire companies are in the firehouse by a citizen knocking on their door or while the fire company is working in the district performing fire-prevention activities, training or doing inspections. Policies should also spell out instructions for companies that encounter emergencies while en route to another call.

Calls for assistance are generally classified as nonemergency or emergency.

Nonemergency calls

When answering nonemergency calls, you should answer the phone promptly and professionally, following your department's guidelines for identifying emergency and nonemergency calls. You should be immediately prepared to take messages and record all pertinent information, including date, time, name of caller, caller's contact information, and the message.

Always record your name with the message and transfer the information promptly to the person or department it is intended for. It is sometimes appropriate to refer a caller to the appropriate authority or agency, according to your department's referral guidelines. Good customer service would suggest that an appropriate referral be made.

Emergency calls

SKILL DRILL

Procedures must be in place for handling emergencies reported directly to personnel at a fire station or on apparatus. For telephone calls, the process begins by identifying the agency and station being called, and the individual answering the phone. The same information that a dispatcher requests should be obtained to include the address, the nature of incident, and the identification or contact information for the reporting party. Members of the company or station must be alerted, and the dispatch center should be notified immediately that the company or units are out on service dealing with an incident reported directly by the public.

The National Emergency Number Association (NENA) has a model policy, NENA Call Answering Standard/ Model Recommendation, which offers guidance for Public Safety Answering points, which is applicable to fire service dispatching.[3] Associated Public-Safety Communication Officials also provides training, job aids, and certification in fire and EMS dispatch (apcointl.org).

Persons reporting an emergency may be stressed, and need support to properly relate the nature and location of an incident. In these cases the firefighter must act as a "telecommunicator" and capture the necessary information to assure an appropriate and timely response. This information must include:

- Location
- Nature of the problem
- Caller's name
- Contact phone number
- Condition at caller's location

FIRE DEPARTMENT COMMUNICATIONS EQUIPMENT

SKILL DRILL

FFI 4.2.3 This section of the chapter will address the "hardware" aspects of fire service communications. We will consider radio systems, system features, user equipment, and ancillary hardware.

Public safety radio systems

Public safety radio systems can be described as systems of radio transmission and reception infrastructure, mobile and portable user equipment, and dispatch or base station location integrated on a common set of frequencies for a defined functional purpose. This definition, while specific, does not address the numerous variations possible in public safety radio networks.

Understanding radio networks is important, although from the standpoint of a firefighter, we do not necessarily need to be concerned with all of the technical specifications, but rather the capabilities and intended purpose of the system and its components.

Radio operations for the fire service are regulated by the Federal Communications Commission (FCC) (https://www.fcc.gov/public-safety-and-homeland-security). The FCC regulates the design, installation, and operation of the typical two-way radio systems used by fire departments throughout the United States. The FCC has developed regulations and guidelines governing the assignments of radio frequencies in order that all fire departments will have adequate access to needed transmission capabilities. The FCC is the assigning agency and licenses fire departments on one or multiple specific frequencies.

Radio systems can be characterized by the location of dispatching personnel, radio infrastructure, channel characteristics, number of channels, and the manner of operation. Dispatching and radio infrastructure can be centralized or dispersed; one or more frequencies may be in use; and the network may have multiple channels available.

Public safety radio systems operate on several different frequency bands:

- Very High Frequency—Low Band: 33 MHz–46 MHz
- Very High Frequency—High Band: 150 MHz–174 MHz
- Ultra High Frequency-UHF: 450 MHz–460 MHz
- 800–900 MHz

Following the scrutiny of public safety communications after the 9/11 attacks, and in response to the shortage of available frequencies for emergency response agencies, some new spectrum is being made available. In addition to these frequency bands listed above, the 700 MHz band is scheduled to become available for pubic safety as television stations (the current users) move to other frequencies. The 4.9 gigahertz (GHz) band is available now, primarily used for data communication in support of public safety.

Each channel can be set up as **simplex** or **duplex** operation. Simplex operation means that only one frequency is used for both transmission and receiving signals (fig. 4–5). Duplex means that one frequency is used for transmitting while a second frequency is used for receiving. Duplex capability can mean that dispatchers may be able to hear transmissions from the field even while they transmit.[4]

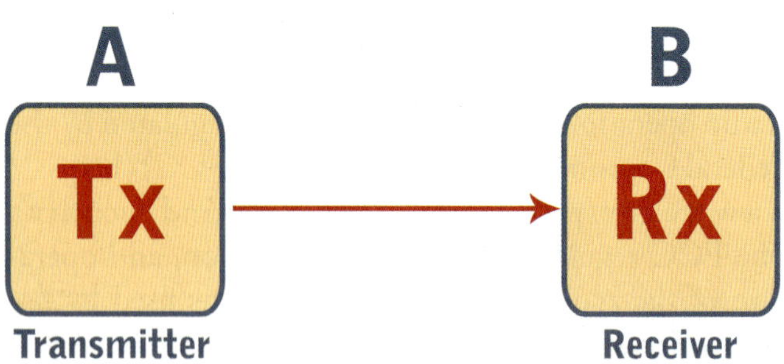

Fig. 4–5. Simplest form of radio communication is a simplex system using only one frequency to transmit and receive messages.

One desirable capability included in radio systems is unit identifiers, which display on the dispatcher or telecommunicator's console and allow the dispatcher to know what unit is calling. Both mobile and portable radios can be equipped with this capability. Another function is the ability to remotely control a radio-coded squelch, which limits the ability to hear interfering transmissions from other organizations sharing the frequency. Distress or Mayday signals can also be sent from some radios. These capabilities are examples of tone encoding. These tones can be subaudible, meaning that users monitoring the channel cannot hear them.[5]

Radios have two distinct functions: transmitting and receiving. In mobile and portable radios, both functions are combined, which is why these radios are called transceivers. Transceivers have circuitry to accomplish both functions, and an antenna that works both in transmit and receive mode. When the transmit button is pressed, the antenna is dedicated to the transmitter. Likewise, when the transmitter is not active, the antenna works in the receive mode.

A major distinction in radio systems is the **conventional** versus **trunked** systems. A conventional system is what commonly comes to mind when describing basic radio system function (fig. 4–6). In a conventional system, channels are selected by the user, and switching of frequencies is not usually necessary to complete a conversation with another user. Channels are designated and function for a known purpose. For example, channel 1 might be Fire Dispatch, channel 2 is Fireground Operations, and channel 3 is Mutual Aid.

Fig. 4–6. Components of a conventional radio system

In a trunked system, the user selects a talk group, and the system selects the channel to be used for any given conversation. Over the course of a conversation between users, it is possible that multiple channels could be used (fig. 4–7). This would be done automatically and would not be apparent to the users.

Trunked radio systems offer the ability to better utilize a limited number of frequencies by aggregating a large pool of users with the anticipation that peak demand for all user groups will occur at different times. Priorities are assigned to users, and, when a user keys their microphone, the system must give them access before they can transmit. Users are organized into **talk groups** that are usually organized by function and geography. For example, in a fire department covering a large area, talk groups may be assigned based on geography. Likewise, there may be talk groups for administration, fire prevention, and suppression. From the user perspective, each talk group behaves as though it was its own separate radio system. Users moving from a conventional to trunked system will require training in use of the system, particularly with regard to switching talk groups.

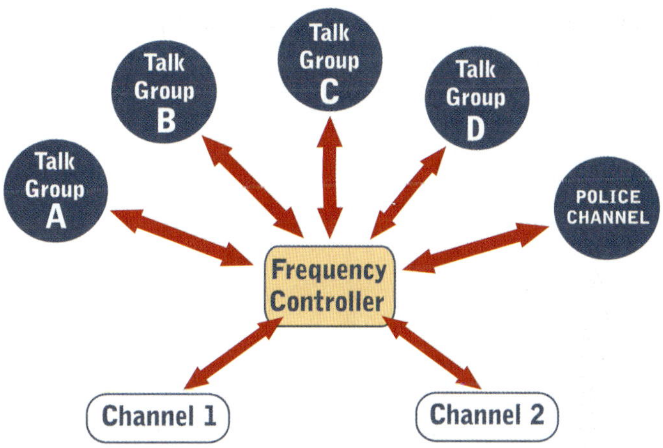

Fig. 4–7. Talk groups that have the proper talk group channel can have communications with any radio within another talk group, including those from different agencies.

Popularity of trunked systems grew out of the limited radio spectrum available, and the desire of some jurisdictions to create large, unified radio systems with the capability of linking multiple agencies. Because of their expense and sophistication, they are normally found only in large municipalities or County or metropolitan departments.

The other major distinction in radio systems is the nature of the signals being broadcast. Most systems use what is known as **analog technology**. This means that the radio signal being sent between parties on the system is the user's voice sent over the airwaves. Like an analog recording, noise or other issues can diminish the quality of the signal. The newer alternative is known as **digital technology**. In a digital system, the user's voice is converted to a digital data then converted back to voice when received. Digital signals have the capability for better transmission range, and are less susceptible to noise being introduced during the transmission process. However, software **algorithms** or programs are used to encode and decode the voice signal, and these algorithms can be affected by background noise when a user is transmitting. Microprocessors control digital systems, which require synchronization of transmitters and receivers.

Digital radio systems have a small time delay between the transmitting party's voice and the receiving party as encoding and decoding take place. This is typically not a major concern, but is perceptible when users are standing in proximity to one another. For fireground applications, there is considerable controversy regarding the applicability of digital voice radio technology because multiple users attempting to transmit are denied simultaneous access to the system. In analog systems, personnel receiving transmissions can detect multiple overlapping transmissions, which may occur, especially during critical fireground situations. For this reason, many organizations rely on direct (non-repeated), analog transmissions for tactical users.

There are numerous technologies available and widely incorporated into radio systems to increase or improve coverage. While range of a base station can be from 10 to 40 miles (16 to 64 km) depending on antenna height, topography, buildings, and output power, the transmitting power of mobile and portable radios is typically much smaller. Mobile radios typically operate in the range of 40–50 watts, while portable radios operate at 2–5 watts (figs. 4–8 and 4–9).

Multiple receivers may be distributed throughout the service area to enable lower-power **mobile** (vehicular mounted) or **portable** (hand held) subscriber units to be heard. This is especially important in fire service applications where transmissions from portable radios operating inside buildings are the norm. Radio systems use various technologies to select the best receive site to send the signal to the dispatcher.[6]

Repeaters are employed to receive transmissions from field units and rebroadcast them at higher power on the receive frequency of user equipment.

In systems covering large geographic areas, multiple transmit sites may be employed. These sites are managed using multicasting or simulcasting. **Multicasting** uses different frequencies on adjacent towers to increase range without interference. **Simulcasting** relies on a common frequency on multiple transmit sites under computer control to ensure that their transmissions are synchronized in order to avoid conflicting with each other.

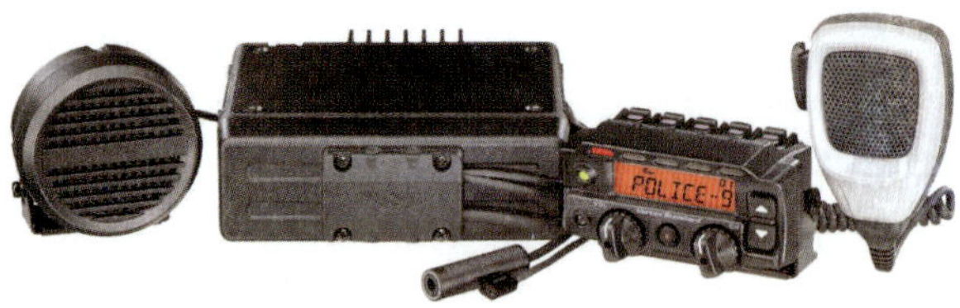

Fig. 4–8. Typical portable radio. Courtesy of PMC Associates.

Fig. 4–9. Typical mobile radio. Courtesy of PMC Associates.

Base station radios are more powerful than both the portable and mobile radios. They are designed for fixed mounting inside a communications center or fire station. The base station radios, with their antennas mounted to a radio tower, enable the radios to have a wide area of coverage. There can be multiple base radio stations located in remote locations to help support large numbers of mobile and portable radios. As more channels are added to the system, it is essential that the number of radio structures increases to support the increased radio traffic. Improved base station technology is an absolute must to ensure fast, accurate communications between the various agencies for a good event outcome.

Other communications technologies

Mobile computer terminals (MCTs) are gaining in popularity in the fire service. While widely used in law enforcement, MCTs and many of their benefits are equally applicable to the fire service (fig. 4–10). They are becoming central to a number of technologies and software applications that are useful in the fire service.

From a dispatch perspective, CAD software can send messages directly to the MCTs, allowing them to be used for dispatching units. With incident information and the address on the screen, this eliminates the need to confirm the address over the radio. **Unit status** can be updated from virtual "buttons" on the MCT, again eliminating the need for verbal radio transmissions and improving accuracy of response times. Unit status includes messages useful to the dispatcher, such as "on scene," "in service," "on the air," and "on scene." This feature is critical to many busy departments, both to conserve radio channel availability and dispatcher workload in manually entering unit status changes into a CAD or other system.

Fig. 4–10. The mobile computer terminal (MCT) installed in the cab of a fire vehicle.

As a functioning computer, the MCT has numerous capabilities, including storage of preplan information, reference material and contact information, incident command software, and related applications. Communications between the dispatch system and MCT is accomplished via radio, over commercial wireless networks, or on dedicated wireless broadband networks. A predecessor to the MCT was the **mobile data terminal (MDT)**, which is a "dumb terminal" that could only display unit status, incident, and other textual information. MDTs communicated almost exclusively over the local public safety radio system (fig. 4–11). As the need for bandwidth has increased, many agencies have turned to commercial wireless data networks to handle their data communication needs.

Fig. 4–11. The mobile data terminal (MDT) is a simple computer that can provide communication and transmit textual information, they are becoming less popular.

Automatic vehicle location (AVL) uses satellite or wireless phone signals to produce an approximate location of a vehicle or person. Early systems were constructed from dedicated infrastructure installed by the local community. AVL, when paired with geographic information system technology, can permit display of a map in the vehicle showing the path of the vehicle, and even recommendations on the route to take to a reported emergency.

AVL is especially valuable in dispatching. The ability to see units located on a map of the service area can add to safety, enhance selection of the closest unit to alarms, and manage system-wide deployment in real time. The provision of AVL is becoming standard in busy systems or systems serving large areas.

Pagers have been used for many years for emergency alerting. They remain in use, often as a backup to other systems, or for use by volunteer or on-call personnel who respond form home. They can also be used to notify others within the organization of any requests for them to respond to an event or to make contact with another person (fig. 4–12). Some pagers are activated by dialing a phone number. Other pagers are activated by an **encoder**, typically from the dispatch center. Following an alert tone, pagers send information in one of two methods: a voice message or a written display (text) message. As opposed to one-time message delivery, there are some fire departments that use pagers that have the ability to monitor radio traffic on their dispatch frequencies at all times. These same pagers are used to dispatch volunteer, on-call personnel, and on-duty staff. These pagers can be used 24 hours a day, so that personnel can be in constant contact with fire administration, operations, and communications.

Fig. 4–12. Firefighter pagers are used to notify firefighting personnel of emergency and non-emergency events.

With the increase of cellular telephone service on the rise, most pagers have been replaced or are used in conjunction with the cellular phones. There is still a need for pagers, especially for those organizations that have minimal budgets, poor cellular service, and where the pager could also be used as a dispatching tool. Cell service can become overloaded during an emergency, and this limits their reliability as a primary means of alerting.

Station Alerting Systems describe a number of systems and technologies in place that link the communications center to the fire station. These systems can vary in sophistication from simple encoded alerts that turn on speakers monitoring the dispatch radio channel, to hard-wired communication systems that may be linked to bells, horns, or electronic notification devices installed throughout a fire station. Newer systems can send alerts specific to the type of alarm, or the units that may be needed to respond.

In some older cities, the alerting may go to a "watch desk" staffed by a firefighter who is charged with monitoring the alert (via phone, radio, or printer) and subsequently notifying the appropriate unit(s) within the firehouse, often via a local public address system. The "watch" system may persist out of tradition, with a firefighter assigned to answer business phone lines, and greet the public during the day.

Communication facilities

What is a public safety communications center? A communications center facility is a unique and complex operation. A public safety communications center is a facility that operates 24 hours a day, 7 days a week, under all conditions, and during any circumstances. It must be built to conform to NFPA standards, specifically NFPA 1221, *Standard for the Installation, Maintenance, and Use of Emergency Communications Systems,* pertaining to how buildings are designed and how notifications are to be sent.[7] During an emergency, the center is responsible for taking the information of the emergency, including type of incident, location, and information about the person reporting the incident. That information is then used to alert the appropriate resources to respond to and mitigate the emergency. The communications center functions much as a quarterback who leads the team, provides the necessary plan or resources, and makes adjustments as needed to achieve the desired outcome.

Communications centers (fig. 4–13) come in many shapes and sizes. In small towns with small populations, it is not uncommon to find a one-room work area with a few dispatch positions and a small staff to handle the workload. In larger areas, the communications center may be a multistory building with 20 or more positions, requiring a large staff and training facilities. A communications center often covers only a local area of jurisdiction. However, some communications centers provide service for multiple agencies and jurisdictions. There can be great variation in organization and practice among centers.

Fig. 4–13. Communications centers come in different shapes and sizes.

It is important to understand some key terms and concepts in relation to communications centers and facilities.

Public Safety Answering Points (PSAP) is used to describe any center designated for direct receipt of 9-1-1 calls. PSAPs often are hosted and managed by law enforcement or independent agencies.

In the United States there are over 8000 primary and secondary PSAPs.[8] Typically, the difference between a primary and secondary PSAP is the ability to handle law enforcement events or incidents. For example, a citizen dials 9-1-1 from a residential telephone, reporting a grass fire or a person needing medical attention. The call is initially routed to the primary PSAP which, in most instances, will be a law enforcement agency. The law enforcement agency determines what the emergency is. If the event does not require a police presence, the primary PSAP will then transfer the call to a secondary PSAP, which handles the dispatching responsibilities for fires and medical-related incidents. If the call requires law enforcement, the police will respond while transferring the other segments of the incident to fire and/or medical services.

Each communications center ensures that the information has been appropriately transferred to the respective agency and that the telecommunicator receives the information of the reported emergency. The telecommunicator then takes the reporting party's information and initiates an incident case file to track the history of the emergency and determines the resources to respond to the incident. Units are then alerted and dispatched to the emergency scene. It is important to be aware of arrangements for dispatching, as any step in the process takes time, and introduces the possibility for error.

Telecommunicators

Telecommunicators are unique, highly skilled individuals who have the ability to multitask. The position of dispatcher has evolved from an individual on light duty who was assigned to the communications center until medically cleared to resume firefighting duties. Today, a telecommunicator must be able to use a **computer-aided dispatch (CAD)** system, operate an extensive telephone system, obtain information from citizens, and alert the appropriate resources. Whether an agency is large or small, high-tech or paper-based, the field units must rely on the professional telecommunicator, not only for incident information and updates but also for their actual safety. The term telecommunicator is preferred to "dispatcher" because it is more accurate, and reflects the professional role, training, and diverse skills required of the position.[9]

Mobile communications vehicles

Many larger organizations operate some type of mobile communications facility incorporated into a vehicle. For small organizations, this may be as simple as a work surface and radio with mutual aid frequencies incorporated into a chief's vehicle. For larger incidents and in extreme climates, more elaborate vehicles are used for this purpose. These vehicles can range from converted cargo vans or ambulances to custom-equipped and constructed trucks or RV-type units.

One important distinction is to keep the concept of a communications vehicle distinct from the notion of an incident command post. While they may be collocated, on larger or long-duration incidents, the communications vehicle will probably not be the best place for the incident commander and supporting staff to be working, as they will interfere with communications functions.

These vehicles have grown in sophistication as technology has advanced, and now such units should include radios, a link to the CAD system, mapping software, capabilities for receiving television and radio broadcasts, digital recording capability, telephone and fax capability, and channel-logging recorders for on-scene communications. Satellite communications, wireless broadband connections, ability to receive video feeds from helicopters, and computers are becoming common as well. Mechanically, these vehicles should be equipped with generators and portable lighting, and they should be capable of operating for extended periods (fig. 4–14a and b).

Figs. 4–14a, b. The Westchester County (NY) communications vehicle includes capabilities of receiving video feeds from helicopters, transmitting video, CAD interface, and ability to replicate the county government's network and computer applications.)

Fire Department Communications

FIREFIGHTER I

Chapter 4

The future direction of these units is to be able to replicate the computer experience of being at a desk of a high-level administrator, with real-time access to enterprise software, e-mail, and data otherwise available in the office.

TRANSMISSION OF ALARMS AND ALERTING

In departments with multiple facilities, or on multi-jurisdictional dispatch centers, there is a need for a more elaborate method of directing units to reported emergencies. Depending on the nature of the emergency or its location, decisions must be made about which stations, units, and personnel to alert. Early municipal fire alarm box systems, which predated radio communications, transmitted a coded signal to the communications or dispatch center. These alarms were often monitored in fire stations as well. Personnel knew the location of each "box" and would know to respond to the appropriate location, where they would hopefully be met by the person who transmitted the signal. In volunteer organizations, these systems were often tied into large air-powered horns so that signals could be heard throughout a community. They were a very efficient way of alerting personnel because they both alerted and provided location information.

Because there were numerous "boxes" in most communities, a term known as a "box" card was used to describe both the physical location of the alarm box and the apparatus or resources that were assigned to respond. In smaller communities, the listing may correspond with mutual aid departments that would be called if necessary. In larger departments, elaborate lists of units assigned to respond, usually based on distance from the box, would be developed. Commonly, box cards contain lists of resources through several alarms.

As municipal fire alarm systems were replaced, and computer-aided dispatch (CAD) systems were developed starting in the 1970s, specific locations throughout a community were selected and predetermined apparatus response was designated for each such area. Even in communities that did not have such systems, areas created were referred to as "box areas." This information is incorporated into modern CAD systems to this day, as each address is associated with a geographic identifier or "phantom box" that contains a unique assignment of apparatus or resources at multiple levels of response.

Figure 4–15 shows a diagram of a "typical" "box" or "run" card. Even the most modern dispatch facilities maintain a collection of physical cards to be used for this purpose in the event that their computerized system fails or is disabled.

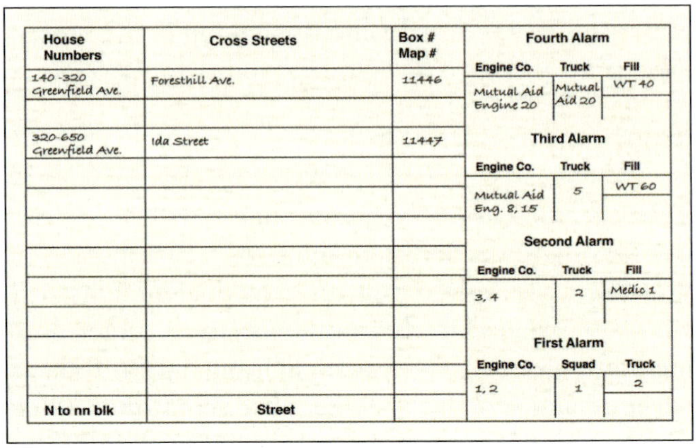

Fig. 4–15. A "box" or "run" card.

The number of units sent on an alarm is a function of local departmental policy. In smaller organizations, all available personnel and apparatus may respond on every alarm, regardless of type. As departments become larger, and alarms more numerous, specific numbers of units may be designated to respond based on the nature of the reported incident. This is necessary to conserve resources for additional alarms that may be reported, and to reduce unnecessary responses.

For example, emergency medical assists may receive a response from as little as one unit with a few personnel, while reported fires in buildings receive the highest level of response. Usually, a minimum of 2–3 engine companies and a ladder or service company are sent to any reported fire in a building. Structural firefighting requires a complement of personnel be on the scene, and if local resources do not meet the need, then other resources should be utilized through mutual aid, or preferably, automatic aid.

CAD systems make it possible to assign resources to specific addresses. Risk-related factors such as the presence of fire suppression systems, hazardous occupancies, or lack of equipment such as hydrants or standpipes can be used to trigger changes in the number and type of units that are sent to an alarm.

After a dispatcher determines the nature of an emergency and the appropriate resources to send, the next decision is how they will be notified. The most common means for alerting of fire service resources is over radio. Often audible or inaudible tones are used to activate speakers, pagers, and audible warning devices in fire stations. These tones, when audible, also serve to alert users monitoring the frequency that an alarm is being dispatched. These systems can be manually activated by the dispatcher, or can be automatically integrated into CAD systems. Other systems used included dedicated voice alerting systems that operate over leased or private lines or microwave links. A typical dispatch work station is shown in figure 4–16.

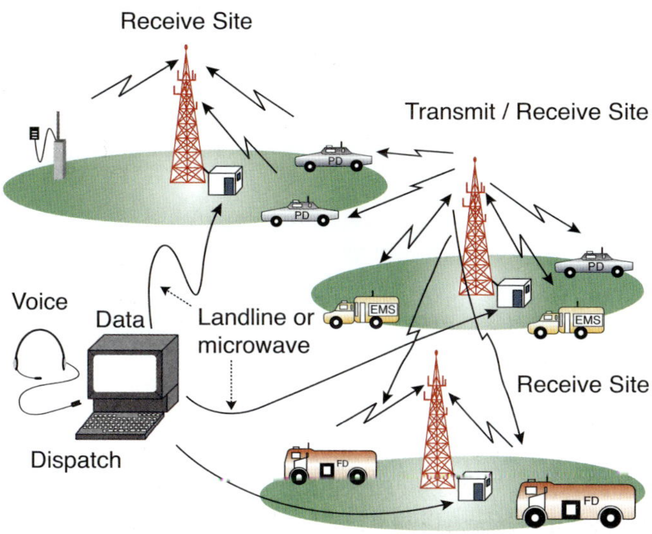

Fig. 4–16. Typical public safety dispatch position. Screens are for 9-1-1 telephone, cad system, radio system, and unit status. Other screens monitor camera systems and administrative uses.

FIRE DEPARTMENT COMMUNICATION DEVICES

Radios

SKILL DRILL

Fire departments use three different radios for service-related activities. The radio used by fire company officers and individual fire personnel to communicate is a fire department–issued portable radio. This **handheld radio** is powered by a battery that has a limited capacity, depending on its use time. At intervals, the radio's battery must be recharged or replaced with a fully charged battery. Most portable radios have the capability of changing channels and even the capability to change talk groups. Most radios have a built-in function to eliminate **squelch,** which is high noise level when no or very weak signal is present. By manipulating a built-in function knob, any weak signals that may be interfering with radio reception are eliminated. These radios provide the users with the ability not only to talk to each other and their communications centers but also to communicate with other agencies.

The portable radio is designed to be carried in the hand or stored in a radio pocket of personal protection equipment (PPE) or even in a holster secured to the belt or on a radio strap around the neck. Newer models are required to have operating knobs and buttons that are distinguishable by feel, and operable with protective equipment worn, such as gloves.

Mobile radios are more powerful units that are permanently mounted on fire department apparatus (fig. 4–17). They do not require the use of external batteries because they are powered by the electrical system of the apparatus. This type of radio is equipped with at least the same features as a firefighter's handheld portable radio. An added feature of the mobile radio is that the radio speakers are mounted outside the mobile radio in a convenient location inside the apparatus. There may also be a speaker located on the outside of the vehicle. This allows the firefighting personnel to be in radio contact while outside the vehicle (fig. 4–18). A microphone is attached to the mobile radio by a coil cord and can be held away from the mobile unit while transmitting information. Many fire departments are now using a tool that incorporates the speaker and microphone into one headset; the external microphone attached to the headset allows the person wearing it to simply push an attached button to talk. The advantage of using a **radio headset** is that the crew always have direct verbal contact with one another. Noise reduction with the headset ensures better communication. This headset also provides hearing protection against loud environments.

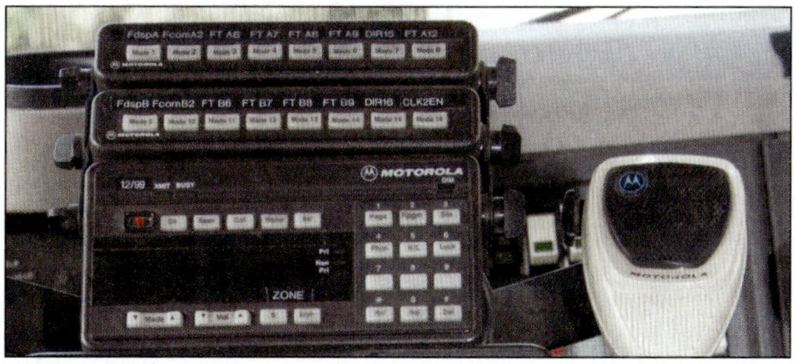

Fig. 4–17. Mobile radios are more powerful units permanently mounted on fire department apparatus.

Fig. 4–18. Firefighting personnel using headset for radio contact while outside the vehicle

EMERGENCY INCIDENT SCENE COMMUNICATIONS

FFI 4.2.3 When communicating via radio, you should be familiar with your department's equipment. Components of the radio itself may include an antenna to receive and transmit radio signals, a push-to-talk switch, and a built-in speaker and microphone. There are also control knobs for volume, channel selection, squelch control, and a personal emergency button.

The best position for a handheld portable radio is out of its storage device and held in the vertical position with the antenna pointing in upright direction (fig. 4–19). Hold the radio at a 45-degree angle, 1–2 inches from your mouth, and speak clearly and calmly. Do not shout or use profanity. Wait until the air is clear before your next transmission, and avoid unnecessary transmission that can clutter the channel. Know what you intend to say *before* keying the microphone to speak. Good **radio discipline** is critical to effective operations.

Fig. 4–19. Firefighters must hold the portable radio in the proper position to be effective.

Follow your department's operating procedures for emergency traffic and emergency evacuation signals and communications.

FFI 4.2.3 Radio communication by fire departments is an everyday occurrence and is normally completed on a routine basis with nonemergency traffic filtering through the airways. Everyday business activities include training, fire prevention, and inspections, along with other miscellaneous information transmitted among fire companies, fire inspectors, and administrative staff.

Routine message transmission makes up the highest proportion of daily communication activity across the radio airways. "Fire Dispatch, Engine 1 is available on radio businesses inspecting." Or you may hear "Fire Dispatch, Battalion 1 is available at Station 2 conducting a vehicle inspection." The situation changes when emergency traffic is required. When communications centers are dispatching apparatus and staffing to emergency incidents, then all normal communications must cease, and emergency traffic procedures take precedence. If you were listening to a fire department radio at this time, you might hear something like this:

"Fire Dispatch: Engine 5, Medic 1, respond to a reported vehicle fire at the intersection of 24th and Maple Street. Reported vehicle accident with injuries."

The next transmission would be by the first-arriving unit to the scene with a transmission back to the communications center, saying.

"Engine 5 on scene; we have a two-vehicle accident, no fire, and checking for injuries; will advise further."

As the incident progresses, additional communication between the units on the scene and communications center will continue until the incident is shut down by the officer in charge.

"Fire Dispatch, Engine 5 is available on radio."

These everyday communications of the dispatch center and the units responding to calls are very similar. Telecommunicators assigned to communications centers dispatch fire units from the district to any request for service. This communication process consists of the following:

- Receipt of a request for service
- Dispatch
- Information delivery to the responding units
- Delivery of any supplemental information while en route
- Arrival information of the first company
- Size-up
- Additional resource request based on needs
- Progress reports during the incident
- Shutdown or termination of the incident
- Availability of unit

SPECIAL CODES AND CLEAR LANGUAGE COMMUNICATIONS

FFI 4.2.3 Communications centers may use what is referred to as plain language dispatching, clear text, or clear speech as opposed to a numeric code like a **10-code** system (numeric codes used to assign messages to represent common phrases). In voice communication, instead of saying in clear language what is meant, a special code is used, such a 10–7 for "out of service." This type of code, along with other numeric coding systems, has been used for many years in the public safety arena. As with any form of communication, there are both advantages and disadvantages. Numeric codes of communication are brief and so use less air time and free up the radio channel for other transmissions. The clear speech or clear text type of communication eliminates the need to remember special codes, which can have multiple meanings. Clear language enables the speaker to say exactly what is meant.

Even though the U.S. Department of Homeland Security's **National Incident Management System (NIMS)**, due in part to the incident command structure and widespread exchange of information, states that the recommended language should be in a clear language format12; there are issues with this form of communication that all users need to be aware of. There are many words that sound alike. "Can" and "can't" are often confused, so "unable" is recommended. There are preferred words to be used for clear language communications. Some choices are affirmative instead of yes, negative instead of no, and standby instead of wait. Other examples include policies that require the person at the destination to repeat the message back to the information source, or requirements that dispatchers "spell out" numbers to avoid confusion (apartment "sixteen" versus apartment "6T" can be aided for example, by saying "one-six or six-"thomas"). The idea is to avoid any possibility that messages are misunderstood.

Clear language is the only means of information transfer authorized by the Department of Homeland Security. If a fire department is using any special codes that are unique to that department, then care must be taken when working with other agencies who may have their own numeric codes but with different meanings. Departments that retain use of codes or "slang" should be alert to the possibility for misunderstanding, and should coordinate with regular mutual aid or automatic aid partners to avoid confusion by adopting the same meanings. The easiest form of communication is simply to say what you mean.

INCIDENT SCENE EMERGENCY COMMUNICATIONS

FFI 4.2.4 There are times when the communications between the telecommunicators and firefighters can literally mean the difference between life and death. Both ends of the communication link must be aware of the protocols and procedures to expedite all information required to stabilize the incident and assist those in distress. Firefighters at the scene of the emergency must be aware of the conditions surrounding them and know how to request assistance, especially emergency assistance. The everyday emergency call that is handled with the utmost efficiency and professionalism is not the most stressful for firefighters. It is the crisis that may occur in the middle of one of the "everyday emergency calls." Emergency announcements that put firefighters on highest alert are the "emergency traffic" callout and, for some fire departments, the callouts "urgent traffic" and "Mayday."

All three of these callouts or messages are used for assorted reasons, but all three are designed for the same function: to signal the immediate need to clear the radio and transmit critical information. Some departments use only the emergency traffic callout. Other departments or jurisdictions may use at least two of the callouts or a combination of all three.

For fire departments using emergency traffic, the radio transmission would be:

- "Incident Command Division 2, emergency traffic."
- Normal radio traffic immediately stops, and Division 2 will proceed with their traffic.
- "Incident Command Division 2, we have a firefighter down inside the building requesting assistance on the second floor east side."

The fire department with the same scenario utilizing the Mayday callout would ask for emergency traffic to clear the airways and then declare a Mayday for a firefighter down and in life-threatening danger. The only time the Mayday callout is used is in a situation in which there is an immediate threat to the lives of firefighters.

The last callout is "urgent transmissions." This type of callout is used in the following situations:

- A fire department member suffers an injury that is not life threatening but requires medical attention and hospital care.
- There is a change in operations.
- A structural problem is discovered, indicating the danger of collapse.
- There is a loss of water, which would endanger members.
- Fire is discovered by firefighters entering an exposure.
- Any delay occurs that would increase the fire problem.
- An urgent transmission might sound like this:
- "Incident Command: Urgent traffic, Division B, be advised, fire is extending to the exposure attic east of the building."

Whatever the method used by the various fire organizations, the callouts are a valuable tool to quickly clear the airways and broadcast notification of the emergency information to all involved.

Personnel accountability system (PAS)

SKILL DRILL

FFI 4.2.4 As part of the evaluating and monitoring process, NFPA 1561, the *Standard on Emergency Services Incident Management System*, requires that all fire departments establish standard operating procedures for a **personnel accountability system (PAS)**. It must also provide for the tracking and inventory of all members operating at an emergency incident.

The PAS sometimes uses Velcro®-backed plastic cards attached to various apparatus. They hold the name tags that identify and account for each member (fig. 4–20). Name tags and status boards are used to track the assignments of the divisions, groups, companies, teams, and individuals at an emergency incident. Fire department personnel must be familiar with the procedures to be implemented and the steps required to ensure an accurate accounting of personnel.

Fig. 4–20. Passports hold name tags that identify and account for each member.

The PAS is designed to be flexible and allows for expansion or contraction as the events or incidents dictate. There are different levels at which the system activates.

Level 1, the most basic activation, takes place at shift change. It is then that each firefighter attaches his personal ID tag to the apparatus accountability passport. At Level 2, activation is in place for all working fires, where more than one piece of apparatus is responding, as well as in hazardous situations. Often a department will assign one of the apparatus officers to document the tracking of the personnel. In extreme situations or incidents involving multiple companies, at Level 3, an accountability officer will take over the passport system. This officer will set collection points for the passports. The officer will also assess personnel entering hazardous areas to ensure proper protective clothing and proper use of the firefighters' personal alert safety system (PASS). PASS is a personal device worn by every firefighter that when activated sounds an audible tone that alerts others that a firefighter is in distress.

The easiest way to understand the use of a PAS is to be aware that all firefighters are issued a set of identification name tags that they retain with their protection equipment. As they report for duty, one of these tags with their personal identification is removed from their personal protective equipment and placed with other crew members' identifiers on a status board on the apparatus. A firefighter's identifier is not to be removed until the individual is no longer with the apparatus. If a firefighter leaves the fire station (or apparatus at the scene) for any reason, he or she must remove the identifier from the status board. If the firefighter does not remove his or her identifier and the company responds to a fire, the firefighter would be assumed lost, and a search would be initiated. Rescue companies would be put at risk because a firefighter did not remove his or her identifier tag from the apparatus. The PAS is a life-saving tool that must be used correctly. Misuse or noncompliance is not an option.

Other PAS systems may use cards, tags, or other systems.

Personnel accountability report (PAR)

Personnel accountability report (PAR) is a polling system used by the incident commander or accountability officer to ascertain that all personnel operating at an emergency incident are safe and accounted for. The following trigger points are generally used to determine the need to conduct a PAR:

- There is a report of missing or trapped firefighters.
- Operating modes are altered: offensive to defensive.
- Dramatic changes have occurred at the incident scene.
- Emergency traffic has been requested due to fireground evacuations, building collapse, or Maydays signaled for lost or trapped firefighters.
- Replacing companies are on the scene.
- At a prearranged time during an active event: Some organizations have a built-in PAR every 20–30 minutes into an event.
- Any time the incident commander deems one necessary.

Personnel accountability is accomplished by an accounting of each firefighter or other personnel to the accountability officer on scene. Company officers must verify that the personnel assigned to them are visually accounted for. The last step in the process is for all division and group officers to confirm that all crew members attached to those companies are accounted for. If during the accounting process an individual or company is not responding, an emergency traffic or Mayday will be requested, and emergency traffic procedures implemented, along with departmental operational strategy and tactics for the missing firefighters.

Building or scene evacuation signal

SKILL DRILL

There are times when the decision is made to evacuate the firefighting or rescue operations inside a building because of excessive danger to the firefighters, or the scene has become too dangerous to continue any tactical operations. Throughout the country, fire departments have worked and trained for the time when it is necessary for the emergency **evacuation signal** to be broadcast. The training ensures that everyone who hears the signal will understand what needs to be done. The following example of an evacuation signal is the one used in California by all the fire agencies throughout the state. The advantage is that during emergency incidents using mutual aid, all the agencies have been trained using one evacuation signal.

Upon declaration of an Emergency Evacuation, Dispatch shall sound the Emergency Evacuation (high/low) tone on the Fireground Channel and announce: "All companies operating in (community) at (incident name) are to evacuate the building immediately!" This shall be repeated three (3) times. The important thing to understand is that the type of emergency traffic communications used by fire departments must be thoroughly understood by all personnel, and procedures constantly trained for, so that when the need arises, they can be used to save lives.[10]

Although it is not your primary responsibility, every firefighter must be aware of emergency and high priority communications. Knowledge of equipment, procedures, and awareness on the incidents scene are just as important as knowledge of breathing apparatus, ladders, and hose practices.

BIBLIOGRAPHY

Desourdis, Robert I., Davis Smith, William Speights, Richard Dewey, and John DiSalvo. *Emerging Public Safety Wireless Communication Systems*. Boston: Artech House Publishers, 2002.

Fire Department Communications Manual: A Basic Guide to System Concepts and Equipment. FA-160. Federal Emergency Management Agency, U.S. Fire Administration, 1995.

McMillan, J. Rhett. *The Primer of Public Safety Telecommunication Systems*, 3rd Ed. Daytona Beach, FL: Association of Public-Safety Communications Officials, Inc. 2000.

Singer, Edward. *Land Mobile Radio Systems*, 2nd Ed. Upper Saddle River, NJ: Prentice-Hall, 1994.

NOTES

1. National Emergency Number Association Fast Facts, http://www.nena.org.
2. Federal Communications Commission. 911 and E911 Services. https://www.fcc.gov/general/9-1-1-and-e9-1-1-services.
3. National Emergency Number Association (NENA). NENA Call Answering Standard/Model Recommendation. NENA 56-005.1 Revised 8/31/2017. Retrieved September 14, 2017 from http://www.nena.org/?page=911CallAnswerStnd.
4. McMillan, J. Rhett. *The Primer of Public Safety Telecommunication Systems*, 3rd Ed. Daytona Beach, FL: Association of Public-Safety Communications Officials, Inc. 2000. p. 77; Singer, Edward. *Land Mobile Radio Systems*, 2nd Ed. Upper Saddle River, NJ: Prentice-Hall, 1994, pp. 72–73.
5. *Fire Department Communications Manual: A Basic Guide to System Concepts and Equipment*. FA-160. Federal Emergency Management Agency, U.S. Fire Administration, 1995, appendix C.
6. Desourdis, Robert I., Davis Smith, William Speights, Richard Dewey, and John DiSalvo. *Emerging Public Safety Wireless Communication Systems*. Boston: Artech House Publishers, 2002, pp. 259–262.
7. NENA Call Answering Standard/Model Recommendation.
8. Essex County Fire, Standard Operating Guideline 602.00 *Accountability*. Retrieved September 14, 2017 from http://essexcounty-fire.org/FIRE/wp-content/uploads/2013/08/SOG-602-00Accountability.pdf.
9. McMillan. *Primer of Public Safety*.
10. Desourdis, Smith, Speights, Dewey, and DiSalvo. *Emerging Public Safety*.

QUESTIONS

1. Another name for a telecommunicator is ________________.
2. The E in E9-1-1 stands for ________________.
3. What is the first step in the dispatch process?
4. What critical information needs to be gathered by an emergency operator or 9-1-1 dispatcher?
5. The deaf and the speaking worlds use what types of devices to communicate with each other?
6. What federal agency regulates public safety radio communications in the United States?
7. In a simplex radio system, how many parties may speak at one time on one channel?
8. A repeater device is used to do what to a radio system?
9. How should you speak when communicating with a radio system?
10. Explain how PAS works in firefighter safety.
11. Passports are used to identify and _____ for each member?

Fire Behavior

by Sean Gray

This chapter provides required knowledge items for the following NFPA Standard 1001 Job Performance Requirements:

FFI: 4.3.11

FFI: 4.3.12

5

OBJECTIVES

Upon completion of this chapter, you should be able to do the following:

- Describe the parts of the fire triangle and fire tetrahedron
- Identify basic measurements of heat and temperature
- Identify the characteristics of the three physical states of fuel
- Describe the combustion process
- Define the term flammable (or explosive) limits
- Identify four types of heat sources
- Describe three types of heat transfer
- Identify the three methods of heat transfer
- Describe heat release rate
- Define the term thermal conductivity
- Identify the three components of smoke
- Identify the four stages of fire development
- Define the terms thermal layering and thermal balance
- Describe the three phenomena of a fully-developed fire
- Describe the two general types of explosions
- Describe the concept of flow path and how it impacts fire growth

INTRODUCTION

Fire has been described as unpredictable, magical, and even a living thing in Hollywood movies. The ancient Greeks believed that air, earth, water, and fire were the four elements that made up the universe and that a god controlled each element. Fires occurred when the God of Fire became upset. To put the fire out, the Greeks prayed to the God of Water for rain and the fires would be extinguished. Even back then it was as simple as putting water on fire. Our knowledge of fire is based on fact, and through research, we are learning more about the principles of fire dynamics and the behavior of fire.

Overview of why it is important to study fire behavior

We have moved into a time when our education about the fires that we respond to, and how those fires react to the tasks that we perform becomes more attainable through research. The importance of understanding fire behavior is an ever-evolving subject that will allow us to progress with smarter and safer firefighting tactics. In an effort to become smarter firefighters, it's imperative to evaluate the research and apply it to fireground operations. The body of knowledge that has been created by researchers for the fire service can only make us better as a profession. Our goal as professional firefighters should be to close the gap between the researchers and the fire service in order to work harmoniously together as a team. Firefighters are viewed as strong, blue-collar workers that heroically show up to mitigate emergencies. We are a profession that continuously trains and educates ourselves in order to master our craft. Understanding fire behavior is a core principle to becoming a proficient firefighter.

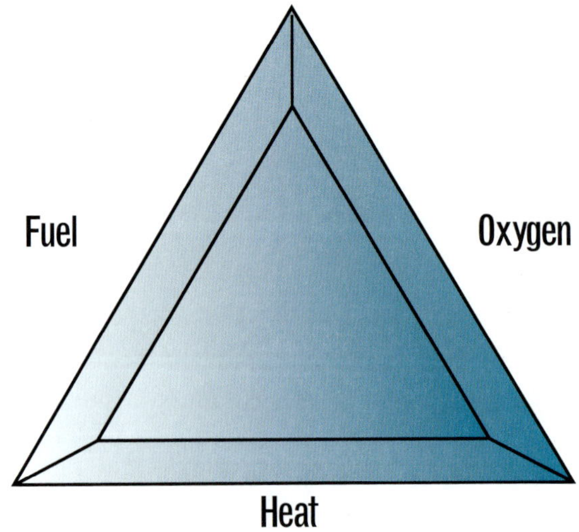

Fig. 5–1. The fire triangle

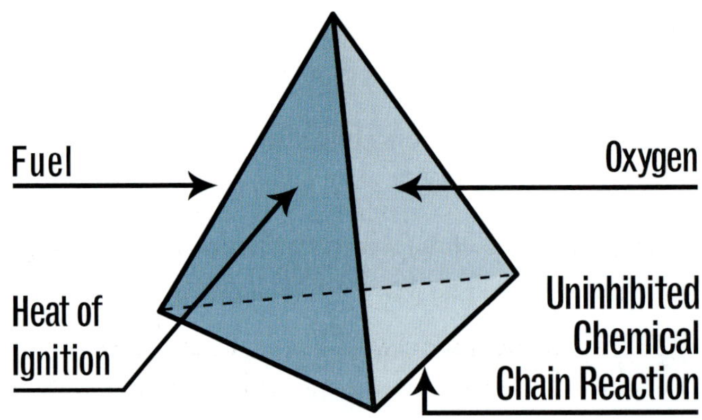

Fig. 5–2. A four-sided tetrahedron, the three-dimensional fire triangle illustrating the interrelationship of the chemical chain reaction in the combustion process.

DEFINING FIRE AND THE FIRE TETRAHEDRON

Fire has been described in many ways. It can be defined as a rapid oxidation chemical reaction that usually results in the creation of heat and light. In the case of a typical fire, rapid oxidation means that a gaseous fuel is chemically combined with oxygen (from the air we breathe in most cases). The flames we see and the heat we feel from a fire are produced by this rapid oxidation.

With this definition in mind, we can identify the building blocks of a fire, the components of creating a fire. A "fire triangle" helps us illustrate this concept—the sides of a fire triangle are fuel, heat (ignition source), and oxygen. The three-sided fire triangle has been used for many years to train new firefighters (fig. 5–1). Several years ago, the fire triangle was modified to include a fourth side, making it a fire tetrahedron. The fourth side, called the uninhibited chain reaction, was added to highlight the important fact that fires are actually a very long continuous series of chemical reactions that keep the fire burning (fig. 5–2). So, for a fire to occur, all four sides of the tetrahedron must be in present.

Fire can take place only when these four components are present: oxygen in some form, fuel (material) to combine with the oxygen, heat sufficient to maintain combustion, and the uninhibited chain reaction. Removal of any one of these four components will result in extinguishment. Breaking the tetrahedron by removing one component will extinguish the fire, and blocking one component from joining the other two will prevent a fire from starting.

A piece of paper and a match are a good example of a combustible material and a source of heat. The molecules within the piece of paper that are close to the heat source begin to vibrate at an enormously increased rate and, almost instantaneously, begin to break apart. In a series of chemical reactions, these fragments of paper continue to break apart, producing carbon and hydrogen that combines with oxygen in the air. This combination releases energy in the form of heat and visible flames. This process of burning, also called combustion, will continue until fuel is exhausted, oxygen is excluded, heat is dissipated, or the chain reaction is disrupted.

TYPES OF FUEL

Fuel is, of course, the material that is consumed in a fire. Fuels can take the form of a solid, liquid, or vapor/gas. Within each of these three categories, there is an endless list of different kinds of substances that will burn. On the other hand, there are some materials that will not sustain combustion.

All matter exists in one of these three forms (also called "states"). Matter can also change from one state to another by heating or cooling, applying pressure, or releasing pressure. When matter changes phases directly from a solid to a gas, this process is called sublimation.

Matter is further classified as organic or inorganic. Organic matter, by definition, consists of those substances that contain carbon. Inorganic matter refers to matter that does not contain carbon. Since all life on earth is based on carbon, generally speaking everything that was living or was once living will burn. Lumber in a house comes from what was once a living tree. The tree has been cut down and the wood has been shaped into varying dimensions. Even though the tree is no longer alive, the wood from the tree will still burn.

This chapter focuses on what fire is, what causes fire to start, how fires grow and spread, and, using that knowledge, what methods can be used to control and extinguish fires.

Solid

When matter exists in a solid physical state, the molecules of the substance are thought to be closely or tightly arranged. The tighter arrangement and greater quantity of molecules relates to the density and the mass of a substance. For a solid fuel to ignite and burn, the solid must go through a transformation into a gaseous state, also known as **pyrolysis**. This transformation requires the solid substance to be heated to its **ignition temperature**. Ignition temperature is the minimum temperature at which combustion can be initiated under specified test condition. The ability of a solid to absorb heat is often dependent on the physical arrangement of the substance.

An issue that has a direct impact on the time to ignition of a solid is whether the substance is considered **thermally thin** or **thermally thick**. Thermally thick means that when a material is exposed to a heat flux on its front face, appreciable temperature rise has not yet occurred on its back face. On the other hand, thermally thin means that at a given instant the material's back face is at a temperature close to that of the front face.

The greater the surface area to mass ratio, the easier it is for the solid to absorb the heat, reach its ignition temperature and ignite. As an example, take a block of wood. If a heat source such as a candle is applied to the block of wood, a considerable period of time will pass before the block absorbs enough heat to ignite. If the same block of wood is converted to wood shavings, the substance is still a solid but has a greater surface area to mass ratio, allowing the wood shavings to reach the wood's ignition temperature and ignite. Finally, if the block of wood is converted into wood dust, the substance now has an even greater surface to mass ratio, and the ability of the dust to absorb heat has increased dramatically (fig. 5–3). And if the dust becomes suspended in air, the ignition process can occur so quickly that an explosion takes place. Yet in all three cases, the substance was a solid at the beginning of the process. From this we learn that not only is knowing the physical state important, but knowing the physical arrangement of the substance is equally important with respect to ignition.

Fig. 5–3. Each of these items is wood

Liquid

Some liquids are capable of burning under normal temperatures and pressures while others are not. The liquids that will not burn can absorb heat, boil, and evaporate but never burn.

Sometimes, these liquids can be used as extinguishing agents, such as water. The liquids that burn are referred to as flammable and/or combustible depending on the flash point of the liquid and the naming convention followed by certain entities. The NFPA uses a break point of 100°F (38°C) to differentiate between flammable liquids (liquids with a flash point below 100°F [38°C]) and combustible liquids (liquids with a flash point at or above 100°F [38°C]). The NFPA further breaks liquids into classes and subclasses. In Europe, the differentiation point is 140°F (60°C). To avoid confusion, the more commonly accepted term for all liquids that burn is ignitable liquids. Flash point is the minimum temperature at which an ignitable liquid gives off sufficient vapor to form an ignitable mixture with air near the surface of the liquid or within a test vessel.

For an ignitable liquid to burn, the liquid must be at its flash point so that when fuel vapors are mixed with it, air will burn. There is a mistaken belief that an ignitable liquid must be suspended in air, or atomized, to burn. Atomization is a process that breaks a liquid into a mist; yet its physical state is still a liquid even though the liquid is in finely divided particles. However, when an ignitable liquid is atomized, there is maximum surface-to-mass ratio that permits rapid vaporization as the liquid is heated. All that is now needed is for an ignition source to be introduced to the ignitable liquids (fig. 5–4) The fire point of an ignitable liquid is that temperature at which sufficient vapors are present to ignite and have sustained combustion.

Fig. 5–4. When an ignitable liquid is atomized it becomes easier to vaporize and ignite.

Gas

Just as there are liquids that burn or don't burn, the same holds true for gases. When subjected to cooling and pressurizing, many gases become liquids. What is key to remember is that by their nature flammable gases in most ambient temperatures and pressures have already vaporized and only need the introduction of an ignition source to burn if they are within their flammable limits.

If there is too little gas, the mixture is too lean and ignition will not occur. If there is too much gas, the mixture will be too rich and will not ignite.

Flammable and explosive limits

FFI 4.3.11 In order for ignition of a vaporized fuel to occur, the mixture of oxygen and fuel vapor must fall into a range commonly called the flammable or explosive limits. While some texts refer to an air/fuel vapor mixture, firefighters need to remember that if the ambient atmosphere is oxygen-depleted, no ignition can occur. Generally speaking, the air at sea level has approximately 20.95% oxygen, which is more than enough to support ignition and combustion.

The flammable or explosive limits have a lower point and an upper point that represent the percentage of fuel vapor. If the amount of fuel vapor is below the lower explosive limit (LEL), the mixture is classified as too lean and will not ignite. If the amount of fuel vapor is above the upper explosive limit (UEL), the mixture is classified as too rich and will not ignite. When the fuel vapor percentage is between the lower and upper flammable or explosive limits, ignition will occur. And if the volume of fuel is significant (as in a house filled with natural gas) the ensuing ignition will have explosive consequences.

The difficulty in dealing with flammable limits is that each fuel gas has a distinctly different range. Additionally, while some gases have a narrow range, such as gasoline (LEL = 1.5%, UEL = 7.6%), others have a very wide range, such as carbon monoxide (LEL = 12.4%, UEL = 74%).

When a flammable mixture is present but below the lower explosive limit, the easiest way to prevent ignition is to ventilate and supply large quantities of fresh air to further dilute the mixture and keep it from reaching the lower explosive limit threshold point. However, if a mixture is encountered where the mixture is fuel rich, then the introduction of fresh air with adequate oxygen may dilute the mixture enough to cause it to fall into the mixture's flammable range. If this occur, the slightest spark may be enough of an ignition source to wreak catastrophic results. In such cases, ventilation and ignition source suppression must occur simultaneously to avoid endangering firefighters.

Technology has advanced to the point where many departments now carry meters and instruments to take air samples to tell them if specific flammable gases are present and, if so, at what percentage. All firefighters would be well advised to learn how to use these devices to protect them from the inadvertent ignition of a flammable gas mixture.

The danger to firefighters is when additional air is introduced to a rich mixture that results in a mixture that falls within the flammable limits. It is also important to remember that even an arc from an electric switch opening or the striking of a match has sufficient temperature to ignite a fuel/air mixture within its flammable limits.

CHEMISTRY AND PHYSICS RELATED TO FIRE

A fire is a complex process with a combination of chemical and physical events. To understand how a fire burns requires knowing some basic principles of chemistry and physics. Firefighters need to understand these basic principles in the course of their careers in the fire service, so the following terms are defined to provide a knowledge baseline.

Specific heat

Specific heat is that amount of heat that a substance absorbs as the temperature of the substance increases. Specific heat is expressed as "the amount of thermal energy required to raise unit mass of a substance by one degree, and its units are J/(kg·K) (Joules per kilogram-Kelvin) in the metric system and BTU/(°F·lb) in the English system.

Latent heat

The thermal energy absorbed when a substance is converted from a solid to a liquid or from a liquid to a gas is known as latent heat. The amount of heat absorbed by a liquid that passes to a gaseous form is called the latent heat of vaporization. Water has an extremely high heat of vaporization, which makes it an ideal extinguishing agent. When heat is absorbed by water as it converts to steam it causes the surface of the burning solid object to cool. The word latent means "hidden," and in a way the heat of vaporization is hidden. Water absorbs a great deal of heat, but upon reaching its boiling point of 212°F (100°C), it turns to vapor (steam). In reverse, when a gas changes to a liquid or a liquid changes to a solid, heat is released and the temperature drops. When its temperature reaches the freezing point of 32°F (0°C), water becomes ice. More information on the latent heat of vaporized water when it is used to extinguish fire is provided in chapter 16, Fire Streams.

Density

The ratio of mass to volume of an object or substance is known as **density**. The greater the density of an object or substance, the more heat energy is needed to cause ignition.

Specific gravity and vapor density

The specific gravity of a substance is the ratio of the weight density of the substance to the weight density of another substance, usually water. All liquid substances have different specific gravities. Water is the benchmark that other liquids are measured against to determine if a liquid is heavier (which will sink) or lighter (which will rise), Assigning water a value of 1, all liquids with a specific gravity of less than 1 will float on water. All liquids with a value greater than 1 will sink below water. Table 5–1 provides the specific gravities of common liquids.

Vapor density is the term for comparing the weights of vapors and gases with the weight of air.

The terms *vapor density* and *specific gravity* should not be used interchangeably, as one measures vapors and gases and the other measures liquids. Some substances will actually have different values for vapor density and specific gravity, depending upon the physical state of the substance (solid, liquid, or gas). Be careful to use the proper terminology when researching a substance.

Air is the standard against which all other vapors and gases are compared and is assigned a vapor density value of 1. Vapors and gases with vapor densities greater than 1 will tend to drop to the ground. Gases and vapors with vapor densities less than 1 are lighter than air. There are only 13 gases known to be lighter than air. All other gases and vapors are heavier than air. The 13 gases lighter than air and their respective vapor densities are listed in table 5–2.

Table 5–1. Specific gravities of common liquids

Liquid	Specific gravity
Acetone	0.787
Alcohol, ethyl (ethanol)	0.787
Alcohol, methyl (methanol)	0.789
Alcohol, propyl	0.802
Carbon disulfide	1.265
Carbon tetrachloride	1.589
Castor oil	0.959
Coconut oil	0.927
Cottonseed oil	0.929
Crude oil	0.876
Formaldehyde	0.815
Fuel oil	0.893
Gasoline	0.739
Hexane	0.657
Kerosene	0.820
Linseed oil	0.932
Mercury	13.633
Milk	1.035
Napthalene	0.963
Olive oil	0.703
Toluene	0.865
Turpentine	0.871

Table 5–2. Vapor densities of gases lighter than air

Gas	Vapor density
Hydrogen	0.070
Helium	0.140
Hydrogen cyanide	0.930
Hydrogen fluoride	0.901
Methane	0.550
Ethylene	0.968
Diborane	0.960
Natural gas (composed primarily of methane)	0.600
Carbon monoxide	0.970
Acetylene	0.900
Neon	0.967
Nitrogen	0.970
Ammonia	0.590

HEAT, HEAT TRANSFER, AND HEAT RELEASE RATE

According to NFPA 921, *The Guide to Fire and Explosion Investigations*, heat is a form of energy characterized by vibration of molecules and capable of initiating and supporting chemical changes of state. In other words, it is the energy needed to change the temperature of an object. It's simple: add heat, temperatures increase; remove heat, temperatures decrease. Heat energy can be measured in a multitude of ways. In the past, had been commonly measured in Joules (J), Calories, or British thermal units (BTUs). While flame and gas temperatures are important measurements, understanding heat and how quickly it is created (heat release rate) are key concepts to understanding fire behavior. Another term commonly associated with heat is the word temperature. Temperature is the actual *measurement* of heat energy, Temperature is a measure of the degree of molecular activity of a material. Temperature is commonly measured in degrees Fahrenheit. Modern fire research often expresses temperature in degrees Celsius.

TEMPERATURE EXAMPLES

- Normal body temperature = 98.6°F (37°C)
- Human skin begins to feel pain = 111°F (44°C)
- Human skin begins to develop a 1st degree burn = 118°F (48°C)
- Human skin begins to develop a 2nd degree burn = 131°F (55°C)
- Human Skin begins to develop a 3rd degree burn = 140°F (60°C)
- Human skin is destroyed = 162°F (72°C)
- Water boils and produces steam = 212°F (100°C)
- Polycarbonate melting temperature (SCBA mask) = 446°F (230°C)
- Charring of modern protective clothing (firefighter PPE) = >575°F (>302°C)
- Post-flashover conditions = >1,100°F (>593°C)

Source: Lawson, J. Randall, "Fire Facts." National Institutes of Standards and Technology Special Publication 110 (Aug. 2009).

Sources of heat energy

There are four commonly accepted sources of heat energy: chemical, electrical, mechanical, and nuclear.

Chemical heat sources. As described earlier, fire itself is a chemical reaction, and one of the products of a fire is heat. So, the heat generated from a fire is often the heat of ignition that causes an uncontrolled fire to occur. As such, combustion is a form of chemical heat energy.

Spontaneous ignition is the initiation of combustion of a material by an internal chemical or biological reaction that has produced sufficient heat to ignite the material. This is observed by firefighters when rags and clothes made of natural fibers such as cotton become soaked with oils (such as linseed oil). The oil attacks the natural fibers, breaking them down. As the fibers break down, a process takes place where heat is generated. For example, when the rags are stacked on top of each other, such as in a pile, the heat of the reaction is not readily dissipated. The piles themselves allow the generated heat to be contained inside, sometimes causing them to ignite. This is the reason fire code regulations call for such oil-soaked rags to be stored in metal containers.

Electrical heat sources. Electricity is an extremely common energy source that is used by most people in everyday life. Electricity is used in residential, commercial, industrial, and institutional applications. A building or area without electricity is an anomaly.

Electricity is used to provide light, power appliances, cook food, and provide heat and air conditioning. Electric power flows from a generating facility, usually called a power plant, through transmission and distribution conductors, usually called power lines, to a service cable and into a distribution panel. This panel is generally equipped with a main or service disconnect and either fuses or circuit breakers that function as overcurrent protection devices.

From the distribution panel, electric current flows out to circuits where the electricity is used.

Some circuits are switch-controlled, meaning that power will not flow to the outlet or electric device unless a switch is turned on. A common example is a switch-controlled light. Other circuits have electricity that is at the receptacle or outlet. As soon as a plug is inserted, electrical power flows into the appliance or device.

The flow of electrical current can generate heat. In the simplest of terms, electricity is a flow of electrons from a negatively charged location where electrons are abundant to a location where electrons are less abundant, known as a positively charged location. This flow of electrons usually occurs along a conductor. The flow of electrons occurs as these electrons move from one atom to another along the conductor. As this occurs, electrons collide with molecules, causing the molecules to break apart, which then results in the liberation of heat energy. The quantity of heat generated depends on a number of variables, including the amount of electricity moving along the conductor and the density of the conductor. Typically, the denser a substance is, the better conductor it is. For example, **metals** are generally good conductors of electricity. Aluminum conducts electrical current (a good reason to be very careful with aluminum ground ladders around any wires), and copper is a better conduz

As the quantity or volume of electrical current (measured in amperes, or amps) flowing along a conductor increases, the potential for molecular collisions increase. This difficulty is sometimes referred to as resistance and is expressed in ohms. As resistance and quantity of electrical flow increase, the amount of heat generated increases. This is why electrically powered appliances that require large amounts of electricity to run need to have larger wires for the circuits. Specific wire diameters (wire gauges) are rated for maximum number of amps it can carry safely; the smaller the diameter, the smaller the amps it can carry without overheating. Toasters with very small diameter wires inside that glow when energized are a good example of this phenomenon.

Electrical heat energy can also be generated when a flow of electricity is suddenly interrupted by the separation of the conductor. A common example is when a switch that is closed (in the on position) is opened up (put in the off position) as current flows. The resulting arc is a release of electrical heat energy. Although these arcs are typically small in size, as is the amount of heat generated, the temperature that is generated can be substantial. Some arcs have been recorded at 2,000°F (1,093°C). Obviously, a sustained arc from a downed overhead wire can generate a continuous high temperature, high heat release event capable of igniting a building on fire.

Lightning is another example of electrical heat energy release. Clouds in the sky may be at differing levels of electron charging. As these clouds connect, allowing electrons to flow, one can observe what some people term **cloud-to-cloud lightning**. When a cloud and the earth have differing charges, we witness a discharge between the earth and the cloud.

Static electricity exists as the name implies, statically. Electric charges collect on the surface of an object. They are not flowing. There are a number of situations or conditions that will cause a static electricity discharge. The more common are the following:

- Contact and separation between dissimilar solids
- Flowing powders
- Flowing liquids or gases

An example that many people can relate to is sliding a foot across a carpet and then touching a metal object. The resulting shock that is received is a static electricity discharge. Another example is the electric charge created when certain flammable liquids or gases flow, particularly at an elevated pressure. The static electric charge can become the ignition source, such as when acetylene gas escapes rapidly from a pressurized gas cylinder. The gas can be ignited as it flows forth from the cylinder without the introduction of any other ignition source.

Induction heating is the process of heating an electrically conducting object (usually a metal) by electromagnetic induction, where currents are generated within the metal and resistance leads to heating of the metal. A microwave oven appliance cooks or heat food by dielectric heating, a method of heating non-conductive materials. A microwave is a radio wave that is between one millimeter and one meter in wave length. Microwave radiation is used to heat water and other polarized molecules within the food.

Mechanical heat sources. Heat of ignition can also come from mechanical actions that, developed at a sufficient temperature, will ignite flammables or combustibles. Mechanical heat is usually developed by friction when two items are rubbed together. The easiest example that demonstrates this is to rub your hands together. The faster you rub, the greater the heat produced.

In that same vein, metals that rub together will also generate heat. Lubricants reduce friction but do not absorb heat, particularly if the two metals rubbing together are part of a piece of machinery or a process. If the lubricant is insufficient or breaks down, the potential exists for enough heat to be generated to ignite surrounding combustibles, including the residual lubricant.

Friction can also create sparks where there is a lack of any lubricant. Visualize the metal edge of an axe coming in contact with a grinding wheel used to sharpen the axe head. As the edge of the metal touches the wheel, a shower of sparks flies out. Although seemingly small and insignificant, these sparks carry enough heat to ignite flammable vapors, finely divided dust particles, or wood shavings.

Compression can also generate heat, particularly if the compression is done under pressure and at a rapid pace. If a compressed gas cylinder, such as a self-contained breathing apparatus, bottle is filled too rapidly, the bottle itself will heat up as the bottle absorbs the mechanical heat energy created by the rapid compression of the breathing air. As the bottle cools off, the compressed gas, in this case air, will be subject to a lowering of the available pressure and thus the volume of air in the bottle. This is why all compressed gas cylinders should be filled slowly.

Nuclear heat sources. The two forms of nuclear heat energy are fusion and fission. Nuclear heat energy is generated when atoms are either split apart, which is called fission, or combined, which is called fusion.

Nuclear material is radioactive and unstable. These materials are constantly breaking down during a molecular process where they seek to become stable. During this process, energetic particles are spontaneously emitted by the disintegration of the radioactive material's atomic nuclei. As this fission process occurs, heat energy is released. In a controlled setting, fission is used to heat water to drive turbines and produce electricity. It should be noted that at the current time nuclear fusion cannot be controlled and has no commercial use, although its use is being explored in some experimental environments.

Regardless of how or why radioactive material is released, fires involving radiation emitting elements should only be fought by individuals specially trained and properly protected against radiation exposure. Regular firefighting personal protective equipment (PPE) offers absolutely no protection to the firefighter from any radiation that may be released. Under no circumstance should firefighters in structural PPE attempt to fight a fire involving radioactive material, no matter how small the quantity, until the radiation exposure hazard has been abated.

Heat transfer

FFI 4.3.12 When anything burns, there are several by products that are released. As discussed in the previous section, one of those by-products is heat. The heat that is generated is transferred to other objects. Depending on a number of factors, the heat that is transferred can cause the fire to spread and grow. These factors include the amount of heat being generated over a particular period of time, the item or substance being heated (combustibility and physical state), the distance between the fire and the item being heated, and the ability of the item being heated to retain the heat and reach its ignition temperature.

There are three modes by which heat is transferred from one substance to another. They are **conduction**, **convection**, and **radiation**.

Conduction. Conduction occurs when heat is transferred by direct contact between solid objects or within a solid object (fig. 5–5). One example is a steam pipe from a building's heating system that comes into contact with wooden structural members. The heat from the steam is transferred to the metal pipe that carries the steam to the heating system, providing heat for the building. The metal pipe absorbs this heat and transfers it to the wooden structural member, such as a ceiling joist in a basement that the steam pipe is in contact with. The constant heat being applied to the wood does several things.

Fig. 5–5. Conduction

The heat dries the wood and causes the cellulose in the wood to break down. When the wood has been subject to enough heat and the wood has reached its ignition temperature, the wood will ignite and begin to smolder. This decomposition—breaking down of the wood or other fuels under heat is also known as **pyrolysis**.

There are several important points to remember regarding conduction. The heavier and denser a material is, the better conductor of heat it is. So metals are excellent conductors of heat. The second point is that ability to transfer heat by conduction is directly related to the mass of the object and the quantity of heat being released. A single candle burning will not generate the same amount of heat as a large bonfire. If you put the end of a 6-in. deep metal I-beam into both fires, the I-beam exposed to the candle would barely warm up whereas the I-beam exposed to the bonfire would quickly heat up and could through conduction from the end immersed in the fire to the other end ignite combustibles away from the initial source of the fire, in this example the bonfire. This happens in real building fires when a steel beam passing through a wall between two rooms is heated by a fire in one room and ignites objects in the other room through conduction (fig. 5–6).

Fig. 5–6. A steel beam exposed to fire in one room can ignite materials in an adjacent room.

All matter is thermally conductive and will absorb, hold, and transmit heat. This ability is generally dependent on the density of a substance, which is the ratio of its mass to volume. If the substance is very dense, its thermal conductivity is high, whereas if the substance has low density, its thermal conductivity is low. Substances that have low thermal conductivity are good insulators since they inhibit the transfer of heat. Examples would include fiberglass and mineral wool. Substances that have high thermal conductivity are thought of as poor insulators and will allow the transfer of heat, such as metals.

Convection. Convection is the transfer of heat via a fluid medium, either liquid or air (fig. 5–7). Flames and heated gases/smoke passing over a material's surface will transfer heat through convection. Many objects are ignited in this way (fig. 5–8).

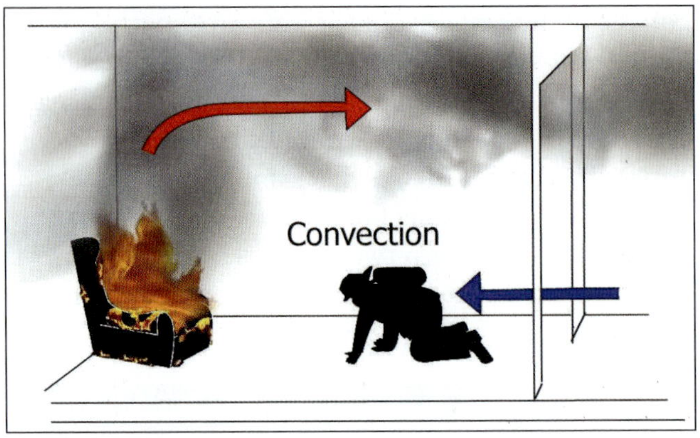

Fig. 5–7. Convection

Fig. 5–8. As flame passes over the surface of a material, it can ignite the material.

An accepted principle of physics is that warm air will rise and cooler air will fall. This is because, as air is warmed, the density of the air is reduced making the air lighter and allowing it to rise. As the air rises in the atmosphere, heat is given off to the surrounding air. When the heated air reaches the same temperature and density as the surrounding air, upward movement stops since equilibrium has been achieved. As the air's density begins to increase, it now begins to fall back down to the surface. This can sometimes be seen when smoke rises out of a smoke stack. It may eventually stop rising at a certain level, when it reaches the temperature of the surrounding air.

For example, smoke from the fire floor in a multistory building enters the enclosed stairwell through a fire door partially opened to allow a hose line to pass through. At first the smoke rises up the stairwell because it is warmer than the surrounding air. As the smoke rises, however, it loses its heat to the point that it can stop rising and not reach the top of the stairwell (the term *smoke stratification* is used to describe this phenomenon where the smoke stops moving vertically).

The same principle applies to liquids. When a liquid absorbs heat, the volume of the space occupied by the liquid expands. When the liquid reaches its boiling point, vapor is given off. Since the density of this vapor is lower than that of the surrounding air, the vapor rises just as heated air rises. If the heated liquid is in a closed container, then pressure will build up inside of the closed container until that pressure is released. Sometimes this pressure release is planned and regulated, such as when boiling water in a tea pot is poured out and the external heat source is removed. However, sometimes the release is not planned or regulated and the release is violent. An example would be a fire impinging on a cylinder or container containing liquefied flammable gases such as propane. A failure of any portion of the cylinder/container would be followed by a violent rupture and along with it the release and ignition of the now-boiling liquefied flammable gas. This type of event is commonly known as a boiling liquid expanding vapor explosion or BLEVE.

Radiation. Radiation is the transfer of heat through space by light waves (fig. 5–9). These light waves range from ultraviolet to infrared and contain electromagnetic energy that travels outward in all directions. This energy is absorbed by objects that are remote from and in a direct line with a fire.

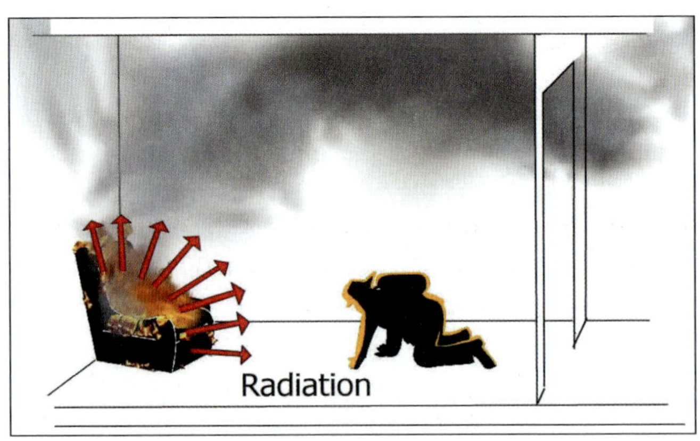

Fig. 5–9. Radiation

Radiation is able to travel through space, including vacuums, as well as through transparent items such as clear glass and water curtains. As the target objects absorb the radiated heat, they in turn begin to give off radiated heat. This is why radiation is now viewed as even more of a hazard than conduction or convection with respect to fire spread (fig. 5–10).

The most common example of radiation heat transfer is the sun warming the earth's surface.

Another example would be the warmth felt on a person's face while looking at a fire in a fireplace. If you turn and look away, your face feels cooler immediately because it is no longer exposed to the direct electromagnetic energy being given off by the fire.

Fig. 5–10. Vinyl siding melted by radiant heat

When there is a fire in a building, radiated heat energy is emitted from the fire in a direct line to all objects surrounding the fire. As the items heat up, they in turn reradiate to other objects. The heat release of the original fire and the ability of the target object to absorb and retain this heat will dictate how quickly or how slowly the fire may spread. One thing is certain: this mode of heat transfer will facilitate flashover more than the other two.

Heat release rate

The previous section discussed common sources of heat: chemical, electrical, mechanical, and nuclear. When we are discussing a fire's **heat release rates** (HRR), we are referring to the amount of energy or heat released over a period of time. We commonly discuss a 100 watt (100 joules-per second) light bulb, a 100 megawatt nuclear reactor, or a 10,000 BTU/hr air conditioner. All are examples of energy being released or used over time. As firefighters, we use metric units exclusively to express heat release rates, typically as kilowatts or megawatts.

For example, a 1 meter by 1 meter (about a square yard) diesel fuel fire will have about a 2 megawatt heat release rate. As long as diesel fuel keeps pouring into that one square meter area, the fire will keep burning at a steady heat release rate (the fire would burn out in a short time if only a thin layer of oil was ignited and the fuel was not replenished). See table 5–3 for other examples.

Table 5–3. An example of the heat release rates of different types of common objects

Heat Release Rates (HRRs) for Common Items

	Mass lbs (kg)	Peak HRR (kW)
Wastebasket, small	1.5–3 (0.7–1.4)	4–50
Trash bags, 11 gal (42 L) with mixed plastic and paper trash	7.5 (2.5)	140–350
Cotton mattresses	26–29 (12–13)	40–970
TV sets	69–72 (31–33)	120 to over 1,500
Plastic trash bags/ paper trash	2.6–31 (1.2–14)	120–350
PVC waiting room chair, metal frame	34 (15)	270
Cotton easy chair	39–70 (18–32)	290–370
Gasoline or kerosine in 2 ft^2 (0.2 m^2) pool	42 (19)	400
Polyurethane mattress	7–31 (3–14)	810–2,630
Polyurethane easy chair	27–61 (12–28)	1,350–1,990
Polyurethane sofa	113 (51)	3,210
Wardrobe, wood construction	154–267 (70–121)	1,900–6,400

Source: NFPA 921, *Guide for Fire and Explosion Investigations*; Quincy, MA, 2017.

Heat flux is the rate of heat energy transferred per surface unit area. For example, radiant heat coming from a hot layer of smoke and gases at the ceiling level down on to the combustible materials at floor level is heat flux, expressed as kilowatts per square meter (kW/m^2). A heat flux of 20 kW/m^2 means every square meter of the floor below the hot ceiling smoke/gas layer is receiving 20 kW. The following are some examples of heat flux.

EXAMPLES OF HEAT FLUX

- Sunny day = 1 kW/m^2
- Typical exposure to a PPE-protected firefighter in a fire = 2.5 kW/m^2
- Pain to skin within seconds = 3–5 kW/m^2
- Threshold flux to floor during flashover = 20 kW/m^2
- Thermal protective performance test (TPP) = 84 kW/m^2
- Flames over surfaces = 60–200 kW/m2

Smoke

As identified earlier in the chapter, smoke is a product of combustion and unburned fuel gases or unburned pyrolysates. According to NFPA 921, smoke is defined as the "airborne solid and liquid particulates and gases evolved when a material undergoes pyrolysis or combustion, together with the quantity of air that is entrained or otherwise mixed into the mass." Carbon particles (soot) is the predominate smoke particulate. Liquid particulates include condensed tars and even water vapor. A multitude of gases are produced including carbon monoxide and hydrogen cyanide (the predominant substances that cause the most deaths in fires).

While the specific health hazards of smoke are discussed in greater detail in chapter 10: Self-Contained Breathing Apparatus and chapter 14: Ventilation, it is important to understand the role smoke plays in the development of a fire in a compartment. As will be seen later in this chapter, the hot smoke at the ceiling of a room on fire will play a crucial role in the development of flashover. Smoke may also become explosive under the right conditions, resulting in a backdraft (smoke explosion).

COMPARTMENT FIRE DEVELOPMENT

FFI 4.3.11 All fires follow a distinct pattern in their development. The four stages of fire development are ignition, growth, full development, and decay (fig. 5–11). And before any fuel can be ignited, the fuel must be heated to a point where the fuel emits ignitable vapors. Obviously these vapors exist if the fuel's state is gaseous, and in some cases, ignitable liquids vaporize at low temperatures (gasoline's flash point is –45°F [–43°C]) while others have to be heated, such as fuel oil, motor oil, or vegetable oil. Solids must also be heated to point where ignitable vapors (pyrolysates) are emitted.

Fig. 5–11. The four stages of fire development (top to bottom): ignition, growth (pre-flashover), fully developed (flashover), and decay (post-flashover)

Ignition

Ignition occurs at the instant when an ignition source unites with an ignitable vapor in an oxygen -sufficient environment, resulting in a chemical chain reaction, and a fire begins to burn. If the ignitable vapor is already present, ignition occurs almost instantaneously. If the fuel is a solid, it may take a few seconds or longer to produce the ignitable vapor. But once the vapor starts to burn, ignition has taken place. In most cases, the fire is small and limited, sometimes called an *incipient fire*. The exception, of course, is where there are massive amounts of flammable gas capable of being ignited simultaneously.

Growth

Once ignition occurs, the fire must grow or it will die. The heat generated from the ignition now begins to cause heat transfer to spread to surrounding combustibles. Once these surrounding combustibles emit sufficient ignitable vapors, they in turn ignite, causing the fire to grow in size and spread.

There are several factors that will affect the fire's ability to grow and spread. The first factor is the amount of available oxygen. Once a fire begins to burn, available oxygen is consumed in the combustion process. If the amount of available oxygen falls below 16% of the atmosphere, the fire has a more difficult time burning. The fire will create even more carbon monoxide when the oxygen level is below 16%. If the oxygen level drops below 8%, the fire will not be able to sustain itself. Air (oxygen) is drawn into a fire through a process called **entrainment** (fig. 5–12).

Fig. 5–12. Fire development showing air entrainment

The second factor is the amount of fuel that is available to burn. A good example is a fire in a fireplace. Once the logs are consumed, the fire goes out for lack of fuel. Conversely, if there is an ample volume of fuel such as a structure full of combustibles, the fire will grow substantially. The physical state of the fuel package and its HRR will determine how fast the fire will grow and spread.

The third factor is the relative size of the space where the fire is to the involved fuel and its distance from other surrounding combustibles. If the space is large and high, the heat generated by the fire will be distributed throughout the space and make it difficult for other combustibles to be preheated. On the other hand, if the space to the ceiling and walls is small, the heat that is absorbed by the ceiling and walls can be radiated back to the fire, allowing it to intensify, which means that the fire can grow larger and faster.

The fourth factor is the insulating value of the ceiling and walls. The better the insulating value, the less heat will pass through. This heat will be radiated back to the fire and the surrounding combustibles. If the insulating value is low and allows heat to pass through easily, that heat will not help the fire to grow.

Thermal layering and thermal balance

FFI 4.3.12 As a fire grows, the hot gases generated by the fire rise until they become obstructed, usually at first by the ceiling. The hot gases will spread out in all directions seeking a new path to rise. Once the hot gases reach this path, they will continue to rise. Conversely, if the gases meet a vertical obstruction such as a wall, the hot gases will begin to bank down until the hot gases find an opening that will allow them to rise again, such as a doorway, open window, or stairway. Some of the heat energy from the heated gases is transferred to the ceiling and walls in a room fire, allowing some of the heated gases to cool slightly. These cooler gases are pushed down in the room's atmosphere as the hot gases that continue to generate rise above the cooler gases. This phenomenon is known as **thermal layering**. Because each portion of heated gases stratifies as equilibrium is achieved. The hottest point will be at the ceiling, while the coolest layer will be near the floor. The temperatures between the ceiling and floor will be layered. Firefighters sometimes refer to this as **thermal balance**.

The flaming gases at the ceiling will be distributed throughout the room or compartment causing the temperature at the floor level to rise quickly. Any firefighters or trapped occupants will be rapidly exposed to excessively high temperatures. This is why it is critical for ventilation to be closely coordinated with fire attack and the application of fire streams.

Fully developed compartment fire

Once the fire has involved an entire compartment or space, the fire is considered to be fully developed. If this is an outside fire, all of the combustibles in a given area would be on fire. The fire would be free burning and has plenty of oxygen to continue burning. A compartment or space fire could be a room, an apartment, a floor, or an entire building. One of the hallmarks of a fully developed compartment fire is that the rate of burning inside the compartment is limited by the amount of ventilation that the room is receiving. This type of fully developed fire is sometimes called a **ventilation-controlled** fire (fig. 5–13a). Fires that have plenty of oxygen (ventilation) but have limited access to fuel are termed **fuel-controlled** fires (fig. 5–13b).

Fig. 5–13. a) Ventilation-controlled fire and b) fuel-controlled fire. (Courtesy of Dustin Hughes)

Rollover

As a fire burns and consumes a room's contents, the heated fire gases rise up to the ceiling. These heated gases and smoke are fuel and under the right circumstances will ignite. Flames will appear at the ceiling level and will travel over the heads of any one in the room who is staying low. This is known as **rollover** (also sometimes called **flameover**). The danger associated with a rollover is that a fire can travel overhead and get behind firefighters, blocking their egress path, allowing the fire to spread, and making subsequent investigation into a fire's origin more difficult by making it appear that there is more than one point of origin. In large rooms with large fuel loads this can be a particularly difficult issue.

In this case pyrolysates that are hot enough to ignite will not be able to burn until they "find" enough oxygen. Usually this occurs when the gases exit (roll) through a window or a door.

Flashover

During the development of a fire, heat is generated. Some of this heat is in heated smoke and fills the atmosphere of the involved room or compartment, while heat is also transferred to the structural elements and contents. The temperature of the contents rises as the heat is transferred. If the contents are ignitable liquids, the liquid will change to vapor once the liquid's flashpoint is reached. If the contents are solids, ignitable vapors will be given off when the ignition temperature is reached.

Obviously, all contents have different ignition temperatures and different physical states, and their ability to absorb the transferred heat is determined by numerous factors. However, because a fire in a confined area such as a room or compartment can result in the generation of significant heat levels, it is possible for all of the contents to quickly reach their ignition temperature. At this point, all of the contents are giving off ignitable vapors (pyrolysates). As these vapors ignite, all of the contents of the entire room or compartment become involved. This is known as **flashover** (fig. 5–14).

Fig. 5–14. Flashover

There are several ways to define flashover, the deadliest of fire "events" (building occupants will not survive the effects of a flashover; firefighters wearing full protective equipment have only a few seconds to reach safety outside the flashover environment before they will succumb—even then, they will likely suffer injuries). One is that flashover occur when a fire changes from becoming a fuel-controlled fire to a ventilation-controlled fire. Another is that flashover occurs when the upper thermal layer reaches a temperature of about 1,100°F (600°C). However you define it, this sudden change can take place in a matter of seconds and create a life-and-death situation for firefighters. It is critical that all firefighters learn to identify the warning signs of an impending flashover and take immediate measures to reduce or eliminate the potential for flashover.

Backdraft

FFI 4.3.11 A fire needs oxygen (or an oxidizing agent) in order to burn. Normally, this is not a problem since the atmosphere contains sufficient oxygen to allow a fire to burn (approximately 20.8%). However, if the fire is burning in a closed or confined area or space, the fire will consume the available oxygen and generate large amounts of carbon monoxide along with an assortment of other fire gases. As the oxygen level drops below 16%, visible flames start to diminish because the fire is being starved of oxygen. The danger is that the compartment is charged with superheated gases and smoke. The sudden reintroduction of oxygen to the compartment or space will literally breathe new life into the fire. There will be a rapid influx of outside air as if the fire is sucking the oxygen in. This will be followed by an even faster reappearance of flames in an explosive manner that has the power to blow out windows and doors and even push out walls. This is known as *backdraft*, sometimes called a *smoke explosion* (fig. 5–15). The violent explosive force of a backdraft has been known to lift buildings off of their foundations and hurl people across streets.

Fig. 5–15. Backdraft

Decay

Once the fuel package has been mostly consumed and the HRR drops significantly, the fire will diminish in size. This is the point at which the fire is in the decay stage. Small pockets of fire may still exist as the fuel slowly moves to the point of total consumption. Glowing embers will be the final indicator that the fire is about to cease because all of the available combustibles have been consumed by the fire. During a fire, all four phases may be seen simultaneously, albeit in different locations. A unique example is a high-rise building fire, especially when the involved area is above the reach of fire department hose streams. As the fire burns a floor, the floor above is preheated. When sufficient ignitable vapors are emitted, ignition occurs on the floor above while the fire is burning. The fire in the floor above, grows until the entire floor becomes involved, which is now preheating the floor above. Concurrently, as the fire now consumes all available combustibles on the original floor of involvement, the fire in that floor enters the decay stage until the fire ultimately dies out from a total lack of fuel.

Explosions

During your firefighting career, you will likely respond to the scene of an explosion after it has occurred. You may even be a witness to an explosion taking place. Explosions are not common events, but is important to understand the basics of these potentially deadly events. According to NFPA 921, an explosion is "the sudden conversion of potential energy (chemical or mechanical) into kinetic energy with the production and release of gases under pressure, or the release of gas under pressure. These high-pressure gases then do mechanical work such as moving, changing, or shattering nearby materials." Explosions occur very quickly and usually have devastating results. Explosions will be covered in some subsequent chapters.

There are two types of explosions: **deflagrations** and **detonations**. The difference is the speed in which they occur: the "combustion zone" (blast wave) moves slower than the speed of sound in a deflagration, while the combustion zone in a detonation moves faster than the speed of sound.

Natural gas explosions and many dust explosions are examples of a deflagration (fig. 5–16a). The use of an explosive such as TNT (trinitrotoluene) results in a detonation (fig. 5–16b). Generally speaking, deflagrations, while destructive and potentially deadly, do not have the same "shattering effect" of the more powerful detonation.

Fig. 5–16. Results of a) gas explosion (courtesy of James O'Connell) and b) TNT explosion (courtesy of David Forward)

Flow paths and today's compartment fires

You may have heard others say, "Fires burn hotter in today's structures." Well, that's not exactly the case. Fire temperatures increase at a faster rate causing fires to reach flashover temperatures sooner. Commonly in fires, flashover occurs before the fire department has arrived. For example, a candle flame burning in 1776 is the same temperature as a candle flame in 2018. But one of the keys in today's fires is the heat release rate rather than the temperature. A visual that is often used is one candle vs. 10 candles—the same flame temperatures but 10 times the heat release rate (fig. 5–17).

Heat transfer is critical to understand because heat is always transferred from the hotter object to colder object. As described earlier, there are different types of heat transfer: conduction, convection, and radiation. Conduction is the heat being transferred to the fires immediate surrounding area. Convection is when heat is transferred from moving gases or liquids and radiation is the heat transferred to you by the wave lengths—the closer you get the hotter it feels. Heat can be transferred by all three mechanisms to the ceiling, walls, and floor of the fire compartment.

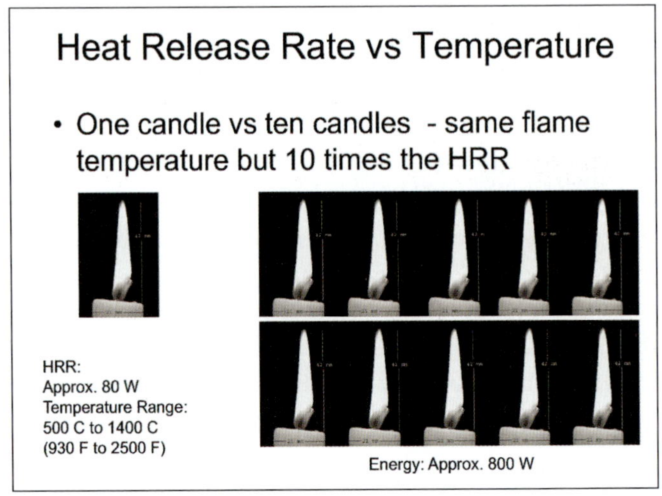

Fig. 5–17. One candle has the same flame temperature as 10 candles, but 10 candles have 10 times the heat release rate. (Courtesy of UL Firefighter Safety Research Institute)

Pressure

Fires create pressure and pressure moves from high pressure to low pressure (fig. 5–18). This is an absolute. The high pressure is the fire room and is being created by the fire. The low pressure is any opening that is created at a distance away from the fire. When such a pressure differential is created, we have the making of a **flow path**. High-pressure combustible gases (i.e., smoke) are forced across the upper parts of a compartment and the fire is drawing in the lower pressure air from a low-pressure opening. If we stop or control the air from being drawn into the fire, we reduce the temperatures and slow the combustion process. Think about the slowing of the combustion process, fire reduces in size along with the amount of combustible gases being produced. Interior visibility will worsen, unless its already zero-visibility, but the rest of the interior conditions improve with lower temperature and less or equal combustible gases.

A good example to illustrate a flow path is the breaking (venting) of an outside window of a home which is burning. As the window is broken, hot smoke will pour from the upper portions of the window while cool outside air enters the lower part of the window. The fire will be provided with oxygen from the low-pressure area at the window, causing the fire to grow in intensity. It is for this reason that we must ensure coordination between firefighters conducting ventilation and other firefighters attacking the fire with a hose stream—we don't want to provide any additional oxygen to the fire before we have hoselines in place to extinguish the fire. Prematurely breaking a window or even indiscriminately opening doors can lead to rapid growth of the fire without having a hoseline in place to suppress it.

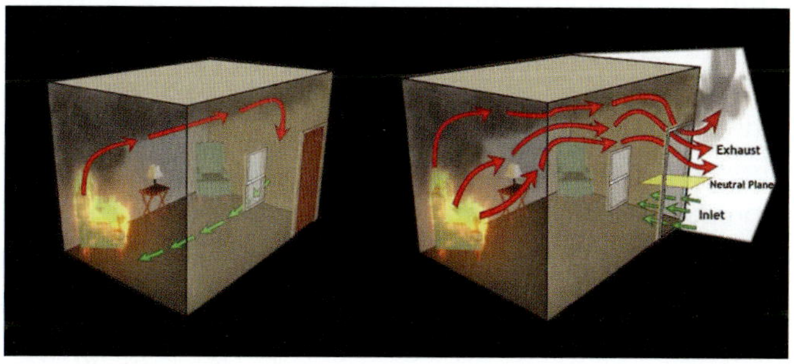

Fig. 5–18. Fire travels from high pressure to low pressure. When a vent opening is made it can create a flow path.

Flammable limits, described above, are important to understand in terms of today's fires. Similar to combustion of automobile engine, smoke burns rich and lean as the fuel/air mixture occurs. This is the flammable Range. Upper explosive limit (UEL) is rich and lower explosive limit (LEL) is the lean. The pressure is at its highest point if the fuel/air mixture is allowed to go above the LEL and gets closer to the ideal mixture. At this point with an ignition source it's possible to see a significant force of pressure during a fire event. As the flammable range goes past the ideal mixture and towards the UEL (rich) is the pressure drops because it's too rich to burn. One of the interesting things about flammable limits is that the fire cannot burn if it's above the UEL or below the LEL, because the mixture would either be too lean or too rich.

Fuel-controlled fire development, also known as **legacy fire development** (fires of years past), was often related to fuel-limited fires. The legacy fire wasn't limited by a lack of oxygen. It kept growing as long it had fuel to burn and once all of the fuel was burning the fire was at its fully developed stage. As the fuel burns down the legacy fire begins to decay, but it still has enough air to mix with gases and complete its combustion (fig. 5–19).

The **ventilation controlled** model commonly relates to ventilation-limited fires. Early on there is enough air to mix with gases and create complete flaming combustion of those gases. Some would consider this flashover but as the fire starts to run out of air, its starts to decay, and temperatures decrease. But then firefighters arrive and open a door to gain access, which gives the fire the air that it is needed to rapidly grow, and likely flashover into a fully developed compartment fire (fig. 5–20).

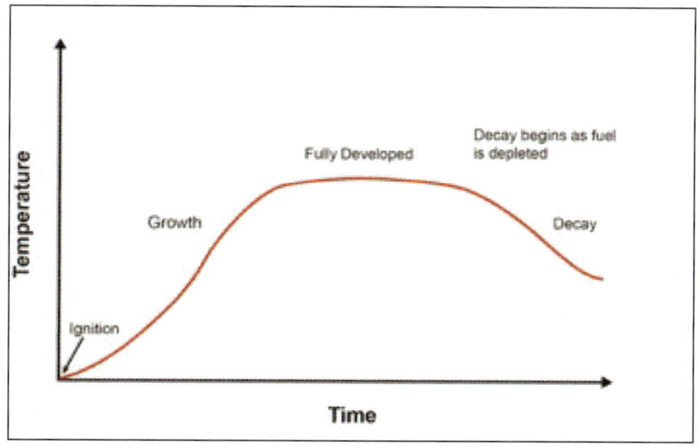

Fig. 5–19. Legacy Fire development graph. (Courtesy of UL Firefighter Safety Research Institute)

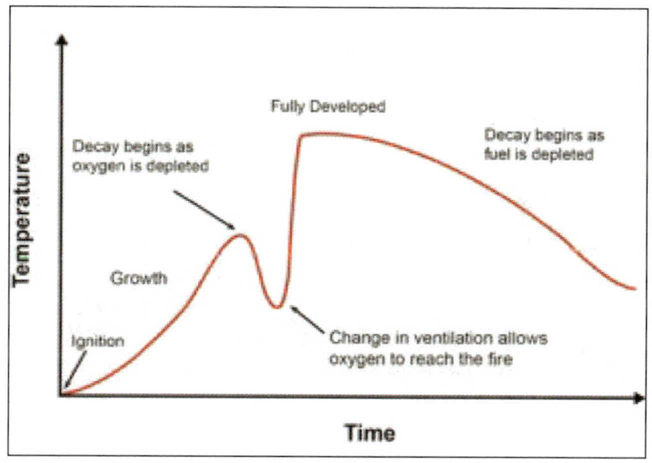

Fig. 5–20. Fire development graph. (Courtesy of UL Firefighter Safety Research Institute)

Flashover is the dangerous transition between the growth stage and fully developed stage. It is predictable and will cause serious injury or even kill firefighters if the precursors go unrecognized on the fire ground. When fire gases completely ignite and the radiant heat from the initial growth stage of the fire ignites the room contents, rapid-fire growth spreads within the compartment. Depending on the compartment size and the amount of pressure that builds prior to flashover, there could be a pressure wave occurring with the flashover, causing witnesses to describe such as a backdraft. While this phenomenon has become more common, it's likely been around many years. The old definition of backdraft from our older basic firefighter textbooks described backdraft as an event that occurred at extreme temperatures greater than 1,600°F, and exhibiting smoke stained windows, yellow-gray smoke, etc. When fires become ventilation-limited the temperatures decrease significantly. However, by the older definition- the temperatures are greater than 1,600°F causing there to be conflict in the definition of the ventilation-limited blasts that we are seeing in today's fires. The current definition of backdraft in NFPA 921 is "a deflagration resulting from the sudden introduction of air into a confined space containing oxygen-deficient products of incomplete combustion." This can occur in attic spaces or *knee walls* as found on many documented fires that have injured firefighters across the country. On these incidents, the fire remains hidden and ventilation limited until a firefighter creates a hole to find the fire and gives it just enough air to rapidly ignite, catching the firefighter off-guard and possibly causing injuries. It could be something as simple as poking a small hole in the ceiling to check for fire extension in a concealed (hidden) "void" space. There are times when creating such holes is necessary. Be aware, however, that introducing oxygen into these voids can lead to rapid fire development. Have a path of escape identified before making the opening in case conditions deteriorate rapidly.

In the above scenario there can be a time delay from the initial opening created by firefighters until the flashover occurs. While this time delay varies, it needs to be recognized by firefighters. The old saying "where there is smoke, there is fire" applies to these situations. If a small opening is made, water should be applied to the smoke and into the compartment as quickly as possible to reduce the temperatures and stop the occurrence of a fire growth. Again- firefighters must be aware of the consequences when creating openings within a structure fire.

Controlling the ventilation openings is critical on the fireground because of the need to avoid creating a flow path, until necessary for rescue or extinguishment. When this occurs the flow path should be recognized and controlled if possible. Recognizing fire behavior as it changes and reacts to our tactics is possibly one of the most underutilized skills for firefighters. Not only should we recognize the fire behavior that occurs from ventilation openings that are made by firefighters, but we should also recognize window or door failures caused by fire conditions. These window and door failures have caused firefighter injuries and deaths. These openings are unpredictable but a solid size up should recognize imminent failure to a window or door.

Interrupting the flow path involves tactics that are innovative and sometimes counterintuitive, but fire research shows that these methods can reduce heat within the structure, minimize the potential for flashover and greatly enhance the safety for firefighters. While this can vary depending upon location and size of the fire of today's structure fires, the time between tactical ventilation and flashover could be two minutes, compared to eight minutes in a traditional fire, giving firefighters only 25% of the time they previously had to recover from poorly timed ventilation. We need to formulate safe and effective ventilation tactics; it is imperative to better understand flow paths. The flow path is the space through which fire, heat and smoke progress, moving from an inlet and the high-pressure fire area to where the fire moves toward lower pressure oxygen sources, outlet areas such as door and window openings. Based on varying building configurations, there may be several flow paths within a structure. Firefighting or rescue operations conducted in a flow path can place firefighters at significant risk. Tests conducted by the National Institute of Standards and Technology and Underwriters Laboratories have advanced the understanding of fire behavior relative to flow paths and helped validate new, more effective tactics to enhance survivability (fig. 5–21). We discovered that increasing ventilation by opening doors, clearing windows, or cutting the roof in a ventilation-limited structure fire may lead to a rapid transition to flashover. In addition, anyone in the exhaust path—between the fire and its direction of travel—is in danger. Conversely, interrupting the flow path by closing doors or windows can limit fire growth and reduce temperatures.

Fig. 5–21. Testing by UL advances the understanding of fire behavior to validate new ways to enhance survivability. (Courtesy of Underwriters Laboratories and Firefighter Safety Research Institute)

The smoke and fire gases are what cause fire spread. Many times, we get focused on the flames. The flames are just a light show. The smoke is what we should be keeping our eyes on. Smoke volume, velocity, density, and color are essential components to understanding fire behavior. Recognizing smoke behavior can help firefighters recognize imminent flashover and predict fire spread. Controlling the ventilation openings is vital to stopping smoke and fire spread if the fire has not been controlled with water.

QUESTIONS

1. Describe two ways fire can obtain oxygen in order to support combustion.
2. What is the definition of fire?
3. What is the process of sublimation?
4. What is specific heat and its unit measurement?
5. Describe the process of latent heat of vaporization.
6. What effect does density play when it comes to ignition of an object or substance?
7. List the four common acceptable sources of heat and energy. Give examples of each.
8. Two types of reactions are ____________________ and ____________________.
9. List three physical states of a fuel.
10. Define ignition temperature.
11. Describe surface to mass ratio and its effects on fuel characteristics.
12. What is flash point?
13. What is fire point?
14. The four stages of fire development are ________________, __________________,______________, and __________________.
15. List four factors that affect a fire's ability to grow.
16. Can all four stages of a fire be seen at once? If so, describe how this can occur.
17. List the warning signs of a potential backdraft.
18. Describe how breaking a window affects the flow path of a fire.

Fire Extinguishers

by W. Jack Miller

This chapter provides required knowledge items for the following NFPA Standard 1001 Job Performance Requirements:

FFI: 4.3.8

FFI: 4.3.16

This chapter contains Skill Drills. When you see this icon, refer to your Skill Drill book for step-by-step instructions. SKILL DRILL

OBJECTIVES

Upon completion of this chapter, you should be able to do the following:

- Define the five classes of fires
- Describe fire extinguisher design and operation
- Describe the different extinguishing agents
- Explain the fire extinguisher rating system
- Have a general knowledge of inspection, maintenance, hydrostatic testing, and recharging of fire extinguishers

INTRODUCTION

Even small fires endanger lives and property. Building owners and managers have a responsibility to protect their tenants and the public by maintaining a fire-safe environment. All public buildings and workplaces need an effective fire protection program that includes strategically-placed portable fire extinguishers.

Portable fire extinguishers, when handled properly by trained and confident personnel, are useful for fighting fires before they spread—provided they are appropriate for the fire being fought and in proper working condition. To choose the right portable fire extinguisher you need to match the extinguisher to the job. This chapter provides the information to help you do just that. It covers the five classes of fire (A, B, C, D, and K) as well as the numbering system that determines the firefighting effectiveness of the different extinguishers. This chapter also reviews the classification of hazards, extinguishing agents, extinguisher design, and operating procedures.

FIRE CLASSIFICATIONS: WHAT IS BURNING?

FFI 4.3.16 The first question to ask is, "What type of fire is most likely to occur?" Fires are classified according to their fuel—that is, what is burning. Some fire extinguishers are more appropriate than others for fighting certain classes of fires. To be effective and safe, the extinguisher being used must match the class of fire to be extinguished. The proper extinguishing agent will put the fire out more efficiently. In some cases, it is dangerous to apply an incorrect extinguishing agent to a fire. The two best examples involve common water extinguishers, which should never be used to fight a fire involving combustible cooking media because water splatters grease and spreads the fire. Also, if you discharge a stream of water on live electrical equipment, you risk electric shock and establishing a short circuit that could damage equipment.

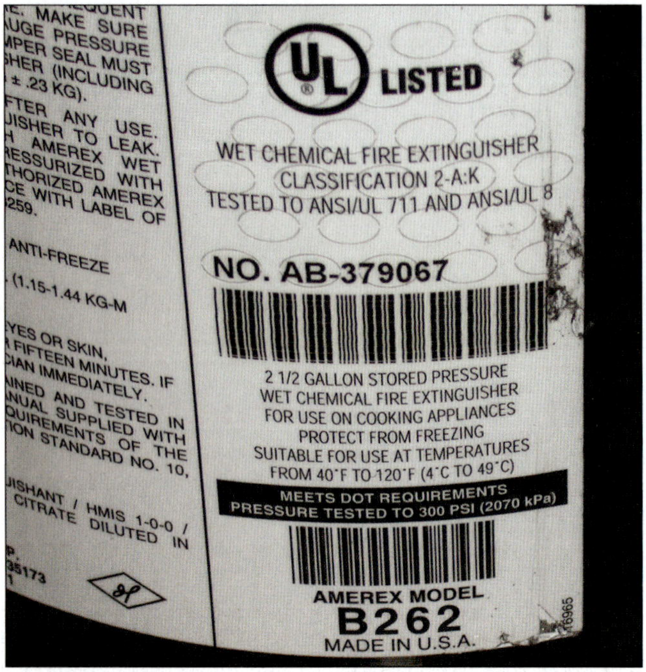

Fig. 6–1. This label from Underwriters Laboratories (UL) certifies that this fire extinguisher has been tested to their rigid standards.

Underwriter Laboratories Inc. or other independent testing laboratories rate fire extinguishers according to the type of fires they can extinguish safely. All fire tests are conducted by trained and experienced personnel in a controlled environment. Extinguishers that are tested and have been found to meet test standards are said to be listed and bear a label from the lab that conducted the testing (fig. 6–1).

Classes of fires

Class A: Ordinary combustibles such as wood, paper, rubber, fabrics, and many plastics.

Class B: Flammable liquids and gases such as gasoline, oils, paint, lacquer, and tar.

Class C: Fires involving live electrical equipment.

Class D: Combustible metals or metal alloys.

Class K: Fires in cooking appliances that involve combustible cooking media such as vegetable or animal oils and fats.

Figure 6–2 lists the classes of fire and their picture symbols.

CLASSES OF FIRES	TYPES OF FIRES	PICTURE SYMBOL
A	Wood, paper, cloth, trash, & other ordinary materials.	
B	Gasoline, oil, paint, and other flammable liquids.	
C	May be used on fires involving live electrical equipment without danger to the operator.	
D	Combustible metals and cumbustible metal alloys	
K	Cooking media (Vegetable or Animal Oils and Fats)	

Fig. 6–2. The five major classes of fire are represented with pictures so that inexperienced users can quickly identify the correct fire extinguisher to use.

TYPES OF EXTINGUISHERS

SKILL DRILL

Fire extinguishers are rated for, among other things, the classes of fire they can extinguish. Note that in the following list of extinguisher types, some extinguishers can be used on more than one class of fire. Do not attempt to use an extinguisher on a class of fire that it is not rated for.

Class A (fig. 6–3)

Class A:B

Class A:B:C (fig. 6–4)

Class A:C

Class B:C (fig. 6–5)

Class D

Class K (fig. 6–6)

Fig. 6–3. A pressured water extinguisher can be used on Class A fires only.

Fig. 6–5. A carbon dioxide extinguisher is common for use against Class B and C fires.

Fig. 6–4. Some dry chemical extinguishers are rated for fire classes A, B, and C. These extinguishers are commonly found in homes and commercial properties.

Fig. 6–6. A Class K fire extinguisher can resemble a Class A pressurized water extinguisher, however, has a special nozzle and water solution designed for cooking oils and greases.

Fire extinguisher design and types

All portable fire extinguishers use pressure to expel their extinguishing agents, but there are a few different designs (fig. 6–7). Many extinguishers rely on a pressurizing gas to expel their agents. Some models store the pressurizing gas with the agent in the cylinder and others store the pressurizing gas externally in a separate cartridge. Some extinguishing agents, such as carbon dioxide, create their own pressure and do not require the addition of a pressurizing gas to expel. Other extinguishers rely on pressure from a hand-operated pump.

The majority of extinguishers in use consist of six basic parts: cylinder, handle, lever, nozzle or horn, locking mechanism, and a pressure indicator (gauge).

The **cylinder** contains the extinguishing agent. Stored-pressure extinguishers store their extinguishing agent in a cylinder with a gas that provides the pressure to operate the unit. Cartridge- and cylinder-operated extinguishers use an external cartridge of a pressurizing gas that feeds pressure into the agent cylinder to expel the extinguishing agent.

Fig. 6–7. Pressure can be generated for fire extinguishers through storing the pressure in the main cylinder, having an outside cartridge, or through a manual pump.

The **handle** is the means by which an extinguisher is carried. The actual design of the handle varies from model to model. In most cases, the handle is located just below the lever that you squeeze for discharge.

The **lever** is what is squeezed or depressed to discharge the fire extinguisher's agent. In cartridge-operated extinguishers, the discharge is accomplished by releasing the pressurizing gas into the cylinder. On most extinguishers, the lever is located above the handle.

The **nozzle** or **hose/horn** is the device through which the agent is expelled.

The **locking mechanism** is a simple, quick-release device that secures an extinguisher's lever to prevent accidental discharge.

The **pressure indicator**, more commonly known as the **gauge**, indicates whether the extinguisher has adequate pressure to operate properly.

TYPES OF EXTINGUISHING AGENTS

FFI 4.3.8 The following extinguishing agents are the most often used in firefighting operations.

Water

Water is an excellent extinguishing agent suitable for use only on Class A (ordinary combustibles) fires. It extinguishes a fire by cooling fuel to below its kindling temperature. Never use water-type extinguishers on fires involving live electrical equipment or for fires in cooking media. The water could flash to steam as a result of high temperatures and cause serious burns.

Foam

Aqueous film-forming foam (AFFF) is made mostly of fluorochemical **surfactants**. As the bubbles break, drain, and lose their water, they form a film on the surface of the fuel that moves easily across the surface and prevents vapors from escaping.

Alcohol type AFFF is similar in base to AFFF with the addition of a **copolymer**. The copolymer reacts with **water miscible** fuels (**polar solvents** and hydrocarbon/polar-solvent mixtures containing 10% or more polar solvents) and reacts with the polar solvents to create a polymeric membrane.

Film-forming fluoroprotein foam forms a film over a flammable liquid surface much like AFFFs. (See fig. 6–8.)

Wet chemical

Wet chemical agents are solutions of water mixed with potassium acetate, potassium carbonate, potassium citrate, or combinations thereof. They are specifically designed for Class K fires, but they have demonstrated superior effectiveness on Class A fires when compared with plain water.

Fig. 6–8. A foam extinguisher

Wet chemical extinguishers work on Class K fires through two methods. The solution is alkaline in nature and therefore reacts with the fatty acids in the cooking medium to form a soapy foam on top of the burning material. This secures the vapors and cools the cooking medium as the foam drains and converts to steam. This reaction is known as **saponification**. In addition to saponification, the agent is discharged as a fine mist that does not submerge below the surface of the cooking medium (preventing a steam explosion), but rather it converts to steam on the surface pulling heat out of the material. Cooking-medium fires must be cooled below their **autoignition temperature** to be successfully extinguished.

Water mist

The **water mist extinguisher** uses deionized water that is discharged as a fine spray onto the burning material. It is designed as an alternative to halon in areas where contamination must be kept to a minimum without the expense of halon substitutes. Using deionized water causes little damage to most electronic circuits and is the safest agent available for use on humans (such as in operating rooms). The wand and wide spray with fine droplets gives a soft and controlled discharge pattern. This extinguisher has passed the Underwriters Laboratories Inc. test for electrical conductivity, allowing it to be listed for Class C applications (fig. 6–9).

Fig. 6–9. Water mist extinguisher

Dry chemical

Dry chemical extinguishing agents have been used since the early 1900s. Early in the development of these agents, sodium bicarbonate was found to have greater effectiveness on flammable liquid fires compared to other chemicals being used at the time. They are still widely used. In the 1960s, major developments in dry chemical agents led to the introduction of potassium bicarbonate (Purple K), and monoammonium phosphate (ABC). Dry chemicals are nonpoisonous, but either the acidic-based (ABC) or alkaline-based (Regular or Purple K) chemicals could be an irritant if inhaled. Dry chemical is not recommended for fires in delicate electrical equipment or aircraft because use of these agents may extinguish the fire but damage the equipment beyond repair.

Sodium bicarbonate

Sodium bicarbonate, or **BC**, is also known as regular dry chemical. In addition to its effectiveness on Class B and C fires, it has some effect on the flaming stages of a Class A fire but no effect on the embers or deep-seated Class A fire. Sodium bicarbonate is alkaline in nature and will not cause corrosion during normal use.

Potassium bicarbonate

Potassium bicarbonate, or **Purple K** dry chemical, was developed by the U.S. Naval Research Lab precluding the use of the term "Purple K" as a trade name. It was discovered that potassium salts are far more effective on flammable liquid fires compared to sodium salts.

Monoammonium phosphate

Monoammonium phosphate, **ABC**, or multi-purpose dry chemical differs from potassium bicarbonate and sodium bicarbonate in its acidic nature. In addition to similar effectiveness on Class B and C fires when compared to sodium bicarbonate, monoammonium phosphate has unique effectiveness on Class A fires. When it contacts the burning surface of an ordinary combustible, a molten residue (metaphosphoric acid) is formed. This residue coats the burning ember and excludes oxygen. Monammonium phosphate does not saponify when used on hot cooking grease and causes corrosion if not thoroughly removed from most hot surfaces.

Dry chemicals are nonpoisonous, but either the acidic-based (ABC) or alkaline-based (Regular or Purple K) chemicals could be an irritant if inhaled. If any physical discomfort is experienced, contact a physician immediately.

Dry powders for Class D fires

Combustible metal fires represent a special hazard unlike Class A, B, and C fires. Extinguishing agents used on all other classes of fires have no success when used on Class D fires. In fact, some agents, such as water, react violently with the burning metal. Most agents used in extinguishers for these fires are proprietary in nature and are classified according to the manufacturer of the dry powder or the brand name of the agent. Each agent has different limitations regarding the type of metal (magnesium, titanium, lithium, etc.,) that it may be used on and the form the metal is in (molten, casting, turnings, or fines). The following NFPA codes, standards, and recommended practices should be consulted:

1. NFPA 400: *Hazardous Materials Code*
2. NFPA 484: *Standard for Combustible Metals*
3. NFPA 472: *Standard for Competence of Responders to Hazardous Materials*

Most Class D agents are applied generously to the burning material, often requiring up to 15 pounds of agent per pound of burning material. The extinguishing agent usually excludes oxygen and performs as a "heat sink" to absorb the thermal energy and cool the material. Caution must always be used when applying special agents to combustible metal fires because these fires react with any moisture in the ground and surrounding materials.

Class D agents

Graphite (carbon) is a finely divided powder that must be applied with a dry scoop or shovel. The powder conducts heat away from the material, reducing its temperature below the point that combustion may be sustained.

Sodium chloride-based powder forms a crust on the burning material, excluding oxygen is available in a 30-pound extinguisher. It also helps dissipate heat from the burning material.

Pure copper powder was specifically developed for use on lithium fires.

Halogenated agents

Halogenated agents have been used for firefighting since the early 1900s. Of the 10 halogenated agents that have been used, two have been the most common since the early 1970s: Halon 1211 and Halon 1301. Halogenated agents suppress fire by interrupting the chemical chain reaction in the combustion process, working in the fire chemically instead of physically. Exactly how this chain-breaking process works is not completely understood. It is generally agreed that bromine is released from the agent as it decomposes in the fire, carrying away the free radicals that cause the combustion and releasing more bromine to continue the chain-breaking process. Halon 1211 and Halon 1301 are ozone-depleting agents subject to control under the *Montreal Protocol on Substances That Deplete the Ozone Layer* and other federal requirements. The primary advantage of halogenated agents has been the lack of cleanup required after using the agent. In some environments (such as electronics, data processing, jet engines, and high-tech equipment), discharging other extinguishing agents such as water or dry chemical could damage more property than the fire itself.

Halon alternatives

Since the ratification of the Montreal Protocol, alternatives to Halon 1211 have been sought. Various proprietary blends of gases are available. For streaming agents (those used in portable fire extinguishers), there are two popular agents in use: FE-36 and Halotron I. Both agents extinguish fires primarily through cooling.

Neither Halotron I nor Halon 1211 should be used in confined areas smaller than indicated on the extinguisher nameplate, food-preparation areas, or the presence of people with cardiac problems. If problems occur, quickly remove the person from the area where the gas is present, apply artificial respiration, and transport to a medical facility.

Carbon dioxide

Carbon dioxide, also known as **CO_2**, is an odorless and colorless gas that does not conduct electricity. It is stored in extinguishers as a liquid under pressure and when expelled turns into a snow (dry ice) on contact with atmospheric moisture. Carbon dioxide displaces oxygen and care should be exercised in confined spaces. Avoid skin contact, which can cause cold burns.

Things to remember

Never discharge a fire extinguisher into anyone's face. Never throw an extinguisher into a fire or leave it unattended if the fire is not out. (Pressure build-up from even a partially full extinguisher can cause an explosion.)

UNDERWRITERS LABORATORIES FIRE-EXTINGUISHING RATING SYSTEM

FFI 4.3.16 The numerical portion of Class A ratings of extinguishers is based on comparative fire tests using various sizes of wood-crib and wood-panel fires (table 6–1).

The numerical portion of Class B ratings of extinguishers is based on fire tests using square steel pans in specific size increments and a flammable, liquid test fuel similar to unleaded gasoline. The fire extinguisher classification is equivalent to 40% of the area of fire extinguished twice by an expert operator.

Class A fire tests

For ratings Class 1A through 6A, the extinguisher shall be tested on the appropriate wood-panel and wood-crib excelsior fires. For a 10A rating, the extinguisher shall successfully extinguish the 6A wood-panel fire plus the appropriate 10A wood-crib fire test. For a 20A rating or higher, only the appropriate size wood-crib fire is required (fig. 6–10).

The minimum allowable discharge time for extinguishers rated 2A and higher is 13 seconds. During the test, the extinguisher must be in the full open position under continuous discharge.

Fig. 6–10. Establishing the rating of an extinguisher often involves the use of wood crib test fires. (Courtesy of Tom Jenkins)

A pan containing commercial-grade heptane is placed beneath the wood crib and is used to ignite the wood members. The heptane is allowed to burn between 2 and 4 minutes, after which it is removed and the crib is allowed to burn until 45% of its weight is lost. After the preburn period, the operator attacks the fire from the front at a distance of at least 6 ft (2 m). The operator may then shorten the distance and attack the fire from all sides except the back of the crib.

For wood-panel test fires, the panel is first placed in a horizontal position, and a pre-determined volume of No. 2 fuel oil is applied to the panel, which is then mounted vertically on a steel frame. A row of excelsior is placed at the base of the panel. Three additional rows of excelsior are placed at the base of the panel. Three additional rows of excelsior are strategically spaced on the floor of the test room. Using heptane, the first row of excelsior at the foot of the panel is ignited. At 45-second intervals, the three remaining rows are pushed to the base of the panel. At 3 minutes and 20 seconds after ignition, all remaining excelsior is cleared from the base of the panel. The fire then burns vigorously for 4 to 5 minutes, at which time the horizontal furring strips located between 6 and 30 in. (150 and 760 mm) above the floor, burn through and begin to fall away from the panel. Within 5 seconds of this observation, the initial attack with the fire extinguisher is made from at least 10 ft (3 m) from the face of panel. The operator must attack this fire using two horizontal sweeps across the bottom of the panel and then may use the technique of his or her choice, provided the extinguisher remains under continuous discharge in the full open position until emptied.

Class B fire tests

Class B fire tests are conducted using a square steel pan at least 8 inches in depth. The test fuel is to consist of at least a 2-in. (51-mm) layer of Heptane, the surface of which is to be located 6 in. (152 mm) below the top edge of the pan. Water may be added to establish the required 6-in. (152-mm) freeboard (table 6–2).

For B ratings up to and including 20B, the fire test is conducted indoors in a large–volume, draft-free room.

Table 6–1. Class A Fire Tests

CLASS A FIRES: Ordinary Combustible: Wood, Paper, Textiles					
Numerical rating	Excelsior fire test	Wood-panel fire test	Wood-crib fire test Overall dimension, quantity, and size		
1A	6 lb/34×68 in. (2.7 kg/86×172 cm)	8×8 ft (2.4×2.4 m)	20×20×15 in. (51×51×38 cm)	72 pcs 12 layers of 6	2×2×20 in. (5×5×51 cm)
2A	12 lb/48×96 in. (5.4 kg/122×244 cm)	10×10 ft (3×3 m)	25⅝×25⅝×19½ in. (65×65×50 cm)	112 pcs 16 layers of 7	2×2×25⅝ in. (5×5×65 cm)
3A	18 lb/59×117½ in. (8.2 kg/150×298 cm)	12×12 ft (3.7×3.7 m)	30¾×30¾×21 in. (78×78×53 cm)	114 pcs 18 layers of 8	2×2×25⅝ in. (5×5×65 cm)
4A	24 lb/72×128 in. (10.9 kg/183×325 cm)	14×14 ft (4.3×4.3 m)	33⅜×33⅜×22½ in. (85×85×57 cm)	180 pcs 20 layers of 9	2×2×33⅛ in. (5×5×84 cm)
6A	36 lb/83×167 in. (16.3 kg/211×424 cm)	17×17 ft (5.2×5.2 m)	33⅜×33 ⅜×25½ in. (85×85×57 cm)	230 pcs 23 layers of 10	2×2×36 in. (5×5×92 cm)
10A	N/A	17×17 ft (5.2×5.2 m)	47½×47½×28½ in. (121×121×72 cm)	324 pcs 27 layers of 12	2×2×43 in. (5×5×109 cm)
20A	N/A	N/A	62¼×62¼×36½ in. (158×158×93 cm)	256 pcs 16 layers of 16 on edge	2×4×55 in. (5×5×140 cm)
30A	N/A	N/A	74⅝×74⅝×36½ in. (190×190×93 cm)	324 pcs 18 layers of 18 on edge	2×4×64 in. (5×5×163 cm)
40A	N/A	N/A	87⅛×87⅛ in. (221×221 cm)	400 pcs 20 layers of 20 on edge	2×4×69 in. (5×5×175 cm)

For B ratings in excess of 20B, outdoor fire tests are conducted under conditions of essentially still air: steady between 3 to 8 miles (4.8 to 12.9 km) per hour with gusts not greater than 10 miles (16 km) per hour and no precipitation. After a 1-minute pre-burn, the operator attacks the fire but is permitted to do so only from one side. The operator is prohibited from extending any part of his or her body past the edge of the test pan while fighting the fire.

FIREGROUND NOTE

Never use ABC dry chemical, Halon 1211, or Halotron I fire extinguishers on fires involving oxidizers (pool chemicals). A violent explosive reaction could occur with the mixture of chemicals.

Table 6–2. Class B Fire Tests

CLASS B FIRES Flammable Liquids: Oils - Greases - Paints			
Flammable liquids and gas rating	Minimum effective discharge time (seconds)	Pan size (inside) sq ft	Heptane used (approximate) U.S. gallons
Indoor Tests:			
1B	8	2½	3¼
2B	8	5	6¼
5B	8	12½	15½
10B	8	25	31
20B	8	50	65
Outdoor Tests:			
30B	11	75	95
40B	13	100	125
60B	17	150	190
80B	20	200	250
120B	26	300	375
160B	31	400	500
240B	40	600	750
320B	48	800	1,000
480B	63	1,200	1,500
640B	75	1,600	2,000

CLASS C FIRES Electrical Equipment	
No fire test	Extinguishing agent is tested for electrical non-conductivity. If acceptable, "C" Symbol is added.

FLOW RATES/APPLICATION RATES

Every agent, regardless of the mechanism it uses to extinguish a fire, has a critical application rate. The **critical application rate** is the minimum rate at which the agent must be applied to the fire and is usually expressed as a ratio of quantity of agent (in pounds or gallons) per area unit (usually square foot) per time unit (either seconds or minutes).

Under most circumstances, particularly when using equipment that does not involve a fixed system (hand hose lines or extinguishers), a rate of application in excess of the critical rate must be used. Other factors involving the fire and extinguishment may cause problems. Wind direction and speed, weather conditions, length of pre-burn, application technique, and fuel sources all affect the critical application rate. Using a higher application rate will help negate some of these factors and decrease the time required for extinguishment. There is a point at which a further increase in the application rate will not result in a further decrease of the time to extinguishment.

Application rates for dry chemicals are theoretical and often expressed in pounds of chemical per square foot per second. Tests done in the 1960s involving flammable liquid in depth and spill fires using a variety of fuels established an average critical application rate of less than 0.01 lb of chemical per second per sq ft. These tests used fixed systems, thus eliminating any operator error. They are not intended for use in the designing systems or choosing hand-operated equipment.

Flammable liquid fires involving fuel escaping under pressure, **three-dimensional**, running, or gravity-fed fuel generally require higher application rates than spill or fuel-in-depth fires. Listed hand-portable dry chemical fire extinguishers may have a flow rate ranging from 0.31 lb (0.14 kg) per second to 2.5 lb (1.13 kg) per second. The flow rate of different fire extinguishers may be established as a requirement for a higher UL listing (to meet minimum discharge times) or to meet a special purpose.

In practical applications using hand-portable or wheeled fire extinguishers, it is not possible to determine the critical application of the dry chemical for every set of circumstances. The advantages in using higher application rates (extinguishers with faster flow rates or multiple extinguishers in unison) are faster knockdown, greater protection from radiant heat (heat shield), faster extinguishment on fuel in depth fires, and greater chances of extinguishing pressure, obstacle, and three-dimensional fires. Disadvantages of using higher flow rates are wasting chemical and having less discharge time.

Personnel trained during live-fire exercises with dry chemical extinguishers are often able to put out larger and more complex fires with extinguishers that have a faster flow rate (and a lower UL rating). This is partly because their application technique wastes less chemical (negating the need for a longer discharge time), has better protection against radiant heat, and allows firefighters to use the faster knockdown to their best advantage. A 20-lb (9 kg) fire extinguisher with a rating of 40 B:C and a flow rate of 1.33 lb (0.6 kg) per second is likely to have far greater extinguishing capabilities in the hands of even a moderately trained operator than a 20-lb (9 kg) fire extinguisher with a rating of 120 B:C and a flow rate of 0.69 lb (0.31kg) per second. The faster–flow, 20-lb (9 kg) extinguisher can easily extinguish a larger UL test fire than its rating implies, but the required discharge time is insufficient to meet the UL standard.

Two operators applying dry chemical to the fire at the same time can theoretically double the flow rate of each of their extinguishers, thereby decreasing the time to extinguishment and providing better safety to the operators and greater chances of success. These are reasons to train personnel to fight fires as a team rather than as individuals.

When conducting training sessions, it is important that the same flow rates are used in training that are available "on line" in the facility. If personnel are trained with extinguishers flowing chemical at 1.77 lb (0.8 kg) per second, and the extinguishers in the facility have a flow rate of 0.68 lb (0.31 kg) per second, then training will not help personnel learn application techniques for their equipment. Application techniques for fast-flow equipment are different from those used with extinguishers having low flow rates and long discharge times. Using old extinguishers for training may have the same result. Many older fire extinguishers have much lower ratings (and faster flow rates) than newer extinguishers of the same agent capacity. For example, a 20-lb (9 kg) extinguisher made in 1972 may have a UL rating of 60B:C, while the same model made in 1992 has a UL rating of 120B:C.

Higher UL ratings do not equate to greater firefighting capability. Many factors, including training, type of fire incident, fuel, and application technique affect any attempt to extinguish a fire. NFPA 10, *Standard for Portable Fire Extinguishers* recognizes this issue and requires the use of fast-flow equipment based upon the possibility of obstacle fires, pressurized fire, and three-dimensional fires.

FIRE EXTINGUISHER OPERATION

SKILL DRILL

FFI 4.3.16 These general instructions are intended to familiarize you with the basic operating techniques of a stored-pressure, hand-portable fire extinguisher. Because there are a variety of extinguishers, the nameplate must be consulted for specific procedures and starting distances.

1. Hold the extinguisher upright and pull the safety pin to break the plastic seal.
2. Stand back from the fire (minimum distance is stated on the nameplate) and aim at the base of the fire.
3. Keeping the extinguisher upright, squeeze the handle and lever together to discharge and sweep from side to side. Move closer as the fire is extinguished, but not so close as to scatter the burning material or liquid.
4. When the fire is out, back away while checking for possible re-ignition.
5. Evacuate and ventilate the area immediately after use. The smoke and fumes from any fire may be hazardous and deadly.

Remember the acronym **PASS** (fig. 6–11):

- **Pull**
- **Aim**
- **Squeeze**
- **Sweep**

Any firefighting effort, first and initial response, or organized attack should minimize risk of injury and death. Once a decision has been made to fight a fire, it is important to remain calm and keep a clear perspective concerning safety. All things of a material nature can be replaced or rebuilt, but human life and health cannot.

Humans, like other animals, will react with either "fight" or "flight" when scared. Fire will cause one of these reactions to occur, sometimes involuntarily. No two people react exactly the same way, and no person is consistent from one fire to the next. Even the most experienced and trained firefighter makes bad decisions regarding safety in the heat of the battle.

Fig. 6–11. The acronym PASS is used to remind untrained users of fire extinguishers to pull, aim, squeeze, and sweep the hose back and forth.

Application techniques

SKILL DRILL

The most important aspect of any fire extinguisher application technique is operator safety. Techniques that needlessly jeopardize the safety of the operator are unnecessary and should never be taught or demonstrated.

It is advisable that all fires should be fought from the upwind side in order to stay out of the smoke, flames, and radiant heat. While this is preferable, it is not always practical. Regardless of wind direction, fires should only be fought from a side that allows a clear path of escape. Never place yourself on the opposite side of a fire from the path of escape in order to take advantage of the wind direction.

Foam application techniques. When applying foam to a fire, it is important to stay back during the initial attack. Unlike dry chemical or water fog patterns, foam streams do not give good protection against radiant heat. Once control of the fire starts to be established, the operator can safely move closer.

Care should always be taken to ensure that the foam is not submerged into the fuel and is applied gently with minimal disturbance to the fuel surface. Methods to do this include deflecting the stream off the back edge, bouncing the stream off of the front edge, deflecting it off an object in the center, or allowing the foam to gently fall down on the surface (known as the "rain down" or "snowfall" technique).

Faster extinguishment can be accomplished if the stream is moved about the surface rather than holding at one application point and waiting for the foam to move across the fuel surface. Once control has been established, shut down the nozzle and see if the foam will seal up, completing extinguishment. As long as the stream is disturbing the fuel surface, the foam cannot seal across the surface and complete extinguishment.

If water-miscible (polar solvent) fuels are involved, even in a small percentage, an alcohol-type foam must be used. When using an alcohol-type foam, it is important that the fuel surface not be disturbed so that the polymeric membrane can form on the surface.

No firefighting foam will secure a fuel surface indefinitely. Reapplication is necessary over a period of time to prevent reignition. Firefighters should never drag hoses through a **foam blanket** or step into a fuel source with a foam blanket on it without having specific equipment and protection on to do so. Water-miscible (polar solvent) fuels such as alcohol are particularly dangerous since they burn so cleanly that the flames may not be visible.

Dry chemical application. Unlike foam on a Class B fire, dry chemical has no securement capabilities. All of the fire must be extinguished or none of it will be extinguished. Because dry chemical uses a chain-breaking action to extinguish a Class B fire, no physical properties of the fire are affected. Oxygen is not displaced, little cooling is accomplished, and the fuel is not removed. Depending on the type of materials exposed to the fire (such as steel, gravel, combustibles), the chances for a reflash usually are present. The operator should anticipate the possibility of a reflash and never turn his or her back on a fire. There is never a reason to step into a flammable liquid source when using only fire extinguishers.

With the exception of fighting fires inside structures or underground mining operations, the operator should remain upright for the best and safest mobility. Bending over or running may cause the operator to trip.

Always start the approach at a safe distance. Humans were made to walk forward well, but running backwards is never safe.

Be aware of what is on the side of the fire opposite the operator discharging. The flare-up phenomenon can catch other fuels or combustibles on fire at a great distance from the dry chemical discharge.

Whether one operator is involved or several, keep alert, communicate loudly, and always have an open escape route in case you are not successful.

Dry chemical and fuel in depth. Class B fires involving fuel in depth present unique problems for the application of dry chemical. **Fuel in depth** implies that the flammable liquid is confined to a certain area either by a vessel, berm, or dike. Only the surface is of concern. It does not make a difference from an application standpoint whether the surface area encloses 5 gal or 500 gal of fuel. Because the fuel is contained to a specific surface area, it is important that the operator does not get too close and splash fuel, causing the area to increase.

Ideally, this fire should be fought upwind, but if the upwind approach would cut off a safe escape route, it can be fought with a crosswind or diagonal wind. The operator should start discharging the extinguisher from a comfortable distance using the dry chemical stream to "lay down" the fire and push it back before stepping closer.

If the fire surface is rectangular in shape, a slow, deliberate cut across the leading edge will get chemical into the **freeboard** area between the lip of the containment and the fuel surface. This is best accomplished by splitting the stream of dry chemical on the lead edge. Generally, the stream should be placed at least 6 in. (152 mm) in front of the lead edge and the sweep or cut should begin at least 6 in. (152 mm) beyond the side of the fire and continue past 6 in. (152 mm) beyond the other side of the fire.

If, after the initial cut has been made, fire is seen rising immediately on the lead edge right after the dry chemical stream moves away, the aim is too high, or in the case of a deep freeboard area, the angle is too flat. If fire immediately reappears on one side or the other, the cut or sweep is not being extended far enough to that side.

If the fire surface is in the shape of a circle, as in a round pan, start the discharge at the closest point, splitting the stream on the lead edge. Hold for a count of one to two, then begin sweeping side to side. As with a rectangular pan, each sweep must extend beyond the edges on both sides.

Once the fire is moving backwards, the operator can take slow steps forward, sweeping side to side until the fire is completely out. During these sweeps side to side, each sweep must extend at least 6 in. (152 mm) past each side. Unless it is necessary because of nozzle range, it is advisable to always back away from the fuel source. Should a sudden wind shift occur, the attempt at extinguishment fail, or a new source of ignition be introduced, the entire surface area of fuel will quickly ignite.

If for any reason it is necessary to be close to the fuel source, the speed of the sweep action must be increased in order to keep from splashing fuel.

Do not chase the flare-up. Keep the stream down on the surface of the fuel. The flare-up will burn out quickly.

Once extinguishment has been accomplished, stand by waiting for reflash, and then back carefully towards the escape route.

Spill fires. Spill fires by themselves are most likely the easiest of all flammable liquid fires to extinguish with dry chemical. There is no "lead edge" or freeboard area to conceal burning vapors. The same basic technique that is used for fuel in depth applies to spill fires, with one extra caution. With a lot of dry chemical flying about, the exact spill area may be hard to define, so it is important that the operator must stay back as far as possible to avoid inadvertently stepping into the fuel source.

Three-dimensional fire (gravity fed). This type of fire involves flammable liquids in motion, usually running from a source through several levels and pooling on the ground and other areas. Often, obstacles are involved, and, at the least, areas where fuel will flow around structures are involved. For these reasons, more than one dry chemical extinguisher operator should be present at the same time.

Initial approach and application are the same as for a spill fire. The operator(s) should open up the stream(s) at a safe distance, keep a clear exit path, and then, once the fire is moving, move forward and take care to stay out of the fuel spill.

Application must begin with the ground fire and then systematically continue extinguishment up to the source. Everywhere that fuel is flowing over hot metal surfaces or wrapping around an obstacle, the dry chemical stream should be held briefly to allow chemical to build up at that spot before continuing.

If ground fire re-ignites at any point during the application, the dry chemical stream must be brought back to the ground fire in order to be successful.

Applications with more than one operator. Using extinguishers in teams can be the safest and most successful way of extinguishing fires in their incipient or early stages. Communication is the key to having a safe, coordinated effort. All other basic rules apply regarding safe exit, distance, staying clear of the fuel source, and individual application technique. Each operator must speak loudly and clearly to each other. Dry chemical streams must be opened at the same time, and each operator should know what the other's plan is going to be prior to the attack.

At no time should the operators be at direct opposite sides of the fire. This will obscure vision and spray fuel and dry chemical on each other. Each operator should be able to see the other at all times and know where each team member is located.

If one operator runs out of dry chemical before the other, that person should say loudly and clearly, "I'm out," and then back away to safety along the predetermined path. This will let the other operator know that it is time to back away and cover both of their exits by discharging dry chemical between them and the fire and creating a heat barrier.

If one operator spots fire that is being missed by the other operator, that person must point it out immediately to the other. While each person may have a specific task to do, or a specific part of the fire to handle, it is important for each person to keep looking at all aspects of the fire. Other operators may be unable to see fire lingering in a small area because their vision is obscured by the dry chemical discharge.

If the fire is successfully extinguished, one of the operators must call the fire out. Both operators should then stop discharging, stand by, look for a reflash, and then back away to safety following the same route used to advance on the fire.

Spill or fuel in depth with an obstacle. An obstacle inside a fuel in depth or spill fire will shield the dry chemical stream from covering the entire liquid surface. At least two operators must be used when obstacles are present.

First, operators must agree on a plan of attack. (Who goes which way and covers what part?) Standing side-by-side, start discharging dry chemical together with each covering two-thirds of the surface area. Once the fire is moving, they then advance and split to each side until they can reach the back of the obstacle from an approximate 45-degree angle. When the fire is out behind the obstacle, they sweep the rest of the surface fire out. Once extinguishment is accomplished, they stand by looking for reflash, and then back away along the same path used to advance.

Fires with flammable liquids under pressure. Generally, the best way to extinguish flammable liquids under pressure is to shut off the source of supply. This is not always practical and dry chemical can be very effective on these types of fires. Usually it is necessary to place the dry chemical stream right at the point of escape and hold at that spot until the fire goes out. Then, continue to put out any ground fire that is burning. Fast-flow equipment with high flow rates must be used on this type of fire since the dry chemical flow rate must match the flow of fuel escaping.

Using dry chemical with handlines. Dry chemical can be applied to fires from behind fog patterns with great success. The fog pattern protects the fire fighters and aids in cooling the fire. The dry chemical can be directed from behind through the pattern without allowing the fire to come back on the operator. When using this technique, it is still necessary to apply the dry chemical using the same application techniques that are used without a fog pattern. Common mistakes with a combined water and dry chemical attack involve just directing the dry chemical into the water stream (this will have little effect on the fire) or sticking the dry chemical nozzle into the fog pattern (thus breaking the pattern and allowing flame to come back onto the hose team).

Dry powder on Class D fires. Applying dry powder to class D fires generally involves a soft discharge pattern that will completely cover the burning material under a mound of dry powder. The operator should be careful that no moisture is present in the surrounding area that may contact the burning metal. Once the material is completely covered, the operator should stand by and watch for "hot spots" that start to burn through the powder and reapply as necessary (fig. 6–12).

Inspection, maintenance, recharge, hydrostatic testing, and obsolete extinguishers.

Fire extinguishers are to be inspected at a minimum of 30-day intervals. They should be inspected more frequently when circumstances require. Per NFPA 10, 7.2 Procedures, the following shall be checked:

1. Pressure gauge reading or indicator in the operable range or position
2. Fullness determined by weighing or hefting for self-expelling extinguishers, cartridge-operated extinguishers, and pump tanks
3. The discharge outlet is not blocked
4. The seal is not broken
5. The operating instructions are clearly visible
6. They are located in a designated place

Fig. 6–12. Applying dry powder to a Class D fire is risky and requires the fire extinguisher operator to monitor the situation for "hot spots" that might flare up.

When an extinguisher reveals a deficiency, it shall be subjected to applicable maintenance procedures.

- **Annual Maintenance.** A more complete inspection of the extinguisher that shall be done by trained, certified persons having the appropriate service manuals, tools, manufacturer replacement parts, and recharge agents specifically listed for use in the extinguisher.
- **6-Year Maintenance.** Every six years, extinguishers requiring a 12-year hydrostatic test shall be emptied and subject to thorough examination of mechanical parts, extinguishing agent, and expelling means. When applicable maintenance procedures are being done during periodic recharging or hydrostatic testing, the six-year requirement will begin from that date.
- **Hydrostatic Testing.** This involves subjecting the cylinder (known as the pressure vessel) to water pressure or pressure applied by some other noncompressible fluid, usually through the cylinder's valve opening, using a specially designed test coupling. This testing should not be performed using compressed air or any compressed gas, and all air and other gases must be vented from the cylinder before you apply any fluid pressure. Hydrostatic testing should be done by trained, certified personnel using suitably equipped testing facilities. How often hyrdotesting should be conducted depends on the type of extinguishing agent used. Examples are water, foam, wet chemical, and carbon dioxide, which require a 5-year test. Dry chemical, whether stored-pressure or cartridge-operated with mild steel cylinders, require 12-year testing.
- **Recharge.** Recharge shall be done by trained, certified personnel immediately after use. The extinguisher must be recharged with the extinguishing agent listed on the nameplate. Substitutions could cause damage or injury. ***WARNING:*** *Always use a regulated pressurizing source set to no more than 25 psi above extinguisher operating pressure.*
- **Obsolete Fire Extinguishers.** Per NFPA 10 section 4.4, the following types of extinguishers are considered obsolete and shall be removed from service:
  1. Soda acid
  2. Chemical foam (excluding film-forming agents)
  3. Vaporizing liquid (e.g., carbon tetrachloride)
  4. Cartridge-operated water
  5. Cartridge-operated loaded stream
  6. Copper or brass shell (excluding pump tanks) joined by soft solder or rivets
  7. Carbon dioxide extinguishers with metal horns
  8. Solid charge-type AFFF extinguishers (paper cartridge)

9. Pressurized-water fire extinguishers manufactured prior to 1971

10. Any extinguisher that needs to be inverted to operate

11. Any stored-pressure extinguisher manufactured prior to 1955

12. Any extinguishers with 4B, 6B, 8B, 12B, or 16B fire ratings

13. Stored-pressure water extinguishers with fiberglass shells (pre-1976)

Further, 4.4.1 states: Dry chemical, stored-pressure extinguishers manufactured prior to October 1984 shall be removed from service at the next six-year maintenance interval or the next hydrotest interval, whichever comes first.

Section 4.4.2 adds: Any fire extinguisher that can no longer be serviced in accordance with the manufacturer's maintenance manual is considered obsolete and shall be removed from service.

LESSON FROM THE FIREGROUND

On arrival the fire in the basement of the 2½-story single family colonial style house was lapping up the wall and starting to spread across the ceiling to the stairway. A firefighter in full personal protective equipment and wearing self-contained breathing apparatus positioned himself at the exterior doorway on the side of the house by the stairway. There he used a 2½-gal (9.5-L) water extinguisher from a fire chief's car to hold the fire in check by placing a fingertip over the discharge point of the nozzle and spraying intermittent bursts of water. The judicious application of water prevented the fire from extending up the stairway and into the house. Shortly thereafter, the attack hoseline was in place and quickly extinguished the fire in the basement. The deployment of the 2½-gal (9.5-L) water extinguisher was a crucial factor in minimizing the damage to the house.

QUESTIONS

1. Explain the differences between applying an extinguisher to a fuel-in-depth fire in a rectangular and a circular container.
2. The rating system for a Class A fire extinguisher is based on what?
3. Explain the difference between dry chemical and dry powder extinguishers.
4. Describe the two ways that wet chemical fire extinguishers work on Class K fires.
5. List examples of the types of fires that halogenated extinguishing agents would be more beneficial than other agents.
6. List the criteria for selecting the proper fire extinguisher.
7. What is the best way to extinguish a fire involving a flammable liquid under pressure?
8. List four fire extinguisher types that are considered obsolete by NFPA 10.

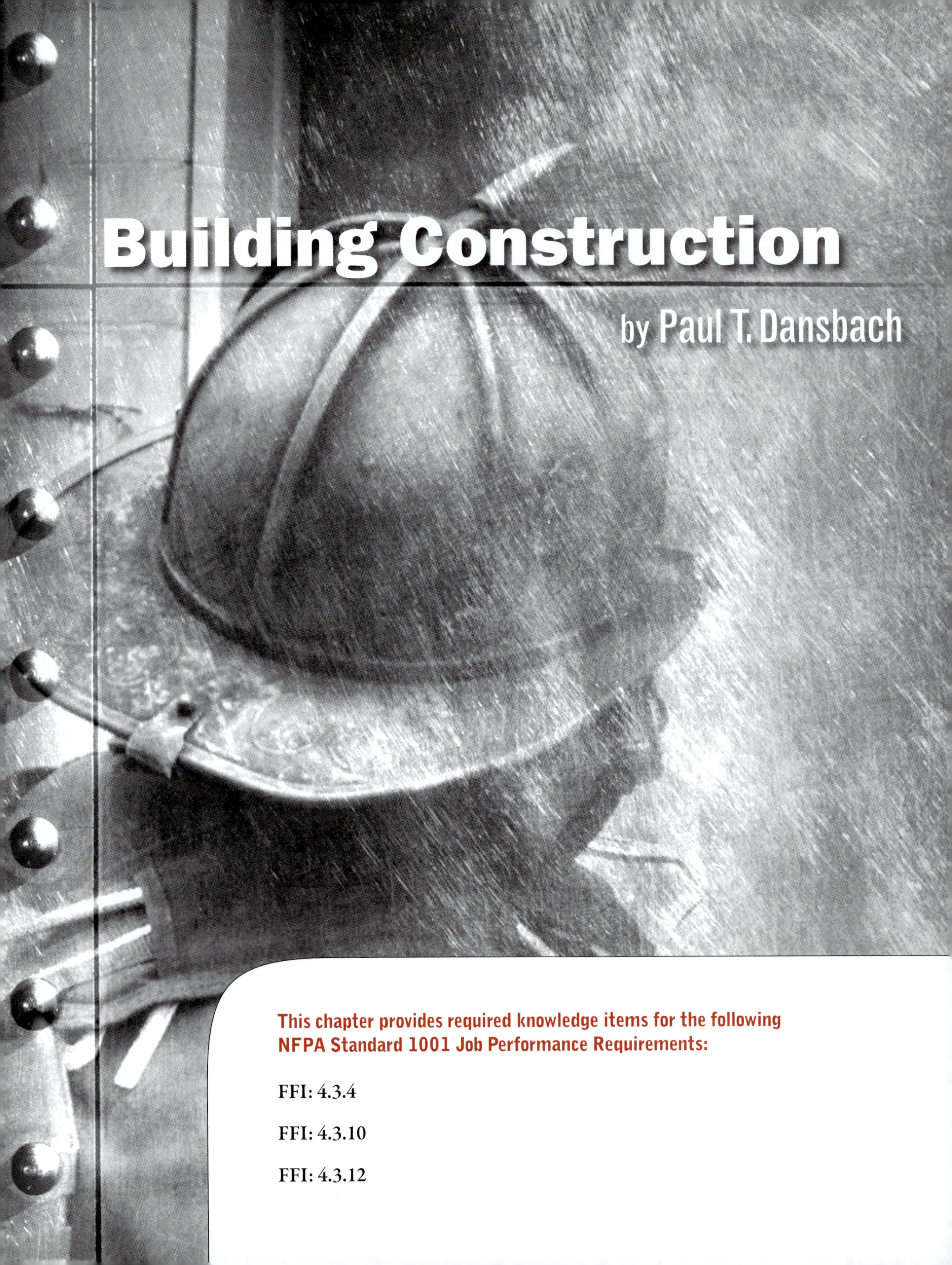

Building Construction

by Paul T. Dansbach

This chapter provides required knowledge items for the following NFPA Standard 1001 Job Performance Requirements:

FFI: 4.3.4

FFI: 4.3.10

FFI: 4.3.12

7

OBJECTIVES

Upon completion of this chapter, you should be able to do the following:

- Identify the types of forces and loads that impact building construction materials
- Identify and describe the basic types of building materials
- Identify and describe the five common types of building construction
- Identify the basic structural components in a building
- Identify factors affecting structural stability of a building
- Describe the hazards associated with light-weight and truss construction
- Identify the indicators of collapse
- Describe the impact of new building construction technology on the fire service
- List actions to take to limit injury or death to personnel in the event of a structural collapse

INTRODUCTION

This chapter begins your study of building construction (fig. 7–1). By the end of this chapter, you will understand the basic concepts of building construction as they relate to firefighting. To be successful as a firefighter or fire officer, the study of building construction must be an ongoing and perpetual learning experience. From the beginning of firefighting, firefighters have been called on to extinguish fires in buildings. Much of what you learn in your formal Firefighter I and II training programs centers on safe and efficient fireground operations. Fighting a fire in a building is inherently dangerous: The building and its structural integrity is attacked and weakened as the fire burns. Much has changed in firefighting since the beginning of organized firefighting, but one aspect remains the same: To successfully extinguish most struc-

Fig. 7–1. A wood frame building

ture fires, a crew of firefighters must enter the building, locate the fire, and extinguish the fire by directing a stream of water from the nozzle onto the burning material. Your success and safety largely depend on your knowledge and understanding of building construction.

FORCES AND LOADS ACTING ON A BUILDING

Forces and how they apply to a building

A **force** is any action that maintains or alters the position of a structure or a part of a structure. Essentially, it is a push, pull, twist, or a combination of these exerted on an object or a structural member. Generally, there are four types of forces that can be applied to a building or its components:

1. **Compression**: the action of squeezing or pushing of a component, a **load** or weight imposed on a structural column that result in a compressive force when applied to the column (fig. 7–2). A compressive force may shorten the structural component.

Fig. 7–2. The studs supporting the floor joists in this wood-frame wall are in compression.

2. **Tension**: the action of stretching or pulling of a component. The threaded rod in fig. 7–3, which supports the stairway's intermediate landing, is in tension. Many members, which are in tension, are supported by attachment to a structural member above (fig. 7–4). A **tensile** force applied to a structural member may lengthen or stretch the member in tension; members in tension often fail by breaking apart or failing at the connection points.

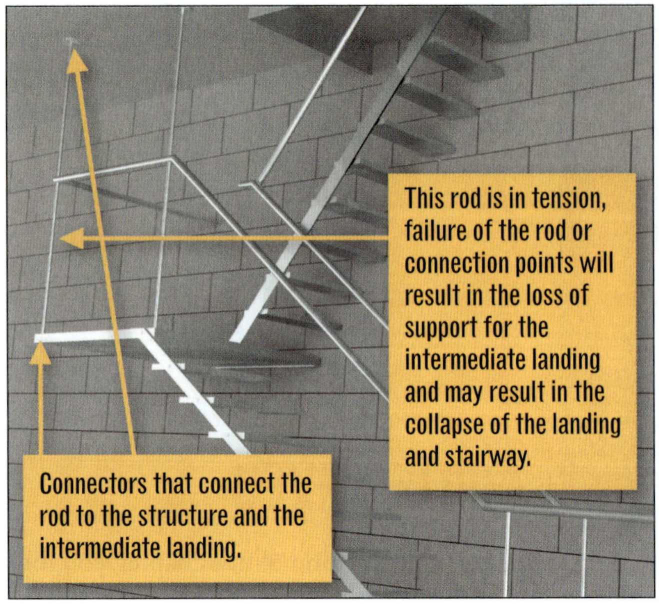

Fig. 7–3. Tensile force applied to a structural component

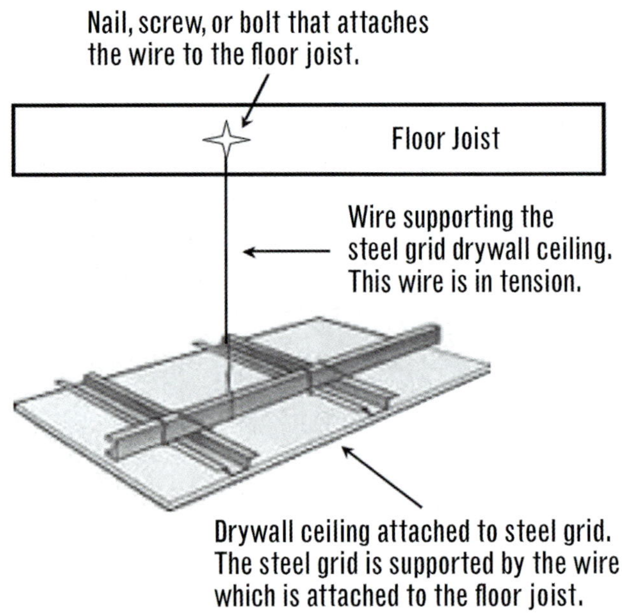

Fig. 7–4. The wire that supports the drywall steel grid is in tension and is attached to the floor joist above.

3. **Torsion**: the action of twisting a building component such as a nut on a bolt. Torsional force is commonly expressed in terms of **torque**.

4. **Shear**: a condition or force causing two structural members to slide past each other in opposite directions (away from or toward each other). Significant shear forces can cause structural members to fail at their connection points (fig. 7–5).

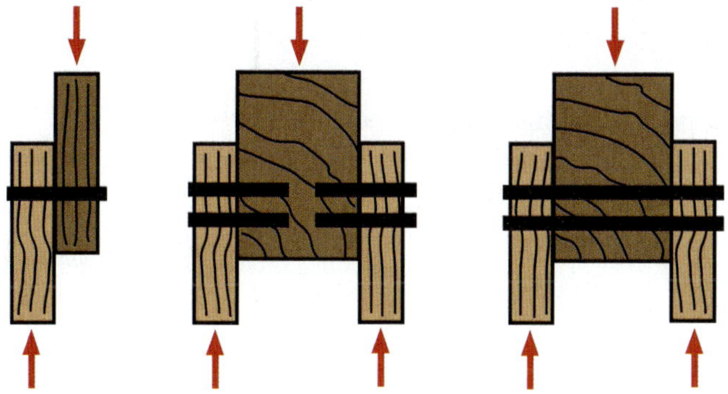

Fig. 7–5. These connections are typical connections in a wood-frame building: a shear force is applied to the connection. If the connection fails, the parts of the connection will slide past each other.

Loads applied to a building

Loads are basically the forces applied to a building from a variety of sources, such as the weight of the contents of a building (occupants, their possessions, and equipment), as well as **external forces** such as environmental effects (weather, earthquakes, etc.).

Loads may be imposed on a building's structural members in one of three different ways: axially, eccentrically, or torsionally. **Axial loads** pass through the center of a particular section or supporting member at a right angle to the cross section of the supporting member (fig. 7–6).

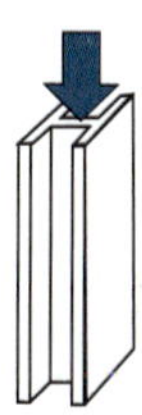

Fig. 7–6. An axial load is imposed on the column as the load passes through the center of the column.

Eccentric loads are imposed on a structural member at some point other than the center section of the supporting member (fig. 7–7).

Finally, **torsional loads** are parallel to the cross section of the supporting member, typically a column that does not pass through the long axis of the structural member (fig. 7–8).

Fig. 7–7. The parapet wall is eccentrically loaded with a cornice (a decorative wall projection attached to the top of many apartment buildings.) This load does not pass through the center of the wall.

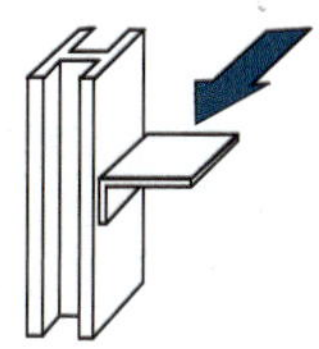

Fig. 7–8. Torsional loads are delivered at a right angle to the supporting member and usually result in a twisting of the member. Torsional loads are common in collapses when the designed structural function of the structural members is upset.

The following terms are also used to describe specific types of loads:

A **designed load** is a load that the building designer or structural engineer has calculated into the building based on sound engineering principles. The structure can carry or withstand all the design loads that are applied to a given building or structure. Several of the loads explained in the following text are easily calculated and required by code to be calculated to ensure the building can withstand the various loads.

Design loads vary based on geographic regions. For example, a roof structure designed for a building in a New England state should be able to carry a much greater load, to support the weight of snow, than a roof structure in southern California, where it does not snow. A building in Florida is designed to withstand a much greater wind load, to endure hurricanes and tropical storms, than a building in a northwestern state, where such weather events are unlikely. A building in San Francisco is designed to withstand an earthquake; however, a building in Washington, DC, is not designed to the same standards, where the threat of an earthquake is much less.

Undesigned loads are loads that a designer or structural engineer did not anticipate or calculate. Undesigned loads are often the result of unauthorized construction, such as when a building owner fails to obtain the necessary approval and permits, or work performed by untrained or unscrupulous contractors. Through the history of fire service, firefighters have been injured and killed in buildings that were renovated or altered in this manner. An undesigned load could be the placement of heating, ventilation, and air conditioning (HVAC) units on the roof of an older building without considering the additional load on the roof support structure (fig. 7–9).

Fig. 7–9. The installation of air conditioning units on a roof without consideration of the extra weight being added to the roof is an example of an undesigned load.

A **live load** is a nonfixed (sometimes moving) variable load that is added to the structure, including people, materials, and other transportable items.

A **dead load** is the load of the building itself—the structural components, building utilities, interior finishes, and any other components that are built in or permanently attached to the structure. The dead load may change through the life of the building as the result of renovations and alterations to the structure.

Environmental loads include snow, rain, wind, and earthquakes. Code requirements for these loads vary based on geographic region (fig. 7–10).

An **impact load** is a force delivered in motion such as a blow or force delivered by an object in motion striking a fixed object (e.g., an axe striking a roof). Impact loads are the opposite of a force applied slowly and maintained over an extended period of time.

A **static load** is a relatively unchanging load, which may be delivered over an extended period.

Dynamic loads display significant dynamic effects or are in motion when applied to a building, including strong earthquakes, gusting winds, and loads applied by the fast movement of water or waves.

A **concentrated load** is applied to a relatively small area, such as a steel column bearing on the building's **footing** or an air conditioning unit on the roof of a building.

Fig. 7–10. An unusually heavy snow load caused the collapse of this bowstring truss roof. The fire department was dispatched to this collapse as an activated fire alarm. When the collapse occurred, the building's automatic sprinkler piping was broken and the resulting water flow tripped the fire alarm system. Note the large area of collapse as the result of the failure of only one of the bowstring trusses.

A **distributed load** is a load distributed over a large area of the supporting surface, supporting a uniform load over the area (e.g., ballast stone applied on a rubber **membrane roof**).

A **fire load** is the total amount or quantity of combustible material stored or used in a building, including combustible building contents and furnishings not built into the building (fig. 7–11). Previously, this load was expressed in pounds per square foot of combustibles or in the combustible's heat energy potential by British thermal units (Btu). Today, however, fire load is more commonly measured in terms of the heat release rate of the building's furnishings and contents (refer to chapter 5, Fire Behavior for a description of heat release rate).

Fig. 7–11. Fire loads will vary from building to building. Flammable liquids, which are used or stored in the building, will present a higher fire load. The wooden structural members of the building also contribute to the fire load.

HOW COMMON BUILDING MATERIALS ARE AFFECTED BY FIRE

The various building materials used to construct a building are affected by fire in various ways. Some materials burn and lose mass as they burn, and others, which are noncombustible, lose their strength as they are heated. Table 7–1 details how common building materials are affected and how the materials react to the heat from the fire.

UNDERSTANDING STRUCTURAL COMPONENTS

FFI 4.3.4, 5.3.10 To understand building construction features and hazards, the firefighter must understand the buildings components, the names given to the components, and what function each component serves in the structure. The following text reviews some of the most common building components and their function.

Arch

Arches are constructed in two basic shapes: round or curved arches and flat arches. The most common arches are round or curved and usually carry forces in compression in one direction only. Arches can fail when the adjoining wall, which helps the arch resist horizontal forces, fails and flattens out. Wooden arches may fail when they cannot support the imposed load because of the loss of mass or the burning away of the arch.

Beam

A **beam** is a horizontal supporting member that transfers weight from one location to another and from one structural component to another. A beam may support floor joists and transfer the weight of the floor joists in a horizontal direction to an exterior wall. A simple beam is supported at two points, whereas a continuous beam is supported at three or more points. Wood beams fail as the result of loss of mass; the beam burns away to a point where the beam can no longer support the load imposed on the beam and a collapse occurs. Steel beams fail because of the loss of strength from the exposure of the fire's heat, and they may sag or warp and drop the beam's load. Steel beams elongate from exposure to the fire's heat. When the steel beam is restrained in a supporting wall, the elongation may push the supporting walls out of plumb and cause the outward collapse of the supporting wall; then the load supported by the beam collapses. A 100-ft-long (30.5-m) steel beam heated evenly to 1,000°F (538°C) elongates 9 in. (229 mm).

Cantilever beam. A **cantilever beam** is supported at one end only (e.g., a balcony, which is supported by attachment to the structure at one end). A wood-cantilevered beam may fail because of loss of mass after being exposed to the fire, or the connections may fail, releasing the cantilevered beam from its support.

I-beam. An **I-beam** gets its name from the cross section of the beam, which is shaped like an I. Today, I-beams are known as **wide flange beams**, because the dimensions of the top and bottom flanges have increased in horizontal dimension; however, it is still called an I-beam in the fire service. The I-beam consists of a top and bottom flange separated by a vertical component called the **web**. I-beams are most commonly found in the form of steel and are used in steel frame buildings to support floors and roofs of these buildings. Lightweight wood floor joists are also manufactured in the shape of an I-beam and are often referred to as engineered wooden I-beams.

Girder

A **girder** is a structural support member usually found in the horizontal position. A girder is supported by a wall and/or columns and supports beams or joists.

Lintel

A **lintel** or **header** is a horizontal beam that usually supports the wall above an opening in a wall such as a window or door. It may be wood, steel, or masonry.

Column

A **column** is a vertical structural element that transmits the load imposed on the top or side of the column to another structural element at the base of the column. The load imposed on the column causes a compressive force on the column. The column may impose a concentrated load on the member supporting the column. Shorter columns tend to fail by crushing, whereas long, slender columns tend to fail by buckling. If connections are made to the column along the vertical length of the column, the connections increase the load-carrying ability of the column and reduce the potential for the column to fail.

Table 7–1. Effect of heat and fire on common building materials [FFI 4.3.12] [FFII 5.3.2]

Material	Common uses in buildings	Reaction when exposed to fire and heat
Wood	Often used in wood frame buildings as the primary structural elements (exterior and interior wall studs, floor joists, and roof rafters), ordinary-constructed buildings (interior wood studs, floor joists, and roof rafters), and buildings constructed using mortise and tenon connections (post and beam buildings).	Wood loses mass as the material burns, and the loss of mass weakens the wood member until it fails. The mass of the wood member is reduced to make the connection, when these connections are attacked by fire, they are usually the point of failure.
Structural steel	Structural steel is used in many forms. The common structural elements include columns, beams, and bar joists. It is the primary structural element in noncombustible buildings. When used as a structural element in fire-resistive buildings, the steel must be protected from the fire's heat. Structural steel may also be found in different forms in wood frame and ordinary construction.	As structural steel is heated, the steel loses strength and expands (lengthens). Once the structural member reaches the yield point (because of high temperature and resultant loss of strength), the structural element fails. The strength of structural steel varies from piece to piece, depending on the age of the steel and the ingredients added to the steel. Steel manufactured early on in the steel age is weaker than steel used today. Industry advances have lead to the development of stronger steel by using different compounds in the manufacture of the steel.
Cast-in-place concrete	Cast-in-place concrete is used in footings, foundations, and grade beams, as well as cast-in-place floors, walls, and columns.	Cast-in-place concrete is a great insulating material, because it is a dense noncombustible material. As with other masonry materials, the cast-in-place concrete is subject to spalling when exposed to high temperatures. If the cast-in-place concrete is a reinforced concrete element, and the spalling exposes the steel reinforcing bars or cables, the steel loses strength and the entire slab of concrete may be in danger of collapse.
Concrete block	Concrete block is found in foundation walls, interior and exterior bearing walls, and interior nonbearing walls.	A great insulating material, it is often used in building to construct fire walls or fire separation wall assemblies, because concrete block resists the passage of fire. As with other masonry materials, the concrete block is subject to *spalling* (fragments of concrete dislodged under the heat of a fire) when exposed to high temperatures.
Cast iron	Structural columns that support floors and roofs are made of cast iron, as are building facades. Cast iron is found in buildings constructed in the late 1800s through the 1920s.	Cast iron may fracture or spall when exposed to high temperatures or when heated and cooled rapidly, such as if cooled by a hoseline operating on the fire.

Truss

A **truss** is a structural component using one or more triangular units to attain stability. The truss consists of **top** and **bottom chords** and **intermediate web members** that connect the top and bottom chords of the truss. Three common types of trusses are available: heavy timber wood truss, lightweight wood parallel-chord or peaked truss, and steel **bar joist** truss (figs. 7–12 and 7–13). Trusses are used to create large open floor areas where the use of columns or interior walls is undesirable. All trusses have large **surface-area-to-mass ratio** that leads to poor performance and early failure of the truss under fire conditions. When a truss fails, the resulting collapse area is large because of the span between the walls supporting the trusses and spacing between the trusses.

Floor truss. A **floor truss** is a lightweight wood parallel-chord truss used in wood-frame construction (fig. 7–14). Trusses are manufactured using 2 × 4-in. (51 × 102–mm) lumber as the top and bottom chord of the trusses. The chord may be spliced from smaller lengths of 2 × 4-in. (51 × 102–mm) lumber and spliced together using a splice plate or finger joint to piece the chord together.

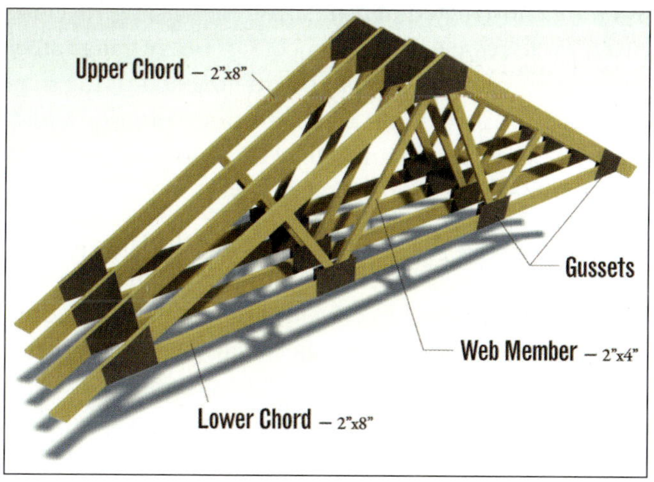

Fig. 7–12. Components of a lightweight peaked-roof truss.

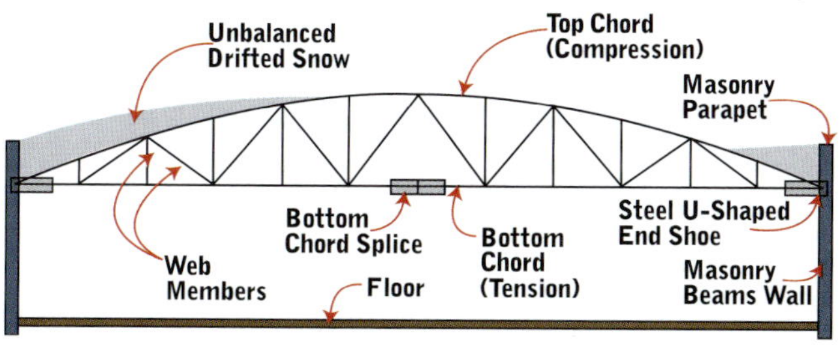

Fig. 7–13. Components of a heavy-timber bowstring roof truss.

Fig. 7–14. Use of lightweight wood parallel-chord truss members in a floor system. The use of floor trusses creates a large combustible void space. Buildings framed with dimensional lumber, which use 2 in. x 10 in. (51 mm x 254 mm) floor joists, will not have this combustible void space.

The web members of the truss are also made from 2 × 4-in. (51 × 102–mm) lumber that separates the top and bottom chord with geometric shapes; web members are connected to the top and bottom chord with gusset plate connectors or finger-joint glue connections. Theses members are used in a series to support floors and ceiling loads.

Wooden I-beam

Wooden I-beams, sometimes referred to as **engineered wood joists**, are engineered floor joists manufactured with a top and bottom chord and a web member between the chords, separating them. The top and bottom chords are made of laminated lumber and are typically 2 in. × 3 in. (51 mm × 76 mm) or 2 in. × 4 in. (51 mm × 102 mm) in dimension. A **dado** or groove is located in the center of the long axis of the top and bottom chord, and the web member is glued in place between the top and bottom chords (fig. 7–15).

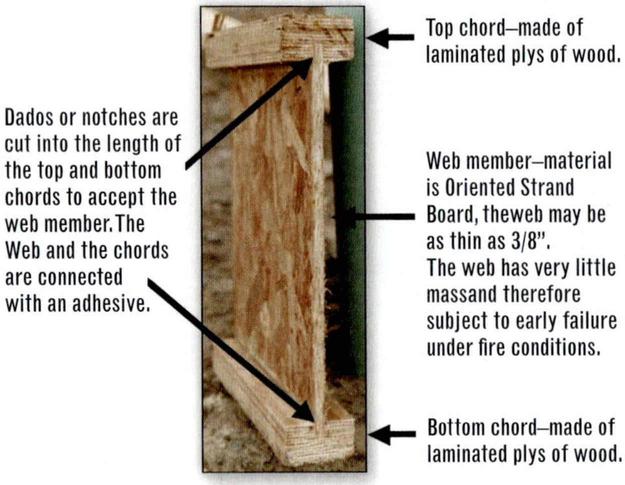

Fig. 7–15. The components of a wooden "I" beam. Math has replaced mass in the manufacture and use of engineered wood structural components. The 3/8" web on a wooden I-beam has greatly reduced the burn time for this component before collapse occurs. The fire service is experiencing a greater number of floor collapses that are injuring and killing firefighters.

The web member is a piece of **oriented strand board (OSB)**, which is typically between ⅜ and ½ in. (10 and 13 mm) thick. Theses members are used in a series to support floor and ceiling loads.

Joist

A **joist** is a parallel framing member used in a series to support floor and ceiling loads. Joists span between and are supported by bearing walls or **girders** or beams.

Rafter

Rafters are a series of sloping roof framing members that support the roof sheathing and roof covering. A rafter is supported by the outside walls at the low point of the roof and is connected to the ridgepole at the uppermost point of the roof. Rafters in buildings constructed with a flat roof may also be referred to as **roof joists**.

Ridge beam or ridgepole

A **ridge beam** or **ridgepole** is the uppermost horizontal framing member of a roof system located at the peak or uppermost point of the roof. Rafters are connected to the ridge beam or ridgepole on opposing sides.

Wall types

Load-bearing wall. **Load-bearing walls** are commonly constructed of wood, brick, structural steel studs, concrete block, and poured-in-place concrete. A load -bearing wall supports its own weight, the floors above it, and the walls or floors above, and sometimes the roof of the building. Load bearing walls can be exterior and interior walls.

Non–load-bearing wall. A **non–load-bearing wall** may be used to separate an area of a building into smaller rooms or individual spaces. A non–load-bearing wall supports no load other than itself and the weight of the finish materials attached to the wall. Non–load-bearing walls may be constructed from common materials including but not limited to wood-frame members and steel studs with an interior finish of drywall or concrete.

Curtain wall. A **curtain wall** is an exterior non-load-bearing wall of a building. The curtain wall is attached to the building or structure by its floors or columns. Curtain walls pass in front of the building **floor slab**. The space between the curtain wall and the floor slab may be a path for fire extension if the space is not properly **fire-stopped** (sealed with a fire-resistant barrier). Curtain walls are constructed of different materials; lightweight curtain wall panels are constructed of **heavy gauge steel stud** with a form of gypsum board attached to the steel stud and finished with a cementitious exterior finish material. These wall panels may contain some form of combustible foam insulating material.

Another form of a curtain wall panel is the large precast concrete wall panel with a decorative finish on the exterior exposed surface. Aluminum frame window systems are usually installed between the curtain walls. These **precast panels** may be attached to the steel frame of the structure by mechanical fasteners such as bolts or may be welded in place. The alternating horizontal bands of curtain wall and window systems that wrap around buildings usually identify office buildings. Residential and hotel occupancies usually have less window area than office buildings.

Parapet wall. The parapet wall is the part of an exterior wall that extends above the roofline of the building. It may or may not be on all sides of the building. Typically, if the parapet walls are part of the original construction, they are constructed of the same materials as the exterior walls of the building. A common alteration to change the façade of a building is to add a parapet wall on the front of the building. Often, the addition of a parapet wall on an existing building with exterior masonry walls uses wood or steel studs. The most dangerous type of parapet wall is a freestanding parapet on only one side of the building that is constructed of masonry (fig. 7–16). Masonry firewalls that separate buildings or parts of buildings that extend through the roof of a building are also parapet walls. Identifying this wall on the roof is a way of locating firewalls in the building.

Fig. 7–16. Freestanding parapet walls are the most dangerous parapet walls, as they do not intersect at any corner. If the roof of the building were to collapse during a fire, the freestanding portion of the wall may become greater and therefore less stable.

Roof types

Roof shapes are known by common names, with which firefighters should become familiar. More importantly, firefighters should learn the operational and collapse hazards as well as the fire spread potential of each roof type (fig. 7–17).

Flat. On a **flat roof** the members supporting the roof decking are horizontal or have a slight pitch. The building may have a cockloft, which is the space between the ceiling and the underside of the roof deck, or the ceiling interior finish material may be directly attached to the roof rafters. In buildings where an interior finish was unnecessary, the roof structure is open to the interior of the building.

Shed or single pitch. A **shed** or **single-pitched roof** is built in a single sloped plane that slopes from between parallel walls of different heights. The interior features of the roof are very similar to those found with a flat roof.

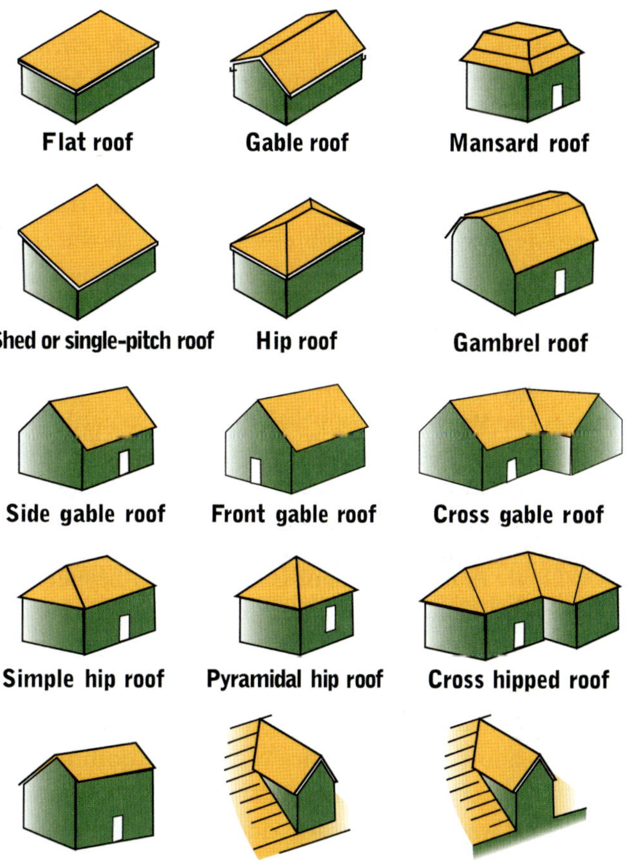

Fig. 7–17. Different types of roofs

Gable. Gabled roofs are identified by the straight slope falling from ridgepole at the highest horizontal point on the roof to opposite sides or walls of the structure. The roof rafters rest at the low end of the roof on an **exterior bearing wall**; the point at which the roof lands on and extends past the exterior bearing wall is known as the **eave**. Gable roofs are most commonly found on wood -frame buildings or ordinary-constructed buildings. The roofing material is typically some form of asphalt or fiberglass shingle. The roof shape creates a peak or triangle on the side walls or the front and rear walls of the building. Wall areas between the slope of the roof are known as the gable walls.

Hip. **Hip roofs** have three or more sides that slope to the exterior walls of the building. The hip roof may or may not have a ridgepole, depending on the size and shape of the building. Most commonly, hip roofs have sloping roof sections on all four sides of the roof. Any side of the building that has a sloping roof and an eve around the roofline is a bearing wall that supports the roof structure. Hip roofs have hips and maybe valleys. The hip is formed where two sides or parts of the roof meet and form an outside or external corner. The intersection where the sides of the sloping roof meet at an inside corner is known as the **valley**. Both hips and valleys typically start at the roof eve and continue up to their ridgepole.

Gambrel. A **gambrel roof** is similar to a gabled roof in that it is made up of two sides that slope away from the ridgepole; however, a gambrel roof begins at the ridgepole with a slight pitch and changes to a steeper pitch. The point where the roof changes slope is parallel to and lower in elevation from the ridgepole. The gambrel roof design allows for greater habitable space or floor area in the story directly below the roof structure. It is commonly found on barns and is sometimes called a **barn roof**. The walls to which the roof slopes are bearing walls that support the roof.

Mansard. A **mansard roof** is similar to a hipped roof in that the roof structure has at least three sides that slope to the exterior walls of the building. A mansard roof has two separate and distinct pitches. The slope or roof pitch that starts at the ridgepole or the flat part of a roof has a low or slight slope. Where the roofline intersects the exterior wall, the roof slope changes to a steep slope and vertically extends down along the exterior wall of the building, perhaps as much as two stories. The construction of a mansard roof may create a combustible void space where fire may extend to spread rapidly both vertically and horizontally. This roof style can also create a tremendous collapse potential because of how the roof structure is attached to the exterior walls of the building.

Saltbox. A **saltbox roof** is a form of a gabled roof where the sloping sides of the roof are asymmetrical. This roof style is used in buildings when the roof must end at two different stories of the building. For example, on a two-story building, one side of the roof slopes to and is supported by the two-story side of the building, and the opposite slope of the roof slopes to and is supported at the first story. The ridgepole on a saltbox roof is parallel to the supporting walls but may not be centered on the building. The walls to which the roof slopes are the bearing walls that support the roof structure.

Dormers. **Dormers** are small structures that rise up out of the roof to provide light, ventilation, and increased usable floor space on the story directly below the roof. Dormers are classified by their roof shape (shed, hipped, gabled, flat, etc.). Dormers are further classified into two classes based on their locations and means of support. A roof dormer extends up from the main roofline, like a small house with its own walls, roof, and window. Only the rafters or other roof structures such as lightweight wood roof trusses support the roof dormer. A wall dormer extends up from the roofline at the roof-to-wall junction and is supported partly by the exterior wall of the building, the same exterior wall supporting the main roof structure.

STRUCTURAL HIERARCHY AND FIREFIGHTER SAFETY

Certain structural components are more important than others. The term **structural hierarchy** determines which structural components are most important, factored with locations of the structural members within the building (fig. 7–18). Columns are some of the most important structural members in a building, and the column's vertical location in the building is also a factor. For example, in a four-story building, a column that fails in the basement affects the entire column line up through the building. In turn, this affects the beams on each floor supported by the column line, so the resulting area of collapse is likely to be extensive. In the same building, if a column fails on the uppermost floor of the building, the resulting area of collapse is likely to be contained to the top floor and the roof area, resulting in a much smaller area of collapse. These principles do not apply to buildings constructed with any form of trusses, because the collapse hazard associated with trusses is more severe.

Firefighters should study the collapse of Boston's Vendome Hotel, which occurred on June 17, 1972, causing the deaths of nine firefighters. The Vendome Hotel was under renovation to change the hotel into a multifamily dwelling when the fire occurred. The fire was under control and firefighters were overhauling and securing equipment when several floors and walls of the seven-story hotel collapsed without warning.

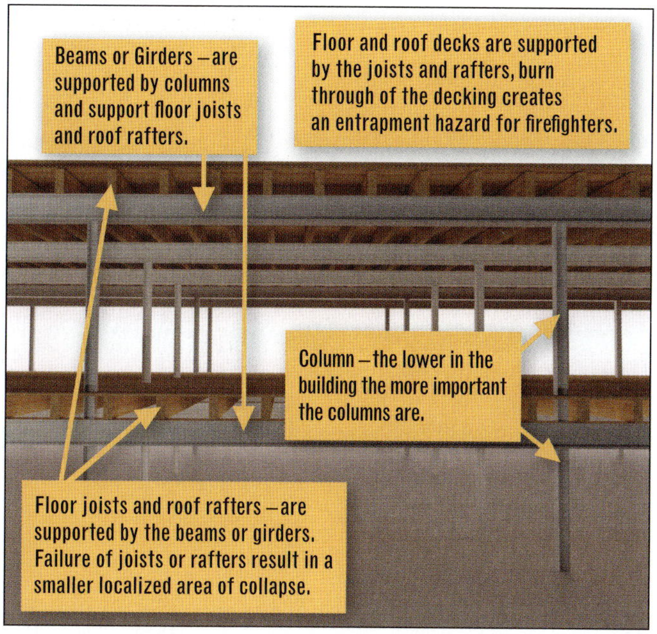

Fig. 7–18. The components and the importance of the components in terms of structural hierarchy

Factors affecting structural stability

FFII 5.3.2 Several factors affect the structural stability of the building. The building's structural stability is the building's ability to resist collapse. Remember, a collapse event may be a minor collapse caused by the loss of structural stability of one or more structural members.

The condition of the structural components of a building may vary and are sometimes deteriorated due to exposure to weather, lack of maintenance, or a combination of factors. For example, deterioration can be found when the ends of a wood timber truss rot because of water leaks in the roof system over the years. The wood slowly deteriorates, and this condition goes unnoticed. When a fire occurs in the building, the truss ends fail quickly as the fire burns through the deteriorated wood.

The mass of the structural elements will also be a significant factor in the structure's ability to resist collapse. Older buildings, especially wood-frame buildings and mill buildings, were constructed with many wooden structural parts. The beams, studs, and joists in the buildings had a fair amount of mass or cross-section value. In other words, there was some **fat** built into these buildings, and the fat of the **dimensional lumber** (2–4 in. nominal thickness) allowed for a certain amount of burn time before the building sustained a structural failure. In today's buildings, especially the lightweight wood-frame buildings that employ floor and roof trusses and engineered lumber such as plywood I-beams, the fat has been removed, resulting in the elimination of any burn time for the building. Another way of looking at the issue is that through engineering and mathematical calculations, the building industry has been able to eliminate the mass of the structural components. Math has replaced mass, making lightweight wood-frame buildings much more dangerous for firefighters. As an engineered beam is built to a certain load tolerance under optimum conditions, a beam exposed to fire, water, or other damage must be considered compromised and unable to withstand its engineered load. Thus, it is not to be trusted. Engineered structural components use the least amount of material to carry a predetermined load over a maximum specified span or at the least expense to the developer or property owner. Many firefighters have lost their lives in buildings constructed of engineered structural components.

Another factor affecting structural stability is the fire loading of the building. A rapidly developing fire with high heat output damages the structural components faster than a small fire with a lower heat output. The type of construction plays a major role in this equation. For example, consider two fires in a noncombustible constructed building with a steel bar joist roof. The first fire occurs in an occupancy involving the storage of baled rags and cardboard. The bales are tightly bound, and the fire has a slow rate of heat release that produces a considerable smoke condition. The low rate of heat release prevents the bar joist roof system from collapsing or failing as quickly as a fire in the same building involving the ignition of flammable liquids. The fire involving the flammable liquids rapidly develops and releases a significant amount of heat that causes early failure of a bar joist roof system. Determining the fire load and fire behavior at all fire incidents is an important part of an ongoing size-up.

CONSTRUCTION CLASSIFICATIONS

NFPA Standard #220, *Standard on Types of Building Construction* and the *International Building Code,* classifies buildings into five classes based on the materials used in a given building and the fire-resistiveness of the structural members of the building. The following text details the five construction classifications, materials used to construct these buildings, common occupancies found in the construction classes, and hazards associated with the construction classification. The classifications are as follows:

Type I:	Fire-resistive construction
Type II:	Non-combustible construction
Type III:	Ordinary construction
Type IV:	Mill or heavy timber construction
Type V:	Wood-frame construction

Fire-resistive construction

In **fire-resistive construction**, the structural members including columns, beams, and sometimes floor slabs are protected from the fire's heat to prevent the loss of the structural element's integrity. The level of fire resistance, measured in an **hourly fire rating**, is high compared to Type II noncombustible construction (described in the following text), which has little or no fire resistance.

Early fire-resistive buildings were constructed primarily of masonry materials, most commonly poured or cast-in-place concrete. Buildings constructed of **steel skeleton frame** used a variety of materials to encase or protect the structural steel from the heat of the fire. These materials include cement plaster, clay tile, **asbestos**, layers of **gypsum board**, and spray-on fireproofing materials (fig. 7–19).

Fig. 7–19. This photo details the use of a spray on fireproofing material on structural steel columns and bar joists. The material is a mixture of an insulating material and cement. The cement binds the insulation together and ensures that the fireproofing adheres to the building component. This material is easily removed by other construction trades and results in a lapse of protection.

Buildings or occupancies constructed of fire-resistive materials include high-rise buildings, mid-rise residential structures, hospitals, old-style warehouses, or cold storage buildings (fig. 7–20).

Fig. 7–20. A poured-in-place fire-resistive constructed building. This construction is also sometime called monolithic construction as the poured concrete forms into one piece. Detailed here are the building columns and floor assemblies, poured into the concrete columns and floor steel reinforcing bars. (Courtesy of Glenn Taldelore)

Pre–World War II buildings of fire-resistive construction tended to have smaller floor areas compared to modern high-rise buildings. These buildings relied on openable windows for ventilation and had fewer accommodations for heating, ventilating, air conditioning (HVAC), and technology. Many pre–World War II high-rise buildings were constructed with true fire tower stairs that vented to the exterior of the building, which effectively prevented smoke from entering the stair towers. The floor areas were smaller and there was greater **compartmentation** of the floor areas in the forms of rooms and offices. This resulted in slower fire spread and smaller fire areas. Pre–World War II high-rise buildings were constructed with heavier materials, which created mass within the structure.

Post–World War II high-rise buildings started using more modern, often lighter building materials. This in effect reduced the mass of the structure. More modern high-rise buildings are often constructed with the lightest weight materials available. The features of modern high-rise buildings include the use of steel bar joist floor systems and curtain wall systems (fig. 7–21).

The buildings are constructed with tremendous accommodations for technology-based business systems and rely on the building's HVAC systems to heat and cool the building, because the buildings do not have any openable windows. Eliminating the need for openable windows allowed the size of the floor area to increase tremendously.

Fig. 7–21. A curtain wall system being installed on a building. The wall panel passes in front of the floor slab and does not support any part of the building other than itself. Curtain walls may be responsible for fire extension between floors if the wall space is not properly fire-stopped at each floor level.

Modern high-rise office buildings are typically constructed using a **center core floor plan**, in which utilities, elevators, bathrooms, and exit stairways are built near the center of the building and are typical on every floor. This method of construction provides the greatest flexibility for the building owner in terms of making the floors available for lease. The floor areas of modern high-rise buildings have much less compartmentation than pre–World War II high-rise buildings (fig. 7–22).

Open office plans using **low-rise** office partitions or cubes create large mazelike fire areas on the floors of these buildings. The open floor plan also allows for faster fire development and a much larger fire area. There have been several notable fires in modern high-rise buildings such as the First Interstate Bank in Los Angeles, California, and the One Meridian Plaza building in Philadelphia, Pennsylvania. As a student of the fire service, take the time to research some of these fires; the articles, reports, and other documents are invaluable to you in your studies. See table 7–2.

Fig. 7–22. Fire-resistive constructed hotels will have the greatest amount of compartmentation because the size of the hotel room will limit the fire area. In this photo the space between the block walls will be further reduced by a non-bearing wall sub-dividing the space into two hotel rooms between the block walls.

Noncombustible construction

In **noncombustible construction**, the structural members are of **noncombustible materials** and have little or no fire resistance. The materials are typically structural steel, concrete block, poured concrete, and cold-formed steel structural elements. This class of construction may also be known as **limited combustible construction**. A major difference between fire-resistive construction and noncombustible construction is that in fire-resistive construction the structural steel must always be protected from the fire's heat. During the early stages of construction, you may be unable to determine the class of construction until the fireproofing material is applied to the steel. The collapse hazard of these buildings is due to the lack of structural steel protection. As the structural steel is heated, the steel loses strength; and when the steel member reaches the yield point, it cannot support its load and a collapse occurs (fig. 7–23).

Table 7–2. Fire spread potential of fire-resistive buildings

Fire-resistive buildings' fire spread potential based on occupancy and construction features (by increasing fire spread potential)	
1 High-rise hotels	These occupancies have the least fire spread potential. The rooms are typically the smallest fire areas found in any high-rise residential occupancy (fig. 7–22). The wall that separates the hotel room from the corridor and the adjoining rooms usually is a fire separation wall that contains the fire to the hotel room of origin.
2 Mid- and high-rise age-restricted and public housing multiple dwellings	These occupancies contain the smallest square footage of multiple dwelling buildings. Typically, the dwelling unit or fire area is between 800 and 1,400 sq ft (74 and 130 sq m). The walls separating the dwelling unit from the corridor and the other dwelling units are usually concrete block; or they may be other fire-rated walls that create a fire separation wall, containing the fire to the dwelling unit. The greatest threat of fire extension in these buildings is the dwelling unit entry door. If the door is left open when the occupants flee, the fire extends into the corridor. This may prevent other occupants from exiting and create high heat and dense smoke conditions in the corridor, making entry onto the floor by firefighting crews more difficult.
3 Mid- and high-rise multiple dwellings	The dwelling units many range in between approximately 1,000 and 2,000 sq ft (93 and 186 sq m) The fire separations and fire spread risk is the same as described above in #2.
4 Mid- and high-rise luxury multiple dwellings	These buildings and dwelling units have the greatest fire spread potential of all multiple dwelling buildings for two reasons. First, they contain the largest square footage within the dwelling unit. The dwelling unit may be 2,000 to several thousand square feet. Additionally, the dwelling unit may also have two levels, meaning there is an interior stairway that leads to the second floor of the apartment within the apartment. The dwelling unit is separated from the corridor and other dwelling units as described in #2. The two-level dwelling units also have the potential for vertical extension to the second floor within the dwelling unit. In some cases, the only access to the second floor of these dwelling units is via the interior stairway within the dwelling unit. This type of dwelling unit presents many different operational concerns to firefighters.
5 High-rise office buildings	The office occupancy has the greatest potential for fire extension. The fire area is defined by the size of the tenant space. The stairways are separated by fire-rated walls, but the tenant separation walls and the walls that separate the tenant space from the corridor may not be fire rated. If one tenant occupies the entire floor area, the fire may spread throughout the floor.

Fig. 7–23. A one story, non-combustible warehouse under construction. The roof system is supported by unprotected steel columns and bar joist trusses. These buildings are very prone to early collapse as the heat of the fire attacks the steel and the steel reaches its yield point and fails.

Common occupancies and buildings made of noncombustible construction are office buildings up to five or six stories high, large single-story warehouse buildings, and smaller two- and three-story storage buildings, such as the self-storage buildings that have become popular over the past 20 years or so. Noncombustible buildings have a variety of exterior wall systems. The exterior walls may be concrete block. The concrete block wall is not a bearing wall; it is only used as a wall in-filled between steel columns and beams and is a nonbearing wall that supports only itself. Another wall is called a **curtain wall**, because the curtain wall panels hang by attachment to the building's steel structure (fig. 7–24). The curtain walls may be made from various materials, such as concrete slabs that may weigh several tons. Wall panels are constructed of **cold-formed steel studs** to which a gypsum material is applied, followed by a stucco-type finish. The space between the curtain walls is then in-filled with windows or glass mirror panels. Steel panels or steel siding is a common exterior wall material that is often found on warehouse buildings.

Fig. 7–24. This office building is a non-combustible constructed building with a glass curtain wall system. This wall system makes identifying the number of floors in the building extremely difficult if not impossible from the exterior.

Another common wall system on warehouse buildings is a **tilt-slab** wall system (fig. 7–25). These wall systems are dangerous under fire conditions. In most structural systems, the walls or columns support the roof of the building. When a building is constructed with tilt wall systems, a skeleton frame of structural steel is erected. The wall panels (usually equal to the full height of the building) are poured either on- or off-site and delivered to the site to be erected. The panels are supported on the bottom by the building foundation or footing and attached to the steel skeleton frame at the roof level to hold the walls in their vertical position.

If a fire occurs, the connections that attach the walls to the steel frame at the roofline may fail and release the wall panels. *The stability of the walls is dependent on the roof structural system.* These panels are then likely to fall out away from the building, but experience has shown some fall into the building. There is little predictability. These walls will always collapse at a 90° angle. Therefore, firefighters must maintain a proper collapse zone that exceeds 100% of the height of the building when operating at buildings with tilt wall panels.

Typically, the roofs of noncombustible buildings are flat with steel decking supported by steel beams or steel bar joist trusses. The roofing system consists of a rigid or semi-rigid insulating board and some form of membrane roofing material, such as built-up layers of asphalt roofing paper applied with hot tar, or a rubber membrane roof system that is sealed at the seams with an adhesive material. The roofs may be covered with small stones known as slag that are raked into a layer of hot tar. For a rubber membrane roof, the roof covering may contain larger rounded river stones that range in size from approximately a golf ball to a baseball. The river stones provide ballast to the membrane roof and add weight to the roof.

The lightweight structural systems add steel bar joist roof systems to the collapse dangers associated with tilt wall panels. Such systems make noncombustible buildings some of the most dangerous buildings the firefighter encounters.

Fig. 7–25. Exterior tilt-slab wall of a noncombustible warehouse occupancy. In tilt slab construction, the exterior wall system relies on the roof for support. If a fire causes the roof to collapse, there is a tremendous likelihood that the walls will collapse next.

Ordinary construction

Ordinary construction consists of exterior masonry load-bearing walls and wood joist floors and roof (fig. 7–26). The interior bearing walls may be either wood frame, or, when required, a firewall constructed of masonry. This class of construction may incorporate some type of roof truss. Typically, buildings along old, downtown Main Street USA are made of ordinary construction. These buildings typically range from one to four stories in height.

Fig. 7–26. Ordinary-constructed buildings are made of masonry exterior walls, wood joist floors, and a wood roof structure. When the building has a storefront, the opening for the storefront windows is created with columns and beams. The steel columns and beams may be steel or cast iron. In the photo, the columns and beams support the total weight of the second and third floors, the roof structure, and the weight of the exterior masonry walls. (Courtesy of Christian P. Dansbach)

The term **taxpayer** has been applied to the one- and two-story buildings that line the streets of some older commercial districts; these taxpayers are cheaply constructed ordinary buildings. The "taxpayer" definition can vary among different regions. In some areas, it means a building erected on a property by the property owner for rental purposes, to generate a revenue to pay the property taxes. In others areas, taxpayers are a general classification of mixed-use buildings that contain a business on the first floor and apartments or living spaces on the upper floors.

The modern **strip mall** of retail stores has replaced its ancestor, the taxpayer (distinctive differences between the two are discussed further in chapter 31: Advanced Fire Attack). Strip malls are also often of ordinary construction with lightweight wood roof trusses and concrete block walls (other strip malls that use steel bar joists for roofs and concrete block walls are usually classified as Type II noncombustible construction).

Common uses of occupancies with this type of construction include older multiple dwellings, garden apartments, as well as commercial and manufacturing buildings. Manufacturing occupancies in ordinary -constructed buildings should not be confused with mill or heavy timber buildings. Ordinary-constructed buildings housing manufacturing occupancies usually have fewer square feet and stories than mill or heavy timber buildings.

Ordinary-constructed buildings that require large open floor areas may be constructed with timber truss roofs (fig. 7–27). This construction feature creates a combustible void space where the fire can intensify without being noticed by firefighters. A large volume of fire may be present in the void space created by the trusses, while conditions on the floor below the truss can be habitable with a light smoke condition. Firefighters must be able to recognize this construction feature and operate safely in these buildings. This construction feature may lead to conflicting reports between the interior crews and the roof crew. The roof crew is in a much better position to identify the fire conditions in the truss space; so when conflicting reports arise, the roof report is the better report to judge fire conditions.

Fig. 7–27. Common ordinary-constructed buildings that are constructed with truss roofs include bowling alleys, automobile dealerships, gymnasiums and churches. In this photo, you can identify the strong areas of support in the brick columns built into the exterior walls. (Courtesy of Christian P. Dansbach)

In multistory multiple dwellings, there are several construction features that permit rapid vertical and/or horizontal fire extension. These construction features include light and air shafts, dumbwaiter shafts and vertical openings created by plumbing waste and vent lines, and openings for steel channel rails in larger buildings.

Light and air shafts extend from the roof down through the building and terminate at the first or second floor level. The light and air shafts allow natural light and ventilation into rooms and apartments near the center. The shaft contains windows; if a fire in an apartment on one side of the shaft extends into the shaft, then due to the proximity of the windows in the shaft, the heat may cause the windows to fail and allow the fire and smoke to extend into the nearby apartments. An **open shaft** is open to one side of the building and easily identifiable from that side of the building; a **closed shaft** is enclosed on all sides and cannot be identified from street level. A fire extending in the shaft and showing above the roof may be mistaken for a fire burning through the roof. A report from the roof crew or identification of the shaft from another side of the building is necessary to determine if the fire is through the roof or extending from the light and air shaft.

Vertical fire extension in any building is always a paramount concern to firefighters. The location of the plumbing waste and vent lines in multiple dwellings creates significant potential for vertical fire extension. The buildings are constructed using typical floor plans, placing the bathrooms and kitchens on each floor in line vertically and resulting in the plumbing waste and vent lines serving all the apartments in that line. The openings created to install and maintain these lines produce a void space running the vertical distance of the building from the basement to and through the cockloft to the roof of the building (fig. 7–28).

Fig. 7–28. There is no fire-stopping or blocking installed around the bathtub waste line on the floor above. From the floor below, we can see the bottom of the bathtub. Fire that starts or extends to the combustible ceiling/floor space will rapidly extend vertically through the building.

Other structural features in some ordinary-constructed buildings are channel rails, which are used in buildings requiring larger square footage. The channel rails form an interior steel frame that supports the floors of the buildings. To determine if channel rails have been used in the construction of a building, an examination of the unfinished areas of the basement of the building will reveal the channels. The steel columns and horizontal beams are often exposed. The space around the channel rails may create a void space for fire to travel, which is especially concerning if the fire originates in the basement of the building.

Dumbwaiter shafts are another means of rapid vertical fire extension that firefighters should be aware of when operating in ordinary-constructed multiple dwellings. Dumbwaiter shafts contain either a mechanical or a manual lift intended to move material from one floor to another floor so the occupants need not carry materials up and down the stairs. The dumbwaiter shafts open on each floor at varying locations. The opening in the basement or the first floor is in a common area accessible to all the tenants, whereas on the upper floors the opening may be in the public hallway or in the dwelling unit. A small enclosure housing the operating mechanism of the dumbwaiter is located on the roof directly above the shaft.

In many older buildings, dumbwaiters are no longer used and may have been sealed up, concealing the original location of the shaft. The housing on the roof is sometimes removed and the roof sealed over the top of the shaft. The abandoned dumbwaiter shafts have been used over the years to run new building utilities such as new electric wiring, telephone and cable wiring, and even new plumbing lines. The new utilities easily run the shafts because the shafts are void of any fire-stopping, making the utility run of several stories an easy task for the contractor. It is for the same reason that rapid vertical fire extension is possible through the abandoned dumbwaiter shafts. For example, the old dumbwaiter shafts landed in the kitchen of each dwelling unit. The old doors have been removed and a single layer of ½-in. drywall has been secured over the door opening. If a fire occurs in a kitchen on the lower floor and extends into the abandoned dumbwaiter shaft, the fire rapidly extends vertically to the kitchens sharing the shaft as well as the cockloft of the building.

Heavy timber or mill construction

Heavy timber or **mill buildings** were designed and used as manufacturing buildings during the industrial revolution era for the manufacturing of products and goods. These buildings may have changed in use and occupancy or have been renovated during the life of the building, but the underlying support structure remains the same (fig. 7–29). The buildings were constructed of exterior brick walls, wood joisted floors, and massive interior wood columns and beams (more than 5 in. [127 mm] in any dimension).

The buildings have large open floor areas, even up to an entire block long. If the building floor areas are subdivided, they are subdivided by firewalls that separate the floor areas of the building. Openings in the firewalls contain rolling horizontal or roll-up vertical fire doors for protection. The firewalls may be easily identified from the exterior of the building by the protruding brick wall that extends above the roof of the building (fig. 7–30).

Fig. 7–29. A typical floor of a mill or heavy timber building; notice the large cross-section dimensions of the columns and beams, the wide-open area that permitted the movement of goods and materials. This area was also the top floor of the building as evidenced by the rows of skylights that allowed natural light to enter the building. (Courtesy of William P. Dansbach)

Fig. 7–30. Mill buildings may have large open floor areas and firewalls that sub-divide the floor areas of the building. In the photo, you can see the firewalls extending above the roof of the building. Mill buildings typically had many large openable windows to allow for natural light and ventilation. (Courtesy of Courtney A. Dansbach)

The buildings contain large open areas to facilitate the movement of raw materials used in manufacturing products. Multistory buildings contain a large freight elevator used for the movement of the raw materials and finished products. The freight elevator shaft represents significant firefighting challenges. The shaft is open, allowing the rapid vertical movement of smoke and heat as well as vertical fire extension. Firefighters operating on floors above the fire should identify the open shaft and be aware of rapidly changing conditions. The open elevator shaft also creates a fall potential for firefighters operating in the building. Even simple tasks conducted in light or limited smoke conditions in low lighting can be deadly if a firefighter falls into the shaft.

Through the years as the manufacturing operations have moved out of these buildings, the buildings have been converted to different uses. The newly established uses in these building have changed the hazards faced by firefighters. Common uses include conversion to residential occupancies, mercantile shops and restaurants, and warehousing ranging from large commercial warehouse space to self-storage facilities. In the conversion to residential occupancies, the building layout and arrangement drastically changes as many rooms are constructed and the entrances to the buildings and dwelling units are altered. Additionally, many concealed spaces are created as chase walls, and new ceilings are installed to conceal the various utilities installed to serve the residential occupancy. As with all void spaces, this alteration creates an avenue for fire and smoke to spread through the building (fig. 7–31). This concealment is not permitted by most building codes but still occurs.

Fig. 7–31. In this photo, a mill building is being renovated and a window is being added in the exterior wall. Notice the thickness of the exterior wall, the size of the interior column, beam, and floor joists. The original building had no ceiling other than the underside of the floor above. If during the renovation a ceiling is added under the floor joists (this is not permissible under most building codes), a combustible void space will be created. (Courtesy of William P. Dansbach)

Conversions to warehouse space may present challenges to the building's automatic sprinkler system if the system is not modified to meet the fire load and wall arrangement presented by the new storage occupancy. Failure of the automatic sprinkler system to control the fire requires the firefighters to manually extinguish the fire. The restaurants and mercantile shops mean higher occupant loads where the original designs did not intend occupancy by the general public. Sometimes, the conversion of a mill or heavy timber building may illegally create combustible void spaces. Modifications to such buildings often result in the rearrangement of the exits and access points to the building, and mazelike conditions from the floors of smaller tenant or occupant areas.

Wood-frame construction

This section of the chapter reviews **wood-frame construction**. Four separate and distinct classes define the features, methods, and hazards associated with each class of wood-frame construction. The four classifications of wood-frame construction are as follows: **post-and-beam** or **braced frame**, **balloon frame**, **platform frame**, and **lightweight wood frame**.

Post-and-beam. This is wood-frame construction in which the frame of the structure is formed by vertical wood posts (located at each corner of the building) connected to horizontal beams (located at each floor level) that support the floors (fig. 7–32). The walls between the beams are vertically filled in by 2 × 4 wood studs. This method of construction uses **mortise and tenon connections** to connect the beams to the posts and the floor joists to the beams.

Fig. 7–32. This photo details a post and beam building. Notice the vertical post at the corners and the horizontal beams at the floor level. (Courtesy of Tim Hetzel)

A mortise and tenon connection consists of tenon (tongue) cut onto the end of the beam that is inserted into the mortise (hole) cut into the post (figs. 7–33 and 7–34). This type of connection removes a significant amount of the cross-section value or mass of the structural members. Up to two-thirds of the beam's mass may be removed to create the tenon. As with any wood structural member, fire resistance and collapse resistance is a function of mass. Simply put, how much wood is there to burn before collapse occurs? The more wood the more burn time before collapse occurs. The mortise and tenon connection at each floor level makes post and beam construction susceptible to an **inward** or **outward collapse** when the first floor of the building is involved in fire. When the mortise and tenon connection at the first-floor ceiling level and second floor's floor is attacked by fire and can no longer support the weight of the floor or floors above, an inward or outward collapse occurs.

Fig. 7–33. This photo details the mortise and tenon connection where the post is connected to the beam, the connection is secured with dowel-like pegs called *trunnels.* The mortise and tenon connection is the weakest point in the framing system as 2/3 of the mass is removed from the beam and 1/3 of the mass is removed from the post to create the connection. (Courtesy of Tim Hetzel)

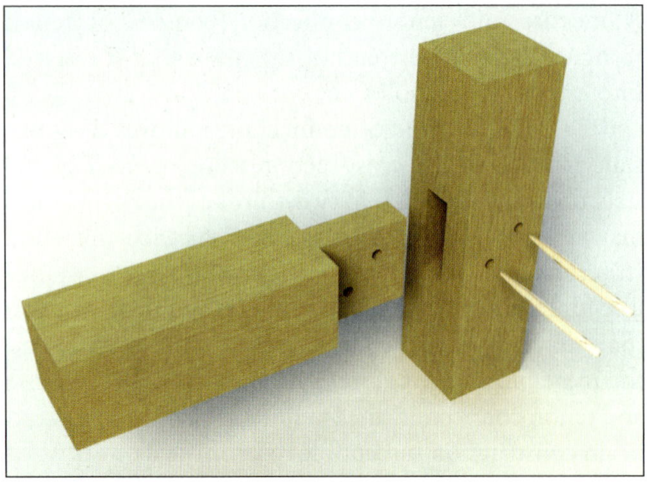

Fig. 7–34. This diagram details how a mortise and tenon connection is constructed. Notice the amount of mass removed from the beam to create the tenon. This makes this part of the connection the weak point in this connection. A wood peg or trunnel is driven into holes to secure the connection.

Fig. 7–35. In a building that is balloon framed, the exterior wall studs run from the basement to the attic without fire stopping. This photo looks up the wall as it passes the second floor joists; note the lack of fire stopping and the ease with which the duct work was run in the wall. This vertical void space becomes a chimney for the fire to travel from the basement to the attic.

Balloon-frame. Balloon-frame construction was a method employed from the mid-1880s to World War II. This type of construction employed the use of 2 × 4 wood framing members for the exterior and interior walls. The major fire problem with this type of construction is that the exterior wood stud walls are supported in the basement on the foundation plate, and the opening between the stud bays is an open vertical channel. The studs extend the full height of the exterior wall, terminating in the attic or top floor of the building (fig. 7–35). The void created by each stud bay allows for rapid vertical fire extension, especially when the fire originates in the basement (fig. 7–36). As a fire starts and continues to grow, two basic laws of physics create conditions for this rapid fire extension: The heat produced by the fire is more buoyant that the surrounding air; and whenever there is an increase in temperature in a confined room or space such as a basement, pressure increases. These factors allow the heated products of combustion, which are naturally rising to the ceiling, to be forced up through the void spaces created by the balloon framing and result in rapid vertical fire extension. Additional void spaces are found in balloon framing. Both load bearing and nonbearing interior walls may be balloon frame. Since the point in the framing system where the floor joists connect to the exterior and interior walls is not firestopped, this opening allows any fire running vertically in a wall to travel the horizontal concealed space created by the flooring and the interior finish applied as a ceiling.

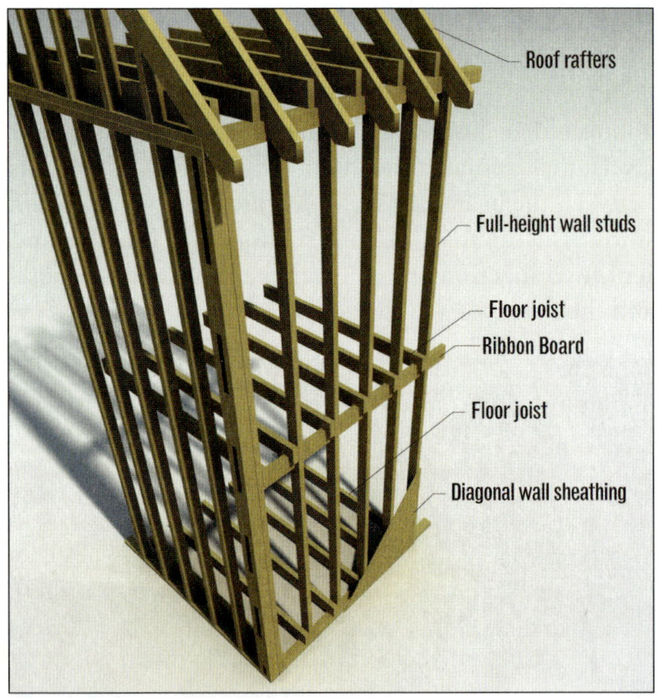

Fig. 7–36. The void spaces created by the 2 x 4–in. (51 x 102–mm) wood studs that run from the basement to the attic. These void spaces result in rapid vertical fire extension.

Platform frame. This method was introduced in the post–World War II construction boom and is still employed today. Traditional platform-frame construction used dimensional lumber such as the 2 × 4–in. (51 × 102–mm) for the interior and exterior frame walls, 2 × 8–in. (51 × 203–mm)for the floor joists, and at least 2 × 6–in. (51 × 152–mm) for the roof rafters. This construction method greatly differs from balloon frame in that the buildings are constructed one story at a time, so the construction method provides **inherent fire-stopping**. After the foundation is completed, the first floor framing is erected and the floor joists extend to the outer edge of the foundation. The subfloor is laid on top of the floor joists, completing the first-floor system. The next step constructs one story of 2 × 4–in. (51 × 102–mm) walls, a bottom 2 × 4–in. (51 × 102–mm) plate is laid horizontally on the subfloor, the walls are erected, and a double 2 × 4–in. (51 × 102–mm) top plate is run horizontally at the top of the first story of studs. This method is employed for interior and exterior walls. The floor of the next story is then constructed on the top of the first floor framing system, extending to the outer edge of the exterior wall framing. Each subsequent floor is constructed in the same manner (fig. 7–37). This framing method therefore provides inherent fire stopping because this construction method has eliminated combustible void spaces that penetrate the floors (fig. 7–38). An area of concern for firefighters is where building utilities (e.g., plumbing drain and vent pipes, HVAC ducts, and electric and communication cable) are run through the floors and walls. The installation of these utilities often results in removing parts of inherent fire stopping in the form of the top and bottom framing plates to run the various utilities. An avenue of fire extension from one floor to the floors above is created if the utility runs are not properly fire-stopped after the installation of the utilities. Experience has shown that in most instances, these penetrations are not properly fire-stopped. Firefighters may identify areas of potential vertical fire extension by locating vent pipes that penetrate the roof and subsequent vertical locations down through the building where fire extension is likely.

The interior finish of a platform-frame building is likely to be some form of drywall or gypsum material. In the beginning years of platform-frame construction, a material known as **rock lath** was used as the interior finish material. The rock lath was a gypsum board material, 16 in.- (406 mm-) to 2 ft- (0.6 m-) wide boards that had holes every couple of inches. The rock lath was installed on the walls and ceilings, and one or more coats of plaster were applied to the rock lath to complete the interior finish. This material, or a form of this material, may still be found in homes where the owner or designer prefers a plaster interior wall finish. There are many different types of drywall installed in wood-frame buildings. The most common form of drywall installed in dwellings is easily breached and removed by firefighters.

Fig. 7–37. In a building constructed of platform-frame construction, each story is built on top of the floor below. Notice the floor joists of the first and second floors extend to the outer edge of the framing, creating inherent fire stopping. (Courtesy of Robert Moran)

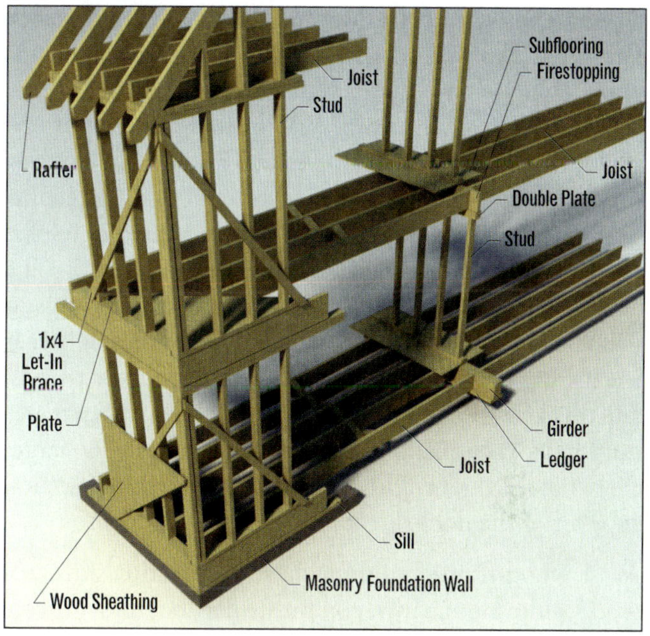

Fig. 7–38. The platform-frame method of construction. Note the inherent firestopping built into the building when one floor and one story of walls are constructed at a time.

Lightweight wood-frame construction. This is any wood-frame construction using lightweight engineered structural components, including lightweight parallel floor trusses, lightweight peaked-roof trusses, and engineered wood I-beams (fig. 7–39). These structural components have been used in the construction industry since the late 1960s; but the proliferation of townhouse and condominium construction in the 1980s saw a tremendous use of lightweight wood structural components. This will continue as the construction industry seeks ways to build more efficiently and economically (but not more fire safe), and natural resources such as wood becomes less available. These structural components have been engineered to carry the greatest load over the greatest span using the least amount of material necessary to carry the imposed load.

The connections in trusses make them dangerous under fire conditions. The trusses are connected with a steel gusset plate that only penetrates the wood between ⅜ and ½ in. The heat from the fire is conducted into the wood through the steel gusset plate and begins to decompose the wood fibers, which affect the connection's strength. Additionally, the steel gusset plate will be deformed and may warp from the heat of the fire, pulling out of the wood truss, and causing the connection and possibly the entire truss to fail (fig. 7–40).

The engineered wood I-beams are constructed of manufactured lumber; the top and bottom chord are laminated plies of lumber that create the top and bottom flange. The web of the beam is OSB, which begins to fail when exposed to the fire's heat as the resins and glues vaporize and allow chips of wood in the OSB to delaminate. From the exterior, a lightweight wood-frame building may be difficult to identify, because the exterior may appear as any other wood-frame building. Lightweight wood-frame buildings are constructed in the same manner as platform-frame buildings: one story at a time. The major difference that makes these buildings so prone to early collapse and so dangerous to firefighter is the use of lightweight wood, parallel-chord floor trusses, and peaked-roof trusses. Using these elements creates numerous combustible void spaces that can be large; there the fire can extend, grow, and destroy the structural support of the floor or roof system.

Most wood-frame buildings have an exterior siding of wood, aluminum, vinyl, composite hardboard or stucco, brick veneer, or another form of a **cementitious** finish. Many older wood-frame buildings may have **asphalt siding** (essentially, thin asphalt shingles that often look like bricks), sometime referred to as **gasoline siding** because of the combustibility and rapid fire development, which occurs over the surface of the siding. In closely built neighborhoods, fires involving this type of siding have been responsible for fire extension to the exposures. Often, the asphalt siding has been sided over with a more modern siding material. This can be a hazard when the fire extends to the siding material and the newly installed material burns away, exposing and igniting the asphalt siding.

Fig. 7–39. This photo details the many combustible void spaces that will be found in buildings constructed with lightweight parallel-chord floor trusses and peaked-roof trusses. Note the floor system in the center public corridor is a different floor system and may be dimensional lumber such as 2 x 8–in. (51 x 203–mm) floor joists which are attached to the ends of the floor trusses and separates the floor systems on each side of the public corridor. (Courtesy of Courtney A. Dansbach)

Fig. 7–40. The lack of penetration of the teeth on the gusset plate allows it to warp and twist out of the truss. Loss of strength in the connection will lead to the failure of the truss.

Building alterations

Many buildings are altered at one time or another. In many cases, buildings have been renovated more than once, and each renovation increases the potential for hazards to firefighters. Common firefighting problems created by renovations include the creation of combustible void spaces, changes to the structural hierarchy of the building, overloaded structure, and changes to the exterior of the building. A common practice during renovation is to create combustible void spaces by lowering ceilings, furring out walls to facilitate the installation of building utilities from floor to floor, and the creation of combustible void soffits to facilitate the installation of buildings utilities horizontally through the structure (figs. 7–41 and 7–42).

Changes to the structural hierarchy of the building often result when the building is renovated (fig. 7–43). The removal of bearing walls or columns and replacement of the bearing walls and columns with other load-carrying components may result in a design incapable of carrying the load brought to bear on the new structural member. A common renovation involved removing a masonry load-bearing wall from the first floor of a multistory ordinary-constructed building and replacing it with a series of columns and beams to support the load of the floors and walls above. The new columns and steel beams may not be capable of supporting the load if a fire occurs, and the steel beams and columns are affected by the heat from the fire. Failure of the columns and beams in this situation may cause the bearing wall and the floors and roof it supports to collapse.

Fig. 7–41. A kitchen renovation that results in the creation of combustible void spaces when the framing for the cabinet soffit has been installed without first installing drywall on the wall and ceiling behind the soffit. Should the fire extend into the combustible void space created by the soffit, the fire will extend horizontally throughout the soffit and vertically in any non-fire-stopped void spaces in the wood framing.

Fig. 7–42. A dropped ceiling of 2 x 4s and drywall has been constructed below the original plaster and lathe ceiling. When checking for fire extension, firefighters must recognize this feature and continue to open the ceilings until they have exposed the underside of the floor or roof above. In this photo, the task was not completed as only the dropped ceiling has been removed leaving the plaster and lathe ceiling intact.

Fig. 7–43. A building in which the first floor exterior masonry wall has been removed and set back into the building from its original location. This renovation has changed structural support of the second floor and roof and the second-floor brick exterior wall. These components of the building had been supported by a masonry wall and are now supported by steel columns and beams. This renovation has changed the structural hierarchy of the building and has changed the collapse potential of the wall. The steel beam will be heated and may fail much sooner than the original masonry wall.

NEW TECHNOLOGY IN BUILDING CONSTRUCTION

One of the greatest challenges facing the fire service is the technological advances in the construction materials. The changes are the result of less raw material being readily available such as lumber and steel; the economic desire to construct more cost-effective buildings, lighter materials, and man-made materials have reduced the cost by decreasing the time to complete a given part of the building.

The advancements in building construction and materials technology have come at great cost to the fire service (fig. 7–44). The changes in construction materials have generally resulted in use of lightweight structural elements in the construction of buildings since the mid-1970s. Materials and structural components, such as lightweight wooden floor and roof trusses, mean a developer can expend less material to support existing loads using structural members capable of spanning greater distances than standard dimensional lumber. This is an example where math, the engineering of wood trusses, has replaced mass, the amount of wood in dimensional lumber (a 2 × 10–in. [52 × 254–mm] floor joist). Using or adding engineered structural components to the buildings has resulted in a considerably reduced period of safe operating in these buildings during firefighting operations.

Fig. 7–44. The lack of quality control has lead to the truss being delivered to the construction site with one entire row of the connecting teeth of the gusset plate not being secured into any part of the wood truss. The heat from a fire will quickly attack these connections; this connection being less secure than the remaining gusset plates is likely to fail very early in the fire.

Another example of new material is the increased use of OSB (previously discussed); OSB is used in place of plywood in many applications such as exterior wall sheathing on wood-frame buildings. OSB is also used as a web member in wooden I-beams, which have replaced standard dimensional lumber in most wood-frame buildings as floor joists and roof rafters. The fire hazard of wood I-beams is again the lack of mass to the structural member. The OSB web of the wooden I-beam of at least 1½ in. (38 mm) is only ⅜ in. (10 mm) thick; compare to a 2 × 10–in. (52 × 254–mm) joist that has a nominal thickness. This results in a difference of 1⅛ in. (29 mm) in the mass of the structural member. Another factor affecting the burn time is the use of adhesive or bonding agents in the production of the OSB. These bonding agents are typically resins and wax; the greater part of the components are resins, and the wax is added to provide a degree of water resistance to the material. When the OSB that is the web of a wood I-beam is exposed to the heat, the resins and wax vaporize; as the materials vaporize, the OSB begins to degrade, delaminate, and lose structural integrity and the resins and wax add fuel to the fire. There are many documented cases where the collapse of lightweight wood floor and roof systems have killed and injured firefighters.

Another example of building designers using less and lightweight materials in Type II buildings is the replacement of steel I-beams with lightweight steel trusses (also known as **steel bar joists**). Trusses are examples of math replacing mass: geometric shapes in the truss design are mainly responsible for the support of a given load. The replacement of steel I-beams with trusses produces a building that is lighter weight. The structure's footings need not support the load of the steel I-beams, only the weight of the trusses.

Many advancements have been made in the manufacture of **gypsum board** or **drywall**. For many years, all-purpose, moisture-resistant, and fire-rated drywall were the most common types a firefighter would encounter. These types of drywall were easily opened up or removed when used in the building. Fire-rated assemblies that employ multiple layers of drywall to construct a fire-rated assembly require significant effort to breach or open up. Drywall is now available that has been reinforced with materials that make it difficult or impossible for firefighters to breach or remove the walls. **Impact-resistant drywall** is manufactured with fiberglass and other materials that make the finish of the drywall harder and therefore more difficult to open or breach. Drywall manufacturers also laminate drywall with other materials such as OSB or sheets of Lexan® plastic to make the drywall impenetrable. These different types of drywall may be found in various occupancy types. Impact-resistant drywall is found in schools and other locations, where durability of the wall is desired, and the use of impact-resistant drywall may also reduce cost by replacing a concrete block wall with a steel stud wall with impact-resistant drywall as the interior finish. **Laminated drywall** may be found on the walls of high-value occupancies such as jewelry stores in malls to prevent unauthorized entry into the occupancy by breaching the walls. Laminated drywall may be found in any location where security is a concern to the owner or occupant. Firefighters should be aware of occupancies that have been constructed with these types of drywall, because opening or breaching walls with these types may be difficult or impossible.

The exterior finishes on many recently constructed buildings appear to be a stucco or cement plaster finish, when in fact the exterior wall finish is a system known as an **exterior insulation and finish systems (EIFS)**. These appear to be typical cement plaster finishes used as an exterior finish material for many years. The EIFS is a lightweight synthetic wall cladding that includes foam plastic insulation and thin synthetic coatings that produce a finish that appears to be a stucco finish. The EIFS may be installed over a concrete block wall or over-sheathing that has been applied to wood or steel studs. The application of any stucco-like wall finish makes identifying the type of construction more difficult from the exterior of the building. The foam component in the exterior wall system is likely to be combustible; and should the fire extend to or involve the wall system, firefighters may be faced with a challenging fire because it may extend through the wall system and require significant overhauling to be controlled and extinguished.

Another part of the building that has seen changes is the roof covering. For many years, flat roofs were covered with a **built-up roof** covering. Built-up roofs employed a fiberboard insulation panel, a base sheet of roofing materials and several layers or roofing felt (paper). The roof felt is applied and lapped over each other to provide multiple plies of roofing. The roofing felt was applied by hot asphalt tar mopped out onto the roof surface and then was rolled out into the hot asphalt. More recent roofing materials include synthetic rubber membrane that provides a watertight roof covering. One form of the synthetic roof coverings is an EPDM (ethylene propylene diene M-class) rubber. The seams in the rubber membrane are sealed with a flammable adhesive.

The roof covering may be fully adhered to the roof, held in place by **mechanical fastening**, or covered with stone to act as ballast to prevent the roofing material from lifting in high winds (fig. 7–45). Firefighters may have to remove the ballast stones to access the roof covering. Additionally, the ballast stone will add a significant weight to the roof structure.

Firefighters should be constantly aware of changes in construction materials and methods. They should also know how these changes affect firefighting in the building as they conduct pre-incident inspections of buildings under construction or renovation in their jurisdiction.

Fig. 7–45. The ballast stones are placed on the roof to prevent the roof covering from lifting and are not secured to the roof. These stones make footing on these roofs difficult. The stones also add weight to the roof structure and create surface tension that slows down the drainage capability of the roof.

INDICATORS OF COLLAPSE

FFI 4.3.12 FFII 5.3.2 An important aspect of the study of building construction is to understand the indicators of collapse and the collapse potential of the various types and forms of construction. There are numerous visual indicators (identified in the following text) of an impending collapse that could cause failure of a building or part of a building. Sounds—creaking or moaning of the building—are also signs of impending collapse. In all cases, immediately report these indicators to your superior officer; rapid evacuation is usually crucial.

Major collapses at fires are uncommon, but firefighters must be aware of the smaller, more localized collapse, which can be just as deadly. Table 7–3 details some of the collapse indicators and collapse potential based on the construction type or building component.

When a building or part of a building is in danger of collapse, **collapse zones** must be established and all firefighters must remain clear of these zones. Each collapse zone has a height and width; the height of the building, or part of the building in danger of collapse, must be transferred horizontally to the ground. The collapse zone away from the building must be at least the height of the building or part of the building in danger of collapse plus a **safety factor**. The safety factor is necessary because parts of the building may bounce outward when they hit the ground (such as bricks). Because measuring the height of the building and subsequent transfer of that dimension to the ground is an estimation, always add or estimate on the high side. Policies on establishing collapse zones may vary slightly from department to department, but the policies must always include a safety factor. The safety factor should be calculated by adding several feet, an additional story, or even as much as an additional 50% to the estimated height of the building. The collapse zone must also include the width of the building or part of the building in danger of collapse; a safety factor should also be added to this distance or dimension (fig. 7–46). Firefighters should know their department's policies and procedures for the establishment of and maintaining collapse zones.

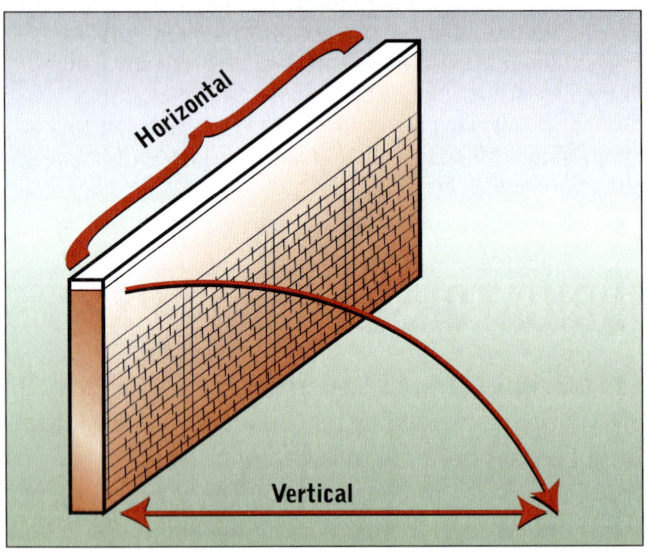

Fig. 7–46. The collapse zone must include the horizontal width of the wall and the height of the wall plus a safety factor.

There are three basic types of collapses that firefighters should understand. A **tent-type** and a **lean-to type** collapse leave void spaces in the aftermath of the collapse, where building occupants who may have survived the collapse may be located. The third type of collapse is a **pancake-type** collapse. (These types of collapses are covered in greater detail in chapter 35: Support of Technical Rescue Teams.) The pancake-type collapse leaves little if any void space for the occupant to survive the collapse.

A paramount concern to firefighters operating at any collapse incident is the potential for **secondary collapse**. Initial fire department operations at any collapse incident should be limited to the following: size-up, removing occupants from known locations that do not require technical expertise, incident stabilization, such as securing utilities from a safe location, and evacuation of adjoining structures. Firefighters operating at collapse incidents that require structural shoring or entry into collapsed areas of the structure should only operate under the direct supervision of fire officers specifically trained in collapse rescue.

RECOMMENDED STUDIES

This chapter begins your study of building construction. Each firefighter must continue to study building construction by a variety of means. Consider continuing your education by attending formal training classes on building construction, conducting walk-through drills of buildings under construction or renovation in your jurisdiction, or in other ways. During the walk-through drill you can (with the owner's or contractor's permission) take a series of digital photos, which can be made into a PowerPoint presentation. This type of presentation can be viewed during company drills and will reveal parts of the building that are no longer visible once it is completed. This serves as a refresher to some firefighters and is an invaluable tool to future generations of firefighters serving your department.

Building construction can be somewhat geographic; firefighters should constantly be aware of new trends in construction methods and materials being used in your response area. Strive to understand how these methods and materials react under fire conditions. When you experience a fire with an unusual outcome or occurrence, share this information as soon as possible with other firefighters in your department, your surrounding departments, and throughout the country. This is one benefit of the Internet! Study fire incident reports that are published regarding major fire incidents, particularly those involving building collapse and serious firefighter injuries and deaths. These reports contain valuable information regarding construction features that may have affected the incident as well as lessons learned from the incident. Each firefighter should seek to understand these lessons learned to become a safer and better firefighter. The funny thing about building construction is that the more you learn, the more you realize there is more to learn!

Table 7–3. Indicators of collapse

Construction type, building component, or feature	Collapse indicator or collapse potential
Wood-frame buildings	Sagging roofs and leaning exterior walls are indicators of collapse.
Wood-frame buildings constructed with lightweight structural components	Fire burning in the void space created by the floor or roof assemblies presents the real potential of collapse.
Freestanding masonry parapet walls	Any masonry parapet wall that is exposed to heavy fire conditions or in which the surrounding roof has burned away or any parapet wall that is out of plumb is in danger of collapse.
Ordinary-constructed building with a heavy timber truss roof	Collapse is a danger when fire is in the truss loft or where the truss has been exposed to fire. Buildings in which the truss loft is a concealed space are dangerous because the true fire conditions are concealed from the firefighters by the ceiling material.
Noncombustible-constructed buildings with a steel bar joist roof system	Steel bar joists are susceptible to collapse even after short-term exposure to fire; they have poor surface-to-mass ratio, and the heat from the fire causes the steel to reach its yield point after short exposure to fire conditions.
Ordinary-constructed flat-roof buildings with rafters of dimensional lumber	This roof system creates a redundant support system and will sustain fire conditions for a significant period of time. When the roof does fail it may do so slowly, and the initial collapse is not easily noticed by firefighters operating on the roof. Indicators of collapse for this type of roof include plumbing vent pipes that start to grow or extend up or reveal what appears to be fresh tar or roof cement around the pipe or areas along the parapet walls that appear to have fresh tar or roof cement.
Walls, floors, and roofs of all buildings	Any wall that appears to be out of plumb is in danger of collapse. Sagging floors or roofs are also indicators of collapse.
Chimneys	A chimney where the adjoining wall or roof has burned away should be considered in danger of collapse. The adjoining wall or roof provides lateral support for the chimney; and when the adjoining wall or roof burns away, the chimney is unsupported for a greater vertical distance.
Wood-frame, ordinary, and noncombustible-constructed buildings	When these buildings become well-involved in fire and the fire has taken hold of multiple floors, they are in danger of a major collapse.

QUESTIONS

1. What are examples of undesigned loads with regards to building construction?
2. Describe the main differences between live loads and dead loads.
3. A balcony that is supported at only one end is a(n) ________________.
4. Many structural columns that support roofs and floors are made of cast iron. When under a fire load, what happens to these columns?
5. In building construction, what are I-beams mostly used for?
6. Describe the factors that affect structural stability of buildings?
7. NFPA ______________ groups buildings into five classes.
8. Briefly describe each of the classes of buildings.
9. What is the main reason for center core floor plans in high-rise buildings?
10. Tilt wall construction poses safety hazards for firefighters when they are under fire conditions. Describe how the construction works and the safety hazards that exist.
11. Name the four classifications of wood-frame construction and describe them.
12. What is the significance of balloon frame construction in fire suppression strategy?

Ropes and Knots

by Mark A. Sulcov

This chapter provides required knowledge items for the following NFPA Standard 1001 Job Performance Requirements:

FFI 4.3.20

FFI 4.5.1

FFII 4.5.4

This chapter contains Skill Drills. When you see this icon, refer to your Skill Drill book for step-by-step instructions.

SKILL DRILL

8

OBJECTIVES

Upon completion of this chapter, you should be able to do the following:

- Identify the materials fire service rope is made from
- Identify and describe the terms and parts of the rope when tying knots
- Identify the basic knots used in the fire service
- Describe and demonstrate the methods used to tie fire service knots and their primary uses
- Describe the inspection process of ropes, how to maintain them, and the standard methods of storing them
- Recognize the difference between life safety ropes and utility lines
- Identify methods used to secure and hoist various fire service tools and equipment
- Understand why fire service rope is downgraded and/or put out of service

INTRODUCTION

Rope work is an important part of the fire service. The firefighter needs to understand the importance of rope selection based on construction, strength, and what it will be used for. The firefighter also needs to understand the basic elements of knots; how to tie knots for fire service use; and the proper inspection methods, storage, and maintenance procedures for ropes.

It is important to note that knots can be tied in more than one way. This chapter provides specific ways to tie knots; your instructor may show you other ways to tie a particular knot. For example, being left-handed or being right-handed affects the way people tie knots. With practice, you may be able to tie a knot with your left or right hand. In addition, people who are left-handed will take this as a personal challenge. Relate this to having more tools in your tool box. There are many tools that do the same job, but do the job differently. In your career, you will find that there is no one tool that fits all situations. Have a selection available.

ROPE MATERIALS AND CONSTRUCTION

Rope is made from either natural or synthetic material. Each rope has different characteristics that affect its use in the fire service (table 8–1). The first ropes used by humans, made of natural fibers such as live vines, were used to construct rope bridges. In the fire service, the natural fiber material first used was manila hemp (essentially dead vines).

Natural fiber ropes were in service in fire departments until the 1980s, when they were replaced by synthetic materials (fig. 8–1). In addition, a disastrous, fatal firefighting rope rescue failure in New York City in 1980 focused attention on fire service ropes and the need for a national standard on their design and use. In 1982, the National Fire Protection Association (NFPA) began the process of developing a standard that was first issued in 1985 and is known today as NFPA 1983, *Standard on Life Safety Rope and Equipment for Emergency Services.* It covers ropes and their uses, that is, life safety versus utility, harnesses, and hardware.

Natural materials

There are three basic natural materials that are used for rope: manila, cotton, and sisal. All of the natural ropes suffer from the same problems:

- Mildew
- Rot
- Deterioration
- Poor **abrasion** resistance
- Size-to-strength ability (natural fiber rope needs to be thicker to be stronger)
- Little predictability in terms of expected strength

Fig. 8–1. Natural fiber rope is made from organic materials and isn't generally used for fire or rescue purposes.

Manila. Manila rope is made from fibers from plants. These small fibers are twisted together (not knotted) to form longer fibers. The twisting of the fibers creates longer and longer strands that are then twisted with more strands to make the natural fiber rope. There is a definite relationship of size (diameter) to strength. The thicker the rope, the stronger it is.

Note: When looking at manila fiber rope, you can see numerous fiber ends sticking up out of the rope. These are the fiber ends that are exposed or have broken due to either stress on the rope or abrasion.

Cotton. Cotton rope is made with the same cotton seed fibers that cotton thread is spun from. Because of the very short cotton fibers, the rope is soft and pliable, but this results in very low strength and makes the rope susceptible to damage by abrasion.

Sisal. Sisal is another natural fiber that comes from plants. It is a very weak fiber that has less strength than cotton. Thus it is only used to make very small diameter rope. Cotton and sisal fiber ropes are used less today in the fire service because of the emergence and widespread acceptance of synthetic fiber ropes.

Table 8–1. Rope material characteristics

	Manila	Polypropylene	Polyethylene	Nylon	Polyester
Loss of strength, °F (°C)	180 (82)	200 (93)	230 (110)	300 (149)	300 (149)
Melting or charring temperature, °F (°C)	375 (191)	275–300 (135–149)	285 (141)	400–500 (204–260)	450–650 (232–343)
Loss of strength when wet	50%	0	0	25%	Minimal
Float	No	Yes	Yes	No	
Acid resistance	None	High	High	Low	Limited
Elongation under loads	Yes			Yes	Low

Natural material ropes are not used for *life-saving* purposes—they are never used to support a human being. When used, they are used for *utility* purposes only. Downgraded life safety rope should be labeled as utility rope.

Synthetic materials

The most widely used synthetic materials in the manufacturing of ropes are nylon, polyester, polypropylene, and polyethylene (fig. 8–2). These materials have excellent resistance to rot and mildew. Deterioration due to age is not considered a factor. Synthetic ropes are also more resistant to physical and abrasion damage, but are still susceptible to damage/destruction from the heat of a fire. The biggest advantage of synthetic materials is that, because they are manufactured, each fiber is the same and so are their properties. Each continuous fiber of the same length is just as strong as the others. In contrast, natural fiber is made up of short pieces, and thus it is weaker. This increased strength factor makes synthetic fiber rope preferable for fire service use.

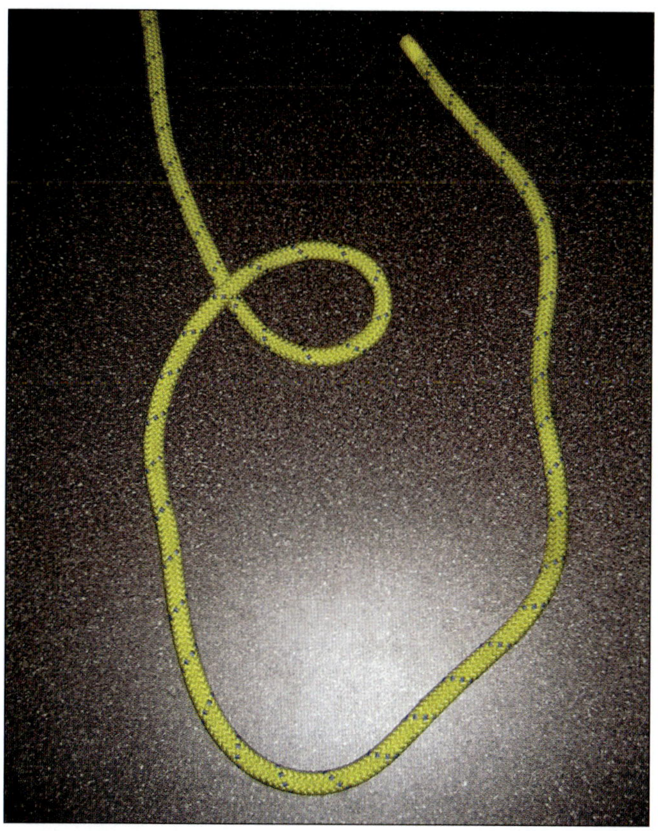

Fig. 8–2. Synthetic rope is used in a wide variety of applications in the fire service, including technical rescue.

Nylon. Nylon was invented in the 1930s by the Dupont Corporation. First used to make stockings for women, nylon was used during World War II for parachutes and cords due to the shortage of raw materials. One nylon strand is called a **filament**. Rope is made from multifilament lines, either twisted into strands or covered with sheathing. The biggest disadvantage with nylon is that it is damaged by acids.

Polyester. Polyester material, unlike nylon, is resistant to most acids. Polyester also has relatively low stretch under a load, but does not hold up well if a shock load is encountered.

Polypropylene. **Polypropylene rope** is primarily used for water applications, that is, water rescue or any area with a high degree of moisture, because of its ability to float. Polypropylene also resists rot and mildew and has a strong tolerance to chemicals and acids. Polypropylene will break down easily from sunlight, heavy loads, and abrasion.

Polyethylene. **Polyethylene rope** is nearly identical to polypropylene, also floats, and comes in many colors. It is weaker and stiffer than polypropylene, but does have better abrasion resistance.

Construction

There are two types of rope used in life safety applications: dynamic and static. **Dynamic rope** is high stretch and very flexible. It is used mostly in rock climbing to absorb falls and put less stress on the anchoring system. **Static rope** is low stretch and is used in most fire department applications, namely, rescue, hauling, and rappelling. It has very little stretch.

Laid (twisted) rope. The laid or twisted method of construction is for natural or synthetic fiber rope. The yarn is twisted into strands, and the strands—typically three—are twisted together to make the rope (fig. 8–3).

Fig. 8–3. Laid or twisted rope construction involves the twisting of three strands of natural or synthetic rope.

Note: The quality of the fiber and the tightness of the twist determine the specific strength of a particular rope. A twisted rope leaves the three strands exposed to all the elements, leaving it open to damage, especially by abrasion. You can easily inspect twisted rope by looking at all surfaces for cuts and abrasions; chemical damage; or embedded dirt, metal, or glass pieces. **Laid rope** can be easily twisted back to expose the inside area and checked for debris that can damage the rope. If any damage is found, the rope should be taken out of service.

Braided. Braided ropes are made from natural and synthetic fibers. The most common and the only one the fire service will use is made from synthetic fibers. The rope is made by braiding the strands together uniformly, just like braiding hair. Braided rope reduces twisting when used for rappelling. The rope is very smooth to the touch, but like laid rope, the majority of the fibers are exposed and subject to damage (fig. 8–4).

Fig. 8–4. Braided rope is commonly found in synthetic rope used by the fire service. A drawback of this type of rope is that all strands are exposed to damage from abrasion.

Braid-on-braid. Braid-on-braid is constructed by covering a braided rope with another braid (some manufacturers call this *jacketed*). The outer braid usually has a distinctive design pattern (fig. 8-5). This type of rope is very soft, and just like braided, it has the same poor characteristics. The inspection process for braided and braid-on-braid is mostly the same. We should look for cuts, abrasions, or excessive picks (fuzz) or deformities in the rope (soft spots). In braid-on-braid, we must also look for sliding of the inner and outer braid. If this is found, the rope must be taken out of service.

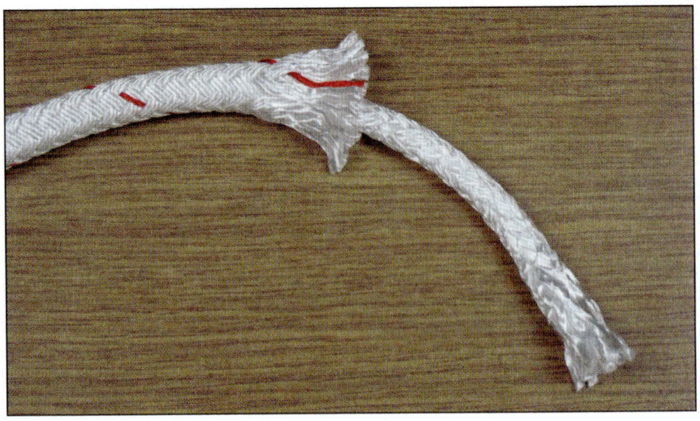

Fig. 8–5. Braid-on-braid rope has many of the same drawbacks as simple braided rope, except that the outer core creates a protective "jacket" for the rope.

Kernmantle. **Kernmantle rope** is composed of two sections: The *kern* is defined as the inner core of the rope, and the *mantle* is the outer cover. There are three types of kerns: twisted, braided, and continuous filament fiber. The first two types create dynamic rope, and the continuous filament fiber creates static rope. To put it simply, the mantle is a tightly woven outer shell that protects the kern. In all cases, the kern carries the majority of the load (strength), normally 75%–80%. The strength of the mantle is 20%–25%. This type of rope is primarily for life rescue (fig. 8–6).

Fig. 8–6. Kernmantle rope is the most common fire service rope since it is designed for life safety operations.

Webbing. **Webbing** is not really rope, but rather woven fabric that is either flat or tubular (fig. 8–7). It is often used for slings and anchoring. Some firefighters carry small-width webbing instead of personal rope. Webbing can be used to aid in securing loads, dragging trapped people, and lowering loads. One of the advantages of webbing is that it loses very little of its strength when bent correctly around a carabiner.

Fig. 8–7. Webbing can be utilized in complex rescue rigging arrangements or carried by firefighters to aid in maneuvering hoselines or carrying tools.

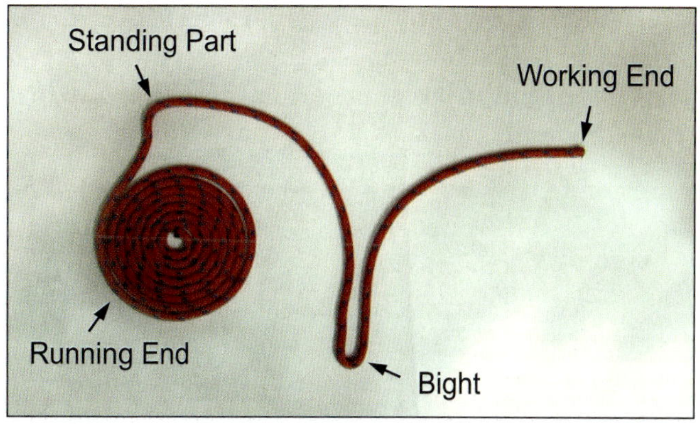

Fig. 8–8. The working end of a rope is the area used to tie a knot, while the running end is used for hoisting and pulling.

ROPE NOMENCLATURE

Primarily rope falls into two categories, utility and life safety. Utility rope is used for such tasks as hoisting and lowering of tools, securing objects, and many other nonemergency uses. Rope used for search and rescue and other non-life-threatening uses also does not fall into the life safety category. Note that these are normally **downgraded** life safety ropes and need to be marked accordingly and stored in a compartment remote from any life safety rope.

Before we can start tying knots, we must first understand what the different parts of the rope are called and what makes up a knot. First, the rope itself can be divided into three sections:

- **Working end.** This is the part of the rope that is used to tie the knot (fig. 8–8).
- **Standing part.** This part of the rope can be above or below the knot, but must be between the knot and the remainder of the rope.
- **Running end.** This is the rest of the rope used for the work of the rope such as hoisting, pulling, and so forth.

Elements of a knot

For a knot to be used in the fire service, the knot must be easy for you to tie and untie while wearing firefighter gloves. This is where repetition and practice come into play. Someone may be able to show you a knot and how to tie it. But you must practice. No one else can learn it for you.

A **bight** is made by bringing the rope back along itself side by side. A **loop** is made by placing a twist in the rope and having the standing part of the rope continue in the same direction. A **round turn** is made by making a loop and then having the standing part go back in the direction from which it came.

KNOT TYPES AND USES SKILL DRILL

FFI 4.3.20 The fire service family of knots consists of the overhand (also known as a safety), half hitch, clove hitch (open and closed), becket bend, bowline, figure eight, figure eight on a bight, figure eight follow-through, water, and handcuff. The tying instructions for these knots are located in your Skill Drill book.

Note: In the emergency services, there are numerous ways to tie various knots. The tying instructions in your Skill Drill book are only demonstrating one way. This should not stop you from learning other ways to tie knots. Try to think of it as putting another tool into your tool box, thus having a selection of ways to accomplish the same goal no matter what obstacle is in your way.

Overhand (safety) knot. This is used as a backup to the primary knot. The overhand knot prevents the loose end of the rope from walking out of the primary knot, so that it does not untie (fig. 8–9).

Fig. 8–9. An overhand safety knot is used to provide a margin of safety with other knots.

Half hitch. Always used in conjunction with another knot, the half hitch is effective at securing and hoisting long tools, such as a pike pole. The half hitch keeps the tool stabilized as it is lifted to an upper floor window (fig. 8–10).

Fig. 8–10. A half hitch being used to hoist an axe

Clove hitch (open). Basically, the clove hitch is two half hitch knots. The clove hitch secures tools or posts with the hitch, using the open end of the rope (fig. 8–11a).

Fig. 8–11a. An open clove hitch

Clove hitch (closed) knot. This is the same as the open version, but used when there is no open end of a rope. The closed clove hitch should always be backed up with a safety knot (Fig. 8-11b).

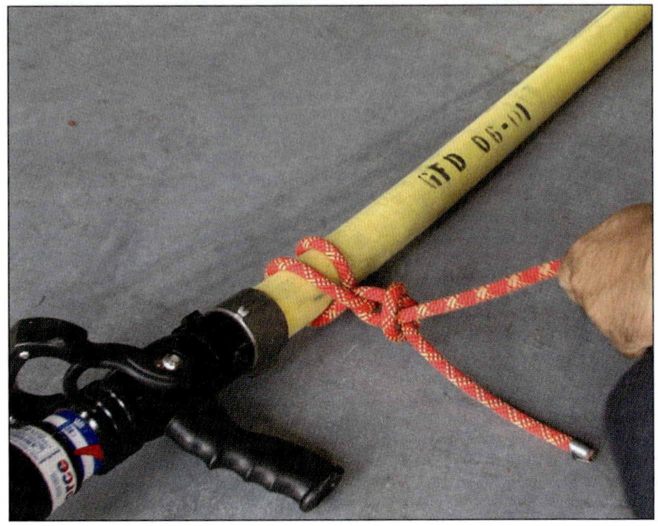

Fig. 8-11b. A closed clove hitch

Becket bend. With the becket bend, two ropes of unequal diameter are tied together. This knot places a bight in the larger-diameter rope to join to another rope or to a chain. The becket bend should not be used for life safety procedures (fig. 8–12).

Fig. 8–12. A becket bend can be used to tie two ropes together.

Bowline. Use a bowline knot to form a secure loop that will hold up to tension and not constrict the object it is placed around. This is one of the most widely used knots in the fire service (fig. 8–13).

Fig. 8–13. A bowline forms a secure knot when tension is applied to the rope.

Figure eight. This knot is used to form several other fire service knots, including the figure eight on a bight and the figure eight follow-through. A figure eight knot will effectively stop a rope from slipping through a pulley or through the grommet of a rope bag (fig. 8–14).

Fig. 8–14. A figure eight knot

Figure eight on a bight knot. This knot forms a secure loop at the open end of a rope. The figure eight on a bight is a good life-safety knot. It is also good for securing the rope to an anchored piece of equipment(fig. 8–15).

Fig. 8–15. A figure eight on a bight

Figure eight follow-through knot. This knot is for tying a figure eight on a bight through an object (fig. 8–16).

Step 1. Make a basic figure eight, leaving sufficient rope on the working end to go around the object you're securing the rope to.

Step 2. Wrap the rope around the object to be tied.

Step 3. Now follow the rope back through the knot in the opposite direction.

Fig. 8–16. A figure eight follow-through

Water knot. This is the most effective knot for joining two pieces of webbing to make a longer piece, or for forming a single piece of webbing into a loop. The water knot is similar to a figure eight. Unlike a knot made with rope, the water knot must lie flat (fig. 8–17).

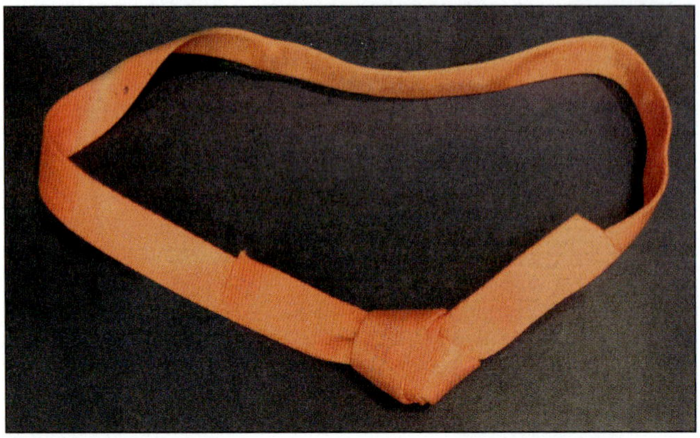

Fig. 8–17. A water knot is typically used to tie ends of webbing together.

Handcuff. This is a life safety knot with two adjusting loops that is useful for hoisting a firefighter or victim from a lower level. As the rope is pulled taut, the loops close up, forming a handcuff, which allows the person to be pulled upward. It is simple to tie, and very strong (fig. 8–18).

Fig. 8–18. A handcuff knot can be used to rescue and then drag a downed firefighter.

PROPER USE OF ROPE

There are some important points to remember when dealing with rope. A firefighter should think of these as the list of do's and don'ts with respect to rope.

1. When using a rope, always find a way to protect the rope over sharp edges. Use a rounded object such as a hose roller, ladder rung, or handle of a tool. Don't let the rope run over construction materials that have a 90° angle, like the edge of a roof or window. If possible, use **edge rollers**, which are made of metal and can be placed over edges and sharp corners to protect a rope while in use (fig. 8–19). They also reduce the amount of friction. Hose rollers are an ideal way of preventing abrasion to a hose.
2. Be aware of the effect that shock loads have on rope, and avoid creating these shock loads.
3. Regardless of the rope material, ropes and fire don't mix well. The heat from a fire can seriously damage a rope. Never let the rope pass through flames or heated gases while it is being used, regardless of the purpose. The rope could fail suddenly and without warning.

Fig. 8–19. Hose rollers can be used to protect hose and rope when drawn over sharp edges.

HOISTING TOOLS AND EQUIPMENT

One of the primary functions of rope is to hoist equipment to an elevated location. The utility rope can be used for this job. All equipment being hoisted should have a **tag line** on the ground. This line is used to keep your tools and equipment away from the structure or obstructions like branches, overhangs, or anything else it could get caught on. Remember that all knots should have safeties.

Axe or Halligan tool

SKILL DRILL

Place an open clove hitch over the handle or an axe or Halligan tool and slide it all the way to the head of the tool and tighten (fig. 8–20). Remember to leave enough rope past the hitch to later tie a tag line to. Now make a bight around the head of the tool and bring the rope back up the handle, and secure with an underhanded half hitch.

Fig. 8–20. Using a clove hitch and a half hitch, along with a tag line, you can successfully raise a Halligan tool to an upper floor.

Pike pole

SKILL DRILL

Like all tools, the pike pole is also tied with the heavy end down (the hook). This method will give greater control at the head of the tool to clear any obstructions. Place an open clove hitch over the handle of the tool and tighten by the hook (fig. 8–21). Remember to leave enough rope for the tag line. Take a bight around the head of the tool and bring up toward the handle. *Rule of thumb*: For every length of an axe in the handle, there should be that many half hitches.

Fig. 8–21. Usually one half hitch and a clove hitch can secure a pike pole to be hoisted.

Exhaust fan

SKILL DRILL

Tie a bowline around the top two handles of an exhaust fan (fig. 8–22). Try to get all three sides of the triangle to be equal. It will look like a triangle when tension is placed on the line. The tag line is simply a closed clove hitch.

Fig. 8–22. A bowline tied around the top of an exhaust fan, along with a clove hitch for a tag line, can allow the fan to be hoisted to an upper floor.

Hoselines, charged and uncharged

SKILL DRILL

It is preferable to hoist both charged and uncharged hoselines the same way to keep it simple. First tie a closed clove hitch behind the first coupling. Then take a bight in the rope and pass it through the bale and over the nozzle. By pulling the rope, the bale is guaranteed to stay shut (fig. 8–23). *Note:* No tag line is necessary, so you can use the hose itself as a tag line. Also, tying the clove hitch just below the coupling eliminates stress on the coupling.

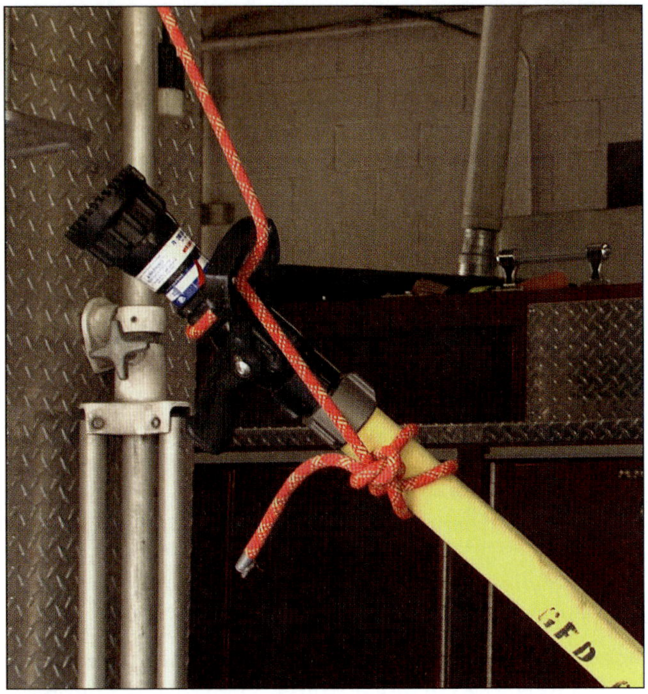

Fig. 8–23. A hose line can be hoisted by tying a clove hitch below the first coupling.

Hoisting a ladder

SKILL DRILL

Using enough rope to make a 3-ft (1 m) loop, tie a bowline. Place the bowline under the top rung, over the second rung, and under the third rung. Now pull the rope toward the top of the ladder and place the loop over the beams of the ladder. Pull up on the rope to take up the slack while sliding the loop down the ladder. Attach a closed clove hitch as the tag line to the bottom rung. This will control the ladder while being hoisted (fig. 8–24). A figure eight knot may also be used to hoist the ladder instead of the bowline.

Fig. 8–24. Firefighters using a bowline and a clove hitch to hoist a ladder.

Tying a rope between objects

SKILL DRILL

This is also known as a chimney or taut line hitch (fig. 8–25). Secure the rope around an upright object (closed clove hitch). Now bring the rope around the other upright object and back approximately three quarters of the way. Pass the working end under the right-hand line; this looks like the number 4. Bring the working end into the loop, pull taut, and make another number 4 with the working end. Bring this end into the loop between the first one you made and pull taut. This last time, hold the rope again as a number 4. Bring the working end into a new loop in front of the other by itself and secure tightly against the first two. If you release tension, the rope gets slack in it and people will be able to pass. Pull taut and the line will be tight again.

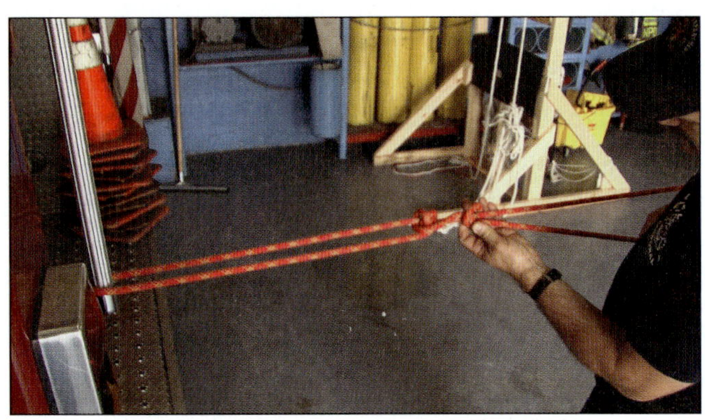

Fig. 8–25. Securing a rope between two objects

Using rope or webbing to secure objects

A rope or webbing can be used effectively to secure an object in place. Some firefighters carry a short length of personal rope so they can control a door that has to be forced open just in case there is a fire or severe heat condition behind the door being forced (fig. 8–26). Once opened and firefighters are advancing, the same short piece of rope can be used to hold the door open. Ropes can also be used to secure objects that are moved out of the way of firefighters, such as fencing that may be cut and rolled back, tree limbs and branches, as well as gates. In warehouses and distribution centers where most of the storage is on racks, ropes can be used to secure boxes stored on racks to prevent them from falling out of the racks and injuring firefighters. Ropes are used as a last resort for securing a motor vehicle when cribbing or blocking are not available.

Fig. 8–26. Firefighters often carry smaller lengths of rope or webbing with them to secure and control doors while performing searches.

ROPE MAINTENANCE, INSPECTION, AND STORAGE

Inspection of rope

FFI 4.5.1 **FFII 5.5.4** Ropes need to be inspected just like any other piece of equipment in the fire service. In addition to a visual examination after each use, ropes should be inspected as part of the regular routine of inspection for all tools and equipment. A weekly inspection is a particularly good idea when the rope is used on a frequent basis. For those departments that use their ropes less frequently, monthly inspections are acceptable. Regardless of the frequency, the inspection should be documented in writing (fig. 8–27). There are commercially available rope inspection forms, or a department may elect to create its own form.

The inspection report should identify what rope was used and when the rope was used. This report should also include information on how the rope was used and what type of exposure occurred to the rope. The date of the inspection, the person conducting the inspection, and the condition of the rope should also be noted.

The rope should immediately be removed from service if the rope is found to be damaged or suspected of having been damaged. Follow your department guidelines on how many picks or yarn pulls from routine wear are considered enough damage to remove from service.

When conducting an inspection, the rope should be examined along its entire length and not just the section that is thought to have been used. The person conducting the inspection should look for damage to the rope from a cut such as a laceration (cut into the rope); abrasion (cut on the surface of the rope from friction); fraying; charring; melting (especially true of synthetic fiber ropes); and damage from exposure to chemicals such as acids, corrosives, and petrochemical products.

The rope should be passed through the hands so that the person conducting the inspection can feel changes to the rope as well as examine the rope visually. These changes include finding a spot that has a different feel (hard or soft spot vs. the regular feel); a slippery feeling; or foreign objects embedded in the rope including grit, dirt, stone, wood or metal pieces. The inspection should also tell if the rope appears to be coming apart, which could indicate some type of interior damage (fig. 8–28).

The best time to conduct one of these inspections is after the rope has been cleaned and thoroughly dried out. The presence of moisture will allow mold or mildew to form, which in turn will cause the rope to rot if it is composed of natural fibers. Synthetic ropes can also disintegrate if they have been exposed to certain chemicals, and those chemicals have not been fully washed off the rope and the rope dried out.

Some people prefer to inspect a rope as they are putting it away in a rope bag or coiling it. Others prefer to tie the rope off, then slowly run their hands over the rope while the rope has some tension on it. Regardless of the method selected, the important thing to remember is that even a minor imperfection could be a tip-off of a major problem when the rope is placed under stress.

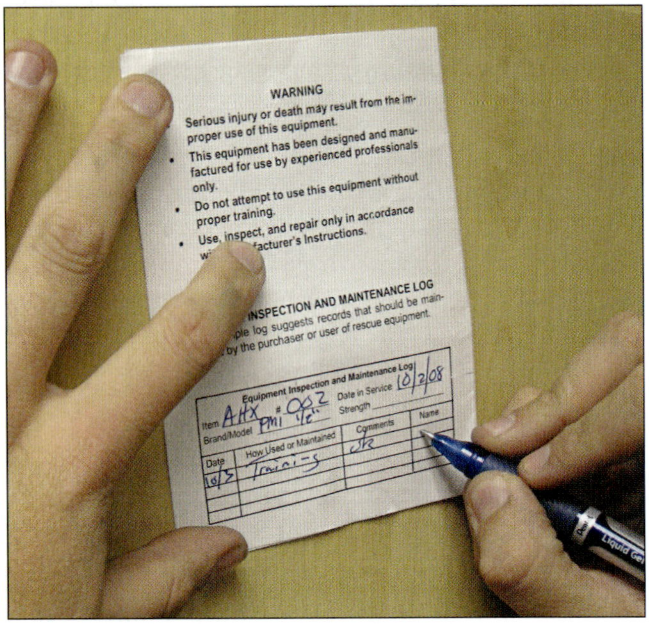

Fig. 8–27. Rope inspections must be documented and records kept in a conspicuous and accessible location.

Fig. 8–28. When damaged rope is located during inspection, it should be removed from service.

Ropes that are laid or twisted together can be untwisted to reveal the inside portion of the rope during an inspection (fig. 8–29). This is not true of braided, braid-on-braid, or kernmantle type ropes, because there is really no way to twist the rope apart to reveal the inner portion of the rope. The key with inspecting these ropes is to feel for differences when running the rope through your hands. Synthetic ropes are especially good candidates for conducting a tactile inspection while putting the rope under slight stress or tension. Any irregularities in the feel should be noted, and the rope removed for more thorough inspection. Whether the situation is one braid running over the other, flat spots, bunching, stiffness, or a different feel, these ropes could be damaged and in need of replacement.

Fig. 8–29. Untwist the strands of twisted or laid rope to inspect.

Proper maintenance of life safety and utility ropes is required to ensure that they are ready for service. Most fire service ropes come with instructions on how to properly clean, dry, and store the ropes.

Cleaning

SKILL DRILL

FFI 4.5.1 **FFII 5.5.4** Please follow the manufacturer's instructions. If your rope did not come with care instructions, then follow the general guidelines for cleaning, drying, and storage of ropes.

Natural fiber. A major drawback to the use of natural fiber rope is that it loses strength when it gets wet. And natural fibers do not regain lost strength. That means that using water to wash or clean a rope made of natural fibers simply cannot be done. The only way to clean a rope of natural fibers is to brush off the loose dirt and foreign material with a stiff broom or brush. This is another reason why ropes made of synthetic fibers are growing in popularity in the fire service.

Synthetic rope. Synthetic material can be cleaned in a number of ways. Again, remember that it is best to follow the manufacturer's instructions. The following is a general procedure to follow:

- Only use cold tap water, not hot water.
- Only use a mild detergent that is well-diluted. Do not use bleach or soap with bleach.

The cleaning and inspecting of kernmantle rope is difficult and time consuming because the damage might not be obvious. This inspection is done by touching all the rope as it passes through your hands. You are feeling for hard and soft spots. Note that some spots may have been made by a tight knot and are only temporary. This could indicate damage to the core, but on closer visual inspection, if the outer mantle is not damaged at that spot, then the fibers in the core may be temporarily misaligned and will relax. If there is any damage to the outer mantle, then the rope should be taken out of service. You should also look for cuts, odd shapes or weave, discoloration from chemicals, glass, metal, roughness (this may occur when a rope moves over another, stationary piece of rope), or an excessive amount of picks (fuzzies). If this or any condition presents itself, and you're not sure whether the rope is really damaged, then you should seek out someone with the training, knowledge, and experience to determine if the rope should be downgraded to another use. All ropes should be inspected after every use.

Hand washing. This is the most basic way of cleaning. Naturally, there are several methods that can be employed. One is to use a large sink or bucket and let the rope soak a few minutes, then move it around with your hands and scrub the rope with a small brush. Although some people suggest using your bare hands to wash the rope, the safer practice is to use a brush in case there are any sharp pieces of wood or metal embedded in the rope. Another method is to lay the rope out on the floor of the apparatus room. Since the apparatus bay is usually a fairly large area, the rope can be laid out in its entirety. Wet the rope with a hose and then scrub the rope with a stiff brush. If there is need to add detergent, you can place the cleaning solution in a bucket and dip the brush, then scrub.

One thing to remember regardless of the method selected to wash the rope is that if a detergent is used, the rope must be thoroughly rinsed of all detergent residue or cleaning solution (fig. 8–30). You can use a bucket of freshwater, pass the rope under a running faucet, or hose it off. However, if you use the hose, remember to turn the rope during the rinse process to get to the soap that is hidden between the rope and the floor.

Rope washer. A commercially available mini-version of a hose washer is the rope washer. The difference here is that the rope washer allows you to attach it to a threaded faucet or hose bib. You then run your rope through it, back and forth. The rope washer should have arrows that indicate the end into which the rope is placed and the end from which the rope emerges. Multiple small jets of water are directed at the rope as it is pulled through the rope washer. If connected to a faucet, the discharged water falls harmlessly into the sink. Using a hose bib may create a slip-and-fall hazard. The major downside to using a rope washer is that you cannot use any detergent in the process.

Washing machines. Only front-loading washing machines should be used. Top-loading machines should be avoided because this type of washing machine has an agitator that may damage the rope. Ideally, the rope should be placed in a mesh-type bag so that the rope does not get caught on anything inside the washing machine. If a mesh bag is not available, you can tie the rope in a **daisy chain** and place it inside the washing machine (fig. 8–31).

You may also want to select a front-loading washing machine that has a window so that you can visually check from time to time to make sure that the rope does not get snagged or damaged during the washing cycle. Add the proper detergent and make sure that all of the dirt and residue are rinsed out. One caution should be kept in mind. If the rope has been exposed to any type of chemical, you should avoid using a washing machine. The rope should be washed by hand or sent to a commercial vendor that specializes in decontamination of equipment when exposed to chemical environments.

Fig. 8–30. When washing rope, it is important to remove all detergent, if used.

Fig. 8–31. If you don't have a mesh bag, tie the rope in a daisy chain before placing it into a washing machine. See your Skill Drill book for steps to create a daisy chain.

Drying ropes. Before the rope can be stored it must be dried. The preferred method of drying a rope is to lay it flat over a ladder so air gets all around the rope. Never expose rope to direct sunlight. Hanging rope to dry is another method. The rope can be hung in a hose tower or suspended between or below bar joists or trusses. Extra attention must be taken not to get any contaminants onto the rope.

Rope storage

SKILL DRILL

It is imperative that rope be stored in a manner that keeps it clean and readily identifiable (length, size, and task), so a firefighter can quickly locate and deploy the proper rope for the task. The location on the apparatus where rope is stored needs to be a dedicated area that keeps the ropes from being buried under other equipment or being exposed to battery acid, fuel, fumes, or sunlight.

Bags. Rope storage bags will keep rope clean, neatly stored, and quickly deployable (fig. 8–32). Depending on the bag construction, some are better than others. Specially designed nylon or canvas bags are best for this purpose. A mesh bag is not. Nylon and canvas bags are best for protecting ropes from dirt, liquids, sunlight, and abrasion. A properly bagged rope is easier to deploy than a coiled rope.

Fig. 8–32. Rope bags help protect rope from abrasions and chemicals while being stored in fire apparatus.

Placing ropes into a bag. The purpose of the rope determines how the rope should be put into the bag. If placing in a **search bag**—used for carrying and deploying rope during a search—then place the rope through the grommet in the bottom of the bag, and tie a figure eight on a bight. This will give you a hand-hold to pull the rope from (bag stationary), or you can tie the bight off to the outside area and carry the bag deploying the rope behind.

If the rope is for a **drop (or life safety)** bag—used for raising tools, search or escape, and other fireground functions—then start out with a figure eight knot on both sides of the bottom grommet. This will keep the rope from being snagged and inadvertently being pulled out the bottom. Then simply load and coil the rope into the bag by starting around the inside edge and filling up the middle of the bag. You should tie a figure eight at the end of the rope and place it on the inside top of the bag.

Note: When deploying the bag, open the top of the bag drawstring and remove the figure eight, making sure that the eight has not pulled through any loops. This would cause multiple knots in the rope and a failed deployment.

Coiling and uncoiling rope

SKILL DRILL

Coiling a rope using the end of a ladders is a common practice that aids in keeping rope organized and easy to deploy (fig. 8–33). Steps for coiling and uncoiling a rope are found in your Skill Drills book.

Fig. 8–33. The end of a ladder is a useful tool for creating an organized rope coil.

OTHER EQUIPMENT

There are several tools and other pieces of equipment that are used in conjunction with a rope. It's important that you as a firefighter have at least a basic knowledge of these items should you be asked to retrieve them or provide them for technical rescue team members at a fire or other emergency incident.

Carabiner

A **carabiner** is a loop-type device with a gate-type opening on one side that opens by depressing it (fig. 8–34). This allows you to slip a rope into the carabiner or to attach the carabiner to an object. It also allows for the carabiner to be attached to a stationary object or to a piece of tubular webbing. A carabiner allows for the rope to change direction with minimal loss of strength. Carabiners are also used as part of any firefighter escape devices that you may be provided with.

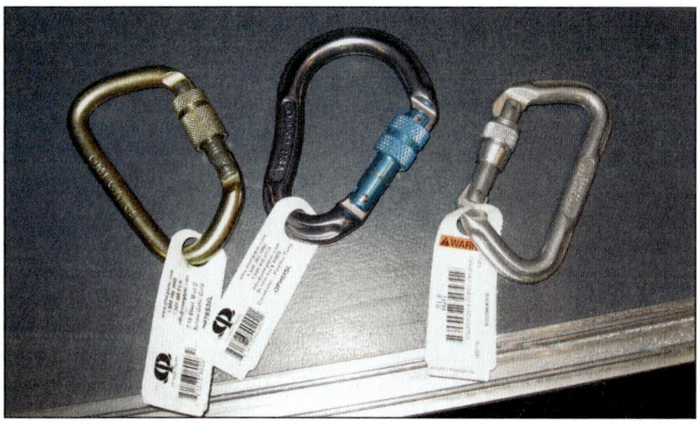

Fig. 8–34. A carabiner is the most common rope tool used in the fire service. When used in conjunction with a length of rope or webbing, it can create an emergency firefighter escape system.

Figure eight plate

A **figure eight plate** is used to assist in descending (fig. 8–35). As the name implies, the plate resembles a number 8. One ring is small, and the other is larger. The smaller ring is attached to a carabiner that a firefighter wears. The larger ring is for the rope. Some figure eight plates now come with ears that stick out at the bottom of the larger ring to prevent the rope from creating a girth hitch, which would essentially stop the firefighter from descending.

Fig. 8–35. A figure eight plate assists with descending on a rope.

Single-sheave pulleys

A **single-sheave pulley** has one sheave that allows a single rope to pass through it (fig. 8–36). Single-sheave pulleys are used to change the direction of pull on a rope. Sometimes the sheave has pivoting sides that allow the rope to be placed into it without having to start from one end or the other and can be inserted anywhere along the length of the rope.

Double-sheave pulleys

As the name implies, a **double-sheave pulley** is similar to a single-sheave pulley except that this device allows for two ropes to be inserted (fig. 8–37). A double-sheave pulley can be used to help create a **mechanical advantage** for raising or lowering a load.

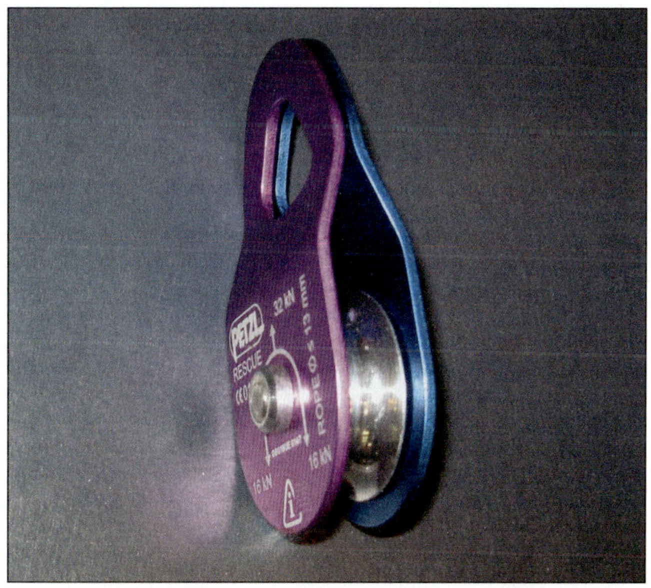

Fig. 8–36. A single-sheave pulley

Fig. 8–37. A double-sheave pulley

QUESTIONS

1. When should a rope be inspected?
2. What are the advantages of synthetic ropes over natural materials?
3. What do you look for when inspecting rope?
4. When should a rope be removed from service?
5. What are the differences between kernmantle, laid, and braided rope?
6. Which would you choose for life safety applications and why?
7. Describe hoisting a hoseline.
8. What is the difference between a bight, loop, and round turn?

Personal Protective Equipment

by Tim Pillsworth

This chapter provides required knowledge items for the following NFPA Standard 1001 Job Performance Requirements:

FFI 4.3.2

FFI 4.5.1

This chapter contains Skill Drills. When you see this icon, refer to your Skill Drill book for step-by-step instructions. SKILL DRILL

OBJECTIVES

Upon completion of this chapter, you should be able to the following:

- Describe the different styles of personal protective equipment (PPE) worn by firefighters
- Describe the relationship of SCBA and PPE
- Describe the purpose and operation of a personal alert safety system (PASS) used by firefighters
- Describe the six components of structural firefighting gear
- Describe the proper method of donning PPE
- Describe the proper method of doffing PPE
- Identify to procedures for inspecting and cleaning PPE

INTRODUCTION

Your **personal protective equipment (PPE)** is your first line of defense when operating at alarms, but it should not be considered the last line of defense. What does this mean? Your PPE is designed to protect you from the day-to-day risks: the mechanical, thermal, and biological risks you may encounter at any alarm. Your PPE has limitations. The best protection that you have is to understand the fire/rescue environment and not put yourself or your team in a position in which your lives depend solely on your PPE.

Whether a structure, vehicle, or brush fire; motor vehicle accident (MVA); medical call; or any one of the calls for help you receive each and every day, when worn correctly, your PPE offers you the best protection against harm from the forces and elements that surround you.

Once you complete this chapter, receive instruction, and train in the correct use of PPE, you will be able to protect yourself; and by protecting yourself, you protect your partners (brother and sister firefighters) and the general public. Your PPE training does not end with this chapter. You must continue to practice donning and doffing your PPE on a regular basis.

Although all PPE ensembles have some similar features and can be cross-used at times, they should only be used for their designed purpose. Structural turnout gear is the garment system or ensemble most commonly used. We will discuss this gear last because of its importance, and because it is the PPE we are all issued and use at most alarms. But the other forms of PPE must be covered to understand their use, function, and limitations.

There are several types of PPE, including the following:

1. Station wear
2. Wildland or brush
3. Proximity
4. Medical
5. Technical rescue
6. Hazardous materials
7. Ice rescue
8. Self-contained breathing apparatus (SCBA)
9. Personal alert safety system (PASS)
10. Structural

Each system/ensemble of PPE is covered under a National Fire Protection Association (NFPA) standard. As your career progresses, you should become familiar with the requirements in the regulations so you can better understand why and how the systems are designed, tested, and maintained. This knowledge will be important if one day you must specify PPE for your department. The garments, devices, and other items described show one or a limited number of manufacturers for PPE. For PPE requirements issued in your department, complete additional training with your training officer to ensure you know and understand the correct use and function of the equipment.

TYPES OF PPE ENSEMBLES

FFI 4.3.2 The following are the standard PPE ensembles worn for most department activities and incident responses.

Station wear

Station wear PPE is covered under NFPA 1975: *Standard on Station/Work Uniforms for Emergency Services*. This clothing is worn while on duty in the fire station, awaiting a response (fig. 9–1). Station wear serves as a "work" uniform and is designed to be worn under other protective clothing such as structural firefighting gear (wearing station wear alone is not appropriate for structural firefighting, as they are not considered primary protective clothing).

Fig. 9–1. Station wear

This gear, at a minimum, meets specific ignition and flame spread testing criteria to avoid injury during firefighting activities. In the past, some fire departments issued non-fire-resistant station wear made of synthetic materials. Such practices sometimes led to burn injuries of firefighters. With certified station wear, such problems have been greatly minimized.

Station wear may also be "dual certified" to meet the requirements of a variety of other specific PPE standards such as NFPA 1951, *Standard on Protective Ensembles for Technical Rescue Incidents*; NFPA 1977, *Standard on Protective Clothing and Equipment for Wildland Fire Fighting*; NFPA 1994, *Standard on Protective Ensembles for First Responders to CBRN Incidents*; and NFPA 1999, *Standard on Protective Clothing for Emergency Medical Operations*.

Wildland or brush PPE

Wildland or brush PPE fall under NFPA 1977, *Standard on Protective Clothing and Equipment for Wildland Fire Fighting*. Wildland or brush gear is designed to protect the firefighters from flash fire and abrasion during wildland fires while allowing the greatest amount of mobility and heat recovery possible (fig. 9–2). Fighting a wildland fire may expose you to extended hours of operation including walking and hiking on uneven terrain, high heat and low humidity, and even the possibility of snake or other small animal bites. The wildland PPE offers limited thermal protection, but is light, less restrictive to movement, and will release body heat more readily for prolonged operations. Materials such as Nomex® and treated cotton are used in pants and coats, shirts, or overalls (jumpsuits) with common closures, which are worn much like jeans and long-sleeved winter shirts. Often reflective stripping is added for safety and visibility. Head protection can be offered by a standard structural helmet with eye protection and earflaps or rated as a wildland fire helmet, which are similar in appearance and to construction hard hats. With the wildland fire helmets, hoods for neck and ear protection and goggles are required for full head protection. Protection for the foot, ankle, and lower leg, wildland boots are similar to hiking and construction work boots. Leather boots have hardened toes and shank for protection from rocks, punctures, and tools used during wildland operations. Using steel for toe and shank protection reduces the amount of heat buildup from walking on embers. Typically, this style of PPE is only purchased and issued to departments dealing with many large and or prolonged wildland fires. Most departments with limited wildland fires will use their issued structural gear for this purpose. The level of protection offered by the structural gear will be far above the wildland gear, but it will be much heavier and can cause higher levels of heat stress for the firefighter.

Although wildland PPE offers a lighter easy breathing garment, in the event of a fire entrapment situation such as a **blow-up** (a sudden increase in fire intensity), the gear will *not* protect you from the extreme heat. With each set of wildland gear, a **personal fire shelter** should be issued. This is a self-deployable shelter is constructed of an aluminized material to reflect the extreme heat during a fire roll over. This shelter is used as a last line of defense for survival.

Fig. 9–2. Wildland PPE. (Courtesy of Andoni Kastros)

Proximity PPE

Proximity PPE is covered under NFPA 1971, *Standard on Protective Ensembles for Structural Fire Fighting and Proximity Fire Fighting*. The proximity gear's outer shell has an aluminized surface. The outer **shell** reflects rather than adsorbs the vast radiant heat created from aircraft and petroleum-based fires (fig. 9–3). With an aircraft fire, a single plane can hold 20,000 gal (75,708 L) or more of aircraft fuel. Although this form of gear offers high levels of protection from radiant heat, the outer shell is extremely expensive and cannot withstand the daily punishment from normal municipal firefighting. The thermal and moisture layers (described in detail in the structural firefighting section later in this chapter) within the proximity gear can be constructed from the same materials found in structural gear. Both the helmets and boots for the proximity gear must meet the same standards as the coat and pants. The boots will typically have an aluminized coating applied, which is similar to the material found on the coat and pants. This coating reflects the high radiant heat. The head protection comes in two forms: a standard structural helmet with elastic covers made from the same material as the outer shell as well as a full-head hood system. To protect the face and the SCBA mask, full-face shields must be incorporated. The shields have a gold tone coating to reflect the radiant heat while still offering visibility. Without the shield, the high radiant heat levels burn the firefighter's face and damage the SCBA masks.

Medical PPE

Medical PPE is covered under NFPA 1999, *Standard on Protective Clothing for Emergency Medical Operations*. The use of medical gear has been increasing in many departments that provide medical responses in their jurisdictions. The use of dedicated medical PPE preserves body substance isolation and will greatly reduce the possibility of cross contamination (fig. 9–4). The PPE is much lighter and less cumbersome than structural gear and, in many ways, appearance; and style looks much like wildland fire gear or technical rescue gear. The reduced weight and bulk make for less fatigue while operating at medical calls and MVAs. Where the medical gear differs is its protection from bloodborne pathogens (BBPs). BBPs are microscopic organisms that, if introduced to the body, can cause illness or even death in the most extreme cases. The medical gear has a moisture barrier, which will prevent any liquids such as blood or vomit from penetrating the PPE and reaching your body. It is important to protect your hands with single-use medical gloves that meet NFPA standards. However, because these medical gloves can be damaged under normal working conditions, it is best to carry an extra pair in your gear at all times. Be aware that some firefighters have latex glove allergies. If you have such an allergy, ensure that you are issued non-latex gloves.

To protect your face from airborne fluids, a shield or goggles that meet the fluid-borne pathogen resistance standards must be donned. When it comes to foot protection, many departments issue **day boots** that meet the standards for slip-resistant soles, hardened toes and shank, and body fluid–borne pathogen resistance. The boots will appear and feel much like normal work boots.

Fig. 9–3. Proximity PPE

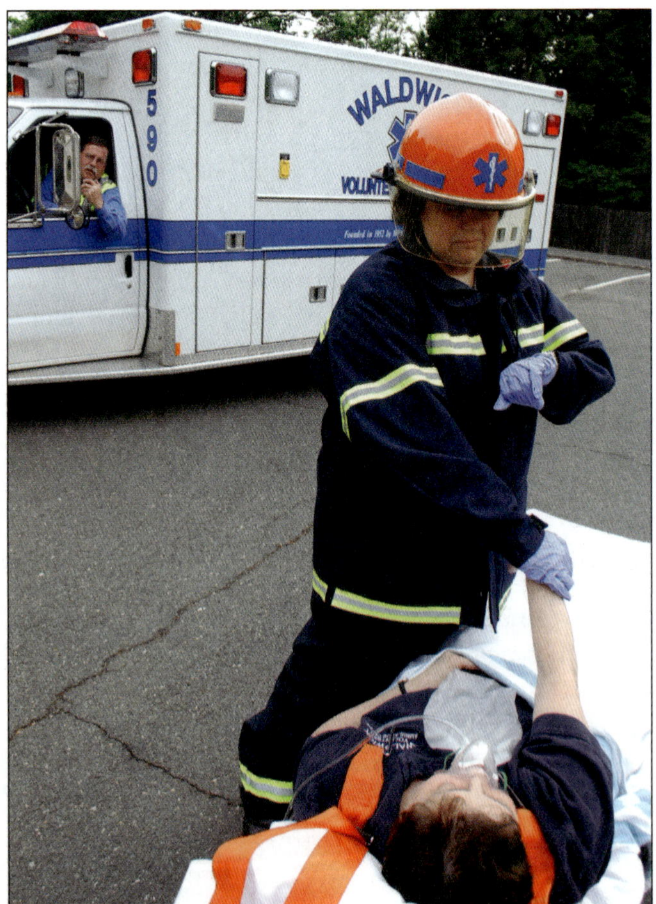

Fig. 9–4. Medical PPE. (Courtesy of Jeanette Kendall/State Line Fire and Safety)

Fig. 9–5. Technical rescue PPE

Technical rescue PPE

Technical Rescue PPE is covered under NFPA 1951, *Standard on Protective Ensembles for Technical Rescue Incidents*. The ensemble incorporates lightweight flexible PPE to offer protection during special rescue operations (fig. 9–5). Whether it is a high angle, trench, or collapse rescue situation, the PPE is designed to offer protection to the head and body. The PPE can come in the form of a coat and pant or a jumpsuit. They both must offer abrasion, liquid, and flame resistance. Once again the boots are similar to medical boots with toe, foot, and general penetration protection. To protect your head, special *technical rescue* helmets are employed. They typically have no rear brim like a structural firefighting helmet; each has chinstraps, sweat bands, and an internal suspension system. In addition to the PPE, rescue harness, rope, and rope accessories can be included.

Hazardous material PPE

Hazardous material PPE is covered under NFPA 1991 and 1992, *Standard on Vapor-Protective Ensembles for Hazardous Materials Emergencies* and *Standard on Liquid Splash-Protective Ensembles and Clothing for Hazardous Materials Emergencies*, respectively. The most technical form of PPE is the equipment used at hazardous material (hazmat) incidents. With the exception of the SCBA, this equipment is vastly different from any other from of PPE. The protection during a hazmat incident can vary from Level A (the highest level of protection) to Level D (the lowest level of protection). The Level A suit is designed to protect you from liquid or vapors and requires full SCBA (fig. 9–6a). Level B will only protect you from liquids but once again requires full SCBA to be donned (fig. 9–6b). Level C has the same level of skin protection as Level B but does not require the use of a SCBA (fig. 9–6c). The lowest protection, Level D, would be covered by your issued structural PPE (fig. 9–6d). The level of required protection is governed by what materials are present, the form they are in (vapor, liquid, particle), their concentration, or if the material is even known. To use this type of PPE, numerous advanced training courses are required and will not be covered in this chapter. This equipment is discussed in further detail in chapter 25: Implementing the Planned Response to a Hazardous Materials Incident.

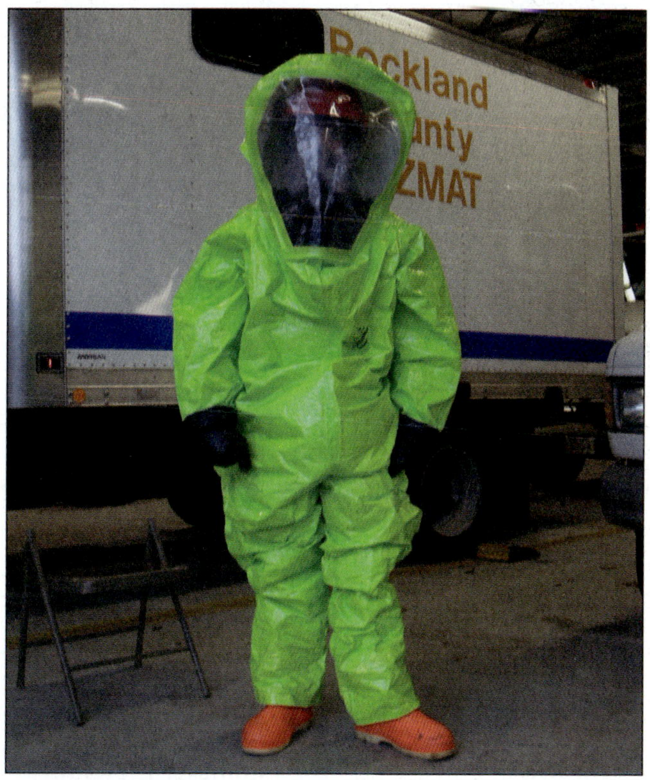

Fig. 9–6a. Level A hazardous material PPE—This is the highest level of protection, with integrated gloves and boots. They come in different materials for protection against different chemicals. SCBA is donned under the suit for respiratory protection. (Courtesy of Jerry J. Knapp)

Fig. 9–6b. Level B ensemble

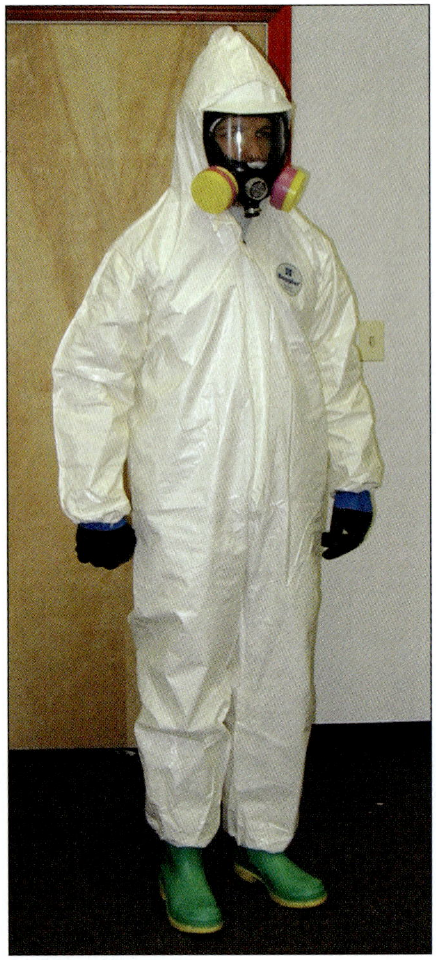

Fig. 9–6c. Level C ensemble

Fig. 9–6d. Level D clothing

Ice rescue PPE

Design of **ice rescue PPE** is not governed by a specific standard (but ice rescue and PPE is covered by NFPA 1670, *Standard on Operations and Training for Technical Search and Rescue Incidents*). Many think that ice rescue is only on ice, but it is considered a cold-water event in any water less than 70°F (21°C). Therefore, the rescuer must be protected from hypothermia. Water has the ability to remove heat from the human body 32 times faster than air. The heat transfer can be so rapid; a person can be hypothermic in only a few minutes. The ice rescue ensemble will seal your body from the water and the only area exposed is your face. The ice rescue ensemble consists of the suit, helmet, and harness (typically integrated with the suit itself). The suit is one piece, much like coveralls with attached boots, gloves, and hood (fig. 9–7). The integrated gloves and boots are formfitting and do not require the need for additional internal apparel for use. Once donned and closed, the hood creates a tight seal around the face to keep water from entering the suit. Additional ice shoes, which are like sandals with spikes, make maneuvering on ice quicker and safer. A helmet much like the technical rescue helmet is required for protection in case of a fall; remember, ice is slippery. The chest harness is for both your safety and for recovery. It will act as your lifeline but will assist in retrieving the victim from the ground crew.

SCBA as part of PPE

SCBA use and care are covered under NFPA 1981, *Standard on Open-Circuit Self-Contained Breathing Apparatus (SCBA) for Emergency Services*. To protect your respiratory system, the use of a SCBA is required. SCBA will be covered in great detail in chapter 10 in its use, care, and procedures. This chapter will cover and show how it interfaces with your PPE garments to complete your full PPE envelope. Without both the SCBA and your PPE garments working together properly, your body will not be completely protected.

Fig. 9–7. Ice rescue PPE

Fig. 9–8. Stand-alone PASS—The stand-alone PASS is still in compliance but must have the key or clip for activation. Many people forgot to turn on the older switch-style PASS units before entering the IDLH area. (Courtesy of A. Zytowski)

PERSONAL ALERT SAFETY SYSTEM DEVICES

PASS is covered under NFPA 1982, *Standard on Personal Alert Safety Systems (PASS)*. The PASS is a device that can be either removable (clipped on a strap) or integrated as a component of an SCBA. When in alarm the PASS will notify firefighters working in and around the scene that a firefighter is down, trapped, or in some life-or-death emergency. The PASS will activate in two different modes. The first is the manual alarm activated by you. You can activate the PASS by turning a switch or pushing a button on the device. The second is by the use of the motion detector within the device. If you are motionless for 30 seconds, the PASS will alarm. In the automatic mode, the PASS will pre-alarm to warn that the full alarm is imminent if there is not motion or a button or switch is not adjusted. Once in full alarm, the only way to deactivate the alarm is to manually turn off the alarm by either a button or switch on the PASS itself. For many years the PASS was a separate unit typically attached to the waist strap of the SCBA and turned on by the firefighter for operation (fig. 9–8).

More recently, most SCBA manufacturers have designed integrated PASS alarms into their SCBAs (fig. 9–9). The PASS will operate in the same manner, but the device will be turned on automatically when the SCBA is turned on. This solved the problem of firefighters not turning on their PASS before entering an **immediately dangerous to life and health (IDLH)** area. To have the PASS taken seriously as a safety device, it must be used correctly, which requires a lower number of false activations. With its correct operation, there will be less chance of a false alarm. To ensure that no PASS is left on, practice turning it off after you doff your SCBA or return to the apparatus.

The aforementioned ensemble of PPE and the structural gear, which will be discussed at length in the following text, will not offer any protection to your respiratory system. The products of combustion found in the smoke from all fires will cause damage to your lungs, mouth, nose, and the remainder of the respiratory track. To protect yourself, learn and understand the use and care of the SCBA. This will be covered in depth in chapter 10. But as part of the complete PPE envelope, it will be covered as it interfaces with your helmet, hood, and coat. Without all the parts of the PPE worn correctly, you will not be protected.

Fig. 9–9. SCBA PASS—Today most newly purchased SCBAs have integrated PASS alarms to insure they are activated when the unit is turned on.

STRUCTURAL FIREFIGHTING PPE

Structural PPE is covered under NFPA 1971, *Standard on Protective Ensembles for Structural Fire Fighting and Proximity Fire Fighting*. Structural PPE is also known as **turnout** or **bunker gear**. All the pieces of your turnout gear working together (helmet, coat/jacket, pants, gloves, boots, hood, SCBA along with miscellaneous personal tools) create an ensemble to protect you from the dangers of all operations on the fireground. To know your total structural ensemble, you will need to understand each of its parts.

Working from the bottom to the top, the boots will be the first donned (fig. 9–10). To meet the NFPA 1971 standard, the use of the hip boots (also known as ¾ **rubber boots**) are not allowed. However, they can serve an import role. In the event of water emergencies from the flooded basement to minor street flooding, they can be used to keep you dry and protected. The hip boots must not be worn for any rescue or fire operation, they offer little thermal protection and do not cover the waist area.

The boots used today are typically called **bunker boots**. They can be pull-on leather or rubber boots, which are typically 14–16 in. (36–41 cm) high; or they may be laced/speed laced boots, which are typically 10 in. (25 cm) high. Boots are typically placed within the pants *before use* for efficient donning. Each have their own pros and cons; and depending on your department PPE specifications, you could be issued one of multiple styles. All offer toe protection by hardened toe, foot bottom protection by a shank, aggressive tread, and ankle support.

Fig. 9–10. Structural firefighting boots—Boots have changed over the years. The tall rubber boots are no longer appropriate for structural firefighting, but are very good for water emergencies. Rubber and leather pull-on bunker boots are the most popular, while the leather zip-up short boots are used in some areas.

Many leather bunker boots offer additional shin guards and greater ankle support. Also, moisture barriers are now found in many boots for better water resistance and BBP protection. Even with the additional lining, with prolonged exposure to water, the boot will be penetrated. The leather and more recently designed boots of leather and high-tech materials are much lighter, offer a tighter fit, offer better foot and ankle support than traditional rubber boots.

The low-cut laced structural boots, sometimes referred to as *day boots*, offer the highest level of foot and ankle support; and they look and feel much like a construction work boot. They are typically lower than the pull-on boots (10 in. vs. 14–16 in.), therefore any deep water will more readily overtop the boot.

Working up from the boots, pants are next in line. The layered construction of the pants and coats being the same, the system will be discussed together. The combination of the coat and pant creates a system designed to work together; they are constructed from three separate layers of materials, which protect you from heat, water, and abrasion (figs. 9–11 and 9–12). Mixing and matching different styles of gear, even from the same manufacturer, will place you at risk of injury and void any testing and certification the manufacturer has on the gear. The coat and pant is one system or ensemble and cannot be mixed and matched.

Fig. 9–11. The three layers of structural PPE—Thermal liner with the moisture barrier and outer shell of a structural coat

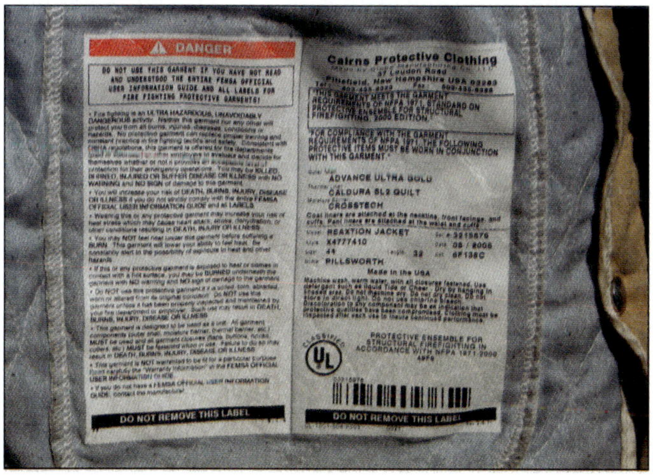

Fig. 9–12. The required NFPA label with the garment's information

The outer shell is the most durable layer and will protect you from flame and abrasion. It can be constructed of Nomex®, PBI®, or Advance Ultra®, just to name a few materials, and can be produced or dyed to a variety of colors. Each material will offer different levels of protection and cost. Many outer shell materials used today are highly heat resistant and required to self-extinguish once the ignition source is removed. When dyed, damage can be noticed by the removal or discoloration of the dye. Although this might not be a sign of damage, it does show a heat event happened to the coat.

The middle layer is the moisture barrier that protects you from water, common liquids (gasoline, battery acid, hydraulic fluid, and chlorine solution of 65%), and BBP; it can be constructed from Crosstech®, Gore®, and Stedair® to name a few. The moisture barrier will also prevent any common liquid from entering the thermal layer of the garment and reaching your body while allowing perspiration and body heat to be released. The moisture barrier is the thinnest layer of the garment and is only seen when the inner liner is removed from the outer shell. This layer can be damaged from heat and small punctures with little to no noticeable damage to the stronger outer shell.

The innermost layer is the thermal liner and will protect you from convected and conductive heat and control the moisture (sweat) that you generate. Common materials are Caldura®, Aralite®, and Synergy®. The liner is woven layers of thin materials quilted together, which trap air in between each layer to create the desired thermal protection. Air is a poor conductor of heat and the small air pockets will not allow the heat to transfer toward your body. This is one of the reasons why such a thin layer of material can protect you from such high heat. The innermost surface will typically have a smooth finish for comfort and easy donning. The thermal and moisture layers are attached to the outer shell by a combination of snaps, Velcro®, and zippers.

Structural PPE is required to meet many performance criteria. The **thermal protective performance (TPP)** test, which is essentially a test to measure the time it takes for convective and radiant heat to penetrate through the three layers of the complete PPE system—the outer shell, moisture barrier, and thermal liner—and injure the wearer. A PPE must receive a minimum TPP numerical rating of 35, which is equivalent to receiving a second-degree burn after 17½ seconds under flashover conditions.

The TPP is a tested by all three layers of the garment been tested at once by applying heat from a burner and a radiant heat source on one side of the test sample, and a sensor to measure the heat transfer on the other. The test results in a time and temperature graph. Therefore, if your gear has additional insulation, the TPP could increase to 40, which would only increase your time to burn by 2.5 seconds. With an increase in TPP, the garment may become heavier and will most likely reduce your body's ability to release heat generated from activity. This test does not mean you cannot get burned while your gear is donned. This is a test for use for comparison and minimum values; it is not actual fire conditions.

The second test is the **total heat loss (THL)** test, which measures the ability of the PPE to allow heat to pass away from the body through the three layers that make up the garment. The minimum required heat loss is 205 W/m^2 (watts, a measurement of heat energy, are described in chapter 5, Fire Behavior). The three layers working together will create air gaps between the layers and have synergistic effect on the values of the tests. This means that the value of the group is higher than the sum of the individual components.

The importance of the removal of heat created from your body (reflected in the THL value) during any alarm is critical. The buildup of body heat within your gear will reduce your efficiency and the total time you are able to perform tasks. If your PPE will not dissipate body heat, the possibility of heat stress and heat stroke increases. The extreme heat stroke can cause permanent brain damage and even death.

In the vulnerable areas of your body and the garment (cuffs, knees, elbows, and shoulders), additional layers of material can be added to increase the thermal and abrasion protection as well as offer padding for comfort. The materials can vary from the outer shell and thermal liner materials as well as high abrasion-resistant materials. Over time leather has been proven to be one of the strongest material for abrasion resistance; but it is heavy, becomes stiff with age, and is prone to absorbing water and fluids, causing problems with decontamination. Many of the new high-abrasion materials such as Arshield® and Dragonhide® offer high levels of abrasion protection, will not stiffen over time, and resist absorbing water and other fluids, thereby reducing the difficulty of decontamination.

For nighttime operations, general visibility and identification NFPA 1971 requires a combination of **retro-reflective** (a characteristic in which the light received by the reflective surface is reflected directly back at the actual source of light, such as a car's headlight) and fluorescent surfaces on the coat and pants. The reflective material must be placed in such a way that it is visible from 360°. Many departments add additional reflective material to the arms and legs as well as letters and numbers for department and personal identification, which add to the level of protection.

With the general knowledge of the makeup of the layers of the coat and pants, now the properties of each can be described. The pants will offer protection from the ankle to the hips and lower back if so equipped. For comfort and protection from heat, additional padding is typically added to the knees. At the fly and waist, there is a positive fastener system to secure the pants to your waist. The suspenders should be worn to ensure that the pants stay at the correct location for the knee pads and ensure they stay on in the event that you need to be rescued. If you go down, your bothers will use every method to pull you out, even pull you by your pants. The suspenders will keep them on.

The coat will protect your torso, chest, back, shoulders, and arms much the same way the pants protect your legs. A fastening system of zippers, Velcro®, snaps, hook and eye, or combinations form a positive seal once donned (figs. 9–13 and 9–14). The collar's neck flap works with the hood and ear flaps to protect the neck and the back of the head. There are two general types of cuffs, **standard** and **wrist guard**. The difference in the cuffs allows different styles of gloves to be worn. The longer cuff with either a thumbhole and palm protection or thumb loop allow the use of **gauntlet gloves** (an **interface component** that extends from the *end of the glove itself* and provides limited protection to the area where the coat and glove meet). The standard cuff requires **wristlet** style (*interface component*, which is typically a specially designed piece of fabric that extends from the *end of a coat sleeve* that provides limited protection to the area where the glove and coat meet). The **drag rescue device** (DRD) is webbing system within the coat to offer a safe and efficient rescue method for a downed firefighter. The DRD is a webbing loop that wraps the back and shoulders when deployed (fig. 9–15). To deploy, lift a flap located on the top back of the coat or within the back of the collar and pull the webbing outward; and you will have 2 ft (61 cm) of a loop to pull your partner to safety. This DRD is designed to allow leverage and distance to pull a firefighter along floor or ground (fig. 9–16). It is not designed to be attached to a rope and used as a lifting or lowering harness or attached to some form of mechanical pulling device (winch). This can cause grave injury.

Fig. 9–13. Coat closure, zipper, and Velcro®

Fig. 9–14. Coat closure, hook & dee, and Velcro®

Fig. 9–15. Stowed structural coat drag rescue device (DRD) under outer shell

Fig. 9–16. Deployed DRD—While short, it is usable for a firefighter rescue in a horizontal direction only.

The structural gloves are required to offer a minimum TPP value of 35, the same requirement as your coat and pants. Typically, the outer layer is made from some form cow or elk leather or kangaroo hide (fig. 9–17). But many gloves made with the newer materials in conjunction with the leather to create a more comfortable and user-friendly glove. Thin layers of thermal protection and a moisture barrier under the outer layer complete your protection (fig. 9–18). One of the largest complaints of fire gloves is the loss of dexterity with the gloves donned. The solution to this compliant is twofold. First, get fitted for the correct size gloves. A properly fitted set of gloves not only offers you the best protection, it allows for the greatest level of dexterity. Second, wear your gloves, all the time, at all alarms and drills. With high level of use, you will gain muscle memory making the tasks of starting saws and operating tools easier.

Fig. 9–17. Structural gloves

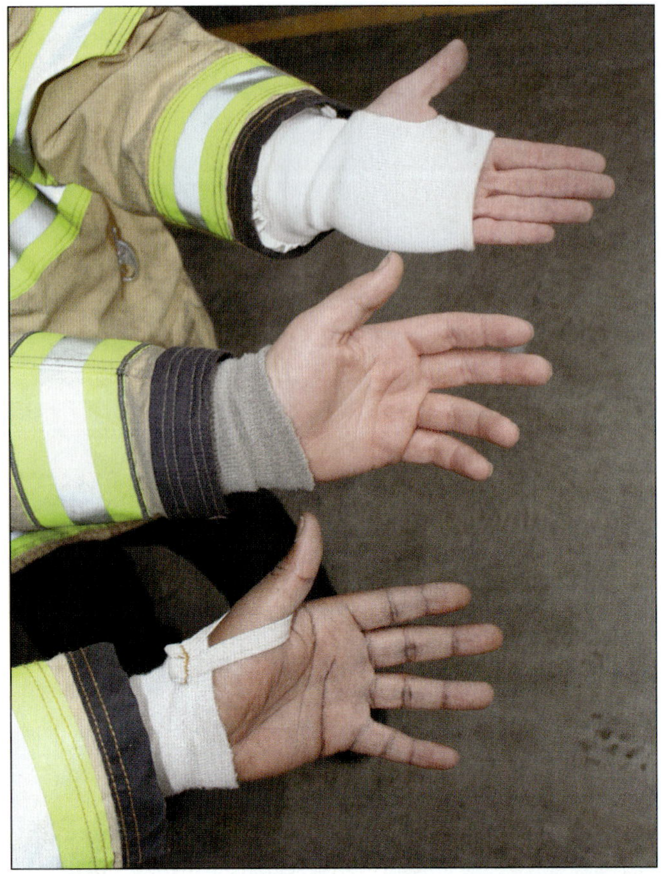

Fig. 9–18. Structural coat wristers

A structural hood is worn under your helmet and over the SCBA facepiece netting to protect your ears, neck, and any exposed skin not covered by the helmet, facepiece, or collar. The structural hoods come in various styles and are constructed of Nomex®, PBI®, and Carbon-X®, and when donned correctly with the helmet and facepiece, there will not be any exposed skin visible. The hood covers the most important and most difficult area of your body to protect. When donned correctly the hood is tucked into the coat so it will lay flat along your head, neck, and shoulders. To have it tucked into your coat it should be donned before the coat.

The interface with the SCBA facepiece is completed by sliding the hood over the back of the head, then donning the facepiece, which is covered in depth in chapter 10. Once the facepiece is in place, you can slide your thumbs around at the base of the head, hook the hood, and slide over the netting. Once complete, adjust any areas to ensure complete coverage. If correct, no visible skin will be exposed and the entire head net of the facepiece will be covered (figs. 9–19a, 9–19b, and 9–19c).

Fig. 9–19a. Hood donned and tucked under coat

Fig. 9–19b. Hood donned correctly

Fig. 9–19c. Hood donned incorrectly—Old, stretched hoods leave open gaps that will allow the super-heated atmosphere to reach the sensitive skin of the face and ears.

Last, let's cover what protects the thing you should use the most on the fireground: your head. Helmets come in many different shapes and styles and are constructed from many different materials.

From traditional to modern styles, from leather to plastic, the structural helmet can take different forms and materials. With all their visual differences, each has the same components (eye protection, suspension system, ear flap, chin strap, identification shield, reflective material, and protective shell), which meet or exceed the same NFPA standards (figs. 9–20a, 9–20b, 9–20c, and 9–20d).

All PPE manufacturers must submit samples to undergo extensive testing, as required NFPA 1971, before it can be used by firefighters. Each PPE component is thoroughly analyzed. For example, structural firefighting helmets must undergo several rigorous tests. These tests include an impact resistance test, a corrosion resistance test, an electrical insulation test, a retention system test for chinstraps, a suspension system retention test, a shell retention test, a retroreflectivity and fluorescence test, a radiant heat resistance test, and flame resistance tests.

To have the helmet protect your head, it must be worn correctly, every time. The **ratchet adjustment** (a changeable headband inside the helmet) will allow you to make changes to the fit, having the helmet donned with or without the SCBA facepiece and hood. Adjust the ratchet for a snug comfortable fit, and use and adjust your chinstrap every time. With both in use and adjusted correctly, your helmet will stay in place offering you the best protection. If your helmet falls off, it can be lost in a smoke-filled room and you will have lost your protection from both heat and impact. *The use of the chin strap is essential at all times; it is critical for your safety if you should fall and strike your head.*

Either a **Bourke shield**™ or face shield can offer secondary eye protection. Both styles will protect your eyes from an object from the straight-ahead direction, but not from spray or an object bouncing under the face shield. If a SCBA facepiece is not donned, to meet the primary level of eye protection, goggles or safety glasses must be used, especially when using tools.

Frontpieces, also known simply as *fronts* or shields, display your fire company identification and rank on the front of your helmet.

DONNING AND DOFFING PPE

Donning your PPE

SKILL DRILL

Now with your newly gathered knowledge of what and how your structural PPE is constructed, you will need to learn how to don it correctly, and in under 1 minute. It might seem impossible to do now, but with practice and some basic tricks, it will be an easy task. But to do this, you first need to have your gear set up for donning (ready for the next alarm).

Place your boots in your pant legs and pull the pants down around the boots, leaving the tops of the boots and bootstraps exposed. Place the suspenders over the tops of the boots so they are visible and remain untangled. Place your hood in an accessible location so it can be donned before your coat; on top of the helmet or on top of the boots or pant pocket works well for many. Have your coat open when it is hanging up and your helmet in close proximity. Your gloves can be in the coat pocket or kept with a glove keeper. Many departments have gear racks, which allow the gear to be stowed in a convenient and orderly manner (fig. 9–21).

Donning is in the following order:

1. Clear the suspenders from the tops of the boots, and step in/pull boots on.
2. Pull up your pants from the waist, pull the suspenders over your shoulders, and fasten the waist closure.
3. Gather your hood and pull it over your head.
4. Don your coat and fasten the closure. Close the coat fastener system and collar.
5. Slide your hood off your head and down to your neck.
6. Don your hand light and radio if so equipped and required.
7. Don your helmet with chinstrap.
8. Don your SCBA.
9. If needed, don your facepiece, pull the hood over the netting, and replace your helmet and chin strap.
10. Gather and don your gloves.

Fig. 9–21. Gear properly stored on rack—Your gear can be stored in an open locker, apparatus compartment, or gear room.

Fig. 9–20a. Structural helmets

Fig. 9–20b. Rear of structural helmets

Fig. 9–20c. Structural helmet suspension—note all approved helmets have attached ear protection and chin straps.

Fig. 9–20d. Structural helmet eye protection

Once complete, all your skin will be covered by one or more layers of protection (fig. 9–22). All pieces of the PPE, when sized correctly, will allow for maxim movement without being overly large or binding.

Doffing your PPE

SKILL DRILL

Doffing (removing) your gear is an important task rarely talked about. Under normal circumstances, the most import thing to remember is to remove it in the order it was donned and stow your gear the same way each time. By placing your hood and gloves in the same location, and stowing your gear the same way each and every time, you will be able to don it faster at your next alarm. Stopping to search for your gloves or hood will be slow and inefficient.

Fig. 9–22. Correctly donned structural PPE

While doffing, perform a quick inspection of any component you feel you might have damaged or contaminated. Before stowing your PPE—whether in a locker, bag, or compartment on the apparatus—make sure all components are completely dry. Wet gear will not dry when packed away, which promotes mold growth. Additionally, donning wet gear is more difficult and very uncomfortable. Make sure the coat cuffs are pulled out and the tops of the boots are clear with the handles exposed from the pants for easy donning for the next alarm.

Under most conditions, doffing your PPE will not take much attention and become a routine after every alarm, with one exception: doffing superheated gear. Injuries from superheated gear have been on the rise for the last few years. This is because of the high level of protection from your PPE. If you are operating within a superheated environment for a prolonged time, all your gear will become superheated. The temperature will be high enough to cause burns if you do not doff quickly and correctly.

A signal that your gear is superheated is the **off-gassing** your PPE exhibits; the gear is literally *smoking* when you leave the building (not to be confused with steam coming off your PPE when leaving the fire building and going into a cold environment). This process is your gear releasing heat absorbed from the fire. If this happens you will feel the heat on your body and at best are very uncomfortable or feel pain from the burning heat. This is not the time to apply water or pat down your gear in an attempt to cool off or *put out* your gear. This will cause the air pockets within the gear to be filled with steam or collapse, keeping the superheated gases closer to your body. The discomfort you were feeling will be greatly increased.

If you are in superheated gear, follow the following steps to limit any burns you might receive. This set of procedures was developed by Firefighter Patrick Brown of the Chicago Fire Department:

1. Keep your gloves on and remove your regulator from your facepiece.
2. Loosen the shoulder straps to your SCBA.
3. Open the collar tab and closure from the top down.
4. Once at your waist strap, open the strap and open the top of your pants.
5. Open your coat as wide as possible and roll the coat and SCBA out and off your shoulders and let them slide to the ground.
6. Use your feet step on your coat and help pull your arms out of the coat.
7. Undo your suspenders and let the pants fall down and step out of your boots.

This procedure can be done by yourself or with the assistance of a firefighter on the scene. If there is a member there to assist you, tell him or her that you have superheated gear and need assistance; talk him or her through the process.

If this does happen, your gear must be removed from service and completely inspected for any damage. There will be a good chance that the gear has been damaged and will require repairs or replacement.

AUXILIARY PPE

Many items may not be required in the NFPA PPE standards, but they do offer you convenience and safety while working on the fireground. Some items may be clips, tabs, glove keepers, and pockets on your coat and pants or the personal tools and equipment you carry in them to make your work safer and easier.

Pockets will allow you to carry tools, radio, and safety equipment at all times while keeping your hands free, reducing entanglement issues, and protecting the tools from heat and loss. Typically, bellow-style pockets that expand for storage are found on each side of the coat and each pant leg. A radio pocket can be added for radio storage, protection, and convenience. Remember that all the tools you might put into your pocket will be there all the time. They are additional weight and possibly snag or become entangled. If possible, place all auxiliary PPE inside a pocket to avoid entanglement hazards. Anything on the outside of your PPE will become an entanglement hazard (e.g., hand light, radio strap, mic cord, rope).

Whenever you don your PPE, have at least one hand light, even in the middle of the day, and a radio. If the handlight or radio has a shoulder strap, wear it under the SCBA straps to reduce the chance of entanglement.

Personal hand tools such as pliers, knife, wire cutters, window punch, shove knife, screw driver (combination flat/Phillips to reduce number), door wedge (two in a pocket, one on helmet), webbing, hose strap, spanner wrench, rope, and extra medical gloves are important to carry (fig. 9–23). Place the life safety tools (knife and wire cutters) in a place that can be easily accessed when full PPE is donned, including SCBA, working in zero visibility. Lives could depend on the life-safety tools, so keep them accessible. Your pant pockets may offer the best location for these tools, because your SCBA straps will interfere with access to them in a coat pocket. The remaining tools can be kept in a comfortable safe location. To keep the tools from damaging your gear, try to wrap them in a tool roller or inside an old glove, or place them in an old section of hose. The sharp points and edges will damage the expensive PPE and your body if you do not protect them.

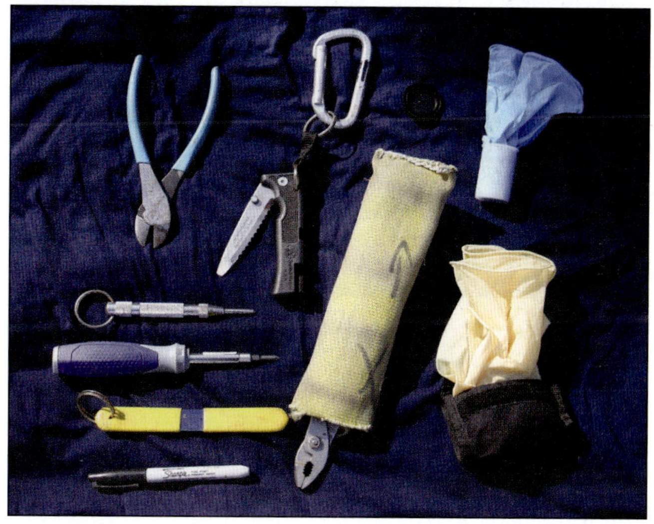

Fig. 9–23. Personal tools—your personal tools should include some combination of extra medical gloves (in a holder or even a 35 mm film case), wire cutters, screwdriver, pliers, center punch, and a knife. They can be used while performing your duties, or even to rescue yourself in an emergency.

PPE INSPECTION AND MAINTENANCE

SKILL DRILL

FFI 4.5.1 Care, inspection, and maintenance of firefighting PPE is covered in NFPA 1851, *Standard on Selection, Care, and Maintenance of Protective Ensembles for Structural Fire Fighting and Proximity Fire Fighting.*

NFPA 1851 requires all PPE to be evaluated after each use, fully inspected and cleaned as required. This will be for the day-to-day exposures to the variety of contaminates that we will encounter. Methods such as dry sweeping, spot cleaning, and water rinsing on the scene or upon return to the station to remove contamination is required to keep the PPE in proper order. Additionally, a complete advanced cleaning is required at a minimum once a year or as needed. Advanced cleaning of your PPE should be conducted by trained members in your organization, manufacturer, or an independent service provider (ISP). At no time shall your PPE be taken home to be cleaned. This will expose your family to the contaminates within your PPE (fig. 9–24).

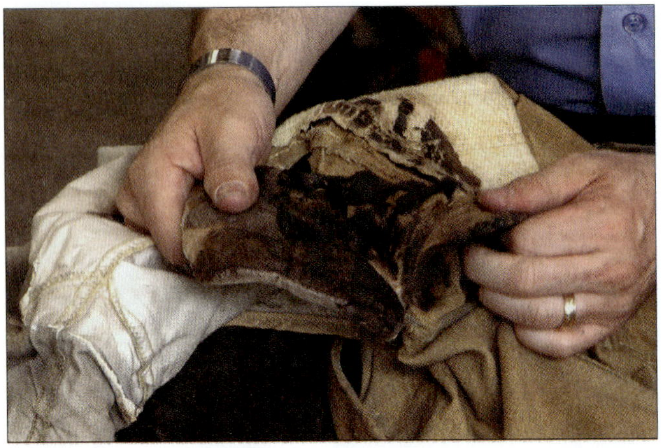

Fig. 9–24. Damaged gear being inspected

Over time, contaminates will be trapped in your PPE which degrade the materials, conduct electricity more readily, can cause illness, and may reduce your overall level of protection. By no means the inspection and cleaning take place only once per year as described in NFPA 1851, it should be completed as often as required to insure your PPE is neither damaged nor contaminated. The idea that having the dirtiest, most worn gear in the station makes you look seasoned or macho is long past. The contaminants found in the smoke—such as hydrocarbons, grease, and oil from MVAs, BBPs, and even carcinogens will penetrate the PPE and must be removed from your gear to ensure it is safe and ready for use.

Inspections should be done to some extent after every Inspections should be done after every alarm as well as prior to and after cleaning. As you are doffing your gear, closely look at it. Does it have a hole, burn, cut, or a seam opening? Is there any reflective striping missing or hanging? Does one area of your body get wet quickly while the remainder stays dry? These are all part of normal wear-and-tear damage that happens to all gear over time. These are all signs of damage to one or more of the layers of your PPE.

A full inspection is completed when the gear is up for its advanced cleaning cycle, or after any event when you feel there was the possibility of damage.

In order to properly clean and inspect the PPE, it shall be separated. Each component, outer shell and inner liner (moisture and thermal layer) needs to be visually inspected. Cleaning the layers separately is required to reduce cross contamination between the layers and to allow for the proper cleaning methods and products. The same layers can be cleaned together if desired and allowed by your department's SOGs and manufacturer recommendations. The outer shell should not have any holes, cuts, separated seams, or missing reflective striping. All snaps, Velcro, and fasteners must be in working order.

The outer shell will be able to withstand damage from heat better than the moisture barrier, and the damage to the moisture barrier might not be visible from the until it is separated from the outer shell. The moisture barrier should be uniform in color. All seams should be intact. There should not be any signs of abrasion (typically in joint and pressure point areas such as the knees, shoulders, and elbows). When the moisture barrier has sustained heat damage, the area will become brown or black and can crust or char.

The thermal liner is the innermost layer and is next to your body. The sweat from your body will be absorbed into this layer, possibly causing bacterial growth. Look for signs of staining, worn seams, and quilting becoming unwoven. Some firefighters may spray air fresheners, odor removers, or other materials to remove any body odors; but this should not be done unless the products being used are approved by your PPE manufacturer. Using non-approved sprays can result in degradation of the materials.

If any damage is found, the PPE needs to be taken out of service and repaired by the manufacturer, certified repair facility, or replaced. If you find or suspect any damage, report it to the station officer in charge of PPE.

For the annual advanced cleaning (or when it is required due to use), there are three methods of approved cleaning, by the manufacture, ISP, or trained member within your department. As noted, never bring your PPE home to clean it!

Whether your PPE is cleaned within your department or by ISP, there are specific requirements within NFPA 1851 that are to be followed. Please review the inspection and care requirements from your PPE manufacture prior to any cleaning. At a minimum, the general NFPA requirements are as follows: layers of the PPE are to be cleaned separately, water temperature not to exceed 105°F (40.6°C), cleaning detergent pH range between 6.0 and 10.5, maximum drum g-force of 100 g for spin cycles, and drying temperatures shall not exceed 105°F (40.6°C) or air drying.

If your department sends your gear to an approved cleaning facility, contact the officer in charge of PPE and follow the department protocols for sending the PPE out. It will most likely be out of service for 3 to 7 days. Once it is returned, inspect it much like you would inspect for dry cleaning. Ensure it is your PPE, it is complete, and there was no damage from the cleaning process.

NOTES

NFPA 1951 *Standard on Protective Ensemble for USAR Operations*

NFPA 1971 *Standard on Protective Ensemble for Structural Fire Fighting*

NFPA 1976 *Standard on Protective Ensemble for Proximity Fire Fighting*

NFPA 1977 *Standard on Protective Clothing and Equipment for Wildland Fire Fighting*

NFPA 1999 *Standard on Protective Clothing for Emergency Medical Operations*

NFPA 1981 *Standard on Open-Circuit Self-Contained Breathing Apparatus for Fire and Emergency Services*

NFPA 1982 *Standard on Personal Alert Safety Systems (PASS)*

NFPA 1991 *Standard on Vapor-Protective Ensembles for Hazardous Materials Emergencies*

NFPA 1992 *Standard on Liquid Splash-Protective Ensembles and Clothing for Hazardous Materials Emergencies*

Recommended additional reading material

Manufacture specifications and product information for the PPE that you are issued from your department. Include all aspects of the PPE you are issued, helmet, hood, coat, pants, boots, gloves, and eye protection. Read and understand the design and construction on the PPE that will be protecting your body at your next alarm.

QUESTIONS

1. What is the best protection you have on the fire scene?
2. What level of hazmat protection does structural PPE provide?
3. The PASS will activate in what two modes?
4. NFPA _______ covers structural PPE.
5. What are the advantages and disadvantages of leather bunker boots?
6. What are the advantages and disadvantages of rubber bunker boots?
7. What are the three layers of structural PPE and what do they protect the wearer from?
8. What is the minimum performance criterion for thermal protective performance for structural PPE? What is this number equal to on a fire scene?
9. What is the minimum performance criterion for total heat loss for Structural PPE? What does it tell us?
10. How is the drag rescue device (DRD) designed to be used?
11. Does the face shield or flip down on a helmet provide primary eye protection? Why or why not?
12. Structural PPE should be donned within _____ minute(s).
13. NFPA requires all PPE to be formally inspected, cleaned, and repaired, if needed, a minimum of _________ times per year to prevent the buildup of contaminates.
14. What are some contaminates that can get on your PPE?
15. When should a full field inspection of your PPE take place?
16. During an inspection, what are you looking for in the outer shell? The moisture barrier? The thermal liner?

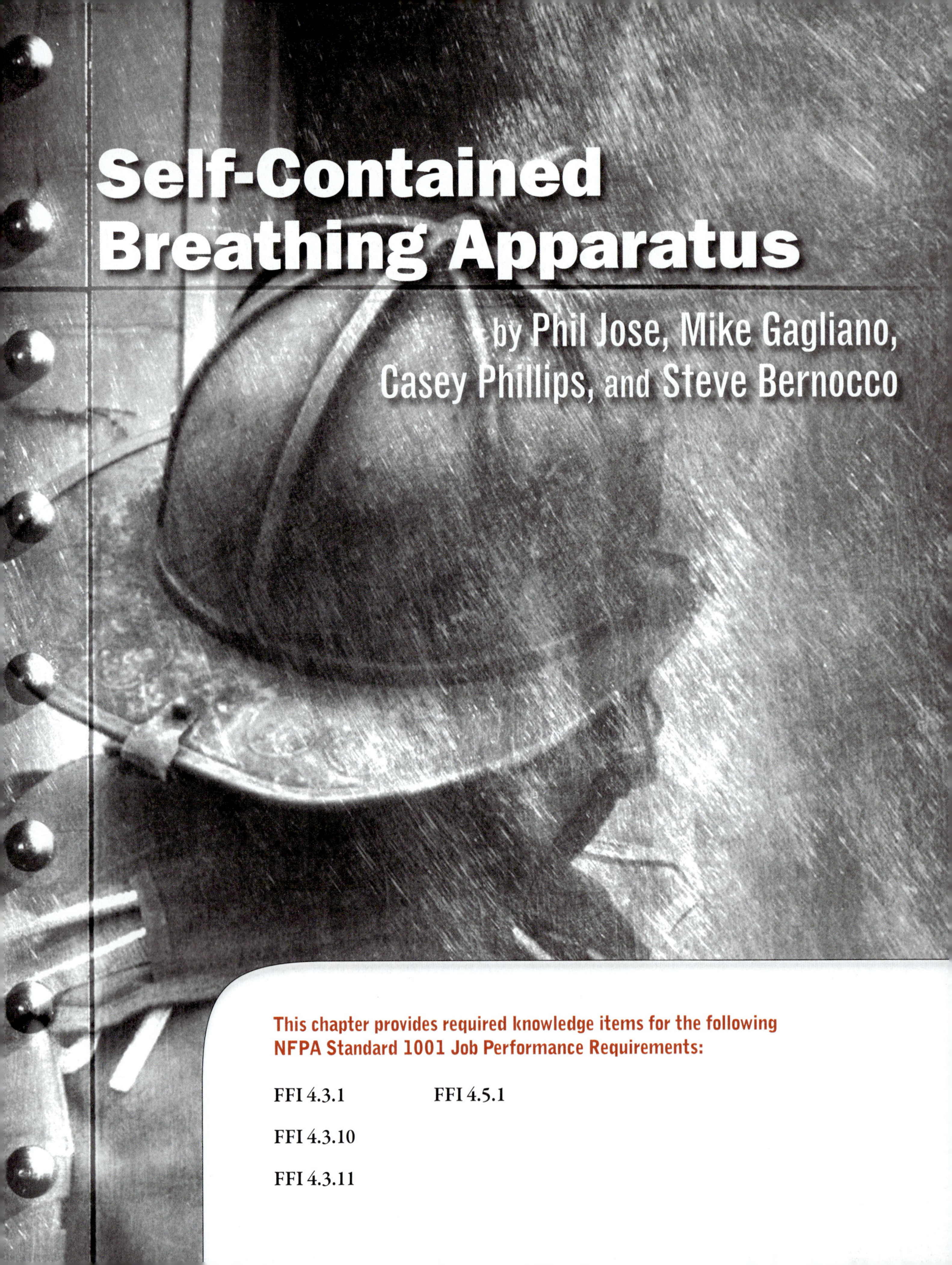

Self-Contained Breathing Apparatus

by Phil Jose, Mike Gagliano, Casey Phillips, and Steve Bernocco

This chapter provides required knowledge items for the following NFPA Standard 1001 Job Performance Requirements:

FFI 4.3.1
FFI 4.3.10
FFI 4.3.11
FFI 4.5.1

This chapter contains Skill Drills. When you see this icon, refer to your Skill Drill book for step-by-step instructions. SKILL DRILL

10

OBJECTIVES

Upon completion of this chapter, you should be able to do the following:

- List types of respiratory devices for fire and non-fire applications
- Describe how self-contained breathing apparatus (SCBA) protects the user at an emergency incident
- List the physical requirements needed by a user of SCBA
- List and describe the major components of an SCBA
- Describe and demonstrate the approved methods of donning and doffing SCBA
- Describe and demonstrate installing an SCBA regulator
- Describe and demonstrate changing SCBA cylinders and filling them
- List three activities accomplished by firefighters while wearing SCBA
- Describe how air consumption rates impact the use of SCBA at an emergency incident
- Describe the procedures for exiting a hazardous area while on emergency reserve air supply
- Describe the proper emergency measures to take when there is a failure of the SCBA
- Describe and demonstrate the proper inspection and maintenance procedures for SCBA
- Describe and demonstrate periodic inspections of SCBA unit
- Describe and demonstrate cleaning an SCBA

INTRODUCTION

This chapter introduces you to the open-circuit **self-contained breathing apparatus (SCBA)**, and includes the necessary information to meet the requirements for National Fire Protection Association (NFPA) Firefighter I and Firefighter II from NFPA 1001, *Standard for Fire Fighter Professional Qualifications*; components of NFPA 1500, *Standard on Fire Department Occupational Safety and Health*; and components of NFPA 1981, *Standard on Open-Circuit Self-Contained Breathing*

Apparatus (SCBA) for Emergency Services. These standards require firefighters to be trained in accordance with NFPA 1404, *Standard for Fire Service Respiratory Protection Training.*

OVERVIEW

Fire departments provide firefighters with many types of respiratory protection so firefighters can operate safely in all kinds of hazardous environments. Protecting the respiratory system is critical to the short- and long-term health of firefighters, and there are various levels of protection available from the minimum protection offered by paper dust filter-style mask to the full respiratory system protection offered by the SCBA. This chapter focuses on the SCBA as the tool of choice for respiratory protection while performing duties at a structural fire. It also provides a brief introduction to three types of respiratory protection for special applications ranging from fire investigation to confined space entry.

RESPIRATORY PROTECTION FOR NONFIRE APPLICATIONS

Fire departments often have air purifying respirators (APRs) for use by fire service personnel, like those worn by fire investigators (fig. 10–1). APRs have particulate filters, chemical filters, or combination filters to clean the air breathed by the user. The particulate or chemical filters in an APR must match the specific hazard from which the wearer would otherwise be exposed. Although APRs do not supply breathing air from a known source, they filter the ambient air surrounding the user. For this reason, they cannot be used in oxygen-deficient atmospheres (less than 19% oxygen).

Fire departments may also use the powered air purifying respirator (PAPR), which has the same limitations as the APR (fig. 10–2). The primary advantage of the PAPR over the APR is that the PAPR unit provides a blower that supplies air to the user. This supplied air reduces the effort needed by the user to breathe, thereby reducing overall fatigue and increasing the work time of the user. Neither the APR nor the PAPR is approved for use by structural firefighters while operating at the scene of a working fire.

The supplied air respirator (SAR) is another type of respiratory protection that is available for fire department operations (fig. 10–3). The SAR provides breathable air to the user from a remote source through an air supply line. Fire departments use the SAR primarily for confined space operations. The SAR can be used safely and effectively by trained personnel in known and properly controlled operational environments. The SAR, when used in fire department operations, includes some type of escape cylinder that is carried by the user in case of failure of the air supply line or the need for the user to immediately exit the space without time to manage the air supply line during exit. The SAR is not approved for structural firefighting and requires additional training.

Fig. 10–1. Sample of an APR often used during fire investigations. (Courtesy of Mine Safety Appliances)

Fig. 10–2. A PAPR with the battery pack attached. (Courtesy of Mine Safety Appliances)

Fig. 10–3. This sample of a SAR shows the air supply line attached. You can also see the small escape cylinder on the waist belt. This cylinder is used for emergency egress only. (Courtesy of Mine Safety Appliances)

SELF-CONTAINED BREATHING APPARATUS

SCBA is the firefighter's tool for breathing air in structural fire environments (fig. 10–4). SCBA is a time-tested and reliable method for firefighters to have safe, breathable air while operating in the most extreme of environments. Modern SCBA are designed and stringently tested to meet the requirements of NFPA 1981 *Standard on Open-Circuit Self-Contained Breathing Apparatus (SCBA) for Emergency Services.*

Fig. 10–4. In some cases, it is necessary to use SCBA outdoors when smoke conditions dictate.

USE OF SELF-CONTAINED BREATHING APPARATUS AT EMERGENCIES

FFI 4.3.1 SCBA provides an excellent level of respiratory protection for firefighters engaged in operations at structural fires. Firefighters arrive with a known quantity of breathable air to operate in immediately dangerous to life and health (IDLH) or potentially IDLH environments. The IDLH environment at structural fires includes gases produced by the combustion of the building and its contents (fig. 10–5).

Fig. 10–5. This firefighter is preparing to make entry into a known IDLH environment. Respiratory protections protects this firefighter, allowing the completion of assigned duties safely and effectively. (Courtesy of Lt. John Lewis, Passaic, New Jersey, Fire Department)

FFI 4.3.10 **FFI 4.3.11** The gases produced during a fire are toxic. The gases are carcinogenic. The combination of toxic and carcinogenic is IDLH for the firefighters. A common example is carbon monoxide (CO), which is present at all structural fires. CO has a much stronger affinity to red blood cells than oxygen (O) and works to displace the blood's ability to carry oxygen. The effects of CO can be seen in (table 10–1). Breathing the gases created during a fire can cause serious injury or death, even from a single exposure. The combination of toxic and carcinogenic is eventually dangerous to the firefighter's life and health. Low-level exposures to fire gases over the long term can cause serious health problems, including cancer (fig. 10–6).

Fig. 10–6. Before SCBA were readily available, firefighters were exposed to toxic gases while operating on the roof of the fire building. Firefighters from this era are at elevated risk for cancer due to their exposure. There is no excuse for this type of behavior on the modern fireground. (Courtesy of Seattle Firefighters Local 27)

Firefighters often underestimate the long-term damage produced by small exposures to toxic gases. Some departments maintain an unhealthy cultural bias against comprehensive use of the SCBA. The reality of the modern smoke environment requires the firefighter to use the SCBA whenever any smoke is present. If you want a long and healthy career followed by a long and healthy retirement, do not breathe smoke.

The following is a brief description of a few fire gases and their effect.[1]

The deadly duo: polyvinyl chloride and hydrogen cyanide

Two of the least recognized components of the modern smoke environment are **polyvinyl chloride (PVC)** and **hydrogen cyanide (HCN)**. The former is a product present in large quantities at most fires today, whereas the latter is a deadly by-product that is the silent killer of the fireground.

PVC, a type of plastic similar to that used in plumbing piping, is likely to be found in more abundance than any other product in today's homes. It is devastating to the firefighter no matter where they are in the smoke as its deadly fumes are emanating from materials at every level of structure. Its "emissions during fires of benzene, chlorinated dioxins, and dibenzofurans, known carcinogens, appears to explain the high frequencies of leukemia, laryngeal and colon cancer, and of rare soft tissue cancers found in many firefighters at relatively young ages."[2] PVC fires also include the presence of the highly acidic gas hydrogen chloride and other gases that impact respiratory and circulatory health.

Hydrogen cyanide is a colorless, odorless gas that emanates from both natural and synthetic sources (fig. 10–7). While carbon monoxide is well known to firefighters, in the last few years hydrogen cyanide (HCN) has been recognized as another deadly component of fire smoke. Thanks to the efforts of groups such Firesmoke.org, HCN is assuming a more prominent place as a cause of fire fatalities. HCN was the gas of choice for Hitler's death camps (Zyklon B) and the method the terrorists of the Aum Shynrikio religious cult attempted to use in their attack on the Shinjuku Station in Tokyo (May 5, 1995).

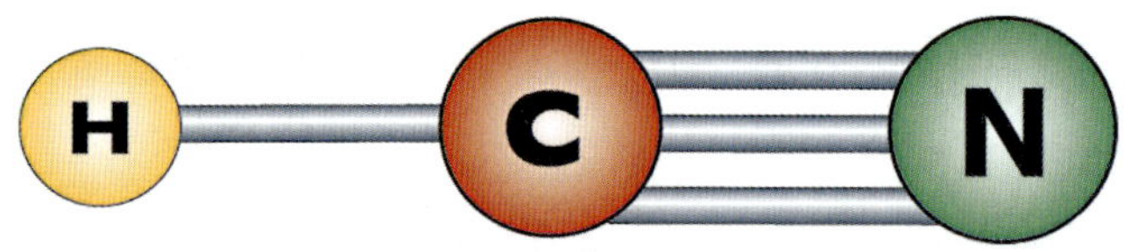

Fig. 10–7. Hydrogen cyanide molecule

A more comprehensive list of chemicals can be seen in table 10–1. Firefighters should understand that these gases are present at every fire. These gases can cause short- and long-term health problems. Furthermore, they can cause the immediate death of a firefighter from a single breath. Firefighters should not breathe smoke. The correct tool to prevent the exposure of firefighters is the SCBA. There is no better way to protect the health of the firefighter than a well-maintained, properly worn, SCBA.

Table 10–1. Toxic effects of carbon monoxide

Carbon monoxide (CO) (ppm)	Carbon monoxide in air (%)	Symptoms
100	0.01	No symptoms-no damage
200	0.02	Mild headache; few other symptoms
400	0.04	Headache after 1–2 hours
800	0.08	Headaches after 45 minutes; nausea, collapse, and unconsciousness after 2 hours
1,000	0.1	Dangerous; unconscious after 1 hour
1,600	0.16	Headache, dizziness, nausea after 20 minutes
3,200	0.32	Headache, dizziness, nausea after 5–10 minutes; unconsciousness after 30 minutes
6,400	0.64	Headache, dizziness, nausea after 1–2 minutes; unconsciousness after 10–15 minutes
12,800	1.26	Immediate unconsciousness, danger of death in 1–3 minutes

Table 10–2. Fire gases and their effect on the human body

Table of primary fire gas toxicities

Gas	Assumed LC50 (ppmv) 5 min	Assumed LC50 (ppmv) 30 min
Acetaldehyde	—	20,000
Acetic acid	—	11,000
Ammonia	20,000	9,000
Hydrogen chloride	16,000	3,700
Hydrogen bromide	—	3,000
Nitric oxide	10,000	2,500
Carbonyl sulfide	—	2,000
Hydrogen sulfide	—	2,000
Hydrogen fluoride	10,000	2,000
Acrylonitrile	—	2,000
Carbonyl fluoride		750
Nitrogen dioxide	5000	500
Acrolein	750	300
Formaldehyde	—	250
Hydrogen cyanide	280	135
Toluene diisocyanate	—	100
Phosgene	50	90
Perfluoroisobutylene	28	6

THE SCBA IN THE POSTFIRE ENVIRONMENT

In the postfire environment, the contents in the structure begin to cool. These contents continue to release toxic gases at lower concentrations. Many gases are the same as those produced during the active fire stage of the incident. Firefighters must be aware that, although a postfire environment may appear clear, toxic gases such as hydrogen cyanide and carbon monoxide and cancer-causing agents such as benzene are still released into the atmosphere (table 10–2). For this reason, firefighters should continue to wear SCBA after the fire is out and during all overhaul activities.

Many fire departments use some type of gas detector to identify the level of CO present in the postfire environment. This is a weak attempt to identify that the air is safe enough to breathe to permit firefighters to operate without SCBA. However, this practice is strongly discouraged. There is ample evidence that low CO readings do not indicate the absence of other toxic and cancer-causing gases in the atmosphere. Keep in mind that CO readings do not provide information about particulate matter, such as asbestos, suspended in the air that will be inhaled by firefighters not wearing SCBA (fig. 10–8). The authors advise a thorough study of the UL-FSRI research project on smoke particulates: https://ulfirefightersafety.org/research-projects/firefighter-exposure-to-smoke-particulates.html.

Fig. 10–8. Firefighters are exposed to toxic and cancer-causing gases when conducting overhaul operations without SCBA are exposed to toxic and cancer-causing gases. Firefighters should always wear respiratory protection in the postfire environment. (Courtesy of Seattle Firefighters Local 27)

The postfire environment, while not necessarily an IDLH environment, is a time when firefighters without SCBA are exposed to chemicals that are known to cause cancer. There is no reason to knowingly expose firefighters to these dangerous gases and chemicals when the SCBA can provide excellent protection. The SCBA provides protection for the user's respiratory tract (fig. 10–9) from toxic and superheated gases. The SCBA also provides a significant level of protection to the user's face and eyes. However, the SCBA does not protect the user from exposure to chemicals absorbed through the skin. There are literally hundreds of **polycyclic aromatic hydrocarbons (PAHs)**, which can be absorbed through dermal contact. These chemicals cause numerous cancers and have been determined to have an affinity for sticking to our bunking gear. They latch on and can be transmitted to other areas such as the seats in the firetruck, equipment, and back at the firehouse.

Many SCBA are approved by the National Institute for Occupational Safety and Health (NIOSH) to provide respiratory protection in **chemical, biological, radiological, and nuclear (CBRN)** environments. NIOSH performs extensive testing of SCBA assemblies for compliance with a variety of federal standards. As those standards apply to firefighter and CBRN qualification, here is what NIOSH reads about their testing:

> **To Protect Emergency Responders Against CBRN Agents in Terrorist Attacks**
>
> *The U.S. Centers for Disease Control and Prevention's (CDC) National Institute for Occupational Safety and Health (NIOSH) has a program to approve self-contained breathing apparatus (SCBA) for use by fire fighters and other first responders to terrorist attacks. NIOSH approval under the program signifies that an SCBA is expected to provide needed protection to first responders in situations where an act of terror has released harmful chemicals, pathogens, or radioactive materials into the air. Approvals are based on positive results from rigorous tests on sample units submitted to NIOSH by manufacturers, and from stringent evaluation of manufacturers' quality-control practices, technical specifications, and other documentation. To ensure protection in CBRN environments, the firefighter should look for a CBRN-NIOSH Agent Approval label on the SCBA assembly. If there is no CBRN Agent Approval label, or it has been removed or destroyed, the SCBA is not approved for use by firefighters in the environment.*

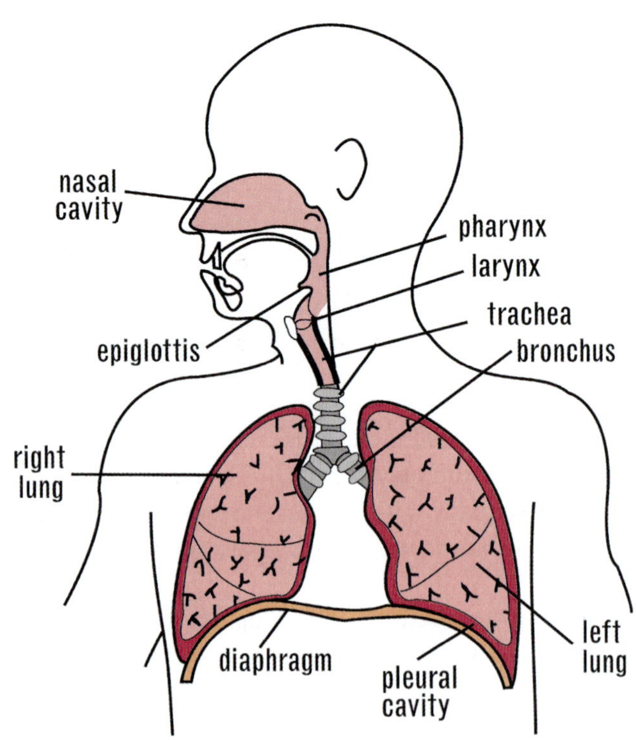

Fig. 10–9. Respiratory tract

PHYSICAL REQUIREMENTS OF THE USER

Firefighting is an extreme physical activity. Firefighters must use SCBA during all phases of firefighting operations in areas that are or could become IDLH. Fire departments should ensure that any person trained to use SCBA for emergency operations receives appropriate medical screening. Firefighting is an ultra-hazardous occupation, and the leading cause of death for firefighters is cardiac arrest. This fact alone should provide ample incentive for fire departments to ensure that all personnel engaged in emergency operations in IDLH environments receive medical screening that complies with all applicable standards. Medical screening should be performed before SCBA training and should continue to be performed annually as long as the firefighter is responding to emergency scenes.

In addition to appropriate screening, the properly trained firefighter must understand that there are multiple layers of regulations regarding the use of the SCBA in IDLH environments. These include local, state, and federal regulations. The basis for most fire department regulations related to SCBA and IDLH environments comes from the **Code of Federal Regulations (CFR)**; specifically, CFR 29 section 1910.134 sets the requirements for many parts of the respiratory protection program including the following:

- Selection of respirators for use in the workplace
- Medical evaluations of employees required to use respirators
- Fit testing procedures
- Procedures for cleaning, storing, inspecting and maintaining respirators
- Procedures to ensure adequate air quality, quantity and flow of breathing air

In addition, CFR 29 1910.134 requires fire departments to meet the minimum requirements of what is commonly referred to as the "two in—two out" rule. This rule requires that firefighters enter the IDLH as a team, the "two in." These two must be supported by at least two firefighters outside the hazard area, the "two out", who can help if the "two in" experience an emergency. The standard reads as follows:

Procedures for interior structural firefighting

In addition to the requirements set forth under paragraph (g)(3), in interior structural fires, the employer shall ensure that:

1910.134(g)(4)(i)

At least two employees enter the IDLH atmosphere and remain in visual or voice contact with one another at all times;

1910.134(g)(4)(ii)

At least two employees are located outside the IDLH atmosphere; and

1910.134(g)(4)(iii)

All employees engaged in interior structural firefighting use SCBAs.

As a Federal regulation, CFR 29 1910.134 applies to all fire departments. The rule above does not prevent firefighters from taking immediate action, when necessary, to save a life. This rule is also supported by other documents such as NFPA 1500, *Standard on Fire Department Occupations Safety and Health Program.*

COMPONENTS OF THE SCBA

FFI 4.3.1 The major components of the SCBA are the facepiece, regulator, harness, and cylinder (figs. 10–10 and 10–11). The cylinder holds compressed breathing air for use by the firefighter. Cylinders come in a variety of construction types and capacities. Cylinders are made of steel (not typically used today), aluminum, composites such as carbon fiber (fully wrapped), and fiberglass or Kevlar (hoop wrapped or fully wrapped). Although there are exceptions, fire service cylinders are commonly grouped into two pressure categories: low-pressure cylinders, which operate with an upper limit of 2,216 psi (15,279 kPa), and high-pressure cylinders that operate up to 5,500 psi (37,921 kPa). Regardless of the pressure rating of the cylinder, firefighters should focus on the volume of air contained within. The amount of air in the cylinder is the limiting factor for a firefighter operating in an IDLH environment.

Fig. 10–10. Components of the SCBA

Fig. 10–11. SCBA bottle valve assembly

Common terminology for fire service organizations is to refer to cylinders by their rated service time (fig. 10–12). There are three common cylinder sizes in the fire service (fig. 10–13):

- 30-minute/1,200 liter/44 cubic feet (cu ft)
- 45-minute/1,800 liter/66 cu ft
- 60-minute/2,400 liter/88 cu ft

The 30-minute cylinder is being phased out in most areas of the United States and has largely been replaced by the 45-minute cylinder for firefighting operations. The 60-minute cylinder is used primarily for special operations including hazardous materials or rapid intervention team assignments.

Firefighters must understand that the minute ratings are inaccurate for fire service applications. Cylinders hold enough air for a firefighter to use at a standard rate of 40 liters per minute (lpm)/1.4 cubic feet per minute (cfm) for the rated time (table 10–3). However, firefighters must understand that air consumption rates (ACRs) during firefighting operations far exceed the 40 lpm/1.4 cfm standard and can easily reach over 100 lpm/3.5 cfm. For this reason, firefighters should not expect the air in the cylinder to last for the rated service time. In a study conducted by Phoenix Fire Department, and published in the article "Rapid Intervention Isn't Rapid" by Steve Kreis (*Fire Engineering*, December 2003), a "30-minute cylinder" will last a firefighter between 16.5 and 18.5 minutes. Firefighters should refer to cylinders by their standard volumes so there is no confusion between rated service times and actual fireground use. Air management and air consumption techniques are addressed later in this chapter.

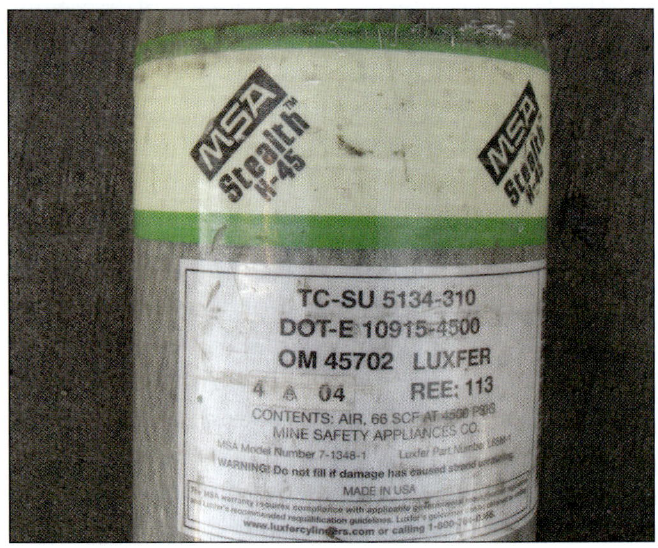

Fig. 10–12. Cylinder label that shows capacity and minute service time

Fig. 10–13. Different types of cylinders

Before each use, the cylinder should be visually inspected to ensure that it is full, has not been damaged, and has a current hydrostatic test sticker (fig. 10–14). All cylinders have a cylinder valve and pressure gauge assembly that should be checked to ensure that the cylinder is full. Although 100% of the rated capacity is desired at all times, fire departments should establish a written policy addressing the minimum acceptable capacity for a cylinder to remain in service. The user should ensure that the cylinder is above the minimum guidelines set by department policy or the SCBA manufacturer.

Table 10–3. Standard SCBA cylinder specifications

Approx. duration (min.)	PSI	Weight (lbs)	Weight (kg)
15	3,000	3.3	1.5
30	2,216	6.7	3.04
45	4,500	9.3	4.23
45	4,500	9.4	4.26
60	4,500	12.2	5.53

Fig. 10–14. This firefighter is checking for a current hydrostatic test label before filling the cylinder at an emergency incident. (Courtesy of Seattle Firefighter Local 27)

Most manufacturers mandate exchanging a cylinder at 90% or less of the rated capacity for a cylinder. Any cylinder that does not meet the requirements must be exchanged for a full one.

Visual inspection for damage to the cylinder should also be performed by the user on a daily basis. Cylinders contain an incredible amount of stored energy, and damaged cylinders should be removed from service. There are different levels of damage (fig. 10–15). Small nicks and scrapes from normal wear and tear are the least problematic and are considered Level 1 damage, which should be noted but does not require action by the user. Level 2 damage includes significant scraping or otherwise missing layers of the clear-coat protective cover. Level 2 damage requires that the cylinder be removed from service and repaired. Level 3 damage is anything greater than Level 2 and includes such items as damaged carbon fiber wrapping or significant damage caused by striking another object. Firefighters should be trained to recognize all three types of cylinder damage. Training should indicate the levels of damage that can be anticipated to distinguish normal wear and tear from more significant and dangerous damage. Check that the cylinder has been hydrostatically tested according to recognized national standards. Essentially, cylinders are subjected to specified high pressures using water as the pressurizing material rather than air. Cylinders require hydrostatic testing at 3- or 5-year intervals. Firefighters must be familiar with the requirements for the SCBA cylinders they are assigned to use. Cylinders that are beyond the accepted testing interval should be removed from service and identified for repair according to department standard practice.

The cylinder is held to the SCBA by the harness assembly. The backpack harness assembly has been the fire service standard for more than 50 years and has an excellent service history. Modern harness assemblies are rated to survive extreme conditions without failing. The back plate of the harness holds the cylinder in an upright position. A high-pressure line connects to the cylinder valve assembly by a threaded coupling to permit air flow from the cylinder to the regulator assembly. This connection includes an O-ring assembly and is particularly vulnerable to leaking and failure. Pay careful attention when connecting and disconnecting this fitting to ensure that the O-ring is present (fig. 10–16).

The harness assembly has waist and shoulder straps that connect the harness to the user like a backpack. These straps should be secured and appropriately adjusted whenever the SCBA is used. The harness assembly is designed to effectively transfer the load of the cylinder and backpack assembly to the body of the user. In addition, many harness assemblies, when fully connected, make excellent handholds to extricate unconscious firefighters from fire environments. Firefighters who operate without the waist belt or shoulder straps connected or properly adjusted increase the potential for negative consequences on the emergency scene (fig. 10–17). Although department policy should state that firefighters wear the SCBA properly, it is ultimately the responsibility of individual firefighters to operate safely.

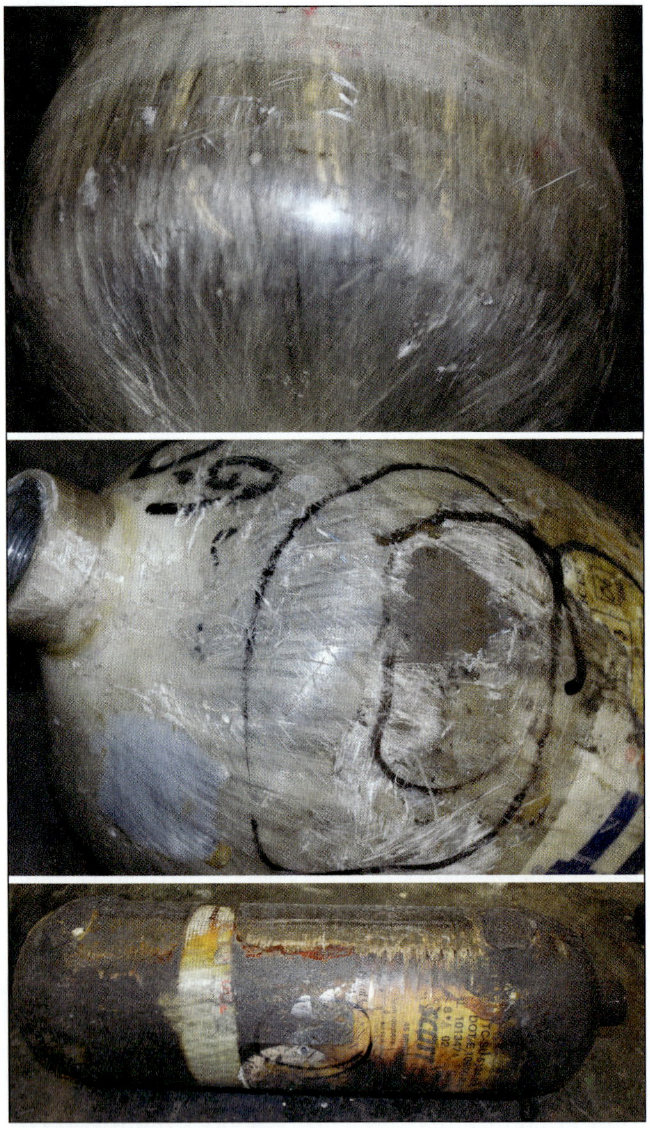

Fig. 10–15. Examples of Level 1, Level 2, and Level 3 damage

Air is provided to the user through the **regulator assembly** (fig. 10–18). First-stage regulators reduce the high cylinder pressure to approximately 100 psi (700 kPa) at the second-stage regulator. The second-stage regulator then reduces the pressure to just above atmospheric pressure. Air is provided to the facepiece at slightly more than atmospheric pressure to provide a positive pressure inside the facepiece. This ensures that the firefighter is not exposed to products of combustion in case of a facepiece malfunction or inadequate seal against the face. One style regulator is belt mounted with a low-pressure tube connecting the regulator to the facepiece. The other more common style is a **mask-mounted regulator (MMR)**, where the regulator connects directly to the facepiece. Firefighters should be thoroughly trained on every type of SCBA that they may be expected to use (fig. 10–19).

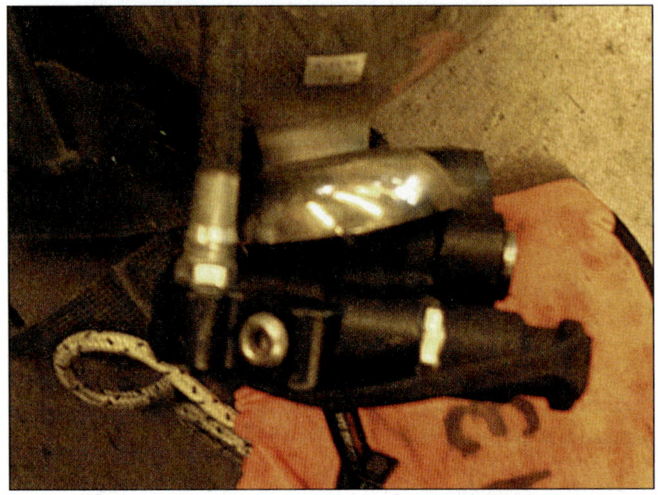

Fig. 10–16. Here you can see the high-pressure hose connection to the neck of the cylinder. The low-pressure alarm and the rapid intervention connection (RIC) are also visible.

Fig. 10–17. This firefighter does not have all of his SCBA straps properly connected.

The regulator assembly has a **main line** valve and a **bypass valve.** The main line valve is designed as the primary method of providing breathing air to the facepiece. The main line valve allows air to flow into the regulator at a high pressure, which the regulator then reduces to just above atmospheric pressure, providing positive pressure in the facepiece. Main line valves on belt-mounted regulators are generally manually operated, while on MMRs the valve automatically opens when the regulator is placed inside the facepiece and the user inhales (fig. 10–20).

Fig. 10–18. Non-mask mounted regulator

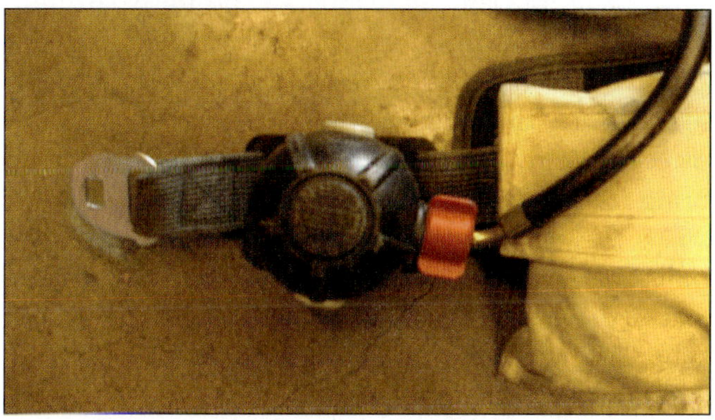

Fig. 10–19. The MMR shown here is properly stored. When placed in service, the regulator is attached to the facepiece.

Fig. 10–20. The mainline valve on this MMR is internal and operates automatically. The bypass valve is red and can be seen mounted on the left side of the MMR.

The bypass valve allows air to go around the moving parts of the regulator and flow at increased pressure directly into the facepiece assembly. Bypass valves are usually red and are manually used only during SCBA malfunctions. Firefighters who must use the bypass valve should notify their supervisor immediately and seriously consider calling a Mayday.

The facepiece is the final component of the SCBA. The facepiece is individually sized to the user, with most manufacturers offering small, medium, or large sizes. Most facepiece assemblies offer optional equipment such as electronic voice enhancement for improved communication and **Heads-Up Display (HUD)** to assist in air management. Each firefighter must receive a fit test before initial training as well as annually to ensure proper fit. Fit testing is required by NFPA 1404 and other national standards. Fire departments should maintain a record of the annual fit test.

A **low-air alarm** is provided on all SCBA (fig. 10–21). Some take the form of a light-emitting diode (LED) lamp on an MMR and vibrating facepiece. Some SCBA manufacturers provide other types of visual and audible alarms. It is important that you determine the type of low-air alarms provided on the SCBA you will be using.

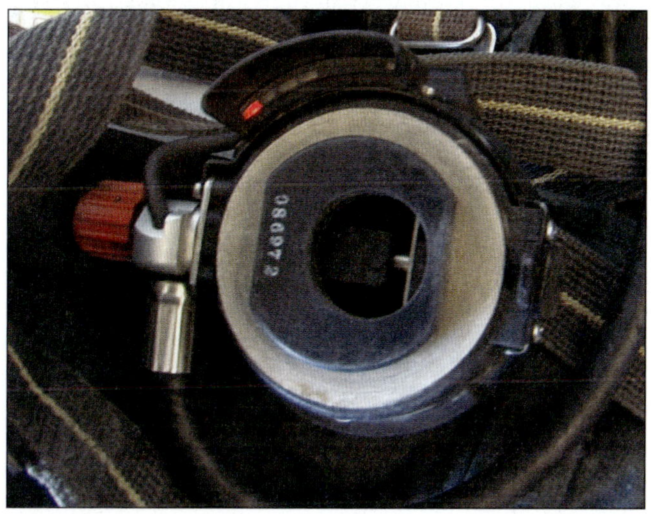

Fig. 10–21. Low air alarm on a HUD

The HUD is a great addition to the modern SCBA. The HUD allows firefighters to get information about air volume by looking at the HUD in the facepiece. While manufacturers place the HUD in different places, every HUD has a combination of lights that indicate the volume of air remaining in the cylinder by one-fourth volume increments. A full cylinder will be indicated by four green lights, 75% by three green lights, 50% by two yellow lights, and 33% by one red light. The HUD can assist firefighters in determining when they should begin to exit the IDLH environment. It is a good idea to consider beginning your exit when the 50%, or two yellow lights, marker is reached. Good air management practice will allow the entire team to exit the IDLH environment before any member reaches the 33%, or one red light, marker.

One critical aspect of effectively using the HUD is understanding the ranges the indicator lights represent. The lights can be deceiving if firefighters are not fully aware of what they mean and adapt their air management decisions appropriately. The common misconception, when seeing three green lights, is that the bottle is ¾ full. This is an understandable mistake as the three lights do seem to be saying just that. It is critical to remember that the three lights mean you have passed by the ¾ full mark and are steadily heading toward ½ a bottle. Those three lights merely tell you the remaining air is somewhere between ¾ and ½. That can be a very big distinction when deeply committed to a structure or working at a heightened respiratory rate due physical exertion. It is always recommended to verify your air with a quick reading of the gauge.

Facepiece assemblies provide air to the user from the regulator. The most common process is for fresh air to flow from the regulator through the inhalation valve across the lens to keep the lens clear of fog. The air then flows through the nose cup and is inhaled by the user. Exhaled air flows through the exhalation valve and out into the atmosphere. A properly fitted facepiece ensures that no leakage occurs around the face. The facepiece should be cleaned by the user after each use, and the lens should be inspected regularly for clarity of vision and safety. The facepiece is held to the head with a weblike harness normally consisting of three or five straps. Firefighters should be thoroughly trained to effectively and efficiently place the facepiece on the head to obtain and maintain a seal. Firefighters who use corrective lenses should consult with the SCBA manufacturer for information on placing corrective lenses inside the facepiece.

Many modern SCBA also have a built-in personal alert safety system (PASS) device, an alarm device that activates when a firefighter becomes motionless. While some firefighters still use individual PASS devices, which attach to their coat, in the past it has been shown that many firefighters failed to turn their PASS devices on. PASS devices integrated into SCBA prevent this problem: when the SCBA is turned on, so is the PASS device. PASS alarms are discussed in greater detail in chapter 9: Personal Protective Equipment.

DONNING AND DOFFING THE SCBA

FFI 4.3.1 The proper use of SCBA by a firefighter is a critical function on the fireground. If not donned properly, the unit may malfunction at a critical time. The two most common methods of donning are over-the-head and the coat methods. For both it is important that the firefighter's protective clothing be properly worn and that the SCBA be fully charged.

Over-the-head donning

SKILL DRILL

The firefighter must fully open the main cylinder valve and make sure it matches the regulator gauge. With the unit being held in front of the firefighter and the cylinder valve facing away, the firefighter will lift the SCBA unit over their head and slide the unit down their back (fig. 10–22). The shoulder and waist straps are adjusted for comfort. The firefighter will then pick up their face piece, set their chin in the chin pocket of the mask, and bring the straps/webbing over their head, adjusting the straps for proper fit and comfort. Their hood is then pulled into place and their helmet is placed on their head and secured with the chin strap. The regulator/breathing tube is then attached to the mask, the gloves are put on, and the firefighter is ready for assignment.

Fig. 10–22. Over-the-head donning

Coat method donning

SKILL DRILL

The firefighter must fully open the main cylinder valve and make sure it matches the regulator gauge. With the SCBA unit in front of them, the cylinder valve facing away, and the straps to the sides, the firefighter will place their dominant hand to the opposite shoulder strap near the backplate and lift and swing the unit over their dominant shoulder (fig. 10–23). The other hand and arm will be slid between the cylinder and corresponding shoulder strap. The shoulder and waist straps are adjusted for comfort. The firefighter will then pick up their face piece, set their chin in the chin pocket of the mask, bring the straps/webbing over their head, and adjust the straps for proper fit and comfort. Their hood is then pulled into place and their helmet is placed on their head and secured with the chin strap. The regulator/breathing tube is then attached to the mask, the gloves are put on, and the firefighter is ready for assignment.

Fig. 10–23. Coat method donning

Doffing the SCBA unit

SKILL DRILL

This is a reversal of how the unit was donned. The firefighter removes his gloves, disconnects the regulator from the mask, and removes their helmet and hood. The mask is removed in the following manner: Loosen head strap, if available, temple straps, chin straps, and remove the face piece. Unbuckle the waist strap and loosen the shoulder straps. While holding the shoulder strap where the regulator is attached, slide the opposite shoulder strap off your shoulder and lower the unit to the ground. Shut off the main cylinder valve and bleed of any air in the regulator by opening the purge valve. Then turn off the PASS device if it is integrated into the unit.

Installing the mask mounted regulator

SKILL DRILL

To place the MMR in service, the firefighter shall pick up their face piece and set the chin in the chin pocket of the mask. They will then fit the face piece to their face and bring the straps/webbing over their head. To adjust the face piece straps the firefighter will do so in this order: First adjust the chin straps, then the temple straps and, if available, the head strap. They will make sure the straps are pulled straight back, not outwards. They will then check for proper seal. They will then pull their hood over the straps/webbing, making sure not to dislodge the face piece. Lastly, the helmet is placed on their head and secured by the chin strap. The regulator will then be attached to the face piece as per manufacturer's recommendations.

Replacing an SCBA cylinder

SKILL DRILL

The firefighter will place the SCBA unit with the depleted air cylinder on the floor and turn off the cylinder valve. Any remaining air will be bled off using the purge valve. Disconnect the high-pressure hose from the cylinder valve and inspect the O-ring for damage. Release the cylinder from the backpack and remove the cylinder from the unit. Then place it by the refill station and take a full cylinder from the refill station. Make sure the cylinder is full by looking at the cylinder valve. Place the full cylinder in the unit's backpack and lock into place, and connect the high-pressure hose to the cylinder. *Hand-tighten only!* Then slowly open the cylinder valve and listen for leaks. Once the cylinder valve is fully opened, check the cylinder gauge with the regulator gauge to see that they match.

Replacing an SCBA cylinder being worn by a firefighter

SKILL DRILL

Have the firefighter lean over forward and remove their helmet and hood. Disconnect the regulator from the mask. Turn off the cylinder valve on the depleted cylinder and bleed the remaining air in the hoses using the purge valve. Then disconnect the high-pressure hose from the cylinder valve and inspect the O-ring for damage. Release the cylinder from the backpack, remove the cylinder from the unit by sliding it forward and away from the backpack, and place by the refill station. Take the new full cylinder and slide into the backpack, avoiding hitting the firefighter in the head. Lock the new cylinder in place, reconnect the high pressure hose to the cylinder, and open the cylinder valve and make sure the cylinder valve matches the regulator valve. Notify the firefighter that the cylinder exchange is complete and they may return to service.

All modern SCBA also have a built-in PASS device, which activates when a firefighter becomes motionless. While some firefighters still use individual PASS devices that attach to their coat, many firefighters in the past have failed to turn their PASS devices on. PASS devices integrated into SCBA prevent this problem—when the SCBA is turned on, so is the PASS device. PASS alarms are discussed in greater detail in chapter 9: Personal Protective Equipment.

ACTIVITIES WHILE IN SCBA

SKILL DRILL

The SCBA is worn by firefighters at all structural fires and many other types of emergencies including hazardous materials, confined spaces, structural collapses, and some emergency medical alarms. It is important for firefighter trainees to have the opportunity to operate in the SCBA in enough nonemergency capacities to prepare for whatever an emergency scene may present. Thorough training includes practice in advancing hoselines, climbing ladders, crawling through windows, and many other activities that the firefighter can be expected to perform on the fireground.

Firefighters are often called on to perform a variety of tasks. To perform these tasks, the firefighter must be competent enough in the operation and use of the SCBA so that the firefighter can concentrate on the other skills. Take the example of a **primary search.** If a firefighter enters the structure from a ladder, the ladder must be taken from the apparatus, carried to the fire building, spotted and raised, and finally climbed by the firefighter to make access through a window (fig. 10–24).

The firefighter then must perform a search of the fire room or fire floor, communicate within the crew, and perhaps even talk on the radio. If the firefighter is accessing the building through the front door, the firefighter may need to perform forcible-entry skills, assist with the hand line advancement, and then proceed with the search. In both examples the SCBA is expected to be a part of the firefighter's protective ensemble. The firefighter must be comfortable and confident enough with using their SCBA to safely and effectively perform the tasks. This level of SCBA comfort and confidence comes through regular and realistic training exercises. This level of SCBA comfort and confidence is maintained over a career through regular and realistic training exercises.

Fig. 10–24. Firefighters putting up a ladder at a fire have many things to consider. Wearing the SCBA should not negatively affect the ability of a well-trained firefighter to perform basic tasks.

AIR CONSUMPTION RATES AND FACTORS THAT AFFECT THE DURATION OF THE AIR SUPPLY

Firefighters operating in an IDLH environment are limited by the amount of air they bring with them in their SCBA. Firefighters must be trained to exit the IDLH environment before the activation of the low-air warning alarm. To do this effectively, each firefighter must be aware of the amount of air they use in a given time or their **air consumption rate (ACR).** The ACR will have a big impact on the duration of the air supply and can help firefighters and fire officers balance the needs of the incident or assignment against the ACR and other factors. Some of the factors that must be considered are outlined in the following text:

- Fitness:
  - The size, weight, and overall aerobic fitness of a firefighter affects the duration of the air supply and the ACR. As a rule, smaller firefighters use less air in a given time than larger firefighters given the same workload. Aerobically fit firefighters generally use less air than the same size firefighter who is unfit.
- Nature of the work being performed:
  - The harder the work performed, or the more physical exertion required by a firefighter, the more air the firefighter uses. This applies independent of other factors. Given an assigned task, there are often higher workload positions within a crew. Understanding the ACR for individual members of the team can assist the company officer in making assignments to maximize team effectiveness related to the ACR.
- The environment where the work is being performed:
  - Hot, dark, smoke-filled rooms create physical and emotional stress on the human body. Although some improvement to the ACR is expected in firefighters with experience, it is important to understand the physiological and psychological stressors that apply in IDLH environments. Firefighters should be aware of their own physical reaction to environmental stressors and train themselves to control their respiratory rate and emotional state during times of high stress.
- The effects of protective clothing and repeated work cycles:
  - The modern PPE, or bunker gear, provides a great degree of protection from fire and other thermal effects. The downside to this level of protection is the encapsulation of the member within the *greenhouse* created by the sweat and heat of their own activity. Fire departments should address the ratio of work cycle to rest cycle requirements for firefighters operating in full PPE and SCBA.
- The duration of a firefighting operation inside an IDLH environment:
  - The amount of air inside the cylinder when entering the IDLH environment affects how long you can operate. The more air you start with, the longer your operating time will be. Fire department policy should address the minimum pressure, or volume, allowed before committing to an IDLH environment. In no case should this be less than 90% of the rated capacity of the cylinder. Firefighters have died after entering an IDLH environment with a cylinder that was less than the minimum required.
- How far firefighters travel inside the IDLH environment and the distance and time required to reach an area of safety:
  - Because a firefighter is required to exit the IDLH environment before activation of the low-air warning alarm (33% of the capacity of the cylinder), the firefighter must be aware of how long it takes to reach the area of safety, how much air that travel time uses, and how this affects the assigned activity. Having a clear understanding of how and when you will exit the IDLH environment allows the firefighter and the team to perform effectively in the firefight. In addition, the company officer can more easily identify whether the team can complete their assigned objective.
- Company officers coordinating the activity of the team:
  - Getting the team to complete the assigned task effectively, exit in a coordinated and safe manner, and maintain an appropriate margin for safety are necessary components of the company officer's responsibility. To do this effectively, the company officer needs information and interaction with the members of the team. The company officer also coordinates with the incident commander to arrange for a crew to relieve the team inside the IDLH environment if the team cannot meet their assigned objective and safely exit the space. This responsibility requires the company officer to know and understand the ACR for each team member and how this applies to the current team assignment. Improved understanding of the ACR as well as the factors that affect the duration of the air supply lead to improved situational awareness for all members of the team.

The ACR is a fluid factor for a given firefighter or team within an incident and over a career. For example, a veteran generally experiences less stress than a rookie on the fireground. Understanding how the ACR, incident, and team members interrelate is learned and maintained through regular and effective training exercises with the SCBA. Well-trained firefighters, working in well-led teams, perform more safely and make their air last longer.

THE AIR MANAGEMENT PROGRAM

SKILL DRILL

FFI 4.3.1 Fire departments are required by NFPA 1404, *The Respiratory Protection Training Standard,* to have an **air management program (AMP)**. The AMP is the foundation for understanding the relationship between the ACR, the factors that affect the duration of the air supply, effective operation of the firefighting team in an IDLH environment, and the safe exit of the team prior to the low-air warning alarm activating.

All SCBA are equipped with a low-air warning alarm. This alarm is designed to indicate to the wearer that they have used all of the air designated for operation in the IDLH environment and are now consuming the emergency reserve. Low-air warning alarms are required to activate when the cylinder reaches 33% of the rated capacity. Manufacturers have created various types of low-air warning alarms: from a ringing bell, to a vibrating alarm, to an electronic beep or whistle. Firefighters must be able to recognize the sound of the low-air warning alarm for each type of SCBA they may be required to use or hear at an incident scene. A firefighter's survival may depend on their ability to recognize, or have others recognize, the activation of the low-air warning alarm on the incident scene.

Each department is required to have an AMP that should include the following three directives for firefighters using SCBA:

1. Firefighters must exit the IDLH environment before consumption of the emergency reserve begins.
2. The low-air warning alarm indicates that the firefighter is using the emergency reserve.
3. Activation of the low-air warning alarm is an "Immediate Action Item" for the individual and the team.

Another way to look at this requirement is to simply follow the **rule of air management (ROAM)**. The ROAM states:

> "Know how much air you have, and manage that air, so that you leave the hazard area BEFORE the low-air warning alarm activates."

- Know what you have.
- Manage it as you go.
- Exit before the alarm.

The firefighter functions as a member of a team while operating in and SCBA. The team leader must understand the components of the AMP and how each applies in different situations. In all cases the team leader is required to ensure that the team follows the ROAM and exits the IDLH environment prior to the activation of the low-air warning alarm. To accomplish this, the team leader must coordinate the activity of the team in conjunction with the ACR and the air management policy. The team leader may consider assigning roles on the team based on the individual ACR (fig. 10–25).

Fig. 10–25. Company officers of the roof ventilation team quickly review assignments before going up the ladder. Understanding the high and low workload positions during a roof operation increases the effectiveness of the team. (Courtesy of Yukari Horikawa)

The team leader may rotate personnel from positions of high work to positions of low work to ensure that air use is equalized within the team. Team members should understand how to use **controlled breathing** techniques. Team leaders must have an area of safety identified (the exit out of the hazard area), estimate travel time and distance to the area of safety, and determine when the team must begin to exit to reach the area of safety before the activation of the low-air warning alarm.

Breathing Techniques

These techniques are advised to slow down your breathing, get your emotions under control and conserve valuable air. There are many reasons these may be needed, including simply to calm down in normal operations, conserve air due to being lost or trapped in a fire, or giving a RIT time to get to you for rescue. Firefighters do not agree on which is the best technique and have found value in one or more of these, often used in combination depending on the circumstance. You must practice these for them to usable in stressed situations.

The default breathing method that usually occurs when the body is physically challenged or stressed is to breathe heavily and rapidly through the mouth. This is inefficient and can quickly deplete air supply. Determining to bring your breathing under control can best be done by forcing yourself to slow down and breath through your nose. It may take some time to do this when you are breathing heavily but the goal is get under control.

1. **Reilly Breathing Technique (RBT)/hum breathing.** This is simply a normal intake of air and on exhale you slowly hum the air out. This has proven to greatly extend air levels and does not have the CO_2 buildup problems that can occur with techniques like counting or skip breathing. See the "Out of Air" video from firesmoke.org for fuller description.
2. **Box breathing.** This technique moves in a linear fashion that has you breathe in for 3 seconds, hold for 3 seconds, breathe out for 3 seconds, and hold again for 3 seconds prior to repeating. This method is used by the Navy Seals BUD/S program and has proven to have positive impact on blood pressure, heart rate, and respiratory rate.
3. **Belly breathing.** Used by Marine snipers to calm breathing, this technique focuses on using the diaphragm to extend the stomach and retract the same without using the chest. It reduces the amount of air intake and has the same homeostatic effects on vitals as box breathing.
4. **Skip breathing.** The simplest of techniques and most widely used in the fire service, this is simply taking a breath and holding it instead of breathing the next breath. You skip over one subsequent breath each time to help slow yourself down and get control. While this has been effective for some, there is research that indicates this may cause a buildup of CO_2 and disrupt the normal breathing drive. For shorter periods of use the technique has proven effective.
5. **Count breathing.** This is a technique that can be varied depending on the individual and the degree of control needed. You simply breathe in and count to a certain number before exhaling. A derivative of skip breathing, this gives you something to concentrate on if you are trying to get your anxiety down, and the counting seems to provide that distraction for some people.

There are other techniques that are simply variations of these and that work for some firefighters. Your task is to practice them during drills and determine which will best suit you. You'll want to know that prior to actually needing to use them successfully. (Special thanks to Ric Jorge of Tactical Resiliency Training LLC.)

Fireground operations are inherently dangerous, with overexertion, strain, slips, trips, and falls being the two leading causes of injury. Overexertion and strain cause 26.5% of fireground injuries. Further, nearly 23% of fireground injuries are the result of a slip, trip, or fall. These injuries often occur while or following firefighting activities and may often be a result of the fatigue those activities have induced in the firefighter.

Firefighters' SCBA protect the respiratory system during firefighting but increases the physiological burden. Fortunately, the SCBA is very effective at protecting the respiratory system when properly worn. But the weight of the SCBA adds to the work that a firefighter must perform and because of the way it sits on the back it changes the center of gravity and may increase the risk of overexertion strain injuries or slips, trips, and falls. A trend in the fire service toward extended duration SCBA (more than 30 min.) may further increase the physical demand on the firefighter. Extended duration SCBA are typically heavier and may reduce the time before the firefighter becomes fatigued.

In a recent study completed by the research program at the Illinois Fire Service Institute (IFSI Research), 30 firefighters were recruited to take part in study to examine the effects of SCBA and duration of work cycle have on physiological strain, balance, gait, and safety of movement. Firefighters completed seven different conditions with various SCBA (30, 45, and 60-min. standard cylindrical SCBA and a low-profile 45-min. prototype) and durations of simulated firefighting (one or two bouts) in a heated environmental chamber. Four activities were performed (stair climb, hose advance, secondary search, and overhaul) on two-minute work-rest cycles. Subjects also completed an obstacle course designed to test their gait and functional balance prior to, and immediately after the simulated firefighting activities.

Following firefighting activity firefighters had elevated heart rates and core temperatures. The firefighters also generally performed worse in the obstacle course. Contrary to the initial hypothesis, the size of the SCBA had a minimal impact on the firefighters, though there was an important reduction in performance on a functional balance test with the largest SCBA. However, during a second evolution, work output significantly declined while heart rates and core temperatures were higher than the first bout of work. In fact, 37% of the subjects were unable to complete the second bout of work in at least one of the trials that included a two-bout work conditions. Importantly, the firefighters who could not complete the work had lower fitness and higher body mass than those who completed all assigned tasks. This data highlights the importance of the SCBA, and extended use of SCBA, in work performance and safety measures, and it reinforces that firefighters must possess adequate levels of physical fitness in order to be able to perform strenuous or prolonged firefighting activity.

—Dr. Gavin Horn, director of the Research Program at University of Illinois Fire Service Institute and Dr. Denise Smith, professor for Health and Human Physiological Sciences Department/director First Responder Health and Safety Laboratory

Exiting the hazard area prior to activation

The requirement that firefighters be trained in the AMP and exit the IDLH environment prior to the activation of the low-air warning alarm is now the accepted standard across the nation. This change in firefighter behavior, which began with the adoption of the 2006 version of NFPA 1404, is directly responsible for the significant decrease in out-of-air emergencies and close calls. This practice is designed to improve the safety of firefighters wearing SCBA and provide a margin for safety in case of an accident inside the hazard area. Managing air and exiting the IDLH with a margin for safety can help prevent firefighter emergency and Mayday situations. Although rapid intervention teams (RITs) are a required and important fire scene resource, prevention of Mayday situations is always preferable to relying on intervention capabilities at the incident scene.

To exit the IDLH safely, firefighters must operate as a team. The foundation for an effective team is training before the run (event). When operating, always ensure that other members of your crew and other firefighters are not endangered by your activities. Work within the team and respond to the company officer's direction to make the team successful. Maintain your own situational awareness, including air supply and ACR. Maintain team accountability, including the location of, and distance to, your exit and the amount of air required for the team to get there safely. Complete the assigned activity and recognize when your air supply or other factors may prevent you from completing your assignment. Communicate within the team and with the incident commander whenever you have completed your task, are unable to complete your task, recognize that the fire conditions are changing, or identify fireground tasks that need to be completed. Begin exiting the hazard area so you reach the area of safety before the activation of the low-air warning alarm.

On occasion, a firefighter or team may stay too long in the hazard area and activate a low-air alarm on the way out. These situations require immediate action by the individual and the team. At a minimum, department policy should require a firefighter with a low-air alarm to contact the incident commander; identify the team; communicate the situation, the projected exit time, and location; and advise if any additional assistance is needed. Be realistic about your situation. This is no time for bravado. You may literally be betting your life on this decision. If there is any chance you will not exit before your air is exhausted, request activation of the rapid intervention team (fig. 10–26). The low-air alarm emergency should be included in fire department policies addressing firefighter emergency, urgent, or Mayday calls.

Fig. 10–26. Incident commanders can only provide assistance when they know you have a problem. Communicate the Mayday early so the appropriate resources can be sent. (Courtesy of Seattle Firefighter Local 27)

SCBA FAILURES

SKILL DRILL

The modern SCBA is a reliable piece of safety equipment when it is properly maintained. Regardless of the reliability, situations occur when the SCBA partially malfunctions or fails outright. The potential for SCBA failure is one reason firefighters must maintain a margin for error with effective air management technniques while operating in an IDLH environment.

One common cause for equipment failure is the abuse or misuse of equipment. Many photographs or videotapes of fire scenes across America show firefighters who are abusing their SCBA by wearing it improperly. The waist belt holds the regulator assembly for most SCBA. Wearing the backpack without the waist strap attached causes the regulator to receive more abuse than with the waist belt attached. In addition, the failure to connect the SCBA's straps and belts decreases overall safety and should not be allowed. Many photographs show firefighters who have removed the MMR from the facepiece but have not stored it on the waist belt. This practice allows dirt and moisture to enter the regulator and increases the potential for failure (fig. 10–27). In addition, it is advisable to use designated training SCBA when training is likely to result in damage to the backpack or the cylinder. Damage is typical when training to rescue firefighters using firefighter rescue drags or specialty training in confined spaces. Masks that will sustain regular damage through training should be considered expendable, are meant to take a pounding, and should have empty cylinders. These conditions will ensure that, if they fail, the wearer is not exposed to any risk. This is not to be confused with regular training in the SCBA for typical fireground situations or firefighter emergencies, which is critical to firefighter safety and effective fireground operations. As much as possible, keep your front-line SCBA free from damage or damaging impacts that are not directly related to true emergencies.

Unapproved equipment attached to the SCBA can cause the device to fail. First among the list of unapproved equipment is the **cheater** that has been used by firefighters for years. The cheater is an unapproved modification to the SCBA that consists of a tube attached to the inhalation valve on the facepiece. The cheater allows a firefighter to choose between breathing the clean air provided by the SCBA or breathing the contaminated air of the fire scene. This modification violates all safety related to the SCBA and cheats only the firefighter and the firefighter's family. In this way firefighters cheat themselves out of a long life. In this way firefighters cheat their daughters out of a walk down the aisle. In this way firefighters cheat their grandchildren out of a good spoiling. Unapproved equipment should *never* be attached to the SCBA, and fire departments should deal with cheaters effectively. The firefighter should never consider attaching the cheater to their SCBA or their reputation.

Fig. 10–27. Two SCBA temporarily stored on the ground with the regulator exposed. Improper storage can result in water or debris entering the regulator assembly and causing a malfunction. Always store the SCBA with the regulator protected.

The practice of **buddy breathing** is another way the user can cause the SCBA to fail. The practice of buddy breathing should not be confused with using the RIC fitting or other devices attached to the SCBA for the purpose of providing emergency assistance to a firefighter experiencing a low-air or out-of-air emergency. Buddy breathing in this context is the practice of sharing the SCBA facepiece with a civilian or other firefighter. This practice should not be condoned or practiced in any way. Firefighters must maintain their respiratory protection to ensure that they can guide an affected civilian safely from the fire building. Proper air management practices can prevent situations where another firefighter experiences a low-air or out-of-air emergency.

Proper maintenance including appropriate service at regular intervals and daily testing is the best way to ensure that your SCBA will provide excellent and reliable service. Training, record keeping, and effective leadership on and off the fireground prevents SCBA failure as well as the attendant urgent, Mayday, low-air, or firefighter fatality situation (figs. 10–28 and 10–29).

DAILY CHECKS – EVERY DAY

Cylinder:
-Cylinder gauge reads above 4050.
-Cylinder with less than 4050 will be replaced.
-Check condition of the cylinder.
-Damaged cylinders shall be removed from service.
-Cylinder band is fully tightened and latch locked.
-Rubber bumper is in place.

Low Pressure Audi Alarm:
-Coupling nut is hand tight at the cylinder valve.
-Bell clear of obstructions, dirt or debris.
-Quick fill cover is in place.

Pressure Hoses and First Stage Regulator:
-Check for tight fit at the bell and the regulator.
-Check for physical damage to the hose.

Second Stage, Mask Mounted Regulator:
-Follow hose checking for physical damage.
-By-pass knob closed finger tight.
-Regulator free of moisture and debris.
-Store in belt mounted holder.

ICM 2000 / Pressure Gauge:
-Open Cylinder Valve fully to charge system.
-Listen for Audi-alarm bell and ICM 2000.
-Listen for air leaks.
-Check that cylinder gauge and ICM 2000 pressure gauge read within 10% psi of each other.
-Close Cylinder Valve and bleed pressure from system using bypass until low pressure bell sounds at 1175 psi.
-Turn off alarm by pressing yellow switch twice in succession.
-Check that ICM 2000 connector nut is tight at swivel.
-Quick-fill cover is in place.

Backpack Harness and Carrier:
-Check that straps are fully extended and buckles work.
-The quick-fill hose with protective pouch in place.
-Check for cracks or deterioration in the frame.

WEEKLY – AFTER YOUR FOUR OFF

Face piece:
-Lens clean and free of cracks.
-Four adjustable head straps have not deteriorated.
-The nose cup is in place and free of debris.
-Nose cup inhalation valves installed properly.

Pass Device and ICM 2000 Check:
-Don SCBA backpack.
-Open Cylinder Valve fully to charge system.
-Stand motionless until warning tones sound, move PASS to cease tones and assure proper operation of motion sensor.
-Remain motionless and let system go into full alarm.
-Reset by pushing yellow button twice in succession.
-Manually operate ICM 2000 Alarm by depressing red emergency alarm button. Reset with yellow button.
-Check operation of backlight by pressing green button.
-Operate Bypass and listen for air flow, then close.

Operation and System Check:
-Don facepiece.
-Check facepiece seal by holding palm of hand over inlet connection. Inhale, holding breath for 10 seconds.
-Keep inlet covered, breath out to test exhalation valve.
-Connect MMR to facepiece and inhale sharply to start air flow.
-Turn on amplifier and verify operation. Turn off.
-Verify HUD operational with 4 green lights.
-Disconnect MMR.
-Close cylinder and check that the alarm pressure gauge does not show a reduction in pressure, for 30 seconds, due to leaks.
-Bleed off pressure, holding face piece close to HUD sending unit and verify that HUD and ICM 2000 low air lights flash at 1175 psi. then reset ICM 2000.
-Place backpack and facepiece on vehicle ready to respond.

SUNDAY
-Lubricate second stage regulator, belt mounted storage unit and face piece with Christo-lube lubricant.

TRAINING
Video instructions for performing the checks on this sheet are available on the in-web under Training/videos.

DATE	REMARKS – REPAIRS	DATE REPAIRED

FORM 029 Page 2 (10/2004) O:dept\forms\form029

Fig. 10–28. A sample of an SCBA maintenance document. This side of the document lists the minimum daily and weekly checks that must be performed and repair notes.

DAILY AND WEEKLY MASK CHECK

DATE	MASK NUMBER							DATE	MASK NUMBER						
1								16							
2								17							
3								18							
4								19							
5								20							
6								21							
7								22							
8								23							
9								24							
10								25							
11								26							
12								27							
13								28							
14								29							
15								30							
Company No.______ Month______ 20__								31							

Form 029 (9/2003) Indicate "OK" or "R" for repair

Fig. 10–29. This side of the document records the identity of the person who completed the daily or weekly check.

FIREGROUND NOTE

Cheating is an act of lying, deception, fraud, trickery, imposture, or imposition. Cheating characteristically is employed to create an unfair advantage, usually in one's own interest and often at the expense of others.

EMERGENCY PROCEDURES FOR SCBA FAILURE

SKILL DRILL

FFI 4.3.1 **FFI 4.3.5** Firefighters should be trained in and regularly practice for failures of the SCBA. The SCBA is a piece of mechanical equipment. No matter how well built, maintained, or checked, failure is not only possible, *it is predictable.* Although SCBA failures are rare, they do occur. Firefighter training should include the potential causes for SCBA failure and specific procedures to follow for different failure situations. Each type of SCBA should include specific emergency procedures in the owner's manual. Fire departments should ensure that each firefighter can demonstrate all emergency procedures associated with each SCBA that may be used. Although it is impossible to address all of the emergency procedures for all the manufactured styles and types of SCBA, some failure considerations are identified in the following text and a basic procedure for a cracked facepiece can be found in fig. 10–30.

One category of problems can be attributed to operating the SCBA in low temperature situations. Firefighters should be aware that low temperatures can cause freezing and icing problems. Icing can occur on the cylinder. The release of compressed air from the cylinder causes the remaining air to cool. The cooling properties of expanding air can also cause icing to occur on the hoses connecting the cylinder to the regulator assembly. In some SCBA, this icing can negatively affect the low-air warning alarm and cause a partial or complete malfunction. Such a malfunction can cause a firefighter who is not practicing air management to run completely out of air in the hazard area and may lead to significant injury or death. Mechanical failure like this is another reason firefighters must manage their air.

MINOR CRACK OR LEAK IN FACEPIECE

- Don't PANIC. The benefits of remaining calm cannot be emphasized enough when dealing with an emergency procedure.
- Leave facepiece on. The facepiece continues to provide protection to the lungs, eyes, nose, mouth, and skin. With a minor leak, leaving the facepiece on provides the best option.
- Place hand on facepiece/regulator and press against face. A leak from the facepiece from a minor crack will act similar to an inadequate fit of the facepiece seal. Due to the positive pressure of the system, excess air will leak out from within the facepiece. Pressing the facepiece against the face helps to alleviate the option that air is leaking between the skin and the seal. It helps to ensure that a leak is coming through the facepiece itself.
- Conserve air by covering crack with hand. The air coming out of the regulator is slightly above the atmospheric pressure outside of the facepiece. This helps to ensure that there is a positive pressure within the facepiece and thereby neglecting contaminated air from entering through a small hole or crack. However, if the crack/hole is not covered, the air in the cylinder is depleted at a faster rate.
- Notify officer and leave with another member. Consider your department's policy on accountability and Federal or State OSHA laws, when a member enters or leaves an IDLH.

MAJOR CRACK OR LEAK IN FACEPIECE

- Continue to cover damaged area.
- Press the manual shutoff after each breath. If air still seems to be leaking out, excess air is being lost while a hand covers the crack/hole. Engage the manual shutoff after each breath. This will thereby limit the amount of air from the cylinder that will be lost due to the positive pressure of the system.
- If manual shutoff will not release on inhalation, control air flow using purge or bypass valve. If a gloved hand over the leak does not cover enough of the crack/hole, the inhalation may not be strong enough to activate the flow of air out of the regulator. Therefore, continue to have the manual shutoff engaged and utilize the purge or bypass valve to allow enough air to enter the facepiece. A partial opening of the purge or bypass valve may be enough for a breath, then close the valve after each breath. SCBA manufactures have their purge or bypass valves on different sides. Check yours to ensure what hand will cover the leak and what hand will operate the valve.
- Notify officer and leave area immediately with a partner. Consider your department's policy on the urgency of these two emergencies in the use of your PASS and the necessity of a Mayday radio transmission.

Fig. 10–30. Cracked facepiece emergency procedure (Jarod Blake, FDNY, FDIC 2007 instructor)

Low-temperature situations can also cause fogging or icing of the facepiece. The design of the positive-pressure facepiece directs the incoming air across the front of the glass lens to minimize internal fogging. The cold air causes the glass to become cold, which can increase the tendency for moisture to condensate on the outside of the facepiece obscuring the vision of the firefighter. Fogging of this type can be easily dealt with by wiping the facepiece with the gloved hand. Firefighters should be aware that this may scratch the glass or deposit debris on the lens. Anti-fogging compounds can be used to minimize the effect of internal or external fogging within manufacturers' specifications.

Low temperatures can also cause high-pressure hose leaks with the contraction of the metal. Firefighters should take extra care to ensure hose connections are tight enough to prevent leaking during low temperatures but still loose enough to operate effectively when warm. A leaking low- or high-pressure fitting can be identified by listening carefully when charging the system with air during the daily check. Leaking SCBA that cannot be fixed should be placed out of service and identified for repair. Unexplained increases in the ACR for a firefighter may also indicate leaking. A leaking hose is cause for immediate exit from the hazard area.

High temperatures can present challenges to the firefighter using an SCBA (fig. 10–31). Although the ambient air may be hot, the breathing air supplied to the firefighter is relatively cool. This occurs because the physical properties of compressed air cause it to cool as it expands. Firefighters must be aware that the temperature around them may be quite high even though they are comfortably breathing fresh air. The facepiece lens is often the weakest link in the SCBA when exposed to heat. Training programs should include information so that firefighters can accurately identify when high temperatures exist in the IDLH environment and how to deal with them effectively when they occur. This training should include situations where rapid temperature changes can be considered indicators of extreme fire behavior, such as rollover or flashover, and the potential for serious injury or death.

Fig. 10–31. Firefighters exposed to high-temperature fire gases while making entry or operating in an IDLH environment are well protected when they use the SCBA properly. (Courtesy of PM Brian Smith, Seattle, Washington, Fire Department)

Many SCBA include built-in or add-on communications devices. Communications devices often improve the ability to talk within the team or over the radio (fig. 10–32). They can present their own challenges and limitations. For example, if many devices are in the same area or the operator is in an area with thick or reinforced walls, the transmission may be compromised. It is also possible for feedback to occur if there are many communications devices within the same small area at a fire. The ability to communicate effectively is critical to fireground safety. All firefighters should be trained and familiar with any communications device that is provided for use during an emergency. Firefighters should also understand alternate techniques to improve communication through a radio microphone. Many firefighters have success by placing the microphone against the throat or against the facepiece glass while talking. One factor that will always improve communication is a calm and steady voice.

Out-of-air situations, although rare, can be fatal. Firefighters should avoid out-of-air situations at all costs. It is possible to prevent out-of-air situations through good teamwork, good situational awareness, good training, and good air management practices (ROAM). Low-air situations should be considered an emergency and an urgent or Mayday should be transmitted. Firefighters' training should include methods to conserve air during low-air emergencies to maximize survival time while finding an exit or awaiting the RIT team. In all cases consider reducing physical activity to reduce respiratory demands. Breathing techniques can be reviewed in the Air Management Program section of this chapter. In the rare circumstance that a firefighter runs completely out of air, the Mayday should already have been called and help should be on the way. Once you are out of air your survivability is in jeopardy. Clear thinking and calm action may make the difference. Remain calm, and do not give up. Whenever possible, leave your facepiece on. This will allow the RIT team to connect to your universal RIC connection and provide breathable air as quickly as possible.

Fig. 10–32. An example of a communication device attached to the facepiece. Firefighters should be well-trained in how to communicate effectively while operating in their SCBA. (Courtesy of Mine Safety Appliances)

Here is a list of last-resort options for dire situations:

- Leave the facepiece on, remove the regulator, and "filter breathe" the ambient air. Use your hood or glove to filter and possibly cool the smoke. Breathe as shallow as possible. Many fire service instructors will advise you to keep low to the ground. However, this may be a bad option if the carpet has been heated to the point that it is off-gassing. If the carpet is off-gassing, the primary component will be hydrogen cyanide. One breath may be enough to incapacitate.
- Attempt to find areas where there is less contaminated air to breathe. These may include toilet tanks, wall or floor void spaces, washers and dryers, windows, ventilation shafts, and sewer vent pipes. It is critical to remember and drill on the reality that there are many sources of air that can be utilized to get a few more breaths. Practice this and embed these ideas in your mind.

Use of universal rescue connection assemblies

SKILL DRILL

All NFPA 1981–2002 approved SCBA come equipped with a URC assembly fitting. The **universal rescue connection (URC)** is a male quick-fill inlet found on the back of the SCBA by the cylinder valve. The URC is for use by RICs for emergency filling operations only. Some SCBA manufacturers also equip their SCBA with a shoulder-mounted quick-fill system.

Fire departments must develop and use standard operating procedures (SOPs) for URC use by RICs or RITs as well as for emergency operations. Firefighters must understand how to use the URC, under what circumstances to use the URC, and when they should not use the URC. One example is the increased existence of firefighter air replenishment systems (FARS), the air standpipe, in larger structures such as high-rise, big box, and tunnels (fig. 10–33). Firefighters are able to use the quick-fill panels of FARS to trans-fill their bottles without removing either their bottle or facepieces during emergencies (fig. 10–34). Of course, a solid hands-on training program focusing on the URC and rapid intervention is critical to the success of firefighters using the URC correctly.

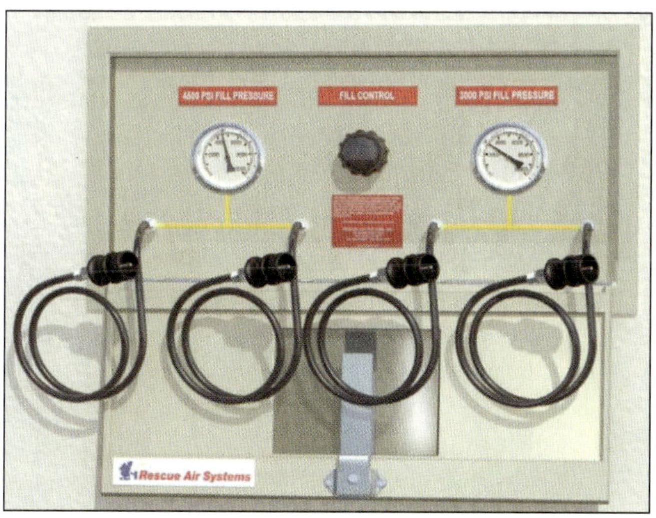

Fig. 10–33. Firefighter air replenishment systems should be in all large structures.

Fig. 10–34. Firefighters are able to use the quick-fill hose and its standard URC connection to refill their bottle without taking off their SCBA.

The following is a step-by-step method for using the URC in emergency situations. It should be noted that RICs must use a separate air supply (a secondary air supply), such as a **rescue air kit (RAK)**, which is a full air cylinder, a quick-fill hose, and an extra facepiece. These kits can be purchased from the SCBA manufacturers. *NIOSH does not approve the use of the URC assembly to transfer air from the cylinder of one SCBA to another SCBA. Failure to follow this warning can result in serious injury or death.*

1. To attach the quick-fill system hose of the secondary source, such as a RAK, to the URC:
   - Turn on the secondary air source, fully opening the RAK cylinder valve.
   - Remove the rubber dust cap from the male inlet fitting on the URC found on the back of the SCBA of the firefighter needing air. Be sure that the firefighter's SCBA cylinder valve is fully opened.
   - Remove the rubber dust cap from the female fitting on the quick-fill hose.
   - Push the female fitting of the quick-fill hose on the male fitting of the URC until it snaps into place. Pull on the hose to be sure the fitting snapped into place. Filling begins when the female fitting is snapped on the URC.
   - After approximately 45 to 60 seconds, pressure equalizes between the secondary air source (such as the RAK) and the SCBA of the firefighter needing air.

2. Compare the SCBA pressure gauge of the firefighter receiving air to the secondary air source cylinder gauge. If the readings are the same, the pressure is equal.
3. Disconnect the quick-fill hose after the transfill is complete.
4. Remove the firefighter who needed the emergency air outside the hazard area as quickly as possible.

INSPECTION AND MAINTENANCE

SKILL DRILL

Basic maintenance on SCBA is very important. If the unit is not checked on a regular basis, there is the chance it will fail at a critical moment on the fireground. To inspect a cylinder, the following is performed. Remove the unit from its holder or container. Check the straps and backpack for any excessive wear or damage. The high-pressure hose between the regulator and the cylinder should be checked for any damage or excessive wear. Next is the cylinder. This outside should be checked for damage to the fiberglass or metal. Make sure that there is no damage around the cylinder valve. The cylinder valve should read full. Open the cylinder valve and check it against the regulator valve. If the unit is an MMR check the low pressure hose from the regulator to the pressure reducing device. Activate the integrated PASS device, if available. Connect the regulator to the face piece and take several breaths. Once this is complete close the cylinder valve, purge the air from the system and make sure the low pressure alarm activates. Close the purge valve and restore to service.

Cleaning an SCBA

SKILL DRILL

FFI 4.5.1 In order for firefighting equipment to work properly, it must be maintained and cleaned on a regular basis. To clean an SCBA unit, do the following. Remove the face piece from the regulator and place on the side. Then remove the cylinder from the SCBA harness. Rinse all parts of the unit with clean water to remove any debris. A soap and water solution should be used to scrub the cylinder and harness with a still bristle brush. Rinse the harness and cylinder off and set aside to dry. To clean the face piece, first mix a solution of mild soap and water in a 5-gal (19 L) bucket. Place the face piece in the solution and allow to soak. Clean the regulator using the solution and a soft-bristle brush, making sure soap doesn't get inside the regulator. Rinse the face piece and regulator with clean water and set aside to dry. When components are dry, reassemble and inspect the unit before placing back in service.

Refilling a bottle with a cascade system and a compressor system

SKILL DRILL

Before any cylinder is refilled, there is certain information that needs to be obtained about the cylinder. The necessary forms to be used should be spelled out in the department SOPs. Information to be filled out will be cylinder type, required air pressure to be placed in the cylinder, cylinder serial number and identification number, the hydrostatic test date of cylinder, date of the refill. Once this is recorded, the operator needs to check the condition of the cylinder, date of cylinder manufacturer, hydrostatic test date. After these items have been checked, the cylinder can be filled. To fill the cylinder, place the air cylinder in the shielded refill station, then fill the cylinder slowly to prevent overheating. Once you are sure the cylinder is full, but not over-pressurized, remove the cylinder from the refill station.

NOTES

1. Gagliano, Mike, Casey Phillips, Phil Jose, and Steve Bernocco. *Air Management for the Fire Service.* Tulsa, OK: PennWell, 2007.
2. Wallace, Deborah. *In the Mouth of the Dragon.* Garden City Park, NY: Avery, 1990.

LESSON FROM THE FIREGROUND

The first arriving company observes heavy smoke pushing from the first floor apartment of a two-story garden style apartment complex as they turn the corner. As the apparatus pulls to the front of the building a woman with a baby in her arms appears in the window of the second-floor apartment above the fire apartment. The officer jumps from his seat on the apparatus and races into the building to make a quick rescue. He thinks he can hold his breath long enough to make the grab and exit quickly. The unfortunate reality of this poor decision is readily apparent as the officer is seen stumbling out of the doorway gasping for air.

One of the firefighters from the first due unit has donned an SCBA while the officer ran into the building. He passes the officer who is stumbling out of the doorway and quickly makes his way to the trapped occupants. The SCBA equipped firefighter is able to quickly remove the woman and her baby to safety. Once outside, the firefighter sees the company officer on the rear tailboard sucking on an oxygen mask and being readied for a trip to the emergency room. The lesson is indelibly engraved in the firefighter's memory—always take the time to don the SCBA.

QUESTIONS

1. Firefighters are provided a known quality of breathable air for operations in IDLH or potentially IDLH. What does IDLH stands for?
2. The IDLH environment is produced by the combustion of the__________.
3. The combination of _____ and _____ substances is IDLH for firefighters.
4. What is likely to be found in abundance compared to any other product in today's homes?
5. What gas is a common cause of death to firefighters?
6. The respiratory tract is protected by SCBA use from toxic and ___________ gases.
7. The leading cause of firefighter deaths is ____________________.
8. When should medical screening start? How often should it be done?
9. When a 2,216 psi bottle reaches ___________ psi, it should be recharged. When a 4,500 psi bottle reaches ____________ psi, it should be recharged.
10. How often do cylinders have to be hydrostatically tested?
11. Fit testing is required by NFPA _____________.
12. Low-air alarms are required to sound when cylinders reach _______ %.
13. ROAM stands for __________ ______ __________ __________.
14. ROAM states that you should leave the hazard area ________________ your low air alarm activates.
15. Skip breathing is when you take a small breath, hold for ___ seconds, take in another breath, hold for ___ seconds, exhale, and repeat.

Firefighting Tools

by Michael N. Ciampo, with Rick Fritz

This chapter provides required knowledge items for the following NFPA Standard 1001 Job Performance Requirements:

FFI 4.5.1

FFII 5.5.4

This chapter contains Skill Drills. When you see this icon, refer to your Skill Drill book for step-by-step instructions. SKILL DRILL

OBJECTIVES

Upon completion of this chapter, you should be able to do the following:

- List and describe general-purpose firefighting tools and equipment
- List and describe the safety considerations for each tool
- Identify tools used during forcible entry
- Identify tools used during overhaul
- Identify tools used for ventilation
- Identify tools used in fire attack
- Identify the different types of saws used in firefighting
- Identify auto extrication and rescue tools used by firefighters
- Describe the maintenance requirements for each type of tool and equipment
- Describe how to clean and inspect each tool prior to and after use

INTRODUCTION

As you can see from your work in the previous 10 chapters, firefighting is a labor-intensive, tool-dependent occupation. Firefighters who master the use of firefighting tools will do well in the profession. Tools accomplish various tasks on the fireground, including cutting and striking, pushing, pulling, and prying. By effectively using tools as extensions of their bodies, firefighters may multiply the force or leverage applied in a given circumstance while quickly and safely accomplishing the task.

This chapter examines the uses of basic firefighting tools during fire responses. You might not use all the tools we describe during the same fire, but chances are good that you will use at least one or more before returning to the firehouse. Many firefighters follow a simple rule: two hands, two tools, meaning if you're getting off the apparatus to perform truck company duties with a set of irons, or a smaller hand tool and hook, you'll be able to accomplish almost anything that you come up against on the fireground. Subsequent chapters in this book will cover the use of these tools in greater detail.

Basic firefighting tools are categorized by what they allow firefighters to do. The first group is hand tools that cut.

TYPES OF HAND TOOLS

Cutting tools

SKILL DRILL

Engines and trucks carry two types of tools that strictly cut: one is a type of axe, the tool most recognized within the fire service, and the other is different-sized bolt cutters.

Pick-head axe. A **pick-head axe** (fig. 11–1) is a steel head single-bit axe, which means it has only one cutting blade attached onto a fiberglass or wood handle. This blade should maintain a sharp cutting surface that is free of nicks and dings. On the opposite side of the blade is a pick. This triangular point can pierce materials such as roofing when it is necessary to create a starting point for cutting. It can also shatter materials such as glass, plastic, and fiberglass.

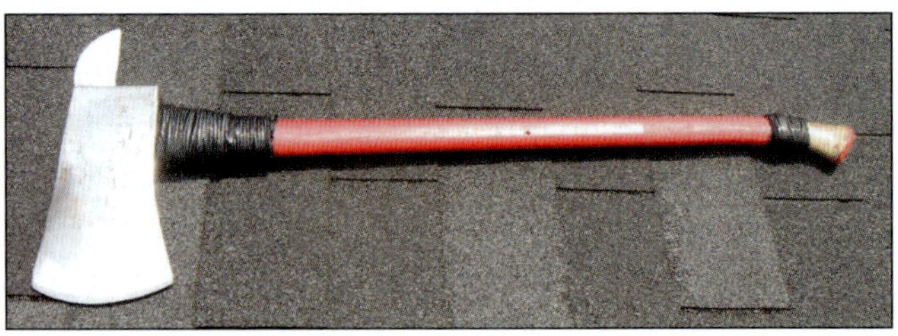

Fig. 11–1. Pick-head axe

FIREGROUND NOTE

These tools are as important to a firefighter as weaponry is to a soldier. Without them, the enemy is going to kill you.

An ideal axe head weighs 6 to 8 lbs (3.6 kg). A heavier head will wear out the user before the task is accomplished, and a lighter head means the user must strike harder, which also will exhaust the user and make the axe hard to control.

Accuracy is the key to effectively using any axe. Swung properly, an axe will cut every time. If a user can't place an axe blade in the same place with each swing, he or she will never open a hole.

To maximize your effectiveness, keep your eye on the area in which you want to drop the axe blade. Cutting with an axe is as much hand-eye coordination as strength. Before using any axe, clear the area around you of any obstructions. Also, it is recommended that the roof shingles be removed from where you are going to cut to speed up the cutting process. This reduces the friction between the roof surface and tar build-up on the axe's head. Remember, axes can severely injure or kill anyone they hit, so always use caution when operating one.

To cut using a pick-head axe, grasp the handle firmly and find a stance that is comfortable and compatible with the surface on which you are working. Try to spread your feet and center your body weight. Flex your knees a little. Hold the axe handle where it feels comfortable, but not close to its head. Your stance should feel good and give you balance. You must be able to rock back and forth on your feet while swinging to maintain your balance. Gravity, the force of the wind, bad weather conditions, a burning structure, and the weight and restrictions of your turnout gear will all work against you. Your hands should go no higher than your shoulders on the upswing. Do not let the axe head go behind your head and shoulders. The axe's weight will pull itself from your hands and create a dangerous situation. Letting the axe drop into place will allow the weight of the axe to do the work. As you swing, slide one of your hands along the shaft of the handle to meet the other hand that is grasping the bottom. Don't try to push the tool or swing with one arm stiffened because you will tire quickly. An axe swung with too much force might slip from your hands and plunge into the hole endangering crews working below you, or you could miss your mark and hit another firefighter.

Concentrate on accuracy. Swing the axe so that the blade strikes the surface you want to cut at a slight angle. A dead-on strike might cause the axe to bounce back without cutting the material. If the blade sticks in the material, don't pull because you'll fall. Work the handle of the axe up and down a few times to release the blade. For pick-head axes that are deeply stuck, release the axe by gripping the handle with one hand and grasping the pick with the other. Work the axe head back and forth to release the blade. Another method is to place your boot, a Halligan bar, or piece of debris under the axe handle near the head to create a fulcrum. Push the opposite end of the handle down to force out the axe head from the material.

The pick side of the pick-head axe allows a user to make a purchase point, or starting hole, in materials that are too hard or that have too much spring to allow immediate cutting. To use the pick, flip the axe over and set your stance.

The axe will swing differently because the mass of the axe head will be at the top of the axe. As you swing as though you were using the blade side, maintain a firm grip on the handle to prevent the tool from rolling to the blade side. Keep your eye on the spot you want the pick to hit, and let the weight of the tool drive the pick.

Like the blade side, the pick might get stuck. To release the pick, don't pull. Push the handle forward a little, and then pull it back toward you. If the pick still won't release, lower the handle parallel to the surface you are cutting and rotate the handle 15 degrees in either direction or place a boot, tool, or piece of debris under the handle as a fulcrum as mentioned above.

It is important to note that a Pulaski axe has a wider pike at one end, but it is actually a hoe. This tool has an axe head at one end and a hoe on the other. It is used during wildland firefighting operations for making fire stops or overhauling. It should not be confused with a pick-head axe.

Bolt cutters. **Bolt cutters** (fig. 11–2) are another cutting tool firefighters use. They provide a quick and relatively easy method of cutting through various types of metal. Although other tools are available to firefighters for cutting materials such as chain, lock shackles, fencing, etc., bolt cutters, when used properly, are fast and efficient. Most important in using bolt cutters on the fireground is having the right set for the material to be cut. Some departments keep different types of cutters on the apparatus to cut materials of various thicknesses. The cutting surface of the cutter may not be suitable for case-hardened material such as some chains and lock shackles, but it is better at cutting fencing, light locks, steel rod or rebar, and small cable.

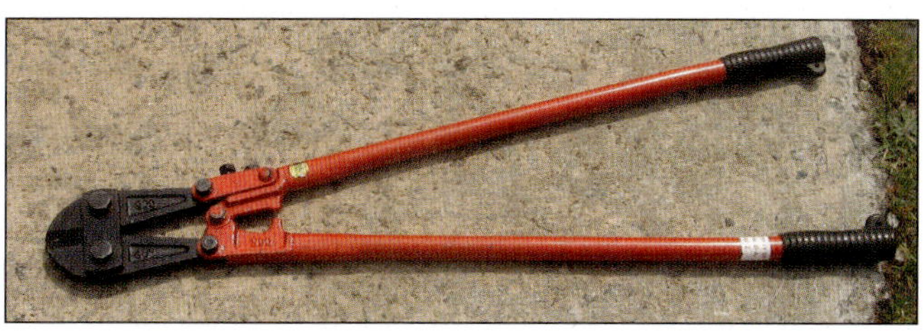

Fig. 11–2. Bolt cutters

As a firefighter, be wary of dielectric bolt cutters and wire cutters, which can be recognized by heavily insulated handles, or they may have wire separating tines attached to their tips. Firefighters don't cut energized electric lines of any type, ever. This includes cutting automobile battery cables of any type, including those on hybrid cars, because the voltage can be lethal. Bolt cutters are made of high-quality carbon steel with long, powerful handles (not fiberglass) with rubber grips. Handles should be long enough to make cutting easy. The shorter the handles, the less leverage firefighters will have for cutting the material.

When using bolt cutters, evaluate the thickness of the material to be cut. Ensure that you have the proper bolt cutters that will bite into and cut completely through the material. Wear eye protection because bolt cutters can launch the cut-off end of chain, bolt head, or whatever you cut. Protect yourself from flying debris. Never intentionally cut a loose end. When cutting cable or some other material, make sure of the end result before you cut. Cutting cables or cords may release an object being held up or in a tension situation, such as a garage door spring.

When performing **forcible entry**, insert the shackle deep into the cutting jaws, then cut one leg of the shackle at a time. Cutting too close to the lock body might jam it if you don't cut all the way through or if the cutters twist. Cutting high on the shackle gives you another place to get a purchase if you can't get the leverage for the first cut. Using a firefighter on each handle is dangerous. The bolt cutter blades might dimple or bend, and the hinge mechanism of the bolt cutters might self-destruct under the tremendous pressure and force. Bolt cutters are not designed to cut case-hardened padlocks found on some security gates. If you encounter case-hardened materials, select another tool, such as a power saw.

Cutting and striking tools

SKILL DRILL

The next category of tools is a step up from single-use tools; these tools allow firefighters to perform two functions: cutting and striking. These tools include the flat-head axe and splitting maul.

Flat-head axe. The **flat-head axe** (fig. 11–3) is a steel head, single-bit axe attached to a fiberglass or wood handle. Its blade should also be maintained with a sharp cutting surface that is free of nicks and dings. Unlike the pick-head axe, the opposite side of the flat-head axe has a flat striking surface that can be used as a striking tool like a sledgehammer. The most common use for a flat-head axe is forcible entry operations. The axe is most commonly "married" to a tool called the Halligan. When firefighters encounter a locked door, the Halligan is inserted in between the door and jamb. It is struck by the firefighter carrying the axe on the command: "hit or strike." During these operations the axe is not swung like it is for cutting roofs. Firefighters should maintain their grip on the tool with one hand on the end and the other about 3 in. to 4 in. down from the head. This prevents them from hitting their hand on the Halligan if they overswing when striking. As they pivot and swing they will strike the Halligan. It is not a baseball swing but more of a controlled short swing. When operating with the axe, only strike when called for; to prevent injury, do not swing the axe or hit the Halligan if the tool is moving because this will prevent injury to either member.

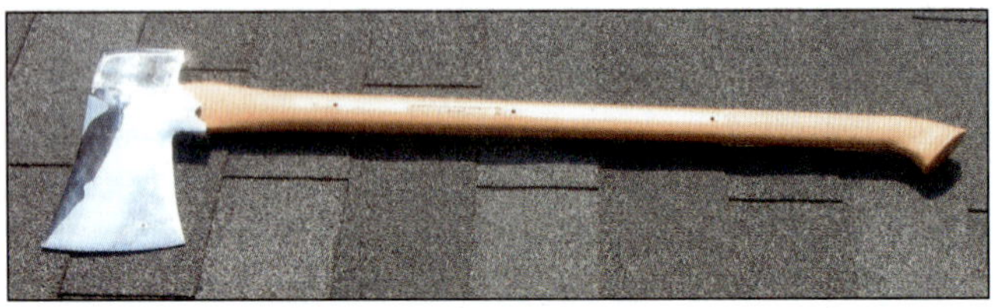

Fig. 11–3. Flat-head axe

To use the flat-head axe as a cutting tool, see the previous directions for the pick-head axe. When using the flat-head axe as a striking tool, allow the weight of the tool to do the work. It is not necessary to swing the tool in a full arc to get the benefit as a striker. Line up the flat striking surface against the tool or object you will strike. Instruct the firefighter holding the tool you will strike to stay still.

Arrange your stance so you can effectively and strongly pivot your hips and hit your target. By putting your body weight behind the flat-head axe, you will strike and efficiently drive the tool into or through the target.

The flat head axe also has the following fireground uses: 1) its flat side can be used to break brick or block when performing ventilation or opening up a sealed building, 2) its blade can be placed behind carriage bolt heads to shear the heads off when struck by another tool, and 3) its blade can be used to cut lightweight metal sheeting or thin metal-skinned door or tin building walls.

Eight-pound splitting maul. The **eight-pound splitting maul** (fig. 11–4) is another tool that will complement your cutting and striking capabilities on the fireground. The maul has a sledgehammer face on one side and a wedge-shaped splitter on the other side. The splitting maul can replace the flat-head axe in some forcible-entry situations. Its weight is effective, it provides cutting and striking surfaces, and it is inexpensive and readily available.

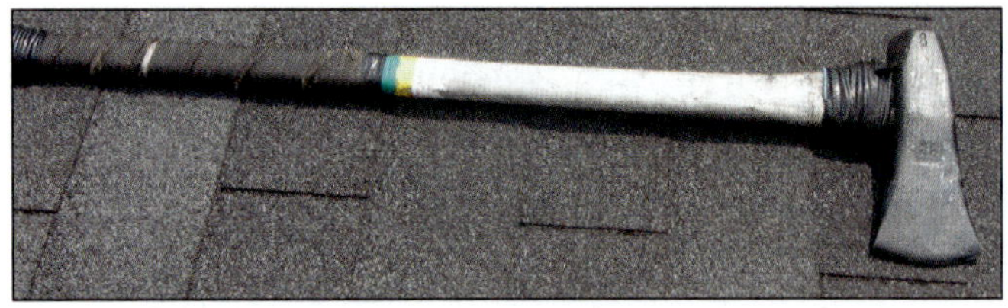

Fig. 11–4. Eight-pound splitting maul

The splitting maul is used primarily the same way as a flat-head axe. It is swung and used the same as a striking tool. The biggest difference is that the maul does not cut. It is designed to split wood, so it is a bit more brutal. The splitting maul makes short work of roof ventilation. Once the site for the vent hole has been selected, a firefighter can smash a hole through the roofing material, rather than cut it. The splitting maul easily splits oriented strand board (OSB) sheeting and even plywood. No shingles have to be pulled, and because of the shape of the head, a simple twist of the handle to the right or left will free the tool head.

Using this tool is much like using an axe; a proper stance must be taken to properly swing it. Hold the splitting maul as you would an axe. The head is a little more off-balance because the bulk of the weight is to the rear of the head.

When using the cutting edge, expect the tool's tendency to invert to the striking side. Hold the handle tightly and prevent it from flipping around in your hands when you swing and strike. Be prepared to shift your weight to accommodate the off-balance head.

As with every cutting or striking tool we have discussed, accuracy is key. Do not swing giant strokes. Slide one hand down the handle no more than shoulder high to meet the other hand that has a firm grip on the bottom of the handle. Allow the weight of the tool to do the work.

When the maul gets stuck, free it by twisting the handle to the right or left about 15 degrees. Pulling will cause you to lose your balance and fall. Remember, this tool is designed to split wood, not cut like an axe. A heavy, deep stroke might bury it into the material being cut and make for difficult removal.

When using the splitting maul or flat-head axe as a striking tool, allow the weight of the tool to do the work. It is not necessary to swing the tool in a full arc to get the benefit as a striker. Line up the flat striking surface against the tool or object to be struck. Arrange your stance so you can effectively and strongly pivot your hips and hit the target.

Striking tools

SKILL DRILL

Types of striking tools. A **sledgehammer** (fig. 11–5) is used for one purpose: to strike an object or another tool. A sledgehammer has striking surfaces on both sides of the head. It is often mistakenly called a maul, which has a striking surface on just one side of the head; there is another tool on the opposite side. The splitting maul is a true maul. The sledgehammer is a big, heavy hammer that comes in numerous sizes and weights (from 8 lb to 20 lb).

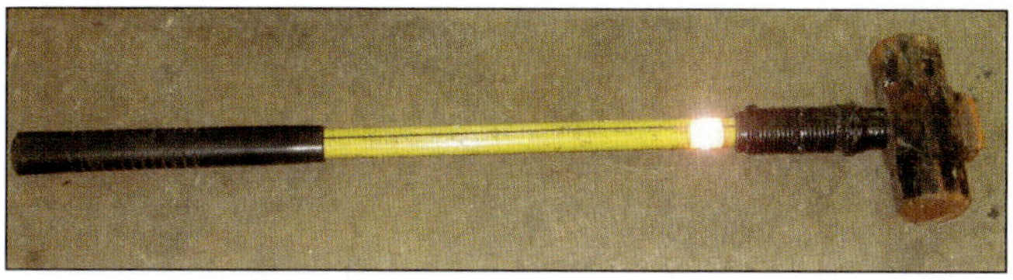

Fig. 11–5. Sledgehammer

When you select a sledgehammer as a striking tool, the weight of the tool is important. Whether driving a stake into the ground to hold a deluge gun, opening up brick or blocked up windows, breaking concrete at a collapsed building or striking a Halligan bar in breaking a lock, you must determine how hard to swing. If you have to swing too hard, get a heavier hammer. Swinging hard is not conducive to swinging accurately. Weight is the key. The mass of the tool determines how effective it is, as well as your ability to swing and strike effectively and accurately. Special care must be taken when you swing a sledgehammer, regardless of its weight. There is no need to swing a sledgehammer over your head and out of sight. Controlling the tool is paramount for your safety and the safety of others.

Miscellaneous cutting and striking tools

Many manufacturers and firefighter's creative innovations and ideas have come together to create combination striking tools. These tools offer some of the same characteristics as pick- and flat-head axes but with special and updated features. A few you may see are the pig, the badaxxe, The TNT tool, and the lock slot 8.

The **pig** has a large metal flat-head striking surface on one side of the head with a thicker pike end on the opposite side. The pike has a slot in it for the forks of the Halligan tool to be married to it. A blade exists under the pike end for cutting materials.

The **badaxxe** has a tapered flat head with a gas valve shutoff built into it. The head has an indentation to hook onto roof sheathing or a structural element to let the tool act as an anchor for firefighter self-rescue tactics. Forged into the tool's metal handle are tie points for a rope assisted self-rescue, a hydrant operating nut, and slots for a carrying sling.

The **TNT Tool** is a combination axe and sledgehammer with the opposite end having a modified pike pole head on it where the point end has been replaced with a wider pry bar end. The axe and sledge end are forged with a "D" handle cut out in it. This allows firefighters to grip the tool there or marry a Halligan tool to it in this spot. The heads are mounted on a fiberglass handle.

The **lock slot 8** is an 8 lb flat head axe that is forged with two slots into the head that allow the Halligan's forks to slide into the grooves and be married to the axe tightly. This allows the firefighter to more easily carry the pair.

The **all axe** is a modified pick-head axe with a metal handle on the opposite end of the head with a Halligan tool's forked end attached to it. The top of the axe head has a welded steel bar on it for use as a striking or battering ram surface. The head also has a metal chain link welded to it for attaching a rope so the tool can be used to break windows from the floor above.

The **pry axe** is a small combination tool often carried by officers. Its head is a combination cutting edge and pike end much like a pick-head axe but on a smaller scale. The handle is a two-piece design, which can act like a slide hammer or be used as two separate tools. Opposite the head of the tool on the handle is a forked end. The tool can also be put together in a 90-degree design for additional uses.

Tool safety

SKILL DRILL

Swinging your tools. You should never lose sight of the tool during your swing. That means no over-the-head or way-around-sidearm, off-behind-you swinging. Keep the tool head in site and maintain good hand-eye coordination to deliver the maximum effective blows on target. Missed blows are just that many more strikes you'll have to make to accomplish the job, and too many missed blows can damage a tool. If you are swinging to cut a roof, let the tool's weight, momentum and speed glide through your hands to assist cutting the material. If you are performing short strokes, use caution and ensure the tool your hitting isn't moved prior to swinging.

Carrying your tools. Safety is a number-one priority at all times on the fireground. Axes, mauls, and sledgehammers are dangerous tools that can do great bodily harm if they are not handled carefully. The safest way to carry an axe, maul, or sledgehammer is to grasp the handle close to the head of the tool with the handle pointing away from you. You can use the handle to gently clear your path without hurting anyone with the steel tool head. You can also use the tool as a search tool when you carry it this way. The handle will act as a blind man's cane, allowing you to extend your reach safely during a search. An axe can be carried with the head handle of the tool tucked under your arm, blade sitting in the firefighter's hand and facing backward, pick facing forward. Hold your hand over the pick. A flat-head axe may be carried under the arm with the blade facing backward, away from the body. If you carry either axe in a belt, always sheathe the axe. Never crawl with an axe in your belt. The weight of the axe head will cause it to slide out or the tool will get hung up on furniture or an object when searching.

FIREGROUND NOTE

Prior to opening up walls and ceilings, ensure that the power has been shut down to the occupancy you are working in to avoid being electrocuted when using hand tools.

Prying tools

Next to cutting tools for getting to a fire, the most important group of tools can pry or increase your leverage. They are pry bars, claw tools, and Halligan-type bars, which are used in forcible entry and overhaul operations, among others.

Pry bars. **Pry bars** are the ultimate in simplicity because they are basic, true levers. There are two types. **Pinch-point bars** (fig. 11–6) have only one beveled side. **Wedge-point bars** (fig. 11–7) have a bevel on both sides of the bar, forming a wedge point. Both bars are available in various lengths and sizes, dimensions, and weights. Both inexpensive pry bars are limited-use tools because many fire service tools work more efficiently. Nevertheless, these ultimate levers should not be removed from service. Learning how to use them will equip you to handle many situations. Their use in collapses, heavy rescues, and some confined-space, industrial rescues cannot be matched by any other tool.

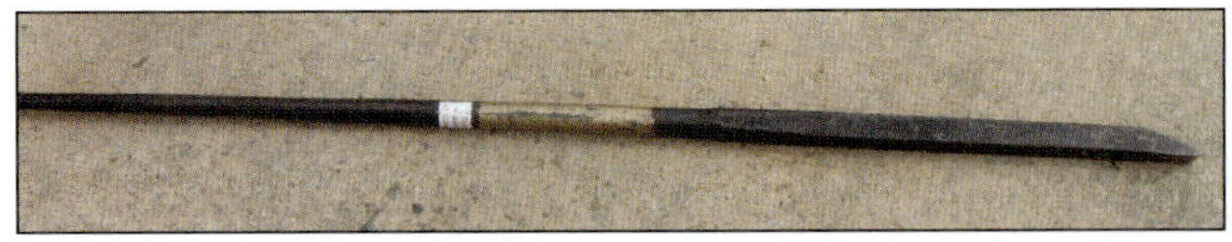

Fig. 11–6. Pinch-point bar

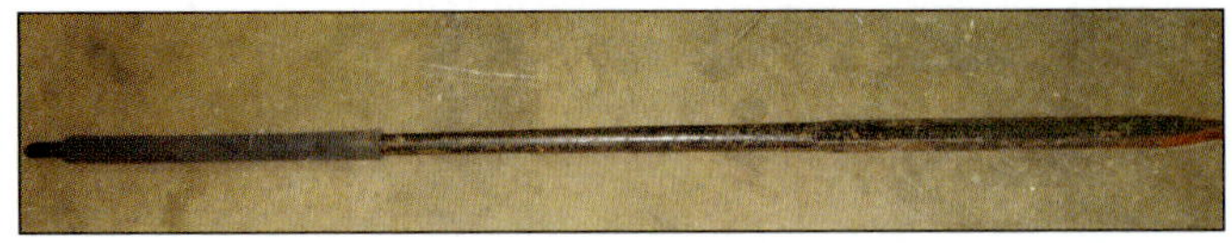

Fig. 11–7. Wedge-point bar

A big advantage pry bars have over other tools is their length and narrow profile. These tools can be used with a proper fulcrum to raise collapsed material, machine parts, automobiles, trees, and other debris that has fallen or blocked a means of egress.

A firefighter may use a pry bar when stabilizing a car with cribbing by using the cribbing as a fulcrum. Firefighters can gently move or lift a vehicle by slipping cribbing materials under the frame.

Pinch-point bars are excellent tools for prying up wood floors. By inserting the tool bevel-side-up into the seam of a wood floor, a user can drive the tool in and lift the first piece of flooring. Then, using the tool as designed, the user can apply leverage and the flooring will come up. Two firefighters, each armed with a pry bar, can quickly remove many floor surfaces this way. Once the first piece is removed, the rest will follow easily.

Pry bars have a variety of uses, none of which has to do with using the tool as it was designed.

To make firefighting operations safer, you may use the tool as a securing post for ground monitors or deck guns. A pry bar can be driven into the ground like a fence post, and a monitor can be secured to it to prevent it from walking. The pry bar must be driven in deeply and could be difficult to remove following a fire, but it provides a secure post to which you may lash a gun.

Pry bars can also be used for securing ladders to windows. Place a long pry bar horizontally inside a window, and secure the ladder rung or beam to the pry bar. Apply sufficient pressure with the rope to pull the pry bar tightly against the interior window framing or walls. Ensure that the pry bar is substantially wider than the window.

Claw tool. A **claw tool** (fig. 11–8) is a step up from a pry bar but doesn't have the versatility of the Halligan tool. It is a multipurpose prying tool with a hook at one end and a fork at the other. It can perform many overhaul functions and some forcible entry tasks, such as padlock removal, on the fireground when used correctly. Its use has declined drastically over the years due to upgraded tools, but firefighters may find these on apparatus.

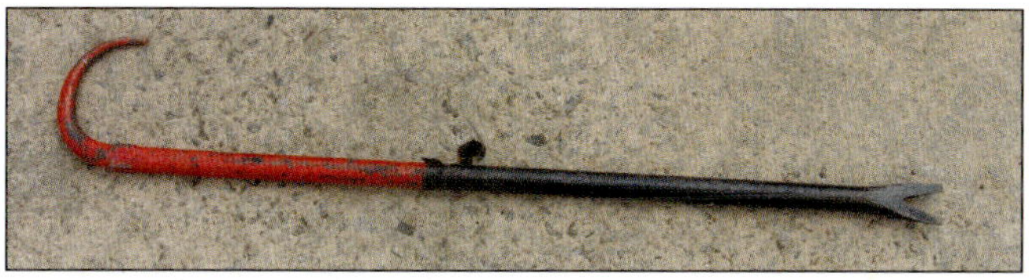

Fig. 11–8. Claw tool

Its biggest disadvantage is the absence of an engineered striking surface on the hook end. When combined with a striking tool, a claw tool can be driven in and conventional, forcible-entry techniques can be used. The tool's length makes it a great lever, but at 42 in. (1,067 mm) long or more and with two sharp ends, it's a tight fit in narrow hallways and rooms. When using the fork, you must pay attention to the hook end. If the claw tool is not well set and it slips, the point of the hook end can rip skin and hook and tear face pieces.

The claw tool has some minor special uses, but they are limited. The hook end is an effective ripper for getting a firm bite into ductwork and other light-gauge, metal surfaces. You may use it to easily perforate and remove ductwork or metal. The curve of the hook end is a great handle when using the claw tool to open walls. Once a hole is created in a wall, slide the tool fork-end-down into the hole. Grasp the hook end and pull toward yourself. The entire length of the tool will pull through the plaster and lath or wallboard, making opening much easier.

Halligan bar. If there has been one fire service-driven innovation that changed the way fire service operations are conducted, it was Chief Hugh Halligan's **Halligan bar** (fig. 11–9). Developed in the 1940s, this bar, although modernized and updated, is still seen as the preferred prying tool of the fire service. With all the upgrades and variations of the tool made by different manufacturers, they are still referred to as the Halligan tool due to its similarity to the original version. These are the most efficient, effective tools available to any firefighter for tasks that require forcible entry, leverage, and overhaul operations. (fig. 11–10).

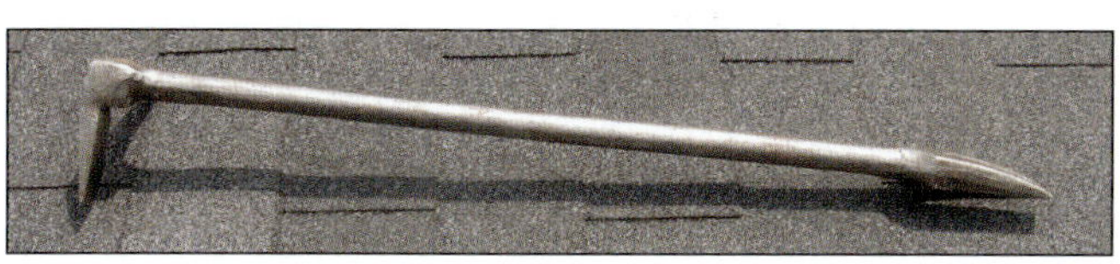

Fig. 11–9. Halligan bar

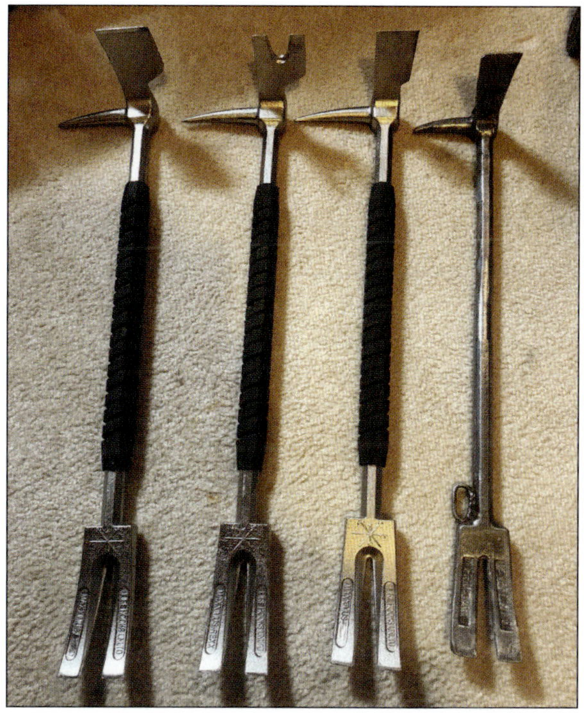

Fig. 11–10. Multiple types of Halligan tools

The many functions of the Halligan bar depend on leverage. The tool's design allows for multiple functions with one tool, but leverage is the key especially in forcible entry operations. Halligans are available in various lengths. The longer the bar, the greater the amount of leverage it is capable of producing. Its use will dictate what length you need. A standard 30-in. (762 mm) Halligan bar is the best bar for day-to-day, conventional forcible entry, ventilation, overhaul, and other standard fireground activities. Shorter and longer bars have their places, but the 15 to 1 ratio of force allow the 30-in. tool to be the best for forcing open doors. Weight is also a consideration, and the 30-in. (762 mm) bar weighs only 9 lb (4.1 kg).

The Halligan bar is a single piece of forged steel that has three workable ends: the adz, the pick or pike, and the fork end. The adz end of the tool is most commonly used to force outward-opening doors or to pull open lath-and-plaster walls. It also pries off door and window molding when performing overhaul operations. At a 90-degree angle to the adz is a hook point known as the pick or pike end of the tool, which is used to start a purchase point in a wall for overhaul, inserted into padlocks and struck to force the lock open, or to remove window and door moldings. Wide at the base where it joins the tool, the pick should also taper and curve, ending in a sharp point. The Halligan tool has four areas designed as striking surfaces: the top of the tool on the flat part of the adz, the side opposite the pick, the side opposite the adz, and the shoulders of the forks. These striking surfaces are designed to receive heavy blows from a striking tool or can be used as striking tools themselves.

The shaft of the tool is usually in a hexagonal shape, which adds strength and rigidity, and the many faces improve a firefighter's grip on the tool.

The fork is broad and tapered and is the primary end to use when performing forcible entry operations on inward opening doors. It can also be used to pry off locks from shackles, cut windshields with a "can opener motion," shut off gas valves, and remove molding from doors, windows, and baseboards. It should have beveled tines that assist in acting as a fulcrum and assist in forcing items open. The tool may or may not have an attached ring just above the fork to which you may snap a utility rope for hoisting and lowering, attaching a carrying strap, or throwing the tool off a roof to ventilate upper-floor windows. The tool can also assist firefighters while searching by simply holding the forked end with the adz and pick set like a tripod. If the tool slides forward and drops due to a hole in the floor or a stair, thus, pinning the firefighter's hand near the fork to the ground, they'll know that a hazard exists in front of them. The Halligan bar is the icon of fire service tools and is mostly seen "married" to a flat head axe or onto a hook to make the quintessential tool couple for firefighting operations.

Push-pull tools

SKILL DRILL

Pike pole. The **pike pole** (fig. 11–11) is one of the oldest tools in our arsenal of weapons to combat fires and, next to the pick-head axe, is the most readily identified tool of our trade. Pike poles are used during most fire incidents to pull apart debris and burning materials, open walls and ceilings, remove trim, pull back roof boards during ventilation, and let firefighters find and extinguish deep-seated fires. Pike poles can also be used to assist firefighters stretch a hoseline up a fire escape or floor-to-floor via a window when the nozzle's handle is placed on the pike pole's hook. (Note: they are often just referred to as a "hook" in firefighting nomenclature.)

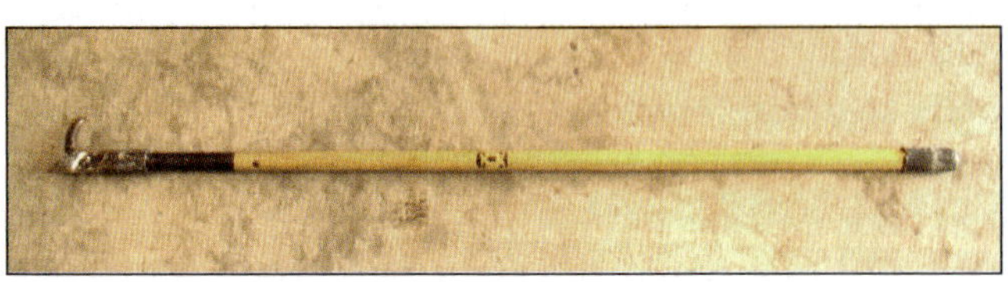

Fig. 11–11. Pike pole

Pike poles come with a variety of pike head styles, handles that make them stronger and more durable, and lengths to accommodate working overhead or below grade made by many manufacturers.

Pike pole handles come in three basic materials: wood, metal, and fiberglass. The shape of the pole can be round, oval, or I-beam. Fiberglass handles can be solid round or oval, hollow core, or solid I-beam. Metal handles can be stainless steel, plain steel, or aircraft steel.

Pike pole length is always a hot topic in firehouses. Many departments use short pike poles (2 ft to 4 ft) often referred to as closet hooks due to their ease of opening up the walls of a closet or working in tight places or when an officer carries them during the fire. Tip: you'll be much more efficient if you use a Halligan bar in place of the short hook due to the ability to use the three sides of the tool.

Pole length should depend on the type of structure for which you use it. Consider taking the standard 6-ft (2-m) pole as the minimum-length pole carried for standard firefighting operations. A 6-ft (2-m) pole will easily reach most parts of a residential structure that you will need to reach. It can be carried into and out of structures safely and can be maneuvered easily inside.

The 8-ft (2.4-m) pole should be the minimum-length pole taken into a commercial structure or light-industrial building, but be aware that a 10-ft pole may be needed.

> **FIREGROUND NOTE**
>
> In commercial occupancies, a drop or false ceiling with ceiling tiles may hide a tin or gypsum board ceiling above it, so longer hooks may be needed to reach the hidden ceiling.

The 10-ft (3-m) pole is normally not suited for use in a typical residential home, but it may help in homes with high cathedral ceilings.

Some poles become extremely flexible at 10 ft (3 m) or longer, especially oval fiberglass poles. It's like trying to use a fishing pole to pull ceilings: The head end bends and wobbles, so be aware of this issue when opening up. Most poles are stored in rear compartments or on the side of the apparatus in brackets, tubes or compartments (figs. 11–12 and 11–13).

Fig. 11–12. Apparatus tool compartments

Fig. 11–13. Tools hung on the side of an apparatus

Other poles. Other types of poles have been created and used in the fire service over the years and due to departments requiring specific needs. Be aware that other styles of pike poles exist; you might see the following in your department (fig. 11–14 from left):

- The **Boston rake** has a goose neck head mounted onto a pole. It is predominantly used in older buildings for opening up plaster and lath walls.
- The **Chicago pike pole** was created by the CFD shops and is similar to a regular pike pole. Its head has an angular hook and its pike is four sided with a pyramid on the top. The shape allows the tool to have tremendous strength and power for penetrating material and opening up moldings.
- The **Providence hook** has a cutting surface on the backside of the point (pike) and another on the top of the hook. It pushes easily into gypsum and plaster and is great for penetrating plaster and lath and tin ceilings.
- The **Halligan hook**, also known as the multipurpose hook, was invented by Hugh Halligan, inventor of the Halligan bar. With its engineered angles, this hook makes quick work of overhaul and prying roof sheathing off a roof joist. It is also good for opening up ceilings, moldings, baseboards, and other trim.

Fig. 11–14. Other pole styles

- The **San Francisco hook** has a chisel point on the top of the pike that allows for quick insertion into a ceiling and efficient trim removal. The hook end is set at 90 degrees to the handle of the tool and has serrated teeth on its underside, giving it the ability to bite into material as it is pulled open. This tool was created in the San Francisco Fire Department to assist in opening up the lath and plaster city.
- The **Falcon hook**, a modified Halligan hook with broadened hooks and a curved cutting blade at the top, is a great hook for lightweight, metal buildings, tin ceilings, ductwork, and other hard-surface materials.
- The **New York roof hook** (fig. 11–15), official hook of the FDNY, is similar to the Halligan hook, but one side of its head is flat, so it can be slid up along the wall to remove a skylight or flashing. The head design allows one end to pry up roof boards easily when placed onto or next to the joist. The opposite side of the head allows lath and plaster to be pulled and knocked away from the firefighter during the pulling motion. The opposite end of the handle has a pry-bar end for assisting in forcible entry, gaining initial purchases in ceilings and removing baseboard molding.
- The **Gypsum board hook** (fig. 11–16), often called the sheetrock rake, is a clawed rake that can cut and open up gypsum wallboard with little or no effort on your part. The angled jaw has three teeth to grab the material, and the cutting blade perforates the gypsum so it can be pulled open in sections.

- The **L.A. trash hook** or **rubbish hook** (fig. 11–17), often referred to as an arson rake or trash hook, has a wide metal head with two long prongs that bite into material that is being pulled. Often used to tear open burning trash, it also is a tremendous asset in roof work, used to sound the roof or, when its tines are thrown into the roof, to make a work platform when working off a roof ladder. Its large head surface pushes down large sections of ceilings from the roof and allows large sections of ceilings to be opened up.
- **D-handle poles.** Many pike poles will come with a D-handle on the opposite end of the tool to assist firefighters in using the tool in a pulling motion. One hand remains on the handle with the opposite hand pulling on the D-handle. Often this tactic might be used if a double thickness ceiling is encountered when opening up. Always use caution that the hook's head isn't caught on something like an electrical cable, water pipe, or structural member. In cases when "throwing" the head of the hook into the ceiling, firefighters must resist the urge to hold the tool's D-handle. If they were to strike a ceiling joist or structural beam, the hook won't enter the ceiling and the repercussion will be felt all the way down into the firefighter's wrist and hand.
- **Double-end poles.** Because firefighters often face varying circumstances when on scene, many firefighters prefer to carry hooks that have two workable ends. Many manufacturers are making tools that allow firefighters to carry one handle with two different tool heads or ends. The NY roof hook has a pry bar end to assist in prying opening items and removing baseboard and crown molding. Other hooks such as: the Colorado Hook has the NY Roof hook's head on one end and the L.A. trash rake on the other. The Tri-State hook has the NY roof hook and the Boston rake on opposite ends. It's important to learn that these hooks exist and could help you on the fireground depending on which type of structures and construction features that are prevalent in your fire alarm district.

Your department may or may not have any of these hooks in addition to a standard pike pole. Mutual aid departments might have them. The more you know about tools in the fire service, the better firefighter you will be.

Fig. 11–15. New York hook

Fig. 11–16. Gypsum board hook

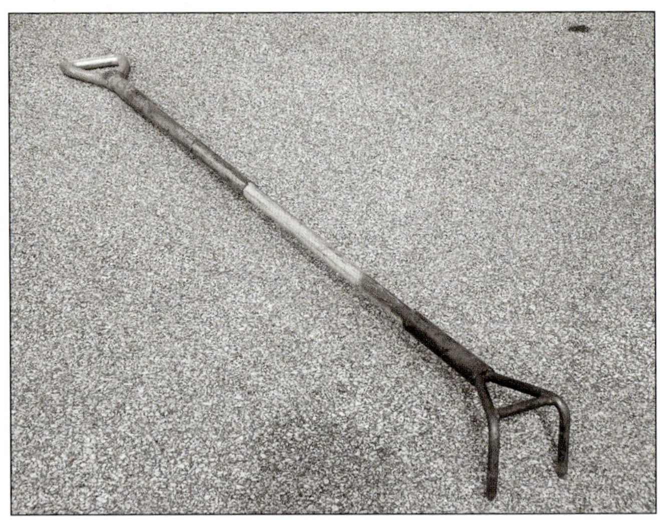

Fig. 11–17. L.A. trash or rubbish hook

Forcible entry tools

Many apparatus and fire departments will have many types of tools that will assist firefighters in specific forcible entry tactics. The following lists some specific types your department may carry or that you'll encounter on the fireground.

The **K-tool** is a piece of metal channel tool used to pull lock cylinders when firefighters perform through-the-lock forcible entry techniques (see chapter 12). The face of the tool has two replaceable beveled blades that are positioned and resembles a large K. The back side of the tool has a bracket for inserting a pry type tool to assist in pulling the lock cylinder from the door. The tool is usually carried in a pouch with assorted picks and or keys to move the levers inside the pulled lock. The K-tool is used often at commercial strip malls and aluminum-framed single-locked doors.

The **duck billed padlock breaker** is a long metal tapered wedge tool used for inserting into the shackles of a padlock. It is struck and driven downward to drive, break, or release the shackle from the body of the padlock. The tool must be inserted properly, and firefighters should not place it where it would wedge the shackles outward.

The **hydraulic forcible entry tool** is a hydraulic tool that is used for forcing inward opening doors or side sliding elevator doors. It is comprised of a hydraulic piston inside a housing that, when pumped with the tool's handle, opens jaws that force the door open when inserted between the door frame and jamb. It is a tool that works under high pressure and must be used properly to avoid a mishap. The original tool was of two-piece design known as the rabbit tool. Updates and revisions have made the tool a "one piece" or one firefighter operating tool, which can be used in zero visibility situations.

The **shove knife** is a small forcible-entry tool made of spring metal or plastic from a detergent container. Its head has some various indentations and cut outs in it to slide open the locking latch on doors or on double-hung windows. It is normally carried with the K-tool kit or by individual firefighters.

Aluminum door wedge/chocks resemble a regular wooden door chock. Wedges are being manufactured to allow firefighters the ability to chock or maintain a purchase when forcing inward and outward opening doors. The wedges are either smooth, solid, or made with grooves and have a hole in them for the pike of the Halligan tool to be used with the wedge. The metal wedge can also be driven into the tight space between the door and jamb to open it up for tool insertion. These wedges are allowing firefighters an easier time when forcing a door in a one-firefighter operation.

Firefighters will be called upon to remove people from stalled elevator cars. **Elevator keys** are small metal cylindrical rods with modified ends that come in a variety of sizes, shapes, and styles with probably the most common being the drop key for sliding elevator doors. Specific safety guidelines must be followed in elevator removal operations, and firefighters must know how the elevators keys work for specific elevators in their response district.

TOOL MAINTENANCE

FFI 4.5.1 **FFII 5.5.4** Don't be too disappointed if the tools at your department could use a little work. Basic hand tools sometimes get overlooked for simple maintenance. Maintenance does not take long and is an outward sign of great company and departmental pride. You will also find that a little effort in the firehouse yields tremendous advantages on the fireground. Well-maintained tools look good, function well, and are safer than neglected tools.

Note: Do not perform any maintenance on your company's tools without authorization.

Maintenance of metal-headed tools

When metal-headed tools are not being used on the fireground, they are usually covered with dried or clumped pieces of plaster, gypsum dust, or even roof tar. Firefighters usually attempt to wash them with water prior to putting them back on the apparatus, creating more of a mess. Upon returning to quarters, these heads can be cleaned with soap and water and a wire brush, steel wool, or a wire wheel on a bench grinder to remove the buildup of material. After cleaning the debris off the heads, always inspect the tool for damage. If the head is cracked or broken, it should be put out of service. File down any small burrs or mushrooming ends on the head with a mill bastard file. This will keep a sharp edge and prevent burrs from detaching while the tool is being used or becoming a flying projectile that could cause injury to a firefighter. Using a grinder can extend the serviceability, strength, shape, and longevity of your tool, but can also quickly diminish the size, shape, or strength of the head. Use caution. When done cleaning the tool a light coat of oil applied to the head will reduce the chances of it becoming rusty and will keep the tool ready for service.

Maintaining tool handles

Handles may be washed with soap and water after use. They may also be heavily caked with debris and need to be washed with a scrub brush upon returning to quarters. They should always be thoroughly inspected after use. Many striking tools or pulling tool handles will show the most damage near the head. Axes and sledgehammers will usually have damage near the base of the tool head due to overstrikes, the handle hitting the tool being struck. The same holds true for longer handle hooks because their handles will see more damage near the head from hitting the material being opened up. Many axes and sledgehammers now come with a rubber protector or overstrike protector just beneath the tool head.

Wood. After using the tool always inspect the handle for damage such as cracks, chips, or splinters. If the crack is severe, place the tool out of service. If there are chips or minor damage to the handle, sand the bad portion down with sandpaper and then wipe off the dust. After cleaned, a light coat of linseed oil can be placed onto the handle and rubbed into the tool to maintain the handle.

Fiberglass. After using the tools always inspect the handle for damage such as cracks, chips, or splinters. Fiberglass has a tendency to create small shards, sometimes invisible to the eye, so use caution with your hand when checking the shaft. If a crack is severe, place the tool out of service. Be aware that fiberglass handles should also be checked for any signs of melting or burned surfaces. Firefighters have been known to misplace tools during a fire and damage has occurred, so inspect all tools thoroughly. Smalls dents, divots, or splintering can be sanded down with sandpaper and to remove them, and a thin layer of polyurethane can be applied to the handle to keep the surface smooth.

Metal-handled tool should always be inspected after use. The most common occurrence is burring, dents, or chips on the handle. These can cause cuts to the hands and should be filed down if found.

Fluid levels

Firefighters will use tools that run on gasoline, a fuel mix of gasoline and oil, or an ethanol-free fuel. After each use fuel levels should be filled to capacity to ensure the tool will run to its full potential the next time it is called upon. In addition, tools that operate with hydraulic fluids should be checked periodically and according to the manufacturer's guidelines for proper fluid levels. Many tools in the fire service operate with specific fluids and can't be intermingled. Always ensure you use the proper fluid for the proper tool.

Specialized tools and equipment

Firefighters will be called upon not only to extinguish fires but also when people are in need in a variety of situations such as automobile accidents, stalled elevators, and building collapse. It is important to know that other tools and equipment also assist us in performing our tactical operations at fires and emergencies.

A **thermal imaging camera (TIC)** is a durable, portable camera device that reads heat sources through smoke or inside of walls using infrared radiation. Many of the cameras are handheld and operate on rechargeable batteries, but helmet- and facepiece-mounted cameras are available. The tool is invaluable to firefighters when performing search operations, including locating the fire and its extension along with locating an unconscious victim inside a smoke-filled atmosphere. The TIC can also be used at hazardous materials incidents to spot a level inside a tank of material. Proper training in the use of this equipment is imperative.

FIREGROUND NOTE

Firefighters using thermal imaging cameras must maintain their bearings using the wall as a reference point when searching. It's easy to just look at the TIC screen and lose your point of reference.

Smoke ejector is a term used for a larger-than-normal ventilation fan used by the fire service. The fan is a heavy-duty design and comes in various sizes and is powered electrically from a generator, a gasoline engine, or even a portable battery. It is normally used to clear smoke from a building and can also provide fresh air into the structure or into a confined space rescue. As with any ventilation operation, communication with units operating inside the structure should be in progress to determine when and where ventilation efforts are needed.

Brush brooms are specialized brooms used by the fire service. They have a head made out of a bamboo-type material that holds up to the wear and tear of overhauling or extinguishing along with water at grass, brush, and wildland fires. These heads are mounted on wooden handles.

Extrication tools and equipment

Hydraulic rescue tools for auto extrication are commonly referred to as the jaws of life. There are numerous manufacturers of these tools, and most come with a spreader, cutter, or extending rams. The tools work off a hydraulic pump that is either powered by a generator or by gasoline. There are also battery-operated hydraulic tools that work on the same principle but don't have the hydraulic lines running into them like the older tools do. Over time the manufacturers have also found ways to combine some tools, and you will find a spreader and cutter in one. Firefighters must use caution when using these tools and must attend training classes to understand auto extrication practices, techniques and tactics.

An **air chisel** is an air-powered tool shaped like an oversized gun that accepts different styles of bits. These bits can be used to cut metal, shear off bolt heads, and break apart objects such as concrete, tile, or block. It was predominantly an auto extrication tool until the arrival of the hydraulic tools but is still carried and used by many departments for intricate rescues, such as a person stuck in a machine. Some larger air chisels are often called air hammers and look like a rifle assembly.

Air bags have layers of neoprene with aramid fiber reinforcement that, when filled with air, are capable of lifting heavy objects. Bags come in different sizes and shapes and are capable of: lifting different objects and weights, moving a load or shifting objects, and even can be used to stop a leaking tank car. Firefighters must attend training to learn the capabilities and tactics that must be mastered for using these bags that are capable of lifting numerous tons.

Rescue struts or **stabilizers** are metal extending supports that can be made in a circle or tubular design that can extend and be used to support or stabilize loads such as a vehicle on its side. The struts are normally extended by pulling a pin and manually extending them or by the use of a hand crank. There is another larger type of these struts powered by air and used to support panels in trench rescues or stabilize unstable buildings. Proper training is required to know how to use these devices.

Bottle jacks are hydraulic jacks that are small cylinder objects with a center yoke that extends under hydraulic pressure to lift an object. They come in numerous sizes and can weigh from 5 tons up to over 50 tons. The jack works by placing it under the load and moving the lifting mechanism with a handle. To release the jack, a release valve on the lower portion of the jack is opened with the same handle and lowers the jack. Once again, training is required to understand the lifting tactics and the safety issues that go with this operation.

POWER SAWS, CHAIN SAW, RECIPROCAL SAW

Rotary saws

SKILL DRILL

Rotary saws play a big part in firefighting tactics and procedures. These saws are commonly used to assist firefighters in performing forcible entry, ventilation, breaching, and demolition operations. The rotary saw is commonly called a cut-off saw, demolition saw, or rescue saw (fig. 11–18). Some firefighters also associate a specific brand name or model number and often call the saw by that nomenclature.

Fig. 11–18. A rotary saw

A rotary saw is a portable saw with a **two-cycle engine** that normally runs on a gas-and-oil fuel mixture or with an ethanol-free fuel liquid made by specific manufacturers. The engine's driveshaft powers a clutch assembly that operates a drive belt that is driven over two pulleys that spin the blade. As the blade spins, a specific material is cut. The speed of the saw is controlled by the firefighter as he or she applies pressure on a throttle lever in the saw handle. For the firefighter to control the saw during cutting, his or her hand that is not controlling the saw's speed must hold onto the carrying handle.

The rotary saw can be equipped with a variety of blades to cut a range of materials. **Carbide-tip blades** are normally used for roof ventilation and cutting wood flooring (fig. 11–19). These blades cut through tar-covered roofs, Lexan®, acrylic glass panels, and light-gauge metal (i.e., **Q-decking**, metal, aluminum, and vinyl siding). These blades should not be used to cut case-hardened locks, roll-down security gates, or structural steel components because the tips could snap off the blade causing a severe injury to a firefighter. Always check the saw manufacturer's recommendations on the type of blades that can be used on their saws.

Fig. 11–19. Carbide tipped blades

For a rotary saw to cut heavy-gauge metals, it must be equipped with the proper blade (fig. 11–20). Rotary saws equipped with **aluminum oxide abrasive blades**, sometimes referred to as discs, will cut these metals (fig. 11–21). These blades are made up of metallic and composite materials held together by an organic bonding agent. These blades may also be reinforced with a fabric or fiber base that prevents the blade from breaking apart during cutting operations. The blades cut through material by friction and heat the metal being cut, disintegrating it into sparks, chips, fragments, and dust. The life span of these blades is relatively short if a large amount of metal material must be cut. The aluminum oxide blade can also cut through wood, although it's not the blade of choice. It actually burns through or grinds up the wood as it is cut.

A rotary saw with a **silicon carbide blade** or abrasive disc can cut through concrete, mortar, brick, block, tile, and stone. The blade closely resembles the aluminum oxide blade but is made of different composite materials that when applied to materials heats it through friction, causing the material to turn into chips and dust. To distinguish the two blades, some fire departments label or paint identification labels on them.

Another blade commonly used in the fire service is known as a multi-functional, multi-purpose blade or **diamond-cut grit blade** that will cut materials such as steel, concrete, and wood. These steel blades closely resemble carbide-tip blades. The major difference is that they are covered with industrial diamonds and grit held together by a bonding agent located toward the tips of the blade. These blades normally don't need lubricating fluids or water to assist the blade in its cutting efforts. The only minor downfalls are that these blades are expensive and once the diamonds become worn, a blade will not cut well.

A chain saw rotary blade is another blade that is available to the fire service. The chain saw blade is permanently attached to the metal circular blade. As it spins the teeth of the blade cut the roofing material. The chain does not run around the blade like a chain does on the bar of the chain saw. Always follow manufacturer's guideline on materials these blades cut; many are just for roofing materials (wood and shingles), while some other hydraulic chain saws can cut concrete with their blades.

Fig. 11–20. An irons saw

Fig. 11–21. The aluminum oxide and silicone carbide abrasive blades are similar in appearance, some departments paint the blade yellow to distinguish the two.

Saw safety. Firefighters must learn to operate a power saw under adverse fire conditions to reduce the risk of a severe injury or accident. Firefighters should follow the safety items listed to operate safely and efficiently with a running saw.

- Full personal protection equipment and eye protection must be worn when operating any saw.
- Prior to beginning a cutting operation, the operator should formulate a plan of action that includes the sequence of cuts, wind direction, and travel to a safe mean of egress.
- Firefighters should make all attempts to stand clear of debris thrown from the rear of the saw. Sparks, splinters, and chipped-off teeth can fly off the blade and injure an unassuming firefighter operating toward the rear of the saw.
- All firefighters should maintain at least 10 ft (3 m) from the saw during cutting operations. This circle of danger should encompass the saw and its operator.
- Once the saw is started, the operator should lift the carrying handle and push down with the hand holding the control handle. Following this procedure allows the saw blade to rise off the ground first.
- The operator should maintain constant contact with the saw using both hands when performing cutting operations. The saw can vibrate and move when cutting and easily slip or slide from an operator's hands if they aren't maintaining a grip on it. In addition, the saw can be awkward to operate and lift in overhead-cutting operations, so both hands should be used.
- When cutting, the operator should keep clear of the blade's path and discharging material thrown from the saw. When cutting metal overhead, the operator should hold the saw so that discharging sparks fly away from the operator and others.
- If manpower allows, a firefighter should be assigned to act as a guide person during cutting operations. This firefighter can monitor conditions around the saw operator and guide his or her movements when working in hazardous conditions and positions.
- Many times as a result of saw noise, communication between firefighters is not understandable or heard. To alleviate this, some fire departments rely on hand signals or taps/slaps on the operator's back to permit nonverbal communication. The following are the meanings of taps or slaps to the back:
  - One tap = stop cut
  - Two taps = cut
  - Three taps = shut down the saw
- Saw operations should be conducted in a well-ventilated area to prevent carbon monoxide buildup from the saw's exhaust.
- Saw should not be started or operated in suspected explosive atmospheres.
- Whenever a saw blade is removed from a cut, the operator should stop the blade on the material so it stops spinning. It's common for firefighters to mistakenly lift the saw and spin it to their side when they remove a saw blade from a cut, which could injure a nearby, unsuspecting firefighter.
- When cutting during dark hours, firefighters should provide adequate lighting to the scene.
- If manpower and conditions permit, a firefighter should be positioned away from the loud saw operations to monitor radio communications.
- If roofs covered with gravel must be cut, a path should be made in the gravel to decrease the chance of sending gravel flying like a projectile.

- When firefighters with a running saw must travel to a new location on a roof, they should stop the blade from spinning and roll the saw on its blade to the new location.
- A running saw should never be refilled with fuel. Firefighters must use caution when refueling a saw that has been running. The engine housing is hot, and any spillage of fuel mixture could ignite and cause an injury. Use a funnel in refueling operations.

Starting operations. Prior to starting any saw, a firefighter should ensure that the saw is in proper working order and equipped with the right blade for the task. There should also be sufficient room around the saw to ensure that the blade is able to rotate when started. Because most saws run on an oil-and-gasoline mixture, shaking a saw prior to starting it ensures the gas and oil in the fuel tank are properly mixed (fig. 11–22).

Fig. 11–22. Starting a cold engine

Starting a cold engine.

1. The firefighter checks that the stop or kill switch is in the off position.
2. The firefighter engages or pulls the **choke lever**, switch, or handle.
3. The firefighter then engages the **throttle handle trigger** and **throttle lock**.
4. If the saw is equipped with a **decompression button**, the firefighter presses the button to reduce the pressure in the cylinder and make the saw easier to start.
5. The firefighter places either boot into the saw handle while bending over to grab the control handle with the left hand. With the right hand, her or she grips the starter cord handle.
6. The firefighter then slowly pulls the **starter cord** until he or she feels resistance. Then, he or she pulls quickly upward with the right arm.

Note: *Do not attempt to pull the starter cord completely out of the saw or too forcefully because the cord could snap or damage the internal coil spring. Damage could also happen if a firefighter were to let the extended starter cord snap back into the starter mechanism.*

7. Once the saw starts, the firefighter applies full throttle to disengage the throttle lock. This stops the saw from running at its full **revolutions per minute (RPM)**.
8. If the saw starts and putters or runs for a few seconds, the firefighter may attempt to push in the choke lever to keep it running. Often, this doesn't happen and the saw stalls. Prior to restarting the saw, the firefighter pushes in or shuts off the choke lever. Many times, there is sufficient gas-and-oil mixture in the **carburetor** chamber to permit starting on the next pull.
9. Now that the saw is running, the firefighter ensures that the blade has stopped rotating and lifts the saw upward toward the cutting task.

For the firefighter to stop the saw, he or she must use a finger to engage the stop button, lever, or switch. Some firefighters refer to this switch as the kill switch.

Starting a warm engine. When a saw has been running, the starting sequence may differ from starting a cold saw. Usually, there is no need to choke the saw again. The saw's engine is warm, and sufficient fuel is in the fuel lines and carburetor chamber. Firefighters should follow the same sequence as above other than choking the saw. If after two pulls the saw shows no signs of starting, then it may have to be choked.

Simple saw starting. A simple procedure to assist firefighters in saw starting was developed by Lieutenant Michael Ciampo of the Fire Department of the City of New York. All the previously mentioned mechanical procedures are followed in reference to the saw's physical positions for starting the saw. The change is in the firefighter's body position. The firefighter positions himself or herself behind the saw in a balanced squatting position resembling a catcher's stance. The right knee exerts downward pressure onto the saw's air cleaner or upper housing (fig. 11–23). The firefighter extends the right arm forward and holds onto the control handle of the saw, keeping it balanced. The left arm pulls up on the starter cord to start the saw.

This method of saw starting has a few advantages:

- It puts the firefighter's hands closer to the choke and throttle levers once the saw is started.
- It shortens the pulling distance of the starter cord, which reduces the chances of damaging it or the saw's **recoil spring**.
- It allows the firefighter to lift the saw upward using leg and back muscles together, reducing the chances of an injury.
- It can be used on different types of saws.

Fig. 11–23. An alternative method for starting an engine

Traveling with a live saw

SKILL DRILL

When a saw is running, it is commonly referred to as a *live saw*. Firefighters must ensure that whenever changing locations or positions, the saw blade has stopped rotating. A rotating blade can injure an unsuspecting firefighter. To combat this problem, firefighters learn to roll a live saw on its blade when traveling. Rolling a saw to its next location decreases the chances of a rotating blade injuring a firefighter (fig. 11–24). It also can assist firefighters working on a smoke-covered rooftop. Pushing a saw in front of them while moving can inform them of any holes in the roof, shafts, setbacks, or edges of the roof.

Fig. 11–24. Rolling a saw

Carrying a saw in a sling

Carrying a saw up a ladder in a sling enables a firefighter to maintain hand contact with the ladder (fig. 11–25). Most manufacturers have made renovations to their particular saws that enable **saw slings** to be attached. If a saw doesn't have specific brackets or attachment rings, the saw sling can still attached to it. Never carry a live saw up or down a ladder.

A **choker hitch** can be made around the control handle of a saw and then the other end looped around the blade and down onto the **arbor**. This permits a firefighter to place the sling around one shoulder and or over the head to carry the saw.

Fig. 11–25. Using a sling to carry a saw up a ladder

Operating with a saw

SKILL DRILL

The work area should be as stable as possible so a firefighter can work in a balanced position while operating with a saw. Obstructions should be removed and all hazards in the cutting area identified prior to cutting operations. The firefighter operating the saw should have a plan of action and relay it to anyone operating within the vicinity of the saw.

Because firefighting is dangerous and conditions can change in a moment, saw operations are often performed in a two-person team: the saw operator and a backup, safety, or guide firefighter. The operator is responsible for the safe operation of the saw while the safety firefighter is responsible for the movement of the pair, giving hand signals, and watching and monitoring conditions around them. This procedure is normally used when performing certain types of roof ventilation. For other simple situations, such as cutting a lock off of a gate, there may be no need for a backup firefighter.

When cutting a roof for ventilation, vinyl siding with wood sheathing, or wood flooring, the saw must be run in a specific manner. Carbide-tip blades or multipurpose blades should be brought up to full rpms prior to engaging the material to be cut. If the blade were not spinning at full rpms, the blade could stop or become wedged in the material being cut.

When the saw is parallel to the material being cut, the saw's **foot bracket** should rest on the material. This will act as a pivot point. Once the saw is brought up to full rpms, the throttle handle is lifted upward and the saw pivots on the foot, bringing the blade closer to the material being cut. The firefighter must maintain control and balance of the saw while operating the throttle with one arm and holding the control arm with the other (fig. 11–26).

Fig. 11–26. One firefighter cuts a padlock while another firefighter holds the padlock with a pair of channel locks.

Once the saw is cutting, the firefighter maintains full rpms while he or she drags, pulls, or walks the saw back toward himself or herself. The saw blade spins toward the firefighter and cuts in a backward direction. Firefighters must not pull fast because this can cause the saw to stall. They must ensure that they are cutting the material during the first attempt and monitor the cut for material being thrown. Sawdust and tar chips are tell-tale signs that the roof covering and roof boards are being cut in one operation. To assist the saw during the cutting operation, some firefighters gently rock the saw up and down by first lifting on the control handle and pushing down on the throttle handle and vise versa. This often assists the saw in throwing debris out of the blade guard and also increases the depth and width of the cut line, commonly referred to as the **kerf** of the blade.

Note: *Once the saw is removed from the cut, the firefighter should release the throttle trigger and slowly place the blade into the material being cut. This stops the blade from spinning and decreases the chance of hitting anyone with the blade.*

When cutting metal with an aluminum oxide blade or brick with a silicone carbide blade, the initial cutting operation is a little different. When the blade first comes in contact with the material to be cut, it should rotate at a slow to moderate speed. This enables the blade to start a groove in the material to be cut. Then the saw can be brought up to full rpms to complete the cut. This also works for multipurpose blades when cutting metal or concrete.

Often during firefighting operations, the saw must be operated in precarious positions and a firefighter finds it difficult to balance or operate the saw. It may be necessary for the saw to be supported in a few manners. The saw can be supported by a firefighter's boot when he or she may have to cut at a low level. The firefighter can also use the upper leg to support the saw when cuts need to be made below the waistline. If the cut is chest-level, sometimes a firefighter may be able to press his or her chest into the saw's housing to support the cut. Cutting overhead is the most difficult, and firefighters must use upper-body strength to support the saw during these cuts. When performing cuts in precarious positions, it may be necessary to move the saw's blade guard to protect the firefighter from flying debris from the cut.

Whenever cutting with any type of blade, firefighters must remember the following:

- Limit side pressure on the blade and try to cut in a straight line. Side pressure or twisting of the saw can bind the blade in the cut or damage the blade.
- Firefighters should cut only as deep as necessary through materials; cutting through structural members can weaken a structure. A firefighter must learn how to listen to the saw's cutting sound and how it feels cutting through materials.
- Firefighters should not play with the saw's throttle, commonly referred to as **gunning the saw**, while waiting to perform another cut. This can damage the saw or injure another firefighter.
- Firefighters should operate to side of the saw in a well-balanced **boxer's stance** while cutting on a flat surface. This helps keep the firefighter clear of the debris being thrown from the blade.
- When firefighters cut metal, they should stand out of the flying sparks at all times. These sparks and molten metal have burned or melted the rubber on SCBA facepieces.
- Do not start or use saw in the presence of flammable gases or vapors.

Refueling the saw

SKILL DRILL

If the saw needs to be refueled during cutting operations, it must be shut down and moved to a safe location. Extreme care must be taken because the saw's exhaust is extremely hot and any spillage of the gas and oil mixture could ignite if the two come in contact. A small funnel should be used for all power saw fueling operations to prevent spillage or injury (fig. 11–27). Ensure that you use the proper type of fuel mix based on engine type (two-stroke vs. four-stroke engine, for example) or an ethanol-free fuel according to the manufacturer's recommendations.

Prior to taking off the fuel cap, ensure that it and its surrounding are clean of sawdust, dust, or debris. Debris that gets into the tank can harm the saw. Also, always open the fuel cap slowly to release any over pressure that exists in the tank. After fueling, always tighten the fuel cap carefully, ensuring that the threads line up and the cap and gasket fit tightly onto the saw housing.

Fig. 11–27. Be careful when refueling a saw.

Maintenance

SKILL DRILL

Saw maintenance is a large factor in determining the life expectancy of a saw. There are specific instructions set forth by the saw manufacturers for specific saws. Firefighters must read and be accustomed with their particular saw-maintenance procedures. Firefighters may perform some maintenance. Qualified technicians must perform other procedures. Firefighters must check the following items regularly and after each use:

- The saw and saw guard blade must be clean and free of debris that could affect proper operation. This includes removing any tar or metal slag from inside the guard, a build-up of material can slow a spinning blade down.
- The saw's throttle components, choke, and shutoff must operate freely and correctly.
- The saw's starter cord and handle must be free from defects.
- The saw's air intake fins must be free of debris.
- All bolts and nuts must be tight and secure (fig. 11–28)
- The fuel tank must be full and the cap must be secured.
- The clutch assembly and belt must be adjusted properly.

Depending on use and maintenance requirements, saws also need firefighters to do the following:

1. Change the air filter(s).
2. Change the fuel filter.
3. Change or clean the **spark plug**.
4. Adjust the belts and clutch assembly.
5. Check all cables and electrical connections.
6. Fix any deficiency with the saw.
7. Clean the cooling fins from dust and debris.
8. Check that the muffler is securely attached.
9. Check that the carburetor functions properly.

Fig. 11–28. Make sure all nuts and bolts are secure.

Miscellaneous rotary saws

There are also many types of saws powered by hydraulic fluid, lithium batteries, electricity, and water in a rotary design (fig. 11–29). These are not commonly used in everyday firefighting operations, but firefighters use them during collapsed-building operations and vehicle extrications. The saws run on the same principle as regular power saws but are constructed for adverse conditions. Specialized training may be required.

Fig. 11–29. Various rotary saws

Chain saws

SKILL DRILL

Chain saws in the fire service are normally used for ventilation efforts at structure fires or for cutting trees to provide a fire break in wildland firefighting operations. They also assist in removing trees from structures and automobiles during weather-related emergencies. Some hydraulic manufactured chain saws can cut through concrete, **reinforced concrete**, bricks, and cement block and are useful at building collapses and for breaching walls and floors (fig. 11–30).

Fig. 11–30. Different types of chain saws

A chain saw is a portable saw with a two-cycle engine that runs on a gas-and-oil fuel mixture. The engine's driveshaft powers a sprocket assembly that operates a chain that rotates over a metal alloy guide bar. As the blade rotates across the guide bar, its teeth chip and cut a specific material. A firefighter controls the saw's speed by applying pressure on a throttle lever in the handle. Some saws come equipped with a **safety brake handle** or lever that prevents the blade from rotating across the bar as the throttle lever is engaged. It also allows the operator to apply the brake, which immediately stops the blade from rotating in an emergency. To control the saw during cutting, the hand not controlling the saw's speed must hold onto the carrying handle.

A chain saw's cutting chain resembles a bicycle chain. Individual links are riveted together, and teeth on the top cut and chip the material being cut. These teeth can be carbide-tipped and can cut a multitude of materials. Some specialty blades reduce the chances that the blade will hook into the material, protect the carbide, and actually file through a material instead of cutting it. A metal finger on the underside of the chain attaches to the link, enabling the chain to ride inside the **guide bar**, carry oil to lubricate the cut, and engage in the sprocket to allow chain rotation around the bar.

The life span of a blade depends on how well it is lubricated throughout the cut and how much material there is to cut. Chain saws are equipped with a reservoir that holds bar oil. A built-in pump disperses the oil across the bar and into the bar's channel, which lubricates the chain throughout the cut. For long operations, firefighters must ensure that the reservoir always has oil. A sharp chain throws shavings and chips and requires little force to cut. A dull chain throws dust and requires a lot of pressure applied to the saw to cut.

Numerous brands and types of chain saws are used throughout the fire service. Some of the most common saws have **chain guards** and cutting **depth gauges**. A gauge set to the depth of the cut helps prevent the blade's teeth from cutting through the roof joists (fig. 11–31). These saws are normally used for roof ventilation. Some chain saws with specially designed guide bars are shaped differently than most common guide bars. Firefighters must know their departments' specific saw specifications and operating procedures.

Fig. 11–31. A chain saw with a depth gauge

Chain saw safety. Many safety items mentioned in the rotary saw material apply to chain saws, with a few additions:

- Never lean over or straddle a moving chain saw blade.
- Position yourself left of the chain line, with the left hand on the control arm and right hand on the control handle.
- When performing prolonged saw operations, wear approved hearing protection.
- Find an alternative to cutting overhead with a chain saw.

Traveling and carrying a chain saw. A running saw is a "live saw." Firefighters must ensure that the chain has stopped rotating when changing locations or positions. A rotating chain can injure an unsuspecting firefighter. To combat this, firefighters should engage the **chain brake** (if the saw has one) before moving to a new location. Firefighters can also roll a live saw on the chain when traveling.

If the saw isn't running, firefighters should carry the saw with the chain facing the rear. If they were to trip or fall forward, they would not land on the blade. In addition, it would less likely strike another firefighter on the fireground.

Chain saws can also be carried using a saw sling. Most manufacturers have procedures for carrying saws. Firefighters must be accustomed to their departments' saw procedures.

Starting a chain saw. Rotary saw starting procedures apply to the chain saw with one additional safety factor: If the chain saw is equipped with a chain brake, it should be engaged before starting the saw. This eliminates the chance of the blade's rotating around the bar once the saw is started. If the saw does not have a chain brake, the firefighter must be aware that the blade will begin to rotate around the bar once the saw is started. Firefighters must remember that a rotating chain can injure. All safety procedures must be followed.

Operating with a chain saw: ventilation. The work area should be as stable as possible so a firefighter can work in a balanced position while operating a saw. Obstructions should be removed and all hazards in the cutting area identified before cutting begins. The firefighter operating the saw should relay his or her plan to anyone operating within the saw's vicinity.

Because firefighting is dangerous and conditions can change, saw operations are often performed in two-person teams: the saw operator and a backup, safety, or guide firefighter. The operator is responsible for the safe operation of the saw while the safety firefighter is responsible for the movement of the pair, giving hand signals, and watching and monitoring conditions around them. This procedure is normally used when performing certain types of roof ventilation.

Because many chain saws used for roof ventilation lack a safety guard, the safety or guide firefighter should place himself or herself slightly behind and to the opposite side of the saw when cutting is in progress. The possibility of the saw kicking back out of the cut line or the operator lifting it up could cause the saw to kick toward the rear and cause an injury. Some fire departments do not allow a guide or safety firefighter to be near a saw with no safety guard. Firefighters must understand their fire departments' standard operating guidelines (SOGs).

When cutting a roof for ventilation, vinyl siding with wood sheathing, or wood flooring, a firefighter must run the saw in a specific manner. Normally, only the tip of the saw and a few inches of the bar are used to cut through a material. Burying the whole saw's bar and blade into the roof could cause the saw to cut right through the structural joist and cause a collapse. Saws that have depth gauges and guards should be set to cut only through the roofing material. Firefighters can cut a small inspection hole, kerf cut, or cut line to verify roof thickness and then set the depth gauge before cutting operations begin.

When saw cutting begins, the chain must be plunged into the roofing material to start the cut line. The saw should be brought up to full rpms before engaging the material to be cut. If not, the chain could stop or become wedged into the material. If a chain saw has a depth gauge or safety guard, the underside of the guard at the tip can be placed onto the roofline to assist in making the initial cut. The saw can pivot on the guard as the firefighter brings the saw up to full rpms and then rotates the tip into the roofing. Once the saw plunges through the material, the saw cuts back toward the operator and should be held almost vertical and run at full speed during the entire cut. The operator should watch the debris being thrown to judge the cut's progression.

Note: *As the saw is pulled from the cut, the operator should engage the brake lock, which will stop the chain's rotation. This prevents injuring anyone when the saw is moved. If the saw does not have a brake, the blade should be kept near the cut until it stops rotating and then be moved to the next cut.*

Chain saws assist firefighters in ventilating hurricane windows (see chapter 12, Forcible Entry, for more information on these windows). Firefighters should be equipped with proper breathing protection during cutting operations. Before beginning the cut, ensure that a purchase point for the saw's chain exists. Cutting these windows can be dangerous and all safety procedures should be followed.

Operating with a chain saw: tree cutting. Cutting trees with a chain saw is different than cutting a ventilation hole in a roof. Different procedures must be followed. Normally when firefighters cut trees or limbs, their chain saws lack a safety bar or depth gauge. The entire bar and chain is exposed during cutting, which increases the risk of injury if safety is not practiced at all times.

Prior to cutting any trees, limbs, or branches, firefighters must perform a scene size-up to ensure they are not in any contact with electrical lines. The size-up should include the position in which the tree is resting or leaning. In instances such as a crushed automobile with a trapped victim, it may be necessary to lift the tree with firefighting equipment instead of cutting it. Another danger is when a tree is leaning on a structure and firefighters begin to cut. Once a few of the larger branches are removed, sometimes the trunk may actually spring up and back into position because the weight is released, or fall farther down into the structure. Firefighters must use caution when performing these tactics. If the situation has no immediate rescue or life hazard, the area should be secured or stabilized and left for the proper attending authorities.

To begin saw operations, firefighters place the saw against the tree with the **bumper spikes** in contact with the tree.

The saw is held with the bottom of the chain just off the tree. Once the throttle is fully engaged, the saw can pivot on the spikes and the chain can cut into the tree. The saw must be run at full speed during the entire cut. Once the chain has cut through the material, the throttle trigger should be released and the operator should engage the chain brake to stop the blade from spinning. The chain saw should not be forced downward into the cut. The firefighter should allow the saw's chain to cut the tree. The saw can be slowly rocked up and down to assist in the cut.

Firefighters must remember that any cutting with the tip of the bar and chain can cause the saw to kick back toward the operator. Using the tip is dangerous and should not be done.

There are numerous cutting variations and situations when cutting trees and dismantling limbs. Wedges and blocks assist firefighters in many of these procedures. While cutting, firefighters should see large wood chips flying from the cut and not just sawdust. This shows that the blade is sharp and still good to use. When cutting, if the blade begins to get dull, the chips usually reduce in size and the blade begins to smoke. Chain replacement would be recommended and the bar oil level should be checked to ensure that it is full and lubricating the bar and chain. Firefighters must always follow specific manufacturers' guidelines when operating their departments' saws.

Maintenance. Chain saws require more labor-intensive maintenance than rotary saws because of all of its parts. Firefighters must know their departments' saws and follow specific manufacturers' procedures. The following are some of the most common post-cutting maintenance requirements for chain saws:

- Clean and inspect the guide bar for wear and damage. It is also recommended to run a narrow object around the entire channel in which the chain rides to remove debris.
- Maintain the saw with a sharp, properly adjusted chain.
- Ensure that the oil discharge ports are clear and oil flows freely through the bar.
- Ensure that the bar oil reservoir and fuel tanks are full.
- Clean and inspect the air intake and remove any dust or debris from the openings.
- Ensure that the air cleaner is free of dust and debris.
- Clean and remove any sawdust or debris from around the sprocket and inside the saw's sprocket housing cover.
- Ensure that the depth gauge or chain guards are in good working order.
- Wipe off the saw housing and any areas that may have excess oil or cutting debris.
- Ensure that the chain is installed properly.

Periodic maintenance should also be conducted.

Reciprocating saw

SKILL DRILL

Reciprocating saws are commonly used for vehicle extrications, machinery extrications, and building collapses. These portable saws are lightweight, small, and easy to transport (fig. 11–32). They are capable of operating on battery power with no need for an electrical cord. Some reciprocating saws operate on air supply and normal, household electrical current. They are commonly called sabre saws or by the brand, Sawzall®.

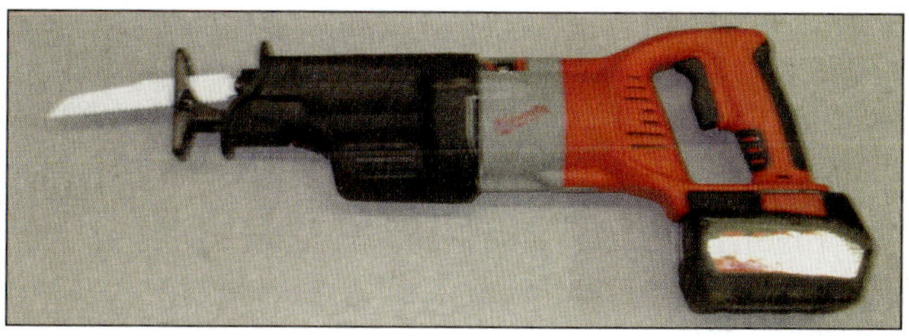

Fig. 11–32. Reciprocating saw

The reciprocating saw uses an alternating-direction motor (reciprocating motion) to move a blade back and forth to cut through material. Many of these saws offer a variable speed control option that works in conjunction with the amount of pressure applied to the trigger switch by the firefighter. The saw can be equipped with a multitude of blades that can cut wood, metal, plastic, automobile windshields, and drywall (fig. 11–33). Many of these blades are capable of cutting a few types of material and may be labeled as **bi-metal**, meaning they can cut metal and wood products. The blades are inserted into a holder and kept in place by a set screw or a manual locking device. They also can be inserted with the teeth of the blade facing either up or down to offer a different angle or plane of cutting. Lubricating some of these blades while cutting may also increase the blade's life span.

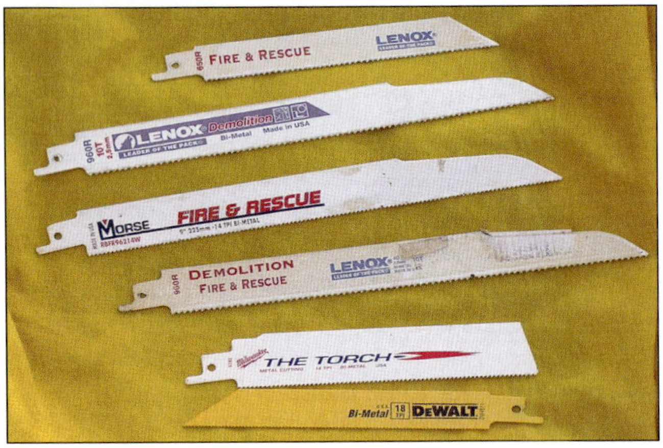

Fig. 11–33. Reciprocating saw blades

Maintenance. After each use, the saw should be cleaned of dust, dirt, and debris, especially around the blade holder. If the saw is cordless, it needs its batteries recharged. A new blade should also be installed even though the blade may still have some life left in it. The fire service must keep its equipment fully ready for service at all times. The old blade can be used for drill periods.

Safety

- Prior to cutting, the firefighter should be wearing all personal protective equipment and safety glasses.
- Anything being cut that has a power supply should be shut down prior to cutting.
- The area or material being cut should be surveyed to ensure other or structural members aren't cut accidentally.
- Firefighters must wear gloves. Blades are hot after cutting and can burn.
- When necessary, clamp or secure the item being cut so it doesn't move during or after cutting.
- Maintain a well-balanced position while cutting, and don't overreach with the saw.

Cutting operations. The firefighter first ensures that the saw is equipped with the proper blade for the cutting operation. He or she holds the control handle in one hand with the index finger on the speed control trigger. The other hand holds the saw near the blade around the tool's shaft or grip handle. Now, the firefighter holds the blade slightly off the material with the safety shoe against the material and slowly engages the trigger to make a cut line in the material. Once the blade bites into the material, the firefighter either operates the saw at full speed or a slower speed depending on the cutting operation.

As the saw cuts, the firefighter should watch the dust and debris thrown by the saw and the depth of the cut. He or she also listens to the saw and blade cutting through the material. If the saw sounds like it is laboring, the firefighter is placing too much pressure on the material being cut, the saw needs to run at a higher speed, the speed of the cut must be slowed to enable the blade work, or the blade may need replacing. In some cutting operations, the saw can be pivoted on the safety shoe in an up-and-down motion to assist the blade in cutting the material.

LESSON FROM THE FIREGROUND

A tool's efficiency depends on you. Size-up is critical to tool selection, and it should be made long before you respond to an alarm. Knowing the types of building construction in your response area is an absolute must. Know your town's history. By studying the history of your community, you will learn what types of buildings you may encounter and what building techniques were used. It is very important. Germans built buildings differently from the Swedish, who built differently from the French, who built totally differently from the colonists. Our own fire service history has played an important role in the way buildings are constructed. Huge conflagrations and high death tolls brought about the enactment of building and fire codes. As a firefighter, you are responsible for every structure in your community, not just those under construction today, but every building ever built and still standing.

Become familiar with what types of security devices are being sold and used. A trip to the local home-improvement store will answer a lot of questions about what types of tools you will need to bring with you to the next fire. Do prefire inspections. Don't just look for code violations. Look at how the occupant secures the building at night. How do you get in? What tools do you need?

MY POINT

Standing in front of a locked door with the most sophisticated tools in the world won't open that door. Know the capabilities and limitations of the tools you are holding in your hand.

QUESTIONS

1. What tool's specific purpose is to pull lock cylinders from doors?
2. What are three different types of fuel used with rotary and chain saws?
3. Describe the maintenance needs for fire service axe heads.
4. According to the text, name some tools that are available for firefighters to strike other tools?
5. Carbide-tip blades on a rotary saw are used to cut what materials?
6. Describe the most common lengths for pike poles and their uses.
7. Describe the maintenance procedures a firefighter should carry out after using a chain saw.
8. Describe the three distinct parts of a Halligan tool and some of their uses.
9. Describe the primary disadvantages of fiberglass poles used on pike poles and other push-pull tools.
10. Which general variety of tools is used for vehicle extrications?
11. Describe the necessary actions to start a cut in a roof with a chain saw.
12. Describe the signal system recommended in the text for physical communication between a saw's operator and the safety guide during roof operations.
13. What is the recommended minimum distance other firefighters should maintain from a firefighter operating a saw?
14. What is the proper method for starting a cold engine on a rotary saw?
15. What is the most important safety action a saw operator must take after cutting?
16. How should a saw in use be refueled?
17. When using a saw to ventilate a roof, what is the recommended method for moving to start a new hole?
18. An ideal axe head should weigh how many pounds?
19. What is considered the ideal length of a Halligan bar?

Forcible Entry

by Dan Sheridan

This chapter provides required knowledge items for the following NFPA Standard 1001 Job Performance Requirements:

FFI: 4.3.4

FFI: 4.3.9

FFII: 5.3.2 (Specialized forcible entry tools)

Quint. The **quint** is a fire department vehicle equipped with a permanently mounted fire pump, water tank, hose storage area, and aerial device with a permanently mounted waterway and complement of portable ladders. Many of these apparatus are constructed on a short frame assembly and resemble an engine company apparatus with a mounted aerial device. They can also be constructed on larger ladder company apparatus. In numerous fire departments with limited manpower or resources, this apparatus performs more than one function and is a fireground asset (fig. 13–5).

Fig. 13–5. Quint

Portable ladders

Straight ladder. The **straight ladder** is a lightweight, single-section ladder with a fixed length, usually 12–20 ft (3.7–6 m), commonly called a **wall ladder**. A drawback with this ladder is that its length can't be adjusted. Normally, these ladders can reach only windows and roofs on certain one- and two-story structures.

Roof ladder. The **roof ladder** is a form of a straight ladder also commonly called a **hook ladder** because of curved metal hooks permanently attached to its tip. When these spring-loaded hooks deploy, they allow the ladder to bite into the ridge of a roof, which provides a stable platform for a firefighter working on a steep roof (fig. 13–6).

Double hook roof ladder. These ladders are roof ladders with a set of hooks on the tip and butt of the ladder as well as butt spurs on each end. They are made to prevent firefighters from having to helicopter a roof ladder around when needing it on the opposite side of the roof. You simply pull the ladder up at the tip, close that set of roof hooks, and slide the ladder down the opposite side of the roof. Once ready to be lowered to the roof's ridge, open the other set of roof hooks so they bite into the roof and secure the ladder. These ladders may commonly be found on the inside of the fly section of an aerial ladder to help firefighters perform roof operations more safely.

FIREGROUND NOTE

If a firefighter encounters a high or decorative ridge vent cap, make two holes with the Halligan tool's pike in the opposite side of the roof. Place the hooks of the roof ladder into these holes, which will secure it substantially to the roof.

Fig. 13–6. Roof ladder with hooks deployed

Roof ladder safety devices. Many upgraded, modified, or newly created devices originate from firefighters who experience situations on the fireground and felt that there could have been a better or safer way to perform a tactic. Firefighters along with manufacturers have created roof safety devices that attach to roof ladders to assist firefighters in ventilation efforts. These devices can attach quickly to the roof ladder for storage and can deploy rapidly for use as a safety step off the side of the ladder. They give firefighters a safety platform to work from, especially in peaked roof ventilation tactics.

Extension ladder. The **extension ladder** is adjustable with two or more sections. This ladder is more practical than a straight ladder because it offers a range of heights (fig. 13–7). An extension ladder has a **bed section** and one or two **fly sections** that rise as a firefighter operates the halyard. As the fly section rises, the ladder's **locks**, also called **dogs**, travel over each rung and make clicking noises. When the ladder locks into a rung just after the "click" of the lock, the fly and bed section rungs align. In an extension ladder with one fly, the ladder rises in 14-in. (356 mm) increments. With a two-section fly, the ladder rises in 28-in. (711 mm) increments. This occurs because both flies rise as a firefighter pulls the halyard. With two fly extension ladders, firefighters must ensure both sets of dogs are engaged prior to climbing. If these locks aren't engaged, a climbing firefighter can descend rapidly as the ladder retracts when climbing on the unsupported section. Firefighters have been severely injured when these ladders aren't properly locked into position. Firefighters must use caution when placing their hands and feet near these ladders when raising and lowering because there are a lot of moving parts and firefighters have injured their fingers or feet when improperly placed supporting the ladder. Some larger extension ladders are equipped with **tormentor poles** to maximize stability during raising and lowering. These ladders are commonly called **Bangor ladders** or **pole ladders**.

Fig. 13–7. An extension ladder and a roof ladder

Folding ladder. The **folding ladder** is narrow, collapsible, and folds into itself for transportation through small, narrow spaces. It is also called an **attic**, **scissor**, **suitcase**, or **closet ladder**. It provides inside access to attic hatchways, some of which are commonly inside closets. One also can be used for: removing occupants of a stalled elevator through the roof hatch and out of the hoist way shaft; confined-space rescues; and climbing to a bulkhead structure to vent a skylight (fig. 13–8).

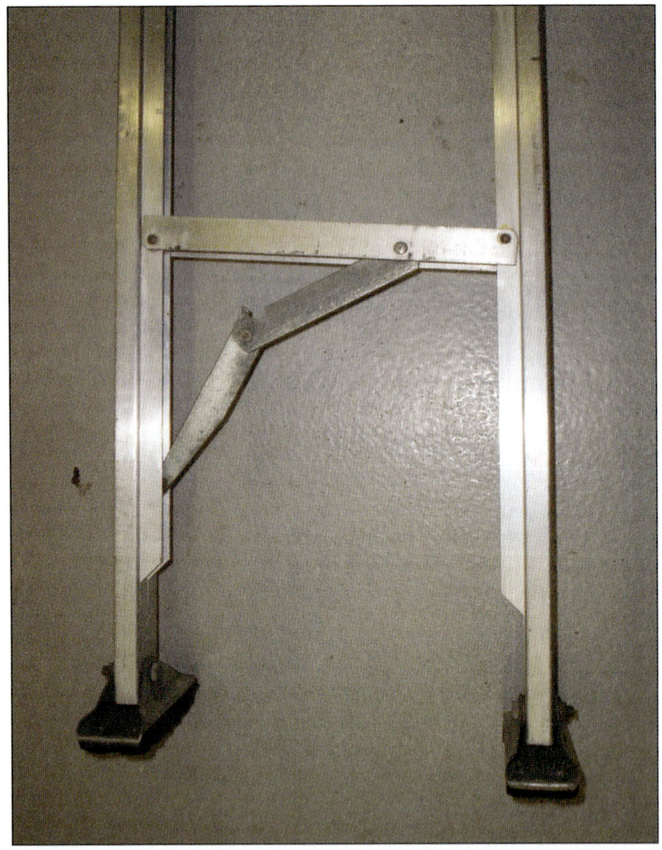

Fig. 13–8. Folding ladder

A-frame ladder. The **A-frame ladder**, also called a **combination ladder**, that when closed resembles a small extension ladder. The main difference, however, is that it does not have any halyard or pulleys. It has two pins and two receiver brackets mounted at the ladder's tip. Most of these ladders have two stationary ladder locks that hook onto a rung to keep them closed or locked together when extended manually. This ladder can be used in the A-frame position or as a small extension ladder. Because of its size, it can be transported conveniently throughout a structure (fig. 13–9).

Fresno ladder. The **Fresno ladder** is a narrow extension ladder with no halyard or pulleys. This ladder also provides access in narrow areas. It rises manually when the fly section is pushed up and locks in position as the spring-loaded dogs pass over the rungs. It is commonly called a **two-section attic ladder**.

Pompier ladder. Also called a **scaling ladder**, the **Pompier ladder** is historic in the fire service. It is no longer recognized by the National Fire Protection Association, but some fire departments and training facilities still use it to increase a firefighter's trust in tools and equipment. In the early years of the fire service, this ladder enabled many dramatic rescues. Its construction includes one center beam with rungs attached to each side. A large, solid, forged hook at the top of the beam was used for breaking windows and then could be hooked onto a window's ledge. Then, a firefighter could climb the ladder and proceed up the side of a building floor by floor, if needed (fig. 13–10).

Miscellaneous ladders. Many fire departments carry portable ladders such as the collapsible combination ladder on their apparatus, while others carry various sizes of the common stepladder. Although a stepladder in the open position resembles an A-frame ladder, a regular stepladder cannot convert to an extension ladder. It is common to find many types of collapsible and telescoping portable ladders on ladder company apparatus (fig. 13–11).

Fig. 13–9. A-frame ladder

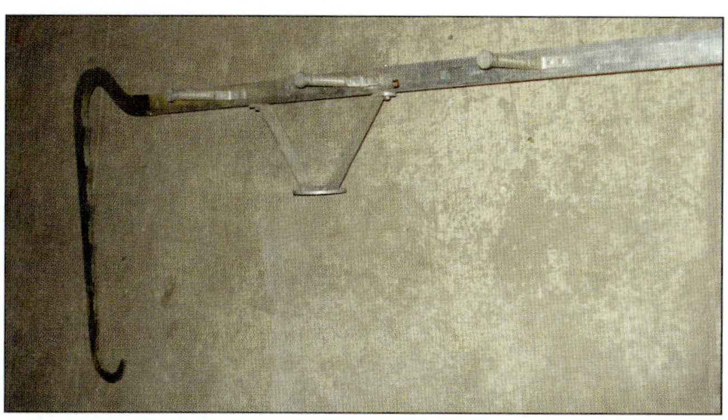

Fig. 13–10. Pompier ladder

Fig. 13–11. Combination ladder

STRUCTURAL COMPONENTS

FFI 4.3.6 The main structural components of portable fire service ladders are aluminum, fiberglass, wood, or a combination of these materials. In addition, lightweight metal assemblies and parts allow ladders to function properly. Most fire departments rely on lightweight aluminum and fiberglass ladders in lieu of heavy wooden ladders with many maintenance requirements. Wood and fiberglass ladders are often referred to as **nonconductive** ladders, meaning they do not conduct electricity. Any damp or wet ladder, however—even wood and fiberglass—can conduct electricity.

Although there are many types of ladders, almost all have the same structural components. Firefighters must know and understand ladder names and parts to operate safely and efficiently on the fireground.

- **Beam:** the main structural component of a ladder that supports a firefighter's weight and transfers it from the rungs to the ground. The beams run the long sides of the ladder and support the rungs at 14-in. (36 cm) intervals. There are a few types of beam construction used in the fire service:
  - **Trussed beam:** a trussed-beam ladder has a top and bottom rail joined together with gusset

COMPONENTS OF A LADDER

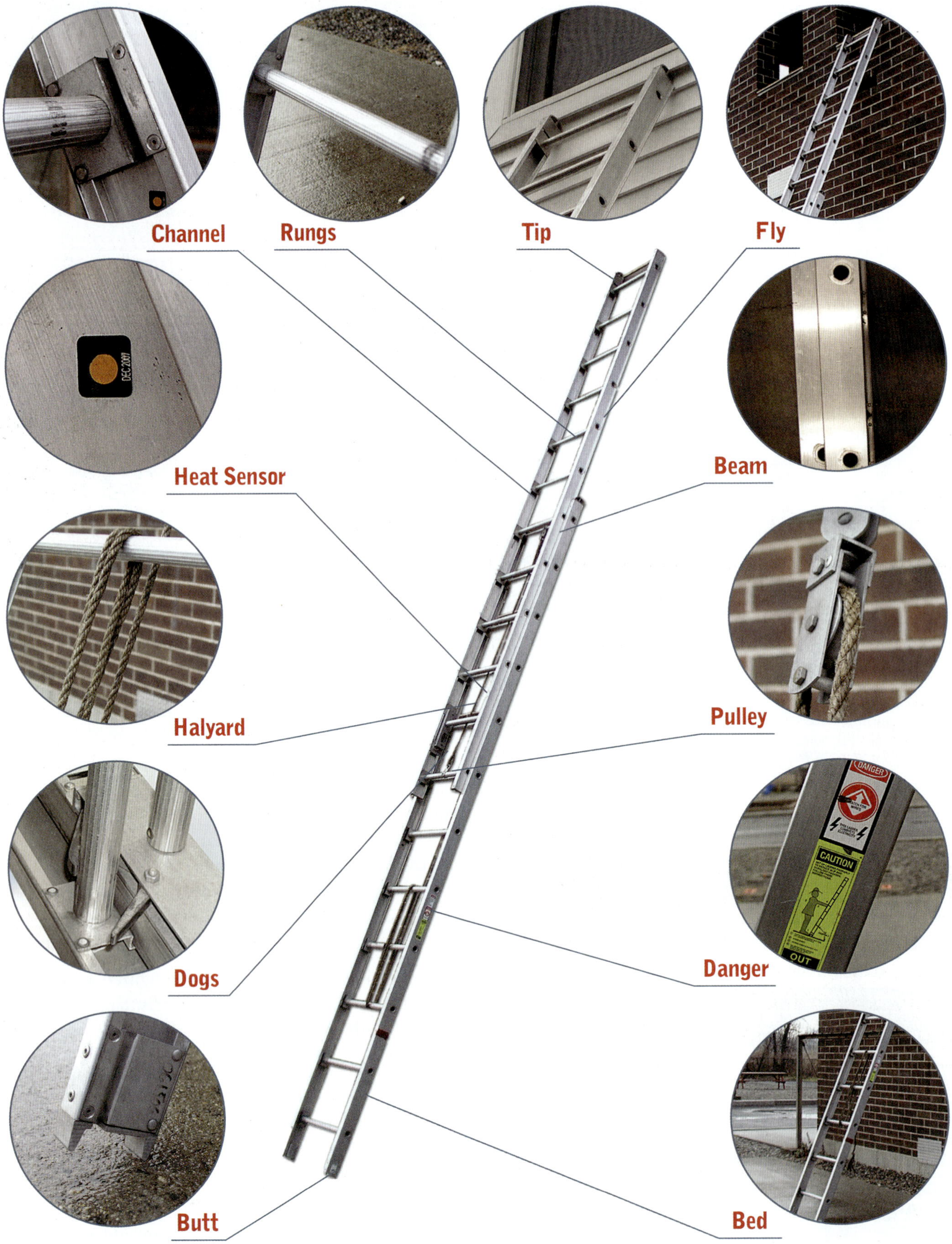

plates, usually at the rung position. It is of open construction and usually made of wood or aluminum. Most have spikes or permanent cleats mounted at the base.

- **Solid beam:** a solid-beam ladder has side components that are solid core as in wood ladders or constructed with a rectangular tube design as in aluminum ladders.
- **I-beam:** the beams of these ladders are shaped in an *I*-like configuration, with the rungs attached to the beams. Fiberglass and common homeowner ladders often carry this design.

- **Balance point:** the spot on a ladder where it balances evenly when lifted. It may not be the true center of the ladder's length because other mechanisms attached to the ladder may render the balance point more toward one end (figs. 13–12 and 13–13).

Fig. 13–12. A sticker marking the balance point

Fig. 13–13. Holding a ladder at the balance point

- **Bed section:** the bottom section of a ladder that remains in touch with the ground or apparatus. The bed section is normally the widest section because upper or fly sections retract into it. The bed section is normally the only section with a designed foot or butt attached to it.

- **Butt:** also called the **heel** or **base** of the ladder, it is located at the end where it contacts the ground when positioned against a structure. In a truss-construction ladder, spurs or spikes are usually mounted onto the ladder at its base. In other types of ladders, a permanently mounted precast aluminum foot, butt, or shoe attaches to the base. Some ladders have a slip-resistant combination safety shoe, a rubber nonslip pad that attaches to the base with a pivoting hinge. Often, these devices have spurs or teeth forged into an end for use on soft surfaces (fig. 13–14).

Fig. 13–14. A pivoting combination safety shoe with slip-resistant rubber

- **Channel guide:** a section, channel, or slot in a ladder that supports and interlocks with a corresponding section of a ladder as it is raised.

- **Dogs:** spring-loaded, mechanical locking devices at each end of the beam on the fly section that engage on each rung as an extension ladder is raised. They are also called **pawls**, **rung locks**, and **ladder locks**, and they are usually encased in a protective metal box shield (fig. 13–15).

Fig. 13–15. A locked "dog"

- **Danger/electrical/angle stickers:** stickers attached to each side of the base section of a portable ladder between the fourth and fifth rungs. They warn of electrical hazards overhead, question whether a ladder is in a correct climbing angle, and remind the user of the proper direction the fly section should face (fig. 13–16).

Fig. 13–16. Caution sticker

- **Fly section:** a section of a ladder that can extend from the bed section to gain height and distance. In many applications, if a ladder has more then one fly section, both fly sections move when the ladder is raised. The fly sections are normally narrower to fit into the bed section. The fly attaches to other ladder sections by fitting inside a channel or guide section of the corresponding section of ladder.
- **Halyard:** a manila or nylon rope used to extend or retract the fly section(s) of a ladder out of the bed section. The halyard runs through pulley(s) for ease of raising and lowering. In three-section ladders, a wire cable runs in conjunction with the halyard to assist in raising and lowering the two fly sections.
- **Heat-sensor labels:** usually small, round, heat-sensitive, orange stickers with expiration dates attached below the second rung from the top of each section of a ladder and on each side of the beams. They warn users when a ladder has been exposed to enough heat to damage it. When this occurs, a sticker turns black to warn firefighters that the ladder has been subjected to more than 300°F (149°C). Ladders exposed to high heat must be placed out of service and tested prior to further use (fig. 13–17).

Fig. 13–17. A heat sensor sticker notifies firefighters when the ladder has been exposed to high heat.

- **Pulley:** a small wheel with a grooved channel attached to a ladder by a bracket. The grooved channel allows the halyard to pass through it, reducing friction on the rope and allowing for easy raising of the ladder.
- **Roof hooks:** permanent curved, spring-loaded, metal hooks that secure the tip of a roof ladder to a pitched roof (fig. 13–18).

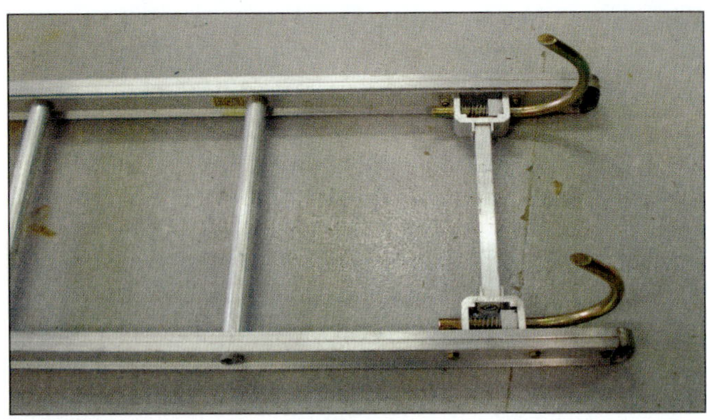

Fig. 13–18. Roof hooks

- **Rungs:** circular, horizontal cross members of a ladder that tie two beams together. The rungs serve as footrests for climbers and transfer one's weight into the beams and then into the ground. In aluminum ladders, rungs may have raised extrusions or tops of rungs may be flat to allow more traction for climbing. There are 14 in. (356 mm) between rungs.

- **Tie-rods:** normally found only on wooden ladders, these metal rods run under rungs and through both beams. They help secure both of the beams together with the rungs.
- **Tip:** the top or top few rungs of a ladder when it is propped over the roofline of a structure.
- **Tormentor poles:** metal poles attached to the bed section of a ladder with a swivel connection. These poles help stabilize the ladder as it is raised and lowered. They also have a single spur at their bottoms to assist in footing the ladder (fig. 13–19).

Figs. 13–19. Tormentor poles used to raise a ladder. (Courtesy of Jerry Naylis)

> **FIREGROUND NOTE**
>
> A 24-foot extension ladder in the closed position is the same length as a 14-foot straight ladder but offers additional lengths.

INSPECTION, MAINTENANCE, AND CLEANING

FFI 4.5.1 Portable ladders are subjected to harsh conditions, including the weight of several firefighters removing a victim, physically venting a window with a ladder, and using a ladder over a burned-out stairway. Firefighters must regularly clean, inspect, and maintain ladders for them to remain in service as beneficial tools.

Inspection

Portable ladders should be visually inspected monthly, after each use, and prior to being placed back on an apparatus. If they are dirty, they should be cleaned prior to inspection. Caked-on dirt and grime can hide defects on ladders. The National Fire Protection Agency has created standards on portable ladder design, use, maintenance, and testing. Knowing these standards and those of a ladder's manufacturer simplifies inspections. The following are some general guidelines for portable ladder inspection:

- Check the halyard for any fraying, kinking, and cut fibers. Ensure it operates smoothly through the pulleys. Halyards often have to be replaced during the ladder's life.
- Check the rungs to ensure none are bent, loose, cracked, or gouged. Also check their ends where they are pressed into the beams for looseness. If they are welded at the end, check the welds for cracks.
- Check that all rivets, nuts, bolts, and plates are secure.
- Check main beams for straightness or bowing, cracks, splintering, and obvious signs of excessive heat exposure.
- Check that dogs or ladder locks and spring assemblies function properly.
- Check that the ladder's feet or spurs are secure and/or pivot freely. If they have rubber cleats on the bottoms, make sure they are secure. If they are cast aluminum, check that they are not cracked and the nut and bolt are secure.
- Check that the ladder slides smoothly through the guide rails and stops at the ladder stops.
- Check that pulleys are secure in their brackets, operate freely, and are not out of round.
- Check the tips of the ladder. If they have cast aluminum rail caps, search for cracks and ensure the nut and bolt are secured.
- Check that hooks on roof ladders operate freely, ends are sharp, and the hook's shape is

not distorted. Also verify that tension springs controlling the hook's movements are well-lubricated and operate properly.

- Check that heat-sensor stickers are in place and haven't been subjected to excessive heat. Also check that they are within their expiration date.
- Check that wire cable halyards are taut and not fraying or kinked. Ensure their cable-tie clips are secure.
- Check all surfaces for gouges, chips, dents, and sharp edges that could injure a firefighter or place the ladder out of service.
- Check that any additional add-on or after-market items (ladder light beacons, adjustable beam legs) are in place and functional.

If any defect or deficiency is noted, a ladder must be placed out of service for repair. Trained firefighters can perform most minor repairs and regular maintenance. Major repairs to a ladder's structural components must be performed by the ladder's manufacturer or at a certified repair center.

Certification/serviceability testing

Annual service and load testing must be performed on all portable ladders to ensure structural soundness for firefighting. Any ladder that has been dropped, overloaded, exposed to high heat, repaired, structurally damaged, or if its serviceability is in question, should be tested. A **horizontal load test** determines serviceability of a ladder by placing weight on a horizontally supported ladder. The amount of deflection or bending determines the ladder's strength and serviceability. Tests are also conducted on roof hooks, ladder locks, and mechanisms of ladders.

Maintenance and cleaning

All firefighters must perform routine maintenance on portable ladders. Regular, preventive maintenance increases the lifespan of a ladder and its parts and can pinpoint a defect prior to a ladder's malfunction or failure. Following the manufacturer's recommendations is the best way to keep a ladder in service. The following are some general guidelines for ladder maintenance:

- Ladders must be cleaned regularly and after each use with water and a mild detergent. Wash ladders with a soft-bristle brush and rinse with fresh water, remembering to flush out any residue inside the rungs. If melted tar covers a ladder, a degreaser or solvent approved by the manufacturer may be necessary.
- After cleaning, lubricate the ladder locks or dogs, slide rails, and the roof hook's spring assemblies and pulleys with lubricants prescribed by the ladder's manufacturer. Normally, this consists of applying either candle wax or paraffin to the slide rails and friction areas. Other recommended lubricants may be applied to the springs of roof hooks and spring assemblies.
- If the halyard is frayed, kinked, or has cut fibers, or if the wire cable is worn, stretched, or kinked, replacement is necessary.
- If there are any sharp edges on a rail, beam, rung, or foot, it may be necessary to file each down with light-grit sandpaper or a metal file.
- If the ends of the roof hooks become dull, they may be filed to a sharp point and painted.
- Replace any heat-sensor sticker that is falling off and any that have reached their termination date. If a sensor shows that the ladder has been exposed to high heat, place the ladder out of service for testing.
- When a ladder is being inspected, also inspect the ladder brackets that secure the ladder (figs. 13–20 and 13–21).

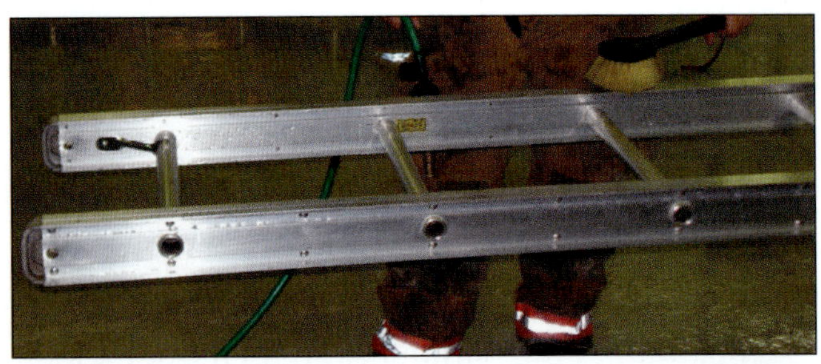

Fig. 13–20. Ladders must be cleaned after each use.

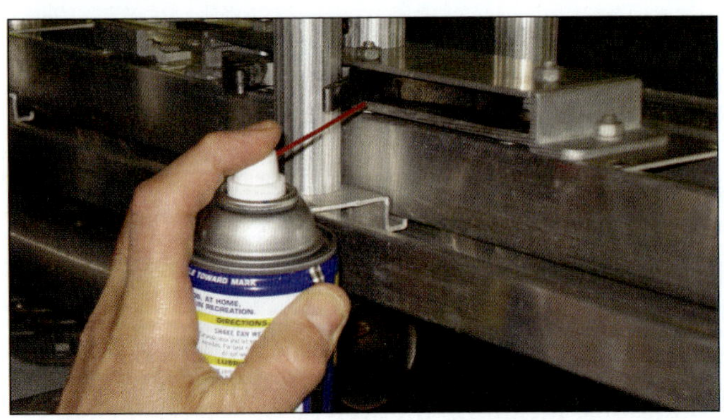

Fig. 13–21. Lubricating the dogs

Fiberglass ladders

Fiberglass ladder manufacturers recommend other maintenance requirements. After a ladder is washed in a mild soap-and-water solution with a sponge or rag, it may be waxed with regular car wax to restore the surface. Small nicks and scratches may be lightly sanded and covered with a thin coat of clear epoxy glue. Once it has dried, the epoxy should be sanded smooth. If there are any doubts about the severity of a scratch or dent, the ladder should be placed out of service and tested.

FIREGROUND NOTE

Keeping a maintenance log for each ladder can assist fire departments with record keeping and data.

Wood ladders

Wood ladder maintenance requirements differ from those of aluminum and fiberglass ladders. They may be cleaned with a sponge or rag in a mild soap-and-water solution, but soaking a wood ladder in water damages surface varnish and allows rungs and joints to absorb water and swell the ladder. When a wood ladder dries, it constricts. These actions damage the rungs, rails, and joints and force additional maintenance.

After cleaning the ladder, inspect it for any varnish damage. Varnish on wood ladders helps keep them nonconductive. Remember, wood ladders can conduct electricity, especially if they are wet or damp. In addition, some wood ladders have aluminum rungs, metal tie-rods, and steel roof hooks attached to their ends, all of which conduct electricity. Spot repair any minor scratch, chip, or dent with a light sanding to the bare wood and reapply varnish to the area. Wood ladders may be waxed after being cleaned and inspected.

Wood rails and rungs also expand and contract in response to humidity and temperature changes. The rungs and metal tie-rods must be checked regularly for tightness in wood ladders because they uphold structural integrity of wood ladders.

MARKING LADDERS

Ladders must be marked with their proper numerical length so firefighters can quickly identify them by size. Normally, all ladder manufacturers label both beams of a ladder near the base with permanent, size-designation labels. These labels allow for quick identification, but viewing them can be difficult when they are stored on an apparatus. To solve that problem, firefighters commonly use paint markers or self-adhesive stickers to display a ladder's size on the base of the ladder's butts for easy identification (fig. 13–22). They also label the ladder's bedded length next to the overall size. With straight ladders, manufacturers commonly attach identification labels at the base and near the tip or roof hooks on both beams.

Fig. 13–22. Stickers display each ladder's size.

Fire departments commonly color code the bases of their ladders to show their sizes. For easy identification, it is also good to put the size of the ladder with the color code. Other departments allow individual companies to color code or paint their company identification colors or numbers on ladders to reflect company ownership. Whichever manner a ladder is marked, it should be large enough to allow easy visibility.

Firefighters also mark the **balance point** of a ladder to permit ease of operations. This proves especially useful when a firefighter must remove a ladder off the back of an apparatus or lift one by himself or herself. Finding this spot is easy. A firefighter can pick up the ladder in the suitcase-carry position and find the spot where the ladder balances itself. At this location, paint a line, stick a piece of colored electrical tape, or label the beam with the company identification numbers. Some balance points on solid-beam ladders fall directly between a set of rungs, so fire departments paint the inside and outside of both beams. When they remove the ladder from a vertically stored ladder rack, firefighters easily recognize the balance point. This is referred to as **painting the box**. A ladder's rungs must never be completely painted.

The tips of many fire service ladders are often brightly painted or covered with reflective tape to enhance their visibility in smoke and low illumination. Remember, ladders should not be painted entirely to improve their appearance. Painting ladders entirely can hide dangerous defects. For instance, painting an aluminum ladder's entire rung can create a slipping hazard.

Fire departments can purchase small lighting or beaconing devices that attach to the tip of a ladder or end rung to improve the ladder's visibility. When activated, the device sends a strobe light or flashing signal that directs firefighters to the ladder.

Another application in marking straight or roof ladders after locating their balance points is to mark their **hoisting points**. Often on the fireground, a ladder needs to be raised with a piece of rope. Marking the rung two up from the center rung with a small paint mark or piece of electrical tape near the beams or in the center of the rung will assist in placing the rope for the lift.

Engine companies commonly mark their ladders' beams or rails with paint or tape to indicate where the ladders sit in holding brackets on their apparatus. This is beneficial for post-fire operations, and the ladders must be put back in their proper places. Often, ladders on these apparatus interfere with compartment openings or the stretching of hoselines if placed in a wrong position.

OPERATING SAFELY

FFI 4.3.6 Operating ladders safely and properly on the fireground is important for all firefighters. Numerous firefighters have died or have been severely injured as results of ladder mishaps during training and on the fireground. The following are some general safety guidelines for working with ladders:

- When possible, work with another firefighter to reduce the risk of straining or injuring yourself and others. Ladders are cumbersome and heavy. Moving or placing them on the fireground can be difficult.
- Most importantly, look for overhead utility wires before raising any ladder. Most departments use aluminum ladders, which are good conductors of electricity. Wood and fiberglass ladders also can conduct electricity because they have metal assemblies and get wet. Maintain a minimum distance of 10 ft (3 m) when operating near overhead power lines. Electrical arcs from overhead power lines have hit ladders operating in their vicinity.
- Prior to resting a ladder on a structure with aluminum siding, release the ladder as it makes contact with the structure to avoid being electrocuted. Electrical feed lines have shorted out and charged siding on numerous occasions.
- While making overhead visual inspections, look for tree limbs and branches, overhangs, canopies, and elevated decks and platforms that can affect the extension or placement of a ladder.
- Survey the ground where the ladder will be placed. Look for a solid, level foundation for the ladder's base.
- Resist overreaching when working on a ladder. Instead, reposition the ladder or use a longer tool. Use one of the leg-lock or arm-lock maneuvers whenever working from a ladder.
- Choose the proper ladder for the job. Extension ladders offer variable heights, whereas straight ladders do not adjust.
- Use caution in placing ladders. A ladder placed in front of a lower-floor window or door is not always advisable. Fire may or can erupt out of the window, cutting off the escape route of a firefighter. A ladder positioned by a door can be knocked off its base by a charged hoseline or unassuming firefighter.
- Always face the rungs when ascending and descending from a ladder. If you were to slip, you could grab the ladder for support. A slip while not facing the rungs could cause you to fall off the ladder and be injured.
- When lifting a portable ladder lying on the ground, use your leg muscles. Just bending over to pick the ladder off the ground could cause a back injury.

- When working off an aerial ladder or inside a tower ladder bucket, firefighters must wear ladder safety belts or installed safety belts. Firefighters will be working at higher than normal heights. Slipping, falling, or being thrown out or off the ladder or bucket can be fatal.

These are a few safety guidelines that must be followed to prevent injury and death. In following sections of this chapter, equally important safety guidelines are presented.

SELECTING THE PROPER LADDER

SKILL DRILL

Firefighters must be familiar with the apparatus-mounted ladder and the capabilities of all portable ladders on the apparatus. Ladder selection often must be made in a split second, and choosing the wrong ladder can have severe consequences. Prior to choosing a ladder, ask yourself where the ladder must be placed, what length is needed, and what its purpose is. An upper-floor fire in an eight-story building might require an aerial ladder or tower ladder, whereas a portable ladder could be used faster than a mechanical device for second-floor fire at the same building.

Knowing a building's characteristics can also assist in choosing an appropriate ladder. Normally, most residential structures measure approximately 8–10 ft (2.4–3 m) from floor to floor, whereas a commercial structure may measure 10–12 ft (3–3.7 m). A general rule passed through the fire service is that taking the first number of a ladder's length determines what floor it will cover. (For a 24-ft (7.3 m) portable ladder, two is the first number, so the ladder will cover the second floor). Exceptions should be noted. A 28-ft (8.5-m) or 35-ft (11-m) extension ladder can reach the next higher floor's window if set up on a level surface. Again, knowing the capabilities of your apparatus' ladders and your response area characteristics will assist you in choosing the proper ladder.

Of course, other factors will influence your choosing a ladder at a fire. Often when operating short staffed, firefighters choose straight ladders because they are easy to maneuver around the fireground single-handedly. If the straight ladder were being placed only to a low porch roof for access, it might be a wise choice. If the ladder were needed later for a task at a higher elevation, it might not reach. An extension ladder offers more variables. Remember, straight ladders offer only one height and cannot be adjusted.

REMOVING LADDERS FROM THE APPARATUS

SKILL DRILL

Following sections of this chapter discuss how to store and remove ladders from different fire apparatus, including apparatus positioning and effects on the removal of portable ladders. Often on the fireground, one apparatus pulls close to the rear of another apparatus and prohibits the removal of a portable ladder. Apparatus operators must leave sufficient clearance to the rear of an apparatus to permit ladder removal. Some fire departments offer apparatus operators tips in this area. If the largest ladder on a ladder apparatus with rear storage compartments is a 20-ft (6-m) roof ladder in nested position, the apparatus must leave at least a 20-ft (6-m) clearance at the rear. Other departments instruct operators to park at an angle to the fire building to expose the rear ladder compartment. If an apparatus has side-mounted ladders, allow enough side access for ladder removal.

Engine company apparatus

Because of the numerous styles and designs of fire apparatus, portable ladders are stored or mounted in many places. In years past, engine company apparatus had portable ladders mounted in shoulder-height brackets on the sides. As fire departments kept pace with society, the need for more compartments was necessary and ladder locations changed. It is common to find engine company ladders mounted on hydraulic or electric lift assemblies that lower themselves from a horizontal position over a hosebed to a vertical, shoulder-height position on the side of the apparatus. Other lift assemblies raise only the ladder over the highest compartment on the side of the apparatus. Note that the controls to lower the ladder may be in the cab of the apparatus or near the ladder itself.

Once either assembly is in the down position, the rung-locking clamps must be released for ladder removal. Normally, the engine company's extension and roof ladders are nested inside one another. It is a good rule to mount the extension ladder so it can be the first ladder off the apparatus. Once the locking clamps are released, two firefighters can lift the ladder off the rack and proceed to where it is needed.

If the roof ladder must be removed to reach the extension ladder, place it out of the way. Carrying it to the rear of the apparatus and sliding it under the apparatus could cause you to interfere with the hose when it is stretched off the back step. Plus, it puts the ladder out of the firefighter's sight, and it might be problematic to retrieve it for other uses. Laying the ladder up against the tire in the standing position can also be problematic. It could fall and strike a firefighter in the leg, or it could become heated from apparatus exhaust and suffer structural damage. Putting the ladder back on the rack and out of the way or taking both ladders with you can be solutions.

On some engine company apparatus, ladders are stored in a vertical or horizontal compartment under or alongside the hosebed. If the hose hasn't been stretched, it is easy to remove the ladders by sliding them out of the compartment. If the hose has been stretched, it might be necessary to lift loose ends back into the hosebed to prevent a tripping hazard or the ladder from snagging onto the hose. If the hose has been charged with water, it might be necessary for one firefighter to lift up on the hose as another firefighter slides the ladder from the compartment (figs. 13–23, 13–24, 13–25, and 13–26).

Fig. 13–23. Ladders mounted on a lift. (Courtesy of Bill Marshall)

Fig. 13–24. A vertical ladder-storage compartment. (Courtesy of S. K. Willis)

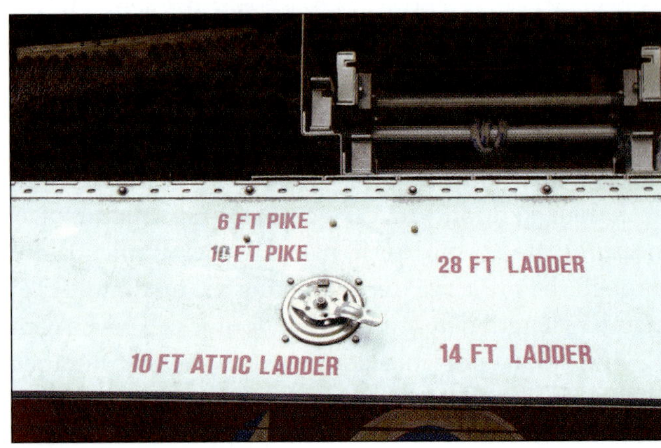

Fig. 13–25. A horizontal ladder-storage compartment

Fig. 13–26. Side-mounted ladder

Ladder company apparatus

Normally, there are three ways to store portable ladders on ladder company apparatus. Two involve the ladders being stored in a rear compartment or trough, which permits storage under the main ladder assembly. Firefighters might have to use caution when working near the rear of some ladder apparatus. Some of these apparatus can have a rear overhead obstruction such as an aerial ladder or tower ladder bucket. Firefighters must wear head protection when working in these areas. These compartments or troughs may be an enclosed or an open-style compartment (figs. 13–27 and 13–28). The ladders in this compartment sit in a vertical or horizontal position with tips facing forward and bases toward the rear of the apparatus (figs. 13–29 and 13–30).

Fig. 13–27. Ladder compartment

Fig. 13–28. Ladder trough

Fig. 13–29. Horizontal ladder storage. (Courtesy of S. K. Willis)

Fig. 13–30. Vertical ladder storage. (Courtesy of S. K. Willis)

There might be exceptions. Some fire departments place roof ladders with the butt end inside the compartment. In this manner, a roof ladder's hooks are closer to the outside of the apparatus. This way there is less chance of the roof hooks snagging the other ladder's halyards in the compartment as the ladder is removed. Plus, when the ladder is carried to the structure, the hooks and tip of the ladder are forward, ready to be raised onto the roof without spinning the ladder around (fig. 13–31). The third way portable ladders are stored on ladder company apparatus is in compartment racks on either side of the apparatus (fig. 13–32).

Removing ladders from the rear compartment is easy for two firefighters. Open the compartment door(s) and/

or release the ladder lock mechanism or stop bracket assembly. Then position yourselves on the side of the ladder that enables you to keep the fire building in sight. Often, this is referred to as **body to the building**. You will be able to remove the ladder and watch changing conditions on the fireground.

Fig. 13–31. Ladder with hooks toward the rear of the compartment

Fig. 13–32. Ladders stored on side compartment racks

After choosing the appropriate ladder and checking that no one is in the way, one firefighter pulls the base of the ladder to slide it out of the compartment. This firefighter should maintain his or her position at the base of the ladder to prevent it from striking anyone as the ladder is removed from the apparatus. He or she should proceed slowly and steadily as the ladder nears the end of the compartment. Once the second firefighter supports the ladder at the opposite end and it is clear of the compartment, this firefighter can turn into the direction of deployment (figs. 13–33 and 13–34).

The second firefighter, also facing the fire building, can now assist in removing the ladder by slowly sliding the ladder out of the compartment. As the end of the vertically stored ladder nears the end of the compartment, the second firefighter can place his or her shoulder into one of the rung spaces near its end. If the ladder is coming out horizontally, it must be grabbed with both hands and then be positioned in a shoulder- or suitcase-carry position for transport. Verbal communication between the firefighters makes this operation smooth. If a third firefighter is available, he or she may be positioned midway on the opposite or same side of the ladder to assist in the ladder's removal and transport.

Fig. 13–33. One firefighter pulls the base of the ladder from the compartment.

Fig. 13–34. The second firefighter supports the ladder at the opposite end.

One firefighter may perform this a few ways, but it is more difficult. One technique involves pulling the base of the ladder almost completely out of the compartment. As the ladder is suspended tilting toward the ground, or its base touching the ground, the firefighter may proceed to the balance point and remove the ladder from the apparatus in a well-balanced position. Another common method involves the firefighter's sliding the ladder from the compartment while maintaining a position near the back of the apparatus. As the ladder's balance point comes out of the compartment, the firefighter either steps into the rung spacing or grabs the ladder and continues walking the ladder out of the compartment. Again, the ladder is well-balanced and can be transported by a single firefighter. A safety concern when using this method is that the ladder's butt is unprotected as it slides out of the compartment and could strike an unsuspecting firefighter (fig. 13–35).

Fig. 13–35. Find the balance point and carefully remove the ladder.

Side-mounted portable ladder storage. Numerous ladder company apparatus have portable ladders mounted on sides of their apparatus. These ladders can be mounted vertically or horizontally in compartments or on mounting brackets. Tractor-trailer-type aerial ladders often have a portable ladder mounted on one or both sides of the trailer apparatus. Some rear-mount aerial ladders are constructed with the turntable mounted lower in the rear. This reduces the overall height of the apparatus and ensures it will fit through the apparatus bay doors of older fire stations. Because the turntable is mounted lower, there is reduced ladder storage in the rear and more side storage. Some rear-mounted tower ladders have a large frame and brace assembly in the rear of the apparatus with reduced ladder storage in the rear compartment. Ladders may be stored on the sides of the apparatus.

For firefighters to remove ladders off the side of fire apparatus, there must be sufficient clearance between any object and the apparatus. Removing ladders off the side of the apparatus is easy for two firefighters. The ladder securing bracket(s) or locking mechanism must first be released, and then both firefighters can position themselves evenly along the ladder. After releasing the holding mechanism, lift the ladder off the bracket or slide it off the side of the apparatus and transport it to the proper location. Sometimes sliding or lifting the ladder off the apparatus from the side is difficult because the ladders are stacked high. If they are, the shorter roof ladders are normally stacked toward the top.

When removing ladders from an apparatus side on a narrow street with parked cars, sometimes it is necessary to carry the ladder to the front or rear of the apparatus to gain access to the fire building. Lifting the ladder over parked cars might be an option, but one that must be done slowly to allow both firefighters room to walk between the cars.

In a one-firefighter evolution, lifting the ladder off the side of the apparatus is difficult when the ladder's balance point is not marked. Many times firefighters guess the balance point, and when they lift or slide the ladder off the side of the apparatus, it tilts toward the heavier end. Sometimes in this situation, firefighters quickly compensate for the tilting with shear muscle, which can cause a pulled or strained muscle. The firefighter should transport the ladder to the front or rear of the apparatus, then try to lift it over a car and maneuver it when working alone.

Roof ladders mounted on the side of the bed section of an aerial ladder. Many of today's newer aerial apparatus are equipped with a roof ladder mounted to the side of the bed section or on the inside of the fly section of the aerial ladder. The roof ladder is usually a 14- or 16-ft (4.3- or 4.9-m) ladder mounted in a holding bracket assembly. Prior to lifting the ladder out, a firefighter must release the hold down mechanisms. If any of these ladders are needed on the ground for portable ladder operations, they must be handed down off the side of the apparatus prior to raising the aerial. If a ladder is transported up a ladder for use as a roof ladder, there are a few options.

If only one firefighter is available to lift the ladder out of the brackets, he or she should find the balance point and lift the ladder out of the brackets once the aerial ladder is positioned on the roof. Depending on the incline of the ladder, he or she has a few choices on transporting the ladder up the aerial ladder. If the ladder is on a slight incline and no obstacles are present to either side of the aerial ladder, the roof ladder may be laid flat across both rails of the aerial ladder and slid up the ladder as the firefighter slides his or her hands up the rails. The firefighter will be in good control of the ladder and the ladder will be well-balanced. Another way to transport the ladder up a slight incline is to place the ladder on one of its beams and slide it up the rungs of the ladder as the firefighter climbs. Some firefighters prefer to open the hooks of the roof ladder prior to sliding it up the ladder because it enables them to quickly secure it to the ridgepole of the structure once they reach the top of the aerial ladder. Remember that the open hooks can strike an obstacle or part of the ladder and cause a firefighter to slip. No matter how slight of an incline, firefighters must remain in constant contact with a ladder at all times.

On a steeper incline, the firefighter must secure the ladder in a shoulder-carry position. Then the firefighter should climb the ladder using a hand-over-hand motion on the rungs. Climbing at steep angles is dangerous, and with a roof ladder, it becomes more dangerous. At steeper angles, it is often better to perform this tactic with more than one firefighter or it might take more than one trip to transport the ladder and tools.

Roof ladders mounted inside the fly section of an aerial ladder. Some apparatus have roof ladders mounted in holding brackets on the inside of the aerial ladder's fly section. When the aerial ladder rises to its desired location, the roof ladder also rises mechanically and is easily accessible. Mounting the ladder in the fly section eliminates having to carry the roof ladder up the aerial ladder. The ladder securing brackets should be released, and now the ladder may be lifted out of its brackets for use on the roof (fig. 13–36).

Figs. 13–36. Roof ladder mounted on aerial

Use caution when climbing an aerial ladder with a roof ladder mounted in the fly section. The roof ladder takes up space inside the fly section, decreases the space between the rails, and makes climbing more difficult. This is especially true if other tools are mounted on the opposite side of the fly section, the ladder is at a steep incline, or a firefighter is carrying up tools or a saw. Roof ladders mounted inside the fly section can make it difficult to slide a Stokes basket up or down the aerial when performing a technical rescue operation. Removing the roof ladder from the fly section before the rescue might be necessary.

Folding ladder mounted inside the tip of an aerial ladder. For buildings with high, front parapet walls (the front section of a building's wall that extends above the roofline), folding ladders mounted inside and near the tips of aerial ladders are common. To use these ladders for roof access over a high or steep parapet, a firefighter must first release the ladder from its holding brackets. At a high elevation or steep incline, a firefighter must lock into the ladder with an approved safety belt prior to releasing the folding ladder.

After releasing the ladder, the firefighter opens it and slides it down onto the roof. As the ladder comes in contact with the roof, the firefighter should **sound the roof** with the ladder to make sure the roof is stable. Making the transition from the aerial ladder to the folding ladder can be dangerous at high elevations. Always use caution and keep one hand on the aerial ladder prior to transferring yourself to the folding ladder (fig. 13–37).

Fig. 13–37. Folding ladder mounted inside aerial

Collapsible ladders mounted inside a tower ladder bucket. To address high parapet walls, many tower ladder companies carry some type of collapsible or folding ladder mounted inside their buckets or the fly section of the main ladder assembly. Some tower ladder apparatus have a short folding or extension ladder attached to the inside or outside of their bucket doors. Once the apparatus is in position, these roof-access ladders can be deployed (fig. 13–38).

FIREGROUND NOTE

Marking the ladder's balance point can make removal off any apparatus easier.

Fig. 13–38. Collapsible ladder mounted inside tower bucket

PLACEMENT, CLIMBING, AND OPERATIONAL GUIDELINES

Transport

SKILL DRILL

Ladders may be carried numerous ways. A firefighter performing the carry must take command of the operation. Normally, fire academies teach that the firefighter assigned to the butt of a ladder controls the evolution. The butt is normally transported first because it must be properly positioned in line with a window or the objective. Therefore, the firefighter leading the transport should control the overall operation of the ladder.

The firefighter at the ladder's butt must consider ladder length in negotiating turns around corners and obstacles. In some situations, the firefighter at the tip of the ladder must swing out or turn wide. Both firefighters must communicate with one another. "Obstacle to the right," "Swing out," and "Okay, clear," are simple statements that communicate much during transport.

When rescuing a civilian hanging onto a window ledge, firefighters may transport a ladder tip-first. In cases such as this, the firefighter at the ladder's tip takes command. If a firefighter lacks the confidence or skill to control a ladder operation, it might be necessary for the other firefighter to take command

Uneven ground surfaces. **FFI 4.3.6** Uneven ground surfaces are obstacles on numerous firegrounds. Many times, the spot in which the ladder must be placed is not level, and one of the ladder's feet dangles far from the ground. Because the ladder is not supported evenly at the base, it tends to lean toward the unsupported side. One of the easiest ways to overcome this is to find a better location for the ladder. Often this is not feasible and the ladder must remain at this location.

When the ladder is on soft surfaces such as grass, dirt, or even hot asphalt, use the sharp point of a tool or the adze end of a Halligan to dig a trench wide and deep enough for the ladder's foot to touch the ground.

Insert the ladder's foot into the cutout. Both of the ladder's feet should touch the ground, and the ladder should be evenly supported at the base. It might be necessary to lift the base out of the trench and dig deeper or along the area in which the bottom rung sets for the ladder's base to rest evenly on the ground. Remember, climbing a ladder with an unsupported base is dangerous and can cause injuries.

When firefighters face an uneven ground terrain with a hard surface such as a poured concrete walkway or street curb, breaking up the area and digging a trench is not feasible. Many fire departments carry a set of wooden step chocks, angular wood cribbing, and 2 × 4 or 4 × 4 in. (51 × 102 or 102 × 102 mm) cribbing on their apparatus. Using a combination of chocks on hard surfaces levels the ladder base, assists a firefighter in footing, and maintains the ladder's balance. Many fire departments drill a hole at one end of each piece of a 2 × 4, 4 × 4 in. (51 × 102, 102 × 102 mm), and piece of angular cribbing. Next, they place a short section of rope through each hole to make a loop. Now a firefighter can grab the chocks, which come in various heights and shapes, and take them with the ladder (fig. 13–39).[1]

Fig. 13–39. Various chocks. (Courtesy of Nicole Ciampo)

Some cribbing is available in a plastic resin design, but it can be slippery. Ensure a firefighter properly butts a ladder when another firefighter climbs on these hard, uneven, cribbed-up surfaces. It is dangerous, but when life is at stake, it might be necessary.

Another technique passed through numerous fire academies is that a firefighter should survey his or her work area. Sometimes nearby objects such as landscape bricks, blocks, wooden railroad ties, pallets, and lumber can level a ladder's base. Placing a large piece of wood such as plywood on a wet, muddy surface prevents a ladder's base from sinking. When ladders sink into mud, suction develops and it is often difficult to lift them out of the mud. This can cause havoc if a ladder is needed for another fireground task.

Some firefighters place a Halligan bar or other tool beneath the foot of the ladder not touching the ground for support. The Halligan bar, when placed with the adze and point facing down, creates a slope and acts like a wedge with variable height adjustments under the ladder's foot. This creates a few problems: First, on terrain with a slight slope, the metal base of the ladder may slide on the metal tool and cause the ladder to move. Second, why would a firefighter want to leave a valuable tool on the ground when he or she might need it at the top of the ladder or inside a building (fig. 13–40)?

Fig. 13–40. Halligan used as ladder support

Today's technology assists firefighters combat uneven ground surfaces. Some ladder manufacturers make adjustable legs that bolt onto the bed section of ladders near the base. When needed, these legs may be adjusted to level the base.

Proper climbing angle. **FFI 4.3.6** Once a ladder has been removed from the apparatus and transported to its objective, it must be raised and set in a proper climbing angle. Portable ladders must be set in 65–75 degree angles against buildings to ensure the ladders can carry their maximum load capacities. This angle range increases the overall safety for firefighters climbing, descending, and operating ladders. A portable ladder angled lower than 65 degrees increases the probability of the ladder sliding down the building or having its base kicked out; it also reduces the ladder's maximum load capacity. A portable ladder positioned in an angle greater than 75 degrees creates a steep climb and increases the probability that a firefighter will fall.

To attain the proper climbing angle, place the base of the ladder one-fourth the **total working length of the ladder** away from the building. The total working length of a ladder is the distance from its base to the spot of its upper support. For example, the base of a portable ladder raised 24 ft (7.3 m) should be 6 ft (2 m) from the building for a proper climbing angle.

Another quick method of checking proper climbing angle exists. Position yourself in front of a ladder with both feet on the ground. Now, reach outward at about chest level. The ladder's rung should be in a comfortable location for climbing. Although this is not a foolproof way to check a ladder's climbing angle, it is often relied upon during firefighting operations (fig. 13–41).

Also, some ladder manufacturers place stickers on the bed section of portable ladder beams to assist firefighters in placing ladders properly. These stickers resemble a small carpenter's square on a distinctive background. When the bottom line of the sticker is parallel to the ground, the ladder is in the proper climbing position.

Physical ladder position. As a result of fire department ladder construction, most manufacturers insist that when a ladder is raised, the fly section must be in the fly-out position. Another way to describe this is that when the ladder is raised, the fly section must be positioned away from the fire building and on top of the bed section. A sticker affixed to the side of the bed section serves as a reference guide and safety reminder. Note that some ladders can be raised opposite, with the fly toward the building. Firefighters must follow manufacturers' recommendations when using portable ladders.

Fig. 13–41. Quick method to check ladder angle

Raising to the objective

SKILL DRILL

When a ladder becomes vertical, the fly section may be raised to its objective. Some fire departments pre-tie their ladders' halyards or ready them in a continuous-loop design. Either way, a ladder's halyards needn't be untied to raise the ladder. Other departments tie ladders closed with the extra halyard tied or looped around the rungs. In these cases, the halyard must be untied and unwrapped from the rungs to permit raising of the fly section.

Because the halyard is behind the ladder or closest to the structure, the firefighter assigned to the butt is in the best spot to be in control. He or she may easily untie and operate the halyard. Once he or she unties the halyard, the firefighter should place the right or left boot against the ladder's butt to brace it as the ladder rises. Before raising the halyard, the firefighter verifies it is safe to begin operations by rechecking overhead and with the other firefighter. When clear, he or she pulls the halyard down in a hand-over-hand motion. The motion should be fluid and deliberate as the firefighter listens to the dogs clicking over the rungs. If the firefighter were to slip, he or she might be able to lock the dogs on the next available rung or last rung passed. Firefighters should not wrap their arms or hands around the halyard because it is dangerous. Some departments require firefighters to tie off the halyard to a rung after the ladder raise has been completed to keep it from getting in the way of firefighters moving on and around the ladder; other departments just push it to the side, keeping the firefighter who raised the ladder from having to take the time to tie a knot, and as well to allow for the ladder to be quickly raised if needed without having to untie a knot. Determine your department's policy on this issue.

The firefighter in front of the ladder should be well-balanced with each hand on either of the beams. He or she balances the ladder as the halyard pulls downward and the fly section extends. This firefighter also watches the extending fly section and lets the firefighter operating the halyard know when the ladder could hit something such as a gutter or roof overhang and when the ladder reaches the objective. Most ladder operations progress this way. In other situations, firefighters must use a person to act as a guide, watching the tip of the ladder meet its objective. Keep in mind that buildings often have different floor heights—commercial buildings often have higher ceilings than residential buildings.

If a raised ladder lowering into its objective is about to come up short, the firefighters operating it have a few options. A ladder being laid slowly into a structure will hint if it will come up short. Firefighters must bring the ladder back to vertical and raise it to the proper height, then lay it back into the structure. If the ladder is laid into the structure and there is not time for this evolution, the firefighter operating the halyard might be able to pull it, raising it a rung, as the bracing firefighter pulls the ladder off the structure to extend the fly section. In some instances, the ladder is a little long, but bringing it down a rung makes it too short. In such cases, the butt of the ladder might need to be pulled out. Pulling the base too far, in some instances, decreases safety margin and proper climbing angle.

Footing or "butting" a portable ladder

SKILL DRILL

Once a ladder is in a proper climbing angle, it must be secured before a firefighter climbs it. This prevents the ladder's butt end from sliding out while a firefighter climbs or descends. It also prevents the tip of the ladder from sliding or moving along the building, and it stabilizes the ladder as the firefighter climbs. The most common way to secure a ladder is for one firefighter to foot, butt, or heel the base while another firefighter climbs.

The preferred method of footing, butting, or heeling the base of a ladder is for one firefighter to face the front of the ladder in the direction of the climb or the building. This firefighter applies pressure with one boot against one of the bottom rails of the ladder near the base or on the first rung, preventing it from kicking backward. Then, he or she places the other foot behind in a well-balanced position and places both arms on the beams. As one firefighter climbs, the firefighter footing the ladder can apply slight pressure on the rails. This takes any bounce out of the ladder and steadies it.

As the firefighter climbs, the one footing the ladder must watch the climbing motion. If the firefighter were to lose his or her balance, the footing firefighter could quickly climb the ladder while holding the underside of the ladder rails and pin the climber back onto the ladder. The footing firefighter could also climb the ladder quickly and reach up with one arm to support the climber. In addition, if the climber were glancing at his or her tools and didn't notice another issue, the footing firefighter could relay that information.

If a climber were operating off one side of the ladder while in a leg-lock position, the firefighter footing the ladder should place his or her foot on the opposite side of the ladder's base. This prevents the base and tip from shifting and equalizes pressure on both rails, keeping the ladder secure. It might be necessary to place additional hand pressure on the opposite side rail to equalize pressure and secure the ladder.

Some fire academies and departments prefer to foot ladders from the rear. A firefighter positions himself or herself under a ladder with arms mounted on the rails and feet positioned for good balance. Then he or she leans back into the building to keep the ladder in place and remove any bounce during the climb. Problems can occur, however, when this firefighter is behind the ladder. First, if the climber drops a tool, it might hit an unprepared firefighter who was trained not to look up in order to avoid falling dirt from the climber's boot or a shingle. Second, the footing firefighter has no vision of the fire. He or she is oblivious to all conditions behind him or her. Last, if the climber were to lose his or her balance, the firefighter footing from the rear could quickly come around the ladder to offer support, but this traveling momentum could knock the climber off balance (figs. 13–42 and 13–43).

Fig. 13–42. Footing a ladder from the front. (Courtesy of Nicole Ciampo)

Fig. 13–43. Footing a ladder from the rear. (Courtesy of Nicole Ciampo)

If the base of the ladder cannot be footed properly, substantial objects near the ladder such as cars, dumpsters, curbs, other buildings, or even parking lot bumper poles may be used. Normally, this will prevent the base from sliding or kicking out. Also, a firefighter with a piece of tubular webbing pre-tied in a loop may tie the ladder's base to a substantial object with a chocker hitch around the ladder's rung (fig. 13–44).

Fig. 13–44. Using the environment to foot a ladder. (Courtesy of Brian Doyle)

Securing the tip

SKILL DRILL

In the past it was recommended that the first firefighter to reach the objective should secure the tip of the ladder to a substantial object. In theory, this is a good safety tactic. In practical fireground operations, however, it is not always acceptable. If a portable ladder will be in position for a long time for access and egress or at a technical rescue situation, secure the tip. Normally, a hose strap, short piece of utility rope, or piece of webbing on a rung, beam, or both beams can be tied off to a substantial object. Emergencies occur on the fireground and a portable ladder placed at one location might be needed at another. If the ladder were secured at its tip, it could delay its relocation to another urgent matter.

Moving a portable ladder once it is in position

SKILL DRILL

Often, portable ladders raised to a window or roof must be moved for another task or because they are exposed to fire. These ladders can be in either the nested or extended position. In many situations, bringing the ladder vertical and retracting the fly section into the bed section takes too long. Getting the ladder off the building and into a vertical position, then attempting to move it while it is extended is often difficult, especially in inclement weather or on uneven terrain.

> **FIREGROUND NOTE**
>
> Portable ladders with extended fly sections are heavier at the tip than they are when nested. Use care in moving top-heavy, extended ladders.

Some firefighters think sliding the ladder along the building is the answer, but a few things could happen. The ladder's tip could move too fast along the aluminum gutter of the building and slide away, causing a firefighter to lose control. This often causes the ladder to fall, and it could injure a firefighter and damage itself. Also, depending on weather and terrain, the firefighter sliding the ladder could slip and the ladder could slide away. Portable ladders slide quickly across buildings with aluminum or vinyl siding, especially if the siding is wet, icy, or snowy.

Many fire academies teach that two firefighters may position themselves on opposite sides of the ladder, lift the ladder once it is nested in the vertical position, then carry it to the new location. Carrying an extended ladder is difficult because firefighters must maintain an awkward physical position. This becomes even harder in inclement weather and on uneven terrain. In addition, firefighters might have to move the vertical ladder around overhead obstacles.

Rolling a portable ladder, whether it is extended or nested, solves that problem. One firefighter may position himself or herself at the base. If the ladder is on soft terrain and the butt is partially buried in the earth, first a firefighter must pick up the ladder or slide it upward to release it from the earth. This tactic is not necessary, but it does assist in the initial roll of the ladder. Next, the firefighter takes a well-balanced position in front of the ladder with his or her back leaning slightly into the ladder and his or her feet behind it. Now, the firefighter does a push-and-pull motion with his or her arms to roll the ladder to its desired location. When the ladder is being moved to the right, the firefighter's left arm pulls the left beam forward, and the right arm pushes the right beam backward. The motion is opposite when the ladder is moving to the left.

Keep your feet and legs away from the rotating ladder. If a firefighter loses control of the ladder during the evolution, both hands may press it into the building to regain control. It might occur if a firefighter were to roll the ladder quickly along the building. Sometimes it is necessary to take up any slack in the halyard lying behind the ladder and attach it to the bed section before rolling the ladder. If the halyard has too much slack, it might get spun around the ladder during the rotation. When performing this maneuver over a long distance, a firefighter might notice the ladder slide down or lose its proper climbing angle. All a firefighter must do is stop, pick up the base, and slide the tip farther up the building. Then, continue the roll (fig. 13–45).

When operating on a moderate-grade terrain, rolling or lifting the ladder might be difficult. It might be better to take the ladder down, carry it to the new location, and re-raise the ladder.

Fig. 13–45. Rolling a ladder. (Courtesy of Megan Ciamp).

FIREGROUND NOTE

Before moving any portable ladder, talk with the firefighters who used the ladder to determine their positions. If they give permission to move the ladder for another task, you must return the ladder to its original position upon completing the new task. In addition, verbally communicate to the firefighters that the ladder is back in its original position.

Climbing a ladder

SKILL DRILL

Climbing a portable ladder. Once the portable ladder is raised in position, it is almost ready to be climbed. Prior to climbing, the first firefighter should glance up the ladder and make sure all the dogs are locked onto rungs. In three-section extension ladders, it is possible that only the lower set of dogs will lock and the top section's dogs will not engage. Once the firefighter has checked that all the dogs are locked and the ladder is heeled properly, he or she may climb.

The firefighter should climb on the balls of the feet with hands positioned on the underside of the beams and maintain a fluid, deliberate motion while ascending. Climbing this way is safest because it affords the firefighter three points of contact: both hands and a foot. While climbing, the firefighter should keep his or her eyes focused on the destination. An occasional glance at the ladder and surrounding area is okay to check for dangers. The firefighter should not race up the ladder because it could cause a slip or fall. In instances such as rescues, it might be necessary to progress quicker than normal, but firefighters should strive to maintain balance and a fluid climbing motion.

In rain, snow, or ice, the ladder-climbing technique might need to change. The firefighter should place the arch of the boot near the heel next to the rung to permit a better grip onto the ladder. The climbing motion should be slow, deliberate, and consistent with weather conditions. Ladder falls and slips increase during situations such as these.

Many firefighters climb with their hands on the rungs, but this can have bad consequences. First, as a firefighter proceeds up or down the ladder, there is only one point of contact made with the hands. Second, if he or she were to slip, only one hand would be in contact with the ladder and the firefighter could fall. Third, a firefighter climbing with a tool in one hand does not keep contact with the ladder with that hand. While he or she climbs, the other hand would release a rung to reach for the next, leaving a short time of no hand contact with the ladder. Climbing with no hand contact, even for a short period, can lead to disaster.

When more than one firefighter climbs a portable ladder, they should space themselves evenly to avoid interfering with one another. Limit one firefighter to each section of the ladder. When climbing an aerial ladder, firefighters should space themselves at 10-ft (3-m) intervals to evenly displace their weight. During many fireground operations such as a rescue, additional firefighters may be on a ladder. Fire department ladders are made to strict standards. If proper maintenance and inspections are followed, these ladders can withstand grueling circumstances.

Climbing a portable ladder with a tool. Firefighters must be proficient in carrying tools on ladders. It is a common fireground experience. Hoisting tools by ropes may be safer, but it is also more time-consuming. Improperly carrying tools up a ladder reduces a firefighter's grip on the ladder and can lead to a slip or fall. Firefighters must maintain a grip on ladders while carrying tools to prevent any slip, fall, or injury to themselves or firefighters below them.

A firefighter should find a comfortable hand position to climb with a hand tool. Sometimes firefighters prefer to carry tools by their balance points and slide them up the outside of a ladder's beam. If a tool is thin enough, it might be possible for a firefighter to grip the underside of this beam and climb the ladder, allowing both hands contact with the ladder. At all times while climbing, the free hand must maintain a position under the opposite beam and slide up the beam. If a firefighter were to climb the ladder using the rungs with the free hand, he or she would have no hand contact with the ladder for each step (figs. 13–46, 13–47, and 13–48).

When a firefighter climbs a ladder with a hook or pike pole, the tool's head can hook onto a rung, and the firefighter can climb the ladder while maintaining their hand on the underside of the beam. Using this method, the firefighter can climb with two tools: one hook and a hand tool in the opposite hand. Initially, the firefighter reaches up with the hook and places it onto the highest rung within reach. Remember, it is not necessary to reach for the highest rung possible.

Next, two methods can move the hook up as the firefighter climbs. The first occurs as the firefighter passes the top of the hook. Once the hook is about thigh-level, the firefighter can straighten his arm and reach down to grab the hook. Now he or she lifts the hook off the rung and slides it up until he or she can hook it on a higher rung. Again, the firefighter does not need to stretch to place the hook on a higher rung. It should be a fluid motion that does not change foot position.

For the second method, the firefighter stops at the balance point of the hook while climbing. Now, he or she removes the hook from the rung and lifts it to a higher rung. This involves additional times the hook must be moved because the total length of the hook is not used when climbing. This method, however, does not cause the firefighter to bend or lean to one side of the ladder in reaching back for the hook as in the first method.

Fig. 13–46. Slide the tool up the outside of the beam. (Courtesy of Megan Ciampo)

Figs. 13–47 and 13–48. Firefighters should always maintain a grip on the ladder when carrying tools to avoid any slip, fall, or injury. (Courtesy of Megan Ciampo)

Climbing a portable ladder with a saw. Climbing a portable ladder with a saw is difficult, especially if the saw is not equipped with a saw-carrying sling or harness. If a firefighter carries a saw without a sling or harness, he or she should carry only the saw—no other tools. While climbing or descending, the firefighter must maintain constant contact with the ladder by sliding the free hand up the outside or under the beam. The firefighter can place the arm carrying the saw either alongside the opposite beam of the ladder using the ladder as a guide, or he or she can carry the saw alongside himself or herself. The firefighter must use caution. The saw, in this position, has a tendency to hit the firefighter in the leg (fig. 13–49).

Fig. 13–49. Carrying a saw

While climbing an aerial ladder at a low angle, the firefighter might be able to slide the saw on top of the ladder's sections. Use caution when the sections become narrower because the saw changes positions.

When climbing with a saw in a sling or harness, there are a few options. The firefighter may place the sling over the head and let the saw rest underneath the arm and back as he or she climbs. Another variation is to push the saw farther back and up on top of the SCBA's bottle. Doing this can assist firefighters when climbing steep angles and keeps the saw from hitting the side of the ladder.

Some firefighters prefer to climb with the saw sling over a shoulder. The sling can ride down the shoulder as the firefighter's arm position changes. The firefighter must use caution when climbing with a dangling saw because the saw can throw off the firefighter's balance and cause a slip or fall. Firefighters must maintain hand contact with the ladder at all times to prevent this (fig. 13–50).

Climbing a portable ladder with an uncharged hoseline. When an uncharged hoseline must be carried up the ladder, have sufficient hose at or near the base of the ladder to ensure it is deployed smoothly as the firefighter climbs. Insufficient hose anytime during the climb could cause the climbing firefighter to be pulled backward off the ladder.

Fig. 13–50. Using a sling to carry a saw is recommended

Fig. 13–51. Place the hose under one shoulder and over the opposite one.

The firefighter climbing with the nozzle should place the hose under one shoulder and over the opposite shoulder with about 2–3 ft (0.6–0.9 m) of hose extending over the back. A firefighter at the base of the ladder should feed the hose up as the firefighter climbs. If the climb is high, a second firefighter might need to be in the middle of the climb and another firefighter at the base of the ladder to assist in hose deployment (figs. 13–51 and 13–52).

The second firefighter should position the hose over the shoulder on the side the hose is being raised. The hose should have some slack and form a small loop to the side of the ladder. This loop prevents the second firefighter from being pulled off the ladder should the hose become taut or be pulled up during the climb.

The hose must be fed at a steady pace and kept outside the climber's body to avoid tripping and entangling the climber.

Advancing a hoseline up a ladder is easier when the hoseline has not been charged with water. If the hoseline has been charged with water, climbing might be easier if you shut the line down and partially drain it. It might be necessary to place firefighters along the line of the ladder to advance the hoseline. These firefighters should be secured to the ladder by either an approved safety belt or leg-lock maneuver. If that is not practical, send a few firefighters up to where the hoseline is needed. Drop a utility rope down and pull the hose up and over a hose roller to this location.

Fig. 13–52. The firefighter at the base feeds the hose.

Climbing and working on an aerial ladder. An aerial ladder's angle often determines how a firefighter climbs it. If the ladder is at a low angle, the firefighter climbs in a hunched position and must maintain at least one hand sliding along the top of the ladder's rails. If the ladder is not extended far, the rails are wide and the firefighter is more comfortable sliding the rail toward the edge. In this manner, he or she can grip the outside of the rail to maintain grip and balance. Sliding the hand flat on the wide surface of the rails should be avoided.

When the ladder is at a steep angle, it might be difficult to reach the top of the rails and climb, especially at the first few sections. Some aerial apparatus are constructed with wide base sections that make it difficult for firefighters to reach. In these situations, they can use the rungs. Climbing should be done in a fluid, hand-over-hand motion while transferring boots from rung to rung. Steep-angle climbs should be done slowly and safely. A slip or fall could result in a severe injury or fatality (fig. 13–53).

Fig. 13–53. Be careful when performing steep angle climbs. (Courtesy of Andy Moloney)

As with climbing portable ladders, a firefighter climbing an aerial ladder with a tool such as a hook can place it on the aerial's rung and move it as he or she climbs and descends. At some angles, it might be more comfortable to slide the tool along the upper portion of the aerial ladder's rail. When a firefighter climbs an aerial ladder at any angle, he or she must maintain a grip on the ladder. Aerial ladders slide between greased tracks, and small drops of grease on the rungs often create slipping hazards. A firefighter walking down an aerial ladder like it is a tight rope or balance beam with no hand contact is an accident waiting to happen.

When working from aerial ladders, firefighters should never perform any leg-lock maneuver through the rungs or beams. If a ladder were to move or mechanically fail, a firefighter could be injured severely. When performing any type of work from an aerial ladder, a firefighter should lock himself or herself into the ladder by wearing an approved safety belt.

FIREGROUND NOTE

Remember, if a charged hoseline is tied onto a portable ladder with a hose strap, the opening and closing of the nozzle will cause back pressure in the hoseline and possible shifting. An unattended ladder could shift and fall if it is not properly secured.

Climbing a vertical ladder on a fire escape. In some areas with exterior fire escapes, firefighters commonly ascend and descend these appliances. Most fire escapes are constantly exposed to the weather, so they might not be in the best structural condition. Climbing them can be hazardous and difficult. Firefighters must use caution on these secondary means of egress.

The small vertical ladder, or **drop ladder**, that connects the first floor landing to the ground when released and the top vertical ladder, or **gooseneck ladder**, that connects the top floor landing to the roof, are difficult to climb. Prior to climbing any vertical ladder, shake the ladder to check its stability and if it is attached to the building and the fire escape's main structure.

Once a firefighter verifies stability, he or she may climb the ladder. A hook in a firefighter's hand can hook either onto the highest reachable rung or the fire escape landing itself. Then the firefighter can maintain hand position on the rails or rungs of the ladder. With these narrow, straight ladders, it is often difficult to maintain a hand position on the rails. Climbing with hands on the rungs is permissible. A firefighter should climb slowly, placing feet toward the rail on each rung and ensuring with each step that the ladder can maintain his or her weight.

FIREGROUND NOTE

A firefighter should always position him or herself under the fire escape landing when releasing the vertical drop ladder. These ladders have fallen out of their tracks when released. A falling ladder could injure a firefighter in the wrong position (fig. 13–54).

Fig. 13–54. Position yourself under the landing to avoid a falling ladder.

Operating and working from a ladder

SKILL DRILL

Throughout their careers, firefighters work with hand tools on portable ladders. Usually, firefighters place ladders to the side (overhauling window trim) or possibly under the work area (removing window bars). In both cases, a firefighter's hands are on his or her tools and he or she won't be able to hold the ladder properly. The first solution is for firefighters to wear and use certified ladder belts or personal safety belts and hook into the ladder. Hooking a safety belt into the middle of a rung, however, can allow a firefighter to slide. If you hook onto a ladder's rung, position the safety hook near the beam to reduce the risk of sliding. There are a few alternatives for a firefighter not wearing an approved safety belt while working from the ladder. He or she can perform a leg-lock maneuver to secure onto the ladder.

Leg-lock maneuver. The **leg-lock maneuver** (fig. 13–55) was developed years ago when firefighters still used **hip boots**, three-fourths rubber boots, and had more flexibility. Today's firefighters wear bunker pants with large side pockets that can hold numerous items. Performing the standard leg lock is difficult and often firefighters opt for alternatives. Whatever type of lock firefighters choose, they must secure themselves onto ladders.

Figs. 13–55. Leg lock

Firefighters perform the standard leg-lock maneuver on the ladder side opposite the work side. For example, if a firefighter were to ventilate a window on the right side of the ladder, the firefighter's left leg would perform the leg lock. The standard leg-lock maneuver is done in the following manner:

- First, a firefighter reaches the proper working height on the ladder.
- Then, he or she climbs to the next higher rung.
- Next, he or she places the leg opposite the work side through the rung spacing. As the firefighter's knee approaches the rung, he or she can begin to bend the leg back toward the next lower rung. The leg is now inserted into this spacing. Depending on the firefighter's physical makeup, the foot may lock onto the outside of the ladder's beam or onto the next lower rung. Some firefighters even place their boot behind their knee to lock themselves into the ladder.
- Now, the firefighter steps down a rung and places a boot nearest the work side, next to the beam of the ladder. Now he or she is in the original work position on the ladder.

Normally when a firefighter completes the maneuver, he or she is two rungs below the rung through which he or she inserted a leg. Many firefighters can stand on the rung at a desired work location, lift a leg up the two rungs, and perform the maneuver. Others with shorter legs find it uncomfortable. It is fine to position themselves one rung below the bent knee.

Hook-in leg-lock (hill) maneuver. Because now firefighters carry bail-out ropes, pliers, cutters, screwdrivers, chocks, etc., in their bunker gear pockets, performing the standard leg-lock maneuver is a nuisance, takes too long to get into, and is uncomfortable. In addition, many firefighters feel trapped in the ladder because releasing the leg lock takes a while.

To prevent ladder falls by firefighters performing no leg lock, a few leg locks have been modified. The **hook-in leg-lock maneuver**, developed by Lieutenant Michael Ciampo of the FDNY, is relatively quick and simple, and it reduces the bending and twisting of a firefighter's leg. Plus, firefighters of varying shapes and sizes can perform the maneuver.

Firefighters can perform the hook-in leg-lock maneuver with the leg opposite the work side of a ladder or the leg on the same side of a ladder. A firefighter's physical makeup decides which position is suitable. Follow these steps to perform the maneuver:

- Reach the proper working height on the ladder with both feet on the same rung.
- Then, place a boot in the center of the rung with the heel slid into the rung.
- Next, lift the other boot over the rung and through the rung spacing.

Variations can be performed depending on a firefighter's body:

- The standard tactic is to place the heel of the boot that has come over the rung on top of the boot's toe that is positioned on the center of the rung. By maintaining downward pressure on the boot while the thigh and knee are wedged under the rung, you are locked into the ladder. Position the thigh and knee next to the beam of the ladder to prevent sliding. Firefighters with shorter legs might have to turn the bottom boot's tip toward the beam for the thigh and knee to lock securely. Firefighters should not perform this maneuver with the thigh and knee positioned in the center of the rung. Movement in either direction could cause a firefighter to slide across a rung and the ladder or firefighter to fall.
- Some firefighters might be able to place the heel of the lifted boot over the rung onto the rung they are stepping on as the locking device.
- Some firefighters feel more comfortable placing the tip of the boot lifted over the rung onto the rung they are stepping on or over the beam of the ladder as the locking device.

Practicing the hook-in leg-lock maneuver determines which option suits a firefighter's body style (figs. 13–56 and 13–57).

Figs. 13–56 and 13–57. Hook-in leg-lock (hill) maneuver

Hyperextend leg-lock (hell) maneuver. Many times firefighters execute unique, dramatic ladder rescues with no time for safety maneuvers. There are also times when victims hand a child to a firefighter who must release his or her hand grip on a ladder to accept the child. Positioning into any of the leg locks can be difficult and dangerous for the firefighter and victim during such a transition.

Lt. Ciampo has also developed the **hyperextend leg lock** to assist firefighters who might momentarily lose their grip on a ladder, slip, or who perform rescues. Its primary purpose is to help firefighters maintain balance, control, and prevent ladder falls. Although nothing is safer than being belted in or performing a leg-lock maneuver, sometimes those are unachievable. To perform the hyperextend leg-lock maneuver:

- Reach the proper working height on the ladder with both feet on the same rung.
- Then, raise one leg, similar to standing on the tip of the toe or flexing a calf muscle. You may use either leg to perform this tactic and should be well-balanced, not attempting to balance on the toe itself.
- As the leg extends upward, wedge the lower thigh or knee under the ladder's rung. Firefighters with short legs might not reach the rung with a knee. If you can't reach, do not attempt to rise all the way up on a toe, but rather follow the next direction.
- As this is performed, you may press outward with both legs and wedge yourself between the ladder's beams for additional locking support. This tactic might be the only one to follow for firefighters with short legs.

This maneuver should be done with the raised leg near the ladder's beam to prevent sliding and movement, especially if a firefighter must reach off the side of the ladder for a victim. Firefighters should never come up on both legs to wedge themselves under the rungs. This gives them little foot contact with the ladder, and any sudden movement could cause them to slide back or lose their balance on the rung and possibly fall (figs. 13–58 and 13–59).

Fig. 13–58. Raise one leg up on tip-toe.

Fig. 13–59. Press outward with both legs.

Fig. 13–60. Arm-lock

Fig. 13–61. Place your left arm between the rungs.

Arm-lock maneuver: venting windows from a portable ladder. The **arm-lock maneuver** can be performed by a firefighter venting windows from a portable ladder.[2] It is relatively easy and requires minimal time to get into position on the ladder, thus saving time for other tasks. It may be used in conjunction with the leg-lock maneuvers previously described or by itself.

To perform the arm-lock maneuver, follow these steps for ventilating a window to the right side of a ladder:

- Reach the proper working height on the ladder with both feet on the same rung. When venting the window off the side of the ladder, place a leg and foot up against the ladder's beam. This offers more support and balance as you swing the hand tool. It also braces you onto the ladder.
- Then, hold the hook in the right arm and place the left arm between the two rungs directly in front of you.
- Then, place the butt end of the hook behind the ladder and grab it with the left hand.
- Next, position the hook to the window and check that no part of the ladder or structure will interfere with the swinging. Now the window can be ventilated (figs. 13–60 and 13–61).

The hook used for ventilation acts as the safety brace for you. If the hook were to bounce off the window and the recoil were too much, you must only pull yourself back into the ladder to prevent falling. When performing the arm-lock maneuver:

- Don't choke up on the hook. The entire length of the hook must be used to clear the opposite beam of the ladder while swinging the tool. If the tool is not used properly, it could cause the swing to come to an abrupt halt, shake the ladder, and cause the hook not to vent the window.
- You must split the rungs with your arms. Hugging the ladder can cause you to pinch a hand between the hook and the ladder's beam.

- Resist pulling material out of the window. The butt end of the hook could come back and strike you or the ladder.
- You can create a small working space between your body and the ladder that allows for easier swinging of the hook.

Another method to vent a window involves the hook's butt end being placed behind the firefighter's back with his or her arm holding the hook. The firefighter's opposite hand holds the rung of the ladder and secures him or her to the ladder. Next, the firefighter can twist or quarter turn with the upper body. The momentum assists in letting the hook strike and break the window. If the firefighter chooses this method of venting a window, he or she must remain below or toward the side of the window and out of the path of the escaping heat, smoke, and gases (fig. 13–62).

Fig. 13–62. This allows the firefighter's free hand to secure him or her to the ladder.

FIREGROUND NOTE

When you first make physical contact with a victim, ask if anyone is trapped in the building.

PORTABLE LADDER RESCUES

SKILL DRILL

FFI 4.3.6 One of the most harrowing experiences for a firefighter is a portable ladder rescue. A victim waiting for a ladder under smoke, fire, and heat conditions normally is panicked and irrational. A firefighter must be trained in rescue procedures for the ladder rescue to be successful.

Often victims reach for and grab onto ladders being raised in their vicinity. Some have even jumped for ladders being raised toward them. Firefighters must constantly watch victims. Once a victim is on a ladder, he or she might become aggressive and panicked. Firefighters must be prepared for and in control of victims once they reach the safety of ladders.

When first faced with victims at windows, make verbal contact with them. Firefighters must take command and judge victims' conditions and states of mind. A severely panicked victim might lead a firefighter to throw a ladder out of the victim's reach, then roll the ladder to him or her after it contacts the structure. In other instances, a victim might remain calm and self exit with no coaxing when a ladder arrives.

When a firefighter reaches a victim able to self exit onto a ladder at a window, the firefighter should make sure the victim faces the ladder. The firefighter should place both arms around the victim and onto the back of the ladder's beams. If the victim becomes panicked during descent, the firefighter can press him or her into the ladder to gain control or stop the descent.

Firefighters must not place their hands on ladder rungs while descending with a victim. If a victim were to suddenly move, turn, or become panicked while the firefighter changes hand position, both victim and firefighter could slip and fall. During the descent, it might be necessary to reinforce verbal commands or encouragement to keep the victim focused and not panicked.

If there are enough firefighters, another firefighter may climb up behind the firefighter and victim, extend an arm up to the firefighter's back, and offer physical and verbal support as they descend.

If two firefighters are available at the start of the rescue, one firefighter might enter the building prior to the victim's removal. If conditions aren't immediately dangerous, the victim might be able to be sheltered in place. If conditions are intolerable or if the victim is panicked, however, this firefighter can assist in lifting or placing the victim into the window and onto the ladder for the other firefighter to remove.

Semiconscious removals

SKILL DRILL

Victims might become disoriented, semiconscious, dazed, or confused. In a **semiconscious removal**, a firefighter can continue their descent down a ladder much like a conscious removal, with one exception: the firefighter may place a knee into the buttocks of the victim as they descend. As the firefighter goes from rung to rung, he or she may transfer the victim's weight from knee to knee. Some firefighters prefer to use just one knee during the whole operation, but on longer climbs, this leg could become more fatigued. The firefighter must place only the ball of the boot onto the next rung. If he or she places the boot too deeply into the rung, lifting the victim's weight and completing the removal is harder (fig. 13–63).

Fig. 13–63. The semiconscious removal

FIREGROUND NOTE

When faced with aluminum gutters on a structure, a firefighter may gently drop the ladder into the gutter or apply pressure into the ladder with the beam to dent the gutter. A good spot is between the attachment spikes. This prevents the ladder from bouncing or sliding as the firefighter climbs. It also allows more room on the rungs for a firefighter's boot when he or she nears the top.

Unconscious removals

SKILL DRILL

In an **unconscious removal**, a firefighter may place both arms out and onto the beams of the ladder, accepting a victim across both arms. The firefighter can control the descent by pressing the victim into the ladder with his or her chest. Sometimes, the victim's legs can be split to assist in the carry down. Cradling an adult victim can be dangerous because the victim might slip through the arms during the descent.

If additional personnel are on scene, a second ladder may be placed next to the first to assist in the rescue. The ladder may slide up behind the victim on the leg side so two firefighters support the victim.

Children are always carried down ladders. In addition, a rescue is not complete until victims are out of harm on the ground. These techniques may also be used to rescue downed firefighters.

In the case of multiple victims, firefighters must decide who is in the most danger and needs to be rescued first. Normally, the most critical places to address first are adjacent areas, then the floor above. Victims two floors above a fire and a few rooms away might also need to be removed, but they might not be the first priority (fig. 13–64).

Fig. 13–64. An unconscious removal

PORTABLE LADDER PLACEMENT

Access and egress

SKILL DRILL

FFI 4.3.6 Safely placing a portable ladder's base is important, but equally important is resting the ladder's top against a reliable structure. Normally the top of a ladder rests on a roofline, eave line, parapet, or wall. Both beams of the ladder near the top must make contact with the structural support. As a firefighter climbs a ladder in this position, the ladder rests firmly on the support. If one of the beams did not touch the support, the ladder being climbed could twist, bounce, or slide off the structural support. Beware of awnings on the front of a row of stores—they have no structural stability. Placing a ladder against the awning and walking on it may cause the awning to collapse, dropping firefighters to the ground.

As the ladder reaches the reliable structure near its tip, the firefighter should check the support's sturdiness. If the support is questionable, weak, or unreliable, the firefighter should gently press in on the ladder about chest level. This added pressure shows whether the structure can support the firefighter. If doubt still exists, the ladder must move to a reliable support area. If a firefighter were climbing the ladder and the support were to fail suddenly, he or she must attempt to quickly climb down the ladder. Remember, before a firefighter climbs a ladder, he or she must ensure that the base and tip are positioned properly, securely, and safely.

Roof level

SKILL DRILL

Portable ladders providing roof access (fig. 13–65) should extend at least five rungs above the roofline because:

- This increases the ladders' visibility when smoke obscures rooflines and means of egress.
- Extra rungs above the roofline assist firefighters dismounting and mounting ladders when they get onto or off roofs. Firefighters have better hand grips on ladders and don't overreach as they get on or off ladders. In addition, they are in more comfortable, upright body positions to transfer onto roofs or ladders. Ladders that aren't raised high enough make firefighters bend over rooflines to grab the ladders. When firefighters lean forward to grab ladders, they lean off the edges of buildings, which is dangerous. A fall could be fatal.
- If ladders have a light beacon attached to the end of one of their rails, they might be more visible in smoke.
- If a firefighter were to accidentally slide down a pitched roof, more rungs above the roofline could stop his or her descent.
- When placing a roof ladder to a pitch roof, a firefighter at the top of the ladder can lean into the ladder for support as he slides the roof ladder up onto the roof.

In situations such as short porch roofs with varying grades of slope, extending the ladder five rungs above the roofline might interfere with firefighting operations. It might be necessary to decrease the number of rungs extended over the roofline.

A firefighter may perform a simple procedure prior to transferring completely off the ladder. If he or she steps onto a pitched or flat roof, a boot or hand tool may **sound the roof**. This means banging a boot or tool to test a roof for structural soundness. In many instances, this tactic informs firefighters that a roof's underside is weak or burned away.

Firefighters should avoid jumping off ladders onto roofs. The impacts could be a problem, especially on a roof weakened by fire. Firefighters should slowly transition from ladders to roofs. Doing it too quickly sometimes causes ladder tips to slide along rooflines.

Fig. 13–65. The ladder should extend five rungs above roof level.

Windows

SKILL DRILL

A portable ladder should be placed to a window with its tip slightly below or level with the windowsill. A tip extending into a window decreases the window's size, which hampers entering, exiting, and performing victim removals and rescues. In addition, during a headfirst, ladder-slide maneuver, the ladder could be knocked from its position or a firefighter could get caught on the ladder's tip. A ladder at this location provides the safest, most effective window-entry method.

Until recently, firefighters relied on an older window-entry technique. They placed ladders next to windows with the tips even or level with the tops of the window-sills. Once a window had been entirely removed to create an access point, a firefighter stepped from the ladder onto the sill and into the building. This is dangerous for a few reasons. First, the firefighter enters high in the window and exposes himself or herself to exiting smoke, gases, and heat. Second, a slip could mean a high fall. And last, it is common for a ladder's tip to slide once the transition from ladder to window occurs.

There is another danger in placing the ladder in this position. If a firefighter must exit a structure rapidly, it can be a long reach for the ladder under difficult circumstances. During this reach, ladders often slide when firefighters transfer their weight from windows to ladders. Fire debris on firefighters' boots can also increase slips and falls during this evolution.

After the tip of the ladder is in position at the window and the ladder is footed by another firefighter, it is ready to climb. The climbing firefighter approaches the window and prepares to ventilate the window for climbing access. Prior to venting, the firefighter positions himself or herself into a leg-lock maneuver. Now the entire window and sash should be removed for easier entry. After completing window removal, firefighters must remember they've created a flow path toward the open window. Next, they would check the floor conditions with the **sweep-and-sound maneuver** before entering the window. The firefighter keeps his or her head outside the building toward the window's side and uses the structure's wall for protection. Next, he or she sweeps the floor with a hand tool to look for a victim. If the tool meets no resistance, the firefighter sounds the floor with a tool. Sounding the floor ensures that there is a stable floor in the immediate area and moment. Then the firefighter may drop the tool forward and listen for that distinct sound, reinforcing that the floor exists.

Now the firefighter may enter the window, driving the right shoulder into the bottom of the left side of the window frame with his or her face facing outward while climbing toward the top rung. While climbing over the top rung, the right leg enters the window while the left hand maintains a grip on the ladder's rung or beam. Next, he or she slides the buttocks back toward the right side of the window frame. This way the firefighter is low in the window frame—away from escaping heat, gases, and smoke. As the firefighter's right leg enters the room, it again sweeps and sounds the floor. The firefighter's physique and the window's size determine which of two options to select. If the window's opening is small, a typical firefighter lifts his or her head into the room and over the sill, riding down the wall to the floor. The left leg naturally follows as the firefighter enters the room. If the window's opening is larger, a firefighter can bring the left foot up behind and into the window while keeping the head outside. Numerous safety items protect the firefighter performing this maneuver:

- He or she maintains contact with the ladder at all times. If the room were to suddenly ignite, the firefighter could reverse direction and exit quickly.

- A large portion of the firefighter's body is outside the window and limits exposure to superheated gases and smoke that could ignite.
- The firefighter can quickly exit a building in the same fashion as entering. He or she does not have to reach or look for a ladder placed to the side of the window. This also reduces the chances of slipping and falling.
- If a firefighter is trained in self-exiting techniques such as the headfirst ladder slide, the ladder is in position from the start.

In another window entry, the firefighter goes over the sill headfirst after sweeping and sounding the floor. Some instructors teach this tactic because it allows the firefighter to slip under the escaping gases and heat. Many fire academies shy away from teaching this method because of its disadvantages:

- With today's newer **lightweight construction**, a firefighter going in headfirst could travel through the fire-weakened floor itself.
- A firefighter could land on furniture near the window, causing it to tumble over and land him abruptly or farther from the window.
- If a quick escape were needed, the window that the firefighter came through might not be visible. He or she might dive into the wall looking for it on an escape.
- Broken glass on the floor could cut a firefighter as he or she extends the hands during the headfirst descent.
- Firefighters entering head first often do this much faster than when entering leg first. Controlling the complete entry is more practical.

> **FIREGROUND NOTE**
>
> As firefighters leave ladders and proceed along roofs, occasionally they should sound the roofs for weakness with hand tools or their feet.

Balcony, deck, or fire escape

SKILL DRILL

Firefighters face numerous victims trapped or congregating on balconies, decks, and fire escapes. Or, this avenue might be the only access point to a fire building, so firefighters must ladder it. All are exposed to weather and constantly deteriorate. Introducing a fire department ladder with a firefighter's weight could break them. It is common to see bent, rusty, or even missing handrails, railings, and balusters on these structures.

It is best to ladder the side of the balcony, deck, or fire escape and use the building as the ladder's main support. When a ladder is placed to the side of a structure, it should be placed with the tip several rungs above the side of the railing with the ladder's beam next to the railing. After climbing, a firefighter can swing a leg over the railing and maintain a hand on the ladder while checking the structure's stability with his or her leg. It is also easier to remove victims because the ladder is more stable. In addition, the ladder is out of the way and does not interfere with operations on the platform. Plus, it is less likely to slide along the handrail and fall.

If a ladder's beams were placed on and just over the front of the railing when a victim transfers his or her weight over the railing, the ladder might slide along the handrail, which could cause panic. This can also happen when a firefighter transfers his or her weight off the ladder and onto the platform. A firefighter must butt the ladder when any firefighter transfers from ladder to platform. If it is necessary to place a ladder in this position, a firefighter must remember before climbing to forcefully tap the handrail with the tip of the ladder to ensure it is structurally sound and can carry the weight of the ladder and firefighter.

Another option for decks that are lightweight, have minor structural defects, or that can't be laddered on the side is to place the ladder at the front of the deck. Both beams of the ladder should rest on the structure's platform frame or flooring. Now, a firefighter with a hand tool may climb the ladder and remove some balusters, permitting an egress point. Using the deck's structural frame is often better than placing the ladder's beams on the top handrail.

Fire escapes often become overcrowded in an instant because the vertical drop ladder is hard for civilians to descend. To relieve overcrowding and make descent easier, place a portable ladder to the first landing opposite the fire escape from the drop ladder. If the fire escape is severely overcrowded, another portable ladder may be positioned on the second landing on the same side of the drop ladder.

CARRIES AND RAISES

There are many ways to carry, transport, and raise a ladder. Many carries and raises depend on the number of firefighters available. Normally, ladders are taken butt-first toward their objectives. Sometimes it is necessary to carry the ladder with the tip forward. As we often find in firefighting tactics and techniques, each has advantages and disadvantages. It is up to the operating firefighters to choose the best technique for each situation.

Single-firefighter evolutions

SKILL DRILL

Although not the preferred method to transport or raise a ladder, one firefighter sometimes finds himself or herself performing the tasks of two. It is essential to learn single-firefighter evolutions when dealing with portable ladders.

Single-firefighter shoulder carry. A firefighter on one side of a vertically stowed ladder may lift the ladder at its balance point with the butt end facing the front or direction of travel. Then, he or she inserts a shoulder between the rungs and rests the ladder on the shoulder. He or she may place this arm's hand on the bottom of the beam, at the base, middle, or top of the rung, or on the next forward rung. If the other hand is free, this hand can grasp the next forward rung to balance the ladder. As the ladder sits on the firefighter's shoulder, it may be tilted slightly down toward the base to help control the ladder, provide good visibility, and prevent striking another firefighter with the butt (fig. 13–66).

Fig. 13–66. Single-firefighter shoulder carry

Single-firefighter suitcase carry. A firefighter on one side of a vertically stowed ladder may locate the balance point of the ladder and hoist the ladder with one arm. He or she may then carry the ladder like a suitcase with his or her arm dangling to one side. Ladders carried in this position often strike firefighters in the leg if they move quickly or walk over rugged terrain. If a ladder is to be picked up off the ground, a firefighter should bend his or her legs to prevent a back injury. This carry is practical for short distances and small, straight ladders. It also allows a firefighter to carry tools in the other hand (fig. 13–67).

Fig. 13–67. Single-firefighter suitcase carry

Single-firefighter ladder drags. During fire operations, firefighters might find themselves shorthanded or alone when tactics must be performed simultaneously. To assist a single firefighter in transporting ladders, firefighters have created ladder drags. These drags may be used on all surfaces during any weather. Both drags incorporate dragging tools and two ladders to a destination in one trip, which benefits short-staffed firegrounds.

The **Swick method** was created by Robert Swick of the Fairborn (Ohio) Fire Department who was assigned to a one-man truck company for 17 years. To perform this drag, a firefighter lays an extension ladder fly-section-up on the ground. Then, he or she places a roof ladder with open roof hooks on top of the extension ladder. The hooks bite into the top rung of the extension ladder, connecting the ladders. Next, the firefighter lays his tools on top of the ladders. Now, he or she goes to either side of the ladder nearest the tip, crouches, and places his or her arm downward to pick up the extension ladder's bottom rung. The firefighter stands up to lift the ladder with the leg and back muscles. It is more comfortable for some firefighters to place the bottom rung inside the elbow and place the hand under the ladder's rail. The ladder sits higher and into the firefighter's chest area. Now the ladders may be dragged to their destination (fig. 13–68).

Fig. 13–68. The Swick method

There are pros and cons. When a firefighter turns a corner with a long roof ladder on top, the longer ladder can catch on uneven terrain and spin slightly off the top of the extension ladder. In this case, the firefighter must only slide the roof ladder back on the extension ladder and continue dragging to the destination. Another drawback is that the ladder's tips or spurs can scratch during dragging over pavement or concrete. If this happens, small damages may be filed or sanded.

One other thing that bothers firefighters is the uncomfortable drag. The ladder has a tendency to rub the turnout coat under the armpit, and firefighters with shorter arms don't have the reach to carry the ladder in this position. When firefighters come across a problem or situation, they quickly find a cure.

The **Creed method** of dragging a ladder evolved after Creed McClelland of the Orlando Fire Department performed the Swick method at a fire conference. After pondering a few minutes about the drag, he came up with a variation. A firefighter may place a roof ladder flat on the ground with its open hooks pointing up. Now, he or she slides an extension ladder bed-section-down on top of the roof ladder and into the hooks. The hooks act like a stop and permit the extension ladder to ride on top of it. Now, the firefighter may come to the foot section of the ladder, lift the first rung, and drag the ladders to their destination. When using a longer roof ladder on the bottom of the drag, a firefighter has more rungs accessible near the base of the ladders, making it easier to drag. This method also removes the possibility of the ladders separating during the maneuver (fig. 13–69).

Fig. 13–69. The Creed method

FIREGROUND NOTE

The Stokes basket may be placed into the rungs of the roof ladder, filled with tools and equipment, and dragged to the fire scene for rapid intervention team operations or during inclement weather (Fig. 13–70).

Both drags take little time to set up and are easy. They allow a firefighter two ladders and tools in one trip, conserving energy for other tasks.

Fig. 13–70. Using a roof ladder to transport a Stokes basket

Single-firefighter flat raise. Although it is preferable to raise a ladder with another firefighter, it is not always possible on the fireground. All firefighters must know how to raise a ladder with one person and practice the skill. After a single-carry transport technique, a firefighter must check for overhead obstructions such as tree branches and electrical service lines, then place the ladder's butt end against a structure for bracing. Now, the firefighter kneels or bends at the tip of the ladder and lifts the tip off the ground using the leg muscles. He or she then walks the ladder up with hands sliding up the rails of the beam and eyes focused on the ladder rising overhead. As the firefighter slides hands up the rails, the fluid motion should not include a bouncing rhythm, which puts undue strain and stress on the firefighter's body (fig. 13–71).

If the firefighter uses a hand-over-hand motion on the rungs to raise the ladder to vertical, a few things may occur. First, the ladder balances on only one arm as it rises. If a loss of balance occurs, the ladder could drop. Second, ground conditions determine the possibility of slipping, and balancing a ladder with one arm might be difficult for the strongest of firefighters. Those who use this method often proceed quickly, which has caused lost grips on the rungs and ladder drops. When using the hands-on-the-rail method to raise the ladder, a firefighter slipping or losing balance could push the whole ladder away to avoid injury.

Fig. 13–71. Single-firefighter flat raise

Single-firefighter high-shoulder carry and raise. Before becoming firefighters, many of us worked on job sites with lightweight, easy-to-maneuver ladders not built to fire department standards. Some fire academies instruct firefighters on a high-shoulder carry and raise that evolved from other occupations. The cons outweigh the pros. There are better, safer ways to perform a single-firefighter carry and raise evolution.

As a firefighter removes a ladder from an apparatus, he or she carries the balance point of the lower beam on top of the shoulder or places the shoulder inside the beams and rungs near the balance point. Then, the firefighter raises the other hand to grab the upper beam or a forward rung to balance it. He or she then transports the ladder, tilting the base toward the ground to avoid striking another firefighter (fig. 13–72).

As the firefighter approaches the proper distance from the fire building, he or she plants the base of the lower beam in the ground to act as the foot. Now, the firefighter uses upper-body strength to push the ladder up and walk it up in a hand-over-hand motion and beam-raise position. In other situations, a firefighter may place the base against the base of the fire building and then raise it upward. This carry and raise puts a lot of stress on the firefighter's upper body. Time and conditions permitting, it is better to transport the ladder to its location in the high-shoulder carry than to perform the single-firefighter flat raise.

Fig. 13–72. High-shoulder carry and raise

Two-firefighter evolutions

SKILL DRILL

When two firefighters begin ladder work, they must perform safely, communicate effectively, and use teamwork. One firefighter must lead the operation. The firefighter at the ladder's butt normally leads and gives commands. This firefighter leads the operation because the butt of the ladder arrives to the structure first with this firefighter in the lead. Once the ladder rises, he or she controls the halyard and raising the ladder.

Two-firefighter shoulder carry. Two firefighters face the same direction on the same side of the ladder, one at the butt, and the other at the tip. Then, both squat, grab a rung, and with their leg muscles lift the ladder onto their shoulders. As they place the ladder onto their shoulders, each firefighter's same arm can enter between rungs and grab the next forward rung. The butt firefighter should be inside the first or second rung spacing to protect the butt and use the other arm to warn approaching firefighters, move obstacles, and open doors and gates. The firefighter at the tip following the butt firefighter steers the ladder around obstructions and corners (figs. 13–73 and 13–74).

The two-firefighter shoulder carry is for removing a ladder from an engine or ladder truck and for farther distances than the suitcase carry. If a third firefighter is available, he or she goes in the middle. A very tall third firefighter will throw off the ladder's balance. He or she might have to keep the shoulder free from the ladder and use only arm strength.

Figs. 13–73 and 13–74. Two-firefighter shoulder carry

> **FIREGROUND NOTE**
>
> Firefighters must attempt to stay at least 10 feet from electrical power lines when raising portable ladders. In some instances, electrical power has jumped from power lines to ladders.

Two-firefighter suitcase carry. Two firefighters position themselves on the same side of the ladder and face the same direction. One firefighter is at the butt of the ladder and the other is at the tip. Both firefighters squat, grab the beam of the ladder, and prepare to lift it. Using their leg muscles, both firefighters begin to stand. Their arms and bodies raise the ladder to a vertical carry position that resembles a suitcase. The butt firefighter must maintain a position to protect the butt of the ladder from striking another firefighter or an object. If a third firefighter is available, he or she may go to the center of the same side (fig. 13–75).

Fig. 13–75. Two-firefighter suitcase carry

Remember, the suitcase carry is designed for carrying ladders a short distance. Note a few disadvantages:

- The ladder often strikes firefighters in the lower leg, making it a nuisance or injuring firefighters.
- The ladder can strike uneven terrain. In addition, if a firefighter were to lose balance or stumble, the ladder increases the risk that the firefighter would fall if the ladder were to strike his or her leg.
- Carrying large ladders is difficult and can injure and tire a firefighter.

Two-firefighter beam raise. After the ladder reaches the desired location, the butt-end firefighter places the ladder the proper distance from the building with the fly section out and, if possible, out of the way of overhead obstructions. Remember, this should be about one-fourth the working distance of the ladder so the ladder won't need to be repositioned once it is vertical. The firefighter places the ladder with one beam planted on the ground so it is in the vertical position on one beam.

The firefighter at the ladder's butt or foot is commonly called the footer. His or her main objective is to apply body pressure to the ladder as a counterweight to help the other firefighter raise the ladder. The footer places one foot forward and onto the ladder's butt end on the ground. The other foot goes comfortably behind the firefighter. This leg must not be directly in line with the other leg and the ladder because it could get struck as the ladder rises. The footer's hands go onto the beam. The arm that's on the same side of the body as the leg on the ladder's butt extends forward. The back arm now goes on the beam near the ladder's butt. Do not place this hand under or on the foot or spurs of the ladder's base. If the raise were performed rapidly, this hand could be pinched between the ladder and the ground. As the firefighter at the tip raises the ladder, the footer slowly leans backward and pulls slightly with his or her arms, reducing the weight of the ladder and assisting in the raise.

The firefighter at the tip rechecks the overhead clearance before raising the ladder. Next, the firefighters position themselves inside the ladder or nearest the building, with the ladder sitting on the opposite shoulder. By doing this, they are out from under the ladder and have a constant view to the objective. They raise the ladder by walking hand over hand, glancing up at the objective, and watching for overhead obstructions or changing conditions. If an obstruction such as a tree limb were problematic during the raise, the firefighter could pivot the ladder on its butt as it raises to avoid the limb.

Now that the ladder is vertical, both firefighters assume positions on opposite sides of the ladder with one foot against the butt to steady it and the other leg behind them to balance. Now they can untie the halyard and extend the ladder. The firefighter pulls the halyard down in a hand-over-hand motion and resists wrapping his or her hand or arm in the rope as it is pulled. If the rope were to snap, break, or slip without warning, it could injure the firefighter. As the halyard is pulled down, the other firefighter positions both hands outside the ladder's beam to steady it as it rises. Placing hands on the outside of the beams reduces the chance of an injury should the fly section suddenly retract. This firefighter has the best view of the raise and can give verbal commands to stop the raise. When the ladder reaches the desired height, firefighters slowly lean it into the building. They place one foot against an opposite foot on the ladder and use their arms to control the ladder's lowering into the building.

The two-firefighter beam raise is preferred over the two-firefighter flat raise because:

- The ladder is in a reduced-profile position because it is more narrow than when raised flat. In this position, the ladder is less likely to hit overhead obstructions. In addition, it is easier to pivot around overhead obstructions because only one foot is secured to the ground.
- Both firefighters can maintain good, balanced positions throughout the raise. The footer does not have to bend over and get onto the base or bottom rung of the ladder, which is a difficult position to assume and maintain. It also removes the possibility that the footer could slip off the ladder's base or rung as it is raised.

- The beam raise is the preferred portable ladder raise in a narrow alley or between overhead obstructions. The ladder's narrow profile allows for use in small, narrow areas (figs. 13–76, 13–77, 13–78, and 13–79).

Figs. 13–76 and 13–77. Two-firefighter beam raise

Fig. 13–78. The firefighter at the tip positions himself between the ladder and the building.

Fig. 13–79. The firefighter at the foot acts as the counterweight.

Two-firefighter flat raise. Once the ladder reaches the desired location by the footer, it can be raised flat. The ladder is positioned flat with both feet lying on the ground. The footer positions his or her feet onto the bottom rung or onto both rails of the ladder. Then he or she crouches with arms extended to grab a forward rung of the ladder. A firefighter's physical makeup determines which rung is feasible and comfortable to grab. As the ladder is lifted and raised, the footer maintains the position and leans slightly back to counterweight and assist in lifting the ladder. As the ladder approaches the vertical position, the footer removes one foot from the ladder and places it behind him or her for balance.

The firefighter raising the ladder checks again for overhead obstructions, then raises the ladder. He or she walks at a steady pace sliding hands up the ladder's rails to raise it to the vertical position. When walking the beams of the ladder up, the firefighter slides his or her arms up the ladder and uses leg muscles to drive the ladder upward. Firefighters should not bounce the ladder up while sliding the rails because a slip could mean an injury.

Many firefighters prefer to walk the rungs up to raise a ladder. This has a few problems:

- As the firefighter proceeds up the rungs, for a moment only one hand balances the ladder, which introduces a tendency to twist. If the firefighter were operating on uneven terrain, a slip could cause a loss of grip and injury.
- A firefighter walking up the rungs normally watches his or her hand position on the rungs and takes his or her eyes off the objective and tip of the ladder.
- When the ladder rises rapidly, it often bounces and twists as a firefighter proceeds up the rungs (figs. 13–80, and 13–81).

Now that the ladder is vertical, both firefighters assume positions on opposite sides of the ladder with one foot against the ladder's butt to steady it and the other placed behind them for balance. They can untie the halyard now and extend the ladder to its objective.

Fig. 13–80. The footer reaches the bottom rungs.

Fig. 13–81. Slide your hands up the rails of the ladder.

Two-firefighter flat-arm carry and raise. Some U.S. fire departments must adapt daily to peculiar hazards such as catenary wires strung across roadways and intersections for light rail train cars. These overhead obstructions make raising ladders difficult because they carry a lot of voltage. Fortunately there is a method for dealing with these situations.

Because ladders can be stored horizontally on some ladder company apparatus, the firefighters can position the ladder between them for removal. As the balance point approaches the firefighters, they grab the ladder and begin walking to the rear of the apparatus. Once they remove the ladder from the apparatus, the firefighters bend their arms and carry the ladder flat to its objective. If the ladder were not stored flat, the firefighters could pick it up, rotate it into the flat position, lift it at the balance point.

As the firefighters approach the objective, they raise the tip with the forward hand or the one nearest the structure. The back hand grips the ladder firmly and drives the ladder forward and up to assist in the lift. Some firefighters perform a short dip with the tip as they begin the raise, which gains momentum. As the ladder reaches the building, the firefighters bring it to the vertical position, brace it, and prepare to extend the fly if needed (figs. 13–82 and 13–83).

Fig. 13–82. Both firefighters are positioned at the ladder's balance point.

Fig. 13–83. Raise the tip toward the objective.

The flat-arm raise is difficult with large ladders and must be practiced before fireground use. It can cause back injuries and its use should be limited. The flat-arm raise is extremely valuable for rescuing a victim hanging to a windowsill by his or her arms when firefighters arrive. Getting the ladder underneath the victim provides support and allows the victim to assist in the rescue.

Wheelbarrow raise. With overhead obstructions such as wires or tree branches, it may be very difficult to perform a flat, beam, or flat arm raise and throw as mentioned previously. There is another method of raising a ladder under an overhead obstruction often called the wheelbarrow raise, as the firefighter on the butt end resembles someone pushing a wheelbarrow. In order to perform this raise, place the taller firefighter at the tip of the ladder just behind the overhead obstruction. They'll now place the ladder over their head while the firefighter at the butt uses an underhand grip on the bottom rung. As they begin to raise the ladder, the tip firefighter may walk a few steps backward to keep the angle of the ladder from hitting the overhead obstruction. The tip firefighter will bounce the ladder in their hands while it is being raised under the obstruction. Meanwhile, the butt firefighter walks hunched over with arms outstretched, keeping the ladder low with their eyes on the tip of the ladder. Their speed may have to be adjusted as the ladder is raised from beneath the overhead obstruction to the structure. It is very important that the butt firefighter doesn't stand up; the ladder's angle should be kept upward toward the building to get under the obstruction. If the ladder's position becomes too horizontal, if will most likely strike the side of the building. They must also focus on the tip of the ladder and not put their head down; they don't want to come in contact with the obstruction during the raise.

Three-firefighter evolutions

SKILL DRILL

When three firefighters begin ladder work, they must perform safely and with proper communication and teamwork. One firefighter must lead the operation. The firefighter at the ladder's butt normally leads and gives commands when performing ladder evolutions. He or she leads because the ladder's butt reaches the structure first. Once the ladder is raised, he or she controls the halyard and raising the ladder.

Three-firefighter suitcase/shoulder carry. The methods to transport a ladder described in the two-firefighter carries also apply to the three-firefighter carry, with one exception. The additional firefighter goes on the same side of the ladder at the center. To perform the shoulder carry when firefighters differ in height, the tallest goes at an end, not in the center. This way the ladder's weight distributes equally among all three firefighters (figs. 13–84 and 13–85).

Fig. 13–84. Three-firefighter suitcase carry

Fig. 13–85. Three-firefighter shoulder carry

Three-firefighter flat-arm carry. Three firefighters can also carry a ladder in flat position with their arms. Two of them go on the same side of the ladder at the tip and the butt. The third goes on the opposite side in the center. All three firefighters crouch onto one knee. On command, they lift the ladder off the ground using their leg muscles. They can place their hands on either the rung or beam, whichever is more comfortable and helps maintain their grip while they lift and transport the ladder (fig. 13–86).

Fig. 13–86. Three-firefighter flat-arm carry. (Courtesy of Joseph Alvarez)

If the ladder were removed from a ladder apparatus with rear horizontal storage, the two firefighters would position themselves on one side of the ladder with the third on the opposite side. Remember, a ladder should be carried butt first to the fireground and tip first upon return to the apparatus. This allows it to slide back into the apparatus storage compartment easier.

FIREGROUND NOTE

Firefighters may place their tools on top of the flat ladder during transport to the fire building.

Three-firefighter flat-shoulder carry. The three-firefighter shoulder carry's steps resemble that of the flat carry, except the ladder rides on top of the firefighters' shoulders during transport. The positioning is the same as the flat-arm carry, with two firefighters on one side and the other at the center of the ladder. All three lift the ladder from the ground to waist level, then continue lifting upward. Now, they can use both arms in lifting and placing it onto their shoulders (fig. 13–87).

Some firefighters learned to face the direction opposite where the ladder is needed (face the tip because the butt goes to the fire building). When the ladder reaches about chest level, the firefighters pivot and rotate by turning into the ladder and placing a shoulder under it. Now they face the butt ready to transport. Each method is acceptable, so firefighters should learn and practice both.

Fig. 13–87. Three-firefighter flat-shoulder carry. (Courtesy of Joseph Alvarez)

Three-firefighter beam raise. The same method of raising the ladder described in the two-firefighter raise applies to the three-firefighter raise, with one exception. The additional firefighter, depending on physical size, either walks the ladder up in front of or behind the firefighter raising the ladder to the vertical position. The tallest firefighter goes behind the other firefighter or toward the tip. This way, both firefighters can reach the ladder and raise it in unison. If the taller firefighter were in front during the raise, the smaller firefighter could be unable to reach and assist in the lift.

Another option is for the additional firefighter to go in the middle of the ladder and face the tip. When the ladder rises, this firefighter pulls the ladder toward them, self-assisting in the lift. As the ladder passes his or her position, this firefighter turns toward the base and either maintains a position on the outside of the ladder assisting in balancing the ladder or follows the raising firefighter assisting in the raise (fig. 13–88).

Fig. 13–88. Three-firefighter beam raise. (Courtesy of Joseph Alvarez)

Three-firefighter flat raise. The same method of raising the ladder described in the two-firefighter raise applies to the three-firefighter raise, with one exception. The additional firefighter goes at the ladder's tip alongside the other firefighter. They raise the ladder using either a hand-over-hand motion with their inside hands on the rungs or they slide the rail of the ladder on their shoulders, using their leg muscles to assist them. Both firefighters must use caution while raising the ladder. They could strike the other's arm or bump one another. Either action could cause a slip or mishap. Many fire departments rely on this raise for the three-section, 35-ft (10.7-m) ladder becoming more prevalent on fire apparatus (figs. 13–89 and 13–90).

Four-firefighter evolutions

SKILL DRILL

When four firefighters begin ladder work, they must perform safely as a team and communicate properly. One firefighter must lead the operation. A firefighter at the ladder's butt usually leads and gives commands because the ladder's butt reaches structures first. He or she controls the halyard and raising the ladder. Firefighters should "step off left"—start on the left foot when commanded to move to the building by the ladder.

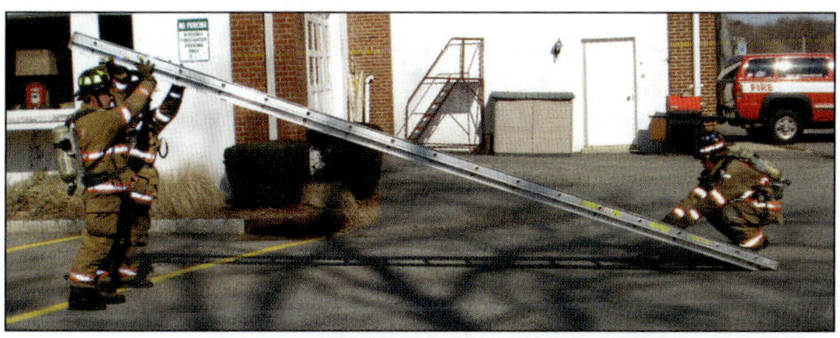

Figs. 13–89 and 13–90. Three-firefighter flat raise. (Courtesy of Joseph Alvarez)

Four-firefighter flat-arm carry. Four firefighters can also carry a ladder in the flat position with their arms. Two firefighters go on each side of the ladder, two at the tip and two at the butt. All four crouch onto one knee. On command, they lift the ladder off the ground with their leg muscles. They may place their hands either on the rung or beam, whichever is more comfortable and helps maintain a grip on the ladder during lifting and transport (fig. 13–91).

Fig. 13–91. Four-firefighter flat-arm carry. (Courtesy of Joseph Alvarez)

If the ladder were removed from a ladder apparatus with rear horizontal storage, two firefighters would go at the ladder's tip, and the other two would go at the butt. Remember, the butt should lead to the fireground, and the tip should lead on return to the apparatus. This makes sliding back into the apparatus storage compartment easier.

Four-firefighter shoulder carry. The four-firefighter shoulder carry's steps resemble those of the flat carry, except the ladder goes on top of the firefighters' shoulders during transport. The positioning is the same as the flat-arm carry with two firefighters at the tip and two at the butt. All four lift the ladder from the ground to waist level, then continue lifting the ladder upward. Now, they can use both their arms in the lift and placing the ladder on their shoulders.

> **FIREGROUND NOTE**
>
> It might be necessary to push down and dent ornamental or sheet metal ridge vents for a roof ladder to sit properly on a roof. If this doesn't work, use a Halligan tool to make two holes on the opposite side of the roof's ridge so the roof ladder's hooks can secure onto the roof.

Some firefighters learned to face away from where the ladder is needed (face the tip because the butt arrives first at the fire building). When the ladder reaches about chest level, the firefighters pivot and rotate by turning into the ladder and placing their shoulders under the ladder. Now they face the butt ready to transport the ladder. Each method is acceptable and should be learned and practiced (fig. 13–92).

Fig. 13–92. Four-firefighter shoulder carry. (Courtesy of Joseph Alvarez)

Multiple firefighter ladder operations

The church raise. The church raise, also called the auditorium raise, is used when there is no wall to support or rest the tip of the ladder. It is commonly used in an open area inside the structure, such as a church or auditorium, hence the name. The raise is not commonly performed on the fireground but is one that firefighters should learn and be able to perform for unusual situations they may encounter.

Prior to raising the ladder, utility ropes should be tied to the tips of the ladder. If using two ropes, find the center of the rope and that will be secured to the ladder. When using four ropes, one end of each rope will be tied to the ladder. Before securing the ropes, extend the ladder by two rungs to expose the tip of the ladder; this makes securing the ropes easier. Next tie the ropes so they are splitting the second rung from the tip and around the rail with a substantial and approved knot. These secured ropes are not for raising or lowering the ladder, they are to keep the ladder taut and secured when it is in the vertical position. The ropes will be held by four firefighters standing at angles to the ladder in a square pattern to keep the ladder supported at all times.

Once the ropes are tied and secured to the ladder, the ladder can be placed with the butt at the desired position and the ladder raised to the vertical position, using a beam or flat raise. Now the halyard can be pulled and the ladder extended to the desired height. Once the ladder is set in place, the ropes can either be tied off to a substantial object or, if none exists, four firefighters will remain on the ropes securing the ladder with their bodies' weight and position. Firefighters holding the ropes should not tie the rope around themselves; they should maintain a good stance with the rope in one hand and running down along their back and then around their body held with the opposite hand. The ladder will move somewhat while a firefighter climbs it and operates from it, so it's very important that the rope firefighters maintain their hold to keep the ladder steady and supported.

The firefighter climbing the ladder must do it steadily and slowly; the vertical ladder is one of the hardest ladders to climb. When the firefighter reaches the tip to perform whatever task needs to be completed, they must perform a leg lock and/or lock into the ladder with an approved safety belt or personal harness.

Seesaw ladder removal. In some situations, removing an unconscious victim or downed firefighter from the first floor could be difficult. If a portable ladder is nearby, it can assist in making the rescue or removal easier. Simply insert the straight ladder or closed extension ladder into the first-floor window. The end going into the structure only needs to be inside the building about 4 to 5 feet. The opposite end of the ladder remains outside the window, resting on the windowsill. The sill acts as fulcrum, and the ladder will be in the seesaw position, with the higher end outside the building. Once the victim is placed onto the ladder, the end outside is pulled downward, lifting the victim off the floor. Depending on the height of the ladder outside the building, a hook might be needed to pull the ladder downward. Always pull from the last rung to increase the leverage and make it easier to perform the tactic. Once the ladder is in the horizontal position, the two firefighters stationed on each side of the ladder next to the window will slide and pull the victim outward and maintain their physical position near the window. The firefighter who pulled the ladder down will walk backward and pull the ladder out of the building. When removing the victim walk slowly and do not pull rapidly, ensuring that the victim or firefighter's equipment does not get hung up on the windowsill. The firefighters operating inside the structure can assist in the lift and guide the individual out of the window during the removal. Once the victim is outside and clear of the structure, the ladder can be lowered on the victim's end. The opposite end of the ladder is maintained in an upward position at waist level so the ladder pivots more easily to the ground.

Bridging with a portable ladder. There will be instances when firefighters may have to use portable ladders as an escape mechanism or for an unusual rescue. Portable ladders can be used as bridging devices between building roofs or even across narrow shaftways to windows on the opposite side. Normally an extension ladder is used because of its ability to maintain its shape and strength when weight is added to it in the closed position. If a straight ladder or roof ladder is used, firefighters can expect to see some flex in the ladder as a firefighter or civilian transverse the ladder to safety. When traveling over the ladder, you should always crawl and not walk over the rungs.

There are many ways to set the ladder in place to bridge between buildings. If the distance is short, the ladder can manually slide over to the other objective. If the distance is longer, a rope can be tied off to the tip of the ladder (if using an extension ladder tie off both sections to avoid the ladder from extending). Once the ladder is tied off, throw the rope over to the opposite side and have those firefighters pull and lift the rope if needed as the ladder is slid over to the objective. On rare occasions, the extension ladder can be extended horizontally to the opposite side to make the rescue. This should be a last resort.

ROOF LADDER OPERATIONS

Firefighters often ventilate structures with steep roof angles and ones covered with ice or snow. Working on these structures without a roof ladder can be dangerous and should be avoided. If there is no way to work from a roof ladder, position an aerial or tower ladder to the roof for a work platform.

A **roof ladder** is a straight ladder with solid aluminum or steel hooks attached to the tip. The hooks are mounted in a small bracket with a tension spring wrapped around the base of the hook and a square stop. To release the hooks, a firefighter must apply moderate downward pressure then turn the hook 90 degrees either away or toward himself or herself. This motion resembles pushing the hook down toward the base of the ladder and spinning it a quarter turn. Firefighters usually hold the ladder about waist level to release the hooks, but a firefighter can also perform this tactic while the ladder is flat on the ground. Applying pressure to the hooks with the inside portion of a boot might release them. Once the hooks are turned, pressure is released and the hooks spring up into a locking position inside their holding brackets. Roof ladder hooks secure the ladder to a structure's roof by allowing the hooks' tips to bite into the roof on the opposite side of the roof's ridge (fig. 13–93).

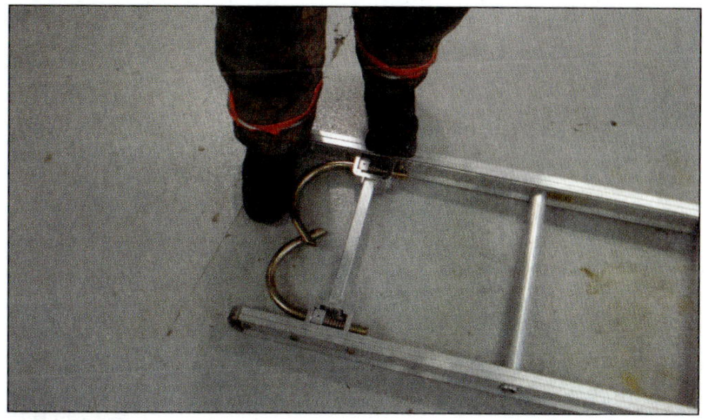

Fig. 13–93. Roof ladder hooks

There are a few methods to raise a roof ladder to a fire building's roof. The method depends on the size of the roof ladder and the type, angle, and length of the roof.

Single-firefighter evolution

SKILL DRILL

It might be possible to raise an 18- or 20-ft (5.5- or 6-m) roof ladder from the ground to the roof of a one-story structure with a steep pitch. A firefighter transports the ladder to the structure, lays it down, and extends the roof hooks to the open position. Next, the firefighter performs a single raise. When the ladder leans onto the roofline in the proper climbing angle, the firefighter lifts up the ladder's base and slides the ladder onto the roof and over the ridge. He or she must pull down on the ladder to ensure the roof hooks have bitten into the roof and will support the firefighter's weight.

Another ladder is brought to the base of the building and raised next to the roof ladder. Now the firefighter climbs the ladder and transfers to the roof ladder at roof level. With this method of sliding the roof ladder upward, firefighters often prefer to slide it flat so both rails ride up the roof. They can use both arms and push as the ladder slides upward.

When tall structure must have an extension ladder to reach the roof, another method must be used. After the extension ladder reaches proper climbing position, the firefighter opens the roof ladder's hooks and the ladder rises next to the extension ladder. Next, the firefighter may place the roof ladder on one beam and next to the extension ladder with the roof hooks facing away from the extension ladder. Now the firefighter climbs up the extension ladder. When he or she reaches the roof ladder's balance point, he or she places a shoulder between the rungs. The ladder now balances on the firefighter's shoulder as he or she begins to climb.

As the firefighter reaches the roofline, the roof ladder begins to pivot over the roof's edge. The firefighter pushes the roof ladder up the roof either on the beam or laid flat. Once it goes over the roof's ridge, the firefighter pulls back on the ladder, sinking the roof hooks into the ridge. Care must be taken while pulling back because the ridge could have a **ridge vent** made of fibers and shingles that tear. Firefighters must ensure the hooks bite into the roof and will be secure so the roof ladder can carry the weight of firefighters on the roof.

Two-firefighter evolution SKILL DRILL

The roof ladder must be raised after an extension ladder is in position. One firefighter climbs to a position near the top of the extension ladder that allows him or her to transfer the ladder from the extension ladder to its locked position on the roof. This firefighter locks into the ladder with a leg-lock maneuver or safety belt. Now the firefighter on the ground raises the roof ladder to the firefighter on the ladder.

There are two common methods: Have the roof hooks open or have them closed. There are pros and cons to both, but both are acceptable. Raising the ladder from the ground to the firefighter on the extension ladder may be done a few ways. The firefighter on the ground can remain on the ground and transfer the ladder up using his or her arms. Or, the firefighter on the ground may first hand the roof ladder up to the firefighter on the ladder. Then, the firefighter on the ground may climb up the extension ladder and assist with the roof ladder's raise and balance.

Roof ladders are removed in the reverse fashion. Now and then roof hooks set deeply into a roof and are difficult to release. Use caution when trying to release them. Apply steady pressure to the ladder. Striking the ladder with a hand tool to release it could damage to the ladder.

VENTING A WINDOW WITH A LADDER SKILL DRILL

It is not the preferred method to vent windows, but venting a window with a portable ladder is often necessary because of manpower shortages, a lack of time, or numerous windows needing to be ventilated. There are a few methods, all of which start with raising or positioning the ladder's tip in contact with the top third of the window. Striking a window high has two advantages. First, the window might break easier as a result of being weakened or distorted by heat. Second, heat, smoke, and gases rise and collect at the ceiling and mushroom downward. Breaking the window high can help release these elements quickly (fig. 13–94). Remember, all venting operations should coincide with the advance of the hoseline. Communication between units operating inside and outside the structure should be paramount when venting windows. Ventilating windows creates a flow path condition in which escaping heat, gases, smoke, and fire can be drawn to the opening.

Fig. 13–94. Venting a window using a ladder

Extension ladder throw

The extension ladder can be used in the beam or flat position to break a window. Some firefighters think the weight and structural makeup of throwing the beam into the window works better. Others think the flat position covers more surface area and breaks more glass. Both techniques work.

To begin, a firefighter raises the ladder vertically to the proper level near the top third of the window. Then the firefighter at the ladder's base forcefully drops or pushes the ladder into the window. As the ladder strikes the window, the firefighter removes his or her hands from the ladder to avoid glass that could slide down the beams. His or her hands remain in the vicinity of the ladder in case the ladder bounces off the window. This way, he or she can grab the ladder's rungs or beams to maintain its stability. Done properly, this often breaks the window's center sash and bottom windowpane as well. Sometimes it takes one or two attempts to break a window sufficiently for ventilation.

After ventilating windows with a ladder and completing fireground duties, the firefighter inspects the ladder for embedded glass shards, structural damage, and thermal damage from vented heat or fire.

Roof ladder throw

A firefighter may also use a roof ladder to ventilate a window. The roof ladder's two large roof hooks attached to its tips resemble firefighting hooks and pike poles. To perform this tactic, a firefighter first opens the roof hooks and places the ladder at the proper height. Then, the firefighter at the ladder's base throws or pushes the ladder into the window.

After the ladder breaks the top pane, the firefighter has a few options. If the window frame is wood, the firefighter places the ladder's hooks onto the center sash. Next, he or she picks up the base of the ladder and gives it a slight tug or pull backwards or out of the window. Usually this additional pressure lets the hooks break the bottom pane of glass and removes the center sash of the window. If the ground's terrain is bad, the firefighter avoids this method and instead drops the ladder into the window again to complete ventilation of the lower panes of glass.

Remember, firefighters must use all eye protection and personal protective equipment when performing these tactics.

FIREGROUND NOTE

If there are not enough firefighters to perform the flat raise with the Bangor ladder, consider using the building to physically butt the ladder as it is raised.

ADDITIONAL LADDER TYPES

Three-section extension ladder

Increasing numbers of fire departments are placing three-section extension ladders onto first-line apparatus for many reasons. Most departments put them on apparatus to meet standards and guidelines for ratings, while others encounter manpower shortages. Those departments think having a large ladder on the apparatus on scene immediately lets them handle any situation. Plus, many bedroom communities are experiencing growth in new and higher townhouses and condominiums.

Firefighters must be aware of three-section ladder characteristics before using it on the fireground. Most noticeable about the short-looking, three-section ladder is how heavy and bulky it is with the added fly section. Also, pulling down the halyard raises both fly sections and raises the ladder 28 in. (71 cm) compared with the 14 in. (36 cm) of regular extension ladders. Before climbing the ladder, a firefighter must ensure both fly sections are locked in place. Sometimes these ladders lock only one fly and require halyard adjustments to lock the other fly. Firefighters have suffered severe injuries using this ladder when it was not properly locked.

Firefighters may raise the three-section ladder in a beam- or flat-raise evolution. Three firefighters should be used if they are available. Once the ladder is vertical, two firefighters position themselves on the ladder's sides or beams. They place both hands out of danger on the back beam. Remember, this ladder has two moving fly sections that shouldn't be held during the raise. The firefighters place their feet behind them, not near the front base of the ladder, because the fly sections might strike the ground if something were to go wrong during the raise. The third firefighter goes behind the ladder at the base to control the halyard. He or she ensures both sets of dogs are locked once the ladder rises to the proper position.

Bangor or pole ladder

The **Bangor** or **pole ladder** is a large extension ladder that has a permanently attached support pole hinged and mounted to each side of the bed section. Firefighters use these poles to assist in raising, lowering, and stabilizing the ladder. These ladders are normally raised only in a flat-raise evolution because of the position of the support poles. In emergencies, the ladder may be raised with the beam-raise evolution, but only one support pole can be used. The ladder should be raised flat and then pivoted to its desired location once the ladder is raised.

The support poles on the ladder's sides are usually attached to the beams with a pivoting bracket near the top of the bed section and either a clevis or cotter pin and mounting rod at the base of the bed section. A firefighter must remove the pin and lift the pole from the rod to release it. Next, he or she reattaches the pin to the rod to reduce the chance of slipping on it. These pins are often attached to a short chain or cable so they remain attached to the ladder and won't become lost.

Once the support poles are released, the firefighters at the base of these poles remove them. Next, they bring them to the vertical position if room allows or bring them up and around in a semihorizontal position to two firefighters near the center of the ladder. Once these firefighters have the support poles, they walk back past the ladder's tip to the ends of the poles. These two firefighters at the base of the poles are ready to assist in the lift by walking forward and pushing upward with the poles. They must keep the spur attached to the base of the poles between their fingers and palm up as they raise the ladder to avoid injuries (figs. 13–95 and 13–96).

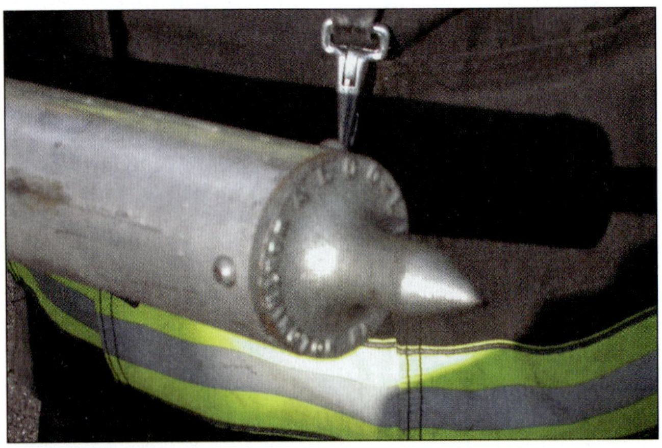

Fig. 13–95. There is a sharp spur at the base of the poles.

Fig. 13–96. Keep the spur between your fingers as you raise the ladder.

With two firefighters at the ladder's tip, one at the base, and two on the support poles, the ladder is ready to be raised. The firefighter at the base foots the ladder and gives the command to raise it. In most instances with the flat Bangor raise, this firefighter positions himself or herself on the ladder as a counterweight. The two firefighters at the tip begin to walk forward, use their arms in a hand-over-hand motion, and raise the ladder. Most of the time, these firefighters position one hand on the rung and the other on the beam.

The two firefighters on the support poles are positioned to the outside of the poles with one hand holding the butt end as described above and the other arm extended up onto the pole at a comfortable position (fig. 13–97).

Fig. 13–97. Two firefighters at the tip, one at the base, and two on the poles.

As the ladder rises, the firefighters push the support poles forward and walk with the poles. They must advance at the same rate as the firefighters raising the tip of the ladder, not running or gaining too much momentum. If they go too fast, the ladder can get away from the firefighters at the tip and cause the ladder to become unstable. If the ladder is being raised properly, all the firefighters should feel about the same pressure or weight throughout most of the lift. Initially, the firefighters on the support poles don't feel a lot of pressure or weight, but this changes once the ladder begins to gain height and the angle increases.

When the ladder is vertical, one firefighter on the support poles proceeds to the side of the ladder with the pole. The other firefighter moves more toward the center of the ladder in line with the beam with the support pole. Now both firefighters are stabilizing the ladder at a 90-degree angle. The firefighter on the side of the ladder controls the ladder's side-to-side movement while the firefighter in front of the ladder controls the front-to-back movement (fig. 13–98).

As the ladder is being raised vertically with the halyard, the pressure on the support poles increases and the firefighters holding the poles maintain their positions and balance the ladder. Next, as the ladder lowers into the building, the firefighter on the front support pole leans back into the pole and assists in the ladder's speed and descent. Now that the ladder is positioned and needs no further adjustments, the two firefighters with the support poles walk toward the building. They place the poles toward the back of the ladder and rest them on the ground.

Fig. 13–98. One support pole to the side and another toward the center stabilize the ladder.

The support poles are not designed to act as an outrigger and carry the ladder's weight. Use caution when positioning the poles because poles angled far from the ladder can cause problems. An unsuspecting firefighter could walk into the pole or an advancing hoseline could knock it out of position.

Folding ladder

The **folding ladder**, commonly called the **closet**, **attic**, or **scissor ladder**, can do more than provide access. If a firefighter were trapped or injured in a basement, the ladder could be lowered through a small cellar window and function as a stretcher. The firefighter would roll onto it and be attached with tubular webbing or hose straps. Then the ladder could slide out of the basement. Sometimes it is necessary to remove an injured firefighter's self-contained breathing apparatus from his or her back and put it between his or her legs. Remember, the face piece must remain attached to the firefighter to deliver fresh air. Repositioning the SCBA should allow enough clearance to slide the firefighter through the small window (figs. 13–99 and 13–100).

For a firefighter facing burned-out floors or a firefighter trapped in the floor, a closet ladder might be able to span the joists to reach the firefighter or distribute the weight of the rescuing firefighters. In addition, if a firefighter were to fall through the floor and into the floor below, firefighters could bring a closet ladder through the building and lower into the hole to reach the firefighter. Once the closet ladder is open, it does not matter which end enters the hole. The ladder usually has adjustable feet at one end and a square end design at the tip. This enables the ladder to sit evenly on either end when placed on a level surface. Plus, once weight is put on the rungs, the ladder can support it in either position, upright or upside down.

Figs. 13–99 and 13–100. Using a folding ladder as a rescue device

MISCELLANEOUS LADDER USES

There are numerous other uses for ladders on the fireground, and often firefighters develop ways to use them while operating at a fire or emergency scene. The following is a short list of ways ladders can be used:

- An exhaust fan can attach to a ladder outside a window to assist in ventilation efforts.
- A ladder can be placed over a burnt out or a stairway of questionable stability for access to the upper floor.
- A ladder covered with salvage tarps, runners, or plastic can function as a chute to divert water during salvage operations.
- Ladders placed on stanchions can serve as barriers.
- Two short ladders placed on opposite sides of a fence provide access to both areas.
- A ladder can prop open or act as a safety bracket for large roll-up and garage doors.
- A ladder can be used as a hoisting point or high anchor point for technical rescue situations and firefighter removals. (See RIT in chapter 17.)
- Lights attached to ladders add additional overhead lighting.
- Ladders can latch together to form a square frame for salvage covers to become a collection area for hazmat operations or a water-holding tank.
- Two ladders on opposite walls with a plank supported by each ladder's rungs can span the distance to create a work platform.
- A ladder can be used out of the tower ladder's bucket to extend the reach of the tower ladder's height.

For firefighters to operate safely on the fireground, they must comprehend and understand the evolutions and tactics described in this chapter. Firefighting is a hazardous, unavoidably dangerous activity, and operating unsafely with ladders increases the risk of a serious injury or death. It is each firefighter's responsibility to operate within safety limits during any ladder activity.

NOTES

1. Mathew Rush, "Ground Ladder Chocks," *Fire Engineering*, March 2007.
2. Michael Ciampo, "The Arm Lock Maneuver," *Fire Engineering*, March 2006.

LESSON FROM THE FIREGROUND

Over the course of the years, having worked in two "big city" fire departments in some very active units, I've seen that learning from actual rescues and firefighting experiences is a large part of the educational process of becoming a seasoned firefighter. During some of these experiences some minor mishaps or "close calls" were averted, mostly due to luck and training. After experiencing some of these ladder rescues and close calls as well as reading and hearing about others who experienced similar mishaps during ladder evolutions, I felt there was a need for additional training and tactics on ladder operations.

Not long after presenting a "Street-Smart Ladders" hands-on training seminar which reviewed some of the procedures in this ladder chapter, I received one of the most rewarding gifts of my career, a manila envelope in the mailbox. I opened it and sat in awe as I read about the heroic act the letter described and the photos that accompanied it. Briefly, it read:

"As the first due units arrived, they were greeted with heavy fire on the 1st and 2nd floors that was rapidly extending to the 3rd floor of a wood frame dwelling. In addition, seven occupants had already jumped from the 2nd floor and were lying on the ground with a multitude of injuries. As firefighter Kevin Galligan proceeded down the narrow alleyway, a distraught woman was dangling an infant out the window on the 3rd floor. After verbally communicating with the woman, he began to raise a 35-ft (10.7-m) ladder with the assistance of Lieutenant Daniel Santry. Firefighter Galligan proceeded up the ladder and removed the infant, handing her down to the Lieutenant, and then entered the dwelling and removed her mother, helping her safely to the ground. Once they got to the ground the woman informed the pair that two of her teenage children were still trapped inside. Without hesitation, the two then ascended the ladder in search of the children. The first child was quickly located and removed and placed into the bucket of Tower 2. As the two searched, they located the second child and dragged her back to the window, just as their air tanks were depleted. Now the duo had to exit and allow Firefighter Eonas to enter the window in order to remove the last child."

After reading the unit citation report, a personal letter followed. The letter briefly stated:

"As a fellow instructor and newly promoted Lieutenant, I would like to inform you that the information you were instructing was accurate and very useful to me at this fire. The methods I found particularly helpful were the modified leg lock (HELL), the arm lock maneuver, Body to the Building while raising the ladder, and victim removal operations. These methods made a lot of sense to me and utilizing them assisted me in completing the successful rescues and may very well have saved some lives. From one firefighter and instructor to another, thank you. The methods and tactics you are teaching are right-on and street smart knowledge that firefighters need. I appreciate all that you're doing for the fire service."

This letter is one that any fire instructor would hope to receive and one that encourages me to keep on teaching ladder techniques to the fire service. Hopefully, some of the information in this ladder chapter will also be able to help you in your firefighting career.

QUESTIONS

1. Identify the primary functions of portable ground ladders.
2. What is the minimum distance that firefighters should maintain from overhead electrical wires when raising a portable ground ladder?
3. What is the main function of a heat-sensor label?
4. A firefighter should hold onto what part of a portable ground ladder while climbing and descending to maintain three points of contact?
5. Generally, there are four classifications of portable ground ladders. A roof ladder is classified as what type of ladder?
6. When working from the side of a ladder to remove burned siding, a firefighter should perform a leg-lock maneuver. Which leg does the firefighter use to perform this tactic if he is working from the right side of the ladder?
7. Prior to climbing an extension ladder, the first firefighter to climb the ladder should ensure that what are in the locked position?
8. What should a firefighter do while cleaning an aluminum ground ladder to ensure that soot, ash, and debris are removed from the inside of each rung?
9. What are the horizontal cross members of a portable ground ladder called?
10. How frequently should portable ground ladders be inspected?
11. Why is it important to regularly check the rails and rungs of wooden ground ladders?
12. Identify and explain a critical safety factor when placing an aluminum ground ladder against a building with aluminum siding.
13. What main function is served by a firefighter footing or butting a ground ladder?
14. Prior to placing a ground ladder in position, what should a firefighter do?
15. What quick reference can a firefighter use to identify the proper length of ground ladder to reach a desired floor level?
16. What part of a ground ladder should be carried first to the positioning point and why?
17. What steps should a firefighter follow to release the hooks of a roof ladder for use?
18. Which firefighter gives the commands when carrying or raising a portable ground ladder?
19. Identify the normal range of a ladder's angle when the butt is positioned one-fourth the distance from a building.
20. When a portable ground ladder is raised to a flat roof, how many rungs should there be above the roof surface?
21. When two firefighters are performing a suitcase carry, where should the portable ground ladder be positioned?
22. When raising a portable ground ladder equipped with tormentor poles, where should the spike on the end of the tormentor pole be placed?

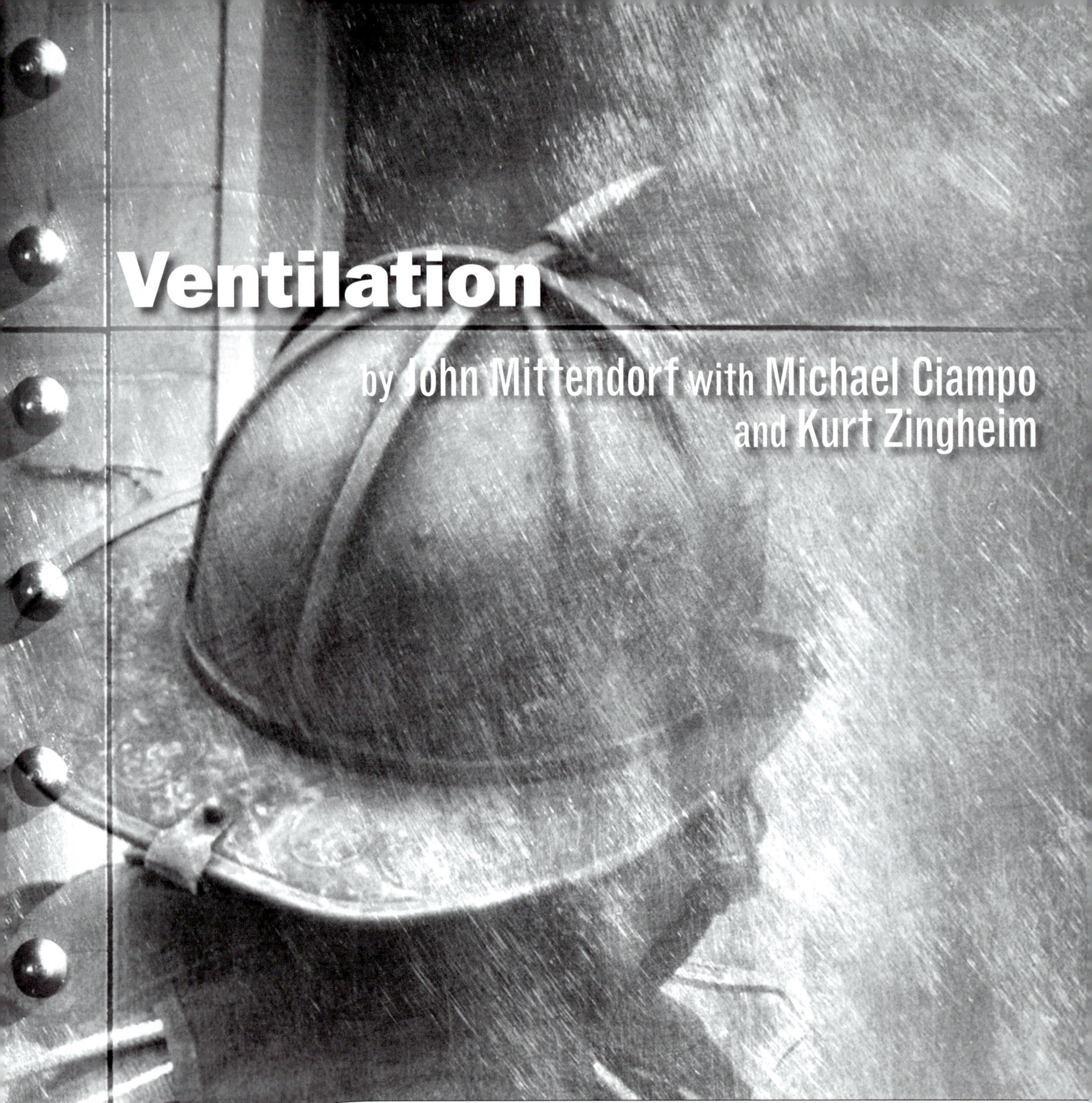

Ventilation

by John Mittendorf with Michael Ciampo and Kurt Zingheim

This chapter provides required knowledge items for the following NFPA Standard 1001 Job Performance Requirements:

FFI 4.3.11

FFI 4.3.12

FFI 4.3.20

14

This chapter contains Skill Drills. When you see this icon, refer to your Skill Drill book for step-by-step instructions. SKILL DRILL

OBJECTIVES

Upon completion of this chapter, you should be able to do the following:

- Define the term ventilation in regard to fire service operations
- Describe principles of venting for fire and venting for life
- Describe the importance of smoke and fire gases in relation to ventilation
- Identify those items needed to perform an effective ventilation size-up
- Identify and describe seven types of ventilation
- Describe the term *flow path* and understand its relationship to ventilation operations
- Identify common construction features in a building that aid in ventilation operations
- Describe the proper methods for venting a window
- Identify the equipment used to perform ventilation
- Identify the firefighting tools used to vent windows and roofs
- Describe the proper procedures for venting flat and pitched roofs
- Identify safety procedures to use when performing a roof ventilation
- Identify the type of roof cuts used in ventilation
- Describe the procedures for performing a "trench/strip cut of a roof"
- Identify the hazards associated with venting truss constructed roofs
- Identify tactical considerations for venting five types of occupancies

INTRODUCTION

Similar to the term firefighting, the word ventilation has several definitions and may be successfully performed using a variety of operations and techniques. In spite of its potential complexity, the word **ventilation** is easily defined and applied to firefighting operations as follows: "Procedures necessary to effect the planned and systematic redirection and removal of fire, smoke, heat, and fire gases from a structure." However, this definition may be too simplistic when we consider that safe, timely, and effective ventilation operations are often performed in hazardous atmospheres and under dangerous conditions.

FFI 4.3.11 Ventilation does not put out fires. However, safe, timely, and effective ventilation is integral to control and extinguish structure fires. According to the International Society of Fire Service Instructors (ISFSI) 2013 *Position Statement on Fire Dynamics Research,* "Fire departments should manage and control openings to the structure to limit fire growth and spread, and to control the flow path of inlet air and fire gases during tactical operations." Controlling ventilation reduces fire development as well as firefighter risk from increased heat release rates.

To understand how fireground conditions dictate the need for ventilation, consider a simple example: An unconfined fire will draw cold air into the bottom of a fire as heated smoke, gases, and air rise vertically. Because the fire is unconfined, it can draw as much cold air (as compared to the heated air from the fire) as necessary to sustain combustion. Any fire must have heat, fuel, and oxygen in the appropriate ratio for combustion.

As our fire utilizes the correct amount of oxygen, heat, and fuel, the result is a free-burning fire. Now, transfer our free-burning fire to a confined room in a building. Initially, the fire has the proper ratio of fuel, heat, and oxygen, which is about 21% in air. As this process continues, the quantity of hot gases and smoke in the room increases, and available oxygen necessary for combustion is reduced. As the oxygen content is reduced below 21%, the burning rate will decrease and the production of smoke and fire gases will increase. The increased smoke will fill the interior of the room and try to exhaust through any available openings.

As the fire continues to burn, the oxygen content will be reduced to about 17%, levels of carbon monoxide and other products of combustion will increase, and temperatures can exceed 1,300°F (704°C). At this point however, another condition will rapidly become significant. As the fire continues to burn, the ratio of heat, fuel, and oxygen becomes unbalanced, with levels of oxygen approaching 13%–15%, which is not adequate to support a free burning fire. When the percentage of oxygen reaches 13%–15%, the fire will begin to smolder due to insufficient oxygen.

Several interesting events will have taken place simultaneously with the fire smoldering and the subsequent reduction of oxygen in the room. First, fire gases will rise vertically until they reach the highest level in the room and then spread horizontally across the underside of the ceiling. Secondly, if this process continues, the structure will become filled with heated smoke and fire gases, offer poor visibility, and develop a level of heat within the structure that can be intolerable for occupants and suppression personnel. At this point, if fireground personnel must enter this environment to conduct a search and/or extinguish this fire, then they are faced with a choice:

- Enter the structure without ventilation and encounter a hot, flammable environment that is dangerous to suppression personnel.
- Use horizontal, pressurized, and/or vertical ventilation to improve visibility, reduce the dangerous concentrations of heat, smoke, and fire gases, and allow suppression personnel to effect a timely and safe extinguishment.

This simple ventilation example points to a common problem. Whether a fire is small or large, smoke, fire gases, and other products of combustion are by-products of any fire and can be expected to be encountered by building occupants and/or suppression personnel.

VENTILATION PRINCIPLES

Some fire departments classify ventilation operations into two basic principles: venting for fire and venting for life.

Ventilation and flow paths

We've been able to learn more about the effects of ventilation on the size and growth of a fire in today's modern fire environment due to recent scientific studies performed by numerous agencies. In years past, we considered all ventilation efforts as the best option to perform by simply ventilating a structure to release the build-up of heat, smoke, and gases. Unfortunately, with newer home furnishings, building products, and synthetic materials, we are seeing different fire conditions than in the past. We need to understand that when we create an opening in a structure (opening a window or a door) we are not only ventilating, but also adding oxygen into the area and allowing the fire to intensify in size and speed. In addition, we are creating or altering a flow path for the fire to travel to our location or into other areas not involved in fire. The amount of energy released by today's furnishings can quickly consume the oxygen in a room. Fuel-rich smoke—lacking oxygen and increasing in temperature—will seek more oxygen to continue to burn. In that manner, the smoke (which is unburnt fuel) will move from areas of high pressure to low pressure, normally from the ceiling to the floor and then to other areas in the structure if the door to a room is open.

When we perform ventilation, we must coordinate it with the engine company's advance with a hose stream on the fire. The fire will want to seek cooler air and oxygen and move in the direction of any opening we make. Indiscriminate ventilation at various locations in a structure can pull the fire or create the flow path for the fire to travel to the ventilation opening and increase the size and intensity of the fire. It is important that members performing ventilation efforts communicate their intentions and the location of the efforts with the interior crews in order not to pull fire toward the units operating inside the structure. Remember, an opened door used for entry is also a ventilation point, and controlling the door to limit the air intake will help prevent increasing the size of the fire and pulling the fire toward the opening.

Firefighters must use caution when performing ventilation operations. Today's fires have heavier fire loads and are burning with more speed and intensity and reaching high heat levels rapidly when we introduce more oxygen. Indiscriminate window ventilation is a thing of the past. As firefighters, we must limit creating a flow path for the fire to grow or push back on interior crews or be pulled into different areas. Always communicate and coordinate ventilation efforts at all structure fires to allow an easier advance for the engine company to extinguish a fire. "Ventilate where they'll operate" should be a mantra for the fire service based on information found through scientific research. Firefighters must coordinate horizontal and vertical ventilation tactics and do them concurrently as the engine company is advancing and operating the hoseline on the fire.

Venting for fire

Venting for fire assists the engine company's advance and extinguishment of the fire (fig. 14–1). Normally, this type of ventilation is performed and coordinated in unison with the engine company's advance on the fire. Proper communication must be coordinated between the outside and inside firefighting teams when performing this type of ventilation. Venting prematurely or venting prior to the engine company having water and preparing to attack the fire can increase a fire's size and intensity. In addition, it can create a flow path and pull the fire in many directions or toward an engine company without water or firefighters searching for victims. Always communicate before you ventilate!

Venting for life

Venting for life is accomplished to assist a firefighter's movement into an area where there is a known or suspected life in danger. There is a calculated risk of pulling the fire, smoke, heat, and gases toward this opening by the path the firefighter created; but it is performed to help firefighters reach trapped victims as soon as possible. Again, proper communication must be coordinated between the outside and inside firefighting teams when performing this type of ventilation.

Fig. 14–1. A firefighter using a 6-ft. (1.8-m) hook to vent a window from the outside. Stay upwind and lower than the window using the reach of the tool to avoid fire, smoke, and gases.

SMOKE

FFI 4.3.11 From a simplistic perspective, **smoke** is nothing more than by-products from a fire. Therefore, as the burning substance changes, so does the composition of smoke. This is best illustrated by considering the basic elements of smoke. Smoke is a mixture of three basic elements: solid particulates, liquid particulates (aerosols), and fire gases. Unburned particulates are primarily comprised of the materials that are burning and carbon. These particulates can burn if their ignition temperature is sufficient.

Fire gases are a combination of various gases released by burning materials, and are also capable of burning if their ignition temperature is sufficient. However, fire gases are much more complicated than particulates, as burning materials can yield a wide range of fire gases.

Here are common examples of fire gases that are released during the combustion process:

- Carbon monoxide
- Sulfur dioxide
- Hydrogen fluoride
- Hydrogen chloride
- Hydrogen cyanide
- Aldehydes
- Benzene
- Acrolein
- And the list goes on . . .

Firefighters should be aware that all of the aforementioned fire gases can be harmful or fatal to any human. However, if asked to choose the most common and deadly fire gas, an average firefighter would most often choose carbon monoxide. Carbon monoxide is well known for its ability to be present in smoke and cause death by asphyxiation if inhaled in sufficient quantities. However, hydrogen cyanide may be the most dangerous fire gas encountered by fireground personnel. It is estimated that hydrogen cyanide is more than 30 times more toxic than carbon monoxide!

This fact was underscored in 2006 when a firefighter at a "routine" fast-food restaurant fire was hospitalized and diagnosed with toxic levels of cyanide. Interestingly, over the next 14 hours, two additional fires in the same area were responsible for four more firefighters being hospitalized. They were also diagnosed with toxic levels of cyanide, and one of the firefighters suffered a heart attack. These incidents have generated a new look at this common combustion product that may be present in high quantities at structure fires. Let's take a brief look at hydrogen cyanide and its inherent risk to fireground personnel.

Hydrogen cyanide can be found in either a liquid or gas form and is a powerful poison that reduces the capacity of blood to carry oxygen. Bodily organs that are vulnerable to cyanide poisoning are the brain, heart, and central nervous system. The toxicity of hydrogen cyanide is emphasized by its use in gas chambers and as a chemical warfare agent. Unfortunately, hydrogen cyanide is found in common materials such as wool, paper, wood, and cotton. However, it is also found in elevated levels in synthetic materials such as foam, pesticides, plastics, synthetic fibers, polyurethane, fiberglass insulation, and other similar modern materials.

These materials are common in the typical residential structure fire as well as other types of structure fires, resulting in hydrogen cyanide likely being present in readily detectable amounts in smoke from smoldering fires, free-burning fires, and overhaul operations. Stated from another perspective, hydrogen cyanide is an extremely toxic gas that is released during the combustion process from any material that contains nitrogen, which is commonly found in structural occupancies. Although there are numerous other toxic gases released during the combustion process, carbon monoxide and hydrogen cyanide are prime examples of why ventilation should be an initial consideration at structure fires, and a primary reason why fireground personnel should not breathe smoke. That statement applies to structure fires, automobile fires, dumpster fires, and other similar incidents. Toxic gases that are inhaled by fireground personnel give credibility to the phrase "when firefighters breathe smoke, they can die."

Ventilation, therefore, is critical for the survival of both building occupants and firefighters.

GENERAL VENTILATION PROCEDURES

Ventilation operations should be governed by a set of standard operating procedures (SOPs), the values of which determine the course of a ventilation operation. Personnel must know how to accomplish the various types of ventilation operations, tactics, and techniques for a specific fire scenario. Other than actual fireground conditions, there is no substitute for training to help determine how to accomplish these types of operations. Discuss and practice ventilation on a frequent basis. If you do so, basic operations will be automatic at a fire, giving personnel a greater opportunity to focus on the specifics. The fireground provides the opportunity to hone operational skills.

Initial ventilation size-up

When sizing up for ventilation operations, firefighter must ask the following questions:

- What is the type of building and the time of day?
- Is there a fire and, if so, what is its severity?

- Is it a small food-on-the-stove fire that may only require opening a few windows to vent the area or a room-and-contents fire that will need immediate ventilation to assist in firefighting operations?
- What are its extension or auto-exposure possibilities before venting?
- Are any occupants endangered?
- Is immediate ventilation needed?
- Is horizontal or vertical ventilation feasible?
- Will a flow path be created and pull a fire in a specific direction or increase its size or intensity prior to the hoseline being in place?

It is crucial to take a few seconds and determine the size and location of the fire and the building type before haphazardly venting.

FIREGROUND NOTE

Communicate before you ventilate? Always communicate and coordinate ventilation efforts at all structure fires to allow an easier advance for the engine company to extinguish a fire.

Building type, age, and characteristics

What type of building is on fire? Suppose it is a multistory hotel. Depending on conditions, personnel may consider venting the stairs and hallways to aid in evacuation. In some larger multiple dwellings, it may be important to ventilate the roof's **bulkhead**, an extension of the stairwell shaft above the roof, or **scuttle**, a hatch in the roof of a building, to alleviate the mushrooming effects of the smoke and permit residents a means of egress down the interior stairs. It is also wise to consider the age of the building. Does it appear to be an older style of construction? If so, it could be balloon frame construction, and introducing forced air into the building could rapidly extend the fire. In newer contemporary homes, firefighters often see numerous skylights on the roof, which aid in vertical ventilation. Often, factory windows are covered with security mesh or grating, which may delay ventilation operations. It is important for firefighters to be familiar with the basic types of construction and buildings in their response district to help them determine appropriate ventilation operations when a fire occurs.

Location, size, and extension of fire

First, what are the fire's location, size, and extension possibilities? If there is a small rubbish fire on the first floor of a five-story apartment house, going to the roof to cut ventilation openings into the roof may not be vital. It may only be necessary to open the bulkhead or scuttle, relieving the stairwell of smoke and gases, and vent the windows in the immediate fire area. However, if the fire is on the top floor of a five-story apartment house, venting the top floor windows first and then cutting the roof open over the fire is an appropriate tactic (fig. 14–2). Whenever we size up the location and size of the fire we should always consider what ventilation operations and efforts might do to increase the size or intensity of the fire or if fire will be drawn to other locations.

Fig. 14–2. Fire and smoke pushing from the top floor of a five-story building

Type of roof

It is important for firefighters to be familiar with the different types and styles of roofs and construction in their response district (fig. 14–3). In some situations it is vital to know the types of construction on which personnel are working.

Fig. 14–3. a) Clay tile roof; b) slate roof; c) wood shingle roof

Fig.14–3. (continued) d) asphalt shingle roof; e) built-up roof

Inverted roofs (flat roof construction with a relatively slight pitch that allows water to flow to a roof drain) are normally *spongy*. Firefighters must know how to use the roof construction to their advantage when cutting ventilation openings, and also the approximate time available for ventilation operations. Firefighters must also be familiar with types of roof coverings such as clay tile, slate, wood shingle, and asphalt shingle on a peaked roof of a single-family home or a "built up" roof consisting of successive layers of asphalt and felt paper on the flat roof of a multiple dwelling.

A relatively new type of roof surface is the "rubber roof," a membrane made of synthetic rubber or certain types of plastic. This material is used on flat roofs to provide a more leakage-resistant barrier. They are often laid down using a torch; sometimes the rubber roof itself is ignited, causing a rapidly moving fire with heavy black smoke.

When ventilating a rubber roof, it is beneficial to remove the rubber first before cutting the roof deck below, if possible. This will make the saw more efficient and speed up the ventilation process if using a rotary saw, as the rubber membrane can become entangled in the guard.

However, membranes can be easily cut without entanglement if using a chain saw with a sharp chain run at full rpm.

Ladders

Determining the type of building and the roof's pitch helps firefighters determine the type and length of ladders needed for roof access or horizontal ventilation. In some instances, an aerial ladder, a tower ladder, or portable ladders (extension and roof) are needed in ventilation efforts.

Hazards

Often, hazards stand in the way of ventilation operations. For example, consider the problems and solutions necessary if electrical wires or razor wire security measures block ladder placement to a roof (fig. 14–4). Also once on the roof, firefighters must use caution and be aware of any air and light shafts or slipping off of a high pitched roof. In one recent incident, a firefighter removing a scuttle cover received numerous electrical shocks. The occupant of the building was tired of break-ins from the roof, so he or she created a homemade security device. Another hazardous condition may be a chain-link fence or steel fences blocking emergency egress to the adjoining roof.

Fig. 14–4. Caution should always be used when working around electrical lines. They are just one of many hazards on rooftops that can have fatal consequences.

TYPES OF VENTILATION

Firefighters should begin ventilation operations by identifying the direction that smoke, heat, and fire gases will travel to exit the structure. Once the direction of travel has been determined, the appropriate method of ventilation should be used. Ventilation can occur both horizontally and vertically. The two methods most commonly used are natural and mechanical (or forced) ventilation. Normally, both methods are used for vertical or horizontal ventilation of a structure. Whenever we perform either type of ventilation, we should be concerned with creating a flow path to properly vent the structure in coordination with fire extinguishment. Premature venting can create the fire intensity to increase or, if done in the wrong location, the fire can be pulled to the ventilation opening created.

Natural ventilation

Natural ventilation at most fire scenes consists of opening up doors, windows, scuttles, and skylights to let the by-products of combustion (smoke, heat, and gases) rise naturally and escape to the outside atmosphere (fig. 14–5). Unfortunately, natural ventilation can be enhanced or limited by prevailing wind currents or atmospheric conditions. This method of ventilation works well when there is a light smoke condition and time to vent the structure out. Small fires such as food on the stove, light ballast odors, or a small rubbish fire may be areas to use natural ventilation procedures.

Fig. 14–5. Natural ventilation

Another process for providing natural ventilation is to cut a hole in a roof to relieve the building of smoke, gases, and heat (fig. 14–6). These elements naturally rise up and out of the ventilation hole. Although the means to cut the hole might be mechanical (power saw), the natural tendency is for smoke, gases, and heat to follow the path of least resistance to the outside atmosphere.

Fig. 14–6. A firefighter venting a roof with a saw.

Mechanical ventilation

SKILL DRILL

FFI 4.3.11 Firefighters often rely on mechanical means to ventilate a structure. The items or systems to provide mechanical ventilation can include an electric, gas, or hydraulic powered exhaust fan or blower; a hoseline to provide hydraulic ventilation; a building's heating, ventilation, and air conditioning (HVAC) system; or a building's smoke management system (covered in more detail in chapter 30, Fire Protection Systems).

These types of ventilation practices are also referred to as pressurized ventilation. This type of ventilation can be used to assist or replace natural ventilation operations at a fire (fig. 14–7). Mechanical ventilation can enhance horizontal and vertical ventilation operations, but must be used when fireground operations favor its implementation.

Fig. 14–7. Mechanical ventilation

Pressurized ventilation is normally classified into two different types: **negative pressure ventilation (NPV)** and **positive pressure ventilation (PPV)** (fig. 14–8). The following material describes the difference between the two types of ventilation.

Fig. 14–8. A firefighter has set up a PPV fan and checks to ensure that the entire doorway is covered by the flow of air into the building.

Negative pressure ventilation. **FFI 4.3.12** To better examine both negative and positive pressure ventilation, assume that the room in fig. 14–9 will be ventilated. In this example, the room is filled with the various by-products of combustion. The warmer gases rise toward the ceiling, and the cooler gases settle toward the floor (thermal layering), assuming that the door and window are closed. In fig. 14–10 the room is now being ventilated by means of negative pressure. The door has been opened and a blower has been placed inside the window frame, or attached to a ladder on the outside of the building. The blower will be turned on to exhaust the contaminants in the room, coaxing them to the exterior by creating suction within the room.

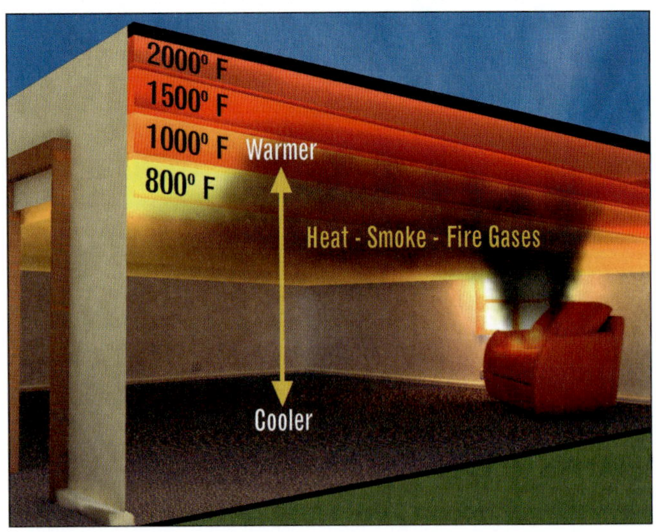

Fig. 14–9. When contaminants fill a room, thermal layering causes the lighter, hotter gases to rise to the top, and the cooler, heavier gases settle to the bottom.

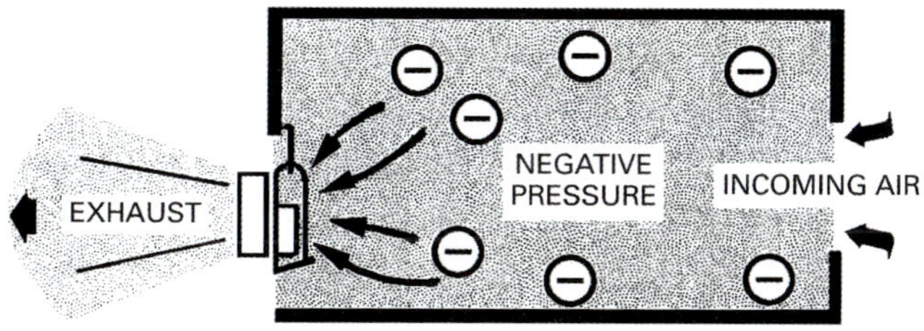

Fig. 14–10. Negative pressure involves placing a blower inside a room to exhaust the contaminants to the exterior.

Although NPV can prove satisfactory, it has several disadvantages, not the least of which is that personnel may be exposed to hazardous contaminants when positioning the blower. Also, when the blower is placed in the window, the suction pulls in air from the room and recirculates some of the exhausted air and gases if the opening around the blower hasn't been sealed with a salvage tarp or plastic. Another problem is that the contaminants from the fire are drawn through the blower, creating a need for additional equipment cleanup and maintenance. Smoke from a wood fire always contains some unburned gases and a fog of unburned tar-like liquids. When plastics burn in a fire, they release hydrocarbon particulates and heavy dense black smoke particulates. When these materials and particulates come in contact with the blower's mechanisms, they stick and form a film that is often difficult to remove. A buildup of materials or a dirty blower passageway can restrict air flow and prevent the unit from functioning up to its capabilities.

Remember, blowers placed in doorways, windows, or hallways can block pathways in and out of the building and firefighters must use caution to avoid striking them with their heads while operating on the fireground. To position blowers effectively, you must often suspend them in doors or windows using straps, ladders, or other accessories. Blowers placed inside buildings can add to the overall noise and confusion, hindering communications. Gasoline-powered blowers should not be put inside a structure because of the addition of more carbon monoxide into the area. Finally, interior blowers aren't totally efficient at removing all the contaminants from the tops of rooms. Air follows the path of least resistance, normally a straight line, from the fresh air inlet to the blower, which can limit the flow of air in the areas near the ceiling.

Positive pressure ventilation. Over the course of the years, with the continual improvement of firefighting equipment such as portable fans and blowers, positive pressure ventilation (PPV) has been an increasingly used tactic on many firegrounds. In fig. 14–11, the same room is pressurized by a blower placed on the outside of the structure to create PPV. Firefighters using this method of ventilation are forcing fresh, pressurized air into the room, creating a positive pressure within the entire room. As in a balloon, the pressure is equal at the top, bottom, and corners of the room. When the window is opened, it becomes an exhausted opening and the contaminants from all parts of the pressurized room ventilate to the exterior. Remember, implementing PPV effectively depends on controlling the flow of pressurized air between an entrance point and exhaust opening.

Compared with negative pressure ventilation, PPV has some distinct advantages in certain fire situations. First, personnel aren't normally exposed to hazardous interior contaminants while positioning exterior blowers. Second, the contaminants aren't drawn through blowers, thereby reducing the need for cleanup and maintenance. Also, doorways, windows, and halls don't need to be blocked by blowers, reducing the chance of a firefighter striking their head on suspended blowers. Some exterior blowers aren't dependent on additional equipment or accessories for setup, because they are self-propelled and self-sufficient for ventilation operations. Placing these types of blowers into service on the fireground requires fewer personnel to deploy them. Exterior blowers may have a minimal effect on the interior noise levels, but many add to the noise level on the outside of the structure. In addition, if gas blowers are not equipped with the proper exhaust flow extensions, the blowers may add to the carbon monoxide level in the structure.

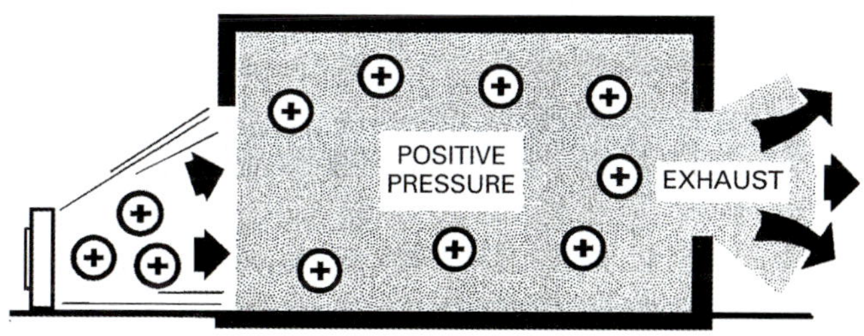

Fig. 14–11. Positive pressure ventilation involves placing a blower outside of a room to force pressurized air inward, thereby driving contaminants to the exterior.

Compared with negative pressure ventilation, PPV is approximately twice as efficient at removing contaminants from a building as long as specific requirements are met. First, exhaust openings should be selected to provide either horizontal or vertical channels for escape of the by-products of combustion. *The exhaust openings should always be created prior to pressurizing the structure or area to be ventilated.* If this is not followed, the pressurized airflow and by-products of combustion have no means of escape from the structure.

Second, it is imperative that the flow and path of pressurized air between an entrance and exhaust opening is controlled to achieve maximum ventilation. If pressurized air is directed from an entrance to an appropriate exhaust opening without being diverted to other openings, contaminants are removed in a minimal amount of time. Simultaneously opening unwanted windows and doors won't facilitate successful PPV operations.

As with many tactics and techniques used on the fireground, a proper size-up must be made to choose the right tactic for the right situation. There are times when PPV may not be a viable option to use on the fireground:

- **Positive pressure ventilation** is used by many departments both just prior to fire attack and for smoke conditions after the fire has been knocked down. Some fire departments, however, use PPV *during* fire attack; this tactic can have deadly results if not conducted properly. If used during fire attack, PPV should only be conducted if the location, extent, and spread of the fire has been established. Additionally, PPV should only be conducted during fire attack if hoselines are in place and operating. The use of PPV during fire attack—which does not meet this criteria—can drive fire at trapped victims and firefighters operating in the building.
- **Balloon construction:** Fire traveling up the walls of these structures could rapidly spread if PPV is initiated before a sufficient exhaust opening is opened. Fire in such structures is difficult to chase if it is in the walls, so use caution in feeding the fire with forced air if fire is in more than one location. In balloon-frame construction, only use PPV if the exhaust opening is opened first.
- **Vent, enter, isolate, and search (VEIS):** Creating an opening to perform VEIS can become an exhaust opening for PPV and possibly make the area to search untenable for the firefighter.
- **Location of the fire:** If the location of the fire is unknown, carefully analyze the consequences before putting PPV into service. A large structure with a complex layout may make PPV operations unsuccessful if it has multiple exhaust openings.
- **Search and rescue:** While these tactics are deployed, the incident commander must evaluate and size up the fire prior to performing any PPV operations.

For more tactical considerations and PPV operations, refer to the end of this chapter.

EQUIPMENT FOR VENTILATION OPERATIONS

Portable fans, blowers, and smoke ejectors

Although natural ventilation is a viable method, it can be assisted or replaced with portable exhaust fans, blowers, or **smoke ejectors** (figs. 14–12a and 14–12b). These fans and blowers can rapidly force the by-products of combustion out of a contaminated area. They also allow you to use openings remote from heat, smoke, and fire, directing such contaminants through preselected, controlled openings or openings not normally viable by natural means. Portable fans, blowers, and ejectors can overcome interior and exterior temperature differentials, as well as the effects of humidity. Overall, they reduce the time necessary to ventilate a building as compared with natural ventilation.

Fig. 14–12a. Various tools may be used to perform both positive and negative ventilation, such as fans, blowers, smoke ejectors, and fog streams.

Fig. 14–12b. A smoke ejector placed in a doorway to remove smoke via negative pressure ventilation

Hydraulic ventilation

SKILL DRILL

Fog streams from hoseline can also be used for ventilation purposes; this is called **hydraulic ventilation**. Simply direct a fog pattern out of a window so that the resultant venturi action expels the contaminants to the outside atmosphere. You can obtain maximum effectiveness by holding the nozzle a few feet back from the door or window. Direct the fog pattern toward the window so that it almost fills the entire opening, leaving some room around the edges. Fog stream effectiveness depends on the size of the opening, the area to be ventilated, the size of the nozzle and hose, the quantity of smoke present, and the discharge pressure. Although some textbooks advocate that hydraulic ventilation must be performed within a few feet of a window (after having first placed the nozzle outside the window, opening it to the correct pattern, then moving it inside the building), it can be applied from a further distance away from the window. If the room in front of the nozzle has burned out floor boards or is possibly compromised, the nozzle can be directed from the door opening to the window frame. The resulting pressure from the stream will pull smoke and contaminants from the structure but may take more time to ventilate the structure. It is also important to note that when firefighters are fog venting, they must remain aware of the fire reigniting or **lighting up** behind them, because more air is introduced into the area.

If a smooth bore nozzle is used for initial fire attack, it can still be used to fog vent out the window. The firefighter can crack the nozzle open slightly to make a coarse pattern that fits inside the window frame. This stream will also ventilate the structure but possibly not as good as a fog nozzle.

Hydraulic ventilation is normally accomplished by a firefighter standing in a contaminated environment and using a spray/fog nozzle to direct contaminants from interior to exterior through available openings such as windows. Although this method can be beneficial, a drawback is that it requires a firefighter to stand in the contaminants while they are being exhausted to the exterior.

For more information on fog venting, see chapter 16, Fire Streams.

HVAC systems

Many multistory buildings use **heating, ventilation, air conditioning (HVAC) systems** to control the environment inside the structure. Such systems draw in outside air, change it to the desired temperature, circulate it, and exhaust it to the exterior. It is possible to use an HVAC system to rid a building of contaminants; however, doing so depends on the capabilities of the system and the expertise of the personnel operating it. It is imperative that the capabilities and operational techniques of a particular system be thoroughly understood, because an HVAC system is also capable of spreading contaminants and fire to uninvolved portions. Initially, during a fire, firefighters must determine if the system was automatically shut down by its safety overrides. All other operations should be handled by the appropriate building engineer or other similar responsible person in conjunction with firefighters.

VENTILATION PRIORITIES

FFI 4.3.11 Integral to any ventilation operation is determining appropriate priorities. The top priorities in any ventilation operation are improving the interior environment to aid in evacuation and increase a victim's chance of living and enhancing the working environment for suppression personnel. Ventilation operations also reduce fire loss and damages to the structure and its contents. The priority of specific ventilation operations is predicated on fireground conditions. Normal ventilation operations are usually performed using a horizontal or vertical ventilation method.

Horizontal ventilation

FFI 4.3.11 For reasons of ease and safety, horizontal ventilation is the most popular method. It is often accomplished simply by opening doors and windows.

Using horizontal ventilation to remove heat, smoke, and fire gases can be effective; however, its effectiveness depends on the size of the opening and its proximity to the contaminants. Horizontal ventilation does not take advantage of the natural vertical path of heated gases and smoke. It also has a minimal effect on the hottest portion of a room (ceiling), where most flashovers originate. Therefore, horizontal ventilation will ventilate that portion around the opening (open door or window) but could have a minimal effect on a flashover.

Before using horizontal avenues, consider these factors regarding wind and fire location:

1. Determine whether the direction of wind could carry contaminants to uninvolved areas of the fire building or exposures. If the building has multiple stories and windows that will be opened on the lower floors, will the direction of smoke then contaminate the upper floors?
2. Consider the speed or force of the wind. When windows are opened on the windward side, will the wind accelerate the fire and enhance extension?
3. Open the windows on the leeward side of the building first. Windows on the windward side can then be opened to allow the wind to force contaminants from the structure (fig. 14–13).

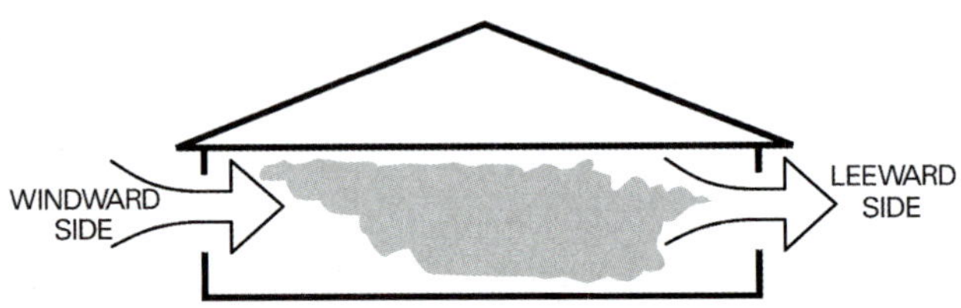

Fig. 14–13. When the wind can be used for horizontal ventilation, open the leeward windows before the windward windows.

4. Opening windows close to the seat of the fire is a top priority. This provides ventilation to the fire area and allows the expanding gases and steam from fire-suppression operations to escape. It also reduces the chances of the fire, heat, and gases from lighting up overhead or wrapping around the nozzle team, because they are escaping through controlled openings.
5. Remember, prematurely opening windows in a fire area can increase the supply of oxygen and cause the fire to accelerate. Therefore, coordinate communication between the hoseline's advance and ventilation operations to maintain a safe working environment.

Once windows in the fire area have been opened, open others away from the fire area if doing so ventilates other contaminated areas without enhancing extension. Firefighters should remove blinds, curtains, and shades when opening windows to ensure that they won't restrict the removal of contaminants. In addition, it is vital to ensure that screens are removed, because they can restrict the movement of air by at least 50%.

When firefighters open **double-hung** or **casement windows**, there are normally two rules to follow. If there is a light wind blowing toward the ventilation opening, the window can be opened in the following manner: two-thirds down from the top and one-third up from the bottom (fig. 14–14). Doing this allows bidirectional flow, which allows cool air to enter the bottom of the window while the hot, contaminated air exits at the top. When there is no prevalent wind, the windows may only need to be pulled down from the top to allow the room to ventilate. Pivoting windows must be opened according to their design. If a window can't be opened, you may need to break or force it open.

Fig. 14–14. A double-hung window opened two-thirds down from the top and one-third up from the bottom

Horizontal ventilation can be enhanced by mechanical means. If multiple areas need ventilation, ventilate each area by using doors as partitions to reduce the size of each. This maximizes the flow of fresh air through each area to be ventilated. When opening these doors, remember to wedge or chock them open so that they don't inadvertently close.

Vertical ventilation

FFI 4.3.12 Depending on conditions and building construction features, vertical ventilation can be the preferred method when the heated products of combustion rise and collect at the highest levels of the structure (fig. 14–15). Vertical ventilation takes advantage of the natural travel path of heated contaminants and ventilates the area that needs it most—the ceiling area, which prevents mushrooming. This is the common term for when contaminants collect at the highest level of a structure and deflect downward, or bank down, onto lower floors.

Firefighters should try to open a vertical artery to displace these contaminants. This decreases both the amount of explosive gases, heat, and smoke collecting at the structure's upper levels and the chances of fire extension. In addition, the release of these by-products also pulls in and introduces cooler fresh air at the lowest levels. This can improve conditions at the floor level and assist firefighters in their fire attack operation.

Vertical ventilation permits firefighters to proceed and perform a more thorough search for victims and fire extension on the floors above the fire. It also relieves the interior hallways and stairwell of smoke, which can assist in the evacuation process.

Note that not all fires require vertical ventilation; a small rubbish fire in a structure doesn't require immediately cutting the roof open to provide vertical ventilation. Firefighters must use an appropriate size-up to decide which ventilation tactic to perform. Vertical ventilation takes additional time, staffing, and equipment for completion, and can also place ventilation personnel in a dangerous location (over a fire).

Fig. 14–15. Vertical ventilation takes advantage of natural convection currents.

VENTILATION USING NATURAL CONSTRUCTION FEATURES

It will always be easier for firefighters to use the natural construction features of a structure for initial ventilation operations. Building features such as windows, scuttle covers, bulkhead or penthouse doors, and skylights can be a fast and efficient means of ventilating a structure. There are a variety of tactics to follow when opening the following construction features for ventilation operations:

Air shafts. **Air shafts**, also know as **light shafts** or **light wells**, may be found in many different types of structures, such as row frames, multiple dwellings, tenements, and hotels. They are intended to provide air and light to inner rooms of a structure. In a situation where there is a small fire and no concerns of a direct flame exiting the structure, an air and light shaft can be used for ventilation purposes by opening windows within the shaft to rooms that need ventilation (fig. 14–16). However, firefighters must close windows within the shaft as appropriate to prevent fire extension into the shaft or autoexposure to the floors above or an adjoining building. In some instances, a skylight is at the top of the shaft to prevent inclement weather from entering it. If the shaft is covered, the skylight may have to be removed to provide ventilation.

Fig. 14–16. An exterior light well may be used to aid in ventilation provided upper story windows are closed prior to venting.

Skylights. **Skylights** in dwellings are normally located over a stairwell or hallway. In commercial occupancies, larger skylights are often placed over manufacturing areas. Depending on the type of structure, individual skylights may also be placed over stairways, light shafts, and air shafts. If a building has an attic or cockloft, the area below the skylight is normally boxed off from the attic/cockloft space. Therefore, when the skylight is opened, the interior of the building is ventilated but the attic space is not. These small walls that extend down from the inside of the skylight are commonly referred to as the returns. When checking for fire extension inside the attic or cockloft, firefighters can open up the returns to size up the conditions. In some instances, firefighters should not open up returns that are remote from the original fire because it could possibly pull the fire in that direction and assist in horizontal spread.

Skylights may be opened for ventilation purposes by removing the entire assembly, removing the glass panels, or breaking the glass panels (fig. 14–17). However, always communicate when taking the skylight so we can time ventilation efforts and not create a flow path that will pull the fire toward the ventilation opening we've created. If the glass panels are removed individually, additional time may be required and broken glass may fall into the building. When a firefighter isn't equipped with a portable radio and is going to break a skylight, a good procedure to follow is to break a small section of it first. This should give warning to the members operating below that ventilation operations are going to take place overhead.

Fig. 14–17. A skylight

Once any skylight has been vented, the firefighter should take a hook and reach down into the hole to ensure a **draft-stop** (a piece of glass, Lexan®, plastic, or drywall installed at the ceiling level, preventing the loss of heat up into the skylight—not to be confused with a draft stop in an attic) isn't secured below and interfering with ventilation operations. If conditions below reveal a fire is evident under the draft-stop, a hand tool tied to a utility rope can be dropped down to vent the area. This safety measure keeps the firefighter a safe distance away from the escaping flames exiting the skylight.

Glass skylights. The most effective method to remove the skylight assembly is by completely prying it up from its attachment points or tipping it onto the roof. This takes minimal time and reveals the entire opening. It also allows you to replace the skylight later.

Skylights can also be opened as follows: If it appears that tar hasn't been used around the edges of the glass and the panels can be quickly and easily removed, then remove the metal tabs or metal stripping along the bottom edge of the panels. This will allow them to slide out. If tar has been used as a sealant and the glass panels can't be removed easily, then attempt to remove the entire skylight's frame attached to the roof if possible. If these attempts do not work, break the glass panels (fig. 14–18). Remember, if you break the glass panels, the resulting shards of glass within the building will be a hazard to interior operations.

Fig. 14–18. Older skylights are often sealed with tar, making them difficult to remove.

It is important to note that on many bulkheads or penthouses, roofs are slanted and dangerous to climb up on and operate on these spaces. Breaking the glass may be the only safe option for firefighters to perform.

Plastic skylights. Newer buildings have plastic skylights, usually a plastic bubble in a metal frame attached to a metal riser. First, try to pry the skylight away from the roof (fig. 14–19). If this can't be done easily, you may have to cut the plastic around the junction of the metal and plastic. Use an axe or power saw to make the cut. If conditions on the roof necessitate operations must be done quickly, attempt to break the plastic bubble with a hand tool for rapid ventilation.

Fig. 14–19. Plastic skylights can often be removed by prying upward.

Often, because of water leakage or to beef up security, roof skylights have been replaced with plywood/oriented strand board (OSB) and lightweight support beams or just tarred over. Thus, firefighters should avoid walking on any skylight or alteration. In situations such as these, firefighters may have to use a power saw to open up the roof or skylight area. Remember, once a skylight is removed, it should be placed near the ventilation opening to warn firefighters of an open shaft on the roof.

Bulkheads and penthouses. Opening the door to a **bulkhead** or **penthouse** can provide significant ventilation to the interior stair shaft and any hallways open to it. When venting the bulkhead, be aware that civilians overcome by smoke could be exiting there. Also, in some areas the bulkhead has been known to be a sleeping area for homeless individuals, so a quick search of the area should be a main focus. Remember with opening the bulkhead in the right conditions, we could create a flow path and pull the fire toward the operating companies below. Always communicate your intentions and actions with the interior teams so they are not overcome by conditions caused by premature ventilation. Because of the importance and simplicity, opening it up should be a primary consideration. If it has a lower door at the bottom of the stairs, verify the conditions on the other side before opening it. Firefighters proceeding down the interior stairs to open this door could be subjected to severe smoke and fire conditions behind it, so they must use caution.

Depending on the type of door (wood or metal) located on these structures, firefighters can use normal forcible entry tactics to open them up. Usually, these doors are outward opening because they are an exit door. In some cases they may have added security such as a slide bolt or eye bolt and hook, and if they are difficult to force, firefighters should size up the hinge size of the door. As always, remember our cardinal rule of forcible entry: **try before you pry**.

Another typical feature of these structures is that they often have a skylight mounted on top. This skylight provides additional vertical ventilation. In some instances, the skylight has a draft stop mounted on the ceiling of the bulkhead. If the skylight is vented and no smoke issues from it, check for the presence of the draft stop. It may be possible for the firefighter to strike the draft stop from the door of the bulkhead with a hook to ventilate it.

Often, firefighters encounter a bulkhead or penthouse with high walls, making it difficult to climb in bunker gear and reach the skylight mounted above. If an aerial ladder is placed to the roof and it has a folding, attic, or scissor ladder mounted at its tip, it can be removed and used to access this skylight. If not, there are a few other options.

A firefighter can remove the bulkhead's door and lay it on an angle on the wall of the structure. If the firefighter has two tools, one tool punctures the roof at the base of the door to act like a brace, preventing the door from sliding outward as they climb. Once on top of the bulkhead, vent the skylight (fig. 14–20).

If the firefighter has a hook and Halligan tool, he or she can place the Halligan alongside the bulkhead's wall with the fork pressed into the roof and the adze resting on the wall. This tool creates a step or small platform. The hook's head is placed on the top edge of the bulkhead's wall and the handle of the tool hangs downward. Now the firefighter can step on the Halligan and use the hook to pull him- or herself upward while climbing.

Fig. 14–20. Sometimes we need to be creative when gaining access to vent.

It is *extremely* important that firefighters enter and exit the roof of the bulkhead in the same location. Many bulkheads are aligned next to a shaftway; and exiting in the wrong place can cause a firefighter to fall many stories, possibly suffering a severe injury.

Remember, any time a firefighter opens one of these doors and is met with a heavy smoke condition, he or she should still probe inside the landing with a tool to check for victims. Also, opening the door to reveal little or no smoke may indicate that the fire apartment door hasn't been opened yet. Also, don't forget to chock the door open to prevent it from relocking if it accidentally closes.

There are a few variations in the type of bulkheads prevalent on buildings across the country (fig. 14–21). Many have two doors and are a larger sized bulkhead that provides access to the roof in two separate locations. Also, it isn't uncommon to find the elevator motor room located inside the larger sized bulkhead. Firefighters must be careful when working around the **walk-through bulkhead**. This bulkhead normally cuts the roof section of a building into two sections and has doors at both ends. Firefighters must resist the urge to transverse through the interior of the bulkhead under heavy smoke or fire conditions, as this places them in peril. Access to the other side of the roof should be done by aerial ladder, tower ladder, or the top of the bulkhead if sufficient room exists.

Fig. 14–21. Bulkhead and penthouse

Roof scuttle. As shown in fig. 14–22, a **roof scuttle** is a small, covered opening providing access to the roof. Some scuttles consist of a wood cover over wooden risers on the roof. The cover is normally protected by a composition covering. Unless the cover has been attached to the risers, it can be removed by prying it up with an axe, Halligan, or similar tool. Newer scuttles consist of metal risers with a spring-loaded metal door that is fastened and locked from the inside. These are difficult to open and may take more time and effort to open. Sometimes it may be easier to cut the scuttle with a saw in these situations. If a scuttle is found tarred over, it also may be easier to cut the scuttle cover open with a power saw to provide vertical ventilation.

In many buildings with scuttles, an interior vertical ladder is mounted inside the scuttle's opening, providing access to the roof (fig. 14–22). These are normally located in the top floor hallway or in a closet in the hallway. Once the scuttle cover has been removed and no smoke is present, position yourself so that only your arm is placed over the open scuttle; and reach down with a hook to probe the walls, looking for a door. Use caution and do not subject yourself to climbing down the narrow scuttle opening and on the steep vertical ladder to open this door. That is a dangerous practice and could seriously injure you if there is fire behind the closed door.

In many of the commercial occupancies, for increased security, these scuttle covers are made of heavy steel plating with heavy security locks. A rotary saw and metal cutting blade may be the best tool to provide rapid ventilation. As in all situations, notify the interior teams of your intentions when venting a scuttle so you don't create a hazardous condition for units operating below.

Fig. 14–22. A roof scuttle with a fixed access ladder

Elevator house or bulkhead. Structures with elevators may have an equipment or motor room located on the roof. **Elevator houses** vary in size according to the number of elevators they service. They are above the elevator shafts and open, by varying degrees, to the shafts. Opening the door and skylight can provide ventilation to the elevator shafts, reducing concentrations of heat and smoke within the structure.

Additionally, the equipment in an elevator house supports the full weight of the attached elevators. Therefore, any fire that has extended to this area may weaken the structural integrity of the elevator equipment supports, possibly causing collapse of the equipment and elevator cars. The resultant hazards and damage that this might cause cannot be overemphasized. Firefighters must use caution if they enter into these rooms; there are holes in the floor that are normally covered with steel grating to allow air currents in the shaft to flow while the cars move up and down. It is common to find these gratings in poor shape or missing.

Ducts. Heating, air conditioning, and other **ducts** of various sizes are often found on roofs (fig. 14–23). Because these passageways may lead to the interior of a building, they must be checked for heat and smoke. If heat and smoke are present, open the duct to provide some ventilation to the interior of the building, as well as to check for extension (fig. 14–24). Also, when firefighters are faced with a grease fire in ductwork, it may be necessary to do the following:

- Tear open the top of the ductwork to allow proper ventilation
- Open the roof near the ductwork to ensure that no fire has left the duct at its seams and extended into the cockloft or roofing materials

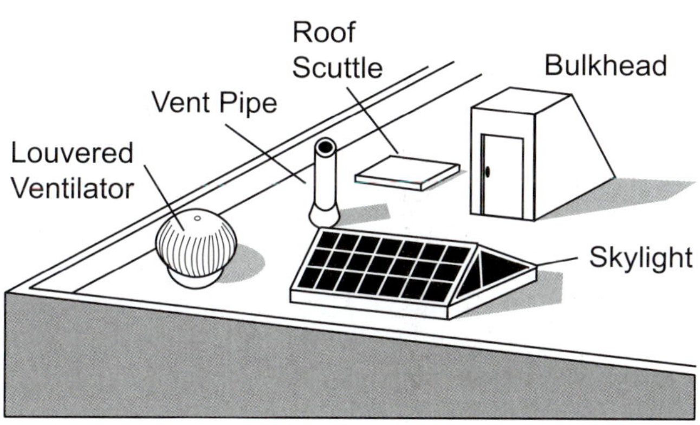

Fig. 14–23. Roof ventilation personnel should be able to recognize and use various types of natural construction features to assist ventilation.

Fig. 14–24. A rooftop duct

Dumbwaiter bulkheads. Some older multiple-dwelling structures may still have their **dumbwaiter shaft's** bulkheads on the rooftops. Although the shafts are now illegal to use in most parts of the country, they haven't been removed from the structure. Most shafts served more than one apartment and were located in the kitchen area so tenants could send down their trash to the basement for the maintenance people to remove. Often fires in the kitchens or apartments can extend into the unused shaft and upward (fig. 14–25).

Fig. 14–25. A dumbwaiter shaft door inside an apartment.

Venting these shafts allows the fire to vent vertically and possibly prevent it from extending to other apartments. Often, fires in these types of buildings have smoke showing from the dumbwaiter bulkheads on the roof (fig. 14–26). Venting these bulkheads is important and must be a quickly accomplished task. The bulkhead may or may not have a skylight on the top and can easily be ventilated by prying up on its sides and hinging it open. Also, the dumbwaiter bulkhead normally has an access door on it, providing access to the pulley and gears located at the top of the shaft. These doors are now normally sealed up and covered with roof tar.

Fig. 14–26. There are many avenues for fire to travel horizontally, as well as places to perform ventilation, such as this dumbwaiter bulkhead.

The doors open outward and can be forced open using outward-opening-door techniques. One of the best methods is the baseball swing with the Halligan tool. The point of the tool can be embedded in the door, opposite the side of the hinges. Now the tool can be pulled or pushed toward the door's hinge side to force it open. In some instances it may take a few attempts to force the door open in this manner because of the magnitude of tar that has sealed the shaftway.

Vent pipes and soil pipes. Plumbing **vent pipes** extend through a given roof at various locations and sometimes can indicate the locations of bathrooms and kitchens. Although they aren't considered ventilation avenues, they do travel vertically within a building. Therefore, they are capable of spreading heat and fire. Smoke issuing from acrylonitrile butadiene styrene (ABS) plastic vent pipes is an excellent indicator that the pipe is burning somewhere within the building. ABS plastic pipes can burn readily and serve as extension avenues wherever the pipe travels. After a firefighter feels the base of a metal or plastic vent pipe and notices signs of heat or smoke, the roof must be opened and checked. Additionally, a vent pipe that appears to be growing upward is an excellent indication that the roof is sagging and possibly about to fail.

Ventilators. Many types, sizes, and shapes of **ventilators** may be found on roofs. Ventilators can be classified as either venting a particular area within a building or providing some type of ventilation for a device or appliance, such as kitchen appliances. The size and shape usually reveal the type; or if it is covered with grease or oil residue, it can inform the firefighter that it is connected to a cooking device. Ventilators for heat-producing appliances generally have a serrated top suspended over the pipe, whereas those that vent the attic or top floor usually have a capped top or the familiar turbine.

Smoke or heat coming from a ventilator indicates that the area it serves is affected by smoke or a fire, or a fire exists in the ductwork itself. If a ventilator is equipped with a screen or cover, it should be removed, because it restricts the efficiency of escaping smoke. Don't remove the top of a rotating ventilator, because the ventilator stack can easily be damaged and restricted. Rotating ventilators are 30% more efficient when the turbine is operational, compared with when the turbine has been taken off. Be aware that these ventilators can detach and fly off the pipes with fire venting from it. If these items are the only ventilation objects on the roof and a firefighter has to check the cockloft for smoke and fire conditions, remove them. It is easier to remove them for a quick size-up of the conditions in the cockloft and the construction features.

WINDOWS

Firefighters encounter numerous different types, sizes, and shapes of windows on the fireground. It is important to learn and identify the basic concepts of windows and how they operate for ventilation tactics to be performed rapidly. Most windows in residential structures have **double hung** (upper and lower sash), **casement** (hinged window, pivots to one side), and **solid frame** (non-opening).

These windows are also constructed in a wood, metal, or vinyl frame with single, double, or triple pane glass. In addition, in areas near the ocean, a new laminated or hurricane-resistant glass is becoming a regular feature. It is also common to find Lexan™ and other heavier gauge plastic laminates used for windows, because of their high security features.

Again, it is important for firefighters to size up the fire conditions and then decide or be told to open a window or vent a window with a hand tool. Remember, it is important to communicate with the interior teams and coordinate ventilation efforts with extinguishment operations. When venting windows, a common theme exists in the fire service: Remove everything and make the window turn into a door. Remove the screen, shades, blinds, cross members, sash, glass, and any other obstacle. This allows unimpeded access for firefighters to enter or exit the structure without becoming hung up on anything.

Venting the windows

SKILL DRILL

Prior to venting windows, all firefighters must consider the effects that ventilation will have on a fire. Premature venting of windows can increase the size and intensity of the fire. It can also draw the fire to the ventilation opening. Size up the conditions and always communicate before you ventilate so other firefighters can monitor the conditions around them. All firefighters should size up the window for any cracks from the heat or if the glass is discolored. These conditions could be a good indication that fire exists directly behind the window, so firefighters must use caution when venting begins. When firefighters ventilate a window with a hand tool on a level surface, they often maintain a safe distance away from the window and position on the windward side of the window (wind at their backs) at a level lower than the window. Firefighters often crouch when ventilating windows at this level. Crouching ensures that the firefighter remains low and out of the flow of hot air, gases, smoke, and possibly fire venting from the new ventilation opening.

Using a tool to its maximum length or choosing a tool with a long handle helps firefighters maintain a safe distance away from the window. It is important to maintain a safe distance away from a window because the escaping by-products of combustion could ignite or explode when escaping to the fresh air. Now that a safe distance has been attained, the firefighter uses the tool to strike the upper sash or pane of glass to vent the window. The upper pane is struck first because it may already be compromised by the heat generated by the fire and break easily. It also doesn't really mater what part of the tool strikes the window with single pane glass; a metal tool is often no match for glass. When firefighters are faced with double or thermo-pane glass, it may be more advantageous to strike the window with the sharp end or point of the tool to break the glass.

As the firefighter strikes the upper pane of glass, it is usually in a downward motion; and the center sash and bottom pane of glass may also break with the downward swing. If so, once both panes are broken, the center sash can be pulled out and then the window can be trimmed.

Trimming a window means that the firefighter goes around the outside of the frame with a hand tool and removes the leftover shards of glass from the window (fig. 14–27). This allows a safer means of egress and access if the window must be used by the firefighter. Once the large glass shards are removed by striking or trimming, the firefighter must ensure the blinds, curtain, shades, and screens are removed.

Figure 14–27. Trimming the window by removing remaining glass shards with a tool.

When operating on a level surface, it is okay for a firefighter to pull out any remaining glass, curtains, and blinds from the inside of the structure to assist ventilation efforts. Dropping them inside the structure could cover a victim. Once ventilation efforts are concluded on a window, the firefighter should always sweep the floor just inside the window to check for a victim who may have attempted to reach a point of safety.

Ventilating windows with a portable ladder

SKILL DRILL

The extension ladder can be used either in the beam or flat position to break a window. Some firefighters believe that the weight and structural makeup of throwing the beam into the window works better than flat, whereas others believe that the flat position covers more surface area and breaks more glass than the beam throw. Either technique can be used to vent the window.

The ladder is raised vertically to the proper level, near the top third of the window. Then the ladder is dropped or pushed into the window with some force by the firefighter positioned at the base of the ladder (fig. 14–28). As the ladder strikes the window, the firefighter should remove his or her hands from the ladder, in case glass slides down the beams of the ladder. Hands should be in the vicinity of the ladder in case it bounces off the window; this way they can grab the ladder's rungs or beams to maintain its stability. It is very important to note that firefighters performing this tactic must have the proper eye protection and wear all personal protective equipment. Also, sometimes it may take one or two attempts to sufficiently break out the window for ventilation. If this ladder evolution is properly executed, often it also breaks the window's center sash and bottom window pane.

Fig. 14–28. A ground ladder venting a window on the second floor

Roof ladder's hooks used to ventilate a window

SKILL DRILL

Another method of performing window ventilation is to use the roof ladder. The roof ladder has two large roof hooks attached to its tips that resemble the firefighting hooks or pike poles firefighters use daily. To perform this tactic, the roof hooks are placed into the open position. Then the ladder is placed at the proper height and thrown or pushed into the window by the firefighter positioned at the base of the ladder.

After the ladder has broken the top pane, the firefighter has a few options. If the window has a wood frame, the ladder's hooks can be placed on the center sash. Next, the firefighter can pick up the base of the ladder and give it a slight tug or pull backward or out of the window. Normally, this additional pressure lets the hooks break the bottom pane of glass and remove the center sash of the window. If the ground's terrain is in bad condition, avoid this method and drop the ladder into the window again to complete ventilation efforts of the lower panes of glass.

Remember, all eye protection and personal protective equipment must be used when performing these tactics.

The arm-lock maneuver—venting windows from a portable ladder. The arm-lock maneuver is a tactic that can be performed for venting windows while working off a portable ladder. It is relatively easy to perform and requires minimal time to get into position on the ladder, thus saving critical time for other important tasks. It can be used in conjunction with the leg-lock maneuvers previously described, or it can be used by itself (fig. 14–29).

To perform the arm-lock maneuver, the firefighter should use the following steps for ventilating a window to the right side of the ladder:

- A firefighter reaches the proper working height on the ladder with both feet placed on the same rung. Note that when venting the window off the side of the ladder, place the leg and foot up against the ladder's beam. This offers more support and balance when swinging the hand tool. It also can help brace the firefighter onto the ladder.
- The hook is then held in the right arm, and the left arm is placed between the two rungs directly in front of the firefighter.
- The butt end of the hook is now placed behind the ladder and grabbed with the left hand.
- Next, position the hook to the window, checking that no part of the ladder or structure will interfere with the swinging motion. Now the window can be ventilated.

For more in-depth information on the ladder tactics mentioned, see chapter 13, Ladders.

Fig. 14–29. A firefighter demonstrates hand position for the arm-lock maneuver.

Ventilating top floor windows. Ventilating windows at top floor fires is a vital tactic to follow because it allows fresh air to enter the structure and helps the superheated gases and smoke escape through the roof ventilation hole. If the top floor windows can be reached using a hook, position above and slightly to the windward side of the window. Swing the hook inward toward the window to ventilate it (fig. 14–30). Once the glass breaks, pull the body backward and away from the rising smoke, heat, and gases. Remember to coordinate communication with the inside teams prior to ventilating any windows.

Fig. 14–30. Breaking a window with a hook

In some instances, a hook might not reach the window; however, with a piece of utility chord and a hand tool such as a Halligan bar, a different tactic can be used. The first step is to secure the tool with a knot and a safety to one of its ends. Next, the tool is lowered to the window to be ventilated. Secure the rope by using the foot to step on it at the roof level; maintain this position for the method to work effectively. The rope and tool are then pulled up to the roof. Finally, toss the tool out and away from the building, causing it to arch into the window with sufficient force to break it (fig. 14–31).

Whenever this tactic is used, firefighters operating in the vicinity must use caution. Flying glass drops from the window, and occasionally utility ropes, are severed by the glass and the tool flies into the structure or down to the ground. Unfortunately, this tactic does not clear all the glass, screens, or curtains from the windows; but it can afford some ventilation efforts when there are limited means to do so.

This technique works well in many fireground situations, and some fire departments have made adjustments to their tools and tool inventory to accomplish the task. In addition, some tool manufacturers have made tool modifications for the utility rope to *clip* to the Halligan bar. Also, the hook and Halligan can be secured together by a short chain and safety clips that allow this tactic to be performed.

Fig. 14–31. A Halligan bar secured with a rope can be an effective ventilation tool.

Ventilating windows covered with security grating or steel mesh. Many factories place metal grating or mesh over their windows to prevent vandalism and illegal entry into the building. These security devices are normally comprised of a steel frame with the grating or mesh welded inside the frame, which is then attached to the building by lag bolts or through bolts. The following are a few options to choose from when encountering these windows that need to be ventilated:

- A forcible-entry saw can be used to cut a section of the grating or mesh out so that a hand tool can be used to ventilate the window.
- A forcible-entry saw can cut the brackets where the security frame attaches to the building (fig. 14–32). This releases the frame from the building, exposing the windows for ventilation operations. Use this method on small frames, but use caution if the frame is large. A large falling frame could injure unsuspecting firefighters operating around the building.
- Some fire departments have created a homemade hand tool to combat these windows. The tool is a piece of ½-in. (13-mm) metal rod or rebar about 4 ft (1.2 m) long, with one end bent downward to form a handle and the other end left blunt. The tool is inserted between the grating, and the glass is punched out. The method works well but requires additional time because it must be inserted and removed numerous times to ventilate the windows.
- In some instances, the distance between the grating is large enough to insert the butt end of a hook to ventilate the windows.

Fig. 14–32. Firefighters use a power saw to cut attachment points of mesh grate to ventilate windows at a one-story commercial building. The tower ladder bucket facilitates the operation.

Note that operating a power saw on a portable ladder is a difficult tactic to perform; use caution when doing so. If possible, working out of a tower ladder's basket can make this operation easier and safer.

Ventilating windows protected by security bars. Many structures around the country have steel security bars placed over their windows for security and protection. Normally, these bars are either attached to a frame or individually embedded into the bricks of the building. There is usually sufficient space between the bars to insert a hand tool to ventilate the windows. Once the windows have been ventilated, remove the bars to permit a secondary means of egress for firefighters operating inside the structure.

After a complete size-up of the installation and types of the security bars, firefighters can chose a particular method of removing them. There are a few ways to attack the bars, and each situation may require a different method (fig. 14–33). One method is to attack the ends of the bars, where they are embedded into the brick walls. Breaking the bricks may release the bars. This may also work when the lag bolts are set into brick or blocks. Another method is to use the forcible-entry saw and cut the bars for removal. A hydraulic forcible-entry tool may also help: Place the jaw behind the bar frame and push the lag bolts out of the building's walls, releasing the bars or their frame. In other instances, cutting the bars with a rebar cutter or torch may also work. Many of these installations can be defeated with a set of forcible-entry tools; they may be pried off the building, forced off, or even sheared off.

For more information on removing security bars, see chapter 12, Forcible Entry.

Fig. 14–33. Window bars vary. Some are flush mounted, some are recessed. Look at your district and consider which tools and techniques can negate these security devices.

Ventilating hurricane resistant glass windows. Hurricane resistant glass windows are a common construction feature in areas where hurricanes are prevalent. Because they are difficult to penetrate and offer increased security, they are becoming more widespread and found in all types of communities. These windows are designed to take an impact load from the outside and can hamper ventilation efforts severely.

The windows are constructed with either two or three layers of glass that are laminated together with a polyvinyl-butyrate and argon gas. They are attached to either a metal or vinyl frame, which is securely mounted into the building. Their locking devices are heavy-duty mechanisms and help prevent intrusion from the outside.

Attacking these windows for ventilation operations can be difficult. They normally cannot be broken by a firefighter swinging a hand tool. When faced with these windows, firefighters must revert to using power tools in order for rapid ventilation operations to be successful. The chainsaw with a carbide-tip blade is perhaps the quickest tool to deploy against these windows. A rotary power saw with a carbide-tip blade also works well. Tests have shown that a reciprocal saw can cut these windows, but the blade can quickly dull. Some firefighters have had success using an auto extrication hand saw or pruning saw when faced with these windows. These saws can cut the material but it may take considerable time to do so.

It is of utmost importance that all personal protective clothing and eye protection be used when cutting these windows. Small glass shards and fragments can cause serious injury to firefighters that aren't properly protected.

Because these windows are designed to take the brunt of a load from the outside, they are weaker and sometimes easier to remove from the inside. Any time firefighters come across these windows on the fireground, the incident commander should be notified immediately.

For more information on these windows, see chapter 12, Forcible Entry.

Ventilating Lexan™ windows. Firefighters often encounter Lexan™ windows on commercial occupancies. Lexan™ is a high-grade plastic material that is resistant to impact loads. It is often chosen for windows in these occupancies for increased security measures. When Lexan™ is exposed to high heat conditions it melts, creating its own ventilation process. Unfortunately, it is no easier to strike and remove the Lexan™ once it has melted and cooled; it still retains its impact resistance.

Firefighters can ventilate these windows with either a rotary or chainsaw. Full protective equipment and eye protection must be used to prevent an injury. Cutting Lexan™ is dangerous because fragments can become projectiles, so use caution when performing these tasks. If firefighters don't have the necessary power tools available, they can attack the frame and molding and attempt to remove the Lexan™ in this manner. If Lexan™ windows are present in a structure, communicate to the incident commander by way of a radio.

Ventilating blocked-up windows. Frequently, firefighters face fires in vacant structures that have had their windows blocked up with either cinder or concrete blocks or bricks for security measures. In other instances, factories block up large windows to reduce energy costs. Ventilating these structures often requires more time than normal ventilation operations. Firefighters must use hand tools to remove the brick or blocks to ventilate the structure.

If firefighters are operating on ground level, they should attempt to break the brick or block slightly higher than waist high. Here they have a good baseball swing and can use the momentum of their body to help break out the brick. It is important to attempt to swing in the same area and penetrate a brick or block with a few blows to make an entry point. Once the entry point is made, begin to work either next to it, above it, or below it, knocking out the next brick.

Also, when the brick or block is broken just above waist level, it leaves much of the upper portion unsupported. Many times the weight of the brick or block allows it to collapse when it is no longer supported from below, or a few simple strikes with a hand tool cause the remainder of the brick or block to tumble down. Firefighters often leave this unsupported piece in place as they complete knocking out the bottom portion of the brick. Leaving in the top section at this point sometimes keeps the area free from smoke and allows the firefighter to see the work area in front of him or her. Firefighters must keep the upper portion under constant watch; knocking out the lower section can create movement and vibrations, and the unsupported top section could come crashing down without notice. Clearing the entire window allows a better means of egress and escape if it is needed later in the operations. This tactic of opening up brick or blocked windows reduces the time and energy the firefighter must use to ventilate these windows.

When faced with these windows at upper levels, the easiest way to ventilate them is from the safety of a tower ladder's bucket. The bucket is placed slightly below the intended work area and the firefighter swings a hand tool to knock out the brick. The area below must be safeguarded by a firefighter to prevent anyone from possibly being struck by a falling brick or mortar. When working out of the tower ladder's bucket, the firefighter must remain cognizant of the fact they could strike their arm or hand on the bucket's safety rails while swinging. When knocking the blocks out in this scenario, start at the top course and work downward. Do one row at a time so the entire blocked up area does not collapse and injure firefighters operating in the vicinity below.

Ventilating wired glass windows. Many windows in walls that require a level of fire resistance—exterior walls near a property line or near other buildings—have glass with wire safety mesh embedded in its layers. These windows are also found in high-traffic areas of structures such as hallways and stairwells. Ventilating these windows often takes a little longer than regular windows because of the wire mesh holding the glass together.

The glass can be struck with a hand tool and broken, but it may only shatter. Firefighters may need to strike it again to break a section out or create an entry point.

Once an entry point has been made, a firefighter has a few options:

- A hand tool can be inserted into the entry point and can cut the glass and wire mesh with its sharp end. Once a substantial section has been cut into the glass, it can be pulled in one direction to ventilate the window.
- A firefighter can punch out a *U*-shape around the edge of the glass and then pull it in the direction of the bottom of the *U* to release the glass.
- In some instances it may be easier to attack the mullions around the window's frame and release the entire piece of glass to ventilate the window.
- Punching out the center and pulling the glass often takes numerous attempts to ventilate the window. It often leaves exposed wire and glass dangling from the window that could cause an injury.
- Some firefighters smash the perimeter section of glass out around three sides of the window. Next they cut the wire mesh with cutting pliers and pull the glass toward the attached side, ventilating the window.

Ventilating windows covered with plywood/OSB. Many vacant structures have their windows protected by plywood/OSB that has been nailed or screwed into their window frames to prevent intruders and protect the vacant structures from vandalism. Most of the time, these sheets of plywood/OSB are easily removed by a firefighter prying it off with a hand tool (fig. 14–34). Some firefighters prefer to start at the top and release those fasteners and then pull the entire sheet down and out of the window. Others prefer to work around the base of the window, pry outward to release it on both sides, and then lift the plywood/OSB sheet up to gain access to the window. In both situations, firefighters must use caution, because a sheet of plywood/OSB can be heavy and cause an injury when it is dropped or released.

Firefighters can also use a baseball swing technique with a Halligan tool, pick-head axe, or hook's head to pry these sheets off. Simply position the tool about 6 in. (150 mm) from the edge of the plywood/OSB and drive the point through the sheet. Next, pry back on the tool's handle and push it toward the sheet of plywood/OSB, pulling the screws or nails out of the window frame. This technique may only release one side of the plywood/OSB, and the firefighter must perform this same technique on the opposite side of the window.

Fig. 14–34. Use a prying action with the Halligan to remove plywood/OSB that has been nailed or screwed into window frames.

These types of security measures are also being modified to make it even more difficult to enter the structure. Often the plywood/OSB is now covered with a thin metal mesh and then a **skim-coat** of concrete mastic is layered over the plywood/OSB. This step reinforces the plywood/OSB and makes removal more difficult. The same method used previously can be attempted, but it may be more advantageous to take a rotary power saw with a carbide-tip blade and cut along the outside edges of the plywood/OSB sheet to release it. If the plywood/OSB sheet is recessed into the window frame, the operator may have to cut a few inches in for the saw blade's assembly to access the plywood/OSB. Firefighters *must* use caution and not plunge the saw's blade entirely into the plywood/OSB sheet. This could cause the saw to hit the glass, window sash, or another obstacle such as security bars behind the plywood/OSB, which could cause damage to the saw and injure the operator. Plunging a chainsaw through the plywood/OSB sheet can also be dangerous for the same reasons, so firefighters should avoid doing this. In addition, if another firefighter or victim comes to that window from the inside, a serious injury could occur.

Ventilating windows covered with HUD window coverings. Firefighters may encounter vacant buildings that have been covered with a HUD-style window covering, a means of sealing up vacant buildings accepted and approved by the U.S. Department of Housing and Urban Development (HUD). Although they are less prevalent than in years past, they are still present in numerous areas.

The **HUD window,** as it is called by firefighters, is made up of a sheet of plywood/OSB that is attached to a window frame by four cross members: 2 × 4–in. (50 × 100–mm) pieces of wood (fig. 14–35). Two smaller pieces of this material can be seen from outside the structure, whereas the other two longer pieces cannot. These two pieces are supported on the inside of the building, spanning the window frame on each side, adding strength and rigidity to the device. All the pieces of wood are connected by through bolts or a piece of threaded metal rod and become secure when they are tightened up. Normally, there are two bolts on each of the upper and lower cross member supports for a total of four. These devices can be installed with the windows opened and left in place, or the entire window and frames may have been removed prior to their installation.

Fig. 14–35. Notice the 2 x 4 cross members that tip off the firefighter that these are H.U.D. window security devices.

To ventilate these windows, there are a few options to consider after sizing up the material. If the wood shows signs of being compromised or rot, it may advisable to attack it. A hand tool such as an axe or Halligan tool can be used to split the outside 2 × 4–in. (50 × 100–mm) cross members, exposing the four metal through bolts. The bolts can then be pushed or struck into the structure to release the inside cross members. Firefighters must use caution because the sheet of plywood/OSB can now become loose and swing down and out from the window, and the cross members are now unsupported. Firefighters must maintain some type of grip on the plywood/OSB to prevent it from falling away from the building and causing an injury. One way to tackle this is to drive a hand tool into the plywood/OSB and maintain a hold of the tool. This controls the sheet of plywood/OSB once its cross members are released.

Another method to remove these devices if they are present on the upper levels of a building is to operate from the safety of a tower ladder basket and use a forcible-entry saw or rotary saw blade capable of cutting the metal through bolts. Once the tower ladder's bucket is placed slightly below the window, place the point of a hand tool through the plywood/OSB as a safety hold; this *must* be maintained the entire time to prevent the plywood/OSB sheet from falling from the building. Another firefighter then uses the saw to cut the upper and lower head of the bolts on one side of the window. The saw's blade cuts through the 2 × 4 at an angle and cuts the bolt head and rod at one time. Next, they cut the top and bottom bolt on the opposite side of the window. Firefighters can now either strike the bolts back into the building, releasing the device, or they can pry up the outside 2 × 4s. These actions normally release the inside cross members and the sheet of plywood/OSB. If no window is present, the sheet of plywood/OSB can be put back into the fire building rather than dropping it down onto the fire scene and possibly causing an injury. If it can't and there is room inside the tower ladder, it may be placed inside the bucket rather than dropping it to the ground.

Some firefighters prefer another method of removing these devices. They use a chainsaw and cut a small triangle in the middle of the plywood/OSB sheet. If fire conditions behind the sheet are severe, they can move onto another window. If they aren't, the chainsaw can be inserted into the hole (once it is safe) to cut the upper and lower inside cross members, releasing the tension of the device. In some instances, when performing this tactic, the plywood/OSB may have to be pried toward either side to remove the inside cross members from the window frame after the cuts are made. When performing this tactic, firefighters must remember to maintain a grip onto the plywood/OSB to prevent it from falling to the ground and causing an injury to an unsuspecting firefighter operating below.

It is also important to assert that plunging a live chainsaw through a HUD window can be a dangerous evolution. This could cause the chainsaw to hit the glass, window sash, or other obstacle such as security bars behind the plywood/OSB, which could damage the saw and injure the operator. In addition, if another firefighter or victim comes to that window from the inside, a serious injury could occur.

Metal casement windows. Many older structures have single-pane glass set in metal casement window frames (fig. 14–36). The glass panels are usually 8 × 8 in. (200 × 200 mm) or 12 × 12 in. (305 × 305 mm) and are easy to ventilate with a hand tool. The windows may or may not have a hinge section(s) that pivots and opens outward with the turn of a handle. If these windows need to be removed for the rescue of a trapped civilian or firefighter, it can be time consuming using hand tools. It often requires numerous strikes of the cross members and uprights to remove them if that's even achievable. Sometimes all they do is bend and warp from the hand tool's impact. A much faster way is to use a rotary saw equipped with a metal cutting blade to remove the uprights and cross members. This reduces the firefighter's exertion level and requires less time to make an opening in the window frame for the rescue.

Post fire control window ventilation. After the fire has been extinguished and the structure is still full of toxic smoke, gases, and heat, the building still must be ventilated. Firefighters should resist the urge to break windows after the fire has been extinguished; this looks unprofessional and increases the workload of those performing salvage operations. Firefighters should size up the windows and see if they can be opened or removed to assist in ventilation efforts.

Fig. 14–36. These casement windows on this Tudor-style multiple dwelling will be difficult to force.

Many newer types of double-hung windows have release clips mounted on the sash that allow the windows to be removed to help clean. Firefighters can release these tabs, then lift one side of the window up and the other side slid downward and the window comes out of the tracks. In some instances, the tracks are spring loaded and slight tension or compression on one side prevents the window from releasing from the track. By performing this technique, we're reducing fire damage and looking more professional (fig. 14–37).

Fig. 14–37. Taking the time to open a window instead of breaking it during post control ventilation helps to reduce fire damage and makes us look more professional.

VENT, ENTER, ISOLATE, AND SEARCH

The **vent, enter, isolate, search (VEIS)** technique of search and rescue is used by some fire departments in certain fire situations. Most commonly, it occurs when access to a trapped victim is unattainable through the main entrance of a structure. This technique allows a firefighter to access the alleged location of the victim by another avenue. In some fire departments around the country, the VEIS technique is used daily in their firefighting operations, while in others it is only used for specific rescue operations. Firefighters must understand and learn this concept prior to operating on the fireground.

Prior to discussing the VEIS technique, it is important to note that firefighters must understand and grasp the meaning of two ventilation concepts: venting for life and venting for fire. Venting for life is performed to permit firefighters access into an area where there is known or suspected life hazard exists. There is a calculated risk of eventually pulling the fire toward the area to be searched with this tactic, because a ventilation flow path opening has been created to release the by-products of combustion. However, these calculated risks are performed to reach potential victims as soon as possible, and firefighters will reduce the chances of creating the flow path by performing a tactic of closing the room's door to isolate the area.

Venting for fire is performed to assist the engine company's advance into the fire area to extinguish the fire. The venting is usually performed once the engine has water in the hoseline and prepares to move in and attack the fire. This venting assists the by-products of combustion and steam from the hoseline's water to escape from the building, lessening the burden on the firefighters operating inside. These two concepts *must* be understood, because premature ventilation can increase the fire's intensity, speed, and location as well as decrease the chances of a trapped victim's survival. Remember, all types of ventilation efforts must be coordinated between firefighters operating inside and outside the structure.

Normally, the VEIS tactic is performed by one or more firefighters going through a remote opening, such as a window to access a suspected or known life hazard area. Prior to window entry from a ladder, porch roof, fire escape, deck, or platform, the firefighter should size up the entry point (fig. 14–38). They should recognize the amount of smoke, heat level of the smoke, and its condition (puffing, rolling under pressure, or a haze) prior to opening the entry point. Once the window is opened, firefighters should position themselves away from the escaping heat, gases, and smoke in case the induction of fresh air or oxygen ignites the escaping by-products. Firefighters should then use the sweep-and-sound method of entry by first probing inside with a tool with a sweeping motion to see if an unconscious victim is on the floor and then tapping the floor forcefully to check its structural stability and if it will hold the firefighter's weight.

Fig. 14–38. A firefighter venting from an extended roof, preparing for entry

Once the floor has been checked and found to be clear, the firefighter should vent the window with a hand tool and remove any sash in between the clearing. They should also remove any screen, drapes, blinds, or curtains to promote ventilation efforts and prevent these items from entangling the firefighter entering the area. Next, the firefighter enters the structure with one leg first and then the other, rather than going in headfirst. The weight of a firefighter diving into a room can cause an impact load to the floor and a possible collapse if the floor is of questionable stability. Also, if a firefighter *dives* into the window and hits a piece of furniture, he or she could go flying across the room and lose all sense of direction and whereabouts.

Once the firefighter has entered the window, they should be aware they've created a flow path for escaping heat and gases, and closing the room's door should be the first priority. As the search begins, there are two basic concepts firefighters may use:

1. Firefighters can probe the area with a hand tool looking for a victim while keeping one hand on the wall to maintain his or her bearings while searching the room.

2. Firefighters can use a hand to search for a victim while using the tool to maintain contact with the wall to keep their bearings.

Following the wall, proceed across the room to shut the room's door. With a 6-ft (2-m) hook in one hand, you may be able to close the door to the room from your initial entry point. The hook can be run along the midsection of the walls feeling for the door, so use all your senses and listen for the sound of the door closing if it is struck with the hook. If you hear the distinct sound of glass, do not immediately break it because you feel it will vent the room more quickly. Remember where you are operating. Premature venting could allow autoexposure or pull the fire into the area by creating a flow path and draw the fire right into the search area. Remember that the sound of glass in the room could also be a mirror, television set, or even a fish tank. Your hands and senses can tell you what you're feeling in a severe smoke condition.

Closing the door to the room can help in a few ways. It decreases the chances of pulling the fire into the room entered and eliminating the flow path. Closing the door may increase visibility in the room and make finding a victim easier. Finally, it creates a safety factor with the door holding back the fire and gases while searching the room.

If a victim is found in the room prior to finding the door, you must decide if conditions allow for the door to be closed or if you'll have to immediately retrace your way back to the entry point with the victim in hand. Remember, if the victim is large and you need time to remove him or her, closing the door to the room increases the overall safety margin of the rescue operation. If the fire has been controlled below, the victim may also be removed through the interior of the structure.

If no victim has been found, searching firefighters can leave the room's door closed and exit the room to safety. Or they can size up conditions out in the hallway to see if an interior search of other rooms may be feasible; if not, a portable ladder can be used to gain entrance to a window in another section of the structure to perform the VEIS tactic. VEIS can be a dangerous tactic to perform, so firefighters must operate in pairs and in conjunction with the tasks of the interior teams to make this a safe evolution. When performing this tactic with two firefighters, it may be performed in two ways:

1. One firefighter enters the room while the other stays at the window as a safety firefighter, coaching the interior firefighter and using the thermal imaging camera to assist the search. Remember the camera cannot locate a victim under a bed or one who may have fallen off the bed and is blocked by it.
2. Another option is for both firefighters to enter the room and search in opposite directions, which works well in larger homes with large rooms. If done properly, this tactic can save numerous lives.

WINDOW GATES AND CHILD BARS

In some situations firefighters find window gates or scissor gates behind a window to protect the occupants from intrusions (fig. 14–39). These gates are normally a hinge-style gate or in a slide track and both locked from the inside. In newer installations, a release button is mounted on the wall to unlock these gates; the button can be located at the base of the wall or somewhere on the wall next to the gate. Unfortunately, these devices can delay VEIS operations and can be difficult to force open from the outside.

Fig. 14–39. Although typical in residential multiple dwellings, scissor gates can be found in commercial occupancies as well, and will delay egress through these windows.

If firefighters encounter these devices while trying to make entry, they should first attempt to see if they are unlocked and can be pushed or slid open. If they are not, they must begin forcible-entry procedures on them. For scissor window gates in a slide track, it may be easiest to work on the stationary side and insert a hand tool behind the frame to pry it out and away from the wall. These installations are normally screwed into the framing and pull out with the use of leverage. If smoke is issuing from the window, pull down the top sash first and pry the top of the gate off first. Then push the window back upward to cut down on the smoke exiting and use it for a protection measure. Next, lift up on the lower sash and attack the gate at the bottom. Once the gate is released from the stationary wall, slide it in the track toward the lock side or hinge it and push it inward. Remember, if this window is going to be used for VEIS, it should be cleared entirely out.

In heavier installations with steel bars, the same method of attacking the gate may be feasible. In some situations it may call for additional tools such as a forcible-entry saw or hydraulic forcible-entry tool to gain entry. In some installations it may also be possible to pry on a set of bars to make an opening to insert a gloved hand into the building. Next, attempt to lift the lock guard and then turn the locking mechanism to open the gate.

Some cities require that windows have child bars on them to prevent a child from falling out of a window. These bars are different from regular security bars; they are a two-piece system that includes interconnected bars and uprights that normally run horizontally across the bottom of the window (fig. 14–40). They are screwed into each side of the window frame and, if struck with a hand tool on one of the center uprights, can be pulled out from the side of the frame. The frame then can be pushed inward or struck in the other direction to remove the opposite side, permitting entry into the building.

For more information on these devices, see chapter 12, Forcible Entry.

Fig. 14–40. Child bars are installed with good intentions

ROOF VENTILATION

Tools used for roof operation

When firefighters are performing horizontal or vertical ventilation operations, the following tools or compliment of tools (some of which are shown in fig. 14–41) can be used to perform the ventilation tactics:

- Full personal protective clothing with eye protection
- Breathing apparatus
- Axe (pick-head or flat-head)
- Pike pole, rubbish hook, or other suitable tool
- Power saw (chain or rotary saw)
- Communications (portable radio)
- Halligan tool
- Utility rope
- Flashlight

Fig. 14–41. These are various tools that can be very useful for roof ventilation operations. It is important to understand the strengths and weaknesses of various ventilation techniques.

Pick-head or flat-head axe. The axe is an indispensable and dependable tool for ventilation operations. The pick-head axe is versatile, because the pick can be used for a variety of tasks:

- Driving the pick-head into the roof and using the axe as a foot brace if cutting with a saw
- Prying off scuttle covers or lifting up skylight frames
- Making quick inspection holes in the roof and tearing off roof coverings

Also, when traversing a pitched roof, firefighters should keep the pick-head axe ready. The pick can be used to pierce the roof and act as an anchor in case you slip or fall.

Most training material regarding axes suggests cutting at a 60° angle to the roofing material. Although this can provide acceptable results, it has several drawbacks:

- Not all of the energy expended for cutting is effectively used.
- It is an inefficient way of cutting plywood/OSB and difficult to cut through shingles and sheathing.
- The axe head may plunge beneath the surface being cut and become trapped.

These problems can be overcome by bringing the axe downward in a smooth arc motion so that the cutting edge is 90° to the work surface at contact. This results in more cutting power due to the velocity and natural swing of the axe requiring less effort. Remember it's very important to position your feet properly apart so that the axe doesn't accidentally strike your leg or foot. If the head of an axe becomes trapped in or below the work surface, simply use the toe end of a boot as a fulcrum at the junction of the axe head and the handle to lever the axe out again (fig. 14–42).

Fig. 14–42. An axe head that is trapped below roof decking can be levered out by placing the toe of your boot between the roof and the junction of the head and handle.

When firefighters are cutting a roof with an axe, they should try and determine the location of the rafters by sounding the roof with the head of the axe. To sound the roof, the firefighter bangs down on the roof with the top of the axe and listens to the sound it makes. A solid thump can alert the firefighter to the position of the joist. A hollow sound means an area between the joist. Cuts should be made as close to the rafter as possible, providing a firmer foundation for the cut.

Although less versatile than the pick-head axe on the roof, a flat-head axe can also be used to cut open roofs. The axe is swung and used the same way as mentioned above. In addition, if the roof sheathing is made up of plywood/OSB or pine boards covered with shingles, another method of using the axe is available. Firefighters can first sound the roof and attempt to cut alongside a joist. Next they'll use the flat end of the axe and swing downward and strike the roof. Each strike makes a cut into the roof; and when these cuts abut each other, they make a cut line. Using this method it is easy to make a louver in a roof. Some firefighters use the maul or a splitting maul to perform the same tactic of opening the roof.

Pike pole or hook. Firefighters must bring a pike pole or hook with them when reporting to the roof for roof ventilation operations. Once the roof has been cut open, the roof sheathing must be pulled up or louvered to create the roof opening. To pull up the sheathing, the hook's head is placed into the knockout and then the sheathing is pulled upward away from the roof (fig. 14–43). If a knockout was made on both sides of a cut, two firefighters can pull up the roof sheathing. If the sheathing or roof planks are difficult to remove by pulling upward, the hook can be used as a giant pry bar. Simply place the hook's head on the joist and pull it so that it bites under the roof planks or sheathing. Now simply push up and away from your body to pry the sheathing upward and off the joist.

Fig. 14–43. Pulling roof boards with a Halligan hook

Once the roof planks or sheathing is removed, the ceiling must be pushed down to ventilate the room and area below. If it is not, the cockloft or attic is the only space ventilated. Using a hook rather than a shorter hand tool allows the firefighters to position themselves away from the ventilation opening. This helps them keep out of rapidly escaping smoke and gases that could ignite without warning. In some buildings with deep cocklofts, hooks longer than 6 ft (1.8 m) are needed on the roof to reach the ceiling below. The hook can also be used to sound the roof as a firefighter proceeds across the roof to check its stability. Also, the hook can used as a safety brace when operating on a peaked roof.

Rubbish hook. The rubbish hook was originally designed as a companion to the pike pole to move debris during rubbish fires, during overhauling operations, and to strip shingles from roofs. Although the pike pole is versatile, the rubbish hook also benefits firefighters performing roof operations. The hook can be used as a sounding tool as firefighters transverse the roof. Firefighters can use the wide heel of the hook's head to strike and sound the roof as they proceed. The width of a rubbish hook's head also simplifies pulling up cut portions of a roof. Firefighters can place the head of the hook over the joist, with both hook tines on either side of the joist; this creates equal pressure on both sides of the sheathing when it is pulled open by the firefighter. The head's large size also assists firefighters in pushing down the ceilings once the roof is open. Remember, it is possible for a hook to catch onto wiring, cable, or ductwork when pushing down the ceilings.

Power saws: rotary and chainsaw. Power saws are versatile tools that can simplify ventilation operations by reducing the time it takes to open a roof with a hand tool. Using power saws also reduces the physical exertion level on a firefighter, increasing his or her effectiveness to perform other tasks once the roof is open. Although power tools are beneficial, they do have some drawbacks (bulky, cumbersome to operate at some angles) and can break at a moment's notice (the saw's starter chord can snap or tear out of the saw), requiring firefighters to use hand tools to provide ventilation.

Rotary power saws are excellent tools to use for cutting the thick or heavy roof composition usually found on older roofs. The saw's rotating blade is very good for cutting through thick tar and expelling the material through the blade guard and out of the saw. For these reasons, the rotary saw is better equipped to handle these types of roof operations on these structures.

As with a rotary saw, a chainsaw can simplify numerous ventilation operations to a significant degree. The chainsaw can be effective on roofs with asphalt shingles and wood plank decking or plywood/OSB sheathing. Depending on the type of chain used, the saw should be able to cut directly through the shingles and sheathing in one operation, enabling the roof to be opened in less time.

FIREGROUND NOTE

When faced with thick tar roofs, firefighters often attempt to pull the whole cut section of tar off the roof first. Then they attack the roof planks or sheathing to complete the roof opening.

Using a rotary saw or chainsaw to perform roof ventilation tactics presents a variety of hazards to firefighters. All safety measures must be followed and all firefighters must be aware of the dangers when a live saw is operating.

For more information on operating saws, see the saw sections of chapter 11, Tools.

Radio

Firefighters performing either vertical or horizontal ventilation efforts must carry and use a radio to communicate with interior teams on ventilation tactics. This is extremely important when performing horizontal ventilation, because it must be coordinated with the hoseline's advance on the fire. Firefighters must keep this in mind when ventilating top floor windows from the roof. It is also vital for firefighters to communicate when performing roof ventilation operations to prevent creating a flow path or pulling the fire toward firefighters operating on the top floor.

A firefighter opening up a roof's bulkhead door when a fire is in a large multiple dwelling can simply relay to their officer that the roof is going to be opened, meaning that initial vertical ventilation over the stairwell is in progress. This informs the officer that the firefighter has gotten to the assignment and the mushrooming effects of the smoke and hazardous and explosive gases are in the process of being reduced or eliminated from the interior of the building. If the officer in the fire area notices that the fire is being pulled in their direction, they can have the bulkhead door closed or the fire apartment door closed to decrease the flow path or pull of fire toward operating crews.

Another simple communication statement should be made to units operating on the fire floor once the roof has been cut open. Prior to the ceilings being pushed down, the roof firefighter should quickly announce over the radio that it is going to be performed. By stating this message, firefighters operating on the floor below can prepare for a falling ceiling to crash down. A simple safety measure is for a firefighter to be positioned in a doorway, thus protected by the door frame when a ceiling is pushed down from above.

It is of the utmost importance for all firefighters to relay any pertinent information (heavy loads on the roof such as HVAC units, etc.) or delays that may affect ventilation operations to the incident commander.

Halligan tool. Firefighters often carry a Halligan tool to the roof for roof operations. The tool has a number of uses and is an excellent tool to force open doors on bulkheads, elevators, and penthouses. It also helps firefighters open up scuttle covers and make a safety step for peaked roof operations.

For more information on the Halligan tool, see chapter 11, Basic Firefighting Tools.

Rope: utility rope and life-saving rope. FFI 4.3.20 Firefighters should carry some type of utility rope; it is another tool on which they often rely. During roof operations the rope can be used to tie a hand tool to ventilate top floor windows. It is also good for hoisting additional tools and equipment to the roof. Firefighters can also use this rope to section off a dangerous section of the roof, preventing a fireground mishap.

In some areas it is necessary to carry a lifesaving rope to the roof. This rope is not to be used for ventilation efforts; its intended use is to save any life that may be trapped by the fire and to provide an emergency escape device for firefighters operating on the roof if no other avenue of escape is accessible.

Access to the roof

SKILL DRILL

Firefighters must have a pre-formulated plan on how to reach the roof of the fire building. For many private dwellings, access to the roof is by portable, aerial, or tower ladder. In many instances, because of the roof's high pitch, a roof ladder is needed as well. For larger multiple dwellings, firefighters should communicate with the inside teams to see if there is an alternate enclosed stairwell other than the attack stair that has access to the roof. If another stairwell is not available, firefighters should use the following:

- The adjoining building. Cross over at the front of the roof if the buildings align. Remember, if they are on a slope or hill, it is always better to choose the higher building and drop down to the fire building than it is to try and climb up to the fire building from the building below.
- An aerial or tower ladder to gain roof access to the fire building. If necessary, use one of these devices to reach the exposure, and use a portable ladder for access to the fire building.
- A fire escape that provides roof access (fig. 14–44). The fire escape should be the last resort because of their age and maintenance issues. Also, they can be narrow and be difficult to maneuver on, often requiring firefighters to perform the reduce profile maneuver.

On row-frame dwellings, using an adjoining building is not always wise, because of the **common cockloft**. A firefighter could get injured climbing up an interior scuttle ladder if the fire breaks out while he or she is climbing. A safer means of travel is to use a portable, aerial, or tower ladder to gain access to these structures' roofs.

Fig. 14–44. At times, the quickest way up is via an exterior fire escape.

Encountering obstacles during roof access and operations. Access to the roof to perform ventilation tactics can often be hampered by a variety of obstacles, such as trees, overhead electrical wires, and street lights, to name a few. These obstacles can often delay placement of ladders to the roof and overall roof operations, because a firefighter may have to find an alternate way to the roof. Many times, using the adjoining building for roof access can be the quickest and simplest method available. Firefighters should notify the incident commander whenever they encounter obstacles that severely delay their work assignment.

Security fences. On many structures that have adjoining roofs, security fences have been placed on the adjoining parapet walls to prevent rooftop intrusions. These fences may be chain link, steel bars, or ornamental aluminum fences. In some instances, the top of these fences have razor wire or barbed wire strung across them for increased security. Also, many of these fences jut out over the front and rear cornice or parapet walls. This prevents access around the ends of the fence.

For access through a chain-link fence, a firefighter should locate the small wire tie bands that wrap around the piping of the fence. These ties secure the chain-link fence to the pipe structure. A firefighter should attack these ties starting at the bottom of the fence. Inserting the point of a hand tool into the chain link next to the wire band and prying in one direction normally pops the tie or pulls it off the fence. Some firefighters carry small wire cutters, which can also cut the wire ties off the fence.

As the bottom ties are removed, the firefighter cuts or releases a few vertical ties on the vertical pipes of the fence. Once a small area has been released, the firefighter can lift up the bottom of the fence and place a hook under it to keep it suspended. Now he or she can transverse to the other rooftop for roof ventilation operations.

When steel bar fences are in between rooftops, access to the adjoining roof can be more difficult. Attacking the welds or trying to bend the individual metal bars is often time consuming and still does not permit access to the adjoining roof. If these fences are visible from the street level, a firefighter should take a forcible-entry saw to the roof. Cutting the steel bars of the fence provides a quick means of egress to the adjoining roof for ventilation operations.

Cell phone sites. Many buildings are now leasing their rooftop space and **air rights** to cell phone service providers. The roofs of such buildings are now loaded with new structures that increase the weight load on their structural components and presents other problems for roof operations (fig. 14–45). In numerous installations, wire conduit and pipe chases provide power and signal service for equipment, and antennas run along the rooftop surface. These conditions can restrict the area available for cutting the roof for vertical ventilation.

Fig. 14–45. Cell site equipment introduces new hazards to roof operations.

It is also important to note that these antennas emit microwaves and are harmful to the health of human beings. If firefighters are within 10 ft of an operational cellular antenna when it is transmitting, serious or fatal damage can occur from radiofrequency (RF) radiation. When operating on the roof, firefighters must restrict the time they must operate in front of these antennas, and stay as far outside a 10 ft radius from a cell tower as possible (fig. 14–46). This reduces exposure levels to the microwaves and increases overall safety while working on the roof. In some installations, natural gas lines are run along the roof from an auxiliary generator in case of an electrical malfunction. These gas lines can become severe exposures and a hazard if the roof is engulfed or compromised due to fire.

Fig. 14–46. Cell site antennae add more risk from RF wave radiation, which can be dangerous or fatal if standing within 10 ft while transmitting. Look for cell site installations on tall buildings or buildings built on high ground in your district.

Remember, inform the incident commander whenever a **cell phone site** is found on the roof or tied into the roof's parapet walls.

Multiple roofs. Multiple roofs are common on many renovated buildings. When the building was retrofitted instead of removing the original roof, a new roof was just placed on top of it (fig. 14–47). If firefighters perform roof ventilation operations and find a second roof below the original roof, they should immediately inform the incident commander. Judging from the fire conditions below, the incident commander may have firefighting operations on the roof cease or attempt to cut the roof below the original roofline. Remember, the addition of the new roof also adds a significant weight load to the structure.

Fig. 14–47. During some building renovations, peaked "rain roofs" are sometimes added over existing flat roofs, which makes ventilation operations difficult.

Determining feasibility. **FFI 4.3.12** Beyond reading the status of the roof, personnel must determine whether or not ventilation operations on it are feasible. This entails knowing such factors as the location, size, and extension of fire and the type of roof construction present. The roof may be constructed of 2 × 12 in. (50 × 305 mm) joists/rafters and 1 × 6 in. (25 × 150 mm) sheathing, 2 × 3 in. (50 × 75 mm) wooden I-beams and ⅜-in. (10-mm) plywood or OSB, heavy wood timber beams, lightweight wood truss systems, or open-web steel bar joist construction covered by corrugated metal.

All these roofs may look similar from the top, yet each reacts differently when exposed to fire. Before committing personnel onto an unknown roof, determine whether it is structurally sound and whether it provides adequate time to perform the intended operations. The roof must also be sized up for items on the roof such as air conditioning units, HVAC units, billboard frames, cell phone sites and antennas, and any unusual conditions. These items can increase the weight load on the structural joist and supports and could cause an early collapse in adverse fire conditions. Remember to communicate the location of these things to the incident commander.

The type of construction can determine the location of a ventilation opening. For example, conventional construction obtains its strength from the size of its structural members. This often allows ventilation openings to be made over or near the fire if the roof is deemed safe. The key ingredient is the time available between impingement by flame and structural collapse. Conversely, lightweight construction is subject to fast failure rates and may not allow adequate time for these operations. Therefore, any ventilation operations on lightweight roofs should focus on uninvolved portions. While working on a roof, consider the potential of fire underneath. Start ventilation cuts in the weakest portion of the roof (near the fire) and work away toward the strongest portion. Spend as little time as possible on the weak portion. Also, work away from the fire and toward the means of egress.

To underscore an important point, consider the following about roof construction: Roofing material is usually made up of an under-laminate or sheathing, corrugated steel decking covered by composition-type materials (insulation board, polystyrene [Styrofoam]), or shingles (asphalt or cedar). It is important for firefighting personnel to be familiar with a particular building and roof and the structural members because sheathing or plywood/OSB decking can weaken or burn away without burning through the composition covering. Therefore, although a roof may look normal from the top, it may not support any weight.

Once firefighters have vented the natural roof openings and the top floor fire or attic area needs additional ventilation, the longer task of cutting the roof open begins. When the first firefighter assigned to roof ventilation accesses the roof, he or she should have a tool to sound the roof as they proceed. The firefighter is also trying to determine if the roof is strong enough to support the weight of personnel or if the decking has been burned away under the roofing material. The strength of a roof can easily be determined by sounding it with an axe, pike pole, rubbish hook, or other such tool. Always evaluate the integrity of a roof before stepping onto it (fig. 14–48).

Fig. 14–48. Always sound the roof ahead of you to evaluate its integrity.

The firefighter on the roof should also use his or her feet in conjunction with the sounding tool. Being sensitive with the feet can reveal a roof's bounce, flex, or sponginess. Sounding a roof also provides an additional benefit; firefighters can often determine the location of its supported and unsupported sections, particularly on lightweight roofs. In addition, firefighters can always cut inspection cuts into the roof to determine roof construction and roof joist layout.

When the first ventilation member sounds a roof to determine its safety, that member also determines the path of travel for other firefighters. Inexperienced members shouldn't determine the path of travel across a roof. A good rule to follow on flat roof structures is to walk along the building's perimeter to access the area of the fire. The roof beams are normally supported at the ends of the building; and in many cases if a firefighter must seek an alternate means off the roof, he or she may be able to cross over to safety on the next building. When operating on peak roofs, the firefighter should work on a portable ladder, on an aerial ladder, or from the safety of a tower ladder basket.

Before venturing off the ladder, personnel must take the appropriate time to read the roof. What is the location of the fire? Is it burning in a specific location, or is it extending from its original location? How long has the fire been burning? Is fire showing through the roof? Is a portion of the roof sagging? Does the roof have ventilators, and are they issuing smoke or fire? Consider evaluating the building as a building under demolition. What type of roof is it? Is it of conventional construction? Is the roof covered with slate or tile, making it difficult to traverse? Such materials can be deceptive, changing the way the roof feels underneath you. Do you know where your means of egress is located? Know where and how to exit the roof and whether or not you have an alternate means of escape (fig. 14–49). If roof ventilation personnel discover that skylights are available and appropriately located, they may be able to accomplish ventilation without cutting the decking.

Fig. 14–49. A good practice is to set up an additional ladder as a second means of egress from the roof in case access to the primary ladder is compromised.

Determining where to cut. Normally, firefighters are directed to cut the roof directly over the fire when performing roof operations (fig. 14–50). Sometimes it may be difficult to determine where to cut the roof. Firefighters can look for certain clues to assist them in choosing a cutting location. The following items should be surveyed for clues:

- Melting or bubbling tar
- Asphalt shingles curling upward
- Wood shingles smoldering or smoking
- A wet roof with a dry area
- A snow- or ice-covered roof with a melting area
- Location of the fire area gathered from a perimeter search
- Checking the base of soil or vent pipes for heat
- Radio communication from the interior firefighting teams

Although cutting directly over the fire is a good theory for firefighters to follow, sometimes it needs adjustment. If a roof is not structurally sound, firefighters should not position themselves over the fire to cut it; moving back to a safer location may be more beneficial. *Firefighters should never attempt to ventilate a roof of lightweight wood construction involved in a fire.* Also, if a fire is already ventilating out of the windows, firefighters should position the cut a little further back, not directly over the fire. Because the fire is already venting to the outside from this specific location, it may be advisable to cut a little further back to assist vertical ventilation. This channels and ventilates the heat, gases, and smoke upwards and also relieves the punishing conditions in the immediate area of the fire. This can assist the engine company's advance into the fire area with their hoseline.

Fig. 14–50. Use visual clues and communication to determine where to cut.

Making the cut. When possible, firefighters should plan ventilation cuts so that the wind is at their backs, with smoke and heat moving away from the cut. They should also attempt to make the long side of the ventilation cuts parallel to the roof joist. Following this method allows more joist channels to be ventilated. When cutting the roof decking open, don't make the cuts deeper than the thickness of the decking, this can cause the saw's blade to hit a joist and possibly damage a structural member.

Remember, to adequately ventilate any building, make the opening commensurate with the amount of heat and smoke inside. Observe the pressure of the venting smoke. If the contaminants are coming out lazily, the opening probably doesn't have to be enlarged. However, if the contaminants are venting under significant pressure, the opening is too small. Keep increasing its size it or create additional openings as necessary.

Openings made in roof decking involve specific considerations to make them quickly and efficiently. This doesn't imply that the openings should be small and made in haste. They should be easy to open and of a size appropriate to the needs of the incident. Therefore, when cutting and removing roof decking, don't cut a ventilation opening that cannot be easily and completely opened. A common recommendation for roof ventilation openings is 4 × 4 ft (1.2 × 1.2 m) or 3 × 6 ft (1 × 2 m) for dwellings and 8 × 8 ft (2.4 × 2.4 m) for commercial properties. Remember these are only guidelines and some fire departments have changed SOPs in roof cut's sizes because of the fire load of today's fire environment.

Often firefighters face stone or gravel placed over the roof membrane on the structure's roof. Placing a rotating saw blade into these materials can create flying projectiles and injure a firefighter. It is important for firefighters to use their boot and sweep or drag the material away from the cutting line. Another option is to use a shovel or brush broom to remove the material from the cut line.

After the cut: pushing down the ceiling. Once the cut in the roof has been made and the roof sheathing is pulled and detached from the roof joist, ventilation operations are not completed. Just cutting and pulling up the roof sheathing only affords ventilation of the cockloft or attic space. To properly ventilate the fire room or rooms below, the ceilings must be pushed down from the roof (fig. 14–51). A good firefighting tactic to perform is to verbally announce this over the radio to inform the interior firefighting crews of the impending ceiling situation.

Fig. 14–51. After cutting the roof and removing the sheathing, the task is completed when the ceiling is pushed down.

When firefighters are going to push down the ceiling from the roof, they should position their bodies to the windward side of the ventilation hole. They should be in line or parallel with one of the sides of the hole and only expose very few body parts over the hole to push the ceiling down. In this situation, if they were to slip on the roof surface or loose their balance and fall forward, they could keep from falling directly into the ventilation hole. Using this technique also keeps them out of the direction of escaping heat, smoke, and gases that could ignite once they escape to fresh air. Firefighters should not bend over and square themselves up to the hole and push downward with a hook. If they slip, they might fall directly into the ventilation opening and be injured.

If another firefighter is assisting with roof operations, he or she should hold onto the bottom of the firefighter's coat or hold onto the safety harness for support while the firefighter pushes down the ceiling. If operating on a roof ladder, the firefighter pushing down the ceiling can also hook into the roof ladder with his or her safety belt to ensure he or she stays on the roof.

When using a hook to push down the top floor ceiling, some fire departments encourage their members to push the ceiling down with the butt end of the hook's handle. By using this technique, the hook's head is less likely to hit and snag on any wire, cable, or ductwork in the attic or cockloft space as it is pulled upward. Some fire departments use larger hooks such as rubbish hooks; this increases the area of the ceiling the hook's head can penetrate. Another method to use for pushing down the ceiling is to use some of the roof planking that was pulled up off the joist. The roof planks are wider than standard roof hooks and can also push down a greater area of ceiling. In addition, if the plank slipped out of the firefighter's hand, he or she would not lose the tool.

Safety considerations for roof operations

FFI 4.3.12 In most cases, these safety considerations apply to both flat and peaked/pitched roofs, except where peaked/pitch roofs are specifically identified.

- When faced with severe smoke conditions, firefighters should crawl on the roof with a tool out in front of them, probing the area for any open shaftways or hazards. Many firefighters have accidentally walked off roofs in smoke.
- Initially, when firefighters arrive on the roof, they should size up an alternate means of escape in case of an emergency retreat. Placing at least two ladders to a roof can provide a secondary avenue of escape.

Ladders placed to each side of a building can ensure an avenue of escape to all sides.

- Cutting an inspection hole in a roof can reveal the type of construction of the roof's framing system.
- When firefighters proceed with a live rotary saw, they can push the saw across the roof like a wheelbarrow, eliminating the chances of the blade striking another firefighter.
- As firefighters proceed across a roof, they should sound the roof with a tool to ensure its stability. Skylights may have been removed and covered up with skimpy materials, which could collapse under a firefighter's weight. It is also recommended that firefighters proceeding across a roof with a large area make inspection holes/cuts if the roof is of questionable stability.
- If time and conditions permit, firefighters can cut a rubber membrane covering off a roof prior to cutting it (fig. 14–52). This inhibits the membrane from gumming up the blade and interior of the blade guard of a rotary saw, but more importantly it prevents the membrane from catching on fire. This can create a flash fire condition and fire extending across the roof's surface rapidly, making a firefighters retreat a priority. *Note:* Membranes can be easily cut without entanglement if using a chainsaw with a sharp chain run at full RPM, rather than a rotary saw.
- When a roof cut is pulled open or skylight removed, the section or skylight should be laid on the roof in the vicinity of the cut or opening. This piece of material should serve as a warning to firefighters that a hole or opening exists in the roof.
- Firefighters operating on peaked or pitched roofs should work from the safety of a roof ladder, tower ladder, or aerial ladder. Wearing an approved safety belt or harness is recommended.
- If footing on a pitched roof is questionable, the firefighter can place the point of the Halligan bar or the blade of an axe into the roof to make a safety step (fig. 14–53).
- For additional safety considerations, see the "Size-up" section in chapter 31, Advanced Fire Attack.

Fig. 14–52. A firefighter using a knife to remove roof membrane to avoid having rubberized material bind the rotary power saw

Fig. 14–53. A tool such as a pick-head axe can be driven into the roof as a safety step while working on a peaked roof.

FIREGROUND NOTE

Many of today's roofs are insulated with thick materials that also create a pitch in the roof for drainage. These thick materials can prevent the rotary saw's blade from reaching the roof sheathing in the initial cut. If an inspection cut reveals thick insulation, firefighters can use a knife to cut off the roof covering, expose it, and cut the insulation away from the sheathing. This helps overall roof operations.

- In some instances it may be necessary to make a hole in the peaked roof on the opposite side of the ridgepole for the roof ladder's hook to sit in. This ensures the roof ladder's hook bites into the roof, securing the ladder over a high ridge vent.
- Prior to transferring from a ladder to a roof ladder, a firefighter should pull down on the roof ladder to ensure it is secured to the roof.
- If when cutting a roof a white/gray powdery residue is found, firefighters should relay this information to the fireground commander and evacuate the roof. This indicates a lightweight gypsum roof, which is prone to collapse.
- When operating at attached row frames, brownstones, or multiple dwellings, firefighters should cross over at the front of the buildings. Open-ended shaftways between buildings can exist toward the rear, and a firefighter could fall.
- If climbing up onto a bulkhead to vent its skylight, always get on and off the bulkhead in the same location. One side of the bulkhead could face an open shaft between buildings.
- Prior to removing slate or tile from a roof, advise the fireground over the radio that this tactic is being deployed to reduce the risk of injury from flying debris.
- Wearing an approved safety belt or harness when operating on the roof ladder can allow the firefighter to work with both hands while properly attached to the ladder.
- Prior to getting to the roof, firefighters should have a preplan regarding what they will be performing and a back-up plan in case the first plan of action cannot be achieved or fails.
- Once on the roof, the presence of any heavy equipment such as air conditioning units, billboard frames, and cell phone sites should be relayed to the incident commander. These things can lead to an early collapse of the roof under fire conditions.
- If a fire wall is separating a common cockloft or adjoining buildings and there is fire inside the cockloft space, consider cutting an inspection hole into the building not on fire. The adjoining building's joist pockets may have deteriorated over the years and fire could spread horizontally, despite the fire wall.
- If a life-saving rope is part of a firefighter's standard equipment, it should be placed near a substantial object on the roof, where it would be tied off to perform an **emergency escape slide** or a **rescue pickup evolution**.
- Because of the complexity of roof operations and their danger levels, firefighters should always operate in teams of two. Whenever a saw is removed from the roof cut, the saw operator should engage the chain brake, if using a chain saw, or place the blade of a rotary saw onto the roof to stop it from spinning. Engaging the chain brake or stopping a blade from rotating reduces the chances of striking and injuring another firefighter.
- Whenever noise levels on the roof make it difficult to monitor radio communications, one firefighter should move to a quieter location to monitor the radio. This firefighter should be in line of sight of the cutting operations, in case they must relay information to the saw team.
- Firefighters should not use satellite dishes mounted on the roof for foot support; they may be insufficiently attached to the roof.
- If a firefighter must use a fire escape for roof access, he or she should always shake the gooseneck ladder before climbing to ensure it is attached to the building's wall and roof.

- Firefighters operating on the roof should always perform a perimeter search, looking for trapped or fallen victims who may be in shrubbery or shafts. They should also check for the fire's location, extension, and autoexposure possibilities.

Inspection holes

SKILL DRILL

When firefighters operate on roofs of burning structures, it is vital that they operate safely. An important safety tactic to follow is to cut **inspection hole(s)**. Inspection holes provide a firefighter with a lot of information pertaining to the conditions below a roof covering if the firefighter is uncertain about the extent of fire in the roof structure. These conditions may include the following:

- The nature of the smoke (i.e., color, volume, temperature, pushing under pressure)
- The volume of fire, direction of fire travel, and the fire's extension
- The type, size, and run of the structural elements supporting the roof

The three most common types of inspection holes firefighters' use on a variety of roof types are a **punched hole**, a **kerf cut**, and a **triangular cut**.

When making an inspection hole, make it small enough that a firefighter cannot accidently step in it and break an ankle when visibility is low.

Punched hole. It is normally relatively easy to punch a hole in most roofs. Firefighters can forcibly drive the following tools through the roof to create an inspection hole:

- Pick end of a pick-head axe
- Sharp point of a hook
- Pry bar end of a hook
- Pick end of the Halligan bar

Once the tool has been driven into the roof surface, the firefighter can pry the tool back and forth to expand the hole; this also helps release the tool from the roof. Once the tool is released, the firefighter can judge the smoke and fire conditions below (fig 14–54).

The punched hole inspection method has a drawback: The hole that the tool creates is not that large (fig. 14–55). This can cause a firefighter to miscalculate; the actual fire or smoke conditions under the roof covering and construction components. The punched hole inspection hole does offer the firefighter an additional safety benefit at fires. As a firefighter proceeds across a roof, he or she should sound the roof with the sharp end of a tool or its butt end. In this manner, if the roof's stability is questionable or the roof's tar is melting, the tool may perforate the roof covering, which can alert firefighters to the dangerous condition below or help them find a good area for roof cutting operations.

Fig. 14–54. A firefighter punching a hole in a roof to make a quick inspection hole using a hook.

Fig. 14–55. An inspection hole punched in an asphalt shingle roof.

Kerf cut. The kerf cut is a quick and easy inspection hole to use during roof operations; it is also commonly referred to as a plunge cut. Firefighters can use an axe, power saw, or chain saw to cut one blade's width into the roofing material. If a power saw is used to make the cut, the firefighter should bring the saw up to full rpm prior to contacting the roofing material. This will prevent the blade from binding into the roof or the saw from stalling. As the blade is pulled out of the roof, it creates an opening the width of a single blade (known as the kerf). This kerf allows the firefighter to judge the conditions below. The kerf cut is often used when a firefighter is quickly trying to determine the boundaries of the fire's extension. The downside of the kerf cut is that it is also small, and often on membrane roofs or rubberized roofs the kerf can close up if the roof is subjected to heat from the fire (fig. 14–56).

Fig. 14–56. The Kerf cut is a quick way to make an inspection hole during roof operations.

Triangular inspection hole. The **triangular inspection hole** is also commonly referred to as a tepee cut or an A cut because of its resemblance to them. The cut is far superior to the other types of inspection cuts and holes because it is larger and gives the firefighter better visibility into the attic or cockloft space. This helps firefighters quickly identify today's modern lightweight construction features. The only setback in making this cut is that it takes a little longer time to cut the three legs of the triangle.

To complete the triangular inspection hole, stand in a well-balanced position, with feet spread open and aligned. Next, flex and bend at the knees, so the saw touches the roof surface. Hold the saw out in front at all times, performing the cuts in the leg space area (fig. 14–57).

Fig. 14–57. An inspection hole can be used to indicate the type of roof, size of structural members, thickness of the roofing material, and the presence of any fire or smoke. Limit the size of the inspection hole to avoid stepping in it.

The sequence of cuts should be preplanned and become second nature to all firefighters. The first cut should be diagonal. The firefighter starts at the center, and cuts down toward the right boot. The second cut is vertical, starting at the top of the first cut and cutting down toward the center of the firefighter's stance. It is extremely important that all cut lines overlap, so the roofing material is cut all the way through. The last cut is the easiest cut to make. It is performed from left to right, overlapping both cut line and forming the bottom leg of the triangle.

Once the cuts are completed, the roofing material can be removed by striking it downward, pulling it out of the cut; or it may fall into the hole once the cut has been completed. In some instances, the roof material may hinge on a joist once the cut has been made.

Roof ventilation cuts

SKILL DRILL

Pullback method. The **pullback** method of roof ventilation uses three cuts in the roof deck for removal and is designed to work on a roof sheathed with roof boards or planking of 1 × 6, 1 × 8, or 1 × 12 in. (25 × 152, 25 × 200, or 25 × 305 mm). It cannot be used successfully on roofs with plywood or OSB decks, as it is too difficult to pull. Once the cut is made it resembles an upside down U.

To perform this cut, first make a head or top cut across three joists/rafters. If on a pitched roof, make the head cut approximately 1 ft below the ridge. Avoid cutting into the rafters by rolling the saw up and over when it makes contact with a rafter. Use chain saws in the vertical position (at a 90-degree angle) for the head cut.

- Then make two downward cuts (the sides of the U), starting at the head cut and moving toward the bottom of the roof, parallel with the rafters between the center and outside or between both outside joists/rafters. Remember to overlap the cut lines with the saw to permit the roofing to be pulled. These joists/rafters are on 24-in. (610-mm) centers.
- To make pulling up the roof boards easier, make a simple crossover or knockout cut in the top corner of the cut. This allows a tool to be inserted under the sheathing to assist in removal. Remove the top board using a hook to pull upward, which separates the board from the joist. When performing this on a roof with joist spacing on 16-in. (406-mm)

centers, two firefighters may find it easier to release the roof boards with hooks at each end.

- Continue to remove the decking material with a steady hooking motion adjacent to the center rafter. If necessary, roof covering can be scored (a light cut in the surface material only) at the bottom portion of the opening to ensure a clean and easy break.

Some benefits of using this method are as follows:

- The head cut reveals the location and spacing of the roof joist.
- A ventilation opening can be initiated near the ridge and opened toward the point of egress.
- A section of decking to be removed is only nailed to a single rafter, or possibly two.
- The decking material doesn't drop into the structure, possibly injuring operating members or trapped civilians.
- The cutting process is a quick method, reducing time for other important tasks to be accomplished.
- Firefighters can use this method with or without a roof ladder on pitched roofs, depending on the pitch.

Some fire departments have another version in their arsenal of cuts and one referred to as the *tepee cut*. Normally, this cut is used in forcible entry on roll-down gates but can be used in roof ventilation. It is much like the pullback method of roof ventilation; instead of cutting an upside down U, the firefighter cuts a tepee design with one crosscut. This cut also reduces the time it takes to cut open a planked roof and only employs three cuts to be performed. The tepee cut can be performed on plywood/OSB-sheathed roofs, but it may be difficult to remove the sheathing once the cuts are made. It is also easy to perform the teepee cut out of the tower ladder's bucket or off of an aerial ladder.

To perform this cut, make one diagonal downward cut from the ridge about 3 to 5 ft (1 to 1.5 m) long. Next put a small crosscut about 8 in. (200 mm) down from the peak; this is a knockout for tool insertion to pull the roof. Ensure that the crosscut overlaps and goes across both downward cuts. The last cut is the second diagonal downward cut. Now that the cuts are complete, a hook can be inserted into the knockout and the roof planking pulled off the roof (fig. 14–58). Remember, once the roof planks have been removed, the top floor ceiling must be pushed down to ventilate the top floor (fig. 14–59).

Fig. 14–58. Roof sheathing is removed after a teepee cut has been made.

Fig. 14–59. Knowing how to use your tools can get the job done faster and easier.

Dicing. The **dicing** method of roof ventilation consists of making multiple downward cuts between and with the direction of the roof joist. This method can be used on roof planking systems and sheathed roofs. A head cut may not be required if there is a ridge vent, but it is required when there isn't one present. Dicing the roof is performed by following these three steps:

1. Make a head cut perpendicular to and across the joist, measuring the desired length of the ventilation opening. This cut should be located near the ridge if on a pitched roof. It is also good practice to make a knockout or crossover cut to allow the roofing to be opened up.

Note: After the head or top cut is made, the firefighter can look into the space and see where the joists/rafters are physically located. Also, visible rafter tails present under the eaves of the structure can indicate where to make the next series of cuts.

2. Make two downward cuts on either side of a single rafter. The length of these cuts should be less than

the length of the tool used to remove the severed section. This ensures that personnel stand on uncut portions of the roof while they remove the decking.

3. Step 2 is continued as necessary to achieve the desired length of the ventilation opening. Firefighters must remember to overlap all cut lines to permit the roofing to be pulled.

The cut sections of decking are then removed by beginning near the start of the head cut. Loosen the first section of sheathing to be removed by hitting the end of the first-cut board with an appropriate tool if no knockout cut was made. Then with the same tool, use an upward pulling motion to remove the decking. Firefighters should work down and across the slope of the roof until the entire cut portions of the roof have been removed.

Some benefits to using this method are as follows:

- The head cut can reveal the location of the joist and their spacing.
- Firefighters can remove the decking away from the fire and toward their means of egress.
- Removing decking that is nailed only to a single rafter can be quick and easy.
- A large section of roof is opened with minimal cutting operations.
- The decking material doesn't drop into the attic or into the structure.

If the need to increase this opening is necessary, firefighters should have the foresight to remove the pulled up roof decking a sufficient distance away from the bottom legs of the cut. Doing so allows the cut to be quickly expanded downward.

Louvering. The **louvering** method (also known as the butterfly or hinging method) can be effectively performed on most types of roofs and roof decks, often more quickly than the coffin cut described below. A power saw and the principle of leverage do most of the work. The following five steps are followed to louver a roof:

1. Make a head cut perpendicular to and across the joist, the desired length of the ventilation opening. This cut should be made near the ridge if performing this tactic on a pitched roof. Remember, when making contact with a joist, roll the saw up and over it to prevent doing structural damage. Chain saws must be used in the vertical position (at a 90-degree angle, perpendicular to the roof). Typically, chain saws are more effective than rotary saws for this operation.
2. Next look into the kerf of the head cut and spot the joist's location. Then make downward cuts between the centers of the joists/rafters. The length of these cuts is determined by the size of the ventilation opening; however, 4 to 6 ft is an approximate rule of thumb.
3. The last cut is a parallel cut made along the entire bottom of the cut, similar to the cut in Step 1.
4. Remember where all the cuts intersect; they must continue to cut several inches past the intersections to ensure that the decking has been completely severed. Remember to avoid compromising structural members.
5. The cut section of decking can now be louvered on the rafter to complete the opening. Use an appropriate tool to push down or pull up on one side of the panel to open it. It may be possible to use one tool to push down and another tool to pull up on the opposite side. The object is to open the roof and not push the material into the hole. The material remains attached to the center of the joist and will be in a *hinged* position.

The benefits of using this method are as follows:

- The head cut can reveal the location and spacing of the rafters.
- Louvered decking that is nailed only to a single rafter can be opened easily, because personnel can use the weight and leverage of the panel against itself.
- Firefighters can louver the decking away from the fire and toward their point of egress.
- The decking material isn't usually dropped into the structure. This point depends on how the covering is secured to the base material. Composition material shouldn't slide into an attic. It must be removed from the decking and placed on the roof.

Note: A cut section of decking must be removed if it cannot be louvered because of contact with sprinkler pipes, bracing, conduit, or other obstacles. Remember, this method is faster and requires less effort than the pullback or dicing method of roof ventilation.

FIREGROUND NOTE

If it is difficult to remove or pull up on roof boards or sheathing, remove them using a hook. Place the head of the rubbish hook/hook onto the roof joist and slide it under the roof decking. Next pry the hook's handle upward and away from the body so that the hook acts as a giant lever to remove the roof decking material (fig. 14–58).

Coffin cut or expandable cut

SKILL DRILL

When a serious fire occurs on the top floor of a structure or in the attic or cockloft space, it is important to provide ventilation. Initially, windows, scuttle covers, bulkhead doors, and skylights should be ventilated first, prior to the slower task of opening up a roof with a saw. When the roof is going to be **opened up**, firefighters should have a preformulated plan or sequence of cuts to follow. A good plan of action to follow is to make an expandable opening or coffin cut in the roof.

The **coffin cut** is a relatively easy cut to make and for firefighters to remember the formula: the 7, 9, 8 cut sequence. Normally, the hole's size runs somewhere between 3 × 6 ft (1 × 2 m) or 4 × 8 ft (1.2 × 2.4 m). Prior to cutting, a firefighter should attempt to determine the run of the joists/rafters. Sometimes this requires making an inspection cut, whereas at other times it may be figured out by looking at the size of the building. For example, a row frame dwelling or narrow building may have the joist run side to side; the distance is short and the joist will rest on the outside bearing walls.

Once the run of the joists/rafters has been determined, the long leg of the coffin cut runs perpendicular to the joist; this enables ventilation of more area of the attic or cockloft space. As with all cutting sequences, firefighters try to have the wind blowing at their backs. In this manner, once the cut is made, the wind drives the smoke and gases away from the firefighters. This increases visibility and also puts the firefighters in a clean atmosphere. Unfortunately, many times this may not feasible because of a blowing or swirling wind condition, so firefighters must use the best approach they can.

Firefighters should perform the following sequence of cuts when cutting the coffin cut (fig. 14–60):

1. Cut line #1 is the top of the seven, approximately 3 to 4 ft long.
2. Cut line #2 is a knockout cut, which allows for tool insertion to pull the roof boards.
3. Cut line #3 is approximately 6 to 8 ft.
4. Cut line #4 and #5 is approximately 3 to 4 ft, making the number 9.
5. Cut line #6 and #7 is approximately 3 to 4 ft, making the number 8.

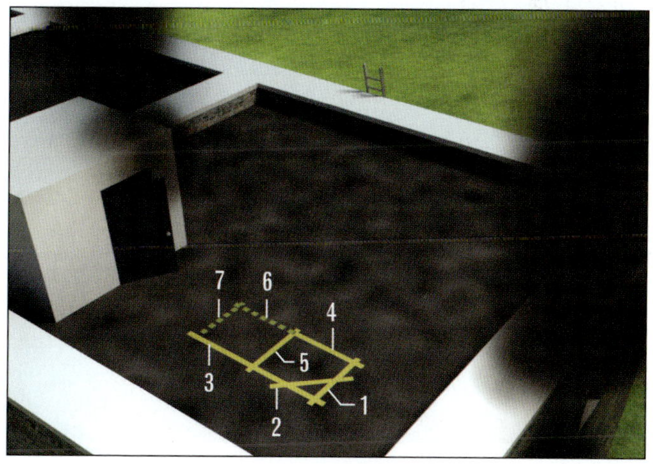

Fig. 14–60. The sequence of cuts needed to make an expandable coffin cut.

If there is time, a second knockout cut can be made just before making cut #4. This enables both sides of the cut to be pulled open by two different firefighters using hand tools.

It is extremely important that the roof cut is not pulled open until all cuts are complete. If conditions on the roof begin to deteriorate before the whole cut is complete, the first box section of the coffin can be pulled open for ventilation efforts prior to firefighters retreating from the roof. If the opening needs to be enlarged, the coffin can be continued in either direction of the rectangle. In addition, it is extremely important that, once the roof has been cut and pulled, the top floor ceilings are pushed down to permit ventilation of the top floor. Also, the top floor windows must be ventilated for a draft of air to flow into the floor and out through the roof opening. By following this procedure, firefighters decrease the chances of horizontal fire spread in the attic or cockloft space.

Trench cut/strip cut. Often in the fire service, firefighters face a large fire in control of an attic or cockloft space. Many fires occur in large buildings that may be shaped in a wing-type design, such as *H*, *U*, *E*, *L*, or even a double *H* type. It is common in such structures to lose or give up a wing to protect the rest of the building. A **trench cut** (also known as a **strip cut**) is a defensive operation that helps firefighters prevent the extension of fire in the attic or cockloft space and reduces the chances of loosing a large portion of the building to fire (fig. 14–61).

Fig. 14–61. A trench cut must extend from one side of the building to the other side and be wide enough to stop the spread of fire.

Firefighters must remember that before any trench cuts are made or opened, a sufficient ventilation hole should be made over the fire. A trench can be cut while a vent hole is being cut, but it should not be opened (if possible) until there is an adequate opening over the fire. One of the main ideas of the trench is that it introduces air into the cockloft space from this remote opening and allows the initial vent hole to pull a draft toward its opening. This reduces the chances of horizontal fire spread in the cockloft and ventilates the area of fire, gases, heat, and smoke.

The trench cut should be made about 20 to 25 ft (6 to 7.6 m) from the initial vent hole or in an advantageous position on the roof. This may be achieved by using the narrowest available roof section or taking advantage of other building characteristics such as the throat section between wings, bulkhead, elevator bulkhead, skylights, or outside walls of the building. The cut must run from wall to wall or another suitable fire stop and be approximately 3 ft (1 m) wide or wide enough for a firefighter to safely transverse it (fig. 14–61). While cutting across the building, a firefighter can make crosscuts and knockout cuts, similar to those used when making a coffin cut, to assist in opening up the roof. The initial position of the trench should also be chosen in an area that isolates the fire to a certain section of the building and allows the firefighter sufficient time to make the cut, before the fire overruns his or her position.

Inspection cuts can be made on the fire side of the trench, and these holes should be monitored for smoke and fire conditions below. Normally, these holes are located about 5 to 8 ft (1.5 to 2.4 m) in front of the trench. Once the fire has reached these holes, pull open the precut trench. In some cases, inspection holes can be made on both sides of the trench. These additional holes inform the firefighter if the fire has overrun the trench while pulling it open or if there is other fire extension.

If firefighters are overrun by fire while pulling open the trench, they must retreat and possibly begin a second trench in a safe area, if conditions warrant. A charged hoseline positioned on the roof allows firefighters to protect themselves and the trench opening. An additional hoseline positioned at the top floor and the ceiling, removed where the trench is located, can also help prevent the fire from extending across the trench.

Performing a trench cut is a time-consuming tactic, so the need for additional saws and personnel on the roof is of paramount importance. If at any time during the cutting operations firefighters feel that this cut may not stop fire extension, they can pull back to a safe position and begin a second trench.

The ridge vent: ridging the roof. On many new homes, condos, and townhouses, the construction can and normally has a ridge vent incorporated into the peaked or pitched roof. The ridge vent is an area at the peak of the roof, where a space is made between the sheathing and covered with a ridge vent (decorative aluminum vent) or roofing material that allows the roof to vent the hot air from the attic space. Many times the ridge vent is just a mesh fiber covered with shingles; firefighters can easily pull this up with hooks to ventilate the attic space. Unfortunately, the ridge vent space is narrow and permits inadequate ventilation for fire conditions.

However, buildings with ridge vents can aid firefighters in roof ventilation. A common technique used by firefighters is known as **ridging the roof**. Prior to describing the technique, it is important for firefighters to understand a common construction practice. When a building is constructed and the roof is being sheathed with 4 × 8 ft (1.2 × 2.4 m) material, roofers start at the bottom of the roof and work upward, with each course of sheathing staggered. When the last course of sheathing is laid in place, it is normally a smaller section of sheathing and can measure anywhere from 8 to 24 in. × 8 ft (203 to 610 mm × 2.4 m) (in some instances it may be wider).

When firefighters reach the roof, they can pull off the ridge vent and attempt to pry up the last sheet of sheathing from the roof. In many instances it can be pried up and allow a larger ventilation hole. In some instances it may be difficult for it to be pried up in 8-ft (2.4-m) lengths. In these cases a saw or axe can be used to cut the sheet in half and then pry it upward to vent the roof. Firefighters can work right along the peak of the roof and continue opening it up in this manner to allow vertical ventilation.

Another method of ridging the roof is to use a saw and cut downward on both sides of the ridge vent. The length only needs to be as much as the last piece of sheathing applied to the roof. This distance can easily be figured out by a firefighter removing the shingles from the roof in a small area to size up the distance needed to cut. Then relief cuts are made about every 4 ft (1.2 m) to make it easier to remove the sheathing on both sides of the roof. Now a sufficient roof ventilation hole has been made for a fire that involves the attic space or cockloft. Note that if a top floor must be ventilated in this fashion, it may take a long hook to push down the top floor ceilings from this higher position on the roof. Also, when performing this tactic, it is very wise to have a ladder or means of egress off the roof on both sides of the building. A firefighter doesn't want to be cut off from his or her access if the fire lights up out of the ridge vent.

Ridge venting is a newer technique used in firefighting and also performs well when working out of a tower ladder or off an aerial ladder when working on buildings with lightweight truss construction.

Cutting peaked or pitched roofs. When firefighters conduct ventilation operations on roofs that are slightly sloped, walking, operating, and maintaining a solid footing on them are normally minor concerns. However, as the pitch of a roof increases, the firefighter's balance, footing, and stability decreases. The importance placed on maintaining stability during ventilation can't be stressed enough. When faced with a high pitched roof, firefighters must properly use a roof ladder or cut from the safety of a tower ladder or aerial ladder to conduct a safe and efficient roof ventilation operation.

Unfortunately, because of overhead obstructions such as wires and trees, aerial or tower ladders may be unable to reach the roof on some structures. If this is the case, use an extension ladder and a roof ladder. Place the roof ladder next to the section of roof to be ventilated on the windward side, allowing the escaping smoke to blow away from the operator. The roof ladder's base should be near the top of the extension ladder in case the firefighter must retreat from the roof. When the roof ladder is attached to the ridge, the firefighter must ensure that it *bites* into the ridge of the roof. Normally, this is done by the firefighter pulling back down on the roof ladder to ensure it is attached.

Once the roof ladder is secured into the roof, a firefighter can proceed up the ladder with a power saw or hand tools. Most firefighters will bring either an axe or Halligan tool up the ladder with them; the tool will be used to make a safety brace into the roof so the firefighter can place one foot on it for balance while making the cut. Remember, a safety brace can assist in maintaining a safer work environment but if the attic is well involved with fire, it can release smoke and pressurized gases around the tool.

If a second firefighter is available, he or she should proceed up the ladder behind the first firefighter, be equipped with a hook to help remove the roofing material, and push down the interior ceiling once the cut has been completed. While the first firefighter cuts, place the hook on a ladder rung, keeping it out of the way, or its head can be driven into the roof across from the work location. The hook can act like a safety brace in case the upper firefighter slips and loses balance.

The following method is presented as a simple and effective method of working off a roof ladder with two firefighters when cutting on a pitched roof:

- To facilitate working away from the roof ladder, insert the pick end of the axe or the point of the Halligan tool into the roof about 2 ft (0.6 m) from the ladder and below the work area.
- The firefighter operating the saw can now use the embedded tool as a foot brace.
- The second firefighter can now act as a safety person, enabling the saw operator to cut away from the ladder.
- The saw operator can now cut either one or two louver panels, depending on his or her reach and

position from the ladder. The firefighter may also cut by using the pullback method of roof ventilation if faced with roof planking.

- Once the cuts have been made, the saw operator should shut down the saw or engage the safety brake and then reposition onto the roof ladder.
- Once on the roof ladder, the firefighter can open up the roof section by louvering or pulling it. Remember, the top floor ceilings should be pushed down once the cut is made.

Using this method results in the following:

- The firefighters maintain stability and efficiency while operating on the roof ladder.
- Decking that is nailed only to a single rafter is easily louvered.
- Decking material should not drop into the building.
- A ventilation opening is easily enlarged as necessary.

Another technique to use for a foot brace and for the saw operator to maintain balance is for the backup firefighter to position the head of a rubbish hook into the roof as a foot brace. The head is placed into the roof out from one side of the ladder, and the butt end of the hook is supported on the roof ladder and braced by the backup firefighter who stands on it. The handle of the hook is now a few inches above the roof and can be used as a support for the saw operator's feet as he or she cuts.

Personnel can also use two roof ladders on either side of an area of roof to be ventilated. Although this is a safe and proven method, its effectiveness is minimized by the number of personnel required and the time necessary to raise two roof ladders, which may be long enough to negate the effects of timely ventilation. Additionally, one of the roof ladders may be in the the path of escaping heat and smoke, depending on the direction of the wind.

When using this method on tile or slate roofs, remember that you may have to remove ridge tile to increase the stability of the roof ladder and ensure positive engagement of ladder hooks into the ridge. You must remove tile to embed an axe and before using power saws or axes to cut the decking. Tile and similar materials can be heavy and may reduce the structural integrity of a roof under fire conditions. In addition, they can cause injuries to members below if they are being removed and slide off the roof.

Operating on a severely pitched roof is a dangerous activity; add some rain, sleet, or snow and it becomes even more dangerous. Firefighters should attempt to work on these roofs from the safety of a tower ladder's bucket or from an aerial ladder. If using the tower ladder, place the bucket below the intended cutting area at roof level. The saw operator can now open one door of the bucket, place one foot on the roof, and keep one in the bucket. Remember, whenever operating in the tower ladder bucket, wear the installed safety belts or an approved belt tied into the bucket. The saw operator can perform either the louver or pullback cut from the safety of the bucket.

The aerial ladder can be used in place of the roof ladder if it can be placed onto the roof for ventilation operations. These same procedures mentioned earlier can be used on the aerial ladder. In some instances, the aerial ladder accesses the roof, and the roof ladder must be used off the aerial ladder.

FIREGROUND NOTE

In some communities, homes with high peaks are spaced closely together. Firefighters may be able to place a 35-ft extension ladder in line with the pitch of the roof and footed by the adjoining home. Performing this task reduces the time it may take to ventilate the roof, because the roof ladder evolution did not need to be performed. The extension ladder acts as the roof ladder in this scenario.

Another tactic to use is to place the tip of the aerial ladder about 4 ft (1.2 m) back from the intended site to be cut open. The saw operator can kneel on the rungs and lean forward and cut the roof with the saw. The firefighter should be wearing a safety belt and can hook onto the rail behind him or her for added support while cutting. The Charlotte Fire Department developed a method of sending a second firefighter up behind the saw operator. This firefighter ties the safety belt into the belt of the saw operator. The saw operator's belt should be placed with the safety ring on the back side. This way, while they are tied together, the saw operator has more reach with the saw. This method of roof cutting works well with practice and is achievable when cutting open truss roof construction.

Gable end venting. On many homes with attic spaces, **gable end vents** are placed into the exterior walls to permit air flow. Usually, these vents are placed in between the wall studs and near the top peak of the side of the home. The vent is usually either an aluminum louver vent or plastic decorative louver vent and can be easily removed by a firefighter with hand tools for ventilation efforts.

When firefighters are faced with severe weather conditions such as ice, snow, or rain and it may be difficult to perform roof operations for an attic fire, they can attack the gable vent (fig. 14–62). Working from the safety of a tower ladder's bucket or off an aerial ladder is the preferred method of performing this tactic. Once the device is placed to the side of the home, just below the vent a firefighter can quickly remove the louver with a hand tool. The firefighter performing the operation must position himself to the side of the louver so as not to be directly exposed to the venting smoke and fire. Once the louver is removed, the firefighter can use a power saw and cut downward along both sides of the vent, removing the exterior sheathing of the home. If the interior wall is finished, the hand tool can be used to knock in the wall and complete the ventilation tactic.

Fig. 14–62. A gable vent being opened for a concealed space helps the ventilation process.

Although most walls in homes are constructed with the wall studs being on 16- or 24-in. (406- or 610-mm) centers, this type of ventilation doesn't provide an excellent means of smoke and gas removal; but it works well and doesn't chance a firefighter from falling off the roof. In some instances, once firefighters have opened up one bay using this technique, they'll open up the adjoining bays to provide more ventilation to the attic space.

Miscellaneous types of roofs

Open-web steel bar joist. Open-web steel bar joist construction usually supports a metal deck roof or a metal built-up roof that may be covered with multiple layers of insulation material, tar, and composition (fig. 14–63). The building industry considers the metal built-up roof to be an efficient form of construction because the materials are widely available, the cost is less than that of a comparable wood roof, and it is an easy roof to install. Unfortunately, the steel bar joist and metal deck roof can be dangerous when exposed to fire. Steel loses a large portion of its strength at 1,000°F (538°C), and such roofs have a quick failure rate with minimal warning, resulting in an unannounced collapse. In addition, the joists can be spaced from 2 to 8 ft (0.6 to 2.4 m) apart, leaving a large section of roof unsupported between the joists. These roofs demand specific ventilation techniques after a comprehensive size-up of the fire conditions below. It is imperative that safety be the number one priority when deciding to vent a metal deck roof. Moderate or heavy fire conditions in the building below will dictate that firefighters will not be positioned on the roof because of the substantial collapse potential.

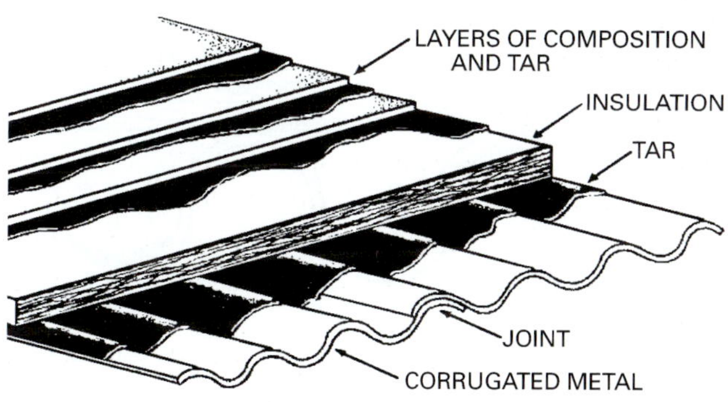

Fig. 14–63. The decking of an open-web bar joist roof normally consists of corrugated metal covered with multiple layers of insulation, tar, and composition.

In many instances these roofs are equipped with glass or plastic skylights, permitting easy initial ventilation tactics to be performed after a detailed size-up. When faced with severe fire conditions in this type of operation, roof operations may be considered too dangerous to perform. In many instances firefighters must rely on horizontal ventilation operations: opening up windows and garage doors when faced with steel bar joist construction.

When dealing with steel bar joist construction, firefighters may face a specific type of fire that does not involve the interior of the building. Many times, vents extending through the roof from grease duct equipment can ignite a fire in the roofing materials. These fires can often spread between the metal deck and the composition covering. Cutting these roofs with a roof saw can be achieved to limit the extension of fire and extinguish the fire. When faced with a fire that only involves the roofing material, firefighters can cut the roof tar, membrane roof, and composite coverings while avoiding the metal decking. Once the cuts have been made, the material can be pulled away from the steel decking.

Remember, when corrugated metal is heated, it expands and the tar liquefies and then gasifies. The gases cannot escape upward, so they travel between the corrugations in the steel decking and burn (fig. 14–64). As the gases burn, they generate more fuel. Thus, the fire can be self-sustaining and independent of the original fire in the room below. This type of fire can exhibit several hazards. For one, the insulation becomes a source of fuel. The fire is difficult to access, and it can cause or contribute to roof collapse if the fire should extend rapidly. Such a fire also creates large volumes of dense smoke, which may hamper suppression efforts.

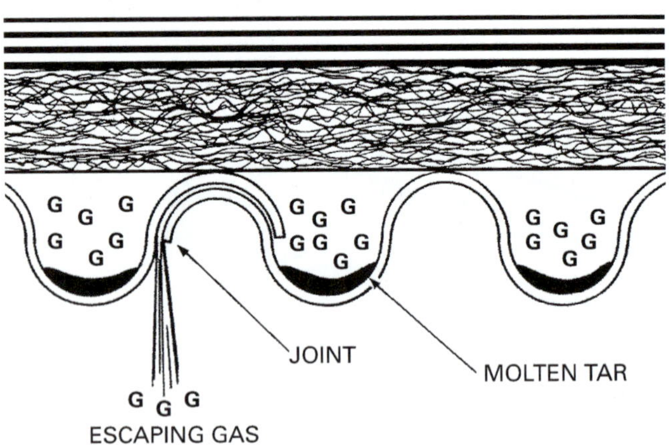

Fig. 14–64. When corrugated metal is heated, the tar liquefies and then gasifies. These gases can easily travel between the corrugations in the steel decking and burn, creating a self-sustaining fire that is difficult to reach.

Lightweight wood truss roofs. Unfortunately, because of costs, many of today's homes are constructed with **lightweight wood truss** and even lightweight wooden I-beams (fig. 14–65). The truss components are normally connected by thin galvanized metal gusset plates that only penetrate into the wood about ¼ to ⅜ in. (6 to 10 mm). Under moderate to severe fire conditions, these metal plates can start to expand and pull out of the wooden truss, which could cause a collapse of the roof. Firefighters must make an initial inspection hole into a roof to size up the roof's construction features. In truss construction, there is no **ridgepole** and normally a ridge vent exists. Firefighters can simply remove the ridge vent to size up the type of roof construction.

Fig. 14–65. Lightweight wood truss roofs make up most of the new construction today. (Courtesy of Emergency Training Solutions)

Working on or under roofs with a severe fire can be a dangerous and deadly practice (fig. 14–66). Firefighters should strive to make a ventilation hole in these roofs by operating out of a tower ladder's bucket or off of an aerial ladder. Working off of a roof ladder should not be an option because there is no ridgepole to support the roof ladder, and the ladder cannot bite into the roof.

Fig. 14–66. There is considerable danger involved in working on or under roofs with severe fire.

In many cases it is preferred to move to a safer location—surrounding areas not involved in fire—and make a trench cut to stop the horizontal spread of the fire. Draftstopping (vertical non-fire-rated barriers such as gypsum board or plywood/OSB installed to compartmentalize an attic space; draftstopping is often compromised when holes are made in them for cable TV, electrical wires, etc.), if properly installed, will help slow the spread of fire in the attic. Remember, firefighters must be deployed to the interior of the structure below the trench to open the ceiling with the protection of a charged hoseline. After notifying the incident commander, the hoseline may be used into the attic or cockloft space to stop the horizontal spread of the fire. Safety is again the number-one priority—if the fire is quickly "running" the attic and is spreading faster than firefighting forces can be deployed, firefighters should not be committed to the building.

All operations in lightweight wood truss roofs must be evaluated prior to employing interior operations. *Never trust a truss.*

Bowstring truss roofs. Perhaps one of the most dangerous roofs for firefighters to face is the **bowstring truss roof** (fig. 14–67). This roof is commonly used on large structures where a wide open floor space is required. The most telltale sign of spotting a truss roof is its two large humplike ends. A comprehensive size-up must be performed before committing any firefighters to roof operations. These trusses have failed without warning and killed numerous firefighters over the years.

Equally dangerous are other truss roof types such as gable truss, bridge truss, or the tied truss (which is the most dangerous of all). The differences are in the roof shapes, but the construction is the same.

It is recommended that if any fire involves the truss area or fire is impinging on the truss, no firefighters should be committed to roof operations. The failure of one truss can result in a very large collapse area since they are spaced far apart.

Fig. 14–67. Bowstring truss roofs can be dangerous for firefighting operations. (Courtesy of ETS)

If ventilation must be performed, a tower ladder can be used to access the humplike ends of the roof. Here, firefighters can cut a ventilation hole into the roof from the safety of the tower ladder basket. Some of these roofs also have skylights in them. Again, after firefighters perform a comprehensive size-up, the decision must be made on whether or not it is safe to let a firefighter operate on the roof to vent these skylights. Many firefighters have been killed in buildings in which bowstring roofs have collapsed. Six firefighters were killed in a Brooklyn, N.Y., supermarket in 1978, and five were killed in a Hackensack, N.J., auto dealership in 1988.

Gypsum roof decking. Gypsum concrete is a prefabricated material made up of calcined concrete and wood chips or shavings milled together. The material is then molded and laminated into planks, which are normally 2 in. (52 mm) thick, 2 ft (0.6 m) wide, and 8 ft (2.4 m) long. These planks are supported by bar joists and subpurlins and then grouted together at all of their ends. Unfortunately, the ends of these planks are often unsupported and end in an area between joists. Once the grout material dries, a roofing material such as a rubber membrane covers the planks. **Gypsum roofs** are vulnerable to moisture and conducive to an early collapse under fire conditions.

As with any type of roof installed over steel metal bar joists, firefighters must use caution and perform a comprehensive size-up before attempting roof operations. When a roof is cut with a power saw and a white or gray powdery residue emits from the saw, a gypsum roof is present. Many fire departments inform their members that if a gypsum roof is found while ventilating the roof with the saw, they should notify the incident commander and exit the roof.

Slate roofs and terra-cotta slate roofs. Many dwellings have roofs covered with slate. The slate is nailed to the roof and laid in rows over each other. The slate is slippery and dangerous to work on without the use of a roof ladder, aerial ladder, or tower ladder. Firefighters operating around the outside of the fire building must be aware of the **slate roof**. These slates can detach easily from each other, slide down the roof, and severely injure or cut a firefighter operating below.

One of the easier tactics to follow for venting these roofs is to use a maul or the pick of a pick head axe to strike the slates. This causes the slates to crack, break, and slide down the roof, exposing the roof decking. Normally, this decking is either plywood/OSB sheathing or wood planking and can be cut with a roof saw with one of the tactics mentioned above.

New materials are always created, and it is common to find slate or terra-cotta made of a plastic material and attached to the roof. From the street level it resembles slate, but closer inspection reveals that it is a type of man-made plastic. The same procedure for venting these fake slate roofs can be performed. Firefighters may also cut right through the plastic slates with a roof saw to perform roof ventilation.

C-joist roof construction. Pre-engineered metal **C-joists** are another product used to support roofs. Like steel bar joists, they are directly affected by fire and heat and prone to a quick collapse under severe fire conditions (fig. 14–68). After an inspection hole is made in the roof and C-joist are determined to be supporting the roof, the incident commander must be notified. A comprehensive size-up determines if roof operations should be performed or firefighting forces pulled off the roof. If the roof needs ventilation, performing this tactic from the safety of a tower ladder's bucket or aerial ladder is recommended.

Fig. 14–68. A "C" joist

Lightweight steel-tile roofs or tin roofs. **Lightweight steel tile** or **sheet tin roofs** have become popular, because they can be installed over tar and gravel, wood shakes, shingles or composition, fiberglass shingles, plywood/OSB- type materials, or open rafters. Installation often involves placing the panels on wood battens that have been nailed to an existing or new roof (fig. 14–69). Although these panels result in an attractive roof, they can be difficult to remove by hand and cause severe cuts to the hands, even through gloves. In addition, when pulling on the light-gauge steel with firefighting hand tools, it often tears and makes ventilation efforts more difficult.

When you encounter this type of roof and ventilation operations are necessary, an opening is best created by using a carbide-tipped power saw. Simply simultaneously cut the steel panels, the batten substructure, and the original roof decking. One roof cutting technique mentioned earlier can be used. Remember, it is important to cut deep enough to sever the various layers of materials to permit roof ventilation.

Fig. 14–69. Lightweight steel-tile roofs are often comprised of stone coated interlocking panels of 26-gauge steel that are nailed to 1 x 4 in. (24 x 102 mm) wood battens nailed to an existing or new roof. (Courtesy of Jim Plaster)

VENTILATION OPERATIONS FOR VARIOUS TYPES OF STRUCTURES AND FIRES

The preceding material presented in this chapter is the foundation of ventilation theory, tactics, and techniques. The following section of this chapter reviews the ventilation concepts and principles in relationship to structure fires as they pertain to certain types of fires in residential, commercial, and high-rise type occupancies.

Basement fires

Because of their subterranean location and lack of openings such as windows and doors, providing either vertical or horizontal ventilation to a basement can be a difficult process (fig. 14–70). In addition, because of the pressure of the smoke and gases created by the fire, they travel into as many cracks, crevices, and voids as they can. Fire, smoke, and gases can fill up an entire building and expose its occupants to these dangers.

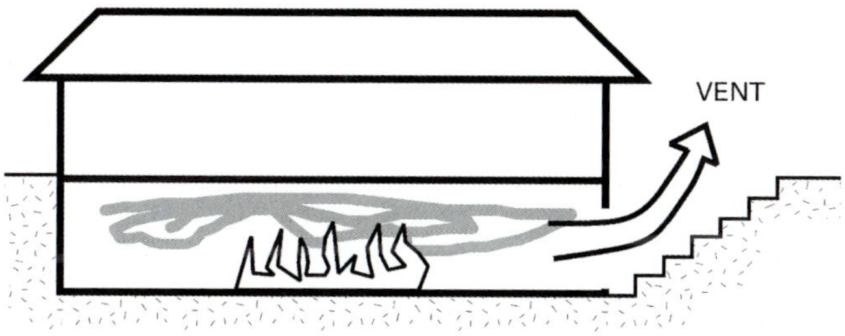

Fig. 14–70. In a basement fire, often the only available ventilation opening is the route that will be used by personnel.

Also, many basements may have limited access and are partitioned off into many rooms with a high fire load because they are used for storage. Partitioned rooms present a problem for searching for the seat of the fire and can make ventilation operations difficult. Another hazardous condition inherent during basement fires is exposed beams, which can lead to a collapse situation when exposed to fire.

Tactics

Timing. Initiate timely and coordinated ventilation of a basement fire to permit the operating and flowing of the attack lines, to reduce vertical extension to other portions of the building, and to minimize rapid development of the fire by giving the fire more oxygen. Normally, venting operations should be performed in conjunction with the attack lines' initial advance and progress whenever possible.

Openings. **FFI 4.3.12** To ventilate the basement, use any of the available openings to the area, but keep in mind that we are trying to limit the induction of fresh air or create a flow path prior to the hoseline operating on the fire: doors (outside entrance doors to the basement, garage doors, **Bilco-type fold-up doors**, old coal chute hatches or doors), windows, and dead lights. In some instances in the larger structures, it may be important to open the elevator bulkhead or dumbwaiter shafts at the roof level that begin at the basement level to provide additional ventilation.

If the initial attack line is entering the outside entrance of the basement to attack the fire, opening and controlling the interior basement door at the top of the stairs for additional ventilation is necessary. Remember, this door should only be opened if it does not expose firefighters to punishing conditions, and first-floor windows must be ventilated for the smoke and gases to escape. A hoseline positioned at the interior door can protect the first floor and the firefighters operating there.

If there is only one route into the basement and no openings to ventilate, the hoseline uses this avenue for fire attack, and ventilation is still necessary, consider opening the floor over the fire, adjacent to exterior windows or doors. (For more information on basement fires, see chapter 20, Basic Fire Attack.) This may be a time-consuming operation, but relief of the heat and gases in the basement is needed to assist the firefighters operating in there. Always ensure that the first floor is well ventilated, vertical passageways are controlled, and a charged protective hoseline is in place. Remember, vertical passageways begin in the basement and can travel to the top floor of a multistory building, requiring search and ventilation of the upper floors and attic area. These conditions appear in balloon-framing structures and many larger multiple-dwelling buildings.

Always keep in mind that, from a simplistic viewpoint, the floor over a basement (normally the grade floor) is also the roof of the fire. So, until the fire in a basement is extinguished and/or the integrity of the floor is verified, remember where you are standing.

Pressurized ventilation. Pressurized ventilation can be used effectively for basement fires if there are openings in the basement and/or first floor to channel the heat, gases, and smoke out of the building. Remember, unlike platform construction, balloon construction can allow fire to travel unrestricted from the basement into the attic. PPV can rapidly enhance the fire if an exhaust isn't opened prior to the pressurization. If an exhaust opening is created prior to PPV, most of the pressurized air flows toward the opening and has little affect on the spread of fire in open walls. Therefore, when you encounter fire extension in balloon construction, the key is to create an exhaust opening before initiating positive pressure. For more information of basement fires, see chapter 20: Basic Fire Attack.

Single-family dwellings (single story)

Single-family dwellings (sometimes referred to as ranch or shotgun homes, bungalows, or single-frame dwellings) may be ventilated by using either vertical or horizontal ventilation tactics. Because most homes are usually accessible from the exterior, horizontal ventilation operations should be the firefighter's first consideration. Remember, just because a window is in reach of a hand tool, indiscriminate ventilation should not be performed, we must communicate before we ventilate; proper and initial ventilation should be initiated as close to the fire as possible while the engine company begins its attack on the fire (fig. 14–71).

Fig. 14–71. Single story, single-family dwellings are usually accessible; however, indiscriminate horizontal ventilation should be avoided. Always size up and follow your plan.

Tactics

Horizontal ventilation. Ventilating the appropriate windows can improve the interior environment of a structure, aid in search and rescue, and speed the advancement of attack lines (fig. 14–72). Normally, the venting and removal of windows close to the fire or fire area should be sufficient. Ventilating these windows provides an exit for the expanding gases when water from the hoseline is applied. However, just opening windows generally only ventilates the bottom (moderate heat) and central (high heat) portions of a structure and has little effect on the upper (extreme heat) regions. In addition, with fire studies conducted, this ventilation without stream extinguishment allows fresh air to enter the fire area, increasing its intensity, so ventilation must be timed and coordinated. Remember that the upper portions are where the highest concentrations of heat accumulate, supporting flashover. These areas must be cooled down by the attack hoseline and can be ventilated to the outside once fire-suppression is completed.

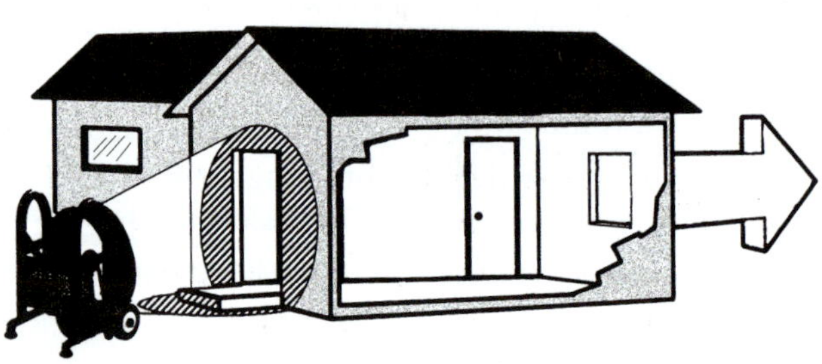

Fig. 14–72. Air blowing in one side and out the other is a form of horizontal ventilation.

It is important to note that opening the wrong windows or premature ventilation can supply additional oxygen to a fire, causing it to accelerate. Numerous broken windows in unfortunate areas can also reduce the effectiveness of pressurized ventilation. Windows away from the fire area should only be broken if they cannot be opened easily or are being used for VEIS operations. Remember, breaking windows may put broken glass inside the structure, hindering interior operations and causing an injury to a firefighter. If security bars are evident on the windows, inform the incident commander, and attempt to remove the bars if firefighters will be operating inside the structure.

Vertical ventilation. If a fire in these structures is extensive and backdraft conditions are present or a fire is present in the attic space, firefighters should consider vertical ventilation of the roof after a proper size-up (fig. 14–73). Although horizontal ventilation operations should be performed first, the need for vertical ventilation may be necessary. Ventilation of the attic can improve the interior environment of the structure, particularly at the ceiling and upper levels of the structure. The roof openings should be located over the fire and at the peak of the roof. Attic windows or louvered vents can also be removed as appropriate to accelerate venting or if weather conditions don't allow roof operations. Firefighters must use caution when opening an attic vent or a ventilation hole at the far end of the structure; it may draw fire throughout any uninvolved portions of the attic.

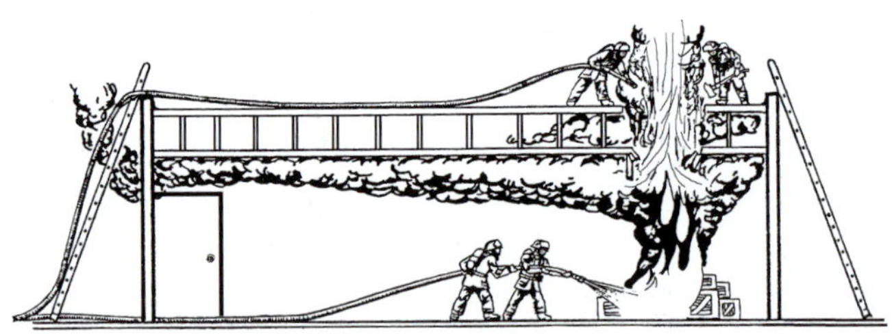

Fig. 14–73. An opening above the fire that allows fire, smoke, and gases to rise and exit the building is vertical ventilation.

Pressurized ventilation. Pressurized ventilation can be used effectively for fires in these types of structures (fig. 14–74). Remember that the exhaust openings should always be created prior to pressurizing the structure or area to be ventilated. If this is not followed, the pressurized airflow and by-products of combustion have no means of escape from the structure. In addition, it is also imperative to control the flow and path of pressurized air between an entrance and exhaust opening to achieve maximum ventilation. If pressurized air is directed from an entrance to an appropriate exhaust opening without being diverted to other openings, contaminants are removed in a minimal amount of time. Simultaneously opening unwanted windows and doors does not facilitate successful PPV operations.

Fig. 14–74. The cone of air pushed towards a building opening during PPV operations should cover the entire opening.

Single-family dwellings (multistory)

These structures are commonly two-, two-and-a-half–, and three-story dwellings, each of which is normally inhabited by a single family. Firefighters should not be surprised to find single-family dwellings turned into multiple-family residences or containing a separate apartment. The age of these dwellings is an indicator of potential vertical extension avenues. Old construction may conceal balloon construction, which provides open passageways within the exterior walls from the lowest floor to the highest. Newer construction uses open space floor plans, which can have features such as fascia, zero-clearance fireplaces, ductwork, balconies, and cathedral ceilings, which can allow fire to spread rapidly. These structures come in a variety of shapes, sizes, and designs and firefighters must become familiar to those structures prevalent in their response areas.

The location of the fire determines where the firefighter must perform ventilation operations. A fire on the first floor of a multistory dwelling calls for coordinated horizontal ventilation, whereas fire on the top floor may place initial emphasis on horizontal ventilation followed by vertical ventilation operations (fig. 14–75) Firefighters must remember that depending on the height of the building, access above the ground floor requires ladders before ventilation operations can be initiated. Firefighters can ventilate upper windows by using the portable ladder throw, roof ladder throw, or the arm-lock maneuver with a hand tool (described earlier in this chapter).

Fig. 14–75. Single-family multi-story dwelling. A fire on the first floor of a multistory dwelling calls for coordinated horizontal ventilation, whereas fire on the top floor may place initial emphasis on horizontal ventilation followed by vertical ventilation operations. (Courtesy of Mike Blatchly)

Tactics

Ventilation. If a fire is only on the first floor, the first priority should be to ventilate this floor horizontally. The immediate fire room and area should be the first priority, in coordination with the attack line's advance. Firefighters searching on the floor above the fire must use caution and not indiscriminately ventilate any windows they come across. Doing so could cause an autoexposure problem by pulling the venting fire into the floor they are searching. Firefighters operating on the outside of the structure must use caution and not ventilate windows on the upper floors if they are in jeopardy of fire entering them from the floor below. Remember, in most instances performing vertical ventilation for a first-floor fire could draw the fire and contaminants to the upper floors.

If extension above the first floor has occurred, horizontally and vertically ventilate the second floor. This reduces the heat and smoke in the building and minimizes mushrooming on the top floor. If the fire is on the top floor, perform horizontal ventilation and then begin vertical ventilation operations. For fires involving the attic space, perform vertical roof ventilation or remove the attic vents at the gable end of the structure.

In newer or contemporary homes, removing skylights can benefit ventilation operations of the upper floors. Many such homes have high pitch and irregularly angled roofs, on which it can be dangerous to operate; firefighters should use roof ladders or work from the safety of an aerial or tower ladder. Accurate roof ladder placement depends on determining the location of the fire and the appropriate location for the opening. Firefighters should place these ladders on the windward side when possible, directly adjacent to the likely roof opening. This allows firefighters to ventilate with the wind at their backs, and the by-products of combustion escape in the opposite direction.

Pressurized ventilation. If a multiple-story dwelling needs ventilation, always start at the lowest level with PPV and ventilate toward the top. To ventilate the first floor, ensure that all the exterior windows on the upper floor are closed or that a stairwell door to the upper floor is closed. Position a blower at an appropriate entrance opening, and then ventilate the contaminated areas on the first floor to provide maximum pressurized air for ventilation on the first floor and no flow of air on the second floor (caused by a lack of an exhaust opening). To ventilate the second floor, leave the blower in the same position and ensure that all the exterior windows and doors have been closed on the first floor. If a stairwell door has been closed, open the door and sequentially ventilate the contaminated areas on the second floor.

Multiple dwellings and center-hallway buildings

Multiple dwellings, multiunit apartments, and some condominiums are normally designed with a hallway in the center of the structure and units on either side, which allows for the maximum number of rooms in a building. Normally present in these structures is an open interior stair running from the first floor to the roof. Buildings with a **center hallway layout** may also have one or more skylights, bulkheads, scuttle ladders and covers, penthouses, and fire escapes. Fire doors may or may not exist in the interior hallways of these buildings. A prior knowledge of these types of buildings in a specific district is essential for firefighters (fig. 14–76).

Determining the location of the fire in these structures often indicates the ventilation priorities. The main concept to understand is that vertical ventilation in these buildings is a priority. It relieves the upper floors of smoke, prevents mushrooming, and allows occupants to survive or escape from the building. Opening up the bulkhead or skylight is considered vertical ventilation and allows the gas and smoke build up in the hallway and the stairway leading to it to be relieved. This allows the truck companies to access the upper floors while searching and allows the engine company to advance to the fire floor in moderate smoke conditions. Firefighters providing vertical ventilation in these scenarios must communicate with interior crews so that ventilation operations do not create a flow path and pull the fire toward the operating crews. At the same time, if interior crews notice the conditions deteriorating or the fire is being pulled toward them, requesting the bulkhead door to be closed may alleviate those conditions.

Fig. 14–76. Multiple dwelling and center-hallway buildings. A prior knowledge of these types of buildings in a specific district is essential for firefighters to learn. (Courtesy of Ed Evers)

The stairs and public hallways must be checked for fire, heat, and smoke, as well as any victims overcome by smoke. The presence of smoke and its quantity in the public hallways is an excellent indicator of the fire's extension and location. The presence and status of fire doors can dramatically affect this; closed fire doors limit an area and the smoke, whereas open doors allow the smoke to enter the entire public hallway. Opening a bulkhead door ventilates the stairway and public hallways while indicating the smoke level in the stairway. Firefighters must remember that if there is a light smoke condition or no smoke condition in the hallway, the fire apartment's door may still be closed and preventing any smoke from entering the stairs or hallway. It is imperative that vertical ventilation still be completed. Although smoke conditions could be light, the buildup of carbon monoxide could be high and overcome individuals or firefighters.

Tactics

Horizontal ventilation. If the fire is on a floor below the top floor, use an appropriate combination of horizontal, vertical, and pressurized ventilation. Firefighters must communicate coordinated ventilation operations and ensure they do not create an autoexposure situation when performing horizontal ventilation.

If horizontal ventilation is to be performed off of the fire escape, the firefighter should size up the situation first. If any occupants use the fire escape as a means of escape, no windows should be vented; occupants could be overcome by smoke or burned if fire vents from the windows. If the windows are to be ventilated, the windows farthest from the fire escape should be broken first and then those nearest the fire escape.

In some fireground situations, the aerial ladder may be used for ventilating windows and firefighters must adhere to their fire department's SOPs in these matters.

Vertical ventilation. Vertical ventilation operations should focus on three basic priorities: stairwells, skylights, and, for a top floor fire, cutting a hole in the roof over the fire. Opening a bulkhead door or skylight provides initial ventilation of the stairwell and public hallway but may need to be controlled in order not to permit a flow path. In some cases the skylight may have been removed and covered up with a new roof. When faced with these conditions, firefighters can use a power saw to cut an opening where the skylight was, which enables stairwell ventilation. If there are operable sprinklers in the skylight openings, they can form a water curtain and hamper ventilation. If ventilation is necessary under these conditions, consider eliminating the flow of water in the appropriate skylight opening by wedging the sprinkler head. Remember, only stop the flow of water from the sprinkler if it has controlled the visible fire below. If fire reappears and takes control of that area below, remove the chock with a hand tool to avoid a burn injury.

If there are doubts of fire in the attic or cockloft space, the **returns** in the skylight should be opened for a fast and easy way to check for extension. Firefighters should avoid opening up returns in skylights remote from the fire, as they could draw the fire toward the open returns.

For fires on the top floor, it is still important to ventilate the bulkhead or skylight and then provide initial horizontal ventilation to the top floor windows. Ventilating these windows in conjunction with cutting a hole in the roof provides vertical ventilation and extension into the cockloft. Venting the top floor windows creates an influx of fresh air and allows that air to be drawn into the room and up and out of the ventilation hole in the roof. The process of cutting some of these roofs may take some time if they have been built up over the years.

Firefighters should cut a hole in the roof over the fire. They must also size up the building's features and what the fire is doing. If a fire is venting out of two windows of one room, they should not cut directly over the room, because it is already self-venting. The cut can be pulled back a few feet to attempt to cut over two rooms and assist the engine company's advance into the fire area.

Condominiums and townhouses

Condominiums and townhouses are a common type of structure scattered across the country. Condominiums can have an interior or exterior stairway that serves the individual units. **Townhouses** are normally one, two, three, or even four stories, constructed side by side in a row with an open interior hallway within the unit. Both occupancies normally consist of a common attic that may have draftstopping or fire walls constructed within them and are supported by dimensional lumber or prefabricated truss roof systems (fig. 14–77).

Fig. 14–77. Condominiums and townhouses. Both of these occupancies normally consist of a common attic that may have draftstopping or fire walls constructed within them and are supported by dimensional lumber or pre-fabricated truss roof systems.

Ventilation operations in condominiums should focus on improving the interior atmosphere in the public hallway and stairwell to facilitate the evacuation of occupants and entry by firefighting personnel. Many such structures have a skylight or scuttle over the stair enclosure, which permits rapid vertical ventilation. Horizontal ventilation should be performed and used for fire in a unit. If the unit is two stories, they can be ventilated by both horizontal and vertical means. It is extremely important to check the attic space for fire and extension throughout the operation.

Ventilation in townhouses may be of a different strategy than condominiums. Although horizontal ventilation is performed for most fires, the townhouse unit should be checked for a skylight. Ventilating the skylight can relieve the open interior stairs of smoke and gases, making the interior tenable for firefighting operations. Townhouses may also have offset windows on the end units; these windows normally provide light into the stairwell area. They are good windows to ventilate to assist in vertical ventilation efforts.

The attic or cockloft space is a main concern with fires on the top floor of these units. Firefighters must check this area for fire extension and involvement during the course of the fire. A visible inspection of the roof may reveal that a fire wall extends through the roof, separating the individual units. Unfortunately, because of builders running wires, ductwork, or plumbing through this space, the fire walls can be compromised. Firefighters must not rely on fire walls alone. A simple procedure for checking the cockloft is to find a hatchway in the unit providing access to the space. It is normally located at the top of the stairs, on the top floor hallway, or in a closet on the top floor. If it takes too long to find, inspection holes can be made in the top floor ceiling with a hook to check on conditions above.

Garden apartments

Garden apartments lack several features that are typical of the center-hallway dwellings. They lack an interior stairwell and enclosed hallway. Many have an unenclosed exterior stairwell serving a few units and exterior balconies on the upper floors. The layouts of the units are less complex in design and have units that open directly to the exterior (fig. 14–78). These apartments also have common attics that run the length of the entire building. Typically, the roofs have a low or zero pitch and rafters that support the roof decking, making roof operations feasible.

Fig. 14–78. Garden apartments. Garden-type structures, whether residential or commercial, have units that open directly to the exterior.

Tactics

Ventilation operations. The floors below the top floor must be horizontally ventilated. This is easily done by natural or pressurized ventilation of an individual unit. A fire on the top floor can be ventilated by horizontal or vertical ventilation operations. A size-up of the fire and its extension possibilities can help decide whether or not to open the roof. Sometimes it is more feasible to open the top floor ceiling to check the cockloft area for extension and then cut the roof open. If a fire has extended into the attic or cockloft space and a ventilation hole has been made, it may be necessary to cut a defensive trench or strip cut in the roof to suppress the travel of the fire.

Commercial occupancies

Strip stores, taxpayers, shopping centers, and mini-malls include multiple occupancies attached to each other that may share a common attic or cockloft space. The cockloft area can be as small as a few inches high to as large as a few feet high. It also provides an open channel above the ceilings where fire can rapidly spread undetected. Many older buildings may have old gas lines that once provided lighting to the store, running through the cockloft. The risk of escaping gas can also lead to increasing fire conditions in this space.

When entering the structure, immediately check the cockloft for conditions above the firefighters. It may also be feasible for firefighters to determine whether the space is common to the entire structure or only a portion. Firefighters operating on the roof should check for division walls, fire walls, smoke from vents, and other such indicators.

Such a structure may have a fascia attached to the front, which may hide the actual age of the building (fig. 14–48). Consider a fascia to be an additional common attic built onto the exterior wall of the building. In some older buildings, ventilating the front fascia exposes the joists and the cockloft and allows access into the cockloft for the penetration of a master stream device for fire attack. Newer buildings are most commonly built of lightweight construction (steel bar joist, lightweight trusses, C-joist construction), and firefighters should make inspection holes into the roof to check for construction features. On these newer buildings, consider a fascia to be an ornamental feature and an additional common attic running only on the exterior of the building (fig. 14–79).

Fig. 14–79. Many commercial occupancies have fascias attached to the front of the building and it may hide the actual age of the building. Consider a fascia to be an additional common attic built onto the exterior wall of the building.

Tactics

Horizontal ventilation. Horizontal ventilation is generally limited to ventilating the front store windows once the engine company has water in the attack line and is ready to proceed into the fire. Firefighters must use caution in ventilating these heavy plate-glass windows, as they can break in large sections, fall outward, strike the firefighter, and cause a severe injury. Firefighters should attempt to use a long hook to ventilate the window from the side so that they don't place their bodies in front of the opening. Rapidly expanding smoke, gases, and heat exit the space and could ignite, causing a serious injury to a firefighter.

It may also be feasible to use PPV, pressurize each contaminated unit, and use the rear door as the exhaust port for minor fires inside the store. Many such structures have rear doors but no windows for security purposes and are perfect for PPV. A fire involving the cockloft area should not be fed by the forced air from a PPV operation.

Remember, fires in these types of occupancies are prone to backdraft conditions. Vertical ventilation can be accomplished before horizontal ventilation is performed.

Vertical ventilation. If possible, vertical ventilation operations should begin over the seat of the fire, reducing the chances of horizontal extension into the common cockloft. Initial operations should be to ventilate any skylights, scuttles, or other openings present on the roof. Cutting the roof open is a longer, more involved process; and these openings take less time to open up, thus possibly preventing flashover conditions inside the building.

When roof cutting operations are to begin, firefighters should cut over the store involved in fire. Unfortunately, this operation may not be possible because of the fire's progress; it may be difficult to conclude where the point of origin is, what store is on fire, or where is the largest body of fire located. Communications with the inside crews can relay the fire's location to the outside crews to ventilate the roof in the proper location.

If horizontal extension within a common attic is verified, it may be possible to cut a trench or strip ventilation ahead of the fire as long as the roof area isn't too large and the fire isn't rapidly spreading. Although strip ventilation is extremely effective, it is time consuming. When selecting a location for strip ventilation, cut the initial strip between the fire and the longest portion of the uninvolved attic to protect the greatest portion of the uninvolved building. Also, leave enough room between the strip and the fire to allow enough time to complete the strip. Two strips may be necessary to cut off an attic fire emanating from a single source but extending in opposite directions. In this case, protect the greatest portion of the building first and the smaller portion second. Strip ventilation should not be used in lieu of a primary ventilation hole over the fire, if possible.

Vertical ventilation for basement fires may have to be accomplished by cutting opening the first story's floor beneath a skylight, cutting open the floor near the front showroom windows, and in some cases removing the stepped-up displays or platforms in the front of the store to help ventilate the basement. Remember that a charged hoseline should be near these openings for protection.

Fascias. Fascias in these structures can readily promote fire extension under their sheathing. If the fascias are a separate decorative structure added to the building at a later date, any fire involved in the space may not be as important as if it had access to the cockloft. If opening a fascia to check for extension or to ventilate, consider these three methods. Openings underneath or on top of the fascia can be dangerous to personnel, because they could pull material on top of themselves and be injured. It also puts the firefighter in a dangerous area, operating under a portion of a structure that could collapse if it was fire ravaged. Openings can easily be made by cutting the vertical projection above the roof if the sheathing extends above the roofline. In some instances, this fascia ties into the brick parapet wall, and access with this method will not work. Using this method, a fascia would collapse away from personnel on the roof. Opening this fascia up from the safety of a tower ladder's bucket is another proven method.

Note that commercial structures often have heavy equipment such as air conditioning units, HVAC, billboards, and cell phone sites directly on their roofs or supported by steel I-beams above the roof. These features can add a tremendous weight load to the roof; and if fire involves the cockloft, the weight of these items can lead to an early collapse of the roof. In some instances, these frames may be tied into the parapet walls and can cause the failure and collapse of these walls under fire conditions. Firefighters should relay all pertinent information on the whereabouts of these items to the incident commander (fig. 14–80).

Fig. 14–80. Heavy equipment on the roof—this information should be relayed to the I.C.

Factories and warehouses

Many different construction methods are used to build industrial structures and complexes. Often, multiple manufacturing processes or tenants are housed in the same building with different hazards. Older buildings tend to have been modified and retrofitted throughout the years. In these cases it is common to find older construction in one section of the building and newer construction features in others. Most of the newer buildings are constructed of lightweight materials and are not conducive to firefighting operations like the older buildings.

Industrial occupancies may have large, open floor plans with a large amount of stock. These conditions can rapidly enhance the spread of fire. Although division walls may be present, the fire doors may be inoperative or blocked open, rendering a division wall ineffective and exposing additional areas of the building to fire and smoke. Larger buildings and roofs may require extensive operations to adequately ventilate. Firefighters must always consider the type of roof on a structure before ventilating it. Many industrial buildings may have flat or truss roofs. They may be of lightweight wood, concrete, steel bar joists with built-up metal decking, mill construction, or conventional wooden construction.

Windows of many such industrial structures may have been removed and covered with bricks for security purposes, whereas others be covered with wire mesh security screens. These obstacles may mandate the use of vertical ventilation and increase the time it takes to provide horizontal ventilation. If horizontal operations aren't feasible in some situations, it may have to be limited to the building's entrances, exits, and loading docks.

FIREGROUND NOTE

During a size-up of factories or warehouses, firefighters should also check for items such as the name and type of business on the front of the building. They should look for the National Fire Protection Association 704 hazardous materials placard ("diamond") on the exterior, the size of the electrical service, the type of equipment or machinery on the roof, stock type and supplies stored out in the yard, and whether hazardous materials are present.

Tactics

Horizontal ventilation. Horizontal ventilation operations can often be done by ventilating windows, doors, or larger openings, such as loading dock doors and display windows. Pressurized ventilation can enhance or replace the natural ventilation process.

Vertical ventilation. Initial roof operations should take advantage of natural features such as skylights, scuttle covers, and any bulkheads that may be present on the roof. Roof operations are predicated on the size and progress of the fire and the type of roof. Initial operations should be offensive and an opening made over the fire. If necessary, they should be followed by defensive tactics if personnel and conditions allow. Larger buildings require additional or larger ventilation openings than the standard hole made in dwellings. Because additional equipment and personnel are required, emphasize safety.

Institutions

The location of a fire in institutional buildings must be determined to evaluate its potential effect. Prime locations for fires in these types of buildings are in the kitchens, lounges, and storage areas. Hospitals, sanitariums, and rest homes are characterized by non-ambulatory residents who may need assistance or need to be removed. Jails and other detention facilities have occupants who may need to be released by security personnel. If a detention facility is in your district, plan your responsibilities ahead of time for incidents that might involve your services.

Firefighters must consider the height and number of floors in relationship to the location of the fire. A fire on the lower levels of a multistory institution can expose the floors above. Rescuing occupants on the upper floors may prove difficult. A fire on the upper levels may not expose the lower floors and may easily be ventilated with roof or stair shaft techniques.

Tactics

Single story. Ventilation efforts should initially focus on creating a ventilation opening as close to the fire as possible to reduce the extension of fire elements to other portions of the building. Openings should be large enough to ensure rapid and effective ventilation so as to facilitate the removal of occupants and the deployment of suppression personnel. Additional operations should provide ventilation for hallways and other passageways to minimize horizontal extension within. Pressurized ventilation in conjunction with horizontal ventilation can be effective.

Multistory. Begin initial ventilation operations with the vertical passageways such as stairwells and hallways if they are contaminated by fire or smoke. If a fire is on the top floor, always consider ventilating over the fire to reduce horizontal extension. Likewise, always consider the effect of using an uninvolved stair shaft to ventilate the floors below. Remember, it is crucial for all firefighters to be aware of the attack and evacuation stairwells in buildings with multiple levels. Lower floors can be ventilated by using a combination of pressurized ventilation, horizontal openings, or stairwells. When ventilating the lower floors, particularly through a stairwell, always consider its effect on the upper floors. Regardless of the size or height of the structure involved, always try to compartmentalize and restrict the extension of fire and its by-products.

Note that access into the stairwells of many institutions is restricted to prevent people from leaving or entering a floor. Many times it is more advantageous for a firefighter to get a key than to force the doors.

High-rise office buildings

Of the various types of structures encountered by the fire service, high-rises can be the most challenging in which to perform ventilation operations (fig. 14–81). In the older high-rises, there may be windows that open, but in the newer structures there are none. The newer high-rises are built with curtain wall construction and tempered glass windows, which are designed to be broken by firefighters to remove smoke from the floor. These windows are typically identified by a glass etching or reflecting sticker of the firefighter's Maltese cross (tempered windows break into small pieces as opposed to large, dangerous shards of glass from a plate glass window). The air in these buildings is pressurized and moved by HVAC systems located on various floors throughout the structure. Firefighters must realize that although a high-rise building may have windows, it may not be wise to indiscriminately ventilate them. Fires on upper floors can be affected by the wind currents present at upper levels of the atmosphere. In addition, falling glass from these high levels could injure anyone in its path.

The migration of smoke and toxic by-products throughout a high-rise building can often present a greater hazard to life than the spread of fire itself. During fires, the migration of air throughout a high-rise, frequently by its HVAC system or natural channels, results in the spread of smoke and toxic gases to areas far removed from the actual fire. This can render vertical and horizontal escape routes impassable and hamper fire-suppression efforts. Because smoke normally travels upward, the floors above become prime exposures for stratification, mushrooming, and possible fire extension. Thus it is important to have some type of guidelines for ventilation operations.

Smoke movement is also affected by a phenomenon known as "stack effect." In essence, smoke will be aided in its movement up through a building when temperatures inside the building are warm and the temperature outside the building is cool. Conversely, reverse stack effect will tend to drive smoke down through a high-rise building when temperatures are warm outside and cool inside the building (fig. 14–82).

Fig. 14–81. A high-rise office building. Of the various types of structures encountered by the fire service, high-rises can be the most challenging for firefighters to perform ventilation operations.

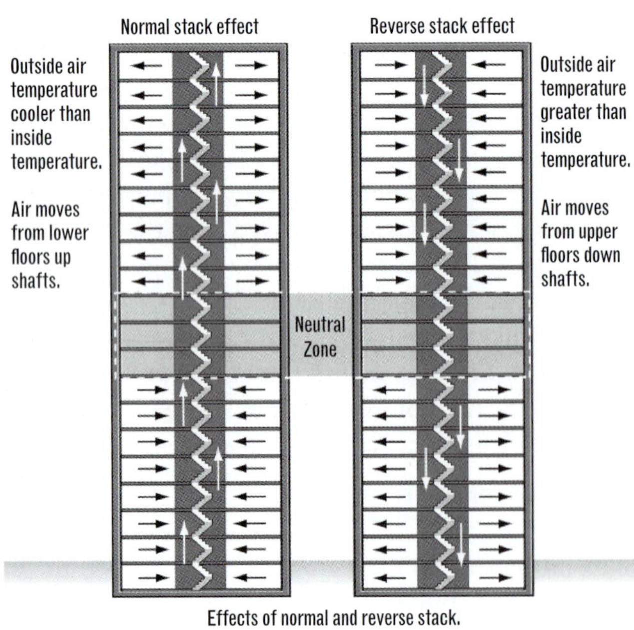

Fig. 14–82. The effect of normal and reverse stack effect in a building

Consider that some buildings will be equipped with smoke management systems. These may protect parts of a building (such as pressurization of an elevator shaft to keep smoke out) or may be installed throughout a high-rise building to exhaust the floor that is involved in the fire and pressurize the floor above and below to keep smoke from migrating to those areas. Most modern high-rise buildings have stairwell pressurization systems that inject air into a stair shaft during a fire to keep smoke out. It is important to identify all smoke management systems and how they work during preplanning operations.

HVAC operating guidelines. The following measures should be taken in regard to the HVAC system when units arrive at a fire in a high-rise:

- Determine the location of a building representative (fire safety director or maintenance worker).
- Determine the status of *all* HVAC systems. Manually shut down any system that has not automatically shut down (including supply and return fans).
- Determine the fire floor or floors.

Once the fire floor(s) have been determined, activate the supply fan in *all* HVAC zones that do not include the fire area. Activating these fans provides fresh air and pressurizes the zones, limiting the spread of smoke and gases. All units operating in the building should report any adverse conditions to the incident commander.

PPV used in high-rise operations. Opening the bottom and top doors of a stair shaft quickly develops a natural upward flow of air. Placing blowers at the bottom of the shaft can augment this flow, regardless of the height of the building. Such currents can be used to keep contaminants from accumulating in a stair shaft (if the roof opening is kept closed) or to remove accumulated contaminants and keep the shaft clear (roof opening kept open). Emphasis is placed on stair shafts that may be used by attack personnel. This creates better visibility in the stairwell, allowing the engine company to hook up to the standpipe system and the truck company to search the floors above and the stairwell. Note that such operations can be affected by the presence of a stairwell pressurization system.

High-rise fires present unique and varied ventilation problems. Firefighters should never attempt a haphazard approach to ventilation without specific plans for the effective use of personnel and equipment. Additionally, when ventilation operations are necessary, it is essential to have a clear understanding of the parameters imposed by high-rise buildings.

High-rise multiple dwellings

In many cities across the country, large high-rises have numerous apartments per floor. These buildings may have windows that open or not and outdoor balconies in some instances. Another common type of these buildings is referred to as projects and are prevalent in lower-income areas. These buildings come in various shapes, heights, and sizes and also have ventilation issues to handle. Recently, many major cities and outside agencies have been studying on the effects of wind-driven fires at these occupancies. Unfortunately, numerous firefighters have been killed in fires in such dwellings over the course of the years. Wind-driven fire events have made conditions unbearable and fatal to many firefighters who were operating in the public hallways or apartments due to premature failure of windows, ventilation efforts by firefighters, and the speed of wind gusts. When wind is a factor all tactics must be coordinated, and in some instances firefighting operations will use other tactics to extinguish the fire, such as a floor-below nozzle or a transitional attack via a pump-mounted monitor nozzle.

When fires occur in these buildings, firefighters should begin with a size-up of the conditions on the upper floors. Either see what the conditions are on the floor below or on the floor above and communicate that information with interior units to formulate the plan of attack. Once a size-up has been taken and wind is not a factor, normal efforts for ventilating the fire apartment windows from the floor above in coordination with the hoseline's fire attack can be performed. The first firefighter who reaches the apartment on the floor above should relay the wind conditions to the incident commander and follow their directions. Ventilation of an upper-floor fire in these dwellings is usually performed by a firefighter reaching downward with a 6-ft (2-m) hook, a Halligan tool attached to the hook, or a Halligan attached to a utility rope to break the fire apartment's windows. The roof's bulkhead is normally opened to clear the stair shaft of contaminants. Opening the door at the base of the shaft also allows a draft to be pulled up the stairs to help clear the stairwell of contaminants.

Unfortunately, the pressure of the wind causes significant problems for firefighters. As they commence their initial fire attack into the fire apartment or public hallway leading to the fire apartment, windows fail and wind currents blow fire and heat back on top of the operating crews. The heat levels become unbearable, because the apartment's fire load burns at a furious pace from the wind, and chase firefighters out of the apartment or hallway. Firefighters have reported that the conditions changed so rapidly that they did not know what happened.

Recent studies have shown that by pressurizing the attack stairwell with large PPV fans, firefighters can lower the hallway temperatures and diminish the effects of the wind-driven fire back onto the fire attack teams. Also, by pressurizing the stairwell at wind-driven fires, the fire attack team can attempt to proceed into the public hallway or fire apartment to extinguish the fire. Another method of controlling the wind at these fires is to prevent the influx of wind into the fire apartment by dropping a **fire blanket** out the window from the floor above (fig. 14–83). The fire-resistive blanket blocks the window channel from the wind, diminishes a flow path of the heat, smoke, and gases toward the advancing hoseline, and allows interior crews to attempt an aggressive fire attack on the fire. This tactic will require additional personnel and time for implementation, and the height and location of the fire will be a major consideration.

Fig. 14–83. A fire blanket or wind control device being placed over the window of an apartment to block wind from entering. (Courtesy of NIST)

QUESTIONS

1. What are the differences between the two basic principles of ventilation?
2. Smoke contains many different gases that are toxic to civilians as well as firefighters. Among those listed in the text, which are the two most common gases firefighters should be concerned about?
3. When performing ventilation size-up of a structure, what are some of the considerations firefighters must ask?
4. Both negative pressure and positive pressure ventilation may use mechanical methods to ventilate a structure. What is the theory behind how each displaces the smoke and by-products?
5. The improper use of PPV during fire attack can have deadly results for firefighters as well as the civilian occupants of a structure. List three conditions that would prevent PPV from being utilized during fire attack.
6. Regardless of type of ventilation used on a fire scene, what is the goal of effective ventilation principles?
7. In relation to ease and safety of firefighters, what is the most popular method of quickly ventilating a compartment of a structure?
8. After removal of skylights to provide ventilation, what must the ventilation crew check to ensure full access of the opening?
9. How do roof scuttles and bulkhead doors differ in the overall effect of ventilation?
10. If presented with multiple panes when ventilating a window, what are the correct steps to open the window?
11. When ventilating windows that are covered with security grating or steel mesh, what are the different methods that may be employed?
12. Lexan windows can present a unique problem to firefighters attempting horizontal ventilation. What steps can be taken when forced to ventilate these types of windows?
13. VEIS techniques are generally used when accessing the alleged location of victims when not accessible by other means. This is known as venting for ___________?
14. What are the proper steps of performing VEIS?
15. Why is closing the door early in the VEIS technique so critical?
16. List some of the typical tools needed for roof ventilation.
17. The Kerf or plunge cut is commonly used to accomplish what task?
18. When using the "pullback" method of roof ventilation, what are some of the advantages?
19. An effective ventilation technique that can be accomplished more quickly than the coffin cut is?
20. When ventilating pitched roofs, what is the primary purpose of the setting a roof ladder?
21. Caution must be used when ventilating open web steel bar joist supported roofs. What are some of the hazards associated with these types of roofs?
22. Identify the most dangerous type of roof and list the reasons why.

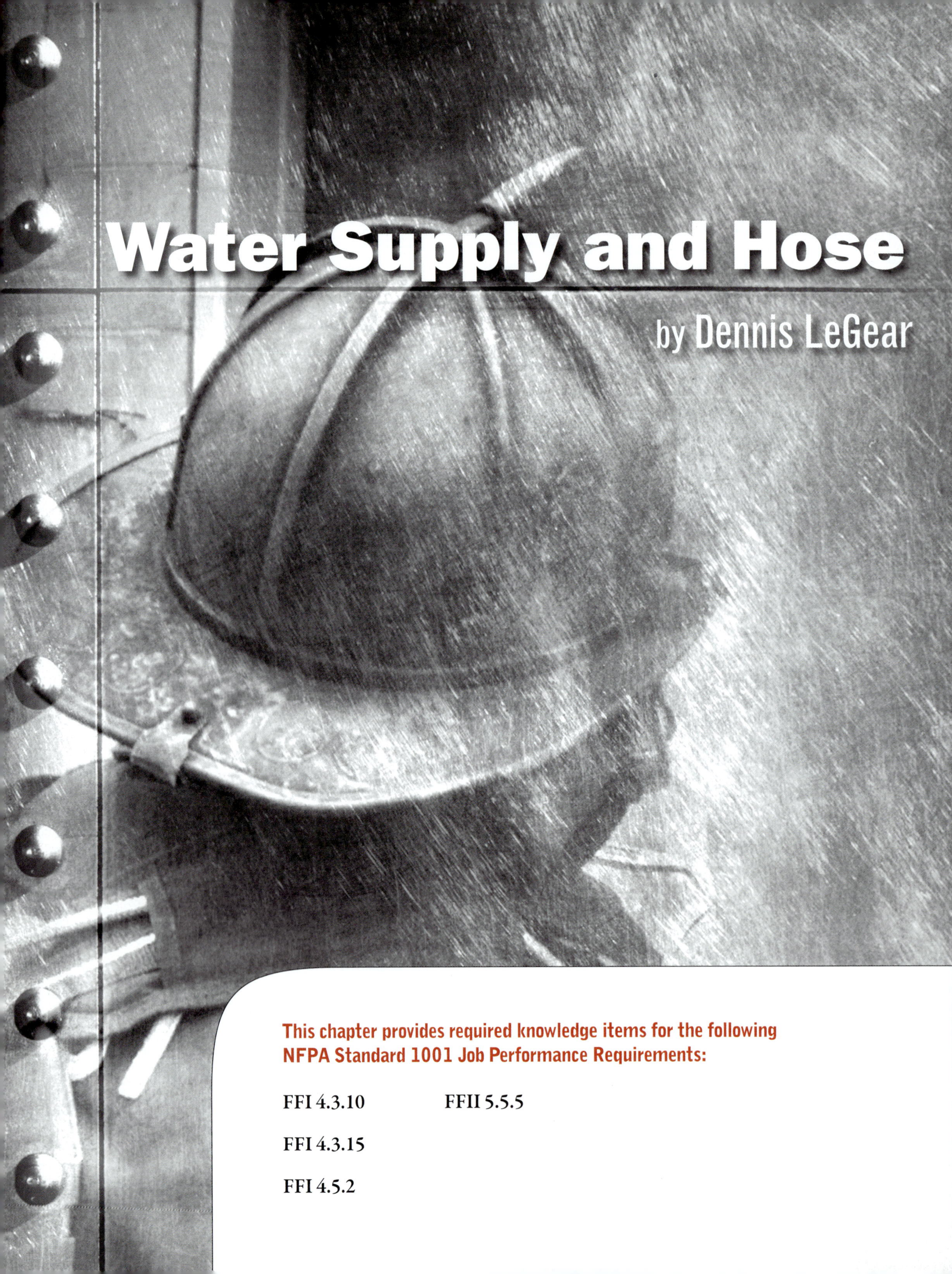

Water Supply and Hose

by Dennis LeGear

This chapter provides required knowledge items for the following NFPA Standard 1001 Job Performance Requirements:

FFI 4.3.10

FFI 4.3.15

FFI 4.5.2

FFII 5.5.5

This chapter contains Skill Drills. When you see this icon, refer to your Skill Drill book for step-by-step instructions. SKILL DRILL

OBJECTIVES

Upon completion of this chapter, you should be able to do the following:

- Identify the basic properties of water
- Differentiate between flow and pressure as they relate to water supply
- Define the term friction loss as it relates to water supply
- Define static and residual pressure
- Identify the common elements of a municipal water system
- Describe the differences between wet barrel and dry barrel fire hydrants
- Describe the proper procedures for opening and closing a fire hydrant
- Identify and describe the proper procedures for inspecting, maintaining, and testing of fire hydrants
- Identify types of rural and auxiliary water supply systems
- Describe the classifications of fire hose
- Identify types of fire hose construction
- Identify the common types of hose and typical flows used by the fire service
- Identify the types of couplings used to connect fire hose together and to various appliances
- Identify the types of hose fittings used by the fire service
- Identify the types of hose appliances used by the fire service
- Identify the types of hose tools used by the fire service
- Describe the proper procedures for testing, maintaining, and cleaning fire hose
- Identify and describe the different methods of storing hose used by the fire service
- Identify and describe the different types of hose rolls used by the fire service
- Describe the different types of hose loads for supply and attack hose
- Describe the different types of hose lays
- Describe the proper procedures for deploying hose at a fire emergency incident

INTRODUCTION TO WATER SUPPLY

Establishing an adequate water supply is the one of the first critical elements of all successful fire attacks. An adequate water supply has to be in place quickly and last for the duration of the incident. Many failed fire attacks can be squarely blamed on a lack of an adequate water supply. Establishing a good water supply is often accomplished under extreme stress, as the firefighter in figure 15–1 can attest. From the firefighter to the incident commander, one critical item must be addressed early on at all fires: water supply. Water is the most common extinguishing agent used by fire departments around the world because is relatively plentiful, absorbs large quantities of heat, is easily transported, and follows a fixed set of physical rules and characteristics.

Fig. 15–1. A firefighter struggles at a hydrant to charge a large diameter supply line. (Courtesy of Danny Barlogio)

Basic water characteristics and terms

FFI 4.3.10 Two of the most important physical rules governing water supply are that water is **noncompressible** (as far as the fire service is concerned) and is capable of being pumped. This means that one cubic foot of water will occupy one cubic foot (1 ft^3) regardless of the pressure.[1] It also means that it can be moved from one location to another. These two basic physical properties allow water to be easily stored, transported, and moved in pipes and hoses by both gravity and pumps.

Two specific terms are used to describe the movement and use of water on the fireground: **flow** and **pressure**. Flow is the volume of water being moved or used and is expressed in **gallons per minute (gpm)** or **liters per minute (L/min)**. Pressure is a force applied over a given area, commonly expressed as **pounds of force per square inch (psi)** or **kilopascals (kPa)**. Pressure (psi or kPa) is the force used to move the volume (gpm or L/min) of water through a hose and out of a nozzle.

As a member of the fire service, you should know a few other basic properties about water. A gallon of water (3.8 L) weighs about 8.33 lb (3.78 kg), and 1 ft^3 of water (12 in. × 12 in. × 12 in.) is approximately 7.48 gallons (gal) (28.31 L). This means 1 ft^3 of water weighs about 62.4 lb (28.3 kg). Water pressure is commonly measured in psi. If we took a 1-in.^{2} column of water (1 × 1 in.) that is 1 ft tall, the weight exerted at the bottom of that column would be measured as 0.434 psi. A column of water 2 ft high would have a pressure at the base of the column of 2 times 0.434, which equals 0.868 psi, and so on. Take a look at fig. 15–2 to help you understand the concept of water pressure. Expressed a different way, "water pressure increases 1 psi for every 2.3 ft (0.7 m) of depth." The actual pressure at the bottom of a 10-ft (3-m) column of water would be 4.34 psi (30 kPa). One can see the advantage of an elevated water tank. For example, if the water level in the tank in figure 15–2 (a and b) was 200 ft (60 m) above the ground, the water pressure would be roughly 87 psi (600 kPa) at the bottom. This force (pressure) is due to one thing: gravity, which acts continuously on all objects on earth. The pressure created by a column of water is called **elevation head**.

Alternatively, the fire service commonly uses a rough estimate of 5 psi (34 kPa) for the amount of pressure that is needed to move (force) water up every 10 ft (3 m) of building height, the approximate height of a building story. While we *gain water pressure* (force) at the bottom of an elevated tank (column of water), we must *apply pressure* (force) to move the water up to the top of that tank (water column). Fire department pumpers (engines) are used to provide the pressure to move the water through hoses and up into the upper stories of a building that is on fire.

In addition to gravity, we must consider the force of **atmospheric pressure**. That force is literally the weight of the air that makes up the earth's atmosphere that is exerted on objects on the earth. At sea level, a common place to measure atmospheric pressure, the atmosphere produces a pressure of 14.7 psi (100 kPa). Let's take a look at a bottle of liquid (fig. 15–3). The surface of the liquid in the bottle is under the pressure of the atmosphere above it. At sea level that pressure is 14.7 psi (100 kPa). If a straw were placed in the bottle of liquid, the level of liquid inside the straw would be the same as the liquid level surrounding the straw in the bottle. This is because the air in the straw has the same atmospheric pressure, 14.7 psi (100 kPa). However, when you drink from the straw, you lower the pressure inside the straw to below 14.7 psi (100 kPa). The liquid is then forced up the straw because the liquid surrounding the straw is still under the greater atmospheric pressure. If you lowered the pressure in the straw by 1 psi to 13.7 psi (94.5 kPa), the straw could be as long as 2.3 ft (0.7 m) and still force the liquid to the top of the straw. If you were to remove the entire atmosphere (14.7 psi) from inside the straw at sea level—creating a perfect vacuum of 0 psi (0 kPa) inside the straw—it would function to a height of 2.3 ft (0.7 m) × 14.7 psi (100 kPa), or about 33.9 ft (10 m)!

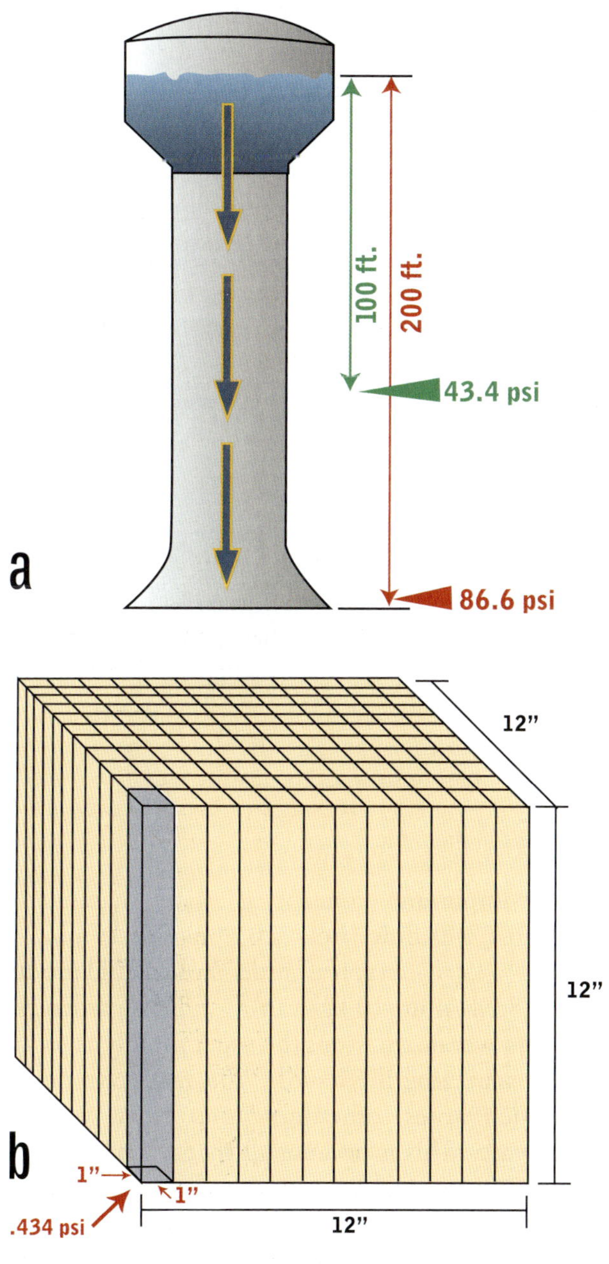

Fig. 15–2 a, b. Understanding water pressure is important to effectively use your water supply on the fireground.

Fig. 15–3. A simple way to demonstrate atmospheric pressure is with liquid and a straw.

To measure pressure, fire department pumpers use pressure gauges. They are mounted on the pumper's **pump panel**, the location where the pumper's operator (chauffeur, motor pump operator, or driver) controls the pumps. All water pressure gauges in the fire service read in **psig**; the "g" stands for gauge, which means that the reading has been corrected to zero psi after taking into the account the weight of earth's atmosphere.

Now that we understand the physical properties of water better, let's move on to how this knowledge is applied in the fire service. For any water supply system to work, one has to be able to move water from one point to another. As we have seen, pressure can be created by gravity, pumping, or the combination of gravity and pumping together. Pressure is used to move the water from short to very long distances. Because pressure is a form of energy, some of this energy will be dissipated (lost) when water is flowing. This loss of pressure (energy) is called **friction loss.** Friction loss is created as water flows through a hose and literally rubs against the inside of the hose, losing pressure all along the way. Friction loss increases as more water is flowed through a given size hose (gpm or L/min); friction loss can be decreased if the size of the hose is increased with the flow kept the same.

For example, a 1¾-in. (45-mm) hose flowing 175 gpm (662 L/min) loses more pressure due to friction than a 2½-in. (65-mm) hose of the same length and flow. Since nozzles and other appliances need certain minimum pressures to operate properly (covered in chapter 16, Fire Streams), friction loss is a very important factor to consider.

There are two other terms used to describe types of pressure: **static pressure** and **residual pressure**. Static pressure is the amount of pressure in a hoseline or water main in a street while *no* water is flowing. Residual pressure, on the other hand, is the pressure remaining in a hose or water main while water *is* flowing.

Let's take a look at how a common garden hose is used to define some of these concepts (fig. 15–4).

Fig. 15–4. When the garden hose is not flowing, but is charged, it is a closed system, and the water pressure will be equal at all parts along the hose.

Most of us take the common garden hose with an attached spray nozzle for granted. We use it to wash our cars and to water our plants without wondering how it works. Let's take a closer look at what makes this possible. Once you open up the hose bib, water travels down the hose and stops at the closed nozzle. If you bled the trapped air off by cracking open the nozzle momentarily and then took a pressure reading at the closed nozzle with no water flowing, you would get the static pressure. Imagine that the static pressure reading was 50 psig (350 kPa). If you then went back to the hose bib and took another pressure reading, again with no water flowing, it would read the same 50 psig (350 kPa). The same would be true if you took a third reading halfway between the nozzle and the hose bib. The fact is that if you took a pressure reading anywhere along this charged garden hose with no water flowing, you would get 50 psig (350 kPa). This 50 psig (350 kPa) reading is an example of static pressure; in this example, it represents the **potential energy** of the water in the garden hose waiting to be released.

Now let's take a look at the garden hose and the effect on pressure once we open the nozzle (fig. 15–5). Notice that with water flowing, the pressure is highest at the hose bib and lowest at the nozzle. The 45-psig (315 kPa) reading on the gauge at the house near the hose bib is called the residual pressure. The 45-psig (315 kPa) residual pressure represents the extra capacity left over in the house's water supply system when the garden hose is flowing water. This extra capacity allows other people in the house to still wash the dishes, take a shower, and so on. The 35 psig (245 kPa) on the middle gauge represents the fact that as the water travels down the garden hose, energy is lost to friction. This drop in pressure represents a loss in the kinetic energy available in the hose. In this case the friction loss is 10 lb (4.5 kg) in the first half of this hose. Now take a look at the gauge located at the nozzle; it reads 25 psig (175 kPa), which means the water leaving the nozzle is at 25 psig (175 kPa). This 25 psig (175 kPa) is the energy used to force the water out of the nozzle and throw the stream a distance away from the nozzle. In this example the remaining 25 psig would be called the nozzle pressure (common nozzle pressures are covered in chapter 16, Fire Streams).

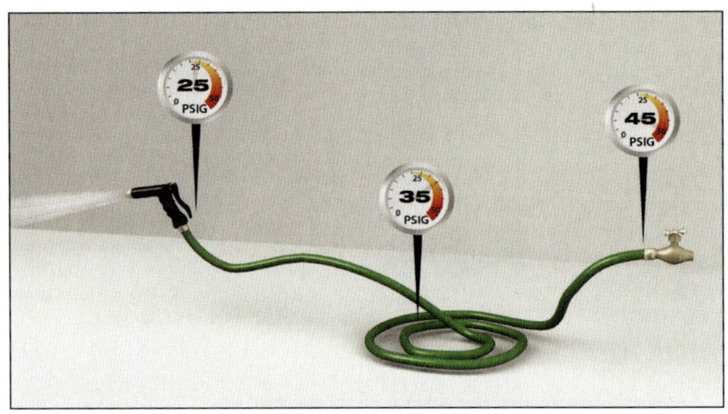

Fig. 15–5. When water is flowing, the residual pressure represents the extra capacity available in the system. The 25 psig (274 kPa) represents the pressure of the water leaving the system. The 20 psi (239 kPa) difference is the friction loss in the system.

Another important concept is water hammer. A garden hose will flow about 10 gpm (38 L/min). At 8.3 lb/gal (1 kg/L), a typical garden hose would produce a flow of 83 lb/min (10 kg/min). A typical garden hose squeeze sprayer nozzle stops the water rapidly as soon as you stop squeezing the handle. You may have noticed how garden hoses sometimes jump when you stop flowing water. This is because all that water in motion was abruptly stopped by that closed nozzle. Since we know water does not compress, that energy must be dissipated in one way or another. This violent dissipation of kinetic energy that occurs when flowing water is stopped rapidly in a closed system is called **water hammer**. In this example, the garden hose stretches and moves, and the pipes in the house shake. This small 10-gpm (38 L/min) flow is not enough to cause damage; however, in the fire service where typical flows are 150 to 1,000 gpm (568 to 3,785 L/min), rapidly stopping water by slamming shut intakes, discharges, or nozzles can result in major damage. This includes damage to nozzles, fire hose, pumps, and water mains. The best practice to prevent water hammer is by always operating valves and nozzles in a slow, deliberate, and cautious manner.

MUNICIPAL WATER SYSTEMS

Now that you have a basic understanding of some of the physical concepts and characteristics regarding water, we will discuss larger water supply systems. Most communities have a water supply system that serves both its drinking water needs and fire-suppression flows. This type of **municipal water system** is designed to provide sufficiently clean water for consumption while providing adequate residual capacity of that same water for fire suppression. The modern water supply system is usually run by a public or private water department. The water company bills consumers for their consumption and in return administrates the system under heavy local, state, and federal regulations to ensure that both drinking water standards and minimum fire-suppression capacities are met. Municipal water systems are typically made up of the following critical elements:

- Sources of raw water
- Treatment and storage facilitates
- Distribution system
- Valves
- Hydrants
- Hydrant inspection, maintenance, and testing

Sources of raw water

Typical sources of water are broken into two distinct groups: **groundwater** and **surface water**. Groundwater is stored beneath the ground in naturally occurring aquifers. These groundwater aquifers are exploited by drilling wells. Communities are often supplied by groundwater wells alone, and substantial sustainable amounts of water can be acquired by drilling wells and pumping the water out. Groundwater has accumulated in these aquifers for thousands of years through rainfall in their watersheds, which then percolates down into the aquifer. We can think of large aquifers as lakes underground.

Most of us are also familiar with surface water sources, of which the most typical types are freshwater sources. Some examples of surface freshwater sources are lakes, creeks, rivers, deltas, and engineered freshwater storage such as large reservoirs (fig. 15–6 a and b). Some water departments have now started to use saltwater sources like bays, oceans, and brackish areas of estuaries to augment their fresh surface water sources. Most water districts or departments usually use only one source of raw water; however, as water becomes more valuable and water resources more scarce, many water systems are now taking advantage of multiple sources.

Fig. 15–6 a, b. A man-made reservoir and a nearby creek

Treatment and storage facilitates

After possible sources of water have been identified, the water has to be transformed from raw water to treated, **potable water**. This process is done at a water treatment plant. The raw water is transported by gravity, pumping, or both to the treatment plant. Sometimes the direct source for a treatment plant is a small-terminal, raw-water-storage reservoir. This type of terminal reservoir acts as a buffer to the main source of water, so that in case of an earthquake, drought, or other natural disaster, the water district or department would still have a limited supply of water for treatment and distribution (fig. 15–7).

Water treatment can be very complex and can have many steps. The amount of water treatment that is required is directly related to how clean the raw water entering the system is. Treatment can include any of the following processes: desalinization, aeration, coagulation, flocculation, sedimentation, filtration, fluoridation, ozone disinfection, and ultraviolet light disinfection. Some of these processes are mechanical, and others are chemical. Some of these treatment processes remove substances and contaminates from the water, while others add things. However, the main goal of water treatment remains the same no matter what processes are employed. That goal is to ensure that the water is safe for human consumption before it enters into the distribution system. Ideally, these water treatment facilities should be designed to handle emergencies like power failures and other situations, to ensure that both adequate drinking water and fire capacity can be maintained through most foreseeable emergencies.

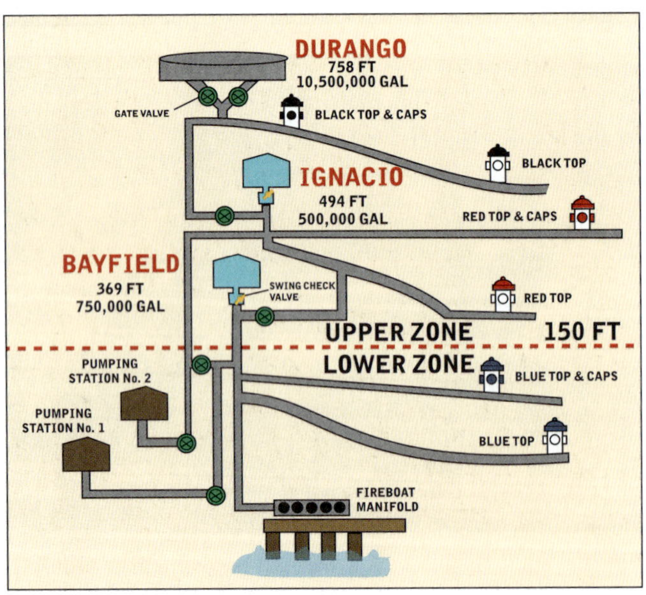

Fig. 15–7. An example of a high-pressure system that is fed with non-potable water and is only for firefighting

Distribution system

The **distribution system** is the most important part of a municipal water system for fire departments. It is the system of treated water storage reservoirs, tanks, pumps, and water mains or pipes that supply the fire hydrants with water. The system should be designed to provide the best flow of treated water throughout the municipality, maintain safe water quality, and meet the service area's minimum standard fire flow. Figure 15–8 shows a basic water system, where a treatment plant receives raw water from a lake, treats it, then pumps the water into a distribution system of several treated water storage tanks in the community. Using the force of gravity, these storage tanks then provide pressurized treated water to the community's water main system. This pressure is created by the elevation head, discussed earlier in this chapter. Some distribution systems have no stored treated water in elevated tanks or reservoirs and rely on a system of pumps to maintain adequate pressure in the system. During large-flow incidents such as large fires using many master streams, more pumps will be automatically called into service by the drop in residual main pressures caused by fire department suppression operations. One example of this is municipalities that operate water main systems from several large water wells. Backup power is always provided in these systems for the distribution pumps. A distribution system can also commonly rely on a combination of both pumps and tanks to ensure adequate fire flow potential.

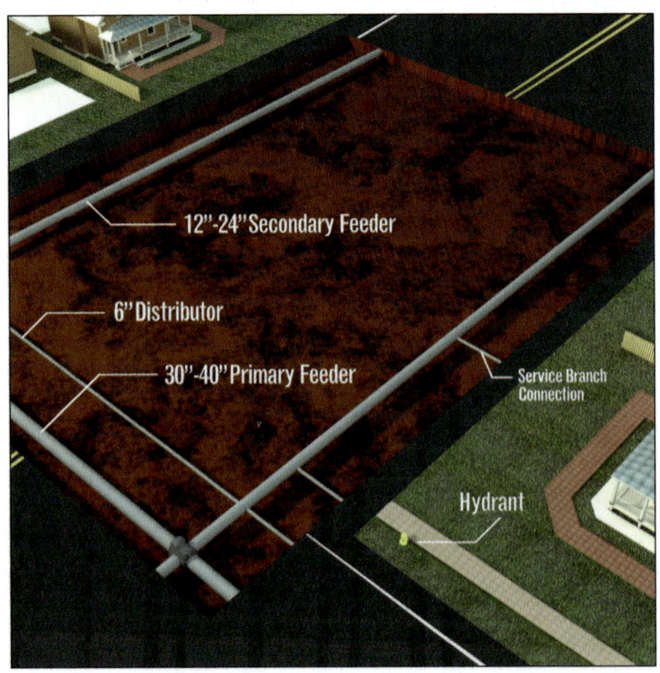

Fig. 15–8. A small water main system

These larger treated water storage tanks ensure sufficient stored treated water capacity for the water main system. This extra capacity could be used to maintain the flow at a large fire. Figure 15–9 (a and b) represents two different examples of treated water tanks that supply the water main systems for the neighborhoods below and around them. Both of them are using gravity to supply their water main systems. Most likely, they are pumped full at night to replace the water used during the day. They are usually refilled at night because demand for water in most communities is lowest at night.

Fig. 15–9 a, b. These treated water tanks use gravity to supply the municipal water supply.

The water main system is very important to the fire department. It is ideally designed to have low friction loss and good residual pressure during normal use. This is done in several ways. First, by laying out the water distribution system in a grid, water is received at most locations from multiple directions and from different sized mains. This efficient design lowers friction loss because no one pipe has to flow all the required water being supplied to a hydrant. Remember, as flow increases in a given size pipe, so does the friction loss. Friction loss can be reduced by increasing the size of the pipe or reducing the flow, which is what happens when flows are split in a grid pipe network (fig. 15–10). Another way water departments and districts lower friction loss in the water main system is to have a series of large-diameter pipes reduced to medium-diameter pipes, and then reduced even further to smaller-diameter pipes that are laid out in a gridlike fashion.

Fig. 15–10. Split flow in a pipe grid is a great way to combat friction loss in a water supply system.

Throughout the water main system, **water main valves** are located at frequent intervals. Water main valves are used by the water district to shut down certain areas of the system or grid for issues like repair and maintenance. Most water main valves are always left in the open position. Water main valves can also be kept in a closed position to separate grids or pressure zones. They are commonly spaced at 500 ft (152 m) and 800 ft (244 m), so only a small section of water main would need to be turned off for repairs, leaving other areas still supplied.

- Large-sized water mains are sometimes called **treated water aqueducts** or **primary feeders**. They can be very large: some are 20 ft (6 m) in diameter in cities with heavy water demand like New York. However, in most cases they are between 30 in. and 48 in. (76 and 122 cm) in diameter. Typically, no hydrants are connected to these primary feeders.
- Medium-sized water mains are commonly referred to as **secondary feeders** and are generally sized from 12 to 24 in. (31 to 61 cm) in diameter. These secondary feeders are common throughout most municipal water main systems. Many fire hydrants are connected to these secondary feeders.

- Smaller-sized water mains are usually referred to as **distributors**. These pipes make up the bulk of most water main systems and generally range in size from the outdated 4-in. (10-cm) main to 12-in. (30-cm) mains. Because of their small size and high friction loss, 4-in. (100-mm) mains are not commonly used anymore; 6-in. (150-mm) pipe is the current minimum size pipe used to supply hydrants. Hydrants are primarily installed on these small distributors.

Hydrants

Fire hydrants are the main access points for the fire department water supply in municipal water systems. Hydrants are connected directly to the water main system by a pipe. This is typically a 6 in. (150 mm) diameter pipe with a gate **valve** (also called a roadway box valve) located in the street between the hydrant and water main. This valve allows the hydrant to be shut off using a special long wrench called a **hydrant key or tee wrench.** It is sometimes necessary to shut off a hydrant at the gate valve for maintenance or if a hydrant is knocked over during a vehicle accident. Ideally, hydrants are supplied by a water main from two directions. Sometimes it is not possible to easily supply a hydrant from two directions; for example, when a water main goes down a dead end court or cul-de-sac street. This is called a dead end water main, and the hydrants on it are called dead end hydrants. Dead end hydrants have less capacity, as they do not have the same redundant supply as those located within the grid where they are supplied from multiple directions. Some water districts mark these with symbols like an arrow showing the direction of water flow in the main and a vertical line denoting that it is a dead end hydrant.

Typically, you will encounter two types of fire hydrants: wet barrel and dry barrel. **Wet barrel hydrants** have water inside right at street level and are usually used in nonfreezing climates. In fact, wet barrel hydrants were first developed around 1900 in the San Francisco Bay area and are sometimes called California hydrants. A typical modern wet barrel hydrant will have more than one outlet. Each outlet has its own operating stem and valve. Figure 15–11 (a and b) shows a picture and a diagram of a typical wet barrel hydrant installation. Notice some of the terms used in the diagram:

Wet barrel hydrants typically have all of these components: stem guide seal, body, stem, stem O-ring, valve rubber, retainer, O-ring seat, retaining nut, cotter pin, stem guide O-ring, stem guide, carrier, valve seat, cap, cap washer, and chain ring.

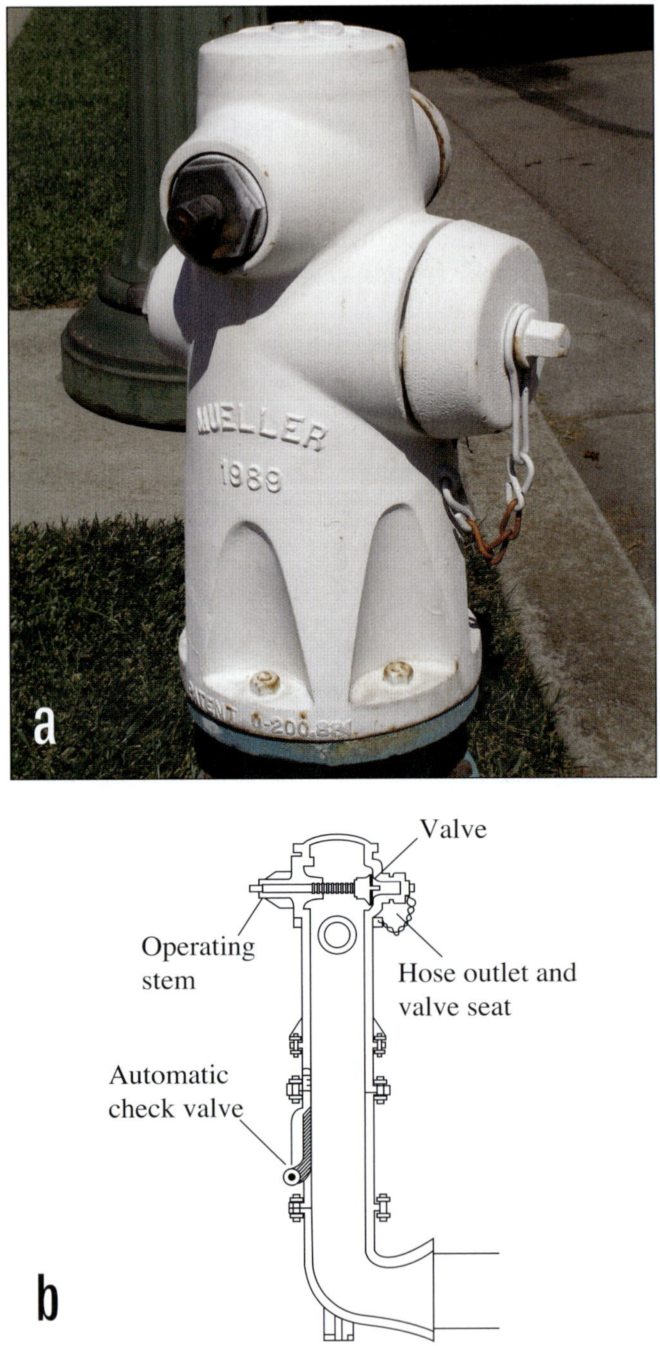

Fig. 15–11 a, b. Wet barrel hydrants are common around the world. Water resides inside the hydrant at street level.

The **dry barrel hydrant** is typically found in freezing climates. However, dry barrel hydrants can be installed in nonfreezing climates as well. This means that you may find dry barrel hydrants, which are specially designed to work in freezing climates, in warmer climates. The main difference between the two designs is that the dry barrel hydrant is just that: dry. In a dry barrel hydrant, the water is held in the pipe serving the hydrant safely below the frost line by a valve that is controlled by an operating nut at the top of the dry barrel hydrant. Figure 15–12 (a and b) contains a picture and a diagram of a typical dry barrel hydrant installation.

a

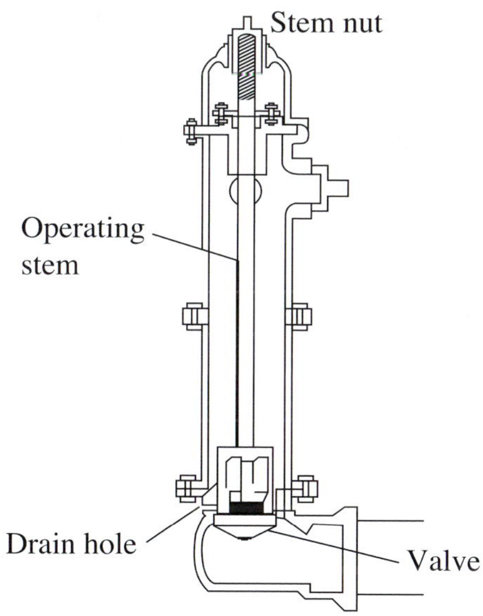

b

Fig. 15–12 a, b. Dry barrel hydrants are common in freezing climates. The water resides safely below the frost line.

Outlets on both wet and dry barrel hydrants are threaded. The current national standard is **National Standard Hose Thread (NST)**, sometimes known as National Hose Thread (NHT). Most fire hose in the United States uses this standard, although there are a number of cities with their own hose threads. The number and size of hydrant outlets vary, but generally include one or two 2½ in. (65-mm) outlets and a single **steamer** connection (also known as a **pumper** connection) of 4 or 4½ in. The steamer (or pumper connection) is a critical item to understand, as it by name denotes that the full flow potential of the hydrant is only available at the hydrant outlets. During large fire flow operations using many master streams, it is critical to place a supply engine/pumper on the hydrant and push water to the fire scene, and not wrongly tax the distribution system residual pressure with the job of moving water through the supply hose to a remote attack engine/pumper (100 or 115 mm). Some cities have incorporated a 5-in. (125 mm) Storz connection on their newer hydrants. (Storz connections are described later in this chapter.)

Basic hydrant operation. **FFI 4.3.15** Before we go any further, it is appropriate to briefly discuss basic hydrant operations. These methods apply to both wet and dry barrel hydrants. Hydrant operations begin with sizing up the area around the hydrant and the hydrant itself. It is important to make sure that it is accessible and that there is no exterior damage to the hydrant that would impact its operability. Before charging the hydrant it is important to check the interior of the hydrant for any damage as well. Once it is determined that the hydrant is operational, it is time to charge the hydrant. Prior to charging the hydrant, it is recommended that the hydrant be flushed by removing the steamer cap and opening the hydrant a few turns so that a fair amount of water is discharged from the hydrant. The hydrant is opened by using the hydrant wrench to turn the operating nut. This will vary between wet and dry barrel hydrants. Look on the hydrant for a directional arrow that will indicate the direction to turn the operating nut (clockwise or counter-clockwise). Once it is flushed, close the hydrant and attach the supply hose to the appropriate hydrant discharge opening or nozzle. It's important to note that once the dry barrel hydrant is charged with water, all outlets have pressurized water. Many fire departments will place a hydrant gate valve on unused outlets to ensure additional water can be accessed without the need to turn the hydrant off (fig 15–13). When you are ordered to "charge" or "open" the hydrant, slowly and smoothly open the hydrant fully, usually 10–14 turns in order to avoid a "water hammer," and then turn back the hydrant wrench a quarter turn.

Once the operation is completed and the order is given, the hydrant can be shut down. Shutting down the hydrant is the reverse of the procedure to open it. Place the hydrant wrench on the operating nut and turn in the reverse direction you did when you opened it, or look for the indicator arrow on the hydrant and turn in the reverse direction. As before, close it slowly and smoothly, to avoid a "water hammer." Once fully closed, turn the operating nut a quarter turn in the opposite direction to prevent over tightening it. Once the hose has been drained, disconnect it from the hydrant and replace to hydrant discharge caps by hand tightening.

Fig. 15–13. Understanding basic hydrant operation and hydraulics is crucial to maintaining a proper water supply on the fireground. (Once the yellow dry barrel hydrant is charged, both 2.5-in. outlets caps will have pressurized water behind them. In a wet barrel hydrant each outlet has its own control.)

Hydrant inspection, maintenance, and testing.

A large percentage of all firefighting around the world is made possible by water supplied from fire hydrants. In the United States, hydrant spacing is typically every 300 ft (90 m) in high-fire-load areas (high value), and 500 ft (150 m) in lower-fire-load areas (low value) such as single-family-dwelling neighborhoods. For example, the City of Oakland, California, with a population of 400,000, has roughly 6,500 hydrants located within 56 square miles (145 sq km).[2] The fact is that an incredible number of hydrants in the world, once installed, will probably never be used unless there is a fire. Based on this fact, inspection, maintenance, and testing of hydrants are critical to ensure that fire hydrants remain functional and perform to their fullest capability during an emergency.

In the United States, it should be the goal to inspect and maintain fire hydrants annually. **Flow testing**, a more detailed test, can be done on a rotating time schedule. Flow testing is covered in chapter 29, Pre-Incident Planning. Many jurisdictions mandate that all of their hydrants must be inspected annually, with a certain percentage to be flow tested. For example, if 20% of hydrants were flow-tested during their annual inspection, it would take 5 years to flow-test every hydrant in the jurisdiction.

Minimum annual company district **hydrant inspections** should address the following items: location, clearance, caps/chains/threads, barrel empty (dry), valves, paint, color code, blue reflector, hydrant gate pot, and curb paint. Minor problems can be corrected immediately, and major problems reported for repair. It is more critical that dry barrel hydrants are inspected biannually (fall and spring) because they are more susceptible to failures than wet barrel hydrants. Fall inspections ensure they are dty to prevent freezing, and spring inspections determine if there is plow/car damage. Examples of serious failures in dry barrel hydrants are debris in the barrel and failure to drain properly. Both of these issues could easily lead to an unusable dry barrel hydrant. The following list describes what you should look for when performing an inspection of a wet or dry barrel hydrant:

- Location: Make sure the hydrant is on your map, and, if working from a list, confirm the hydrant number or address.
- Clearance: Ensure that the hydrant is readily visible. Typical fire code provisions require that "a 3-ft clear space shall be maintained around the circumference of fire hydrants except as otherwise required or approved." The hydrant shown in figure 15–14 needs to be dug out and have vegetation trimmed back. If a retaining wall is required to keep it uncovered, it should be reported.

Fig. 15–14. This hydrant needs to be dug out and have the vegetation trimmed to ensure access when needed.

- Caps, chains, and threads: Caps should be free of damage, attach to the hydrant by chains, and turn freely. Remove caps and check for gaskets, because the rubber seals are used to maintain a leakfree condition in unused outlets in dry barrel hydrants when in operation. The other primary purpose of gaskets when hydrants are not in operation is to prevent electrolysis and severe rusting of caps. Ensure that threads are in good condition. The 2½ in. (65 mm) cap nut in figure 15–15 has broken off, and its chain is missing.
- Barrel empty (dry): With the caps off the hydrant, ensure that dry barrel hydrants are empty and free of visible debris.

Fig. 15–15. This hydrant is missing a cap nut and chain.

- Valves: Check for leaks around the operating stem on wet hydrants and for water in the barrel of dry hydrants. Open the hydrant slowly and operate all valves, making sure that they work easily. It is not important to achieve full flow; just make sure the hydrant works. Turn the hydrant off slowly. Ensure dry barrel hydrants drain completely by verifying that the water is draining down the hydrant barrel. If the water only drains down to the level of the open hydrant outlet, it is not draining properly. This can be pumped out with a dry barrel hydrant pump (fig. 15–16). If the hydrant is not draining properly, it must be repaired. Dry barrel hydrants with frozen water in them will not flow correctly, if at all. When operated, dry barrel hydrants should be fully opened. When fully opened, water is stopped from flowing out of the drain hole. Partially opened dry barrel hydrants can undermine themselves if pressurized water flows out of their drain holes.

- Paint: Check the condition of the paint. It should be in good condition, but can have minor cracks and light rust. If the hydrant is severely rusted and needs a complete repainting, report it. The hydrant shown in figure 15–15 should be repainted.

Fig. 15–16. Dry barrel hydrant pump

- Color code: Many jurisdictions use **color codes** from NFPA 291, *Recommended Practice for Fire Flow Testing and Marking of Hydrants*, located on the bonnet indicating the hydrant's available flow at 20 psi residual pressure. If so marked, have the proper paint in aerosol cans for touch-ups (table 15–1).

- Reflector: A roadway reflector is a method for identifying a hydrant's location. If your jurisdiction uses them, check them during the annual inspection and replace damaged ones or install new ones when necessary.

- Hydrant valve cover (roadway box): Ensure that the hydrant gate valve cover is not paved over. Some jurisdictions also paint them white to help locate them during emergencies such as sheared hydrants.

- Curb paint: Make sure the curb is painted red. If the red curb paint is missing or needs repainting, report it.

Table 15–1. Properly color coding the hydrants in your jurisdiction helps to rapidly identify expected flow rates.

Class C	Less than 500 gpm (1,892 L/min)	Red
Class B	500–999 gpm (1,892–3,782 L/min)	Orange
Class A	1,000–1,499 gpm (3,782–5,674 L/min)	Green
Class AA	1,500 gpm and above (5,678 L/min and above)	Light blue

The annual test just described is a good basic inspection for determining serviceability of the hydrant. What it does not determine is the maximum flow of the hydrant, nor does it assess the available capacity of the water distribution system in the area. To attain this important information, one must subject the hydrant to a flow test. Flow tests need to be coordinated with the water district. All information collected should be done under a mutual program with shared data. This is because the information will be used by the water district to determine how their distribution system is performing. Has there been a drop in flow in an area? Is it because the pipes have become clogged? Is there a regulator valve not opening properly? Has a change in the system had a negative impact in an area of the distribution system? Is the hydrant street valve partially closed? The data gathered from annual flow testing of hydrants can be an invaluable tool for both fire departments and water districts.

Usually, large municipal water districts will have a telephone number to call at least 24 hours before flow tests are performed. Proper notification of the water department during flow testing is imperative, so the system operators will understand the sudden spike in usage. It is also good for the fire department to foster a solid relationship with the agency that oversees the water distribution system. Large water districts sometimes have an employee or department that is assigned to manage all the flow testing data and hydrant testing programs. Sometimes they will even provide the training and necessary equipment for flow testing.

RURAL AND AUXILIARY WATER SUPPLY

Many firefighters work in rural areas of the country where large portions of their response area are not served by water main systems with fire hydrants, or the systems in place have inadequate capacity and pressures to supply enough water to mount an effective fire attack. In other cases, some older highly urbanized areas have outdated or obsolete water supply systems that are no longer capable of providing an effective fire flow. Municipal water supply systems become obsolete when the size and age of the water mains become inadequate or when regular demand on the system has grown to a point where it no longer has enough reserve capacity to provide large fire flows. Poor maintenance of a water supply system can also lead to dramatically less water capacity from fire hydrants. Furthermore, many cities in earthquake zones have water main systems that may completely fail during a large earthquake.

In all of these cases, it is still necessary to provide an adequate supply of water for fire control. Fortunately, effective fire control may be accomplished if an adequate water supply is established in a timely manner. In the rural firefighting world, many well-prepared fire departments have identified reliable auxiliary sources of water and have made plans to access them during emergencies. To these rural agencies, alternate water supply operations are considered bread and butter. In the urban world of firefighting, many municipal fire departments are now developing these same types of auxiliary systems to ensure adequate water service in times of disaster. Some fire departments, like San Francisco, have good auxiliary water supply systems already in place. Others cities, for example Seattle and Vancouver, have recently completed a plan that addresses auxiliary water supply.

The following are examples of auxiliary water supply:

- Water tenders and portable tanks
- (Static sources) drafting hydrants, cisterns, ponds, lakes, swimming pools, and jet siphons
- Relay pumping
- Fireboats and dry main systems
- Reclaimed nonpotable water main systems and hydrants (some of these are high pressure)

Water tenders and portable tanks

FFI 4.3.15 The modern fire engine is a **triple-combination pumper**, which consists of three key critical items: a pump, fire hose, and a water tank. The water in the engine's tank usually does not exceed a storage capacity of 1,000 gal (3,758 L). This water carried on the engine is only good for mounting initial attacks on small to medium fires; often, several engines on-scene can back each other up, leading to rapid control of up to medium-size house fires. However, a purpose-built **water tender** apparatus can have a water tank that holds up to 4,500 gal (17,034 L) of water. These specialized units make large-scale **water shuttle operations** possible. Typically, these water tenders have an easily deployable, folding, portable water tank that has a slightly larger capacity than their water tender tank (fig. 15–17).

An efficient water shuttle operation can supply large fire flows (fig. 15–18). A basic water shuttle operation consists of three elements, a **dump site**, a **fill site**, and the necessary fire apparatus. The dump site is located at the fire. It is usually made up of portable tanks taken off water tenders and then filled by those water tenders. The attack engine then drafts from these portable tanks. As the attack engine uses the water in the portable tanks, the water tenders travel to the fill site and fill their tanks. The water tenders must then return to the dump site and refill the portable water tanks supplying the attack pumper. This must all be accomplished in a safe manner.

Fig. 15–17. Most water tenders have an easily deployable, folding, portable water tank.

Fig. 15–18. A simple water shuttle operation can effectively supply water for fireground operations.

The fill site should be located as close as possible to the fire. The fill site should have a capacity of at least 1,000 gpm (3,785 L/min), have good access, and be located on the best available roads that preferably provide a circular traffic pattern between the fill site and the dump site. The fill site does not need to be the nearest good hydrant; a static source of water, such as a lake, river, swimming pool, or bay can also be a good fill site. If your water supply is a static source of water, you must have a means of moving the water. This is usually accomplished by a drafting fire engine, which in turn fills the water tenders.

The most basic of water shuttle operations from a static water source need the following apparatus, equipment, and actions: (1) A dump site by the fire with a portable tank is initially filled by the water tender. (2) A route is established to a nearby water source with the capacity to fill a water tender in a rapid fashion, preferably at least 1,000 gpm (3,785 L/min). (3) Upon arrival at the fill site, the water tender is filled by an engine drafting from the static water source or hydrant. (4) Once filled, the water tender returns to the dump site and fills the portable tank again. This completes one shuttle cycle, which provides a limited capacity of water at the fire scene. Ideally, repeating this cycle will supply a continuous water supply to the attack pumper during the fire until the fire is extinguished.

Fire department personnel involved in water shuttle operations should never try to increase the amount of water that can be delivered by speeding up the shuttle cycle. Rushing these types of operations can lead to serious and sometimes fatal injuries. Firefighters have been hit and run over at fill and dump sites. Water tenders are heavy vehicles that demand respect when operated and must be driven cautiously and in a defensive manner at all times. Many firefighters have lost their lives in vehicle-related accidents involving water tenders. The simple fact is that if more water is required, the water shuttle operation must be scaled up. This means using more portable tanks, possibly multiple dump sites, more water tenders, more routes, possibly multiple fill sites, more attack engines, more fill engines (fill hydrants), better routes, road closures to all traffic but emergency equipment, and so forth. During this whole process, positions must be assigned at critical areas: dump site officer, fill site officer, traffic control officer, water supply group supervisor, and the like. Many safe options are possible to increase capacity and efficiency in water shuttle operations. Speeding up can be fatal and is not an option in water shuttle operations. In the end, fire departments that rely on this difficult operation must train actively and often.

Obtaining water from static sources

FFI 4.3.15 Many fire departments are surrounded by raw water supplies such as lakes, rivers, bays, and even the ocean. Note that saltwater from the ocean or brackish water from a river that empties into the ocean must be flushed from fire department pumpers, hoses, and nozzles because the saltwater can damage them. Cities across the United States are often located by vast sources of water; these bodies of water are used for commerce and transportation. If a bay-front warehouse is on fire, and the old water mains servicing that district of the city are inadequate for the volume of fire encountered, the fire department can draft water right out of the bay. This is a good, solid tactic, and some fire departments have addressed their outdated water systems further in waterfront areas by installing drafting hydrants and providing some engines with jet siphons to assist with waterfront supply issues.

A drafting hydrant, properly designed on a reliable static water source, will easily provide the full-rated capacity of the engine's pump (fig. 15–19 a and b). A typical **drafting hydrant** is a specialty hydrant that consists of the following components: a strainer basket submerged at least 2 ft (0.6 m) at the end of a 6-in. (15 cm) diameter pipe, and a pumper connection with a female swivel pumper connection in easy reach of an engine's drafting hose. These drafting hydrants are just that: ideally, pipes into a static water source, with less than 10 ft (3 m) of lift. Drafting hydrants function as an easy way to draft water from an untreated static water source. If a drafting hydrant is not provided, you can still draft using the **hard suction hose** on the engine (described in detail later). However, this is usually less efficient and can have many disadvantages when compared to using a drafting hydrant. Some of these disadvantages include access issues, lack of access to water, poor bank conditions, leaves and debris clogging the strainer, and frozen surface water during the winter.

Cisterns are large, watertight underground tanks used for storing liquids, especially water. In the fire service, jurisdictions that would like to add emergency stored water capacity for firefighting sometimes build cisterns. In cities, cisterns are typically located below ground at major intersections. These cisterns can hold hundreds of thousands of gallons of water. Cisterns in large cities are usually filled by the nearest hydrant via an access plate in the street. This plate also serves as the drafting port. Cisterns are also commonly found in rural areas. Building codes in rural areas are starting to require property owners to provide water for firefighting, and cisterns can fill this role. Many times on rural properties, instead of a cistern the same amount of required water is stored in an above-ground water tank. It is now possible to buy precast concrete or plastic cistern/tanks kits that have fill ports, vents, and drafting hydrants as part of the overall design. Chemicals are added to prevent the growth of algae and other aquatic organisms. Cisterns, once filled, are basically tanks of non-potable water for drafting during emergencies.

Jet siphons are typically used to move water from one portable tank to another in water shuttle operations. A jet siphon (fig. 15–20) is a device in which water is pumped into a small, 1½-in. (38-mm) intake and then under high pressure, rapidly discharged into a large hard-suction hose. New strainer-equipped jet siphons now can generate up to 700 gpm (2,650 L/min) utilizing regular, soft-suction hose. However, it is always better to try to achieve a true drafting operation using properly sized, traditional hard-suction hose with a total distance of less than 30 ft (9 m) and with no more than 10 ft (3 m) of lift. Drafting in this manner should provide the full rated capacity of an engine's pump.

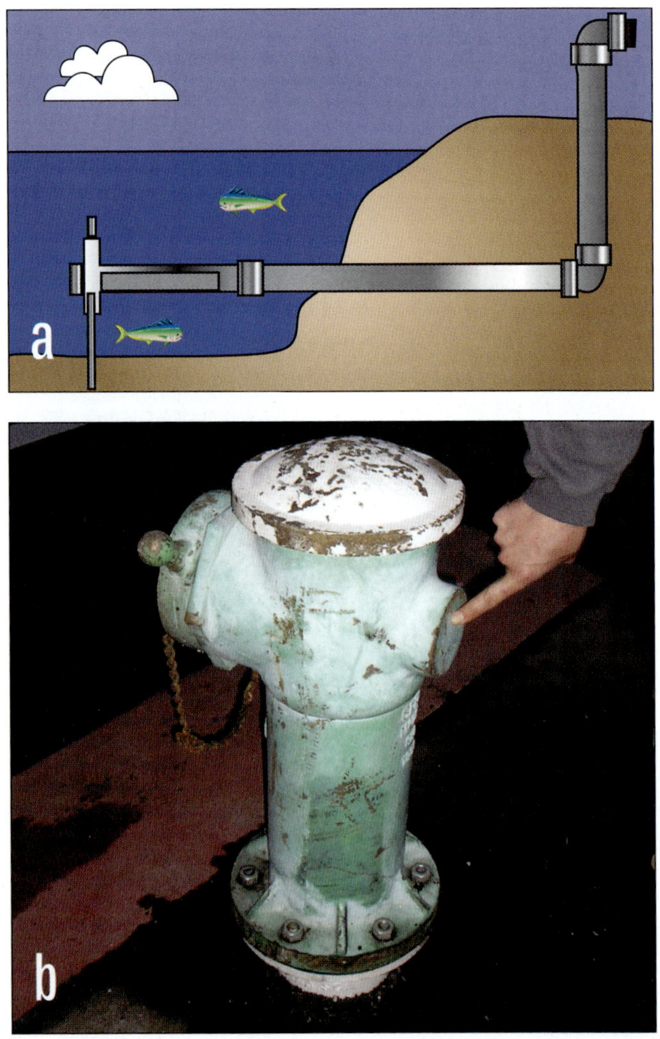

Fig. 15–19 a, b. A drafting hydrant allows the fire service to easily access an untreated water source. Notice that there is no operating spindle on the hydrant.

Fig. 15–20. Jet siphons are typically used to move water from one portable tank to another.

Relay pumping

Relay pumping in its simplest form refers to one pump supplying water through fire hose to another pump (fig. 15–21). Fire departments relay pumps to overcome water supply problems where the fire is at a distance from the nearest water source, but typically not more that a mile away. Pumping water through hose takes energy, and fire hose has friction loss. For example, a 4½-in. hose at 1,000 gpm (3,785 L/min) has 10 psi (69 kPa) friction loss per 100 ft (30.5 m). A mile (roughly 5,300 ft or 1,609 m) of 4½-in. (115-mm) hose would require 530 psi (3,654 kPa) of pump pressure to overcome the friction loss to continuously supply 1,000 gpm (3,785 L/min) through the mile of hose. 530 psi (3,654 kPa) of friction loss a mile in 4½-in. (115-mm) hose at 1,000 gpm (3,785 L/min) is not possible to achieve with the standard fire department hose and equipment. However, to pump a mile of 4½ in. (115 mm) hose at 1,000 gpm (3,785 L/min), a fire department could place a pumper every 1,000 ft (304.8 m), which would bring the required pressure per pumper down to 100 psi (700 kPa) for each 1,000 ft (304.8 m) section of 4½-in. (115-mm) hose it is assigned to pump for friction loss. Friction loss in hose is the main reason fire departments establish relay operations.

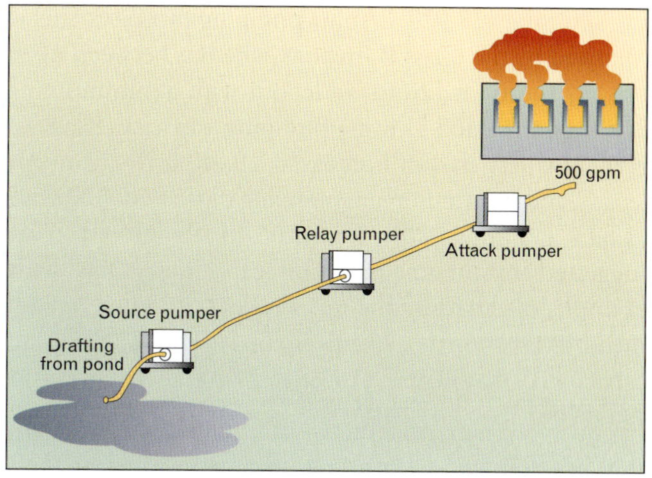

Fig. 15–21. A simple relay operation

For terminology purposes, let's take a look at a three-pumper relay operation. The pumper located at the water source that pumps water to a second pumper is called the **source pumper**. The second pumper, or **inline pumper**, receives water from the source pumper then pumps the water to a third pumper. The third and last pumper in this example is called the **attack pumper**. It supplies the fire attack lines and is located at the fire. The number of required pumpers, size of hose, length of lay, type of lay, required gpm, and elevation change are just some of the factors that have to be considered for a successful relay-pumping operations.

Fire departments that routinely use relay operations should be well practiced and drilled; they should also be aware of the limitations of their relay operations in their training. These fire departments sometimes have special hose wagons and relay valves to help speed the deployment of large amounts of hose. This makes engaging in relay-pumping operations in rapid fashion more feasible. Long relay-pumping operations can be established in a rapid manner by departments that routinely train in this tactic. In some jurisdictions, fire departments have set up preformed relay-pumping task forces that can be special-called by the incident commander. These task forces then establish an entire secondary water supply upon arrival, with a goal of attaining a certain water supply capacity, usually no less then 1,000 gpm.

Fireboats and dry main systems

Many fire departments located near large bodies of water have **fireboats** with pumping capabilities. These fireboats vary in size and use, but we will discuss them based on solely on pumping capacity. The NFPA classifies fireboats as Class A, B, or C. Class A is the largest class of fireboat.[3] It must be 65 ft (20 m) in length or longer, with at least a 5,000-gpm (18,927 L/min) pumping capacity. This is followed by the Class B fireboat, which is between 40 ft (12 m) and 65 ft (20 m) in length with a minimum of 2,500-gpm (9,463-L/min) pumping capacity. Class C is the smallest class of fireboat, is 20–40 ft (6–12 m) in length, and has a minimum pumping capacity of 500 gpm (1,893 L/min). Many fireboats greatly exceed their suggested minimum pumping capacities. One such example, operated by the Los Angles City Fire Department, is a 105-ft (123-m) fireboat with a pumping capacity of 38,000 gpm (143,846 L/min).

Fireboats are amazing pieces of equipment. Typically, they are designed to continuously pump water at their rated capacity for at least 8 hours. Following the terrorist attack of September 11, 2001, on the World Trade Center in New York City, fireboats were used to fight the ensuing fires and supplied a peak flow of 60,000 gpm (227,125 L/min) to the World Trade Center site through fire hose.[4] Fireboat pumping operations continued for days at Ground Zero.

The 1989 Loma Pretia earthquake caused several major fires in San Francisco's Marina District. San Francisco Fire Department's (SFFD) fireboat *Phoenix* (fig. 15–22) pumped roughly 5.5 million gal (20.8 million L) of water over a 15-hour period through a portable water supply system built out of 5-in. (125-mm) hose because water mains in the area had been severely damaged by the earthquake.[5] No hydrants were working reliably in the area of the main fires near the waterfront. During this conflagration, the pumping capacity of the *Phoenix* was largely credited with saving a whole district of the city. Soon after the 1989 earthquake and fire, SFFD purchased a second fireboat to augment their waterfront fire protection and recently purchased a third.

Fig. 15–22. SFFD fireboat Phoenix and waterfront-mounted dry hydrant. (Courtesy of Danny Barlogio)

Some cities with fireboats have wet or dry **manifold systems** connected to special water main systems. This type of water main system carries nonpotable or raw water, which can be pumped into by fireboats or pumping stations that draw water from a static source, fed by large reservoirs under the force of gravity, or a combination of all of these methods. San Francisco's Auxiliary Water Supply System is the best example of this in the United States.

Other cities have separate high-pressure water supply systems—in some cases using water supplied directly from an adjacent river—in their downtown congested commercial districts. These systems are activated during large fires. Vancouver over the last decade has developed a downtown auxiliary water distribution system capable of 300 psi (normal hydrant pressure is 50 to 120 psi) and surviving a 8.3 Richter scale earthquake.

Reclaimed water main systems and hydrants

As drinking water becomes an increasingly scarce natural resource, a new trend has begun on the West Coast: the use of **reclaimed water**. Sometimes called **gray water**, reclaimed water is becoming so regularly used that wastewater treatment plants (sewage treatment plants) are now providing it back to consumers through reclaimed water distribution systems. Gray water is used to water golf courses and flush toilets in high-rises, and it is used in other situations where potable water is not needed. Some fire departments are now requesting hydrants to be installed on these newly installed gray-water main systems and are using these hydrants as auxiliary water supply sources. These gray-water hydrants are painted purple, which is the same color as reclaimed water pipes. Figure 15–23 shows just such a gray-water hydrant. This gray-water fire hydrant is located on a reclaimed water main system that waters a very large city park and school grounds.

As we conclude our discussion about water supply, you now can see just how important an adequate water supply is, for many firefighters the most expensive system you will interact with in your entire career ends at the fire hydrant.. A sufficient water supply is necessary at all fires once an attack has begun, and it needs to last the whole incident without interruption. There are billions of dollars of water supply infrastructure throughout the United States. In addition to the municipal water supply infrastructure, billions of dollars have been spent on fire department equipment to use that available water. In the end, it is up to us, the fire service professionals, to get the most out our available water supply system. As you move forward in your career, it will become necessary to expand your knowledge regarding water supply. This chapter is just a basic introduction to effective and efficient water supply. Remember, poor water supply equals poor service.

Fig. 15–23. Reclaimed water wet barrel hydrant

INTRODUCTION TO HOSE

Now that we have discussed water supply systems, we need a way of moving the water from place to place. The moving of water in the fire service is accomplished with the use of hose specifically designed and constructed for the fire service. The ideal fire hose must be the right mix of the following characteristics: adequate friction lost for flow, durability, weight, and ease of coupling. Fire hose must also be made in a cost-effective manner. Weighing these needs and characteristics, the major manufacturers of fire hose provide a wide array of choices to the modern fire service. In this discussion about fire hose we will be covering the following topics:

- Classification of hose
- Hose anatomy and construction
- Size and type of hose and common flows
- Couplings
- Fittings, appliances, and hose tools
- Basic hose testing and maintenance
- Hose storage
- Water supply evolutions

Classification of hose

The breakdown in classification of fire service hoses is based on its tactical uses. Typically, hose is discussed in terms of drafting, supply, attack, and wildland. **Drafting hose** is a rigid hose that is used by a pump to acquire water from static water sources. **Supply hose** is designed to move larger volumes of water; naturally these supply hoses have a larger diameter than attack hoses. **Attack hose**, sometimes referred to as hand line hose, must have a diameter large enough to move the required flow, which can be up to 325 gpm (1,230 L/min), but must be small and lightweight enough to be easily moved around the fireground by firefighters. **Wildland hose** is a type of specialty attack hose designed to work well for combating vegetation fires. Wildland hose is not produced in diameters greater than 1½ in. Two main criteria for wildland hose are reduced weight and tolerance for higher pressures. One other type of specialty hose used extensively in the fire service is booster line or reel line. It is basically a big rubber garden hose made in either a ¾- or 1-in. (20- or 25-mm) diameter. The booster line typically is only used for nuisance fires, but has its place in fighting both small urban nuisance fires and in the control of small vegetation fires. Booster line has a limited role in fighting fire due to its low flow and fixed length. The full length of booster line is mounted on a hose reel and is usually no longer than 200 ft (60 m) in length. A typical maximum flow from a booster line is 60 gpm (227 L/min). *A booster line is never to be used for fire attack in a building or for a vehicle fire.*

Hose anatomy and construction

Hose anatomy and construction is another area of the fire service that has seen dramatic changes with the help of technology. Advances in materials science have led to fire hoses that far exceed the performance characteristics of hose manufactured as recently as two decades ago. It is common now to have fire hose capable of withstanding continuous pressure of up to 400 psi (2,800 kPa), even though regular fire service pumps are not designed to operate efficiently at pressures above 250 psi (1,750 kPa). This increased capability is a demonstration of just how resistant to high pressure currently manufactured modern fire hose really is.

When fire hose is constructed, the manufacturers try to build it for the task it will be performing. For example, is the hose going to be used for attack, supply, or wildland? What types of pressures will the hose encounter? Will it be exposed to unduly harsh environments inside fire buildings, or will it mainly be used outside? Is it going to be continually exposed to sunlight in open hosebeds or routinely be exposed to moisture for prolonged periods? Is the overall weight of the hose likely to be a big issue? How much kink resistance and flexibility is required? With these things in mind, manufacturers set out to construct hose based on consultations with fire departments and industry experts while meeting written industry standards and laws. The goal is to make fire hose that performs its function well with low overall weight, low friction loss, and reasonable cost, while maintaining adequate durability and service life.

Hose must be resistant to all types of mechanical damage, such as that caused by sharp objects or being regularly dragged over rough surfaces. It must be able to handle long-term storage without deteriorating in a variety of conditions in apparatus hosebeds, compartments, bundles, packs, rolls, and hose racks. Fire hose needs to absorb heat without failing or easily catching fire. Furthermore, it must be resistant to damage from long-term exposure to sunlight. On top of that, modern fire hose will routinely be put away wet, which promotes mildew and mold. Although constructed to overcome these performance demands, fire hose must maintain its ability to flow water while holding its shape and not easily kinking. Fire hose must also fold adequately and store in a space-efficient manner, lying semi- to nearly completely flat when not in use. Facing all of these challenges, the fire service hose manufacturers have stepped into a world of synthetic materials and exotic rubbers to meet these demands. Moving to improve one or some of these characteristics can negatively affect others and vice versa. It is a balancing act where the end users should have a large say based on their needs.

Typical fire hose can be broken down into **lined hose** with either one or two outer reinforcing **jackets**. The linings of the hose can be made out of rubber compounds, thermoplastics, or blends of both, as well as polyurethane or natural rubber latex.[6] This liner must be smooth, watertight, and durable while maintaining a surface that provides an acceptable friction loss based on the amount of water that the hose is designed to flow. A common liner material is an ethylene propylene diene monomer (EPDM) rubber and polyurethane. The inner liner, which is the waterway, is always protected by one or two outer jackets.

The first hose jacket serves many purposes, so it must be tough. The main purpose of the first jacket is to resist the expansion of the waterway, which enables the fire hose to retain its shape and not deform under pressure. Hose jackets or reinforcements are commonly made out of synthetic fiber, natural fiber, or a combination of both.[7] The first, critical jacket allows the liner to operate at high pressures, though usually not more than 250 psig (1,825 kPa) during normal fire service application. In the case of **single jacketed (SJ) hose**, a fire hose designed with only one jacket, the outer sheath must also provide a layer of protection from mechanical, thermal, and chemical damage. The first hose jacket is usually made from synthetic fibers in the form of polyester or polyester/nylon (combination), or in rare cases, cotton fibers. For example, wildland hose is still sometimes single-jacketed hose lined with cotton and rubber. Wildland hose also comes in SJ nylon rubber–lined hose construction. The cotton jacket performs its functions well, especially under heat, but it is susceptible to mold and mildew damage more than the SJ synthetic fiber version of wildland hose. A fire service agency must spec the type of hose they wish to purchase, because there is a wide variety of construction available for similar uses.

Double-jacketed (DJ) synthetic fiber hose is used throughout the United States and is commonly referred to as structure fire hose. DJ structure fire hose is a very widespread specification for urban fire departments for operations at building fires. DJ structure fire hose is commonly purchased in sizes ranging from 1½ to 5 in. (38 to 125 mm). The extra outer jacket is usually made of similar material as the first. It is designed to shed water and resist mechanical abrasive forces, heat, and flame. One added benefit of having this extra outer jacket, which encloses the first or single jacket and liner, is that it provides an extra layer of protection against forces on the fireground that can cause hoses to fail. Both jackets are treated to minimize mold and mildew even when put away wet. In addition, modern hose jackets are usually designed to be resistant to petrochemicals. Figure 15–24 (a and b) shows the difference between SJ and DJ fire hose.

Two other types of hose the fire service uses are drafting hose and booster hose. Drafting hose, also called hard suction hose, needs to maintain a waterway under a vacuum (i.e., negative pressure, at or below 0 psi). Suction hose was traditionally made out of a wired frame wrapped with rubber belting and coated in rubber. This made a hard semiflexible black hose that was very heavy and rigid. Drafting hose is now frequently either made out of an EPDM rubber liner with outer nylon and wire reinforcement or a durable, flexible, lightweight polyvinyl chloride (PVC) plastic (This PVC type should never be connected to a pressurized source.) Drafting hose allows pumps to acquire water from static water sources by providing a waterway in which a vacuum can be created and maintained. Once a good vacuum is attained, water from the static source simply flows up the hose under atmospheric pressure into the pump. This is the same principle that is at work when you use a straw to drink a glass of soda. The straw is basically a drafting hose made of plastic tubing. Typically, drafting hose is made in 10-ft (3-m) lengths, and engines carry two lengths of it. Fire engines carry only 20 ft (6 m) of drafting hose because the laws of physics dictate that the practical limit in the fire service for drafting vertically is around 20 ft (6 m). Drafting hose is manufactured in a variety of diameters; large pumps need large drafting hose. A 1,500-gpm (5,678 L/min) pump requires 6-in. (15 cm) diameter hard suction hose to reach capacity at a 10-ft (3-m) draft. We must strive to remember that we only ideally have 14.7 psi of atmospheric pressure to work with during drafting operations.

Even though the lion's share of all attack and supply hose is either single jacketed or double jacketed, there are other options. One type of uniquely constructed hose being offered by hose manufactures is an extruded hose made out of nitrile rubber or polyurethane with a weaved matrix providing the backbone of the hose (fig. 15–25a). In essence, it is a jacketless hose designed with a nitrile rubber or polyurethane material composing both the waterway and jacket, then the embedded, woven synthetic fiber fabric matrix providing the required strength. Extruded hose is offered both in supply and attack hose diameters. Its main advantage is that it is lightweight while still providing the majority of all required aspects of fire service hose. However, extruded hose is notorious for its tendency to kink. Finally, there is a hybrid double jacketed extruded handline structure hose that has gained popularity over the last 10 years in which the extruded hose construction is cover by a standard synthetic fiber outer jacket (fig. 12–25b).

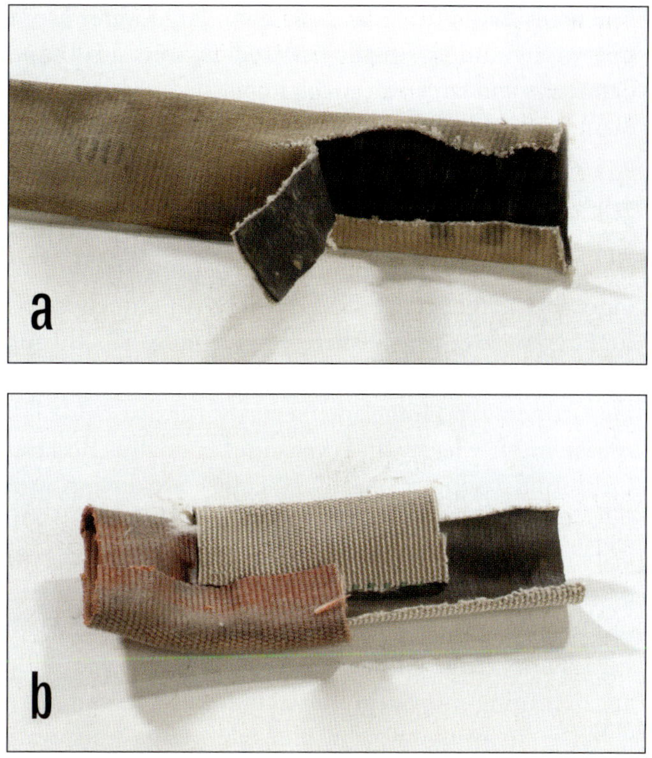

Fig. 15–24 a, b. Top: 1½-in. (38-mm) SJ cotton fiber rubber hose; Bottom: 1¾,-in. (45-mm) DJ synthetic fiber structure hose. Both have rubber liners. (Courtesy of Adam Weidenbach)

Fig. 15–25a. Extruded 1½-in. structure fire hose. (Courtesy of Adam Weidenbach)

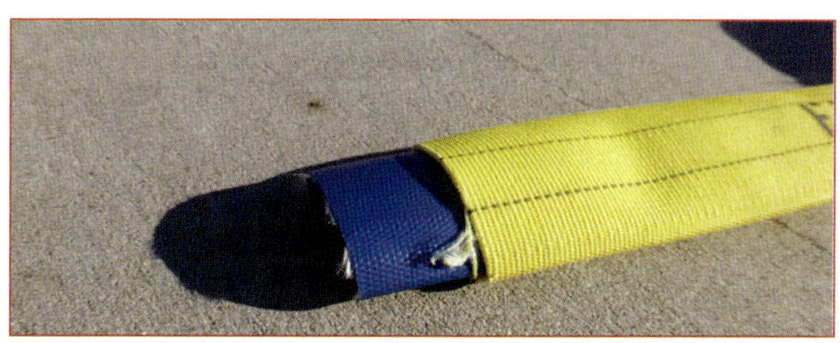

Fig. 15–25b. Hybrid Extruded DJ 1¾-in. extruded inner waterway cover by woven synthetic outer jacket.

The main use of extruded hose is for large water supply operations and is usually ordered in 4-in. (100-mm) diameters and larger, although it is also found in smaller-diameter hand lines. It is now even purchased in diameters of 6, 8, 10, and 12 in. (152, 203, 254, and 305 mm). These diameters are referred to **ultra-large-diameter hose (ULDH)**. This is an exciting new type of hose, which is providing solutions in governmental and big industrial applications. ULDH hose is uniquely suited for large-scale disasters where flows of up to 10,000 gpm (37,854 L/min) through a single hose are required. The Department of Homeland Security funded a ULDH system in New Jersey, and even cities as small as Berkeley, California (population ca. 100,000) are investing in this technology. Berkeley has purchased a ULDH system to combat large-scale water main failures that are predicted by the water district serving their community following a major earthquake on the Hayward Fault. ULDH systems with portable pumping capabilities are becoming more and more common and can act as a buffer to overcome the inevitable large-scale water problems during catastrophic events.

Size and type of hose and common flows

FFI 4.3.10 In this section, we will look at different diameters, flows, and standard lengths of various fire hose. The nice thing about fire service hydraulics is that it is based in math and physics. There is always one manufacturer or another trying to say that their hose flows the best or claiming that their 2-in. (51-mm) hose is as hydraulically efficient as a competitor's 2½-in. (65-mm) hose. Do not fall into this trap. Technological advances are usually applied to all similar types of hose at the same time, no matter who the manufacturer is.

The simple fact is that between two hose brands, if the liners are similarly smooth, the larger the diameter of the hose, the more water it will flow. If both hoses are labeled 1¾-in., one may actually be nearer to 2 in. in actual internal diameter. Much of fire service hose claiming very low friction loss is actually significantly larger than what is printed on the jacket. This difference is allowed under the current construction standard and has caused significant problems, especially in handline hose. This has been such a large problem that in 2012, a major NFPA-funded study Determination of Fire Hose Friction Loss Characteristics attempted to address the "hose diameter creep" problem because it was significantly affecting pumping operations and safety. Some manufacturers have started to produce actual true internal diameters in handline structure fire hose. From the new recruit to fire chief, this is critical to understand to ensure proper flows and hose behavior on the fireground.

Take, for example, a flow of 200 gpm (757 L/min) in a 1¾-in. (45-mm) DJ attack hand line hose. A 200-gpm (757 L/min) flow in 1¾-in. (45-mm) attack hose will require 62 psig (434 kPa) per 100 ft (30 m) of hose just to overcome the friction loss. On the other hand, that same 200-gpm (757-L/min) flow in 3½-in (90-mm) DJ medium-diameter supply hose (MDH) only loses 4 psi (28 kPa) per 100 ft (30 m) due to friction loss. This example clearly demonstrates the need for different hose sizes for different jobs in the fire service. Note that in table 15–2, which is calculated using those actual diameters, the largest attack hand line hose is 2½ in. (64 mm). This is because a 2½-in. (65-mm) diameter fire hose can reasonably move the maximum recognized hand line flow of 325 gpm (1,230 L/min). There is no need for a larger-diameter hand line hose in most circumstances. This is discussed in more detail in chapter 16, Fire Streams. The other type of hose used in supply situations is commonly referred to as **large-diameter hose (LDH)**. Look at table 15–2 to get an idea of the size and type and common flows of these types of hose, attack/hand line, MDH, and LDH. The table includes the dry weight of 100 ft of hose as well as the same 100 ft (30 m) of hose when charged with water. Notice again in the attack hose section the dramatic difference in weight between 2½-in. (65-mm) attack hose and 3-in. (75-mm) MDH. Some of the currently marketed 2.5-in. (64-mm) handline hose is commonly 2.75+ inches nearing, in some circumstances, 3-inches. This leads usually to overpumped handlines, excessive nozzle pressure, kinking issues, and very heavy (more than 100 lb per 100 ft) and difficult-to-move attack 2.5-in. (64-mm) handlines.

The fire service fights more than just structure fires. So far, the hose we have discussed is mainly used to supply and attack fires related to buildings, and thus is commonly referred to in the broadest sense as structure fire hose. Most common structure fire hose is double jacketed in design. If you look again at table 15–2, you will notice that it states that all information is based on nylon double-jacketed structure fire hose construction.

Beyond structure fires, the fire service is responsible for extinguishing all types of fires. One specialized facet of the fire service, wildland firefighting, has its own type of hose, called single-jacketed (SJ) wildland hose. Wildland SJ hose was developed strictly for wildland firefighting and is usually either 1 in. or 1½ in. (25 or 38 mm) in diameter. It is designed to be lightweight and to withstand higher pressures than DJ structure fire hose. In contrast to a

Table 15–2. Size and type of hose and common flows, based on nylon double-jacketed structure fire hose construction

Size	Typical flows, gpm (L/min)	Type	Weight per 100 ft (30 m) empty, lb (kg)	Weight per 100 ft (30 m) full, lb (kg)	Typical length, ft (m)
1½ in. (38 mm)	60–150 (227–567)	Attack/hand line	30 (13.6)	107 (48.5)	50 (15)
1¾ in. (45 mm)	95–200 (960–757)	Attack/hand line	32 (14.5)	152 (68.9)	50 (15)
2 in. (51 mm)	150–250 (568–946)	Attack/hand line	40 (18.1)	176 (79.8)	50 (15)
2½ in. (65 mm)	200–325 (757–1,230)	Attack/hand line	52 (23.6)	264 (119.7)	50 (15)
3 in. (75 mm)	0–500 (0–1,893)	MDH/supply	68 (30.8)	375 (170.1)	50 (15)
3½ in. (90 mm)	0–800 (0–3,028)	MDH/supply	78 (35.4)	493 (223.6)	50 (15)
4 in. (100 mm)	0–1,200 (0–4,542)	LDH/supply	88 (39.9)	631 (286.2)	50 or 100 (15 or 30)
4½ in. (115 mm)	0–1,500 (0–5,678)	LDH/supply	100 (45.4)	787 (357)	50 or 100 (15 or 30)
5 in. (125 mm)	0–2,000 (0–7,571)	LDH/supply	110 (19.9)	958 (434.5)	50 or 100 (15 or 30)

standard DJ structure fire hose with couplings every 50 ft (15 m), wildland hose generally comes in standard 100-ft (30-m) lengths, thus eliminating a full coupling every 100 ft (30 m), which reduces its overall weight. This same goal of weight reduction was applied when the outer jacket was removed from the specification of wildland hose. The main reason for this reduced weight is that wildland SJ hose is usually carried long distances, commonly on foot while hiking. Wildland SJ hose can also commonly be used to form very long continuous progressive hose lays where pressures can become high, based both on distance and terrain. There is even a special lightweight ¾-in. (20-mm) wildland hose. Flows during suppression of wildland fires from a single line rarely are more than 70 gpm (265 L/min) in 1½-in. (38-mm). SJ wildland hose and 30 gpm (114 L/min) in 1-in. (25-mm) SJ wildland hose. These low flows can be complicated by high pressures due to hilly terrain and portable, low-volume capacity high-pressure pumps. These are some of the reasons why there is a special hose developed specifically to combat wildland fires.

As far as booster hose and reel lines are concerned, they typically flow between 30 and 60 gpm (114 and 227 L/min). They are used in both urban and wildland environments. The booster hose is stored on motorized reel line. When the booster hose is in use, you pull out only the required amount. When finished, you can take it up quickly onto the reel. Booster hose should only be used in situations where required and expected flows will not exceed their limited capability, and *never for building or vehicle fires*. Booster line hose is made in 100-ft (30-m) lengths in both ¾-in. (20-mm) and 1-in. (25-mm) diameters. Booster hose is stored on reel lines in coupled lengths of up to 300 ft (91 m). In an urban environment, booster hose is commonly used at small outside trash fires or during overhaul. In the wildland environment, booster hose is commonly used on small grass fires: the vehicle it is attached to slowly moves while firefighters extinguish the burning grass. This type of wildland fire attack is called a mobile attack. These reel lines with booster hose are severely limited to low flows of 60 gpm (227 L/min) and less and have a fixed length usually not exceeding 200 ft (60 m). Some highly urbanized departments do not even use them; the Los Angeles Fire Department (LAFD) is good example of this. Other departments have stopped using booster lines because they are frequently misused due to their ease of deployment. One must recognize that speed is no excuse for bringing the wrong tool for the job. Hoses are tools used to move water. It is imperative that the right size of hose and flow is selected for the job at hand. There will be further instruction on selecting hose in the fire stream and fire attack chapters.

Couplings

SKILL DRILL

Couplings are a key component in making fire hose versatile. The coupling makes it possible to join two pieces of hose together or break apart two joined pieces of hose in a rapid manner. The two most common types of couplings in use today in the U.S. fire service are the threaded coupling (made up of individual male and a female couplings) with NST or a type of sexless coupling called **Storz**. Both types of couplings allow good, strong, watertight connections to be made in a rapid manner. Modern couplings can handle great pressures and stresses developed inside the hose during pumping operations.

Today, couplings are typically made out of hardened aluminum. In the past, brass was a very commonly used material. Some fire hose may still be in service with brass couplings today. However, the vast majority of all fire hose now has hardened aluminum couplings. Hardened aluminum couplings resist all of the following areas of concern regarding fire hose: corrosion, abrasive forces, compressive forces, and expansive forces. This durability makes hardened aluminum ideal for the fire service.

Hose lengths are individually marked for identification purposes. A unique number is stenciled on the hose jacket. Alternatively, a special steel numbering die is used to number a hose coupling on a length of hose when it is received new; for example, the first 2½-in. (65-mm) hose length purchased in the year 2009 is marked 2009-1, the next is designated 2009-2, and so on. Records are kept of each hose length so that the age of a hose can be tracked when it is tested (described later). In addition, in large departments the engine company that receives the new hose may have the couplings painted with a specific combination of colors or provided with another numerical designation in order to return the hose to the proper company after a large fire.

Male and female couplings are one of the most commonly used hose couplings in the fire service. They are attached usually at the factory to lengths of fire hose typically in 50- or 100-ft (15.2-m or 30-m) lengths. Hose commonly comes from the factory standard, with a female coupling on one end and a male coupling on the opposite end. The couplings in figure 15–26a are a male/female set designed for use on DJ 3-in. (75-mm) hose. One way to attach a coupling to a piece of hose is by using what is called a brass expansion ring system. On the right in figure 15–26a, the male coupling has a brass expansion ring sitting in its bowl or shank. Notice that this male coupling has **rocker lugs** extending the full length of the bowl. A lug is a raised part of a coupling that allows a **spanner wrench** to grab onto it. On the left in figure 15–26a, the female coupling has rocker lugs only on the swiveling part of the female coupling and its bowl is smooth in nature. Also notice the female coupling has a gasket inside the female threaded area to allow for watertight connections. Both of these couplings are threaded with, 2½-in. (65 mm) NHT. These couplings are designed to attach to one another with or without the use of spanner wrenches. At times, handline hose may need stretched vertically in either a well hole or up the side of the building in handline sizes 1.5-in. to 2.5-in. (38-mm to 64-mm) hose properly coupled. They can easily be safely put into service at distances of 20 stories (fig. 15–26b, c).

Fig. 15–26a. Male and female set of hardened aluminum coupling with brass expansion rings. (Courtesy of Adam Weidenbach)

Fig. 15–26b, c. Modern expansion ring couplings provide the required strength for substantial vertical stretches, easily surpassing the overhead 4:1 safety to 20 floors.

Shown in figure 15–27a is a 2½-in. (65-mm) NH female coupling designed with a bowl for 3-in. (75-mm) sitting on top of a machine called a hose expander. The removed tail bowl gasket sits on top of the machine, and the expansion ring is visible on the draw bar. This gasket is inside the coupling and is nonremovable. It makes the hose-to-coupling mating water tight. The draw bar is the part of the machine that will eventually expand the brass expansion ring inside the hose and smash the bowl gasket. To get a watertight seal, the brass expansion ring is inserted into a cut piece of 3-in. (75-mm) DJ hose, and then the hose with the expander ring in it is pushed into the bowl of the proper sized coupling. In the bowl, the expansion ring makes contact with a rubber bowl gasket located inside at the end or bottom of the bowl. The assembled hose, expansion ring, and coupling are then pushed as a unit onto the expander (fig. 15–27b). The expansion ring is then hydraulically expanded under tremendous force by the expanding machine. This particular application for 3-in. (75-mm) DJ hose with 2½-in. (65-mm) couplings requires about 1,300 psi (9,100 kPa) of expansive force. The brass expansion ring ends up sandwiching the hose and the tail gasket against the inside of the coupling's bowl, which is ribbed to help the hose remain in place when pumped under high pressures. In this case, you end up with a 2½-in. (65-mm) female coupling joined to a length of 3-in. (75-mm) DJ hose. This is a very strong, mechanically pressed attachment.

At times, handline hose may need stretched vertically in either a well hole or up the side of the building in handline sizes 1.5-in. to 2.5-in. hose properly coupled. They can easily be safely put into service at distances of 20 stories (fig 15–26b).

Fig. 15–27a. Coupling expander

Fig. 15–27b. Hose mounted with coupling expander

In figure 15–28, you can see four different hose sizes, all with male and female couplings. The top example (4½-in. [115-mm] red supply hose) has extended lugs on the female swivel. The next hose down is a 3-in. (75-mm) tan DJ MDH, which has a coupling design to allow for a 2½-in. (65-mm) threaded connection. The blue hose is 2½-in. (65-mm) DJ hand line hose, with 2½-in. (65-mm) threaded couplings. The last example is 1¾-in. (45-mm) DJ hand line hose with 1½-in. (38-mm) couplings. All the hose in figure 15–21 is of DJ construction with hardened aluminum couplings that were attached to the hose using brass expansion rings.

Fig. 15–28. Different hose sizes with male and female couplings

Coupling fire hose. We will now discuss techniques of joining two hoses together using male and female couplings. Anyone who has screwed together a nut and bolt or joined two garden hoses together can understand the basic principles of coupling and uncoupling threaded fire hose. Firefighting is a team endeavor, but there are times when one must do a certain operation alone. This applies to coupling and uncoupling lengths of fire hose. There are various methods that can be used if you are by yourself. Some of the methods are the "foot-tilt method" and the "knee press method." If there is an additional firefighter, both of them can assist in coupling or uncoupling the hose (fig. 15–29). One common method is called a "stiff-arm method." For each of these methods it is important for the firefighter to make use of a thread design incorporated in the male thread of a fire hose coupling, called a "higbee cut." This allows for an easier connection of male and female connections. The "higbee cut" can be located by finding the "**higbee indicator,**" usually located on the coupling lugs. Most times, fire hose can be coupled and uncoupled by hand-tightening the coupling. If they are too tight to be uncoupled in this manner, a spanner-wrench can be used, but only to uncouple, not to couple, as this can damage the gaskets in the female couplings.

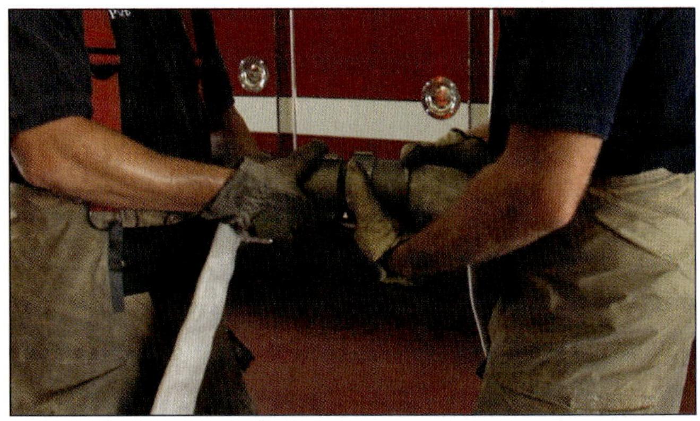

Fig. 15–29. Firefighters utilize various techniques to aid in coupling and uncoupling hose.

The other common coupling that has gained widespread use in the United States is the Storz sexless coupling. It is manufactured for use on all sizes of hose. However, the most common sizes are 4-in. and 5-in. (100 and 125 mm) Storz couplings. These unisex couplings have some advantages over a male/female design. The most obvious is the fact that one does not need any **double male** or **female adaptor fittings**, which are needed when you find yourself in a situation where you are trying to join two hoses together with the same sex threaded coupling. No matter which side of the hose you grab, it will automatically be able to attach to another Storz fitting of the same size. Storz couplings are also fast to connect, needing just a quarter turn to lock together. This can lead to problems if the hose has twisted during operation, because they have been known to uncouple themselves in older hoses without a locking mechanism. (See fig. 15–30 for nonlocking 5-in. [125-mm] Storz coupling.) Modern Storz fittings have a locking mechanism that has solved the uncoupling issue and often swivel on every piece.

Coupling and uncoupling a Storz fitting is basically the same as working with threaded hose. They also tighten to the right or clockwise and loosen to the left or counterclockwise. The same coupling techniques discussed previously—both the single-firefighter and two-firefighter coupling methods—can be applied to Storz couplings. The notable difference is that it only takes a little more than half a turn to completely couple and uncouple a Storz fitting. It is therefore necessary to ensure that when completing a coupling, both locks are secure and the lugs are lined up. This will ensure that the Storz coupling will stay together under pressure. If you work in a fire department still using nonlocking Storz, it is critical to make sure the Storz lugs are lined up and excessive twists are not in the uncharged hose. Not doing so can lead to catastrophic self-uncoupling of nonlocking Storz hose when charged.

Fig. 15–30. Nonlocking 5-in. (125-mm) Storz coupling

The 5-in. (125-mm) Storz coupling in figure 15–31 does not use an expansion ring attachment to join it to the hose. Instead of a bowl, there is a tail piece and an attached external collar or retaining ring. The hose is simply cut and pushed over the tail piece, usually with a **hose gasket** of some sort (fig. 15–32). Then the retaining ring goes over the hose and the tail piece. The retaining ring is then mechanically fastened to the hose with (in this case) four heavy-duty bolts. The retaining ring design also makes a strong watertight connection between the hose and the coupling.

Fig. 15–31. Locking Storz coupling

Fig. 15–32. It is important to check hose gaskets for wear and tear, on a regular basis.

Per NFPA 1962, *Standard for the Inspection, Care, and Use of Fire Hose, Couplings, and Nozzles and the Service Testing of Fire Hose,* couplings should be visually inspected after each use for the following problems: damaged threads, corrosion, slippage on the hose, out of round, swivel not rotating freely, missing lugs, loose external collar or internal gasket, and other defects that could impair operation.

When inspecting couplings for damage, the best tools you have are your eyes and hands. First, look for abrasion damage to the shanks and excessive wear from use. If they are excessively worn from dragging, they should be put out of service. Corrosion is not common because the hardened aluminum is very resistant to corrosive forces. Hold the coupling and pull on the hose, and there should be no movement. Any sign or feeling of movement of the hose from the shank is a failure, and the hose and coupling should be put out of service. Any collar-type coupling should have all of its collar bolts, and there should be no signs of collar slippage or looseness. Any collar problems should cause the coupling and hose to be put out of service. Any female swivel should be checked by freely spinning it. If stiff in motion, the female swivel can be washed in hot soapy water and lubricated with silicone lubricant.

One specific item that is often overlooked, but is easy to address, is the female swivel gasket in the common threaded coupling. If it looks old, worn out, cracked, nonpliable, or smashed, just pull it out and check it. The gasket simply sits in a groove, and it is easily removed by pinching the gasket with a thumb and index finger and pulling it out of the groove (fig. 15–32). If it fails inspection because it is cracked or no longer elastic, just replace it. Once again, pinch it, place it back in its groove in the female coupling, and make sure it is seated all the way. It is good practice to join the hose to another male couple after replacing a gasket to ensure proper working condition. This also guarantees that proper installation has been achieved.

It is obvious that great care must be taken to ensure that couplings work in a flawless manner. This takes discipline in inspection and care of fire hose. Without a properly coupled hose, you do not have a means of moving water efficiently, and without that, you cannot successfully use water to mount an effective fire attack. Couplings are a keystone item found on all fire hose, and they are vital to fire hose function.

Earlier in this chapter, it was mentioned that there was specialty hose called ultra-large-diameter hose, or ULDH. This type of hose can be coupled in many ways, and fig. 15–33a shows an example of a two **Victaulic couplings** joined together with a retaining ring and gasket. Victaulic couplings (mechanical couplings) are designed to be field repairable and are used on very large hose (up to 18 in. [457 mm] in diameter). Figure 15–33b shows a 12-in. (305-mm) coupling with no hose on it. This ULDH special hose is simply cut and pushed onto the tail piece. Then, a series of metal bands are tightened over the hose in the grooves located on the coupling's tail piece to make a watertight connection. This is not a fast method of coupling hose, but when using ULDH in a major emergency, the ULDH hose may be in place for months. Having couplings every 100 ft (30 m) is just not an option, given that the hose lay may be miles long. There are, of course, many other ways to couple hose used in all types of industry. In the United Kingdom, the fire service uses British Instantaneous Hose couplings. Japan has a new sexless coupling system as well. The point is that as a professional firefighter, you must know what your agency uses and also understand that there are other types of coupling systems that may be used in your mutual aid response areas.

Fig. 15–33 a, b. There are various specialty couplings that are utilized in the field, such as those used in ULDH.

Fittings, appliances, and hose tools

SKILL DRILL

Now that we have a general understanding of what fire hose is—a conduit for flowing water—a question comes to mind: How do we get all these different types of hose to work together? How do we join two different sizes of hose together or two different couplings designs together? How do we get a hose to divide into two hoses or combine into one hose? This is where hose fittings, appliances, and hose tools come into play.

First, let's take a look at **hose fittings**. They used to be made out of brass, but today they are made from the same hardened aluminum that couplings are made out of. When looking at a fitting, it is always good to take a systematic approach. One must have a common way of identifying a fitting, so that when you need a fitting, it is easy to describe. It is always proper to start with the female side when describing a fitting. However, if the fitting has a Storz side, start naming with it. Other items that need to be identified are the sizes of both sides of the fitting. After you have addressed those two main steps in naming a fitting, do not forget to identify anything else that is special about it. For example, it could have different types of thread, swivels, or extended lugs. If you do not name a thread type, it is assumed that the fitting is NHT. Take a look at the fittings and naming examples in figure 15–35. All the fittings are named from left to right. Notice that some of these fittings are still made of brass. In the past, all fittings were generically called *brass*.

All the reducing fittings in figure 15–34 do exactly what their names imply: They go from larger to smaller. Most of the time, reducing fittings are just called **reducers**.

Fig. 15–34. Reducers. (Courtesy of Adam Weidenbach)

1. 4½-in. (115-mm) female to 2½-in. (65-mm) male reducer
2. 3-in. (75-mm) female to 2½-in. (65-mm) male reducer
3. 2½-in. (65-mm) female to 1½-in. (38-mm) male reducer
4. 2½-in. (65-mm) female to 1½-in. (38-mm) male bell reducer. (Bell component keeps it from hanging up on corners or objects.)

The couplings in figure 15–35 are all increasing fittings; they simply make it possible to go from small to larger. Most of the time increasing fittings are just called **increasers**.

Fig. 15–35. Increasers. (Courtesy of Adam Weidenbach)

1. 3-in. (75-mm) female swivel to 4½-in. (115-mm) male pin lug increaser. (Notice that the swivel end and also the pin lugs were mentioned.)
2. 2½-in. (65-mm) female to 3-in. (75-mm) male increaser
3. 1½-in. (38-mm) female to 2½-in. (65-mm) female increaser

Figure 15–36 shows all double female fittings. They make it possible to join two males together. These are the most common type, as they join males of the same diameter together. Most double female fittings swivel on both sides. It is assumed they do, so if you have a specialty double female that is ridged on one side, you must state it.

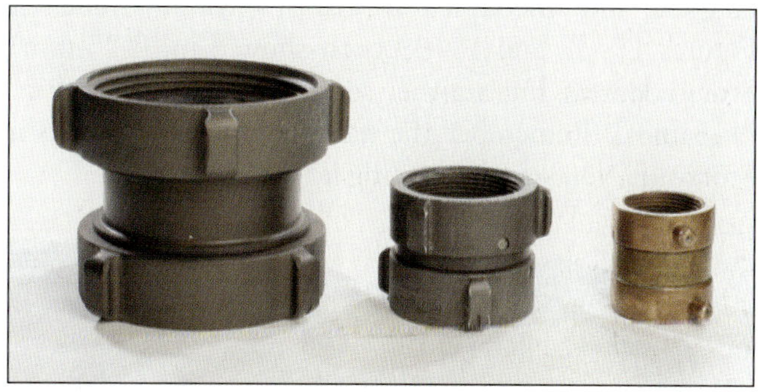

Fig. 15–36. Double female. (Courtesy of Adam Weidenbach)

1. 4½-in. (115-mm) double female (DF)
2. 2½-in. (65-mm) DF
3. 1½-in. (38-mm) DF

Figure 15–37 shows all double male fittings. They make it possible to join two females couplings together. These are the most common type, as they join females of the same diameter together.

Fig. 15–37. Double males. (Courtesy of Adam Weidenbach)

1. 4½-in. (115-mm) double male (DM)
2. 2½-in. (65-mm) DM
3. 1½-in. (38-mm) DM

Figure 15–38 shows a common type of adaptor fitting used in the fire service. This is the first adapter we have looked at that changes thread type. These types of adapters are often needed because many departments use Storz fittings on their large-diameter supply hose.

The two Storz adapters in figure 15–38 are identical and have a locking feature on the Storz side. This safety feature is now required with Storz fittings. Since Storz fittings only require about a half turn to couple, they also only take about a half turn to uncouple. This half turn feature does make coupling Storz fittings easy, but also makes them susceptible to self-uncoupling if the hose twists too violently during operation. This is why modern Storz fittings have locking mechanisms. Storz fittings with locking mechanisms should be able to be coupled and uncoupled by hand without wrenches. Figures 15–38 and 15–39 both show 5-in. (125-mm) Storz adapters, but another common Storz size is 4 in. (115 mm). Remember the naming rules: If there is a Storz side, you start by naming it first.

Fig. 15–38. Storz adapter: 5-in. (125-mm) Storz locking to 4½-in. (115-mm) female swivel, extended lug. (Courtesy of Adam Weidenbach)

Fig. 15–39. Storz adapter: 5-in. (125-mm) Storz nonlocking to 4½-in. (115-mm) male. (Courtesy of Adam Weidenbach)

The fittings in figure 15–40 are examples of fittings that change thread types. In older cities, some existing buildings predate the standardization of thread **pitch** and **count** of the standard NST. Other times, you many need a thread adapter to be able to connect to other types of common thread design used in construction. For example, you may need an adapter such as a 3-in. (75-mm) pipe thread (a common thread used to join pipes) female to a 2½-in. (65-mm) NH male. This adapter would be very useful if you needed to connect fire hose to an industrial pump that had a 3-in. (75-mm) male pipe threaded outlet. Figure 15–40 shows some examples.

Fig. 15–40. Thread adapters. (Courtesy of Adam Weidenbach)

1. 2½-in. (65-mm) Female National hose thread to ¾-in. (20-mm) male garden hose thread (GHT)
2. 1½-in. (38-mm) Female Pacific Coast thread (PAC) to 1½-in. (38-mm) male National hose
3. 1½-in. (38-mm) Female National hose thread to 1½-in. (38-mm) male Pacific Coast thread (PAC)

Notice that the fittings are clearly stamped with their thread designators. The two Pacific Coast (PAC) thread adapters are reverse of each other. Look at their male end and you can see that PAC has a finer thread than the NHT.

The fittings in figure 15–41 are elbow fittings. Elbows are sometimes called drop fittings. Their main purpose is to allow a hard bend from a ridged outlet, so that hose connected to it will not flex or kink. You commonly see them on pump panel discharges, and some departments carry them in standpipe kits to overcome poorly designed standpipe outlets. They can be very useful.

Fig. 15–41. Elbow fittings. (Courtesy of Adam Weidenbach)

1. 5-in. Storz locking to 4½-in. (115-mm) female swivel, 30° elbow
2. 2½-in. (65-mm) female swivel to 2½-in. (65-mm) male, 60° elbow

Figure 15–42 shows a simple cap and plug. A cap covers a male outlet, and a plug fills a female inlet. Most caps and plugs have some sort of chain attached to them. This is so that when the plugs or caps are removed from their corresponding male outlet or female inlet, they will be hanging nearby, so they can be reinstalled after use and are not easily lost.

Fig. 15–42. Cap and plug. (Courtesy of Adam Weidenbach)

1. 2½-in. (65-mm) cap
2. 2½-in. (65-mm) plug

Appliances. Now that we have looked at some common fittings, there is another set of equipment that is used when working with water and fire hose. This general category is called **appliances**. Simply put, an appliance in a piece of equipment that water flows through that is portable in nature. Appliances include gate valves, wyes, siameses, water thieves, water distributors, portable hydrants, and hydrant valves. Water appliances allow water flow to be manipulated in different ways. As you have probably noticed, the fire service names many of its tools based on function and appearance, and appliances follow this same basic principle. Again, we call it as we see it. When talking about an appliance, generally they are described based on the direction of water flow through the appliance. Figures 15–43 to 15–46 demonstrate this commonsense naming practice.

Figure 15–43 (a and b) shows examples of **wye** appliances, commonly just called wyes. A wye is a device in which water enters through a single female inlet and then leaves through two male outlets. For this reason, when naming them, it is just a given regarding the sexes of the one female inlet and the two male outlets. When asking for a wye, it is critical to include the sizes of inlets and outlets, and whether you need a gated wye (a wye with individual control valves for each outlet) or not.

1. 2½-in. (50-mm) straight wye. You need only to name the size since all inlets and outlets are the same; **straight** refers to the fact that there are no valves to control the flow of water.
2. 2½-in. to 1½-in. (65-mm to 38-mm) gated wye. This wye changes size as water flows through it, and it also has quarter turn ball valves. The valves allow the water to each outlet to be controlled individually. The inline handle on the left is an open outlet, and the perpendicular handle on the right is closed outlet.

Figure 15–44 (a and b) shows two appliances called **Siamese.** A Siamese is the opposite of a wye. A Siamese takes two female inlets and turns them into one male outlet. For this reason, when naming a Siamese, it is a given regarding the sexes of the two inlets and one outlets. When requesting a Siamese, it is necessary to include whether or not it is straight or **clappered.**

1. 2 ½-in. (65-mm) straight Siamese.
2. 2 ½-in. (65-mm) clappered Siamese with drain. The clappers, or sometimes a single, swinging clapper, prevents water from flowing out the other female inlet if only one of inlet is being supplied water under pressure.

Appliances are made for all sorts of special purposes. There are pressure reducers, hydrant valves, portable hydrants, ladder pipe assemblies, and ground monitors. Technically, nozzles are considered appliances, but we will not discuss them here because they are covered in great detail in the chapter 16, Fire Streams. Appliances are much like fittings. There are just too many variations to cover them all. Fire departments stock the fittings they use and need, and the same holds true for appliances. Figure 15–45 has one such example: a 5-in. (125-mm) Storz to three 3-in. (75-mm) gated wye with a built-in pressure relief valve. Many appliances that are used with LDH have a built-in pressure relief device to help reduce the chance of water hammer damage. These pressure relief devices have become mandatory in many applications regarding LDH and large flows of water. Before we move on to hose tools, look at some of the Websites maintained by fitting and appliances manufactures. You will see that the options out there are amazingly diverse, and many special appliances and fittings exist for firefighting.

Fig. 15–43a. Wye. (Courtesy of Adam Weidenbach)

Fig. 15–43b. Gated wye. (Courtesy of Chris Martin)

Fig. 15–44a. Siamese. (Courtesy of Adam Weidenbach)

Fig. 15–44b. Clapper Siamese. (Courtesy of Chris Martin)

Fig. 15–45. 5-in. (125-mm) Storz to three 3-in. (75-mm) gated wye. (Courtesy of Adam Weidenbach)

Hose tools. Hose tools aid firefighters in the use of fire hose. Some examples of **hose tools** are spanner wrenches, hose rollers, hose edge rollers, hose jackets, hose clamps, chafing blocks, and hose bridges. Figure 15–46 shows a variety of spanners. Starting at the top of fig. 15–46 is a large spanner with one end for use on couplings between 2½ and 3½ in. (65 and 90 mm) and the other end for 3½ and 5 in. (90 and 125 mm). This large, stamped, aluminum spanner is good to use with Storz fittings as well. Remember that when using a spanner it is always good to grab a set of spanners. If you have two identical spanners, then it is easy to hold both the male and female sides of the hose when you are trying to couple or uncouple the hose. Typically, you can couple hose by hand, but many times when uncoupling hose after use, you many need to use two spanners because the connection has become very tight. The second spanner down in figure 15–46 is a smaller pocket spanner. It is an old design, made to be used on pinhead lugs only. The third spanner down in figure 15–46 is another pocket spanner. Firefighters should have a small pocket spanner on hand at all times. This particular pocket spanner is the Oakland (California) Fire Department's (OFD's) version. If you look at the end, you will see that it can be used with either pin lugs or the more modern rocker lug. It also has a hydrant pentagon and a smaller striking surface. The bottom spanner is the smallest of the four. You can see a small protruding pin at the end of it. This small pin is for recessed designed couplings, which are found on booster lines. This is because booster hose is stored on a reel, and smooth couplings without raised lugs prevent the booster hose from damaging itself when reeled in on top of itself.

Fig. 15–46. Spanners. (Courtesy of Adam Weidenbach)

Hydrant spanners are included in the hose tools section. Hydrant spanners are also sometimes called hydrant wrenches. A hydrant spanner is just a wrench designed to open and close hydrant, as well as take off hydrant caps. The most standard type of hydrant spanner is the one on the top in figure 15–47. This standard type hydrant spanner, as you can see upon closer examination, is fully adjustable by screwing the handle in and out to make the pentagon opening larger or smaller, and it can also be used to couple and uncouple hose. The standard hydrant spanner is shown with a 2½-in. (65-mm) regular hydrant outlet cap. Notice the pentagon protrusion, which is usually the same size as the operating nut on the hydrant it is attached to. On the bottom in figure 15–47 is one type of specialty locking cap hydrant spanner, which is becoming more common because the illegal use and vandalism of hydrants has become common in some areas of the country. Drinking water is very valuable and open hydrants waste a lot of water. For example, a flow of 1,000 gpm (3,785 L/min) in just 1 hour equals 60,000 gallons (227,125 L) of water, enough to fill about three average residual swimming pools. Locking caps, like the one shown with the locking cap hydrant spanner, make it more difficult to illegally turn on a hydrant or gain access to the water. Notice that the hydrant operating nut pentagon is not adjustable, and also the locking cap spanner has no built-in lug grabber for use with hose; this one is truly a specialty tool. The locking hydrant outlet cap is designed to be tamper-proof and requires a special hydrant spanner to remove it. As a firefighter, you should become familiar with all types of hydrants and locking mechanisms used in your jurisdiction, including the types of failures and troubles these systems have. Access to water in a rapid fashion by using the proper tools, such as hydrant spanners, is one of the most basic tasks assigned to a firefighter at a fire.

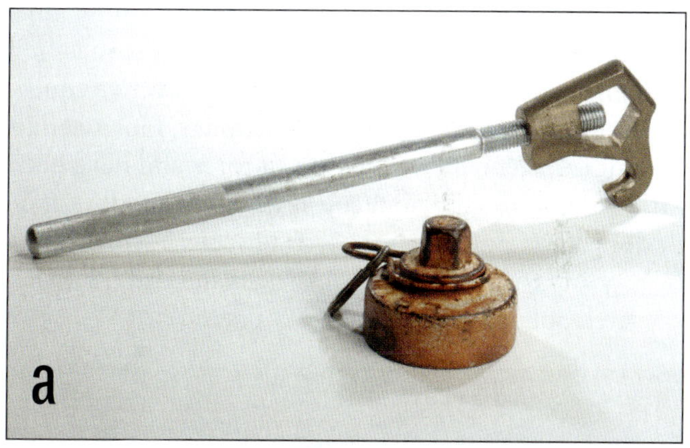

a

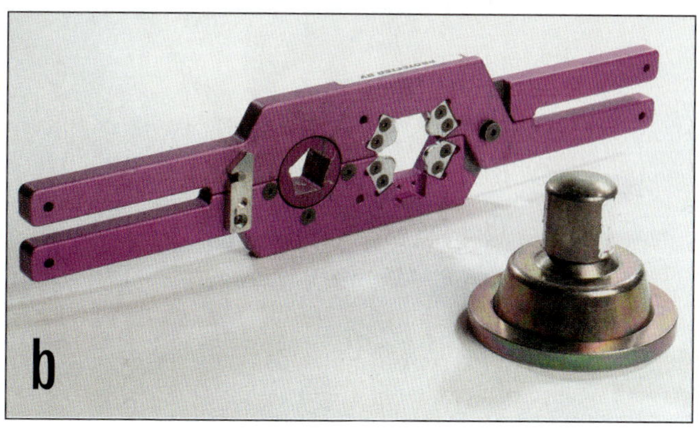

b

Fig. 15–47. (a) Standard hydrant wrench/spanner (b) Locking cap hydrant spanner. (Courtesy of Adam Weidenbach)

Hose clamps are not used today as much as they were years ago. This is mainly attributable to the more common use of LDH for supply operation. Basically, a hose clamp is just a clamp that can be screwed down in design, has lever action, or is even hydraulic. They were commonly used on fire hose before charging. This allowed water to flow into the hose until it arrived at the clamped section. Firefighters working beyond the clamped section then had time to make the necessary connections and hose lays. They would then return to the closed clamp that had stopping water from continuing down the fire hose and carefully open it to allow water to continue to its destination. Hose clamps were primarily used on 2½-in. (65-mm) (when it was still considered a supply line), 3-in. (75-mm), and 3½-in. (90-mm) supply line.

Figure 15–48 shows a common screw-down hose clamp on a length of 3-in. (75-mm) MDH. Notice that it has stopped the water from the hydrant just before the rear hosebed. This is because it was placed by an engineer who called for water before he or she broke the supply line. This maneuver allowed the hydrant firefighter to charge the hydrant and come up to do other tasks. This technique of using a hose clamp to free the hydrant firefighter from the hydrant was once common practice in fire departments that used MDH as supply line. Releasing clamps or applying hose clamps to charged hoseline must be done in full personal protective equipment (PPE) including gloves and helmets. Hose clamps are under a tremendous amount of force and have been known to break spontaneously and/or rapidly come undone. One should never partly unscrew a screw-down hose clamp that is holding back water and then just kick the latch open. This is a dangerous practice.

Fig. 15–48. Screw-down hose clamp

Figure 15–49 shows a press-down type hose clamp, which achieves its mechanical advantage from a level type action instead of a screw. Press-down hose clamps are used the same way as screw-down types; however, they are more effective at stopping flowing water because they can be more smoothly and quickly applied. The screwing action of a screw type hose clamp is a cumbersome endeavor when used against an already charged line. Using a hose clamp to stop the flow of water to replace a burst section of hose is a difficult and dangerous task, and this task is much easier to accomplish by simply closing the proper valve at the pump panel.

Fig. 15–49. Press-down hose clamp

Some departments still use hose clamp, especially departments that use 3-in. (75-mm) supply line; however, it is usually not part of their standard operating procedures (SOPs). Many times these clamps are carried on the apparatus but not regularly used. One place hose clamps are still used regularly is for a wildland hand line evolution call or a progressive hose lay. Figure 15–50 shows a wildland hose clamp designed to be used on 1-in. or 1½-in. (25 or 38 mm) SJ wildland. It has a simple clamping design and is handheld.

Fig. 15–50. Wildland hose clamp

One other technique that should be mentioned is the simple field hose clamp. This action does not require any special tools, just good technique and strength. Also, it is not effective in hose size above 2½ in. (65 mm). Basically, a simple field hose clamp is just a simple double bend in the hose that is then just pressed down on itself. This creates two kinks in the hose in close proximity to each other. The goal is not to stop the flow of water completely, but merely to slow it enough to allow a burst length to be replaced or (usually) to add an extra length of hose. The firefighters in figure 15–51 are performing the field hose clamp maneuver.

Fig. 15–51. Field hose clamp maneuver

Hose jackets are another hose tool that are not in regular use anymore. A hose jacket is a hose tool that is put around a hose that has sprung a leak. They were mainly designed for use on fire hose sizes between 1½ and 3½. in. (38 and 90 mm). While no longer commonly used, they can serve an important function if a department still has one. Some hose jackets were made of a metal clam-type clamp design with a hose orifice in the middle and a rubber seal at each end. This clam-type clamp hose jacket was simply placed over the leak and then clamped into position. Another type of hose jacket was simply made out of leather, straps, and buckles. The use of a leather-type hose jacket is self-explanatory in nature. All hose jackets mainly just slowed a flowing leak and did not function well above 150 psi (1,050 kPa). You can still order them or make them; however, they are not used frequently anywhere. Figure 15–52 shows a metal clam-type clamp hose jacket on a 3-in. (75-mm) MDH. This type of design works on 2½-in. (65-mm) to 3-in. (75-mm) fire hose, partly because of modern hose construction, in which the outer jacket of fire hose is designed in a ripstop fashion so that small leaks usually do not become bigger ones rapidly. Hose jackets also have proven to be ineffective on LDH, mainly due to the difficulty of maneuvering LDH because of its heavy charged weight.

Fig. 15–52. Metal clam-type clamp hose jacket. (Courtesy of Tim Olk)

Hose rope tools serve a very useful purpose. They help firefighters move and secure hose. Some hose rope tools are designed with a cast hook that also serves as a handle with an attached piece of webbing or rope that forms a loop. These types of hose rope tools are used by the fire department to secure fire hose to aerial ladders during ladder pipe operations and to secure hose to ground ladders, fire escapes, and window ledges. Figure 15–53 shows a traditional cast metal hook and rope hose tool, known as a **hose strap**. Notice that the metal hook also has a opening for your hand. One could just loop the rope around the hose and pass the hook through. This creates a **lark's foot** around the hose, which acts like a hand of rope. The cast hook part could then be hung on a ladder fire escape or grand ladder or just used as a handle to help manipulate the charge hoseline. This was particularly useful for large hoseline like 3-in. MDH.

Fig. 15–53. Hose strap

Hose rope tools do not have to be custom-designed, as discussed previously. Some fire departments require firefighters to carry a length of 1-in. (25-mm) tubular webbing tied into a loop with two overhand bend knots. This inexpensive hose tool usually forms at least a 4-ft (1.2-m) loop that can be used not only on fire hose but also on victims and injured firefighters as a rescue aid. This loop of webbing is very strong and can easily be wrapped around the hose and passed through itself, forming a lark's foot knot around a charged hoseline. In this fashion, it can be used as a handle to pull on a large hose or help move it around. Figure 15–54 shows just such a piece of looped webbing lark's foot around an uncharged hose. This 1-in. (25-mm) tubular webbing hose rope tool can also be lark's footed around a charged vertical hoseline below a coupling and then clove-hitched to a banister or railing to prevent the charged hoseline from slipping backward under the force of gravity. The lark's foot and clove hitch (a simple knot covered in chapter 8) are easily tied with gloves on.

Fig. 15–54. Rope hose tool with lark's foot. (Courtesy of Adam Weidenbach)

The final hose tools that will be covered are **hose bridges** and **chafing blocks**. These two items are not used enough in today's fire service. Many departments have stopped their use all together. Both of these tools are used to prevent hose damage. As has been mentioned previously, fire hose has come a long way since the time of simple, single-cotton-jacketed, rubber-lined construction. This has led to some complacency because of the durability of current specification of fire hose. The item in figure 15–55a is a chafing block. This chafing block has an obvious channel for fire hose and straps to attach it to the hose. Chafing blocks should be used especially on supply hose that bends up from street level to a pump inlet connection. This is a point where supply hose is exposed to a lot of vibrations, which cause the hose jacket to wear as it rubs back and forth against the pavement. One of the reasons supply hose is mentioned is that much of it is still of SJ design.

Fig. 15–55. (a) Chafing block and (b) hose bridge. (Courtesy of Adam Weidenbach)

Fire hose should never be driven over, whether the hose is charged or uncharged. If it is necessary to drive where there is fire hose, the hose should be protected using a hose bridge.[8] This commonsense rule is often broken. Although fire hose is very durable, it is not intended to be driven over. In figure 15–55b, the hose tool is a hose bridge used for 3-in. (75-mm) fire hose. The hose bridge in this case is two pieces of triangular shaped wood jointed by two nylon straps. To use it, you simply lift up the 3-in. (75-mm) fire hose and place the hose in the valley formed between the two pieces of wood. This hose bridge is only around 20 in. (508 mm) in length, so you need two hose bridges to make a spot where vehicles can safely drive over. All types of hose bridges are available for fire hose up to sizes of 12 in. (305 mm) in diameter. However, as fire departments have begun using more and more LDH, it has become difficult to find space for efficient storage and easy deployment of hose bridges stored on fire apparatus. The lack of hose bridges on apparatus has led to some poor practices, like running over hose that is not properly bridged.

Basic hose maintenance and testing

Cleaning. **FFI 4.5.2** Fire hose must be properly cared for. The vast majority of fire departments load hose back onto the apparatus at the scene. Before loading hose back onto a fire apparatus after use, one should wash the dirty used lengths with clean water. This can be accomplished by spraying water from a reel line or another type of charged hoseline and a stiff, long-handled broom kept on the engine. Once sufficiently clean, the hose can be loaded back onto the apparatus. Remember that if the fire hose is cotton jacketed, it should be thoroughly dried before being loading onto the apparatus, using a **hose dryer** or **hose tower**. This is done to prevent the growth of mold and mildew. In the case of cotton jacketed hose, just roll it up and thoroughly clean and dry it back at the station.

When loading hose onto a fire apparatus, make an attempt to reload the hose in a different order than the way it deployed. Attack handline lines should be reverse order loaded, nozzle length first then the next length, etc. This prevents 200-foot preconnected lines from having a very worn and overused nozzle length of hose followed by another heavily used section, then a good section to a nearly unused section by the outlet. This method of reloading handline hose is best practice. To avoid permanent folds, try to pack the hose in such a manner that the folds are in different locations than before. Fire hose that has been used must be checked for serviceability, because it could have incurred serious damage during use. One should check for obvious damage to couplings and the outer jacket. It is good practice to check for leaks in attack and supply lines after use by simply inspecting them while they are still charged. Many leaks and other damage to fire hose are found this way; take those particular lengths of hose out of service. This practice prevents damaged or leaking hose from making its way back onto the engine in service at fire scenes. Simple actions such as these can really help prolong the life of hose and also prevent premature failure of fire hose. They also help prevent unexpected fire hose failure at emergencies.

Some fire departments have to deal with extremely cold temperatures. In these cases, fire departments that have this type of severe weather problem usually take wet, used hose back to the station where they can then clean the hose properly and dry it. Sometimes hose used in freezing temperatures has to be thawed. It is best to never let hose freeze completely, especially with water in it. Hose that has frozen completely with water in it needs to be thawed and then be subjected to an in-service hose test before being placed back into service. Dealing with freezing temperatures is one of the toughest weather conditions for firefighters. Some agencies that have to contend with cold weather on regular basis have specialized equipment like mechanical hose washers and dryers. Many of these departments store many extra lengths of hose, sometimes up to entire second hose complement, back at the station.

When cleaning fire hose, it is also necessary to check the couplings for damage to the male threads or to the female swivel. Ensure that there is a pliable rubber gasket in each female coupling. Female couplings that are sticking and not swiveling freely can be dunked in warm water with a mild detergent. Once dry, apply some silicone lubricant to the swivel. Mild detergent can also be used on the fire hose if necessary, for example, if the hose was exposed to some oil. It is not necessary for fire hose to be scrubbed completely clean or beaten into a germ-free state suitable for eating off of it. It is only necessary to keep the hose free of large pieces of dirt and debris that can be easily scrubbed or washed away.

Fire hose is susceptible to gaining a **memory** at the folds. In other words, if fire hose is left for a long enough time, a heavy crease will develop where it is folded. This memory at the folded areas can lead to hose failure caused by damage to the liner. To prevent this failure, "hose shall be removed from the apparatus and reloaded so that the folds occur at different positions with sufficient frequency to prevent damage and the setting of permanent folds in the rubber lining."[9] This is called exercising the hose. All hose should be removed from the apparatus at least four times a year. One of these times will be for the annual in-service hose test. At each of these times, hose should be inspected and then loaded back on the apparatus, using care to make sure the hose is loaded in such a way that the folds end up in different locations.

Fire hose testing

SKILL DRILL

FFII 5.5.5 NFPA 1962, the *Standard for the Inspection, Care, and Use of Fire Hose, Couplings and Nozzles and Service Testing of Fire Hose,* is really the definitive resource for fire hose testing and maintenance. This standard has become more complex over the years as fire hose has become increasingly diverse. Standards like recommended pressures for service testing hose have changed a lot over the years as new hose has become more resistant to high pressure. It has gotten to the point where testing companies exist strictly to test fire hose to the NFPA in-service test standard. Some fire agencies now hire testing companies to complete the annual in-service hose test and help meet some of or all of the requirements of NFPA 1962. These testing companies have specially built hose-testing rigs that test thousands of feet of fire hose at a time. Test procedures will be discussed in the following paragraphs.

Mentioned early in the chapter was the specter of handline hose marked with the same diameter with vastly different flow characteristics. This can lead to substantial varied friction loss in a department hose fleet, and liner elongation and delamination of waterway liner (automatic failure) can affect friction loss as well. System tests shall be conducted at least annually on each preconnected line or any attack line used for interior firefighting operations on a fire apparatus, together with the nozzle or hose-connected appliance it supplies.

In addition to the previously mentioned general hose care, fire hose has to be hydrostatically tested once a year. The annual in-service hydrostatic test is defined in NFPA 1962, chapter 7, Service Testing. Before we discuss the details of the annual service test, it is important to understand that when testing hose, all fire hose is described based on its ability to withstand pressure. For example, a 5-in. (125-mm) LDH marked as "supply hose" by the manufacturer has a maximum operating pressure of 185 psi. However, similarly sized 5-in. (125-mm) LDH marked "attack hose" by the manufacturer has a 270 psi maximum operating pressure. Obviously the 5-in. (125-mm) LDH "attack hose" cannot be used as a hand line. The stenciled "attack line" on this particular 5-in. (125-mm) LDH is referring to its ability to withstand high pressures that are used to supply hand lines, standpipe, and sprinkler systems.

Figure 15–56 shows a 3-in. (75-mm) DJ 50-ft (15-m) length of fire hose. Earlier in this chapter, 3-in. (75-mm) fire hose was described with respect to its use as medium diameter hose (MDH), mainly for supply operations and sometimes as a hand line. We discussed how 3-in. (75-mm) hose is too large to function properly as a hand line and slightly too small to act efficiently as supply line based on modern fire flows. However, pump-supplied 3-in. (75-mm) hose can still move a lot of water. The 3-in. (75-mm) DJ hose shown in figure 15–56 is stenciled by the manufacturer as "Attack Hose" and also "Service Test to 400 psi PER NFPA 1962."

Fig. 15–56. 3-in. DJ Hose

Using the 3-in. (75-mm) DJ hose in figure 15–56, we will define the following terms: **service test pressure**, **proof/acceptance test pressure**, **burst pressure**, and **operating/working pressure**. All of these pressures are defined using the service test pressure as a guide. The service test pressure is required to be stenciled on all new fire hose. It is the pressure used to determine if fire hose is still suitable for use, and it is the pressure the hose is tested to during the annual service test by the user (unless derated by the end user. In figure 15–57a, the service test pressure is 400 psi (2,800 kPa). The proof/acceptance test pressure is the pressure the hose is subjected to by the manufacturer at the factory prior to shipment. The proof/acceptance test pressure is to be no less than twice the service pressure, and in this case that would mean a minimum of 800 psi (5,600 kPa). Burst pressure is tested by the manufacturer when the hose is new, and it is at least three times the specified service test pressure, which in this case is 1,200 psi (8,400 kPa). Operating/working pressure should not exceed 90% of the service test pressure, so if the service test pressure is 400 psi (2,800 kPa), then the operating/working pressure should not exceed 360 psi (2,520 kPa). This 10% safety factor over the service test is designed to ensure that hose that passes its annual hose test will not fail during use. Most fire department pumping operations will rarely exceed operating pressures of 250 psi (1,750 kPa); however, water hammer can expose hose in use to higher pressures.

All fire hose should be kept track of using a written record. This record can be kept at the firehouse or in a central location. It is beneficial to have this information in a computerized database. If this is done, things like common failures or the need to order hose can be identified easily. Hose records should include the following for each length of hose: identification number (usually stamped or inscribed on the male coupling), size, length, type of hose, in-service date, repairs, annual service tests, and service test pressure. Some organizations record more or less information about their fire hose. The details listed here are considered the minimum acceptable practice. This hose record should be updated any time a length of hose is put into or out of service.

Testing process. Annual service testing of fire hose is a requirement of NFPA 1962. This testing should always be reflected properly in the hose records. The following will just cover the basics of the service test. NFPA 1962 should be consulted for definitive step-by-step procedures and requirements. If your department does not follow the NFPA standard, they should follow the manufacturer's recommendations.

Fire hose testing involves three parts. The first two are applied to all hose consist of a visual inspection of the hose and couplings and a pressure test. The third was added in 2013 to NFPA 1962, Chapter 8, "System Testing" in section 8.1.1: "Each pre-connected line or any attack line used for interior firefighting operations on a fire apparatus, together with the nozzle or hose connected appliance it supplies, shall be flow tested as a system at least annually." All three are critical parts of hose testing. Visual inspection of the hose involves inspecting the couplings, gaskets, and the interior rubber lining of the hose. Some damaged parts, such as a gasket, may require replacement. Others, such as significant damage to the inner lining, known as delamination or couplings, will require that the hose be condemned.

If the hose passes the visual inspection, it needs to be identified. Hose will have stenciling on it indicating the manufacturer, type, and size of the hose, as well as service test pressures. Testing of the hose should be in an open area where the hose can be laid out. It is recommended that all personnel wear a minimum of a helmet and gloves during the testing process. The hose can be tested by using a **hose testing machine** (fig. 15–57a). If a pumper is being used, the operator should be fully qualified and certified on that particular pumper.

To conduct the test, lay the hose out straight, with no bends, and coupled together up to 300 feet (90 m). Place a mark with a pencil behind each coupling to determine if slippage occurs during the test. Cap the last male coupling with a cap and bleeder or a fire service nozzle (if using a nozzle make sure it is rated to the test pressure), and secure the hose to prevent it from whipping during the test. Bleed the air out of the hose while slowly filling it with water. When all the air is out of the hose, close the nozzle or bleeder and begin the test. First check for leaks, and then bring the pressure up to the recommended service test pressure and hold for 5 minutes while observing the hose visually from a distance of 15 ft (5 m). If no problems are observed, reduce the pressure in the hose and open the nozzle or bleeder valve to release the pressure, and record the results. If a problem occurs such as slippage, burst length, etc., remove the length from service, mark it as condemned, and record the results.

Hose system testing has become critical for fire departments to ensure that target flow and proper nozzle pressure is achieved using the department's standard pump operating procedures and handline hose fleet. Earlier in the chapter, the specter of handline hose manufactured and marked the same diameter with vastly different flow characteristics was identified; this can lead to substantial varied friction loss in a department hose fleet. In addition, problems such as liner elongation and delamination of waterway liners (an automatic failure) have led to substantial low flow conditions. The opposite is very possible as well; a very hot overpressured nozzle (very high nozzle reaction) at low pump discharge pressure in oversized hose is marked much smaller than its actual internal diameter. This is currently occurring on fire grounds across America, and this is why NFPA 1962 was modified. In a delaminated hose fleet where problems are masked by automatic nozzles, fire departments can barely achieve half their desired target handline flow. Figure 15–57b is an example of delamination failure. Also, a significant issue is liner elongation (fig. 15–57c) where the inner assembly of a through-the-weave type design hose is covered by a woven jacket that becomes dimensionally incongruent with each other through use and the extra liner material bunch at the male end of the hose. Both of these conditions can affect friction loss enough to substantially reduce target flow in handlines. Finally, figure 15–57d shows just how much a hose fleet's friction loss can vary and the substantial impact on pump discharge pressure to achieve proper nozzle pressure.

Proper service testing is the only way we can determine the suitability of hose for fire-suppression use. As of now, there is no recommendation for the maximum life span of hose. Committees of industry leaders and manufacturers have come up with a guideline of 10 years. It is probably more reasonable to state that hose can become unserviceable at any time based on use. Heavily used fire hose can fail an in-service hose test after just a few years. Rarely used hose may pass a service hose test 30 years after its date of manufacture. Generally, hose that passes its service-pressure and system hose test is suitable for use and should be retested every year to confirm this. The maximum life span issue will be more definitively addressed in the near future, but most likely it will be closer to 15–20 years. Hose has a definite cost factor associated with it, and it is very durable. It should not be discarded strictly based on age if in good serviceable condition.

Fig. 15–57a. Proper hose testing is important to ensure proper working equipment on the fireground.

Fig. 15–57b. Delamination failure of liner

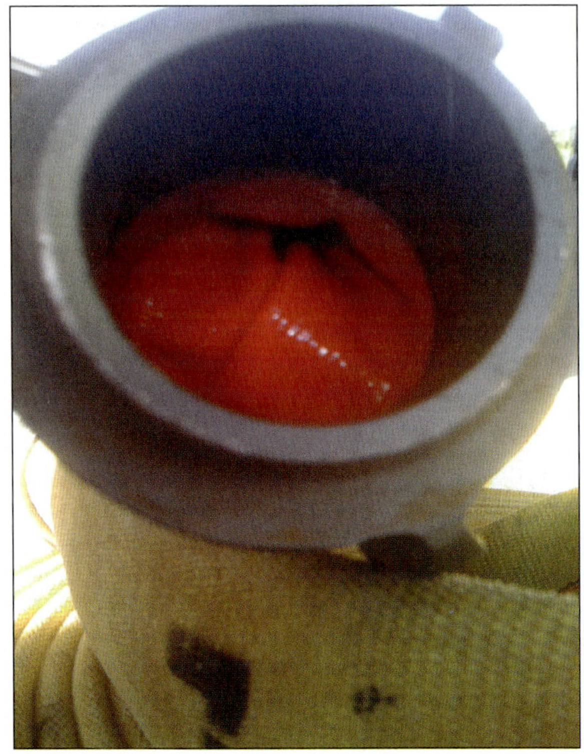

Fig. 15–57c. Elongation condition of liner

LAWRENCE, KS — 1.75" HOSE TESTING — OCTOBER 2014

PSI LOSS IN 1.75" HOSE										
HOSE MANUFACTURER	YEAR	MARKED DIAMETER	HOSE LENGTH	TIP SIZE	OUTLET PSI GUAGE	100' PSI GUAGE	TIP PSI	TOTAL PSI LOSS	LOSS PER 50'	TOTAL COUNT (285/46)
Key	2000	1.75"	200'	15/16"	150	105	50 PSI	100	25	97
Key	2002	1.75"	200'	15/16"	132	95	50 PSI	82	20.5	32
Key	2005	1.75"	Not Tested							3
Key	2010	1.75"	200'	15/16"	138	80	50 PSI	88	22	19
Key Eco10	2013	1.75"	200'	15/16"	135	98	50 PSI	85	21.25	7
Angus	2005	1.75"	200'	15/16"	235	138	50 PSI	185	46.25 (See Note)	17
Angus	2006	1.75"	200'	15/16"	200	124	50 PSI	150	37.5/ 32.5 (See Note)	41
Mueller	2005	1.75"	200'	15/16"	150	110	50 PSI	100	25	24
National	2003	1.75"	200'	15/16"	120	88	50 PSI	70	17.5	19
Firequip 800 (Attack Lite Poly)	2009	1.75"	200'	15/16"	130	92	50 PSI	80	20	18
Mercedes MTSS 800	2012	1.75"	150'	15/16"	110	98	50 PSI	60	15	3
(HR) Mercedes MTSS 800	2011/2012	1.75"	150'	15/16"	125	96	50 PSI	75	18.75	4
(HR) Firequip	2008	1.75"	150'	15/16"	155	104	50 PSI	105	26.25	8
(HR) National	2003	1.75"	150'	15/16"	105	84	50 PSI	55	13.75	30

Note:

- Further testing of different Angus 2006 (FH #: 06129, 06144, 06145, 06153) was done on October 17, 2014 to ensure accuracy due to the amount of Friction loss originally found. 32.5 psi per 50' section was found (180 psi @ Outlet, 120 psi @ 100', 50 psi @ tip). These higher friction loss numbers impact our pump charts significantly.

Fig. 15–57d. PSI loss in 1.75-in. hose. (Courtesy of Nate Jamison)

Hose storage

SKILL DRILL

FFI 4.5.2 Fire hose is typically stored in one of two places, either on the apparatus or in the firehouse. We are first going to address hose stored in the firehouse. Typically, hose stored in a firehouse should be in-service hose. Hose that is out of service or marked for repair should never be stored anywhere near or next to in service hose. Hoses stored in a firehouse are typically rolled in a straight roll. Figure 15–58 is a 2½-in. (65-mm) DJ attack hose in a 50-ft (15 m) length. It is rolled in-service with male coupling at the center of the roll, which protects the NH threads.

Hose that is out of service should be clearly tagged "out of service" and then rolled into a straight roll with the male coupling out. If the firefighter placing the hose out of service knows that the hose is not repairable, both couplings should be cut off the fire hose before discarding it. This ensures that bad hose will not be used in error. In some circumstances, the removed couplings can be reused. Figure 15–59 is an out-of-service straight roll with the male coupling exposed, that is, missing a clear and complete repair tag. You will notice that just by looking at this 3-in. (75-mm) DJ hose you cannot tell what is wrong with it. The repair tag is very important!

Fig. 15–58. In-service straight roll, with male-threaded coupling protected. (Courtesy of Adam Weidenbach)

Fig. 15–59. Out-of-service straight roll, with male coupling exposed, but with no repair tag. (Courtesy of Adam Weidenbach)

Figure 15–60 shows a 3-in. (75-mm) DJ hose rolled out of service with the male coupling removed and clearly labeled with an attached repair tag. Originally, it was a 50-ft (15-m) length of 3-in. (75-mm) hose. The damaged section has been removed, and it is awaiting a new male coupling. The repaired 3-in. (75-mm) DJ will be a 25-ft (8-m) jumper and will need to be service tested after the coupling is installed. This will ensure that it is ready to go back into service as a 25-ft (8-m) jumper, sometimes called a **pony length** (often used to hook up to a hydrant close to the pumper to fill the water tank).

Fig. 15–60. Out-of-service hose, male coupling removed, with properly filled-out and attached repair tag. (Courtesy of Adam Weidenbach)

Finally, fig. 15–61 shows a returned out-of-service hose. This set of couplings is from two lengths of 50-ft (15 m) DJ 1¾-in. (45-mm) attack line. The outer jacket of this hose was severely damaged at a structure fire; however, the hose did not fail. The first jacket and liner remained intact, and the engine company was able to successfully complete its fire attack. You can now see how valuable an extra outer jacket can be. This is an example where both couplings should be cut from the hose, and the hose discarded. The couplings should be inspected thoroughly and may be reused if in serviceable condition.

Figure 15–62 shows an example of clean, serviceable hose stored on a hose rack. Notice that all lengths are in-service, straight-rolled with the male coupling protected. In-service hose should be clean and dry, and it should have passed an annual in-service hose test. Hose stored in this manner should be good for a year. It is important not to hang hose in hose towers for longer than it takes to dry the hose. Excessive hanging of fire hose can lead to extreme drying of the hose liner, which can cause liner failure. Hose is best stored as shown in figure 15–62, where it is kept in a climate-controlled area of a building. Remember, hose that sits can be damaged by passing equipment or environmental degradation, and even attacked by rodents. When a year of proper storage has passed, stored fire hose should again be subjected to an annual in-service hose test.

Fig. 15–61. Severely damaged 1¾-in. (45-mm) DJ hose. Cut the couplings off and discard hose. Do not roll. (Courtesy of Adam Weidenbach)

Fig. 15–62. In-service hose properly stored on a hose rack. (Courtesy of Adam Weidenbach)

Hose loads

SKILL DRILL

FFI 4.5.2 Hose is stored on apparatus in many ways. The most common ways are in rear hosebeds, crosslays (also known as a "Mattydale load," which was invented by the Mattydale, New York, Fire Department), reels, and compartments. Hose can be stored in different types of tactical manners, depending on how it is going to be deployed. This is one area of the fire service where there are many differing opinions. Depending on the type of district a company responds to, it may alter the design and layout of how hose is carried on apparatus. These factors include available water supply, types of buildings, types of hose used, and common overall lengths of hose stretches, to name a few. It is good for fire departments to have a standard hose complement and hose load developed for their particular district or city. This standard load should be the right mix of hose sizes and loads for their particular needs.

Fire hose is mainly carried in one of two ways on a fire apparatus: either preconnected to a discharge or suction inlet or not attached to any discharge or pump inlet. First, we will take a look at attack lines. **Preconnected** attack lines, sometimes called live lines, are generally attack hand lines not usually exceeding 300 ft (90 m) in length. They are usually made with either 1½-, 1¾-, 2-, or 2½-in. (38, 45, 50, or 65 mm) hose with the last female coupling attached to a pump discharge, and in turn, the last male coupling having a nozzle attached to it. The preconnected attack lines in figure 15–63 are called preconnected crosslays (hose stored in a "trough" that runs from one side of the apparatus to the other). They are both made with 1¾-in. (45-mm) DJ hose. The front (left) preconnect crosslay is 150 ft (45 m), or three lengths of hose, and the rear crosslay (right) is 200 ft (6 m), or four lengths of hose. Preconnected hose does not have to be mounted in a crosslay bin. Many fire departments have preconnected attacks lines that lay off the back of the rig as opposed to a crosslay. Once deployed, a preconnected attack line must be pulled off and completely flaked out. After the deployment of all the preconnected hose, the pump operator can then open the discharge outlet that the preconnected deployed hose is already attached to. This is only done after the call for water is made by the members manning the pulled preconnected attack line.

The preconnected attack lines in figure 15–63 are loaded in the simplest manner. Both are loaded in a simple flat-load style. The flat load is the gold standard of loading hose in today's fire service. It is simple, reliable, reasonably deployable, predictable, and easily duplicated and reloaded, even when handled by extremely fatigued members. It requires no special setup or folds; simply connect the female couplings to the proper discharge in the crosslay and then load the correct number of lengths of hose one after another. The final step is to attach a nozzle. The company in figure 15–63 also included some **pull loops** when loading their 1¾-in. (45-mm) flat-loaded preconnects. The loops, in this case, are on both sides and appear to be roughly at every length of hose. The loops are made to help facilitate the deployment of hose as additional members join in the stretching of the selected preconnected line. Remember that, with preconnected or live lines, all of the hose must clear its bed before charging, because premature charging of a preconnected line without complete clearing of its bed leads to catastrophic tangling and delays in proper deployment.

There are other ways of loading preconnected lines. Two very popular methods of loading preconnected lines go by the names **minuteman load** and the **triple fold**. Both of these techniques of loading preconnects come with pros and cons. The first alternative load we will discuss is the minuteman load. Basically, this is a hose load designed for preconnected operation no longer than 200 ft (60 m) and usually using attack hose with a maximum diameter of 1¾-in. (45-mm). The main reason for these limitations is that all of the hose, when initially deployed, is pulled onto a single firefighter's shoulder. Overall lengths exceeding 200 ft (60 m) and hose sizes larger than 1¾ in. (45 mm) invite common failures related to tangles and loss of stability and control during deployment. It is difficult to properly deploy this much hose from a single bundle on a shoulder without practice. With hose larger than 1¾ in. (44 mm), this can cause significant kinking and deployment issues.

Fig. 15–63. Preconnected attack crosslays. (Courtesy of Daryl Liggins)

The minuteman load can be a time and labor saver; however, it is best served with 1½-in (38-mm) (1¾-in. [44 mm] is very common) attack hose, preferably in a maximum length of 150 ft on the shoulder the rest loaded should be dump hose and flat loaded. (45 m). The minuteman load, because of its design, is best suited for rear-facing preconnects in the hosebed. This is because the nozzle sits at the bottom of the load in the direction it must be pulled. Some departments that load their crosslays with minuteman loads have one load facing toward the officer side of the engine and the other facing toward the engineer side.

Minuteman load. A minutemen load is a method of deploying fire hose and a nozzle with minimal effort using minimum manpower. To load the minuteman on a fire apparatus for a 150-ft (45-m) stretch of hose you connect a 50-ft (15-m) length to a pump discharge and make one pass and let the rest sit to the side for later. Connect the other 50-ft (15-m) lengths together and place in the bed with a nozzle on the open male end at the bottom and the rest of the 100 ft (30 m) of hose on top. Then connect the first 50-ft (15-m) length to the 100 ft (30 m) already in the bed and you have a finished minuteman load.

To deploy the minuteman load, grab the entire bundle of hose onto your shoulder with the nozzle on the bottom. Walk away from the pumper and allow the hose to deploy off the load from the top. When you get to your destination you will have all the hose smoothly deployed and the nozzle in your hand ready to be used (fig. 15–64 a and b).

Fig. 15–64. (a) The minuteman load and (b) deploying the minuteman load

Triple fold. Another method of storing and deploying hose for fire attack is the triple fold method. The triple fold method has the hose folded in three layers and then laid in the hose bed in an S-shaped fashion with one layer on top of the other so the nozzle is placed on top. It is designed so one person can easily remove the hose pack. For this hose lay to be used, all of the hose must be removed from the hose bed before deploying the nozzle. It is typically used for 150- and 200-ft (45- and 60-m) hose stretches (fig. 15–65).

Fig. 15–65. The triple fold load

Fig. 15–66. Static attack crosslays. (Courtesy of Daryl Liggins)

Attack hose can also be carried at the rear of the apparatus in the hosebed area, along with the supply hose. In the hosebed, attack hose can either be preconnected to a rear discharge and loaded or be loaded in a static fashion. Figure 15–67 shows two pumpers with two different static attack loads. Figure 15–67a shows a single 2½-in. (65-mm) horseshoe finish, and fig. 15–67b shows both a 2½-in. (65-mm) and a 1¾-in. (45-mm) attack line. Notice that the 1¾-in. (45-mm) attack hose line is filled out with 2½-in. (65-mm) hose on the bottom. This horseshoe finish is sometimes called a reverse horseshoe finish. It is usually just the last length of hose with the nozzle attach fed back upon itself in a U-shaped (thus horseshoe) fashion. This makes a nice 50-ft (15-m) pack of hose to grab when deploying the hose from a static bed. In that case, make sure the nozzle team has at least a 50-ft (15-m) length of working line. Notice also that the hosebed in 15–67b has vertical divider separating the beds; these are called hosebed dividers, and they are adjustable. There are other options regarding finishing static hosebeds to aid in the deployment of hose depending on fire department preferences (fig. 15-68).

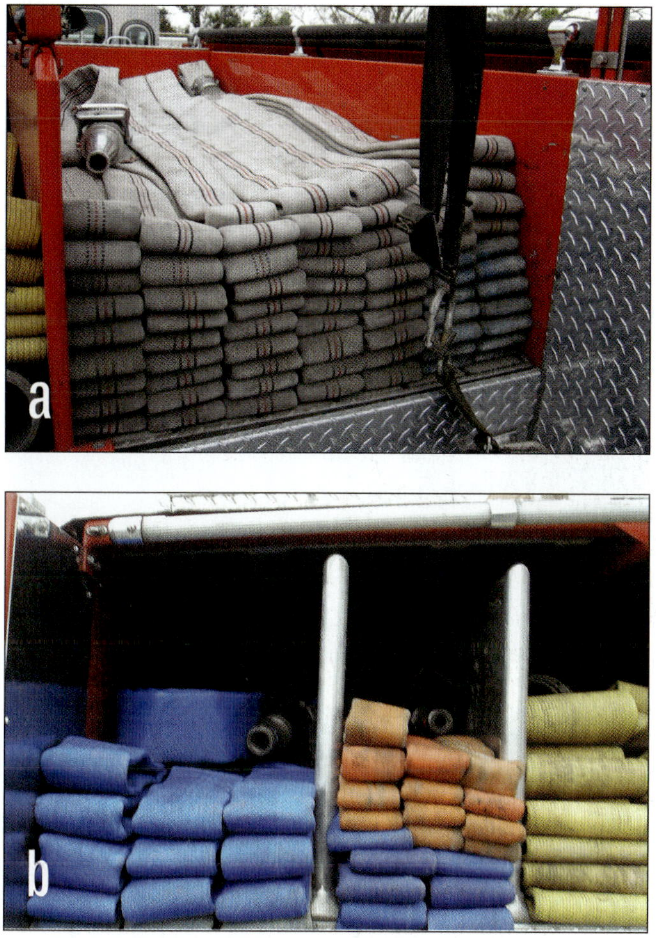

Fig. 15–67 a, b. Static attack lines loaded in the hosebed. (Photos courtesy of Daryl Liggins)

Fig. 15–68. Reverse horseshoe finish

Hose bundles

SKILL DRILL

FFI 4.5.2 We have discussed attack hose loaded on the apparatus in crosslay and at the rear of the apparatus in the hosebed. In both places you can find the attack hose loaded in both a static or preconnected fashion. One other way attack hose is often carried on apparatus is in **bundles** or specialized rolls. Bundles and specialized rolls of hose are commonly kept in compartments or seat-belted onto running boards and fenders. Once again, there are many types of bundles and specialized rolls used by the fire service. Many departments choose the one they feel best suits their operations. Let's take a look at the **horseshoe bundle**. A horseshoe bundle is easy to fold, does not get tangled, and deploys in an efficient fashion. Figure 15–69 (a, b, and c) shows a series of pictures:: a completed 2½-in. (65-mm), 50-ft (15-m) horseshoe pack with a nozzle on it; a firefighter rebuilding a 2½-in. (65-mm), 50 ft (15-m) horseshoe pack using a self-contained breathing apparatus (SCBA) bottle as a guide; and two 50-ft (15-m) horseshoe packs of 1¾-in. (45-mm) joined together at the coupling and then strapped together.

Bundles of hose are necessary for many reasons. For example, if you enter a building to use a standpipe (bundles in this case are called **standpipe packs**), or walk up a hill during a wildland fire, you have an easy way to carry the hose. The bundle should be folded in such a way that it is easy to deploy and couple the hose. Figure 15–70 clearly shows the value of a hose bundle. All of these firefighters are able to climb the stairs of a building while having both hands free. Also note their front and side profiles do not increase much. This will allow them to pass by civilians or others who are attempting to exit the building easily.

In addition, the hose bundle, some specialized hose rolls need to be discussed, including the donut roll, twin donut roll, and self-locking twin donut roll. Hose rolls are used as a means of storing hose on the apparatus or in a hose rack in the fire station. There are several variations of hose rolls: the straight roll, the simple donut, and the twin or double donut. The straight roll is designed for storing hose at the fire station on a hose rack. It is made by laying the length of hose straight out and, starting with the male coupling, rolling the hose so that the male coupling is in the center and female coupling is on the outside.

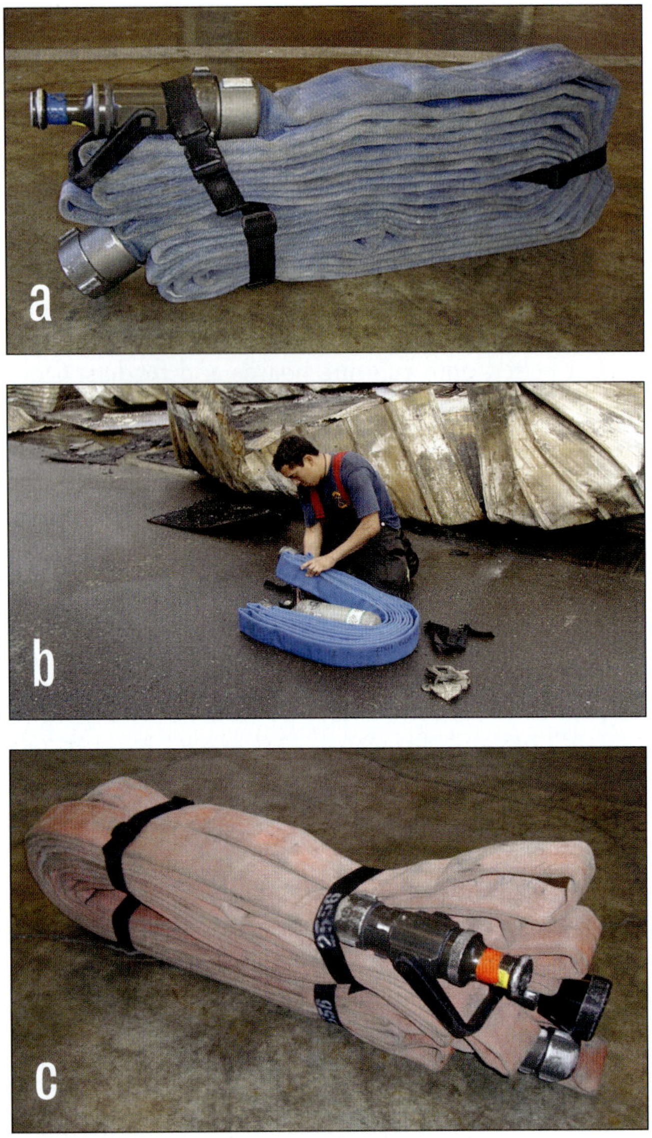

Fig. 15–69 a, b, c. Horseshoe hose bundles of 50 ft (15 m) and 100 ft (30 m). (Courtesy of Daryl Liggins)

Fig. 15–70. Firefighters ready to enter a building with 2½-in. (65-mm) hose in bundles. (Courtesy of Daryl Liggins)

Donut roll. The simple **donut** is made by folding the hose length onto itself so that the male end of the hose is about 3 feet (1 m) in from the female end. The hose is then rolled on to itself so, when completed, the male coupling is inside the roll and protected from damage (fig. 15–71).

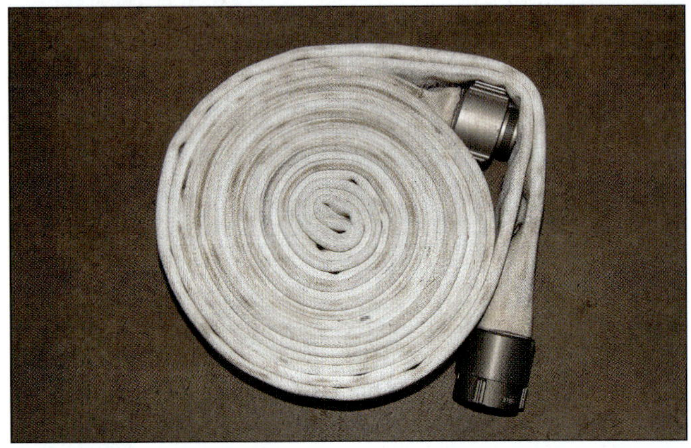

Fig. 15–71. The donut roll

To deploy this roll, you simply grasp the two couplings and roll the hose out from where you are standing or in congested areas hold on to the two couplings and drag the hose with you while the roll deploys behind you.

The donut roll is a staple in the wildland firefighting environment. During progressive hose lays, wildland hose is easily deployed from the donut roll. It is also very useful in supply line pony sections for the pump operator to use to make or finish supply connections at the engine/pumper. It's commonly kept in compartments on the pump panel side of the apparatus.

Twin donut roll. Another popular donut roll is the **twin donut roll** (fig. 15–72). The twin or double donut roll allows for a compact roll that can be stored in a smaller compartment on an apparatus or can be used for applications such as stand-pipe packs or high-rise packs. It is made by laying the length of hose out so that the two couplings are side-by-side. The hose is then rolled so that, when it is complete, both coupling are on the outside and easily accessible. This roll can be held together by using a hose strap or by a variation known as a self-locking twin donut roll. To deploy it, simply remove the hose strap that is locking it, and, by grasping the couplings, roll the hose away from you so that, when deployed, you have both couplings in you hands

The self-locking twin donut roll. This roll is particularly useful because it forms a carrying handle without the aid of hose strap or other securing devices such as hose packs (fig. 15–73).

Fig. 15–72. Twin or double donut roll

Fig. 15–73. Self-locking twin donut roll

There are additional hose bundles and hose rolls that will not be discussed here. There are several types, and they all have their pros and cons. Your department will only use a few of them. Please pay close attention when your instructors go over the bundles and hose rolls because you will be responsible for how to use them properly.

Reel line. The last type of attack hose carried on an apparatus is a form of preconnected attack hose called **booster/reel line**. Earlier in this chapter, we detailed the many downfalls of using a reel line as an attack line. It has a very limited flow and should only be used in situations such as very small outside fires. Most reel lines, even if 1 in. (25 mm) in diameter, do not flow more than 60 gpm (227 L/min). This is a very small flow and is completely inadequate for any interior building operations. As stated previously, some departments no longer spec hose reels for their apparatus. Reel lines can provide a good utility function. A reel line, like all tools, is useful only if used properly. Using a 60-gpm (227 L/min) reel line on an inappropriate fire is not the fault of the reel line, but of the person who selected it.

Water supply evolutions

Supply line is fire hose used to supply water for fireground operations. Supply line is typically deployed in hose-laying operations where the engine is in motion, and the supply hose pays out of the hosebed as the engine is driven. These supply hose-laying operations are typically made either from a hydrant toward the fire, called a **forward lay**, or from the fire to a hydrant, called a **reverse lay**. In this section we will cover forward lays, reverse lays, and also **split lays**, as well as the basic use of hydrants. These are the most common water supply evolutions using supply hose.

Supply line is almost always loaded statically in the hosebed. An exception to this rule is short, preconnected intake jumpers, usually no longer than 35 ft (10.7 m) in length. These short lengths of supply hose are commonly preconnected to the pump's intakes to assist firefighters in making rapid connection to hydrants or already laid or deployed supply lines. LDH supply jumpers are made in various lengths: 15, 20, 25, 35, and 50 ft (4.5, 6, 7.6, 10.7, and 15 m). Supply line in the hosebed may be loaded with the female coupling out, male coupling out, or Storz couplings. Depending on which sex coupling or type of adapter with Storz is the first out of the hosebed, one can determine what tactic a department usually uses during its water supply evolutions. You have probably noticed by now that most of the figures in this book show hose loaded in a flat style. In figure 15–74 (a, b, and c) an engine company loads 4-in. (100-mm) DJ LDH in a flat load, taking care to place the large 4½-in. (115-mm) couplings at the front of the hosebed. Placing the large couplings in carefully is a must, because they should not be allowed to flip during a hose lay operation. When loading supply hose, it is sometimes necessary to take a short fold of hose, not only to ensure that the coupling does not flip over while paying out, but also to allow placement of the coupling in a desirable location in the hosebed. This short fold of hose is commonly called a **Dutchman**. The flat load has become the standard practice for loading hose. In fact, "excessive edge wear can occur when 100% synthetic yarn-reinforced hose is loaded on the apparatus in the conventional manner (horseshoe U-load, accordion, or skid loads). To prevent this edge wear, hose manufactures recommend that if 100% synthetic yarn-reinforced hose is used, it should be loaded on the apparatus in the flat load manner."[11]

Fig. 15–74 a, b, c. Firefighters load 800 ft (244 m) of 4-in. (100-mm) hose in a flat load. This 4-in. (100-mm) NST LDH is set up for a forward lay with the female coupling out.

Some departments may still use a **horseshoe (U-load)** or **accordion load**, both of which require the fire hose to be loaded on its edge. Let's take a brief look at a horseshoe load, as shown in figure 15–75. You can see that the hose is loaded in a U-shaped horseshoe fashion. The original main advantages of this horseshoe load are that it promotes good air flow for cotton jacket hose, minimizes severe bends in the hose, and is very difficult to hang up on laying line. The negatives of the horseshoe load are that all the lengths, if hand-jacked or pulled, are of unequal sections due to the nature of its outside in loading technique, and it also promotes excessive edge wear of the supply hose.

Fig. 15–75. The horseshoe load

The accordion load promotes good air flow for cotton jacket hose, although it can be easily packed too tightly and might hang up upon deployment. The accordion load provides easy shoulder-carrying capabilities with each layer being of the same back-and-forth distance. Firefighters can merely walk up to an accordion load, grab a few folds, and know exactly how much hose they are pulling and then easily place it on their shoulders. Figure 15–76 shows an example of an accordion load. Horseshoe and accordion loads are usually made up of 3-in. (75-mm) MDH, which can be set up in forward or reverse fashion and also in dual beds. Another reason that both the horseshoe and accordion loads have been generally phased out is that they do not work well with any type of LDH supply hose. In addition, either the accordion or horseshoe load's bottommost layer of hose is in direct contact with the metal hosebed bottom on edge. This lower layer of hose on edge is exposed to excessive wear and is not recommended, because with modern synthetic hose, it has been known to cause premature failure.

Fig. 15–76. The accordion load

However, finishing the last length or two of supply hose in a horseshoe or other edge-laid finish to aid in hose evolutions is an acceptable practice, because there is no added weight of hose above the finish, and the finished, on-edge hose is on top of the other hose and not the metal hosebed bottom. Many departments practice this tactic to aid their lead-off or hydrant person in gathering enough hose to complete a water supply evolution.

Supply line is usually no smaller than 3 in. (75 mm) MDH because of the friction loss encountered when moving large volumes of water. The NFPA states that the minimum size supply line is 3½ in. (90 mm) However, many departments still use 3-in. (75-mm) hose. With the trend toward increased hand line flows due to harsher fire conditions caused by modern fire loads, 3-in. (75-mm) MDH is beginning to be too small to appropriately act as supply hose. Some departments use dual beds of 3-in. (75-mm) MDH, which does overcome single 3-in. (75-mm) MDH supply line high friction loss when moving a large amount of water. This is made possible by laying two supply lines from a single engine; hence the name **dual bed**. For example, the friction loss for a 1,000 gpm (3,785 L/min) in a single 3-in. (75-mm) hose lay per 100 ft (30 m) is 80 psi (560 kPa). However, if two 3-in. (75-mm) MDH supply lines are laid side by side and share the 1,000 gpm (3,785 L/min) flow by each flowing just 500 gpm (1,892 L/min), the friction loss is only 20 psi (140 kPa) per 100 ft (30 m). Laying dual 3-in. (75-mm) supply lines is the hydraulic equivalent to a single 4-in. (100-mm) LDH supply line.

Modern fire departments that continue to use a 3-in. (75-mm) MDH supply line are usually heavily urbanized cities with dual 3-in. (75-mm) MDH supply beds that also have a strong water supply infrastructure that supports its use. This infrastructure includes close hydrant spacing of no more than 300 or 500 ft (91 or 152 m) and good water pressure, as well as using the tactic of placing their pumps at hydrants to provided pumped 3-in. (75-mm) MDH supply. These characteristics and tactics allow for continued use of 3-in. (75-mm) supply line in some cities. Supply line is like any other tool in the fire service: there is a vast array of choices and selection is based on many variables, including available water supply infrastructure and tactics.

Forward lay. A forward lay is just what it sounds like. An engine spots a hydrant near the fire and then lays a supply line towards the fire building. Most fire departments that use a forward lay water supply tactic use a single bed of supply hose, usually in any of the following sizes of LDH: 3½, 4, 4½, and 5 in. (90, 100, 115, and 125 mm). Figure 15–77a shows an engine performing a forward lay evolution by laying a single supply line from a hydrant to a residential building fire. Fire departments that utilize a forward lay should load their LDH supply line with the female coupling out to provide easy connection to the male hydrant outlets/nozzles.

Forward lay tactic departments that have Storz fittings on their LDH supply line usually pre-attached a Storz to threaded swivel female adapter to speed hydrant connections; the attached adapter is usually sized to the largest common outlet on their jurisdiction hydrants. This tactic eliminates the advantage of the Storz coupling to quickly support reverse, forward, and split lays with no adapters. One option is to keep the LDH Storz-to-hydrant adapter in a hydrant bag. If forward laid supply hose is attached directly to the hydrant outlet, the fire department utilizing this tactic is relying on the residual pressure in the water mains to overcome the friction loss in the supply hose. This limits the practical water capacity of the hydrant. This downside to a forward lay can be overcome using a four-way hydrant valve.

Figure 15–77b, is a Akron four-way hydrant valve being used to augment pressure in an existing 3-in. (75-mm) forward lay supply line. A four-way hydrant valve appliance offers an immense advantage to departments using a forward lay tactic, because if it becomes apparent that due to water supply issues (a extremely long forward lay, for example) placing a pump at the hydrant will significantly increase water supply, one has the option of accomplishing this without interruption of water flow in existing forward laid water supply hose.

Placing a pump at the hydrant is the best way to use all the existing flow capacity in water main system. The engine in figure 15–77b could be called upon to supply more water; it could easily gain access to more water by stretching an LDH to the other non-used outlet on the hydrant and then supply additional water through more hose. An engine assigned to a hydrant that uses all of the hydrant's outlets and then pumps it down to near zero on the compound gauge (70-kPa) has effectively used all of the available water from that hydrant; this is called welling the hydrant. Remember, the hydrant pressure will be well above 10 psi even when you have near zero on the compound gauge, due to friction loss in the hose.

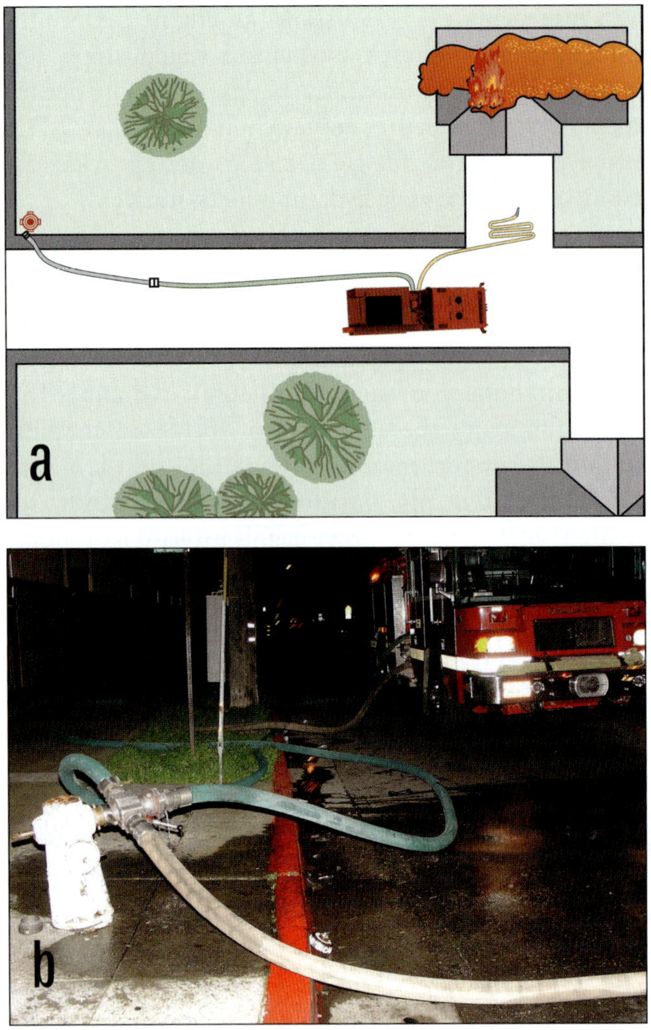

Fig. 15–77. a) Diagram of a typical forward lay. b) Engine pumps through a four-way hydrant valve.

Reverse lay. The reverse lay is simply a water supply evolution where an engine company lays supply line or attack line back to a hydrant (fig. 15–78). This means an engine performing a reverse lay stops at the fire building and then lays hose to the nearest hydrant. The reverse lay is typically used in two different ways: for water supply or fire attack. In the supply scenario there is usually an engine already on the fire scene with attack lines deployed. The attack engine either has no water supply or has an inadequate existing supply. One such example would be an engine operating off its tank. The supply engine that is performing the reverse lay to establish a water supply would pull near the attack engine either by passing it on a wide street or by backing down the street when access is an issue. Once the reverse laying supply engine is in position, the crew would pull the proper supply hose for attack pumper and then release the supply engine, whose crew would proceed to lay out to a nearby hydrant. Once at a hydrant, the engineer would then connect the pumper to the hydrant and the reverse laid supply hose to the engine's discharges. The supply engine at the hydrant must strive not to block access for arriving ladder companies.

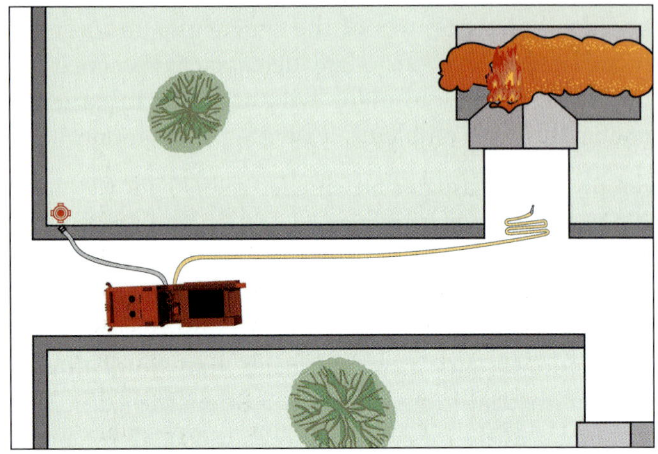

Fig. 15–78. Reverse lay

The other type of reverse lay is an attack line operation. In this case the engine pulls up to the fire building and the crew gets out. The engine crew then proceeds to make a line selection (typically either a single attack line or both an attack line and a backup line are pulled). These lines are located in the hosebed and must be static loads. The company must select the proper size of attack line and estimate how much hose is needed at the fire building to make an effective attack. At the same time, they must strip any equipment they need. Some examples of this would be SCBAs, lights, and hand tools. Sometimes a gated wye or manifold is attached to the engine's supply hose; later, the attack lines, usually removed from engines in bundle configuration, will be connected. Once all of that is accomplished, the engine is then released by the company officer, laying supply hose as it travels. The engineer then drives to the nearest hydrant, leaving the company behind at the fire scene. As the engineer makes the necessary connections to the hydrant and engine, the rest of the company at the fire building starts to flake their attack line and get into position. Once a call for water is made by the attack crew, the engineer charges the proper attack line. Reverse lays for attack hand line operations are typically done in large urbanized areas where hydrants are closely spaced. This leaves the front of the building open for extensive use of aerial devices.

Many fire departments use reverse lays as a water supply evolutions for apparatus located near the fire building. Figure 15–79 depicts the Milwaukee (Wisconsin) Fire Department's (MFD's) rear hosebed and a diagram of a supply reverse lay. Notice that at the time this photo was taken they still used 3-in. (75-mm) MDH as a supply line in the two center hose bins. Both 3-in. (75-mm) MDH supply beds have the male coupling facing out to facilitate reverse lays. Each hose bin has 750 ft (230 m) of 3-in. (75-mm) hose, in 50-ft (15-m) lengths. One 3-in. (75-mm) MDH supply bed has a red hose strap, and the other 3-in. (75-mm) hosebed has a tan hose strap. The MFD can either reverse lay a single or dual 3-in. (75-mm) lay. Pumped supply of 3-in. (75-mm) MDH in this configuration can easily supply 1,000 gpm (3,785 L/min) even if the full 750 ft (230 m) is used, as long as a dual 3-in. (75-mm) hose lead is taken. This leaves a 500-gpm flow in each 3-in. (75-mm) lead at 20 psi (140 kPa) per 100 ft (30 m) of friction loss. The MFD also benefits from optimal hydrant spacing of 300 ft (90 m). Such short lays are a good match for 3-in. (75-mm) hose, especially when it is pump-supplied.

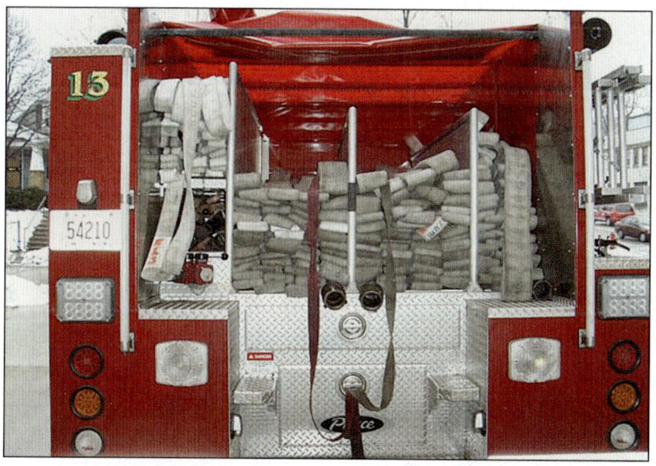

Fig. 15–79. MFD hosebed setup for reverse lays

Another example of a city that uses 3-in. (75-mm) hose is San Francisco. The SFFD benefits from both an excellent water supply system and close hydrant spacing. The SFFD also deals with a densely constructed city, predominately of multiple-story wooden structures. They use dual beds of 3-in. (75-mm) MDH, one with the male coupling out and the other with the female out. This allows them to accomplish a single 3-in. (75-mm) forward lay or a single 3-in. (75-mm) reverse lay with ease. They can also take dual reverse lays or dual forward lays by using either a double female or double male adapter. In figure 15–80, the engine at the hydrant is pumping two 3-in. (75-mm) leads. Notice one of the leads is attached to a pump discharge by a brass 3-in. (75-mm) double female adapter. The engineer operating the engine has welled the hydrant with two suction jumpers. This engine is properly positioned, out of the way, and connected in a manner such that it will supply all the available water from the hydrant. This SFFD engine could easily be called upon to supply more lines until the hydrant flow reached the available capacity. Notice that in the background extensive aerial operations and ground ladder operations are taking place directly in front of the fire building. Although the SFFD uses 3-in. (75-mm) MDH as supply line on their engines, they also maintain a small fleet of 5-in. (125-mm) LDH hose tenders. This is a common practice for departments that use 3-in. (75-mm) MDH, given that there is truly no substitute for size of hose when called to move very large amount of water—2,000 gpm (7,571 L/min) and up. These hose tenders are usually for special calls or assigned to certain alarms.

Fig. 15–80. SFFD pumps dual 3-in. (75-mm) MDH lay at hydrant

Split lays

Now that we have discussed the water supply evolutions of both forward and reverse lays, we will describe another useful operation called a split lay. The split lay is basically where two engines are used to complete one water supply line evolution from the hydrant to the building fire. The split lay is not commonly done; however, split lays can be extremely effective in the right circumstances. For example, the first due engine arrives at an extremely long alley way or private drive with a building fire located at the end of it. This engine in figure 15–81 came across just this circumstance. The company officer in this case decided to use a split lay by dropping a supply line at the beginning of the private drive and continued into the fire building. The first due company officer then radioed the second due engine company to pick up their dry lay. This second due company now was then tasked with finishing the water supply line. The second due company stopped and attached their supply line to the dry supply line laid by the first due company.

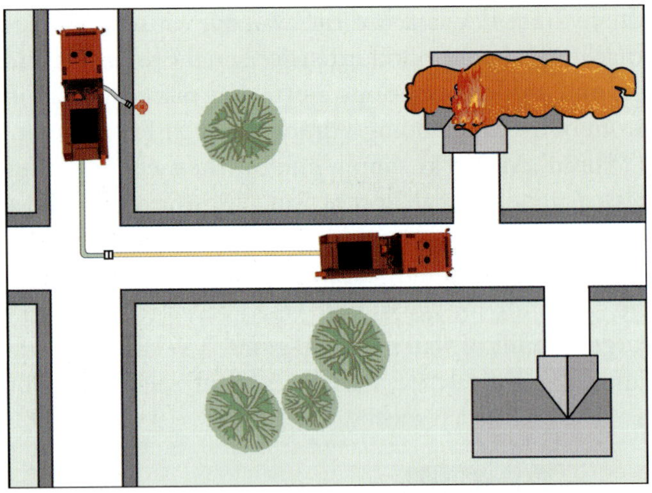

Fig. 15–81. Diagram of a typical split lay

The second-due company's engineer then drives back to the nearest hydrant. Once at the hydrant, all necessary connections were made. Then engineer ensured that the middle connection or **split connection** had been completed and the attack pumper was ready for water. Once all of this was confirmed, water was pumped from the hydrant through the now completed split laid supply line to the attack pumper.

Let's say in this case the total length of the split lay worked out to 1,000 ft (305 m) of 5-in. (125-mm) hose, and each engine in the example only carried 800 ft (245 m) of 5-in. (125-mm) LDH. This would mean that due to the long driveway and distance to the nearest hydrant, it would have been impossible for either company to have made this supply operation on their own. It is not necessary to only consider split lays if the amount of hose required is more than one company's complement. Split lays are sometimes performed for access issues alone.

Drafting operations

SKILL DRILL

In rural settings, drafting operations are commonplace. This may involve the use of a dry hydrant connected to a lake, a cistern, or a swimming pool, as described earlier. Perhaps the most common operation is simply the use of a rigid (hard suction) hose placed in a body of water.

Once the engine has been placed at the edge of the source of water (lake, pond, river, etc.), the two hard suction hoses are removed and connected. The strainer is then attached.

A **floating strainer** is typically a square-shaped device attached to the hard suction hose that floats on the surface of the water, maximizing its ability to avoid underwater debris and to draft from shallow water depths. It can simply be placed on the water, and the pump operator can begin to pump after establishing a vacuum (fig. 15–82).

A traditional, round strainer (a circular tube with many holes cast into it) is deployed into the water using a rope to keep it off the bottom of the lake, pond, or other body of water. The rope is then tied to the engine, and the pump operator can initiate creating a vacuum and then begin pumping operations (fig. 15–83).

Fig. 15–82. Floating strainer in operation

Fig. 15–83. Traditional barrel strainer

HYDRANT APPLIANCES

Departments across the country use different **hydrant appliances** to accomplish a variety of tasks. Figure 15–84a depicts an LDH four-way valve attached to a hydrant, which provides the option to pump at the hydrant after water flow is established through the initial supply line. Many departments use four-way hydrant valves. In figure 15–84b, the dry barrel hydrant with two 2½-in. (65-mm) outlets has a hydrant gate valve attached to it and a single 3-in. (75-mm) MDH supply line. The gate valve allows the use of the other hydrant outlet without turning the hydrant off.

The yellow hydrant in figure 15–84c has a 5-in. (125-mm) lead attached to it, which is using a Carlin automatic hydrant valve.

Fig. 15–84 a, b, c, d. From left to right: a four-way hydrant valve, a gate valve, a Carlin automatic hydrant valve, and a Gleason pressure-reducing valve.

The last picture on the right in figure 15–84d shows a large high-pressure hydrant. The water available from this hydrant is under such tremendous pressure that it is necessary to put a pressure-reducing valve on it. This type is called a Gleason valve, and it is used by SFFD on their high-pressure auxiliary water supply hydrant system. Boston is another city that has a unique hydrant system; one part is a hydrant stem system located under manholes that require a Lowry hydrant. The Boston Lowry hydrant is carried on the engine and is attached directly to a connection in the street. The City of Boston has three hydrant systems; the Lowry system is located mostly in downtown. Today, five engine companies still carry Lowry hydrants.

In summary, many fire departments use different hydrant appliances. Some of these appliances are very simple (like gate valves), and others are rarely used and complex (like the Carlin automatic hydrant valve). You must be familiar with your own fire department's operations as well as the water supply operations and capabilities of your surrounding jurisdictions. The Carlin automatic hydrant valve is used by a jurisdiction near the fire department of one of the chapter authors. Therefore we should know how it works, even though we think it is very impractical.

Basic hydrant use

SKILL DRILL

The ability to use a hydrant correctly is a critical skill that all members of the fire service should not only master but also routinely practice. Engine companies should routinely drill with fire hydrants. It is a bread-and-butter operation that needs to be second nature to all firefighters. All common problems related to hydrants in your particular jurisdiction should be preidentified, and solutions should already be in place. Very few fire departments have the luxury of hydrants that are perfectly maintained and work properly very time. Many departments face significant hydrant problems from vandalism and poor maintenance. I strongly recommend that fire departments outfit all of their engine companies with hydrant bags. A hydrant bag should have all of the necessary equipment to make a good connection to a hydrant, even when problems are encountered.

Take a look at the hydrant bag in figure 15–85. Fire departments can encounter significant problems with their hydrants based on a variety of factors. This hydrant bag was design by a department that use wet barrel hydrants. It includes a pipe wrench for stripped spindles, a **cheater bar** to apply extra force (avoid breaking the hydrant stem connected to the operating nut by applying too much force), a 3-lb (1.4-kg) maul to knock loose rusted or stuck caps; a Hydra-Shield® locking cap spanner to deal with locking caps; a 2½-in. (65-mm) female to 4½-in. (115-mm) male increaser, in case the steamer outlet is nonfunctional; and a regular screw-type hydrant wrench. Fire departments that access water with dry barrel hydrants should consider these additional items to aid in hydrant connections: extra hydrant caps (in case one or more are missing), outlet plugs or blocks (in case of damaged outlet threads), and different hydrant gate valves.

Fig. 15–85. An example of tools in a hydrant bag

Hydrant bags are especially useful for departments that rely on forward lay water supply evolutions. During a forward lay, the engine company leaves a firefighter at the hydrant. This firefighter has only the equipment that has been removed from the engine, to obtain a good water supply by rapid connection of the supply line to the hydrant. By placing all of the necessary equipment in a hydrant bag, the lead-off firefighter should have all of the tools required to "catch the hydrant." After checking the PPE and SCBA, one of the first things the lead-off firefighter at shift change should check is the hydrant bag. Hydrant bags are also useful for departments that use reverse lays. In case of a reverse lay, the engineer is able to grab one bag to locate all the tools necessary to make a good hydrant connection.

Let's walk through hitting a hydrant while performing a forward lay.

The forward lay is used when the water source, static or municipal, is located before the structure on fire. In a forward lay, the pumper approaches the water source— in this case a hydrant—and stops. A firefighter, sometimes known a "hydrant firefighter," gets off the pumper with the needed equipment: hydrant wrench, spanner, gates, hose traps, etc., and removes a sufficient amount of supply hose to "wrap" the hydrant. As mentioned previously, the hydrant firefighter checks the hydrant for damage and access, and then proceeds to secure the hose (wrap) around the barrel of the hydrant to anchor it safely in place. The pumper then proceeds to the fire scene. The hydrant firefighter then checks the hydrant, removes the caps and flushes the hydrant. The hose and gates are then connected to the hydrant, and the hydrant firefighter awaits a signal from the pump operator or fire officer to open or charge the hydrant.

Once water is received at the pumper, then those necessary functions (fire attack, exposure control, hydraulic ventilation, overhaul, etc.) can be accomplished at the fire scene to control and extinguish the fire. This is usually beyond the suppression capacity of the initial engine's tank water, which commonly charges the first handline. It is important to have this evolution down pat, as firefighters often work from tank water until it is completed.

Once the fire is extinguished and all overhaul functions have been accomplished, the process of shutting down the operation can begin. At this point the incident commander notifies the pump operator to begin shutting down attack lines and the hydrant firefighter is notified that they can begin closing the hydrant, as discussed in a previous section. Once the hydrant has been shut down and the pressure bleeds off the hose, the gates can be disconnected. The hydrant caps are replaced and the supply hose is drained of excess water and prepared to be repacked on the pumper.

NOTES

1. Bachtler, J. R., & Brennan T. F. (Eds.). (1995) *The Fire Chief's Handbook* (5th ed.). Tulsa, OK: PennWell.
2. Oakland Fire Department. (1997). *Water Supply Reference Course*. Oakland, CA: Oakland Fire Department (OFD).
3. FEMA U.S. Fire Administration (2003). *Fireboats Then and Now, Special Report*, Retrieved November 2007 from http://www.fireboat.org/FEMAfire-boatsthennowMay2003.pdf.
4. FEMA U.S. Fire Administration (2003).
5. FEMA U.S. Fire Administration (2003).
6. NFPA 1961. (2007). *Standard on Fire Hose*. Quincy, MA: National Fire Protection Association.
7. NFPA 1961.
8. NFPA 1962. (2008). *Standard for the Inspection, Care, and Use of Fire Hose, Couplings, and Nozzles and the Service Testing of Fire Hose*. Quincy, MA: National Fire Protection Association.
9. NFPA 1962.
10. NFPA 1962.
11. NFPA 1961.

LESSON FROM THE FIREGROUND

The fire department arrived at a well-involved two-story garden apartment fire. The fire was burning throughout a second floor dwelling unit and the fire coming from the windows had spread the fire into the combustible attic space. The buildings were situated between 1,000 and 1,200 ft (305 and 366 m) off the main road. There was a hydrant in the complex but the fire department knew this hydrant was fed by a dead end 6-in. (150-mm) water main. The first arriving pumper (1,500 gpm [5,678 L/min] with 500 gal [1,892 L] tank) stopped at the hydrant at the entrance to the complex and dropped a 5-in. (125-mm) supply line.

The volume of fire dictated the use of heavy caliber streams. A deck gun was ordered to hit the main body of fire. The pre-piped 1¾-in. (45-mm) smooth bore tip master stream was charged. A 2½-in. (65-mm) hoseline with a one and ⅛-in. (3-mm) smooth bore tip was ordered to the adjoining second floor apartment where the fire was now spreading. A second two and ½-in. (13-mm) hose line with the same nozzle was stretched to combat the fire in the attic, which was now rapidly involving the attic space over the entire building.

As the hand lines were charged, the pump operat realized the supply line was collapsing. A second supp line was called for and an additional 5-in. (125-mı supply line was stretched from the dead end hydra using two 100-ft (30-m) lengths of hose. Although tl improved the situation slightly, there was still a proble supplying adequate water to the single deck gun and tv large hand lines.

The immediate assumption was there had to be a proble with the hydrant and the public water supply. T initial static pressure was 70 psi (490 kPa) but quicl bottomed out. An examination after the fire showed tl there was plenty of water. The problem was a basic pur operations flaw. There was too much friction loss creat when flowing the large volume of water through the fi inch supply hose from the hydrant. The simple answ would have been to have another pumper at the hydra to overcome the friction loss problem.

QUESTIONS

1. What is normal atmospheric pressure and how does it register on a standard fire service gauge?
2. Explain the effect friction loss has on flow pressure in a fire hose.
3. In the water distribution system, on what type of water main are hydrants normally found?
4. Why are dry barrel hydrants typically found in climates where the temperature can be expected to drop below freezing?
5. What items should be included in a hydrant inspection?
6. What color should the bonnet of a hydrant capable of flowing more than 1,500 gpm (5,678 L/min) be painted?
7. When setting up a water shuttle, what conditions make for a good fill site?
8. When is relay pumping typically needed?
9. How would a 2½-in. (65-mm) hoseline be classified? Why?
10. What is the advantage of double-jacketed hose when compared to single-jacketed hose?
11. Can large-diameter hose (LDH) be used to draft water from a static source? Why or why not?
12. Why are Storz connections commonly used on large-diameter supply hoses?
13. Prior to reconnecting sections of hose together, what should the couplings be inspected for?
14. What is the definition of a hose appliance?
15. Explain the difference between a gated wye when compared to a clappered Siamese.
16. Why should a fire hose not be folded or reloaded in the same order it was prior to being used?
17. What does it mean if a fire hose is stamped "400 psi service test"?
18. What advantage do attack lines stored preconnected on the pumper have over attack lines stored in a dead load?
19. Laying a supply line from the hydrant to the fire is referred to as what?
20. In what circumstances might a split lay be utilized?

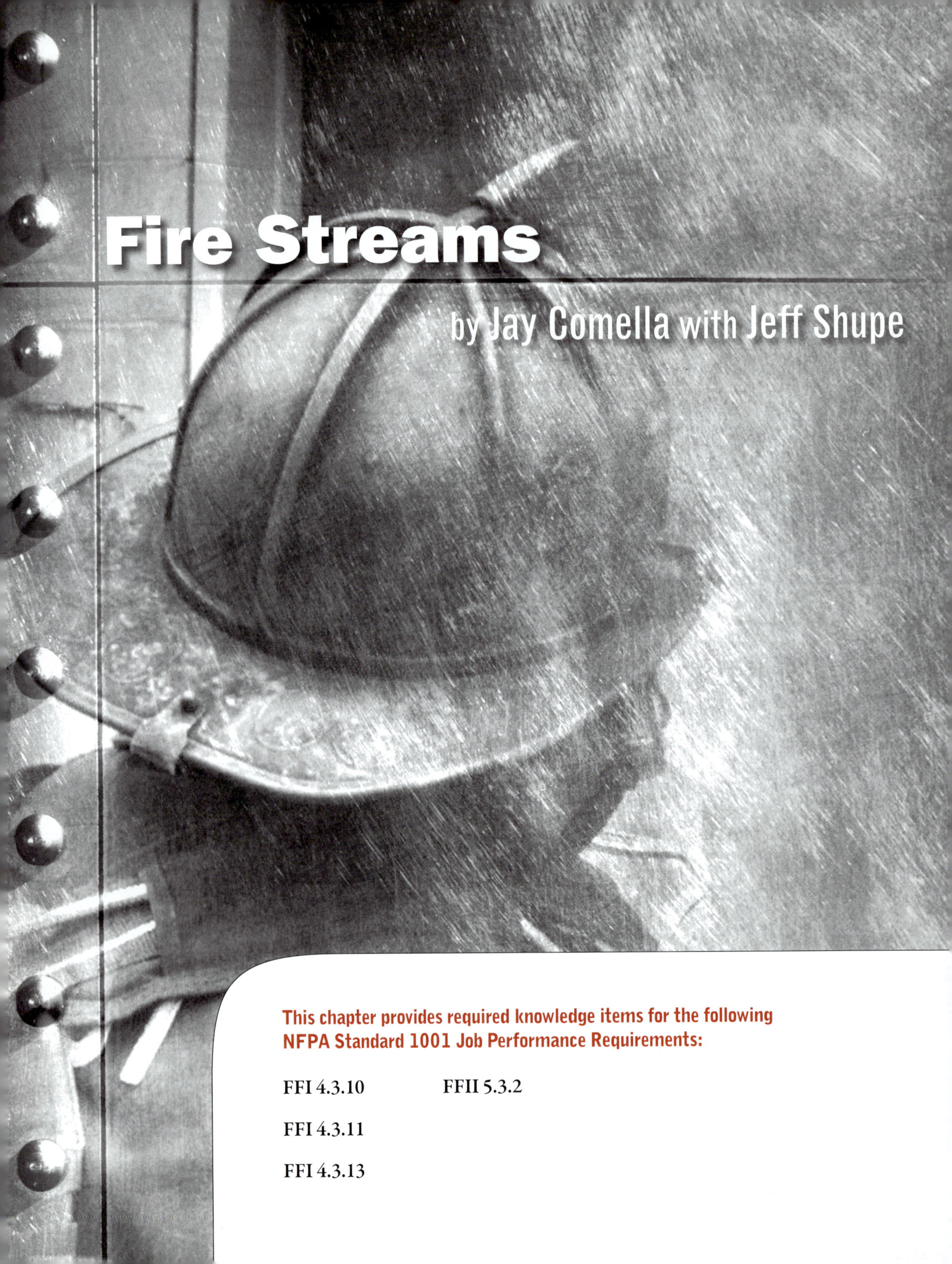

Fire Streams

by Jay Comella with Jeff Shupe

This chapter provides required knowledge items for the following NFPA Standard 1001 Job Performance Requirements:

FFI 4.3.10

FFI 4.3.11

FFI 4.3.13

FFII 5.3.2

16

Fire Streams

FIREFIGHTER I

OBJECTIVES

Upon completion of this chapter, you should be able to do the following:

- Identify the advantages of water as an extinguishing agent
- Identify the factors that affect the firefighter's ability to handle fire hose
- Describe the characteristics of the four nozzle types
- Identify the mnemonic device "ADULTS" as it applies to using a fire hose and nozzle
- Describe the six different methods of fire attack
- Identify the three types of fire streams
- Describe the proper procedures for advancing a charged hose line
- Describe the proper procedures for advancing an uncharged hose line
- Describe the responsibilities of the three positions on a charged hoseline
- Describe the proper procedures for using a hoseline to perform hydraulic ventilation
- Describe the proper procedures for using hoselines during exposure and overhaul operations
- Identify the different types of master stream devices used by the fire service
- Identify the types of specialty nozzles used by the fire service
- Describe the basic care and maintenance of fire service nozzles

INTRODUCTION

Over the years the leading instructors of the fire service have repeatedly made statements such as "The greatest lifesaving action on the fireground is the proper placement of the proper size initial attack hoseline," and "As the first line goes, so goes the fire."

FFI 4.3.10 Such statements denote the critically important actions involved in the proper selection, deployment, and use of **hoselines**, **nozzles**, and **fire streams**. When applied properly, they diminish or extinguish the fire completely, reduce the temperature of burning materials, and stop the spread of fire. They directly affect lifesaving by extinguishing fire and inhibiting the products of combustion. They also allow all other fireground lifesaving functions to proceed more quickly, efficiently, and safely. The entire job revolves around the acts of stretching and advancing hose and operating the nozzle to extinguish fire. The members who perform this work are the tip of the spear of the fire service.

WATER AS AN EXTINGUISHING AGENT

Today's fireground is a much more volatile environment than that of the past. The flow rates of 95–125 gpm were deemed adequate at a time when fuel loads were lighter and comprised of so-called ordinary combustibles, such as wood, paper, and cloth (**cellulosic materials**). Quantities of combustibles have dramatically increased. Fuels are heavier and largely **hydrocarbon**-based (plastics); plastics are petrochemical products that behave like **solid gasoline** and generate large quantities of thermal energy. Higher heat release rates associated with plastics (discussed in chapter 5, Fire Behavior) bring a room to flashover more quickly. Couple these factors with better insulated buildings that inhibit fire from self-venting (tight building syndrome), and today's engine company most definitely faces a much more dangerous enemy than in the past.

FFI 4.3.10 Because the enemy has become much more dangerous, the weapon used to combat the enemy must be upgraded accordingly. Akin to the police evolving from the 38-caliber revolver to the 9 mm semiautomatic, the fire department also must make a more intelligent weapon selection. The hose and nozzle system is the engine company's weapon for attacking the fire. Most of the American fire service now considers 150 gpm (568 L/min) to be the minimum acceptable flow rate for interior structural fire attack. Many fire departments use a target flow rate of 180 gpm (681 L/min) to ensure an added margin of safety.

In his brilliant treatise on the art and science of applying water on fire,[1] the late Andrew Fredericks (a New York City Fire Department [FDNY] firefighter who was killed at the World Trade Center on 9/11), the foremost expert on engine company operations in modern times, further states that in addition to 150 gpm (568 L/min) being the minimum acceptable flow for residential fires, 250 gpm (946 L/min) is the minimum acceptable hand line flow for operations in commercial occupancies.

The establishment of robust, occupancy specific, minimum flow rates is in effect an extension of the Powell Doctrine to the fire service. The Powell Doctrine is the culmination of General Colin Powell's many years of battle experience, training, and study of the military arts. The doctrine is a set of guidelines meant to ensure the highest probability of success in the conduct of military operations. The essence is that once combat is joined, one must bring overwhelming force to bear upon the enemy in an extremely rapid manner (shock and awe) to ensure the highest likelihood of victory in the shortest duration. This in turn reduces the overall depletion of one's resources, both material and human.

The outcome of fireground operations depends on the outcome of the battle between the water the engine company delivers (gpm) and the fire's heat release rate. The flow at which the engine company can win the battle and kill the fire is defined as the critical flow rate. If the critical flow rate is not met, the battle will be lost. This dictates that the single most important characteristic of a hose and nozzle system is water flow capability. The water the engine company delivers must not merely meet theoretical flow rates; it must be sufficient to expediently overwhelm and kill the fire. Maneuverability of the hose and nozzle are important factors, but sacrificing flow for ease of use has proved to be suicidal in too many instances.

Water is an ideal fire-extinguishing agent. Besides the fact that it is readily available and inexpensive in most locales, it is efficient in terms of its fire-extinguishing capabilities. Water extinguishes a fire primarily through cooling, reducing the temperature of the burning fuel and the fire gases. In addition, water applied to unburned fuel surrounding the burning materials wets them, making it difficult, if not impossible, for the fire to spread.

Water has a high **specific heat** compared to other materials (pound for pound, it absorbs more heat than many other substances). Thus it takes more heat energy to raise the temperature of water compared to other materials. It takes one **British thermal unit (Btu)** of heat energy to raise 1 lb water 1°F. (You may be familiar with Btu in the context of air conditioners and the amount of heat energy they are capable of handling, it is a measurement of their **cooling power**.)

For every 1°F, 1 lb water is raised by a fire, it absorbs 1 Btu of heat energy. When the water turns to steam at 212°F, it absorbs an additional 970 Btu, called the **latent heat of vaporization**. In fig. 16–1 for example, 1 lb water at 55°F heated until it turns to steam absorbs

$$157 \text{ Btu } (212 - 55) + 970 \text{ Btu} = 1{,}127 \text{ Btu}$$

Taken a step further, because 1 gal water weighs 8.3 lb, 1 gal water at 55°F turned into steam absorbs

$$8.3 \times 1{,}127 \text{ Btu} = 9{,}354 \text{ Btu}$$

When liquid water turns to steam, it expands approximately 1,700 times its original volume (see fig. 16–2). Although smothering by steam is a useful firefighting tactic in unoccupied buildings and equipment, it is neither safe nor desirable to generate large quantities of steam because of the danger to occupants and firefighters. In typical compartment (room) fires, firefighters must avoid creating steam by applying water properly and in the right quantities. Creating larger quantities of steam is dangerous and can lead to **steam burns**. This issue is discussed in the following text in terms of nozzle selection and fire streams.

Although an adequate flow rate cannot be sacrificed for ease of use, handling characteristics cannot be completely overlooked, either. The **nozzle operator** must exert enough force to resist the nozzle reaction. Nozzle reaction is measured in pounds of force and is a function of two factors: flow rate and **nozzle pressure (NP)**. An increase in one or both factors results in increased nozzle **reaction force (RF)**. The higher the nozzle RF, the more difficult the nozzle is to control. Because adequate flow rate is the ultimate goal of a well-conceived hose and nozzle system, the logical way to keep nozzle RF within the manageable range is to keep nozzle pressures low and avoid sacrificing flow. More than 75 lb (34 kg) RF is considered to be too much reaction force for a hand line. However, RF less than 45 lb (20.4 kg) is considered a sign of an ineffective stream.

Hoseline handling characteristics are a function of the following factors:

1. Flow rate
2. Hose size
3. Friction loss
4. Pump discharge pressure

Hand line maneuverability is determined by the pressure at which a given size line must be pumped to attain a desired flow rate. If hose size remains constant and flow is increased, pump discharge pressure must be increased to account for greater friction loss. This reduces maneuverability as the line approaches the stiffness of a pipe. Conversely, if hose size increases while flow remains constant, pump discharge pressure may be reduced because of lower friction loss requirements. This results in improved maneuverability because the line becomes more bendable.

The aforementioned parameters lead to certain conclusions about what constitutes a well-planned hose and nozzle system for residential fires. The hose should be capable of flowing between 150 and 180 gpm (568 and 861 L/min) with relatively low friction loss. The nozzle should have similar flow capability at a nozzle pressure that will maintain reaction force in the range of between 45 and 75 lb (20 and 34 kg).

Because of the pressures required to account for friction loss, the practical flow limit for 1½-in. (38-mm) hose is 125 gpm, whereas the practical flow limit for 1¾-in. (45-mm) hose is 200 gpm (757 L/min) (fig. 16–1).

Fig. 16–1. The 1¾-in. (45-mm) hose allows significantly higher flow than the 1½-in. (38-mm) line, yet size and weight differences are nominal.

NOZZLES

FFI 4.3.10 **FFII 5.3.2** The tool at the very heart of the entire fireground operation is the nozzle. It is the weapon with which members enter into close-quarter combat with the enemy. If the nozzle malfunctions or is improperly used, all other tools and tactics on the fireground are likely to become quite limited in their effectiveness in saving life and protecting property. All kinds of nozzles perform their all-important mission by providing some rather simple, uncomplicated, albeit incredibly necessary functions. They control flow, create shape, and provide reach. Because the functional requirements for a nozzle are relatively simple and yet immensely important, it intuitively makes sense to select the nozzle with the least complicated design and the fewest moving parts. The most low-tech choice in nozzle selection ensures the greatest degree of durability and reliability. Simple, durable, and low tech are all qualities that contribute to low initial and long-term costs. More importantly, these qualities lead to reliability, which leads to increased safety. The use and care of nozzles is covered under National Fire Protection Association (NFPA) 1962: *Standard for the Inspection, Care, and Use of Fire Hose, Couplings, and Nozzles and the Service Testing of Fire Hose.*

In the early days of fire service, hoses were leather and the first nozzle was nothing more than a piece of pipe on the end of the hose. The addition of a controlling device, or shut-off, between the male hose butt (coupling) and the piece of pipe was the genesis of today's fire nozzle. The bore of the pipe, or **smooth bore nozzle (tip)**, was eventually tapered to improve hydraulic efficiency. The controlling device consists of a shut-off valve and a handle by which to control it. Over the years various valve and handle sizes and types have seen use. Most configurations fell by the wayside as the shut-off evolved into the modern incarnation: a quarter turn ball valve with a 1⅜-in. (35-mm) waterway activated by a bale-type handle.

The rule of thumb for smooth bore tip orifice size is that it should be one-half of the inside diameter of the hose. This equates to a ⅞-in. (22-mm) tip for 1¾-in. (45-mm) hose and a 1¼-in. (32-mm) tip for 2½-in. (65-mm) hose. However, many fire departments have had great success with slight variances from this rule. The 15⁄16-in. (24-mm) tips for 1¾-in. (45-mm) hose and 1⅛-in. (30-mm) tips for 2½-in. (65-mm) hose are the most common sizes, and practical experience proves they deliver efficient and effective fire streams.

The first **fog nozzle** was developed in 1863 by Charles Oyston of Little Falls, New York. Fog streams are fog nozzles, and fog streams went relatively unnoticed and had little effect on the fire service for quite some time. It was not until the post–World War II period that fog nozzles gained widespread use. This increased favor within the fire service was a result of wartime experiences gained by the naval services who successfully used fog streams to control shipboard fuel oil fires in confined spaces. The intervening years have seen various combination nozzles, such as variable flow nozzles, constant gallonage nozzles, adjustable gallonage nozzles, and constant pressure nozzles (also known as automatic nozzles) all come into being.

After a relatively short-lived duration, the variable flow nozzle fell into disfavor. As its name indicates, it delivered varied flows by design. As the pattern selection changed, so would the flow. It soon became obvious that tying one's ability to achieve critical flow rate to stream selection was a distinct disadvantage in the design of the variable flow nozzle. This characteristic caused its use to decline and then cease. Variable flow nozzles flow less at straight stream than at full fog.

The three other types of combination nozzles mentioned, constant gallonage, adjustable gallonage, and constant pressure, all remain in present day use.

Nozzle characteristics

Fog streams are characterized by small droplets of water in a dispersed pattern compared with the tight, compact stream of a smooth bore nozzle; the distinct droplets of water in a fog stream evaporate more readily, generating steam (figs. 16–2a and 16–2b). Combination nozzles can be adjusted to discharge stream patterns differing along a range from straight (very narrow fog) streams to wide fog pattern streams. The kinds of nozzles available today, in descending order of simplicity and durability, are smooth bore, constant gallonage (single-gallonage) fog, adjustable gallonage fog, and constant pressure (automatic) fog (fig. 16–3). Combination nozzles are often referred to as fog nozzles and sometimes called **spray nozzles.**

Fig. 16–2a. A firefighter using a combination nozzle discharging a fog pattern

Fig. 16–2b. A firefighter using a combination nozzle discharging a straight stream

Fig. 16–3. Left to right: 15⁄16-in. (24-mm) smooth bore nozzle (180 gpm at 50 psi [681 L/min at 45 kPa]), constant gallonage nozzle (150 gpm at 50 psi [568 L/min at 345 kPa]), and an automatic nozzle (50–300 gpm [189–1,136 L/min]). (Courtesy of Jerry Knapp)

Smooth bore nozzle

SKILL DRILL

The smooth bore is the most low-tech of all nozzle designs. It consists of a ball valve shut-off device onto which is threaded the smooth bore tip, which is basically a piece of tapered pipe. Together, the shut-off and tip present a compact (7¾ in. [20 cm]) and lightweight (2½ lb [1.1 kg]) package. Genius lies in the simplicity of its design. It has only one moving part: the ball valve (fig. 16–4).

It is difficult to clog a smooth bore nozzle. It requires the least maintenance of any nozzle type and has the longest service life.

Smooth bore nozzles are by far the least expensive kind to purchase and maintain. Of all nozzles, the smooth bore requires the least amount of training for pump and nozzle operators to become proficient (fig. 16–5). The incredible reliability of the smooth bore nozzle is a significant safety feature. Because you can produce only a solid stream with the smooth bore nozzle, its use ensures that members and victims are not being exposed to the potentially debilitating or lethal effects associated with introducing a fog stream into the fire area.

The smooth bore nozzle is a safe and efficient weapon for combating interior structural fires. Therefore, it is the preferred nozzle to be taken into the most hostile work environment on the face of the earth—the interior of a burning building. Fog nozzles are appropriate for other uses, such as flammable-liquid fires, outdoor rubbish fires, vehicle fires, and the like. However, some departments use fog nozzles as their primary nozzle for **interior fire attack**. Fog nozzles used during interior fire attack must use the straight stream pattern to avoid the possibility of creating large quantities of steam from the vaporization of the fog droplets, scalding firefighters, and trapped occupants.

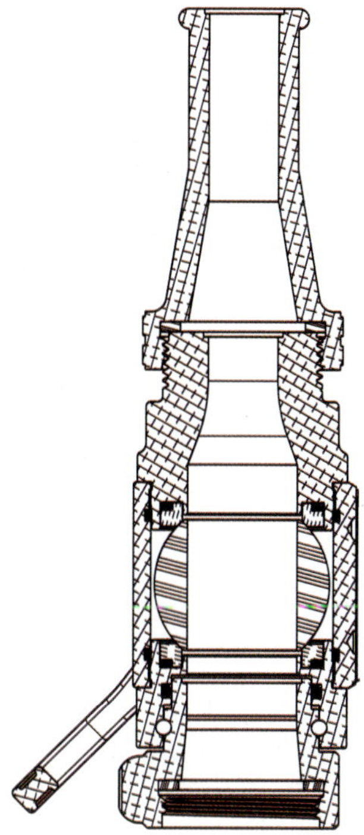

Fig. 16–4. Cut away of a smooth bore nozzle. (Courtesy of Elkhart Brass Mfg. Co.)

Fig. 16–5. California's Oakland Fire Department (OFD) members operate a 2½-in. (65-mm) hoseline with a 1⅛-in. (30-mm) smooth bore nozzle (266 gpm at 50 psi [1,007 L/min at 345 kPa]) during testing and evaluation. (Courtesy of Daryl Liggins)

Constant gallonage combination nozzle

The **constant gallonage combination nozzle** is the simplest, most reliable, least maintenance-intensive and, hence, safest member of the fog nozzle family. Of all fog nozzles, this type requires the least training. It does, however, require somewhat more training than the smooth bore nozzle. Constant or single gallonage indicates that this nozzle is designed to flow a specific gallonage when operated at the specific pressure for which it is designed, such as 150 gpm at 100 psi NP (568 L/min at 700 kPa). In addition to the 100-psi (700-kPa) model, constant gallonage nozzles also come in 75-psi (525-kPa) and 50-psi (345-kPa) models. The nozzle is 12¼ in. (311 mm) long and weighs 6.1 lb (2.8 kg). As the name suggests, there is the distinct possibility of introducing fog stream into the fire area. This has the potential to turn a still-tenable environment into one that is untenable in short order. As with all fog nozzles, when the water flows from the hose—through the shutoff, into the tip, to be broken into a spray stream—a clog point exists. Constant flow nozzles flow the same gallons per minute through the range of patterns from full fog to straight stream.

Adjustable gallonage combination nozzle

The **adjustable gallonage combination nozzle** takes fog nozzle technology to the next level of complexity. It has more moving parts and is more maintenance-intensive than the constant gallonage nozzle and, hence, has an increased potential for nozzle failure or malfunction. Using a flow-selection ring, the nozzle operator can choose a desired flow. This operation requires an increased level of training for both nozzle and pump operators. If the nozzle operator changes the flow setting, the pump operator must be informed so he or she can adjust pump discharge pressure to the appropriate level for the selected flow. It is possible to put the flow-selection ring on the wrong setting, so the nozzle's flow is less than the desired amount of water. So, in addition to possibly introducing a dangerous fog stream into the fire environment, there is a great potential for producing a flow that is less than the acceptable minimum. The adjustable gallonage nozzle is 12¼ in. (311 mm) long and weighs 5.6 lb (2.5 kg).

Automatic combination nozzle

Automatic nozzles originally were designed in the late 1960s by Chief Clyde McMillan of the Gary Fire Task Force, an auxiliary unit of the Gary Fire Department in Indiana. One role of the task force was to respond to large fires and put master streams into operation. Often, initial water supply was inadequate when transitioning to defensive operations. McMillan set about designing a master stream appliance nozzle that would produce a stream with good reach, even at the low flows available during the transitional phases of operations. He also wanted that same nozzle to be appropriate for the high flows achievable after augmentation of the water supply.

The automatic nozzle is also called the **constant pressure nozzle**. Constant pressure refers to the fact that the nozzle produces a stream of reach and appearance consistent with 100-psi (700-kPa) tip pressure regardless of the pressure actually coming into the base of the nozzle. This is accomplished by a baffle and spring arrangement. As a given amount of water enters the nozzle base, it puts the spring under a given amount of compression. This, in turn, moves a baffle that changes the nozzle's orifice size. As the amount of water flow fluctuates, so does the orifice size. The orifice is thus maintained at a size that, for the given amount of water, provides approximately 100 psi (700 kPa) nozzle pressure (NP). This allows the nozzle to create a visually attractive stream with good reach over an extremely wide range of flows. Often, the stream produced by the automatic nozzle looks good but doesn't contain much water. Remember, volume of water is critical to fire extinguishment.

The automatic fog nozzle is bulky (length: 13¾ in. [350 mm], weight: 6.5 lb [2.9 kg]). It is at the high-tech end of the spectrum of fire service nozzles (fig. 16–6). To be used properly, it requires more training for both nozzle operators and pump operators than any other nozzle type. It has the most complicated design of any nozzle and the most moving parts. Many automatic nozzles come with a shut-off that has a slide valve as opposed to a ball valve. A ball valve (on a smooth bore nozzle) has a completely unobstructed waterway, while the slide valve can become a significant clog point. Given the potential for clogging, the nozzle is equipped with a protective screen at its base to keep debris from entering the slide valve. Clearing the clog necessitates shutting down the hoseline, resulting in problems during a fire attack. The automatic nozzle is the most maintenance-intensive nozzle type.

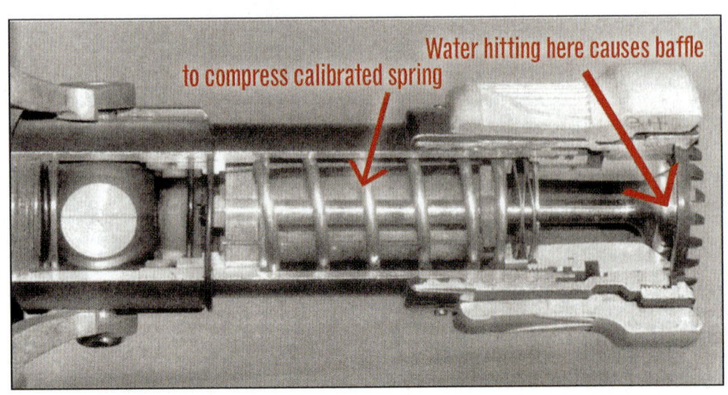

Fig. 16–6. Cutaway illustration of automatic nozzle.

Operating characteristics of smooth bore and combination (fog) nozzles

SKILL DRILL

Smooth bore nozzles are simple to operate. To flow water, firefighters need only pull back on the bale and water discharges from the nozzle. The bale can be adjusted between the fully closed (the **bale** is in the forward position, farthest away from the firefighter) and the fully open positions (the bale is in the rear position, closest to the firefighter). Positioning the bale between these two extremes allows the firefighter discharge smaller amounts of water. To open or close the bale, move in a slow, deliberate manner (figs. 16–7a and 16–7b).

The bale of fog nozzles operates in the same way as a smooth bore nozzle. Typically, the nozzle has a rotational control to adjust the pattern from straight stream to the wide-angle spray also known as a **wide fog** (characteristically, a 100-degree angle). Turning the adjustable rotational control from left to right changes the pattern from a wide-angle fog to a narrow fog, and then finally to a straight stream (and shutoff in some cases) (figs. 16–7c, 16–7d, 16–7e, and 16–7f). Some fog nozzles use a lever-type control to adjust the pattern and open or close the nozzle. Some fog nozzles have a **flush feature** that allows small pieces of debris to be discharged.

Fig. 16–7a. Smooth bore nozzle in open position

Fig. 16–7b. Smooth bore nozzle in close position

Fig. 16–7c. Fog nozzle in open position

Fig. 16–7d. Fog nozzle in closed position

Fig. 16–7e. Fog nozzle adjusted to straight stream

Fig. 16–7f. Fog nozzle adjusted to wide fog

SELECTING THE PROPER HAND LINE SIZE

FFI 4.3.10 **FFII 5.3.2** The first step in planning a hose and nozzle system is to establish the needed flow for the occupancy type in question. The flow requirement is derived by determining the flow at which the engine company most often will overwhelm the heat generated by the encountered fuel load. To deliver the desired volume of water, parameters for hose selection are based on flow and friction loss characteristics. Parameters for selecting a nozzle to couple to the business end of that hose are based on flow and reaction force characteristics. Actual friction loss in hose varies with the specific brand (manufacturer), grade, and age of the hose. It is important for firefighters to understand the actual friction loss characteristics of the hoses used by their own department. This holds true for residential occupancies and for fires in commercial buildings.

Under most circumstances, a 1½-, 1¾-, or 2-in. (38-, 45-, or 50-mm) hand line suffices for a typical room and contents fire in a low-rise residential buildings and the **bread and butter** structures of the fire service: single- and two-family homes. This rule of thumb does not apply to high-rise residential structures or heavily involved residential structures. The company officer (following the particular fire department's standard operating procedures) decides what size hand line to use in a particular building.

As mentioned earlier, the minimum acceptable hand line flow for operations in commercial occupancies (including industrial and institutional occupancies) is 250 gpm (946 L/min). For this type of flow, 2½-in. (65-mm) hose is the line of choice. Friction loss at 250 gpm (946 L/min) is 10 psi per 100 ft (70 kPa per 30.5 m) of 2½-in. (65-mm) line. For the same flow in 2-in. (51-mm) hose, the friction loss is 50 psi per 100 ft (345 kPa per 30.5 M). Although a 2½-in. (65-mm) line is a substantial piece of equipment, it is not too heavy to aggressively advance as a hand line, as would be the case with 3-in. (76-mm) hose.

The key to efficiently using a 2½-in. (65-mm) line is proper nozzle selection. The 100-psi (700 kPa) combination nozzle effectively removes the 2½-in. (65-mm) line from many a fire department's arsenal of offensive weaponry because of the very high nozzle reaction force of 126 lbs (57 kg) while flowing 250 gpm at 100 psi (946 L/min at 700 kPa) nozzle pressure. Low-pressure nozzles (50-psi [350-kPa] tip pressure) impart significantly less reaction force."

Many departments successfully employ a 1¼-in. (32-mm) tip. Its 324-gpm (1,226-L/min) flow technically classes it as a large-caliber stream, making this size tip possibly better suited for use with master stream devices. A far greater number of departments use the 1⅛-in. (30-mm) tip. With a flow of 266 gpm at 50 psi (1,007 L/min at 350 kPa) nozzle pressure, it has a reaction force of 95 lb (43 kg). Although it is still crucial to keep nozzle reaction force low, it would be impractical to try to apply the previously cited 75-lb (34 kg) cap to flows from large-caliber hand lines. Recent advancements in the fire hose industry are serving to significantly lessen the workload involved in deploying and operating 2½-in. line. Changes in materials and hose construction techniques combined with a return to a true 2½-in. inside diameter (as opposed to larger dimension hose being labelled as 2½ in.) results in a much more manageable large-caliber hand line.

Paired together, the 2½-in. (65-mm) line and the 1⅛-in. (30-mm) tip create a user-friendly, offensive, large-caliber weapon. Fredericks states the following:

> *No combination of smaller hand-lines can duplicate the volume, reach, and pure knockdown power of a single, well-placed 2½-in. line. In addition to its high-volume flows (between 250 and 320 gpm) and long stream reach, 2½-in. hose provides the following benefits when used with a 1⅛-in. solid stream tip:*
>
> - *Low friction loss per 50-ft length (only about 5 psi at 266 gpm).*
> - *Exceptional penetrating power due to hydraulic force of the stream.*
> - *Little premature water vaporization in highly heated fire areas.*
> - *Easy reduction to smaller hand line(s) after knockdown, and much better maneuverability than 3-in. hose (sometimes used as a hand-line) or portable master-stream devices.*[2]

Using a 2½-in. (65-mm) line is indicated in situations in which fire conditions are likely to overwhelm smaller hand lines. Fredericks cites the oft-used mnemonic device ADULTS, which refers to scenarios requiring the use of 2½-in. (65-mm) line:

- Advanced fire on arrival
- Defensive operations
- Unable to determine extent (size) of fire area
- Large, uncompartmented areas
- Tons of water
- Standpipe system operations

The ADULTS acronym is reminiscent of an anecdote related by retired Chicago Fire Department Battalion Chief Ray Hoff regarding proper hand line selection. On seeing an engine company stretching a 1¾-in. (45-mm) line toward a commercial occupancy exhibiting a heavy fire condition, Chief Hoff requested, "Would you please put that down and bring me an adult-size line?"

Advanced fire on arrival. When the engine company encounters advanced fire on arrival, the high flow available from 2½-in. (65-mm) hose is needed for rapid control. Even a private dwelling may exhibit a fire condition heavy enough to warrant the quick knockdown power of the 2½-in. (65-mm) line. This is especially true of extensive involvement of the first floor or front porch.

Although using master stream appliances is not recommended for occupied residential buildings, the same cannot be said of 2½-in. (65-mm) hose. The 2½-in. (65-mm) line coupled with the 1⅛-in. (30-mm) smooth bore nozzle is a large-caliber weapon that is aggressive, mobile, and offensive. It can rapidly darken down a heavy fire condition to allow an interior attack. This permits three tactical options:

1. The 2½-in. (65-mm) hand line can be advanced into and through the structure.
2. The attack can transition to the use of a smaller line with the big line left where it is.
3. The 2½-in. (65-mm) line can be reduced down to a smaller line to press the interior attack for final extinguishment.

Defensive operations. Whether operations are initially defensive or transition from offensive to defensive, smaller-caliber hand lines should not be used. The 2½-in. (65-mm) line is a much safer and more efficient alternative. The reach afforded by the larger line allows it to be operated from outside the collapse zone. Once its high-volume stream penetrates the fire area, it has a much greater effect on conditions than does a stream from a smaller line. The 2½-in. (65-mm) hand lines are much more mobile and easier to deploy than master stream devices. This allows streams to be brought to bear from a greater variety of locations.

Unable to determine extent (size) of fire area. If the engine company officer is unable to determine the extent (size) of the fire area, use a 2½-in. (65-mm) line. The high-flow stream allows for unforeseen contingencies. During the course of operations, it may be determined that the amount of fire encountered can be handled with smaller hose. As with the above-mentioned private dwelling scenario, the 2½-in. (65-mm) hose can be reduced to, or replaced by, a smaller line.

Large, uncompartmented areas. Fires in large, uncompartmented areas require levels of reach, penetration, and volume that are beyond the capabilities of smaller hand lines. In addition to wide-open floor plans, occupancies such as supermarkets, bowling alleys, warehouses, theaters, houses of worship, and the like often have high ceilings. High ceilings allow massive amounts of heated fire gases to accumulate. Once these flammable vapors ignite, they may prove to be too formidable for streams from smaller lines. The reach and tremendous cooling power of the 2½-in. (65-mm) line with 1⅛-in. (30-mm) tip allows operation from an entranceway into the rolling flame front of combustible gases beneath the ceiling. Once the hazard in the fuel-laden overhead area has been mitigated, the attack can be pressed deeper into the structure's interior.

Tons of water. At some fires, extinguishment simply requires tons of water. This is often the case for fires in piles of tires, junkyards, garbage dumps, and lumberyards, to name a few. A 2½-in. (65-mm) line with 1⅛-in. (30-mm) tip operating at 50 psi (350 kPa) NP discharges more than a ton of water a minute. The use of smaller lines in this kind of situation would be an exercise in futility.

Standpipe System Operations. Proper consideration for members' safety demands the use of 2½-in. (65-mm) hose and smooth bore nozzles for standpipe operations. For many years, NFPA 14, the *Standard for Standpipe Systems*, was based on the use of 150 ft (45 m) of 2½-in. (65-mm) hose equipped with a 1⅛-in. (30-mm) smooth bore nozzle at a 65 psi (455 kPa) standpipe hose valve outlet pressure. Depending on which version of the standard a given standpipe system was designed under, outlet pressures can be either 65 psi (455 kPa) (pre-1993) or 100 psi (700 kPa) (post-1993). Outlet pressures such as these simply do not meet the friction loss requirements for smaller-diameter hose, especially in conjunction with 75-psi (525 kPa) or 100-psi (700 kPa) nozzles.

Many standpipe systems have pressure-reducing valves (PRVs) that are not field adjustable. This means that no matter what pressure fire department pumpers pump into the system, outlet pressure does not rise above a given outlet's rated pressure. Retired New York City Fire Department Deputy Chief John Norman[3] admonishes that to use anything other than 2½-in. (65-mm) hose and smooth-bore nozzles for standpipe operations is to misuse the standpipe system. Prior to becoming a member of the career fire service, Norman was a fire protection engineer and made his living designing sprinkler and standpipe systems.

Because of design configurations and conditions of standpipe systems, flow and pressure problems chronically plague operations. As stated by Fredericks, "A standpipe system is like a big black box which fire department Members did not design or install, do not maintain, and in most cases do not inspect or test yet Members expect to put water in one end and have it come out at the other end at the proper pressure and flow." Although certainly not an ideal situation, even at a low outlet pressure, the combination of 2½-in. (65-mm) hose and a 1⅛-in. (30-mm) smooth bore tip still can develop a usable fire stream.

In February 1991, the Philadelphia Fire Department had a disastrous experience dealing with a fire in the One Meridian Plaza building. At the time, the Philadelphia Fire Department used 1¾-in. (45-mm) hose and 100 psi (700 kPa) automatic fog nozzles for standpipe operations. Misadjusted standpipe PRVs combined with the smaller hoseline size and pressure-intensive automatic fog nozzles resulted in a weak stream with which to fight the large fire. Three firefighters were trapped in the building and killed. Firefighting forces were pulled from the building in anticipation of possible collapse; the fire consumed several floors until its upward progression was stopped by nine sprinkler heads in the partially sprinklered building.

METHODS OF FIRE ATTACK

FFII 5.3.2 For quite some time, firefighters have debated the various methods of applying a fire stream to attack a fire. For many years, only one method existed by which to extinguish fire and that was to apply a solid stream of water directly to the burning solid fuels, commonly known as the "direct attack." This is one of several types of fire attack; it is described below along with five other types: the indirect, combination, and modified direct attack, blitz, and transitional.

Indirect method of attack

Prior to World War II, direct attack was the long-established mainstay of the fire service. During the war Lloyd Layman, Chief of the Parkersburg Fire Department in West Virginia, became the commandant of the United States Coast Guard firefighting school. His duties included determining the best way of combating fuel oil fires in the confined below-deck spaces of ships. Toward this end he conducted extensive research and testing. A major part of his efforts consisted of a series live-fire tests aboard a decommissioned liberty ship.

Results of Layman's research and testing led to new theories about fire attack and eventually to the formulation of the **indirect method** of fire attack. He found that it was efficient to remotely inject a fog stream into an unoccupied compartment. To use this method, the space must be unoccupied, because it would rapidly become untenable upon the introduction of a fog stream. After the fog nozzle operates for a sufficient amount of time, it shuts down. The hatch through which the nozzle operated is then shut to preclude ventilation and confine the expanding steam. The steam is the major factor in extinguishment, through cooling as well as smothering. When water turns to steam, it expands by a ratio of 1700:1 (fig. 16–8). Water turns to steam at 212°F (100°C). Few fires only reach a maximum 212°F (100°C). What many people fail to realize is that at 1,000°F (538°C), a ceiling temperature easily reached at structure fires, the expansion ratio of water is 4,000:1.

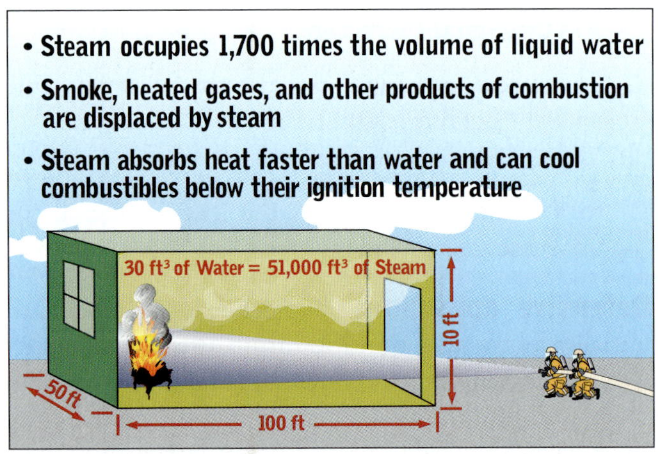

Fig. 16–8. When water turns to steam it expands by a ratio of 1,700:1.

At the war's end Layman resumed his peacetime career as the Parkersburg, West Virginia, fire chief. This is when he began to transfer what he had learned about the efficient extinguishment of shipboard fuel oil fires in unoccupied confined spaces to the suppression of interior structural fires (fig. 16–9). He did, however, maintain the following tenets for the use of the indirect attack for fires in buildings.

- First and foremost, there must be no life hazard within the fire compartment. The term **life hazard** includes both building occupants and fire department members. The application of the indirect method of attack rapidly destroys the thermal balance of the fire compartment, making it an environment both untenable and unsurvivable.

- The ceiling temperature must have reached at least 1,000°F to have an environment conducive to the efficient conversion of the water fog into steam to cool and smother the fire.
- The water fog must be remotely injected into the fire compartment from the exterior, preferably from a window. The fog stream is held in a fixed position, because it is directed into the overhead area of the fire compartment. The injection of water fog into the overhead area results in the fine water droplets, rapidly vaporizing and expanding, which destroys the thermal balance and pushes a high level of wet heat down to the floor. Avoid ventilation to keep the voluminous amounts of expanding steam necessary for efficient extinguishment within the fire compartment.

Layman emphasized that if any of the above-listed tenets could not be fulfilled, a direct attack on the fire is then the proper course of action.

Fig. 16–9. The indirect attack as described by Lloyd Layman

Combination method of attack

The **combination method** of attack was developed by Keith Royer and Floyd W. Nelson during numerous Iowa State University test fires in acquired structures. They expanded on Layman's work and deemed it more efficient to rotate the fog stream in a clockwise direction hitting the ceiling, walls, and floor. They also deemed it appropriate to inject the fog stream through either a window or a doorway. The reason that they stated that the nozzle be rotated in a clockwise manner is that during their many test fires they were able to observe that clockwise nozzle rotation pushed products of combustion and extinguishment away from the nozzle operator. Counterclockwise nozzle rotation, they observed, would draw products of combustion and extinguishment toward the nozzle operator (fig. 16–10).

Like Layman, Royer and Nelson maintained that there must be no life hazard within the fire compartment (again referring to building occupants as well as department members). They also stated that ventilation should be as limited as possible to confine the massive amounts of steam generated by the combination attack to the fire compartment. As with the indirect attack, the combination method of attack required a ceiling temperature of a minimum of 1,000°F (538°C) to take advantage of water's expansion ratio of 4,000:1 at that temperature. Also, like Layman, they maintained that if the aforementioned parameters did not exist, a direct attack on the fire would be the best course of action.

Fig. 16–10. The combination method of attack requires a fog stream injected through a door or window and rotated clockwise to push products of combustion away from the nozzle operator.

Direct method of attack

The **direct method** of attack is simply the application of a solid stream or a straight stream directly on the burning solid fuels (fig. 16–11). This type of attack is used when the fire is relatively small: a single object or a small number of objects are burning and can be easily extinguished by applying the water directly on the burning materials. When several objects or an entire compartment is on fire, then the modified direct attack (described below) is called for.

Larger metropolitan fire departments have a long tradition of aggressive interior direct attack of structural fires. This is because they are most often tasked with the protection of a high life hazard in the form of dense concentrations of people in large buildings. Experience has proved time and again that the best way to protect life is to put out the fire in the most expeditious manner. Once the fire is extinguished, every other fireground activity (entry, laddering, search, rescue, removal, ventilation, etc.) can be performed more safely and efficiently. This results in greatly enhanced probabilities of safe outcomes for both building occupants and operating members.

Prior to the widespread use of breathing apparatus, evidence of the use of such aggressive tactics at more arduous fires was the presence of numerous members prostrate on the sidewalk and incapacitated because of smoke inhalation. It was realized that this risk was worth taking because of the likely reward of saving many lives and thus fulfilling the fire service's custodianship of the populace.

During and after the development of the indirect and combination methods of attack, most large fire departments never varied from their tried and true tactics of aggressive interior direct fire attack coupled with coordinated aggressive natural ventilation and an aggressive primary search. "Natural" ventilation refers to allowing the products of combustion to move *without* the use of positive pressure fans or other mechanical equipment. With the advent and consequent widespread use of more efficient, user-friendly breathing apparatus, many smaller fire departments began to emulate the aggressive interior tactics of larger urban departments.

Along with this seemingly positive development came efforts by many to press into service for interior operations the indirect and combination methods of fire attack. The indirect and combination methods of attack are basically exterior or defensive operations. Combining defensive tactics with offensive interior operations is normally dangerous and counterproductive. Always keep in mind that Layman, Royer, and Nelson—the creators of the indirect and combination methods of attack—strongly stated that the indirect and combination methods should not be employed if a life hazard exists within the fire compartment. Even if the building is unoccupied prior to the arrival of the fire department, once members enter the building to operate, a life hazard exists in the building.

The foremost mission of the fire service is to protect life, which is why operating members enter burning buildings, plain and simple. Because the paramount mission of the fire service is to save life, members must make extreme efforts to access the area of the building near the seat of the fire. That is the area where victims are in the most extreme peril. All efforts must be made by engine company members to aggressively push the initial attack line in to extinguish the seat of the fire. All efforts must also be made by ladder company members to aggressively vent, enter, and search the building as near to the seat of the fire as possible. The parameters set forth earlier indicate a method of fire attack appropriate to deal with the life hazards of occupied buildings. A major facet is that it allows members to enter fire compartments for extinguishment, ventilation, and search. Additionally, the chosen method of fire attack should be that which does the most to preserve the thermal balance of the fire compartment as well as to first preserve and the expediently improve the tenability of the fire occupancy.

Aggressive interior direct attack coupled with coordinated aggressive natural ventilation is best tactic for expediently improving the tenability of a fire compartment without compromising thermal balance or introducing unintended ventilation flow paths. Combined with aggressive primary search, this is the operational doctrine that most greatly enhances the probabilities of survival for both victims and operating members. The synergistic effect of simultaneous operations, a combined arms approach if you will, allows the fire service to most completely fulfill its custodianship of the public's safety.

Fig. 16–11. A direct attack procedure

Modified direct method of attack

The **modified direct method** is a two step attack: the application of a solid stream or a straight stream is directed into the overhead area (the first step), out front ahead of the nozzle team (fig. 16–12). The nozzle must be moved vigorously in a clockwise or side-to-side motion, splattering the stream against the ceiling and upper walls. Breaking the stream up in this way causes large chunks of water to rain down all over the fire area, finding the seat of the fire in the burning solid fuels. The stream is then directed on to the burning solid objects in the room (the second step) to achieve knockdown and extinguishment. (If the stream was to be applied first directly onto the large group of burning solid fuels it would very likely forcefully push fire ahead of itself, into and then up a wall, then sending it rolling back across the ceiling above the heads of the nozzle team.)

It should be noted that some texts refer to the modified direct attack as a "combination attack." This is historically incorrect; the combination attack involves the use of a circling fog stream in an uninhabited compartment, as described above.

Applying the stream first into the overhead area simultaneously does two things. It cools the superheated area near the ceiling where flashover and other rapid fire progress phenomena are born while also dropping water down onto the burning solid fuels. The application of water onto the burning solid fuels quenches the fire at its source, the fuel-flame interface, and quells the production of flammable vapors.

Fig. 16–12. Water is applied first into the overhead, striking and sweeping the ceiling and upper walls, and then onto the burning objects in the room in a modified direct attack. The nozzle operator here is in a hallway door, inside the building. (Courtesy of Sean Gray)

The blitz attack

Developed in the 1970s, the blitz attack is a swift knockdown of a large quantity of fire that is rapidly spreading on the face of the building. This exterior attack applies massive quantities of water to the entire face and windows of the building in order to slow the fire down and attempt to stop its spread both horizontally and vertically. The blitz attack is a very rapid application of massive amounts of water immediately after arrival of the engine, using the water from the engine's tank, applied from a distance. The time-consuming effort of laying of a water supply line for the engine is sacrificed for the benefit of getting water on the fire soon after arrival.

The original form of the blitz attack utilizes a large volume deck gun device (monitor) on top of an engine, which can be aimed and used once the pump operator has sent water to the device (fig. 16–13). In some cities, preconnected large volume 2½-in. handlines with smoothbore nozzles are deployed with the same effect. More recently, lightweight portable master stream devices have been developed for this type of attack. Water is applied from a distance in all cases.

It is important to recognize that blitz attacks are limited by the quantity of water available in the engine's onboard water tank. What this means is that the attack may only last a few minutes. It is important to calculate exactly how long the blitz attack can be sustained given the volume of the tank and the flow rate of the nozzle that will be used.

It is critical that the stream be kept moving across the face of the building to hit all areas to try to have the biggest impact on slowing the fire. Do not spend too much time in one area while fire is still visible in other areas. Remember, the goal is to try to knock down the large amount of fire sufficiently so that hand lines can subsequently follow up and extinguish the remaining fire.

Fig. 16–13. The blitz attack shown here uses a master stream from an exterior attack, however a 2½-in. hand line with a smoothbore nozzle can also be used on a rapidly spreading large volume of fire. (Courtesy of P.J. Norwood)

The transitional attack

The newest type of attack is the transitional attack, developed as a result of the flow path research studies conducted over the last few years. It is an exterior attack using a handline, applying water through an opening (typically a window) directly into a specific fire compartment. The goal is to reduce the temperatures in the fire compartment, while handlines are being deployed to attack the fire using a direct or modified direct attack immediately following the transitional attack. The transitional attack "resets" the fire, forcing it to redevelop to full intensity.

In the transitional attack, the attack takes place at the opening. Using a 2½-in. handline with a smoothbore nozzle, the stream is applied through the lower third of the opening. The stream is angled to hit the ceiling of the compartment, just as described in the modified direct attack discussed earlier. Positioning the nozzle in this manner will allow the products of combustion and steam to exit the structure through the top two-thirds of the opening and allow the water stream to break up at the ceiling and drop onto the fire below (fig. 16–14).

Fig. 16–14. Position the nozzle correctly in a transitional attack, applying the water through the lower third of the opening and with the stream directed at the ceiling. (Courtesy of Sean Gray).

Vigorously rotating a straight or solid stream should never be performed, as it will have the same effect as filling a window with a fog pattern, effectively preventing a bidirectional flow of hot gases and steam from venting out the upper portion of the window and cooler air from entering the lower portion of the window opening. Recognize that the water stream will not suppress fire outside the compartment into which the water is being applied, such as adjoining rooms and hallways.

The transitional attack is used only when fire is issuing from the structure, not when the fire is contained on the interior. First-arriving officers will make the decision when to use the transitional attack.

It is important to point out that the transitional attack does not take the place of an interior attack; the transitional attack is intended to temporarily slow the fire down with rapid follow-up with interior attack. If the fire can be attacked more quickly with an interior attack, then the transitional attack will not be used. A limited amount of personnel available to stretch hoseline is also a concern. Initiating an interior attack should not be delayed in order to conduct a transitional attack when there aren't enough firefighters on the fireground to get a hoseline stretched to the interior.

FIRE STREAM TYPES

FFI 4.3.10 All three stream types (solid, straight, and fog) accomplish extinguishment when applied for sufficient duration and in sufficient volume that meet or surpasses the critical flow rate. In the case of fog streams, however, prior to extinguishment, conditions in the fire area—especially in the fire compartment itself—first worsen before finally improving. Upon initial application of the fog stream, scalding steam is produced because of a large degree of premature vaporization of the water fog. This excessive steam generation destroys the thermal balance of the fire compartment. As a result floor level temperatures increase and visibility is reduced. In addition to the quantitative increase of the heat at floor level, there is also a qualitative change in the heat. Because of excessive steam generation, conditions at floor level change from dry to wet heat, further multiplying the debilitating effects to firefighters and fire victims. All these factors lead to the existence of a period of decreased tenability between the time of first water application and the time of extinguishment and ventilation of the products of combustion and extinguishment (flame, heat, smoke, gases, steam, and air movement). This decreased tenability can be inadvertently caused when using a straight stream on a combination nozzle and changing the pattern to a fog stream.

Much discussion revolves around the tenability of conditions within the fire compartment at floor level. This is because the floor level is where members operate and more importantly where victims wait to be rescued, unprotected and vulnerable. The potential to save human life is the major motivating factor that compels members to go into burning buildings in the first place. Hence, the fire service must focus attention on how its operations either increase or decrease the window of survivability in the area where human life is expected to exist.

This text also touches on the concept of products of combustion and extinguishment. Products of combustion, as well as the various methods by which to combat combustion, have differing consequences. Each method of fire attack and each type of fire stream affects the tenability of the environment in the fire compartment differently. Products of extinguishment include steam generation, air movement, visibility reduction, and temperature change. Hence, members of the fire service must be aware that the choices of method of fire attack and type of fire stream significantly affect the probabilities of survival for the civilians they vowed to protect from fire.

The amount of air, fire gases, and smoke pushed ahead of a fog stream is so voluminous that it quickly overwhelms the likely ventilation openings ahead of the nozzle in a residential setting (e.g., windows). Only so much product of combustion and extinguishment can be forced through the window(s) at any given moment. The significant portion of products that cannot be vented when forcefully pushed forward by the fog stream then impact the wall surface and roll back over and around the nozzle team. This phenomenon has been observed in innumerable building fires and tested at the Rockland County, NY Fire Training Center. The phenomenon known as **fog nozzle ricochet** is a major contributor to the destruction of thermal balance, excessive steam generation and accumulation, and loss of tenability. This phenomenon is not caused by a fog nozzle, but by the untrained use of a **fog stream**.

Another attribute of a hose stream that greatly affects both speed of extinguishment and member safety is the reach of the stream. A stream with greater reach allows members to apply water to the seat of the fire sooner and from a greater distance than does a stream of lesser reach. Ability to engage the fire sooner leads to quicker extinguishment and hence greater overall safety for everyone on the fireground. The greater distance also enhances safety; members need not be in as close a proximity to the seat of the fire to initiate extinguishment. The closer the nozzle team must approach the fire to initiate and accomplish extinguishment, the greater the compromise to their safety (fig. 16–15). A close approach to the fire in order to achieve extinguishment is dictated by the limited reach of a fog stream (the wider the fog, the closer that you have to get to the fire to extinguish it).

Fig. 16–15. The limited reach of a fog stream

The superior efficiency and efficacy of solid streams versus fog streams is further evidenced in tests performed by William Clark, former FDNY battalion chief and director of Wisconsin's fire training programs. Clark's test results showed that at equal flow rates and with equal fuel loads:

- Solid stream extinguished the fire faster than did the fog stream.
- Solid stream required less total water to achieve extinguishment than did the fog stream.
- Solid stream led to less measurable runoff water than the fog stream.

Thus, it is not a great leap in logic to conclude that a solid stream applied at a high rate of flow causes flame, heat, smoke, and gas production to cease in the shortest amount of time, accomplished by using the least amount of water. Thus, the solid stream is the most efficient and effective in the limiting of property damage caused by fire, smoke, and water. Far more important, however, is the fact that it is the most effective, efficient, and expedient method for the fire department to attack and kill a dynamic, out of control, untenable physical phenomenon that is threatening the lives of victims. In this manner fire department operations return the environment to nonthreatening equilibrium in the shortest possible amount of time and preserve to the greatest degree any savable human life.

ADVANCING THE CHARGED LINE

SKILL DRILL

FF1 4.3.10 The efficiency and effectiveness of the advancement of the charged line is in no small way determined by the efficiency with which the dry hoseline is stretched and flaked prior to charging. A major factor in the efficiency of stretching is determined much prior to arrival at the fire building. Engine company apparatus design directly effects the proficiency with which engine company members can stretch and use fire hose. An apparatus design that highlights user-friendly attack hosebeds greatly facilitates the performance of the engine company's primary function, its reason for existence, the stretching, advancing, and use of fire hose to kill fire. User-friendly attack hosebeds are low enough so that even the shortest member of the fire department can reach the top layer of hose in the attack hose load while standing on the ground. They also provide ample room for sufficient lengths of hose to make up static attack hose loads capable of covering the distances of any foreseeable stretch in the response area. Static loads also allow the supply line and the attack line to, in effect, simultaneously stretch via the backstretch as opposed to supply and attack lines stretched sequentially with preconnected attack lines. This also has the advantage of placing the apparatus where the pump is most efficient: at the hydrant. This type of hose load is best situated in a section of the main, or rear, hosebed (fig. 16–16).

Fig. 16–16. A bed of supply hose on the left and three beds of static attack hose loads. Static loads are not preconnected and are also known as dead loads. (Photo by Jay Comella)

In addition to apparatus design, policy makers affect the degrees of efficiency, effectiveness, and safety with which engine company members are able to perform their lifesaving function through the levels of company staffing deemed adequate to protect the public from fire. The quantity and quality of well trained members in a company has an extremely significant effect on the outcome of the company's operation and by extension the outcome of the entire fireground operation. A successful outcome of a fire department operation equates to lives and property saved. Unsuccessful outcomes of fire department operations result in the loss of life and property that was salvageable upon arrival. Fire departments able to stop the shrinking of staffing at six members per company should be emulated in this respect. Company staffing directly relates to the efficiency of the work performed as well as death and injury rates in the fire service. Personnel is a fire department's most valuable commodity with respect to mission capability.

Over the years staffing levels have shrunk in many departments as the budget axe has fallen time and time again. Governmental entities that have repeatedly tasked fire departments to do more with less have effectively increased the risks of injury and death to both department members and the citizens they are sworn to protect. In light of this negative external influence, the fire service's traditional organizational mind-set must change. Long accepted practice in many departments has been for engine companies to compete to get first water on the fire. The approach to a company's role within the entire fireground operation must change so that engine companies work in concert with one another, rather than in competition. Sufficient personnel must be committed to getting the first line to the seat of the fire before addressing the deployment of additional lines. Attack lines are best stretched in series, not in parallel. Even the better staffed, forward thinking departments take the approach that all but the simplest hose stretches or advances are, at a minimum, two-company evolutions.

POSITIONS ON THE LINE

Engine company officer

FFI 4.3.10 Ideally, if staffing levels allow, the engine company officer should not physically take part in handling the line. The officer bears overall responsibility for the operations of the company and member safety. Staffing levels that allow the engine company officer to be a true supervisor positively affect the highest order on the efficiency, effectiveness, and safety of engine company operations. By remaining physically uninvolved in the operation, the officer is better able to direct it. The officer must be able to avoid the potential tunnel vision that can come with being tied into the line and assisting with its movement. The officer's proper position is slightly behind and to the side of the nozzle operator. It is important that the officer is at the point of attack to observe and experience conditions as they are encountered by company members.

When unencumbered by the need to perform the physical labor of moving the hoseline, the officer is at liberty to move out ahead of engine company members to perform reconnaissance regarding the fire's location and nature. This in turn provides the information needed to perform the following:

- Properly direct members to the correct route to the fire, the size of line needed, the number of lengths sufficient to cover the entire fire area.
- Ascertain forcible-entry problems and life hazard.
- Perform search prior to arrival of line to the extent allowed by fire conditions.
- Confine fire by closing door if possible.

Confirming, to the extent possible, the location of the fire is necessary prior to initiating the hoseline stretch. Failure to do so can lead to many unwarranted mistakes:

- Stretching short
- Stretching long
- Stretching to the wrong location on the fire floor
- Stretching to the wrong floor
- Stretching via the wrong stair well
- Even stretching into the wrong building

The consequences of stretching short are obvious: failing to reach the seat of the fire. It is often wise to include one or two extra lengths in one's hose stretch estimate to err on the side of caution. However, realize that as the amount of excess hose increases, the difficulty in managing the hoseline and the potential for kinks also increases. Kinks in the hoseline lead to significantly reduced water flow. Stretching long can also lead to unduly high pump pressures.

Ensure that all doors through which the line is stretched and advanced are positively chocked open. A door closing on the line can severely hamper the stretch and/or advance, sometimes with dire consequences. This is especially true if the door closes over a dry line that is then charged while it is under the door. The door becomes, in effect, a hose clamp and little or no water gets to the nozzle. The charged line becomes an effective door chock that secures the door in the closed position. Those caught on the wrong side of the door in this situation are definitely in dire straits. As is so often the case, in fire department operations, the devil is in the details. The use of an inexpensive wooden wedge properly positioned to securely chock the door in the open position can quite literally mean the difference between life and death. Shakespeare addressed the importance of paying attention to detail thusly, "For want of a nail the shoe was lost. For want of a shoe the horse was lost. For want of a horse the battle was lost."

Hose should be stretched dry as far as is safely possible to avoid expending time and energy on undue labor. The hoseline must be charged before entering the fire area or that which may rapidly become the fire area. Prior to the line being charged, sufficient dry hose must be properly **flaked out** at the entrance to the fire area. Properly flaked hose is laid out so the bight (bends of hose where it changes direction) of the line are open enough to have the least propensity for **kinking** (fig. 16–17). The fire is often referred to as the enemy and the fire building as the battleground. The logical extension of this line of logic is that kinks are collaborators. Kinks rob valuable amounts of water flow from the attack hose stream. It is the duty of all personnel to remove kinks whenever they are found. Removing kinks is such an important consideration that it is responsibility of all personnel on the fireground, whether engine company or ladder company members, the newest probie or the chief of the department (fig. 16–18).

Fig. 16–17. Properly flaked hose greatly lessens the propensity for kinking.

The pump operator must be notified to charge the line only after it has been properly stretched and flaked out. To do so any earlier invariably increases the time necessary to get water on the fire. Whenever a hoseline is prematurely charged, the labor and time involved in advancing the line and removing the kinks increases dramatically.

Fig. 16–18. Kinked lines prevent maximum flow of water to the fire.

Prior to entering the fire area, entrapped air must be bled from the charged line (fig. 16–19). The nozzle must be opened fully to ensure the attack line is supplied with sufficient water flow and pressure before commencing the attack. This is known as **bleeding the line**. At this juncture more than any other, the engine company truly reaps the benefits of having an officer in a purely supervisory role, free to coach, guide, and direct the members through prompt extinguishment and also be a custodian for their safety throughout the process.

Fig. 16–19. Nozzle operators should bleed the charged line prior to entering the fire area.

The officer must strive for the utmost possible awareness of conditions as the company prepares to enter the fire area and throughout the advancement of the line and extinguishment of the fire. Although senses are severely muted, the officer must make use of what little sensory input he or she is afforded. After advancement into the fire area begins, smell is negated with the donning of the self-contained breathing apparatus (SCBA) facepiece. Hearing can be affected by a cacophony of competing fireground noises. Sense of sight is severely diminished in most cases because of voluminous, thick, black smoke. Perhaps most damning is the fact that the total embunkerment of today's fire service severely dulls one's sense of feel and the ability to gauge temperature or increases in temperature.

Important audible stimuli that compete with other fireground noises include victims' muffled coughs, moans, and cries for help; crackling fire; falling or collapsing construction material; breaking glass; forcible-entry procedures; and operating saws. All these things give clues about the conditions that exist, how conditions may change, and how operations should be conducted. All this aural input must be constantly received and analyzed while the officer concurrently maintains the ability to continually monitor the handheld radio for vital communications. The remote microphone/speaker is an invaluable tool for tactical fireground communications.

Smoke impairs an officer's sight while advancing through the fire area. The most significant visual input to consider may, in fact, be that there is no visual input. Prior to entering the fire area, while observing the smoke issuing from it, may be the last opportunity to visually analyze conditions.

The personal protective equipment of today's operating forces has significantly curtailed the ability to feel the level of heat and discern rapid increases in heat in the operating environment. Prior to this development, the ability to receive this physical stimulus on one's skin from the environment was the gauge by which many veteran members of the fire service determined how and how long they could operate in a given environment. The time between feeling a painful heat stimulus and being able to remove oneself from the environment or take action to improve it is called **alarm time**. As the insulation from the environmental heat stimulus increases, the environmental temperature at which one feels heat inside the protective envelope also increases. This delay in becoming aware of high heat decreases alarm time, the window of opportunity one has to improve the chance to avert a catastrophic outcome.

Significant reconnaissance can be gained if the door to the fire area is open and conditions allow approach to the doorway without water application. At floor level one may be able to look under the smoke layer to ascertain the layout of the occupancy, location of victims, location of fire, and presence of potential obstacles in the path of the advancing line. Sweeping the beam of one's hands-free, forward-facing hand light under the smoke layer will aid immensely in this task (this is covered in greater detail in chapter 19, Search and Rescue). Of course, severe conditions warrant immediate stream application. Once the stream is operating, however, visibility is likely nil, even at floor level.

If the door to the fire area is closed, ensure the nozzle team is low and to the side of the doorway. The side of the doorway that is chosen will depend on a number of factors including the expected layout of the fire compartment, the layout of the area outside the entrance to the fire compartment, which way the door swings, and so forth. The nozzle team should be at an angle relative to the doorway that allows stream application, if necessary, when the door is opened.

The door can be opened once the nozzle team is ready with ample charged and bled line and forcible-entry issues have been addressed (fig. 16–20). The nozzle operator must be ready to flow water if conditions dictate. If a fog nozzle is being used it must be adjusted to the straight stream pattern. If possible, take a few seconds to stay low and let it blow. It is beneficial to pause the operation briefly while the energy behind the pressurized, venting products of combustion diminishes somewhat. This is a wiser choice than occupying the doorway as soon as it is opened and being in the path of whatever may vent through it. The doorway used for access of the initial attack hand line may be the first ventilation opening (flow path) to which the fire has access. The initial amount and pressure of the products of combustion that vent through this opening can be significant, and it is unwise to be directly in their path of travel. Of course, if fire venting through the doorway threatens members' position, immediately meet it with copious water from the hand line. If conditions allow a brief delay in water application, a quick look under the smoke layer reaps information beneficial to the operation.

Fig. 16–20. The nozzle team's position at a doorway prior to entering the fire area is critical for a successful operation.

In the past, most members of the fire service were drilled with the mantra, "Don't open the nozzle until you see fire, never put water on smoke." This was fine back when dictum came into being. It was consistent with the type of fire and smoke conditions normally encountered during that time period. With the advent and proliferation of plastics, the amount and type of smoke has drastically changed. At many fires one cannot depend on seeing the fire at all, in which case it is common to use heat to indicate when to open the nozzle. The advent of full personal protective equipment negated that tactic. Now when a member feels heat it may be too late to open the nozzle in time to avoid burn injury. In the modern fire environment, the prudent method of determining when to open the line is to do so based on observation of the smoke condition before becoming immersed in its blinding cloak.

Prior to entering the fire area, the smoke condition must be rapidly observed and analyzed. Thick black smoke is heavily laden with unburned fuel. A compartment whose volume is full or nearly filled with thick black smoke contains a lot of ignitable fuel. Rapidly moving smoke is under pressure, a result of heat in the fire compartment; the higher the heat, the higher the pressure. Thick, black, rapidly moving smoke is a mixture of heat and fuel that only needs oxygen to complete the fire triangle. A fire compartment disgorging a heavy volume of dark black, dense, pressurized, and turbulent **angry smoke**, known as **black fire**, may light up (autoingnite) and potentially bring the compartment to flashover. If smoke conditions at the entrance to the fire occupancy are light enough to not indicate a potential rapid fire progress event, the line can be advanced, without flowing water, until either the fire is found or conditions indicate the need for stream application. Once fire is visually located, or angry smoke conditions are encountered, the nozzle must be opened to begin gaining control of the environment and improving conditions (fig. 16–21).

Fig. 16–21. Angry, black smoke usually indicates unburned fuel which could ignite within the fire compartment.

Let the water do the work. Long before members are ever close to the enemy, decide on proper weapons configuration and selection to ensure they are armed with a powerful, long-reaching, high-volume, hard-hitting attack stream. Open the nozzle as soon as the stream can affect conditions. Members need not be in the close proximity to the seat of the fire required by the short reach of a fog stream. The long reach of a solid or straight stream allows extinguishment to begin sooner and from a farther distance than does the short reach of fog streams. The time and distance aspects associated with solid streams are both factors in increasing the safety of operating members. Even when stream application is at an obtuse angle to the doorway of a room involved in fire, as when approaching the doorway from down a hallway, getting water into that room begins to improve conditions. It is unnecessary to wait until the members of the nozzle team are in the doorway to start getting water into the fire area (fig. 16–22).

Fig. 16–22. A long-reaching stream applied to a fire area allows extinguishment to begin sooner and from a farther distance.

If, on opening the doorway to the fire compartment or approaching the open doorway, a voluminous and angry smoke condition is indicative of a rapidly deteriorating environment, copious amounts of water from the nozzle must be swept back and forth across the ceiling out front and ahead of the nozzle team. A fire exhibiting such characteristics is an ominously volatile environment. The smoke is a large quantity of superheated fuel seeking enough oxygen to become ignitable. Members cannot afford to give any concern in this situation to the outdated cliché, "Don't put water on smoke." In these circumstances the characteristics of modern fuel loading (plastics) and the effects of tight building syndrome are likely combining to create a black fire scenario. Without any visual indicators, black fire has the potential to instantaneously transition into flashover. Water kills flashover. The tool to improve the environment is literally at the nozzle operator's fingertips. For members who have had both visual and heat sensing capabilities taken away, it is highly imprudent to push into this environment without preparing it to have a higher degree of human survivability. Sweeping the ceiling with copious amounts of water out front and overhead through the superheated fuel–laden smoke serves to sever the chain reaction of rapid fire development through the cooling effect of the stream.

The phrase concerning stream application, "sweep the ceiling with copious amounts of water," is purposefully used to convey the need for adequate water to kill the potential for flashover. Another phrase, "pencil the ceiling with the stream," has had widespread exposure in the fire service. **Penciling** has numerous interpretations, one of which is to use limited amounts of water for short durations to delay flashover. When restoring control to an out-of-control environment at a building fire, the goal should not be to delay flashover or to cool fire gases just enough to hold conditions back from crossing the threshold of rapid fire progress phenomena. The greatest contributions an engine company can make to protecting the most savable human life in the shortest amount of time are bringing the highest practicable flow rate to the fire at the earliest opportunity and continuing stream application until extinguishment.

Once a significant amount of water has been applied to the overhead area of the fire compartment, the stream should be swept across the floor prior to advancing any farther (fig. 16–23). This is necessary to cool the scalding runoff created by applying water into the superheated upper areas of the fire compartment where temperatures can exceed 2,000°F (1,093°C). During the majority of the advance and extinguishment, the stream will be operated at an upward angle outward ahead of the nozzle team. However, the stream should also be swept across the floor intermittently throughout the advance. In a further attempt to reduce the potential for knee burns, members must advance so the knees are not in contact with the floor. If this is unfeasible, members must limit the time that knees are in contact with the floor by frequently raising them from the floor to interrupt heating. Knee injuries occur, even through bunker pants, because the insulating layer of the pants becomes thinner as it stretches and compresses when kneeling. Sweeping the deck with the stream can also blow debris out of the path of advance. This can include a wide range of things from hypodermic needles to all manner of feces. It is important to pay attention to the sound of the stream sweeping the floor. If it becomes quieter, there may not be any floor ahead of the nozzle team. There may be hole or an open stairway.

Fig. 16–23. Sweeping the floor with the water stream prior to applying it to the burning objects in the room (for final knockdown and extinguishment) removes debris from the path of the advancing nozzle team.

Nozzle operator

SKILL DRILL

FF1 4.3.10 During the stretch the nozzle operator ensures sufficient lengths of dry hose cover the entire fire area are near the entrance to the fire area (fig. 16–24). This is always a minimum of 50 ft (15 m), or one length of line. The nozzle operator is responsible for carrying one folded length, often called the **working length** to the **drop point**. The drop point is the point near the entrance to the fire area where the working length is flaked out so that slack line can be advanced into the fire area. If more than one length of line is required to cover the fire area, it may or may not be carried by the nozzle operator. However, the nozzle operator must ensure that all necessary slack hose is flaked out as near as possible to the fire area entrance. Having sufficient slack line to cover the entire fire area prior to the advance pays off in the proficiency with which nozzle team advances through and is able to extinguish the fire.

Fig. 16–24. The nozzle operator must ensure that sufficient lengths of dry hose reaches the entrance to the fire area.

The nozzle bale should be a slightly bent arm's reach out in front of the nozzle operator. The line should be on the side of the nozzle operator's dominant arm; however, there is a school of thought that the line should always be on one's right side because of the direction a burst length tends to rotate. Being on the dominant side puts the dominant arm to the rear, and this arm does the bulk of the labor in holding the line. For ease of explanation, further description refers to a right-handed nozzle operator.

The forward, or left, hand controls flow and directs the stream. The forward hand operates the bale (fig. 16–25a). Once the bale has been operated, the hand moves to the hose behind the last male hose butt. The hand must be in an underhand position on the hose (fig. 16–25b). This is the position from which the stream is directed by the forward hand. If the hand were to be left on the bale, nozzle, or hose butt, a hard-to-control kink would likely develop behind the hose butt. A further negative aspect of leaving one's hand on the bale is the likelihood of partially closing the bale and reducing flow. If the hand is placed on the hose in an overhand manner, there is great likelihood of forcing an unwieldy kink into the line behind the forward hand.

Prior to operating the nozzle, the nozzle operator must expect that when the nozzle is opened up, reaction force must be overcome. The body must be in an attitude that reflects this. He should be leaning forward into the direction from which the nozzle reaction force will come, with the upper body positioned so the spine's angle roughly mimics that of the hoseline. In this manner the body is ready to receive, absorb, and overcome the nozzle reaction force once the line begins to operate.

In any fire that is past the incipient stage, advancing the hoseline takes members into a volatile high-heat environment. Members must be able to advance the line while operating in the lowest portion of the fire compartment (i.e., floor level). Using a solid stream maintains the thermal balance and preserves the floor area as the coolest portion of the fire compartment. Down on the deck is the safest place to be in the fire compartment should any rapid fire progress event occur.

The classic duck walking method keeps one low and keeps one's knees off the floor (fig. 16–26). However, this can be an awkward position to maintain and may not be the most efficient for achieving forward momentum. Some variation of the duck walk where one or both knees touch the floor at intervals may be more efficient. It is important to avoid having both knees on the floor all, or even a significant amount, of the time. Not only does it place one at a higher risk for knee burn injuries, but it is also harder to achieve both forward momentum and traction. It is also an unstable body position. To be on one's knees is to place oneself in a subservient position.

Fig. 16–25a. The nozzle operator's forward hand operates the bale.

Fig. 16–25b. Once the bale is operated, the forward hand should be placed under the hose behind the last hose butt.

Fig. 16–26. The classic duck walk keeps firefighters low and their knees off the floor.

It is also important to ensure that the part of the foot in contact with the floor is the bottom of the foot, chiefly the toes and the ball of the foot. Some people end up on their knees with the tops of their feet in contact with the floor. This is a weak and unstable position and should be avoided. It is amazing how much more muscle mass can be brought into play when the bottom of the foot is used. It provides far greater traction, power, and stability. While engaged in active fire-suppression, the nozzle bale should be in one of two positions: fully open or fully closed. Many fires can be extinguished with the hit and move technique where the line is operated from a static position and briefly shut down to ease the effort required to advance. However, fires of greater magnitude and/or intensity require the cooling power of a continuously operating stream throughout the advance. Small incipient fires can be extinguished with the bale cracked partially open.

During most of the advance, the stream should be directed outward and upward at an angle such that water is applied to the overhead area well ahead of the nozzle team. Sweeping side to side or rotating in a clockwise direction are the best patterns of nozzle movement (figs. 16–27a, 16–27b, 16–27c, and 16–27d). The nozzle must be moved in a rapid vigorous manner to physically distribute the water in the form of large drops by splattering the stream off the ceiling and upper walls (figs. 16–28a, 16–28b, 16–28c, and 16–28d). As the stream is worked around in the overhead area, it paints the ceiling and walls with thick coatings of water. This causes a great amount of surface cooling, which contributes significantly to gaseous cooling. Recent extensive research and testing involving water mapping (tests that track the movement of water as it leaves the nozzle and follows it through the compartment and where the it eventually ends up), shows that this process not only takes place in close proximity to the point of stream impact on a given surface. But, due in large part to the water exiting the nozzle at approximately 60 mph and water's surface adhesion qualities, this also takes place well ahead of the points of surface impact as the water forcefully surfs forward along the ceiling and the walls. One would think that all or most of the water directed at the ceiling would splash to the floor, but modern research has proven that a significant amount of water costs and is absorbed by the ceiling. Using the nozzle in this manner is the most efficient for extinguishment, and it also has the tertiary benefit of lessening the reaction force burden that the nozzle operator must bear. If the stream were horizontal, the nozzle operator would receive the full reaction force burden. As the stream is angled upward from the horizontal, a greater and greater portion of the reaction force is directed into the deck. When the nozzle is moved vigorously to break up the stream, it significantly diffuses the energy of the reaction force.

Once the overhead has been cooled, the stream is applied directly onto the burning objects in the room to achieve final extinguishment. This is the modified direct attack.

Fig. 16–27a, b, c, d. The clockwise pattern of stream application is the most efficient method for extinguishment, and rapid vigorous nozzle movement lessens the reaction force burden on the nozzle operator.

Fig. 16–28a, b, c, d. Rapidly moving the stream pattern side to side off the ceiling helps distribute large droplets of water to the fire area.

Backup position

FFI 4.3.10 In comparing the nozzle team to a vessel, the nozzle operator is the pilot and the backup is the engine room. The **backup position** does the lion's share of the physical labor and provides the bulk of the momentum to the nozzle team.

The backup person must be on the same side of the line as the nozzle operator (fig. 16–29). Backup should be like a tractor motor: always in low gear and always providing momentum to the nozzle operator without pushing him or her faster than he or she wants to go. The nozzle operator is an effective heat shield for the backup position and may become aware of extreme temperatures before the backup does. The nozzle operator may also become aware of holes in the floor or other obstructions before the backup position realizes the situation.

Fig. 16–29. The backup position does most of the physical labor in the nozzle team, and provides the bulk of the momentum.

The goal of the backup position should be to make the nozzle operator's job as easy as possible. Ideally, the nozzle operator should only have to think about operating the bale and directing the stream. To relieve the nozzle operator of the bulk of the reaction force burden, the backup must physically back up the nozzle operator. The backup must attempt to maintain physical contact with the nozzle operator throughout the advance and extinguishment. Through physical contact between the nozzle operator and the backup position, the bulk of the reaction force burden is transferred through the nozzle operator's body to the backup.

Another goal of a good backup is to keep the line straight behind the nozzle operator. Keeping the line straight helps make dealing with the reaction force manageable. The backup achieves this by treating the line as the first 5–8 ft (1.5–2.4 m) of hoseline behind the nozzle is a straight piece of pipe. Working together to move this section of line as if it were a straight piece of pipe, the nozzle operator and the backup person should be able to prevent any kinks directly behind the nozzle operator. Any bends in the line should occur behind the backup position. If the line bends between the nozzle operator and the backup, the nozzle operator's job gets much tougher (fig. 16–30).

Fig. 16–30. The backup position should keep a straight line behind the nozzle operator, moving the section like a straight piece of pipe.

The fact that the stream is predominantly directed outward and upward dictates that the hose usually trails from the nozzle at a downward angle. The hose trailing from behind the nozzle operator's armpit also naturally falls at a downward angle. While providing forward momentum, the backup position must also maintain the line's natural downward inclination. Many inexperienced backup people lift up the line behind the nozzle operator (fig. 16–31). This may make it easier for the backup, but it definitely makes things tougher on the nozzle operator. Lifting the line up behind the nozzle operator tends to force the nozzle tip down, making it harder to direct the stream into the overhead where it is needed for efficient extinguishment. Lifting the line while the nozzle is pointed upward also forces a bend into the line directly behind the nozzle operator, making it more difficult to deal with the nozzle reaction force.

Fig. 16–31. The backup firefighter should not lift the hose behind the nozzle operator.

The backup position takes a grip on the line so that the forward hand is underhand and the rear hand is overhand (fig. 16–32). The backup grabs the line where it naturally falls, at downward angle from the rear of the nozzle operator. After the forward hand is wrapped around the line, it is slid up the line fast and hard to make positive contact with the nozzle operator. This brings the forward forearm into contact with the nozzle operator. Leaning forward then brings the shoulder into contact with the back or trailing side of the nozzle operator, further solidifying physical contact. As forward movement occurs, space between the two bodies slightly fluctuates. Simultaneously, at a minimum the backup's forearm must stay in contact with the nozzle operator to physically provide momentum. Depending on the backup's body position, the rear hand can pull the line in against the leg. This applies friction to the line to assist with gripping the line.

Fig. 16–32. The correct position for the backup firefighter behind the nozzle operator.

The backup must be in a low compact position. The downward angle of the line behind the nozzle operator dictates that if the backup can attain and maintain a body position lower than that of the nozzle operator, it benefits the biomechanical efficiency of the operation.

The backup position needs to apply force outward and upward along an angle approximating that of the hoseline and through the body of the nozzle operator, as if trying to push forward a piece of pipe that is screwed into the base of the nozzle. He or she should face forward with eyes scanning to monitor whatever visual clues exist regarding conditions ahead of and around the nozzle team. The backup should lean into the direction needed to apply force, with the spine approximating the angle of the hoseline. The nozzle operator and the backup drive forward with the hoseline, focusing their energy outward along an upward angle similar to the way that football players drive a blocking sled or the way in which rugby players provide motive power to a scrummage.

The backup position is the engine room of the nozzle team. The power plant truly resides in the muscles of his lower body. To make efficient use of the calves, thighs, and gluteus he or she must push off of the floor with the bottoms of his feet. Having the tops of the feet in contact with the floor effectively negates a significant amount of the potential to harness the power of the muscles of the lower body.

If the nozzle operator needs to direct the stream horizontally or downward, the backup can facilitate this by raising the line accordingly so that the first section of hose, from behind the nozzle to behind the backup, is maintained in a straight line (fig. 16–33). If the nozzle operator turns to the left the backup's movement must swing through a wider faster arc. This is so that as the nozzle is turning to the left the hose behind it will move correspondingly to the right to avoid the occurrence of any bends directly behind the nozzle operator (fig. 16–34). If the nozzle operator turns to the right the backup's movement must swing through a wider faster arc. This is so that as the nozzle is turning to the right the hose behind it will move correspondingly to the left to avoid the occurrence of any bends directly behind the nozzle operator (fig. 16–35). If the two members are working in concert as if to move a straight piece of pipe, there will be no bends between them.

Fig. 16–33. As the backup firefighter lifts the hose, the nozzle operator can direct the stream downward.

Fig. 16–34. As the nozzle moves to the left, the back-up firefighter moves the hose to the right.

Fig. 16–35. As the nozzle moves to the right, the backup firefighter moves the hose to the left.

Door position

FFI 4.3.10 Although perhaps not immediately apparent, one of the most important positions on the attack hoseline is that of the door position. The **door position** has two primary responsibilities: to ensure the smooth and sustained movement of hose to the set of the fire and to eliminate any kinks in the hoseline. This is a critical firefighting position that ensures the hoseline is not be starved of water because of a kink and is not hung up under a door or caught on some obstruction. Although this may not be a glamorous position, it is nonetheless an essential job that must be performed.

The term "door position" refers to the physical location of this firefighter as the hose moves through the structure to the fire, which is often in a doorway. During the final push toward extinguishment, the door position is likely to be in the doorway to the fire occupancy or fire compartment. The door position is *not* with the nozzle operator and the backup; this firefighter is at least a length or two behind them. If assigned to this position, it is important to avoid the tendency to *move up* on the hose to the nozzle operator and the backup position. They depend on the door position to stay back and provide them with slack hose to advance with.

Once the hoseline has been initially stretched and charged, the door position must **chase kinks** and take them out of the hoseline. While doing this, the door firefighter positions the hose for advance. The excess hose on the outside of the structure is arranged in large loops on the front lawn or sidewalk to avoid kinks and ease movement into the building (fig. 16–36). The loops help the advance by allowing the hose team to only pull on the last few lengths of the hose as they move in on the fire, not the entire hoseline starting at the engine. The portion of the hose from the engine to the front of the building does not move; only the last few working lengths of the hose move with the final advance on the fire.

Fig. 16–36. Arranging hose in loops in front of the building entrance reduces kinks and allows easier access of the hose into the building.

If the fire is on the first floor of a building, the door firefighter should position the hose in front of the building so that it can move easily through the doorway (fig. 16–37). This entails keeping the hose in a straight path as it enters the building (as it comes out of the loops) and ensuring that it doesn't get hung up on any obstacles or caught under any doors. It may be necessary to also move objects out of the way to avoid entanglements and impediments.

Fig. 16–37. Door position moving hose through a door.

As the working lengths of the hoseline move inside, the door firefighter must assist in its movement. This position takes finesse; it is important for the door position to give the nozzle operator and backup position just enough hose to keep it moving and help them advance at the proper speed to the seat of the fire. Too much **push** on the hoseline creates kinks in the hose because the nozzle team is moving slower into the fire area than the hose is being pushed. Too little push on the hose and the backup position has difficulty moving the hoseline forward, forcing the nozzle operator to pull on the hose instead of controlling the nozzle.

Once the hoseline has been properly positioned and is advancing into the building, the door position may advance forward along the hoseline to help move it through the structure. The door firefighter must remain mobile, moving back and forth between the front door and a length behind the nozzle, looking for kinks or possible hang-up points. The door firefighter's goal is to always provide slack hose to the nozzle team. This can be done by continually placing a horizontal arch or bow of hose on the floor behind the nozzle team. Beware of arranging slack hose vertically against a wall, as it can more easily become entangled in obstacles. Also, beware of placing hose in circular loops. Advancing from a circular loop often causes a cat's eye kink, which in effect places two kinks in the hose.

When advancing a hoseline up a stairwell and down a hallway in larger structures, the door position may help the nozzle operator and the backup position by creating loops in the hose to ease the advance inside the building (fig. 16–38). These loops are usually located a length behind the nozzle so as to not affect the operation of the nozzle.

Fig. 16–38. A firefighter can create loops in the hose to ease advance inside the building.

Advancing hoselines up stairwells inside multiple dwellings and commercial buildings or through large area buildings is covered in detail in chapter 20, Basic Fire Attack. Such hose stretches require precise estimates of how much hose is required to get to the seat of the fire (fig. 16–39). Occasionally, firefighters and fire officers underestimate the number of lengths needed to get to the fire. In such cases, a hoseline may need to be extended.

Many smooth bore nozzles are actually a series of **stacked tips**, essentially a set of decreasing nozzle tips threaded into each other. Some fog nozzles can also be broken down in similar fashion. To extend the hoseline, additional hose and *another* nozzle are brought to the nozzle attached to the end of the hose. With the attached nozzle shut down, the tip or tips are removed until the 1½-in. (38-mm) outlet of the nozzle is exposed. New hose and the new nozzle are attached. When completed and the hose has been flaked out, the original nozzle can be opened, charging the line. The original nozzle now becomes a control valve. Some departments use 2½-in. (65-mm) attack lines to knock down the fire and later bring in 1¾- or 1½-in. (45- or 38-mm) hose as a **mop-up line** for overhaul purposes. In such cases, the same procedure is used to attach the mop-up hoseline to the original attack line (fig. 16–40).

Fig. 16–39. Stretching hoselines into multi-story buildings requires precise estimates of lengths of hose needed to reach the fire.

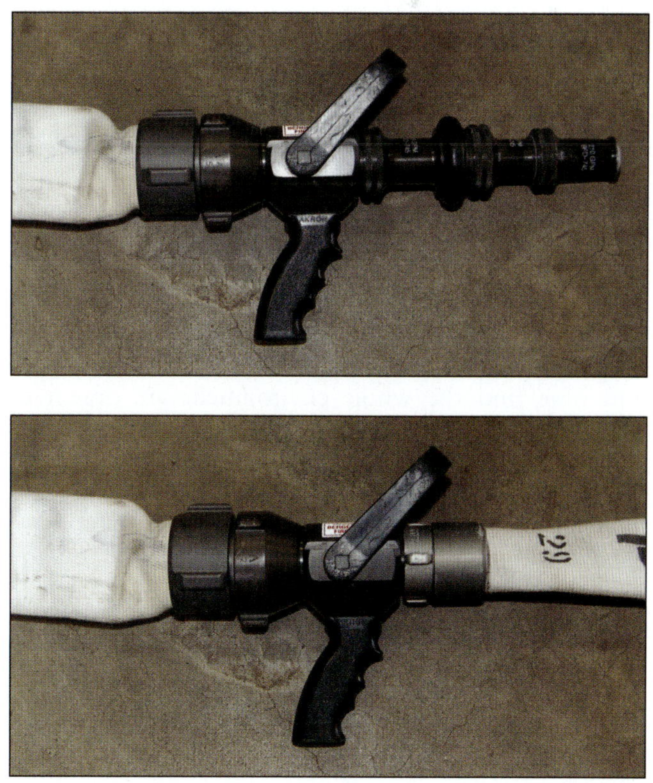

Fig. 16–40. The stack tips on a 2 1/2-in. smooth bore nozzle can be removed allowing a smaller hoseline to be attached. The 2 1/2-in. then functions as a gate valve.

HYDRAULIC VENTILATION

FFI 4.3.11 Some fire departments use fog nozzles exclusively for fire attack as well as ventilation. This type of ventilation process is known as **hydraulic ventilation.** It is where firefighters use the power of a fog fire stream to exhaust heat, smoke, and gases from a room or an enclosure after a fire has been controlled and/or extinguished.

Fog streams are effective for helping to ventilate fire buildings and have been used effectively throughout the fire service for many years. In fact, in some cases a fog stream used for ventilation is more effective than fans because of its ability to move greater amounts of air. This may be because of the location of the fire in the fire building, or wind conditions, or other circumstantial items. It is quick and easy to employ because the attack hose and nozzle is all that is needed and is right there in the fire room or area as soon as the flames are being controlled. After the fire is darkened or extinguished, and the area is checked and ready for ventilation and overhaul, then firefighters need only to find an opening for venting the environment to the outside—usually through a window or door. While ventilation is taking place, overhaul can also begin.

Hydraulic ventilation is most effective with fog streams from fog nozzles. This is because of the ability to change the shape of the stream from a straight stream to a wide-angle pattern that resembles a hollow cone. Smooth bore nozzles can also be used to ventilate fire rooms, but the speed of the water leaving the nozzle is not as fast as that from fog nozzles. The fog stream is able to function as a ventilation tool because of some laws of nature. Because we live in a positive pressure atmosphere (14.7 psi [100 kPa] at sea level), as a fog stream takes on a wider angle, a low pressure point exists where the stream comes out of the nozzle. It is at this point where the room air, smoke and heat, and the whole environment are drawn to that low pressure point and then carried along the jets of water that form the steam's umbrella pattern and are exhausted from the building.

To perform hydraulic ventilation using a fog nozzle, the nozzle operator must ensure that the nozzle is in the straight stream position and aimed at the open window or door. Once the nozzle has been opened, the stream is adjusted so that the fog pattern fills the window or door without hitting the wall and running onto the floor, causing water damage (fig. 16–41). Similarly, a smooth bore nozzle is aimed at the opening and is opened, making sure that the water is leaving the building and not ending up on the floor. Additionally, smooth bore nozzle streams should be rotated *within the inside perimeter of the window or door*, which increases the volume of smoke moved by the stream (fig. 16–42). The smooth bore nozzle can also be partially closed to create a broken spray stream, which causes more air movement.

Fig. 16–41. Venting out a window with a fog pattern

Fig. 16–42. Venting through a window with a broken spray stream from a smooth bore nozzle

In the past, some fire training programs have taught that nozzles must first be placed outside the opening, opened, and then brought back into the building. It should be positioned to ensure that water damage is held to a minimum because there is a possibility of water hitting the interior surfaces of the room. However, such procedures take time to accomplish; if properly done, the procedures described in the previous paragraph ensure minimal water damage.

When a wide-angle fog stream is directed out a window or door, it can ventilate an average size room or two in a matter of a few minutes depending on the volume of smoke inside the building. For the stream to be most effective, the window or discharge opening should be completely free of any glass, debris, curtains or screens so the water can flow freely without anything to stop it.

As a safety precaution when ventilating by fog stream, it is important to watch for changes in smoke and heat conditions in the fire area. In some cases firefighters have knocked down flames and moved quickly to begin hydraulic ventilation before the fire has been controlled. In these situations, the room clears of smoke and heat after a minute or so; but soon after, a rush of fresh smoke and/or heat is drawn to the nozzle, because the fire reignites and grows intensely from being fanned by air currents created by the fog streams. Any undetected fire in walls and ceilings grow in volume and intensity and may burst through and fill the room with flame. This testifies to the adverse nature of a fog stream's ability to push or pull products of combustion (depending on the nozzle's location and what it is doing at the time—either attacking a fire or ventilating), which firefighters must know.

Hydraulic ventilation is not limited to just fire operations. The power of a wide fog stream makes it able to direct vapors away from serious exposures where there may be a life and/or property concern. An example of this could be illustrated where there is a gas leak near an occupied building. The stream may be used to push or force vapors away to make a temporarily safer condition. It can also be used to divert liquids such as in hazardous materials incidents.

Note: It should be emphasized that this type of ventilation is absolutely no substitute for actual ventilation—where there is a need to force entry and open windows, roofs, doors, shafts, or use other means necessary to rid a building of fire products and to create a safer atmosphere inside.

EXPOSURE PROTECTION

FFI 4.3.10 Exterior exposures need to be considered for protection from fire extension. Many exterior exposures become involved where the fire building is heavily involved and there is a tremendous amount of convective heat or radiant heat being given off. Remember, convective heat is the air and atmosphere around a fire. If an exposure building is located within a few feet of a raging fire, the superheated air and gases may be all that is necessary to ignite an exposed building. However, where there is plenty of fire coming out of a fire building, there is plenty of radiant heat; it travels equally in all directions from the heat source, in this case the fire. In the past, many people in the fire service believed that to stop radiant heat from traveling to an exposure, you would direct a fog stream between the buildings for protection. In many of those cases, the exposure buildings burned to the ground! The cause of this was the radiant heat's ability to penetrate a curtain of water fog. Most fire departments that have experienced this have realized that an exposure stream must be played *onto an exposure surface* to keep it cooled below its ignition temperature. Otherwise, there is good possibility that the exposure will reach its point of ignition.

Another point to consider about exposure streams is their size. Smaller hand lines like those used for residential interior firefighting do not provide the amount of water needed to cool the side of an exposure in most cases. Remember, if there is a large amount of flames, there must be a large amount of water used for cooling. Exterior exposure lines of 2½ in. (65 mm) should be considered for maximum cooling efficiency. After the fire has been contained or knocked down, exposure lines can be reduced to more mobile hand lines that handle the lesser amount of fire.

As a matter of safety when protecting exterior exposures, hoselines should be positioned so that in the case of any possible collapse, firefighters are not in any collapse zones. Larger streams have greater reach and allow firefighters to operate at safe distances under heavy fire conditions.

When selecting an exposure hoseline, choose a stream big enough to do the job. If the fire is severe, use an effective volume of water and stop the fire. Don't gamble or believe that a small amount of water from a lighter hand line controls a large amount of fire—it won't work!

HOSELINE USE DURING OVERHAUL

FFI 4.3.13 One of the basic steps of firefighting is overhaul. It generally takes place after the main body of fire has been controlled or extinguished. In this process firefighters use tools to open walls, ceilings, and floors and to examine furniture and anything that looks like it may contain any hidden fire. A hoseline should be present in an area where there has been fire damage and should be in the possession of an assigned firefighter. In many cases overhaul begins quickly with the knockdown of flames. As firefighters are afforded better personal protection from turnout gear, they sometimes approach or move into a fire area quickly without realizing the amount of fire yet to be extinguished. This example and others make overhaul one of the more dangerous times of a working fire, even though the bulk of flames has been knocked down.

FFI 4.3.13 Overhaul operations generally require less water volume and pressure than fire attack operations, therefore 1¾- or 1½-in. (44- or 38-mm) hoselines are usually adequate. Follow your department's guidelines for hose selection with overhaul.

Depending on the amount of damage from a fire, the use of water during overhaul should not be excessive. Realistically, there are fires where a little amount of water is used for actual extinguishment compared to the amount used for overhaul, predominantly because of unusual conditions within the structure. In so many cases, firefighters risk their personal safety to get as close to the fire as possible to extinguish it with a good, quick hit from the hose stream, then shut the line down quickly and move in for the final kill, only to flood the room out 10 minutes later with a deluge after overhaul! The result is little fire damage but great water damage. Water use during overhaul should be guided by more careful consideration.

One of the first things firefighters should look for during overhaul is any structural damage from the fire. A deadly combination is a fire-damaged structural framing system, the added weight of several firefighters working with their tools, and the added live load of water that may have not drained away or has collected in clothing and furniture. This new live load has resulted in structural collapse and firefighter fatalities in the past.

It is important that an attack hoseline is never left unattended while firefighters are actively opening walls, ceilings and floors to search for hidden fire. Many ceilings have been opened or "pulled" and flames have come rushing out and down to the floor, catching unsuspecting firefighters. When this happens there should be someone ready with the hoseline to extinguish the fire. Firefighters have also fallen through fire–weakened floors and have needed protection from any debris burning around them.

During extended overhaul operations where there is plenty of debris, such as plaster and lath, dry wall, or other wall and ceiling materials, the hoseline can become buried; and if there is a need for the line, it may be difficult to locate. Even if the hoseline is found, the nozzle may still be buried and inaccessible in time of need. Again, it is important to keep a firefighter with the hoseline just for any unexpected situations.

MASTER STREAM DEVICES AND SPECIALIZED NOZZLES

Master stream devices

FFII 5.3.2 When an incident commander has ordered firefighting forces to assume defensive operations because the fire has taken possession of the building, **master stream devices** are employed to contain the fire (when there are no living people, including firefighters, inside the building). Master stream devices, by definition, flow in excess of 300 gpm (1,136 L/min). There are a variety of master stream devices used. Although care must be exercised with all nozzles, flowing such large quantities of water makes them dangerous when misused. Master stream devices must never be pointed at people. They must be operated within manufacturer's recommendations; for example, master stream nozzles must not be operated outside their specified range of elevation angle and lateral (side to side) movement. Nozzle reaction is substantial for such devices; and operating them outside the specified ranges can make them unstable, allowing them to move and injure people. They should never be over pressurized. In addition, other dangers of master streams that firefighters must avoid are striking objects such as roof tiles, sending them airborne, lowing large amounts of water into a building causing a floor collapse problem, and the uncontrolled movement of master streams due to nozzle reaction.

Elevated master stream devices include **ladder pipes** used on aerial ladders and monitor nozzles used on tower ladders. Each of these devices is capable of flowing in excess of 500 gpm (1,893 L/min). Their advantage is that they can apply water from above (e.g., through a collapsed roof) or through an upper-floor window.

The ladder pipe is a nozzle attached to the rungs of an aerial ladder. Some ladder pipes are detachable, others are permanently attached, or prepiped. The detachable models use supply hose from engine companies on the ground to provide water. These models also typically use a camlock system for attachment. Always follow manufacturer recommendations for attachment and use (fig. 16–43a, 16–43b, and 16–43c).

Once an aerial ladder has been raised, the ladder pipe is controlled (raised up and down, moved side to side) using halyard ropes or, in the case of newer models, using electronic remote controls. A firefighter should never be positioned on a ladder to operate a ladder pipe, it is simply too dangerous.

Tower ladders typically use mounted **water monitors** (also known as a **water cannon** or **water turret**) controlled from the platform (bucket). These master stream devices have individual control valves that allow them to be opened and closed slowly (fig. 16–44). Before operating these nozzles, make sure that there are no obstructions or dangers in the path of the stream (such as power lines). Powerful streams such as these, flowing at more than 1,000 gpm (3,785 L/min), can knock down walls.

Fig. 16–43a. A permanently mounted ladder pipe

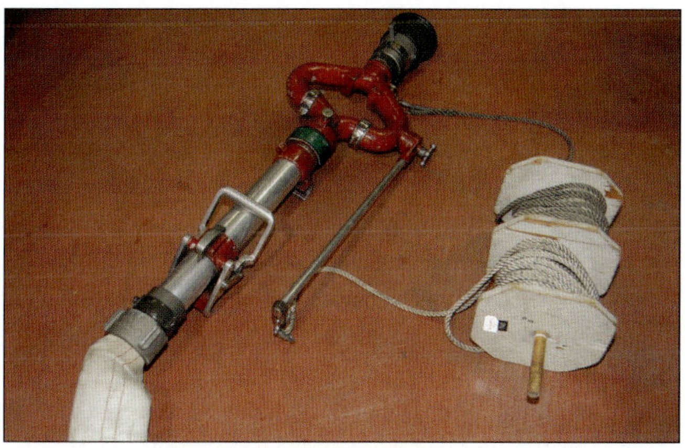

Fig. 16–43b. A detachable ladder pipe

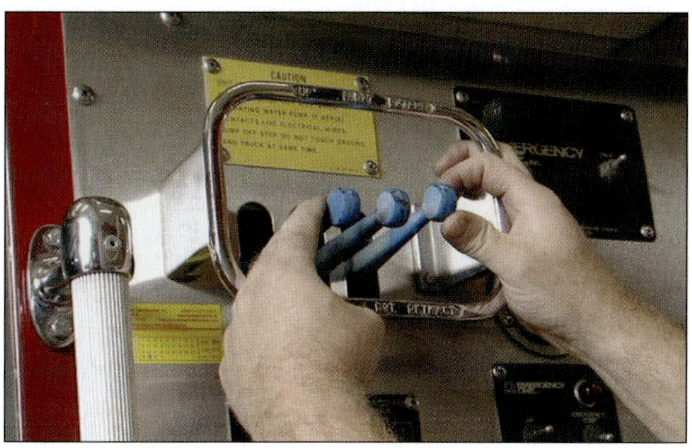

Fig. 16–43c. Electronic controls for ladder pipe

Fig. 16–44. A tower ladder water monitor and control valve

Deck guns are mounted on the top of engine companies. They are used to apply water from the ground onto burning piles of debris, over walls, and sometimes into unoccupied buildings to get a quick knockdown. Some devices are permanently mounted to the engine; others can be detached and moved to a location on the ground and supplied with water at that location (figs. 16–45a, 16–45b, 16–45c, and 16–45d).

Fig. 16–45a. A deck gun on apparatus

Fig. 16–45b. Locking device

Fig. 16–45c. Handle to raise and lower nozzle

Fig. 16–45d. Attaching nozzle to base

Deck guns have a rotating wheel so that they can be adjusted up and down, angle, and swivel so the nozzle can be adjusted side to side. Locking devices secure the nozzle in position. A key safety issue is making sure that the deck gun is securely attached to the engine, particularly if it is a detachable model and has not been reattached properly after use. These nozzles also often have a base, which can be taken off the engine and used on the ground. The base often has spiked feet (to dig into the ground) and a securing chain/strap to attach it to a fixed object. **Deluge guns** are portable monitors that are similar to detachable deck guns except that they are always used on the ground (fig. 16–46). The same safety issues apply here as well.

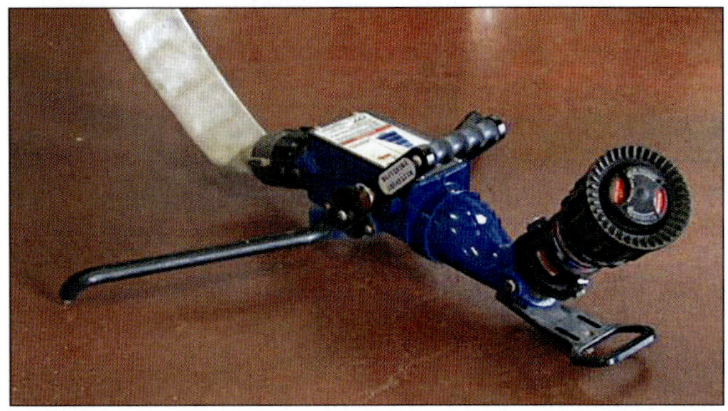

Fig. 16–46. A non-apparatus mounted deluge gun is a portable monitor that is always used on the ground.

Specialized nozzles

Over the years, specialized nozzles have been developed to deal with specific fireground problems. One such problem is that of hidden fire in void spaces and inaccessible areas. **Piercing nozzles** apply water to hidden spaces such as in wall and floor cavities (fig. 16–47). These tapered nozzles have a pointed end that can be driven through a wall or floor surface to penetrate the void space. The nozzle is charged and discharges a fog spray into the burning void space, creating large quantities of steam. This nozzle has been used to fight fires inside attics, walls, and floors.

Fig. 16–47. A piercing nozzle applies water to hidden spaces such as in wall and floor cavities.

The **distributor**, more commonly known as the **Bresnan distributor** after inventor FDNY Battalion Chief John J. Bresnan, is a spinning 1½- or 2½-in. (38- or 65-mm) nozzle with multiple orifices pointed in different directions, spraying water in all directions. (Bresnan also invented the hose roller for passing hose over a roof's edge and improved the **swinging harness** for fire horses who pulled fire apparatus in the 19th century. He was killed at a fire in 1894.) These nozzles are used primarily for basement fires where normal fire attack is not possible, and they have been used on attic fires as well (fig. 16–48). A hole is cut in the floor and the nozzle is lowered into the basement, just low enough to clear any overhead obstructions to get the widest possible distribution. The hose is secured to an object at the point of entry to the floor. In addition, a gate valve should be located between the last two sections of hose before the nozzle (50 ft [15.2 m] before the nozzle) to control flow to the nozzle.

Fig. 16–48. A Bresnan distributor is used primarily for basement fires where normal fire attack is not possible.

The 2½-in. (65-mm) **cellar pipe** (a common model is named after its inventor, Baker) is used in similar situations as the distributor nozzle (fig. 16–49). These nozzles, however, have a directed stream that can be pointed in any direction through the use of a lever at floor level. Some of these pipes have control valves attached to them; if no gate valve is attached, one must be inserted between the last two sections of hose before the cellar pipe. Similar nozzles are used to fight fires under shipping piers as well.

Fig. 16–49. Cellar pipes have a directed stream that can be pointed in any direction underneath a floor.

CARE AND MAINTENANCE OF NOZZLES

Although nozzles are designed to be used in a rough environment, take care to keep them fully operational at all times (fig. 16–50a, b, and c). Nozzles must not be dropped or thrown. After each use they must be cleaned and inspected. Inspection procedures include checking that the waterway is clear of obstructions, the bale works properly, there are no dents or nicks in the tip of the nozzle, and there are no missing parts. Worn out gaskets must be replaced.

A fire stream is the heart of all fire attacks. Knowledge of the correct volume of water, the proper attack method, and the best nozzle to use are essential to become an effective firefighter. Practice what you have learned in this chapter to ensure that you are prepared for meeting the fire head on as the "point of the firefighting sword."

Fig. 16–50a. Nicks and dents on a nozzle

Fig. 16–50b. A blocked waterway inside a nozzle

Fig. 16–50c. Replacing a worn out gasket in a nozzle

NOTES

1. Fredericks, Andrew. "Little Drops of Water: 50 Years Later," Parts 1 and 2. *Fire Engineering*, February and March 2000.
2. Fredericks, Andrew. "The 2½-Inch Hand-line." *Fire Engineering*, December 1996.
3. Norman, John. *Fire Officer's Handbook of Tactics*. Fire Engineering, 1998.
4. Clark, William. *Firefighting Principles and Practices*. Fire Engineering, 1991.
5. Fornell, David. *Fire Stream Management Handbook*. Fire Engineering, 1991.

QUESTIONS

1. Describe why the needed flow rates of handlines have increased over the years.
2. How does the Powell Doctrine apply to firefighting tactics?
3. When presented with various nozzle selections, list some of the advantages and disadvantages of each.
4. List advantages of the 2½-in. (65-mm) attack line.
5. When a life hazard exists in a structure, identify which method(s) of attack are appropriate and why.
6. How does the direct attack method differ from the modified direct attack method?
7. What is a transitional attack? What is a blitz attack?
8. Explain how the indirect and combination attack methods extinguish a fire.
9. According to William Clark's test, which stream was found to be most effective in extinguishing an internal structure fire?
10. How does company staffing relate to advancing hose lines?
11. What is black fire?
12. When positioning the hoseline into operation, what are some of the mistakes that should be avoided?
13. Which position on the hoseline is responsible for ensuring that enough hose is stretched to cover the entire fire area?
14. Describe proper positioning of the nozzle operator while flowing an attack hoseline.
15. What is the goal of the backup position on an attack line?
16. The door position is one of the most important positions on the attack hoseline. What are the door positions responsibilities?
17. How would you perform hydraulic ventilation with either a smoothbore or fog nozzle?
18. When would master stream devices be employed?
19. Bresnan and Baker nozzles are most often used for what type of fires?

Firefighter Safety and Survival

by Anthony Avillo, Frank Ricci, and John Woron

This chapter provides required knowledge items for the following NFPA Standard 1001 Job Performance Requirements:

FFI 4.2.4	FFI 4.3.9	FFII 5.1.1
FFI 4.3.2	FFI 4.3.10	
FFI 4.3.3	FFI 4.3.17	
FFI 4.3.5	FF1 5.3.18	

This chapter contains Skill Drills. When you see this icon, refer to your Skill Drill book for step-by-step instructions. SKILL DRILL

OBJECTIVES

Upon completion of this chapter, you should be able to do the following:

- Identify the reasons for and preventative measures for firefighter injuries and fatalities
- Describe the importance of injury reporting
- Describe the importance of Fire/EMS Safety, Health, and Survival Week
- Identify and explain the 11 rules of engagement for structural firefighting
- Describe the need for a fire department safety committee
- Describe the responsibilities of an incident safety officer
- Define the terms *accountability officer*, *fireground rehabilitation*, and *on-deck air management*
- Describe the purpose of a fire department wellness program
- Describe methods of limiting injuries to firefighters during response and return from emergency incidents
- Describe the role that the Occupational Safety and Health Administration (OSHA) plays in firefighter safety
- Describe the role staffing plays in firefighter safety
- Describe the role the National Fire Protection Agency (NFPA) Standards 1710 and 1720 plays in fire department staffing
- Discuss at least five tactical breakdowns made at an emergency incident that impact firefighter safety
- Differentiate between "top-down" emergency transmission and "bottom-up" emergency transmission
- Identify the required tools for a rapid intervention team (RIT)
- Explain the on-scene actions of a RIT
- Describe the proper procedures for packaging and removing a downed firefighter for a hazardous environment
- Discuss 10 factors relating to poor air management at an emergency incident
- List and explain five types of drags and carries for removing a firefighter from a hazardous situation

INTRODUCTION

The day will start like any other, with thoughts of the weekend or an upcoming vacation with the family. You are dispatched to a working fire, and everything seems routine: light smoke is showing, and *all hands* are working. Then, in a tenth of a second, everything goes wrong, resulting in a significant injury or death.

Due to the inherent dangers of this job, most firefighters will be injured over the course of their career. Each year between 60 and 100 firefighters are killed in the line of duty, with an additional 60,000 injured. We work in a hostile and dynamic environment that can change rapidly (fig. 17–1).

Fig. 17–1. Fire conditions and fire buildings must be constantly monitored for clues to changing conditions. Ignition of heavy smoke may cause firefighters to become trapped and buildings to become unstable. Failure to recognize changes in conditions can be fatal. (Courtesy of Ron Jeffers)

We must control what we know by using safe practices and preparing ourselves for the unknown. On a routine basis, life-and-death decisions will be made based on imperfect information adapted to fit the circumstances at hand. Simple, almost meaningless actions or habits that we should strive to achieve will serve to minimize the risks we all face. Although some failures at fires will be catastrophic, most will happen incrementally until the sum is greater than the parts, resulting in an injury or death. This is commonly known as **drifting into failure**. It is your responsibility to keep yourself safe and to look out for your fellow firefighters. You are no longer an individual. You are a member of a team, and your attitude toward safety will have an impact on that team. By following your department's standard operating procedures (SOPs), training regularly, and developing proper habits, you will strive to become proficient. Every skill that you learn must incorporate safety into the lesson. It is critical to realize that these skills are perishable and must be practiced, always keeping the safety aspect of the task in the forefront. The goal is to operate safely, effectively, and efficiently.

There is no greater priority on the fireground than that of firefighter safety and survival. Safety is a mind-set. The fire department's responsibility is to foster a culture of safety. Safe attitudes and operations don't just magically show up on the fireground. Safe mind-set, attitudes, and procedures come out of the firehouse just like every other tool we use on the fireground. Safety must be a cultivated, enforced, and reinforced discipline. Safety is a 24/7/365 commitment to excellence. A department that does not have a safety mind-set off the fireground will never have it on the fireground. Safety discipline is about operating with a safety mind-set in the 98% of the time we spend off the fireground. If we can do that properly, then the 2% of the time we spend in actual emergencies are likely to be conducted in a safe manner. The safety mind-set must permeate all nonfireground activities, including training and both apparatus and equipment maintenance and operation. The safety mind-set is about doing things safely when no one is looking. Responsibility for a safety mind-set starts with the chief of department and must permeate the organization downward and sideways through every level—right down to the recruit firefighter.

Firefighter safety must be the common thread that runs through three general priorities of fireground operations: life safety, incident stabilization, and property conservation. Utilizing your protective clothing and following established policies, procedures and precautions allows the fire department to conduct aggressive operations while managing the risk. If we cannot make an operation safe, we must seriously consider whether that operation should be conducted at all. This is called **risk versus gain**. If the risk to be taken is not worth the gain to be accomplished, then alternative courses of action may need to be considered. For example, a building that is fully involved in fire does not typically receive a primary search. This is because the likelihood of finding any live victims in this fiery environment is extremely remote, if not impossible. The risk to firefighters far outweighs the gain (fig. 17–2).

These decisions should not be made by the probationary firefighter or even a 3-year firefighter (a rookie by most accounts) but must be carefully considered by more seasoned veterans of the fireground. There is an old adage in the fire service: "If you cannot make things better, at least make them safe." According to the late Phoenix Fire Chief Alan Brunacini, there is another saying that addresses the same concept and deals with the same risk versus gain issues: "Risk a lot to save a lot, risk a little to save a little, risk nothing to save nothing." *Fire Engineering* editor-in-chief Bobby Halton added, "When it is time to risk a lot, we do not just risk a lot, we risk everything." The mark of the true pro is the ability to differentiate when risking everything is justified and when it is not.

Fig. 17–2. Buildings that are fully involved or in danger of collapse present a risk to all responders. The risk taken by entering such buildings far outweighs what could be gained. (Courtesy of Ron Jeffers)

Firefighter safety is rooted in training and awareness. Unsafe conditions or actions are absolutely unacceptable and must not be tolerated at any level. To this end, the chief and company officers, as well as senior firefighters, have the responsibility to set the example with regard to safe operations and actions on the fireground and beyond. No one is tougher than a fire.

This chapter will cover the four critical principles:

- Safety
- Prevention
- Self-rescue
- Intervention

Survival is predicated on training, knowledge, and experience coupled with safe staffing, **situational awareness**, and tactics that occur in unison to and with each other. This chapter will address common tactical breakdowns, driver safety, scene safety, and actions to take to avoid getting jammed up in the first place, while recognizing that even when operations are conducted right, things can still go wrong. It will take the student through tactics to effect self-rescue and provide an understanding of the roles and responsibilities of the rapid intervention team (RIT). Ideally, the RIT should be made up of experienced, well-trained firefighters, but in reality, a recruit may be assigned to rapid intervention on his or her first fire. By no means are these the only methods or evolutions for firefighter removal. It is important to establish a "foundation of safety" during Firefighter I and II training. That foundation must be built upon for the rest of a firefighter's career. By incorporating a safety mind-set and reinforcing it continuously, we can reduce the number of firefighter deaths and injuries.

Statistics

Firefighter fatality and injury statistics. To properly take the steps toward firefighter fatality and injury prevention, we must first analyze the causes of death and injury. As we shall see, with the exception of the past few years, (a most promising trend), on average, there are approximately 100 firefighter **line-of-duty deaths (LODD)** per year in the United States. This is more than two per week! None is acceptable. All are preventable. There is no such thing as an acceptable loss in the fire service (fig. 17–3).

Fig. 17–3. This sight is all too common in the fire service. Prevention of casualties must be the number-one priority of all fire departments. (Courtesy of Ron Jeffers)

Fire doesn't care whether we are firefighters or civilians. Fires do not discriminate between career and volunteer firefighters, old or young firefighters, or superior officers or subordinate firefighters. Fire kills and injures at every opportunity. With that in mind, all fire department personnel need to understand the ways in which our brothers and sisters are becoming casualties so we can begin to take the steps of injury and fatality prevention.

> **FIREGROUND NOTE**
>
> Statistics exist for one reason—so that you don't become one.

Firefighter fatalities and prevention. According to provisional statistics from the National Fire Protection Association (NFPA), there were 69 on-duty firefighter fatalities in the United States in 2016. Five out of the last six years recorded 70 or fewer firefighter fatalities. While that is promising news, it is still more than one per week. That number equals the previous year in 2015, when 69 firefighters also made the supreme sacrifice during the execution of their duties. There were three multiple fatality incidents. Deaths among career firefighters were at their lowest level in 2016 (19), as were deaths on the fireground (15). Seventeen deaths occurred as firefighters responded to or returned from emergencies. Sudden cardiac death and traumatic injuries accounted for the largest share of the on-duty deaths (26 deaths in each category). Of the total number, 19 fatalities were career firefighters, while 50 were volunteer firefighters. The ages of those who died ranged from 16 to 79.

Of these fatalities, the two most prevalent duty types were during fireground operations and while responding to or returning from alarms. Incredibly, 10 deaths occurred during training activities, an environment that is supposed to be strictly controlled and supervised. A further danger of the modern emergency response world claimed the lives of three firefighters: shootings. Additionally, two firefighters took their own lives.

Causes of the fatalities were as follows:

1. Exertion, stress, and medical issues: 29 fatalities; 26 were due to sudden cardiac death.
2. Vehicle Crashes: 17 fatalities were motor vehicle-related.
3. Fatal falls: 7 fatalities; 1 was a result of rapid fire development.
4. Struck by falling objects: 4 firefighters; 3 wildland, 2 struck by trees, 1 struck by falling boulder

Included in these figures are three fatalities that were collapse-related or the result of firefighters getting lost and running out of air in a structure.

Exertion and stress: Sudden cardiac death. Sudden cardiac death, most often as a result of heart attacks, is a leading cause of death for firefighters, accounting for close to half of the LODD per year. Many of the victims have had some type of prior condition.

FFI 4.1.1 Firefighters who follow a proper diet and exercise **regimen** and are part of a department **wellness program** are less likely to fall victim to exertion and stress fatalities. In addition, regular physical fitness checkups, including stress tests, will assist in proactively identifying potential cardiac concerns. No-smoking policies and **smoke cessation programs** for department members are also part of a healthy wellness programs. Departments who practice regular relief and rotation on the fireground, including limiting firefighter work time in immediately dangerous to life and health (IDLH) atmospheres will put less work-related stress and exertion on its members. Fatigued firefighters rapidly become injured and possibly dead firefighters.

Struck by an object or contact with an object. Motor vehicle–related injuries are the second leading killer of firefighters. This category also includes firefighters that have been killed by falling objects such as collapsed walls and roofs. It also includes contact with live electrical equipment.

A portion of vehicle fatalities occurs in the wildland environment where terrain is less than conducive to safe driving. In addition, many of the victims were not wearing seat belts. Contributing factors include excessive speed, medical emergencies, texting, negligence, and intoxication.

Caught or trapped. This is usually the result of either 1) a rapid fire development such as a flashover or backdraft or 2) firefighters losing their orientation with their surroundings, losing contact with their partner or team, or both, and running out of air. Building collapse is also included in this category, as firefighters fall through collapsed floors or are trapped under collapsing ceilings or roofs. Another overlooked issue when firefighters are categorized as being "caught" is when their personal protective equipment (PPE) is not worn properly and becomes a liability, causing them to get caught up on an object, become trapped, and subsequently run out of air. One item that is often ignored as an important piece of gear is the self-contained breathing apparatus (SCBA) waist strap. Failure to properly fasten the waist strap has cost firefighter lives when the straps become hung up on something inside a building.

Prevention

Driver safety. In 2015, an estimated 16,600 collisions involved fire department emergency vehicles responding to or returning from incidents. Injuries, death, and exorbitant sums of money in damages to equipment and property are the consequences of these incidents. Fire departments must have mechanisms in place that create and enforce policies that address, maintain, and enforce safe driving practices at all times. To that end, departments should actively engage members in driver safety courses and driver evaluations on at least a yearly basis. Training should be in accordance with NFPA 1002, *Fire Apparatus Driver/Operator Professional Qualifications*, and NFPA 1421, *Standard for a Fire Service Vehicle Operations Training Program*. Departments must have driver safety SOPs in place and enforced at all times, for both apparatus response and volunteer response from home in privately owned vehicles (POVs). In addition, seat belt regulations should be equally enforced with a zero-tolerance policy, and all accidents must be both investigated and documented. Departments should also see that apparatus maintenance is proactively pursued and there is a mechanism in place to address issues regarding same.

Collapse safety. A comprehensive training program regarding building construction and collapse, as well as a hazardous building marking system and building familiarization program, should be in place. Buildings that are dangerous should be brought to the attention of all department members via department safety bulletins and computer-aided dispatch (CAD) print-outs. On the fireground, emergency transmission protocols should be in place to warn firefighters of dangerous building conditions. Proactive reconnaissance, ongoing awareness, and timely communications should be the cornerstone of the fireground safety plan.

Electrical safety. Departments should conduct joint training sessions with local utility companies. In addition, electrical safety SOPs should be in place, and members should be trained to give all electrical equipment a wide berth. Operations such as overhaul should not be initiated without power shutdown in affected areas (fig. 17–4).

Fig. 17–4. All equipment should be kept at least 10 feet from all electrical lines. Unless power has already been shut off, operations here show an unacceptable disregard for safety. (Courtesy of Ron Jeffers)

Fire behavior safety. The need to train firefighters in fire behavior and to further train them to recognize and properly react to deteriorating fire conditions cannot be overemphasized. Enforcing the proper wearing of turnout gear may save a firefighter from himself or herself. Strictly enforced **mandatory mask rules** for SCBA use as well as training firefighters in proper air management and awareness is a must for safe interior firefighting. Many times, firefighters become lost and disoriented because they are operating outside of the established attack plan; in essence, **freelancing** (described later in this chapter). Structured command and control along with disciplined officers and firefighters operating under the parameters of effective scene assignment SOPs are the best accountability system and the backbone of firefighter freelance prevention. Ongoing training in Mayday policies and rapid intervention procedures also helps prevent casualties.

Firefighter injury statistics and prevention.

FFI 4.3.10 Firefighter injuries often occur for some of the same reasons that fatalities occur; sometimes we are just lucky that we only got hurt instead of killed. Overexertion, lack of awareness (not paying attention), failure to properly wear protective clothing, complacency, freelancing and lack of discipline, and just plain lack of common sense all contribute to firefighter injury.

Injury statistics from the NFPA show that there were more than 68,000 firefighter injuries in 2015. That is more than a 20% reduction since the year the first edition of this book first reported it in 2006. Almost half (43%) occurred on the fireground, that is, more than 29,000 injuries. The leading types of injuries during fireground operations are strains, sprains, and muscular pain (53%), followed by wounds, cuts, bleeding, and bruising (14%). Next are burns (5%) and smoke or gas inhalation (4.4%). Of all nonfireground injuries, strains, sprains, and muscular pain account for more than half the reported injuries (60%). Overexertion and strain were tied for the leading causes for injury with falls, slips, and jumps, both at just over 27% (fig. 17–5).

Fig. 17–5. Fireground injuries account for more than half of all reported injuries. Injury prevention must be a main priority of all fire departments. This injury occurred when an aerial device was accidentally retracted on a captain's leg. (Courtesy of Ron Jeffers)

Other estimates for injuries include over 8,000 exposures to **infectious disease** (HIV, hepatitis, meningitis, methicillin-resistant **Staphylococcus aureus** [MRSA], etc.) and over 27,000 exposures to hazardous materials and conditions such as asbestos, radioactive materials, chemicals, and fumes.

The firefighter injury rate for 2015 represented an increase in 7.5% over 2014 after nearly a decade of yearly decreases in annual injury rates. The Northeast has the highest fireground injury rate, more than twice that of the rest of the country.

Summary of firefighter fatalities and injury statistics. Even a statistic that represents 1% of the fatality totals is significant. Remember that 1% represents someone who left behind a family and friends and whose department and community are no longer recipients of that individual's service, dedication, and talent. Statistics exist for one reason—so that you don't become one.

Types of injuries and exposures

In all cases, the best ways to prevent injuries are proper training and supervision, adopting a safety mind-set, having a philosophy of continuous vigilance and awareness (paying attention to your surroundings), promoting and enforcing anti-freelance policies, maintaining the proper relief and rotation of companies, and wearing PPE properly. A safety-conscious department will experience fewer injuries than one that allows safety violations on a routine basis; thus, as was mentioned earlier in this chapter, safety of personnel begins at the top and must permeate the entire departmental structure. Unsafe actions must be subject to a zero-tolerance policy at all times.

Strains, sprains, and muscular pain prevention. A progressive program in muscular and cardiovascular fitness can minimize strains, sprains, and muscular pain. There are many parallels between firefighting and athletic competition; however, unlike athletic competition, our opponent is not human. Instead, it is fire, smoke, buildings, blood, and other nonhuman hazards. Although we cannot warm up as an athlete does for a game, we can ensure that we use common sense in physical endeavors, such as not overextending ourselves or attempting something alone that we should get help with. Proper relief and rotation is the key to injury prevention, especially in extremes of weather.

Falls, slips, and jumps. Falls, slips, and jumps are also often the result of fatigue. Therefore, proper relief is a key to prevention of not only these but all injuries. In addition, providing proper lighting at night and in reduced visibility conditions will help prevent falls. To prevent slips on ice, all apparatus should carry ice-melting compounds such as rock salt during the winter months. These compounds should be spread around the fireground, on the steps of the fire building, and around the pump panel. There are also traction devices available that fit on firefighting boots to help prevent slips and falls.

Falls from heights have severely injured and killed many firefighters. The severity of a fall is often a function of the height of the fall and how properly the firefighter wore the turnouts. Chinstrap policies are ignored in many departments. Firefighters have fallen from heights as low as 10 ft and have been killed when their helmets fell off on the way down. On the other hand, firefighters have

LESSON FROM THE TRAINING GROUND

In one case, a fire instructor fell approximately 30 ft during a training evolution at a fire academy after the wind blew the ladder over that he was working on. His helmet cracked in half on impact, but stayed on his head on the way down as his chinstrap was properly fastened. The severity of his other injuries, a broken jaw, a cracked cheekbone and orbital, as well as shoulder and hip injuries suggests a more severe head injury would have been suffered had the helmet fallen off on the way down.

LESSONS LEARNED

1. Always wear your gear properly, especially during training evolutions.
2. Just because you are an instructor doesn't entitle you to forego safe PPE practices.
3. Ladders should be properly tied off or footed at all times. This ladder was tied off. The fall occurred as the instructor disconnected the chain that was securing the ladder to the building.

been spared from death because their helmets stayed in place during the fall by virtue of a properly fastened chinstrap.

Conditions that cause firefighters to jump out of a window can come about as a result of a failed size-up, a failure to secure a secondary means of egress, or some type of uncoordinated fire operation. An equally unacceptable failure in recognition of rapid fire development signs has also led to many jumps from buildings. Jumps occur when all other methods of self-preservation have failed. A firefighter jumping out of a building or leaving a building by way of an emergency bailout represents a failure at every level of the fireground organization and fire department, from the firefighter who jumps or bails out to the incident commander (IC) and other commanding officers to the department as a whole. Mechanisms such as effective and enforced scene assignment SOPs, a proper accountability system, disciplined communications, and ongoing departmental training in bailout prevention are all remedies to the jump (and fall) dilemma. Gravity never takes a day off.

Responding and returning injuries. Responding to calls and returning to the firehouse together are the second leading cause of firefighter deaths and are also a major cause of injuries, accounting for nearly 4,000 injuries. There were over 16,000 fire apparatus accidents in 2015. This number does not include accidents involving firefighters' personal vehicles, which accounted for more than 700 additional collisions and more than 50 injuries (figs. 17–6 and 17–7).

Figs. 17–6 and 17–7. Seven firefighters were injured in this squad company versus rescue company versus building collision en route to a reported fire. Seat belts saved their lives. (Courtesy of Ron Jeffers)

As mentioned earlier in response to motor vehicle-related fatalities, training, awareness, and enforced driver safety SOPs will help prevent these occurrences. Firefighters must understand that they do not have the right to circumvent motor vehicle laws such as speed limits and right-of-way just because they have a big truck and make a lot of noise (or they have a personal vehicle with a flashing light on it). Firefighters have been judged to be liable for violations of motor vehicle laws and in some cases have been prosecuted, convicted, and jailed.

Injuries during non-fire emergency calls. Non-fire emergency calls include responses to non-fireground rescue situations, hazardous calls such as spills, and disaster responses. Often the cause of the injury is directly related to incident awareness (or lack thereof) when conducting operations that personnel are not adequately trained for or equipped to handle. When responding to these incidents, firefighters must choose caution over aggression and not get involved in an issue they were not trained to handle. Statistics have shown that in many of these types of incidents, over half of victims are would-be rescuers—many times firefighters who got too close to the problem before they realized the magnitude of the consequences. This issue boils down to discipline. Disciplined firefighters are safe firefighters.

Weather-related injuries. Injuries such as frostbite and heat exhaustion that occur in weather extremes are best minimized by proper relief of personnel, recognition of symptoms, medical monitoring, and providing shelter and other comfort measures based on the weather. For example, getting firefighters out of turnouts, ensuring proper hydration (in all seasons!), and providing cooling equipment like misters and tents will help reduce the effects of heat and humidity on personnel. Speaking of hydration, once you are feeling parched, it is probably too late to sufficiently rehydrate quickly enough to avoid the consequences. When the weather is hot and humid, firefighters should pre-hydrate all day long and stay away from caffeine and heavy foods.

On the opposite extreme, in cold weather, providing shelter, urging firefighters to carry additional dry clothing, and providing early and regular rotation and relief all help minimize the effects of cold on firefighting personnel. In all cases, the answer to proper rotation is to summon additional personnel. In weather extremes, a one-alarm fire may require a three-alarm response to ensure proper relief and rest for weather-fatigued firefighters (fig. 17–8).

Fire departments can be proactive in injury prevention by conducting training on recognition and prevention of heat-related injuries in the spring and cold-related injuries in the fall before the onset of the extreme weather. Training for recognition of symptoms is extremely important because the victim of these injuries is usually unaware that he or she is being affected until it is too late. Thus, early recognition by observant colleagues can be a lifesaver in many cases.

Fig. 17–8. Firefighter-unfriendly extremes of weather require both mental and physical toughness. Firefighters have a personal responsibility to dress accordingly. Command has the responsibility to relieve and rotate personnel. (Courtesy of Ron Jeffers)

Exposures to infectious disease. The department must conduct annual training on the dangers and prevention of bloodborne contamination occurrences. Training must be documented. In addition, there must be in place a comprehensive and enforced SOP regarding operations where bloodborne pathogens may be encountered. Recognition of the potential for exposure and contamination is the first step in safeguarding personnel. Once the potential is recognized, proper body substance isolation (BSI) protection should be available and worn. Policies should be in place for decontamination of personnel and equipment. Departments must have **disinfection** equipment available dedicated to simple decontamination operations of tools and other equipment, as well as procedures for the proper isolation and disposition of contaminated clothing and turnout gear. In addition, as will be mentioned in the next section, all exposures to bloodborne pathogens must begin with a process that includes a trip to the hospital for medical evaluation as well as documentation.

LESSON FROM THE FIREGROUND

At a structure fire in a residential occupancy, a captain and his crew were trapped above a fire. This occurred because the attack teams two floors below in the basement did not know they were up there. When the order to evacuate the building was given by command, the lines were withdrawn, allowing the fire to roar out of the basement and block the stairwell, their main egress point. They realized they were trapped, and they began making their way to the rear and a secondary means of egress. Visibility was extremely poor, and as a result the captain tripped over a footstool and crashed headfirst out a second-floor window and into a *light shaft*. He landed on his head. His chinstrap was in place. Like the fire instructor in the previous lessons learned, his helmet cracked in half on impact, but had it come off on the way down, he would have been killed. His injuries were enough for him to leave the fire service on a disability pension.

LESSONS LEARNED

1. You cannot trip when you are crawling. If you can't see your feet, you need to crawl.
2. Our turnout gear is our first line of defense and is meant to be used as a system. Don't ever compromise that system by cutting corners.
3. Always let someone know when you are going above the fire without a hoseline.
4. Inadequate firefighter management systems and accountability controls often lead to loss of coordination on the fireground.

FIREGROUND NOTE

Red lights, sirens, or blue lights should not be a signal to the public that says, "Watch out for me. I am about to do something stupid."

Exposure to hazardous conditions. Annual refresher training, along with solid and enforced SOPs outlining the parameters of personnel involvement when hazardous materials and conditions are encountered, must be in place. Documentation is also mandatory, especially the completion of department exposure forms whenever any member is exposed to a harmful chemical or agent, including chemicals used by firefighters in the fire station and from equipment at an emergency scene.

Preventing injuries

Preventing training injuries. Many injuries occur during training sessions. Recall that training is often the third highest category of fatalities and injuries. In 2015, there were nearly 7,600 documented training injuries. As such, training sessions should be well planned with safety as the overriding concern. The area where the training will occur should be safe and checked out in advance, especially if a dedicated building is being used. Many states prohibit burning except in a dedicated burn building at a fire academy. For those states that do allow live burning of structures, strict compliance with NFPA 1403, *Standard on Live Fire Training Evolutions*, must be followed. Even when not burning a structure, the parameters of the drill should be explained and an area familiarization should be conducted for the participants. This should include any considerations regarding the dangers associated with the specific training being conducted. A safety officer (SO) should always be assigned, and the span of control of instructor(s) to trainees should be maintained.

Preventing non-fireground injuries. These injuries are classified by the NFPA as "other on-duty" activities. Non-fireground injuries are those injuries that occur at the station, during fire prevention and maintenance duties, while exercising, and during everyday routine activities. These are preventable and many times directly attributed to lack of communication, not cleaning up after oneself, and basically not doing the right thing. The following is an excerpt from North Hudson, New Jersey, Regional Fire and Rescue's *Safety Matters* bulletin. It was put together as a result of a plethora of non-fire duty injuries and is titled, "There's No Fire, But We're Still Getting Injured."

Don't be a slip-and-fall person. Preventive maintenance prevents injuries.

- Report broken steps, cracked sidewalks, leaks, etc.
- If you spill it, clean it.
- If you notice it, see above rule.
- "I" = ownership.
- "I saw it" = "I will take care of it."
- Don't touch things twice: When you are done with something, put it away. (Your mother does not work here!)
- If it's wet, icy, oily, you will slip—unless you dry it, clean it, de-ice it.

Details should not equal disaster.

- Be aware of your surroundings at all times.
- Don't just check out the rig, check out the house.
- Use the same caution, care, and common sense in a different firehouse as you would on a fireground.
- House duties, routine maintenance, meal preparation areas, and work-out equipment and areas differ from house to house—familiarize yourself before using.
- House and apparatus familiarization should be ongoing.
- Check with the "normal" crew about idiosyncrasies before you discover them by "accident."

All rigs are not created equal.

- Doors and compartments open and are secured in different manners—be aware that some also don't secure well at all—and some are as insecure as a first-day probie!
- Warning devices may not be the same as your "home" rig.
- Watch your step: Heights, footholds, and hand grips for mounting and dismounting apparatus can vary.
- Check what's above your head before you step up; ditto for stepping down.
- Overlooking small details can cause big injuries.

Don't be a tool—Let your equipment work for you.

- Some items may have shifted during travel, so use caution when opening overhead compartments.
- Nothing ever falls up, and gravity never takes a day off.
- If it falls out once, straighten it out; if it falls out twice, you need straightening out.
- It's a scientific fact that what you don't secure today will hit you or someone else on the foot or head tomorrow.
- Securing devices need attention too: If you don't check to see that they are doing their job, you are not doing yours.

Respect your back, or you'll wind up on it.

- If it looks heavy, it probably is, so get help.
- We are in a team sport all the time, not just on the fireground.
- When lifting, remember that your knees bend for a reason—bend your knees, not your back.
- Giving someone a hand doesn't mean clapping for him or her—everyone pitches in.
- Try to keep the object you are lifting close to your body.
- The extent of the injury is often in direct proportion to the haphazardly hasty manner in which it is performed. If you are looking to get done quickly, take your time, think it out, do it safely.

Summary

- Pass information on—chances are someone knew about an unsafe condition before it hit you in the head.
- Set the example—don't be the example.
- Your personal safety is your own responsibility.
- Accident prevention is rooted in awareness and begins with you. Although in some circumstances ignorance may be bliss, in the fire service it can be deadly—be aware!
- If you can't make it better, at least make it safe.

Injury management and reporting

Injury reporting is a vital part of the injury prevention cycle. An injury properly analyzed and accurately reported may be an injury prevented in the future. Departments must have a protocol in place for the reporting and management of injuries. It is generally the responsibility of the immediate supervisor to report and document injuries. Other individuals may also be involved, such as the department safety officer (SO) or the department health officer. Regardless of who is responsible for documentation of an injury, one thing must be made clear to all department personnel: All injuries must be reported and properly documented no matter how minor they may seem.

As an example of improper injury reporting that has played out many times in fire departments, suppose Firefighter Jones twists his ankle as he is getting off the apparatus after a run. After the initial pain, he feels it is OK, and he will walk it off. He declines a request from his officer to seek medical attention, and the incident goes undocumented. Two days later, while off duty, Firefighter Jones's ankle has swollen considerably, and he decides to go to his own doctor to have it checked. It turns out that there is a fracture. Firefighter Jones calls on injury leave. The department has no record of the injury and places Firefighter Jones on sick leave instead. Firefighter Jones now has to expend six weeks of sick time, which comes out of his sick bank and affects his payout at separation of service. Because the department has no record of the injury being documented, it is treated as if it did not happen while on duty. In addition, in order for a firefighter to receive workers' compensation for an on-duty injury, he or she has to go to the department's doctors, usually at the time of the injury. Firefighter Jones went to his own doctor, which, if he wants to make an issue of this being an on-duty injury, probably violates department policy in regard to medical treatment for on-duty injuries. This may also result in disciplinary charges for Firefighter Jones and his officer for not following the departmental injury policy. Most departments' insurance policies do not recognize any diagnosis other than that of their own doctors for on-duty injuries. All this hassle, disciplinary charges, and lost benefit potential for Firefighter Jones occurred because of improper documentation of an injury. Document everything!

Injury management and documentation create a paper trail that will follow an individual through the treatment and recovery process, and beyond. Documentation should begin as soon as the injury occurs. Even jotting some notes down on a notepad is better than trying to remember the details of what happened: The palest ink is better than the sharpest memory. Fire departments are required by law to retain personnel documentation for many years after an individual terminates service with the department. Litigation might surface years later from an injury that occurred today, and the department has the responsibility not only to protect the individual but also to protect itself.

Injury management begins as soon as the injury is reported. If the injury requires care at a medical facility, prompt notification of emergency medical services (EMS) should be made. A department representative should accompany the injured member to the medical facility and act as a liaison between the medical authorities and the department. This includes updates, further documentation, and arrangements made for travel back to the department once the individual is released. We cannot just send our injured firefighters to a medical facility and then forget about them! Decisions and notifications may have to be made on the injured person's behalf while at the medical facility. It is a departmental responsibility to provide a representative to accompany and support that member from the beginning of treatment to release from the medical facility. Once the individual is released, the department must follow up on recovery orders and monitor the recovery (fig. 17–9).

Fig. 17–9. Injured firefighters must be cared for properly. Assigning a department representative to accompany the injured firefighter to the hospital is part of that care. (Courtesy of Ron Jeffers)

Causes of all injuries must be investigated. Departments should have forms to document this information, and supervisory personnel should be thoroughly trained in proper completion of these forms. The injured member, his or her superior, and anyone who witnessed the injury should be directed by department protocol to submit reports documenting the incident. Information regarding the illness and injury should be entered into the company's house journal or logbook. Failure to properly document an injury can have severe consequences at a later time, which could result in monetary loss for the department and the individual, caused by either litigation in the case of the department and its officers and proper compensation in the case of the individual.

FIREGROUND NOTE

Firefighters should focus on illness and injury prevention as an alternative to medical intervention.

Another critical area for documentation that is often overlooked is when department members are exposed to hazards during or after an incident. Such exposures may be to bloodborne pathogens, hazardous materials, or other potentially contaminating substances. The department must also document these exposures on the proper reports and ensure that medical attention is made available as required. In the case of bloodborne pathogen exposure, department members exposed to blood or bodily fluids must be transported to a medical facility, and their contaminated gear or clothing must be bagged, properly tagged (documented), and addressed according to department policies for bloodborne exposure. Follow-up treatment and testing must be performed in accordance with the department's bloodborne policy, which is mandated by OSHA (fig. 17–10).

In regard to exposure to an atmospheric contaminant, or a hazardous material or substance, many times it is not clear what the individual was exposed to, such as in the case of a fire where toxic smoke (all smoke is toxic!) has blanketed the area. In these cases, it is prudent to have the air monitored, but even more important to document an actual or possible exposure. There have been cases where firefighters have been exposed to toxins such as carcinogens at some time during their career and later developed cancer, emphysema, or other diseases, but because the exposure was never documented, no compensation was awarded.

Fig. 17–10. Hazardous exposures are not just limited to blood. At this fire in a live produce store, firefighters were exposed to runoff that contained animal waste. A decontamination station was established and exposure forms completed to document this exposure. (Courtesy of Ron Jeffers)

Fire/EMS Safety, Health, and Survival Week. Formerly called the National Firefighter Safety Stand Down, Fire/EMS Safety, Health, and Survival Week was initiated by the International Association of Fire Chiefs (IAFC) in 2005. That year, an estimated 10,000-plus fire departments participated in the event. Safety, Health, and Survival Week usually takes place during the third week of June and is announced on the IAFC Web site (www.iafc.org). Safety, Health, and Survival Week raises awareness and calls national attention to the unacceptable number of firefighter LODD. Information specific to the year's theme and suggested activities can also be downloaded from the site. Each year, a different focus is highlighted. In the year 2017, the theme focused on Mayday, self-rescue, and rapid intervention. The IAFC recognizes that safety initiatives and policies must be instituted at the organizational and administrative levels and filtered down and enforced through the chain of command. If the uppermost levels of the department do not think safety is important and do not create and enforce policy that addresses these issues, it will never be important to the department's members or become a reality on the fireground, on the training ground, or in the firehouse.

The rules of engagement for firefighter survival. The International Association of Fire Chiefs created and adopted 11 Rules of Engagement for Firefighter Survival. They are relatively simple and straightforward, but failure to abide by them has led to both injury and LODD on the fireground. There are also 14 Incident Commander's Rules of Engagement for Firefighter Safety. For this text, we will concentrate on the 11 Firefighter Rules.

1. *Size up your tactical area of operation.* Know where you are going and what you need to do. If you (and your department) did your homework before the incident, you will have vital information going in. Understand that even if this information is available, it is critical that firefighters conduct an initial and continuous size-up of the tactical area. What dangers are present? What is the condition of the structure? Where is the fire, where is it going, and what is it doing to the building and/or its occupants? Answers to these questions will help you determine risk vs. gain when deciding on a course of action to take.

2. *Determine the occupant survival profile.* The question that must be asked is, "Can anyone be alive in this structure?" If the answer is yes, then we must look to engage to the extent it will take to conduct a rescue. If the answer leans in the negative direction, we must be careful about how far we venture into the danger area and to what degree we risk ourselves. Trying to rescue someone who is already dead is not part of the prime directive of firefighting (fig. 17–11).

Fig. 17–11. Can anyone be alive in this structure? The answer is yes, therefore we must engage. The first floor appears clear and is searchable. The rear of the second floor can be searched with a hoseline in place on the second floor between searching firefighters in the rear and the fire. (Courtesy of Anthony Avillo)

3. *Do not risk your life for lives or property that cannot be saved.* There will be times when, due to conditions, we cannot save lives or property. A good working knowledge of building construction and fire behavior will help determine the extent to which we can engage in an aggressive manner.

4. *Extend limited risk to protect savable property.* Notice that there is no mention of savable life here. Property can be restored or rebuilt. Life cannot.

5. *Extend vigilant and measured risk to protect and rescue savable lives.* The ability to measure risk is something that comes along with experience. Vigilant is defined as keeping careful watch for possible danger or difficulties. This watch and evaluation is the barometer by which we extend ourselves in addressing savable lives.

6. *Go in together, stay together, come out together.* Firefighting is a team sport. No one gets left behind and no one operates alone. Fire companies must be assigned tasks as a team and perform them as a team. Department-created, supported, and universally enforced accountability policies, scene assignment procedures and, above all, firefighter discipline play an important role in the maintenance of the "Get 'em in safe, work 'em safe, get 'em out safe" mantra.

7. *Maintain continuous awareness of your air supply, situation, location, and fire conditions.* Your responsibility to those you work with is to keep yourself from becoming a liability. If you fail to maintain your awareness of your equipment, your air supply, and/or current conditions, the potential for the need for additional firefighters to have to risk their lives to save you is increased. You should never put yourself in a situation where others must rescue you, as when this occurs, the fireground dangers increase exponentially.

8. *Constantly monitor fireground communications for critical radio reports.* Proper and disciplined fireground communications provides a panoramic spectrum of what is occurring on the fireground. It fills in the gaps not only for command but for those fire companies working in various areas of the fireground. Proper communications are a vital link in the safety chain. Improper and undisciplined communications can break that link.

9. *You are required to report unsafe practices or conditions that can harm you. Stop, evaluate, and decide.* Vigilant situational awareness provides an opportunity for everyone to take care of everyone else. If you see something, say something. Unsafe conditions are often beyond our control, and identification and reporting of them must prompt command to action. This action can vary from an acknowledgement of the conditions, to an investigation by a chief or safety officer, to a complete withdrawal of personnel from a structure. Conversely, unsafe actions are usually within our control. No firefighter, regardless of rank, should ever

turn his or her head and ignore a safety violation. Safety is everybody's business, regardless of rank. It must be reported and known to be addressed and corrected. Often, failure to adhere to this "say something" mind-set does not have consequences the first time you see it, but unaddressed, that failure sets the stage for an even bigger problem in the future. There are countless examples of fire officers and firefighters who wished they had addressed an issue when they had an opportunity, for that missed opportunity was a direct cause of a casualty later. You never know if you will have an opportunity to fix something again, so say something *now* while you have a chance to make a difference.

10. *You are required to abandon your position and retreat before deteriorating conditions can harm you.* This ties right into the need to not only stay aware of what is happening not only in your position, but to keep an ear to the radio to be aware of the conditions in all areas around the building, perimeter, or wherever the emergency ground happens to be. Oftentimes, a timely radio report will provide a warning to those who cannot see what you see but will benefit from the information as it applies to their position.

    For instance, interior fire companies were operating in an old three-story multiple dwelling. The fire was in the roof space above the top floor. The building occupants were evacuated, and after a primary search, an "all clear" was transmitted to command. Ladder companies were making their way to the roof for ventilation. As the hoselines from the engine company reached the top floor and ceilings were opened, the interior crews observed that the building's roof support system had been renovated with lightweight truss construction and that it was involved in fire. A prompt report to command prompted a complete and immediate evacuation of the structure and roof and a change of operational mode to exterior, defensive positions. Soon after the companies had exited the building, the roof began to collapse, causing a secondary collapse of the top floor. Failure to get that information out to command and the working companies could have caused multiple firefighter casualties. Prompt evacuation of the structure because of a proper radio report made the difference. This scenario showed how all the rules of engagement are often tied together to create a safer fireground.

11. *Declare a Mayday as soon as you think you are in danger.* There is no shame in declaring a Mayday. The "I got into this mess, I can get out of it" mentality in the fire service has often made bad situations worse. Realize that every second you wait to issue a Mayday is at least two seconds that no one is coming to get you. If you think you are in trouble, *you are!*

    Included here is the first rule in the original set of rules of engagement, because no matter how many rules are established, if fire departments and personnel adhere to this rule, the others will fall into place.

    No building or property is worth the life of a firefighter.

    Often, our ability to keep out of harm's way is directly proportional to how well we abide by this all-important rule. Although this does not mean we will operate in a defensive manner and let all buildings burn down once everyone is evacuated, it means that the risk to personnel must be properly managed if we are going to continue to operate inside a structure. It also means, if in doubt, see rule number one.

Risk management personnel

Risk management and casualty prevention begins well before any fire department personnel even set foot near the incident or on the fireground. One of the best ways for a department to prevent an injury or LODD is to be proactive in its approach to the conditions and actions that lead to these events. Dealing with identified hazards by minimizing the chance of injury is called risk management. Fire departments must have an official, written risk management plan in place according to NFPA 1500. This program must cover all fire department policies and procedures including the following areas:

- Administration
- Facilities
- Training
- Vehicle operations
- Personal protective equipment (PPE)
- Operations at emergency incidents
- Operations at nonemergency incidents

Addressing risk management should be thought of as a cycle that repeats itself over and over as new information is discovered, analyzed, and resolved. It will follow four basic components of the risk management plan:

1. Risk identification
2. Risk evaluation
3. Risk control techniques
4. Risk management monitoring and follow-up (fig. 17–12)

Fig. 17–12. This collapse occurred several hours after fire operations ended. Fire watch was still on scene. The risk was evaluated, and control measures were taken by cordoning off the area, which kept personnel away from the danger. No one was injured.

Department safety committees. A valuable tool the department has in its quest to reduce injuries and fatalities consists of its own members. Who better than the major stakeholders to address and correct safety concerns? To this end, many departments have developed departmental safety committees to address risks and identify methods and controls to inform department members of safety concerns and give members an active voice and participation in addressing these concerns. A departmental safety committee should be represented by department members of all ranks and provide an alternate avenue outside of the chain of command to advocate for safety. The committee should be a mechanism by which concerns can be addressed and fixed, thereby preventing unsafe actions and unsafe conditions. The department chief and the administrative staff must also be active members of this committee. In a career department, members of the collective bargaining unit may also be represented on the committee. The members of the safety committee usually meet once a month and discuss issues, both old and new, that have been brought to their attention. Some areas that may be addressed include PPE, station concerns such as diesel exhaust and maintenance issues, and apparatus concerns. Through cooperative efforts, many problems can be addressed and solved when they are small rather than serious problems, causing injury or worse. This committee can also be effective in bringing about change in a progressive manner and potentially saving the department millions of dollars in legal fees by prevention of, rather than reaction to, unsafe conditions or actions. The committee works to ensure compliance with a myriad of OSHA and safety regulations and fosters a positive relationship between labor and management. We all benefit when labor and management work together to increase safety.

The committee can also keep department members informed with newsletters and bulletins, reminding members of safety issues and giving important information about current safety concerns, both at the department and at state and federal levels.

One effective information tool used by the North Hudson (NJ) Regional Fire and Rescue is a newsletter focusing on a specific safety topic, either a current trend in the fire service or one that has affected the department recently. The bimonthly newsletter is called *Safety Matters*. It is written by the platoon SOs on a rotating basis. Anyone, however, can submit information for a topic or for inclusion into a specific portion of the newsletter (fig. 17–13).

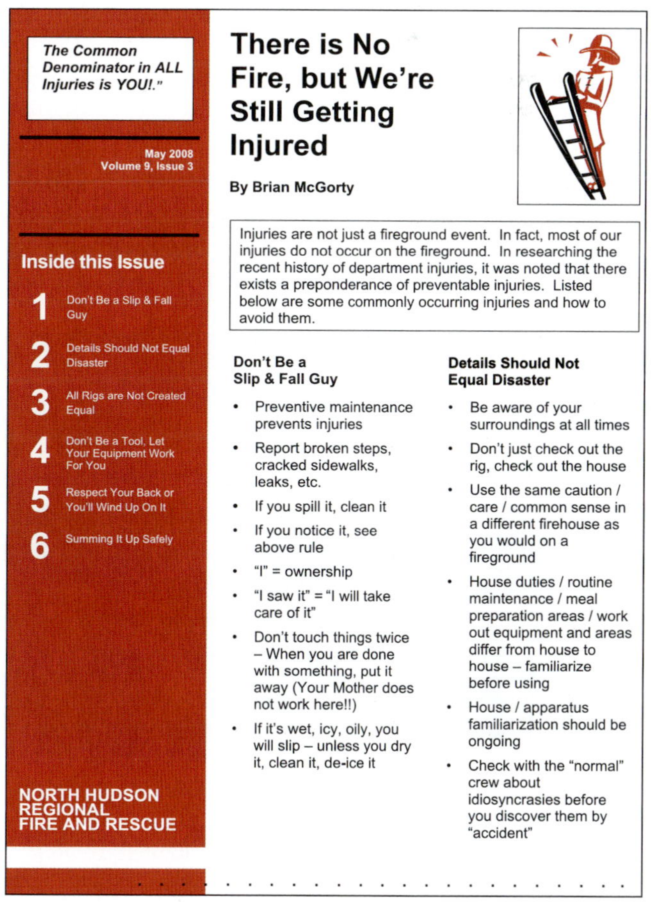

The Common Denominator in ALL Injuries is YOU!."

May 2008
Volume 9, Issue 3

Inside this Issue

1 Don't Be a Slip & Fall Guy
2 Details Should Not Equal Disaster
3 All Rigs are Not Created Equal
4 Don't Be a Tool, Let Your Equipment Work For You
5 Respect Your Back or You'll Wind Up On It
6 Summing It Up Safely

NORTH HUDSON REGIONAL FIRE AND RESCUE

There is No Fire, but We're Still Getting Injured

By Brian McGorty

Injuries are not just a fireground event. In fact, most of our injuries do not occur on the fireground. In researching the recent history of department injuries, it was noted that there exists a preponderance of preventable injuries. Listed below are some commonly occurring injuries and how to avoid them.

Don't Be a Slip & Fall Guy

- Preventive maintenance prevents injuries
- Report broken steps, cracked sidewalks, leaks, etc.
- If you spill it, clean it
- If you notice it, see above rule
- "I" = ownership
- "I saw it" = "I will take care of it"
- Don't touch things twice – When you are done with something, put it away (Your Mother does not work here!!)
- If it's wet, icy, oily, you will slip – unless you dry it, clean it, de-ice it

Details Should Not Equal Disaster

- Be aware of your surroundings at all times
- Don't just check out the rig, check out the house
- Use the same caution / care / common sense in a different firehouse as you would on a fireground
- House duties / routine maintenance / meal preparation areas / work out equipment and areas differ from house to house – familiarize before using
- House / apparatus familiarization should be ongoing
- Check with the "normal" crew about idiosyncrasies before you discover them by "accident"

Fig. 17–13. Safety always matters.

The fireground scene safety officer. SOs monitor and assess safety hazards and unsafe situations and develop measures for ensuring personnel safety. SOs have the emergency authority to alter, suspend, or terminate any operation they deem hazardous to the operating personnel. On the fireground, the authority of the SO is on par with the IC. Because the IC has the responsibility for everyone and everything that happens on the fireground, it is critical that the department or the IC assign someone who is reliable, safety conscious, and rich in fireground knowledge. The IC can delegate the authority for the SO to operate, but the IC must retain the responsibility for any actions of the SO.

The department should adopt SOPs that address the duties, responsibilities, and the authority of SOs so that no ambiguity exists regarding the scope of their duties. The scope of these roles is addressed in NFPA 1521, *Standard for Fire Department Safety Officer*. These are outlined in a course called Incident Safety Officer, developed by the National Fire Academy.

Proper deployment of the SO can assist the IC in running a controlled, safe, and informed fireground. The SO should not be shackled to the command post, but rather assigned as the eyes and ears of the IC. Orders and directives that come from the SO carry the weight of the IC, and personnel should be directed to follow them as such. The SO should be mobile, going where the areas of concern are and furnishing reports back to command, either face-to-face or via radio. This concept, sort of a "roving recon," puts an arm of the IC where it is needed most. Basically, the SO becomes another division supervisor, one with the flexibility to check on all areas of the fireground as the need arises (fig. 17–14).

Fig. 17–14. The safety officer should be in full protective clothing and operate as an arm of command, alternating between areas of concern and the command post. (Courtesy of Ron Jeffers)

Many departments have the luxury of a dedicated SO on duty. Others must assign this at the scene. The position may be assigned to a company or a chief officer. Regardless of who or how the SO is assigned, upon arrival to the scene, the SO reports to the command post to check in with the command. Once checked in, the SO should immediately do a 360° tour of the fireground. After giving a face-to-face report to the command post—or via radio, if something critical arises—the SO begins to move to the areas of most concern to liaison with personnel assigned to that area. This officer provides support from the operational safety point of view and advises as required. Safety reports back to the command post are issued by the SO at regular intervals and are a vital part of command operational evaluation.

All personnel must understand that the role of the SO is not that of a safety cop. The SO should not have to micromanage a fire scene. This means that obvious safety actions, such as wearing PPE properly and operating within the parameters set forth by department SOPs, should not have to be addressed on the fireground. Disciplined officers supervising disciplined firefighters will make the SO's job easier, which in turn enables the SO to focus on greater safety issues, making the fireground an overall safer place. In contrast, how could an SO, who had to spend most of the time telling personnel to get dressed properly, watch for signs of building failure or other dangerous conditions? It would be a crime if the SO was telling a company to put their masks on while the building was collapsing on personnel in another area of the fireground.

In technical rescue situations such as confined space, trench rescue, and high-angle incidents as well as at hazardous materials incidents, the resources designed to handle these exotic incidents require an SO on scene who has a level of expertise equal to those being requested to handle mitigation. In this instance, there would be assigned a hazardous materials safety officer (HMSO) or a technical rescue safety officer (TRSO), depending on the incident. This specialized SO oversees the safety of the immediate incident hot zone, while the department SO usually addresses the perimeter and overall scene safety. Just as it is unsafe for personnel, it is not a good idea to assign SOs to an operation that they have not been trained for.

The accountability officer. The accountability officer serves to coordinate the flow of personnel into and out of the fire area with the IC. The accountability officer should use a command board to track the location and assignments of resources. This officer should also carry a riding list of all personnel on duty for a career department, or of all personnel on scene in a volunteer department. To best do the job, the accountability officer should be at the command post. Mobile human command boards do not work on the fireground (fig. 17–15).

Fig. 17–15. The accountability officer should use a command board to track movement and assignment of companies. A headset will assist in the distraction-free monitoring of communications. (Courtesy of Ron Jeffers)

Further responsibilities of the accountability officer are to continually monitor the radio and to liaison with command about deployment and tactical reserve. A headset is helpful to use in these duties. Some departments also give the duty of the personnel accountability report (PAR) to the accountability officer. Others give that duty to dispatch and require that the accountability officer use the roll call (PAR) to update the command board and ensure that companies are where they should be.

To support the accountability officer, the department should have a strong personnel accountability SOP in place and enforce it at all incidents. Operational safety starts with accountability, and it is everyone's responsibility.

Risk management programs

Fireground rehabilitation. **FFI 4.1.1** Firefighter rehabilitation is vital to the well-being of personnel as well as the continuity of the operation. Policies need to be established and enforced to rotate crews from IDLH areas. Firefighters need to be properly rehydrated and replenish their caloric intake, have their turnout gear removed, have their vital signs monitored, and either be returned to service or removed from service. Improper rehabilitation not only affects air management but also leads to cardiac events, one of the top killers of the members of the American fire service. Fire departments must ensure that when members are removed from the IDLH environment, they report to fireground rehab and are properly cared for, as just described.

A rehab division with supplies appropriate for the season should be present at every major fire and should be at all fires when the weather is extreme. All apparatus should carry water coolers so that personnel can constantly rehydrate themselves during an incident, no matter what season it is or how minor the incident. The water should be changed every day. In addition, ICs must see to it that shelter is provided in the extremes of weather. A warm shelter in the winter where firefighters can change into dry clothes is a necessity. Firefighters should be urged during all seasons to carry extra clothes and especially gloves on the apparatus. In cold weather, while the firefight is continuing, one tends not to feel the cold. But after the fire is knocked down and it is time to pick up equipment, or when faced with a long, drawn-out defensive battle, cold-related injury can occur, especially if the firefighter is wet. In the summer months, provide shade and fans for relief. Large fan-driven misters are excellent for cooling down firefighters during the battle. Cooling towels worn around the neck also provide welcome relief from the heat. ICs who do not provide for their firefighters are not meeting the most important fireground priority, that of providing for firefighter safety (figs. 17–16, 17–17, and 17–18).

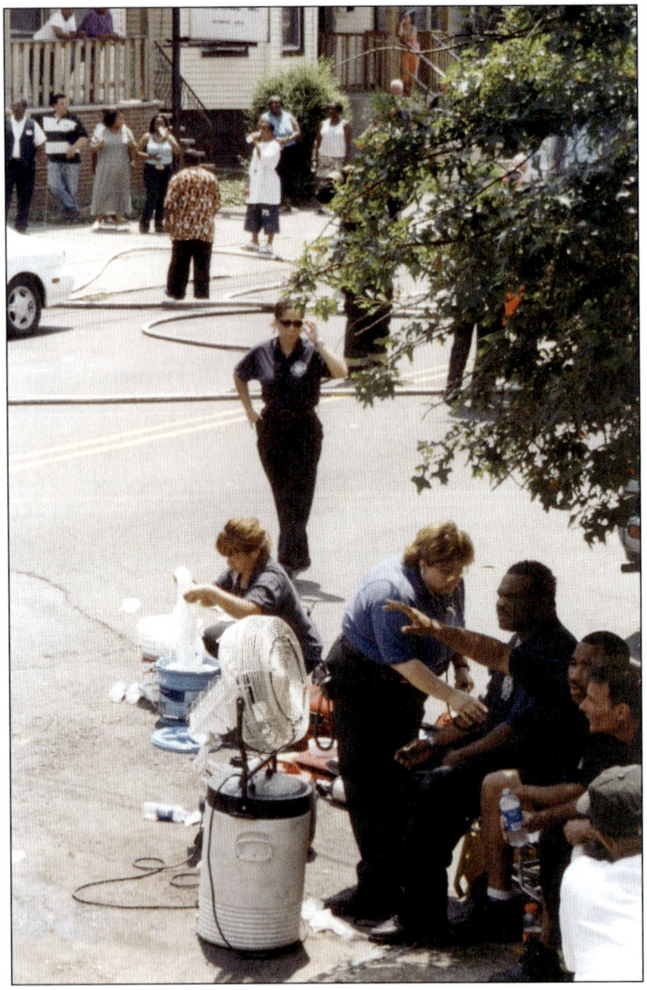

Fig. 17–16. Rehabilitation is a critical element of the fireground safety profile of the incident. Rehab equipment appropriate for the season should be available. (Courtesy of Ron Jeffers)

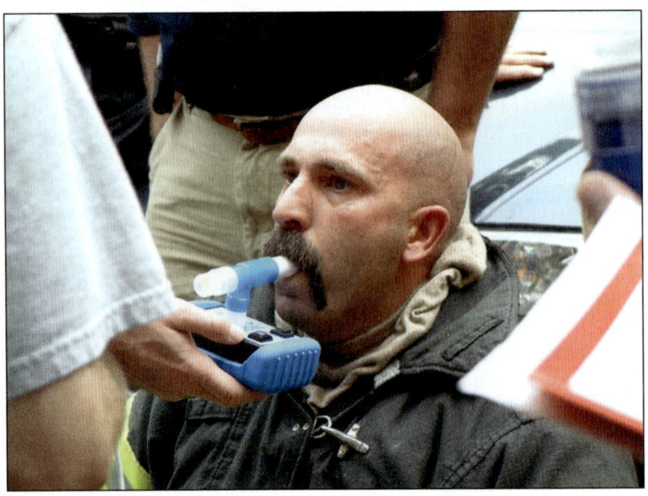

Fig. 17–17. Firefighters should be evaluated for exposure to carbon monoxide as part of the rehabilitation process. (Courtesy of Nancy Ricci)

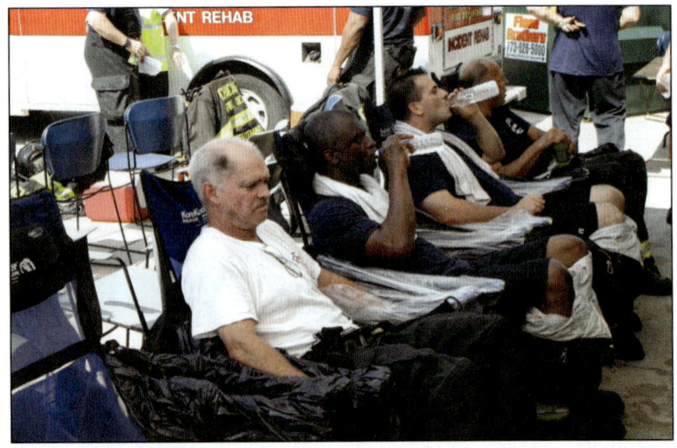

Fig. 17–18. Cooling towels used at a fire scene during a hot day. (Courtesy of Tim Olk)

On-deck air management. On-deck air management refers to having on hand properly staffed companies ready to relieve crews that are nearing their one-fourth reserve air and meeting them inside the structure to give the exiting crews ample time to remove themselves. Many fire departments are utilizing the on-deck system developed by the Phoenix, AZ, Fire Department. This allows the operations section to monitor air consumption levels of the crews electronically. As the crew nears their one-fourth reserve air level, an on-deck crew is sent in to relieve them. This system allows them to exit the IDLH before they begin to consume their reserve air. Once outside the structure, they get a cylinder change and are put back on deck. Based on factors such as available personnel, weather conditions, and fireground operational status, when a crew has used two bottles of air, they are sent to rehab for medical evaluation, hydration, caloric intake, and cooling or heating as needed. If conditions just stated limit working time, one cylinder may be the limit before being assigned to rehab. Through personnel management, the initial attack crews once released from the rehabilitation division are returned to quarters to clean up, change clothes, and be put back into service. Refer to chapter 10 for more information on air management.

Fire department wellness programs. Wellness programs improve the overall health and well-being of our firefighters. The better we take care of ourselves, the better we can take care of each other and the public we are sworn to serve. To this end, firefighters should take advantage of every opportunity to incorporate and bolster their own health and fitness, both physically and mentally. Wellness programs include annual physicals; fitness programs including dietary, flexibility, and exercise programs; rehabilitation of injuries; critical incident stress management (CISM) programs; and employee assistance programs (EAPs) that include trained peer support. The most important part of any department-sponsored program is willing participation by its members and the support of the department administration.

Fire department physicals and ongoing wellness. A firefighter first entering the service should have a physical used as a baseline. The requirements for this physical shall meet the minimum standards set by NFPA. Physicals should follow on an annual basis after that. Annual physicals are of the utmost importance to firefighters. When one considers the frequency with which cardiac sudden death takes the lives of firefighters, it is vital that this be a part of a comprehensive fitness and wellness program. We can relate fire prevention and fire suppression to illness and injury prevention. Our main job is to prevent fires. Only when that prevention fails do we need to suppress fires. In regard to health, we should focus on illness and injury prevention as an alternative to medical intervention once our poor habits have caused a bodily malfunction. To this end, many departments offer incentives for firefighters who pass their annual physicals, and still more offer comprehensive physical checkups to firefighters free of charge. Stress tests, blood testing for cholesterol levels and prostate screening, lung capacity testing, calcium screening to determine the extent of arterial plaque buildup, and electrocardiograms (EKGs) are all ways that firefighters can determine disease potential in a proactive manner and thus begin to take steps to prevent serious illness or injury.

In New Jersey, there is a program called "A Gift from Captain Buscio," which has been endorsed by the unions and the departments alike. It was started by the wife of a captain from Jersey City who died at a very young age of a cardiac sudden death. The program, which is offered to firefighters across the state, consists of a strictly confidential, comprehensive medical evaluation with no out-of-pocket expense to firefighters. The program is also offered to immediate family members as well. All firefighters must take advantage of opportunities regarding fitness and medical evaluations, not just for themselves, but for the loved ones they might leave behind (fig. 17–19).

Fig. 17–19. Medical evaluations provide a baseline in case of a future work-related illness or injury. (Courtesy of Christine Ricci)

Firefighter fitness. Firefighting is one of the most demanding occupations in the world, both physically and mentally. Firefighters in top physical condition can perform their jobs with more efficiency and less risk of injury than those firefighters who do not regularly practice cardiovascular and strength/endurance/flexibility conditioning. The nature of our work demands that any exercise program for the fire service should include constantly varied functional fitness that is geared toward our essential job functions. We must realize that at a moment's notice, firefighters can be thrust into a veritable pressure cooker of stress, strength-depleting manual labor, and split-second, life-or-death situations. These demands on the body and mind require that firefighters be in top physical shape. This includes not only cardiovascular and strength/endurance conditioning but also a major emphasis on flexibility conditioning. Since sprains and strains are the most predominant injury in the fire service, stretching should be part of the pre-and post-exercise routine. In addition, a healthy diet must also be part of the regimen for all firefighters (fig. 17–20).

Fig. 17–20. Firefighters must be prepared both mentally and physically for the many challenges of the profession. This bus versus light rail train collision created additional hazards such as electricity, unstable vehicles, jagged metal, multiple injuries, and bloodborne pathogens. (Courtesy of Ron Jeffers)

Departments should have exercise rooms and encourage personnel to use them. In fact, many departments are mandating exercise and setting aside time so that firefighters can engage in some sort of aerobic exercise, as well as strength/endurance conditioning as part of the daily duties. A note of caution: When in the station's exercise room, never work out alone. Be sure to use a spotter when lifting weights and consult a certified trainer before starting a workout program. Remember to inspect all equipment beforehand, and after use, put all equipment where it belongs to prevent a trip hazard. Clean the equipment with an approved disinfectant. Fire personnel should also be encouraged to prepare healthy meals at the station. In addition, many departments consult with or employ nutritionists, health specialists, and physical trainers to assist their members in developing and maintaining physical fitness and dietary regimens. This is one way for a department to protect and support its greatest resource: its members.

Injury rehabilitation. For years, many firefighters have felt forgotten when they received an injury. Today, many departments have comprehensive programs to aid the firefighters in their return to work. Firefighters who are not ready but return to duty anyway not only endanger themselves but also place their team and the citizens at risk. The Injury Rehabilitation Initiative of New Haven, CT, focuses on three components:

- Clinical pathway training: This training takes doctors and physical therapists through a fire department orientation period that includes riding with the fire department and participating in a full day of hands-on training.
- Physical therapy that is geared to our essential job functions.
- Practical evolutions that allow the firefighter to build strength and confidence.

Departments have found that these programs reduce secondary injury and reduce the number of lost days from work (fig. 17–21). One significant difference between this program and traditional rehabilitation is that once the member progresses, the therapist will start to include the entire body for conditioning.

Fig. 17–21. An occupational therapist guides a firefighter through a practical evolution so that he can aid the doctor in making better return-to-duty decisions. (Courtesy of Mike Cianciulli)

Critical incident stress management (CISM). Behavioral health is just as important as mental and physical health. All fire department personnel, at some point in their career, will be exposed to the more gruesome aspects of the profession, and most, if not all, will experience some effects of critical incident stress. There is a significant difference between experiencing symptoms of post-traumatic stress disorder (PTSD) and being diagnosed with PTSD. As firefighters, you must be prepared to witness and participate in emotionally disturbing events. Although you may be affected by these events, you must keep your composure and conduct yourself professionally on scene. The severity of such effects will vary from person to person and from event to event. Some people will not be affected by a certain event, while others will be. Sometimes a later incident may trigger a reaction that was buried after a traumatic event months earlier. There is no abnormal reaction (fig. 17–22).

Fig. 17–22. The events of September 11, 2001, exposed not only firefighters to critical incident stress but also the entire nation, who were witnesses to the tragic events that day. Countless emergency responders and their families have sought counseling as a result of this incident. (Courtesy of Ron Jeffers)

Critical incident stress anxiety evolves from normal reactions to abnormal situations and events. When the effects of these reactions begin to affect the individual's life to the point where lifestyle, family and peer relationships, and job performance are affected, intervention in the form of peer-focused counseling and possibly further professional counseling may be necessary. Symptoms of critical incident stress can vary in not only severity and scope but also in time. Sometimes, symptoms may not surface until weeks or months after a traumatic event occurrs.

Although the effects of critical incident stress cannot be prevented completely, they can be reduced by the establishment and maintenance of a department support mechanism. Written policies should be in place that include peer support and early intervention. Partnerships should be developed with mental health care facilities to provide the best stress management program for department members. Department members, especially supervisory personnel, must be trained in the recognition of symptoms of critical incident stress.

Some of the signs members should be on the alert for include, but are not limited to: Any change of behavior including use of drugs and alcohol, head and body ache and tremors, difficulty making decisions and concentrating, memory loss (especially of the incident), anxiety, guilt, fear, feelings of abandonment and isolation, emotional outbursts, withdrawal from others, suicidal thoughts, and any significant change in behavior.

Some departments have their own CISM team or peer support teams, and others rely on hospital-sponsored teams or regional teams. In several states, there are teams that respond based on which geographical region the incident occurred in. To best use their services, CISM teams should be requested as soon as the potential for exposure to a critical incident is suspected.

CISM teams usually consist of health care professionals, fire personnel from different departments (peer support), and other support personnel trained in addressing the concerns of those in need. On the fire end of it, many times the members of the team are fire personnel who have been through traumatic events and are now taking an active role assisting others in need. Working with peers who have "been through it" and can empathize with those who are experiencing critical incident stress offers great benefits in the way of support and reduction of the effects brought on by the incident. Contact information for these teams should be kept available and current. Dispatch may be required, at the request of the IC or designee, to request the CISM team or peer support while personnel are still at the scene of an incident. Members must realize after the alarm they can activate peer support on their own.

CISM, in the initial stages, usually takes the form of a defusing or debriefing. Especially at incidents where an emotionally charged environment exists, it is best, if personnel permit, to relieve all on-scene companies and send them to a designated CISM area away from the scene. Medical personnel should also be available in this area. Debriefings are aimed at getting people to talk, although speaking is not mandatory. Some people feel better just listening and identifying with other members' feelings. Intervention by the team is focused on both listening and offering helpful ways in which members can relieve the stress they are experiencing and get on with their lives. There may also be a further and more comprehensive debriefing a few days later, sometimes on the next shift, where all members are invited to participate and continue the process of discussion and healing. It is a very effective process. If further counseling is required, it is usually channeled through the CISM team and network. It is critical to understand that peer support and CISM teams must be trained to be a bridge to definitive treatment if advanced treatment is required.

Employee assistance programs. Employee assistance programs (EAPs) exist for both the physical and mental welfare of firefighters. Your department's EAP should be trained in working with firefighters. They exist to help a department member and/or family to get through troubling times and returning their lives to normalcy. Departments should have written policies outlining the parameters of this assistance as well as the steps involved in accessing and utilizing the available programs. Departments need not have an exclusive program available but must provide access to a program. Some larger departments have a fully staffed employee assistance division, while most other departments offer this service through their health care provider. The programs offered by EAPs encompass a wide variety of help, including smoking cessation programs, weight control programs, family and marital counseling, substance abuse programs, and stress counseling programs. Some also offer financial counseling and legal counseling, such as living will documentation. There may also be affiliations with larger, well-known national counseling groups such as Narcotics and Alcoholics Anonymous, which have vast resources in confronting and conquering both mental and physical issues. These programs are confidential and are usually offered to not only the employee, but also the members of the immediate family.

There may be times, especially in the case of substance abuse, where completion of a program through the EAP is the criterion to retain the job of firefighter. Absent that condition, neither the department nor its supervisors can intervene in the private lives of their members unless job performance has been affected. They can, however, if the members allow, offer counseling to assist the members in addressing a troublesome concern in their lives. The first time your members meet their EAP provider should not be the first time they need them. An EAP provider should be brought into each recruit class and be invited at least annually to a department-wide meeting.

Suicide prevention. As an industry, we are seeing more firefighters die each year from suicide than we are seeing deaths due to fireground operations. We need to do better. As firefighters, we need to know it is OK to ask for help. EAP programs should have 24-hour counselors available for firefighters; but you are not limited to that resource. The National Suicide Prevention Lifeline can be reached at 1-800-273-8255. Expect to reach a short recording before you are transferred to a trained individual who can provide assistance.

Leadership in behavioral health. We all need to take responsibility for behavioral health. While there are no easy answers, we do know that early intervention is the key to having the best chance for a positive outcome. For far too long the fire service has been slow to help our own. At times, due to archaic attitudes, issues have been covered up instead of being addressed. Issues left to fester often manifest themselves into more significant issues that will negatively impact the member's quality of life and job.

Personal risk management

Personal responsibility. The key to the safety mind-set starts with everyday actions and attitude toward work, self, and others. You should embark on this path with the goal of always doing a little more than required. This goes for housework, equipment maintenance, and fireground operations. Laziness and complacency contribute to injuries in the firehouse and on the fireground.

Firehouse and personal safety. Many preventable injures can occur at the stations. The remedy for most unsafe conditions is proper and diligent house work. Make sure that paths to apparatus are clear and wet floors are clearly marked. When cooking, ensure the stove is turned off before responding to an alarm. Fire poles should be safeguarded, and sliding a pole when wet is not recommended. Cleaning supplies should be properly marked and stored in a designated area. Material safety data sheets (MSDS) should be kept in a highly visible place and updated whenever a potential hazardous product is brought into the station or taken out (figs. 17–23 and 17–24).

Fig. 17–23. This right-to-know station on the apparatus floor is clearly marked MSDS.

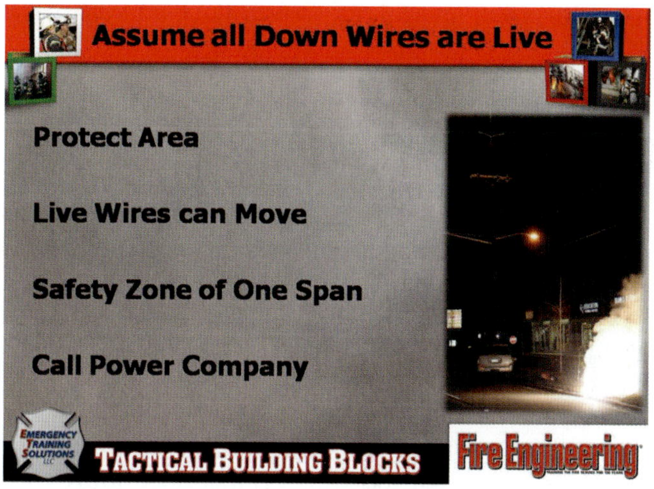

Fig. 17–24. Safety postings throughout the station serve as constant reminders to be vigilant. (Courtesy of Jim Duffy)

FFI 4.3.2 Horseplay can also lead to injuries. Although jokes are part of the fire service, there are several lines that must never be crossed. Any practical joke that contains references to race, sex, or religion, or that involves any safety equipment, has no place in the fire service. These inappropriate actions can lead to a breakdown in the team and have a direct impact on safety as well as invite litigation. Treat others as you would like to be treated. You are now part of a family that must work together (fig. 17–25).

Fig. 17–25. The kitchen table is a great place to learn, and it brings the company closer together. (Courtesy of Nick Ricci)

Responding and returning safety. Responding to and returning from an alarm is the second leading cause of firefighter LODD in the United States. It is also a leading cause of firefighter injury. Driver safety SOPs, seat belt policies, and annual driver evaluations are some of the tools that can lower this figure. Probably the biggest reason the casualty rate is so high is that the rules are not enforced. Also, if a company that arrives before us transmits a report of a working fire, we tend to want to speed the response up. This is counterproductive and contrary to safe response actions. In almost all cases, the first thing that must be changed is our attitude. Once that is on track, we can begin to address these totally preventable tragedies. The issue to be aware of is that if we don't get there safely, we can't help anyone (fig. 17–26).

Fig. 17–26. Response to an obvious working fire is not justification to jeopardize the safety of personnel and civilians alike by driving in an unsafe manner. Arrive safely first, then address the situation. (Courtesy of Ron Jeffers)

Maintenance. Most day-to-day operations will work toward the goal of ensuring operational readiness. When the alarm comes in, there is no time to find out that something doesn't work. The most important tool we have is the one you need right now, so no tool is more important than another, and all must be in a ready condition. Preventive maintenance programs should be incorporated into every firefighter's day. They can be broken down into three sections:

- Apparatus
- Hand tools
- Power tools

Firefighters are often charged with assisting the driver in completing the morning apparatus checks. In volunteer departments, apparatus checks are generally done on a weekly basis. This will include the Department of Motor Vehicles pre-trip. A pre-trip is a systematical check of the vehicle inside and out. Too many firefighters have paid with their lives so we could learn this lesson. Brakes, fluids, gauges, and tires are just a few of the critical items that must be inspected. The pre-trip inspection should be logged in the company log book (fig. 17–27).

Fig. 17–27. A firefighter conducts a pre-trip inspection at the start of his shift to ensure the safety of his crew.

These added responsibilities will be in addition to your position checks. Often, riding assignments will direct what equipment a firefighter is responsible to check. This firefighter must go over the inventory and conduct an in-depth inspection of the tools assigned to that position. If you are assigned on an engine company to the nozzle, your responsibility may be to take the nozzles off, check for debris and any other abnormality that may affect proper operation, and make sure there is water in the booster tank. On author Frank Ricci's first-time assignment to the nozzle ("pipe"), he took off the nozzle to find parts of a plastic bag lodged in the ball valve. Know where all the equipment is located on each apparatus and verify that all is in working order. External equipment must be securely mounted.

When checking hand tools, make sure they are clear of rust, properly marked, and free of damage. Power tools should be started regularly and run till the motor heats up. For a saw, this usually takes 3–5 minutes and will run the fuel through the system. Make sure that blades, chains, and disks are free of damage. If a chain is missing three teeth in a row or a total of five teeth, notify the officer and take the chain out of service. The chain should be sharp and be under enough tension so that if you lifted it out of the track with two fingers it should snap right back into place.

Having a piece of equipment fail on the fireground often leads to unnecessary delays. Because time is a constant factor, safety precautions may tend to be overlooked. If a problem is suspected with your tools or apparatus, it is your responsibility to report it to the driver and officer. There are no time-outs or do-overs on the fireground. The only way we can be sure that our apparatus and equipment will perform properly is by developing good habits and conducting proper maintenance.

Seat belts. Fire departments must have SOPs in place for the use of seat belts for each occupant of any department vehicle. Seat belts must be fastened at all times. It is the responsibility of both the apparatus operator and officer, or senior firefighter if no officer is assigned, to ensure that all personnel are safely on board and belted before the apparatus begins to move. It is also a responsibility of each individual on the apparatus.

Firefighters should don the proper level of PPE for the response prior to the vehicle responding. Drivers should be permitted to wear PPE that does not restrict their driving. At no time should personnel be allowed to remove their seat belts to don PPE or SCBA. If they cannot properly sit in their seats with their SCBA on and use their seat belts, they must wait until they arrive and remove their seat belts to don the SCBA.

Firefighters not riding in enclosed cabs should wear head, eye, and hearing protection. In addition, a ManSaver™ safety bar should be used in conjunction with a seat belt. Tail board riding must be prohibited at all times. In this day and age, it should not even have to be mentioned.

Mounting and dismounting apparatus. **FFI 4.3.2** Firefighters should always use the handrails and steps while mounting and dismounting the apparatus. Use of handrails reduces the chances of falling or slipping. Firefighters should always try to maintain three points of contact with the apparatus. They should face the apparatus when mounting or dismounting. This allows a safe check of traffic or hazards in each direction. It is recommended that if the apparatus comes into contact with live electrical lines, the firefighter should stay on the apparatus until the power company removes the power. In an extreme case where you would have to dismount from the apparatus, you must make sure to jump clear of the apparatus and not touch it and the ground at the same time to avoid electrocution.

Firefighters should also not stand or attempt to dismount the apparatus while the vehicle is still in motion, including when the vehicle arrives at the emergency scene. Trying to save a few seconds by dismounting as the vehicle pulls up is extremely dangerous. One slip can result in a fall that lands you in the path of the rear tires. The best way to ensure that this does not happen is to wait until the apparatus has stopped before unbuckling your seat belt.

SAFETY AND THE EMERGENCY SCENE

SKILL DRILL

Traffic and scene control

FFI 4.3.3 One of the biggest dangers firefighters will be exposed to occurs in the streets in which they respond and operate. Firefighters have been killed or severely injured when stepping off apparatus at scenes where traffic has not been controlled. Firefighters have been struck by civilian vehicles, police and EMS vehicles, and even other fire apparatus. Night is a particularly dangerous time. Hit-and-run accidents, civilian vehicles crashing into stopped fire apparatus, and (even more prevalent, though almost never reported) near-misses all take their toll on firefighter well-being, the ability of apparatus to respond, and the general feeling of security when operating. Firefighters must be vigilant at all times and take early preventive actions to safeguard themselves and other firefighters on the scene from becoming victims. Traffic scene control can be accomplished in many ways by utilizing any and all of the following:

- **Local or state police** should block roads and secure the fireground perimeter.
- **Volunteer fire department fire police** (typically older, retired members who direct traffic at a fire scene) can take proactive steps to assist in establishing a perimeter.
- **Fire department personnel** can perform traffic control duties if no police or fire police are on the scene, until police agencies arrive.
- **Traffic cones** should be carried by all apparatus to identify perimeters.
- **Road flares** can be used at night to alert motorists, but should be used judiciously, because they present an ignition source that might not be desirable at some incidents.
- **Fire apparatus** should be positioned to block traffic, especially for highway incidents. Apparatus should be angled so that operational areas such as tailboards and pump panels are as far away from the traffic flow as possible.
- **Traffic-diverting warning lights** on apparatus should be turned on at all times when companies are operating where traffic is still moving past.
- **Reflective trim on apparatus** assists in making apparatus more visible at night or in visibility-obscured areas.
- **Reflective trim on turnout gear** helps denote presence of emergency responders, but reflective trim will only work if it is kept clean and serviceable.
- **American National Standards Institute (ANSI) Class II vests** can be worn in addition to or in lieu of (if not in danger zone) turnout gear.

The bottom line is that if motorists, including other apparatus drivers, can't see you, they can't stop in time or get out of the way. All illuminating devices and equipment should be used as a first action when arriving at incidents at night or when visibility is obscured or expected to be obscured (fig. 17–28).

Fig. 17–28. Safety vests provide optimal visibility for oncoming traffic when working on the road. (Courtesy of firegroundimages.com)

Roadway emergencies and traffic hazards. Roadway emergencies include car fires, motor vehicle accidents, and hazardous materials responses. These incidents pose many potential problems for responders, including stopped traffic, which can cause a delay in response. When confronted with this concern, companies should relay their predicament to dispatch. They can suggest alternative routes for other responding companies and advise dispatch to alert other responding companies to avoid the congested area. Additional concerns involve access problems and long distances between exits and turnarounds or jug-handles. In these cases, if communications are timely, companies can be directed into an emergency scene in the most efficient manner, avoiding traffic tie-ups and response delays. Failure to communicate properly can cause a delay in scene arrival and loss of operational coordination.

There may be times, especially on a highway, where companies may need to respond against normal traffic flow. This is not recommended unless communications are properly transmitted and the police are on the scene to prevent accidents. This operation must be strictly controlled.

When arriving, fire apparatus must slow down to walking speed and turn off audible warning devices. If backing in, always use personnel as guides. Never back up without personnel in the street guiding you. This rule goes for return to quarters as well. A departmental policy addressing backing up should be in place and enforced. Apparatus illumination and warning lights should stay on to warn others that an emergency operation is in progress.

When on scene, especially an accident scene on a highway, it is critical that at least one lane be closed in addition to the lane the accident is in. This will provide a buffer for personnel operating. This practice is also recommended at other incidents such as structure fires, provided it does not block the movement of apparatus into and out of the scene. For the purposes of safety, additional lanes can be closed. If this is the case, it would be wise to coordinate with local or state police agencies or highway personnel.

For personnel safety, it is a good idea, especially if police are not yet on scene, to use apparatus as a shield from oncoming traffic. Place the apparatus between the scene and the normal flow of traffic. Do not get between apparatus and traffic; firefighters have had their turnout gear snagged by vehicles that passed too close and then dragged them. It is best to place apparatus at a 45° angle with the hosebed in the area furthest from the traffic. This is especially important when packing hose. This is usually the time when the police agencies want to reopen as many traffic lanes as possible. It is critical to maintain this one-lane buffer until you are ready to leave the scene.

When leaving the scene, turn off warning lights so as not to confuse civilian motorists who may think you are on a response. This can cause panic on their part, irrational actions, and further accidents (fig. 17–29).

Fig. 17–29. Apparatus blocking an accident scene with its wheels turned away from personnel working on scene. (Courtesy of firegroundimages.com)

Size-up at a roadway incident scene. Fire responders must size up the scope and severity of the incident upon scene arrival. First responders should determine:

- The magnitude of the incident (number of vehicles, injuries, etc.)
- Utility hazards such as wires down or spilled vehicle fluids
- Proper apparatus placement
- Potential traffic hazards

Regarding traffic tie-ups at the incident scene, as far as the fire department is concerned, the overriding issue is safety, not congestion.

Emergency vehicle lighting at roadway incident scenes. The use of emergency lighting at roadway incidents is essential. This lighting is intended for the safety of both responders and civilian motorists. Emergency lighting provides warning but no traffic control, and it may be confusing or blinding to motorists, especially at night.

According to the Federal Highway Administration (FHWA) emergency lighting may be reduced if proper emergency traffic control procedures are being used. It is safer to divert traffic with advance placement of signs and cones rather than relying on warning lights and vehicles alone. Reduce lighting at the scene as much as possible, without compromising the safety of responders.

Apparatus operators are urged to use caution in the deployment of floodlights at nighttime roadway scenes. Floodlights must be raised and deployed in a manner so as not to blind motorists driving past the incident scene. When floodlights are used, they must be raised to a height that allows the light to be diverted down onto the scene.

Protective clothing for roadway incidents. The reflective trim that is found on most firefighter turnout clothing is insufficient for providing adequate safety on the roadway. The FHWA states that reflective trim must be supplemented with additional garments that make the firefighter more visible. SOPs must clearly dictate that all personnel wear appropriate reflective vests when operating on the roadway. Reflective vests increase worker visibility regardless of the use of turnout gear. The vests must have both retroreflective and florescent properties. **Retro-reflective** material returns most of the light from the light source back to the observer. Florescent material absorbs ultraviolet light of a certain wavelength and regenerates it into visual energy.

Lighting the incident scene

SKILL DRILL

Many emergency incidents require emergency power or lighting. If the incident location has poor lighting or the incident happens in the dark, firefighters must be aware of and have the knowledge to properly set up and use the required equipment to light the emergency scene. Lighting is often brought into a building after the fire has been knocked down. The key to lighting is to make sure all areas are covered, including lighting specific hazards both inside and outside the building.

Power plants. Fire apparatus and other emergency service vehicles utilize several different methods of power generation, including inverters and generators.

Inverters. An inverter is a device that converts the vehicle's 12 or 24 volt DC (VDC) current into 110 or 220 volt AC (VAC) current. Inverters are usually used when small amounts of power are needed. These systems are advantageous for fuel consumption and initial cost, but the amount of power they can produce is very limited.

Generators. Generators are the most common type of power source used to provide emergency power. These systems can be either vehicle mounted or portable equipment.

Portable generators. Portable generators are usually powered by gasoline or diesel fuel (fig. 17–30). They typically have a power output of 110 or 220 V. Smaller systems can be carried by small groups of firefighters, whereas larger units are usually equipped with wheels and handles to transport them. A portable system is essential for areas that a fixed system on a vehicle cannot access. The disadvantage of a portable system is the limited power that can be produced by the unit.

Fig. 17–30. A gas-powered portable generator. (Courtesy of Lou Tibor)

Vehicle-mounted. Vehicle-mounted generator systems usually provide greater power capacity than the portable units. Vehicle-mounted units can be used to power vehicle-mounted lighting equipment along with providing outlets for portable lighting, power cords, or electrical powered equipment.

These systems have power outlets of 100 or 220 V with capabilities up to 60 kilowatts (kW). Some vehicle-mounted systems are very noisy and require the vehicle's motor to be running to power them.

Lighting equipment. Lighting equipment falls into two categories: portable and fixed.

Portable lights. Portable lights are used at night or dark emergency scenes to provide a safe working environment (fig. 17–31). Adequate interior and exterior lighting makes fireground operations safer for everyone involved. Apparatus floodlights and light towers provide visibility for exterior operations. For the interior, portable lights allow firefighters visibility to enter and exit the structure safely.

FFI 4.3.17 These lights are constructed in many ways, and they can range from 500 to 1,500 W. Portable lights can be supplied by either a fixed-vehicle or portable power plant. They are usually attached to a power cord. Portable lighting can be set up as string lights nailed to walls or other supports. Some portable lights are mounted on tripods or telescoping poles. For centralized lighting, an A-frame ladder can set up in the center of a room as a tripod to mount portable lighting. Try not to string lights along the floor so that firefighters will not step on fixtures or trip over lighting cords.

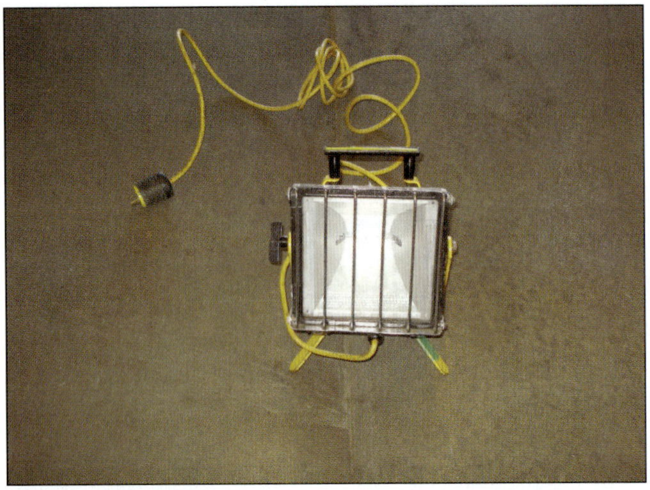

Fig. 17–31. A portable light. (Courtesy of Ken Nolan)

When using portable lighting equipment and power generators on the scene, remember the following safety practices:

- Keep electrical equipment and power cords dry, and position off the ground or away from areas of standing water.
- Inspect electrical cords for damage or fraying before connecting to a power source. Re-inspect cords and equipment for damage after use.
- Keep power cord, string lights, and other equipment out of traffic paths inside the structure or around the fire scene.
- Make sure portable generators and other electrical equipment are properly grounded with ground fault indicators.
- Always make sure the amp requirements for lighting devices match the rated amps of the generator and power lines or junction boxes.

Generators create carbon monoxide and should not be used indoors.

Fixed lights. Fixed lights are mounted permanently on the apparatus (fig. 17–32). They can provide immediate lighting around and under the vehicle. They are wired so that they may be turned on by a breaker panel or switches in the cab of the apparatus. When it is dark out, these lights should be switched on as soon as the company arrives on scene. If a generator is used to power the lights that are operated by the cab switches, they should be turned on during the response. On many apparatus, they too are usually operated from a switch inside the cab.

Another popular configuration is the fold-up or telescoping light mast. This device can contain multiple banks of lights. The system of lights can be rotated and tilted up and down to get the desired effect. A typical bank of lights ranges from 500 to 3,000 watts per light. These systems, along with other anticipated electrical loads, are matched to the vehicle's fixed generator capacity.

FIREGROUND NOTE

Overhead lighting works well for major incidents. Light mast and tower ladders will make it look like daytime.

Fig. 17–32. A fixed light on an apparatus

Auxiliary electrical equipment. Auxiliary electrical equipment constitutes many different devices that can be used in conjunction with power cords, lights, and generation systems to help provide emergency power. This equipment should meet NFPA 70E, *Standard for Electrical Safety Requirements for Employee Work Places*.

Electrical cords. Electrical cords can be rated at 15, 20, or 25 amps, and they come in different lengths. Most fire departments will carry a combination of electrical cords that terminate with twist lock and conventional plugs. This allows a department to have multiple options.

Twist locks and regular adaptors. Most fire department electrical cords have either twist-lock connectors or standard two-prong and ground plug male and female connections (figs. 17–33, 17–34, and 17–35). Twist-lock connectors require the same amperage rating for the male and female connections. You cannot mate a 15-amp connector with a 20-amp connector.

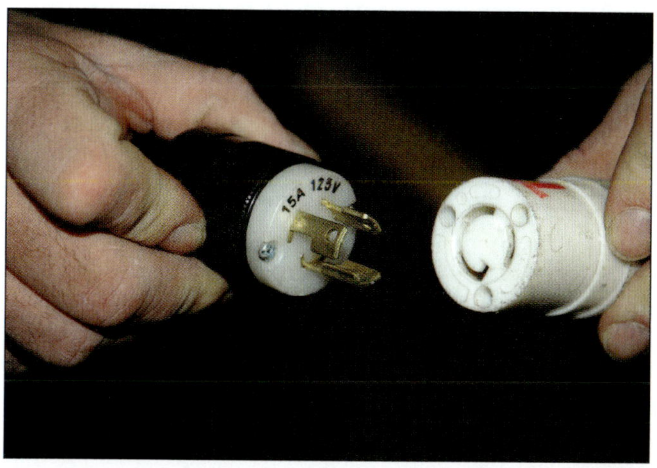

Fig. 17–33. A twist-lock connector. (Courtesy of Lou Tibor)

Fig. 17–34. A three-prong connector. (Courtesy of Lou Tibor)

Fig. 17–35. A two-prong connector. (Courtesy of Ken Nolan)

Electrical cord reels. Electrical cord reels can be either vehicle-mounted or portable units. They are designed to carry a specific amount of electrical cord. The advantage to a portable unit is it can be brought to wherever it is needed. The fixed unit is typically hard-wired to the apparatus and cannot be easily removed. Fixed units hold more cord.

Junction or gang boxes. Junction or gang boxes may be used when multiple outlets are needed (fig. 17–36). They are supplied by one connection back to the power source. They can provide multiple connections for twist-lock or conventional plugs, or a combination of both. The advantage to this device is that multiple cords can be powered from the same location. A variety of electrical adaptors should also be carried. All electrical equipment should also be equipped with ground fault interruption (GFI) capabilities to eliminate the possibility of electrical shock for the users.

Fig. 17–36. A gang box. (Courtesy of Ken Nolan)

Controlling utilities

SKILL DRILL

Electricity. If an emergency incident warrants that the electrical power to a structure be turned off, the local utility should be immediately requested by the IC. In the short term, it may be necessary for firefighting personnel to locate the electrical panel or fuse box and disconnect the main switch (figs. 17–37 and 17–38). Many times, these disconnect areas are in the basement on the wall directly below where the electrical service enters the building. Firefighters should make a mental note if a breaker that they turned off was tripped. This must be documented and reported to command and the fire marshal. Pulling the electrical meter from the socket on the structure is not an acceptable method. This should only be done by qualified electrical personnel. Improper removal of an electrical meter can cause serious injuries or even death.

Fig. 17–37. An electrical panel with circuit breakers

Fig. 17–38. A fuse box

In some commercial occupancies, it may be required to leave the power energized for certain processes or electrical-sensitive systems. The fire department officer must check with plant personnel before randomly cutting the power to a building. Having a person familiar with the building's system to advise you is critical. For example, if tasked with shutting off the power to minimize a possible ignition source, you will have to ascertain if this action will cause a generator to automatically start causing a greater hazard. You may have to de-energize the generator first. Auxiliary power supplies are common and you may also find a building fed with electrical power from two sources or sides. This information must be ascertained beforehand.

FFI 4.3.18 Utility disconnections should be handled in accordance with the safety procedures of your department and any guidelines or training required by the local utility companies for the specific systems. If required to shut off a knife switch, the firefighter should use a closed fist and apply pressure down to close the switch. Electrocution often results with the contraction of your muscles. Having a closed fist will ensure that if electrocuted your hand will remain free (figs. 17–39a and 17–39b).

Fig. 17–39a. A firefighter uses a closed fist to shut off a knife switch

Fig. 17–39b. Using a closed fist ensures that if electrocuted, your hand will remain free. Electrocution often results with contraction of your muscles.

Gas. If it becomes necessary to shut off gas utilities to a building, the local gas utility or propane company should be notified by command. Natural gas utilities usually can be shut off on the street side of the gas meter (fig. 17–40). This can be accomplished by turning the gas cock one-quarter turn into the off position (fig. 17–41). Individual gas appliances should also have an inline shutoff with a one-quarter-turn valve. In rare instances, the gas may have to be controlled by turning off a valve in the street. If the gas cannot be controlled by the fire department, firefighters must take a defensive position until the utility crew arrives. This means protection of exposures while letting the gas burn. Extinguishing the gas flame may cause a gas buildup inside a structure, which can lead to an explosion.

FFI 4.3.18 Once the gas is shut off at the meter, it should not be turned back on until the system undergoes a safety inspection by the utility company personnel or another qualified professional. Only after inspection should gas service be restored and pilot lights relit.

Fig. 17–40. A gas meter with a gas cock

Fig. 17–41. A tool closing a gas cock (perpendicular to flow)

Most propane systems have shutoff valves on the top of the tank and at inline shutoff valves at the appliances (fig. 17–42). Once the gas supply is shut off, the area should be metered to verify the reduction of gas vapors. It is not recommended that the fire department turn the gas back on at the end of the incident.

FFI 4.3.18 Personal protective equipment should always be worn when working with live utility systems in the event of electrical shock or gas explosion.

Fig. 17–42. The valve on a propane tank. (Courtesy of Lou Tibor)

Water. Most water utilities can be controlled by closing the valve where the water enters the structure. Most water utility companies recommend shutting the valves on the building side of the meter, not the street side. If the fire department encounters difficulty finding or shutting down the water service to a facility, they should contact the local water company. At times, it may be necessary to have the water shut off in the street. In multiple occupancies or large buildings, there may be multiple water services that need to be controlled.

PERSONNEL SAFETY

Personal protective equipment

NFPA 1500, *Standard on Fire Department Occupational Safety and Health Program*, under requirement 7.1.3, states, "Structural firefighting clothing shall be cleaned at least every six months as specified in NFPA 1851." (See fig. 17–43.) Chapter 7 of the NFPA 1851, *Standard on Selection, Care, and Maintenance of Protective Ensembles for Structural Firefighting and Proximity Fire Fighting*, addresses the cleaning and drying of turnout gear ensembles. Wet and unclean turnout gear is an important health and safety issue. Wet turnout gear is thermally unstable. Section A.7.4.3(6) of NFPA 1851 states, "Ensembles and ensemble elements should be completely dry before reuse to avoid the potential for steam burns caused by moisture remaining in the layers of the ensemble." Moreover, the storage of wet or moist turnout ensembles can affect the strength of some materials and promotes the growth of mildew and bacteria, some of which can lead to serious medical conditions. Gear that is not cleaned can also contain toxic products of combustion that can result in an exposure. Volunteers should not keep gear in the passenger compartment of their personal vehicle, and they should be sure that gear is stored out of direct sunlight. Also, wet turnout gear can place further stress on the firefighter. It is recommended that a firefighter be issued two sets of gear. Further details on protective clothing is found in chapter 9, Personal Protective Equipment.

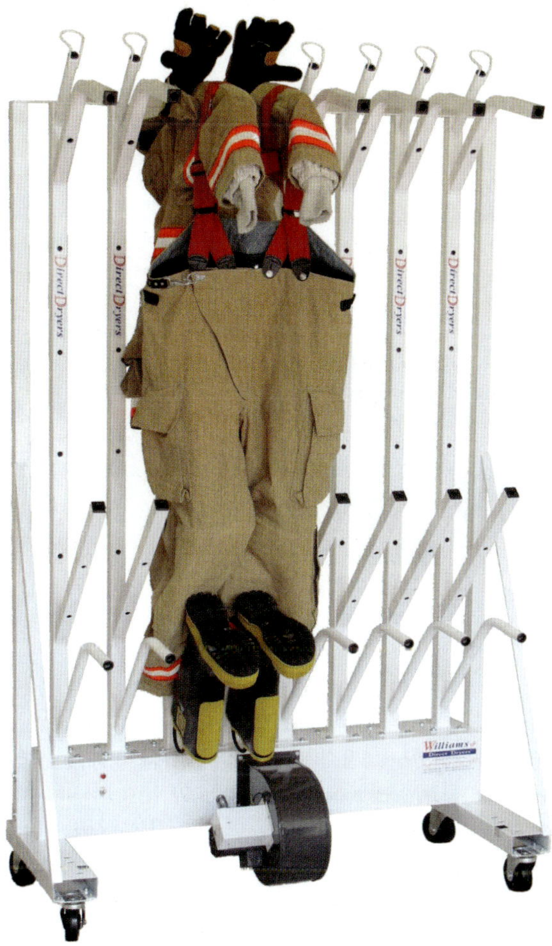

Fig. 17–43. Ideally, gear should be dried in a non-tumble-style dryer with a forced-air dryer around 100°F (38°C). (Courtesy of Williams Direct Dryers)

Cancer

One of the most prevalent occupational hazards to firefighters is cancer. Firefighters have a higher risk of contracting cancer than the general public. The body of science and research on this issue is well documented and growing. Due to the number of toxic containments in the form of gases and particles that are released at a fire, it is not possible for firefighters to be completely protected. The goal is to manage the risk and minimize the impact from the exposure. While use of self-contained breathing apparatus (SCBA) has decreased but not eliminated many of the respiratory hazards firefighters are exposed to, SCBA does not protect your skin.

In addition to the respiratory hazards, firefighters are also exposed to cancer-causing toxins through absorption. Containments are rapidly absorbed through your skin and enter your blood stream. You not only absorb containments while participating in fireground tactics but after an incident is over. You can absorb toxins through our skin anytime you come in contact with any of the toxic byproducts of the fire. This includes the handling of our gear or from gear that is off-gassing in the passenger compartment of a vehicle.

Minimizing the risk. While on scene, gross decontaminate your gear if you have a second set. Note wet gear lacks thermal stability and should not be worn on the next alarm due to the risk of burns. If a second set is not available, doff your PPE and store it in a compartment to be washed at the end of your shift. Remove any potential toxins from your skin. Pay special attention to the area of your neck and wrists. Utilize a moistened wipe to clean your neck, hands, and face. Upon returning to your station shower immediately and change into a clean uniform. To further reduce this intrusive threat to your body and internal organs, fire departments have been conducting testing on the effects of saunas in removing toxins from firefighters' bodies (sweating out the toxins).

Be proactive by including decontamination equipment on the apparatus including your radio. Departments should have a spare PPE for all members to utilize after a fire; ideally each member should have a second set of PPE. This provides all members the ability to wash their gear according to the manufacturer's recommendations after every exposure. Do not forget to wash all parts of your protective clothing including glove, hoods, jacket, pants, boots, and helmets as per the manufacturers' directions.

All firefighters must take the threat of cancer seriously and actively take steps to decrease our exposure. The days of having dirty gear are gone.

Hearing protection

As firefighters, we can be exposed to noise levels above 115 decibels on the fireground. NFPA 1500 and OSHA *Standard on Occupational Noise Exposure* 1910.95 both require that firefighters be provided with hearing protection when charted levels exceed 90 decibels. The standards further state that fire departments shall establish hearing conservation programs. Personal responsibility plays a large role in ensuring that the provided hearing protection is used when responding to alarms and windows are kept closed to minimize the noise from the sirens. It has been well established that when responding to alarms, you will be exposed to over 90 decibels. OSHA does recognize that it is impractical to use hearing protection once on scene because it could interfere with other life-saving equipment. Many departments, however, urge firefighters operating power equipment such as saws and jackhammers to use hearing protection (fig. 17–44). Steps should be put in place to engineer safety controls when practical. For example, having fire apparatus with headsets that increase communication between the crew allows for the firefighter to hear communications through the internal speaker and provides needed hearing protection.

Fig. 17–44. Hearing protection is recommended, although not always practical on the fire scene. (Courtesy of Lou Tibor)

Occupational Safety and Health Administration (OSHA)

FFII 5.1.1 OSHA standards are federal regulations that mandate minimum training standards, equipment performance, administration, and operations. It is important to realize that the fire service is unique in the way we are regulated because of the type of work we must perform. For example, the fire service has no fall protection standard for the fireground. Although it may sound counterproductive, the time it would take to set up protection on a roof would place firefighters at greater risk of falling through the roof. All departments should strive to exceed all OSHA (or state) standards. That can only happen if you commit yourself to safety.

Staffing and safety

FFII 5.1.1 Lack of proper staffing is a leading contributor to firefighter injures and death. The tactics that you have learned cannot occur incrementally. They must take place in unison with each other to allow us to control the building and safely mitigate the incident. At a structure fire, tasks are time sensitive. NFPA 1710, *Standard for the Organization and Deployment of Fire Suppression, Operations, Emergency Medical Operations, and Special Operations to the Public by Career Fire Departments*, mandates that the first-arriving company shall be staffed with four personnel and arrive in less than 4 minutes (fig. 17–45).

Fig. 17–45. Company staffed with four personnel. (Courtesy of Dave Esposito)

The standard also states that the entire first-alarm assignment of 15–17 should arrive in 8 minutes. These requirements are based on a 2,000-sq-ft (186-sq-m) house with no exposures. Numerous studies have found a definitive link between injures and crew size. The International Association of Fire Fighters (IAFF) report, "Analysis of Fire Fighter Injuries and Minimum Staffing per Piece of Apparatus in Cities with Populations of 150,000 or More" (December 1991), found that crews smaller than four almost doubled their percentage rate for injuries when compared with crews of four or more firefighters.

A study by the Providence, RI, Fire Department, analyzed in an applied research project submitted to the National Fire Academy as part of the Executive Fire Officer Program by Curtis Varone (1994), found that "four-person staffing led to a 23.8% reduction in injuries, a 25% reduction in time lost injuries and a 71% decrease in time lost due to injury when compared to three-person staffing." Inadequate staffing contributes to companies having to play catch-up. This increases risk to firefighters and contributes to firefighters taking shortcuts that lead to injures.

To show the effect apparatus staffing has on effectiveness and safety, a landmark study done by the National Institute of Standards and Testing (NIST) showed that crew size has a substantial effect on the fire service's ability to protect lives and save property. The study found that four-person firefighting crews were able to complete 22 essential firefighting and rescue tasks in a typical residential structure fire 30 percent faster than two-person crews and 25 percent faster than three-person crews (NIST, 2010).

Safe staffing requires ESP:

- *Education and emotion:* Be passionate about your job. Take classes, read trade magazines, and continue your education. Take advantage of educational opportunities. At fire prevention talks, explain the role of all the different companies on scene. Participate or conduct the Fire Operations 101 program developed by the IAFF. This program takes elected officials through an orientation program and educates them on the dangers of firefighting. Fire Operations 101 gives them a firsthand view of how difficult fire operations are with limited personnel and builds lasting relationships with officials.
- *Standards and statistics:* You should be well-versed in standards such as NFPA 1710 and the OSHA Two-in, Two-Out Rule (discussed in a later section). Read reports from the National Institute of Occupational Safety and Health (NIOSH), NFPA, and the U.S. Fire Administration to stay informed on safety and current trends.
- *Politeness and political action:* Get involved in your department and your community. Standing up for appropriate staffing is protected speech because it is a legitimate matter of public concern. Note that until you are off probation, you are not entitled to an opinion, so be quiet and learn. You have two ears, two eyes, but only one mouth for a reason.

Volunteers and staffing. The majority of fire departments are made up of volunteers. Commitments to work and family and the fact that many members work out of town have placed a strain on many volunteer departments to respond during daytime hours. NFPA 1720, *Standard for the Organization and Deployment of Fire Suppression Operations, Emergency Medical Operations and Special Operations to the Public by Volunteer Fire Departments*, applies to volunteer fire departments.

Many departments have instituted innovative approaches to increase staffing. Rockville, MD, started a college live-in program in which students receive room and board in return for service to the department. Other departments have started duty nights in which the members commit to staffing the station during a predetermined night. This builds camaraderie and provides many of the benefits of career departments. Departments have also established Fire Explorer programs or Junior Firefighter programs to increase interest in the fire service from younger members of the community. This is a valuable tool that many career and volunteer fire departments have benefited from. Volunteer departments in Long Island, NY, assign members to cover certain types of calls to limit the impact on their family and encourage greater participation. For example, five or six members may be assigned to cover single engine calls for a particular night for such responses as activated alarms and car fires.

FIREGROUND NOTE

The Washington Post Theory:

Conduct yourself so that if your actions were to be published on the front page of the Washington Post, you would be proud.

—Captain Michael Abrashoff
U.S. Navy

Staffing, you, and the budget. You may be asking yourself, "I am new to the fire service. How can I have an impact on achieving or maintaining safe staffing?" As a new firefighter, you can play a key role by interacting with our "customers." Take time to ensure that you treat all citizens with respect and go out of your way to make their interaction with the fire department a positive one. Be friendly; talk to the elderly a little longer; and at the completion of a call, show the kids the fire truck. It is always a good idea, during the conversation or when the children are playing, to mention fire prevention issues (fig. 17–46).

Your department's reputation will play a direct role in the budget process. The budget, whether career or volunteer, has a direct impact on the training funds, equipment repairs, protective clothing, and staffing. Your actions in the field may determine if the community supports funding new apparatus or if they stand with you when need to save a neighborhood firehouse from being closed.

Personnel accountability systems. **FFI 4.1.1** Ensure that you have read and understood your department's SOP on accountability. Accountability is everyone's responsibility and must be strictly adhered to. Crew integrity is not just the officer's responsibility. If you are separated from your crew, radio or report to another officer and let them know your location. Accountability systems make tracking personnel and crews more manageable. First and foremost, they establish who is on scene and what companies those personnel are assigned to. Departments use many different tag systems, boards, or written logs to track crews on scene (fig. 17–47).

Fig. 17–46. Firefighters taking time to interact with the community. (Courtesy of Jacqueline Bender)

Fig. 17–47. An example of a tag system. Firefighters clip their tag into a ring on the apparatus when they report for duty. (Courtesy of firegroundimages.com)

Whatever system is used, it must address the following:

1. Who is on scene
2. What company they are assigned to
3. Where they are operating

Tracking dynamic crews is difficult and requires the human factor to continually update the locations of companies. Manufacturers are working to establish electronic accountability using radio waves and other technology (fig. 17–48).

Fig. 17–48. Accountability officer tracks location of crews on an accountability board. (Courtesy of Ron Jeffers)

Personnel accountability report. The personnel accountability report (PAR) provides a roll call to ensure that all personnel are accounted for. A PAR can be called at any time by the IC. An accountability officer may be designated to track crews. Some departments require the dispatcher to notify them at predetermined intervals (e.g., every 15 minutes) to mark time for command and ensure that PARs are conducted. PARs can also be given after each radio transmission. PARs are usually called for after an evacuation order or a catastrophic event. They should also be requested once the fire is declared under control.

Diagramming the fire. All firefighters should have a radio, but not everybody should talk on the radio. Think before you speak. Ask yourself, "**D**oes **I**t **M**atter **W**hat **I**'m **T**ransmitting?" (DIMWIT). Radio discipline leaves radio channels available for necessary transmissions.

All members must listen to the radio. Time and again, Mayday calls are being missed. Radio transmission is one of the ways in which operating crews are tracked. Officers and firefighters should give their location with every radio transmission when inside a hazard area.

For example: "Truck 4 to command: Second floor kitchen, we have fire running the stud spaces with possible extension to the attic, Truck 4 has PAR."

This information is not only beneficial for command, but it also allows the RIT and other companies that haven't arrived to track crews. If a Mayday is given that is missing all the necessary information, it allows for a starting point to locate the member(s) in trouble. Fire department personnel who are not on the alarm should monitor the radio and try to track crews on a board. This helps develop listening skills and gives multiple ears the task of listening to the radio.

Evacuation order. This is the one order that can never be questioned. As firefighters involved at the task level, you are only seeing part of the picture. Command is basing the decision to evacuate a structure or area on the totality of the incident. Command is positioned to see visual cues such as collapse and should continuously be receiving input from other officers. If ordered out, get out with your crew! Most departments will issue an order and have the closest fire apparatus blast their air horns three times in case a radio message is missed. Emergency fireground transmissions such as withdrawal orders will be discussed later in this chapter.

Crew integrity and hazard area exit. Complete fireground accountability can only be effective with 100% commitment to the team concept. Firefighters should never work alone. If you are separated from your crew, you must immediately call for help. At times, you will lose your crew or get mixed into the wrong crew. If this occurs, notify your commanding officer that you are separated and identify what company you are from. If one member must exit the hazard area, he or she should exit with another team member. When exiting the hazard area and vision is obscured, you must crawl. If you are following a guide line that is taut, grasp the line and slide your hand on the rope. If you are exiting on a slacked line, one method is to kneel on the line and grasp the line with a closed hand, then slide your hand on the rope. Next, move your knee and repeat the process (figs. 17–49 and 17–50). Hoselines may also be used to guide you to the outside by using the hose coupling lugs.

Furniture should not be moved during fire operations. When exiting, the location of furniture will help keep you oriented to your location. Even when using a **thermal imaging camera (TIC)**, move along the walls in a systematic fashion. This method has worked for rats for hundreds of years, and, as with any technology, the camera can fail. When crawling, keep a low center of gravity and always check for the floor. When we crawl, we need to move quickly, but we must also ensure that we will be able to pull back if a weak floor or shaft is found (fig. 17–51). Make an effort to look up and check conditions above you. Periodically stop and listen, as you might be able to hear the fire above you.

Fig. 17–49. This method is slower than crawling along the line, but it ensures you will not get turned in the wrong direction if the guide line crosses over itself. (Courtesy of Kris Sundwall)

Fig. 17–50. If you are lost and you find a hoseline, remember the phrase "bumps to the pumps" as a guide. Move in the direction of the hose coupling lugs. Couplings that are painted identify particular engine companies. (Courtesy of Maureen Duffy)

Fig. 17–51. The shaft that author Frank Ricci fell into when searching for a trapped security guard. Make sure that if you encounter a shaft, you can pull yourself back up.

TACTICAL BREAKDOWNS

This section will cover predictable and preventable events that the fire service has paid the highest price to learn. If you are on the RIT and are activated, it is a clear indication of failures on multiple levels. If you look at the NIOSH LODD reports as a random report card of our service, we are failing in many areas. This is mentioned with the disclaimer that mistakes will be made at every fire. It is not the mistake that you make, it is how you recover from it and the fact that you recognize it, share it, and learn from it. It takes a balance between experience, training, and knowledge to be safe and successful in this field. Experience is not enough. When we develop unsafe habits and they are reinforced with positive experiences (good luck reinforces bad habits), they become the norm and contribute to greater failures later that will culminate with an injury or death. Although most of these things are covered in other chapters, the seriousness of failing to heed these warnings can cause serious injury or death. Let's look at some of these failures.

Command and control

Failure to properly size up the situation. Improper size-up costs lives. Taking a couple of extra seconds to analyze a situation can save you minutes and maybe your life. The New York City Fire Department came up with the acronym, COAL WAS WEALTH (construction, occupancy, apparatus and staffing, life hazard, water supply, auxiliary appliances and aids, street conditions, weather, exposures, area, location and extent, time, and height), to standardize size-up. This is an effective method, but experience has shown that new firefighters often miss the very basic points due to adrenaline. You need to first concentrate on the task at hand. If hooking up the hydrant, then focus on this task and accomplish it first. When you are performing any task that will put you inside or around the building, at least five points should come into your mind:

1. Construction
2. Occupancy
3. Layout
4. Where the fire is located
5. Where the fire is going

This will give you a foundation to build on as your experience increases. Take advantage of routine calls to size up homes and buildings in your district. You will find similarities that will aid you in size-up of other similar occupancies.

Combining offensive and defensive attacks. This action can injure a firefighter on the inside and if done with an improper stream could push the fire due to the movement of air. If you are ordered to extinguish fire that is coming out of a window, and you know that crews are operating inside, direct the line at the eaves of the house or above the window. This will allow the water to cascade down the house and stop the fire from enveloping the exterior of the structure. This action will also prevent a firefighter from being hit with a hose stream. Once the interior hoseline is working the fire, shut the exterior line down. The exterior hoseline always wins in these cases.

> **FIREGROUND NOTE**
>
> Combining offensive and defensive attacks is like ordering artillery on yourself.
>
> —Alan Brunacini

Uncoordinated fireground operations. Another major tactical breakdown, one that has caused countless firefighter injuries and LODDs, is uncoordinated fireground operations. There are three crucial fireground operations that must go hand in hand to keep the operation as safe as possible. These are fire attack, search and rescue, and ventilation, represented by the fireground coordination triangle (fig. 17–52). If any one of these three principles of the fireground coordination triangle is out of sync, missing, or unsupported by the operation, the triangle, like a truss, will collapse. A complete triangle will support the center of the triangle—fireground safety.

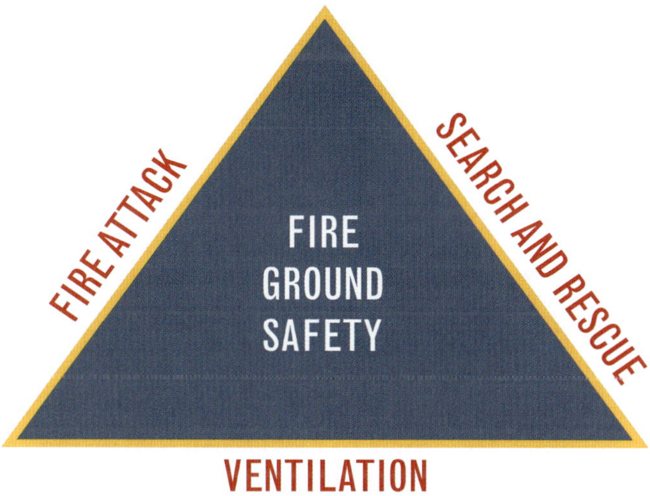

Fig. 17–52. Fire coordination triangle

Failure to recognize when your incident is beyond the scope of your training. Incidents such as technical rescues, biological and hazardous materials incidents, and major utility failure incidents like ruptured gas mains require intervention by specially trained personnel, sometimes from agencies other than the fire department. Firefighters should understand that they are limited by their training and the capability of their equipment. Fire personnel may still operate in a support posture, providing such assistance as evacuation of exposed areas, conducting decontamination operations, and operating exposure lines. They should not get directly involved in these incidents, meaning "hands off" the problem. The mark of the true professional is the ability to recognize that your incident may be best handled by someone else (fig. 17–53).

Failure to pay attention to your surroundings. It has been mentioned several times in this chapter that complacency is a killer. A large proportion of firefighters who become casualties are usually not brand new firefighters, but seasoned veterans who have been allowed by lack of leadership and supervision to become too comfortable in their settings. There is no firefighter tougher than the fire. When working near or above the fire, failure to pay attention to the location of the fire, as well as paths that it can spread has cost many firefighters their lives. In addition, firefighters who become too complacent may fail to recognize such dangers as overhead power lines, dangerous vacant buildings and their hazards, signs of rapid fire development and collapse, and traffic hazards. There is also no firefighter tougher than electricity, a falling building, or a moving vehicle. The fireground should be treated as extremely hazardous from arrival to termination. Firefighters should be continuously watching out for hazards and for each other. Casualties can be prevented by identifying hazards and taking appropriate action (fig. 17–54).

Fig. 17–53. Technical rescue operations such as this trench rescue require expertise beyond most first responders' capabilities. Firefighters must know their limitations in regard to operational capability. (Courtesy of Ron Jeffers)

Fig. 17–54. The truss above this firefighter's head is clearly involved in fire. This is a tragedy waiting to happen. (Courtesy of BC Mike Oriente NHRFR)

Failure to request early assistance. Fires and other emergency incidents are mitigated by people. Not having enough people on scene is a major contributor to incident escalation and firefighter casualties. Additional personnel are required for relief purposes, for reinforcement of operational areas, and to address those little surprises that often pop up on the fireground. When there are enough personnel on scene to address them, small concerns stay small. When you are behind the eight ball in the personnel game, small concerns become big problems and may affect both the strategy and the safety of personnel. As a firefighter, based on department protocols, you may be detailed someday as an acting company officer. If this is the case, you might wind up as the initial IC. The decision to strike an additional alarm upon arrival may be your decision to make. The need for this might not be apparent when you arrive and everyone goes to work. If all personnel are working, and there is no one waiting in the street to go to work, the tactical reserve profile is at zero. The dilemma will emerge when the first-alarm companies begin to exhaust their first SCBA cylinders, when they become fatigued, when someone calls for additional help in an operational area, or when an unexpected concern surfaces. If there is no tactical reserve on the sidelines to put into the game, the game will either be lost or the fatigued firefighters will have to overextend themselves to continue the operation, risking injury. When in doubt, ensure there are enough personnel on scene or ordered to maintain the operation, especially in extreme weather. If you don't need them, you can always send them back.

Water supply and fire attack

Failure to charge hoseline before descending to a below-grade fire. **FFI 4.3.10** The water is your only protection. If a member vents early, before you are at the bottom of the stairs, you may be caught on the stairs by rapid-fire progression and will have no water to protect yourself (fig. 17–55). When ventilating horizontally (for extinguishment), wait until the line is charged and hitting the main body of fire before breaking out any windows. Remember also that the basement stairs may end away from the wall. Bring an extra light to mark the bottom of the stairs as a reference.

Failure to chock an entry door as the hoseline is being stretched. Chocking doors is fundamental. The cheapest and simplest piece of equipment, a block of wood, is often the most important. There is a misconception that if a line charges under an inward-opening door, you will be able to open it easily. This is not true. Having a line charged under a door will block your egress and interrupt the flow of water. Any door that a line goes through should be chocked open (fig. 17–56).

Fig. 17–55. When descending basement stairs, go feet first with your weight to the sides, and ensure you have enough hose to clear the stairs. (Courtesy of Chris Saraceno)

Fig. 17–56. Any door that a hoseline goes through must be chocked open.

FFI 4.3.5 When members are searching without a hoseline, just the opposite is true. Once inside, the door must be closed to provide your team an **area of refuge** (also referred to as **safe haven**) while you search. This is a place that is isolated from the main hazard, where a firefighter can retreat in event of a fireground emergency. By creating an area of refuge, you can reduce the flow path and buy time for the fire to be controlled. Having a contingency plan of where to go if something goes wrong is critical. Any door opened before the line is stretched through it should be controlled and prevented from locking. This can be accomplished with a small clamp. (figs 17–57 a and b.) This allows the search team to start their search and control the flow path, limiting energy from reaching and influencing the fire. With the door closed and the clamp holding it unlocked, when you horizontally vent the windows (provided you are not in the fire area), it will have no effect on the fire. This action will also increase visibility and decrease the toxic environment in the room being searched. If the engine loses water or the fire extends to your location, you will have time to react. When your search has been completed and you are back at the door, be sure to check conditions before moving to the next room.

Figs. 17–57 a and b. Inserting a small clamp will hold the door unlocked, control the flow path and allow the search team to begin its search.

If you are searching the fire room without a hoseline, do not go any farther than 5 ft (1.5 m) into the room. If the room flashes over and you are beyond this "point of no return" distance, you will not be able to retreat fast enough. One safety tactic is to hook your foot around the door frame and sweep with a tool. Once the sweep is complete, close the door and move to an area of refuge (fig. 17–58).

Fig. 17–58. Flashover also indicates that the structure itself is under attack from fire. (Courtesy of Ron Jeffers)

Stretching too many hoselines through one opening. Hoseline management, as a rule, should provide that no more than two lines be deployed through any one opening. There have been numerous cases where firefighters have become disoriented because the hoseline they were following out of a building was tangled with other lines (fig. 17–59).

Fig. 17–59. Stretching more than two lines through one opening makes the lines difficult to advance and manage. (Courtesy of Tim Olk)

Failure to stretch the appropriate hoseline to match the present or potential fire conditions. Improper line size for the volume of fire encountered has resulted in numerous deaths. In some situations, the firefighter will have to determine the size of the hoseline to pull. Understand that proper water volume puts out fires, and fire volume is based on potential fire load. This having been established, stretching a hoseline that is too small and therefore delivering a less than adequate supply of water will not only put all members in the fire building in jeopardy, it will not put the fire out.

Large area buildings, commercial structures, and below-grade areas are no places for anything smaller than a 2½-in. (65-mm) hoseline. Attack teams have been outflanked by fire, burned, killed, caught in collapses, and have caused casualties to other firefighters in the area because small-diameter hoselines have been stretched into buildings that are showing heavy fire or have a fire load conducive to rapid fire spread. Staffing should not be an excuse for stretching an improperly sized hoseline. If it takes two companies to get the first line in place, then that is what will have to be done if the fire or occupancy demands it. The time to stretch the 2½-in. (64-mm) hoseline is with light smoke when it is easier to facilitate a clean stretch. In a commercial building, you will need the reach and sufficient gallons per minute to overpower the fire.

Booster lines should never be used inside a building. They are meant for trash fires only. Small-diameter, 1½-in. (38-mm) lines are similar in effectiveness. They don't belong in structures and are not even particularly effective on car fires. Residential structures require at least a line of 1¾-in. (44-mm) diameter, and high-rises should have at least a line of 2 in. The mantra of "big fire, big water" always applies. If in doubt, stretch the big line. If it is not needed, a smaller line can be stretched, or the larger line can be reduced to a smaller diameter. Attack crews who stretch lines that are too small can be outgunned and overrun by a fire with staggering speed (fig. 17–60).

Fig. 17–60. Firefighters stretching the wrong size hoseline at a commercial structure increase property damage and places firefighters in jeopardy. (Courtesy of Tim Hunt)

Failure to properly flake out the hoseline. Ensuring that the hoselines are properly flaked out is everyone's job. Kinks in a hoseline can reduce water volume up to 50%. One kink can jeopardize the entire fireground. Remember that the hoseline not only extinguishes fire but also protects the primary search team and egress points for both victims and firefighters. Firefighters should never pass a kink (fig. 17–61).

Fig. 17–61. Hoseline management must be a top priority. Water equals protection for everyone. No one should ever pass a kink. (Courtesy of Ron Jeffers)

Failure to back up attack lines. Firefighters searching above the fire and operating in stairwells have been killed when initial attack lines have failed to control the fire or have burst and there was no hoseline in place to reinforce the attack. All attack lines should be backed up by additional lines of equal or larger diameter once the first line is in place. The second hoseline should go to the same position as the first. If the first line is controlling the fire, this second line can then cover adjacent areas and/or the floor above, but only if the first line is sufficient for fire control. Hoselines that are stretched elsewhere in the building rather than as a backup line may be of little use, and their personnel will be endangered if the initial hoseline fails.

Fire attack: the right way

- Must protect the primary search
- Must protect the main areas of egress
- Must be located between the fire and any victims
- Must be properly supported by and coordinated with timely ventilation

Search and rescue

Failure to create barriers during a search operation. This is especially true when searching private dwellings where open interior stairs are present and the building is relatively small. Openings such as those used for vent, enter, isolate, and search operations are essentially vent points and create paths of least resistance for fire spread. It is critical that the first action after entering the room via a window is to close the door to the room. This also holds true for any other type of occupancy, including multiple dwellings. If the fire is nearby, it will move rapidly toward the firefighter-created vent path.

In the Lessons from the Fireground case where the captain and his crew were trapped above the fire (where the captain fell through the shaft window), the main reason the firefighters made it out alive was because they closed the door when they entered the second-floor apartment. They made their way to the door for exit when the evacuation tones were sounded over the radio. When the door was opened by one of the firefighters, he stated later that there were flames in the stairwell and that the heat knocked him immediately to the ground. He struggled but was able to close the door again, most likely because it was an inward-swinging door and all he had to do was push it closed. Had the door swung the other way, he would never have been able to reclose it. Had that apartment door not been closed, they almost certainly would have been overrun by the fire (fig. 17–62).

Fig. 17–62. Fire roaring up the stairs from the cellar trapped firefighters searching above the fire. Only a closed apartment door protected them from burning to death. In private dwellings, this apartment door is not available.

Especially for cellar or basement fires in private dwellings, if you are going to search the upper floor before a hoseline is in place, ensure not only the door to the cellar is closed before you go up, but the door to the area you will be searching must also be closed. Again, it is worth mentioning that some type of notification must be made before personnel go above the fire (fig. 17–63).

Fig. 17–63. Basement fires have limited ventilation points and offer the greatest chance for extension through voids. (Courtesy of Chris Saraceno)

Crowding the hall or stairs. These are paths of egress and should not be blocked. If an evacuation is ordered, it will be difficult to get out or down. If loss of water occurs, a door burns through, or a wind-driven fire is encountered, your escape can be blocked by a parade of firefighters. If you find this condition when entering, don't add to it—find another way. Wouldn't it be great if we had ladders and actually used them! Resist the urge to crowd the hall. Getting trapped in a center hall with no windows is one of the most dangerous positions in a structure (figs. 17–64 and 17–65).

Fig. 17–64. Keep the hallway and stairs clear for a quick egress if things go bad. (Courtesy of Dan Nocera)

Fig. 17–65. Firefighters standing on stairs create an unsafe condition.

Failure to use a lifeline for search. There have been too many cases of firefighters getting lost, then dying because they ran out of air inside large buildings. They usually run out of air because they cannot find their way out of the building. Commercial buildings, below-grade areas such as cellars or basements, and any large area structures such as townhouses or estate homes should only be searched with a lifeline. In any of these areas, any firefighter not on a hoseline should be on a lifeline. A thermal imaging camera (TIC) is also a necessity, but overreliance on this technology, which can malfunction, could leave a team of searching firefighters disoriented in a large area. A **lifeline** does not malfunction or need batteries.

Failure to identify and establish a secondary means of egress. Firefighters entering a building to search should take note of areas of egress before they enter. Egress-friendly areas such as porch roofs and fire escapes must be noted. The same should be practiced for roof access. If you have gotten into a building or onto a roof, you should know at least two ways out.

RIT operations should include laddering all sides of the building and the roof for additional means of egress. These locations should be announced over the radio by command or the RIT officer as soon as they are established, as personnel already inside the building will not be aware of the existence and location of these egress points. Failure to establish additional means of egress leads to an unacceptable alternative: the bailout.

Failure to identify the signs of rapid fire development. To survive rapid fire development, firefighters must be trained in the recognition of the signs leading to these developments. This recognition comes in the form of using the senses to detect these signs. This can be either from visual signs such as the reading of smoke for the signs of a flashover or backdraft or reading the building. Reading and understanding the building and its inherent construction will give clues as to where the paths of least resistance for fire travel will be and where areas of danger from rapid fire development might exist. Rapid fire development signs can come from the sense of hearing as in the case of a fire crackling in the overhead in a smoke-obscured room or above a drop ceiling or in a cockloft. Hearing will also be the sense used when the cues of impending fire development or loss of control of the fire is transmitted over the radio. It can also come from the sense of touch as the heat conditions in a room begin to rapidly increase. This points to one basic need: Be alert at all times (fig. 17–66).

Fig. 17–66. Firefighters must be constantly aware of cues of rapid fire development. Heavy, pressurized smoke is one of those indicators. Things are not getting better when you see smoke like this. (Courtesy of Ron Jeffers)

Search and rescue: the right way

- Must be protected by the attack lines
- Will be made safer by the creation of barriers between the searchers and the fire and its path of spread
- Will be infinitely more effective with and benefit from proper and timely ventilation (fig. 17–67)

Fig. 17–67. A hoseline is stretched through the front door to cut off fire and protect living areas, such as bedrooms. (Courtesy of Kurt Tiedeman)

Ventilation

Improper ventilation. Improper ventilation contributes to our failure to control the building and plays a large role in fire progression. Whether you use positive pressure or roof ventilation, it must be controlled, coordinated, and communicated for firefighter safety. When venting for extinguishment, vent as close to the fire as possible and make sure that you are venting ahead of the line. We are seeing more and more fires where the fire is venting out a window, but further size-up shows that heavy black to brown smoke is pushing down the hall and out the door. This is an indication that additional ventilation is needed. The first priority is to increase ventilation close to the fire after the hoseline is operating on the main body of fire. If conditions dictate, cool the ceiling as you advance to push back the fire and cool the overhead. The old teaching of not applying water to smoke does not hold true anymore. If the smoke is hot and pushing down on you and hindering your advance, then putting water on smoke is appropriate (fig. 17–68).

Fig. 17–68. With smoke pushing out the front door, crews could be caught in the hall by rapid fire progression. This is an indication of the need for more ventilation close to or over the fire. (Courtesy of Chris Saraceno)

The rate of heat release experienced in today's simple house fire results in flashover much more quickly than in the past. Most furniture is made of synthetic materials. The heat release rate of these combustibles approach those of flammable liquid. Conditions can get really bad really fast. Always be aware of your surroundings. The warning signs of rollover, fire propagation down the wall, and everything going dark quickly may all be masked by thick black smoke. The only reliable sign will be rapid heat buildup or high heat. In buildings with high ceilings or drop ceilings, smoke can ignite, changing conditions rapidly.

Failure to sound the floor or roof. Not sounding the floor or roof can result in ending up at the bottom of a shaft or falling through the floor. Entering through a window onto a weakened floor will have a devastating effect. It is important to realize that sounding is not an absolute indication of floor integrity. A hidden hazard consists of windows in two-story foyers that access ledges that homeowners place decorations on. When a firefighter enters, the floor or ledge will be sound, but then, about 2–3 ft (0.6–1 m) in, there is a 15-ft (4.6-m) or more drop to the first floor. When working on the roof, if you cannot see, you must crawl. Beware of light shafts and high parapets. For a commercial building, try to cross in the front where the storefronts line up (fig. 17–69). In multistory multiple dwellings, the shafts can be anywhere, and although their location may be uniform from one building to the next when built in a row, they may not be.

Fig. 17–69. Store depths will vary, making it easy to walk off the roof. Try to cross in the front where they are usually even with the street and sidewalk.

One note of caution when sounding floors: Victims who cannot get to doors often fall victim beneath windows, their next choice for egress. Before bashing your tool onto the floor to see if there is a substantial place in which to enter, sweep first to determine if a victim is below the window. In this way, you will not turn a rescue into a recovery. Blasts to the head with a Halligan tool or axe are not conducive to rescue operations.

Failure to recognize the presence of truss construction. This size-up failure has contributed and continues to contribute to firefighter deaths each year. Truss roofs should be located during preplanning. A building's occupancy will also clue you in to the potential presence of truss construction. Occupancies of suspicion include bowling alleys, auto dealerships, warehouses and "big box" stores, and fast food restaurants. When confronted with any business where a large open space is present or any new construction, especially in commercial or residential occupancies, assume that a truss is present until ruled out. In New Jersey, a system enforced by state law is in place that denotes the presence of a truss. A triangle on the building with an "R" inside it indicates that a truss roof is present. The same triangle with an "F" inside it indicates that a truss floor is present. If both a truss floor and roof are present, an "RF" will be inside the triangle (fig. 17–70). This provides safety warning at a glance. New York State and Florida have also adopted similar guidelines for truss buildings.

Fig. 17–70. Triangle markings indicating the presence of trusses

A truss assembly can fail in 5–8 minutes under flame impingement. When using a TIC, confirm roof construction and report to command. You also will feel the difference when walking on a lightweight wood truss roof versus conventional construction. The truss roof will have more bounce when you walk. You should not operate above or below a truss roof with fire in the roof space (fig. 17–71). The same goes for truss floors and laminated wood I-beam floors. Lightweight construction anywhere in the building must be identified, reported, and checked for the presence of fire. The flow chart in figure 17–72 is a guideline to be used when lightweight construction is suspected or present.

Fig. 17–71. This truss roof in Hamden, CT, failed 6 minutes after the fire department would have arrived on scene.

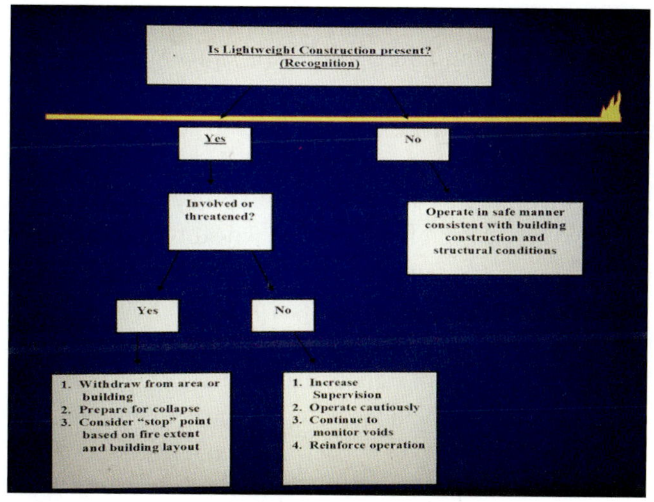

Fig. 17–72. Use this flow chart when dealing with lightweight construction

Ventilation: the right way

- Must be controlled, communicated, and coordinated with the fire attack to help locate, confine, and exhaust the products of combustion in the proper direction
- Will increase visibility, enhancing the safety and expediency of the primary search (fig. 17–73)

Fig. 17–73. Firefighter wearing full PPE while cutting a hole in roof. (Courtesy of the Middletown, CT, Fire Department)

Failure to recognize hazardous exposure potential. Not all deaths are as dramatic as building collapses. As we get older, we are plagued by the cumulative effects of breathing toxic gases that can cause heart attacks, strokes, and cancer. Although it is impossible to completely limit exposure to all the products of combustion, soot on your skin or a shift in the wind can result in a direct exposure. We must do everything in our power to minimize exposure. We often drop our guard at the wrong time.

Overhaul and minor fires pose many hidden hazards. Firefighters should wear breathing apparatus until the environment is metered for the presence of carbon monoxide and hydrogen cyanide. If these gases are present, they will not be found by themselves. Relying on permissible threshold limits will provide a false sense of security. Although it is not practical to meter for all the products of combustion, using these two gases as a benchmark will provide a margin of safety. Masks should not be removed until the atmosphere reads zero. If, after your mask is off, you start to experience watery eyes, it is likely that you are being exposed to acrolein or hydrogen chloride. Place your mask back on or exit the area. Other strategies for limiting exposure include washing gear, especially your hood and taking a shower after a fire (fig. 17–75).

Fig. 17–75. Firefighters should meter the atmosphere before removing SCBA. Remember that the consequences of exposure may, in the long term, result in stroke, heart attack, and cancer.

Freelancing. The most critical tactical breakdown is **freelancing**, which means wandering off to perform tasks on your own without an assignment. Maintaining company integrity and keeping personnel accounted for is the most important duty of the company officer, and maintaining self-discipline is the most important duty of each firefighter. Freelancing is always a mistake, and departments must have a zero-tolerance policy in regard to this dangerous issue. A simple company accountability model explains this very nicely. According to this model, there are only four places a company should be assigned on the fireground:

- At the command post awaiting an assignment
- In a designated staging
- Operating in an assigned area
- In the rehabilitation division

The keyword here is *assigned*. The only time a company deploys without direct orders is when they are one of the first-arriving companies and their actions are being guided by scene assignment SOPs. For example, according to a department's initial scene assignment SOP, suppose the first-arriving engine company is responsible for attack and water supply (and command if they arrive before a chief), the second engine is responsible for a backup line and a second water supply and the first-arriving ladder company is assigned forcible entry, primary search, reconnaissance, and ventilation duties. Unless something out of the ordinary is going on, that is where they should be and what they should be doing. If a later-arriving company self-deploys without orders from command, they are freelancing. This is usually more prevalent with later-arriving companies and additional alarm companies. The moral: *Do not go to work without orders!*

Company accountability model: the right way

Step 1. Companies report to the command post as a unit.

Step 2. Companies are assigned to an operational area, reporting to a division supervisor.

Step 3. Companies operate in assigned division *only*.

Step 4. When relieved, companies report as a unit back to the command post.

Step 5. If reassigned, go back to Step 2.

Step 6. If sent to rehab, when complete, go back to Step 1 (fig. 17–76).

Note that with each assignment and reassignment, including the order to take-up, the company officer must ensure that the accountability officer is aware of that assignment and where that company will be operating. Note also that they always report back to the command post once an assignment is complete, whether it is in a fire building or in the rehab division. The command post should be like an organizational manifold. All activities must pass through it. Anything or anyone who goes around it is freelancing (fig. 17–77).

Fig. 17–76. The accountability model. Note that there are no arrows between the operational assignment (white) areas. The command post should be like an organizational manifold: everything goes through it, nothing goes around it. All changes of assignment must run through the command post. If your company moves from one white area to another without orders or acknowledgments from command, you are freelancing. It is as simple as that.

Fig. 17–77. Companies must report as a unit, receive orders as a unit, and operate as a unit. Freelancing is unacceptable. (Courtesy of Ron Jeffers)

> **FIREGROUND NOTE**
>
> Many firefighters fail to focus on size-up but allow adrenaline to take over.

Emergency bailouts. FFI 4.2.4 An emergency bailout is an absolute failure of the fireground at every level of the game. Firefighting personnel who exit a building any way other than where they entered it have more than likely become disoriented and are often exiting by sheer luck. Emergency bailouts are usually a result of a combination of factors that may include, but are not limited to, the following:

- Failed size-up
- Freelancing
- Failure to properly wear PPE and SCBA
- Communications failure
- Failed command and control
- Failure to follow SOPs and the action plan
- Failure to coordinate fire operations
- Improper and uncoordinated ventilation
- Failure by firefighters to confine the fire
- Failure by firefighters without hoselines to create barriers between themselves and the fire, and/or passing fire with no known exit
- Undisciplined hoseline operations
- Fire streams applied from the wrong place
- Opposing hose streams
- Complacency

Training on the proper procedures to execute an emergency bailout is absolutely necessary, but training in the prevention of the points of failure mentioned above are more important and will serve as bailout prevention measures.

Prevention, not intervention

Tactics are the building blocks that make up our overall strategic objectives of life safety, incident stabilization, and property conservation. It is commitment to the basics—safe staffing and company excellence—that will prevent most tragedies. If we increased physical fitness, healthier lifestyles, and quality training for all ranks, there would be little need for intervention. However, we must acknowledge that we can do everything right and still find ourselves in dangerous situations on this job. Prevention of fatalities and injuries is difficult at best given the places we must conduct operations. We must do our best, and then some, to prepare for working in dangerous buildings through activities such as **prefire planning** and building inspections. Departments must also actively engage in and encourage information exchange between personnel and between shifts. To take this a step further, information exchange is also a concept that mutual aid groups should embrace. How many departments fight large fires alone, especially in large and dangerous buildings? If the only time you see your mutual aid partners is at fires, the potential for uncoordinated operations is off the chart. Information exchange as well as interagency drills with neighboring departments lay the groundwork for a safer, more cooperative fireground. To this end, the following sections outline some ways to share the information, both locally and regionally.

VACANT BUILDING MARKING SYSTEM

The purpose of the hazardous vacant building (HVB) marking systems is to alert responders at a glance to HVBs and their inherent dangers. Marking buildings with orange spray paint (the international color of hazard) is SOP-driven. Buildings are usually marked by the SO on orders of the IC at a fire scene or by the SO when notified by company members that they have come across a HVB, either on a response or on a building survey. The markings are placed in a conspicuous position so they are easily recognized by arriving personnel. This system should be adopted by entire mutual aid groups and be included in dispatch announcements.

Level 1 hazardous vacant building: entry permitted

Vacant buildings that may be entered should be signified by a box painted on the exterior of the building's walls. The department's initials or other adopted distinguishing mark should be written below the box, as should be done with any building hazard notification symbol. This will alert members that the symbol is department-approved.

NHRFR

Level 2 hazardous vacant building: enter with extreme caution

Other buildings may be entered, but only with extreme caution. This condition is denoted by a box with one line cutting diagonally across it (like a spare in bowling). This building may have suffered damage from previous fires, vandalism, and rot due to neglect. Firefighters should not take unnecessary risks in this building. Any attack should be carried out with extreme caution and with a pessimistic prediction of success. This and the no-entry symbol may be accompanied by another symbol, such as the ones that follow.

NHRFR

Level 3 hazardous vacant building: no entry

Vacant buildings that have undergone previous fire damage or other destructive forces should be considered off-limits to all members at all times. A building that has hazards so severe that firefighters should not enter it under any circumstance should be denoted by a box with an X covering the whole box (like a strike in bowling).

NHRFR

Identifying unsafe building features

Some buildings may be safe to enter, but because of some safety concerns, firefighters should use extreme caution in the particular area denoted by the box. The boxes below show an example of a building with a dangerous fire escape on the left and dangerous stairs and/or landings on the right. Departments are not limited to these and may create their own symbols to address a particular hazard. If building owners object to the department putting spray paint on their building, tell them that the fire department objects to their unsafe building. The choice is with the owners: Fix the hazard or have their building marked up.

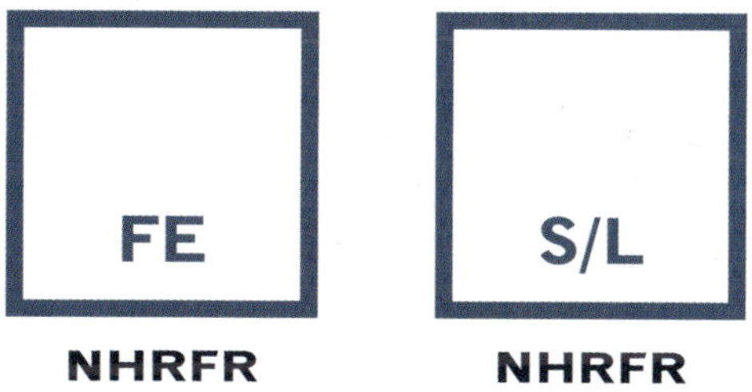

Structural damage (postfire)

Both primary (the damage done by the fire) and secondary damage (the damage done by firefighting) may leave the building in a dangerous condition. Firefighters from other shifts who may have another fire in the same building should be made aware of the hazards. Holes in the roof and/or the floors must be noted. The building should be marked to warn firefighters of these hazards. **Roof open** is denoted by the RO symbol, and **floor open** is denoted by the FO symbol. Floor hole notations should include the floor that is in question. A building with holes in both the floors and the roof may use two separate symbols or they may be placed in the same box. If there are multiple floors with holes, the FO-M symbol may be used. These symbols may be used in conjunction with limited or no-entry symbols.

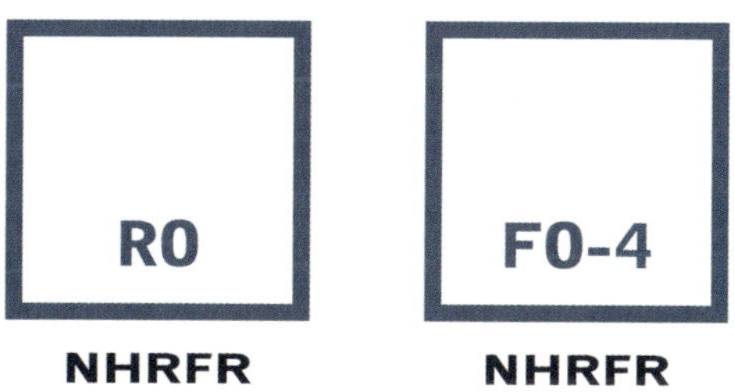

Establishing a building hazard awareness program is an effective way of informing responding personnel about unsafe building conditions. Any discovery of an HVB by any company should prompt the response of the SO or battalion commander. Other notifications must include dispatch and may include the building department, fire prevention, or the health department.

Hazardous vacant building notification system

This system works hand in hand with and is the follow-up system to the hazardous vacant building marking system. It is an SOP-driven system, and once a building is identified as a HVB and marked (fig. 17–78), a paper trail and notification system is set into motion to ensure that all personnel are made aware of the building and its hazards. It begins with the completion of a form (fig. 17–79) that is then submitted through the chain of command where the information is duplicated and disseminated to the companies. The information also goes to dispatch to be included in the CAD system and to the building official of the city. At times, a memo is also distributed with photographs taken by the SO in order for personnel to see the hazards instead of just reading about them (fig. 17–80).

Fig. 17–78. Hazardous vacant building (HVB) markings on this building prohibit entry as well as bring attention to hazards created by an open roof and floor. This is a Level 3 hazardous vacant building. Marking systems such as this allow a fire department to alert its firefighters to the dangers before they enter a building.

FIREGROUND NOTE

Before you put paint on a house, make sure it is vacant!

NORTH HUDSON REGIONAL FIRE & RESCUE
STANDARD OPERATING PROCEDURES

HVB NOTIFICATION FORM

Date:________________ Day of Week:____________________

Platoon:_____________ Safety Officer:__________________

Address:____

Municipality:____

Owner:____

Fire Building ☐ Vacant (Non-Fire) ☐

Permanent ☐ Temporary ☐

HVB LEVEL: Check One

☐ Level 1 Entry permitted ____________

☐ Level 2 Caution Required ____________

Additional Symbols Added: ______ _________ __________

☐ Level 3 Entry Prohibited ____________

Is this an HVB Level / Occupancy Status Update? Yes_________ No________

Comments:____

____

____

____

Signature:_________________________

Fax to:
Deputy Chief
Battalion Chiefs
Headquarters (Chief of Safety)
Fire Control
Building Official of applicable jurisdiction

Hazardous Vacant Building Marking System

Fig. 17–79. Hazardous vacant building form

Office of the Platoon Commander

North Hudson Regional Fire & Rescue

11 Port Imperial Blvd.
West New York, NJ 07093

MEMO

DANGEROUS BUILDING CONDITION: 7301 RIVER ROAD
LEVEL 3 HVB

Side A – marked as level 3 HVB

Side B – note the sagging roof above the Roll-down gate

This splintered truss is beneath the sag. It is supported by a single T-shore
DO NOT OPEN THE ROLL-DOWN DOOR – IT WILL LIKELY COLLAPSE THE SHORE AND THE ROOF
DEFENSIVE OPS ONLY FROM OUTSIDE COLLAPSE ZONES

Fig. 17–80. Hazardous vacant building memo

RADIO OPERATIONS

Urgent call

When an **urgent call** is given on the radio, it is mandatory that all other nonemergency radio traffic stop until the urgent call is cleared by command. This transmission is given over the radio and repeated three times. Some states and/or departments use the term "emergency traffic" instead of urgent. Examples of situations where an urgent or emergency traffic transmission may be required are an electrical hazard such as downed wires on the fire scene, an open shaft, a localized collapse, and a loss of water. The urgent call is an important message that usually will have an impact on tactics. These messages normally follow an unexpected event on the fireground. A proactive urgent transmission may very well avert a Mayday transmission later.

Mayday, Mayday, Mayday

Mayday is a distress call for help and an indication that you or a firefighter you are with is in grave danger of serious injury or death. Like the urgent transmission, this call is given over the radio and repeated three times. This allows all on the fireground to distinguish between the call for help and other units communicating about the Mayday. Although you must call the Mayday as soon as you realize you are in trouble, there will be times where you or your crew will have to take protective measures first. For example, suppose you and your crew are caught in a center hallway with a loss of water. There is fire rapidly progressing down the hall. In this case, you would need to get to an area of refuge first, such as going into a room and shutting the door, before transmitting the Mayday. Although the circumstances will dictate your actions, there are basic guidelines for when to call a Mayday.

- Medical emergency
- Trapped by fire
- Fall through floor or roof
- Building collapse
- Lost and can't find crew
- Lost one of your own crew
- Trapped by debris and can't free yourself on first attempt
- Low on air and not near an exit
- SCBA malfunction

Although this list is not all-inclusive, it all comes down to a simple catch-all rule of thumb: *If you think you are jammed up, you are! Call for help!* Help can always be turned back. We must keep in mind the **zero-impact factor** (the amount of time it will take for a company to make an impact on your situation). Survival is predicated on your ability to remain calm. Don't wait to sound the Mayday. Every second delayed is 2 seconds (at least) that someone is not coming to help you.

FFI 4.2.4 Activate your personal alert safety system (PASS) device intermittently, but not when you are transmitting your call for help. Even if in contact with the RIT officer, periodically activate your PASS device. This will give the RIT a target to locate you. We recommend training using a modified version of the acronym LUNAR (location, unit, name and nature of problem, air supply and assignment, resources needed).

> **LUNAR**
>
> **L**ocation
> **U**nit
> **N**ame and nature of problem
> **A**ir supply and assignment
> **R**esources needed

Location. What is your location? This is where you are now or your last known location: what floor, what division, front or back of the house or building. Can you describe something that will give a clue to where you are?

Unit. What company are you on? For example, Truck 4.

Name and nature of problem. Give your name. Your saving of your own life outweighs any privacy issues. There have been several incidents where the wrong firefighter has been accounted for, resulting in minutes being lost from the rescue. Your rescue will come down to seconds, not minutes. State the nature of the problem. This information may aid the RIT in getting to you quicker, having them take a different route, or other actions. Describe if you are lost, cut off by fire, or trapped in a collapse. Also, just because one member is in distress does not mean others are not with him or her. This may be the case when a team has fallen through a floor, but

LESSONS FROM THE FIREGROUND

A fully developed cellar fire burned through the floor and into a first-floor apartment. Seeing the fire in the apartment, the engine company on the first floor protecting the cellar stairs applied water to stop the extension. In doing so, they unknowingly influenced the flow path through an open door at the rear of the apartment where it roared up a rear stairwell. There was a ladder company in the stairwell, making their way to the top floor to search for a reported missing child. As the fire consumed the stairwell, three of the four members were able to evacuate into a rear yard. The fourth firefighter, who was ahead of them, did not realize what was happening until it was too late to get back down the stairs. The fire and heat were moving so fast up the stairwell that it chased him up to and across the third floor where he had no other choice but to bail out the window. As he was scrambling up the stairs, the force of the flames actually knocked him over and he lost his helmet (no chinstrap!). Already severely burned on his ears, face, and head, he hung onto the windowsill until his hands and wrist almost burned off and then fell three floors. He smashed through a plastic awning on the way down and landed right next to an upside-down picnic table and washing machine. His burn injuries required extensive surgery and therapy. His ears were almost completely burned off and had to be replaced by surgery. He survived and is back on the job.

LESSONS LEARNED

1. Always wear your protective gear properly.
2. Create barriers between yourself and the fire, especially when going above the fire.
3. Communicate your position and assignment to the attack team when you may be opposite the attack.

only one member is capable of transmitting the Mayday. Always ask the member if he or she is alone.

Air supply and assignment. How much air do you have left? If the emergency is that you are lost, time is your biggest concern. This is the hourglass nailed to the table—don't let time be a nail in your coffin. If your normal policy is to be out of the IDLH environment before your low air alarm activates (and you have followed it), you will have more time to find your way out or be located. Assignment will give clues to your likely locations.

Resources needed. What will be needed to extract you from your predicament?

"WHO, WHAT, WHERE": LUNAR made easy

LUNAR is a perishable skill and although it is an accepted practice and many departments use it, the phrases "Who, What, Where?" ("WWW") are more easily transmitted by the firefighter in distress.

- WHO: Who are you?
- WHAT: What is the problem?
- WHERE: Where do you think you are?

Unlike trying to remember an acronym like LUNAR when you are facing your possible demise, the WWW is more of a natural reaction that comes out without having to remember (and try to think of) an acronym like LUNAR. If a firefighter can get this simple information out after calling a Mayday, then command can ask further questions to round out the information required to find and rescue the firefighter. Keep it simple!

Emergency fireground communication

FFI 4.2.4 It is imperative that all fire departments have an emergency fireground communications system in place, including the procedure for transmitting a Mayday message. It is best to develop joint procedures of this nature among those mutual aid departments who work together on a routine basis. When buildings are falling and/or personnel are missing or in distress, SOP-driven control of the fireground will often make the difference between chaos and order.

There are two general types of emergency transmissions on the fireground:

- *Top-down* emergency transmissions (command to operating personnel)

- *Bottom-up* emergency transmissions (operating personnel to command)

Top-down emergency transmissions

There will be times during the operation where the IC needs to transmit a message of urgency to all operating personnel. It may be a warning about a dangerous building condition, a withdrawal from the building, or any other critical message that is essential to the fireground operation. It may even be an announcement regarding the location of secondary means of egress from the building or the roof. This is called a top-down emergency transmission and will usually be transmitted as an urgent message.

All firefighters should be aware of when this type message is about to be transmitted across the air. An emergency transmission should be initiated with a special radio tone dedicated for just that purpose alone. This tone should be tested every day at a specific time so that firefighters become conditioned to immediately stop what they are doing, listen up, and maintain radio silence when they hear it. An emergency transmission has absolute priority over all other transmissions.

In North Hudson Regional (NJ) Fire and Rescue, a system has been in place that has worked well. After the emergency transmission is initiated by someone on the fireground, usually the IC, dispatch transmits a series of distinct tones followed by a boilerplate (standardized) statement announcing the emergency transmission. After the boilerplate announcement, the emergency transmission is repeated several times. Here is an example of an emergency transmission and procedure: (starts with a series of tones) "Dispatch to all companies operating on the fireground, stand by for an emergency transmission. By order of Park Avenue command, all companies operating in the fire building evacuate the fire building immediately; companies operating in exposures Bravo and Delta, hold your position. Repeat (tones again), by order of Park Avenue command, all companies operating in the fire building, evacuate the fire building immediately; companies operating in exposures Bravo and Delta, hold your position. 1533 hours." This message is repeated several times.

FIREGROUND NOTE

It is better to be harassed at the kitchen table about calling for help than having everyone say what a great firefighter you were at your funeral.

—Jim Duffy,
Wallingford Fire Department

Bottom-up emergency transmissions

Just as command has a method for transmitting a top-down emergency transmission, personnel on the fireground and in the fire building must have a way of initiating a bottom-up emergency transmission. This is basically a Mayday transmission, but it can also be used for an urgent transmission. A firefighter who is lost, trapped, or in need of some type of assistance, or who needs to get some information to command or dispatch should have a system that is recognized by all personnel as an emergency transmission, which gets the same priority as a top-down emergency transmission.

FIREGROUND NOTE

A proactive urgent transmission may very well avert a Mayday transmission later.

Many departments have portable radios that have the capability to send an emergency transmission by pushing a button. Called a Mayday or emergency identification (EID) button, it sends a signal to dispatch or command, or both, and informs them by a readout exactly who is transmitting the emergency message. It also emits a distinctive tone over all the radios except for the member who pushed the button. The drawback here is that if the button was pushed accidentally, the member who sent the transmission usually does not know it has been sent until someone comes looking for that individual. The button may be activated accidentally when the radio is inadvertently banged around while operating on the fireground—one very good reason why operating personnel should wear their radios under their coats, with only the collar microphone exposed. There have also been cases where the EID button was activated by getting it wet. This is discussed as a Lessons from the Fireground case study later in this chapter. In any case, it is a method by which firefighters can draw attention to an emergency situation.

In the fire service, there should always be a contingency plan for when something goes wrong or equipment unexpectedly fails. In North Hudson Regional Fire and Rescue, even though the portable radios have the EID button, a backup system has been adopted should the EID button fail. In fact, it is preferred that members in distress transmit an additional bottom-up emergency transmission to better ensure they can be heard amid the normal fireground radio chatter. This procedure is also tested daily over the air. The member who needs to send the message initiates the transmission in the following manner:

Step 1. The member manually activates his or her PASS device, causing the PASS alarm to sound.

Step 2. The portable radio is keyed for transmission.

Step 3. The keyed radio mike is placed next to the PASS device for a period of 10 seconds. The PASS tone is heard all over the fireground, signaling the initiation of a bottom-up emergency transmission.

Step 4. The PASS device is turned back to the arm position.

Step 5. The emergency message is broadcast using the following format:

Mayday, Mayday, Mayday, this is Firefighter ________________.

Mayday, Mayday, Mayday, this is Firefighter ________________.

(Message is then broadcast).

If the emergency transmission is initiated to signal that an entire company is in distress, the company name is substituted for the individual firefighter.

The key to transmitting the bottom-up emergency transmission is to use an audible alarm such as the PASS device or the radio EID button. The aim is to get someone's attention. There have been two cases in North Hudson where Mayday messages went unheard initially because the person sounding the Mayday did not use the PASS device, and thus the Mayday transmission got lost in the fireground communication shuffle. This was before we had radios with the EID button capability. Even a muffled PASS device activation is better than a muffled Mayday transmission without the audible tones. Even though North Hudson Regional Fire and Rescue has new portable radios, each of which has an EID button that sends a tone to the command post identifying who is giving the Mayday, we still test this procedure every day and urge firefighters to use the PASS device over the air as an additional bottom-up emergency transmission.

None of these methods is foolproof, but together they can form an established procedure to allow the IC (top-down) and an individual firefighter or officer (bottom-up) to let someone know that something extraordinary or unusual is occurring. It is hoped that these procedures are never used; however, to be of any use, they must be taken seriously, be no secret to anyone, and be practiced.

RAPID INTERVENTION TEAMS/CREWS

RIT activation—reacting to and managing the Mayday

RIT activation should not cause a crisis at the command post or on the fireground. The RIT operation must be a deliberate and procedure-driven operation that is carefully coordinated and properly carried out. The IC and the officers on scene, both chief and company, have to demonstrate leadership ability, and all firefighters must demonstrate self-discipline to avoid letting the emotional pressure of the situation drive the incident. To this end, whether one is a part of the RIT team, an aide to command, or in a company that is supporting the RIT operation by continuing to fight the fire, every firefighter should understand the process of the Mayday operation.

The first priority is to determine who is missing and their location. Any LUNAR or WWW information missing from the Mayday report must be ascertained to best determine where to start looking. If the identity of the members cannot be determined, a PAR will need to be conducted. If the identity is known, a PAR should still be conducted, but possibly on an alternate frequency so it does not tie up rescue communications. The IC needs to clear the air by using an emergency radio transmission.

Dispatch may be the best tool in accomplishing this, utilizing a dedicated emergency transmission tone, as was discussed in earlier in this chapter. It is advantageous to switch the firefighting operation to another frequency to concentrate on the communication requirements of the rescue operation.

If the distressed firefighter must contact dispatch or command on a different frequency, allow that member to continue to communicate on that frequency and direct the rescue to that frequency. It might also be wise to assign the firefighting operation to another chief officer and for the IC to step back to better manage the Mayday.

Organizationally, the rescue operation should also be delegated to a chief officer designated as the rescue group supervisor. The responsibility of this position is to directly supervise the rescue operations and keep command informed of progress and needs. The officer in charge of the rescue must also become like an air traffic controller and give direction to the officer or firefighter calling for help. An additional SO should be requested as well as additional chief officers. An officer should also be assigned to monitor fire conditions and structural conditions in the area. At least two additional alarms should be struck: one to support the rescue operation, and one for the firefighting effort. These companies should be sent to a Level 2 staging area.

It is critical that firefighting operations continue with a focus on keeping the fire from spreading to the area of the downed firefighter. If conditions require, an additional team may have to deploy a hoseline with the RIT to protect the rescue.

Additional resources must include an immediate request for an additional RIT. If possible, in the interim, one should be used from the tactical reserve for the incident so there is minimal time elapse for positioning a backup RIT. Even if this is possible, an additional RIT must still be summoned.

The original RIT is basically a recon team and should enter the building with minimum equipment to locate the downed or lost firefighter. There is no need to lug all the RIT equipment in at this time because it will slow down the team. Many departments have put together or purchased a RIT pack with all the required tools in one easily transported bag (mentioned in the RIT tools section). If that is not available, at a minimum, a spare SCBA cylinder and mask, a ready-made rescue sling, hand lights, a lifeline, a TIC, and a set of irons (flat-head axe and Halligan tool) may be all that is needed for the initial RIT. After the firefighter is found and assessed, the equipment needs can be relayed back to the rescue group supervisor, where the second RIT can bring it in more quickly. At this point, all they have to do is follow the lifeline to the rescue area. The requirements of the rescue operation will depend greatly on the reports of the first RIT to reach the downed firefighter (fig. 17–81).

Fig. 17–81. RIT team removes a downed firefighter in training.

The IC must ensure that EMS (both basic life support [BLS] and advanced life support [ALS]) are summoned to the command post. They should be positioned in close proximity to the rescue group supervisor where the RIT has made entry. It might be best to have two medical teams standing by because RIT operations often cause injury to the rescuers. A liaison with the police should be established so that there is a lane open for the ambulance to get in and get out without delay. Arranging for and ensuring availability of a police escort may save valuable time during rush hour. If necessary, a **landing zone (LZ)** can be designated in case a **med-evac** is necessary. Med-evac protocols are usually set up in advance and may only require a phone call to be made by dispatch. Where to land the helicopter can be a concern, and the police can be of great help with that. Command should also have dispatch contact the hospital that the firefighter(s) will be taken to so the hospital personnel can make preparations prior to EMS arrival at the hospital.

After the Mayday situation has been stabilized, command should concentrate on further stabilizing the scene. Relief should be provided where needed and the troops kept focused.

All officers, including the SO, should be on the lookout for signs of critical incident stress and address it accordingly. To proactively address firefighter emotional needs, it might be a good idea to request a CISM team respond. Firefighters who were directly involved in the firefighter-down situation should be relieved. These firefighters should be sent to a common area away from the scene, such as a firehouse where CISM counselors are available for consultation and medical personnel are there to evaluate them.

FIREGROUND NOTE

The RIT represents a trained, fresh company with full air. The RIT should work toward proactive safety while being ready to deploy.

Introduction to the intervention dilemma

The RIT dilemma, as we shall call it, plagues the fire service. The RIT response across the country is wide and varied. Some departments assign RIT as an engine company, while others use a squad company, a ladder company, or a rescue company. Some departments respond the RIT with the first-alarm companies, while others dispatch the RIT as soon as a working fire is confirmed. Although this is acceptable and meets most state statutes, other departments either respond with the RIT on the second alarm (which can be a deadly mistake—what if no one strikes a second alarm?), while others respond to it only on the request of the IC (an even more potentially lethal situation—what if the IC forgets?). Some RITs respond from neighboring jurisdictions. This may be an automatic response or it may have to be requested. Other RITs are formed by committee—that is when the IC or SO takes individual personnel as they arrive, and form them into a RIT. Still others designate the RIT on the fly, over the radio, during response.

History has shown that most firefighter rescue situations occur during the first few minutes of the incident when the situation is unfolding and the least amount of information about the building and the fire is available. This is the time when firefighters are actively seeking out the fire and searching for victims, sometimes without a hoseline available, doing it with less than adequate personnel, and usually without the immediate services of a RIT.

Personnel requirements are also wide and varied. Some RITs consist of two members, and some have three, four, or more. Some have officers assigned, and some do not. In New Jersey, the *Model Fire Department Standard Operating Guide* issued by the New Jersey Division of Fire Safety states that the **rapid intervention crew (RIC)** shall be a designated crew of at least four firefighters, optimally six.

Another major mistake can occur when the incident is rapidly escalating, and the first thing the IC does when the RIT arrives on the scene is to put them to work fighting the fire. This is the biggest mistake of all. Firefighter safety is the number-one priority on the fireground. If a building or a whole block burns down, but no firefighters get hurt, lost in the building, or killed, we have met the top priority. Buildings can be rebuilt, firefighters cannot. A RIT is dedicated resource with one focus—firefighter rescue capability. That is all they do. That is all they ever do.

With all these variables, we can state with some validity that the rapid intervention issue in this country is not rapid at all, but often consists of a hodgepodge of directives and agreements between departments that do not always get followed and hopefully will never have to be used. It is a gamble.

To be effective, RITs should not have to be requested, but should be a standard part of the department's response. They should operate under an established set of guidelines as outlined in the department's RIT operation SOP. For operations in large buildings, high-rises, or when exterior defensive operations cover a large geographic area, two or more RITs can be requested and assigned. In high-rise operations, the RIT should respond to the operations division on the floor below the fire where, if needed, they can be rapidly deployed to the area of the distress signal. If they are staged in the lobby, the reflex time for rapid intervention will be unnecessarily increased (and not rapid at all).

Standards and regulations

OSHA 1910.134 is the *Respiratory Protection Standard* that guides fire departments in procedures regarding such areas as respirator (breathing apparatus) use, maintenance, storage, and repair; mask fit testing; record keeping related to breathing apparatus; and procedures for structural firefighting. It is in this last section that regulations regarding the OSHA **two-in, two-out standard** is outlined. This standard is federal law in most states, and all fire departments and fire service personnel should comply with it. The standard states that the employee (the department), during interior structural firefighting, shall ensure that at least two employees who enter the immediately dangerous to life and health (IDLH) atmosphere shall remain in visual or voice contact with one another at all times; that at least two employees are located outside the IDLH atmosphere; and that all employees engaged in interior structural firefighting use SCBA.

The standard also allows one of the two individuals located outside the IDLH atmosphere to be assigned to an additional role, such as the IC in charge of the emergency or the SO, so long as this individual is able to perform assistance or rescue activities without jeopardizing the safety or health of any firefighter working at the incident. The authors cannot think of a situation where using these personnel to conduct a rescue would not further jeopardize the safety and health of firefighters at the incident. Many jurisdictions will also count the two-out as including the pump operator. What this means is that if there is an emergency situation and no one else is available, the IC (the one running the show) and the pump operator (the one controlling the water, which may be only thing keeping the fire away from the downed firefighters) could be used to enter the building and rescue the trapped firefighter.

The standard further states that "nothing in this section is meant to preclude firefighters from performing emergency rescue activities before an entire team has assembled"—which allows us to attempt rescue with inadequate personnel. Studies done by the Phoenix Fire Department have shown that during firefighter rescue operations, the number of personnel needed to accomplish these tasks safely is about 12, and that some of them will also get lost or injured in the process. Having only two firefighters—never mind the fact that they may have further critical responsibilities for the incident engaged in the rescue operation—is hardly an acceptable number. It is the responsibility of the department to have a mechanism in place that addresses this in a safer and more efficient manner.

The other caveat of the standard is an actual dichotomy or contradiction of the intent. It allows for the situation when, if based on dispatch reports, scene reports, or size-up indicators, an imminent life-threatening situation exists where immediate rescue activities may prevent the loss of life or serious injury, the requirements of the two-in, two-out standard may be suspended and individual members or the team may enter to save lives. During the initial phase of the fire operation, we are always in the life safety or rescue mode. Unless the building is fully involved, we must always assume a life hazard exists. Once we determine that there are no victims or they have been removed, we enter the incident stabilization phase and the standard applies.

Although the mandate for the two-in, two-out is a step in the right direction for firefighter safety, other standards such as NFPA 1710, *Standard for the Organization and Deployment of Fire Suppression Operations, Emergency Medical Operations, and Special Operations to the Public by Career Fire Departments* and NFPA 1720, *Standard for the Organization and Deployment of Fire Suppression Operations, Emergency Medical Operations, and Special Operations to the Public by Volunteer Fire Departments*, are the mechanisms by which fire departments can properly staff companies and should also be made law. At this time, these are only suggested standards, as are all NFPA standards not adopted by a state or jurisdiction, where they would then become law. As standards, they have no teeth and do not need to be complied with. It is ironic, though, that these same standards, when a firefighter casualty occurs or other legal issue arises, are the criteria used to show negligence and collect monetary damages from municipalities. Very few departments meet these staffing standards.

Self-rescue

SKILL DRILL

FFI 4.2.4 Although situational avoidance is clearly the best choice, when things do go wrong—and at some point in your career they will—past training, self-reliance, and a proactive approach will give you the best chance of survival. A variety of techniques that every firefighter should know will be described next. Whether you need to help your partner clear an entanglement or seek an area of refuge after a loss of water, you owe it to yourself and your fellow firefighters to be prepared.

Personal tools. Regardless of your position on the fireground, you should always carry tools. One of your hand tools should provide you with a way to breach a wall or force a door. The most practical and versatile tools are the Halligan tool and the flat-head axe. Beware: Never give up your tool. If you do, you will not have it when you need it. We have all seen a firefighter without a tool take the new firefighters' tool. This is unacceptable. For example, if a firefighter or officer grabs your hook or bar, hold onto it, and say, "Where would you like me to pull?" This will ensure that you maintain control of your tool. Smaller hand tools can play a role in clearing wires that you may get tangled in. You should carry a small pair of wire cutters in a pocket that is located opposite the air line on your SCBA. This will allow you to access your cutters without taking your hand off your air supply strap, so you can clear the entanglement even while you manipulate your SCBA to reach the wire.

Area of refuge. Having a contingency plan of where to go if something goes wrong is critical. Continual size-up of your location in the structure will aid you in planning for failure. When in a public hall in an apartment building, first force the door to the apartment next to the fire apartment. This will not only show you the likely layout of the fire apartment, but it will also give you a place to go if there is a wind-driven fire or a loss of water.

When moving down an interior hallway, map it out. Know where all the doors are, what they are constructed of, and which way they open. When inside the structure, a door that opens toward you will either be the closet or basement, which is a change in grade if you are on an upper floor or the way outside on a ground floor. In older buildings, and in some newer townhouses or estate homes, it may be an elevator door. Conversely, doors that open away from you usually open into other rooms. This may be critical if you are looking for a safe haven or to possibly access a window where a fire escape or RIT-placed ladder is located. Understanding the typical layouts of structures can aid you in your size-up.

As a member of a truck company, if you find the fire, you should close the door, ensure your area is safe, and call for a hoseline. If fire is coming down the hall, retreat into a room and close the door. This action should buy enough time for you to exit safely. By closing the door to the room, you may be able to prevent making a headfirst dive onto a ladder or bailing out on a rope. When seeking an area of refuge, never retreat into a bathroom. There may be no windows or only a small window high up on the wall, making escape impossible. Knowing where abutting and porch roofs are can also provide a safe haven. If there is no ladder there, you can exit the window, hook your tool on the sill, using it as an improvised lifeline, and wait for a ladder below any venting fire and heat.

Bailout kits. Theoretically, bailout kits place a ladder at every window. They provide a way out in the most extreme conditions. Although seeking an area of refuge by closing a door may negate the need for such devices, there are circumstances where this will not be possible. You may find yourself in a living room, kitchen, or room without a door, and the bailout may be your only option. There are many different systems on the market. We recommend a device that has the ability to anchor at the window because it is unlikely that under deteriorating fire conditions that you will be able to tie off to a remote anchor. Several devices have the ability to stop you after you exit the window, so you can reposition yourself and then lower yourself to the ground (fig. 17–82).

Fig. 17–82. Bailout kits provide an exit at every window for the times when you cannot close the door. (Courtesy of Coda Ricci)

Wall breaching. **FFI 4.3.9** Breaching a wall may be necessary to exit a hazardous environment or effect the rescue of a civilian or a trapped firefighter (fig. 17–83). Wall construction differs and your size-up will determine if this is a practical option. Brick, concrete, and block walls are not practical for a firefighter to breach for self-rescue and may require special tools. Block walls can be effectively breached for firefighter rescue by employing a sledgehammer. This option is labor intensive, requires room to swing, and is usually performed from the outside in. For self-rescue, the walls most often breached will be lath and plaster, or gypsum board. Even with this type of construction, don't be surprised to find studs that are not conventional in their spacing or a chimney inside that makes breaching impossible. Wall studs are usually 16 in. (41 cm) apart, but this may differ based on factors such as the age of the building. If a wall breach is the action you must take, the Mayday should have already been called. Breaching a wall with a Halligan tool entails the following steps:

- Check for an outlet on the wall. If one exists, move to the next bay.
- Plunge the tool completely through the wall about 3½ ft off the floor to ensure that the back of the wall is clear.
- If you hit something solid (maybe a refrigerator, dresser, or wall unit), move down a few bays.
- Pull the bar back through the wall, drop to bottom, and pull back toward yourself.
- Breach the rest of the wall by pushing the back of the wall down with the Halligan or kicking it out with both feet.

Fig. 17–83. Wall breaching. (Courtesy of Local 825 president Patrick Egan)

- If wires are present, knock them down with the Halligan.
- If time permits, hit a stud sideways at the bottom. This will provide a wider opening. You may then get through without manipulating your SCBA.

Detroit dive

- Breach the wall as previously described.
- Place back to hole.
- Lean back, to allow the SCBA cylinder to clear the hole. Your cylinder gauge must be on the other side of the wall or you will get caught up.
- Rotate slightly to allow one of your arms to be placed on the opposite side of the wall.
- Lean back and rotate on your side toward the arm that went through.
- Note: In all breaches, be sure to check the stability of the floor on the other side to ensure it is actually there and you will not be entering a shaft or two-story foyer.

This method is highly effective, but works best on a gypsum board wall (fig. 17–84). If going through lath and plaster, take an extra second to make sure the bay is completely cleared. In chapter 19, Search and Rescue, the section on "Mask Confidence and Emergency Procedures" covers the reduced profile maneuver and the disentanglement maneuver. Any method that requires you to manipulate your SCBA to self-rescue shall be cause to call the Mayday. If you must use one of these methods to get by an obstruction for fire attack, you should radio a message to command apprising them of the situation and make a careful risk-versus-gain assessment of the situation.

Fig. 17–84. The Detroit dive. (Courtesy of Local 825 president Patrick Egan)

Breathing techniques. Air consumption rates vary between individuals. Factors will include lung capacity, physical fitness, level of exertion, and emotional state. It is critical to conduct comparative testing to evaluate emergency and normal breathing techniques for each firefighter. This is accomplished by synchronizing SCBA bottles with a digital air gauge after the bottles have cooled down. Each bottle must be labeled for the firefighter being tested. Whatever method your department uses, it is not recommended to use skip-breathing or any method that requires you to hold your breath. Holding your breath causes carbon dioxide to build in your bloodstream, resulting in an increase in respirations.

Many firefighters try to breathe in through their nose and out through their mouth until their respirations demand that they take a full breath. You must find a method that works for you. Currently the air gauge works off pressure, not volume. You must be able to judge your available air supply from your pressure gauge, and that takes practice. The following factors should be considered:

- How far did it take you to get to where you are?
- What is your level of exertion?
- How much experience and training have you had?
- What is your emotional state?

Far too often, new firefighters gauge their air supply based on the smokehouse they were trained in. If your first fire is in a commercial building, the smokehouse reference will do you little good. Firefighter I and II classes should be exposed to different buildings in class, and instructors are encouraged to conduct as many evolutions on air as possible.

Reilly emergency breathing technique. As with any method, the Reilly emergency breathing technique is predicated on remaining calm. With this technique, you hum out your breath. This is a low hum and usually cannot be heard over the low air alarm. A firefighter's normal breath cycle is around to 6 to 8 seconds. By extending your exhalations to between 12 and 15 seconds, you can compound the time to escape or wait for rescue. For example, if you had 50 breaths left, and you extended your breathing cycle by just 6 seconds, the compounded time would give you an additional 5 minutes of time on air. In tests, we found this method to also have a calming effect by giving the firefighter something to focus on. If you are moving, it is difficult to use any method. If you feel like you need to take a normal breath, do it, then go back to the technique. The more you use it, the more time you will buy (see fig. 17–85).

DAY 2: Breathing Technique test while firefighter in prone position
500 psi cylinder

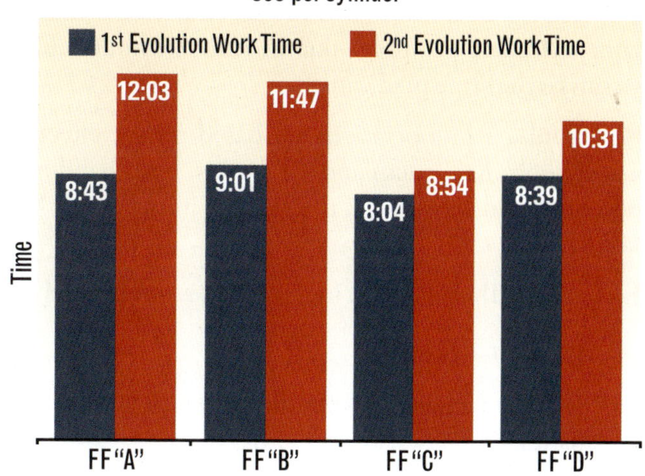

Fig. 17–85. Chart of prone test for breathing technique

Out-of-air emergency. Stay calm and know your breathing apparatus through constant training. For example, one manufacturer's SCBA stops vibrating at 100 psi indicating that the firefighter only has a few breaths left. In this situation, the Mayday should have been called already, and the focus has to be on keeping your facemask in place. On your last breath disconnect the regulator and pull up your hood over the opening. Note this will not block out CO or HCN, however it will help keep soot out of your lower respiratory tract and protect against thermal insult. It will also help facilitate the RIT team's ability to quickly get you back on supplied air.

Intervention teams

The rapid intervention team. Response of the RIT is crucial for firefighter safety. At the very least, the RIT, also known as a Rapid Intervention Crew (RIC) or Firefighter Assist and Search Team (FAST), should respond upon confirmation of a working fire. Second-alarm RITs are not acceptable and, in some states, illegal.

A RIT team should also be dispatched on any special response that is out of the normal response realm, such as hazardous materials incidents, technical rescue incidents, and extrications. RIT response and operation should be part of a departmental SOP, addressed in advance. The duties, responsibilities, and assigned equipment of this team should be addressed in the SOP.

At incidents where there is an unusually large commitment of personnel on the scene, command may consider requesting two or more RITs. RIT members may also be required to don 1-hour SCBA cylinders in a large-area building. The rationale here is that if the firefighters requiring rescue are wearing 30-minute cylinders, and they can't get out, how will the second team with the same cylinders get them out in time without falling victim to the same predicament?

RIT tools. RIT operations are based on the effective use of tools. RIT packs and RIT tools, both hand and power, should be staged at the ready in an accessible but safe area.

RIT pack. The RIT pack includes but is not limited to:

- A 1-hour cylinder
- RIT connections for your department's SCBA manufacturer
- Extra facepiece
- Regulator to RIT pack
- Multiple stand-alone PASS devices
- Wire cutters with reflective handles
- Tin snips with reflective handles
- Small wonder bar
- Two lengths of 15-ft loop webbing
- Two large carabiners
- Small anchor strap
- Two filters

Make sure that there is a way to secure the RIT pack to the firefighter. Firefighters should be trained on the use of all connections to **transfill** the downed member's cylinder. Note that the downed member's cylinder will equalize when transfilled and will not give the downed member all the available air. If you are rescuing a lost company, you must bring more than one pack.

Hand tools. The following is a suggested list of hand tools that the RIT should carry and stage:

Portable radios (headset radio for RIT Officer)

- Spare SCBA cylinder and mask
- Supplied air respirator with spare cylinders
- Automated external defibrillator (AED)
- Ready-made rescue sling
- Hand lights
- Rope-guided search equipment with carabiners
- Thermal imaging camera (TIC)
- Set of irons (flat-head axe and Halligan tool)
- Pike pole or Halligan hook
- Knife
- Target exit device
- Sledgehammer or maul

Based on the building construction and occupancy, other tools may be required. For example, for operations in a high-rise, a hydraulic forcible-entry tool (HFT), also known as a rabbit tool, should be a standard part of the RIT equipment.

Power tools. Power tools are often required when firefighters become trapped under debris or parts of the structure. If the following are available, they should be staged for RIT use.

- Power saw with wood-cutting blade
- Power saw with steel-cutting blade

- Fire service chain saw
- Concrete blade available

Where concrete may be encountered, staging a saw with a concrete blade is recommended. Do not rule out of the use of hydraulic rescue tools as required by the building construction, especially with fire-resistive buildings.

Safe power saw operation. Power saws can end a firefighter's career in a split second. One manufacturer stated that the limb will be gone before you feel the pain. To this end, power saw and all power tool operations should be conducted with extreme caution. The following are some safety guidelines to observe when operating power saws:

- Respect the tool.
- Wear safety glasses or SCBA.
- Ensure all saws are started in a safe area, shut down, and then taken to the area of operation. For example, before going to the roof, start the saw on the ground.
- Always be aware of ignition sources when fueling.
- Be aware of the **circle of danger**—anyone you can reach and touch with the saw (or any tool) is in the circle of danger.
- Always plan the cut and inform others of that plan.
- Never pick up the saw to start it—always start it on the ground.
- Keep good balance, with your feet in a boxer's stance 18–24 in. (46–61 cm) apart.
- When operating a saw with a wood blade (carbide tip), rev the saw to its maximum rpm before sinking it into the work.
- When operating a saw with steel blade, set the saw in the place to be cut before revving the saw. Begin to rev the saw slowly, and when the saw has a bite in the material, then progressively increase the rpm and make the cut.
- Do not store composite blades with hydrocarbons, because this can cause catastrophic failure.
- Position your legs so the plume of sparks is not directly hitting your gear. Turnouts can burn.
- Never step on anything that has been cut.
- Use the **buddy system** when cutting. Always have a guide person behind the saw operator. The guide person keeps an eye on the area, while the saw operator concentrates on the cut.
- Use the "contact-go/no-contact-stop system" when guiding the saw operator.
- As long as the guide person has a hand on the saw operator's shoulder or back, the cutting continues. If the guide person breaks contact, the saw operator stops cutting (fig. 17–86).
- Always control the saw.
- Always be prepared to stop the cut and the saw.
- Never let the saw operate more than 6 in. (150 mm) away from the cut. If possible, let the saw blade "ride the roof" to the next area to be cut.
- Listen to the saw—it will clue you in when to add or take away pressure.
- When the cut is stopped, bury the blade in the roof to stop the blade from spinning.
- Make sure that when the job is finished the saw is cleaned, fueled, and checked for damage.

Fig. 17–86. Nonverbal communication systems eliminate guesswork during cutting operations. This allows the firefighter operating the saw to concentrate on the cut. (Courtesy of Bob Scollan)

FIREGROUND NOTE

There is no politician on scene who is telling command that they can't call for the resources they need to protect us!

RIT on-scene actions—proactive preparation

SKILL DRILL

Prior to standing by at the ready, the RIT can perform specific tasks aimed at increasing firefighter safety. In fact, the more preemptive action the RIT can take to help firefighters inside the building and on the roof to help themselves out of danger areas, the safer the fireground will be. If a firefighter can get to an egress point that is already there and doesn't have to wait for one or create one by force, the job of the RIT is half done. Such actions include, but are not limited to the following:

- *Sizing up entry and exit points from the building.* The RIT should be identifying multiple ways in as well as multiple ways out for all floors or areas of the building. This includes plans for the breaching of walls.
- *Staging building construction-specific tools for firefighter rescue applications.* Without proper tools, the RIT is useless. Some departments use canvas tarps called RIT tarps on which all RIT tools are staged. The essential tools required for basic rapid intervention operations are listed on these tarps. If the building construction or some other condition requires additional equipment, the RIT officer ensures this is secured (fig. 17–87).
- *Ensuring the area is well lit.* Lighting on the fireground is a most basic safety issue. Most new fire apparatus can activate their powerful flood light systems with the flick of a switch. It should be part of the initial scene assignment duties for the ladder at the front of the building. The RIT, as required, can set up additional lighting to illuminate all sides of the fire building and possibly the roof. Illuminated operations are far safer than operations conducted in the dark. In addition, a lost firefighter may find the way to a window easier if the area outside it is lit up.

Fig. 17–87. RIT equipment should be gathered and staged in one place on the fireground. A dedicated RIT tarp is one way of organizing this equipment. Mandatory equipment can be listed on the tarp in permanent marker. (Courtesy of Ron Jeffers)

- *Raising ground ladders.* One of the main tasks of the RIT is to raise ground ladders around the building for secondary and tertiary egress from the upper floors and from the roof. In fact, they should not have to be told to do this. It should be done automatically (and directed by SOP). At one very smoky taxpayer fire, visibility was reduced to less than a foot in the street. When an evacuation of the building and roof was ordered by command, the roof team could not find the aerial in the smoke. Fortunately, the RIT had raised ground ladders to the roof on the C and B sides of the building. The ladder company got off safely without incident. Either command or the RIT officer or both must notify personnel inside the structure and on the roof of RIT-created secondary egress points as soon as they are in place. This can be done via emergency fireground transmission. If companies know the existence and location of additional egress points there before they need it, they are halfway out (fig. 17–88).
- *Removing barriers (window gates).* The RIT should not wait until someone needing help shows up behind window gates. Barred windows hamper both egress and ventilation operations. The removal of bars and properly coordinated ventilation of windows is especially critical at first-floor and below-grade fires (figs. 17–89 and 17–90).
- *Assisting in the set up and monitoring of the command board.* Keeping an up-to-date profile of where personnel are operating allows the team to have a better idea of where to begin. This will include monitoring roll calls (PAR's).

Fig. 17–88. One of the major duties of the RIT is to provide secondary egress routes. Here a RIT is raising a ground ladder for secondary egress from the roof. (Courtesy of Ron Jeffers)

Fig. 17–89. Window bars on wood frame structures are easier to get off from the exterior than from the inside. The firefighter needing to get out may not have the time to remove them from inside the building. Remove them as soon as you see them to avoid critical situations later. (Courtesy of Ron Jeffers)

Fig. 17–90. Metal casement windows pose the same hazards as window bars. (Courtesy of Jim Duffy)

- *Monitoring radio reports.* Many Mayday notifications are made during initial fireground operations when companies are most unfamiliar with the building and the fire. At this communication-heavy time, Maydays are sometimes easy to miss. For this reason, RITs must be vigilant in radio awareness at all times. This includes during the response. Rapid intervention crews must pay close attention to the initial on-scene radio report as well as the assignment of companies. When on scene, they should report to the command post for a briefing and consult and work closely with the accountability officer. It is best that both the command technician (accountability officer) and RIT officer stand by at the command post with headphones on. This helps eliminate external noise and allows them to concentrate exclusively on fireground transmissions.

- *Forcing doors and other egress points not accessed by the initial entry team.* Providing egress helps those inside help themselves to get out faster and more easily. Egress includes rear and side doors, as well as cutting fences and releasing any fire escape drop ladders.

- *Making available a dedicated RIT hoseline and staging it in an easily accessible area.* It may be possible that fire is blocking a firefighter's egress. A quickly deployed hose stream may be all that is needed to allow a potentially trapped member get to safety.

Although these tasks are essential to ensuring the quickest ways of entry and egress are assured, the RIT must not get sidetracked and involved in firefighting. When these safety-oriented tasks are complete, the team must go back to the command post and stand by.

RIT activation considerations. When the RIT is activated, and the team members have a potential focus area to access, they should enter the building to find the firefighter and ascertain his or her medical condition. That entry area should be sufficiently illuminated. When found, one of the first things that should be relayed to command must be the status of the member and his or her predicament. They can also use the LUNAR or WWW method discussed earlier to provide the proper information. We must determine if the firefighter is conscious and assess the degree of entrapment, if any. If the victim is conscious and not trapped, a major concern will be if the victim can walk under his or her own power.

If the answer is no and/or the firefighter is trapped, things will get more complicated as to how to get him or her out and transferred over to EMS. If the firefighter can walk, it may be that he just needs guidance out. If the firefighter was merely disoriented, that firefighter should still be medically evaluated and the entire company reassigned to rehab or released from the scene. Be aware that CISM may still be required and should be made available.

Height may also be an issue when a firefighter gets injured. If so, aerial devices may have to be used by the RIT for access and rescue. At one fire, a firefighter fell into an open scuttle hatch on the roof and dislocated his shoulder. The RIT was activated, and one of the aerials was dedicated to removal of the firefighter. He was helped down by his company on the roof and by the RIT working their way up from the turntable. The fact that he was able to walk helped make his predicament of being in an elevated position less complicated.

Upon finding the firefighter, the RIT should turn off any activated PASS device. This serves two purposes. It will decrease the frenzy and allow rescuers to think better. It will also help determine if there is another PASS device going off that was previously not heard by rescuers.

At the very least, RIT personnel should be trained in the basics of firefighter assessment and rescue. They should know how to check for the ABC (airway, breathing, circulation), and place the firefighter on supplied air, devise various slings for removal, and know how to perform all these skills in the dark, the smoke, and the heat.

One note about the communications in these situations: Be aware of the feedback that is often created when personnel are communicating in close proximity to one another. Often all that is heard at the command post is feedback and squelch.

Air management. **FFI 4.3.9** Poor air management has killed firefighters each year, so it is important to review its implications before we discuss how to remove a downed firefighter. The following factors have contributed to these situations:

Physical limitations. All firefighters have differing amounts of air consumption. Factors include their weight, aerobic conditioning, age, respiratory capacity, condition, and weight of their PPE. Overexertion and stress are also major limitations to air consumption rates.

Environmental issues. High heat and other environmental stressors encountered will affect the firefighter's air consumption. Breathing elevated air temperatures, encountering thermal insult on the body through the facepiece, and retention of heat inside PPE will cause an increase in the core body temperature and an adverse impact on air management.

Physiological issues. Working in certain conditions can cause physiological issues. These conditions include overexertion, issues of competence and/or trust in equipment, and difficulty working in obscured or blacked-out conditions. These lead to such physiological reactions as feelings of claustrophobia, panic, increased respirations, increased blood pressure, and critical incident stress.

Resistance to change. Change is a constant in the fire service and must not cause stress in the ranks. Resistance to change occurs through challenging of the norms set by peer pressure, traditions, and misinformation. One of the best ways to reduce the stress of change is let personnel know that change is coming and to involve those who will be affected by the change. Lack of involvement leads to resistance. Resistance leads to potential injury. Fire departments must be willing to train and enforce current practices and standards. Company officers must be willing to enforce and practice safe use of SCBAs and embrace air management techniques. It is the responsibility of Fire Department management and safety divisions, and all officers, to enforce rules and not overlook violations.

Disregarding low-air alarms. A common problem in the fire service is working under the assumption that it is acceptable to stay in an IDLH environment with your low-air alarm sounding. As mentioned in the NFPA 1404 Standard, "No firefighter will be allowed to operate with less than ⅓ tank of reserve air in an IDLH environment." Fire crews must be trained to consistently monitor their remaining air and match it to the members of their crews.

Tunnel vision. Tunnel vision can cause companies to operate in an IDLH environment without being aware of the constantly changing conditions. This phenomenon, which can have (and has had) deadly repercussions, results from an individual firefighter's lack of training in situational awareness and a weak command structure that puts company-level crews in jeopardy. Some responsibility, however, must fall on the company commander. It is the company commander who, most of the time, is at the root of the issue and, coupled with the exterior size-up conducted by command, can round out critical information on conditions and potential safety concerns.

Lack of RIT. Air management emergencies are often mitigated by proper RIT actions. Problems result from fire departments not utilizing RIT, lack of or improper RIT training, or the RIT being incapable of performing an actual emergency deployment. Too many fire departments staff their RIT with any available firefighters. This group should meet established minimum training standards. They should perform under established SOPs, and have the proper tools, equipment, and knowledge to provide adequate RIT services. In addition, the fire department command structure should have established procedures and training on how to manage a Mayday call.

Poor company-level tactics. There is a national trend of senior fire personnel and officers retiring, which has led to a new generation of younger fire officers and firefighters without the experience of operating in the ever-changing structural fire environment. This lack of experience translates to issues of improper air management that have spiked the number of close-call incidents. The answer here is more realistic and battle-ready training.

Standard operating procedures. Fire departments need to establish and enforce air management SOPs. They must ensure that all personnel are trained in, practice, understand, and use the procedures. The SOPs should be consistently enforced so a consistent behavior is maintained.

SCBA conversion. When a down firefighter is located, it is critical to convert his or her SCBA into a harness. This simple action will greatly assist in a rapid removal of the firefighter. Yes, it takes a few seconds to complete, but it will save you minutes when extracting the firefighter. The conversion can usually be completed while your partner is assessing the firefighter's air supply. A common mistake is for the RIT to try to convert the harness in a confined area. Although this may be necessary, size-up is the key. You will know the terrain from which you just came. If it is clear a few feet away, pull the firefighter out of the debris any way that you can, then convert the SCBA. This will help facilitate a more efficient conversion and will ensure the firefighter has air or can be changed over to a supply of air. The following are the steps to a proper conversion:

- Extend the waist strap while it is still buckled.
- For a very large firefighter, loosen the shoulder straps and pull the SCBA cylinder down.
- Lift one of the downed firefighter's legs over your shoulder.
- Grasp the straps, never losing contact, and move your hands around the firefighter's leg and buckle the straps so that they harness the firefighter's leg that is over your shoulder.

This conversion will allow the firefighter at the top to grab the shoulder straps and pull as a unit. You may choose to keep the leg over your shoulder and push while the firefighter is being pulled (fig. 17–91).

Fig. 17–91. Firefighter converting the SCBA into a harness. (Courtesy of Christopher McGuire)

Firefighter packaging and removal considerations. This section will cover likely RIT evolutions that you may have to face as a member of a RIT team or while helping one of the members out of a structure. These are not all inclusive, but will open your eyes to the necessity of safety and survival training. Removing a firefighter is more labor intensive than removing a victim. Case studies have shown that the team that locates the firefighter often will not be the team that completes the removal. It is important to bring in a lifeline that can be secured to an object to facilitate a quick exit and entrance path for egress, personnel, and resources. Discipline will also play a role. Too many hands can slow the removal. Officers will be needed to manage the rescue, not perform the tactics. This will fall to you. Make sure you are prepared!

FIREFIGHTER RESCUES AND DRAGS

2:1 mechanical advantage system

SKILL DRILL

With the 2:1 mechanical advantage system, a rope and a carabiner is used to help remove an unconscious firefighter in a restricted or cluttered area.

1. One firefighter secures the rope to a substantial object out of the danger area and deploys the rope as he or she searches for the trapped firefighter.
2. Once the firefighter is located, attach the middle of the rope to the top of the SCBA harness, where the shoulder straps are attached, with a large carabiner or snap hook. Make sure the SCBA waist strap is secured between the unconscious firefighter's legs so that the SCBA will not be pulled off.
3. The firefighter backs up in a straight line with the remaining rope, turns toward the unconscious firefighter in a seated position, and pulls the unconscious firefighter from the danger area (figs. 17–92 and 17–93).

Figs. 17–92 and 17–93. The rope is run through a carabiner attached to the downed firefighter's SCBA harness. The firefighter at the anchor point pulls the rope, bringing the firefighter toward him as he pulls. (Photos courtesy of NHRFR)

Drag rescue device (DRD)

SKILL DRILL

NFPA 1971, *Standard on Protective Ensembles for Structural Fire Fighting and Proximity Fire Fighting*, outlines guidelines that requires a device that enables firefighters to be manually dragged horizontally for a quick removal from a danger area. The drag rescue device (DRD) meets this compliance.

The DRD is simple to set up and deploy. It is tucked into the collar of all turnout coats. The harness operates by forming loops that cinch themselves beneath the armpits. The device is a continuous loop, not unlike a rescue sling. It is positioned on the wearer between the exterior portion of the turnout coat (on the inside) and the thermal barrier portion (the shell). The device basically sits wrapped under the wearer's arms between the vapor barrier and the shell of the turnout coat. When deployed, the top loop is released by pulling up on a Velcro® tab on the collar and pulling the top of the device loop upward and outward, taking the slack out of the harness and tightening it around the wearer's armpits, making it ideal for dragging (fig. 17–94).

Fig. 17–94. The drag rescue device (DRD) deployed. (Photo by Jim Ricci, courtesy of Globe)

There are some firefighter-caused limitations with deployment. Firefighters are trained to carry the weight of their SCBA on their hips. This allows easy access to the DRD. If firefighters are carrying the weight of the pack on their shoulders, it is harder to deploy. Some models of SCBA completely cover the DRD even when worn correctly (fig. 17–95).

Fig. 17–95. When SCBA is worn properly, it is easier to deploy. (Photo by Jim Ricci, courtesy of Globe)

The device comes with a user information guide, outlining the manufacturer's recommendations for the device. It is highly recommended that the wearer not only read but become familiar with the layout, setup, and deployment of the harness. After each deployment, the handle (pulling loop) must be reset on the back of the coat in accordance with the user guide instructions. It should also be periodically checked to ensure that it is properly installed and not damaged in any way.

According to the user information guide, the purpose of the firefighter DRD is to aid firefighters with the rescue of an incapacitated firefighter by dragging him or her along a horizontal plane. The DRD is not designed nor tested for use as an escape harness for lifting or lowering a person on a lifeline.

The handle or pull loop, once slack is pulled out of it, is big enough for two firefighters to grip it simultaneously. If the firefighter who is being rescued does not have his or her turnout coat closed all the way or the SCBA waist strap is not in place, the unsecured areas start to separate and the coat begins to ride up. In one case, it almost pulled off.

Belt or harness drag

For a horizontal drag with one firefighter we found this method to be the most effective and the least labor intensive (fig. 17–96). As the rescuer, you would convert your waist strap and send it through the loop of the DRD. This could also be accomplished by hooking into the DRD with the Class II harness. Once hooked in, straddle the downed firefighter, ensuring that you do not dislodge the firefighter's face mask. Now you can crawl with your tool on all fours. The downed firefighter is now dragged in line between your legs.

Fig. 17–96. This photo demonstrates the belt or harness drag; note that your waist strap must go through your legs. (Photo by Jim Duffy, courtesy of Globe)

> **FIREGROUND NOTE**
>
> Preferred firefighter and occupant removal
>
> 1. Interior stairs
> 2. Horizontal exits
> 3. Fire escape . . . (check condition)
> 4. Ladder
> 5. Rope

Stair drags

SKILL DRILL

The DRD is not recommended to be used to move a firefighter up or down stairs (figs. 17–97 a and b). However, when the downed firefighter's SCBA is converted into a harness and the rescuers are trained to pull the DRD in opposite directions under both shoulder straps, the DRD will provide two short loops to grab, giving you both the leverage and control to manipulate the downed firefighter up the stairs. When trying to go

LESSON FROM THE FIREGROUND

The fire was in a two-story wood frame vacant building. It was showing heavy fire and was exposing a wood frame, 2½-story, occupied dwelling on the D side, which was 3 ft (1 m) away. Flames were licking against the soffit of the exposure. The vinyl siding on the other exposure, approximately 20 ft (6 m) away, was already melting. Having been the scene of several previous fires, it was in a state of disrepair and was marked as a Level 3 hazardous vacant building. The first engine on the scene had stretched a line to the rear via an alley where most of the fire was concentrated to try to knock it down quickly. A second line was being used to wet down the exposure on the D side. As it was early in the fire and companies were still arriving or en route, no other extinguishing actions were being carried out. Additional lines would be stretched later to protect the D exposure from inside, protect the other exposed buildings, and extinguish the main body of fire. At this time, however, only two companies were operating.

As the ranking chief arrived and assumed command, it became clear that an EID button had been activated on the fireground. Dispatch confirmed which individual had activated the button. It was the officer of the first-arriving engine company. When contact was attempted over the radio with that individual, there was no answer. Since the second alarm had been struck by the first company on scene, the IC struck a third alarm to address the Mayday. An additional RIT was also requested. The dilemma at hand was that there was no RIT yet on scene. They were still responding and were still a few minutes out. They would still have to park, finish gearing up, and bring tools and equipment to the command post. There was not enough time if someone was in trouble. A battalion chief and a ladder company who had just arrived were sent to find the company. They had none of the RIT equipment that was carried by the dedicated squad companies that responded as RITs, and this meant that some other extinguishment and support functions would not get done, but there was no other choice at the time. All this time, the Mayday transmission was sounding over the radios, but there was no answer from the officer or his company. The firefighter and his company were found in the back of the building. They were operating a hoseline at the rear and were fine. They did not even know that the Mayday was sent. The radio of the officer in had gotten wet and malfunctioned, sending out the Mayday. His collar mike had been knocked loose from the shoulder strap and was not near his face, but was dangling near his knees.

LESSONS LEARNED

1. The person who activates the Mayday button may not know it has been activated.
2. The Mayday tones come out of all radios except the person who activated it. The officer whose Mayday button was activated was never aware of it.
3. Complacency can kill. There can be no such thing as a false Mayday when the tones or a PASS device is heard. All Mayday and PASS device activations must be treated as the real thing until proven otherwise. Waiting to determine the validity of the activation will waste precious time, which is our enemy during firefighter rescue operations.
4. In the early stages of the operation, command must be prepared to take whatever steps are necessary to address a Mayday, even if it means pressing non-RIT companies into service.
5. All companies should be trained in at least simple firefighter rescue procedures, whether they are assigned to dedicated RIT companies or not.
6. RITs should be responding from as close to the incident as possible. The farther away the RIT is responding from, the more the likelihood that on-the-spot, improvised firefighter rescue tactics will need to be implemented.

down the stairs, the head is not supported and will tend to hit the stairs unless the strap is used as just described.

Figs. 17–97 a and b. Both straps are pulled under the shoulder straps, producing two separate short loops. Figure 17–97b shows that after the SCBA waste strap is converted, the shorter loops allow for a safe, efficient and effective pull up the stairs. (Photo by Jim Duffy, courtesy of Globe)

These are just a few of the ways in which the DRD can be used more efficiently in the rescue of an incapacitated firefighter. These are by no means the only ways to accomplish this objective. Like any tool, the skills must be practiced, or they will not be successful at the time they are needed the most.

One of the most valuable attributes that firefighters have is the ability to modify and adapt to overcome and make the job more efficient. That is part of who we are. An old chief once said, "You must adapt to be effective, you must invent to overcome, and you must create to compensate." Remember the time to find out if something works should not be at 3 a.m. when the building is on fire.

One other note about the DRD: In a recent safety and survival class, the strap deployed and got caught, trapping the firefighter as he tried to breach through a wall. Make sure it is repacked according to the manufacturer's guidelines.

Removal of a firefighter down a set of stairs. Convert the SCBA harness as described earlier.

Place the firefighter on his or her side and grasp the shoulder straps.

- Pull and guide the firefighter down the stairs.

It is easier to remove a firefighter down the interior stairs than down a ladder. If the stairs are charged with smoke, ensure that the firefighter's air supply does not become dislodged.

Removal upstairs. Convert the SCBA harness as described earlier.

- The rescuer on top will grasp the shoulder straps and pull.
- As the firefighter on bottom, you will bury your head in the crotch of the downed member and place both of the downed member's legs over your shoulders. Your hands should grasp the downed member's buttocks and push the firefighter up.

Removal will take some coordination between the two rescuers. This process can be enhanced if additional resources become available. Without slowing down the removal, a 2:1 mechanical advantage system can be added to assist the two rescuers (fig. 17–98).

Fig. 17–98. Firefighter being moved upstairs to safety. (Courtesy of Christopher McGuire)

Removal out a window

Upon finding a downed firefighter, the following are several methods of removing the downed firefighter out a window:

- Window cutdown
- Firefighter assist
- Aerial ladder
- Tower ladder
- Ground ladder

As the firefighter is being brought to the window, cut down the window to the floor (basically making a door out of the window) so the firefighter can be removed at floor level. This will keep you from having to lift the firefighter into the window. This requires proper size-up of the inside wall. Once the window has been cut down, the downed firefighter can be removed via ground ladder, aerial ladder, or tower ladder. As the firefighter is removed from the window, he or she should be stripped of the SCBA to ease in moving through the window (fig. 17–99).

Fig. 17–99. The window cutdown allows the downed member to be removed with little effort but takes resources and time.

Denver drill

The Denver drill allows an efficient method to remove a downed member from a confined area with a high window. The development of this technique was in response to a line-of-duty death. Although there are many variations to this drill, we will discuss one method that we have also found that can be used in an open area as well. The downed firefighter will be located under the window:

- Clear window of glass and drapery, blinds, and so forth.
- Rescuer 1 will sound the floor and climb over the downed firefighter.
- Rescuer 1 will roll the firefighter on their back and sit them up.
- Rescuer 2 will enter the window and manipulate their SCBA cylinder so their back will be flat against the wall with Rescuer 2's knees bent.
- Rescuer 1 will lift the downed member and place them on Rescuer 2's knees.
- Rescuer 1 will place the member's legs over their shoulders and coordinate the lift with Rescuer 2 and the outside team.
- Rescuer 2 will push up on the member's bottle while Rescuer 1 pushes the member out the window (fig. 17–100).

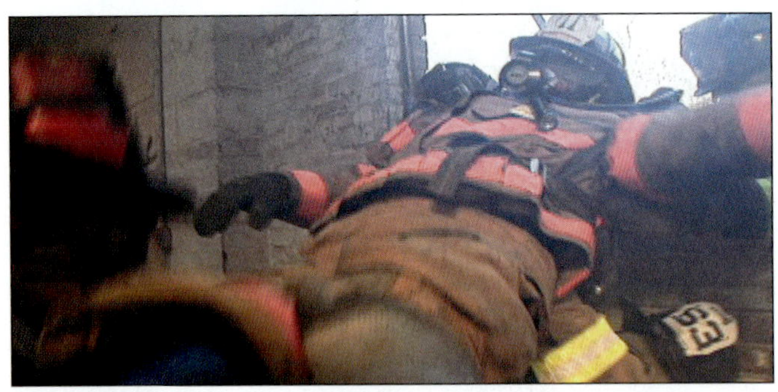

Fig. 17–100. Firefighter removed by utilizing the Denver drill

The outside team should use the building as protection and reach in with the arm that is closest to the window. They will each use only one arm and will reach to grab the downed member. This action will allow the outside team to turn in as the member is pulled out, without switching hands, cradling the downed member, and pulling the member away from the window.

High-point removal

SKILL DRILL

A downed firefighter can be removed out of a window by using a high-point removal system. This can be accomplished by rigging a pulley or carabiner to an anchor strap above the window. A ladder can be used for the **high-point** with a life safety rope run through the pulley or carabiner. A figure eight on a bight with an extra-large carabiner can be attached to the two SCBA shoulder straps on the rear of the downed firefighter and a rated webbing strap that is secured to the firefighter. Most SCBAs are not rated as a harness; a small piece of webbing can be used to make a hasty harness as a redundancy. The *downed* firefighter's SCBA waist strap should be converted as outlined previously. Once connected, the downed firefighter can be assisted out the window and lowered to the ground or into the bucket of a tower ladder utilizing the rope system. The end of the rope will be run under a rung of the ladder to add friction at the bottom of the ladder being used as the high point. Controlling the friction from the off-center side under the ladder is best because the rope also helps to anchor the ladder. This system can also be deployed using a 2:1 system. When using pulleys, keep two rules in mind:

1. If the pulley doesn't move, it only changes the direction of rope travel.
2. If the pulley moves, it will create mechanical advantage (fig. 17–101).

High-point can also be used without the 2:1 simply as a friction-control lowering device.

Fig. 17–101. High-point removal creates friction through the lower rung. Pull in the direction of the building to ensure stability of ladder. (Courtesy of Christopher McGuire)

Firefighter through the floor

Removing a downed firefighter who has fallen through the floor is a high-risk operation. You will be working at the top of a chimney. The most likely situation you will encounter is a firefighter who has fallen into a basement. This complicates removal because of mechanicals (wires, pipes, and duct work) that run along the ceiling. The hole where the firefighter falls will be jagged, may present a secondary collapse area, and may require **shoring**. When making contact with the downed member, and for any pulling operations, work from the corners and size up any obstructions that will need to be cleared. This can also be accomplished by using rope with specific knots.

Utilizing a charged hand line

SKILL DRILL

1. The charged hoseline is pushed down through the opening in the floor. A team of firefighters control one side of the hoseline, while another two rescue firefighters wrap their legs and arms around the hose and slide down it to find the downed firefighter.
2. Once located, the rescue firefighters must assess the air of downed member. Sit the firefighter up and place the charged hoseline across the victim's chest directly under the armpits. The rescue firefighter then opens the downed firefighter's waist strap and places it across the downed firefighter's upper arms (above the elbows) and tightens it.
3. The rescue firefighters pull the hose and downed firefighter up through the hole, working together as the victim comes up.
4. The hose is then pushed back down. The rescue firefighters put the hose across their chests and under their armpits and hold on. The team above then pulls them to safety. This can also be accomplished by standing in the loop and grasping the hose with your hands by crossing your arms (fig. 17–102).

Fig. 17–102. An unconscious firefighter is pulled to safety using a hoseline with no knots. It is unlikely for firefighters to use knots under hostile conditions.

POSTFIRE ANALYSIS

Something of value can be learned at every fire, and thus all fires should receive a critique or what can be better termed an incident evaluation. What is revealed and learned in the evaluation can be of value for the rest of a firefighter's career. Each fire should be a learning experience that should go beyond the kitchen table or bar. The downside of evaluations is that by the time they are held, not everyone is there because of vacation, sick leave, or other commitments. Other members who are present may not remember exactly what happened. Sometimes the best lessons are forgotten forever.

How can the fire department effectively hold an incident evaluation where all the pertinent information is offered? Remembering that the palest ink is better than the sharpest memory, notes should be jotted down as soon as possible after the incident. This will make the evaluation more meaningful. An adapted form will work well and put everyone on the same page during the evaluation session. A form adopted by the North Hudson Regional Fire and Rescue is the after-action report (fig. 17–103). The report is completed anytime a hoseline is charged at a structure fire. The forms are filled out as soon as the company returns to quarters and is forwarded to the battalion commander. That way, if the officer is off the next tour, his or her input will still be included in the incident evaluation. What is likely to happen in this instance is that the acting officer takes the place of the company officer and speaks on his or her behalf at the evaluation, using the form.

Incident evaluation

A specific format should be used for the incident evaluation. It should cover specific areas so that the focus is kept on track. One way to head this off is to conduct a pre-evaluation meeting or a post-incident analysis. This is conducted by the chief officers and the SO as a way to preplan the evaluation and steer it in the proper direction. In the postfire analysis, after-action reports and dispatch tapes are reviewed. SOPs are evaluated and operational strengths and weaknesses are determined. What went right and what went wrong from the command officer's point of view must also be discussed. A strategy on how to best address those issues in the incident evaluation is the goal of the post-incident analysis. If the post-incident analysis is done properly, the incident evaluation should be productive.

The format the incident evaluation takes covers the following areas.

Communications

- Were transmissions prioritized?
- Was proper radio etiquette followed?
- Were personnel on the fireground being "stepped on?" Chief officers should never be stepped on.

Secure a copy of the dispatch tape. It is like the eye in the sky and can tell you a lot about how well things went regarding communications. It doesn't lie. As in football, if the referee says you were holding, you were.

Apparatus positioning

- Were SOPs followed?
- Were there obstacles to response?
- Were apparatus used properly and positioned to take advantage of their capabilities?
- Were there any unusual circumstances, and if so, how were they addressed?

One piece of apparatus out of position can affect the entire fireground. Improper positioning can never be swept under the carpet.

Company integrity

- Was freelancing an issue?
- Were accountability procedures followed?
- How well coordinated was the relief and rehab of companies?
- Were companies broken up as a result of an injury or other unforeseeable occurrence? How was that handled?

Reinforce fireground discipline at this time. Violations of the principle of company integrity are unacceptable and cannot be tolerated at any time for any reason.

Tactics and lessons learned

- Officers discuss their positions and assignments.
- What problems were encountered?
- What solutions were used? How effective were they?
- What lessons were learned from the incident?

IR8.
12/01/08

North Hudson Regional Fire & Rescue
After Action Report

Date: ________ Platoon:_____ Battalion:_____ Company: ______ Officer:______________

Address: ____

Construction: ________________ Occupancy: ___________________

Size-up on arrival, Location of Fire etc.:____

1st Alarm Arrival Sequence: 1st E____ 2nd E____ 3rd E____ 4th E____ 1st L____ 2nd L____

Initial Action: (Circle all that Apply) Fire Attack Water Supply Back-Up Line Exposure Line Outside Stream Ops
Force Entry Locate Fire Primary Search Rescue Ground Ladders Aerial Ladder Ops

Other Action * Ventilation: Horizontal Vertical Mechanical ________________________
Explain equipment used and location

Explain Actions in Detail:

Concerns? / Problems? ____

LessonsLearned:____

Draw Diagram on Other Side

Fig. 17–103. After-action report

Safety considerations—this portion is conducted by the SO

- Was everyone wearing proper protective equipment?
- Were department SOPs followed?
- Was the fire attack coordinated with proper and timely support?
- Did companies stay together, or was freelancing evident?
- Did the operation provide for emergency firefighter egress from the interior?
- Did the operation provide for emergency firefighter egress from the roof?
- Were communications conducive to safe operations?
- Were tools and equipment used safely and properly?
- Was the RIT properly equipped and ready to be deployed?

The real value of the SO's input into the evaluation is that it should be from a point of view of someone that is standing back and taking in the big picture at the fire scene. This point of view should be from a different angle than the operating companies.

The incident evaluation should be a constructive session, aimed at improving the department's performance from a safety standpoint. A safer operation will always translate into a better, more effective operation.

Finally, anyone, regardless of rank, who has something to say should be given a chance to talk about the incident. Allowing all players to have input into the evaluation process makes them more likely to buy into the process.

In concluding the evaluation, together with the SO and the shift officers, the platoon commander should set goals that are both measurable and realistic in regard to a safer and more effective fireground operation. Unsafe practices as well as subpar scene performance should be looped back into the training process and approached in a positive manner. Remember that the further you are from the last "big one," the closer you are to the next.

Postfire analysis in many departments is conducted on scene. Montgomery County, MD, conducts company officer critiques at the end of each fire. This allows each officer to know the actions of every company. After these critiques are complete, the officer or senior firefighter will often walk through the tactics used at the fire with any new personnel. This provides valuable insight to the new firefighter. Even if your department is lacking a formal policy, ask your officer to explain the other company's actions and the fireground action plan. This will allow you to visualize and understand all the tactics that were utilized.

The goal of the incident evaluation is to improve department performance and identify training needs. The overriding concern is always the safety of the participants.

THE 16 FIREFIGHTER LIFE SAFETY INITIATIVES

These 16 initiatives of the National Fallen Firefighters Foundation (NFFF) serve as a blueprint for reducing firefighter deaths and serious injury.:

- Define and advocate the need for cultural change in the fire service.
- Enhance personal and organizational accountability.
- Focus attention on risk management.
- Empower firefighters to stop unsafe acts.
- Develop national training/qualification standards.
- Develop national physical fitness standards.
- Create a national research and data system relative to the initiatives.
- Use technology as it relates to health and safety.
- Investigate all firefighter fatalities, injuries, and near misses.
- Ensure grant programs of safe practices.
- Develop national response procedures.
- Develop national procedures to violent incidents.
- Provide firefighters and their families access to counseling.
- Provide resources for public education in fire and life safety.
- Advocate residential sprinklers.
- Make safety a design consideration for apparatus.

QUESTIONS

1. Safety is the overriding concern of _______________ fireground operations, firefighter training, and station activities.
2. Explain "drifting into failure."
3. Who is responsible for firefighter safety?
4. On average, there are over _________ firefighter fatalities and __________ firefighter injuries each year.
5. Name five causes of firefighter injuries.
6. Injury ___________________ is a vital part of the injury prevention cycle.
7. When should injury documentation start? Why?
8. What on the job injuries should be investigated? Why?
9. List and describe the 11 Rules of Engagement for Structural Firefighting.
10. What role does a fire ground accountability officer play?
11. What is the most important part of a fire department wellness program? Why?
12. NFPA _______________ discuss the recommendations for fire department staffing for fire suppression.
13. What is freelancing and why is it dangerous on the fireground?
14. What is a fireground Mayday? When should it be initiated?
15. What does RIT stand for? When is it used?
16. A ________________________ should be done after every fire.

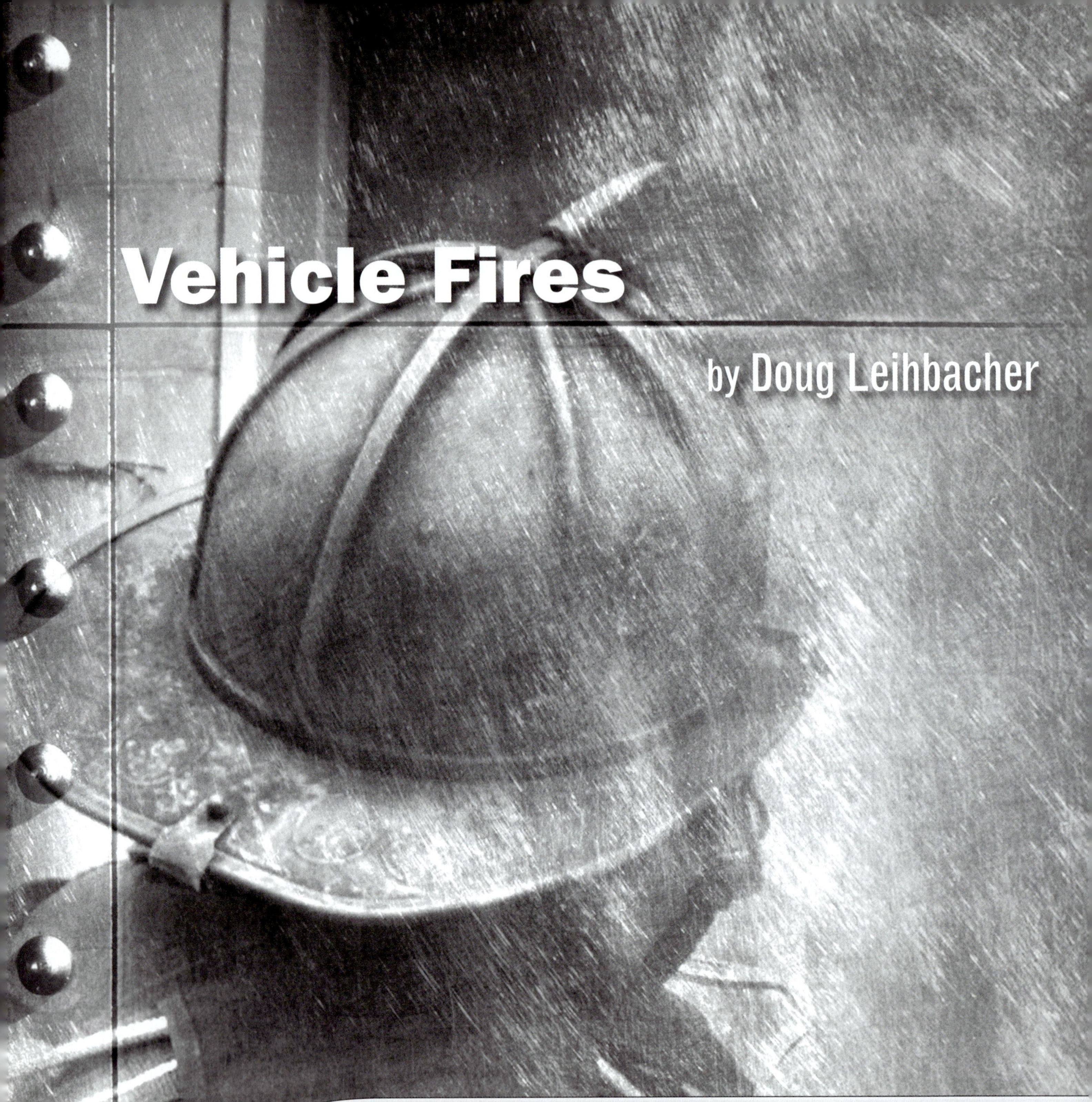

Vehicle Fires

by Doug Leihbacher

This chapter provides required knowledge items for the following NFPA Standard 1001 Job Performance Requirements:

FFI 4.3.7

18

This chapter contains Skill Drills. When you see this icon, refer to your Skill Drill book for step-by-step instructions. SKILL DRILL

OBJECTIVES

Upon completion of this chapter, you should be able to do the following:

- Identify the response priorities for all vehicle fires
- Identify the three compartments where vehicle fires can occur.
- Identify the major safety concerns to emergency responders at vehicle fires
- Identify and describe proper hose selection and attack methods for vehicle fires
- Identify signs that a fire stream has been applied effectively
- Identify and describe the potential hazards of hybrid vehicles and alternative-fuel vehicles
- Describe the functions of the truck company at vehicle fires

INTRODUCTION

According to the National Fire Protection Association (NFPA), a vehicle catches fire every 2 minutes, amounting to 172,000 **vehicle fires** each year (fig. 18–1). These fires kill approximately 500 people and injure more than 1,300 people annually. These numbers represent a downward trend in the number of fires that exceeded 350,000 annually in the 1990s. It is not surprising we experience so many vehicle fires, considering that motor vehicles are composed of complex mechanical and electrical systems involving volatile liquids, internalized combus-

Fig. 18–1. Firefighters take an upwind position and stand clear of the bumpers as they prepare to extinguish an auto fire on the side of a roadway. (Courtesy of Keith Muratori)

tion, many moving parts subject to friction, pressurized fuel lines, wiring, batteries, and extremely hot parts such as exhaust manifolds and catalytic converters.

Vehicle fires can occur when the vehicle is driven, when it is parked and turned off, or as a result of an accident that compromises one of the vehicle's systems. Figure 18–2 provides a breakdown of the common causes of vehicle fires based on the National Fire Incident Reporting System (NFIRS). Note that the number of incendiary fires may be undercounted because some vehicle fires are not investigated for arson.

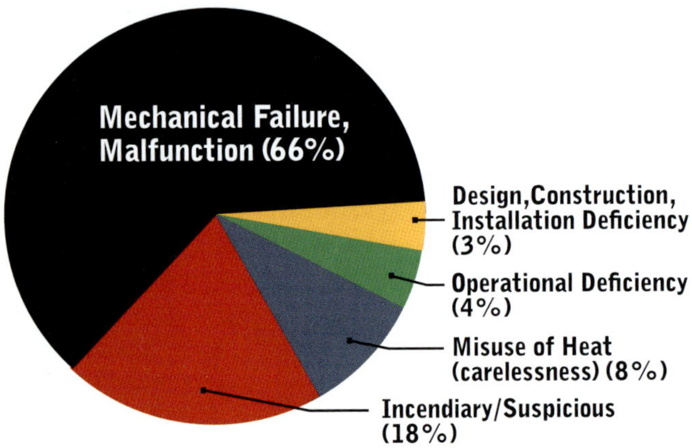

Fig. 18–2. Common causes of vehicle fires

The term **vehicle fire** can potentially refer to any fire that occurs in an automobile, truck, bus, train, plane, or boat. Unless assigned to an airport or a marine fire unit, most vehicle fires to which firefighters respond occur in automobiles. Hence, this chapter focuses on vehicles that travel on roadways.

The priorities at vehicle fires are the same as for all fires:

- Scene safety
- Life
- Exposure protection
- Extinguishment

Notice that extinguishment comes last. At an automobile fire, our first priority is to create a safe working environment for the crew to do their job in. Next is to protect and rescue any victim(s) who may be trapped in or near the vehicle. If there is a victim in the vehicle, steps must be taken to prevent the fire from entering the passenger compartment. Once the life hazard has been managed, our next priority is to protect any exposed property or vehicles by keeping the flames from spreading there. To achieve this, the hoseline should be positioned between the fire and any vehicles or buildings nearby in the direction of fire spread. After taking care of these priorities, firefighters can then extinguish and overhaul the vehicle.

To accomplish these priorities, a response assignment of fire apparatus is dispatched to the scene. Response assignments to motor vehicle fires vary from jurisdiction to jurisdiction depending on available resources, type of roadway, and local protocols. Some departments dispatch a single engine or quint; others two engines; still others, an engine, ladder, and chief. On a **divided highway**, two separate assignments may be sent, one assignment for each direction of traffic flow.

SCENE SAFETY

Regardless of the type of assignment sent, there is only one way to fight a vehicle fire: safely. Extinguishing vehicle fires safely and effectively is a process of controlling an array of hazards that can cause injuries if firefighters are not cautious and alert. Although common, vehicle fires should not be taken for granted. Crew members should expect the unexpected. Addressing these potential hazards begins with taking a look at the big picture as you approach the scene and size up the situation.

The potential hazards associated with motor vehicle fires are influenced by where they occur. There are eight probable locations where a motor vehicle fire is likely to occur. Each of the following can present a unique set of challenges and hazards for firefighters and should be considered while en route and during initial scene size-up:

1. Divided highway: high-speed traffic, limited water supply (no hydrants)
2. City Street: two-way traffic in motion, possible downed wires, onlookers
3. Garage: Gaining access to fire, exposed structure, exposed hazardous combustibles
4. Outdoor parking lot or junk yard: fire spread to nearby autos, limited water supply
5. Accident scene: potential rescue of victims from burning auto
6. Driveway: potential involvement of nearby structure, dropping house service lines
7. Overpass, bridge, or ramp: no shoulder, structural collapse, elevation, dry standpipe operation

8. Automobile repair garage: flammable liquids, structural involvement, possible victims

Responding

Safe operations commence when the crew boards the apparatus and responds to the scene. When responding to a vehicle fire, units should proceed at a safe rate of speed. Unless there are reports of a person trapped in the vehicle, it is foolhardy to put lives in danger by using excessive speed to save property (the burning auto). Instead of racing to the scene, know your district and plan the response route wisely. Consider that the quickest route may vary with the time of day and differing traffic patterns. Blocked or one way streets can cause a deviation from the most direct route. Traffic as a result of the vehicle fire may clog normal approach routes. Two heads are better than one, and discussing the best route with the officer on the apparatus may provide additional response options.

Divided highways

Divided highways can present a challenge to responding firefighters attempting to get to the fire. These high-speed roadways have a limited number of entry ramps where responding apparatus can enter the traffic lanes and, in areas where there is no shoulder, few places for traffic to pull aside.

In addition, remember that cars are traveling at high speeds and may not hear the siren. These factors call for approaching the scene with caution and patience. As mentioned previously, in some jurisdictions two assignments are dispatched to divided highways, one responding from one direction and one from the other; to allow the best opportunity to reach the fire. The preferred approach is for the apparatus to respond in from the same direction that the traffic is flowing. In some cases a safe approach to the burning vehicle is easiest from a service road running along side the highway.

On a divided highway, the fire should only be fought from the same traffic lanes in which the fire vehicle is located. If the fire is in the opposite traffic lanes, firefighters should not stop the apparatus and stretch hose across the guard rail or center barrier unless traffic has been stopped in both directions by police or fire units. Similarly, apparatus should not approach a vehicle fire against traffic unless it is confirmed that traffic has been stopped by the police or other fire department units.

Approaching the scene

As the apparatus approaches the fire scene, all crew members should begin a scene size-up. Are there hazardous materials (hazmat) involved, downed wires, a fuel spill, or icy conditions on the pavement? Is the burning vehicle situated on a blind bend? Is the vehicle a standard automobile or a hybrid? What position is the vehicle in? Is it in the roadway or off to the side? Is it stable or overturned? Is the engine running? Is the driver present? Are there any ambulance personnel, police or Good Samaritans already on the scene? Are there civilians putting themselves in danger trying to extinguish the fire? Almost half of all persons injured in vehicle fires, were injured when they attempted to extinguish the fire.[1]

All scene safety hazards must be addressed before extinguishment operations begin. Firefighters must identify the primary hazards at the scene and manage them first, before taking any other actions. Depending on local protocols and departmental standard operating procedures (SOPs), this may involve directing traffic, placing flares or traffic cones, or establishing a safety officer.

While sizing up the vehicle fire, crew members should consider what stage the fire is in: incipient, developing, or fully involved. Also note where the fire is burning. Is the fire under the hood, in the passenger compartment, beneath the vehicle or confined to the trunk? Most importantly, is there a life hazard? Is someone trapped in the vehicle?

Traffic control

A primary hazard associated with vehicular fires and extrications that occur on high-speed thoroughfares is the danger of being struck by an automobile passing the accident scene at high speed. Firefighters have been severely injured and even killed by passing motorists whose views were obstructed by smoke or terrain, or who became impatient and tried to squeeze by the incident. This danger is increased at dusk, at night, and during inclement weather conditions where visibility is impaired.

Wearing full protective gear with reflective stripes makes a firefighter more visible; but in addition, members must avoid focusing exclusively on firefighting procedures (tunnel vision) and remain mindful of oncoming traffic and other hazards at all times during the incident. This attentiveness begins when you step off the apparatus. Be sure to look before you leap off, and remain vigilant until the entire operation is complete. Often firefighters can be at greatest risk when returning to the apparatus for a tool or packing hose after the fire is out.

Traffic control and apparatus positioning are essential to protect firefighters operating at vehicle fires on highways. Notifying the police department to respond and provide traffic management is good practice. For vehicle fires on city streets, follow the guidelines for fires on divided highways even though motorists may not be traveling as fast. In both cases, vehicles in motion are the primary concern. Overhead wires are another hazard that must be taken into account, especially if the fire is the result of a collision with a pole.

Positioning the apparatus

Fire apparatus must be positioned on the roadway so that all oncoming traffic can see it. If there is a curve or steep hill, the apparatus must be visible from a safe distance. All warning lights should be left on throughout the incident and only shut off after the apparatus has merged back into traffic and is returning to quarters. When the geography makes apparatus visibility difficult, flares, traffic cones, or other appropriate visual warning devices must be set up early in the incident to prevent unsuspecting motorists from colliding with the apparatus.

In addition, police or another fire unit should be requested for traffic control upstream of the curve. Firefighters placing warning devices must be careful to face on coming traffic, stay on the shoulder, and not walk on the roadway except when placing it. The farthest warning device should be placed on the shoulder, then subsequent devices placed further onto the roadway diagonally, until the lane (or lanes) is blocked and a buffer zone is created.

When fuel has spilled at an accident scene, flares should not be chosen over other warning devices in areas where they could ignite a fire. When using flares, the firefighter placing them should carry a lit flare or flashlight to increase visibility to oncoming traffic. To be effective, flares must be placed far enough away from the incident to allow traffic to slow and stop safely.

Table 18–1 lists approximate distances visual warning devices (flares, etc.) should be placed in relation to posted speed limits. This table is based on the stopping distance in feet at a given speed plus the posted speed limit to calculate the approximate placement of **road flares** or other warning devices. It should be used only as a guideline with distances varying to match traffic and weather conditions. This task should be delegated to police personnel when they are present.

When positioning the apparatus, do not place it where it could be exposed to fire spread (fig. 18–3a). Locate the apparatus a minimum of 100 ft (30 m) from the burning vehicle, parked uphill and upwind whenever feasible. Try to position the apparatus so that it shields both the pump operator and crew from oncoming traffic. Depending on the traffic flow direction, this can mean either stopping before the fire or passing it and parking beyond.

The apparatus should be parked on a diagonal across the roadway with the pump panel facing the incident. This creates a traffic-free environment in which both the pump operator and the firefighters can work safely. Parking on a diagonal uses the full length of the vehicle so that it blocks both the lane (or shoulder) where the fire is located and the adjacent traffic lane. This positioning provides a traffic-free zone for members to flake out the hose and work (fig. 18–3 b and c).

Once the apparatus is stopped with the pump panel protected from oncoming traffic (facing the fire), the hose is stretched from the driver's side (the same side of the apparatus that the pump panel is on). This is achieved by using a cross-layed preconnect or by storing the preconnected hoseline on the driver's side fender well or front bumper. Operating on the opposite side of the apparatus from the pump panel may place either the crew or the pump operator in jeopardy if traffic has not been completely stopped (fig. 18–4a). Figure 18–4b shows an aerial view of working from the same side of the apparatus.

Incident commanders should remember to give due consideration to restoring traffic flow at the completion of the incident, while prioritizing crew safety. Coordination with police is important to achieve this. Notify a tow truck early, and pick up tools and equipment promptly after the fire has been extinguished and overhauled. If safety is not compromised, the apparatus can be moved off the roadway and onto the shoulder while hose is being repacked.

Table 18–1. Visual warning device placement

Posted speed limit, mph (kph)	Distance of the farthest warning device, ft (m)
30 (48)	150 (45.7)
40 (64)	220 (67.1)
50 (80)	310 (94.5)
60 (97)	420 (128)
70 (113)	550 (467.6)

EXTINGUISHMENT OPERATIONS

Auto fires have the potential to seriously injure firefighters. If there is no life hazard (in effect, all occupants are out of the vehicle) the fire should be attacked with consideration of the cost benefit analysis. Few motor vehicles are saved or salvageable after a fire, and none is worth the life of a firefighter. Vehicle fires should therefore be fought safely, not perilously.

FFI 4.3.7 Firefighters should wear full protective clothing including self-contained breathing apparatus (SCBA) when operating at vehicle fires. The petrochemical-based parts of modern automobiles are particularly combustible and the smoke they produce is harmful when inhaled. Consider this: At a common auto fire, each breath of smoke a firefighter takes may contain partially combusted products of burning brake fluid, vinyl seat covers, polyurethane foam seats, polyvinyl chloride (PVC) wire insulation, rubber fan belts, and so forth. This is why crew members must use their full protective equipment including SCBA throughout the incident, from extinguishment through overhaul. Here is a list of tools commonly needed in vehicle fire operations:

Tools Required

- Traffic cones/flares
- 100–150 ft (30–45 m) of 1½- or 1¾-in. (38- or 45-mm) hose of with combination nozzle
- Wheel chocks

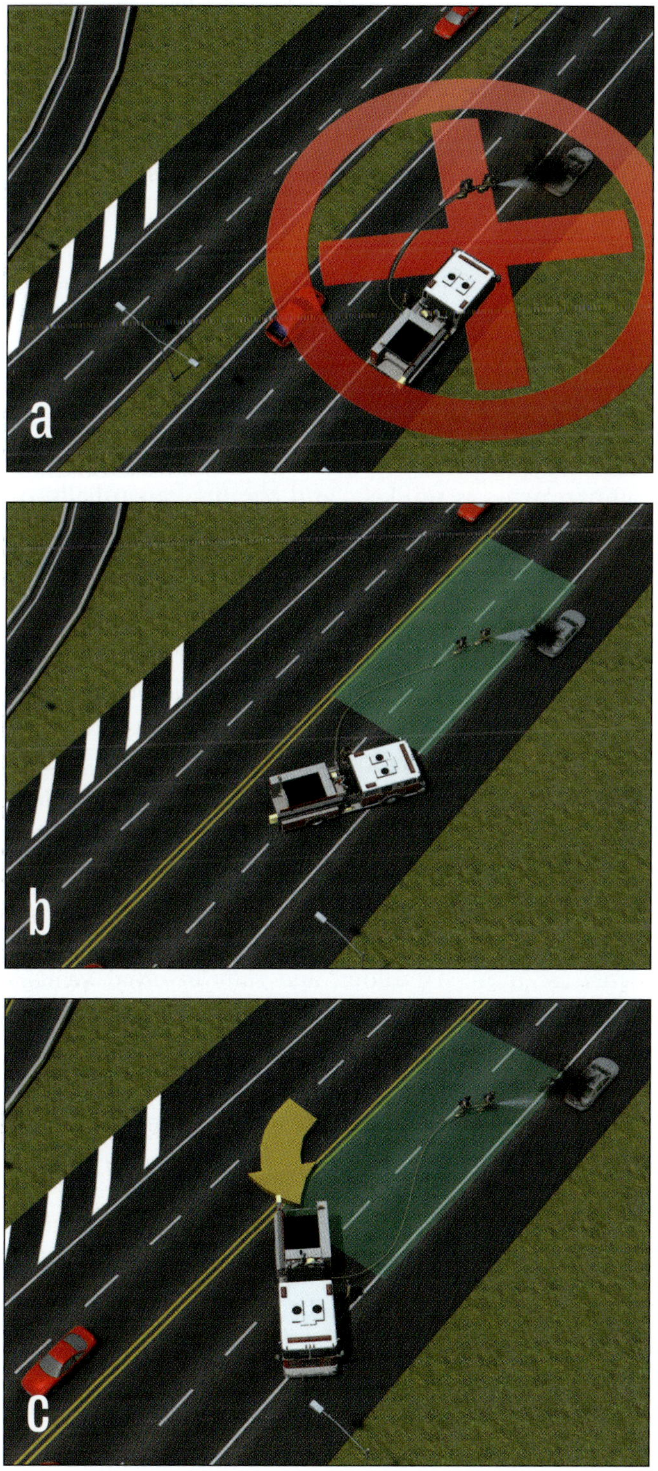

Fig. 18–3. Blocking traffic

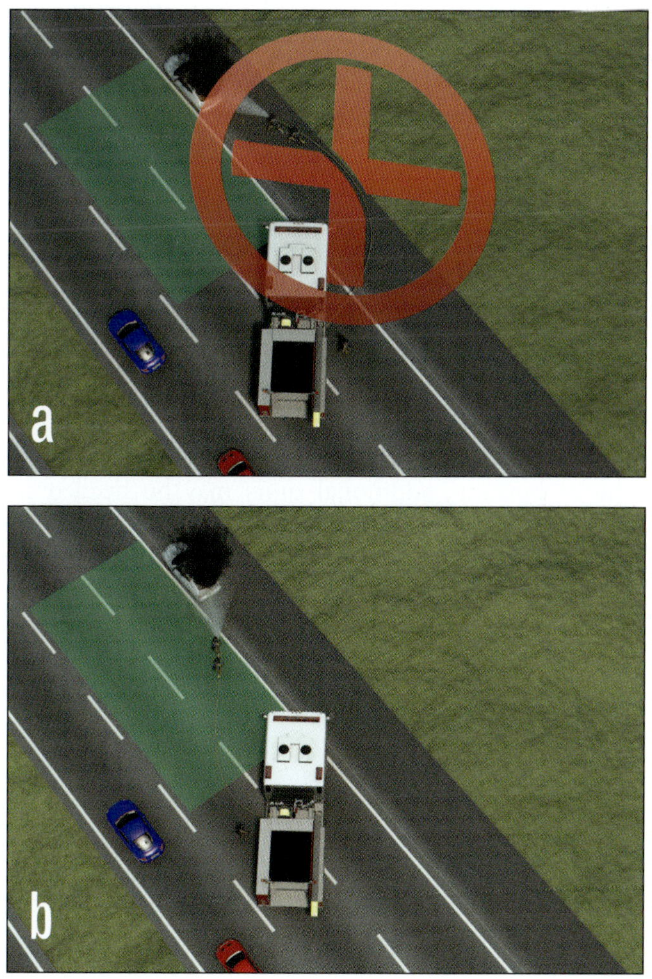

Fig. 18–4. Hose stretch

- Dry chemical extinguisher
- Class B foam and equipment
- Halligan tool or pick-headed axe
- Bolt cutters
- 6-ft (2-m) hook
- Rotary saw with metal cutting blade

Supplementary Tools

- Spring loaded punch
- Piercing nozzle
- Sledge hammer

Water supply

Water supply can be a consideration on divided highways where hydrants are unavailable. Some jurisdictions automatically send two engines to highways for this reason so that the second engine can supply additional water if needed. Most engines carry between 500 and 1,000 gal (1,893 and 3,785 L) of water on board. The pump operator supplying the attack hoseline should monitor the level of water in the tank as the operation proceeds and advise the officer when the tank is down halfway.

At most automobile fires there is sufficient water on board the engine to do the job without connecting to a hydrant. Flowing 100 gpm (378 L/min), for example, yields 5 minutes of continuous water from a 500-gal (1,896 L) tank, assuming that the tank was topped off. The pressure can be lowered after the fire is darkened down to conserve water. However, if a large truck, bus, or electric vehicle is involved, or there are multiple vehicles burning, additional units may be needed to provide extra water and should be requested early if not already assigned. The officer of the first arriving unit should size up the situation and give a radio transmission describing the initial conditions encountered and request additional resources if the incident is beyond the resources of the assigned unit(s).

Here are several circumstances where additional water supplies are likely needed:

- Multiple vehicles involved
- Vehicle fire impinging on a structure
- Fuel spill associated with vehicle fire
- Heavy involvement of a truck or bus
- Master stream foam operations in progress

Having a second source of water is a good practice at vehicle fires so that if something unexpected happens with either the burning vehicle or the pumper, the firefighters can continue operations.

Vehicle size-up and safety

When sizing up the burning vehicle take note of the make, model type, and age. This information can give clues about how the vehicle will burn. Also note what compartment(s) of the vehicle are involved in fire. There are five possibilities:

1. Engine compartment only
2. Passenger compartment only
3. Trunk only
4. Two compartments involved
5. Fully involved

Once the hoseline is charged, any extension into other compartments is unlikely, and the fire can usually be confined to the area of the vehicle involved in fire, especially if the fire is attacked from the unburned side.

Based on where the fire is located, it is possible to quickly assess the potential hazards. For example, at a fire involving the engine compartment, firefighters should be cognizant of mechanisms of injury such as hood pistons, the front bumper, and battery acid. As long as the fire is confined to that compartment, there is no need to worry about injury from an air bag inflator in the passenger compartment or an aerosol can in the trunk. Similarly, a fire confined to the passenger compartment does not affect the energy-absorbing bumpers, and so on. Fires involving two compartments are more hazardous, and fully involved vehicles present the greatest number of hazards and must therefore be approached with the most caution. A more detailed discussion of tactical considerations for each compartment of the vehicle can be found in the following text.

Securing the burning vehicle

An essential preliminary task when extinguishing a vehicle fire is securing the vehicle so that it remains stationary during and after knockdown. Vehicles can start up when burning ignition wires cross. If the vehicle is in gear when this occurs, it can begin to move forward or backward and even drive away. To preclude this the burning vehicle should be immobilized with wheel chocks as soon as it can be done safely (fig. 18–5). When fires result from traffic accidents, the vehicle may be on its side or roof. Such vehicles may have to be stabilized with cribbing or jacks after the fire in knocked down to allow firefighters to overhaul the vehicle safely.

Selecting the right hoseline

FFI 4.3.7 Vehicles burn vigorously. There is usually an unlimited supply of oxygen available to support combustion, and they contain a high concentration of hydrocarbon-based plastic materials throughout. These materials have a high heat-release rate and give off a lot of smoke. The faster the fire is darkened down, the safer the operation will be. That is why a booster line (a rigid hard-rubber hose of ¾- or 1-in. [20- or 25-mm] diameter) that flows only 35 gpm (132 L/min) is inadequate to the task. To provide the needed punch, an 1½- or 1¾-in. (38- or 45-mm) hand line flowing no less than 100 gpm (379 L/min) should be selected for this task.[2]

Fig. 18–5. Wheel chocks. Immobilizing the vehicle is an often overlooked but essential task at vehicle fires. Failure to chock the wheels can result in injuries.

For a 100-ft (30-m) stretch 1¾-in. (45-mm) line, this is a pump discharge pressure of approximately 140 psi (980 kPa) (to supply a 100-psi [700-kPa]–combination nozzle). The pump operator should supply higher pressure for the knockdown but lower the pressure during overhaul to conserve water. In contrast to a booster line, the 1½- and 1¾-in. (38- or 45-mm) lines have the capacity to achieve the quick knockdown you seek.

Note that some pump operators are in the habit of flowing at a rate of 150–180 gpm (568–681 L/min) whenever they charge a 1¾-in. (45-mm) hand line. This is good practice at structure fires; but unless there are unusual circumstances, it is unnecessary to charge the hoseline to 150–180 gpm (568–681 L/min) at vehicle fires. Although more water is usually better, flowing such high volumes of water makes the nozzle harder to control; and more importantly, it drains the tank quickly, potentially leaving insufficient water to complete extinguishment when there is no hydrant available. Things can happen suddenly at vehicle fires and it is important to keep water in reserve in case it is needed.

When there is no life hazard, there is no reason to get close to the vehicle during the initial stages of extinguishment. With a **flash point** of –43°F (–42°C) and an energy-release potential of dynamite, gasoline must always be considered and respected during vehicle fire operations.[3] While flaking out hose and awaiting water, the nozzle crew should remain a safe distance away from the vehicle, out of the traffic lane, uphill, and outside the smoke plume (fig. 18–6).

Fig. 18–6. Firefighters standing at a 45-degree angle and out of the way of the rear bumper use the full length of a straight stream to darken down this jeep. Notice there is fire at road level below the gas tank. It is wise to sweep the street under the vehicle with the hose stream before widening the pattern and advancing closer. (Courtesy of Kevin Imm)

Once water reaches the nozzle and it is bled, the nozzle team should stay back and use the full reach of a straight stream pattern when initiating knock down of the fire (fig. 18–7). To exploit the stream's full reach and maintain visibility, the attack should be initiated from an upwind position if possible, especially if the wind is strong. The nozzle should be kept moving and directed at all sections of the vehicle. Periodically aiming the stream down and banking it off the street allows water to bounce up and cool the undercarriage of the vehicle and gas tank.

Whenever fire is present under the vehicle, it takes priority and should be extinguished before proceeding further. If the nozzle crew is crouching, the stream can be aimed upward and banked off the underside of the roof to create a sprinkler effect on the material burning below. This is preferable to spraying the stream through one window and out the other.

Fig. 18–7. Reach of hose stream. The nozzle team uses the full reach of the hose stream to knock down visible fire before advancing closer to the vehicle.

The right angle of approach

As a rule of thumb, the hoseline should be advanced from the corner of the vehicle at 45°, not directly from the front, back, or sides. This angle of approach keeps the crew away from energy-absorbing bumpers, hood struts, and air bag inflators that can blow during a vehicle fire. The safest direction of approach is from the unburned side at a 45° angle to the corner of the vehicle as illustrated in figure 18–8. Applying a straight stream from the corner at a safe distance allows the nozzle team to sweep two sides of the vehicle before they move in.

Fig. 18–8. Approaching vehicle from safe angle

Energy-absorbing bumpers, pressurized by hydraulic pistons, have caused serious leg injuries to firefighters during extinguishment. Though most were phased out in the 1990s and replaced by polystyrene foam, they are still found on older vehicles and some late models. If exposed to heat, firefighters should not work or stand directly in front of a vehicle bumper until it has been thoroughly cooled (fig. 18–9). Energy-absorbing bumpers can be a greater hazard when the car was in an accident preceding the fire. If the bumper was compressed, it could release without warning.

Fig. 18–9. An older vehicle with an engine compartment fire and fire beneath the front bumper presents a danger to the lower extremities. Do not approach this vehicle from the front. (Courtesy of Chief Michael Urquides, Salinas Rural Fire Protection District)

FFI 4.3.7 The nozzle crew can gauge the success of their fire attack when the flames diminish and the smoke absorbs steam and turns white. When the fire darkens down, and the crew advances, the nozzle pattern should be adjusted from a straight stream to a wider angle and eventually to a medium fog pattern to provide more coverage as well as wider protection to the crew as they advance toward the vehicle (fig. 18–10). A fog pattern also pushes noxious smoke away, redirects flammable liquids, and covers a broader area of the vehicle. The nozzle should be moved continuously so that water is directed over, under, around, and throughout the vehicle giving all involved surfaces a thorough wash.

Fig. 18–10. Wider pattern. As the nozzle crew gets closer to the vehicle, the hose stream pattern is widened.

The crew should take time to continue cooling the vehicle after visible flame has been extinguished and give the undercarriage, passenger compartment, and bumpers a final wash down, before approaching the vehicle closely. The nozzle crew should also direct the stream onto smoldering tires, which can be difficult to extinguish because of the heat retention qualities of the steel belts, which can prompt reignition.

Firefighters often hear loud bangs and pops when fighting a vehicle fire. This is caused by the expansion of liquids or gases in a confined housing. Airbags, tires, steering, air conditioner components, and the like can burst under fire conditions; or one of the numerous struts located throughout the vehicle can rupture when heated beyond its ability to expand. Struts may be found in the wheel wells as well as under hoods, hatchbacks, and trunks (fig. 18–11). These pressurized components can burst or explode when heated. If they contain a hydraulic fluid, it may spray about. If gas filled, the pneumatic strut may launch shrapnel or projectiles that can cause serious injuries. Although most ruptures are confined to the interior of the vehicle, they can penetrate the sheet metal and fly outward with great force, primarily toward the front or rear of the vehicle. Therefore, firefighters should avoid approaching the vehicle closely until it is thoroughly cooled.

Fig. 18–11 Hood strut. This type of pneumatic hood piston, found in a wide variety of passenger cars, can rupture when heated by an engine compartment fire.

Mechanisms of injury in the passenger compartment include seatbelt pretensioners, airbags, and air bag inflators. The number of airbags found in a given vehicle varies with the manufacturer and model. Luxury models often contain more airbags, each of which is activated by a separate inflator. Still, there are other hazards inherent in vehicle fires. Tires can rupture, scattering hot rubber fragments, and should be cooled with the hose stream if on fire. When impinged on by fire, gasoline in a liquid or vapor form can escape under the vehicle from a burst fuel line or loose connection and intensify the fire. This is especially true of fuel-injected engines. Taking the time to cool the vehicle down from a distance prevents these hazards from causing injuries.

Once thoroughly cooled inside and out, the doors can be opened to ventilate the passenger compartment if it is charged with smoke. When the doors are locked and no key is available, a hole punch or the point of a Halligan tool in the lower corner of a side window fractures the tempered glass, allows ventilation, and permits the firefighter to reach in and unlock the door (fig. 18–12).

Members working close to the vehicle should be mindful of hot surfaces and wear their gloves to avoid burn injuries. When venting a vehicle side window, stand to the upwind side and swing the tool to impact the glass in the corner of the window. Once the glass is taken, smoke and flames can push out. Ventilation is frequently unnecessary at vehicle fires, however. Vehicle side windows either self-vent when the fire grows beyond the incipient stage or shatter when the hot glass is hit with the cool water of the hose stream. Never place your head or any part of your body inside the vehicle until it has completely cooled down.

Fig. 18–12. When venting an automobile, the point of a Halligan should be directed into the lower corner of the side window. If the interior is charged with smoke, the firefighter should stay low and to the side when the glass is taken.

The gas tank

Although automobile models such as older Crown Victorias and 1970s-era Ford Pintos (whose fuel tanks were mounted aft of the rear axle) were notorious for erupting into flames following rear-impact collisions, exploding gas tanks are not common during auto fires. This is primarily due to the position of the gas tank. It is mounted on the underside of the car. At automobile fires where the vehicle is upright, the fire is predominantly above the tank with limited heat radiating downward. Nonetheless, whenever flammable vapors are present it is necessary to consider the possibility of a failure and not take careless risks. Hose crews extinguishing vehicle fires should be mindful of the location of the gas tank and protect it from any fire in the rear of the vehicle impinging on it. Flames visible at the rear tires or under the car can heat the gas tank from below and should be doused quickly; and the hose team should keep back a safe distance until the fire under the vehicle is knocked down. **High-density polyethylene (HDPE)** plastic fuel tanks are becoming the norm on newer cars. They are lighter than metal tanks and not subject to rust. However, plastic fuel tanks can fail under conditions of direct flame impingement and release their contents. If this occurs, vapors can flash, or a stream of flaming gasoline can flow out from the vehicle and downhill, spreading the fire to other vehicles or entering a sewer.

A gas tank failure requires the quick application of a Class B extinguishing agent to control. A dry chemical or Class B foam extinguisher should be kept at the ready during vehicle fire operations incase of this eventuality. As mentioned earlier, sweeping and/or banking the hose stream under the vehicle at street level whenever fire is observed beneath the vehicle can impede direct flame impingement and prevent the gas tank or fuel line from failing.

Years ago, many firefighters believed that removing the gas cap could prevent a gas tank explosion by relieving the pressure in the fuel system. This is incorrect and should not be attempted. Although it is true that pressure increases within the tank as it is heated, especially if the fuel tank is nearly empty, the air/fuel mixture in the tank is too rich to burn. Attempting to relieve pressure by removing the fuel cap can release pressurized gasoline vapor that could burn the firefighter. The proper solution is to put the fire out at a safe distance using the full length of the hose stream, and to cool the tank and undercarriage as you do so.

Foam

In contrast to gas tank failures, fuel spills result more frequently from compromised fuel lines. Fuel lines can rupture as the result of an accident or flame impingement from below. They also frequently fail when hot catalytic converters ignite leaves and grass below the vehicle when drivers carelessly park on top of these materials. Using foam or dry chemical on the leaking fuel line is necessary until the fire is out, all sources of ignition are controlled, and the leak can be plugged. Most engine companies carry a complement of foam (in 5-gal [19-L] containers) on the apparatus along with a foam nozzle and eductor. If escaping flammable liquids are involved or there is a spill, the pump operator should get the foam equipment ready for operation.

A growing number of fire departments routinely use **Class A foam** at vehicle fires and report successful results particularly when tires are involved. They use it because they feel it is more efficient than plain water in effecting a quick knockdown. Class A foam is a surfactant that lowers the surface tension of the water molecules allowing superior penetration into dense Class A fuels. When using Class A foam at a vehicle fire, it is important to note that it does not suppress flammable liquid vapors and is ineffective in fighting fuel spill fires sometimes associated with the burning vehicle.

Class A foams do not meet national standards for use with liquid hydrocarbon fires. In contrast to the penetrating qualities of Class A foam, Class B foam floats on top of the liquid **hydrocarbon fuel**, creating a barrier between the fuel and atmospheric oxygen, thus curtailing combustion. A Class B foam blanket must suppress fuel vapors for at least 15 minutes without reapplication. If burning liquids are associated with an automobile fire, Class A foam may achieve an initial knockdown of flames. But beware: In many cases, it does not suppress flammable vapors long enough for complete extinguishment and reignition is possible until an impervious blanket of Class B foam is applied on top of the Class A foam.

When using Class B foam, some departments find that the foam blanket is breaking down at a faster rate than in the past. In most cases this is caused by the new ethanol-blended gasoline that is incorporating alcohol in varying mixtures. The solution is to use alcohol-resistant (AR) foam.

The hood and engine compartment

SKILL DRILL

Opening the hood and checking the **engine compartment** are standard procedure at vehicle fires. Opening the hood is necessary at automobile fires for two reasons:

1. To complete extinguishment of the fire
2. To disconnect the battery cable to prevent a reignition

However, the engine compartment contains several potentially hazardous components and firefighters working in and around the engine must wear full protective gear including SCBA, eye protection, and gloves to prevent burn injuries. Members approaching the vehicle should avoid walking in front of a heated bumper and be cognizant of hot surfaces.

The first hazard of the engine compartment is the battery which can be found in various locations under the hood depending on the manufacturer and model. In addition to an electric charge, it contains sulfuric acid, a corrosive that can cause burns to the skin and eyes. Under fire conditions, this sulfuric acid can produce hydrogen gas as the lead-acid battery is heated. Hydrogen gas is highly flammable and can rapidly accelerate any existing fire conditions. Hydrogen flames burn clear and are easily missed. In addition to the battery, pneumatic hood pistons (or gas struts) located in the back corners of the hood can burst, separate, and become forceful projectiles. Firefighters are at risk of becoming impaled with these pistons when they explode, especially if they walk in front of the vehicle.[4] There are also several flammable/combustible liquids contained within the engine compartment. These include gasoline, engine oil, brake fluid, and transmission fluid which can add to rapid burning and reignition. Radiator and air conditioner coolant also present and may be under pressure when heated. Magnesium engine components intensify the fire when water is applied. For these reasons, members should be safety conscious when working in and around this area. Firefighters should not be in a rush to open the hood if the engine compartment was involved, and be careful where they position themselves when working in this area. Although opening the hood is required, it should be postponed until the engine compartment is thoroughly cooled.

Fires in the engine compartment are difficult to completely extinguish until the hood is opened. Bouncing the stream on the street below the engine compartment or directing it though a headlight opening may affect partial knockdown but rarely produces complete extinguishment. An alternative method of cooling the engine compartment is to use a Navy nozzle with a bent applicator to spray the engine from below. It can also be used to cool the gas tank. When using this technique to cool the engine compartment, the firefighter should stand at the side of the vehicle and position the applicator under the wheel well so that the stream contacts the engine from below. A piercing nozzle, driven through the hood or fender can also be effective. If driven through the fender, be sure that it penetrates not only the outer but also the inner fender wall.

If a piercing nozzle is unavailable, a firefighter standing along side a fender can drive the point of the Halligan into the side of the hood 4–12 in. (100–305 mm) from the edge. Prying up raises a small section of the hood and creates an opening through which water can be directed into the engine area. This operation should be done from the side of the vehicle, not from the front as illustrated in figures 18–13 and 8–14. Only a member in full protective gear with the nozzle alongside should attempt this procedure, because flames can issue out from under the hood as soon as the opening is pried up. Directing the stream through this hole helps cool the hood struts and generates steam while the hood remains closed. It should be left down long enough for the steam to do its job. The advantage of this method is that the line does not have to be shut down to attach a special appliance.

Fig. 18–13. This procedure begins by driving the point of the Halligan into the hood. The point should be set well before prying up the edge.

Fig. 18–14. Once the point is set, the Halligan is pried up to a vertical position. This creates a triangular opening along the edge of the hood through which the hose stream is directed to darken down fire beneath the hood.

Regardless of which method is used to effect knockdown, the hood should remain closed until the engine compartment has been cooled. It is then safe to open the hood to achieve complete extinguishment. Until this occurs, internal wiring and combustible parts such as the air filter element continue to smolder. A layer of insulation affixed to the bottom of the hood commonly drops and blankets the burning engine components, requiring removal before extinguishment can be completed. To do this, the hood must be raised.

Opening the hood. Once the engine compartment and front bumper have been thoroughly cooled, efforts can be made to find the hood release whose cable runs under the hood along the driver's side. Fires that cause heavy involvement to the passenger and/or engine compartments usually damage or destroy the hood release cable, making the job of opening the hood more complicated. An attempt to locate and pull the cable should begin where it is usually located, in front of the driver's seat. If only the plastic handle is damaged, the cable can sometimes be activated by a tug with vise grips. If this fails, an attempt to locate and snag the cable under the hood can be made (fig. 18–15). Remember that depending on the manufacturer, the hood may either open from the front or tilt forward from the back (by the windshield). Autos whose hoods tilt forward include Saabs, BMWs, Corvettes, and some imported sports cars such as the Fiat and Alpha Romeo.

Fig. 18–15. Skillful use of a hooked tool can snag a hood-release cable where it passes along the front of the engine compartment. (Courtesy of Dave Walsh)

On standard hoods, prying from the front is typically ineffective. Leverage is limited in this area because grills are constructed of plastic or fiberglass and collapse long before the latch gives way. Moreover, when operating in front of the vehicle, care must be taken not to inadvertently jam the tool into the radiator or battery in search of the hood latch. Firefighters have gotten hot coolant in their faces when they inadvertently punctured the radiator while attempting to gain a purchase. As an option, the plastic grill can be removed to look for the hood release cable. If the cable can be located along the front of the radiator or driver's side of the hood, the fork of the Halligan can be inserted and twisted until it releases the latch (fig. 18–16). However in many newer vehicles, the cable runs inside the radiator to prevent theft and is not very accessible.

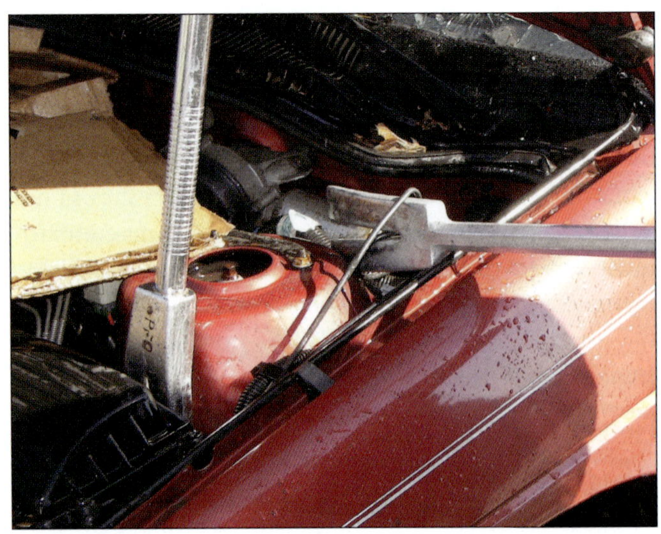

Fig. 18–16. If the fork of the Halligan can be slid under the cable, it can be twisted to release the hood latch

Another option is to snap off the hood latch staple, which is attached to the bottom of the hood. The hood latch staple can be twisted off with the fork of the Halligan tool. This is done by carefully sliding the curved surface of the Halligan's fork along the bottom surface of the hood at its centerline. This is done by a firefighter from a low angle pushing the fork of the Halligan in the direction of the windshield. Once in position, the Halligan is struck with a flat-head axe or sledge by another firefighter until the fork end is set around both the front and back legs of the staple (fig. 18–17). The Halligan is then rotated either clockwise or counterclockwise to snap the staple off.

Fig. 18–17. Instead of prying, the Halligan's fork can be driven over the hood staple and then rotated to snap it off

Another method that many departments have found effective is to use a rotary saw with metal cutting blade to make a V-cut in the front of the hood over the latch mechanism. This cut can be made quickly and takes the latch pin and staple out of the equation (fig. 18–18).

Whenever fire is present in the engine compartment, opening the hood introduces oxygen that could cause the fire to flare-up. Firefighters should thus stay low when lifting the hood, use eye protection, and have the hoseline ready. Once the hood has been raised, it should be propped open with a tool while the engine compartment is washed down (fig. 18–19). Do not rely on heat-damaged hinges or pistons to hold up the hood. Any smoldering material should be removed and soaked.

Finally, the battery cable should be cut at the negative terminal with a bolt cutter. Cutting the cable to the negative terminal does not produce a spark. However, cutting the positive terminal could create a spark that in turn might ignite hydrogen vapors. Similarly, crew members working with metal tools must be careful not to bridge the positive terminal to a metallic surface. This may produce a spark as well. Disabling the battery prevents rekindle, renders the air bags inoperable, and keeps the car from starting.

Fig. 18–18. Firefighter completes a V-cut around the latch mechanism

Fig. 18–19. Once the hood has been raised, it should be propped in open with a tool. A Halligan or 6-ft (1.8-m) hook serves the purpose and keeps the hood from closing while the engine compartment is overhauled and washed down.

The passenger compartment

SKILL DRILL

Separated from the engine compartment by the firewall, the **passenger compartment** burns vigorously once vented. The foam seats, padded dash, vinyl trim, and the wire harness give off a great deal of smoke and produce a substantial fuel load. There are several hazards found in the passenger compartment that can injure you, so avoid placing your head inside the passenger compartment. When heated, air bags can deploy without warning. Similarly, air bag inflators, mounted in steering wheel hubs; dashboards; seats; A, B, and C posts (fig. 18–20); and doors, may explode under fire conditions. These canisters or cylinders vary in size and contain compressed gases (usually argon or nitrogen) under high pressures. Under normal conditions they are designed to discharge and inflate the air bag during collisions. As a safeguard, many compressed gas canisters are designed to activate and release the compressed gas at temperatures ranging from 300°F to 350°F (149°C to 177°C).[5] Thus, if the passenger compartment is well involved, these canisters should discharge their contents before the fire department arrives to make firefighter injury less likely. However, there have been cases where these compressed gas cylinders have experienced a BLEVE (boiling liquid expanding vapor explosion) while the fire department is on the scene, even after the fire has been extinguished. When they fail, they can discharge fragments of shrapnel under pressure. To minimize this risk, it is important to continue thoroughly cooling all areas of the passenger compartment after the fire has been darkened down and extinguished. The cooling hose stream should be directed at the steering wheel hub, dash, side posts, and seats where the driver, passenger, and side-impact air bags are located. Close approach to the passenger compartment should be postponed until all areas have been thoroughly cooled.

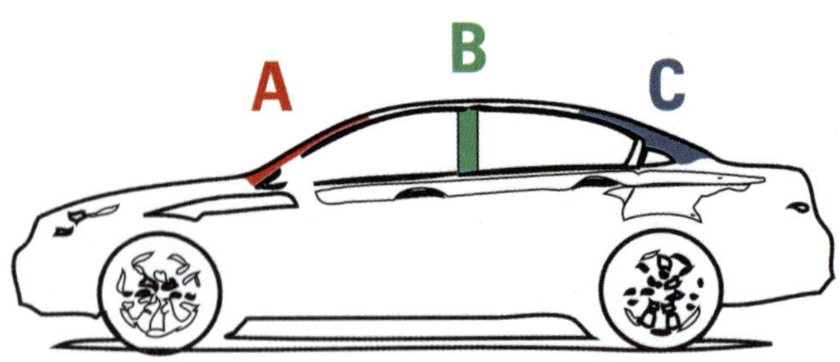

Fig. 18–20. The A, B, and C posts of a passenger vehicle

An occupied vehicle on fire is another matter. Although most occupants self-evacuate at the first sign of smoke, the passenger area should be checked for victims if there is evidence that passengers are still in the vehicle. Victims are more likely to be trapped in a burning vehicle if the fire is the result of a collision. If victims are trapped, immediately direct a hose stream between the fire and the passengers to drive the fire away from them. When there is a life hazard, a second hoseline may be deployed to provide additional protection to victims in the vehicle (fig. 18–21). An immediate rescue operation must be started and coordinated with water delivery. The doors must be opened and the victims removed. The passenger compartment may also have to be vented promptly. Life takes preference over other priorities on the fireground, and a victim trapped in a burning vehicle is a true emergency.

Fig. 18–21. A second hoseline may be deployed to provide additional protection to victims in the vehicle. (Courtesy of Chris Saraceno)

The trunk

SKILL DRILL

When fighting fires involving the **trunk** of an automobile, it is important to cool the gas tank and shield it from heat and flame with the hose stream. Trunk fires can be darkened down by approaching from the front at a 45° angle and directing a hose stream through the back seat if it has been burned away. Trunk fires can also be knocked down by directing the stream through a taillight opening. If fire cannot be controlled through these avenues, the point of a Halligan can be driven into the trunk a few inches from the edge of the trunk lid and pried up in a manner similar to that used with the hood; an alternative to this method is to attach a piercing nozzle to the hose. The nozzle is then driven through the trunk lid, taillight, or fender and directed into the trunk. Before the job is complete, the trunk must be opened, overhauled, and checked thoroughly. However, as with fires below the hood, this must be done carefully and only after it has been thoroughly cooled from a safe distance. The contents of the trunk can cause serious injury.

Motorists transport a wide variety of articles in the truck that can injure firefighters. A person returning home from the grocery store may be carrying a shopping bag containing aerosol cans. Aerosols can explode when heated, releasing flammable or corrosive propellants. A portable gas tank for the lawn mower can rupture and add flammable liquid to the fire. A 20-lb (9-kg) propane cylinder placed in the trunk to be refilled can cause a BLEVE. Beneath the trunk is the gas tank and fuel line. Therefore the trunk should not be approached and opened until the fire has been knocked down, the gas tank has been washed from below the vehicle, and the rear bumpers have been thoroughly cooled. Whenever the driver is present, it is a good idea to ascertain the contents of the trunk before crews get too close.

Opening the trunk can present entry difficulties similar to that of opening the hood. First, try the trunk release, which is usually located next to the driver's seat. If the driver is present, get the key. When these options are unavailable, the trunk lock cylinder can often be pulled revealing either a stem or cam. A long screwdriver or key tool can then be inserted into the locking mechanism and turned to open the trunk using the through-the-lock method. If nothing else is available, a dip stick can be used for this purpose.

Mopping up

Once the fire is extinguished, conduct a thorough search of all areas of the vehicle to confirm that the fire is completely extinguished and make sure a crime has not been committed. If flammable liquids have been spilled, containment efforts should be made to prevent them from entering sewers and waterways. The officer in charge is responsible for recording all information that can identify the vehicle and its owner. The make, model, license plate, and **vehicle identification number (VIN)** must be written down for inclusion in the incident report and given to the police for validation. They are also needed for the NFIRS report. The VIN is commonly listed on the registration sticker mounted on the windshield. If this is destroyed, look for a stamped plate affixed to the front of the dash beneath the windshield on the driver's side. Take care to avoid damaging the contents of the glove box during overhaul. If the vehicle registration, insurance card, and other documents are undamaged in the glove box, give them to the owner or police.

Firefighters should follow their local protocol relative to having a fire investigator examine the vehicle. Some jurisdictions require that all vehicle fires are investigated, whereas others do not. However, if the fire occurred in a remote area after dark with no driver present, the circumstances are suspicious. Correspondingly, if the fire originated in the passenger compartment, a gasoline can is found in or near the vehicle, or the fire otherwise appears to be incendiary in nature, notify a fire investigator to respond. Once an investigator is called, at least one fire department or law enforcement unit must remain on the scene until the fire investigator arrives. As with any fire, the area of fire origin must remain undisturbed as much as possible, commensurate with complete extinguishment.

Magnesium vehicle components

Endeavoring to improve gas mileage and reduce carbon emissions, vehicle manufacturers are beginning to use magnesium in their engine and body components because of its reduced weight (approximately 25% of steel). Magnesium is currently used in valve, cam, and head covers as well as steering wheels, gear boxes, instrument panels, seat components, and wheels. Future uses may include engine blocks, engine cradles, transmission cases, intake manifolds, oil pans, and water pump housings.[6] Vehicles that already have magnesium engine blocks include 1970s-era Volkswagen Beetles, Porsches, and late-model BMWs and Corvettes. Cadillacs currently incorporate magnesium instrument panels, and Audi is integrating magnesium into its valve covers and power train components.[7] Future uses may include more engine blocks, engine cradles, transmission cases, intake manifolds, oil pans, and water pump housings.

When magnesium is involved in an automobile fire, firefighters see a bright white flame and the burning magnesium showers and scatters white sparks when water contacts it. The brilliant flame appears similar to a welder's arc or an arcing electrical wire and can be mistaken for a live wire on the vehicle. Firefighters should avoid staring at the luminous flashes and operate at a safe distance, away from where they may be struck by a piece of molten magnesium.

An engine compartment fire that involves magnesium can stubbornly resist extinguishment with water and flare-up, increasing in magnitude as soon as water contacts it. This is in part because magnesium burns so hot (7,000°F [3,814°C] or more); that it can chemically separate the hydrogen and oxygen in the water, negating its cooling properties and causing the fire to increase in intensity. Industrial magnesium fires are usually controlled with plentiful quantities of a Class D agent such as G-1 powder, soda ash, rock salt, or dry sand, applied in an effort to smother the fire. If a Class D extinguisher is kept on the apparatus, it can be applied to the fire. Only members trained in the application of Class D agents, wearing dry gear, and eye protection should be involved in the extinguishment process. If you do not have sufficient quantities, the best option may be to suspend water application, keep onlookers away, protect exposures, and let the fire burn itself out.

Extinguishment of other sections of the vehicle may continue, with care given not to spray the burning magnesium. Some sources have stated that magnesium fires can be extinguished with copious amounts of water. However, in many cases *copious* means far more water than a 1¾-in. (45-mm) hoseline provides. A hand line is often ineffectual unless the magnesium has been burning for some time and is nearly consumed. Apart from the manufacturers mentioned previously, magnesium auto parts are expensive and not used widely at this time. For the most part, magnesium use is still limited to small parts that are consumed quickly in the fire. Consequently, they have not presented a wide-ranging extinguishment problem to date.

Hybrid electric vehicles (HEV)

Hybrid, plug-in hybrid, electric (EV), and alternative fueled vehicles are becoming more and more common in the United States as automotive technology seeks to increase fuel efficiency and reduce carbon emissions. To date, more than 4 million hybrid electric vehicles have been sold in the United States. They have become common on the roads and are particularly popular in taxi fleets. Tesla, Nissan, and Chevrolet are producing purely electric vehicles, and hydrogen fuel cell–powered vehicles may be on the horizon. Hybrid vehicles are the most common and combine a standard gasoline-fueled engine with a high voltage (HV) battery-driven electric motor that recharges when the vehicle is in motion via kinetic energy or generator. The first generation hybrids featured HV **nickel metal hydride (NiMH)** direct current (DC) dry cell battery packs to supply current to the 165 horsepower electric motors. Newer hybrids have replaced the NIMH battery packs with more efficient Lithium ion battery packs (HV Li-ion) that store energy more efficiently. The more purely electric the vehicle, the larger the battery pack. Although these HV Li-ion batteries have been adapted from those used in cell phones and laptops, to date, they have not shown the same propensity to catch fire as these other devices. Responding units should determine if the vehicle is hybrid or electric during size-up. It should also be noted that hybrid vehicles are silent when stopped. The gasoline engine shuts off to save energy. Therefore, it is not possible to tell if they are on or off simply by listening to the engine. This makes stabilizing the vehicle with chocks all the more essential.

Although the Toyota Prius and Honda Insight have a distinctive hybrid look, many other hybrid models have the same body type as standard vehicles. The only distinguishing characteristic for these hybrids may be a small logo on the vehicle's trunk, hatchback, front fenders, or doors (fig. 18–22).

Fig. 18–22. Placards like this may not be visible if obscured by smoke

To date, hybrid and electric vehicles have not been shown to be more hazardous or fire prone than conventional vehicles. However, high voltage (HV) battery packs do present additional tactical issues if they become involved in fire. Fires in these vehicles can be fought with water by conventional means, but if the HV battery becomes involved, significantly more water may be needed beyond the standard capacity of onboard water tanks, creating the need to secure an additional water source. Hybrid manufacturers have published emergency response guides that can be accessed via the Internet. Firefighters should review these response guides to familiarize themselves with the special systems and hazards associated with these vehicles during extrication and extinguishment operations.

As mentioned, in addition to an internal-combustion engine and standard 12-volt electrical system, hybrids also contain specialized high voltage electrical systems that include HV batteries, HV cables, and generators. The battery packs are usually located in the rear of the vehicle or under or behind the back seat and enclosed in a metal housing. Power is transmitted to the electric motor under the hood through high-voltage cables that run beneath the floor boards. Situated on the underside of the vehicle, these orange high-voltage cables are encased in conduits, many of which are black and either partially or totally cover the cables, making identification based strictly on the orange cables difficult. Do not cut, crush, or touch these high-voltage cables during extinguishment, overhaul, or extrication; and avoid direct contact with the high-voltage battery pack itself.

Although there are high voltages involved, there are no warnings advising against using water to extinguish fires in these vehicles. That is because the voltages are DC and therefore do not travel up a hose stream seeking a path to ground. Several hybrids utilize alternating current (AC) system components as well, but they are protected by ground fault interrupt circuits (GFICs). Nor is there danger of electric shock from touching the car body or chassis even if it is partially or wholly submerged. However, when overhauling or using metal tools, crew members must use caution not to create a bridge between the negative and positive sides of the circuit.

Battery pack fires

HV battery packs can catch fire either by exposure to fire originating in another part of the vehicle, physical impact from a crash, or by thermal runaway within the battery itself due to overcharging. HV battery involvement is often accompanied by a hissing or pressure release sound. A standard fire attack can be made on hybrid and electric vehicles with one caveat: fires involving battery packs will require more water. Applying water is less effective because the HV battery pack shielding inhibits water penetration, even when hose streams are directed through vents. However, the housing and adjacent areas must be cooled with water. This deflection of water will require additional extinguishment time. Patience is also required because HV battery pack fires can be subject to rekindle even after a significant amount of time has passed. Popping or whooshing sounds or sparking indicates reignition is imminent. Hybrid vehicle batteries are typically located behind the back seat and may require opening the trunk to directly apply the hose stream to the battery pack. HV battery packs in electric vehicles are larger and typically located below the floor boards and are therefore more difficult to reach. Once the fire is out, continue to apply water to cool the battery pack inside the housing. Make no attempt to disassemble, pierce, or detach the metal cover to overhaul the HV battery pack. Such action could cause sparking or bring you into contact with high voltage. Puncturing the battery pack can also lead to thermal runaway. Continue to monitor the HV battery pack for signs of reignition. Use of a thermal imaging camera can indicate if the battery pack is cooling or reheating. Continue to protect exposures, and do not have the vehicle removed from the scene until the battery has cooled. As with any vehicle fire, the use of SCBA is mandatory. Burning HV batteries may produce vapors containing heavy metals and can explode or discharge sparks. A fire involving the HV battery may present an environmental hazard and require specialized disposal by a battery recycler.

Summary

1. If the HV battery pack is not involved, handle the fire like a standard vehicle.
2. If the HV battery pack is involved, secure a water supply to avoid running out of water.
3. Consider all HV components to be energized.
4. Avoid contact with any HV cables or HV battery packs, and do not attempt to overhaul them.
5. Allow time for the HV battery pack to cool after extinguishment and monitor it for signs of rekindle.

Alternative fuel vehicles

FFI 4.3.7 Alternative fuel vehicles can present a more serious hazard. In an effort to be more fuel efficient, some municipalities are purchasing fleets of vehicles powered by **compressed natural gas (CNG)**, **liquefied petroleum gas (LPG)**, **liquid natural gas (LNG)** or hydrogen (H). Such vehicles are often found at airports where taxi fleets and airport shuttles are converting to these cost-saving, clean-burning fuels. They may also be found on city streets where city owned vehicles may be powered by alternative fuels. These vehicles can be any type, ranging from small passenger vehicles, to jitneys, to full-size city busses. United Parcel Service (UPS) is powering many of its new delivery trucks with propane. Ford, Toyota, and Honda now offer some auto and truck models with alternative fuel options, the Honda Civic being the most prevalent. These vehicles have the same body type as standard gasoline fuel models, and responding firefighters may not suspect an alternatively fueled vehicle except for a small placard on the trunk lid or rear bumper (fig. 18–23). Aftermarket conversion kits are becoming more readily available to convert gasoline engines to run on alternative fuels such as ethanol, methanol, CNG, and LPG as well; and these my not display a placard at all.

Fig. 18–23. Alternative fuel vehicles display a small placard on the rear bumper or trunk lid. Under fire conditions, it may or may not be visible to responding firefighters.

Because of their chemical nature, these fuels are compressed into cylinders mounted horizontally inside the vehicle with straps in much the same way a propane cylinder is mounted to a fork lift (fig. 18–24). Although not more fire prone than ordinary vehicles, they can explode when exposed to heat, causing catastrophic damage (fig. 18–25). When operating at a fire involving one or more of these vehicles it is important to look for the placard, get information from the driver, and determine the location and extent of the fire during initial size-up. As with all automobile fires, approach the vehicle at a 45° angle and keep a safe distance from the vehicle until the fire is controlled. Although it is impossible to predict with certainty the direction the blown cylinder will travel, it is likely to blow out from the side that is exposed to fire. Thus, for example, if the fire is coming from the front of the vehicle, the front of the cylinder will likely fail first, propelling the cylinder toward the rear, depending on how the retaining straps fail (fig. 18–26).

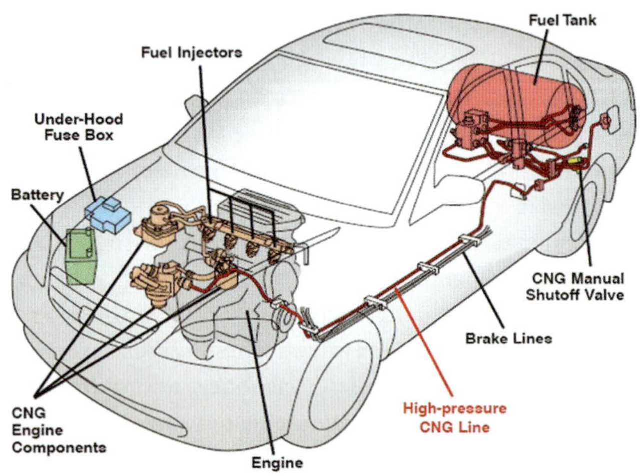

Fig. 18–24. The compressed gas cylinder is mounted in a compartment between the back seat and the trunk in most CNG vehicles. This compartment may or may not contain adequate insulation from fire. (Courtesy of the Seattle Fire Department and Honda Motor Co.)

Fig. 18–25. The aftermath of a fire in a CNG Honda Civic in which the fuel storage cylinder exploded. Note the extreme destruction to the vehicle. Fire spread to several nearby vehicles as a result of the blast. (Courtesy of the Seattle Fire Department)

Fig. 18–26. What remained of the fiberglass wrapped CNG cylinder from vehicle shown in fig. 18–25, which was propelled 100 feet to the rear of the vehicle when it exploded. (Courtesy of the Seattle Fire Department)

The fire attack varies depending on the situation found during size-up. If the fire is limited to a part of the vehicle that is separate from the fuel storage cylinder, an attempt can be made to confine the fire to the area of origin and cool the tank(s) if accessible. Preventing the fire from reaching the tanks prevents the situation from escalating. If the fire is already involving the tank area and the tank(s) is venting, a catastrophic failure may be imminent, and all personnel should withdraw from the area. If it can be done safely, cool the tank from a safe distance, protect exposures, and allow the fuel to burn off. Avoid extinguishing the flames venting from the cylinder relief valve. This could lead to vapor release and explosion. Always consider risk-benefit analysis and avoid approaching these vehicles closely unless a rescue has to be made or they have been thoroughly cooled. The Seattle Fire Department has put together an excellent PowerPoint presentation on the subject of CNG vehicle fires on their Web site.[10]

Ladder company functions at vehicle fires

Although most fire districts dispatch a ladder or rescue to fires involving school buses or larger vehicles, many assign only an engine to respond to standard vehicle fires. Although once thought to be the purview of a single-engine company exclusively, some districts and departments find that the addition of a ladder company or a second engine to perform support duties benefits safety and extinguishment. Whether an additional unit is sent or not, these support tasks must be performed by the crew members on the scene.

On arrival, the initial function of the ladder, second-engine, or other support unit is to provide for scene safety as described previously, by parking the vehicle in a position that creates a safe working area, controlling traffic, and setting cones and other warning devices as needed. On a steep curve or grade, the additional company can be located well upstream of the curve to warn oncoming traffic that might be unable to see the engine. Scene safety also includes lighting the scene, controlling any hazardous material or flammable spills, and keeping crews and the public away from downed wires and other hazards. The ladder can be used to stretch a second hoseline for exposure protection, extinguishment, or crew safety.

When no life hazard is present, the ladder or support company vents the vehicle and provides entry into the hood, passenger compartment, and trunk for overhaul. These operations should be postponed until all areas have been extinguished and cooled thoroughly. Opening up the hood and trunk often requires particular skills, and specialized tools may be needed to gain access to some of these areas. Once the vehicle is opened up, a thorough secondary search is conducted to check for extension or victims.

Vehicle search by the ladder

Most vehicle fires are *not* associated with a collision, and if passengers are in the vehicle at the time the fire occurs, they pull over and self-evacuate. Therefore at most vehicle fires, life is not a factor, there is no rescue emergency and extinguishment is the primary assignment. However, in some instances firefighters may have to initiate a search of the vehicle on arrival at the scene. For example, a truck driver could be trapped in the cab when his 18 wheeler jackknifes. A driver who hit a tree could be unconscious behind the wheel when a spark from the battery cable ignites a fire in the engine compartment. Children could be overcome by smoke from an engine fire in a school bus. Whenever there is a serious accident that results in fire, a rescue emergency could exist. If there are reports of a victim within a smoldering vehicle, or if the scene survey indicates that not all passengers are out of the vehicle, rapid water delivery coordinated with a primary search of the vehicle is necessary.

This task is delegated to the ladder company on the scene while the hoseline is charged. Searching a vehicle is a hazardous maneuver because of the speed with which vehicles burn and the many hazards associated with the passenger compartment of a burning vehicle as described earlier. Speed and water limit the risk. Crew members involved in making a primary search/rescue must be in full protective gear, including SCBA, before initiating a search of the passenger compartment. In an auto, searches are done rapidly. Keep low (below the level of an air bag deployment), open the door, and reach across the seat. Search the front and back seats under the protection of a hose stream. If a victim is in the vehicle, rapid water delivery is vital and the stream must be placed between the occupant and the fire. If the fire has already entered the passenger compartment, a fog stream should be directed there immediately, to lessen the effects of heat and smoke while the victims are being freed. A seat-belt cutter helps free a victim.

Conducting a primary search inside the vehicle is not a routine procedure unless a victim is present. Because of the danger of a sudden air bag deployment, firefighters should keep their head and torso outside the vehicle unless they are engaged in a rescue procedure.

If the fire is associated with a crash or any vehicle occupants cannot be accounted for, initiate a secondary search of the scene on the interior and exterior of the vehicle once the fire is out. At night, a thermal imaging camera can be used to look for injured persons who may have been thrown from the vehicle.

TRUCKS AND LARGER VEHICLE FIRES

Fires in trucks and over-the-road vehicles can present distinctive challenges to responding firefighters depending on the cargo, size, and type of vehicle (fig. 18–27). Fires can occur in the engine, cab, cargo, or wheel areas. A frequent cause of truck fires is overheated brakes or wheel bearings that can ignite tires or grease deposits under the trailer. When responding to a truck fire, it is essential to determine if the truck is carrying hazardous materials before committing resources. When no hazmats are involved, the standard strategy is to rescue any passengers in the cab, confine the fire to the area or origin, and prevent extension to the rest of the vehicle. If the truck has been involved in an accident, or has jackknifed, attention must be given to the saddle tanks beneath the cab. These tanks contain diesel fuel. Although less volatile than gasoline, diesel leaking from a saddle tank should be covered with Class B foam to prevent ignition of vapors.

Fig. 18–27. A truck fire presents a heavier fire load than an automobile fire. This fire involved a pole and transformer. Downed wires are an important size-up consideration at any vehicle fire on a city street.

In general, the same guidelines about protective equipment, apparatus placement, and hose stream application described earlier apply to standard truck fires (trucks not carrying hazmats) with the consideration that more water may be needed to control the fire, requiring additional companies, tanker shuttles, or a potential relay if hydrants are unavailable. Class A foam helps penetrate layers of densely packed cargo. Overhaul can be effort-intensive, when the entire contents of a tractor trailer or garbage truck must be separated and hosed down. Ground ladders and power saws with metal cutting blades may be needed to open access holes in the side of the trailer. Members should avoid working on the interior of semi-trailers or garbage trucks, however. A truck fire is not a place for an aggressive interior attack. In some cases, the burning product may have to be pulled out onto the roadway so firefighters are able to hose it down with water or foam. This is not an ideal situation, because such an operation will tie up traffic for several hours. If it can be done safely once the fire is knocked down, the truck can be towed to a parking area, for overhaul and final extinguishment.

The hazards associated with truck fires vary with the type of cargo the truck is carrying. An immediate effort must be made to determine the contents of the truck. The best source of this information is the driver. Is the truck carrying contents that are hazardous, flammable, or water reactive? Is there a mixed cargo that could interact when exposed to heat? Over-the-road trucks are required by the U.S. Department of Transportation (DOT) to display placards on all four sides of the vehicle if they are carrying hazardous cargo. These placards reveal the general chemical properties of the cargo. Any placarding information should be cross referenced to a hazmat guide on the fire apparatus or with computer-aided management of emergency operations (CAMEO), to determine if any precautions must be taken. The shape of the truck can also tell responding firefighters a lot about what is being carried. Even from a distance, a standard trailer looks vastly different from a gasoline tanker; and a tanker, in turn, is shaped differently than a propane truck, and so forth.

Fires in large recreational vehicles (RVs), motor homes, mobile lunch wagons, and campers should be fought cautiously if there is a significant amount of fire. It is important to establish a sufficient water supply to control the fire. These vehicles may contain standard voltage AC-powered water heaters and refrigeration systems powered either by inverters or, if stationary, external hookups. LPG-powered heaters, refrigerators, and stoves are also found on these vehicles and their cylinders are subject to BLEVE under heavy fire conditions. Standards for design of recreational vehicles can be found in NFPA 1192, *Standard on Recreational Vehicles*.

Gasoline tanker fires

Each day in the United States, gasoline tanker trucks make approximately 50,000 trips delivering gasoline from distribution terminals to local gas stations.[11] Tanker fires can occur on almost any highway in any community, regardless of size (fig. 18–28). Notable gasoline tanker fires have occurred in recent years in as diverse locations as Emeryville, California; Branson, Missouri; Yonkers, New York; Chicopee, Massachusetts; and Washington, DC, to name a few. These fires present a set of challenges that can greatly challenge the capacity of any department.

Gasoline tanker fires can threaten exposures, damage infrastructure, endanger the environment, and create a severe life hazard. Any fire department that has a gas station or interstate within its boundaries should consider developing guidelines on how to handle a tanker fire. Will a tanker fire overtax the capacity of the Class B foam supply? Are water and foam tenders available? Does your department or a neighboring department have a master stream foam capability? If not, is there a mutual aid agreement with a nearby airport or military base? Does the department have enough diking materials to contain a significant gasoline spill?

Fig. 18–28. This fire on an interstate originated in the engine compartment. Firefighters from Washington, DC, and Prince George's County, Maryland, extinguished it with hand lines before it spread to the cargo tank which contained 8,000 gallons of gasoline. (Courtesy of Mark E. Brady)

Gasoline tankers can carry up to 9,000 gal (3,407 L) of fuel when full. Their tanks are cylindrically shaped and approximately 40 ft (12 m) long. The ends are elliptical according to MC-306/DOT—406 specifications (fig. 18–29). Knowing the shape of the various DOT vehicles gives the responding firefighters an idea of the potential cargo from a distance even when the placard is not visible. Gasoline tankers have an aluminum exterior shell and are placarded with the number 1203. Because of the potential hazards, tankers are designed under DOT standards with numerous safety mechanisms designed to reduce the hazard of fire or explosion during transportation. Some of these safety devices include automatic heat-activated vents; vapor-recovery piping; pressure-relief valves; baffling to subdivide the cargo tank into separate compartments; and a reinforced aluminum exterior shell, which melts under fire conditions, thus preventing a BLEVE. Fires in gasoline tankers may occur during filling and offloading operations. However, most fires are the result of roadway accidents that may compromise these safety mechanisms.

Chemical properties. Gasoline is one of our most common fuels. It is refined to perform within the confines of the internal-combustion engine and is not hazardous when contained within properly designed vessels. However, it can become hazardous when it escapes. Gasoline is a petroleum-based flammable liquid whose vapors are 3 to 4 times heavier than air. As such, its vapors are slow to dissipate and instead gather in low areas such as drainage ditches, sewers, and basements where they can travel a great distance in search of an ignition source. Whenever operating at tanker incidents, efforts should be made to prevent spilled gasoline from entering sewers and manholes. If a source of ignition is found, these vapors instantaneously flash back to their source, igniting everything in between. For safety, crew members should remain upwind and uphill when operating at the scene of a gasoline spill, and every effort should be made to suppress vapors with foam application and eliminate potential sources of ignition.

Most people are familiar with the smell of gasoline because it is exceptionally volatile. With a flashpoint of −43°F (−42°C), gasoline generates flammable vapors at any conceivable ambient temperature. That is one reason why it cannot be extinguished with water, because water cannot cool it to a low enough temperature to prevent it from generating flammable vapors.[12] Although heavier than air, gasoline is lighter than water, which is another reason water cannot be used. With a specific gravity of 0.69, if water is applied to burning gasoline, the latter

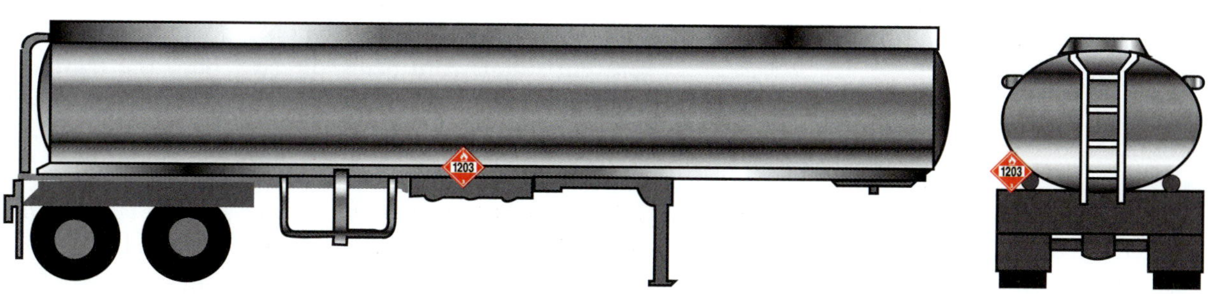

Fig. 18–29. MC 306 tanker profile

will simply float on top and spread about. That property is the basis for the "light water" foam concept in which carbon dioxide foam bubbles keep a water film on top of the gasoline.

On the positive side, gasoline has a narrow flammable range of 1.5% to 6% in air. Any concentration more than 6% is too rich to burn. However, ignition can occur even in diluted states; and when a rich mixture is dissipated, it passes through the flammable range as it is dispersed. The fact is that gasoline is easily ignited, and all ignition sources should be eliminated before venting.

Once ignited, gasoline fires burn with tremendous quantities of radiant heat and copious amounts of heavy, sooty smoke. The heat release rate—the amount of heat energy released in a given period of time—for a small pool of gasoline (2 sq ft [0.2 sq m]) is approximately 400 kW, compared with a heat release rate of 50 kW for a wastepaper basket of the same footprint. The radiant heat makes close approach to a gasoline tanker difficult, but not out of the reach of a properly charged foam nozzle. This high radiant heat also makes exposure protection a primary consideration whenever a tanker fire occurs in a populated area. Traffic must be stopped one-half mile in each direction and structures within this radius should be evacuated to maximize life safety and minimize the impact to the surrounding area. It is not uncommon for surrounding structures to be heavily involved in fire before the arrival of the fire department.

Tanker fire operations

Equipment needed:

- Foam nozzles, eductors, reservoirs
- Class B foam sufficient to the task
- Combustible gas detectors
- Shovels
- Sand truck
- Spill booms and absorbents

Supplemental equipment:

- Master stream foam equipment
- Small excavator
- Sandbags
- Dry chemical extinguishers

There is no single tactical plan that fits all gasoline tanker fires. Although life safety, exposure protection, spill control, and vapor suppression are primary strategies at all incidents, extinguishment procedures vary with the incident encountered based on the scene size-up. Critical factors include incident location, terrain, vehicle position (upright/overturned), the part of the vehicle involved in fire, and whether there is a large discharge of product associated with the fire.

The primary considerations for the incident commander to ascertain while responding are location, degree of involvement, and life hazard. (Is the fire on a remote stretch of highway, on a bridge, under an overpass, or in a crowded neighborhood? Is the terrain level or sloped?) If sloped, are there downhill exposures that must be evacuated? What part of the truck is on fire: the cab or cargo tank area? If the fire is confined to the engine or cab area, can hoselines be directed between the cargo tank and the cab to prevent fire from spreading to it? Has the fire spread to nearby exposures already causing a rescue priority? Is there an immediate life hazard in or near the vehicle? If so, can immediate steps be taken to rescue the driver/passenger before the fire extends to the passenger compartment?

Another important early consideration is the position of the vehicle. Is the vehicle upright or on its side? Is it leaking fuel through the dome covers? Can the leak be controlled and the vapors contained? If the vehicle is upright and the vents are showing fire, sufficient quantities of Class B foam and dry chemical must be brought to the scene immediately, before an attempt to extinguish the fire can be made. An operation that runs out of foam solution while extinguishing a fire burns back to its original size and ends up having no net effect. Sufficient quantities means having enough **foam concentrate** and water to both blanket and reblanket the burning product and to suppress vapors from spills. Water streams can be directed to cool the sides of the tank to keep the aluminum shell from melting too quickly. Cooling the sides of the tanker with water in conjunction with applying foam to the burning liquid can slow the rate of burning somewhat, thus assisting the foam in successfully extinguishing it.[13] When using water, take care to prevent water from being directed onto the flames or entering the tank. This only displaces gasoline and may cause the tank to overflow product, spreading the fire to exposures. The same goes for the foam stream. It must be aimed in an arc so that the foam is laid on top and not driven beneath the surface. Even with careful application, spills and slop over occur and additional foam lines and booms are necessary to control fuel spills.

Life hazard

As previously mentioned, most tanker fires result from accidents in transit. Potential victims include the driver, occupants of a vehicle that collided with the truck, or people trapped in an occupied structure into which the truck collided. Because gasoline fires ignite and spread so quickly, in most well-involved tanker fires, driver survival often depends on the ability to self-evacuate. Few rescues are made, unless the fire is small on arrival, in which case an attempt to extinguish it with Class B foam and/or dry chemical may resolve the problem. However, often by the time the fire department arrives, the level of radiant heat prevents close approach by rescue crews. Nevertheless, in situations where victims are at risk, an attempt should be made to stop the fire from spreading to them with an exposure line, if the fire has not become too large to do so. Aside from those in the immediate vicinity of the tanker, life safety obliges the immediate evacuation of all persons from the area 1,000 ft (305 m) downwind and the eventual isolation of an area of one-half mile in all directions along with the elimination of ignition sources in that area. To accomplish this, a unified command structure must be set up to include police to assist in the evacuation.

Tanker fire strategies

Option A. Defensive strategy: Do not attempt to extinguish the fire. Letting the fire burn off is a legitimate tactic in some cases. If there is no life hazard, available water/foam is insufficient, crews would be placed at unwarranted risk, or the fire is in a remote location with no exposures, the preferred tactic is to stop traffic, set up perimeters, protect exposures and let the fire burn itself out. Even in a moderately populated area, if exposures can be safely protected from igniting, and no critical infrastructure is threatened, the safest procedure is to let the fire consume the product. Even though gasoline is an expensive commodity, the fire should not be fought simply to save the product or the vehicle. The gasoline will be contaminated with foam and offloading it can be a costly and hazardous process. The runoff of foam/gasoline solution into nearby waterways may cause environmental damage far in excess of the value of the saved product. When a defensive strategy is employed, cooling streams may still be directed from a safe distance to cool the outer shell and adjoining tank compartments to impede further involvement.

Allowing the fire to burn off is the only option when insufficient Class B foam is available to extinguish it. Although most vehicle fires, including fuel-oil trucks, can be fought with either water fog or foam; however, a gasoline tanker fire can only be extinguished with Class B foam. Once initiated, foam application must be continuous to be effective; so if the tanker is well involved (one or more tank compartments), the foam operation should be curtailed until sufficient foam can be brought to bear (fig. 18–30). As mentioned earlier, the sides of the tanker shell can be cooled with water streams in the interim, to prevent failure of the aluminum skin and the escape of product. This is a defensive strategy. Whenever water is used to cool the sides of the tank, care should be taken not to direct the stream onto the burning gasoline. If water splashes into the burning tank opening, flaming gasoline can overflow the sides and travel to exposures, particularly those downhill. Similarly, no attempt should be made to direct a hose stream onto flames venting from a pressure reducing valve.

Fig. 18–30. A fully involved tanker can place a heavy demand on foam supplies. A defensive strategy aimed at exposure protection should be initiated until sufficient supplies of foam are on the scene to support a continuous application. (Courtesy of Brent Perkins, Shelby County Tenn. F.D.)

Option B. Mount an offensive attack on the fire. If life safety is at risk, buildings are exposed, infrastructure is threatened, and sufficient foam can be brought to the scene quickly, an offensive attack on the fire can be mounted. Whenever there is life safety involved, firefighters must act to save a life by attempting a quick knockdown (if the fire is small) or by placing a hose stream between the fire and the endangered people (if the fire is too large to extinguish quickly).

A quick knockdown is also advantageous for a tanker truck fire that does not involve the cargo area. A fire involving the engine or cab area of a gasoline tanker is essentially a routine truck fire until the cargo tank gets involved. Quick water application may prevent the fire from spreading and developing into a full blown tanker fire. Handlines should be directed to extinguish the fire and an exposure line should be leveled at the space between the cab and the tank shell to keep it cool. If the saddle tanks below the cab are involved, water fog or Class B foam may be needed to complete extinguishment. Similarly, a fire under the cargo tank (a tire on fire, for example) should be quickly extinguished before the contents become involved.

If the fire is located under a bridge, in a tunnel, or affects a populated building or other infrastructure, every effort should be made to protect exposures and control the fire before it causes significant damage. Flame plumes can extend vertically 50 ft (15 m) or more. Master streams can be directed to cool the steel girders with water, taking care not to displace the gasoline. A larger fire threatening exposures may require water, foam, or a combination of streams to confine the fire.

Because of the intense heat they generate, gasoline tanker fires can be slow to respond to foam application, and close approach can be impeded. Generally, when operating at a tanker fire, nozzles should be located at the furthest distance that will still reach the target. The use of monitors and master stream foam equipment can extend the reach and keep members in safer locations.

Estimating the amount of foam needed. The use of foam, while covered in the context of tanker fires here, is described in greater detail in chapter 25, Implementing the Planned Response to a Hazardous Materials Incident as well as chapter 31, Advanced Fire Attack. Refer to those chapters for foam terminology, foam equipment, and foam application techniques.

In estimating the amount of foam needed, an involved tanker fire can demand the application of several hundred gallons of Class B foam concentrate to bring under control. The application rate must be proportionate to the intensity of fire development. If the tanker is fully involved, several hand lines or a master stream may be called for. Additional foam is also necessary to blanket pools of spilled gasoline in order to prevent ignition. Such operations demand team work and should be practiced beforehand, during training exercises.

When deciding whether to make an offensive attack, it is important to determine if you have enough foam to complete the job. Table 18–2, based on NFPA 11, *Standard for Low, Medium, and High Expansion Foam,* provides a reference that can be used to determine the minimum amount of foam needed. Remember the difference between **foam concentrate,** which typically comes in 5-gal (19-L) pails, and **foam solution,** which is the mixture of foam concentrate *and* water.

Here's an example: Let's say that on arrival, you have a tanker on its side with a burning spill area of approximately 40 ft × 40 ft. How much total foam solution will you need for the initial application period of 15 minutes?

- 40 ft × 40 ft = 1,600 sq ft
- 1,600 sq ft × 0.10 (from the table) = 160 gpm for 15 min
- 160 gpm × 15 min = 2,400 gal of foam solution

Now, to determine how much foam concentrate you need to bring to the scene, you must know what percentage concentrate you are using. So, let's say your department purchased 3% foam. That means that 3% of those 2,400 gal of *foam solution* is going to be *foam concentrate* (with the remaining 97% water).

- 0.03 × 2,400 = 72 gal of *foam concentrate* (15 pails) for initial 15-minute attack
- Additional foam concentrate to maintain foam blanket until picked up by environmental recovery firm (foam blanket will degrade over time)

Table 18–2. Foam application rates for spills, based on NFPA 11

Foam type	Minimum application gpm per sq ft	Minimum discharge time (minutes)	Types of spill
Protein or fluoroprotein	0.16	15	Hydrocarbon
Aqueous film forming foam (A.F.F.F.), Fluoroprotein film forming foam (F.F.F.P.) and alcohol-resistant AFFF or FFFP	0.1	15	Hydrocarbon
Alcohol-resistant foams	Check listing criteria provided by the manufacturer	15	Flammable and combustible liquid spills necessitating alcohol-resistant foam

Controlling runoff. Containment of spilled product is a crucial challenge in tanker fires. Streams of gasoline can escape from the damaged tank structure and quickly find an ignition source. Flowing gasoline travels rapidly and can quickly spread fire to nearby exposures. Dry chemical quickly extinguishes streaming gasoline, but vapors can reignite. Several members should be given dry chemical extinguishers to quickly douse any ignited gasoline on the ground. Gasoline spilling, pooling, or running along the roadway should be contained by a dike and blanketed with foam. Diking can be done with commercial booms, sandbags, or shovels and dirt. For large releases, excavators and earth moving equipment should be requested to form dikes and berms in an effort to control the spillage. Once contained and covered with a blanket of foam, the vapors must be monitored with combustible gas detectors and foam reapplied whenever vapors return. Building a barrier around sewer grates and manholes can prevent gasoline from entering subterranean conduits where it is likely to find a source of ignition and spread the fire. If gasoline does enter a sewer, a unit should be assigned to flood the sewer with foam, and the wastewater treatment plant must be notified. To minimize environmental damage, efforts must be made to prevent **runoff** from entering waterways. If this cannot be avoided, the local Health Department and the Coast Guard should be advised.

Safety management. Clearly, tanker fires require a large number of personnel to control. Multiple alarms or mutual aid are commonly necessary. Many different tasks must be completed simultaneously involving numerous groups and divisions coordinated by means of the Incident Command System (ICS). Essential tasks may include rescue, evacuation, exposure protection, fire attack, foam management, water supply, runoff containment, air monitoring, hazmat control, medical, traffic control, and utilities. The safety officer has a critical role to play in all of this.

Safety of personnel is paramount at these incidents. A rapid intervention crew (RIC) should be established at the beginning of the incident. Firefighters intent on their mission can get tunnel vision and unwittingly find themselves standing in a pool of gasoline or otherwise put themselves in jeopardy. The principle of approaching the vehicle from uphill and upwind is especially important when attempting to extinguish a gasoline tanker fire. Wind can direct the thermal plume toward firefighters advancing from downwind, decrease the reach of the hose streams, and disrupt the application of the foam blanket. Crew members operating downhill are in danger of flowing gasoline or explosive vapors collecting in low areas, and the safety officer must keep crews positioned safely.

Once the fire is extinguished, there may be several hundred gallons of gasoline remaining in the tanker that didn't burn. This product must be off-loaded with specialized equipment before the vehicle is towed away. The trucking company and/or petroleum transporter are responsible for the cleanup and should provide the necessary apparatus for safe removal. The process of removing the remaining product can be time consuming and hazardous. Fire department units must remain on the scene until this process has been completed. Every effort must be made to prevent vapors from reigniting. Combustible gas detectors help determine when the foam blanket must be refreshed to control the vapors.

LPG tanker fires

Liquefied petroleum gas (LPG) tank trucks can carry up to 11,500 gal (45,532 L) of liquefied fuel under pressure when full. Smaller, local delivery trucks known as **bobtails** carry between 2,000 and 3,000 gal (7,571 and 11,356 L) of LPG. Bobtails also have hose reels and pumps to transfer the LPG at delivery points. The fuel is plentiful, with approximately 18 billion gal (68 billion L) are transported and used annually.[14] LPG is used more prevalently in rural areas where gas mains are limited. The tanks on LPG trucks are cylindrical with rounded ends that resemble a SCBA cylinder. They are constructed according to DOT MC-331 and NFPA 58 specifications (fig. 18–31). They have a steel exterior shell that is usually painted white and are placarded with the number 1075. Knowing the shape of the various DOT vehicles gives the responding firefighters an idea of the potential cargo even when the placard is not visible. Because of the possible hazards associated with a spill or fire, propane tank trucks are also designed under Department of Transportation (DOT) and American Society of Mechanical Engineers (ASME) standards that reduce the hazard during transportation. These features include fusible closure valves and remote emergency valve controls.

The most significant safety device is the pressure relief valve located on top of the tank either at the center or toward the rear. To keep the LPG in liquid form, the tanks maintain an internal pressure of about 250 psi (1,750 kPa), and the relief valve is calibrated to open when the pressure exceeds this level by a specified margin. However, if the relief valve is damaged during an accident, it may not work properly. The biggest hazard

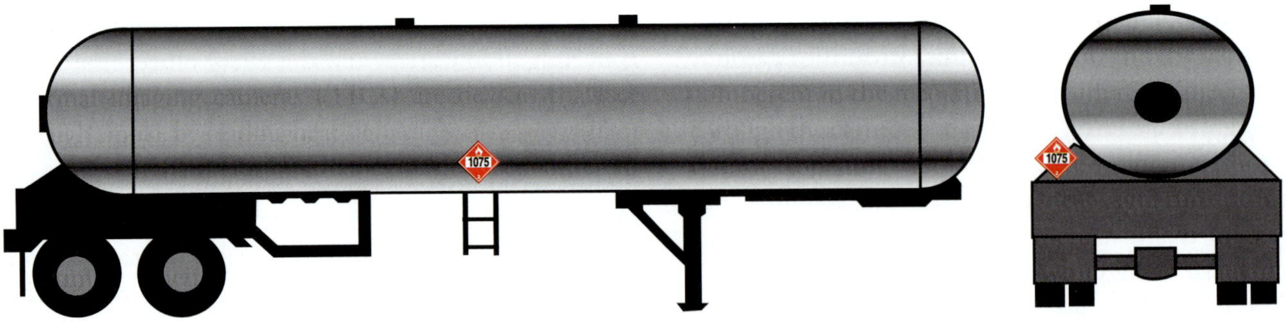

Fig. 18–31. MC-331 tanker

associated with fires in LPG tank trucks is the danger of a BLEVE. A propane cargo tank involved in a fire could cause a violent BLEVE in as little as 5 minutes if flames are impinging on the steel shell above the level of the liquid.[15] This does not give firefighters much time to intervene in the incident and incident commanders must consider this time frame when deciding whether on not to employ an offensive attack.

Chemical properties. LPG refers to a category of petroleum-based fuels that include propane, butane, isobutane, methane, and propylene. Propane and butane are similar in chemical structure and characteristics. The term LPG can mean either propane, butane, or a combination of the two with butane use more prevalent in warmer climates. LPG is used in home heating systems, barbecues, warehouse forklifts, and as an alternative fuel in some motor vehicles. It is also used in stoves found in campers, RVs, and boats; making a BLEVE a possibility at fires in those vehicles. All of these fuels are volatile, highly flammable gases. Like gasoline, vapors that find an ignition source will flashback violently.

Although it is a vapor at ambient temperatures and atmospheric pressure, LPG becomes liquefied when placed under pressure. This allows for easy transport in liquid form. When compressed, propane gas shrinks to $\frac{1}{270}$ of its original volume. Because of its high compression/expansion rate, propane tankers are never filled all the way, but they maintain a vapor space above the liquid to allow for expansion during normal warming. This vapor space is significant to firefighting operations because the steel tank shell above the level of the liquid will heat and weaken much more quickly than below the surface.

LPG is colorless, odorless, and nontoxic. Although invisible, a whitish cloud of condensed water vapor may be visible near the leak site. However, given its expansion rate, an explosive mixture extends well beyond the visible area. Therefore apparatus should be staged well outside the vapor area and only intrinsically safe radios and flashlights should be operated. To make allowance for detection when seeping into the atmosphere, an odorant known as **ethyl mercaptan** is added to the fuel. Ethyl mercaptan gives LPG a pungent aroma that can aid in early discovery and prevention of an explosion.

Propane has a flashpoint of −156°F (−104°C). In contrast to CNG, propane is 1.5 times heavier than air. This property keeps it from dissipating readily into the atmosphere unless there are breezy conditions. Instead, it collects in low areas in much the same way as gasoline; and as such, it is subject to detonation when an ever-present ignition source is found. Its flammable limits are 2.2%–9.5% in air. Although the flammable limit is narrow, only a small amount is needed for ignition. During a leak situation, a combustible gas indicator can be used to determine whether the released product is within the flammable range. However, even when too rich to burn, there is a mixture gradient as the vapors dissipate through which the flammable limit is entered. This hazard can be reduced by the use of water fog. LPG is water soluble. An LPG leak should be handled by evacuating the area, eliminating sources of ignition, and dispersing the vapor cloud with water fog. Members operating the stream should approach from uphill and upwind to stay out of the vapor cloud.

Fire operations. Whenever a propane truck is involved in a fire, it is a serious matter. The primary strategies in a fire situation are to evacuate the area and to cool the tank so that no hot spots develop. Propane tanks are constructed of steel, and steel weakens at 1,100°F (593°C). The combination of a weakening steel shell with an increasing internal pressure from a boiling, highly compressed gas in liquid form can lead to a BLEVE in which the tank separates, explosively propelling flames and fragments over a wide area. Hazmat guides recommend the evacuation of all individuals for 1 mile in every direction when an LPG vehicle is involved in fire. Fire operations involve setting up operations from the sides, not the ends of the vehicle. Although the tank can go in any direction, if a BLEVE occurs the ends tend to travel the greatest distance.

Before initiating an offensive attack on a propane fire, the incident commander should know the length of time the fire has been impinging on the tank and the amount of LPG being carried. If a tire or engine compartment fire can be extinguished expeditiously, a disaster may be prevented from occurring. If the fire cannot be extinguished quickly, or it is unknown how long the tank has been heating, a defensive approach is necessary. Depending on the specific circumstances, the primary tactic may either be to evacuate the area or to cool the tank from both sides. Cooling streams should be directed from unmanned monitors using smooth bore nozzles to obtain the greatest reach possible. The upper part of the tank is the most vulnerable because that is where the vapor space is. Lacking liquid to dissipate the heat, this part of the shell weakens the fastest. Therefore streams should be directed toward the top of the tank, allowing water to roll down the sides (fig. 18–32).

Fig. 18–32. Unmanned water monitors are used to cool a LPG tanker truck.

The emptier the tank is, the faster the steel weakens. If the driver reports that several deliveries have been made and the tank is near empty, the window of opportunity to control the situation is greatly reduced. Sometimes a frost line can be seen on the side of the tank. Formed by condensation inside the tank, this line indicates the level of LPG inside. As soon as the monitors are set up, firefighters should withdraw to a safe location, approximately one-half mile (0.8 km) back. If copious amounts of water are not quickly available, or if flames have been impinging on the tank for an unknown period of time and there are no rescues, crews should withdraw from the area immediately.

Relief valves dissipate internal tank pressure up to their capacity if working properly. On arrival it is important to determine whether the relief valve is operating. If the tank pressure increases faster than the relief valve can disperse it, the internal pressure builds. Cooling the tank can assist the relief valve to do its job. The dispersal of pressurized vapor from the relief valve is accompanied by a rushing sound. If the pitch or volume increases, it is a signal to retreat to a safe location. If flames show from the relief valve, the propane has either found a source of ignition or has reached its auto ignition temperature of 842°F (450°C). Care must be taken to not extinguish flames coming from a venting pressure relief valve. Doing so allows rapid emission of unburned gas into the surrounding area and could lead to an explosion. It is important to note that an overturned LPG tanker's relief valve may be below the liquid level and thus not operable.

Bus fires

During the 5-year period of 1999–2003, fire departments across the country responded to an average of approximately of 2,200 bus fires annually in total. These fires caused an estimated annual average of 3 civilian deaths, 30 civilian injuries, and $24.2 million in direct property damage per year. Bus fires (including school buses) accounted for only 1% of the total reported vehicle fires, 1% of the vehicle fire deaths, 2% of the total vehicle fire injuries, and 2% of the vehicle fire property damage. On average, six bus or school bus fires were reported daily throughout the country.[16] Most bus fires were ruled accidental and attributed to mechanical or electrical failure. Statistics show that most bus fires originate in the engine compartment or wheel/brake area. Engine compartment fires often extend to the passenger compartment when flames spread vertically up the exterior of the bus and compromise the windows above, allowing heat and smoke into the interior (fig. 18–33). Tire fires can burn through the wheel well and enter the passenger compartment from below. According to the National Highway Traffic Safety Administration (NHTSA) less than 0.05% of all reported bus crashes resulted in fire.

Fig. 18–33. Fire at the rear of an occupied bus presents a life hazard to passengers. (Courtesy of Boise, Idaho, Fire Department)

Although they are relatively infrequent, bus fires can present a high life hazard, especially when the passengers are the elderly or school children. One bus fire that involved the tragic deaths of 23 senior citizens received national attention, which may lead to the use of automatic fire detection and suppression systems on buses. The fire occurred September 23, 2005, on Interstate 45 in Texas, when a bus carrying the seniors caught fire while they were being evacuated from Hurricane Rita. Investigators found that the fire resulted from an overheated wheel bearing that ignited the right-rear tire of the bus.[17]

Bus fires present several challenges to responding firefighters. Rescue and evacuation are chief among them. A bus is a large vehicle with a potentially high life hazard that requires more than a single unit response. Bus capacities range from 25 passengers on minibuses, to 66 on intercity motor coaches and school buses. Bus fires spread quickly and may require a significant amount of water to control. Depending on the size of the pumper's onboard tank, a well-involved bus may require far more water than a single engine carries. A task force assignment of two engines, a ladder, a rescue, and a battalion chief would be needed to cover the initial assignments at an occupied bus fire. When the engine compartment is located in the rear of the bus, a fire originating there may call for two hand lines: one brought to the passenger area to protect occupants, and one directed onto the burning engine or tires. Hose crews may find it difficult to penetrate to the interior because the confined nature of the bus with its narrow center aisle and tight, high back seats. Advancing a hoseline is made more difficult when people are still on the bus. In most cases, passengers, if ambulatory, will self-evacuate before the fire department's arrival. Ventilation for life is another important tactic. The thick smoke generated by synthetic seats and trim on the interior can quickly asphyxiate passengers. Horizontal ventilation via the side windows is an immediate priority, and a ground ladder is needed to reach and open the roof hatch. The side widows are generally constructed of safety glass that can be broken with a tool. Most have an emergency release handle. If equipped with two doors, both must be opened to assist with evacuation. Bus doors are opened and closed by a pressurized air system. DOT standards require that each door is furnished with an emergency release valve that will dump the air pressure. These valves are red and located on the interior of the bus next to the door. Their location may vary, so it is a good idea to contact your local bus company and arrange a training exercise.

School buses. School buses are a special class of vehicle designed with the protection of children in mind, in the event of a collision or rollover. Every school day, approximately 450,000 yellow school buses transport more than 24 million children to and from schools and school-related activities.[18] Most up-to-date school buses are equipped with two emergency doors (one in the back and one on the side) and multiple emergency escape windows. The doors have handles on both the inside and outside of the bus. These emergency mechanisms work well; however, if the bus frame is torqued, children may not be strong enough to open the doors. In addition, newer buses have one or more roof hatches that can be opened manually from the interior. These hatches are designed to provide another means of escape if the bus has rolled over. At this time school buses are not equipped with self-opening roof hatches to vent smoke in case of a fire, though the technology exists.

Depending on the bus design, the location of the engine varies (fig. 18–34). In a traditional school bus, the engine is in the front, separated from the passenger compartment. A fire occurring in the engine compartment, if significant, may require the evacuation of the students via the rear and side exit. In modern passenger and school buses, the engine is in the back, beneath the rear passenger compartment. This configuration can affect the passenger area more directly, but passengers evacuate through the front exit where they are used to entering and exiting. The best strategy for an occupied bus is to stretch two hand lines, one to the passenger compartment and one to the engine compartment. Entry to the engine may require a key, or forcible entry through the side access door might be necessary.

Fig. 18–34. Firefighters apply foam to a school bus fire that originated in the engine compartment and spread to the passenger compartment through the fire wall, where it quickly became fully involved. Note the proximity of the flame plume to overhead power lines. (Courtesy of Kyle Hastings, Mendham, NJ, F.D.)

According to the National Coalition for School Bus Safety, there is no federal requirement mandating that school bus seats have a higher horizontal flame spread rating or fire resistance than other vehicles, but some states may require one.[19] Many seats are constructed of urethane foam padding covered by vinyl. They can produce thick smoke and add significantly to the fire load.

A review of school bus fires reveals that many tragedies have been averted by alert, well-trained bus drivers who pulled over and got the kids off the bus at the first sign of smoke. At this time most school buses are powered by diesel fuel, which is mounted between the axles, beneath the passenger compartment in a steel enclosure to prevent damage in the case of a side impact. Some states are now allowing school buses to be powered by propane and other alternative fuels. The use of alternative fuels may affect tactics, and local fire departments should adjust tactics accordingly in areas where alternative fuels are used.

NOTES

1. U.S. Fire Administration, Topical Fire Report Series, Highway Vehicle Fires. 2(4). http://www.usfa.dhs.gov/downloads/pdf/tfrs/v9i1.pdf. October 9, 2008.
2. Gustin, Bill. "New Tactics for New Car Fires." Fire Engineering, April 1996, p. 5.
3. Prather Timothy G. "Storing Gasoline and Other Flammables." National Agricultural Database, Agricultural Engineering Department, Agricultural Extension Service. Knoxville, Tennessee: University of Tennessee. Available at http://www.cdc.gov//nasd/docs/d000701-d000800/d000760/d000760.html/, April 2002.
4. Junkins, Lee. "Is It Time to Change Our Training Yet? Part 2" Available at http://www.firelink.com/training/articles/332-is-it-time-to-change-our-training-yet—part-2/, September 6, 2007.
5. Hollins, Lee. "Airbag Inflator Explosion." Fire Engineering, December 1996, p. 34.

6. Automotive Engineering International Online, "Tech Briefs," http://www.sae.org/automag/techbriefs_06-00/.
7. Automotive Engineering International Online, "Tech Briefs," http://www.sae.org/automag/techbriefs_06-00/05.htm.
8. Emery, Jason D. "Compound Factors." Fire Chief, August 2007, p. 92.
9. Emery, p. 94.
10. www.seattle.gov/fire/publications/CNGAutoFire/CNG%20Auto%20Fire%20May%202007.ppt/.
11. Wilen, Saul B. "Countering Terrorism Threats: Trucking Industry and Multi-Modal Distribution Network Vulnerabilities." San Antonio, TX: International Horizons Unlimited. Available at www.intlhorizons.com/article-trucking.htm/, February 10, 2003.
12. Fried, Emmanual. "Fireground Tactics." HM GINN Corp., 1983, p. 199.
13. Fried, Emmanual. "Fireground Tactics." p. 202.
14. Suburban Propane. Available at http://www.suburbanpropane.com/propane.html/.
15. Hermann, Stephen L. "Haz Mat Studies." Firehouse, March 1991, p. 22.
16. Ahrens, Marty. "Vehicle Fires Involving Busses." NFPA Journal, Sept/Oct. 2006. Available at http://findarticles.com/p/articles/mi_qa3737/is_200609/ai_n17193680/.
17. Levin, Alan. USA TODAY. August 8, 2006. Available at http://www.usatoday.com/news/nation/2006–08–08-bus-fires_x.htm/.
18. School Bus Information Council. "School Bus Information Facts 2004." 2004. Available at http://www.schoolbusinfo.org/School%20Bus%20Fact%20Sheet%202004.pdf/.
19. U.S. Government Printing Office. "Title 49—Transportation." U.S. Government Printing Office. October 2002, CITE: 49CFR571.302, pp. 740–743. Available at http://a257.g.akamaitech.net/7/257/2422/14mar20010800/edocket.access.gpo.gov/cfr_2002/octqtr/49cfr571.302.htm/.

LESSON FROM THE FIREGROUND

A call is received reporting an automobile fire on a city street at dusk. The dispatcher sends the nearest engine and ladder company to the scene. On arrival, the engine officer reports a heavy fire in the passenger compartment of the vehicle which is parked parallel to the curb with no other vehicles near it. As the engine company stretches a hoseline, one member of the ladder "takes the glass" using a 6-ft (2-m) hook while another walks to the front of the vehicle and prepares to force the hood open in order to cut the battery cable and check for extension. What is wrong with this picture so far? Water reaches the nozzle and the nozzle team approaches the vehicle at a 45 degree angle from the side, directing the stream into the passenger compartment and sweeping it beneath the vehicle to cool the gas tank and undercarriage.

Before the flames fully darken down, the vehicle's horn begins to sound, the headlights begin to flash and the sound of the ignition cranking can be heard. Realizing that the ignition wires have shorted and the car could start, the engine officer calls for a chock. A few seconds later, the motor turns over and lurches forward as the firefighter arrives with the chock. Although he places the chock in front of the rear wheel, there is too much momentum and the flaming vehicle rolls over the chock and continues down the street. The truckman at the hood has been knocked down and is lying on his back, his legs under the vehicle holding onto the radiator as the vehicle drags him down the street. If he lets go, he will be run over! If he holds on, he will be crushed against a wall!

A routine car fire has just turned into rather hazardous incident. What happened next? A quick thinking member sees an old, discarded TV on the sidewalk in a pile of refuse, grabs it, and throws it under the front bumper. The TV is too large for a car to drive over. It lodges in front of the front tire on the passenger side, causing the car to come to a stop. The truckman was not seriously hurt, but could easily have been injured or killed.

Lessons Learned: expect the unexpected at auto fires, stabilize the vehicle as soon as it can be done safely, and avoid working in close proximity to a burning vehicle—especially in front of or behind it—until after all visible fire has been extinguished, and all surfaces have been thoroughly cooled.

QUESTIONS

1. Identify the specific automobile components likely to contribute to the ignition of a fire.

2. What are your priorities when a large motor home is heavily involved in fire and parked next to a convenience store? How will you prevent extension of the vehicle fire into the store?

3. Identify the eight probable locations of a motor vehicle fire and discuss the potential hazards associated with each location.

4. As the officer of Engine 7 you are responsible for the safety of your crew. Engine 7 is responding to a truck fire on the expressway. Off in the distance heavy smoke can be seen as you enter the on-ramp. Traffic continues to speed by the truck that is well involved with fire. The posted speed limit on the express way is 70 mph (113 kph). How far from your operation should visual warning devices be placed and why?

5. What are some of the products of combustion coming from today's vehicle fires?

6. Most fire engines carry 500–1,000 gallons (1,893–3,785 liters) of water onboard. Occasionally, additional water will be needed to fight specific vehicle fires. Identify five instances in which you would need to consider additional water supplies.

7. Explain the proper hoseline selection for a small car fire that is fully involved, a full size pick-up truck that is involved in the passenger compartment, and a semi truck burning in the engine compartment.

8. As a rule of thumb, the hoseline should be advanced from the __________________ of the vehicle, at a __________________ angle, and not directly from the front, back, or sides.

9. What is the recommended approach to fighting a fire resulting from a failed fuel tank?

10. Opening the hood and checking the engine compartment are standard procedure at vehicle fires. Opening the hood is necessary at automobile fires for two reasons. What are they?

11. List the differences in the way vehicle fires should be fought in hybrid or alternative-fueled vehicles as opposed to conventional automobiles.

12. Gasoline tankers can carry as much as __________________ gallons of gasoline. What UN number placard would you expect to see on one?

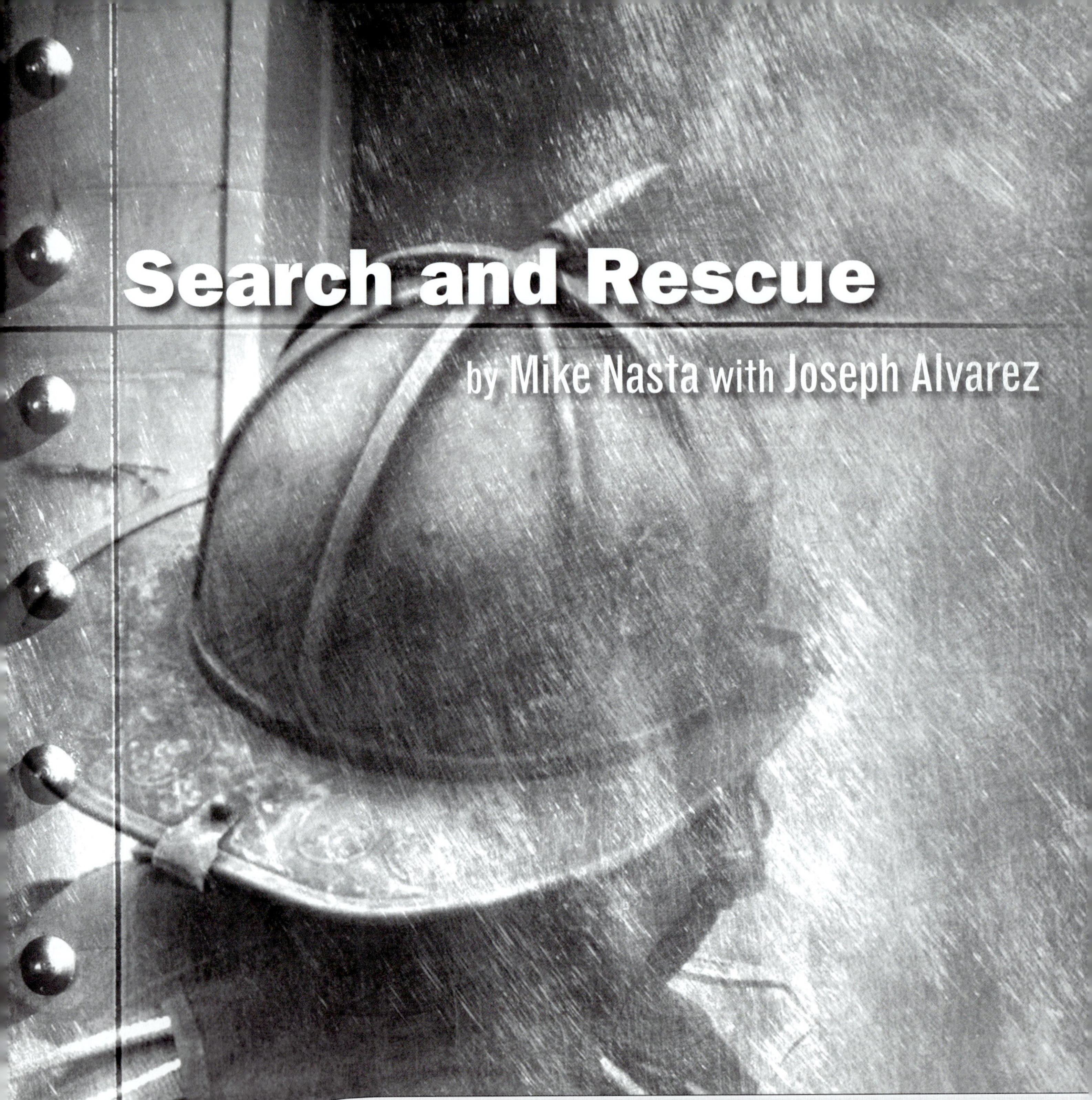

Search and Rescue

by Mike Nasta with Joseph Alvarez

This chapter provides required knowledge items for the following NFPA Standard 1001 Job Performance Requirements:

FFI 4.3.9

FFII 5.3.2

This chapter contains Skill Drills. When you see this icon, refer to your Skill Drill book for step-by-step instructions. SKILL DRILL

OBJECTIVES

Upon completion of this chapter, you should be able to do the following:

- Identify the necessary tools required to perform a search of a building safely
- Identify the five search priorities in order of importance
- Describe a primary search
- Describe a secondary search
- Describe three primary search techniques
- Describe the three emergency procedures for firefighter safety and survival
- Describe six methods of victim removal

INTRODUCTION

The fire department's primary goal is to protect life (including our own) and property. As firefighters we must perform these actions in a safe and responsible manner. The object is not to win medals, but to protect the people we serve and safely return home to our loved ones. We accomplish these lofty goals through tactics that are well thought out and quickly executed. To the untrained eye it may seem that our actions are arbitrary, but in reality fighting fires is a well-orchestrated event carried out by a highly trained team. Make no mistake about it: The procedures and techniques that make up this chapter are critical for the survival of the rescuers as well as the victims. By definition, to search is to look for or seek out. In the world of the firefighter, it means to seek out or look for victims who require our help to remove them from danger, whether it comes from a fire or any other type of emergency to which firefighters may respond during the course of their duties.

Preparing for a search and search tools

Before beginning any search it is imperative that all firefighters are in full protective gear, including self-contained breathing apparatus. In addition, each firefighter must be equipped with a portable radio, not only to transmit information about search results to command, but also to transmit a Mayday message should the firefighter become trapped. All firefighters must be equipped with a tool such as a set of irons (flathead axe and Halligan) or a hook or pike pole (fig. 19–1). Additional tools may include thermal imaging equipment: ropes for rope guided searches, webbing, and a powerful hand light. Tools such as Halligan, axes, and pike poles should be used to extend the reach of the searching firefighters. Great care should be exercised when searching with these tools. Although these items greatly enhance our ability to perform searches, if used improperly they can injure or kill the very people we are searching for. The searching firefighter must swing the "light end" of whatever tool being used (the fork end of the Halligan tool, for example) toward the victim. The tool can also be used to extend your reach under obstacles as well as allow the searcher to cover more area within the space being searched.

Fig. 19–1. Search tools come in a wide variety of shapes and sizes. Many commonly used tools were actually developed by firefighters looking for innovative ways to combine or improve standard tools.

The searching firefighter can also use these tools to sound floors for stability. It should be noted that when sounding a floor in a case such as entering a window to perform vent, enter, isolate, and search (VEIS) tactics, the area should be swept gently first in the event that a victim is in the area that the firefighter is sounding. Forcibly hitting a victim with the tool will injure or kill the victim. No firefighter should ever enter a structure without a tool. Whenever a firefighter enters a building without full personal protective equipment (PPE) and a tool, it is not only that firefighter who is in danger, but all other firefighters on the scene.

SEARCH AND RESCUE

Search size-up and safety

FFI 4.3.9 **FFII 5.3.2** Every firefighter must perform a task analogy or size-up for all events we respond to. These mental notes are critical in keeping us safe. We continually drill and train to hone our physical skills. Although it is imperative to maintain operational readiness, we tend to forget about the mental aspect of our jobs; part of this is taking a critical look at the entire scene as we approach it. Are there any barriers that may impede access or visible signs of structural failure? How well trained am I in building construction? Do I know the signs of failure? What are the fire or smoke conditions telling me? The answers to these and many other questions must be asked and answered long before stepping one boot into the structure.

Upon approaching an event, look for some indications of whether or not the structure is occupied. Some indicators are the presence of cars, time of day, closed window dressings, children's toys outside, air conditioning operating in windows, and the all-important, yet so often overlooked, relevant preplanning information. Every firefighter should have intimate knowledge of the response district. Every day as you travel through your response area and the areas nearby, make mental as well as written notes that you can use if and when the time comes to respond to an incident.

Every firefighter should make mental notes of window and door locations before entering a building. This information may be a lifesaver should that firefighter become disoriented or trapped and suddenly in need of an emergency escape route. The firefighter should be cognizant of how many floors they may have ascended or descended. This information is a must for two reasons:

1. It keeps the firefighter oriented.
2. This information may help a rapid intervention crew locate the firefighter if necessary.

Before entering the structure, pay attention to the smoke conditions. What color is the smoke? Does it contain any indicators that suggest an imminent backdraft or flashover? Is it drawn back into the building as you open the entry door? If so, this may be an indicator of backdraft. Is the smoke violently pushed out of the building as if under extreme pressure? This may be indicative of an ensuing flashover.

FFI 4.3.9 Once you're inside and beginning your search, pay strict attention to smoke, heat, and fire conditions. They can change rapidly, making a survivable atmosphere into a violent and deadly space that you must escape very quickly. Be ever mindful of construction type, layout, and how fire affects that type of construction (e.g., trusses, lightweight structural members, wide open spaces, and confusing layouts). Watch for missing or burned-through floor boards, or building features that have been purposely removed for whatever reason. Use a hand tool to sound the floor as you walk or crawl, to detect sponginess or weakness before placing your weight on it. Before opening a door to an adjacent room, look for signs of smoke or fire coming under the door, and check the temperature of the door. If it is hot there is likely fire on the other side.

Take notice of door swing. The rule of thumb is that doors that open away from you usually lead into living spaces such as apartments or bedrooms. Conversely, doors that open toward you usually lead into service areas such as utility rooms or can be indicative of an elevation change such as steps that lead to a basement or attic. Doors that lead to utility closets may have ventilation louvers built into them. Doors in places of public assembly, such as theaters, usually swing in the direction of egress for the occupants. Before opening a door, always check it for heat by first feeling the door from bottom to top, and carefully control the door as you begin to open it. Door control offers the opportunity to close it quickly should conditions in the room be untenable. Now where do we begin our search or what areas are our top priority?

Search priorities

- Fire room/fire floor
- Room above/floor above
- Top floor
- Floors in between
- Floors below and the building exterior

The most endangered occupants are in and around the fire room and fire floor. Now this may sound like common sense to most; however, firefighters are easily distracted if other victims cry out for help. We may tend to give those victims our attention first when they are not the ones who require our immediate attention. Of course, we don't want to minimize the fact that the victim calling for help needs it, too. However, firefighters must realize that the victims who can't call for help are the ones who are in the most need. So we must either address both victims simultaneously or prioritize and help the ones in the most danger.

The next area that needs attention is the room immediately above the fire. The room above directly receives all the products of combustion as well as a good amount of the heat. Keep in mind that the amount of heat the room and floor above receives depends on the type of building construction. These factors also determine how much danger the rescuer faces when searching the floor above. Obviously, the rescuer faces more danger in a balloon frame tenement than in a fire-resistive building.

Subsequently, the firefighter must address the top floor. The top floor is the next stop for all the smoke and gases rising through convection. The conditions on the top floor are affected by two factors: the location and extent of the fire, and whether the floor has been ventilated properly.

After the top floor, all the floors in between the top floor and the previously searched lower floors must be searched. In buildings that are more than two or three stories in height, searching all these areas will require multiple search teams and a great amount of coordination.

Lastly, and many times overlooked, are the floors below the fire and the building's exterior. Although people on the floors below the fire are rarely affected by the products of combustion, fire departments are always responsible for the well-being of all occupants within the structure. The outside of the building must also be thoroughly examined for victims who may have jumped from the structure. By this we mean all four sides of the structure must be searched to find any victims who may have exited the building by other than normal means. Victims have often been located in shrubbery and in other locations outside the structure.

After entering the room to be searched, there are several areas that the searching firefighter must pay strict attention to. The first area is the normal path of egress, in other words, the areas that are directly in line with the door or hallways. Behind doors is another area where victims are often found; any resistance a search team finds in trying to open the door once forcible entry has been completed is probably an indication of a victim located behind the door. Normally if the path to the door or the hallway is blocked the victim will choose to move toward the window. This fact makes it imperative that we check the area below and around the window. Firefighters that are assigned to search must keep all possibilities of victim location in their mind.

Search categories

FFI 4.3.9 Search can be divided into two different categories: primary and secondary searches. Primary search is a rapid search of high-priority areas. Primary searches are often conducted prior to control of the fire. Secondary search is slower and more methodical. Secondary searches are usually conducted after the fire has been controlled.

Primary search. **Primary searches** are a rapid check of a fire building focused on areas that are normal areas of egress for the occupants. These areas include but are not limited to doorways, windows, and all adjacent areas that lead to these locations. For firefighters to successfully search a structure, they must understand the mindset of the escaping victim. In times of stress, most people resort to leaving the building by the same routes they use during their everyday routine. In other words, if they normally use the side door daily, they are likely to use that same door during a fire. Humans are creatures of habit, which drives us to think in this manner.

Firefighters only have seconds to try to determine what victims may be thinking while they are in a state of stress and disorientation. For these reasons we search high-probability areas first. These also just happen to be the areas that we normally use for access to the building such as the front door, pathways leading to exterior doors, bedroom windows, and the paths leading to those areas. Because of the speedy nature of this type of search, firefighters must cover these areas quickly and aggressively to greatly improve the survivability of the victims.

Secondary search. **Secondary searches** are usually slower, more methodical, and more thorough than primary searches. These types of searches are often done after the fire has been controlled. During secondary searches firefighters must search for victims in every possible spot inside and outside the structure. Areas searched during the primary search are rechecked along with the areas that were not checked. These areas include but are not limited to closets, toy boxes, showers, and under beds or other furniture that could provide a hiding spot for a victim. Hiding spots can differ for adults and children. It is at this point that firefighters must learn a little about how their victims think when they are under the extreme stress of being trapped.

The mind set of adults and children often differs during a chaotic state. Adults try to reason their way out or try to think the problem through; children resort to areas of comfort or protection. Adults may try to escape harm by going into the shower and turning the water on thinking that the water protects them from the fire only to find that the smoke kills them or they wind up steaming themselves to death. Conversely, children try to get to their parents thinking that mommy and daddy can protect them. Parents are often found trying to get to their children. It is for this reason that if a firefighter finds an adult victim that firefighter must also look for any children that may be with that adult. On many occasions firefighters trying to remove the adult victim to safety leave the child behind. In one such case, the child was right underneath the mother. The mother was removed, but the firefighters never saw the child and they left the child behind. So take an extra second and sweep around the found victim just in case a second victim is nearby. Children will also hide from the danger in areas that are comforting to them, such as in toy boxes, under beds, in closets, and any place that provided them refuge in the past.

Children may be frightened by the sight of the firefighter in full protective equipment and hide. The only way for the fire service to overcome these fears is through public education. Your fire department must teach children how a firefighter looks when wearing full protective equipment. Fire departments should go to schools and allow children the opportunity to meet firefighters before they are attired in PPE and then again after that firefighter is fully geared. It allows children to face the person in full PPE.

Search techniques

SKILL DRILL

- Light scan search
- Perimeter search
- Lifeline search
- Vent, enter, isolate, and search (VEIS)

Search techniques are designed to keep a firefighter safe as well as oriented during searches. Searching firefighters are exposed to extreme risk, especially during primary search. Disorientation continues to be a deadly problem for firefighters. Each year many firefighters are trapped and either killed or severely injured during search efforts. Ideally, by using the proper search techniques, we can reduce or even eliminate these tragedies. The following techniques used under the correct conditions are designed to help firefighters efficiently search the building as well as keep the firefighter oriented in the structure.

Before beginning any search, it is imperative that all firefighters are in full protective gear as well as self-contained breathing apparatus. All firefighters must be equipped with a tool such as a set of irons (flat-head axe and Halligan) or a hook or pike pole. No firefighter should ever enter a structure without a tool. By entering a building without full PPE and a tool, a firefighter endangers him or herself as well as all other firefighters on the scene.

Light scan search. This technique can be used if the smoke condition in the room is several feet off the floor, allowing the searcher to quickly scan the room with a portable light to check for any victims located on the floor or on any furniture such as couches, beds, or chairs. To properly perform a **light search**, the firefighter must be in full PPE and equipped with a powerful portable light. The firefighter enters the room by first checking the door from bottom to top using the back of the hand. Once inside the room, the firefighter either crouches below the smoke layer or lies down flat on the floor below the smoke condition and uses the light to completely scan the floor area and any furniture for victims. This technique allows the area to be searched in a very short period of time so the firefighter can continue searching other areas. This technique is usually limited for use in rooms that are smaller in size because of the limitations of the light and how far the firefighter can see across the room (fig. 19–2).

Fig. 19–2. A light search involves using a flashlight to quickly and effectively search rooms with only a light smoke condition.

Perimeter search. It can be said that this search technique is the bread and butter of all search operations in the fire service. This technique lends itself well for use in private dwellings, apartments, and relatively small operational areas. Either a two- or three-person search crew can perform this technique.

Two-firefighter perimeter search. The two-person perimeter search always begins by checking the door in the same manner: from bottom to top with the back of the hand (fig. 19–3). The first firefighter enters the room and immediately places the right hand on the right wall and starts to crawl along the right wall. The second firefighter lies on the floor and performs a quick light scan search of the area. The second firefighter then places the light on the floor at the doorknob side of the door near the door jamb. This light serves as a beacon and additional orientation point for the searching firefighters. (Note: The light should not be placed in the path of the door swing in case another firefighter enters the room and knocks it away from the door.) The second firefighter then places a left hand on the left wall and continues to crawl along that wall until the two firefighters meet.

Fig. 19–3. The two firefighter perimeter search is one of the most common search methods.

Upon meeting, the two firefighters should pause momentarily and listen for any sounds such as a victim breathing, moaning, or crying. After listening for any victims, the two firefighters return to the door by following the beam of the light placed by the second firefighter (fig. 19–4). The two firefighters continue toward the light through the center of the room. Working side by side, the firefighters should stay in physical contact with each other while returning to the door. By following this pattern the firefighters can cover most of an average-sized room.

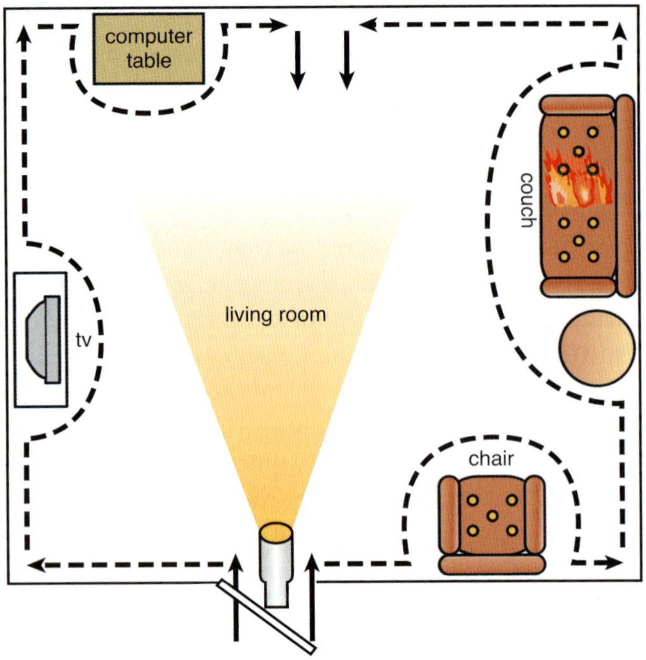

Fig. 19–4. Two-firefighter perimeter search

If the light is not visible to the firefighters from across the room, they should follow one wall back to the doorway. At this point the two firefighters follow the same wall: one firefighter remains in contact with the wall while the second firefighter, keeping contact with the first, stretches out into the middle of the room in an effort to cover as much of the room as possible. While searching, the firefighters should keep mental notes of all doors and windows but should not enter any other doors that would take them into another room. Rooms should be searched one at a time. Closet doors can be quickly swept with a tool to check for any victims. After completing the search of the room, firefighters should pause momentarily at the door of the completed room and quickly discuss any discoveries, such as doorways they may have found or any additional rooms that were encountered. This room is now completely searched and should be marked as such before the firefighters move on to the next room (fig. 19–5).

Marking a room after the search is completed can be done in many ways. For example using a marking pen to mark the door, leaving mattresses folded in half, or any other means dictated by your department's standard operating procedures (SOPs).

Three-firefighter perimeter search. When the situation arises and there are three firefighters available to perform a perimeter search, the procedure changes. After the door is properly checked, the first firefighter enters the room and starts the search by crawling along the right wall. The second firefighter then enters the room and crawls along the left wall. The third firefighter then enters the room and conducts a light scan search; however, this firefighter remains at the door with the light shining toward the middle of the room so the searching firefighters can use it as an orientation point (fig. 19–6).

Fig. 19–5. In order to prevent a duplication of effort, it is necessary for firefighters performing a search to mark the room as "searched" in a conspicuous manner, according to local policy.

Fig. 19–6. When possible, using a three-firefighter perimeter search provides the best combination of firefighter safety and victim search methods.

FFI 4.3.9 Firefighters engaged in perimeter searches should communicate with each other on a regular basis during the search. This communication allows firefighters to keep check on one another and ensure that each firefighter is safe. Communication also gives each firefighter psychological assurance that they are not alone in the room. This assurance will help keep each firefighter calm during the search, as low visibility and extreme conditions can cause disorientation and panic. The communication does not necessarily need to be verbal at all times, the firefighters could track each other by listening for the other firefighters' movement, listening for sound from the other firefighters' breathing apparatus, or by tapping on the floor occasionally with a tool. The firefighters should guard against making noise or talking constantly because constant noise will make it harder for the search crew to hear any noises that possible victims may be making.

Lifeline or rope-guided search. Lifeline or rope-guided searches should be used whenever it is necessary for firefighters to search large areas such as warehouses, supermarkets, or movie theaters. This type of search should also be used in basements or any place with confusing or mazelike floor plans such as office cubicles. Firefighters should not use any of the previously mentioned search techniques in these types of situations. Disorientation in large area buildings is a constant danger to firefighters. All too often a firefighter becomes complacent because most fires occur in dwellings and the main search technique is the perimeter search. Failing to use a lifeline search in a building with a large or confusing floor plan can and often has proved deadly for many firefighters.

Lifeline searches should begin with a fully equipped firefighter (full PPE, self-contained breathing apparatus [SCBA], and tools) and a search rope that is stored inside a rope bag. The average search rope should be approximately 200 ft (60 m) long. The rope may contain indicator knots that can be tied in the rope at 20-ft (6-m) intervals. These knots allow the firefighter to determine how far they have penetrated the structure by simply counting the knots as they move along the search rope (fig. 19–7).

The search rope is tied to an unmovable or substantial object outside the building or fire area, but not a vehicle. Firefighters then enter the area to be searched, and the first firefighter holds the bag and allows the rope to deploy as they move forward into the search area. Any remaining firefighters, whether it is one or more, then guide themselves along the rope, never losing physical contact with the rope. Whenever the search crew takes a turn inside the structure, the rope should be tied off; this keeps the searching firefighters from dragging the rope into areas that the crew did not cover as they move further into the building (fig. 19–8). By dragging the rope across areas not searched on the way in, the search team runs the risk of endangering themselves on the way out. If the firefighters cross areas not covered previously, they may fall into an open shaft, become entangled in furniture, or be trapped by many other obstacles found in burning structures.

Fig. 19–7. Search rope can be invaluable on the fire ground. Besides increasing firefighter safety, using a search rope can allow for a faster and more coordinated search. Indicator knots tied at intervals along a rope can help firefighters estimate their penetration distance into a building.

Fig. 19–8. To control the deployment of a search rope, it is recommended that the rope be tied off to a substantial object outside the structure.

Search crews performing a lifeline search can expand the search area by carrying personal **tether lines** that can be attached to the main search line by carabiners. The firefighter simply attaches the tether line to the main search line and searches off the main line to cover areas such as cubicles (fig. 19–9). It must be noted that tether line must be no longer than 10 or 15 ft (3 or 5 m) in length, thus reducing the possibility of the firefighter getting too far off the main line.

Fig. 19–9. Tether line

Safety is further enhanced by the use of an entry control officer who is positioned at the point of entry. This control officer must be equipped with a clipboard and a timing device (stop watch). The control officer records the names of the firefighters on the entry team, air cylinder duration time (e.g., 30-minute or 60-minute cylinder), and time of entry. The control officer as well as the entry team must be equipped with portable radios preferably on their own tactical frequency. The control officer records the entry time of each member. Following proper air management techniques (described in chapter 10, Self-Contained Breathing Apparatus), the control officer reserves enough time for the team to exit the building. For example, a team equipped with a typical 30-minute rated cylinder is allowed a total operational time of approximately 16.5 to 18.5 minutes to enter the building, conduct search operations, and leave the building. This leaves 25% of cylinder capacity, the level at which most SCBA low-air alarms activate, in reserve for safety purposes. In reality, a 30-minute cylinder does not last for 30 minutes—for the typical firefighter, they last the 16.5–18.5 minutes cited above. Use of longer duration cylinders (larger size cylinder) will increase these times accordingly. Firefighters must not wait until their low-air alarms activate—they need to be out of the building before alarm activation.

Upon being recalled, the firefighters simply reverse their direction on the line by 180° and follow the line out to the starting point. The firefighter maintaining control of the bag leaves the bag on the floor where he or she stopped and follows the rope out, leaving the bag behind. The lead firefighter should make a mental note of the last knot that was past so that he or she can let the control officer know how far into the search area has been searched. The control officer can then use that information to let the second team know where the first team left off, so the second team could proceed immediately to where the bag was left without researching the area that the first team already completed. Finally, lifeline searches are an important tool; all firefighters must be trained to maintain their orientation in large buildings or buildings with confusing layouts (fig. 19–10).

Fig. 19–10. Here firefighters use a search rope or "lifeline" to guide their search of a structure.

VEIS: vent, enter, isolate, and search

VEIS is a high-risk, high-reward and highly effective search tactic that should only be assigned to firefighters with training in this tactic and also experience in fireground operations. In other words, this tactic is not for new firefighters (see chapter 14). VEIS allows firefighters to immediately access the sleeping areas, eliminating the time it might take for rescue crews to work their way to upper floors (where most sleeping areas are found), giving any occupants trapped in these areas a greater chance of survival. However, this tactic exposes the firefighters assigned to this task to great danger by putting them in a position of working right in the path of an advancing fire. This tactic contains several steps, as follows:

Vent. The team starts off by first determining where the bedrooms are located. They then set up a ground ladder to the bedroom window. The first firefighter ascends the ladder and properly vents the window, allowing a few seconds for the hot gases to escape so the room can be searched (fig. 19–11).

Enter. The search team then enters the room, and for the safety of the team must immediately proceed to the isolation step before beginning to search (fig. 19–12 a and b).

Isolate. The first team member must immediately (before beginning the search) proceed to the interior door and make sure that the door is shut.

Note: This step is paramount to the safety of the search team and the survival of any possible victims. By entering the building through a window, the search team is creating a point of least resistance for the products of combustion to follow (flow path), so by closing the door you are isolating the room you entered, protecting the search team and victims from the danger of the advancing fire.

Keep in mind you are counting on getting to the room's door before the fire and betting on your life. Fireground commanders should assign this task only to firefighters with fireground experience.

Search. Once the room has been isolated, the search team completes the search of the room in a timely manner. Remember that time is not an ally here. You are in a very precarious position working above and in the path of an advancing fire. Once the search is completed, the search team should leave the room through the same window and down the same ladder that the team used to enter. Never leave the room via the interior door unless you are absolutely sure that the fire has been completely extinguished and a safe path through the interior is guaranteed.

Fig. 19–11. A firefighter vents the window with a hand tool.

Figs. 19–12 a and b. The firefighter should go into the structure with one leg first and then the other, rather than headfirst.

Thermal imaging equipment

Thermal imaging cameras (TICs) are devices that see through smoke by reading heat signatures, using very slight variations in temperature that exist between different objects to create an image on the camera's screen. In other words, cameras read the temperature difference between a victim and their surroundings. By using this heat differential, the user can view outlines of the objects within the camera's range. Thermal imaging cameras are available in many designs. The camera can be handheld or helmet mounted. However, no matter what their differences, they all work on the principle of heat differential.

Thermal imaging must *never* take the place of any of the tried-and-true search techniques. Although thermal imaging technology has added a new dimension to how firefighters work on the fireground, the fire service must guard against becoming too reliant on this equipment. Remember, thermal imaging cameras can still fail unexpectedly, leaving any firefighter not using proven search techniques lost and disoriented inside a structure. Searching firefighters should use perimeter or lifeline search techniques in conjunction with thermal imaging cameras. Combining both standard search techniques and thermal imaging equipment optimizes the search safety. By doing this, both firefighters and the public will benefit (fig. 19–13 a and b).

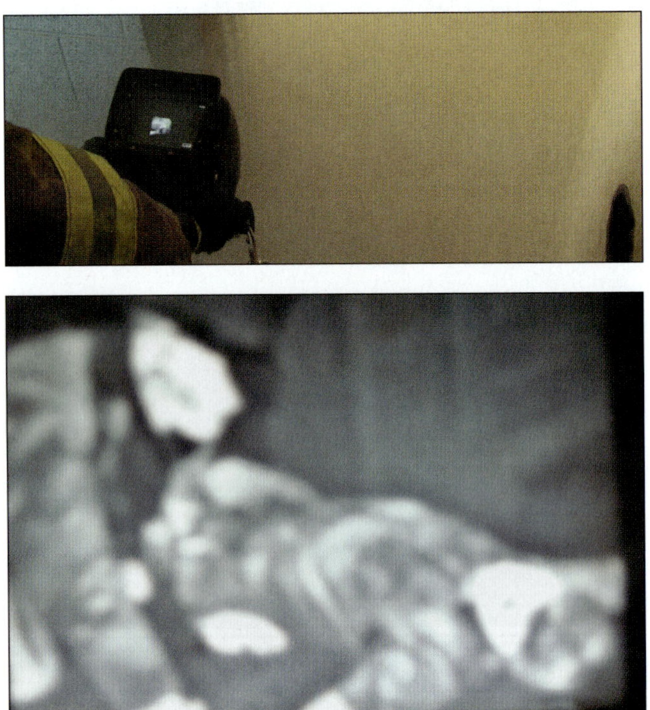

Fig. 19–13 a and b. Thermal imaging camera technology is arguably the most important innovation to search and rescue. The use of TICs has spread to nearly all fire departments, small and large. (Courtesy of Bullard)

Firefighters must know how to operate the controls on their thermal imaging camera and be completely competent in the recognition of different objects viewed through the camera (e.g., what a person, door, or different types of openings look like). The firefighter should be trained in what the different heat signatures look like when viewed through the camera. Some cameras read heat signatures on a black and white screen: the brighter the white, the hotter the object. Newer cameras read heat on a color screen using oranges and reds to indicate levels of heat. No matter how your camera operates, you should be thoroughly trained in its operation before trying to operate it in a hostile environment. The best training combined with the latest technology results in a safe and successful search. When the thermal imagining camera is assigned to the officer that is directing a search crew, the officer can use the advantage of the camera to direct the searching firefighters. The officer can direct them toward any victims that he sees and away from any encumbrances the searching firefighters will encounter. This will greatly reduce the amount of time required to search an area. It should be noted that the searching firefighters still must exercise great care to keep from becoming disoriented.

Searching with a hoseline

A hoseline may be used in conjunction with firefighters' individual tether lines if a rope is not available. One advantage to this method is that the hoseline will protect the firefighters that are searching. A disadvantage is that the search will be limited to the area covered by the hoseline. In addition, a hoseline will be cumbersome and difficult to relocate, thus slowing the entire search operation to a slow crawl. However given a choice between no rope and searching off a hoseline, choose the hoseline every time. It's better to be safe than sorry.

MASK CONFIDENCE AND EMERGENCY PROCEDURES

Mask confidence or emergency procedures are techniques that each and every firefighter must learn and become highly competent in using. Techniques such as the reduced-profile maneuver, swim method, and the quick-release maneuver must be practiced and mastered by each and every firefighter. The use of these methods at a moment's notice may mean the difference between life and death. Every firefighter should also be intimately familiar with the emergency operation of the particular SCBA that they are using. All these procedures and methods must be mastered before a firefighter enters a hostile environment.

Reduced-profile maneuver

SKILL DRILL

The **reduced-profile maneuver** enables the firefighter to move through areas that would normally be too small to fit through, such as a stud wall frame (fig. 19–14). The normal spacing for stud walls is 16 in. (40 cm) on center, leaving the firefighter a little bit less than that amount to squeeze through. The firefighter executes the following steps:

- Start by loosening the waist strap of the SCBA.
- Loosen the SCBA's strap (the strap opposite the unit's regulator) and remove your arm from that strap.
- Reach back with one hand and grab the bottom of the unit's cylinder near the valve. Then turn the SCBA unit in the direction of the strap that is still in use.
- Tuck the top of the air cylinder under your armpit, thus aligning the bottle with the side of the firefighter.
- After the firefighter moves through the restricted area, reverse the previous steps and redon the SCBA.

Figs. 19–14. In emergency situations, firefighters may have to reduce their profile in order to fit through wall studs or other tight spaces.

Swim method

SKILL DRILL

The **swim method** is used when firefighters encounter an entanglement hazard but are not yet entangled. Turn the SCBA unit toward the corner where the wall and the floor meet (fig. 19–15). This allows the air bottle to be directed toward an area and reduces the probability of it catching on the hazard. This also forces you onto your back and in the correct position for the next step in this maneuver.

Extend an arm with the palm facing the floor above the head, similar to the motion a swimmer would use to perform the backstroke. The extended arm then lifts the entanglement above your head as you continue to crawl forward under the entanglement. Continue this maneuver until you clear the area of the entanglement hazard.

Fig. 19–15. The swim method is used to help a firefighter avoid becoming entrapped in an entanglement hazard. It is always preferred for firefighters to learn to avoid hazards rather than have to learn how to recover from them.

Quick release maneuver

SKILL DRILL

To avoid entrapment, use the **quick release maneuver** (fig. 19–16). It should be employed if the firefighter has already become entangled and needs to escape.

1. Backtrack in a straight line. Drop down on your stomach try to move forward again. If you are still entangled, start over.
2. Fully extend both shoulder straps.
3. Slip your right arm through the right shoulder strap.

4. Grasp your left shoulder strap *and regulator hose* with your left hand.
5. Disconnect waist belt buckle with right hand.
6. Turn 180° *to the left* to face entanglement, maintaining *grip with left hand.*
7. Free yourself from the entanglement.
8. Redon unit.

Fig. 19–16. Entanglement hazards have caused serious injuries and deaths to firefighters over the last several decades. Because of this, it is essential that all firefighters learn basic disentanglement techniques.

VICTIM REMOVAL

Drags and carries

SKILL DRILL

Victim removal is an important part of our job, which is often not practiced enough. Keep in mind that when performing these techniques, you must use your leg muscles, *not* your back, to prevent injury to yourself.

The goal of victim removal is to prevent further injury to the victims and remove them as quickly and safely as possible. It is easy to practice these techniques in a controlled environment; however, it is extremely challenging when firefighters are in a stressful situation with zero visibility, high heat, and limited knowledge of the layout of the building. Victims that are located by searching firefighters should be removed to a clean area or removed to the outside of the structure as soon as possible. When removing said victim, the firefighters should make every effort to keep the victim's head as low to the floor as possible, thereby keeping the victim closer to the cleanest air.

FFI 4.3.9 The practice of a firefighter removing his/her face piece to give it to the victim is strongly discouraged. The reason for this is that the victim might not want to give it back and then the firefighter would be in a life-and-death struggle to get the face piece back from the victim. In addition, firefighters should try to remove victims to the exterior of a building by the path that gets the victim safely out of the "immediately dangerous to life and health (IDLH)" atmosphere to medical treatment as quickly as possible. For example, when a fire is located on the first floor of a single family home and has not yet been knocked down, it is usually better to remove a victim through a second-floor exterior window rather than down the stairs to the first floor and then to the exterior. Similarly, it may be advisable to take a victim to the public hallway in a multiple dwelling *when the hallway is relatively clear of smoke.*

There are multiple procedures for removing victims from an IDLH atmosphere and many carrying techniques, depending on your environment. Some carries are more feasible than others, depending on the conditions. These include the following:

- The blanket drag
- The seat carry
- The firefighters carry
- The extremity carry

The blanket drag. If the victim is laying on his or her back, spread the blanket next to the victim, fully opened. Roll the victim toward the rescuer and roll them back onto the blanket. Roll the blanket around the patient, supporting the head and neck. The patient should now be dragged head first, with head and shoulders slightly off the ground (fig. 19–17).

The seat carry. Two firefighters are required for this carry. The rescuers are to face each other. With your left hand, grasp your own right forearm just around your wrist. When both have done this, face each other. With your right hand grasp the other firefighter's left forearm firmly, and he or she should grasp yours. This forms a square seat (fig. 19–18 a and b).

The firefighter carry. The firefighter carry can be performed by one firefighter on either a conscious or unconscious patient (fig. 19–19).

This carry is not recommended for a smoke-filled environment because you may subject yourself and the victim to extreme heat and toxic gases. Firefighters performing this carry should make sure that the victim weighs less than their own weight.

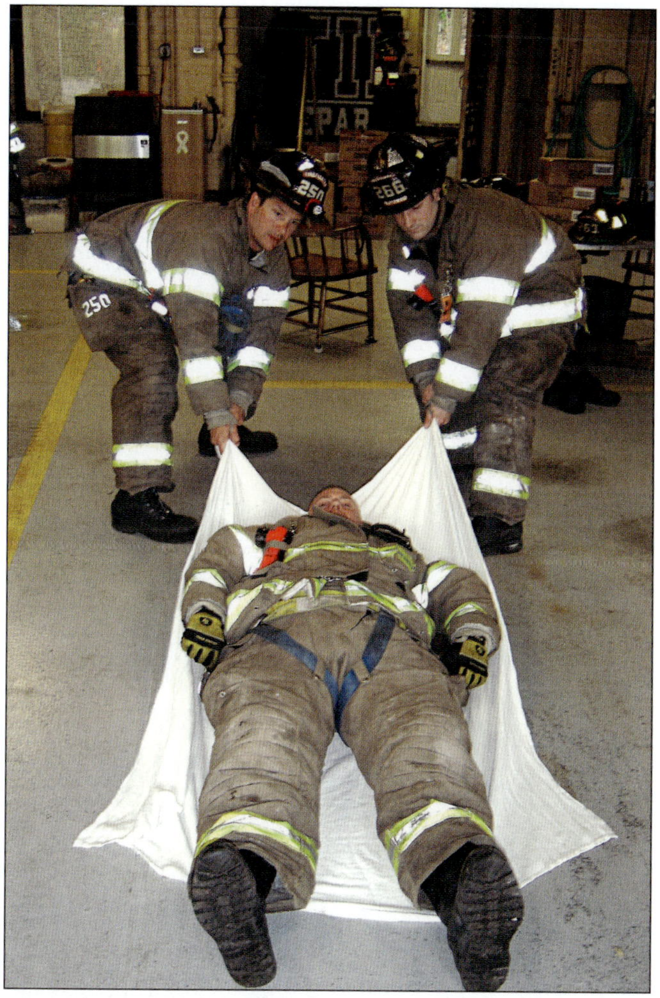

Fig. 19–17. The blanket drag can be used to move unconscious or heavy victims. This maneuver can be attempted by a single firefighter, but is most effective with two firefighters.

Fig. 19–19. The firefighter carry can be used on a conscious or unconscious patient, but should *not* be used under smoke or fire conditions.

Fig. 19–18 a and b. The seat carry is accomplished with two firefighters on a conscious victim.

The extremity carry. The extremity carry requires two firefighters and uses victims' extremities to carry them to safety (fig. 19–20).

Fig. 19–20. The extremity carry

The next set of carries should only be done when conditions warrant it:

- Clothing drag
- Webbing drag
- Downed firefighter drag
  - Using SCBA
  - Using turnout coat harness

Clothing drag. The firefighter should squat or kneel at the head of the victim grasping firmly on the clothing on each side of the victim's head. Drag the victim across the ground. Support the victim's head on your forearms when using this technique. It is important to stay as low to the ground as possible so that the victim's head does not push downward toward the chest and cause the victim to have difficulty in breathing (fig. 19–21).

Fig. 19–21. Firefighters using the clothing drag to move a victim must monitor and support the victim's head.

Webbing drag. The webbing drag can be performed by either one or two firefighters. One-inch tubular webbing is one of the most important tools a firefighter carries. This webbing should be joined by way of the water knot to make a complete loop of the webbing and should be carried this way (fig. 19–22).

Steps for these and other drags and carries are in your skill drills book.

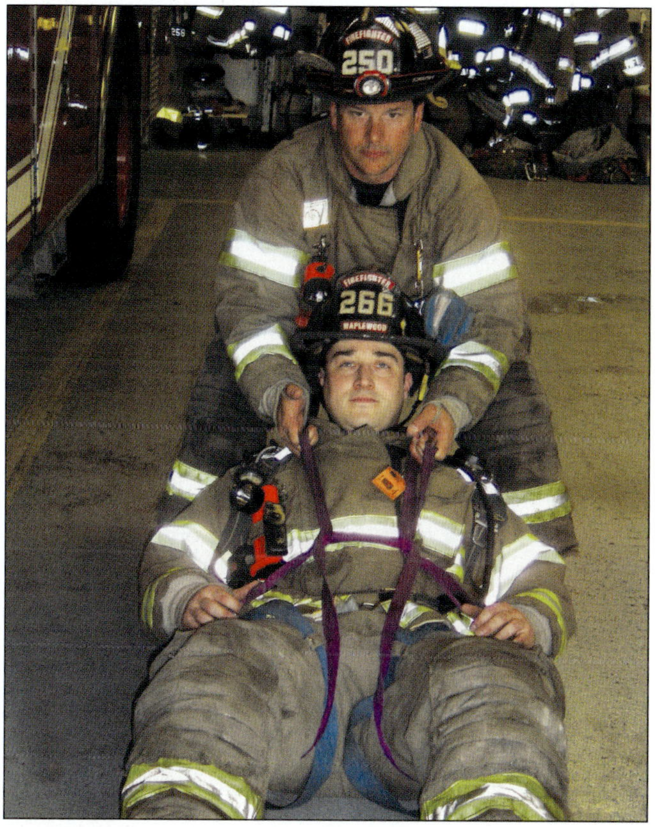

Fig. 19–22. Using 1-in. (25-mm) tubular webbing, one or two firefighters can assemble a makeshift harness and drag a victim to safety.

Drag rescue device. A system that was placed into service in all turnout gear, the drag rescue device (DRD) became a mandatory feature of firefighter protective clothing as part of the 2007 edition of NFPA 1971, Standard on Protective Ensembles for Structural Firefighting and Proximity Firefighting.

The purpose of the DRD is to aid firefighters in the rescue of an incapacitated firefighter by dragging him or her along a horizontal plane. The DRD is not designed or tested for use in vertical rescue operations.

1. Locate the DRD handle beneath the flap on the upper back of the coat shell (just below the collar).
2. Insert a hand though the DRD handle and grasp firmly.
3. Pull the handle to eliminate slack in the DRD.
4. Drag the downed firefighter from the life-threatening environment.

Once deployed, the DRD provides mechanical leverage for dragging a downed and incapacitated structural firefighter from a life-threatening environment.

Note: Do not use the DRD for lifting or lowering a person on a lifeline or emergency escape. Using the harness for unintended purposes may result in bodily injury or death.

FIREGROUND NOTE

Search quickly; search thoroughly! We owe this to the citizens we are sworn to protect (if you forget this, refer back to the oath you took when you were sworn in). Remember to protect yourself, stay oriented, and, most of all, keep your head in the game. If you lose your cool, you may next lose your life!

LESSON FROM THE FIREGROUND

The time is 0230 hours. Your ladder company is dispatched to a report of a dwelling fire. As you step on the rig and try to shake the cobwebs from your head, you begin to run through a million scenarios. The radio breaks in, stating that a patrol officer is reporting a confirmed structure fire with possible entrapment. Your ladder company is riding with one officer and three firefighters. Per SOP, you and your partner are assigned to roof ventilation and the officer and remaining firefighter are to provide forcible entry and begin the primary search. As you approach the scene you find a three-story multiple dwelling with fire showing from the second floor and heavy smoke showing from the third floor and cockloft.

You and your partner position the aerial and begin to ascend to the roof position to open up the scuttle and provide much-needed relief to the firefighters trying to make the stairs. Suddenly the radio comes alive—it's your company officer ordering you to change tasks and place the aerial to the front bedroom window in order to vent, enter, isolate, and search for missing 7-year-old twin boys. You obey the officer's request and remove the bedroom window, allowing the smoke to vent. You sweep under the window before sounding the floor, just as you were trained to do. You enter the room, climbing over a box located under the window, and immediately proceed to the bedroom door and close it. As you begin the search, the heat becomes unbearable, so you quickly perform a perimeter search and exit back out to the aerial just before the room flashes. As you push yourself out the window onto the aerial you knock over the box that you climbed over on the way in—something that will change your life forever.

The engine company makes the room shortly after your exit and quickly knocks the fire down. As they enter the room, they find victim number one partially hiding under the bed. Remember the box you knocked over? A second victim is found under the box. It was a toy box that the child used to hide from the fire. You are still unaware of these tragic deaths as you return to the roof and complete your original assignment. Upon exiting the roof you hear the tragic outcome and begin to second-guess every move you made during that search. You wonder what you could have done differently. Did I search thoroughly enough in the amount of time I was in the room? Did I give those children every chance to survive that I could have? This author was that firefighter. This is an event that I will question for the rest of my life. Remember: You will not be able to save everyone, but you must give every victim the best chances possible.

Remember: Search quickly, but search thoroughly. We owe that to the people we protect and to the oath we took. My hope is that after reading this chapter, you will never have to wonder if you did all you could.

Stay safe, search well, and Godspeed.

QUESTIONS

1. When performing a size-up, what are possible indicators that could lead an emergency responder to believe a structure is occupied?

2. When searching a multi-story building involved with a fire, why would searching the floors above the fire be of a higher priority than the bottom floor?

3. When performing a primary search, it is helpful if rescuers are aware if they are looking for an adult or a child. What are some different reactions that adults and children tend to demonstrate?

4. Secondary searches are usually slower, more methodical, and more thorough than primary searches. Why are these attributes not exhibited during the primary search?

5. Before entering into a room or structure with a closed door, why is it recommended that the rescuer check the door prior to entering with the back of a closed hand?

6. How can door construction and direction of swing provide clues to the type of space located on the other side?

7. When performing a search for victims, why is it important to search the outside of the structure as well?

8. How do the stages of fire within a structure affect your choice on the technique to be used to perform a search?

9. What are some of the advantages for rescuers to occasionally pause when performing search operations?

10. Why is it important that marking procedures showing when a room search is/has been conducted are dictated in the standard operating procedures?

11. What are some of the risks of using a thermal imaging camera that a rescuer is unfamiliar with?

12. How does the fire stage affect the choice of drag or carry when removing unresponsive victims for heated atmospheres?

13. How can the dangers of using lifeline searches increase or decrease when searching large structures?

14. Why is it important for rescuers to train in using mask confidence or emergency procedure techniques before rescue situations?

Basic Fire Attack

by Jerry Knapp with Christopher Flatley

This chapter provides required knowledge items for the following NFPA Standard 1001 Job Performance Requirements:

FFI 4.3.5

FFI 4.3.8

FFI 4.3.10

FFI 4.3.11

FFI 4.3.12

FFI 4.3.19

FFI 4.5.1

20

This chapter contains Skill Drills. When you see this icon, refer to your Skill Drill book for step-by-step instructions. SKILL DRILL

OBJECTIVES

Upon completion of this chapter, you should be able to do the following:

- Define *strategy, tactics,* and *size-up* as they apply to firefighting
- Describe the 10 steps for a successful fire attack
- Describe offensive and defensive strategies use at a fire incident
- Describe the importance of command control and communications at a fire incident
- List the common responsibilities of engine company personnel at a fire incident
- List the common responsibilities of truck company personnel at a fire incident
- Describe the key considerations for hoselines, water supply, forcible entry, ladders, ventilation, and search at a fire incident
- Describe improvements to traditional fire attack procedures based on current research studies on residential fires, most notably the research conducted at Underwriters Laboratories.
- Describe the importance of an after-action review of a fire incident

INTRODUCTION

This chapter brings together all the knowledge, skills, and abilities you have gained in previous chapters. It will provide you with a basic overview of what is required for a successful fire attack operation at a structure fire. As a Firefighter I, you will be given specific tasks to perform by your company officer or a senior firefighter. You will execute your assignment as part of a team, your fire company. It is the coordinated efforts of multiple fire companies working together that result in a successful fire attack (see fig. 20–1).

You might be part of an engine company, delivering water, deploying hoselines, or executing other tactics to extinguish fires. Or maybe you are a truck company member, assigned to place ladders; perform search and rescue, ventilation, or forcible-entry operations; secure utilities;and perform salvage or overhaul.

While you are performing your particular duties, it is important to understand the overall "big picture." Your understanding of how all the pieces of fire attack fit together will help you see the importance of your role. While not every task may be as dramatic as rescuing a child or being on a nozzle and knocking down a rapidly spreading fire, every task at a fire is critical to overall success. Your proficiency in carrying out your assigned tasks will directly affect the outcome of the fire attack.

Fig. 20–1. Aggressive interior firefighting saves lives and property and is at the heart of what we do best. (Courtesy of Tom Bierds)

It is important that you understand two important terms: strategy and tactics. While used interchangeably by civilians, they mean two different things. **Strategy** is the development and implementation of a plan, incorporating specific goals, to bring a fire under control. **Tactics**, on the other hand, are specific actions taken to accomplish the goals identified in an overall strategy.

Take a fully-involved car repair shop, for example. The strategy may be to take a defensive stance and protect exposures with the specific goal of preventing fire from spreading to an adjacent exposure. Tactics would involve placing a hoseline between the fire building and the exposure flowing water directly on to the exposed structure.

In order to understand all the elements of a basic fire attack, we will use a series of scenarios to illustrate the important details. The first scenario will be relatively simple. After you gain an understanding of the basics, we will advance into other, more complex, scenarios.

It is important to note that this chapter will cover only the basics of a fire attack operation, from pulling up on the scene to returning back to quarters after your job is done. The first scenario will be an uncomplicated fire but will require techniques, skills, tactics, and strategies covered in almost every other chapter.

At the end of this chapter the firefighter trainee will understand the following concepts and, more importantly, how they fit together for a successful fire attack operation:

- Strategy: An overall plan, including goals, of a basic structural fire attack
- Tactics: The specific firefighting procedures performed by company officers and firefighters to meet the overall strategic goals
- Size-up: An evaluation of what is happening
- Doctrine for an aggressive interior fire attack: Critical steps used across the country for structural fire attack
- Type of fire attack operations: Offensive, defensive, defensive/offensive, offensive/defensive (and the conditions that cause the transition from one to the other)
- Command, control, and communication: On the fire scene and why they are important
- Fire-extinguishment methods: Direct, indirect, and combination methods of fire extinguishment
- Coordination of engine and truck company functions

This scenario is based on a house fire because house fires often require all of our knowledge, skills, and abilities to be successful and safe. The house fire is also a good example because it is the most dangerous and deadly type of fire for both civilians and firefighters (fig. 20–2).

- U.S. fire departments responded to an estimated average of 358,300 home structure fires per year during 2010–2014, which represents three-quarters of all structure fires.
- Home fires caused an annual average of
- 2,560 civilian fire deaths, or 93% of all civilian structure fire deaths,
- 12,720 civilian fire injuries, or 87% of all civilian structure fire injuries, and
- $6.7 billion in direct damage, or 69% of total direct damage in structure fires.

Fig. 20–2. 2010–2014 Home Fire Data. (Source: NFPA U.S. Home Structure Fires Fact Sheet)

Communication

FFI 4.3.5 There are a few basic levels of communication that will occur at a house fire. Communication is the lifeblood of any emergency operation. Without effective communication, the operation is doomed to failure.

First, the dispatcher will provide information about the location, type of alarm (automatic, telephone, witness, etc.), the alarm location, and other important information the dispatcher may have (fig. 20–3).

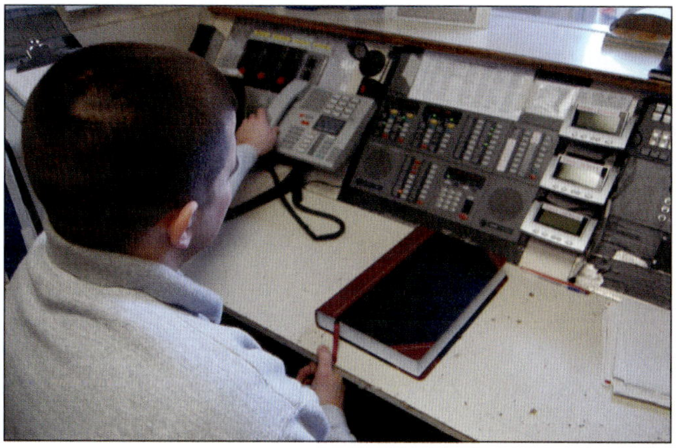

Fig. 20–3. A dispatcher

Messages received from dispatchers often hold clues to the fire condition. Words like "across from" or "next to" will indicate that the fire may not have been reported by someone inside the structure but by a passerby.

Generally, chief officers will arrive before the apparatus (though not always) and give a quick "on-scene report" confirming the address, providing a brief description of the fire, and giving orders for the responding units (fig. 20–4).

Fig. 20–4. Chief officer giving orders

Let's take a look at a typical communication from a first arriving chief at a house fire. A typical initial fire report would be: "Battalion 23 on scene, 20 Main Street, a 20×40, 1½ story single-family home fire showing on A and B side, report of victims trapped." This report provides valuable information to responding firefighters and confirms the correct address and gives dispatch an initial report.

Important points you should gain from the initial report:

1. Confirmed address—you are going to the correct place.
2. Type of occupancy—residential (general knowledge of the typical hazards).
3. Size and height—type of ground or aerial ladders needed.
4. Location of fire in building and severity (may determine tactics).
5. Report of victims reportedly trapped (clearly this is the priority). When a life hazard exists, all on-scene operation support the search and rescue operations.

The on-scene report paints the picture for incoming units so you can begin to consider what priorities, actions, and tools will be necessary at this call.

Command will communicate with engine and truck officers to give assignments, direct search and rescue, coordinate ventilation and fire attack operation, monitor progress of assigned crews, and ensure the safety of all firefighters on the scene.

Unit leaders (engine and truck officers) will report back to command via radio on the progress of their assigned tasks and other valuable information such as fire location, the location and status of victims, and other safety considerations.

Interpersonal communication among the company members and with the company officer is critical for success and safety. The company officer will give initial orders, for example, "stretch the 1¾-in. attack line to the front door," or "we have a report of persons trapped on the second floor." Company members will have to coordinate actions to ensure that the orders and tactical assignments are executed properly. The nozzle operator may ask the backup position to flake out the line or ask for more effort to pick up the nozzle reaction. Firefighters will use their portable radios to communicate with one another during the entire fire attack operation.

COMMON TERMINOLOGY

Firefighters use common terminology to describe different areas of the fire building. This allows firefighters to communicate quickly about strategy and tactics and to give orders. On larger fires it is a way to divide up areas of responsibility. For example, a chief officer may assign a subordinate officer to be responsible for a particular side of the building. That officer will coordinate all actions necessary for the operation on that side or area and report to and work for the incident commander (IC).

FFI5.3.10 It is important to clarify the location terminology here. In National Incident Management System terminology, we would use the terms *Division A, Division B, Division C*, and *Division D* (fig. 20–5). These are essentially geographic areas or parts of the entire fire area. The division chief (in the past, this position was termed the "sector" chief") is responsible for all activities in that area and how those activities impact all other **sectors**. These latter designations also help orient all on-scene firefighters to particular locations in a building where things may be happening: spreading fire, a trapped victim, or a localized collapse, for example. Exposures are designated by letters corresponding to the side of the fire building they face: exposure D is the building that faces the D side of the fire building. Exposure D1 is the next furthest exposure on the D side of the fire building (fig. 20–6).

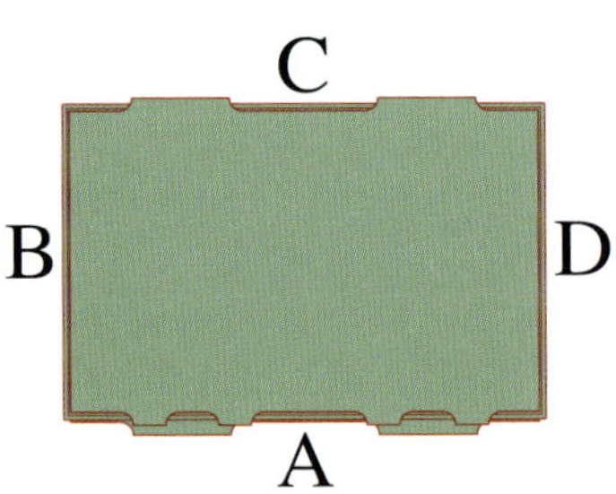

Fig. 20–5. Building sectors

The A side of the structure is usually the side of the building with the main entrance that faces a street, although not always. Sides B, C, and D are the other three sides assigned clockwise around the structure. Divisions can also refer to floors, such as floor 2, division 2. It is up to the incident commander to make the division designations.

It is important to note that your department may use a similar but different number or letter system to identify buildings and floors or areas of the fire building. We encourage you to learn your department's system. Clearly identifying locations in and around a fire building is critical to operational effectiveness and firefighter safety.

Structures that are built on a grade may have what appears to be a ground or first floor in the front and the back if the land slopes downward from front to back. This seemingly simple situation has caused a great deal of confusion on the fireground and has led to firefighter fatalities. The Cherry Road fire in Washington, D.C., in 1999 (www.niosh.com) is an example of this tactical confusion. The location and timing of ventilation is critical to fireground success; in the Cherry Road fire, uncontrolled ventilation of the lower floor of this caused a ventilation induced flashover on the first floor living area. Three firefighters on the first floor were attempting to find the basement stairs and advance hoselines into a fire in the basement. An open front door and uncontrolled ventilation from the rear (floor below) caused a flow path, providing the oxygen the fire in the basement and first floor needed for flashover. Two firefighters were burned to death and a third was seriously burned. Firefighters may report to command via radio that they are on the first floor in the rear when they are actually on the floor below the first floor from the front, where command will likely be. This leads to obvious strategic, tactical, and often fatal confusion and errors.

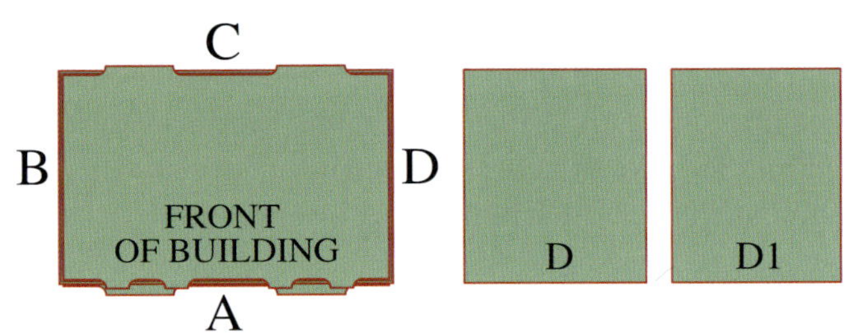

Fig. 20–6. Sector view with exposure designations

STEPS TO A SUCCESSFUL FIRE ATTACK

1. Size-up: the ongoing process of evaluating the situation before arrival, upon arrival, an during the incident
2. Develop a strategic plan (the incident commander's job)
3. Establish a reliable water supply
4. Force entry or exit into or out of the structure
5. Conduct search and rescue operations
6. Stretch and operate a hoseline; confine and extinguish the fire

7. Ventilate the fire building
8. Protect exposures
9. Fire control and overhaul
10. Pick up and return

YOU ARE HERE

1. **Size-up**
2. Strategic plan
3. Establish water supply
4. Force entry/exit
5. Search and rescue
6. Fire attack
7. Ventilate
8. Protect exposures
9. Fire control and overhaul
10. Pick up and return

Size-up

SKILL DRILL

Step 1 in any emergency situation, size-up is critical for gaining awareness to develop an action plan. The ISFSI's 2013 *Position Statement on Fire Dynamics Research* states "Size-up must occur at every fire... A tactical plan for that fire must be developed, communicated, and implemented." Size-up is the ongoing evaluation of what has happened, what is likely to happen, and the available resources (in what priority order) needed to resolve the situation. Everyone on the fireground does an initial size-up and a continuing size-up as the situation changes.

Therefore, you must conduct your own size-up: evaluating the situation, determining what you know and what you don't know, and formulating what you think the situation really is, despite several unknown factors. As a Firefighter I, you will be given specific orders. However, you must continually conduct your own size-up of the situation to ensure your safety, to be sure you are performing the correct tactics, and to ensure the operation is having the desired result.

Dispatch time. Here is an example of how to use size-up and existing information. Assume that your dispatch time is early in the morning. Would you expect to find people sleeping? Where would these people be? In this case they are likely to be in the bedrooms in the rear of the house. Consider also that the victims may have attempted to escape and have collapsed by windows or on usual exit paths within the house.

You might think that the time of day of this call makes it unlikely that anyone is awake. Consider that people may work nights and be coming home or just going to work. Was someone cooking, and was this the cause of the fire? Don't discount any bit of information; you may not initially see how it fits in the overall understanding of the situation.

On-scene information. Bystanders, civilians, and police officers may have valuable information about the fire or occupants. A neighbor may report that the family is all away on vacation, or report that one person was home alone. Not all civilian reports are reliable. No bad intentions intended, but they may be swept up in the excitement of the moment and provide bad information. Conversely, police officers may have good information. They may have been on the scene before the fire department and have had a few minutes to gather and possibly verify that information.

Location of fire building. From the dispatch information you know that you are not responding to the industrial side of town. The fire alarm was in a new development. Developments are usually built with the same or similar style. This fact may help you understand the floor plan of the house, even if you have never been in it.

Floor plan. Have you been in the houses before? Did you inspect them when they were being built, or does your neighbor live in one just like it? Knowing the layout of the fire building is one of the most important factors in a successful operation. For example, you can move quickly through the structure toward the sleeping areas when you know where to go.

If this were a two-story home, what rooms may you expect to find on the first floor? Kitchen, living room, dining rooms, and maybe a utility room if there is no basement. What does that tell you bout what type of fire you may be encountering? What and where are the utilities? Electric, natural gas, or is it bottled gas or propane? Will that affect operations?

Fire development and fire flow paths. The size and location of the fire are some of the more critical parts of your size-up. Let's look at two examples: if you have a basement fire in a home and the occupant left the interior basement stair door open then ran out of the house and left the front door open, the fire will likely race up the interior stairs. The fire will then quickly involve other fuels on the first floor and show out the front door of the home as the first-floor furnishings begin to burn. This is a simple explanation of a flow path. The fire is going to go from an area of high pressure (near the fire) to an area of lower pressure (up the stairs). Air is being drawn into the fire along the floor flowing down the interior stairs and increasing the fire's rate of heat release, increasing the pressure and driving the fire and smoke up the stairs. Smoke and heated fire gases will be coming out the top of the door and fresh air in the bottom; this is a bidirectional flow path, air in the bottom and hot products of combustion out the top. Your size-up should include noting this bidirectional flow path. This indicates there is not another intake vent from which the fire can get air. An intake vent in a house fire can be another door or window that was failed by the fire or by uncoordinated ventilation. Intake vents allow oxygen into the fire and speed its growth and intensity.

If the fire causes the basement window to fail, this will allow the fire to have a fresh air intake. The home now can be compared to a wood stove or barbeque grill with the vent at the bottom open and the vent at the top open. The heat release rate will increase quickly and the fire will surge both in the flow path and to other areas of the home, just as it would in your wood stove. The flow path is fresh air in the basement window to the fire and up the stairs out the door. The exhaust vent, if it is the front door, will likely be completely filled with fire and smoke pushing out for the entire area of the vent.

Here's another example: Your size-up upon arrival from the front of a single-family 2½-story home shows heavy fire on the first floor. You are assigned to VEIS a second-floor bedroom. You ladder the window, take the window, sound the floor, and immediately after entering the room, close the bedroom door. Why close the door? To control the flow path. Did the interior search team or engine company bring the hoseline in leave the front door open? If they did, this is the inlet for the flow path. By closing the upstairs bedroom door you did not give the flow path an exhaust opening and did a great deal to slow down the fire's progress up the stairs.

Construction hazards. If the housing development where the fire is located is described as new, how new is it? The words "new construction" now almost always means "lightweight" construction. Lightweight construction has been around long enough to almost not be considered new. If this home is built with trusses, laminated beams, or other composite materials, those factors will hasten fire spread and early collapse.

Lightweight construction is now very prevalent in our country. It is critical to understand that lightweight construction (trusses, laminated beams, metal studs, etc.) often has little or no fire resistance. Expect early and catastrophic structural failures that will cause floors to collapse, sending firefighters into the burning basement; second and third floors to drop firefighters a floor or more, trapping those below; and roofs to collapse, trapping firefighters in burning rooms on upper floors. Prefire intelligence and planning are your best survival tools.

Resources available. You also know what other resources are responding with you. You know that another engine, a truck, and chief are on the way. Is this the typical response to a fire that is reported by telephone? Based on the report of a confirmed fire, the dispatcher may have dispatched additional units. The police department, whose unit may be on the scene before you, may also provide valuable information either directly or through dispatch.

You know what the staffing is on your engine; do you know what your assignment will be when you pull up? The ride to the scene is a time to make those assignments if they are not dictated by department policy.

Weather may also play an important part in your size-up. Will very high or low ambient temperatures limit the time firefighters can work? Will the doors and windows be closed or locked, delaying entry, confining the fire? Is there a high-wind condition that could drive fire quickly or unnaturally through the building immediately after the ventilation is completed?

The dispatch information reported a fire on the first floor. Did the caller give the address or say "next to" or "across from"? This may indicate that the call was made by someone not in the house but a person passing by or nearby. Could this indicate the possibility of an advanced fire on arrival? Are you receiving several calls reporting the same fire? Does this indicate that the fire is large enough to be seen by others? Or does it confirm a working structural fire?

At this point in your career many of these strategic decisions are not made by you. They are offered to give insight into what you will come to understand later in your career. You will do well if you control your anxiety and follow orders. When you put on your facepiece, try to control your breathing and keep your head.

Size-up upon arrival. The following are seven specific size-up points that must be considered to complete your size-up upon arrival:

1. Location of the fire and type of construction
2. Life hazard
3. Fire control
4. Forcing entry

5. Ventilation
6. Stretching considerations
7. Water supply

Size-up begins with receipt of the alarm and is a continuing process until you leave the scene. Everyone on the scene must conduct a size-up of the initial situation and a continual size-up of the fire scene. Everyone on the scene must conduct individual size-ups of the initial situation and continue to conduct size-ups, since incidents change rapidly.

YOU ARE HERE

1. Size-up
2. **Strategic plan**
3. Establish water supply
4. Force entry/exit
5. Search and rescue
6. Fire attack
7. Ventilate
8. Protect exposures
9. Fire control and overhaul
10. Pick up and return

Developing a strategic plan

Although it is the chief or first-in officer who will determine the overall strategy based on his or her size-up, it is important to understand the four basic fire attack strategies.

Strategy is the overall rescue and fire attack game plan for this call. Strategy is determined by command level officers. One possibility is to conduct a rapid search and rescue operation, supported by forcible entry and ventilation, then rapid deployment of the initial hoseline to protect the rescue crews and to attempt to control and or extinguish the fire.

Tactics, on the other hand, are the supporting tasks necessary to accomplish the overall strategy. It will be up to the company officer (or, in some cases, a senior Firefighter II) who has to make some tactical decisions (what room to hit with the line during the advance, where to cut the roof, etc.). It will be the firefighter who will execute the tactical decisions made by the company officers.

Offensive strategy. Offensive strategy is used when:

- Size-up indicates there are salvageable human lives in the building and by aggressively searching or attacking the fire we have a reasonable chance of saving them.
- There is significant property at risk and by offensive operations, with a manageable risk to firefighters, we can minimize the loss to that property.

Offensive, interior operations must be balanced against the risks to firefighters, the probability of success, and of course consideration of civilian lives and the value of property we are trying to protect. Offensive operations are generally interior operations from the beginning and place both the search and rescue teams and the hose line teams into the building.

Aggressive operations may have some of the following factors present:

- Lives can be saved by rapid interior search and fire attack operations.
- The building has a limited amount of fire and has areas that are salvageable.
- The building is safe for firefighters to operate in for a short period of time or until complete extinguishment (fig. 20–7).

Fig. 20–7. Size-up of this fire reveals that the fire appears to be contained to one or two rooms on the second floor. An offensive attack could result in saving the building and contents with a reasonable risk to firefighters. However, if this building was built with lightweight construction materials, it could be a deathtrap for firefighters. Incident commanders must realize these dangers to firefighters. (Courtesy of Brian Duddy)

Offensive transitional attack. Offensive operations can begin with application of water from the outside of the building before both search and suppression teams enter. Research using fully instrumented full-scale live burns in private dwellings have proven this to be a valuable tactic available to firefighters.

Example: Let's assume you have just arrived at a daytime fire, and as always you are low on personnel. You and your officer have stretched the first line, thinking you can get inside, up the stairs, and get a quick knock down on this fire. The door is a high-security door, there are bars on first-floor windows, and you cannot get through the door to get your line upstairs. The fire has control of two rooms on the second floor, may spread to other rooms on the second floor, and will be in the attic soon.

Underwriters Laboratories (UL) research has proven the advice of experienced firefighters: apply water to the fire as quickly as possible. Use the reach of your hose stream to put water through the front or side windows to control the fire. Once you have darkened down the fire and likely prevented it from going to flashover, get inside and finish the job. If you were delayed for any reason, the fire will continue to grow unchecked. The option of not applying water allows the fire to spread unchecked further endangering any occupants and spreading fire throughout the structure. Water on the fire, even from the outside, has proven to both slow or stop the fire growth, reduce temperatures, and generally increase the chances of survival of trapped and unconscious occupants. This is called a transitional attack.

Many firefighters have been told this tactic will push fire through the home into other rooms that were not involved. The research and numerous instrumented live burns in residential structures show this concept of pushing fire not to be possible using proper nozzle techniques. In this case the stream is directed to bounce off the ceiling at a steep angle. It is important not to move the nozzle and block the exit of fire gasses from the window. It is important to note that the research as of this printing is limited to residential occupancies.

Defensive strategy. Defensive strategy is employed when it is determined that some or all of the following factors are present (fig. 20–8):

- There is no saveable human life present; it is not worth risking firefighters' lives for a futile rescue attempt.
- The building is so heavily involved in fire that it is not salvageable (there is nothing left to save, neither contents nor the building itself).
- The building is unsafe due to heavy fire involvement.
- The building is unsafe due to a fire load on the lightweight construction components.

Fig. 20–8. This fire calls for a defensive strategy. There is no salvageable human life inside and no saveable property, therefore no need to risk firefighters' lives.

LESSON LEARNED

Rock Hill Missouri Fire Chief John Kriska had preplanned all commercial structures in his first due area. Heavy fire greeted the engine company. The Engine captain, following the preplan, established a defensive attack. Moments later, the metal bar joist (lightweight construction) failed under the fire load, sending roof HVAC units crashing to the floor, exactly where the engine company would have been if an offensive attack would have been chosen.

Defensive/offensive strategy is used when the firefighter cannot immediately enter the building because of an extremely heavy fire involvement or other hazards prevent you from getting close enough for an offensive attack. Defensive/offensive operations are employed when some or all of the following factors are present:

- Heavy fire upon arrival
- Hazards that must be controlled before firefighters can gain entry
- When the fire is knocked down quickly, allowing the IC to consider changing to an offensive operation

A common method to employ this strategy is to use the **blitz attack**. The blitz tactic uses heavy streams from a safe distance until the fire is controlled enough to safely approach for complete extinguishment (see fig. 20–9).

Fig. 20–9. Here, downed electrical wires prevented firefighters from making an offensive attack.

Offensive/defensive. When possible, we try to mount an offensive strategy to make an aggressive interior fire attack in an attempt to save trapped occupants or the building or contents. Sometimes we must fall back to a defensive strategy, giving up the interior fire attack for the following reasons:

- This could have been the intention of the incident commander: Mount an interior attack until a search can be completed, then back out due to the severe risks to firefighters.
- No headway has been made on the fire (fire not darkening down), and extinguishments with interior line will not be possible (fig. 20–10).
- If the fire is progressing faster than our ability to put water on it, we may need to shift to a defensive position (proper use of a limited water supply).

Fig. 20–10. a) This fire started out as an offensive attack. b) Heavy fire conditions inside combined with a large amount of accumulated materials inside the house prevented the interior attack lines from advancing and thus forced firefighters to change from an offensive operation to a defensive attack. c) The fire was contained with no injuries to firefighters.

YOU ARE HERE

1. Size-up
2. Strategic plan
3. **Establish water supply**
4. Force entry/exit
5. Search and rescue
6. Fire attack
7. Ventilate
8. Protect exposures
9. Fire control and overhaul
10. Pick up and return

Establish a reliable water supply

There is one inescapable truth in the fire service. To extinguish the fire, you must apply more gallons per minute to overcome the total heat the fire is producing to ensure extinguishment. To accomplish this, one of the first priorities is to establish a reliable water supply (figs. 20–11 and 20–12). The first-in engine company usually has two main options to establish a reliable water supply: forward lay from a hydrant or reverse lay to a hydrant. Fires in a rural setting will require drafting from static source such as a swimming pool or lake or using water from tenders. Tender shuttles may be required for large fires in a rural location. (See chapter 15, Water Supply for details.)

Fig. 20–11. Establishing a reliable water supply

Fig. 20–12. Establishing a reliable water supply is a critical step in the extinguishment process. (Courtesy of Brian Duddy)

YOU ARE HERE

1. Size-up
2. Strategic plan
3. Establish water supply
4. **Force entry/exit**
5. Search and rescue
6. Fire attack
7. Ventilate
8. Protect exposures
9. Fire control and overhaul
10. Pick up and return

Forcible entry

Firefighters conducting forcible entry are usually the first team to the building because they have only hand tools (Halligan, axe, hydraulic door opener, etc.) with them, as opposed to the engine company, which is slowed by stretching the hoseline.

Common tools carried by the forcible-entry team include the flat-head axe, Halligan, and/or a hydraulic door opener. The flat-head axe is often replaced with a maul or sledgehammer as a tool to drive the Halligan tool. However, the occupancy and characteristics of the response area will dictate other appropriate tools. For example, if the area has a lot of industrial occupancies tools selected, maybe a rotary saw and even a hydraulic spreader could be used on bars or reinforced metal doors with multiple locks. Metal foldout gates or roll-down security shutters also require a rotary saw for rapid forcible entry.

The forcible-entry team will make their size-up of the building and choose the most appropriate place to enter. At house fires, this is usually the front door (fig. 20–13). Choosing where to enter the building is an important tactical decision. In this chapter we will discuss residential fires, and in *Firefighter II* we will discuss other types of occupancies. It is important to note that where you enter the building may have significant consequences in terms of your effectiveness and safety. This is significant for several important tactical reasons:

- The front door is generally the fastest and most used way into and out of the home.
- Victims may be found near the door if they were overcome and unable to escape.
- Opening the door may allow victims to escape.
- Opening the door will allow the engine company members to move quickly to the seat of the fire.

Fig. 20–13. Firefighters from the truck company force the front door. Note that the officer is holding a piece of webbing on the door handle to control the door once it is forced open. (Courtesy of Jerry Knapp)

There are three important safety factors to consider when conducting forcible-entry operations.

First, forcing the front door of a home or other building that has a fire in it is a ventilation opening. This large opening may make conditions very near you better by lifting the smoke and heat but you must consider the overall effect on the fire. Did you create a flow path for the fire? Did that flow path either on the floor near you or in a remote area or upstairs or in the back of the home or near the seat of the fire endanger other firefighters or civilians? Did the air you let in cause the fire to dramatically increase in size and intensity? If it did, your ventilation opening may have been more of a hazard than a help to the operation. Ventilation does not equal cooling unless the fire is being or has been suppressed. Ventilation means more air for the fire which is exactly what this ventilation limited fire needs to become a monster.

Second, control the door. When forcing the door to the fire area, always be able to control that door with a rope, piece of webbing, or tool. The sudden inrush of air could cause the fire to flashover or backdraft, trapping you or firefighters in the building. Many firefighters have been killed by rapid/instant ignition of smoke and products of combustion in a public hallway, or above the floor the door is on or on the floor when fire overruns their position. Often, as in the case of the infamous 1999 Cherry Road (Washington, DC) fire, firefighters did not have time to open the nozzle to save their own lives. Consider the hazard of rapid fire development as a result of inappropriate ventilation.

Third, consider forcible exit, that is, how are you going to get out of the area you just forced your way into if conditions suddenly worsen? Is there a window you can get out of? Is there a ladder at that window, or are there bars on the window? In a home or apartment, can you breach the gypsum board wall in time to save yourself and crew? If you are operating in a multistory apartment building and plan on using your bailout system, find your anchor point as soon as you enter the area. Trying to find it while the place is lighting up around you will be too late!

YOU ARE HERE

1. Size-up
2. Strategic plan
3. Establish water supply
4. Force entry/exit
5. **Search and rescue**
6. Fire attack
7. Ventilate
8. Protect exposures
9. Fire control and overhaul
10. Pick up and return

Conduct search and rescue operations

The most important strategic and tactical objective on the fireground is search and rescue. Saving the occupants of the building is always our first priority. Controlling or suppressing the fire may be the best way to accomplish this task, but the point is all operations support the search and rescue operation until that objective is met. We must remember that when we enter the building, we are equally valuable occupants. When a search and rescue operation is in progress at a scene, all operations must support the search and rescue. For example, the first hoseline must protect the means of egress for search team members. If the fire can be extinguished by this line, that's fine, but the primary purpose of the line is to protect the means of egress and the crew from advancing fire.

We must remember that search and rescue are two separate events. Entering the structure and searching for victims (and the location of the fire) is a dangerous task. When an entry team finds the fire, that information must be communicated to command and the engine, so they can bring the line to that location and extinguish the fire. When the team finds a victim, that ends the search and begins the rescue operation for that victim. This does not mean the search is over; other victims may still be in the structure. Rescue, in this case removing the person from the building is a physically and emotionally taxing task. Consider how difficult it is to move an unconscious human weighing 200 lbs. The victim maybe burned and hence very slippery and none of us are equipped with carry handles.

When the victim is found, that information must also be communicated to the IC. Command will then be sure that medical treatment is available as soon as the victim has been removed from the structure. The egress route the team will be taking should also be communicated to other members operating in the area so they can clear the path to the exit.

This also allows the other fireground activities that are occurring simultaneously to support the rescue effort. When the rescue status is known, it greatly reduces the chances that the fire will be pushed on the search team by an advancing line or inappropriate ventilation. Ventilation will be performed to lift the heat and smoke and increase visibility. Forcible entry will be performed to remove obstacles and clear the path to the exit.

Search and rescue tools. It is important to select the proper tools to take with you for a search and rescue mission. Obviously, full personal protective equipment (PPE), including self-contained breathing apparatus (SCBA), are required. If you are going inside to work at any structure fire, a portable radio and handlight are also critically important. A thermal imaging camera will make your search much more rapid, effective, and—when used properly—safer for you and your crew.

Hand tools should be selected based on your department's SOPs. If the responsibility of the search and rescue crew is also forcible entry, then the Halligan tool and an axe are good choices. Your partner will have a 6-ft hook and possibly a water extinguisher, which are good complements to your search and rescue tools. Situations dictate specific actions, but this combination of tools provides a variety of options and good use of these tools to complement your search effort.

The Halligan and flat-head axe can be used in combination to force doors to gain entrance to the building. (See chapter 12 on forcible entry). The axe can be used to chock a door open after you force it. The Halligan can be used to sweep the floor as an extension of your arm to increase the efficiency of searching larger open areas (fig. 20–14).

Fig. 20–14. Sweeping with a tool

The 6-ft (2-m) hook can be used also as a search tool. It can be used to take windows, if necessary (fig. 20–15). The power extinguisher can be used to hold a fire or extinguish a smaller fire found during the search.

Fig. 20–15. Using a 6-ft (1.8-m) hook to open a window panel

When selecting tools, consider the number of personnel available and what specific task you will be accomplishing with that tool. Be wary of "do everything" tools because they may not function in any purpose with the effectiveness you expected. Also be suspicious of hooks shorter than 6 ft (2 m). Although seemingly a great idea, you selected a hook to provide an extended reach, and 3 ft (1 m) is not much of an extended reach. The standard time-tested tools described above have been successful in the fire service for many years. Learn to use these basic hand tools effectively and with mental agility, and they will assist you in the search and rescue mission.

LESSON LEARNED

We pulled up at about 0530 hours one morning, and the police officer met us as we stepped off the rig. He said there was a victim in the back bedroom, just below the window. When I asked him how he knew that, he said he saw him standing there, then collapse due to the smoke. Lesson learned: The police department maybe on the scene before you in enough time to collect valuable information about the status of victims or not inside the house.

The search and rescue plan. A standard search and rescue plan for single-family dwellings uses the two-team approach (fig. 20–16). The two-team search plan at this fire will include one team entering the front door and the second team attempting to enter nearest to likely victim location. In two-story homes this will be the second floor for nighttime fires. The outside team does not necessarily have to enter through the second floor. One team forces the door at the most common entrance to the home, and the one most likely used by the occupants (usually the front door). This opens a door the escaping occupants might be using. The second team makes their way from the outside to rooms or areas where people are most likely be found. They may make use of portable ladders or porch roofs. You can see where knowledge of the layout of the house would be a huge benefit in this case.

Outside search teams use VEIS. They vent the window (working off the porch roof or ladder), enter the room and immediately close the door to that room to isolate that room from approaching fire. They then conduct the search of that room. If they leave the door to the room on the second floor open, this will likely create a flow path fire to travel up the stair or into the now open room from rooms involved on the second floor.

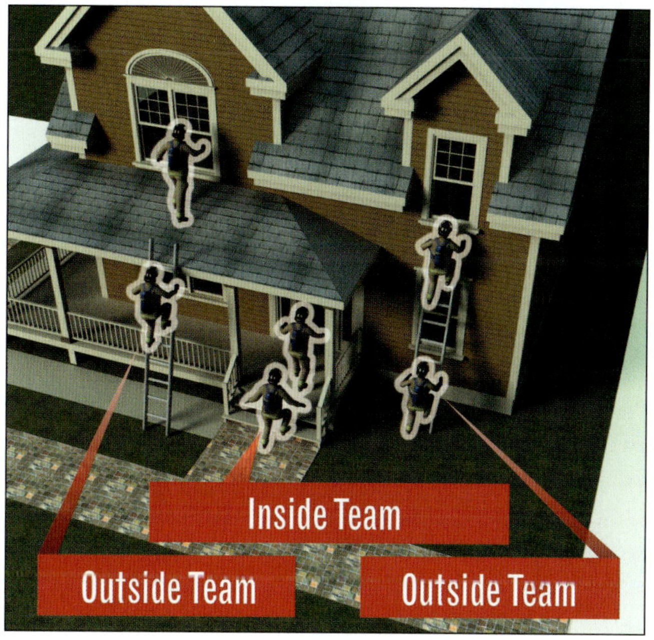

Fig. 20–16. Two-team search plan. The second team can either enter above a porch or through a second story window.

The preferred means of removal (as taught by the New York City Fire Department [FDNY]) is to remove victims by interior stairs first, then through some horizontal exit or possibly an adjoining building, then fire escapes, then ladders (with an aerial being preferred to a portable ladder), and last being by a life-saving rope. All of these may not be an option in a private dwelling. Use the most practical method based on the situation and the best method to do no further harm to the victim.

The condition of the victims will have a great deal to do with how they are removed. With unconscious individuals, you may have limited options. Victims with burns and severe injuries may have to be moved as quickly as possible, risking other injuries to provide the best chance of survival.

Persons with lesser injuries may be "sheltered in place" until conditions improve to move them out of the building. A rescue should not place an individual at a greater risk than the current conditions. Removal of victims must consider the fire flow paths in the building. Obviously it would not be in the victims' best survival interest to take them down a stairwell that was turned into a flow path by improper ventilation.

LESSON LEARNED

I had the nozzle at a working house fire one night, and we were stopped at the locked front door. For some reason the truck company was delayed, so I decided to conduct the forcible entry operation with my foot. I leaned back and planted my size 10 boot on the door, missing the frame of the door and hitting a center wood panel. My foot went right through, and there I was with a charged line, fire blowing out, with my foot stuck in the door! Heavy fire or a large dog behind the door could have made this a painful lesson. Lesson learned: Use the forcible entry tools; the door provides protection from fire and other hazards if you control it!

An engine company with a charged hoseline can do more to effect the rescue of trapped or injured persons than any other small group of firefighters.

It is important to note that there are a variety of effective ways to protect lives of both civilians and firefighters. What does not change is that saving and protecting lives is always our first priority. Four of the most common ways lives can be protected are the following:

1. Immediate removal
2. Placement of hoselines to protect means of egress
3. Protection in place in areas of refuge
4. Fire extinguishment (if it can be achieved very quickly)

When a search and rescue operation is required at a fire, all other operations (fire attack, ventilation, forcible entry, and overhaul) must support the rescue operation. Putting the fire out and removing the bodies later is not a successful outcome. Hoselines can push fire onto victims or search crews. Inappropriate ventilation tactics can cause the fire to light up and trap and burn firefighters. It is up to the IC and the IC's division officers to ensure that these inappropriate and dangerous acts do not happen. All fireground actions must support the search and rescue.

Size-up is an important part of the search and rescue plan and operation. Evaluation of what you see during your size-up for your search operation will provide you with many clues that will help make your operation effective and safer for firefighters. It is important to note that these are only clues, not definitive answers. For example, does five days' worth of newspapers on the front porch mean that no one is home? Of course not; maybe the occupant is sick or simply has not picked them up. Here are few other considerations in your size-up for search and rescue:

- Does the house look occupied?
- Are there multiple mailboxes, utility meters, and so forth, indicating a multiple dwelling?
- Are there bikes or kids' toys outside the house?
- Are there cars in the driveway?
- Are TVs, air conditioners, or heaters running?
- Are there ramps that might indicate the presence of disabled people in wheelchairs?

Don't base your entire search and rescue strategy on one bit of information.

LESSON LEARNED

Peck street: Upon arrival, civilians told us that a person was trapped in this house. Firefighters entered the building and left the front door open, thus allowing the fire unlimited oxygen, which caused it to go to flashover very quickly, nearly killing three of the search and rescue crew. The victim was at a local bar. Lesson learned: Information from civilians at the scene may or may not be correct. Always ask, "Is everyone out?" But carefully evaluate the answer! Ask, ask again, and look for inconsistencies. Evaluate the answer, interrogate if you must, and remember that you are placing the lives of your firefighters on the line based on what people tell you!

YOU ARE HERE

1. Size-up
2. Strategic plan
3. Establish water supply
4. Force entry/exit
5. Search and rescue
6. **Fire attack**
7. Ventilate
8. Protect exposures
9. Fire control and overhaul
10. Pick up and return

Fire attack

SKILL DRILL

This is the heart of what we do. Decisions will have to be made as to the size of the hoseline used in the attack, the positioning of the hoseline(s), and coordinating the hoseline attack with ventilation operations. It is not your responsibility to make these decsions—these decisions will be made by officers. You will, however, be expected to carry out the tasks dictated by these decisions (fig. 20–17).

Fig. 20–17. As the search and rescue team enters a structure, the open door provides the fire with needed air that allows the fire to continue in the free-burning phase.

A chief officer will coordinate the ventilation and fire attack operations. When the hoseline is in place, the officer will direct the ventilation team to break out or **take the windows** ahead of the hoseline, then direct the hose team to advance to the seat of the fire. These operations are important for the following reasons: Brother and sister firefighters have entered this building to conduct a primary search and rescue operation. The fire now can grow unrestricted until the engine company gets water on the fire. Controlling the fire minimizes all other problems and most other hazards on the fireground. Fire growth in compartment fires (inside a building) is often determined by the amount of oxygen available. The open door we enter provides an excellent source of the oxygen required to light up the building or room we are in.

Ventilation, entry, and search make up the fire service doctrine in North America. Although this doctrine provides the greatest opportunity for firefighters to make a successful rescue, it is very dangerous for firefighters. Rapid and effective engine company (fire attack) operations greatly reduce these hazards to firefighters that we have intentionally put in harm's way. Rapid and effective fire attack operations are the responsibility and mission of the engine company. It can be argued that this operation is the most important of all we do in terms of firefighter safety.

Knowing where to stretch the line is as important as what line to stretch. There will be times in your career where you will pull up to a fire and need a larger line than the standard 1¾-in. or 1½-in. (45 mm or 38 mm) hand line.

Large-caliber streams are produced by deluge guns, either apparatus-mounted or ground-based, ground-based rapid attack monitors, or the use of a 2½-in. (65-mm) hand line with a large tip to generate flows of 250 gpm (946 L/min) or larger.

For large fires, the FDNY has developed the acronym **ADULTS**. This is easy to remember by the phrase "ADULTS use big hose."

The **A** in ADULTS is for a fire that is *advanced* on arrival. Any fire that has surpassed the extinguishing power of your standard (1¾-in. [45-mm]) attack line calls for a larger attack line. Two ineffective hand lines will not have the reach and penetration as one large-caliber line. If the water cannot reach the burning material, it will have no extinguishing power.

The **D** is for the *defensive* positions. This means any line stretched for outside exposure protection should be capable of a large-caliber stream. This will allow the water to be applied alternately between the fire building and the exposure, providing fire-killing stream reach and penetration and good reach to provide excellent exposure protection, thereby minimizing damage to both with one stream (fig. 20–18).

The **U** is for fires where you are *unable to determine* the size of the fire area. If you can't tell how much is burning or how large the space it is in, you should not be going in without the protection and capability of a large line. This is not the firefight to be caught short of water!

Fig. 20–18. This is a good example of a fire that needs large lines. There is a greatly advanced body of fire, necessitating a defensive operation, and at this point it is still undermined how much of this building the fire will involve. The high heat production will require tons of water to extinguish the fire. (Courtesy of Brian Duddy)

The **L** is for *large* and uncompartmentalized fires. For example, the sales floor of a department store is no place to be without a large volume of water. The fire spread in a situation like that could easily overwhelm a small hand line.

The **T** is for the times when you need a *ton* of water to cool large amounts of burning material (fig. 20–19). For example, a lumberyard fire or building that has a lot of combustible contents, such as furniture. Another possible scenario is a flammable gas tank with flame impingement. For larger tanks you need a "ton" (500 gpm [1,893 L/min] minimum) of water to prevent a boiling liquid expanding vapor explosion (BLEVE). The much longer reach of the larger line keeps you much further away from the tank if it does explode.

The S is for *standpipe* operations. The FDNY has a policy that all lines stretched from a standpipe will be 2½-in. (65-mm) hose with a 1⅛-in. (30-mm) tip on a smooth bore nozzle. Most standpipe systems were not designed to supply the high pressures (100 psi [700 kPa]) required by a combination fog or automatic nozzle. The smoothbore nozzle and large-diameter hose will deliver superior flow at low (design) pressure.

Fig. 20–19. Fires in industrial occupancies require the reach, penetration, and cooling effect of large lines. Here firefighters stretch a 5-in. (125-mm) line to a ladder truck. (Courtesy of Brian Duddy)

Stretching. We will now explore the stretching responsibilities of each position on an attack hand line. These positions include the officer, the nozzle operator, the backup position, and the door position. Each position is critical for a successful hose stretch (figs. 20–20 and 20–21).

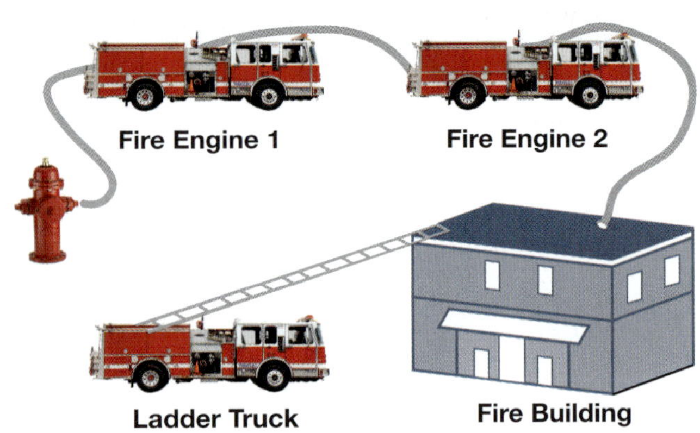

Fig. 20–20. Engine apparatus is positioned to establish a water supply and get the first hoseline in operation. Truck companies are positioned for search, rescue, ventilation, and other truck functions.

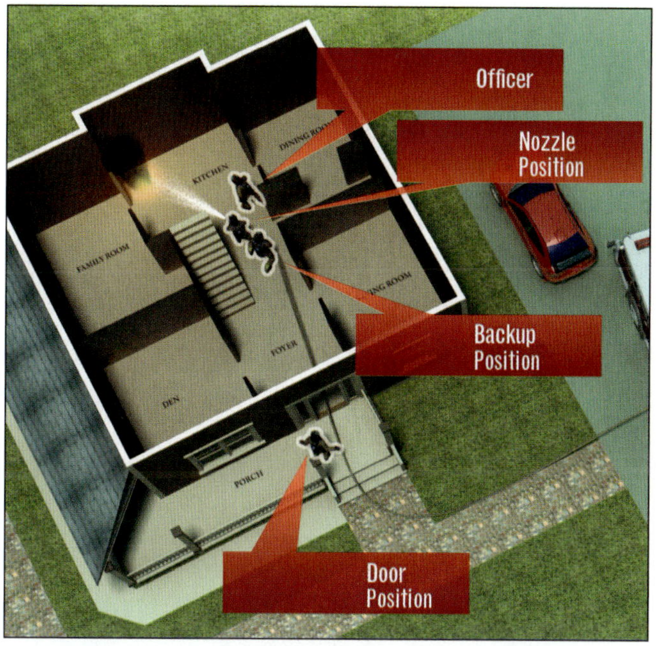

Fig. 20–21. Inside the fire building, each firefighter has a specific assignment and responsibility.

Officer duties in the stretch. The engine company officer (and, in some cases, a senior Firefighter II) is responsible for getting water on the fire as quickly as possible. Generally this officer will take direction from a chief officer on the scene. However, if the engine arrives before the chief, the engine officer will have to make some very important decisions and take decisive action. The engine officer can then pass command to any incoming officer or unit if appropriate. There are several subtasks the officer must accomplish to ensure water gets on the fire as quickly as possible.

1. Size up the fire (i.e., the size and number of lines).
2. Clearly order changes from the SOP.
3. Locate fire and tell the crew the best route to it (fig. 20–22).
4. Determine goal of line: exposure protection, fire attack, personnel protection, etc.
5. Be sure the line will not endanger others.
6. Keep members out of doorway "flue."
7. Keep members at floor level, below dangerous heat levels.
8. Coordinate timing of ventilation.
9. Be the eyes and ears of attack team.
10. Supervise the nozzle operator.
11. Call for relief when necessary.
12. Be responsible for the lives of the team.

Fig. 20–22. Here the engine officer directs the nozzle operator.

Nozzle operator duties in the stretch. There is a variety of ways to pack hose and an equal variety of methods get hose off the engine and to the fire. Described here are basic, universal skills that can be slightly modified to meet your department's specific attack line hose load. These are skills that will serve you and your company and department well if you learn, practice, and maintain them at a high skill level. In fire department terminology this is called **stretching the line**.

Nozzle operator duties

1. Carry the nozzle and the first 50 ft (15 m) of hoseline.
2. Work under command of the officer.
3. Know your department SOPs.
4. Get information from the officer.
5. Stop and wait for the backup position to shoulder the assigned hose.
6. "Bleed" air at door, check the flow, don facepiece.
7. Communicate with the officer and backup position during stretch and advance.

The nozzle operator is responsible for the first 50 ft (15 m) of hose, that is, to remove it from the engine or pumper and carry it to the point of attack (fig. 20–23). This length of hose is called the working line or lead length. It is the hose you will advance into the fire area with to kill the fire.

The nozzle operator shoulders the first 50 ft (15 m) of hose and carries it toward the fire. While moving toward the fire, an important step is to stop and wait for the backup position to shoulder his or her amount of hose. The nozzle operator should stop about 40 ft (12 m) away from the engine and wait just a couple of seconds for the backup position to grab the hose and carry it from the engine (fig. 20–24). Now moving together, you and your backup are stretching the initial attack line! If this is a 200-ft (60-m) preconnected hoseline, then you are carrying 50 ft (15 m), there is about 50 ft (15 m) between you, the backup is carrying 50 ft (15 m) and he or she is dragging the remaining 50 ft (15 m) off the engine. Properly executed, you two can stretch 200 ft (60 m) of hoseline (fig. 20–25).

The **point of attack** is the point at which you as the nozzle operator will drop your working length, flake it out, and request to your officer that the line be charged with water. This is termed **calling for water**.

You will have to use your size-up observations to determine where to drop your working length. In this case, because you don't know exactly where the fire is behind the front door, you will charge the line just outside the door.

Fig. 20–23. The nozzle operator has the working length on his shoulder and is moving away from the engine.

Fig. 20–24. The nozzle operator must wait a few seconds for the backup position to shoulder the load of hose. These few seconds will save a lot of time in the overall stretch.

Fig. 20–25. The door position clears the bed of any remaining hose, or if the load is not preconnected, will determine how much hose is needed and connect it to the pump discharge.

Preparing to charge your line requires you to flake out your working length (fig. 20–26). Dry hose can be literally thrown to allow water to flow through it easily and to minimize kinks that can reduce water flow and pressure.

It is at this point that you don your facepiece from your SCBA (figs. 20–27 and 20–28) and await the water in the line. When the hose is being charged, open the nozzle to allow the air that is trapped in the hose to bleed out. When the line is solid and fully charged with water, flow the nozzle fully to ensure that the pump operator has given you adequate pressure and volume of water. If you only open the line a small squirt, that is all you know you have, a small squirt. Flow the line for about 10 seconds to be sure you have full flow, correct pressure, and pattern selection for combination nozzles.

All the while you are stretching, flaking out, and advancing the line, you should be listening for a variety of things on the fireground. First, you should be listening to the radio to hear that coordination has been done for ventilation and that it is being planned or accomplished. Listen for the sound of glass breaking and/or the ventilation saw running. Good, timely ventilation will reduce the possibilities of flashover and backdraft that could injure or kill you. Listen also to reports from the search and rescue teams. Ensure that you will not push fire or superheated products of combustion onto them, causing injuries, burns, or death.

Listen for the sound of the line as you operate the line inside. The sound of the line will indicate if the stream is bouncing off the walls or ceilings. The sound will also indicate openings like doors and windows and also holes in floors.

Fig. 20–26. The nozzle operator and backup firefighter flake out their line in the direction of their advance.

Fig. 20–27. Firefighters don their hoods and facepieces. Note they are holding their helmets between their knees so they don't get kicked around or lost.

Fig. 20–28. Firefighters must be able to don SCBA facepieces and hoods with gloves on. This reduces the preparation time needed to attack the fire and prevent injuries. Firefighters must learn to conduct all necessary actions (full escape, reduced profile maneuver, etc.) with gloves on. Burned hands cannot help you save your own life.

Backup position's duties in the stretch

1. Second firefighter in stretch.
2. Carry second 50 ft of line and drop it at appropriate time and place.
3. Communicate with nozzle operator.
4. Relieve the nozzle reaction for nozzle operator during the advance.
5. Watch the nozzle operator's body and head positions during the advance.

FFI 4.3.10 The backup is the second firefighter in the stretch, responsible for carrying 50 ft (15 m) of hose and dropping the hose as the line is laid out (fig. 20–29). The door position drops his or her line first. Like the nozzle operator, door position must continually be evaluating the position of the engine relative to the fire building location. From this observation, door position will determine where and when he or she will drop the length and assist in the stretch.

In our example, the point of attack is going to be at the front door. The backup firefighter will drop the hose and flake it out in a position such that it does not kink and allows for an easy advance into the fire area if the 50 ft (15 m) of working hose the nozzle operator has is not enough. Once the backup position has flaked the hose out so that it will not kink, he or she should quickly move toward the nozzle operator and assist him or her in flaking the line and preparing to attack the fire.

The primary responsibility of the backup firefighter is to relieve all the nozzle reaction from the nozzle operator. As water is discharged and flowing from the nozzle, an equal and opposite reaction is realized by the nozzle operator. This is called **nozzle reaction**. Generally, the higher the nozzle pressure and greater the flow, the greater the nozzle reaction. During the advance into the fire area, the backup position must push the hoseline forward to counter this nozzle reaction so the nozzle operator does not tire and continues to advance and attack the fire.

Fig. 20–29. The backup position, moving in conjunction with the nozzle operator, carries one length of hose.

Each member should be able to carry approximately 50 ft (15 m) of hose and drag the same amount, allowing them both to efficiently stretch 200 ft (60 m) of hose. In our example, the point of attack will be at the front door. The backup will drop the hose and flake it out so that it does not kink. This will allow for an easy advance into the fire area if the nozzle operator's 50 ft (15 m) of working hose is not enough (fig. 20–30). Once the backup's hose is flaked out, the backup should quickly move toward the nozzle operator to assist in flaking that line and preparing to attack the fire.

Fig. 20–30. The backup position drops the hose first, allowing the nozzle operator to continue the advance to the fire.

Door position duties in the stretch. The door position is the third firefighter in the stretch. In our example, the door position is responsible for making sure that the preconnect bed is clear of hose, following the line quickly to remove any kinks, and feeding hose to the nozzle operator and the backup firefighter.

As the officer, nozzle operator, and backup position prepare to advance into the fire, it is important that the door position be very mobile (move back and forth along the line) and ensure that there are no kinks in the hoseline and that he or she can quickly supply the team with the amount of hose they need to advance to the seat of the fire.

The third person in the stretch clears the preconnect bed (fig. 20–31), carries the last 50 ft (15 m) of line toward the fire, chases kinks, and feeds the line from edge of the fire area outside the "immediate danger to life and health (IDLH)" atmosphere. They can relieve the nozzle operator, backup firefighter, or officer, because the third person has conserved air by being outside of the IDLH environment. The third firefighter also has a safety link to outside, and knows how far and where the team has advanced.

Fig. 20–31. The door position makes sure the hosebed is clear of all hose during the first part of the stretch.

When the nozzle operator advances into the smoke, the door position must be careful to push just the right amount of hose to him or her. Since the line is rigid, the door position can push the nozzle operator into areas he or she may not want to go or push the nozzle out of his or her hands. A good technique is to push up a bow of hose onto a door frame or wall and allow the backup position or nozzle operator to pull it in as needed. This causes less physical exertion on them and provides hose only when needed, and doesn't push into dangerous areas.

The door position is responsible to see that every foot of hose needed by the nozzle team is provided to them so they can control their advance (fig. 20–32).

Fig. 20–32. One of the door position's responsibilities is to keep the hose advancing. This may require him to move hose around objects that snag the line. These include car tires, fence posts, trees, door frames, and saddles.

Advancing charged hoselines

Officer duties during the advance. The officer will be the eyes and the ears for the nozzle operator. Often the stream blocks the view of the nozzle operator. The company officer can be off to the side of the nozzle operator and provide a better view of the progress of the fire attack, for example, if the stream is hitting the desired target. As the officer controls and monitors the progress of the company, the officer has two hands free to send important radio transmissions and other important tasks. The officer will monitor the level of fatigue for the company and request additional firefighters, companies, or hoselines to assist if required. The officer's other duties include pulling the company out when possible as they become less combat effective. Tired firefighters are dangerous firefighters and are prone to injuries. In the end, it is the company officer who is responsible for the overall safety and accountability of the company members.

Nozzle operator techniques during the advance. There are a variety of ways to position yourself when you advance. There is no right way, and it is important to note that you may use many or all of these during your advance into the fire area until final extinguishment. Fatigue, speed of fire control, and other conditions will determine your speed of advance and subsequent positions while operating or backing up the nozzle operator.

In figure 20–33, the nozzle operator demonstrates good technique, with the nozzle out in front at about arm's length where he can rotate it around and distribute the water into the fire room. He has it under his armpit, using the large muscles of his shoulder to hold it. His right elbow is braced against his knee to help counter the nozzle reaction.

Fig. 20–33. Proper nozzle handling technique while stationary

Another nozzle operator position that provides a great deal of stability while advancing is demonstrated in figure 20–34. It allows the nozzle operator flexibility to direct the stream in almost any direction and to move in quickly when necessary. This position provides a great degree of overall flexibility, allowing one leg to be out in front confirming that the floor is safe. The knees can be switched to allow the bunker pants insulation to cool and reduce the chance of burns. Having only one knee on the floor instead of both reduces, by 50%, the chance of contacting nails, hypodermic needles, and dangers such as broken glass on the floor.

The two-knees-down method has the following disadvantages: It does not allow for rapid movement of the nozzle inward toward the fire, or for the firefighter to feel to see if the floor is intact ahead. Also, the thermal insulation in the knees of the bunker pants is compressed and can lead to painful burns if the floor is hot, if the floor covering is melted or was burning, or if hot water soaks through. Also note that the backup position is holding the hose too high, causing the nozzle operator to fight the spring action of the charged line (fig. 20–35). Firefighters should consider the charged line a solid pipe from the nozzle back 12 ft (3 m). Keeping this "pipe" straight will reduce fatigue on both the nozzle operator and the backup position.

Fig. 20–34. Proper nozzle handling technique while advancing

Fig. 20–35. Improper two-firefighter hose-handling technique while advancing

The backup is a key position in the aggressive interior fire attack (fig. 20–36). The backup takes a firm grip on the hose and pushes forward to relieve all the nozzle reaction. This makes the nozzle operator's job much easier and more effective. The backup position can also use an alternate grip to relieve the nozzle reaction to assist the nozzle operator (fig. 20–37).

The position in figure 20–38 allows the most rapid advance and provides a high degree of safety for the nozzle operator because both knees are off the floor. It puts the nozzle operator in a little higher position, which helps detect high heat levels that may precede flashover, thereby preventing advance into deadly conditions. Being higher also allows the backup to be in better position to relieve the nozzle reaction. This position is favored by experienced nozzle operators.

Fig. 20–36. Proper two-firefighter hose-handling technique while advancing

Fig. 20–37. Alternative two-firefighter hose-handling technique while advancing

Fig. 20–38. Alternative nozzle position hose-handling technique

As the advance begins, the nozzle team must remain low in the door (fig. 20–39). This will allow the smoke and heat to vent over their heads as they enter. Some experienced firefighters have professed that you should wait to see what the smoke does when the door is initially forced. The theory is that after the door is forced, if the smoke gets drawn back in, this may be a warning sign of backdraft. This is potentially difficult to see as it may be caused by the natural convection of air into the fire area. Remember the warning signs of backdraft, and look at what the smoke is doing before you force the door. Smoke is always a reliable indicator of the fire condition when you learn to read it.

Fig. 20–39. The hose team needs to stay low as they make entry to the fire building.

As you enter the structure, move into the room and out of the door opening. When you operate the stream, rotate the nozzle and listen to the sound of the water splashing off the ceiling, walls, and the floor. The sound will indicate any openings, including windows, doors, and holes in the floor (fig. 20–40).

It is important to remember that the hoseline will be charged with 50–100 psi (350–700 kPa) of water pressure. When the line is first charged, it will quickly straighten out. Hold it firmly and crack the nozzle to release the air. Flow the nozzle at full volume to ensure that you have proper water pressure and volume. If you crack the nozzle and release a squirt, you only know that you have a squirt, but you need to know you have full volume.

Fig. 20–40. Sweeping the floor with the line will wash debris away from you; this can be any type of hazard, whether biological, sharp, or other. Swooping the floor will also cool the floor, reducing the potential for burn injuries.

Water hammer. The hoseline is pressurized, and when the nozzle is open the water has significant velocity as it flows through the line and nozzle. Always open and close the nozzle slowly to prevent water hammer. Water hammer is pressure that is created when a flow of water is rapidly stopped and the pressure is redirected to the path of least resistance. Water hammer can increase pressure by as much as six times the pressure contained in the hose. This obviously can cause a burst length or damage to the pumper and other appliances.

Door position duties during the advance. The door position is the third member of the engine company and has an important role in the advance of the hoseline (fig. 20–41). The door position remains at the door during the advance and is responsible for chasing kinks from the door outward and supplying hose as needed by the nozzle operator and backup position.

The door position must be careful not to push in too much line because it will kink or could push the nozzle team into areas where they do not want to go.

Figures 20–42 and 20–43 show good techniques that the door position can use to supply the appropriate amount of hose to the nozzle team.

Fig. 20–41. The door position assists in the advance by feeding line in from just outside the fire area. He tries to maintain a full tank of air so he can assist and relieve the nozzle team when they become fatigued.

Fig. 20–42. Making a loop of hose on the floor allows the nozzle team to advance quickly and easily.

Fig. 20–43. Putting a loop of hose up on the wall allows the nozzle team to very easily pull the hose in as they move down the hallway. This is not always possible but is an excellent method.

This position has many important duties that are critical after the engine company has made it into the fire area. All of these duties relate to the safety of the crew inside.

- Some department's procedures direct the door firefighter to close the door as much as possible after the hoseline enters. This is designed to limit the amount of air entering the fire area and preventing a flow path from developing and increasing the fire.
- Monitor the fire behavior using the smoke conditions as an indicator using the acronym VVDC (volume, velocity, density, color). Volume of smoke: more smoke indicated the fire maybe getting larger. Velocity: speed of the smoke indicates the amount of heat and pressure driving it. Water on the seat of the fire should cause the smoke to reduce velocity and pressure. Density: smoke from uncontrolled combustion can be very dense. After water application slows or ceases the combustion process, density should decrease. Color: as combustion is controlled, unburned bits of fuel that make up the visible part of the smoke should decrease and white steam will start to overtake the original color of the smoke. Although this is generally true, sometimes depending on the fuel that is pyrolizing the smoke and products of combustion can be brown, light or white in color as well. If the smoke was dark when you arrived, water is on the fire and steam is being produced, you can assume the attack is being successful.

- The door firefighter has the extreme advantage of being able to see into and monitor what is going on inside the fire building and communicate easily with command and outside firefighters. This is a huge safety benefit to those inside. If firefighters inside cannot call for help, the door firefighter can be their link to positively impact the crew safety by communicating with command. A handlight positioned by the door firefighter near the door facing inward will act as a safety beacon under the smoke layer showing the location of the door.
- The fireground can be a confusing place, especially when there is an intense operation such as a search for civilians or a rapid intervention team deployment. The door firefighter can be a practical and immediate accountability position.
- As the title implies, the door firefighter must insure that the door locking mechanism does not trap firefighters inside. You can do this by disabling the lock with a tool or placing a rubber inner tube around the handles and latch. An alternative method is to use a spring clamp and attach it to the door. Latching the door is not a problem if there is a hoseline through the doorway. If a hoseline is stretched uncharged and slips under a door and then is charged, it can force the door shut and very effectively trap firefighters inside. An important task for the door firefighter is to prevent this.

The well hole stretch. This is a special type of stretch commonly used in older apartment buildings. It can be used whenever you encounter a well hole in the stairwell. The well hole is the space between the stairs and the edge of landing of the public hall in a multiple dwelling. Using the well hole will reduce the number of lengths required in the stretch, speeding the stretch and reducing friction loss (fig. 20–44). On average, one length of hose will reach from the base of the stairs to the fifth floor.

After dropping the folded hose at the base of the stairs, the second firefighter will proceed up the stairs, assisting with the stretch. When sufficient hose has been hauled up to the point of attack, the line must be secured by hose straps to the railing before the line is charged. When the line is charged, there is a possibility that the weight of the water could cause the line to fall back down the well. Having sufficient line at the point of attack cannot be overstressed. When the line is secured and charged, it is extremely difficult to haul up more line.

Additional firefighters can assist in the stretch, making sure that enough line has been brought into the building and properly flaked out. When the line is charged, they can chase kinks as they move up on the line.

If the well is wide, the nozzle operator may opt to carry a length to the drop point before hauling the line up the well. If a second line is required to be stretched up the well, the nozzle operator has no option but to take only the nozzle up the stairs. When advancing the line up the stairs, the nozzle operator must pass the second line around the first to prevent the two lines from becoming twisted around each other.

Often companies will store utility ropes in empty bleach bottles or other plastic jugs to facilitate rope stretches or hauling tools to the roof. If the well is wide enough for the jug to fall down it, a utility rope stretch can be used to get the first or the second line up the stairs.

Standpipe stretch. Standpipe stretches are another type of important stretch. Standpipe systems are covered in greater detail in chapter 30, Fire Protection Systems, but is also covered here as an introduction.

Simply put, a standpipe system is a fixed set of water supply piping with hose valves that is typically installed in high-rise and large area buildings. Standpipes avoid the necessity of stretching hose all the way from a fire engine in the street to a remote part of a building; firefighters simply attach their hand line to a hose valve close to the fire area to attack the fire. The standpipe system is usually pressurized with water (but not always); a fire department connection (FDC, also known as a "Siamese") at street level allows firefighters to pump water into the system to supplement it. In a high-rise, hose valves are most often located in the stairwells.

Operating hoselines from standpipe systems can be challenging. It requires close coordination on many levels to be not only safe but also effective. It requires the pump operator on the street to supply the correct pressure to the system for the floor the nozzle will be operating on. Simple friction loss calculations don't apply. It is important to coordinate the efforts of the firefighter at the floor outlet where the line is hooked up who maintains the correct pressure and an experienced nozzle team that ensures the nozzle is flowing properly.

Whenever a hoseline is stretched from a standpipe outlet, a pumper must supply the FDC to ensure an uninterrupted water supply. The hookup location must be out of the IDLH environment and never run the risk of being exposed because if there is any problem with the line, the firefighter controlling the outlet must be able to make corrections without the stresses of heat and smoke. The closest location that satisfies both

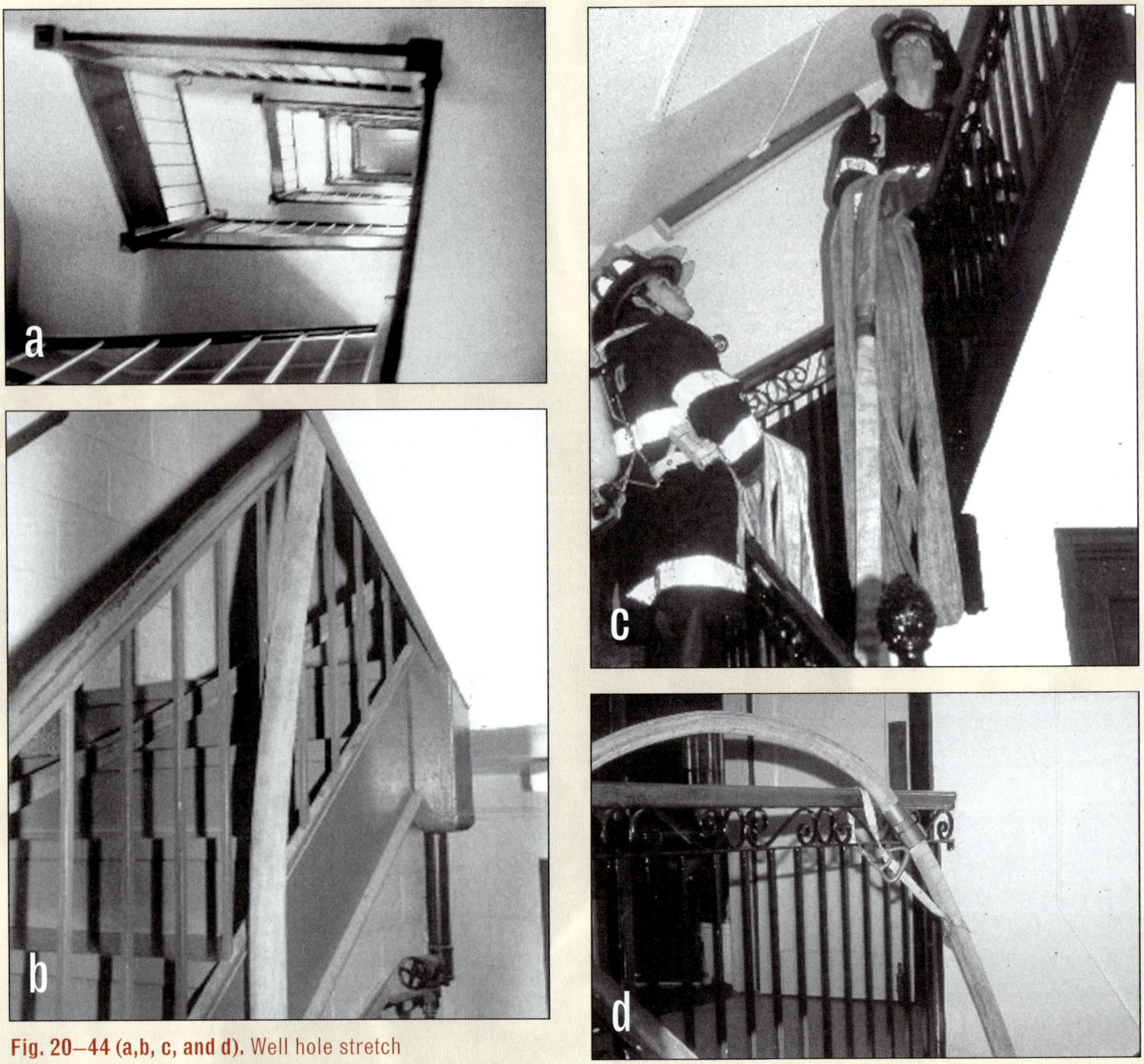

Fig. 20–44 (a,b, c, and d). Well hole stretch

criteria is often an enclosed stairway on the floor below. That stairway should be designated the **attack stair**, and all firefighting efforts should be started from that stair. If building occupants can be alerted of this stair's designation, escaping occupants can be directed to an **evacuation stair** and thus will not be subjected to heat and smoke.

When the hookup site is selected, based on fire conditions, closest to the fire and providing access to the fire area, sufficient hose must be brought to the location. In addition to lengths of hose, a standpipe kit is required. This kit must consist of items to facilitate the hookup with some contingencies. When operating from standpipes, the apparatus can be many floors away, so a complete kit can save a great deal of aggravation.

Some equipment that may be necessary in the kit includes, but is not limited to, a pipe wrench large enough to remove stubborn caps or pressure-reducing devices; spanner wrenches to tighten hoselines, to prevent excess water from accumulating on floors and creating slip-and-fall hazards or unnecessary water damage; and any specific adaptors that may be required for your location (fig. 20–45). Often adaptors are required to convert from pipe thread to hose thread or between municipal threads. Door chocks are always useful when stretching lines, and they take up little room and add little weight to the kit.

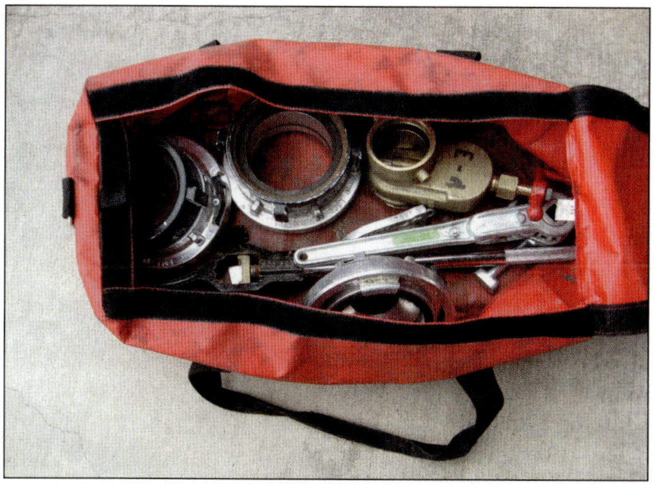

Fig. 20–45. Standpipe kit showing different components

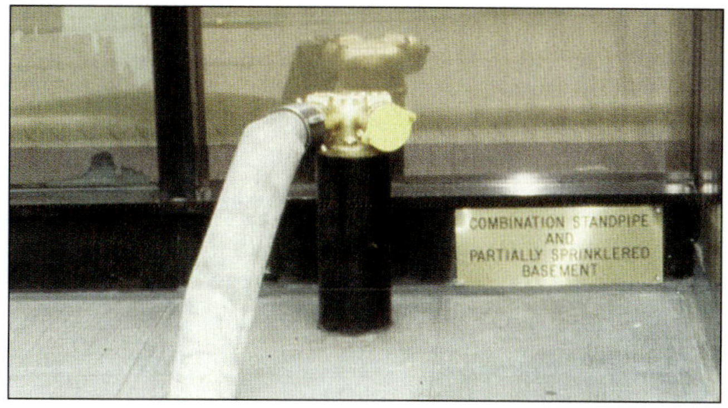

Fig. 20–46. Hoseline attached to a standpipe Siamese connection

Every standpipe kit must include either a pressure gauge or flow meter to ensure that the correct pressure or flow is being supplied to the line. A firefighter must hook up the gauge after the standpipe has been tested, so that any debris flushed from the pipe does not damage the gauge. A firefighter must be positioned at the outlet when the attack begins to make any corrections to pressure or flow when water is flowing. If available, this firefighter should remain in this position in the event of an emergency, burst length, and so forth.

Once the hookup has been made and the fire attack is to begin, charge the line and bleed the line as you would at any other fire operation (fig. 20–46). But when stretching from a standpipe, a final check of the stairs must be made before the door to the fire area is opened. Occupants' floors above the fire will be exposed to the toxic smoke. *A fire attack must never be commenced with civilians above the fire floor in the stairwell.* If a firefighter is available to be posted as a sentry on a floor above the fire floor that provides access to the other stair, this firefighter can remove any occupants from the attack stair to the evacuation stair.

When the door opens and the line is advanced in to the fire area, firefighters will be needed to advance the line down the hall. Flaking the line up to the next landing in the stairs will aid in the advance due to the weight of the line being pulled down the stairs.

Buildings that require standpipes are typically office buildings, high-rise apartment buildings, and "projects." They are built of steel and concrete, and the combination of the two will produce some of the hottest fires you will ever feel. Exacerbating the situation, these buildings have limited means of ventilation. These "brick ovens" will require a skilled team of firefighters with knowledgeable leadership to extinguish the fire and reduce injuries.

Methods of aggressive interior fire attack. There are six methods of structural or aggressive interior fire attack: combination, indirect, direct, modified direct (as described in chapter 16, Fire Streams), blitz, and transitional. Note that before an interior attack is initiated, several factors must be considered. Here are a few basic thoughts to consider:

Is the building safe to enter? Advances in lightweight construction have resulted in buildings that simply will not tolerate a fire load and will collapse rapidly, trapping and killing firefighters.

Heavy fire in any building should be carefully analyzed before firefighters are committed to entering the building and attempting an aggressive interior fire attack.

Are there salvageable human lives inside the building? "Risk a lot to save a lot; risk a little to save a little" is something a wise old fire officer once said. If fire is blowing out of every window in the house and through the roof, the odds of someone surviving are near zero. Firefighters are occupants also, so we must be careful not to unreasonably endanger the "new occupants" (firefighters) if the old occupants are already dead.

Similarly, if there is a reasonable chance that we can make a rescue, and we are not assured that everyone is already out of the building, a search and rescue operation is the key event at this fire. It is of the utmost importance to recall that trapped occupants cannot wait until the fire is extinguished to be rescued. After the fire is out, this operation is in fact a recovery (of a dead body), not a rescue operation.

Is there any property left to save? We don't have to get inside on every fire. Consider the outcome of many fires: heavy fire upon arrival, firefighters make an aggressive attack, one or more firefighters die or are injured, the fire is extinguished, and the insurance company pays huge sums of money so the building is built back better than it was before!

The point is this: We must take risks on the fireground to save lives and property. We must ensure that these are manageable and reasonable risks for ourselves and our members.

Recall that the doctrine for aggressive interior fire attack generally follows this sequence: Truck company members or other firefighters will force entry into the building to conduct search and rescue operations. Engine company members will establish a reliable water supply and stretch a hoseline to fire building to the seat of the fire or between the victims and the fire. Other truck company members will ventilate the fire building just before the engine crews push in to attack the fire. Officers will plan, direct, and coordinate these actions. Because firefighters will enter the building before water is applied, and the fire is at its most dangerous time (it now has an unlimited amount of oxygen causing it to grow unchecked), rapid, aggressive, and proper fire attack operations often will make the difference between life and death for other members inside. It can be argued that nothing is more important than extinguishing the fire because it minimizes all other hazards on the fireground.

Modified direct method of attack. The modified direct method of fire attack is the most common type of attack used. It is used when the fire has gone through the flashover phase or extended beyond one room to several rooms or fire areas, heavy smoke has obscured visibility, and/or conditions are impossible to make a direct attack (fig. 20–47). The room may literally be filled with fire from floor to ceiling. A good sign that these conditions are present is fire pushing out a door or window or heavy smoke throughout the fire area (fig. 20–48).

Fig. 20–47. This room has gone to flashover. The chair on the left and sofa on the right are fully involved almost to the floor level, and flames fill the entire room from floor to ceiling. Ceiling temperatures reach 2,000°F (1,093°C) or more.

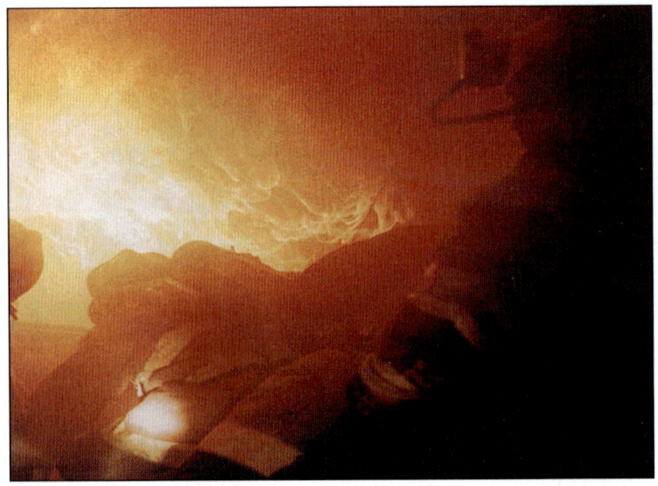

Fig. 20–48. Heat from flashover will force firefighters to the floor and make conditions impossible for firefighters to operate in until the fire attack begins. Turnout gear will become saturated with heat and not protect you from flashover conditions. The modified direct attack will help cool the atmosphere and extinguish the fire, allowing firefighters to enter, attack the room, and move through to the next room or fire area.

At the direction of the officer, the nozzle operator opens the nozzle fully and discharges water into the fire area. The nozzle operator will move the nozzle in an O, N, T, or Z motion to distribute water in the fire compartment to get water on the ceiling, walls, contents, and floor. Sweeping the floor with the nozzle will push burning debris out of your path. If the building happens to be infested with drug paraphernalia, sweeping the floor will push used needles and other biological hazards away from you. Listening to the sound of your stream can tell you if there are holes in the floor and where windows and doors maybe located across the fire room. Most fires, especially residential fires, will darken down in a few seconds if you are flowing decisive amounts of water. Fires with a high load of fuel such as warehouses and stacked material will take several minutes to darken down even if your stream can reach the seat of the fire.

When these conditions are present, the modified direct method of fire attack (as described in chapter 16, Fire Streams) is the most appropriate. This is an aggressive interior fire attack. The nozzle operator operates the nozzle in a straight stream or "zero degree fog" position if using a fog nozzle. Preferably, a smooth bore nozzle would be used for this type of attack.

The nozzle operator directs the stream in circles, starting at the ceiling level and moving the nozzle in a clockwise or side-to-side motion. The water bounces off the ceiling and is broken up and distributed in large drops of water that hit the burning fuel and extinguish it. The nozzle is rotated around, and the fire stream will directly hit the burning fuel (direct attack effect). This water will also cool the ceiling and walls. This reduces the heat that is radiated back into the room. It also helps cool the upper atmosphere in the fire area.

Note that cooling the fire area does not extinguish the fire. By cooling the room, in effect, you are treating the symptom (heat) and not the disease (fire). The large water droplets that are bounced off the ceiling and land on the burning fuel are large enough to extinguish that bit of fire, which contributes to the overall extinguishment process.

The nozzle operator uses the reach of the stream to put water on the fire, not only to first control some of it, but to see how the fire was going to react to the application of water. Live electrical circuits, flammable liquids, and **water-reactive chemicals** in the room may react violently with the application of water. Better that this happens when the nozzle team is outside the fire area rather than in it!

FFI 4.3.12 There are several important facts to understand when using the modified direct method of attack. According to The United States National Institute of Standards and Technology (NIST) report NBSIR 80-2120, *Fire Development in Residential Basement Rooms,* tests that were conducted utilizing 16 full scale burn tests of living room fires resulted in ceiling temperatures reaching 2,000°F. The room you are entering to extinguish the fire has very dangerous products of combustion near the ceiling that you do not want near you. It is therefore vital to maintain the thermal balance of the fire room during the attack (fig. 20–49).

Fig. 20–49. Thermal layering simply means that the hottest air is in higher levels of the room, and cooler air is in the lowest level, near the floor.

Nozzle choice for modified direct attack. The most proven fire stream (and the safest for firefighters) for interior fire attack is a solid stream from a smooth bore nozzle or a combination nozzle set to a straight stream pattern from a fog nozzle.

Because of the dangerous conditions inside during a modified direct attack, it is important to choose the correct nozzle and stream pattern. Since we are inside and close to the fire during the modified direct attack, it is important not to disturb the thermal balance that the fire has established. In essence you want to leave the bad (superheated) air up and away from you and keep the good (coolest) air down near you at the floor level.

The modified direct method of attack pits us against a fully involved fire inside the fire building where it is the most dangerous. Using a straight stream from a fog nozzle or a solid stream provides us the most fire-killing power and the least danger as proven by both experience and laboratory testing.

All fire streams will move air into the fire area as a result of their water movement. Firefighters usually enter through a door directing the stream into the fire area. Thus there is an unlimited amount of air that can be pulled into the fire room or area.

From actual air measurements, we know that a straight stream or solid bore nozzle will move about 725 cubic feet per minute (cfm) (20.5 cubic meters per minute [cmm]) into the fire area. This small volume of air can be adequately vented via a window or door ahead of the nozzle team. The result is that the thermal balance in the room where firefighters are operating is maintained as much as possible. The hottest air remains at the ceiling, and the coolest remains at the floor.

Maintenance of the thermal balance allows firefighters to move through this room that has just been extinguished and into other fire involved areas.

Fog streams: combination and indirect attacks. If you choose a combination nozzle for a direct attack and select a spray or fog pattern, huge amounts of air will be moved both inside the fire room and possibly drawn into the fire area. In tests conducted at the Rockland County, NY, Fire Training Center, air volumes of 2,000 cfm were measured when a 150 gpm fog stream was used. Estimates of peak air movement were as high 10,000 cfm. As a result of this air movement superheated gases from near the ceiling can be pushed around and drawn down on the advancing engine company. To put air flow in perspective, a 12 × 16 × 8 ft. residential room contains 1,536 cubic feet. In about 10 seconds a fog stream has enough energy to move this entire volume of air. During a fire attack with a fog stream the air movement results in what we call "Firemen's Soup" of which we are the main and painful ingredient. The engine company maybe enveloped in heat and steam. This technique will generally work for a one room fire but due to the disturbed thermal layering it is usually impossible for the engine company to move thru this room to other areas involved in fire.

A similar and equally dangerous effect created by fog nozzles when used inside a fire building is that if a nozzle team inadvertently moves down the hall past a room of fire, then operates a fog stream, it is very likely that the fire and superheated products of combustion will be drawn back onto the attack team, creating the potential for injuries and cessation of the fire attack (fig. 20–50).

Fig. 20–50. Firefighters have disturbed the thermal balance with a fog stream inside causing superheated steam to envelope them, creating the potential for injury and stopping the fire attack in the first room.

Further confusing the issue is that a fog stream appears to be a very successful technique for firefighting, especially from the nozzle operator's limited point of view. Holding the fog nozzle, you have a wall of water in front of you. The water spray leaving the nozzle, usually at 100 psi (700 kPa), makes considerable noise and nozzle reaction. Essentially the nozzle operator is fooled into thinking there is a significant strategic weapon in his or her hands.

Additionally, when demonstrating the use of fog nozzles in a parking lot, the fog stream appears to provide protection from the fire by the massive amount of air that is moved past the nozzle operator toward the fire area, thereby "pushing" all the dangerous heat and smoke away from the operator. This "parking lot experience" is in fact exactly the opposite of what happens when a fog nozzle is used inside, as described earlier in the nozzle selection section.

LESSON LEARNED

We responded to a garden apartment that had fire blowing out one window. The assistant chief told me to take the nozzle and move into the left. After several attempts at operating a fog pattern in a circular motion, and with the heat increasing each time, we retreated. We found that we were drawing superheated air from the room adjacent to us. We did not make the sharp turn into the fire room, and our fog stream pulled fire onto us from the next room.

A myth developed that you can "breathe off the nozzle" if you put your face close to the stream. Fog nozzles were used frequently before SCBA became a part of our personal protective equipment; air was available here because it was being drawn in from a clean air source behind the nozzle team. Some firefighters even incorrectly believed that oxygen was being broken off the water molecules. If this were true, the remaining hydrogen atoms would have created a very explosive condition in the fire area.

The fog stream in reality has little reach and penetration to reach the seat of the fire. Many a nozzle operator and backup position have been burned by using a fog stream inside a fire room.

It is important to understand the limits of a fog hand line and its dangers for interior firefighting. Do not base your fire attack strategy and techniques on your experience with one-room fires or parking lot experience.

The fog streams of the indirect and combination methods of fire attack (described in chapter 16, Fire Streams) often prevent firefighters from entering the compartment/building for search and rescue. Recall that the indirect method of attack was developed for shipboard firefighting where no one was occupying the cargo hold areas. The combination method is an extension of the indirect method, with a rotating fog stream in a compartment. The indirect and combination methods doom anyone still inside to certain death by scalding steam and toxic products of combustion. It also allows fire that the steam cannot reach in void spaces to continue to grow and spread unchecked.

All fire streams, including smooth bore and straight streams off combination nozzles will move air as a result of their water discharge. A low-pressure area forms near the nozzle, and air entrained by the water pushes nearby smoke and air away from the nozzle. From recent testing by Underwriters Laboratories we know that in general a 150 gpm straight stream or solid bore nozzle will entrain about 700 cfm of air if the nozzle is held in a fixed position. If, however, the nozzle is moved in the O, N, T, or Z motion, the air movement increases to generally around 7,000 cfm, which is obviously enough to disturb the thermal layering. Properly timed ventilation, just after or just before the engine company has applied water to the seat of the fire, will result in a successful coordinated fire attack. UL research on residential fires has shed considerable new information on the importance of carefully coordinating ventilation with the interior fire attack.

YOU ARE HERE

1. Size-up
2. Strategic plan
3. Establish water supply
4. Force entry/exit
5. Search and rescue
6. Fire attack
7. **Ventilate**
8. Protect exposures
9. Fire control and overhaul
10. Pick up and return

Ventilate the fire building

FFI 4.3.11 Ventilation is often described by experienced firefighters as the key to success for any aggressive interior fire attack operation. If ventilation is performed in the right place at the right time, it allows the engine company to rapidly advance on the fire and extinguish it, which will minimize all other problems on the fireground. Ventilation, or opening up of the fire room or building allows the products of combustion, heat, and smoke to exit the building, raising the thermal layers, raising the smoke layer, increasing visibility, and decreasing the potential for flashover. See chapter 14, Ventilation, for a detailed explanation of ventilating different types of buildings. There are four types of ventilation: horizontal, vertical, positive pressure, and hydraulic ventilation.

Ventilation: when and where. Numerous recent research studies by UL (previously mentioned in this chapter) have provided a wealth of knowledge regarding both the positive and negative effects of ventilation in residential structures (single story ranch homes and 1½-story homes).

These studies have shown that improper ventilation (time and/or place) can have disastrous results on the success of your fire attack and literally kill the crews inside.

When. Ventilating before there is water on the fire does not make it cooler. It actually may make it cooler immediately near you, if for example you take a window during a search. After all, you are letting a lot of heat out of the building. However, data from the live burn research shows that because there is so much fuel in typical homes (plastics, foam, other synthetics) the air you allowed in actually increases the heat release rate and total heat produced. It is the same for smoke production: more air equals more fire equals more smoke and more heat. Most structure fires are fuel rich and ventilation limited; they need air to grow. Ventilation before fire control will increase the fire and its toxic products of combustion and rate of heat release.

When do you ventilate during the fire attack process? In general, just before or just as water is being applied to the fire. This is a critical part of a coordinated fire attack. Do it right and it turns into an easy operation. Do it wrong and it will be a long and possibly painful operation.

Where. As we saw in the previous examples, ventilation can provide flow paths for the fire to extend along and into other parts of the structure. Fire will grow based on the conditions such as fuel and oxygen. Smoke that is so dense and oxygen poor and is beyond its flammable limits when mixed with air from improper ventilation can produce volumes of fire remote from the seat of the fire. Location of ventilation openings along with timing are critical to successful ventilation support for the fire suppression crew.

Generally speaking, you ventilate ahead of the fire attack team that is attacking the seat of the fire. This may be a horizontal vent opening such as a window and, of course, may create a flow path for the fire. Water on the seat of the fire should prevent a flow path from forming and increasing the fire intensity. Additionally, as the attack team moves in, some departments close the door (behind the attack team) as much as possible to delay the inrush of air that may create a flow path to or from the open window. Let's look at the two main types of ventilation: horizontal and vertical that are most often used during a fire attack.

Horizontal ventilation. As its name implies, horizontal ventilation consists of using or making horizontal openings (doors or windows) to ventilate the fire area or building. Horizontal ventilation will provide more efficient venting when doors and windows are opened opposite the advancing hoseline. The hoseline can drive smoke and heat from the building through the horizontal openings. A single firefighter with a tool (hook, pike pole, Halligan, or portable ladder, depending on whether it's a first- or second-floor fire) can create a great deal of ventilation (fig. 20–51).

Vertical ventilation. Vertical ventilation is the opening of a hole in the roof to allow the products of combustion, smoke, and heat to leave the building via the newly created opening. As with horizontal ventilation, vertical ventilation will allow the products of combustion, heat, and smoke to exit the building, raising the thermal layers, raising the smoke layer, increasing visibility, and decreasing the potential for flashover. The purpose is the same; the tactic is different (see fig. 20–52).

If fire has penetrated the attic space due to failure of the gypsum board ceiling or walls, opening the roof will draw the fire through the opening instead of allowing the closed up roof to cause the fire to mushroom and spread horizontally inside the attic (fig. 20–53).

In figure 20–54, truck company members have done an excellent job by cutting the roof allowing smoke and heat to escape. This prevented the fire from spreading into the main part of the house. When possible, ladders should not be positioned over windows. If fire erupts out these windows, firefighters could be trapped on the roof. In this case firefighters could escape safely with a short jump to the ground level.

Communications. Fireground communications are critical to this ventilation. The firefighter assigned to vent should be listening for the engine officer to order the hoseline charged. This should serve as a warning that they will be ready to move in shortly. If any doubt exists about when to vent, ask the officer directly. The FDNY has a written procedure that only the ladder company officer can order horizontal ventilation of the fire area. This ensures that members in an apartment are not exposed to the increase in fire from premature venting.

Fig. 20–51. a)Horizontal ventilation is most often accomplished by removing windows. b) When ventilating a window, remove all glass, screens, curtains, shades, and blinds. Remove the window sashes as well, so firefighters can escape quickly if necessary (make a door out of a window). (Courtesy of Tom Bierds and Brian Duddy)

Fig. 20–52. Vertical ventilation

Fig. 20–53. Fire drawn up through vent hole. (Courtesy of Tom Bierds)

Fig. 20–54. Firefighters exit the roof after successfully venting the fire.

YOU ARE HERE

1. Size-up
2. Strategic plan
3. Establish water supply
4. Force entry/exit
5. Search and rescue
6. Fire attack
7. Ventilate
8. **Protect exposures**
9. Fire control and overhaul
10. Pick up and return

Protecting exposures

FFI 4.3.10 Confine, then extinguish is the overall strategy to limit fire spread. Protecting exposures is the first step in confining and limiting the spread of fire, and ultimately extinguishing the fire. Confining the fire essentially means cutting off the fire spread and not allowing it to extend to other areas or buildings.

Exterior exposures. An **exterior exposure** is a building that, left unprotected, will be ignited by radiant, convected, or conducted heat from the original fire building or room.

When a fire attack operation is going to be defensive, one of the most common goals of a hand line is exposure protection. Remember that this goes back to the main goal of firefighting—to confine and extinguish the fire. By confining the fire to the building on fire, you are protecting the exposure. Protecting the exposure may involve placing the hand line between the fire building and the exposure, in this case, the neighboring structure (fig. 20–55).

Fig. 20–55. A hoseline is positioned between the fire building and the exposure building to protect the exposed building. (Courtesy of Brian Duddy)

When officers order a line to be stretched for exposure protection, they must make it clear what the goal of the line is going to be. A large hand line or ground-based monitor will probably be in a fixed position outside. A large hand line may be advanced into the exposure to direct their stream on the fire building. These are important distinctions that must be clear to the firefighters on the line.

An important note: When selecting the exposure that needs protection, the rule is to protect the exposure with the greatest value, not necessarily where the fire is going to spread.

To properly protect this exposure, a large-caliber hand line should be positioned between the fire building and the exposure. In this way the line can be used alternately on the fire building and the exposure. Water is applied *directly* to the exposure; water should be applied at the top of the wall so that it will cool the surface of the wall as it runs off.

It was believed that nozzles called water curtains that were similar to today's portable monitors were effective in protecting an exposed building from advancing fire. In truth, the water "curtain" or spray that is projected between the exposure and the fire building is not dense enough to absorb a significant amount of radiant energy to delay the ignition of the exposure. It is very likely that the reason these nozzles provided some degree of protection was simply because much of the water ended up on the exposed building. Put water on the exposed surfaces.

Further exposure protection may involve getting into the building and removing any curtains or drapes from the windows and any flammable materials that could ignite from the radiant heat. Operating a hoseline from the exposure is often a tactic. But in this case, because they are so close, that may not be appropriate since it might only allow smoke and heat to damage the structure.

YOU ARE HERE

1. Size-up
2. Strategic plan
3. Establish water supply
4. Force entry/exit
5. Search and rescue
6. Fire attack
7. Ventilate
8. Protect exposures
9. **Fire control and overhaul**
10. Pick up and return

Fire control and overhaul

FFI 4.3.10 Once the fire has been knocked down (i.e., when all visible fire has been extinguished), it is important to let it blow. Letting it blow means allowing the heat and smoke vent from the building so that any unextinguished fire can be discovered. If a hoseline is operated too long, it will overcool the area and the smoke will not lift. When to shut down the line is generally an officer's decision. If the timing is left to the nozzle operator, be alert for changing conditions and the need to open the line again.

This may be a time when a hoseline can be used for ventilation. A hoseline set to a fog pattern, held 4–5 ft (1.2–1.5 m) back from the window and directed out, will move the heat and smoke out. The same may be accomplished with a solid stream nozzle; by taking the tip off and using a broken stream from the shutoff, similar effects can be felt. This technique is often called hydraulic ventilation and can be very effective in immediately moving large amounts of air directly after fire suppression. This provides immediate relief for the attack crew. However, it is important to recall that you selected a nozzle for its ability to extinguish the fire, not ventilate the building

At this point the members assigned to overhaul should be allowed into the fire area with their tools to open up the area and examine for extension.

The process of opening up means that the interior wall, ceiling, and other burned surfaces must be removed. The idea is to expose the concealed spaces behind them until the unburned areas are exposed (fig. 20–56).

Fig. 20–56. Opening walls and ceilings after the initial fire attack must be done quickly to search for hidden fire. Fire can travel undetected in void spaces and is a serious threat to firefighters. (Courtesy of Tom Bierds)

Once the walls, floors, or ceilings are open, the engine company can come back in and wash down any hot spots. The hoseline should be directed on any charred surfaces until they are cooled. Pay particular attention should be paid to the areas where structural members are attached and where floors lay on beams and where boards are nailed together where heat can be held and continue to smolder. These members should not be removed, but instead the stream should be directed to drive the water between them until completely cool. Some of this work may be performed by relief crews from other companies.

LESSON LEARNED

We responded mutual aid one night to a fully involved house. The chief met me on the front sidewalk of the involved home and directed my line to put water on the exposure. My line operating off tank water, which saved the home next door and was not wasted on the fully involved, nonsalvageable home.

Often during overhaul, because of smoldering fuels, high levels of carbon monoxide and other toxic gases are produced. Although it seems safe to take off your SCBA because the majority of the visible smoke has ceased, many invisible toxic gases will remain and pose a threat to you both in the short term or show as various types of cancer in the long term. Wear you mask during overhaul to help reduce your odds of becoming a firefighter-cancer statistic.

YOU ARE HERE

1. Size-up
2. Strategic plan
3. Establish water supply
4. Force entry/exit
5. Search and rescue
6. Fire attack
7. Ventilate
8. Protect exposures
9. Fire control and overhaul
10. **Pick up and return**

Pick up and return

Generally, by the time we have made an aggressive interior attack on a working fire, we are all pretty well exhausted. Repacking hose and replacing ladders and tools now seems to take on a low priority. Resetting for our next alarm is, however, very important.

How well we reset and recover our equipment from this alarm may well determine how our company, or maybe even another group of firefighters, will perform next time. Like packing your own parachute, take time to do it right because lives—firefighters' lives—will depend on it.

When repacking hose, inspect it for damage. Worn jackets can indicate that the hose is compromised and may fail when pressurized. All firefighters should be familiar with their department's procedure for reporting damaged hose. If there is any doubt as to the serviceability of any length of hose, it must be tested before it is returned to service. Steps should be taken to prevent a damaged length from becoming mixed with serviceable ones.

Hose should be repacked when it is clean and dry. Mild soap and water is usually enough to clean the hose. Follow the manufacturer's recommendation and department SOP.

When the hose gets repacked, the important thing is not that it goes on but how it will come off. This has been compared to packing a parachute.

Tools. **FFI 4.5.1** Preparing your tools for your next alarm is also very important. Hand tools should be inspected for damage, cleaned, and placed back in their proper places on the rig according to your company and department policies. There is nothing that looks less professional and competent than a firefighter going from cabinet to cabinet looking for a tool at an emergency.

Power tools such as saws need to be refueled, oiled, and sharpened if necessary. All blades must be inspected to determine whether the blade is the correct one for the saw, has the required teeth, and is in generally good condition. Damaged or worn blades and saw chains can be swapped out to put the saw back in service quickly.

SCBA also needs to be inspected, serviced, cleaned, and tested before you put it back on the rig. Remember that this is life-saving equipment that you or other firefighters will depend on at the next alarm. Be sure to read and comply with manufacturers' instructions for cleaning and servicing. Harsh cleaners and cleaning methods should never be used. Simply reading the directions and properly maintaining SCBA will save your life one day. Improperly maintained SCBA may result in your death.

After action review. Before you leave the scene, maybe even before you recover your equipment and repack hose and ladders, your officer should conduct the **after action review (AAR)**. The purpose of this informal review, often held on or near the back step of the rig, is to capture any lessons learned from this fire. It is a time to review what SOPs you used or deviated from, how successful the operation was, what actions you want to sustain because they were successful, and what actions you want to improve or change.

It is important to do the AAR as soon as possible because we all want to "sanitize" our memory of the fire. That is, we want to dwell on what we did well and forget what we did that was not so effective or sometimes even just incorrect. Remember, this is not a critique of the fire or firefighters! It is a time to review the scenario you pulled up to, the size-up (was it really what you thought?), the tactics and strategy used, and what unknowns made this fire unique. It is also a great time to praise actions that were appropriate or outstanding.

A good technique to conduct the AAR is to look for three things you did well and three things that you would change. Ask firefighters first for three "ups"—what they did right. Then ask for three "downs"—what they did wrong. Additionally, a great question to ask is: What would you do differently at the next similar fire? Often you will get answers such as: "Well, I did this and it worked OK, but this other technique may have been better." This kind of discussion is noninflammatory and fosters the concept of what could we have done better, not what did we do wrong.

Above all, remember this is an after action review, not a critique of personal actions. In the end, it should help share information and develop the team, not criticize individual actions.

SCENARIO: FIRST FLOOR FIRE IN A ONE-AND-A-HALF-STORY HOUSE

Here are the details of your fire scenario:

This scenario assumes that you are familiar with your rig or apparatus (engine or truck) and its hose loads, ladders, hand tools, and all other equipment. This scenario is not intended to replace your department's standard operating procedures (SOPs) but to offer a good introduction to fire attack. As always, operate according to your department's guidelines.

It is 0223 hours, and you are dispatched to a one-and-a-half-story house fire with reported fire on the first floor (figs. 20–57 and 20–58). It is a prosperous neighborhood, a relatively new development with a lot of young families with children. To provide an overall understanding, we will keep this scenario simple and understandable: Only four units are responding: Engine 1, (your unit), Engine 2, Truck 1, and a battalion chief on the first alarm. The engine companies have a driver, an officer, and two firefighters. Truck 1 has a driver, an officer, and two firefighters.

Description of scenario: Heavy fire out kitchen window, heavy black smoke out front door, fire out side door, rear windows seeping smoke, and heavy smoke from the gable vent at the roofline. Occupants outside in their pajamas report that their teenage son did not get out.

To conduct a safe and effective response to this alarm, we will individually examine the necessary steps for success. In this scenario, as in most structure fires, these are the large steps or **benchmarks** that must be considered and dealt with. It is important to note that these steps may be accomplished concurrently or in sequence or not at all depending on the specific situation. Firefighting is not a cookbook type operation. It is one-third training, one-third experience, and one-third art and science.

Firefighters and fire officers must be mentally agile to adapt to the situation and conduct the most critical operations and prioritize others, if personnel is insufficient to do all of them at once. For the purpose of this scenario, we will have adequate personnel.

So let's go step by step at this fire.

Fig. 20–57. The house

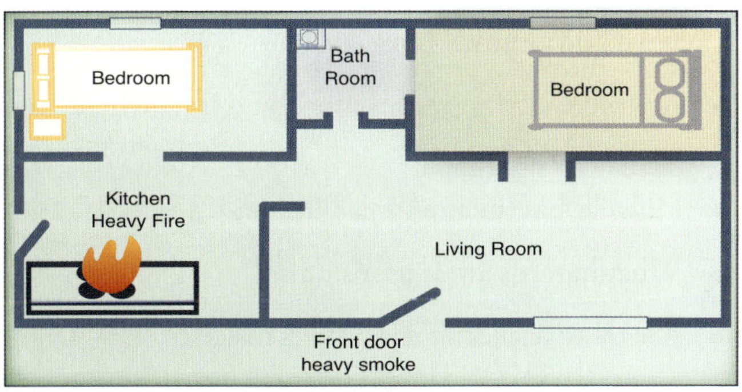

Fig. 20–58. Diagram of scenario

Size-up

Let's consider size-up in the context of our 1½-story house fire scenario and how it will affect the tasks you may be assigned:

1. *Location of fire* and *type of construction*. Fire is on the first floor. It will be easily accessible to both search and rescue crews and engine company members. The fire is exposing the unfinished attic (this is the uninhabited half story), but vertical extension is not a top priority right now. If this house was built with lightweight construction, wooden I-beams or trusses collapsing would always be a concern. Collapsing roof rafters, wiring, and attic contents have killed firefighters inside. Since this is a house, it is likely fully open and unenclosed inside.

2. *Life hazard*. You have reliable information that at least one person is still in the home. Immediately this puts emphasis on the need for search and rescue operations. The last reported position of the victim was in his bedroom, but the smoke alarms may have awakened him, and he could be anywhere in the home. Based on this information, you will start the search in the most likely place to find the victim, then move on to the most dangerous place (closest to the fire), and finally, if there were upper floors, above the fire (which is not the case in the particular fire since there are no occupied floors above the first floor).

   Due to the confirmed life hazard at this alarm, truck company emphasis will be on forcible entry, search, and ventilation. Since the aerial device will not be used due to the low height of the building, the truck company members may be assigned like this:

   The inside team: Two members with forcible-entry tools are assigned to force the front door if necessary, enter, and search. The outside team: Two members are assigned as a second search team and will attempt to gain access via a different manner from the first team. The remaining two truck company members will perform ventilation from the outside of the building to vent hot gases and smoke to make the engine company's push-in easier, safer, and faster.

3. *Fire control*. The fire building is a home, and we are making an aggressive interior fire attack so 1¾-in. hoselines will be adequate because the fuel load is not excessively high. A good general rule is two rooms of fire per hoseline. Even though your size-up reveals that one line should be sufficient, always stretch a backup line. Engine 1 has laid a supply line from the nearest hydrant, following their SOP. The driver has stopped the engine just past the building, leaving room for the ladder truck. Engine 1 has left one member at the hydrant who will charge the line when the driver is ready for water. Engine 1 officer has received orders from the chief on the scene and has told the crew to stretch an attack line to the front door and make a modified direct attack (described below) on the fire. Engine 1 officer has requested Engine 2 officer to send personnel to Engine 1, stretch a second attack line as a backup line to the first attack line, and assist with advancing the initial line.

4. *Forcing entry*. This is a home in a good neighborhood, so forcible entry should not be a problem. It is likely that there will not be excessive locks or security devices. Homes generally have wooden doors and frames and can be forced easily and quickly with hand tools. This function will be taken care of by the truck company crew who forced entry for their search mission and at the same time paved the way for the engine company to stretch and advance the attack line.

5. *Ventilation*. Well-timed ventilation will greatly assist the search team and the engine company pushing the line in. Windows provide excellent sources of horizontal ventilation openings. This important function was taken care of by two members from the truck company. Their initial actions will be horizontal ventilation to support the search and fire attack, then vertical ventilation of the roof if necessary. Horizontal ventilation will be conducted upon order from the incident commander. The IC will be sure the engine crew has water on the fire before he orders ventilation.

6. *Stretching considerations*. Stretching at this fire will be relatively easy. The first line will go to the front door across the lawn from the engine parked just past the fire building. Stretching off the rear of the engine is unobstructed by fences, hedges cars, and so forth. The working length of line at the front door should be enough hoseline to reach all parts of this building. A four-person engine company can establish a water supply and rapidly stretch the first attack line. The second engine company can assist with the first line if necessary and add a second backup line.

7. *Water supply*. Your size-up shows active fire that may extend into the attic of the structure. Where is your hoseline's water supply coming from? Are you operating solely with water from your engine's booster tank, or is the water supply coming from a hydrant, drafting source, or tanker shuttle? Standard operating procedures vary from department to department, but a good choice here is for the driver to provide water to the single attack line from the 500-gal (1,893-L) booster tank. This gives the driver about 2 minutes to hook up the supply line and get a reliable water supply from the hydrant.

Strategic plan

In our 1½-story house fire scenario, we have chosen the offensive strategy for the following reasons:

- The building is safe for us to enter.
- There is or may be a need for search and rescue operations.
- Properly placed hoselines can cut off the fire spread.
- Property can be saved with a manageable or reasonable risk to firefighters.

It is very important to note at this point that we know several factors, but several others we do not know.

What we know (or think we know):

- Where is the fire?
- What and how much is burning?
- What will it take to extinguish it?

What we don't know:

- Are there victims in the house, and if so, where are they?
- How far has the fire spread inside?
- What hazards await us inside?
- Are there backdraft conditions inside?

Establish a water supply

Your engine company has chosen to lay a supply line from the hydrant (forward lay) to conduct the first phase of your mission: Establish the reliable water supply. Engine 1 stops near the front of the building but leaves the actual front for the ladder company.

In rural settings it may be necessary to establish a tanker shuttle or drafting operation. For urban fires there may be a hydrant at the corner of each block. No matter where you are, a high priority is establishing a reliable and adequate water supply.

Force entry/exit

Forcible entry is usually accomplished by the search and rescue crew as they enter the building. In this case, firefighters from Truck 1 will be conducting this important task. However, engine companies should assign someone to bring forcible-entry tools in the event the truck's arrival is delayed. Controlling the door after entry and shutting it to prevent a flow path from developing should be considered. Forcing and chocking this door open must be considered ventilation and we must consider its effect on the fire growth.

Search and rescue

In our scenario, members of Truck 1 will force the front door and search for those in the most danger first, usually closest to the fire or above it (is it possible the teenager has a bedroom in the unfinished attic?). A second team will attempt entry (VEIS) into a bedroom window in the rear of the house, near the Division B and C corners. This provides them direct access to a bedroom where there is a good chance the occupant will be found and provides an escape route for firefighters if they are trapped by worsening conditions.

How does this affect the engine company with the hoseline? Consider all the moving parts. Members of the truck company are out ahead of you searching; members of the outside team are in remote areas, possibly out of sight, moving through rooms for victims as a result of VEIS tactics. How can you protect them with your hoseline? There are three critical options for the engine company:

1. Place the line between the truck team and the fire.
2. Protect the means of egress (and interior stairs).

Fire attack

The engine company is responsible for stretching the attack lines to the seat of the fire in this offensive fire attack. Getting water on the fire quickly is their most important task. Consider what has happened to the fire environment. As the search crews entered the 1½-story single-family home, they moved through an open door or window. This has allowed an unlimited amount of air for the fire. Most structure fires are controlled by the amount of air they have available. More air means more free burning fire.

In our example, we are conducting an aggressive interior fire attack based on the officer's size-up and strategy. The heart of the fire attack is stretching and operating the first hoseline. Although other hoselines will be stretched for this fire, let's focus on the all-important first line. As you approach the scene, your Engine 1 officer turns and tells you the company will be stretching a line directly to the seat of the fire.

This is a residential fire, so a 1¾-in. (45-mm) hose will provide the required water volume to begin our attack and possibly extinguish the fire. The recommended water flow is 180 gpm (681 L/min). This target flow provides for a margin of error incase of kinks in the line, inadequate pump pressure, or other real-world issues that happen on the fireground. This flow will also provide some extra flow to knock the fire down quickly and provide for an adequate amount of water for possible encounters with special hazards such as a broken gas main, leaking propane tank, or leaking and burning fuel oil tank. Using a 1¾-in. (45-mm) line is the SOP for your department for residential fires. Your officer should have sized up the fire and determined that the best route to the fire is through the front door. Your line will go directly to the fire or fire room and conduct a modified direct attack. Your officer's size-up will continue and will do his or her best to determine that this line will not push fire onto other members or into unburned areas

After the line is stretched to the front door, you or the engine officer will call for water, and the line will be charged. At this time this officer will call command and advise that the line is in position; command will direct that ventilation be accomplished so you can push in. Also, among the engine officer's many responsibilities is to verify that the company members are down near the floor below the heat level and out of any door ways during the attack.

A modified direct attack is selected for this fire. The nozzle operator directs the solid/straight stream into the fire area from outside the room to darken down the fire. The nozzle operator will use the reach of stream to begin to extinguish the fire. This firefighter will then move into the room or down the hallway to complete extinguishment.

Ventilate

In our 1½-story home fire scenario the proper ventilation to support the fire attack is horizontal ventilation of the windows in front of the hoseline. In reality, it is very likely that windows in and near the fire area or room will be broken out by firefighters when ordered by command or their company officer. Other windows in areas that are not severely exposed can be opened normally by firefighters. Often sashes and storm windows can be removed, limiting damage to the property. However, as previously stated, will timed and effective ventilation is often the key to success at fire attacks, resulting in rapid fire suppression and saving significant property. Conversely, by not taking the windows, the push-in by the engine company due to untenable conditions results in excessive dangers to firefighters (flashover, backdraft, burns) is delayed, and often the building dies a slow death, consumed by fire in void spaces concealed by smoke and steam. Coordinated ventilation and engine operations are key.

In our example, vertical ventilation of the roof is likely not required by first-in companies. Assuming the fire has not breached the walls or ceiling and has not reached the upper level of the home (the uninhabited ½ story attic), cutting a hole in the roof to vent out fire gases will do no immediate good. It will cause additional damage to the building that may not be necessary.

If, however, fire has reached the unfinished attic above the first floor, it will be necessary for firefighters on the roof to cut the roof shingles and sheathing, opening up the roof. The ideal place to cut the roof, of course, is directly over the fire. Timing of this operation with the operation of hose lines is of course critical as well.

Protect exposures

In our scenario, we have interior exposures, adjoining rooms, adjacent rooms, and rooms above the fire room if this house had second or third floors. Hoseline placement was designed to limit the fire to the room of origin and prevent it from extending to exposed areas outside that room. There are no traditionally described exterior exposures since the home is not close to adjacent buildings. The exterior exposure could be described as the soffit and attic if fire extends out the windows. Additionally, there has been a large body of research on combustible siding and insulation used on modern homes. It is far too voluminous to detail here but like all other previously mentioned studies is available on the web.

Fire control and overhaul

Since we were able to contain the fire to the kitchen and adjacent bedroom area, we can focus our efforts on checking for extension in these areas. We can use a hydraulic vent out the kitchen window with our hoseline to clear the area of smoke, exposing any pockets of fire. We must also open up the walls and check for extension into the unfinished attic. A thermal imaging camera can be used to "see" any hot spots. If there is a scuttle and stairs into the attic, the area above the fire must be checked closely and thoroughly for fire extension.

Pick up and return

While this was a "bread and butter" fire, fought routinely by any department, there are always things to learn. What did you learn at this fire? Did any unusual things happen? How well-coordinated were the firefighting operations? Your officer will often take a few moments after picking up to go over the fire and discuss any important issues. Take this time to ask questions while the fire is still fresh in your mind.

Ensure that all of your tools and equipment have been picked up and returned to the apparatus. When you get back to the fire station, perform any equipment maintenance (including SCBA) or tool refueling tasks so they are ready for the next fire.

ADDITIONAL FIRE SCENARIOS: SIZE-UP CONCERNS

Now that you have thoroughly analyzed a typical first floor fire in a 1½-story home, we will move on to other scenarios. The following examples of alternative locations for a fire in a single-family home (in the basement, for example) and other types of occupancies build on your understanding of basic fire attack. Each of the succeeding examples reviews the size-up issues that may be encountered. However, only the size-up concerns that change are identified for brevity purposes.

Basement fires in a private dwelling

Basement fires provide some unique challenges to firefighters.

Location. The location of the fire makes this not a favorite among firefighters. For basement fires we often have to work above the fire, which is one of the most dangerous places. The key to safe and successful response to basement fires is to get below it or at least be on the same level with it during the fire attack. Attempting to push a hose line down the interior stairs is like trying to climb down the chimney in an active fire place.

The way to do this is to gain access to the outside doors if available and control the fire from that position (fig. 20–59). Attacking the fire by pushing the attack team down the interior stairs is the least viable option and the most dangerous. Consider the condition of interior basement stairs before the fire and after they have been exposed to fire. They are usually, weak, narrow and difficult for firefighters to maneuver in full turnout gear and SCBA.

Fig. 20–59. Access a basement fire from outside. (Courtesy of Ryan Hart)

It is often difficult to determine where the fire is in the structure since smoke may be rising and filling the entire structure. A good way to determine if the fire is in the basement is to take out a small basement window with a tool. You will immediately be able to determine if the fire is down there or not.

Once you have determined that the fire is in the basement, check to see if it has extended to the first floor or attic. If this is a balloon frame home (see chapter 7: "Building Construction"), a fire in the basement will rapidly extend to the attic and often other floors. If the fire is extending vertically for any reason (pipe chases, balloon construction, floors burned through, etc.) you must get lines into position on these areas rapidly.

Before committing firefighters to these dangerous areas, always check for structural integrity. If trusses or wooden I-beams were holding up the first floor and the basement fire has involved them, expect an early collapse and use defensive tactics (fig. 20–60). There are several recent studies based on live burns that show how quickly wooden I-beams and truss supported floors will fail under a fire load from a basement fire. Additionally, there are several NIOSH reports detailing LODD that you should read to help prevent you from becoming another victim of these dangerous fires.

Utilities in the basement may provide a heavy fire load and rapid fire development. Oil tanks, natural gas lines, and meters will result in heavy volumes of fire if they are involved (fig. 20–61).

Further complicating a basement fire may be the huge fuel load from stored materials. Simply put, more stuff equals more fire.

Fig. 20–60. Wooden I-beams

Fig. 20–61. A gas meter inside a building

Life hazard. Basement fires can threaten occupants of the building in ways described above. In apartment buildings, upstairs occupants may not be aware that there is a fire in the basement, which will complicate your search and rescue operations. With real estate prices escalating across the country, basement apartments have become more and more common.

It is very important to remember that we are occupants too! It cannot be overemphasized, if the first floor is being supported by lightweight construction (trusses, glue-lam beams, etc.), keep members out from underneath and from on top of this very dangerous type of construction.

Fire control. Controlling the fire in a basement is an extreme challenge to firefighters because of its lack of accessibility. The most desirable method is to apply water into the basement thru the basement windows. Often, your fire stream will go directly on the burning fuel and provide quick fire knock down. It was believed that we should never apply water through windows because it would push the fire thru other areas of the building. Extensive testing by UL has proved this to be incorrect. Even for a basement fire with the interior stair door open, a hose line (straight stream) operated onto a basement fire will not push fire, heat or smoke up into the first floor.

The least desirable method of accessing the basement is by using the interior set of stairs within the home—this point of access is like descending a fireplace chimney. Some basements have exterior access through a set of "clamshell" bulkhead doors or a set of exterior stairs in a well—these are the more desirable ways of accessing a basement fire because it minimizes the punishment firefighters must take while descending the stairs (fig. 20–62). Additionally, these stairs are usually shorter, sturdier and wider than interior stairs. A hose line can be operated directly into the basement from the bottom of these stairs in a relatively safe position with an immediate access to safe exit.

Fig. 20–62. Clamshell doors provide exterior access to basement.

If a very large body of fire exists in a basement (heavy fire is showing at most or all of the window/door openings), and it has been determined that no one is in the basement or there is no savable human life there, consider putting the nozzle in the window you broke to see if there was fire in the basement. Often knocking down the fire this way prevents firefighters from having to enter this very dangerous fire building which may be on the verge of floor collapse (particularly for exposed wooden beams, and especially for lightweight wooden trusses or wooden I-beams).

Because basements are frequently not finished, exposed wood joists and subflooring will be involved in fire, creating more of a fire load. Also remember to shut off utilities from the outside before firefighters enter for any fire attack.

Force entry. Remember that the goal is to get even with or below the fire if possible. The outside doors to the basement often provide this access. If there is a full-size personnel door in a well, by all means use it, forcing with standard techniques. Be prepared for a face full of fire when you open this door, because you are on the fire floor—it is now just below grade.

As described above, basements often have clamshell doors leading down a few steps that end at the basement floor level. These doors cannot generally be forced by standard means and require cutting with metal-cutting saws.

Gaining access to the basement is only part of the issue with basement fires. Planning your secondary means of egress is the more important part. Generally, for many types of occupancies, there are two ways into and out of the basement: the interior stairs and, if present, the outside stairs. If you make your advance through the exterior stairs and it becomes blocked, your only other egress option is via the interior stairs. Attempting to conduct an emergency egress via the interior stairs has the following problems: First, is the door locked from the first floor? If it is, forcible-entry tools will be necessary. It is important to note that the area at the top of the stairs is usually very constrained space and may not allow forcible entry tools to be used, causing you to become trapped. Interior stairs are generally not well maintained, may be weakened by rotting due to previous or exiting water in the basement, and may not have handrails or other safety items.

Since this area is the highest part of the fire area, it will collect the most heat and densest smoke and will likely flashover first; certainly this is not a good place to be, especially during an emergency egress! If you entered the clamshell doors and now are exiting the interior door to the basement to the first floor, you may have created a flow path providing oxygen to the fire. This may make the previously non0burning (ventilation limited) fuel rich smoke flammable causing rapid fire development and/or flashover. This may prevent your successful escape. Consider reviewing the fatal fire report from the Cherry Road fire in Washington, DC.

Windows to basements are generally small, too small for a firefighter to exit, and often placed high in the exterior walls to allow some light and ventilation. Although windows may be present, they are of little value as a secondary means of escape.

Ventilation. Ventilation of basement fires requires special tactics. First, use what the building has made available to you, the windows above the exterior grade, high in the foundation. These are in a very good place, high in the fire area, and will be a good first step. Usually these windows are small and therefore provide limited capability to release heat and smoke. Often basements have low ceilings, further trapping heat and smoke, so ventilation may be a critical step in gaining access for firefighters.

An alternative tactic for ventilation of basement fires is to cut a hole in the floor above and vent smoke out through the first floor then out a nearby window. Be wary of fire-weakened floors when performing this tactic. Be especially careful in newer buildings of lightweight construction; it is important to recall how rapidly trusses and laminated wood joists fail under even moderate fire loads. Working above or below a floor constructed with these lightweight members must be performed with extreme caution—they are easily capable of quickly killing firefighters when they collapse.

Stretching considerations. Basement fires present unique challenges to engine company members as well. Because you are in a confined space with the fire, you must ensure you have enough line to fight your way all the way to the seat of the fire. Delays and interruptions in flowing water will cause the fire to light up, possibly trapping you and your team. By the same token, basements are generally cluttered and mazelike areas, so too much hose will cause problems like kinks that will delay or cease your attack, again endangering you and your crew. Using the floor above to lay out hose is a good tactic to help you advance.

No matter which access to the basement you are using, interior or exterior, remember that your advance will have to be rapid as heat and smoke will be pouring out of the upper levels of that door way. Get below it fast.

Water supply. A basement is the worst place for a fire in the building because it will expose and threaten the entire building above it. When considering water supply, simply have enough in case the fire finally involves the whole structure. If the fire building becomes fully involved, make sure you have enough water to protect exposures as well.

FIREFIGHTING CONSIDERATIONS IN OTHER TYPES OF OCCUPANCIES

Now that we have an understanding of the basic fire attack principles using our single-family dwelling example, let's move on to fires in other types of buildings. We will not examine every phase, but rather focus only on the different and specific size-up challenges faced by firefighters in these common structures. In this "basic" *Firefighter I* chapter, we examine these different types of building occupancies from your perspective as the firefighter. As mentioned at the beginning of this chapter, you, as a Firefighter I, will be asked to perform a specific task, under an officer or senior Firefighter II's direction. Firefighter I training instructs you *how* to perform these specific tasks.

In the future, if you enroll in a Firefighter II training program with this textbook, you will use these same set of buildings and occupancies to examine the fire problems from a leader's perspective. Firefighter II training emphasizes the "decision-making" aspects of firefighting—specifically *where* and *when* to ventilate, search, and attack the fire.

Wood frame garden apartments

Garden apartment type buildings have become common in suburban and even rural areas. They are economical to build and provide residences for substantial numbers of people that can either rent or own them if the co-op ownership method is used (fig. 20–63).

Fig. 20–63. Garden apartment. (Courtesy of Tom Bierds)

In a sense, garden apartments are simply a group of single-family homes put together both horizontally and vertically, and many of the same issues must be dealt with by firefighters in much the same way as detached single-family homes. Let's take a look at some of the important differences for firefighters.

Location of fire and type of construction. Garden apartments are usually set back off the road or parking lot. This causes firefighters to have to carry ground ladders, tools, and hoseline greater distances than to fight a single- family dwelling. It is important to pace yourself in walking to the fire building so you are not too fatigued to fight the fire when you get there. Resist the temptation to run to the scene. When you don your mask to enter the building, you will quickly deplete the air supply because of the stress of the run from the rig to the building.

Construction of the garden apartment will affect the overall strategy. Lightweight construction often has no level of fire resistance; no protection is provided for the structural members holding the building up (in the form of gypsum board covering the individual floor and roof beams, for example). This lack of fire resistance will often cause incident commanders to quickly abandon offensive operations and move to defensive operations for the safety of firefighters. Defensive operations use larger hoselines, portable monitors, and water tower operations.

Life hazard. As a result of the dense population in garden apartments, search and rescue operations often require many more firefighters than a single-family dwelling. A fast-moving fire, especially in a lower unit, will rapidly endanger people in units above and adjacent to the fire apartment. Simultaneous search and rescue will often be necessary to save lives. For firefighters, occupants rescued from apartments and brought out to the public hallway will need to be carried or at best assisted down the stairs and out of the building.

Fire control. Fire control can be very challenging in garden apartments simply because there is so much to burn. Lots of people mean lots of furnishings, beds, clothing, computers, and the like. Also, items such as bicycles and outdoor toys are stored in living areas. Firefighters on hoselines can get caught and trapped and die during what was thought to be a routine fire attack in a couple of rooms of fire.

In addition to all the furnishings and property inside a garden apartment, there is a tremendous amount of fuel in the wood framing, flooring, sheathing, and interior trim of garden apartments. Since the entire building and contents are combustible, often a huge fire load results in a huge fire. When fires in these buildings gain control of the building and involve the wood structure, it is difficult to get enough water on the fire quickly enough to stop it. When attacking fires in garden apartments, consider the use of big lines quickly.

A good technique for fire control if the engine company cannot get in the fire apartment is to breach the wall in the adjoining unit or next to the front door and apply water through this hole. This firefighting technique may hold the fire to a room and contents fire, thus preventing it from extending to the void spaces, which would lead to eventual destruction of the building.

Forcing entry. Gaining access to a residence is generally not a problem unless it is in an urban area or one plagued by crime. In areas where residents have fortified their apartment doors, firefighters should come armed with more effective and possibly hydraulic tools to speed entry.

Ventilation. Ventilation of garden apartments is very similar to that for single-family homes. Taking windows in front of the hoseline is always a good technique. This may be a bit more difficult because windows can be two or three stories above the ground. Use of ground ladders (carried around the building because of the set back) will be required.

Roof ventilation will be necessary for fires that have entered the attic void space. Often the attics are not divided and provide a rapid lateral and vertical fire spread opportunity. Since the building is usually wood frame and the roof is sheathed with plywood, firefighters can use rotary or chain saws effectively.

Stretching considerations. Stretching hoselines at garden apartments can be very challenging. In order to make the units appealing and supply the garden appearance, U-shaped or zig-zag layouts are used in the design. This means that hose stretches have to be very long. If your engine company uses preconnected hoselines, these will almost always be too short.

Many engine companies that have garden apartments in their fire due area use a thief line. A **thief line** consists of 200–300 ft (60–90 m) of a 2½-in. (65-mm) hose, with a wye or water thief at the end to reduce the friction loss and supply the required volume. Smaller-diameter attack lines can be attached to the wye or thief and used for interior fire attack. Preconnected lines can be detached from the engine, shoulder carried, and used; or apartment packs can be carried in canvas bags or tied together with webbing or quick-release straps.

Water supply. Multiple reliable and high-volume sources are necessary for these types of fires. Experience has shown that if fire walls or interior gypsum board partitions and ceilings are breached by fire, it will turn into a fast moving and large body of fire requiring huge amounts of water.

Taxpayers

Location of fire and type of construction. Taxpayers are a general classification of buildings that contain a business on the first floor and apartments or living space on the upper floors. These are typical Main Street–type buildings.

Compared with strip malls (below), these buildings are usually of ordinary construction and are relatively fire resistant. However, basement fires in these occupancies often present a problem because of stock or storage items from the business.

For firefighters, access to fires in taxpayers are relatively easy. Large front windows and front doors for customers provide good access directly off the street. Beware, however, of the parapet (a decorative set of brick or other masonry unit that is constructed on top of the front wall of the taxpayer), as it is often not well supported laterally and can easily fall during a fire (fig. 20–64).

Fig. 20–64. This taxpayer's parapet is leaning and may be unstable.

Life hazard. Life hazard in the storefront is usually minimal. Store patrons or employees are awake and alert when the store is occupied. Residential occupants upstairs can be a real problem. Frequently a door to the side of the storefront leads up a set of stairs to the upstairs apartments. In older buildings, these stairs are steep and narrow. Fires originating down below can cut off access to these stairs, trapping occupants on upper floors.

Fire control. Controlling the fire can be a challenge in taxpayers as the storefront may contain anything from a candy store to a hardware store, a pharmacy and general goods, a toy store, or an auto parts store. Firefighters should consider using large, 2½-in. (65-mm) high-flow hand lines for these fires. Ladder towers placed near the front of the building with buckets placed near the ground can direct master streams effectively up into the ceilings of the stores (to get at hidden fires) with excellent results. The large-caliber streams make fast work of the fire and prevent the hazards to engine company members from making an interior attack.

Forcing entry. Forcing entry can be a very challenging process. Roll-up gates, sliding gates, and other heavy security measures often are present. Rear doors may have been bricked up or, at a minimum, contain very strong security measures. Heavy-duty tools and power tools will be required to gain access to these doors, which are critical as a secondary means of egress for horizontal ventilation.

Ventilation. Since these stores are usually side by side, ventilation in the front and rear is essential for first floor fires. Display windows can provide large openings in the A side of the building. These are readily available at ground level.

Large fires in one-story taxpayers may extend in to the **cockloft**, the combustible void space above the ceiling of the store. In such cases, vertical ventilation will be necessary to prevent the fire from moving in this concealed space and spreading to other parts of the building above the other stores. This will typically require a large primary vent hole directly above the store, and, in some cases, a trench cut to cut off the fire's horizontal movement in the cockloft. It is important to remember that if flammable smoke has accumulated in the cockloft, ventilation may add oxygen to this ventilation limited fire causing rapid fire growth possibly extending downward onto firefighters opening the ceiling from below. Get water on the fire using small holes (just big enough for the nozzle before opening up large holes that will allow massive inflow of air).

Stretching considerations. Although hoselines can be easily stretched to the front of the building, and large windows provide excellent access if the first floor is involved, stretches to the rear can be difficult. A good firefighter decision is to stretch the hoseline through an adjacent building to gain access to the rear of the fire building.

Fire escape stretches, rope stretches, and other advanced methods may be required to get hoselines in proper position.

Water supply. As with any nonresidential structure, always stretch big lines right from the start. Big lines obviously require large amounts of water, so a reliable and redundant water supply is important.

Strip malls

Location of fire and type of construction. Strip malls are the modern version of the taxpayer (fig. 20–65). They are constructed cheaply and quickly of materials that do not maintain structural integrity under a fire load. Like taxpayers, they contain stores on the first floor. They may be one or more stories and are subject to early and deadly collapse from burning contents (fig. 20–66). Some strip malls are of non-combustible construction with metal bar joists and metal deck roofs. Other strip malls are of ordinary construction with wood joists or (more commonly) lightweight wood trusses or wooden I-beams.

Fig. 20–65. The front of the strip mall shows it contains a variety of different occupancies.

Fig. 20–66. This strip mall has two floors on part of the building.

Often, security issues that create limited access (fig. 20–67) or cost considerations during construction, such as leaving structural members unprotected (figs. 20–68 and 20–69), factor into firefighter safety concerns.

Fig. 20–67. The rear of the building provides limited access and usually high security. Note the roof-mounted HVAC units. (Courtesy of J. Knapp)

Fig. 20–68. Unprotected steel structural members will fail rapidly under a fire load.

Fig. 20–69. Note the lightweight and unprotected C channel members holding up the roof.

The location of the fire in the structure is of major significance to the firefighter. Fires are often in the rear of these buildings in storage or work areas. This means firefighters must push through the building, generally from the front to rear to attack the fire or conduct search operations. These firefighters are directly under the collapse zone of the unprotected steel trusses and beams supporting the HVAC units and the dead load of the roof itself. High security of rear doors ,if they even exist and are not bricked up, usually prevent rapid attack or access from the rear.

Life hazard. The life hazard of strip malls can vary depending on the occupancy.

Fire control. Fire control operations should be attempted from outside the structure if significant fire has attacked the unprotected structural steel components. Because of the small front-to-back distances of these structures, firefighters can use the reach of the stream to extinguish heavily involved areas. This tactic keeps firefighters out of the most extreme danger zone. Staying out from under this building and using the reach of the stream may be a life-saving tactic for you and your crew one day.

Force entry. Again, like taxpayers, strip malls may have extreme forcible-entry challenges, especially in bad neighborhoods. High security on doors and windows may necessitate cutting locks and gates with rotary saws. Barred and or bricked-up doors and windows in the rear are formidable challenges. Consider breaching interior walls as an alternative, especially if life safety is an issue. Breaching an interior wall to escape dangerous conditions maybe more effective than attempting to force high security doors, window bars or roll down gates.

Ventilation. It is very tempting to vent the roof of these buildings. The roof is easily accessible by ground ladder, usually flat and inviting to the aggressive truck company. A "traditional" roof of solid sawn wood joists offers the possibility of vertical ventilation, if necessary. However, other types of roofs are dangerous for firefighters attempting vertical ventilation. Firefighters must be aware of these critical dangers:

- In the case of non-combustible construction, the structural steel joists are usually unprotected (bare steel without any fire resistance). Weakening and failure of the steel joists is a distinct possibility with a moderate or large fire below. Cutting the metal deck itself between joists can leave the portion of the deck where the firefighter is standing unsupported, resulting in the deck bending and sending roof firefighters into the fire. Ventilate using the front and rear doors and windows if possible. Stay off of the roof.
- In the case of lightweight wooden trusses and wooden I-beams, firefighters must stay off the roof. Failure of one component of the truss causes the truss to lose all its structural strength. In addition, a fire in the wood trusses or I-beams themselves results in the very real potential of collapse onto firefighters below. This happened in 1989 to two firefighters in Orange County, Florida, who were killed when the wooden trusses collapsed. The smoke condition in the stores was light while there was a raging fire above their heads in the truss void. Use hooks to open up the ceiling and thermal imaging cameras to determine the extent of fire above the ceiling.

Stretching considerations. It is tempting to use small interior attack lines for fires involving these structures. The fire appears small because stores in strip malls are often small square footage. What firefighters must recall is that there is likely a very high amount of combustible material, which will require a high flow. Stretch and use a 2½-in. (65-mm) hoseline for these fires. The reach, penetration, and flow will make quick work of the fire from a position of safety outside the collapse zone.

Water supply. In the case of ordinary construction, a large combustible cockloft/void space will add tremendously to the fuel load in addition to the combustibles in the store itself. A large volume of water is necessary. Another problem with the Orange County, Florida fire cited above was the lack of a continuous water supply; the firefighters were relying solely on the water in the fire engine's tank rather than being connected to a fire hydrant. Fire can quickly get above the ceiling and spread laterally.

Big box retail

Location of fire and type of construction. Big box retail buildings are very much like a large strip mall that is undivided and has a higher ceiling (fig. 20–70). Many types of businesses may occupy these type buildings: pharmacy, toy store, hardware, housewares, grocery, and other retail businesses. The construction weaknesses under a fire load are similar to strip malls, though the size, square footage, and height vary.

The location of the fire plays a key role in firefighter safety and responsibilities. Fires may be in high-rack storage, hazardous materials may be stored in the building, or HVAC units may be on top of or under the roof surface. All of these are very dangerous type fires.

Fig. 20–70. Note the lightweight construction, unprotected steel, and large size of this building.

Life hazard. Big box stores may contain a relatively high life hazard. Occupants may not evacuate if there is not a perceptible danger (fig. 20–71). Fire alarms often go unheeded. Fire may grow in storage areas or void spaces, then flashover and spread, rapidly trapping or killing occupants who did not escape.

High ceilings in these buildings allow combustible gases to accumulate high above firefighters' heads, possibly igniting the underside of the roof deck. Use the thermal camera to check over your head to be sure you are not in danger from a fire in the roof itself or accumulating hot gases that could lead to flashover. As with all suspended ceilings, lift a ceiling tile immediately inside the door you enter to check for fire above you.

Fig. 20–71. Here a big-box-type construction building houses a retail outlet.

Fire control. These are generally large unenclosed (inside) buildings. Hose streams may not reach into all areas from outside the building. Consider using master streams from tower ladders or deck pipes off engines to obtain the necessary reach. Roof ventilation like the strip mall is very dangerous. Roof vents and scuttles can be used if available for ventilation. Sometimes these buildings are protected by automatic sprinklers; if they are, by all means use and support the systems.

There are lots of products to burn in these buildings. Always use a 2½-in. (65-mm) hoseline if you are going inside and have any significant fire. This size line will provide the reach and flow you will need to extinguish the fire.

The good news is that many of these buildings contain sprinkler systems. Sprinkler systems (discussed in greater detail in chapter 30, Fire Protection Systems) are your best ally. They work in the dark, don't need an SCBA, extinguish fire only when necessary, and function until we shut them off.

Force entry. Generally front entrances are the easiest to force. Large windows may be present for ventilation or entry. Rear access may be limited by security concerns similar to strip malls.

Ventilation. Retail stores make a profit by selling a volume of products. One advantage available to you may be a large loading dock door. This may provide excellent horizontal ventilation.

Stretching considerations. Parking lots usually provide good access to the front of the building. Stretching will be relatively easy until you get into the store. The building may be very deep from front to rear and require a long stretch similar to garden apartments. Always consider using big lines at these fires.

Water supply. Since this is a big box building that has big volume and big contents, a robust water supply must be established. If the building has a sprinkler system, be sure not to steal water from the system by using a nearby hydrant that may be trying to supply the sprinkler system.

Office buildings

Location of fire and type of construction. The location of the fire in office buildings is important because these buildings often are divided up into small offices or cubicles (fig. 20–72). The floor layout can become a deadly maze for firefighters. If it is an open office type floor plan, fire can spread rapidly from cubicle to cubicle quickly.

Office occupancies can occupy a variety of construction types, too many to be discussed here. Refer to chapter 7, Building Construction, for specific construction hazards.

Fig. 20–72. Office building

Life hazard. Typically occupants will be awake and alert in office buildings. Accounting for occupants can be easy if there is a fire plan and they are well schooled in its use. However, this is more the exception rather than the rule, so plan for an extensive and extended search and rescue operation. Use of life lines and large area search procedures may be necessary.

Disabled persons working in the office may provide special challenges to firefighters. Many multi-story office buildings have designated "areas of rescue assistance" or "areas of refuge" for disabled people (fig. 20–73). These rooms and areas are typically separated from other areas by a one-hour fire rated wall (sometimes these locations are actually inside stairwells) and will be provided with a two-way communications system with the building's lobby. These areas are designated with special signage identifying this area and may be equipped with a communications system that connects to a building's lobby. Refuge locations will be a high priority search area.

Fire control. Most modern buildings will be protected with automatic sprinklers and a standpipe system. As previously stated, these are a firefighter's best friend. Support the system with hoselines as early as possible. If the fire building is not equipped with a sprinkler system, plan for long hoselines and high fire volume because of a high fire load.

Force entry. Forcible entry concerns for offices are similar to any other building and are usually dependent on the level of crime in the area and need for security in the particular occupancy. Obviously, banks and financial institutions will have more security than a general-purpose office. Some office buildings may have secure areas with the main office area or complex.

Fig. 20–73. Disabled people who are incapable of evacuating the upper floors of a multistory building on their own may be waiting for you in an area of rescue assistance such as this one.

Ventilation. Ventilation considerations depend on the type of building construction.

Stretching considerations. As previously stated, office buildings and layout can vary. Good preincident intelligence can go a long way in making your stretch successful. Some offices have a large open lobby or waiting areas. These areas will consume a large amount of hose before even reaching close to the fire area. Other offices with cubicles or highly divided floor space will present equally challenging and hose-consuming stretches.

Water supply. As with any structure fire, establish a reliable water supply.

Restaurants

Location of fire and type of construction. This category of building fire is really dependent upon the type of building construction the restaurant is housed in, how big or small the building is, and the type and size of cooking equipment in the kitchen (fig. 20–74). A common important factor you will deal with is something that has little to do with fire attack operations: how often and how well the hood ventilation system is cleaned and serviced. Grease builds up from frying of food in these systems and can trigger fast-moving and dangerous fires.

Fig. 20–74. Restaurant

Fast-food restaurants are special hazards that must be recognized by firefighters. In order to minimize cost and maximize profit, these restaurants are often built of lightweight material, with minimal fire protection systems and with no regard or minimal regard for firefighter safety. This is a class of "disposable" buildings that should never be paid for with a firefighter's life. Sadly, these buildings have taken several firefighters' lives in recent years. It is critical to remember these are designed by the chain owners as disposable buildings, built with the cheapest material and construction methods available solely to sell, sell, sell and maximize profit.

A common hazard in these disposable buildings is a lightweight truss supporting roof-mounted heating, ventilating, and air conditioning (HVAC) units. Under the roof deck are HVAC distribution systems; deep fat fryer duct work; and electrical, gas, and plumbing utilities. From the inside, all this is hidden by a drop ceiling. If fire invades the truss space and has attacked the trusses, expect a rapid and fatal collapse. Obviously, firefighters can be trapped and burned to death in this common fire scenario.

Life hazard. The life hazard can be high during busy times when the restaurant is in operation. However, occupants, customers, and employees will be awake and alert and usually exit the building if they perceive danger.

If the fire occurs after business hours, check for maintenance or cleaning people in the building. However, always consider the risk to firefighters in these dangerous buildings. Another consideration is that you may be able to conduct a "search" from outside the building. Fast-food restaurants often have lots of windows and doors. If you open all these and simply look inside, because of the small floor space, you may be able to visually search (looking under the smoke) the building from outside in a position of safety.

Fire control. Some restaurants will have built-in fire-extinguishing systems, especially in hoods and deep fryers. If these appliances are involved, activate the built-in fire-extinguishing systems if they have not already been automatically or manually tripped.

There is a very good possibility that the fire actually started in the cooking hood and ductwork due to the accumulated grease—these are very hot fires which can spread beyond the sheet metal ducts. Look for fire spread to surrounding combustibles (wood studs, stored paper goods, etc.) with the use of thermal imaging cameras and by opening up suspected fire extension locations.

Force entry. Forcible-entry considerations are similar to other structures. Large restaurants may have exit doors with panic hardware on the inside. After the initial team gains entry, these doors may provide excellent access for other firefighters.

Ventilation. Especially for kitchen fires, consider turning on the hood ventilation system to exhaust smoke from the building. *Make sure, however, that the fire is not in the cooking hood and duct system, a common fire location.* Other than this, usual ventilation considerations apply.

Stretching considerations. It is important to recall that restaurants come in a variety of shapes and sizes. We have all been in the mom and pop restaurant that is squeezed into a long narrow building, sometimes with a dining area up one or more floors. On the other side of the restaurant scale are large catering halls that seat hundreds. Obvious stretching considerations are associated with each of these varieties and everything in between.

Water supply. As always, establish a reliable water supply.

Vacant buildings

A wise old fire officer once asked, "What is in a vacant building to save that is worth the life of a firefighter?" Sure it is a challenge to attempt to put the fire out using an aggressive attack, but what are you saving? Never forget, whenever you enter a burning building, there are a lot of things, some of which we considered here and others, that in the end can kill you. If the building had enough fire in it that you need to aggressively attack it, how much damage is being done to that building, and what will it be worth after the fire? The answer is probably not much. Risk a little to save a little. A vacant is a vacant until the occupants (us) arrive and enter (fig. 20–75).

Fig. 20–75. Vacant building

Location of fire and type of construction. Vacant buildings are very dangerous. For the safety of members you must assume that the structure could be compromised. Here is an example: an old, heavy timber building can withstand a small fire for a relatively long time (due to the massive nature of the wooden structural members) before it becomes structurally unsafe. Simply put, these massive timbers provide us with time to get inside and extinguish the fire. That statement assumes the building is well maintained and the structural members have not been compromised. Since many heavy timber buildings have been around for a long time, they often are not well maintained and have had numerous occupancies during their long and storied lives.

If the building is vacant, how long has the hole in the roof allowed in rain and snow to help rot the wood or rust the connections? Have insects attacked the heavy timber to further weaken it? What types of changes were made by each occupancy over the years?

The point is, even the most fire-resistive structures and structural members may be compromised by forces during the time it has been vacant.

Some other safety concerns with vacants

- Have expensive conveniences been removed? Elevators and dumbwaiters may have been removed or vandalized, leaving open death traps for firefighters in smoky conditions. Did the previous owner leave barrels of leftover hazardous materials or hazardous waste inside?
- Are heavy items left on shelving that may fall on interior teams?
- Were stairs and holes cut in the floor to allow the removal of large machinery?
- Were inexpensive repairs like tar paper put over holes in the roof made by the previous or current owner?
- Are windows and doors boarded up for security, thus limiting egress of the new occupants? (That would be us, firefighters, as we are occupants too!)
- Are there multiple fires in the building? Did the owner want it to burn?

Life hazard. A false report of squatters in an abandoned cold storage building caused the death of six Worcester, Massachusetts, firefighters in 1999. Consider the hazard to firefighters and the lack of reliability of on-scene reports before committing members to search this dangerous type of building. If you have to search the building, use search ropes and thermal imaging cameras. Be aware of the numerous hazards this building presents.

Fire control. Attempt work from the outside. Use the reach of the hose streams to darken down the fire from positions of safety. Use large-caliber streams with high volumes of water and long reach and penetration. Attack the fire if necessary from positions of safety.

Force entry. In vacant buildings, this usually means taking the plywood off the windows and doors. If truck company members get the door open and accessible, wait until they get a couple of other openings available to you to use before making entry. Vacants often have bricked-up or window security devices that are difficult to remove from the outside. Obviously, whatever prevents or slows down the mal-intentioned vandal slows down the well-intentioned firefighter.

Ventilation. Standard ventilation practices apply to vacant buildings—with one important caveat: You never know how sound the structure is, so assume it is not safe; for example, the roof is not safe to vent unless you are supported by a ladder or tower bucket. Consider the use of horizontal ventilation (windows) and not risking firefighters on the roof. Horizontal ventilation may be difficult, as noted previously.

Stretching considerations. This is not an aggressive interior fire attack so leave the 1¾-in. (45-mm) hose on the rig. Bring the big lines, and overwhelm this fire from the outside. Water damage is not an issue. Consider the use of portable monitor nozzles and large-caliber streams. The sooner you put the fire out, the less danger it is to everyone on the scene.

Water Supply. As always, establish a reliable water supply for the size of the fire you are fighting.

EXPOSURE PROTECTION

When protecting exposures, apply water directly on the exposed surfaces, as this is much more effective at absorbing heat and delaying or preventing ignition due to radiant or convective heat (fig. 20–76). Application of water on the exposed building is not only more effective but can be done with less water than other means. Water curtain nozzles were used for a number of years. Recent scientific evidence has determined that these water sprays are not as effective as applying water directly to the exposed surfaces (fig. 20–77). The water curtain is simply not dense enough to absorb all the radiant heat from the source fire. Water curtains have enjoyed success because they inadvertently apply water to the exposed surface. Additionally, these nozzles are supplied with 2½-in. (65-mm) lines, which require significant personnel to stretch and copious amounts of water, which may be in short supply on the fireground.

Fig. 20–76. Water curtains aren't as effective at absorbing heat than direct application of water to the exposed building.

Fig. 20–77. This is a good example of a firefighter with a properly positioned hoseline protecting a heavily exposed building. A good sign an exposure is getting ready to light up is paint discoloring and peeling, bubbling, and beginning to smoke. (Courtesy of Brian Duddy)

"SMELL OF SMOKE" CALLS

Often you will be called to respond to a "smell of smoke in the building" with only this vague information: odor of smoke in the building, 20 Main Street. This call presents unique and challenging situations to firefighters. The most important consideration at these calls is this: is there significant fire in the void spaces of the building that is hidden from you? Void spaces can be the truss space between floors of a building, the space above a drop ceiling, in the walls of a balloon frame home in attic or cockloft spaces, in utility chases or even in soffits above kitchen cabinets. If there is significant fire, follow standard firefighting practices. Always have hose lines in place when opening void spaces where you suspect there is fire. If you are looking for hidden fire and you find it you had better have something to extinguish it with. Firefighters have been killed after they opened a space and heavy fire was driven out by a flashover in that space caused by the newly introduced oxygen. Never underestimate the potential of hidden fire.

After you have determined that there is not significant fire in these void areas, it is often a challenging detective game to find the source of the smell or visible smoke. Ask the occupants obvious questions first: where did you smell it? Where was it the strongest, when did you smell it? What were you doing and what was happening in the building when you smelled/saw it? These questions will help you eliminate some potential sources and highlight potential suspected areas to concentrate your search. As employees enter the building for the first time that day or occupants return home, they may notice the odor or haze of smoke. If you can't find the source ask them to retrace their steps to determine what appliances or circuits they energized and eliminate these as possible sources.

In multi-family residential buildings, a "food on the stove" run is very common. Forgetful people leave a pot of food cooking on the stove after they leave the premises. Working fires can erupt in these situations. Overheated food on the stove has a distinctive odor; you will come to know the smell very well in your career.

Frequently it is electrical appliances or lighting fixtures that are the source of the odor. Always check the breaker panel to determine if an appliance, motor, fan, or lighting circuit has shorted or burned out, tripping a breaker or blowing the fuse. A tripped breaker may point directly to the source of the smoke. In addition, use the thermal imaging camera to scan circuit breakers in a panel—hot breakers (indicating an electrical circuit that is at capacity) will glow brightly.

In occupancies that have fluorescent lighting, check to see if one light fixture is not as bright as others. If you find one of these, the ballast or starter mechanism in the light maybe the culprit and the source of the smoke (fig. 20–78). An overheated ballast has a particular odor; you will recognize it once you've experienced it.

Especially early in the heating season, dust on heating appliances may burn off and cause an odor or some visible smoke. New appliances of any kind may emit odors when used for the first time. Ask occupants if there is anything new in the building.

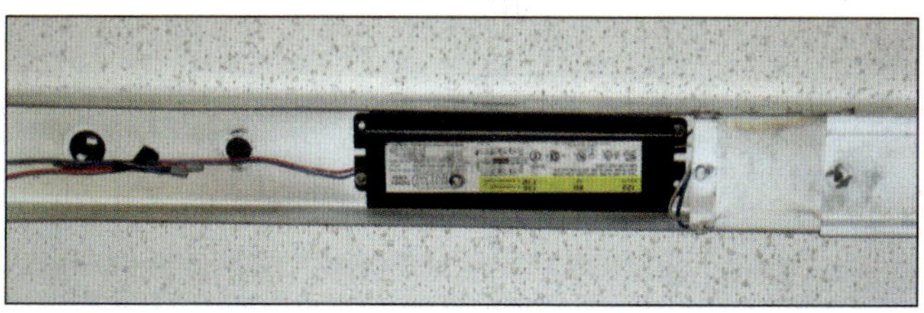

Fig. 20–78. Starter and ballast

Use of the thermal imaging camera can be a great assistance in determining the source of the alarm. Use it to scan suspected appliances, motors, fans, and lighting fixtures you think maybe the cause. This is often the definitive step in determining the cause of the odor or light smoke.

Smoke detectors are another cause of fire department responses similar to odors of smoke. Typically a smoke detector will sound for a short time causing the occupant to call the fire department. By the time we arrive on the scene there is no apparent cause. Again this is a detective game to determine the cause similar to "odor of smoke" responses. Always rule out active or past fire as the first step.

Common non-fire causes of smoke detector alarms are: steam from a shower or other source, dust, cleaning or air fresheners sprayed in or near the detector, insects, or cooking.

In all these cases always remind the occupant to call the fire department again if the odor returns, the alarm sounds again or if they have any reason to suspect a dangerous condition. The last thing we want as firefighters is to respond to a working fire because of a delayed alarm.

OUTSIDE FIRES

FFI 4.3.8 One of the most common types of outside fires is the dumpster or trash fire (fig. 20–79). If these fires are small in size or contained to a small bin, they can easily be extinguished with a booster line or preconnected "trash line." Larger containers, larger piles of trash, and construction debris should be overwhelmed with an 1¾-in. (45-mm) line, or in case of a large debris pile, a 2½-in. (65-mm) hoseline or master stream device. When choosing the size line to extinguish the fire, always remember that you want to overwhelm the fire. A fair fight only exposes us to unnecessary dangers for extended periods of time. Take a cannon to the gunfight and you will win decisively.

Fig. 20–79. Dumpster

Most often with these types of fires, there is nothing to be saved, so no one and nothing should be risked. Use the reach of the stream to apply water to the fire from a distance. Once the fire is knocked down, firefighters will have to pull apart the debris to reach smoldering pockets for final extinguishment. Pike poles, commonly referred to as hooks, are used to pull apart debris to ensure final extinguishment.

Safety at outside fires is often deceiving. The fire is visible and easily accessible, and apparently will be easy to extinguish. We must always remember that closed containers or other hazards may be contained in the trash or garbage container. First, consider that if something started the fire, it could be a hazardous material that may produce toxic fumes. Possibly an arsonist started the fire to cheaply dispose of trash or to dispose of evidence or even a body. Disposal of hazardous waste is expensive. A common practice by unscrupulous persons may be to hide it in common trash. At construction sites or even in household waste, closed containers such as propane torch cans or 20-lb (9-kg) propane tanks may be found. Exposed to fire, they can explode (BLEVE), exposing firefighters to rocketing tanks and lethal shrapnel from tank fragments. Chemical hazards such as unused or outdated pesticides and used or unused cleaning chemicals and construction material can often be found in trash.

At a recent fire in the northeastern United States, a small propane cylinder became a projectile at a dumpster fire. It hit the building with such force that it set off the automatic alarm in the structure.

You never know what is burning in a trash or dumpster fires. For this reason it is important to wear full PPE and SCBA. A shift in the wind or a sudden flare-up of fire or smoke can expose firefighters to dangerous chemicals or other hazards. If you always wear your PPE and SCBA, these are two decisions you don't need to think about on the fireground.

A good tactic for dumpster fires is to fill the container with water. Often dumpsters will have a drain hole in lowest part of the side of the container. Plug this with cone shaped wooden plug by tapping it in with a hammer or flat-head axe, then simply flood the container with water (fig. 20–80). This tactic reduces the time and hazards for firefighters. However, be careful when flowing water into them. Be wary of the weight of water that you are adding and the possibility that the dumpster will move with added weight. Occasionally, it may be necessary to remove the contents of the dumpster to properly extinguish the fire (fig. 20–81).

Fig. 20–80. Dumpster plug

Fig. 20–81. Here a large dumpster had to be unloaded to be extinguished. Trash fires are a common call for firefighters. Here firefighters extinguish a large fire in a container. Remember to always be aware of hazardous materials illegally concealed in trash fire. (Courtesy of Tom Bierds)

Junkyard fires

Junkyard fires are outside fires that pose unique and special hazards to firefighters. Junkyards and scrap metal recycling sites are often large and wandering areas, forcing us to make long supply hoseline lays. Frequently on the outskirts of town or rural areas, water supply may be limited to nonexistent.

Automobile junkyards often stack cars to save space, creating unstable stacks of cars that can fall on firefighters. Obviously, sharp metal, burning plastics, and piles of hot and possibly molten metal dripping down from stacked cars all present us with serious hazards at these fires.

Although these fires may be spectacular, resist the urge to conduct an offensive attack. There is nothing to be saved, so nothing should be risked. Use master streams right away, because they have greater knockdown and can be applied from a distance away, thus providing a margin of safety. It sounds simple but is an often underused method, using the reach of the stream to provide some standoff distance for yourself at these fires (fig. 20–82).

Fig. 20–82. At large outdoor fires, like this one in an automobile junkyard, the use of master streams is required. Note how far the streams reach providing a good margin of safety for firefighters. (Courtesy of Brian Duddy)

Outside tire fires

Rubber automobile tires are sometimes stacked in huge piles, either awaiting disposal or recycling. When these mountains of rubber catch fire, they are very difficult to extinguish (fig. 20–83). They are generally stacked to conserve storage space. This stacking provides just enough space for air and oxygen for combustion, but not enough accessibility for effective stream application. Because they produce huge quantities of heat and black smoke, it is difficult and sometimes impossible to provide and effectively apply enough water to effect extinguishment. These fires often burn for days. Experience has shown that the most effective extinguishment method is to physically separate the unburned tires from the burning tires. This is usually accomplished with contracted or on-site heavy equipment while firefighters control the fire spread for safety. Class A foams have some effect on increasing the ability of water to penetrate these difficult fires.

Fig. 20–83. Tire fire. (Courtesy of Robert Ladd)

Outbuilding fires

Another common type of outside fire is a shed or small storage building (fig. 20–84). These fires can be deceptive and often contain hazards to unsuspecting firefighters. A shed or detached garage may be referred to as an outbuilding, garage, shop, or barn. These buildings may contain products, chemicals, and processes that the home or business owner does not want in his main house or business. Common hazards found in outbuildings include pesticides, pool chemicals, herbicides, flammable liquids, and gases. Recently, a fire in an outbuilding sent eight firefighters to the hospital. The owner ran a seamless linoleum floor business and had a variety of toxic chemicals that were burning. Strategy for these types of fires is to prevent them from extending to surrounding exposures, which may include the main house, business, or storage building, other nearby buildings, cars, trucks, or wildland. Always use full PPE and SCBA.

Fig. 20–84. Firefighters prepare to extinguish a shed fire. Note that they are going to use the reach of the stream to apply water, thereby keeping themselves at a safe distance. There is nothing to be saved, so nothing should be risked. (Courtesy of Tom Bierds)

Groundcover fires

Small brush fires. **FFI 4.3.19** A very common type of response involves small fires in vegetation, including leaves and grass. These brush fires often occur along roadways and railroads, sometimes sparked by improperly disposed cigarettes or hot exhaust systems in cars (a car's catalytic converter can ignite leaves and grass when it is parked on top of these combustibles). Other brush fires are caused by campers who leave campfires unattended. Brush fires can grow in size to become wildland fires, described below.

Small brush fires are usually controllable with a single fire company and small hand lines (1½-in. or 1¾-in. [38-mm or 45-mm]). A key to extinguishing these fires is to determine the direction of spread, anticipating where it is headed. These fires tend to move as a linear front and can be knocked down in a sweeping side-to-side attack along the front.

Be extremely careful of even small brush fires burning in windy conditions. These fires can grow exponentially in size, erupting into a full-blown wildland fire. Also, be wary of fires burning near buildings. In these situations, it is best to call for additional fire companies quickly, before the fire gets out of hand. Since many brush fires occur in locations where there are no fire hydrants, the available water supply—onboard water tanks on the fire engine—may not be enough to deal with the fire if it grows too large.

Wildland fires. **FFI 4.3.19** Large fires involving vegetation have come to be known as wildland fires. Some of these fires occur in the "urban-wildland interface," locations where residential development has penetrated into forests and heavily vegetated areas. Some of these wildland fires involve large responses of hundreds of pieces of firefighting equipment from many states, much more equipment than even the biggest building fires.

While a complete discussion of wildland firefighting is beyond the scope of this textbook, it is important for you to have a basic understanding of the terms and equipment used in wildland firefighting.

The use of proper terminology in wildland firefighting is just as important as it is in structural firefighting. Some of the more important terms to describe a wildland fire are the **head**, which is the portion of the fire with the greatest rate of spread (usually the edge of the fire opposite from which the wind is blowing; a fire may have multiple heads); **fingers**, which are long narrow strips that extend from the main body of fire; **pockets**, which are unburned areas between the main body of fire and the fingers; the **rear** or **heel,** which is the end of the fire opposite from the head; **flanks,** which are the sides of the fire; and **islands**, which are unburned areas within the fire's perimeter (fig. 20–85). **Spot fires** are small fires started by flying embers far ahead of the main body of fire. *Green* is an area that has not been burned (the material may be live or dead vegetation). *Black* is burnt area.

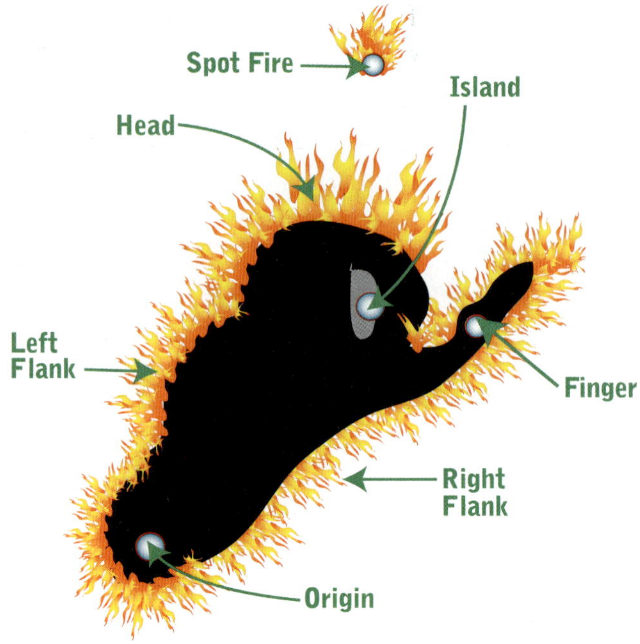

Fig. 20–85. A wildland fire

In terms of fire spread, creeping fires move very slowly. A running fire is one that spreads quickly with a well-defined head. A backing fire moves away from the head, downhill, or against the wind. A **crown fire** consists of serious fires that move across the tops of trees or shrubs, often independent of the fire at ground level (fig. 20–86).

Fig. 20–86. A crown fire in Yellowstone Park

Blowups occur when there is a sustained, rapid increase in the rate of spread, preventing control of the fire. A blowup will often necessitate changing the plan of fire attack. A **flare-up** is any sudden acceleration in the rate of spread for a short period of time.

Firefighters should be constantly aware of ground fire behavior, as well as changing weather and wind conditions when fighting ground fires. Escape routes and areas of safety should be identified and kept in mind throughout the firefighting operation in case of a sudden change in fire behavior.

When fighting a wildland fire, certain techniques are used. **Control lines** are "constructed" or natural fire barriers used to contain the fire. Firelines, on the other hand, are cleared strips or portions of a control line where flammable material has been removed by scraping or digging down below vegetation to mineral soil. An anchor point is the starting location for a fireline. A fire barrier is a safe anchor point. Using anchor points helps reduce the possibility of being outflanked by the fire while the fireline is being constructed. **Slopover** is fire that crosses the control line or natural barrier.

The primary methods for perimeter control of wildfires are direct and indirect attack. Direct attack is achieved through establishing control lines either at or close to the burning edge of the fire. Control lines for indirect attack are established at a distance from the fire, when the fire is too large or intense to use a direct attack.

When fighting a wildland fire, proper wildland PPE is required (refer to chapter 9, Personal Protective Equipment, for specifics on wildland PPE). In addition, the wildland firefighter carries a fire shelter (fig. 20–87). Shelters are used as a last resort when a firefighter becomes trapped and in danger of being overrun by the fire.

Fig. 20–87. Wildland shelter. (Courtesy of Andoni Kastros)

Wildland firefighters also use special tools and equipment. The most common hand tool is the Pulaski (fig. 20–88). It is a combined axe and mattock (a mattock is a chisel-like shape), which is capable of chopping down trees (the axe) and breaking up ground to create a fireline (the mattock). Large wildland fires necessitate the use of heavy equipment such as bulldozers, as well as air resources such as tankers and attack aircraft.

Fig. 20–88. Pulaski tool. (Courtesy of Andoni Kastros)

Specialized training is required for wildland firefighting. Such training includes safety procedures (LCES: lookouts, communications, escape routes, safety zones), constructing and maintaining firelines, wildland fire behavior, and tool use among many other specialized topics.

FIRES IN STACKED MATERIAL

Fires in stacked material, whether inside or outside a building, present two major challenges for firefighters: water application and collapse hazards. Common occupancies where you may find stacked material are lumberyards, junkyards, recycling plants, and warehouse or storage areas. Applying water to stacked material is difficult because extinguishment requires water to be applied on all six sides. Pallets of raw materials, parts, fabric, or other items are stacked on top of another to save space. As a result of this method of storage, fire can be on all four sides and the top and bottom. Bins or pallets of materials usually have access for storage, and retrieval from only the front side and may be stacked several deep. Further complicating water application is the fact that shelving may prevent water from dropping down onto the top of the burning material. In-shelf sprinkler systems are required to apply water to stored materials and are a good option. If these systems are in place in the building, always supply the system with one or more hoselines.

In the case of warehouses with automatic sprinklers, water application can be compromised by shelving in storage racks, blocking water from dropping down onto the top of the burning material. Rack sprinkler heads are required in such situations to apply water directly on to stored materials.

As the packaging material burns away, as in the case of a cardboard box or carboy container, its contents may spill out or fall. Wooden pallets that are the base for many types of container systems burn, and they may lose structural integrity and cause contents to shift and fall onto firefighters making an interior attack.

Falling debris from stacked material presents several hazards to firefighters. Obviously injuries can occur from this heavy material falling on unsuspecting firefighters. The falling debris will not be seen by firefighters below, because it will be obscured by smoke and can be initiated by lateral forces applied by hose streams. Falling debris can cover exit ways and hoselines, trapping firefighters or at the very least making a tactical withdrawal a very dangerous and time-consuming operation. Depending on the type of material stored in stacked fashion, the collapse or opening of containers by physical collapse or burning of containers can result in rapid fire progress, killing or injuring firefighters.

Extinguishing stacked material fires helps create an additional hazard that may kill or injure firefighters. Water will be absorbed by both packing and packaging materials and will therefore increase the weight of the stacked material. The weight of water, at 8 lb per gal (1 kg/L), can cause catastrophic failure of the shelving system. If the material is stacked without shelves (e.g., large rolls of paper or bailed materials such as hay or recycled materials), water can soften it, causing it to sag or become unbalanced and fall on firefighters. Firefighters have been injured and killed in this exact scenario. Remember to stay out of the collapse zone for stacked material.

Foam use on stacked material

Class A foam can be used to increase the effectiveness of water on stacked material fires. (Foams are discussed in greater detail in chapter 31, Advanced Fire Attack.) These foams lower the surface tension of water and therefore make it more effective. The water penetrates baled materials with greater effectiveness because of the reduced surface tension of the water, thus increasing the cooling effect of the water applied. Essentially, the water is allowed to penetrate deep in the material instead of beading up like raindrops on a car hood because the water's surface tension is reduced. This is caused by the foam concentrate, which is a detergent-based synthetic foam. These concentrates are not designed for use on flammable liquids but rather on any Class A combustible.

When **aspirated** much like Class B foams, Class A foam has a cooling and smothering effect on Class A combustibles. Class A foams are generally used at less concentration than Class B foams; 0.1%–1% and can be proportioned a number of ways. Used as a wetting agent, standard solid bore or combination nozzles can be used. When used for pretreatment (before the fire) or as a mop-up tool, air aspirating nozzles or compressed air systems are needed.

NOTE

1. The United States National Institute of Standards and Technology (NIST) report NBSIR 80-2120, *Fire Development in Residential Basement Rooms,* October 1980, (Fang and Breese).

QUESTIONS

1. What is some of the basic information that can be gained from a proper communication report of a fire officer that has arrived on scene?
2. List at least five steps to a successful fire attack.
3. What is the purpose of size-up?
4. There are seven specific points to size up that must be considered upon arrival. List all.
5. What are the primary differences between an "offensive vs. defensive" type of strategy?
6. Which type of strategy is the "blitz attack" method employed?
7. Why are doors the primary entry and egress points to initiate a fire attack?
8. What is the primary purpose of the fire attack line?
9. When performing a rescue of trapped occupants, what is the preferred means of removal?
10. An engine companies operations can do more to effect the rescue of trapped or injured persons than any other group of firefighters. What are the common ways this is accomplished?
11. List the primary responsibilities for the nozzle operator during the initial stretch.
12. List the primary responsibilities of the backup position during the stretch.
13. Which position of the fire attack line is responsible for monitoring fatigue levels of the crew, communicating with command, and requesting additional assistance during the advancement of the hoseline?
14. What type of building will most likely require the use of a standpipe stretch to get the initial attack line into operation?
15. When utilizing the direct or modified direct method of attack, the nozzle should always be set on which pattern?
16. What are the dangers of opening up a fog stream during fire attack?
17. What are some of the basic duties firefighters will perform during the pick up and return part of an incident?
18. What is one of the most challenging types of fires to control, and why?
19. When sizing up a garden apartment for fire attack, what must often be considered when making the hoseline stretch?
20. Taxpayer buildings present a unique combination of occupancy during firefighting operations. What types of occupancies can typically be expected?

21. Big box retail structures should always have what size of line deployed?
22. When sizing up a vacant structure for fire attack, what are some of the general considerations for this type of building?
23. Outside tire fires prove to be extremely difficult to extinguish and often burn for days on end. What is the most effective way to combat tire fires?
24. List some of the hazards firefighters face when combating fires in stacked material.

Salvage and Overhaul

by Jeff Shupe

This chapter provides required knowledge items for the following NFPA Standard 1001 Job Performance Requirements:

FFI 4.3.13

FFI 4.3.14

21

This chapter contains Skill Drills. When you see this icon, refer to your Skill Drill book for step-by-step instructions.

SKILL DRILL

OBJECTIVES

Upon completion of this chapter, you should be able to do the following:

- Define the term *salvage*
- Identify types of damage done by fire
- List the tools and equipment needed to effectively perform salvage operations
- Describe methods of grouping furniture in order to protect them from damage
- Describe two methods of deploying salvage covers
- Describe two methods of storing salvage covers
- Identify two methods of removing standing water from a structure
- Identify two methods of stopping water flow from an activated automatic sprinkler system
- Describe different methods of securing a building after firefighting operations
- Define the term overhaul
- List the equipment needed to effectively perform overhaul operations
- List four ways of detecting hidden fires
- Describe two devices that are used to detect hidden fires
- Identify methods for opening walls, floors, and ceilings
- Identify techniques for overhauling roofs and building contents
- Describe overhaul's role in conducting a fire investigation

INTRODUCTION

Salvage can be defined simply as "the art of saving property." It is a process whereby firefighters try to minimize damage to a structure and its contents when a fire or other emergency takes place. During firefighting operations, the basic way to save anything from harm is to either extinguish the fire and stop its destruction, or remove items of value from affected areas. This effort helps fire departments keep property loss down and achieve the stated goals of the American fire service, which are to save lives *and* property.

Sometimes known as **loss control**, salvage is a component of firefighting that can go a long way toward building good public relations between a fire department and the community it serves. Many fire departments have received praise and accolades for jobs well done at saving someone's property from fire. Some fire departments have even benefited from this financially and in other ways. Even though firefighters learn the importance and value of good salvage techniques early in their careers, it is one of the things that probably receives the least amount of attention on the fireground until after the bulk of the fire has been knocked down. This lack of attention may be partly because it is not truly associated with the excitement of actual fire-suppression efforts or other closely related *action* work at a fire. However, it is truly gratifying when someone acknowledges that you have done a commendable job of saving someone's property from loss.

KINDS OF FIRE DAMAGE

Generally, there are two classifications of fire damage: **Direct** (or primary) and **indirect** (or secondary). Direct fire damage is that which is caused by the actual combustion process, where there is physical destruction from flame, heat, or smoke. Indirect damage is all other damage caused from suppression efforts and other activities. An example of indirect damage is opening up of ceilings and walls to check for fire travel. There may not be any fire, but the job must be done, because if overhaul is not done and fire is left unchecked, the results will show later with more damage and the need for more overhaul.

SALVAGE OPERATIONS AND PROPERTY PROTECTION

FFI 4.3.14 **Salvage** is one of three incident priorities of firefighting. It falls under the heading of *property* conservation. It is also considered one of the basic steps of firefighting. With these points in mind, the importance of saving what we can or what we must save cannot be overstated.

Salvage begins with fire attack. It can take place in many forms, ranging from suppression efforts to the actual removal of property or the grouping and covering of property from threat of damage. An example of salvage from suppression action takes place when firefighters make a quick, aggressive knockdown of a fire followed by efficient ventilation to rid the building of heat, smoke, and gases. The size of attack, the point of approach, and the type of ventilation used in this scenario all affect the total outcome of damage. To illustrate this point, think of a working fire in the living room of an occupied two-story single-family residence. A quick and well-placed attack line, discharging the right volume of water, may be just what is needed for a rapid knockdown of a fast spreading fire while confining it to its area of origin. This effective method of attack can prevent the fire from spreading to uninvolved areas of the house, which will help to limit fire spread and damage. This effort relates to stopping direct and indirect fire damage. Quickly attacking the fire this way can stop greater loss from happening. Other fireground actions such as ventilating the building will also help the salvage effort by relieving the interior of heat, smoke, and gas accumulation.

Planning for salvage

For salvage operations to be efficient, there needs to be prior planning and training. Procedures should be developed to deal with the different kinds of salvage operations that might be needed for the different building types and occupancies found in a particular community. For example, special preplans might be needed for an office building that has especially valuable contents or documents in the offices. Preplans can tell firefighters where these contents are located and what measures should be taken to protect them.

Industrial manufacturing businesses might have materials on-site that are important to the company's survival or may require special handling or salvage techniques. Firefighting history has shown us that some small towns have existed because of their dependence on a large industry or business. Many of the town's citizens were employees of this business, and when a major fire took place, there was not enough saved to keep the operation going. The end result was jobs lost and a faltering local economy that led to scaled-back town services. That loss eventually caused a loss in the quality of life for the community. There could also be a threat to firefighter safety or a much larger problem if fire control in a particular type of facility was not protected properly and created a larger, more complex safety and salvage concern.

Sprinkler systems are a valuable asset to protecting buildings from the threat of fire. Their ability to control or hold fires until the fire department arrives has been proven over time. However, when a sprinkler system is activated, it has the ability to control a fire or hold it in check, but after that, it has the potential to cause extreme water damage. Because of that, fire departments must know from preplans, walk-throughs, and training where sprinkler systems' riser control valves and piping are found in a structure.

Salvage tools and equipment

FFI 4.3.13 Tools and equipment used for salvage should be carried on fire apparatus in compartments where they are easily accessible. In years gone by, ladder or truck companies were generally responsible for carrying this type of equipment and taking care of salvage operations. In larger cities where there were numerous businesses requiring property protection, salvage or **fire patrol** companies were established. These were crews who were paid by insurance companies, and their only job was salvage on the fireground. The equipment carried in their apparatus was strictly for that purpose. These companies no longer exist in our cities, so all salvage operations depend solely on the efforts of the local fire departments.

Nowadays, even though certain units may carry salvage equipment, it is generally accepted that everyone has a hand in salvage operations. Many departments have some basic salvage equipment on all of their fire apparatus. Because of this, every firefighter should know where the tools and equipment are located and know how to use them.

The following is a list of tools and equipment that might be needed to perform salvage duties (fig. 21–1):

Shovels
Brooms
Mops
Squeegees
Pike poles
Floor runners
Polyethylene plastic sheeting (e.g., Visqueen brand)
Salvage covers—canvas type
Roll roofing paper
Buckets
Sprinkler head kit and sprinkler tongs
Wooden wedges of different sizes
Submersible pumps
Water vacuums
Hand tools
Hammers
Electric or battery-operated drill
Supply of nails and screws
Wood lath strips
Duct tape
Assortment of rags and towels
Sawdust—carried in large bags
Debris carryall bags
Polystyrene or wood blocks
Linoleum or carpenter's knife

Fig. 21–1. A wide variety of equipment is used to protect property during salvage and overhaul operations.

Where and when to begin salvage operations

When beginning salvage, first determine where the initial salvage efforts will have the greatest impact on saving property. If operations start in a room below the fire floor, then you will most likely arrange or group furniture into piles. Furniture or items located below a dripping ceiling should be moved to another area of the room or removed from the area all together if necessary. Where there is a tall piece of furniture like a chest of drawers, place it in the center of the pile or group so that when a salvage cover is thrown over everything, it acts as a pole draping the salvage cover over everything while allowing water to run off and away (fig. 21–2).

Fig. 21–2. Vinyl covers protect valuable contents from hot embers and water damage. Grouping contents together allows for the most efficient use of salvage covers.

If a floor is collecting water, then it will be necessary to use blocks or even something like soup cans to lift the furniture or objects off of the floor to prevent water absorption and damage (fig. 21–3). Where fires occur in industrial or commercial occupancies, skids can be used to pile items on and lift stock items off of the floor to keep them away from water pooling.

Fig. 21–3. Sometimes simply lifting valuables up off the ground can prevent them from being damaged from water runoff.

Generally, salvage operations begin on the fire floor or the floor directly below. It is these two areas that are likely to sustain the most amount of damage from fire, smoke, heat, and water. Just after a fire is extinguished, water most likely will collect on the fire floor, then quickly find its way to areas below. If a large amount of water has been used for the fire, then operations should start directly below the fire floor as soon as possible. If the area below the fire floor has ceilings, then firefighters must check them to see if they are accumulating water and might possibly collapse. It may be necessary to puncture the ceiling to relieve built-up water in the ceiling space.

Before water is relieved, it may be necessary to figure the best way to channel water out of the area. Also, water in the ceilings may affect electric wires and fixtures. Perhaps water chutes (see later in this chapter) can be devised to get water out of the building. Floor runners may need to be placed for firefighter traffic and or water droplets falling from above.

As stated earlier, salvage operations should begin as soon as possible. They can begin immediately during a fire attack in an indirect manner or as soon as firefighters are able to initiate a more formalized effort.

Throwing salvage covers

SKILL DRILL

FFI 4.3.14 The most common salvage tool is probably the salvage cover. Typically made of canvas, salvage covers are cut or made into different sizes (fig. 21–4). Some have coatings or treatments applied to them to help shed water. Nowadays, covers vary in size, including 10 × 12, 12 × 12, 12 × 16, and 14 × 18 ft (3 × 3.7, 3.7 × 3.7, 3.7 × 5, and 4.3 × 5.5 m).

Fig. 21–4. While canvas salvage covers are found in most fire departments, they are being replaced by vinyl covers such as these.

If possible, salvage covers should be thrown over furniture piles and tucked in at the bottom of the pile to prevent water splashes from damaging the furniture.

After the salvage covers are used at fires and taken back to the fire station, the maintenance work begins. They are washed and hung up to dry. When they are completely dry and free from any chance of mold developing, they are checked for any holes or tears, then folded (fig. 21–5).

Fig. 21–5. Salvage covers must be properly folded and cleaned after each use.

A new material that is being used for salvage covers is polyethylene plastic sheeting (fig. 21–6), commonly the Visqueen brand. It has several advantages over canvas covers and tarps. One advantage is that it is economical. One roll can be 120 ft (37 m) long by 20 ft (6 m) wide. It is disposable, so there is no maintenance of cleaning, drying, and folding. It is also quick and easy to use on the scene, and it can be cut to fit the situation. The only item necessary to make it work is a utility knife or other type of cutting tool. On an apparatus, it is easy to store and can fit in any apparatus.

Fig. 21–6. Polyethylene plastic sheathing is a popular alternative to traditional salvage covers because of its low cost and the ability to manipulate its size.

Water chutes and catchalls

SKILL DRILL

Salvage covers are primarily used to protect furniture and other valuables but can also serve other uses, such as **water chutes**, where a trough is made using covers, pike poles, and ladders to catch dripping water and divert it away (fig. 21–7). Covers can also be used for holes left in the roof of a building. They can be used as catchalls after the sides have been built up or rolled up, using pike poles once again (fig. 21–8).

Fig. 21–7. Creating a water chute from a salvage cover can be necessary to divert water away from valuable belongings.

Fig. 21–8. Conventional salvage covers can have the sides rolled up to manufacture a catchall.

Removing water and dewatering equipment

Getting water out of a building is often great concern for firefighters. As you know from chapter 15, Water Supply, water weighs 8.34 lb per gal (1 kg per L). Considering that a 2½-in. (65-mm) hand line can flow upwards of 250 gpm (946 L/min), the use of such streams can add more than a ton of weight every minute to the structure. This puts an additional stress on an already compromised building. In addition, we may need to remove water from a structure to aid in a postfire investigation or to find fire victims in a flooded basement. Finally, some cities assist homeowners in pumping out basements after a flood as a public service. Whatever the reason, firefighters are often expected to use dewatering equipment in such situations.

It may be possible to remove water by clearing floor drains, if present. Additionally, it may be necessary to cut holes in floors to allow water to drain to lower levels. Toilets can be removed from their attachment to the floor in bathrooms to allow drainage as well—make sure that the toilet is replaced or the drain is sealed after such actions (fig. 21–9).

Fig. 21–9. To facilitate draining, a toilet can be removed. This tactic can be especially important in multi-story buildings where creating chutes and catchalls isn't feasible.

Although not found too often on fire apparatus, the water vacuum is a useful device for removing water from areas that are not easily accessible with mops or squeegees. Some water vacuums are made like a backpack and are carried on a firefighter's back, making it very portable, but also limiting the amount of water it can pick up (fig. 21–10). However, if there is a large amount of water that needs to be removed and there is limited capability to do it, then the services of professional cleanup crews should be called to the scene because of their larger water vacuum machines.

Fig. 21–10. Using a vacuum to remove water is effective at small fires or incidental discharges of sprinkler systems.

The use of dewatering pumps is sometimes called for. These pumps range in capacity from a few gallons per minute to several hundred gallons per minute. Some are electric powered, while others have their own generators. Ironically, some dewatering pumps use water (hoselines) to power them. A key to the efficient use of these pumps is to select a low point location where water will flow to the pump (fig. 21–11). Ensure that the outlet of the pump is located outdoors in a location where the discharged water will not find its way back into the structure.

Fig. 21–11. Dewatering pumps should be placed at the lowest collection point so that the water will drain toward the pump.

FFI 4.3.13 Safety is of prime importance here. Ensure that all electricity has been shut down before entering an area with standing water. Never stick your hand or any body part near any floor or roof drain; use a tool with a long handle to clear drains.

Stopping water flow from a sprinkler system

SKILL DRILL

FFI 4.3.14 When ordered by the incident commander (and only the incident commander [IC]) once a fire has been extinguished, an operating sprinkler system will be shut down to avoid additional water damage. The firefighter at the riser control valve (most often an outside stem and yoke valve, or OS&Y valve) will shut off the valve (figs. 21–12 and 21–13). After closing the OS&Y valve (the stem is now completely inside the valve), the firefighter can open the main drain and drain out the water from the system (fig. 21–14).

Note: After the water is drained from the system, the sprinkler system will no longer be capable of extinguishing a subsequent fire or rekindle because the water supply will have been shut off to the sprinkler heads.

Fig. 21–12. An outside stem and yoke (OS&Y valve) in the "open" position

Fig. 21–13. An outside stem and yoke (OS&Y valve) in the "closed" position

Alternatively, a firefighter may be asked to insert a **sprinkler wedge** or wood chock into an operating sprinkler head (if it is still flowing water) (fig. 21–15). Firefighters who perform this function will get very wet—a consideration when cold temperatures would cause the firefighter's gear to freeze. The advantage of this technique is that the sprinkler system will remain in service, and the other heads that have not been activated will remain capable of extinguishing a subsequent fire or rekindle.

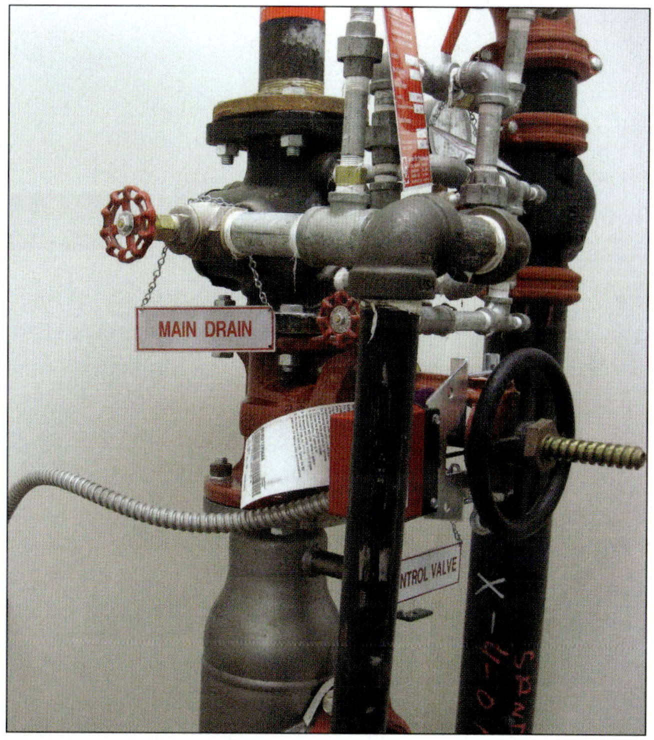

Fig. 21–14. The main drain of a sprinkler system is normally labeled and easy to locate.

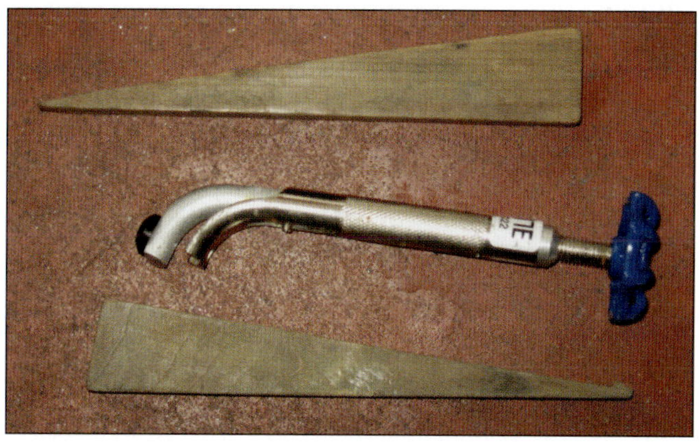

Fig. 21–15. Sprinkler wedges or tongs come in a variety of sizes.

After a fire, most fire departments have their fire prevention bureau follow up to make sure the sprinkler system has been properly restored; some establish a **fire watch** until the system has been restored by a licensed contractor. Other fire departments actually restore sprinkler systems to full operation by replacing heads and resetting **dry-pipe valves** (the alarm valve on a dry-pipe sprinkler system in which the pipes are normally filled with air to avoid freezing) under nonfire conditions. When performing such procedures, ensure that exactly the same sprinkler head is used to replace the head that activated.

For detailed information on total sprinkler system operations, see chapter 30, Fire Protection Systems.

Securing a building after firefighting operations including forcible entry

SKILL DRILL

FFI 4.3.14 In many fire situations, postfire salvage operations require the fire building to be secured before the fire department leaves the scene. During the course of firefighting, there might have been structural damage caused from forcible-entry work to doors and windows. Forcible entry methods for salvage can include anything from forcing doors of rooms adjacent to the fire area to cutting holes for drainage or other types of wall or floor breaches. Firefighters should try to consider the least damaging methods for forcible entry during salvage, to reduce property damage and minimize the tasks of securing the structure before leaving the scene.

If the building is one where it is going to be reoccupied after repairs, then the fire department should close any openings in the building made from firefighting or fire damage. This is to prevent any environmental damage that might occur.

For example, if the fire department had cut a hole in the roof of a house during fire operations, then it should cover the hole with a tarp or other material that will keep the elements out of the building. Generally, canvas tarps will work fine because they are heavy enough to resist wind forces and can keep the elements out also as long as they are secured to the roof. Visqueen or other plastic film sheeting can also be used; however, it may need to be double-layered to resist wind forces. The outer edges of the plastic sheeting should be folded over two or three times to give strength where it will be attached to the structure, so it won't tear from high wind forces. Also, it will be necessary to use long pieces of wood or lath strips to anchor the perimeter of the sheeting to prevent wind forces from getting under the plastic covering and tearing it off the roof decking. Another reason for attaching it this way is to stop rain or other elements from being blown underneath the covering and getting inside the building by that route.

If the roof being covered is a flat roof like that found over a commercial store or "taxpayer," it may be necessary to build a little frame with a pitch to direct the rain off the cover, so the water won't be able to collect and pool in the cover, which would cause the cover to hang down into the vent hole. Firefighters will need to figure out the best direction to pitch the runoff to.

When there has been any substantial fire damage to the first floor or lower level that will cause the building to be vacated during rebuilding, then security will be an issue. If doors or windows have been damaged from forcing, then it may be necessary to cover them with plywood sheeting or oriented strand board (OSB) (fig. 21–16). In many cases, all that is necessary to attach the sheeting is a hammer and nails and pieces of sheeting cut to the size needed to cover the window. Usually, it can be nailed to the perimeter of the window. If the building's exterior walls are of brick, then the sheeting can be attached to the top, bottom, and side rails of the window creating a now secured structure.

Fig. 21–16. Plywood can be used to secure broken windows.

If the building has sustained damage beyond repair and will need to be demolished, then firefighters will need to establish a safe perimeter with barrier tape around the property that will let people know their limits.

Floors with holes in them from cutting operations or from fire damage can be secured by simply putting a door over them and nailing it in place. In extreme cases in which much of the floor is too weak to support anyone, barrier tape over the door can serve as a warning device or nailing the entrance door shut to that area.

OVERHAUL

A simple definition of **overhaul** is "making sure the fire is out and leaving the building in as serviceable a condition as possible." Overhaul is a learned process that is extremely important to the outcome of the incident and should be carried out in a systematic way to minimize the chances of rekindles happening. The process of overhaul generally begins after the fire is extinguished or brought under control, and firefighters are able to move into fire-damaged areas and finish extinguishment by examining burned contents and opening the building's suspected void spaces, typically the walls, floors, and ceiling spaces, to search for hidden fire.

Safety first!

Firefighters must always keep personal safety in mind, even after the bulk of the fire has been extinguished. During overhaul, it is necessary to use complete personal protective equipment, including self-contained breathing apparatus (SCBA), unless the atmosphere has been determined safe to operate in without it (fig. 21–17). A company-level officer should be assigned the responsibility of supervising safety and the overhaul effort in each area where it is performed. The officer should watch for any problems and also monitor all radio communications. If fire is found to be spreading quickly to other areas of the structure, it is that officer's job to alert other crews and the IC about changing conditions or fire travel, or anything that might affect the safety of others working.

Before overhaul begins, it must first be determined if the building is safe to be operating in. In situations where there has been severe fire damage to the building's structural components, it will be the IC's responsibility to determine if it is safe to work in. This decision should be based on reports from company officers, firefighters, and the IC's own personal observations of the structure's condition. Fire service history has taught us that it is this time of a fire incident that is very dangerous and in some cases deadly to firefighters. In other situations where firefighters have had a tough, punishing fire, they can be fatigued from physical stress and effort. It can be times like this when firefighters lets their guard down about the building or fire conditions and just want to get the rest of the job done and get back to quarters. Perhaps there are obvious safety hazards, but they are not being recognized or considered. Here is where we can get into trouble.

Fig. 21–17. Numerous toxic gases are produced during the decay stage of a fire, when overhaul takes place. Therefore it is imperative that firefighters always wear SCBA.

Some points to consider before overhauling a structure are the following:

- Building construction type and type of materials used in its construction: Is there any fire damage to them?
- Occupancy of the building and its contents: Are there any unsafe or hazardous materials involved?
- What is the volume of fire and the length of time the fire has been burning?
- What is the actual damage to the building's structural system?
- Length of time for extinguishment process, namely water: How much was used and for how long?
- Are there any telltale signs of a weakened, unsafe structure that may be a collapse hazard, like any sagging roofs or exterior walls that may be bowing outward?
- Is there any water running out from any part of the building?
- Are there unstable chimneys that may topple?
- Is there enough lighting to see inside the building? Do lights need to be brought in?

If firefighters are already operating inside the structure, are there any reports of water pooling or building up on floors? Do any of the floors feel springy or spongy? Are firefighters reporting any unusual sounds or occurrences within the structure, such as ceilings dropping or wall plaster falling to the floor, or sounds like cracking or stressing noises? Is there a buildup of water on any flat roof surfaces, indicating a weakened roof that is developing a new concentrated, live load? If any of these points is happening, it may be prudent to remove firefighters from the area or building until a further determination can be made concerning the structure's stability. Firefighters noticing any of these signs or other things need to make sure everyone on the scene is alerted to the possibility of an unsafe environment.

Having a plan and where to begin

Just as every fire should have an attack plan, the overhaul process should have a plan, too. Generally, the fire's location within a building is the starting point for the process, and more than likely, that is where overhaul will begin. Remember, that while fires have six sides—front, rear, both sides, top, and bottom—so too does the overhaul process. Firefighters need to remain proactive during overhaul and be suspicious of fire travel. To illustrate this point, picture a fire that has been localized and confined to one or two rooms on the third floor of a four-story garden apartment building. Firefighters need to check if the fire has extended beyond these rooms. Fire remaining unchecked in any wall, floor, or ceiling void spaces or pipe chases could eventually find its way to the cockloft space over the entire structure and cause tremendous damage. The best way to know for sure is to have a crew go above, below, and around the fire area to examine for fire extension. Visual investigations or observations along with good radio reports will help keep everyone aware of the fire's activity—its location, what it is doing, and where it might be going.

If the fire building itself is large and has sustained a substantial amount of fire damage, or if it has a large amount of contents that will require considerable attention, there will probably be extensive overhaul efforts. The IC might consider making a request for more firefighters to the scene to set up and rotate overhaul crews. This move serves to gives those who have already been working at the fire a break, and it helps to spread the workload by bringing in fresh, alert personnel.

Before beginning overhaul, it is important for firefighters to be fully aware of the damaged area(s) and to be looking for possible causes of the fire. In some cases, burn patterns will give clues as to damage and ignition sources. If these are detected, everything should be done to preserve evidence and not destroy or remove any evidence until fire investigators arrive and are able to determine its importance to the fire's cause. In many instances, firefighters begin overhauling and destroying evidence or end up throwing it out of the building.

Attack hoselines

Attack hoselines should not be removed from any areas needing overhaul. Where there is suspected fire travel, hoselines should remain charged and in place until overhaul has been completed and the IC has approved it. In far too many instances, fire departments have placed too much reliance on a single hoseline to extinguish a fire and cover all the potential points of fire travel and extension. After fire has been discovered extending to different areas, the time and effort used to reposition this line might let a fire gain great headway, growing more than the single line of hose can handle. Incidents like this have resulted in greater loss and, in some cases, firefighter injuries. Fire departments should know when and where multiple hoselines should be stretched for efficient coverage. An old adage of the fire service says where fire extension is suspected, don't send a crew to investigate ... send a hoseline and tools!

Tools and equipment

The task of overhaul is generally one of manual labor, meaning hard physical work. The tools used most commonly to accomplish this are hand tools of the pushing, pulling, striking, and cutting variety, but in some cases power tools are used also. Some of the more common tools are axes—both pick-head and flat-head axes; pike poles of different styles and lengths depending on the amount of reach or pulling needed; Halligan and Kelly tools for prying or leverage; shovels; and power saws, both chain and rotary. Other useful tools include rubbish hooks, carryall bags, and even wheelbarrows.

Detecting fire

After the main body of fire or flames have been knocked down or extinguished, it may be hard to actually know if there is any remaining fire due to poor visibility—all that is visible might only be the blackened, charred remains. Many firefighters have been fooled after a fire has been knocked or extinguished and the smoke has been removed from the room, only to have the room fill with smoke and heat minutes later from fire burning behind the walls or ceilings. In these cases, firefighters might have to use their own senses to find hidden flames, such as in the following examples.

Sight. Look for the following:

- Discoloration of paint or wall coverings
- Blistering or bubbling of paint or varnish or shellac finishes
- Heavily damaged or severely burned areas showing extreme amounts and depth of charring
- Smoke coming from around window or door frames or moldings, or registers
- Small sparks or cinders dropping from ceiling areas
- Smoke arising from floors, especially finished floors with subfloors
- Openings (vents, shafts, etc.) near the fire that may be allowed heat and smoke to travel to other parts of the building

Touch and feel. Use the back of your hand to touch suspected hot surfaces, first sensing for heat without pressing your skin to the suspected surface.

Hearing. Listen for the following:

- Any sounds of crackling or snapping as so often happens when wood burns
- Sounds of pressure building from pipes or other vessels

Smell. Use your sense of smell to detect the odor of fresh wood burning or other fresh smoke. If odors are noxious or smoke appears to be building in volume, then redon your face piece for personal safety.

Technology for detecting fire or heat. In the past, firefighters had only their learned principles of techniques and good instincts to help them search out hidden pockets of fire. Many times, firefighters worked extra hard trying to find the source of odors of smoke, only to find nothing, and so they would leave the building to return to quarters, only to return an hour or two later to find a fire roaring out of the windows and through the roof. Nowadays, there is a great tool that employs modern technology that most fire departments have or at least have access to. It is the thermal imaging camera, commonly called TIC (fig. 21–18). Using this device allows firefighters to detect heat or fire in hidden spaces without having to tear a room completely apart to find it. This saves time, effort, and guess work. The TIC is especially useful for the scenarios already described in this chapter, and it has many other uses where any amount of heat greater than the ambient area temperature can be identified through observation on the camera screen.

Fig. 21–18. Thermal imaging technology has allowed the overhaul process to be less damaging to property and more time efficient.

Opening walls, floors, and ceilings

SKILL DRILL

Before any wall, ceiling, or floor enclosure is opened, first determine if there is a real need to do so! In far too many cases, firefighters have caused needless damage to occupied buildings during overhaul by opening areas where there was no fire, nor was there any threat of fire extension. This makes firefighters and their fire departments look less like the professionals they want to be. And their community will perceive that, too. However, when done properly, the overhaul process will leave the building with no fire remaining, and a structure that will sustain minimal damage.

Ceilings need to be opened whenever there is a concern of hidden fire. In obvious situations where there has been substantial fire damage to a room, there will be no question about the need to open the ceiling, especially above the area where the fire burned. When ceilings located below cockloft or attic spaces are opened and there is a large volume of fire found, the burning area should be opened completely for extinguishment and further examination. If the fire is found to be extending horizontally, firefighters may need to get ahead of the fire's direction of travel and begin pulling down the ceiling in that area to cut off or limit the fire's spread. Once the fire is stopped, the entire ceiling will now need to be opened because of the threat of sparks or embers remaining. When fire is found in a ceiling space, the firefighters who found it should make sure this condition and the need to get above it are communicated immediately to everyone.

A special technique for opening walls and ceilings has been developed by Lt. Michael Ciampo of the New York City Fire Department (FDNY). Called the punch technique, the firefighter uses a hook on ceilings and a Halligan tool on walls to create a series of initial perforations in the gypsum board to precut the surface, before removing the gypsum board. Once the perforations have been made, the firefighter can easily and effectively remove the gypsum board in large sections (figs. 21–19 and 21–20).

When it is time to open walls, it should be done completely from stud to stud and floor to ceiling (fig. 21–21). Doing this will expose the channel(s) and allow firefighters to see if there has been any charring inside the wall space. Remove the baseboard to expose the bottom of the wall and edge of the floor. If fire is found in several wall bay spaces, they must be opened completely. A concern here is if the fire in the wall spaces is extending either up or down vertically, especially in balloon frame construction. If a wall is opened and fire jumps out, it should be extinguished by the hoseline, and then the nozzle should be pushed into the wall space to direct water up and down the space to knock down any possible extension. Once again, the officer supervising should contact everyone by radio or verbally to let them know about extension and that crews should be ordered to check for it. Fires in outside walls will need to be checked on the outside of the structure, also. Fire traveling up an outside wall can extend to the upper parts of a building and may reenter the structure.

Fires hidden in floors or floor spaces can require extra effort, depending on the type of floor. In many residential structures, the floor is plywood laid over wood joists or trusses. There may be a layer of carpeting or linoleum on top of that. For others, there may be a subfloor of plywood or floor planks, then a hard wood floor on top of that. Still other residential floors have this supporting system with ceramic tiles for the finish. In any event, the firefighter must first understand the type of floor and the strength of its supporting system in order to determine the safest method to open the floor (fig. 21–22). For example, where there is carpeting, it may be best to cut carpet away with a utility knife before any tool use. If it is a wood floor, it may be best to use a power tool for cutting, as an axe may bounce away or glance off the floor boards when struck. Ceramic tiles may need to be struck with a maul or sledgehammer first.

Fig. 21–19. The ceiling punch technique involves using a pike pole to weaken gypsum board so that it can be removed more easily in large pieces.

Fig. 21–20. The wall punch technique works in a similar manner to the ceiling technique. However, the tool of choice for this evolution is the Halligan.

Fig. 21–21. Firefighters must expose the interior of a wall surface to prevent rekindle during the overhaul process.

Fig. 21–22. When opening up floor assemblies, firefighters must be careful not to cut through the beam below, compromising its load-carrying capability.

Roof overhaul

Different kinds of buildings have different kinds of roofing systems. Because of that, it is important for firefighters to recognize different types of building construction and roofs. Structures have different types of roof-supporting systems and roof-covering materials. Anytime fire or heat has reached the roof area, the roof must be examined for fire (fig. 21–23). In many cases this means the roof decking and covering must be removed. Appropriate methods and proper tools should be used, as this task is labor-intensive.

Fig. 21–23. The roof assembly must always be analyzed and opened up if it has been involved in fire.

Some structures have roof systems that are totally combustible, meaning the rafters, ridge boards, decking, and shingles can burn, whereas others may have a metal deck and an insulation covered by roofing materials, finished with an asphalt-type coating. In these cases, hand tools are most likely to be used along with hose streams. Sometimes power saws can be used to cut away portions of the roof to make less work. If necessary, the job can be done from a roof ladder, aerial ladder, or tower platform.

Techniques for overhaul

Once the flames are knocked down and overhaul has started, the manual work with hand tools begins. It is best for firefighters to have an assortment of tools with them in the structure to deal with the different overhaul needs. Here is where another old fire service adage applies: "The firefighter should let the tools do the work." Firefighters who do not have the right tool for the job will end up working harder to get things done. Firefighters should first examine the type of structure they will be overhauling and take the right tools in the first time. Many times firefighters have tried to make up for not having the right tool for the job and have worked extra hard to do a task, only to receive an injury for their effort.

An example of proper tools for the job is where fire is found in the first floor of an older two-story multiple-occupancy structure that has 12-ft (3.7-m) ceilings that need to be opened to check for fire extension. Firefighters generally carry 6-ft (2-m) pike poles for residential structures, but in this case, the firefighter needs at least an 8-ft (2.4-m) pike pole or ceiling hook.

The firefighter should hold the hook or pike pole so the hook end faces away. As the firefighter pierces the ceiling with the point, it may be necessary to turn the pole at a quarter turn so the business end punches past the lath or ceiling surface. The pole is then turned so the hook once again faces away. The firefighter should locate the joist to which the lath or ceiling is attached, then position the hook along side of it. As the firefighter pulls downward and to his or her front area, the connections between the two are broken, releasing bigger pieces of the ceiling area. The firefighter should open up as much as is necessary to get to clean wood.

Overhauling building contents

When overhauling the contents of a structure, sometimes all that is necessary is to use a stream of water from a hoseline. But in some situations, other methods are necessary to get the job done. For example, a bedroom has sustained heavy fire damage, and there was some extension into the hallway. There will probably be extensive overhaul needed. There might be clothing hanging in a closet that is still smoldering. There might also be window drapes or shades that need attention to be sure they are extinguished. Instead of using the force and volume from the attack hoseline to do the job, firefighters can pick up the clothing articles and take down the shades and drapes, and remove them all to the outside or into the bathtub for a dousing (if the tub is still in service). There may be piles of clothing in the room also smoldering. In some cases, a bucket of water or a tub nearby with water in it can serve the purpose.

If you find it necessary to remove the contents from the structure, ensure that proper safety procedures are followed. Never just throw things out of a window—make sure the area below is clear before dropping the objects. In tall multistory buildings, never use the elevator to remove mattresses and the like—they have been known to flare up inside the elevator, endangering firefighters. Roll up and tie mattresses with rope before bringing them down through the building stairs.

There will be fires in occupied structures where the owner or occupant has smaller, valuable items found after the fire, such as items on a dresser or shelf. It may be best to get a box or other container to put them in for safe keeping while they are returned to the owner. Care should be taken to minimize any breakage by firefighters. This action can serve three purposes: (1) it helps to save something that might be very valuable to someone; (2) it helps firefighters to minimize property loss; and (3) it can build positive public relations between the fire department and its community and make a positive situation from a bad experience for someone.

When overhauling, there is a natural tendency for firefighters to use water heavily for final extinguishment. Firefighters may be tired and want to get done quickly, especially if there was a well-involved fire situation necessitating a lot of hard, physical work. There is always the belief that insurance is going to pay for the loss to the owners anyhow! If the structure is not going to be rebuilt and will be demolished, then by all means use as much water as needed to make sure a rekindle does not happen. But where fire damage is limited in size and if the building is going to be repaired and reinhabited, then firefighters should be careful about the amount of water used during overhaul. If there is any question in this case, then maybe overhaul efforts were not thorough enough to open all areas of potential fire spread, and water is being used to flood it out. This will lead to unnecessary water damage, and it could lead to an uninhabitable building. Officers and firefighters should always be aware of the amount of damage caused *after* overhaul efforts.

Tips for overhauling

- Determine that the building or structure is safe for personnel to operate in.
- Maintain communications and control of the task (company officers).
- Have a plan on where to start and where to suspect likely fire travel.
- Make sure firefighters use the proper tools and techniques for the task at hand.
- Keep hoselines charged and in place until they are no longer necessary.
- Ensure that holes in the floor or other safety hazards are identified, and proper barriers are placed to protect firefighters operating in the area.
- Perform overhaul in coordination with other operating crews.
- Remove debris to the outside to be completely extinguished.
- Look for any possible signs of how the fire may have started.
- Cover any windows or openings in an occupied structure to prevent any further damage.
- Before leaving the premises for the final time, conduct a walk-through of the structure to ensure there is no remaining fire. You must avoid an embarrassing and dangerous **rekindle**—a situation in which a hidden smoldering fire reignites after your departure, setting the building on fire again. Use all of your senses and the thermal imaging camera to search for any hidden fire.

Preserving fire scene evidence and arson

FFI 4.3.13 During the entire incident—from dispatch to picking up and returning to quarters—firefighters should be aware of any unusual signs or clues that might lead to an explanation of what may have started the fire. (Fire investigation is covered in detail in chapter 32.) During overhaul, firefighters are likely to come across things that might point to where the fire started and what may have caused it. Signs of origin can include:

- Location of fire or point of origin
- Single or multiple fire locations

- Direction of fire spread
- Faulty wiring or electrical equipment
- Signs of accelerants
- Cans or container
- Unlocked doors or signs of forcible entry

When signs are noticed, firefighters should be careful not to destroy any evidence or throw anything out that fire investigators might find useful in their probe. If possible, notify investigators and stop the overhaul process as long as the fire is under control. Should firefighters discover the cause or perhaps find something that might point to a crime, do not disturb any evidence if possible—notify investigators immediately and protect the scene.

Arson is the intentional setting of fire to a building or object for the purpose to destroy it. There are many reasons why someone would perform this crime, and that is why it is so popular. Arson can be a very costly crime when all things involved are considered, including firefighter health and safety. Sometimes a building is rigged to burn quickly, and that could lead to a structural collapse that might injure firefighters.

Because firefighters never know when an arson fire will happen, they must be alert to signs that point to arson. For example, if you are a member of the first due company on the scene of a structure fire that is well-involved, and your response time was very good, that might be a clue. If you are stretching in your initial attack line, and it is an occupied house, but the owners are not on the scene and the side door was found open, there might be another sign. As you have knocked the fire down and are beginning overhaul and you smell an odor like that of gasoline or another flammable liquid, then you very possibly have an arson fire to deal with. All this information should be given to the IC, who will then notify the fire investigators. During your overhaul, be on the lookout for anything unusual or any container or flammable liquid stains. Be careful not to disturb anything that looks out of place in a normal living environment. Remember not to destroy this crime scene.

QUESTIONS—SALVAGE

1. How do you think you and your department can be prepared for preventing extensive water damage to property with minor fire conditions?
2. In a multi-story occupancy, on what floors will the salvage operations generally begin?
3. When the fire has been knocked down and the fire floor is measuring inches of water, what are a few techniques learned in this chapter that you as a firefighter can use to alleviate the excessive weight of the water condition?
4. Discuss some ways you can cover holes in the floor or warn others of potential risks resulting from fire suppression efforts.
5. Why is it important to secure a building or door after firefighting operations have been concluded?
6. List some tools you believe would be essential in the preservation of people's property after the main body of fire has been knocked down.
7. Why is it important to follow through with the shut-off of all the utilities before entering an area with standing water?
8. How would you clear a flat-roof drain to allow water to run freely and remove the added weight caused by water on the roof?

QUESTIONS—OVERHAUL

1. When searching for extension of fire in an adjoining room or attached structure and a minor fire has been knocked down, why is the use of thermal imaging cameras a good practice to follow?
2. In a multi-story building, do you think it is safer to remove smoldering mattresses down the interior stairs, creating a tripping hazard to firefighters, or to quickly bring it down an elevator?
3. Is it helpful to the fire investigators to remove suspicious items or appliances that may have caused the fire?
4. If you are responding to a commercial occupancy, which size of pike pole would you take? Answer the same for a two-and-a-half-story private dwelling.
5. How would you prepare to check for extension in the floor before actually getting down to the plywood for various flooring types, and what tools would you use?
6. When the main body of fire is knocked down by the firefighters, should the hoseline stay in place while overhaul takes place, or should it get out of the way of members performing overhaul? Why?
7. What must be done next if fire is found in wall bay spaces?
8. When is it safe to remove SCBA during overhaul?
9. If you can summarize the definition of overhaul, what would it be?

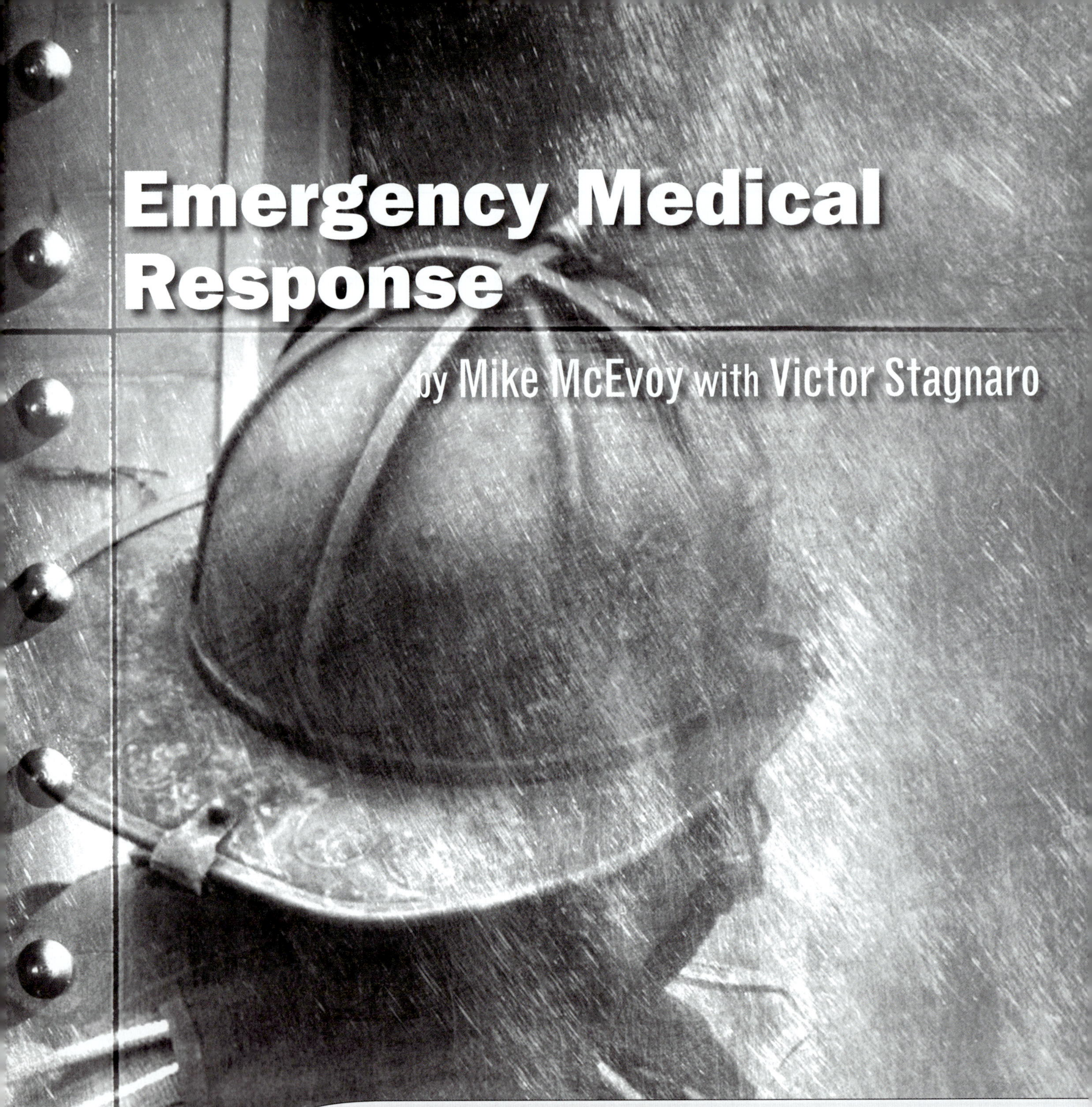

Emergency Medical Response

by Mike McEvoy with Victor Stagnaro

This chapter provides required knowledge items for the following NFPA Standard 1001 Job Performance Requirements:

FFI 6.1

22

This chapter contains Skill Drills. When you see this icon, refer to your Skill Drill book for step-by-step instructions. SKILL DRILL

OBJECTIVES

Upon completion of this chapter, you should be able to do the following:

- Describe the role of the emergency medical system in the fire service
- Describe the role of the emergency medical service as it relates to firefighter safety
- Identify the training requirements for emergency medical care in the United States
- Describe medical-legal concerns associated with emergency medical care
- Discuss scene safety concerns related to emergency medical responses
- Describe the basic rules of initial patient assessment
- Identify precautions used to prevent disease transmission during patient care
- List and describe the common types of medical emergencies encountered by firefighters
- Demonstrate proper techniques for control of severe bleeding
- Describe proper procedures for assessment and treatment of patients with burns and smoke inhalation
- Demonstrate proper assessment and treatment for injuries to the head and spine
- Identify the symptoms of shock and other medical emergencies as well as the associated treatment procedures for each
- Describe the proper procedures for the transfer of patient care

INTRODUCTION

The public has come to expect many things of firefighters. When they don't know who to call, they call the fire department. From bringing a cat down from a tree to having a ring stuck on a finger, it is common for the fire service to respond to all types of emergencies.

Whether or not the fire department where you work or volunteer is fully integrated with an emergency medical service (EMS) system, it surely interacts as first responders work side by side with emergency medical technicians and paramedics. Whether called to vehicle crashes with injured people or an unconscious or unresponsive person at a local mall, firefighters are often the first to arrive and are expected to begin basic medical treatment until the arrival of personnel with appropriate training and expertise. Most fire department statistics show that 70%–80% of what they do is EMS related. Those are some staggering numbers; and as the baby boomer generation grows older, those numbers are more likely to increase. Firefighting is also a dangerous activity with high potential for injury. Firefighters must have a minimal knowledge of emergency medical care in order to provide immediate, lifesaving care to another firefighter should an injury or sudden cardiac event occur during training or firefighting operations. This chapter provides an introduction into the EMS service and covers some basic skills firefighters need to protect themselves from injury and disease as well as begin appropriate care of sick and injured patients. (Note: This chapter does not cover how to perform cardiopulmonary resuscitation [CPR]; such training is typically provided in a standalone course). Your fire department will establish the level of emergency medical services that will be provided in your community. Just like in any emergency, the goal is to prevent further damage or injury and to stabilize the situation. Further training should be sought for more advanced care. Firefighters should *never* attempt to do more than their training has prepared them to do.

THE EMERGENCY MEDICAL SERVICES SYSTEM

EMS in the United States is an organized system designed to provide emergency medical care to people who become ill or injured outside of hospitals or health care facilities. Care is provided by trained and certified or licensed providers acting under the supervision and license of a physician medical director. Calls for EMS are typically routed through a 9-1-1 system, and the response models vary widely. 9-1-1 centers typically utilize medical dispatch protocols to prioritize EMS calls, dispatch appropriate resources, and provide emergency medical dispatch (EMD) instructions to callers prior to arrival of responders. Fire departments may provide basic life support (BLS) or advanced life support (ALS) first response and/or BLS- or ALS-level transport. Firefighters are trained to the level required by the type of EMS service their departments provide, which may range from firefighters assisting emergency medical responders (EMRs), emergency medical technicians (EMTs), advanced emergency medical technicians (AEMTs), or paramedics to firefighters who are trained at any of these levels themselves. The role of the new firefighters on an emergency medical response is to learn from the more experienced firefighters and officers, but be careful to avoid causing more harm. The rewards of being a firefighter most often come from using your skills and training as well as seeing people's lives and property saved. Not everyone's belongings can be salvaged and not every life will be saved, but many are given the chance to survive through the efforts of the fire department and EMS. As part of an EMS system, firefighters have the opportunity to positively affect lives.

FIREFIGHTER SAFETY

SKILL DRILL

As in all aspects of firefighting, safety should be the highest priority of every crew member. A firefighter would never enter a burning building without proper protective clothing and self-contained breathing apparatus (SCBA). Likewise, there are standard precautions and personal protective equipment (PPE) needed when providing emergency medical care. Familiarity with disease transmission helps firefighters to protect themselves, their brothers and sisters, their families at home, patients, bystanders, and hospital staff from exposure to commonly encountered infections. Under federal (and sometimes, state) regulations, every fire department, whether they provide EMS or not, is required to have an exposure control plan. Exposure control plans mandate that members use certain standard precautions and specify what engineering and work practice controls the department will use to prevent exposures. Should an exposure to blood or an infectious disease occur, the exposure control plan specifies what needs to be done. Departments must provide PPE such as gloves, gowns, masks, mouthpieces, and resuscitation bags as well as replace these when necessary. They also must provide annual influenza vaccines and immunization against a number of other common communicable diseases. Cleaning and decontamination of equipment also must be outlined in the exposure control plan. Annual training of all members on the elements of the plan is mandatory. Firefighters should be familiar with their department's exposure control plan and the PPE commonly used as well as how to clean and decontaminate EMS equipment following each use.

The five most important elements of a good infection control program include:

- Handwashing

- Gloves and other PPE
- Decontamination of EMS equipment
- Vaccination
- Common sense

Several of these are elements of standard precautions, a term that replaced what was previously referred to as universal precautions. **Standard precautions** are a set of basic infection control principles used for all patient encounters. A tenant of standard precautions is the presumption that all blood, body fluids, and wounds are potentially infectious. Handwashing or hand hygiene means cleaning your hands with soap and water or an alcohol-based hand cleaner. It helps to prevent spreading germs to patients and lowers the risk of firefighters carrying infectious germs on their hands that could cause an infection in the firefighter or in others they have contact with. The critical times when hand hygiene should be practiced are:

- Before eating
- Before and after any direct contact with a patient's skin (taking a pulse or blood pressure, performing physical examinations, lifting a patient off the floor)
- After contact with blood, body fluids, non-intact skin, or wound dressings
- After contact with surfaces (including EMS or medical equipment) in the immediate vicinity of a patient
- After glove removal
- After using a restroom

Gloves and other PPE are an important part of standard precautions and should be worn whenever touching non-intact skin (wounds or rashes), the mouth, nose or eyes, or when there is potential for contact with body fluids. Gloves should be put on just before patient contact and removed immediately after patient contact and well before touching or handling any medical equipment (EMS bags, stretchers, radios, etc.). Firefighters sometimes inappropriately don gloves well before encountering a patient, touching equipment, doors, and other items that contaminate their gloves prior to patient contact. This defeats the purpose of preventing spread of infection to patients. Additionally, firefighters often fail to remove gloves after patient contact, spreading germs from the dirty gloves to their uniforms, equipment, radios, computers, and vehicles. Gloves are only for patient contact. They should be donned in front of the patient and doffed before moving on to any other tasks. Hand hygiene should be performed before and after use of gloves. Other PPE such as masks, goggles, and gowns should be utilized based on common sense and department exposure control plans.

EMS equipment should be considered contaminated after each patient use. Disposable supplies should be discarded and nondisposable equipment cleaned with an approved disinfectant.

Vaccinations should be kept up to date for all members, regardless of whether they provide EMS care. The purpose of vaccinations for common diseases such as measles, mumps, tetanus, chicken pox, and whooping cough is to stop the spread of these diseases. The vaccines serve to protect firefighters to some degree but do considerably more to protect vulnerable populations that firefighters come into contact with such as their own children, elderly parents, spouses, and patients they care for who may have compromised immune systems. Firefighters interact closely with many sick patients every year and, in doing so, become carriers of many vaccine preventable diseases. While firefighters themselves are unlikely to become ill from these organisms, they can and do transmit them to others. Vaccination breaks that chain and protects others. Every year in the United States, some 36,000 people die from the flu and more than 200,000 are hospitalized. Those numbers would likely be smaller if all EMS and healthcare providers received an annual flu vaccine.

Scene safety

One of the most dangerous EMS environments is at the scene of an incident on a roadway. Use of blocking apparatus, strategic positioning of EMS units, donning high visibility safety vests, and assigning someone to serve as a spotter are all wise safety practices when operating on a roadway. Effective traffic incident management (TIM) practices serve to further decrease the risks to responders, motorists, and others involved in or nearby traffic incidents.

An appropriate scene size-up should not be limited to fires. EMS-related size-up includes potential hazards and scene safety warnings provided at dispatch, knowledge of the neighborhood, past experience with known patients, determining where the exits are, establishing the best means for other providers to approach the building, and ensuring a safe working area. When there is a concern about scene safety, law enforcement should be requested.

Having police on a scene in no way makes the scene safe. Continued and ongoing situational awareness is imperative. EMS providers should receive training in communications skills and de-escalation tactics. Individuals present on the scene of 9-1-1 calls are never having a good day. Often, the stress of an emergency brings out the worst in people. If at any time the safety or welfare of firefighters is in danger, members should retreat from the scene and request law enforcement support.

TRAINING AND STANDARD OF CARE

A single chapter on the role of a firefighter in EMS cannot provide enough training to cover all needed aspects of EMS care. Regardless of what level of EMS training a firefighter may have, there are several legal principles that apply to providing patient care.

Consent is required before providing care to any person. They have the right to consent to or to refuse any treatment. In EMS, written consent is not usually obtained, but firefighters should introduce themselves and offer care in a fashion that allows the person to accept or refuse. When patients are not able to give consent for care such as when they are unconscious, consent is considered to be implied as long as the EMS provider believes that the care would be in the patient's best interests. Some individuals such as children or persons with mental incapacity may require consent from a parent or guardian for treatment. Firefighters should be familiar with their state laws regarding consent for EMS and medical care.

Duty to act refers to the responsibility of a firefighter to provide medical care in an emergency. In most locales, an ordinary citizen has no responsibility to stop and render assistance to someone who has fallen, crashed their car, or is otherwise in need of medical help. Firefighters have a duty to act by virtue of their being on duty in an organized response system. There is an established relationship between a citizen calling 9-1-1 for help and the emergency responders that community dispatches. While the legal requirements vary by location, firefighters should be aware that, while on duty, they have a legal duty to respond when called. Additionally, once a firefighter initiates care, that duty continues until care is no longer needed or the patient is turned over to another provider. Breach of the duty to act can leave members open to liability for failure to act and/or for **abandonment**.

Expectations for the provision of emergency medical care comprise the **Standard of Care**. The U.S. Department of Transportation develops the national EMS educational standards that serve as a blueprint for EMS care. State, regional and local jurisdictions adopt protocols and patient care standards. EMS care is delivered under the license of a local physician medical director. The expectations of the community, coupled with the many published sources of guidance for EMS care serve to comprise the standard of care, which changes and evolves continually over time. Breach of the standard of care may lead to a claim of negligence. To help providers remain current on the standard of care, EMS systems operate quality improvement (QI) programs. QI utilizes review of patient care reports, real time observation of patient care activities and ongoing educational programs to help continually improve the quality of EMS care delivered by a fire department. The physician medical director should be actively involved in QI activities.

Negligence is the failure to adhere to the standard of care. To prove negligence, four elements are needed: duty, breach, causation, and damages. We have already noted that firefighters have a duty to provide EMS care when called (in most circumstances). They are also expected to provide care according to the standard of care, which is often defined as what another firefighter with the same level of training and experience would do in a similar situation. When it can be shown that a firefighter either failed to meet the standard of care or did something far outside of the standard of care, they can be considered to have breached the standard of care and hence, their duty to the patient. Causation and damages imply that the breach actually resulted in some sort of damage to the patient and that damage can be directly connected to the actions (or inaction) of the firefighter. In some cases, a firefighter may be negligent but statute provides immunity from liability. Ordinary citizens are often protected under Good Samaritan statutes when they voluntarily stop rendering care. Immunity protections vary from state to state.

Confidentiality is a core duty of anyone providing medical care. Patients routinely share their personal health information with EMS providers with the expectation that it will be kept private. The **Health Insurance Portability and Accountability Act (HIPAA)** was enacted in 1996 to establish a national standard for protecting health information privacy. Often used as a shield to inappropriately prevent disclosure of information that should otherwise be shared, HIPAA only applies to departments that are billing for EMS services. If you work for a department that bills for care, you are required to have regular training on HIPAA requirements and this information must be provided on scene to every patient.

A patient's health care information should never be shared with the public or the media or discussed afterwards in a location where people not involved in patient care might overhear you. Patient care reports and records should be maintained in secure locations and copies released only with written consent of the patient. As a firefighter who treats injured and ill patients, it is important to understand local and state privacy laws and become familiar with your department requirements for confidentiality.

When considering the legal ramifications and responsibilities of providing EMS care, the best way to avoid litigation is simply to be nice. Treating every patient, bystander, friend, and family member as though they were a member of your own family will assuredly work to have your customers perceive you as a competent, capable, and compassionate firefighter.

REPORTING ABUSE

Federal and state laws require firefighters report certain cases of suspected abuse, maltreatment, or neglect to local or state authorities. These laws specifically permit disclosure of confidential information and often require firefighters to make reports. Reportable circumstances include child abuse, elder abuse, domestic violence, and sexual abuse. Often, there are penalties for failing to report actual or suspected cases. Reporting should not be done in front of the patient, caregiver, or parent. Be familiar with your state requirements and department policies.

PATIENT ASSESSMENT SKILL DRILL

Patient assessment begins with scene size-up as described previously. Once you have evaluated the scene for potential hazards, considered the dispatch information, and made sure you and your crew are safe, you can begin to consider what occurred prior to your arrival and what care you might need to provide. Following scene size-up, you introduce yourself and your crew to the patient, bystanders, and family; obtain consent for care; and begin a primary assessment. Following a primary assessment, you obtain a history and, if time allows, begin a secondary assessment.

The purpose of a primary assessment is to determine and treat any immediate life-threatening conditions. This begins by forming a general impression using the information you have already collected on scene and from dispatch and your initial view and conversation with the patient and those nearby. An initial impression might be, "an 87-year-old female complaining of lower back pain for two days who met us at the door carrying her suitcase and appears to be in mild pain," or "a 25-year-old male thrown from motorcycle after striking a tree, unconscious with snoring respirations." Your initial impression describes the situation, establishes the seriousness of th event, and determines the resources needed.

Part of forming a primary impression includes determining the patient's level of consciousness. Patients can be conscious or unconscious, and when they are conscious, can be alert or not alert. This assessment is described in CPR textbooks as an element of initial patient assessment. When the patient has signs of life, additional assessment is usually ordered in a sequence referred to as the ABCs:

- A = Airway
- B = Breathing
- C = Circulation

The airway of a patient, especially when unconscious, can be obstructed if the head is tilted downward. There are two different techniques used to open the airway: one is for patients without a traumatic injury and the other is for a trauma patient. For a patient without a traumatic injury, the provider places his or her hand on the patient's forehead and the fingers of the other hand underneath the chin. Placing slight pressure on the hand and the fingers, the provider pushes the patient's head back to open the airway (fig. 22–1a).

If a neck or spinal injury is suspected, the provider uses the jaw-thrust method to open the airway. The jaw-thrust is performed by placing the provider's hands on each side of the patient's head, with the fingers along the jaw bone, pushing the jaw open, which in turn opens the airway (see fig. 22–1b). If unsuccessful, revert to the head-tilt, chin-lift method previously described. Opening the airway is more important than the potential risk of worsening damage from a spinal injury.

Once an airway is established, it is necessary to ensure that the patient is breathing. One way to check for breathing is for the provider to place his or her ear near the patient's mouth and look, listen, and feel. The provider looks for chest rise and fall, listens for breathing coming from the patient's mouth, and feels for the breath on their ear. If the patient is not breathing, the provider should be prepared to deliver rescue breathing.

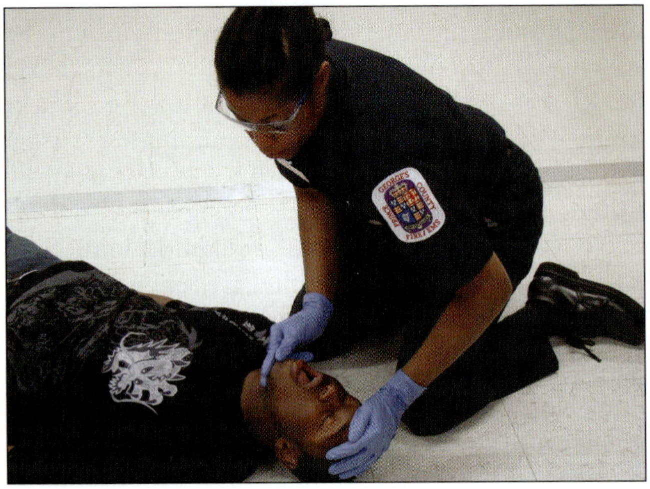

Fig. 22–1a. The provider places his or her hand on the patient's forehead and the fingers of the other hand underneath the chin.

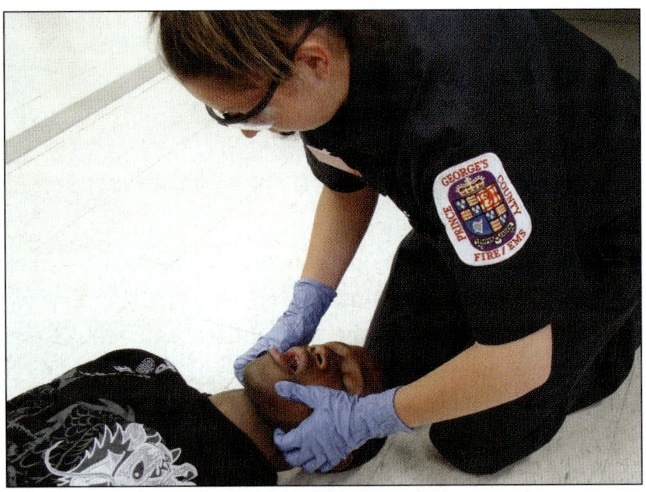

Fig. 22–1b. The provider places his or her hands on each side of the patient's head, places the fingers along the jaw bone, and pushes the jaw open.

After airway and breathing are evaluated, the provider should check for circulation. For an unconscious patient, the provider can check for a pulse by placing his or her index and middle finger on the patient's carotid artery in the neck. This should be done simultaneously with looking, listening, and feeling for breathing (fig. 22–2).

For conscious adult patients, the radial pulse in the wrist is most often used to determine circulation. The brachial artery in the upper arm is used for infants younger than one year. In addition to checking for breathing and circulation, knowledge of respiratory and pulse rates can help the first responder determine treatment options. The Maryland Medical Protocol lists some average heart and respirations rates (table 22–1).

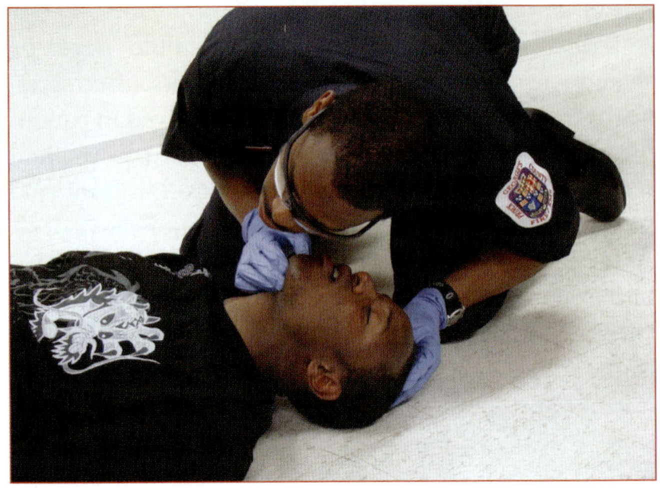

Fig. 22–2. The provider is looking, listening, and feeling for a breath.

Table 22–1. Average heart and respiration rates

Age	Estimated Weight (lb)	Estimated Weight (kg)	Heart Rate	Respiratory Rate	Systolic B/P
Premature	<7	<3	160	>40	60
Newborn	8	3.5	130	40	70
3 mo	13	6	130	30	90
6 mo	18	8	130	30	90
1 yr	22	10	120	26	90
2 yrs	26	12	115	26	90
3 yrs	33	15	110	24	90
4 yrs	37	17	100	24	90
6 yrs	44	20	100	20	95
8 yrs	55	25	90	20	95
10 yrs	77	35	85	20	100
12 yrs	88	40	85	20	100
14 yrs	110	50	80	18	110
Adult	<100	>50	80	18	120

Once the patient's ABCs have been established and any serious findings treated, the provider should conduct a rapid primary patient assessment. This includes a rapid head-to-toe exam to identify any major life threats such as major bleeding, serious deformities, or cyanosis (bluish discoloration of the skin due to lack of oxygen). Life threats discovered during the primary assessment are treated immediately.

Following completion of the primary survey, it is helpful to obtain some history beginning with the chief complaint and history of the present illness. Generally, history taking can be guided by your department's patient care reporting system. The SAMPLE mnemonic is often used to gather pertinent history:

S—**Signs and symptoms**: When did the problem start? What signs and symptoms are present? Is the patient experiencing any pain?

A—**Allergies**: Does the patient have allergies (medications, foods, insects)? If not, this is referred to as a pertinent negative and should be documented as, "no known allergies" or "NKA."

M—**Medications:** Is the patient currently taking any medications? Note the names, doses, and frequencies prescribed (i.e., metoprolol 25 mg, twice daily). Is the patient taking any non-prescription medications or recreational drugs?

P—**Past medical history**: Has the patient ever been hospitalized? Are they being treated by a health care provider? Have they travelled recently or had any recent illnesses or injuries?

L—**Last meal**: When was the last time the patient ate or drank and how much did they consume?

E—**Events leading up**: What led to the illness or injury? What was the patient doing when they were stricken?

When time allows, and for providers who are trained, a secondary survey is conducted. The secondary survey is an exam focused on the body systems involved in the chief complaint. A full set of vital signs including pulse, blood pressure, respiratory rate, pulse oximetry measurement, pupils, and level of responsiveness are included in the secondary assessment.

TRAUMA

Firefighters commonly see trauma patients who have been involved in wide variety of injury-causing events from motor vehicle crashes to assaults. Trauma resulting from crashes and falls produces multiple injuries, including head and spinal injuries.

Previously, prehospital providers spent considerable time and energy immobilizing the cervical spine (c-spine) of traumatically injured patients. This practice resulted in more harm than benefit to the majority of trauma patients, leading virtually all EMS systems to develop protocols for selective immobilization. In general, spinal immobilization is recommended for patients with significant mechanism of injury who complain of neck pain or exhibit symptoms of spinal injury (neck pain to palpation, paralysis, neck muscle spasm) or for patients who have a mechanism of injury that might have caused a spinal injury but are unconscious or otherwise unable to cooperate with a good exam for pain and symptoms. In circumstances where a c-spine injury is suspected, a rigid **cervical collar** should be applied to restrict movement of the head and neck.

Prior to the placement of the cervical collar, the head is kept from moving by holding the head with both hands, using only enough pressure to keep the head still. If the patient is lying down on his or her back, position yourself above the head or in the backseat of a vehicle if the patient is being extricated from the front seat of a vehicle. Hold the head still, as shown in figure 22–3, until the cervical collar is applied. Talk to the patient and providing reassurance to help keep them calm.

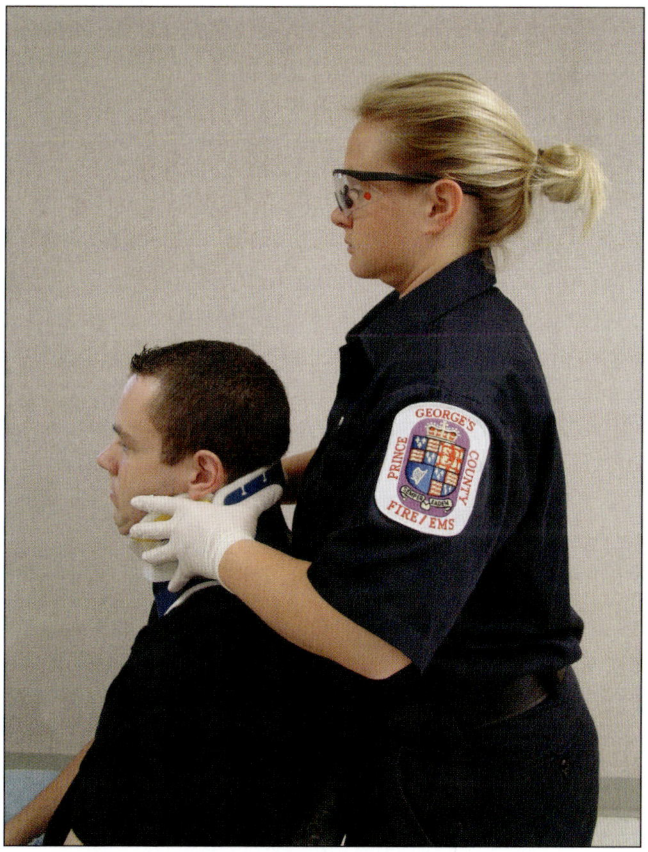

Fig. 22–3. Hold the head still until the appropriate immobilization device is completely applied.

Burns and smoke inhalation

Burns are also classified as traumatic injuries. The majority of burns seen by firefighters will be thermal burns (i.e., caused by heat or flame). Burn survival has improved considerably over the last decade; today, fewer than 4% of patients presenting to hospitals with burns die today. Unlike many other illnesses and injuries, 85% of all burn patients in need of care are seen by EMS. Burns are currently classified by their depth. Of note to firefighters, burn injuries are cumulative, that is, a superficial burn (such as a minor sunburn) usually heals within a week. If, during that time, the burn area is exposed to additional thermal injury, the superficial burn will progress to a much deeper burn. This is important for firefighters, as seemingly minor burns can convert to major burn injuries during routine firefighting operations.

The first goal of treatment for burn injuries is to stop the burning. This is often done with copious amounts of water, keeping in mind that loss of protective skin layers often leads to hypothermia, as patients cannot control their own body temperature. When using water to cool a burn or stop the burning process, flood the area for no more than 3 to 5 minutes. When someone is on fire, never use water to extinguish a patient or firefighter because the water will evaporate into steam, resulting in more serious burns. ABCs should be determined and managed immediately as well. Next, all clothing, jewelry, and other items on a seriously burned patient should be removed. This includes any contact lenses. Next, the patient should be kept warm. Usually, this requires covering the burned patient with blankets. An assessment of total burned surface area and depth should be made along with checking for any associated injuries that might have been missed. The airway should be evaluated for burn injury, which includes checking for burns inside the mouth, hoarse voice, or persistent wheezing. Singed nasal hairs or soot in the nose, which was previously considered a sign of potential airway burns, is no longer considered useful.

Providers should become familiar with estimating the percentage of burned area and know their local protocols for treatment, which are usually based on the total body surface area burned (see fig. 22–4, a diagram for a body surface area rule of nines). Estimates of smaller areas can be done using the patient's entire hand surface, which is equal to approximately 1% of body surface area.

Burned areas should be covered with dry sterile dressings. The patient's head should be kept elevated at least 30° at all times to alleviate swelling in the face and airway. Pain medication should be given if ALS is available.

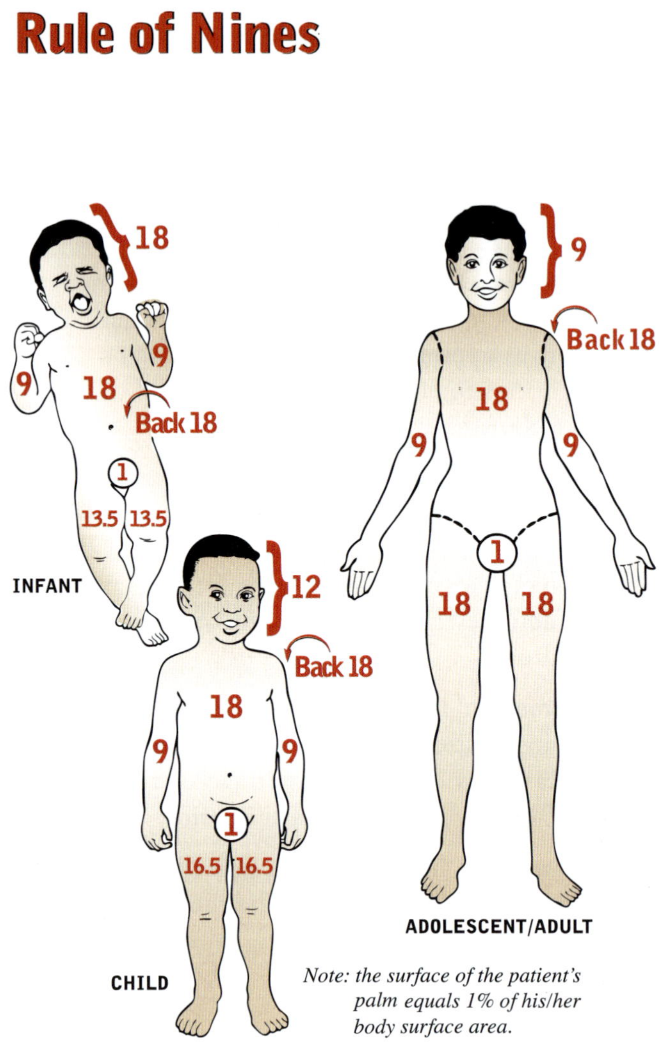

Fig. 22–4. The rule of nines

Smoke inhalation is common in patients with thermal burns, occurring in 17% of all burn patients. It is important for firefighters to consider both carbon monoxide and cyanide poisoning in burn patients and in patients removed from confined spaces where fire smoke was present. Pulse oximeters are available with the capability to measure carbon monoxide (CO) levels in the bloodstream; levels exceeding 10% suggest significant exposure to CO and may require medical assessment and treatment (fig. 22–5).

Cyanide antidotes are recommended in many paramedic protocols for resuscitation of patients removed from fires who are not breathing or are in cardiac arrest. Cyanide antidotes have resulted in survival of smoke inhalation patients who emergency medical and toxicology experts would otherwise have expected to die. Firefighters should learn whether their transport ambulances have access to cyanide antidotes and how quickly they can be deployed if needed for a fire victim or firefighter.

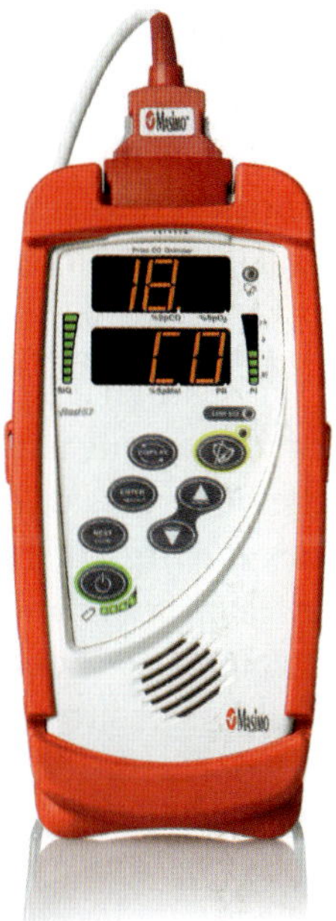

Fig. 22–5. A handheld oximeter measures oxygen saturation, pulse rate and at-a-glance patient status data. (Courtesy of Masimo Rad-57).

Fractures, dislocations, and sprains

SKILL DRILL

It can be difficult for a field provider to tell the difference between a fracture, sprain, or dislocation. Even with the benefit of x-rays, physicians may not be able to distinguish between them. Regardless of the type of injury, the treatment is generally the same: immobilization. When examining an injured area, the provider should look for deformity, swelling, and bruising. It is possible to find a wound that suggests an **open fracture**, which is when a bone is broken and there is a wound through the skin in the same area as the break (often caused by the bone end piercing the skin). The bone may be visible through the open surface of the skin. In addition to splinting, the open area must be bandaged.

The patient will likely have pain during splinting of a possible fracture. Prior to applying a splint, check the extremity for a pulse to ensure that the fracture has not impaired blood flow to the rest of the limb. Lack of blood flow between the injury site and the end of the limb may cause significant nerve or tissue damage. The patient should be transported to an appropriate medical facility without delay. Blood flow can be monitored by checking the pulse, or by checking for capillary refill or by applying a pulse oximeter to a finger or toe further away from the injury. It may be necessary to remove shoes, socks, and gloves to check appropriate blood flow. Capillary refill can be checked by gently pressing down on the fingernail or toe of the injured limb until the skin turns white. After releasing the pressure, the nail bed should turn pink within 2 seconds, suggesting that there is normal circulation. If it takes longer than 2 seconds, it may indicate a problem with circulation. Capillary refill should be checked before and after splinting. It should be noted that capillary refill is generally delayed in cold weather. In extremely cold weather, capillary refill may be an unreliable means of checking blood flow, especially in adult patients.

In addition to checking for a pulse or capillary refill, the provider should check the end of the limb for sensation and movement. Using the dull part of a pen or the eraser end of a pencil, gently rub the bottom of the foot or palm of the patient's hand, checking to see if the patient can feel it. Movement in the fingers and toes is a good indication that he or she has proper blood flow.

Splinting can be done using a variety of devices from cardboard to commercially available splints. Splinting of long bones such as arms and legs that do not involve joints can be a relatively simple task. Splinting shoulders, elbows, knees, and ankles can be complex and require more finesse. Pelvic, hip and femur fractures can be life-threatening injuries. A femur fracture, for example, can result in life threatening internal bleeding that would not exhibit any external signs (such as discoloration or swelling). This type of fracture may require providers to put traction on the injury site and use specialized traction splinting to prevent further injury. Improper splinting (too tight or too loose splints), can cause more damage. Therefore, providers should receive appropriate training in splinting before applying splints to any patient. Unless you have been properly trained in splinting, simply make the patient as comfortable as possible and wait for a higher level of care to arrive.

MEDICAL EMERGENCIES

Medical emergencies comprise a significant number of 9-1-1 responses, and not all of the possible call types can be covered in this chapter. Instead, we'll focus on the more common emergencies encountered and provide some cursory directions on assessment and management. Many medical conditions require additional training to treat. In some situations, the patient may not be able to clearly convey the reason why you were called. They may be unconscious, confused, under the influence of drugs or alcohol, or having a seizure. During scene size-up and primary assessment, look for identification such as a medical alert bracelet (fig. 22–6), necklace, or information sheet inside or attached to the patient's refrigerator. Also, look for bottles of medication that may provide clues to medical conditions the patient is receiving treatment for. Insulin is often refrigerated; checking the patient's refrigerator may reveal that they are being treated for diabetes.

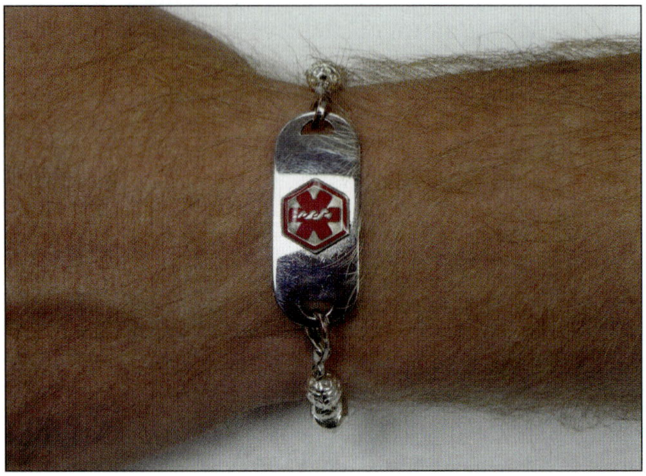

Fig. 22–6. A medical alert bracelet

Any information found during the assessment should be recorded and communicated to incoming EMS units.

Seizures

Seizures are among the more common medical emergencies to which the fire service responds. In addition to monitoring ABCs, attempt to obtain as much information as possible about the patient. Patients who are actively seizing should be protected from injuring themselves by moving objects away from them, placing padding under their head, and sometimes relocating them to a safer location. Do not attempt to restrain seizing patients, and never put anything in their mouths. Seizures can have a variety of causes. Patients with epilepsy who have a history of seizures will often be confused and sleepy once the seizure stops. At times, their confusion may exhibit as hostile behavior. Continue to monitor the ABCs until the transport unit arrives. If patients are unconscious or unable to stay awake, they should be placed in a side lying or **recovery position**.

Shock

Shock is a condition of inadequate oxygen delivery to the body's tissues and organs. Without sufficient oxygen, energy production becomes very difficult, and organs (such as the kidneys and brain) can be damaged. Severe shock states can lead to death. There are three types of shock: hypovolemic, cardiogenic, and distributive. Hypovolemic shock usually results from blood loss; cardiogenic shock results from damage to the heart muscle; distributive shock results from infections in the bloodstream, severe allergic reactions, or damage to the spinal cord.

Symptoms of shock vary, but characteristically a person in the early stages of shock will have rapid heart and respiratory rates as the body tries to increase the supply of oxygen to the tissues and organs. The skin of a patient in shock is often sweaty, cool, and pale. They may become nauseous or vomit. As shock progresses, dizziness can occur, especially while sitting or standing. Some patients may become confused, restless, or argumentative when their brain is not receiving enough blood flow. Low blood pressure is a late sign of shock and is often seen just before the patient dies.

Treatment for shock includes reducing extra demands on the body by allowing the patient to rest in a comfortable position, often lying down. An unconscious patient should be flat or in a recovery position. Elevating the legs can help to increase blood supply to the upper body. Keeping the patient warm by covering with blankets helps to conserve energy. Although oxygen may be helpful to some shock patients although, as mentioned elsewhere in this chapter, recent studies suggest that oxygen may be harmful to many seriously ill patients. Because patients in shock are likely to have nausea or vomit and may require surgery, they should not be given anything to eat or drink.

Chest pain

Another common medical emergency is chest pain. The causes of chest pain are so variable that even hospital physicians have difficulty diagnosing the cause in many patients. For the first responder, the best practice is to treat all chest pain as a potential heart attack or myocardial infarction (MI). Once responders with more training arrive, they can help to narrow down other possible causes. Signs of a heart attack can come on suddenly or may develop over days or weeks. They typically include significant discomfort or pressure in the chest, shortness of breath, sweating, and nausea or vomiting. The longer the symptoms persist, the greater the likelihood of damage to the heart muscle. Women often have heart attack symptoms that are different from men. Women are more likely to have pain in the jaw, back, or neck, lightheadedness, and nausea or vomiting. People experiencing chest pain that results from a heart attack are at risk for sudden cardiac arrest and should be carefully monitored. A patient with chest pain should be made comfortable while awaiting the arrival of a transport unit. If local protocols allow for the administration of aspirin and the 9-1-1 call taker did not already instruct the patient to take aspirin, it should be administered as soon as possible. Some patients have prescribed nitroglycerine and, if local protocols allow you to assist the patient to take their prescribed medication, you should do so.

Anaphylaxis

Anaphylaxis is the name for a severe allergic reaction that occurs within minutes to hours of a person being exposed to something they are allergic to (an allergen). There are many allergens capable of causing anaphylaxis including medications (such as penicillin); stings (wasps, hornets, bees, ants); foods (nuts, shellfish, eggs); and airborne particles (dust, pollen, molds). People with anaphylaxis often have a history of reactions and may have been prescribed an epinephrine autoinjector device for emergency use. This is not always the case—allergic reactions can occur in people with no history. The most common signs of anaphylaxis are respiratory distress (often exhibited by difficulty breathing, wheezing, trouble swallowing, and sometimes swelling of the face, lips, and tongue) and signs of shock such as low blood pressure, dizziness, loss of bladder control, profuse sweating, and fainting. Many patients with anaphylaxis also have skin involvement including hives or rashes (often starting on the chest and spreading to the neck, face, and arms), flushing of the skin, and significant itching. Sometimes anaphylaxis is exhibited by gastrointestinal symptoms such as severe diarrhea or vomiting.

The most important treatment for anaphylaxis is administration of epinephrine. If the patient has an epinephrine autoinjector, and your local protocols allow you to assist with administration, you should. If not, you need to obtain care for the patient from an EMS level of care capable of administering epinephrine. While waiting for assistance, the patient should be kept in a position of comfort.

Stroke (CVA)

Stroke, or cerebrovascular accident (CVA), is much the same as a heart attack but instead involves the brain. Stroke is sometimes referred to as a "brain attack." The same urgency about care and treatment of heart attack applies to stroke, that is, the longer the delay in obtaining treatment, the greater the likelihood of significant damage to the brain. A stroke occurs when blood flow to a part of the brain is blocked. This can occur from a clot or from a ruptured blood vessel in the brain. Recognizing the signs and symptoms of a stroke early can help to get appropriate treatment started sooner. First responders can use the Cincinnati Prehospital Stroke Scale (CPSS) to assess for signs of a stroke. The CPSS has three assessments: facial droop, arm drift, and speech. Any abnormal finding means that the CPSS is positive:

Facial droop: ask the patient to smile.

- Normal: both sides of the face move equally
- Abnormal: one side of the face does not move or moves unevenly from the other side

Arm drift: ask the patient to extend both arms forward, close their eyes, and hold their arms in the air for 30 seconds.

- Normal: both arms remain extended
- Abnormal: one arms drifts downward

Speech: ask the patient to repeat a sentence

- Normal: repeats the words correctly and speech is clear
- Abnormal: uses incorrect words or speech is slurred or garbled

When obtaining history of a patient with a suspected stroke, it is critical to determine what time the patient was last known to be well, that is, when the stroke symptoms started. This often affects the transport destination and the types of treatment that can be used.

Asthma

Asthma attacks are usually triggered by an environmental cause such as an allergen, air pollution, temperature extremes, internal stress, cold, or respiratory infection. The primary dysfunction in an acute asthma attack is spasm of smaller airway passages in the lungs that makes it extremely difficult to exhale air. Patients with an asthma attack should be positioned comfortably. If they are hyperventilating, encourage them to slow their breathing down. Ask the patient if they have a prescribed inhaler or other medication for their asthma. If local protocols allow, assist the patient with their medication as prescribed.

Syncope

Syncope refers to a momentary loss of consciousness, also called fainting, caused by a temporary drop in blood flow to the brain. Like other medical emergencies, syncope has a variety of potential causes, some of them serious enough to require medical follow-up. A person who feels faint should be assisted to a seated or lying position to avoid falling if they suddenly lose consciousness. When assessing a person who had a syncopal episode, it is important to determine whether they fell and if they sustained any injuries when they lost consciousness. Events leading up to the syncopal episode are particularly important as is questioning the patient and bystanders as to whether the patient actually passed out or simply felt as though they were going to lose consciousness. Patients who had some warning that that they were going to pass out may have suffered a seizure or cardiac problem. People who passed out without warning might have internal bleeding from an ulcer or aneurysm or perhaps an ectopic pregnancy. Patients with syncope following complaints of a headache may have bleeding on the brain. Severe allergic reactions (anaphylaxis) can also present as syncope. The history you obtain may help additional responders to focus in on the cause of the problem.

Poisoning and overdose

A **poison** is any substance that causes health problems when it enters or touches the body. Most poisonings are accidental, and virtually all poisonings can be prevented. The majority of poisonings occur at home and typically involve common medications or household products. A system of poison control centers (PCCs) has been established in the United States to deliver advice and treatment recommendations. They are all interconnected to a common 24-hour a day phone number, 800-222-1222. Your 9-1-1 center may have already consulted with the PCC prior to dispatching EMS, as many emergency medical dispatch protocols recommend this for poisoning patients without symptoms. When a 9-1-1 operator already initiated PCC contact, the PCC probably decided that the poisoning was serious and needed care and emergency transportation. If a PCC had not already been consulted prior to your arrival and assessment of the patient, you should make that call once you have an idea of the patient's condition, the name and ingredients of what was taken, what time the poisoning happened, how much the patient was exposed to, and the patient's age and weight.

Most of the care and treatment for poisonings is supportive; meaning that you continually monitor the ABCs and support breathing and circulation as necessary. Prepare for vomiting when poisons have been ingested and assist vomiting patients into a position (usually seated and leaning forward) so they are less likely to choke on their vomit.

An **overdose** is a poisoning caused by taking too much of a medication. Most overdoses occur in people who are intentionally abusing drugs, but they can occur when a drug is misused. Today, EMS responds to a significant number of overdose calls. Many of these involve the effects of alcohol, but more frequently, they involve the use of opioids including heroin, morphine, oxycodone, methadone, hydrocodone, fentanyl, and others. Opioid medications are primarily used to treat pain. Illegal opioids are drugs with no medical purpose and are abused for their euphoric effects. High doses of opioids can lead to unconsciousness and then to respiratory depression which can progress to cardiac arrest. Opioid abuse in the United States has reached epidemic proportions, killing more people each year than motor vehicle crashes. The primary treatment for an opioid overdose is assisting breathing, usually done with a bag-valve-mask device, the use of which is covered in a basic life support (BLS) cardiopulmonary resuscitation (CPR) class. Naloxone is a drug that can reverse the respiratory depression caused by opioids and is widely distributed to patients, families, and emergency responders who might expect to encounter opioid overdoses.

The safest route to administer naloxone is through the nose (intranasal) using an atomizer device. (fig. 22–7) Since a needle is not required, this antidote is safe and easy to use. Many medications sprayed into the nose work well because the nasal cavity has a large surface area that readily absorbs the medication into the bloodstream. Signs and symptoms of an opioid overdose include:

- Unconsciousness (not able to be aroused or awakened)
- Pinpoint pupils
- Not breathing or very slow, shallow breathing (less than 8 breaths per minute)

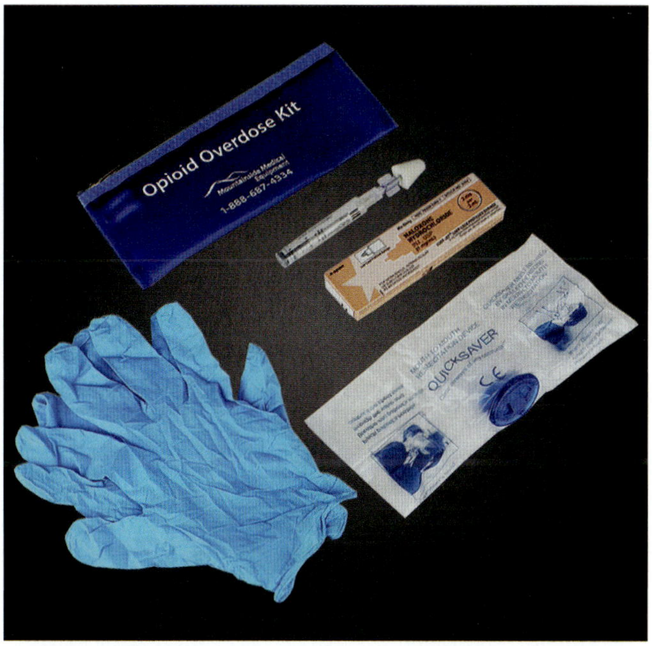

Fig. 22–7. Naloxone can reverse respiratory depression caused by opioids. (Courtesy of Harm Reduction Coalition)

The immediate care for a patient with a suspected opioid overdose is to open the airway and assist breathing with a bag-valve-mask device (ABCs). If the person is not breathing and has no pulse, CPR should be started without delay. If breathing is present but ineffective (unconscious, very slow or shallow, possibly with bluish discoloration to the skin), assist breathing while monitoring a pulse. Unconscious patients who are breathing normally (more than eight breaths per minute) should be positioned on their side in the recovery position.

If firefighters are trained to administer naloxone and authorized under local medical protocols to do so, a dose can be administered (fig. 22–8). The goal of naloxone administration is to restore normal breathing, not to completely awaken the patient. Effects of intranasal naloxone are usually seen within 3 to 5 minutes. Once breathing becomes adequate, the patient should be monitored and replaced in the recovery position. Administering repeated doses of naloxone or repeating administration at more frequent than 5-minute intervals can induce withdrawal symptoms in patients who are addicted to opioids. Withdrawal symptoms can be life threatening and include severe vomiting, explosive diarrhea, violent behavior, profuse sweating, and markedly elevated heart rates and blood pressure.

Media reports of emergency responders becoming ill from contact with opioid analogs such as fentanyl and carfentanil have led to unsubstantiated and illogical paranoia. There is little to no possibility that skin contact with any narcotic or opioid substance could lead to immediate illness or respiratory depression. Common sense should prevail when responding to overdose calls. Avoid inhaling powders, utilize appropriate PPE during patient contact, and remove your PPE when leaving the patient's side so as not to contaminate EMS equipment, personnel, or vehicles. Should providers become ill during or after delivering patient care, consider the usual sources of environmentally caused illness such as carbon monoxide, oxygen deficiency, or chemical inhalation. If skin or clothing become contaminated with liquids or powders during patient care, remove the clothing and wash skin with soap and water as soon as

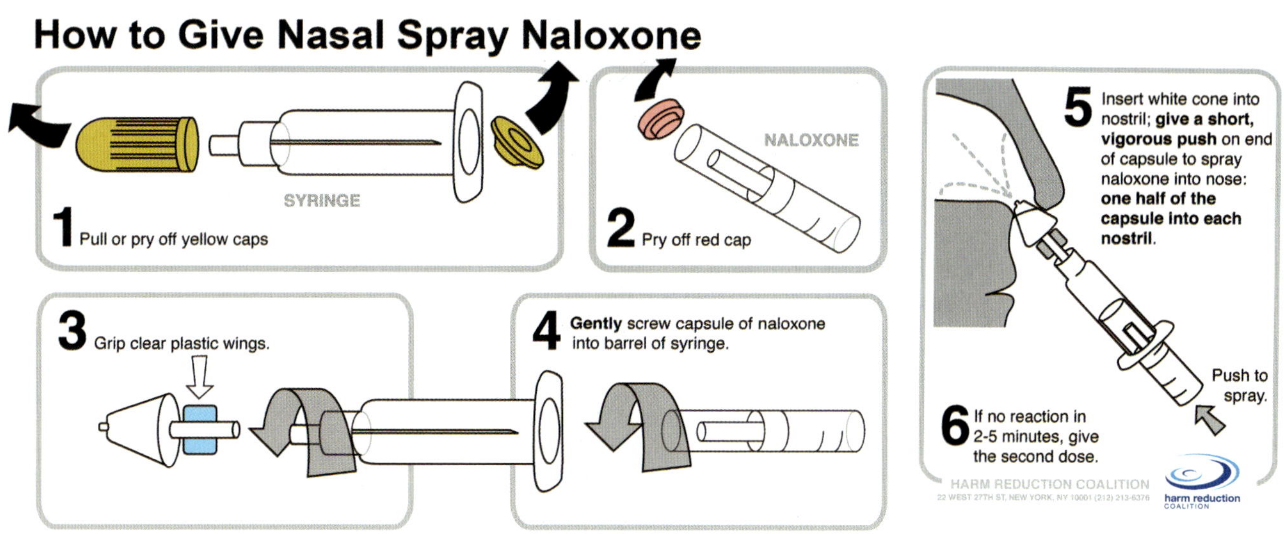

Fig. 26–8. An opioid overdose kit. (Courtesy of Mountainside Medical)

practical. Alcohol-based cleaners should be avoided in such situations as the alcohol may increase skin absorption. Unconsciousness always precedes respiratory depression. Naloxone should only be administered to treat severe respiratory depression (i.e., to patients who are unconscious with respirations of less than eight per minute).

Heat exhaustion and heat stroke

Heat exhaustion can occur in hot environments when the body loses water because of heavy sweating. **Heat stroke** is caused by an inability of the body to release heat and is a life-threatening condition. Even when properly treated, heat stroke is often fatal. Patients who are suffering from heat exhaustion feel light-headed, dizzy, and/or nauseated and they are often sweaty. Heat stroke patients have extremely high body temperatures, their skin is often flushed and dry (because they have lost the ability to sweat), and they may have an altered mental state (confused or combative), or are perhaps unconscious. Heat exhaustion patients should be moved to a cool environment, encouraged to hydrate with oral fluids unless they are vomiting, and monitored until they recover. Heat stroke patients should be cooled immediately. This can best be accomplished by stripping them down to their underclothes and soaking them with water. There should be no delay in transporting a heat stroke patient to an appropriate medical facility.

Hypothermia and frostbite

Hypothermia is a lowering of the core body temperature resulting from cold exposure. Firefighters can become hypothermic when they remove their turnout gear in cooler environments and experience rapid cooling from evaporation of sweat. Hypothermia also frequently occurs in subjects lost in the outdoors and people who accidently fall into larger bodies of water. Initially, hypothermic patients shiver, often uncontrollably, in an effort to generate additional body heat. As body temperature falls, shivering subsides and the patient will often become disoriented or confused. Treatment of hypothermia involves moving the patient to a warm environment, removing any wet clothing, and covering the person to keep them warm.

Frostbite occurs when uncovered parts of the body are exposed to extremely cold temperatures. Frostbitten skin can look white or waxy and may be firm or frozen. Swollen and blistered skin may eventually result. Never rub or massage a frostbitten area. Patients should be removed from the cold and protected from further cold exposures. Rewarming is best done in a medical facility under controlled conditions and should not be attempted in the field unless transport is excessively delayed. Avoid using heating devices or stoves. Patients cannot feel the frostbitten area, and it may cause more damage by burning the area. Initially, frostbite can be treated by using body heat (i.e., putting frostbitten fingers under the arm). Jewelry should be removed; and if the patient is wearing wet clothes, they should be removed as well.

BLEEDING

SKILL DRILL

Virtually all bleeding can be controlled with direct pressure. In the ABCs of assessment, a firefighter should attempt to control any severe bleeding immediately. Using a gloved hand, the firefighter should attempt to find the exact area where blood is coming from. It may be necessary to use a gauze pad to blot the area in order to locate the wound. Once located, direct pressure should be applied initially with the fingers. Once a dressing is available, it should be applied to the wound and direct pressure continued. If the dressing becomes saturated with blood, it is likely that pressure is not being applied directly to the wound. Saturated dressings should be replaced and the firefighter should attempt to maintain pressure directly over the wound. If bleeding cannot be controlled after several minutes of direct pressure, a tourniquet should be applied. Tourniquet application should be done using a commercial tourniquet device such as a CAT (Combat Application Tourniquet) (fig. 22–9).

There may be instances where a firefighter might want to apply a tourniquet immediately rather than attempting to control bleeding with direct pressure. These include multiple casualty situations, firefighter rescues by a rapid intervention team (RIT), while working alone or shorthanded, and when other priorities such as airway management exist. Tourniquets can be safely left in place for several hours. It is important to note the time the tourniquet was applied and be certain that application of a tourniquet is mentioned when handing the patient off to other providers. Tourniquets are initially successful in controlling 82% of bleeding from extremities. If bleeding is not controlled, a second tourniquet should be applied above the first. Applying a second tourniquet controls 92% of bleeding. Tourniquets should be applied as soon as it is recognized that direct pressure is ineffective.

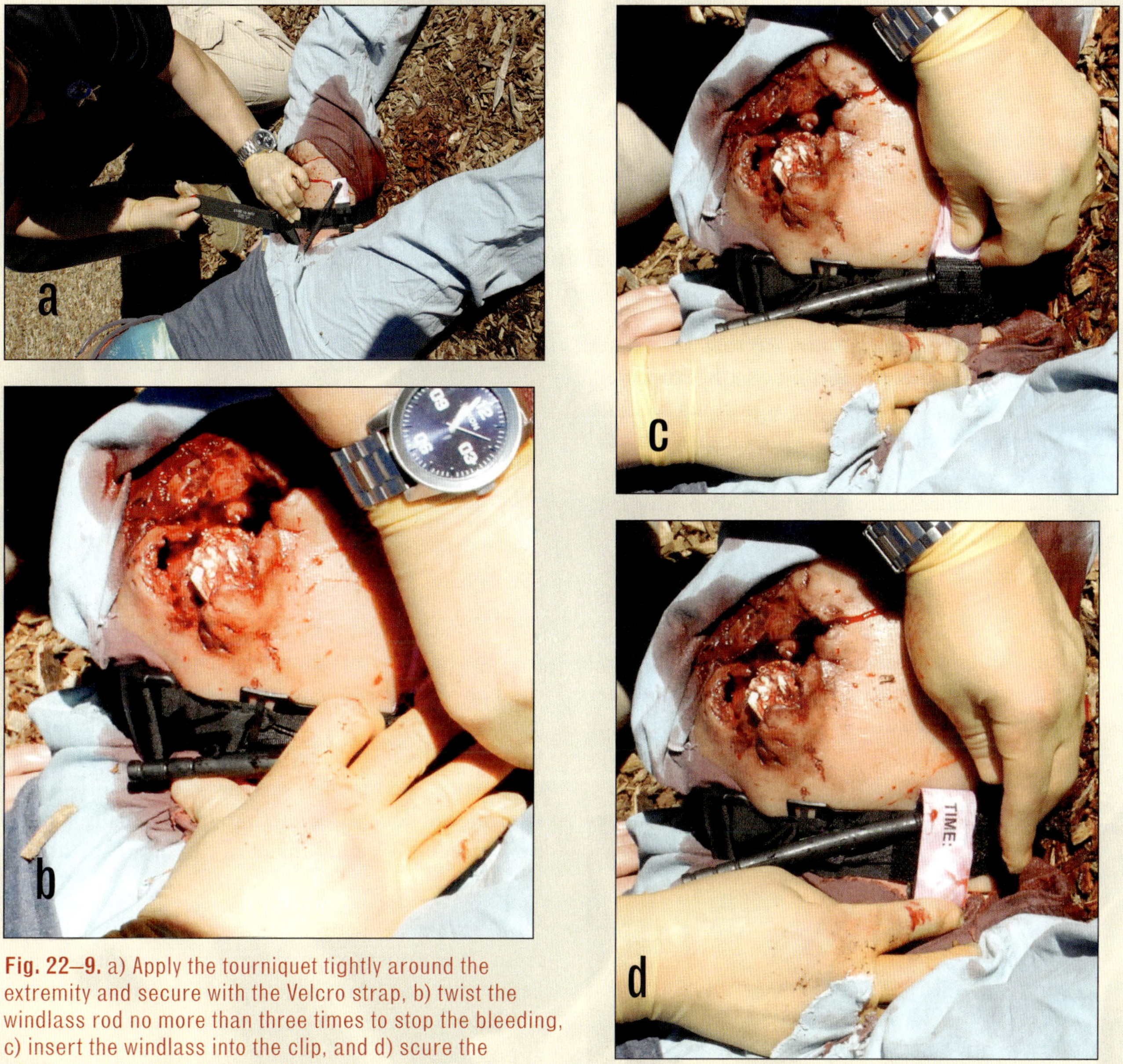

Fig. 22–9. a) Apply the tourniquet tightly around the extremity and secure with the Velcro strap, b) twist the windlass rod no more than three times to stop the bleeding, c) insert the windlass into the clip, and d) scure the windlass using the windlass clip strap.

Waiting for signs of shock before applying a tourniquet significantly increases the chances that the patient will die from their bleeding. It is also important to keep in mind that tourniquets hurt. Patients who require application of a tourniquet will need pain medication. This should be considered when requesting additional EMS resources.

CHILDBIRTH

The fire department may be among the first to assist a mother in labor. Most events are not emergencies, but they may require some initial assistance to prevent a mother or child from becoming injured during birth. Actions should be taken until the arrival of personnel with greater training; transport of the mother may need to be delayed if birth is imminent.

The area where a baby develops in a mother's womb is called the uterus. The baby in the womb receives nutrients from the mother through the umbilical cord attached to the placenta. Attached to the wall of the mother's uterus and the baby inside the womb, the placenta is enveloped in a sac called the amniotic sac, which is filled with amniotic fluid.

There are three main stages of labor. The first stage is when the mother begins to sense initial contractions, indicated by typical labor pains. The provider should note the time and duration of the contractions. When contractions are approximately 1–2 minutes apart, and last for more than 1 minute each, the provider and mother should prepare for the baby's birth.

Sometimes the mother may announce that her "water has broken" meaning that the amniotic sac has ruptured, releasing some amniotic fluid. This is nothing to cause alarm. The provider should look at the mother's vaginal area and determine if the baby is crowning, which is when a part of the baby emerges from the vaginal opening (fig. 22–10). It should be noted that the average duration of the first stage of labor for a first-time mother is more than 15 hours. Therefore, the provider should determine how long the mother has been having contractions, if this is her first baby, or if she has had multiple births. Labor and delivery times can be significantly shorter for a mother who has had multiple births.

Fig. 22–10. A crowning baby

The provider should use a sterile sheet under the mother and prepare to wrap the baby on arrival. Keep the area clear of unnecessary bystanders and unneeded providers. Allow the mother to get comfortable, laying on her back; raise the buttocks, if possible, with her knees drawn up and spread apart. Make sure the mother is as comfortable as possible and ensure the vaginal area is clear of any obstructions, including clothing. Baby delivery is a natural process that can often be done with little assistance. The main role of a provider is to prevent an explosive birth (where the infant rockets out of the birth canal too quickly, tearing the opening of the vagina) and assist in the case of a complicated delivery, which would require immediate transport to an appropriate facility.

As the baby begins to emerge, place one gloved hand below the baby's head to provide support; avoid placing pressure on the baby's head and spread the fingers to avoid pressing down on the soft spots (known as fontanelles). If the amniotic sac has not broken, the provider can puncture the membrane around the baby's head and pull it away from the mouth and nose. Allow the baby to continue to come out naturally, never pulling the baby. When the head emerges, ensure that the umbilical cord is not wrapped around the baby's neck; if it is, gently unwrap the cord from the neck using two fingers to bring the cord forward and over the baby's upper body. After the shoulders and feet emerge, continue to support the baby and check the baby's airway.

Once the baby is completely delivered, place the baby on his or her side with its head slightly lower than its body. Check on the exact time of birth. Once the infant begins to breathe, tie the umbilical cord halfway between the baby and mother to stop blood flow through the cord. Although it is preferable to use a sterile umbilical tie or commercial clamp, a shoelace will suffice in an emergency. Assure that the tie is sufficiently tight to stop blood flow through the cord. Initially, only one tie is needed. Cutting the cord can await the arrival of providers with additional training.

Shortly after the baby is delivered, the placenta emerges from the mother's vaginal area; the placenta should be kept intact and wrapped in a plastic bag if possible. Keep the baby warm and dry and the mother as comfortable as possible.

If any complications arise during childbirth, the mother and baby should be transported to the closest appropriate medical facility. Complications could include significant bleeding, unusual presentations such as a foot or arm instead of the head, or appearance of the umbilical cord in the vaginal opening prior to seeing the baby's head.

OXYGEN

For many years, emergency responders have administered oxygen to patients with various medical problems believing that it would help to improve delivery of oxygen to tissues of the body and relieve shortness of breath. More recently, oxygen has been determined to be harmful to many patients, particularly those encountered by EMS. Molecules called free radicals that arrive on scene after the injury can do damage to tissues in the body from trauma, strokes, and heart attacks. Oxygen is a free radical. Flooding injured tissues with oxygen proportionately increases tissue damage. High levels of oxygen increase brain damage in strokes, heart damage in heart attacks, death rates in trauma patients and distressed newborn babies, and probably is not helpful in a variety of other medical emergencies. Additionally, oxygen does nothing to improve the feeling of breathlessness in patients with respiratory distress unless they have very low levels of oxygen in their bloodstream. Today, administration of oxygen should be guided by measurements of blood oxygen levels obtained using a pulse oximeter. The target oxygen saturation is 94% to 99%. Patients with an oxygen saturation above 94% should not be given supplemental oxygen unless they have been poisoned by carbon monoxide, in which case high-flow oxygen may be helpful. When administering oxygen to patients, the least amount needed to achieve measured pulse oximeter saturation of 94% to 99% is likely to be most helpful and the least harmful.

TRANSFER OF PATIENT CARE

As a firefighter on an EMS call, your primary responsibility is to stabilize the patient until the arrival of personnel with a higher level of EMS training. An assessment begins when the call is dispatched, and evaluation of the scene and the patient starts on arrival. When additional responders such as emergency medical technicians or paramedics arrive on scene, it is your responsibility to provide them with pertinent information. Sometimes this can be relayed via radio or cellular call prior to their arrival.

Information about the current condition of the patient can help incoming EMS providers prepare their action plan to assist the patient. It prompts them to consider what equipment they may need on arrival, particularly if it is different from what the initial dispatch information provided. When they arrive, be prepared to give them a brief report of what you saw when you arrived, what actions you took, and any history of the patient you obtained.

Sometimes, the EMS providers request assistance in getting a patient loaded onto the transport unit. Every firefighter should become familiar with their local EMS transport services, their providers and their equipment. You should be familiar with how to take the stretcher out of the ambulance, and how to put it back in with a patient on it. Often a wheeled chair device (i.e., stair chair) is used to bring patients from indoors to outside a building. There are other devices used to move patients from difficult to access locations such as motor vehicle crash scenes or out of the way locations inside a building. Providers should be familiar with the type(s) of devices used in their local jurisdictions. If used improperly the specialized equipment used by EMS providers can cause injuries; therefore, it is vital to become familiar with them (fig. 22–11).

Fig. 22–11. A wheeled chair device or stair chair

Remember, it is not one person's job or one individual crew's responsibility to provide care. As a firefighter on an EMS call, you are part of a system designed to bring the best possible outcome to a person in need. The transition of care is a vital link in the EMS system. Keep in mind that a leading cause of medical mistakes that harm patients is poor communication during hand-offs, that is, when one provider hands a patient off to another. Your work is not necessarily over when the EMS crew arrives. Ask them if they have any questions or if they need anything before you leave the scene.

QUESTIONS

1. You are a member of an engine crew that arrives first on scene of a bicyclist struck by an automobile. The rider is lying in the road with an obvious open fracture to his left lower leg. What personal protective equipment will you wear while providing patient care, and what standard precautions will you need to consider after the patient has been transported from the scene?

2. Under what circumstances can you as a firefighter disclose information about your patient to others?

3. Discuss why the ABCs of medical care are three of the most important patient concerns you will have while treating your patient.

4. Discuss the procedures for checking a pulse in a patient. Include the differences for an unconscious versus conscious patient and adults versus small children.

5. You are on the first arriving engine to a bleeding call. Upon arrival, you find a 29-year-old male sitting on the house porch with a deep bleeding gash in his left arm just above the elbow. Describe your actions in treating this patient.

6. Your engine crew is the first-arriving unit to an automobile that has crashed into a telephone pole. The only occupant is the 17-year-old male driver still seated in the car. He is conscious and speaking. He has an open fracture below the right knee. Explain the priorities of care and how you will direct other crewmembers to assist.

7. You are assigned to an engine company fighting a house fire on a hot July day. Upon exiting the house and going to the rehab area your partner states he does not feel well. How would you determine if he is suffering from heat exhaustion or heat stroke?

8. You are conducting your morning truck checks in the station when a car screeches to a stop at the station door and a male driver gets out of the car asking for help. He explains to you that his wife is nine months pregnant, her water has broken, and her contractions are less than a minute apart. They were trying to make it to the hospital but the mother believed that birth was imminent. Discuss your actions to help deliver the baby.

9. Upon arriving at the scene of a supposedly unconscious victim you realize the patient has in fact been shot. He is still alive. Describe your concerns from a scene safety standpoint.

10. You are the first arriving unit to an altered level of consciousness call. Upon arrival, you find a 50-year-old woman who is weak and talking nonsense. She is surrounded by several family members who state this is not her normal behavior. What are some of the questions you will ask the family, and what are some of the things you are looking for to determine the needed treatment?

FIREFIGHTER II

Incident Command System

by Forest Reeder and Tim Flannery with Alan Brunacini

This chapter provides required knowledge items for the following NFPA Standard 1001 Job Performance Requirements:

FFII 5.1.1

OBJECTIVES

Upon completion of this chapter, you should be able to do the following:

- Identify the types of command systems
- Identify the five types of incidents developed by the National Incident Management System (NIMS)
- Describe the three organizational levels of command.
- Identify the command and general staff positions within the Incident Command System (ICS)
- Identify incident tactical priorities
- Describe how command is assumed and transferred
- List and describe the three command options
- Identify the ICS functions used at an incident
- Identify types of incident facilities
- Describe the purpose of a staging area
- Demonstrate the ability to function within an assigned role in ICS
- Demonstrate the ability to organize and coordinate ICS until command is transferred

INTRODUCTION

The simple truth is that a strong and functioning **Incident Command System (ICS)** will save firefighter lives. As in the military, organized command and control functions are essential to the success of every operation. Responder safety is always the first consideration, and the use of ICS is key to achieving this goal. In common with all systems or functions, the ultimate reason to use a command system is to provide for the safety of the emergency workers in the hazard zone.

From a single-company operation to a multilevel conflagration or mass casualty incident, a single person must be in charge. This person is identified as the **incident commander (IC)** and is responsible for establishing a set of predetermined outcomes and an overall plan of action to accomplish those goals. This is the guiding principle of incident command.

In this chapter you learn how an ICS functions and why the NIMS was created to provide a standardized-while-flexible ICS framework for emergency incidents across the United States. Standardization is crucial, but flexibility must exist to accommodate a vast array of incident types and levels as well as a broad spectrum of organizational types and makeups.

NIMS

After the 9/11 attacks, the federal government and local emergency response units discovered a disconnect when it came to adequate response to large-scale natural disaster or weapon of mass destruction (WMD) events. President George W. Bush issued **Homeland Security Presidential Directive 5 (HSPD-5)** in early 2003. The HSPD-5 directed the federal government and state and local agencies to work under a National Incident Management System, to be developed and administered by the Department of Homeland Security.

The purpose of NIMS is to create a framework for federal, state, and local agencies to work together effectively as may be required by the size of an incident. NIMS is now the national standard used throughout all 50 states, structured to manage situations of all sizes from large-scale national events down to small local incidents.

The NIMS is based on the principles of *standardization* and *flexibility*. These two components are required to create a scalable system that can be implemented for any incident from a small local fire to a large-scale natural disaster or man-made event.

By standardizing terminology and incident command structure and concepts, NIMS creates operational efficiency and unification among agencies when brought together for an emergency event. At the same time, by using incident command components that can be added, subtracted, or customized to a specific region or event type, NIMS maintains the flexibility to be applied to any incident response situation.

The principle of unified command

Unified command is the structure in which multiple individuals are cooperatively responsible for all strategic objectives of the incident. It is typically used when an incident is within multiple jurisdictions or managed by multiple disciplines.

For instance, a local fire event might require not only firefighters, but EMS and police responders for medical services and area control. Such an event would require a unified command structure in which representatives from each emergency service branch work together as a command team to implement a multi-branch response.

Common NIMS terminology

Some of the most common concepts and terms used in the NIMS framework include the following:

Unity of command. All persons in the operation have a designated supervisor to whom they report at the scene of the incident.

Span of control. The number of personnel one supervisor can effectively manage. The range is two to seven, with five being the optimal number.

Modular organization. An organizational structure where command components are used as needed and expandable according to complexity and size of the incident.

Freelancing. Uncommanded action, or operating outside of the established fire attack plan, operational guidelines, or operating protocols.

Resource management. A process to identify, acquire, mobilize, track, recover and demobilize, reimburse, and inventory necessary resources.

Incident categories

NIMS categorizes incidents as Type 1 through Type 5 (fig. 23–1).

Type 4 and Type 5 incidents represent more than 99% of all fire department incident activity. Each incident type may evolve into a higher, more complicated type. A Type 5 report of a car fire may turn out to be located inside the parking garage of an apartment building, requiring upgrading to a Type 4 response, bringing additional resources. In the same vein, a single localized Type 4 wildfire could grow to become a Type 3 statewide management incident.

Most U.S. fire departments use this type of system for the everyday management of their resources working in hazard zones. You will not typically hear of an incident being labeled as a particular type or classification; these labels are provided to illustrate the complexity of the incident and the resource management needed for it.

HISTORY OF ICS

Incident Command Systems are nothing new to the fire service. Several variations of ICS have existed since the 1970s, but their importance cannot be underestimated because prior to this period, emergency event management and command structure differed from department to department, and sometimes even between chiefs in the same department. It was during the California wildfires of 1970 that the need for a standardized command structure, communication, and coordination became apparent. All of the following systems are varieties of an ICS.

FIRESCOPE

In 1972, the California-based command consortium **FIRESCOPE** (**FI**refighting **RES**ources of **C**alifornia **O**rganized for **P**otential **E**mergencies) was developed to manage firefighting efforts for large-scale wildfires that frequently plague the West coast. It was also later used to manage other large-scale disasters, including forest fires, floods, earthquakes, civil unrest, and so on. This system coordinates massive amounts of necessary resources during incidents that can last from weeks to months, require thousands of incident responders, and generally include federal and state resources.

Due to the effectiveness of the FIRESCOPE ICS, it was revised and adopted on a national level in 1982 to become the **Incident Command System/National Interagency Incident Management System (ICS/NIIMS)**.

Fire Ground Command

Also during the 1970s, a similar command system known as **Fire Ground Command (FGC)** was developed by Chief Alan Brunacini during his service with the Phoenix Fire Department. The FGC was a drastically scaled-down version of ICS used for the typical, fast-moving, high-hazard incidents that make up 99% of American fire service responses. These include fires in houses, apartments, and warehouses, as well as calls involving hazmat, multiple-victim motor vehicle accidents, and so on. FGC was similar to FIRESCOPE in application but used a slightly different organizational structure and terminology.

Incident Management System

Even with the development of FIRESCOPE, FGC, and NIIMS, the need was still apparent for a more common approach to incident command. This led to the 1990 creation of the National Fire Service Incident Management System (IMS) Consortium. The IMS Consortium was made up of fire service leaders and organizations across the nation, as well as FIRESCOPE, the National Fire Academy (NFA), Federal Emergency Management Agency (FEMA), and Phoenix Fire Department's FGC. The Consortium agreed to develop common protocols, pulling from aspects of both ICS and FGC, so it could be applied effectively from department to department in all regions of the United States. The new set of protocols was approved in 1993, and the new system was officially titled Incident Management System (IMS).

The IMS is used to apply local fire department resources to bring under control and mitigate local emergency events. The typical operational period for these incidents can last from 10 minutes to several days. Major objectives of an IMS are hazard zone management and protection of hazard zone workers.

In many areas, the terms *IMS* and *ICS* are used interchangeably to mean a local incident command structure with one individual in charge of the major decisions relating to the incident. Responders must remain current with locally and regionally accepted ICS practices.

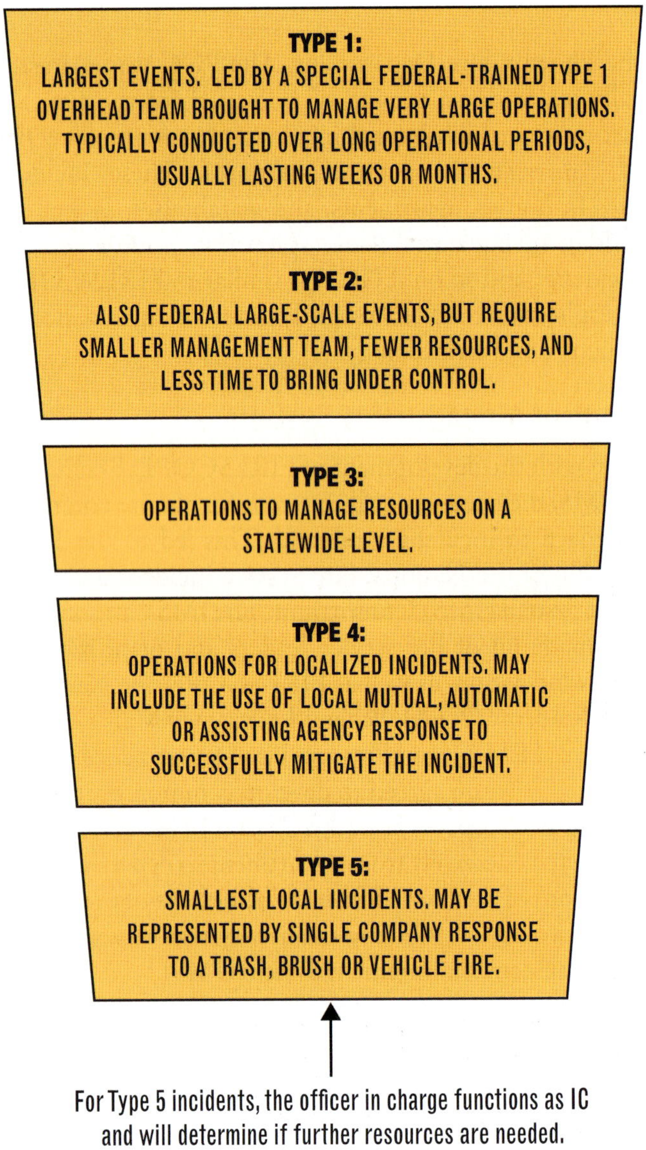

Fig. 23–1. NIMS event categories

The Blue Card incident command system

The Incident Command Certification Program is designed to first instruct, train, and then evaluate and certify fire department officers who serve in the role of incident commanders that supervise and manage emergency and hazard zone operations for every day, local NIMS Type 4 and Type 5 events.

The "Blue Card" training program provides fire departments with a training and certification system that defines the best standard command practices for common, local, everyday strategic and tactical emergency operations conducted on NIMS Type 4 & Type 5 events. This training and certification program produces incident commanders that make better decisions that will potentially eliminate the lethal and costly mistakes that cause injury, death, and unnecessary fire losses in the local response area.

The ICS

The most familiar NIMS component for all incident types is the ICS, which is the basic management and command structure applied to all incident types. A Type 4 house fire will use ICS through the local incident commander and response team. A Type 1 multiagency overhead team uses ICS to manage the goals and objectives of the overall incident. At the same time, hazard-zone workers protect themselves using ICS while using the NIMS ICS structure to report back up into the Type 1 or 2 team chain of command.

The major goals of a local ICS are to ensure initial and ongoing firefighter safety at incident sites and to coordinate all command and operational action to bring incident hazards under control. **Resource management** of crews, apparatus, and other equipment needed to mitigate the problem can be achieved through the use of ICS.

The command system creates an organizational agreement among all incident responders that the incident commander is the overall site manager and that everyone on the scene will play their assigned roles in the IC's plan, reporting up to the IC through their designated supervisors (**unity of command**).

Command procedures give the IC responsibility for overall incident results along with the organizational authority to put responders to work with the same set of tactical objectives. For this basic incident management plan to be effective, responders must follow a set of command procedures that are agreed upon, written (in SOP form), and practiced prior to the event. This is particularly crucial in the typically fast-moving, dangerous incident conditions. *The ICS must be applied to every incident.* Saving the command system for "the big one" is just another way of saying that your department doesn't use, practice, or have an actual ICS. The ongoing everyday/every-event use of the system makes it a regular, habitual way to manage all emergency responses.

The dangers of uncommanded action. Some of the most serious, unsafe conditions on the fireground are those with **no command** or **multiple commands**. Without a strong, central IC from the beginning of emergency operations, the typical incident scene quickly deteriorates into an unsafe, out-of-control situation. These no-command beginnings often lead to uncoordinated action, commonly known as **freelancing**.

Freelancing may eliminate the incident hazard on small incidents, but repetitively practicing uncommanded mistakes creates dangerous habits that can lead to serious injury or death on large, unusual, or time-consuming incidents. When an incident begins with uncommanded, fragmented freelancing, and the team grows as the incident progresses, it becomes impossible for all participants to safely support and coordinate actions to control the incident.

Uncommanded action causes duplicate searches of one area while exposed areas go without primary searches. Attack lines tend to congregate in the same location and leave concealed spaces and exposures uncovered. Salvage occurs after savable property has been unnecessarily damaged by later-arriving freelancers.

The most effective way to avoid these situations, provide firefighter safety, and stabilize an incident is to have a well-trained IC in place from the beginning of operations on *all* events. A strong IC with an operational team that understands and follows basic incident management procedures consistently outperforms a larger force of freelancing firefighters. Habits developed on day-to-day incident operations are the foundation for performance on bigger, longer, and more hazardous operations.

Organizational levels. There are three organizational levels for all incident operations (fig. 23–2):

***Strategic level*:** The IC is responsible for managing the strategic level of the incident organization. The IC sets the overall incident strategy, develops and manages the **incident action plan (IAP)**, and assigns and controls the position and function of all incident resources.

***Tactical level*:** Individual companies perform task-level work. When multiple companies operate at an incident scene, their coordinated tasks combine to make up the tactics. A classic tactical example is a ladder company performing vertical ventilation on a roof so engine companies can safely operate on the interior of a structure.

***Task level*:** The task level is performed by firefighters and is where the work actually gets done. Engine companies stretch and operate hoselines while ladder and truck companies perform support work. These combined efforts complete the tactical priorities of rescue, fire control, and property conservation for the incident (for structural firefighting).

As an incident is managed, it is usually necessary to subdivide or organize the companies working into a more manageable **span of control**, usually no more than five to seven companies. When multiple units are assigned to the same geographic area of the incident scene (side A, the north side of the building, etc.), their actions should be coordinated, managed, and supervised by a tactical-level officer. Designations for these tactical-level company or command officers are sector, division, and group supervisors. In the ICS, divisions represent geographic locations where resources are assigned, whereas groups representing resources fulfilling specific functions don't have to be tied to specific geographical areas. An example is a refueling group that drives around in an ICS sector designation and refuels all of the division's apparatus.

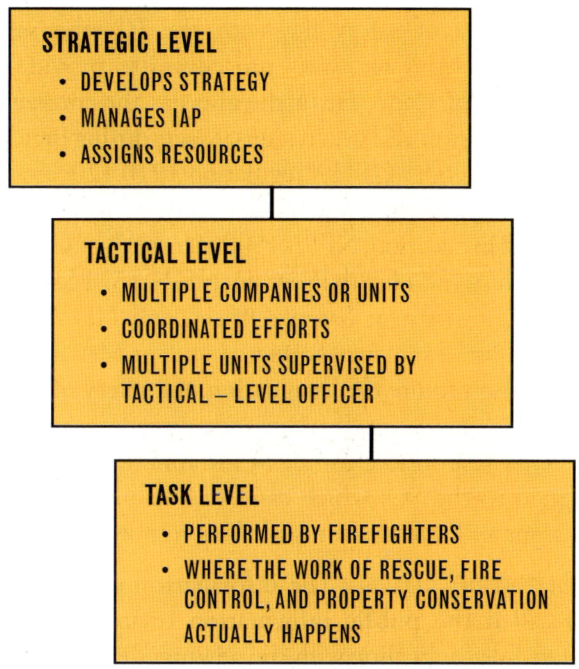

Fig. 23–2. The three organizational levels for all incident operations

Command staff positions

FFII 5.1.1 The command staff answers directly to the IC and provides him or her with essential information and assistance to operate on a strategic level. The IC activates these positions based on the needs of the incident. Only if determined necessary are these positions activated. The people filling these positions are called safety, information, and liaison officers (fig. 23–3).

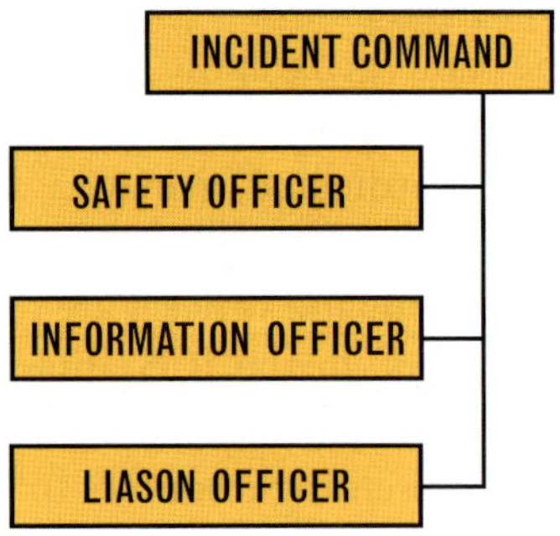

Fig. 23–3. Command staff positions

Safety officer. The **safety officer (SO)** is responsible for keeping the IC advised on all operational safety matters relating to an incident. Although the ultimate responsibility of safety lies with the IC, the safety officer carries out necessary procedures to provide for a safe incident scene. He or she also has the emergency authority to stop and prevent unsafe acts during incident operations. Under OSHA regulation 1910.120, a safety officer is required by law at incidents involving hazardous materials. In addition, NFPA 1500 Standard for Occupational Safety and Health Program, NFPA 1521 Standard for Fire Department Safety Officer, and NFPA 1561 Standard on Emergency Services Incident Management System also require the designation of a safety officer at emergency incidents. If one is not directly appointed, the IC is responsible for all of the functions normally assigned to the SO, which can overwhelm an IC with decisions and information-gathering responsibility.

Information officer. The **information officer (IO)**, also called the **public information officer (PIO)**, is responsible for dealing with the news media and general public. He or she provides the media and public with accurate, complete information about an incident's cause, size, and current situation; resources committed; and other matters of general interest. There is only one PIO per incident, but he or she may have assistants. The IC must approve any information released by the PIO. Not all incidents require a PIO. Reasons for an IC to designate a PIO include working on a highly visible or sensitive incident, media demands interfering with the IC's ability to run the incident, and limiting the possibility of conflicting reports.

Liaison officer. The **liaison officer (LO)** is responsible for coordinating with any outside agencies that respond to an incident. These may be federal, state, or local government agencies; utility companies; and private organizations such as the American Red Cross. This position is established when the number of responding agencies begins to overwhelm the IC.

General staff positions

Typically, four general staff positions can be filled at an incident.

- Operations section
- Planning section
- Logistics section
- Finance/administration section

Sometimes a fifth section for intelligence or investigations may be filled, depending on incident type and scope.

General staff functions are more incident-specific. In other words, a mass-casualty incident may differ from a hazardous-materials incident. The incident drives this part of the command system. Certain functions that apply to a mass-casualty incident do not apply to a hazardous-material incident. Not all functional sections applicable to the general staff may be needed at an incident. The IC may need assistance with logistics, so a logistics section is established, but the IC may still handle the tactical aspects of the incident, making an operations section unnecessary. As the incident expands and the IC realizes that he or she is unable to effectively handle other functions, the IC hands those functions to general staff managers. Anyone assigned to run the operations, planning, logistics, or finance/administration section is called "chief," regardless of his or her official title within an organization (fig. 23–4).

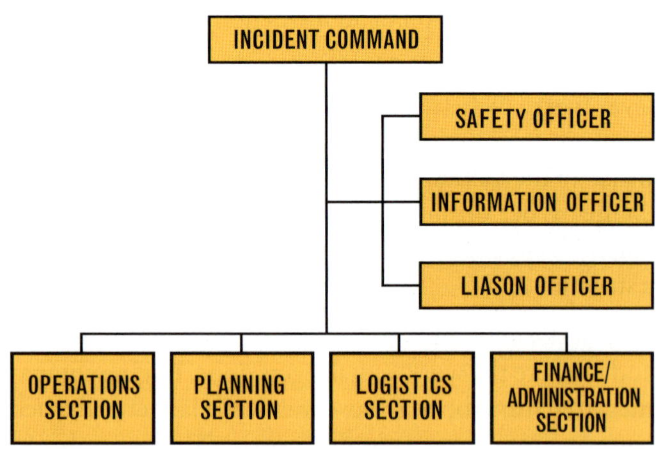

Fig. 23–4. General staff positions

Operations section. The **operations section** is responsible for all tactical operations relating to an incident. Operations at an incident can be set up depending on the following:

- Kind of incident
- Agencies involved
- Objectives and strategy

When the IC determines that he or she can no longer coordinate the strategic and tactical level of an incident, the IC designates an operations section consisting initially of those few resources first assigned to an incident. As additional resources are committed and the incident becomes more complex, a separate operations section may be established. The operations section expands or contracts based on the existing and projected needs of the incident.

The operations section develops from the bottom up by first establishing divisions, groups, and, if necessary, branches. Also, the operations section may have staging areas and, in some cases, an air organization.

The subordinate units of **divisions, groups, branches**, and **resources** within the operations section can be activated based on the needs of the incident (fig. 23–5).

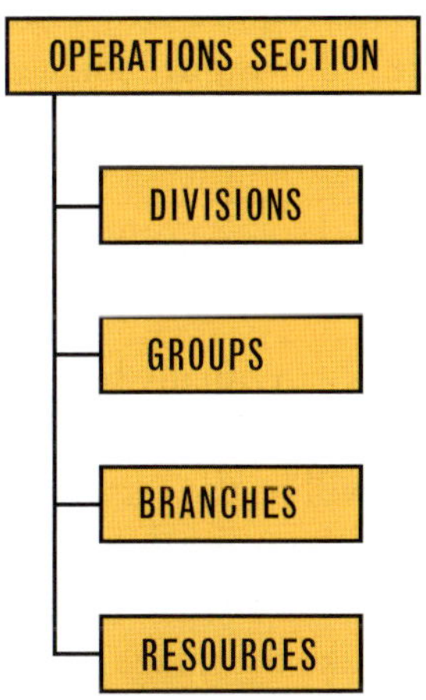

Fig. 23–5. General staff also report directly to incident command, as do command staff.

Divisions. A common method of organizing tactical operations at an incident is for the IC to first establish two or more divisions. Divisions always refer to geographically defined operational areas such as the area around a building, the inside or floors of a building, or an open area.

When defining an incident by division, there is a certain amount of latitude. An incident can be divided using compass points, such as east division, west division, and so on. It can be divided using letter-specific areas, such as side A, side B, side C, and side D. When using this format, the front of a structure indicated by street address is always side A. The other sides are determined in a clockwise pattern around the structure.

An incident may also be divided by floors of a building, such as floor 1, division 1; floor 2, division 2; and so on (fig. 23–6). Initially, establishing divisions may be done to define the incident and may or may not include the designation of separate division supervisors. When the resources assigned within a division exceed, or will soon exceed, the recommended span of control guidelines of one to five, division supervisors should be designated. Division supervisors manage divisions not under the direct management of the IC or operations section chief.

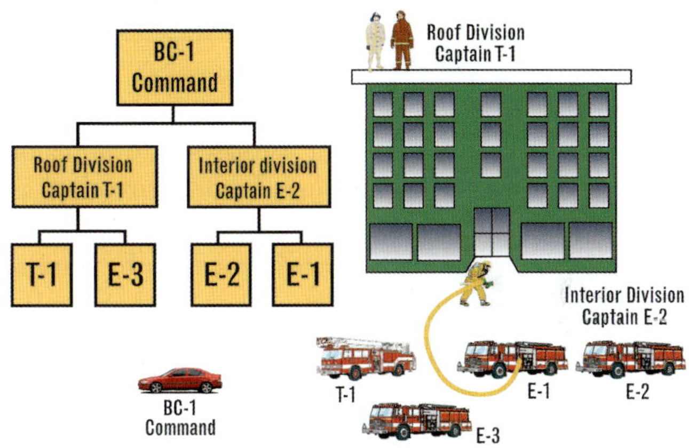

Fig. 23–6. Example of using divisions

Groups. Another common method of organizing operations at an incident is to establish functional **groups**. As the name implies, this form of organization deals not with geographic areas, but with functional activity. Examples of functional groups include fire attack groups, vent groups, search groups, evacuation groups, and so on. Groups, similar to divisions, are managed by supervisors.

A simple way to create groups at an incident is to look at the basic strategies to be used and then create a group to fill those strategies. For example, if the strategies are to search the building, vent the building, and control and extinguish the fire, then the groups become search group, vent group, and fire attack group (fig. 23–7).

Note: Divisions and groups are common designations used on the fireground. Within the command system, they fall between branches and task force/strike team. They are equal in authority, and each is overseen by a person with the title of supervisor. When divisions and groups are utilized at the same incident, the division and group supervisors must coordinate actions together.

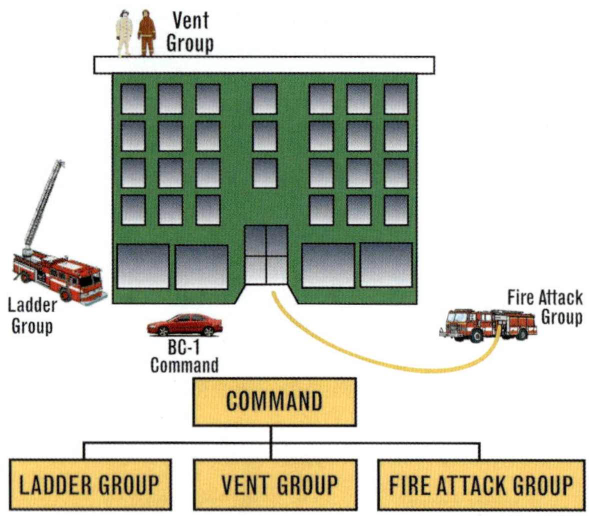

Fig. 23–7. Example of using groups

Branches. As a small incident escalates into a major incident, the span of control may become stretched as more divisions or groups are implemented. In addition, the IC can be quickly overwhelmed with various functions of command. The immediate need of the IC is support. **Branches** may be added to facilitate more effective incident management (fig. 23–8). The person in charge is called the **branch director**. Branches can be geographical or functional, similar to divisions and groups. Reasons to create a branch include the following:

- The span of control for the number of divisions/groups in place becomes unwieldy or unmanageable.
- The incident becomes more complex (e.g., multijurisdictional, worsening conditions).
- The incident has two or more distinctly different operations (e.g., fire, medical, hazardous materials).

Resources. Most incidents start small and grow based on type and severity. In these incidents, most of the resource's apparatus and personnel are committed to the operations section. This section almost always builds from the bottom up. When an incident is small, there are few resources, so there is a small command system. As the incident grows, so does the command system as more resources arrive. There are several ways to organize resources: single resources, task forces/strike teams, divisions/groups, and branches.

- ***Single resources.*** Single resources are individual apparatus or a crew of firefighters with an identified supervisor that can be used in a tactical application on an incident. A single resource is the most common way of initially using resources on an incident. The person in charge of the single resource is designated by unit title, for example, captain, lieutenant, and so on.

  Single resources can be typed to reflect capability. Unless a single resource is typed, its specific resource capabilities may not be clear to everyone. Examples of resources include a single engine company, truck company, rescue company, and so on. Because the operations part of an ICS normally develops from the bottom up (less to more), most of these incidents require only a small amount of resources. The following are some examples:

  - Room and contents fire: two-engine and one-truck response
  - Vehicle fire: two-engine response

  Type 5 incidents can usually be handled by a single resource supervised by an IC. At times, the IC and single resource will be the same: for example, single engine for a dumpster fire, battalion chief called to smell of smoke. In each incident, the command structure looks like the image in figure 23–9.

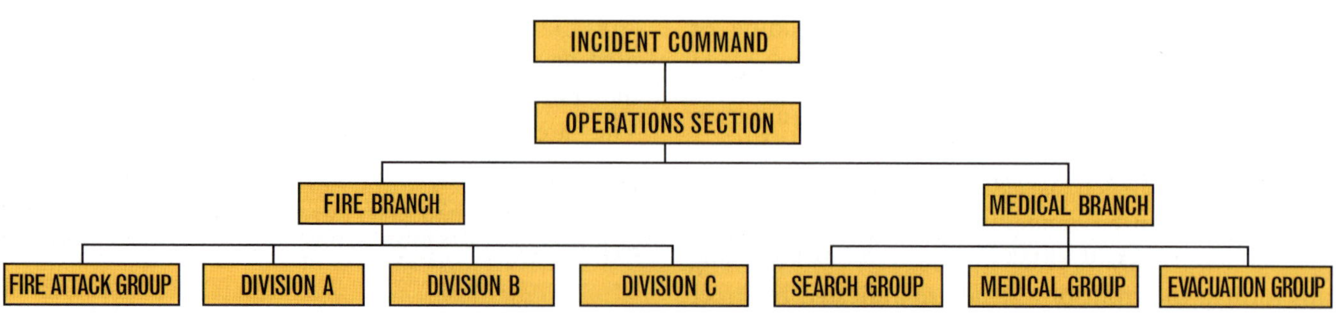

Fig. 27–8. Example organization of operations groups and branches

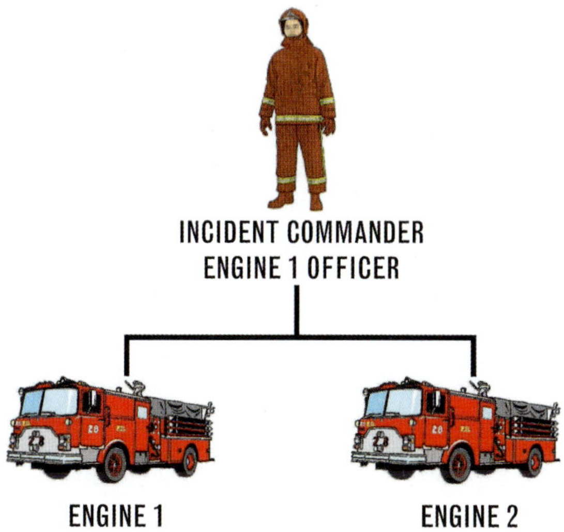

Fig. 23–9. Example of using single resources

- ***Multiple resources.*** One of the concepts under NIMS/ICS is the use of strike teams and task forces. Although similar in function, they are different. The difference is based on the kind and type. A *kind* of resource is what the resources is. Examples are an engine, truck, ambulance, rescue, and so on. A *type* of resource is a designation based on its size, capability, staffing qualifications, and ability to accomplish a task or function. For example, under engines you can have Type 1, Type 2, or Type 3. A Type 1 is larger and has more capabilities than a Type 2, and so on.

Strike teams are composed of resources of the same kind and type (fig. 23–10). They must have a leader and common communications between resources and the leader, have transportation, and operate within span of control limits. Strike team examples might include:

- Search team of five firefighters
- Ventilation team of three firefighters
- Strike team of four Type 1 engines

Task forces are any combination and number of single resources (within span of control limits) assembled for a particular tactical need (fig. 23–11). Task forces may be a mix of all different kinds of resources (of the same kind but different types, or of several resources of one kind mixed with other resources). Task force examples might include:

- Fire attack task force of four pumpers and two aerial ladders
- Water tender task force of five water tenders

As an incident becomes larger or more complex, the IC has additional options to use task forces, strike teams, divisions, or groups to increase the number of resources and still maintain a manageable span of control. For Type 5 and going into Type 4 incidents, an option for using more resources is to combine them into task forces or strike teams. In either of these, the IC can increase the number of resources substantially and still maintain span of control within acceptable limits. For example, one task force or strike team can have up to five resources. So an IC can have three engines and two trucks in a task force and have three of these task forces under his or her control. In reality, the IC has 15 resources under his or her command but only three subordinates reporting directly to him or her, thus maintaining span of control.

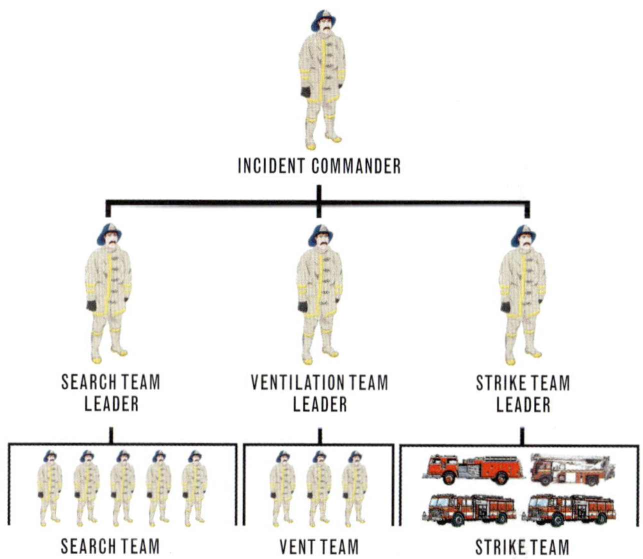

Fig. 23–10. Example of strike team

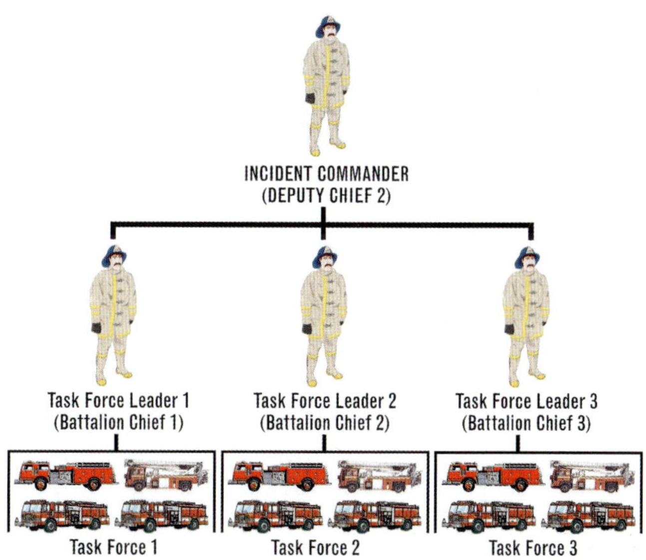

Fig. 23–11. Example of task force

Planning section. The **planning section** is responsible for collecting and evaluating incident situation information, preparing situation status reports, displaying situation information, maintaining status of resources, developing an incident action plan (IAP), and preparing required incident-related documentation. One of the most important functions of the planning section is to look beyond the current and next operational period and anticipate potential problems or events. The planning section consists of four units: resources, situation, documentation, and demobilization units. The person in charge of these subunits is designated as a "leader" (fig. 23–12).

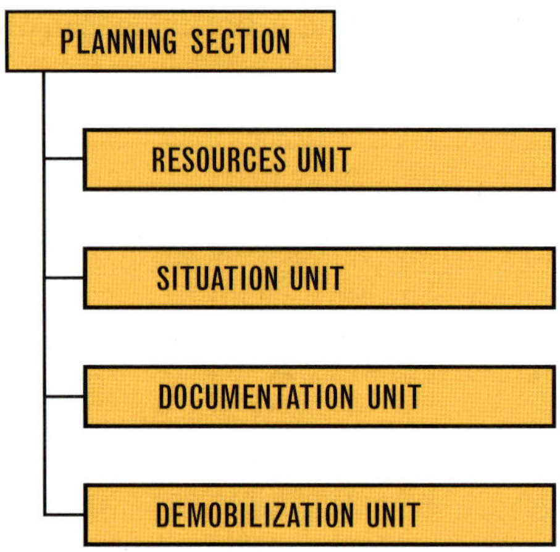

Fig. 23–12. Planning section positions

- ***Resources unit.*** The resources unit is responsible for checking in and maintaining status on all personnel, apparatus, and equipment assigned to the incident. It is the accountability unit for the incident that assists the IC and operations section in tracking all personnel and apparatus so they can determine where they are, if more are needed, and if certain resources can be released or sent home.
- ***Situation unit.*** The situation unit collects all current incident information and, based on this, prepares maps, displays, and projections. This assists the IC in determining future strategies and necessary changes.
- ***Documentation unit.*** The documentation unit is responsible for all documentation related to an incident, including distributing copies of the IAP, objectives, and strategies to all supervisory personnel involved in the incident. This provides responders with the IC's written expectations of them.
- ***Demobilization unit.*** On large or complex incidents, the demobilization unit works with the IC and operations section to develop a plan to release personnel and apparatus no longer needed at an incident, and it ensures that the incident is adequately staffed to meet the IC's objectives and strategies.

Logistics section. The **logistics section** is responsible for providing services and support to meet all incident needs. Early recognition of the need for a separate logistics function and section can reduce the time and money spent on an incident. There are six units within the logistics section: communications, medical support to responders, food for responders, supply, facilities, and ground support. The person in charge of these subunits is designated as a "leader." The logistics section chief may establish separate units for one or more of the logistics support or service activities (fig. 23–13).

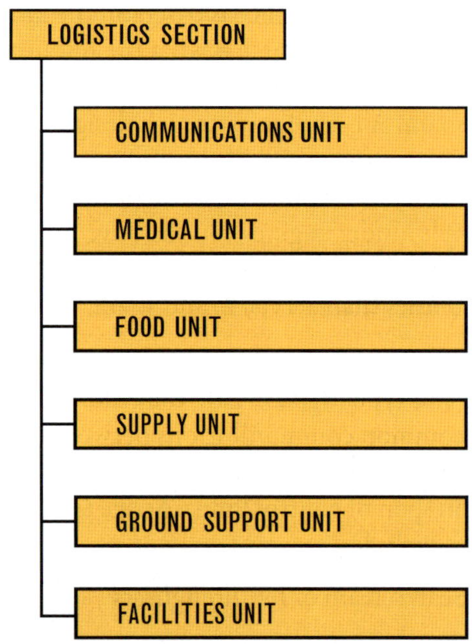

Fig. 23–13. Logistics section positions

Communications unit. This provides on-site communications at large incidents. It also ensures that there are adequate radios and the ability to repair them on scene. The unit is also responsible for developing a communications plan, for example, radio frequencies to be used and their purpose. An example is a mobile communications unit provided by the local, county, or regional emergency management coordinator.

Medical unit. The medical unit's functions include providing medical treatment for incident responders (victims are treated by medical personnel assigned to operations section) and developing an incident medical plan that includes transportation route, available hospitals, and EMS organizations. The unit is also responsible for implementing and maintaining the rehabilitation area at an incident.

Food unit. This unit ensures that incident responders have food and drinking water for a long-term incident. This can be accomplished by local organizations, the Salvation Army, American Red Cross, local canteens, or a ladies' auxiliary. (Note: In April 1995, the Oklahoma Restaurant Association maintained the food unit for the Oklahoma City bombing.)

Supply unit. At large incidents, the supply unit orders personnel, equipment, and supplies. It also stores and maintains supplies, services, and nonexpendable equipment.

Facilities unit. This unit sets up any fixed facilities that may be required for an incident. Typically, this unit is not established except on very large incidents. If a base needs eating and sleeping facilities for incident responders, this is the unit that accomplishes it. NIMS/ICS includes the following types of facilities: **incident command posts**, **staging areas**, **bases**, **camps**, and **helispots.** Other types can be implemented depending on the type of incident. For example, there are specialized facilities for mass-casualty, high-rise fire, and hazardous-materials incidents. A "manager" oversees all of these facilities except the incident command post.

- ***Incident command post.*** The **incident command post** is where the IC oversees the incident (fig. 23–14). Depending on the type and size of the incident, it may be located in a vehicle, trailer, tent, or building. The command post should be where the IC can view the incident, if possible, but not too close to danger. For large and complex incidents, the command post may be comprised of several structures that form a unified command of multiple agencies or jurisdictions. There may also be structures for command staff and communication centers. Examples of incidents and possible command posts include the following:
  - For a two-engine response to a vehicle fire, the command post would be the officer in charge located near one of the engines.
  - For a two-engine, two-truck response to a dwelling fire, the command post may be the back of the responding chief's vehicle.
  - For a mutual aid response to a hazardous-material incident, the command post may be a mobile trailer provided by emergency management.

Fig. 23–14. Internal view of a mobile incident command post

- ***Staging areas.*** Staging areas are where any resources to be used for a tactical assignment by the operations section go and remain until needed (fig. 23–15). The staging area is overseen by a "staging area manager." Depending on the size and complexity of an incident, there may be several staging areas. The staging area is normally under the operations section if one has been implemented by the IC. If not, then it reports directly to the IC. How large the staging area should be depends on the size and complexity of the incident. A staging area can range to two or three fire apparatus parked on a side street near the incident to a large fenced area capable of holding 30 or more pieces of apparatus.

Fig. 23–15. Staging area. (Courtesy of Jocelyn Augustino/FEMA)

Staging can be used several ways. Depending on the type, size, and complexity of the incident, two types or levels can be used. These are level 1 staging and level 2 staging. At a small incident or early in an incident, a level 1 staging may be more appropriate. For an incident that is escalating, it might be more prudent to use a level 2 staging. This concept first appeared in Chief Alan Brunacini's *Fireground Command* text.

For level 1 staging, arriving units radio incident command of their arrival and position. Incident command then determines which units are needed at the scene and instructs the rest to stop approximately one block from the scene in the direction of travel and report their unit number and direction. For level 2 staging, all additional responders other than first alarm report to the staging area and manager until requested by command.

- ***Bases.*** The base is where most of the logistics section and finance/administration functions are coordinated. This includes dispensing of food to responders, medical treatment for responders, and apparatus maintenance and repair. According to NIMS/ICS, there is to be only one base per incident. The person in charge of the base is a **base manager**. In many cases, this base may be located next to or near the incident command post. Most Type 5 and Type 4 incidents do not establish a base, but sometimes they provide logistical functions at a given area. Examples include the following:
  - A fire in a single-family dwelling may have EMS units establish a rehab unit near the scene to monitor responders' vital signs and provide them with fluids and a rest area.
  - A fire in a moderate to large warehouse may have a rehab unit established, the same as at the dwelling fire, and also have an area set up to provide food for responders. Providers may be a local canteen, the ladies' auxiliary of a volunteer fire department, or a local eatery.
  - For a large hazardous materials incident, the functions mentioned earlier are provided, and emergency management provides a communication unit to manage radio traffic. A sleeping area may also be established for responders who go off shift but will be needed the following morning.
- ***Camps.*** These are a carryover from the influence that forestry has had on the NIMS/ICS. These will not typically be used at Type 5 and 4 incidents. Camps are where resources may be kept to support incident operations if a base is not accessible to all resources. Multiple camps may be established at an incident. The person in charge of the camp is a **camp manager**. An example is multiple agencies searching a large area for missing children. If the area covers several square miles, it would make sense to handle functions at a base that may be a distance away.
- ***Helispots.*** These are temporary locations where helicopters can safely land and take off. This is a term used under NIMS/ICS. Another term for it in the fire and EMS service is a **landing zone (LZ)**. This facility is supervised by a helispot manager. Helispots can be used to load or unload personnel, equipment, supplies, water, and so on. Medical helicopters often land at helispots to pick up victims injured at an incident.

Ground support unit. This maintains and refuels any vehicles assigned to the incident. It also provides transportation of personnel to and from the incident, if needed.

Finance/administration section. The **finance/administration section** is responsible for monitoring incident-related costs and administering any necessary procurement contracts. The IC retains responsibility for all finance-related activities until finance/administration units or the section is activated, and on most incidents it is not activated. Four units may be established: time, procurement, compensation/claims, and cost units. The person in charge of these subunits is designated as a leader (fig. 23–16).

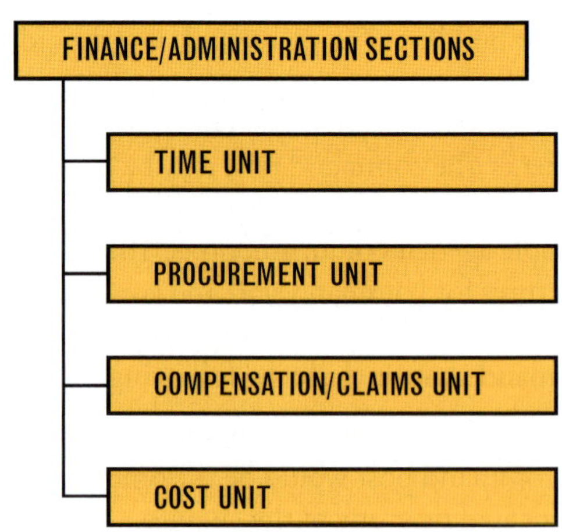

Fig. 23–16. Finance/administration section positions

- ***Time unit.*** This unit keeps track of personnel arrival and departure times at the incident to help with accountability and aids the resource unit in tracking personnel and apparatus.

- ***Procurement unit.*** This unit processes rental and purchase paperwork for specialized equipment to be used at an incident. The unit is also responsible for reporting when the equipment arrives, is used, and leaves.
- ***Compensation/claims unit.*** This unit is responsible for all workers' compensation claims reported at an incident. The unit also files all injury and illness reports associated with the incident.
- ***Cost unit.*** This unit is responsible for collecting all information relating to the cost of the incident and providing cost-savings recommendations.

COMMAND CONCERNS

Role of the incident commander

Successful incident operations require the application of a simple, straightforward management system that is effective for the incident and user-friendly for the responders. This ICS must start at the beginning of every event and only works when a strong IC performs the standard functions of command. The lack of understanding and application of this effective central-command role adversely affects incident outcomes more than any other incident challenge.

Tactical priorities

The IC must manage incidents around completing the basic tactical priorities present at all incidents. The tactical priorities in their completion order are life safety, incident stabilization, and property conservation. These standard priorities create a framework for the tactical operations required to conduct incident operations. Standard operational priorities create an accustomed order for the task-level work that must be done.

Life safety. The IC's primary responsibility on any incident is to ensure the safety of all members working on the incident scene. An effective ICS also serves double duty as the safety system the IC uses to manage firefighter safety for the strategic level of the incident organization. The IC's safety routine becomes the foundation of effective incident operations and customer service.

Firefighter safety. The ultimate test for an ICS is whether the IC can always control the position and function of all operating resources. The IC's ability to change firefighter locations and actions is the most powerful strategic-level safety capability. As an example, when the IC determines that the truss roof attic is well-involved in fire and changes the strategy from offensive to defensive, it outperforms any other safety activity or system. The IC announces the strategy change over the tactical radio channel and orders everyone out of and away from the building. After the building is evacuated, the IC gets a roll call, or personnel accountability report (PAR), to ensure all troops are out of harm. Many times the roof collapses within the next several minutes, showing why the IC must always be in control of hazard-zone workers and bosses.

The IC uses the ICS to act out the standard **risk management plan,** which determines the level of risk firefighters can take. The risk management plan is simple and straightforward:

- We will risk our lives *a lot* to protect *savable lives.*
- We will risk our lives *a little* to protect *savable property.*
- We will *not take any risks* for lives and property that are *already lost.*

The risk management plan is the major tool the IC uses to determine the incident strategy based on the critical incident factors. The IC also uses the risk management plan as the basis for keeping the strategy and IAP current throughout the entire incident operation.

Victim safety. After firefighter safety, the life safety priority extends to the search and removal of viable, endangered victims in the hazard zone and the protection of any people exposed to hazardous elements of the incident. For offensive structural firefights, the life-safety priority is achieved by performing a search in the fire occupancy and any exposures threatened by the fire. The IC uses the standard rescue order to prioritize and manage the search:

- Most endangered
- Largest group
- Remainder of fire area or structure
- Exposures

Offensive structural firefighting operations are initiated with companies advancing attack lines to the interior of the burning structure. These initial lines are directly advanced to the most hazardous area of a building—the part on fire. These initial attack crews search the most hazardous area of a structure and attempt to control and eliminate the fire. The life-safety profile improves for firefighters and victims when the fire is extinguished. Putting out the fire greatly reduces everyone's risks and buys time to complete the search in the remainder of the structure. In these situations, early fire control supports search and rescue.

The IC is responsible for assigning all resources so that a quick, effective primary search of affected structures can be managed. The IC assigns companies to search specific areas. This eliminates duplicate searches while other critical areas go unsearched. The IC manages the initial IAP while getting a primary search completed. The primary search happens quickly during the compressed time periods in the beginning stages of the incident operation. After the fire is controlled, operations can slow down. Smoke, heat, and other hazards that can impede the primary search have been eliminated, allowing companies to perform a more thorough secondary search.

Because defensive fire areas can injure and kill fully-protected firefighters, any unprotected person in these areas is considered lost (see the risk management plan). For this reason, searches are not conducted in defensive fire areas and all-clears will not be given. After a fire has been controlled and the fire area deemed safe to enter, crews can search for victims. This is considered a secondary search and recovery, not a search and rescue activity.

Incident stabilization. For structural firefighting, stabilizing the incident problem revolves around controlling and eliminating the fire. The IC manages this tactical priority by getting companies around all seven sides of the fire and overwhelming it with water. The six sides are the following:

- Top, including ceilings, attics, and floors above
- Bottom, including the floor below, crawl spaces, and basements
- All four exterior sides, including adjacent rooms, occupancies, or other buildings and concealed spaces of all those sides (walls, attics, utility chases, void spaces, build overs, etc.)

(An alternate "seventh side," referring to the interior, is used by some departments in training.)

When the forward progress of the fire is stopped and no other resource is required for fire control, the IC transmits an "under control."

Defensive fire areas have been written off as lost because of fire involvement or some other type of hazard that makes them too dangerous to enter. In these tactical situations, the IC establishes collapse areas, builds an effective incident organization to keep firefighters operating in safe positions, determines cutoff points, gets all-clears in, and protects the exposures. In defensive situations, a fire is declared under control when the main body of fire has been cut off and all savable, exposed property is protected. Defensive fires are typically fought using master stream devices.

Property conservation. After fire control is achieved, all efforts on the incident scene must be directed toward controlling and preventing unnecessary damage to property. Offensive firefighting produces a certain amount of secondary damage. Primary damage results from the fire; secondary damage results from what we do while putting out fire. Ladder/truck companies often must force entry into fire areas, cut holes in roofs to vent smoke and heat from the structure, and open walls and ceilings to check for fire extension. Engine companies stretch attack lines into buildings and operate nozzles. The IC orders and manages these tactics at the correct times, places, and in proper amounts to keep fire from further damaging the structure and its contents.

After a fire is controlled and knocked down, a loss control plan that describes how salvage and overhaul will be performed should be developed. It should include lighting a building's interior to increase safety and reduce damage from firefighters stumbling in the dark. Attention should be given to the customer's personal and irreplaceable possessions, such as pets, pictures, family heirlooms, records, and medications.

All three organizational levels must be constantly aware that our actions are designed to protect savable property. The golden rule of loss control is to treat the customer's property as you would want your family's possessions treated.

Commanding effectively

Assuming command. **FFII 5.1.1** The first unit or member arriving at an incident scene, typically an engine company officer, is responsible for assuming the role of first IC. The standard assumption of command is performed by the first-arriving officer as part of the initial radio report. The initial IC must include the phrase, "I will be assuming [insert street name] command."

The standard initial assumption of command ensures that incident operations begin with a single IC and eliminates multiple, competitive ICs. The IC must assign all later-arriving units' personnel per standard operating procedures (SOPs).

Communications. The IC begins the communication process by transmitting a standard initial radio report. The report includes the following:

- Unit ID
- Arrival to the scene
- Description of the incident problem with action taken
- Declaration of the overall incident strategy
- The assumption of command

An initial radio report for a room and contents fire in an ordinary house would sound something like this:

> "Engine 1 is on the scene. We have a working fire in a medium-sized house. Engine 1 is laying a supply line and attacking the fire with a 1¾-in. attack line. This is an offensive fire. Engine 1 will be assuming Main Street command."

The IC must control communications throughout the incident. The radio is the only tool in the IC's arsenal to stay connected to all incident responders. If the IC gets knocked off the air, incident management goes down the drain.

The assumption of command automatically triggers staging procedures for all responding resources. Staging procedures cause later-arriving units to stop short of the incident scene and announce their staged location. Staging engine companies do not pass their last water source, while staged ladder and truck companies do not pass their last access point to the scene. This places incoming resources where the IC can actually assign them to the incident operation according to the IC's plan. Following staging procedures eliminates freelancing.

Situation evaluation. In most situations, the first-arriving company officer works and operates on all three organizational levels: strategic, tactical, and task. The company officer IC must assume command and perform the functions of command. This includes sizing up the **critical factors** for the incident; developing the incident strategy and IAP; directing and leading the crew to the right places; and ensuring they take correct, safe, and effective actions (fig. 23–17).

Fig. 23–17. Before employing a strategy, the incident commander must size up the critical incident factors.

Critical factors are a list of different aspects of the incident, along with the incident hazard, that have a major effect on completing tactical priorities. The critical factors also have a major impact on firefighter safety. Critical factors vary from incident to incident. They include:

- The size, age, and condition of a building
- Occupancy type
- Fire size, intensity, extent, and location

Deployment procedures describe the cycle companies follow regarding:

- Responding to the scene of an incident
- How they are assigned to positions around the scene
- How long they can safely operate in the hazard zone
- Methods to cycle companies in and out of the hazard zone (work-rest cycles, on deck, etc.)

Transfer of command. The incident command system is designed so the fast-attacking company officer IC has time to evaluate conditions, then effectively assign the next two or three companies to the most critical areas. The command system is supported and reinforced when the first-arriving response chief assumes command from the initial IC (using SOPs) and inherits an effective, safe, and controlled operation. This upgraded IC then sets up standard command functions.

Command is transferred when a higher-ranking officer arrives on the scene and goes through the SOP transfer-of-command process. Depending on your department's procedures, command may be transferred whether over the radio or face-to-face. Face-to-face transfer is considered the most effective, as it allows the most complete incident communication. Regardless of the method, the officer who assumes command should contact the dispatch center and announce that the specific unit is now in command. For example, "Battalion 2 to alarm. Battalion 2 has transferred command from E-1 and is now Main St. Command."

Incident management is designed and applied to support and manage task-level work. The system is built from the ground up. In many cases, the initial fast-attacking IC will solve the incident problem, eliminating the need to transfer command. Command should be transferred before the current IC becomes overwhelmed. Command is typically transferred from a company officer IC to the first-arriving response chief. For many fire departments, this is the only command transfer during an incident. After a strategically-positioned IC is in charge of the incident, command should be supported and reinforced by assembling a command team.

A key to effective incident management is building the properly sized incident organization and support staff. The command team uses the section positions to delegate functional and support responsibilities. This allows the IC and command team to focus solely on managing resources in the hazard zone (fig. 23–18).

Fig. 23–18. A command team consists of the IC, support officer, and senior advisor.

Termination of command. When command is terminated, the incident priorities have been met and companies are typically released from the incident to either return to service or to quarters for servicing. During this phase, final accountability of crews and resources take place, and the status of both are considered. Documentation of incident actions, injuries, and initial stages of critique processes can also occur. In some instances, termination of command may also include a radio message to dispatch advising of the final status of the incident and if any additional priorities are being handled by remaining companies. In this case, it is important to note that if companies remain, someone will be designated as in charge of the remaining support activities taking place.

Post-incident review

Command procedures also establish the framework for reviewing the effectiveness of incident operations. Before decommitting the working units involved in the incident, the IC should conduct a standard post-incident review. Here is a list of basic post-incident items to discuss:

- What were the most significant critical factors?
- Did the strategy and IAP match the critical factors?
- Were there effective incident communications?
- Were there adequate resources (on-scene, staged, response), and were the deployment procedures followed?
- Did the assignments match the IAP?
- Did the incident organization match the problem and the resources required?
- Was the IAP current throughout the entire incident?

The knowledge, reinforcement of good standard practices, and lessons from these post-incident reviews should be integrated back into the SOPs and training to continually improve performance for all emergency incident operations. Opportunities for organizational improvement and safety analysis also exist at almost every incident. Any accident or injury "near-misses" should be discussed as a chance to increase safety awareness in the organization. A national database exists, and individuals are encouraged to submit their reports through an anonymous system at www.firefighternearmiss.com.

FIREFIGHTER'S ROLE IN THE COMMAND SYSTEM

FFII 5.1.1 Typically, most firefighters work on the task level on the incident scene. An effective IMS protects hazard-zone workers and coordinates task-level work that solves the incident's problems. No command system can outperform the nonperformance of the task level. It is imperative that all hazard-zone workers (companies and firefighters) do the following:

- Follow deployment and safety SOPs
- Be assigned and fit into the IC's plan (no freelancing)
- Stay together as a company (voice, vision, or touch)
- Manage their air supply in a way that allows them to exit the hazard zone safely *every time* they enter a hazard zone

The ICS is as much about firefighter safety as it is about effective incident management. The most effective, well-managed incidents are also the safest. Firefighters must understand and fit into ICS to make it work because its goal is to provide the best possible service and ensure that all the workers are protected and go home at the end of their shifts. Understanding some basic management principles such as **unity of command** and **chain of command** helps us know where we fit within the system. Unity of command and chain of command tell us that a firefighter is responsible for reporting to their immediate supervisor for task assignments and who their company report to IMS/ICS. Respect for the chain of command allow for more effective assignments to be made and for safe and effective incident management to take place. Freelancing should not occur during an incident because it poses a risk to individuals and the entire team working in the hazard zone. A firefighter must report observations, considerations, and hazards encountered to their company officer to properly take necessary action and to communicate to the IC.

PUTTING ICS TO WORK

The ICS is part of the daily operations of the American fire service. We should use it on all response types and, as stated earlier, not reserve its use for actual working incidents. Routine responses can be the best places to use the communication processes and principles of ICS.

The principle of modular organization illustrates that the components are used as needed and expandable according to complexity and size of the incident.

Standardizing the functions of command allows for the IC's job duties to be covered initially by the first-arriving company officer, who is then supported by a designated incident commander, who is then supported by a functioning incident management team. If the incident requires multiagency response due to multiple objectives, then the full elements of a command system in a unified fashion are implemented tracking a wide variety of objectives and resources as well as the costs and administrative details needed to manage a massive response of emergency resources.

The individual firefighter's role in ICS is to work within the system as part of the task-level response to the incident objectives. You may be assigned to an engine company, truck company, or heavy rescue, and, at any given time, be responsible for performing tactical duties as varied as fire suppression to ventilation. You will report to and take orders from your company officer who, in turn, takes orders from a sector/division or group officer.

The IC will monitor the completion of the incident objectives and reinforce or modify task-level response based on the analysis of the standard functions of command and critical fireground functions. Maintaining crew integrity and communicating progress reports through the command structure keeps IC and other task-level groups apprised of your status and welfare. Companies needing support or reinforcement should request and receive it through the ICS. The safety officer should constantly monitor your position, condition, and activities, and you should always be aware of those facts as well.

A typical fireground illustration of ICS

A scalable ICS for a structural fire response may build as follows:

A fire department is dispatched to a fire in an apartment building. The first-in company officer arrives on the scene, positions the apparatus according to SOP and delivers a brief yet concise incident report on the mobile radio to inform dispatch and incoming companies of the conditions present, actions being taken, and the needs of the incident action plan. That company may assume a fast-attacking command position, moving to the interior of the building to meet a pressing incident objective, or stay in a command-mode position on the outside of the structure to assign resources as they arrive. Your SOP will determine which is appropriate for your agency.

As a dedicated incident commander arrives, a transfer of command will take place, and critical information including progress and needs are exchanged. An upgrade of the response is requested, and the incident commander will begin to build a command team to successfully manage the quickly emerging incident needs and priorities. Subdivision of the incident into manageable areas of supervision to track crews and assignments will be necessary. IC designates division supervisors in the four quadrants of the structure and assigns companies to them. Groups are formed with specific tasks such as ventilation, rescue, and water supply. A group supervisor may be assigned and could include a company officer to fulfill that objective. Crews will need to be rotated and rehabbed, and reinforcement of tactical positions will take place. The incident management team will monitor where companies are and what their assignments are.

During the course of the incident, the IC or SO will order periodic personnel accountability reports and, when the incident is stabilized, will begin to release resources to their quarters. Crew and resource management allowed this incident to be managed effectively, and the life safety, incident stabilization, and property conservation needs of the occupants were addressed by the correct selection of an incident strategy that was executed through proper task-level or tactical operations.

A statewide or regional incident response

In a small town in the central United States, an engine company is placed on alert for a possible tornado touchdown. The engine company officer responds to a residential area of the community to deal with multiple reports of damage from the storm.

The first-arriving officer considers crew safety and the capabilities not only of his or her company, but then of the department and surrounding communities. As they arrive at the incident scene, they view mass destruction including fire, injuries, and structural damage. The first-arriving engine assumes incident command and begins assessment of the situation. It is clear that this incident is well beyond the response capabilities of almost every fire department.

The company officer has begun this response in Type 5 response and then requests mutual aid from the surrounding cities and counties. This incident has now become a Type 4 response. As the assessment continues, it is necessary to bring in even more assistance to staff an emergency operations center and field command post, and to determine many incident priorities. A Type 3 response is now active, bringing in the combined resources of the entire state and possibly even surrounding states. The incident runs for several days in a high state of operation but slowly stabilizes, and a systematic process of relieving the statewide companies down through the first-in engine company is drawn up and implemented.

Disaster aid and financial assistance from state and federal agencies is now possible thanks to the use of the NIMS, ICS, and all of the documentation of time, costs, and materials used carefully completed. A full incident debriefing can take place to identify and analyze the response to prepare for future deployments across the region.

CONCLUSION

Outside of any federal law or industry standard to use ICS, the responsibility to provide for the safety and welfare of all responders to an emergency incident falls directly to the incident commander. The standardization and modular approach used in ICS provides for a time-proven and effective solution to the safety and survival problems experienced by the fire service. A practiced and full-time utilized SOP on ICS should be a primary goal of every response organization. Each firefighter will work within the system under the direction of their company officer who is given direction and assignment at the task level of a single incident commander. As an incident expands due to size, complexity, or resource needs, the early use of the system makes it a smooth and effective expansion of the modules needed. The goal that *everyone goes home* can be achieved in great part through the effective use of ICS.

RESOURCES AND ADDITIONAL INFORMATION

Completion of NIMS training and certification at the levels of NIMS 100, 200, 700, and 800 levels may be required as part of many state firefighter training certificates. Online course work and information can be found at http://www.fema.gov/emergency/nims/NIMSTrainingCourses.shtm.

QUESTIONS

1. What is the purpose of utilizing an incident command system on everyday/every-event responses?
___

2. List the main difference between unified command and single command, and explain why all the other elements of NIMS/ICS remain the same.___
___

3. As incident operations escalate in time, size, and complexity, the strategic level responsibilities can overwhelm the command team. What can be done to address this problem?___
___

4. List one similarity and two differences between strike teams and task forces.___
___

5. What are the six units under the Logistics section of the ICS?

   i. ___

   ii. ___

   iii. ___

   iv. ___

   v. ___

   vi. ___

6. The size and scope of an incident determines if incident facilities are needed and what kind. List the five incident facilities that are recognized by NIMS/ICS and give an example of their physical locations.

   i. ___

   ii. ___

   iii. ___

   iv. ___

   v. ___

7. The IC's ability to change firefighter locations and actions is the most powerful strategic level safety capability. List one example of this.___
___

8. The IC manages the tactical priority of stabilizing the incident problem of a structure fire by getting companies around all seven sides of the fire and overwhelming it with water. What are the seven sides of the fire?

   i. ______________

   ii. ______________

   iii. ______________

   iv. ______________

   v. ______________

   vi. ______________

   vii. ______________

9. When the IC determines that he or she can no longer coordinate the strategic and tactical level of an incident, what should he or she do?____

   ____

10. In this chapter, there are seven post-incident items to discuss. List at least three of them, and give short examples of each.____

    ____

    ____

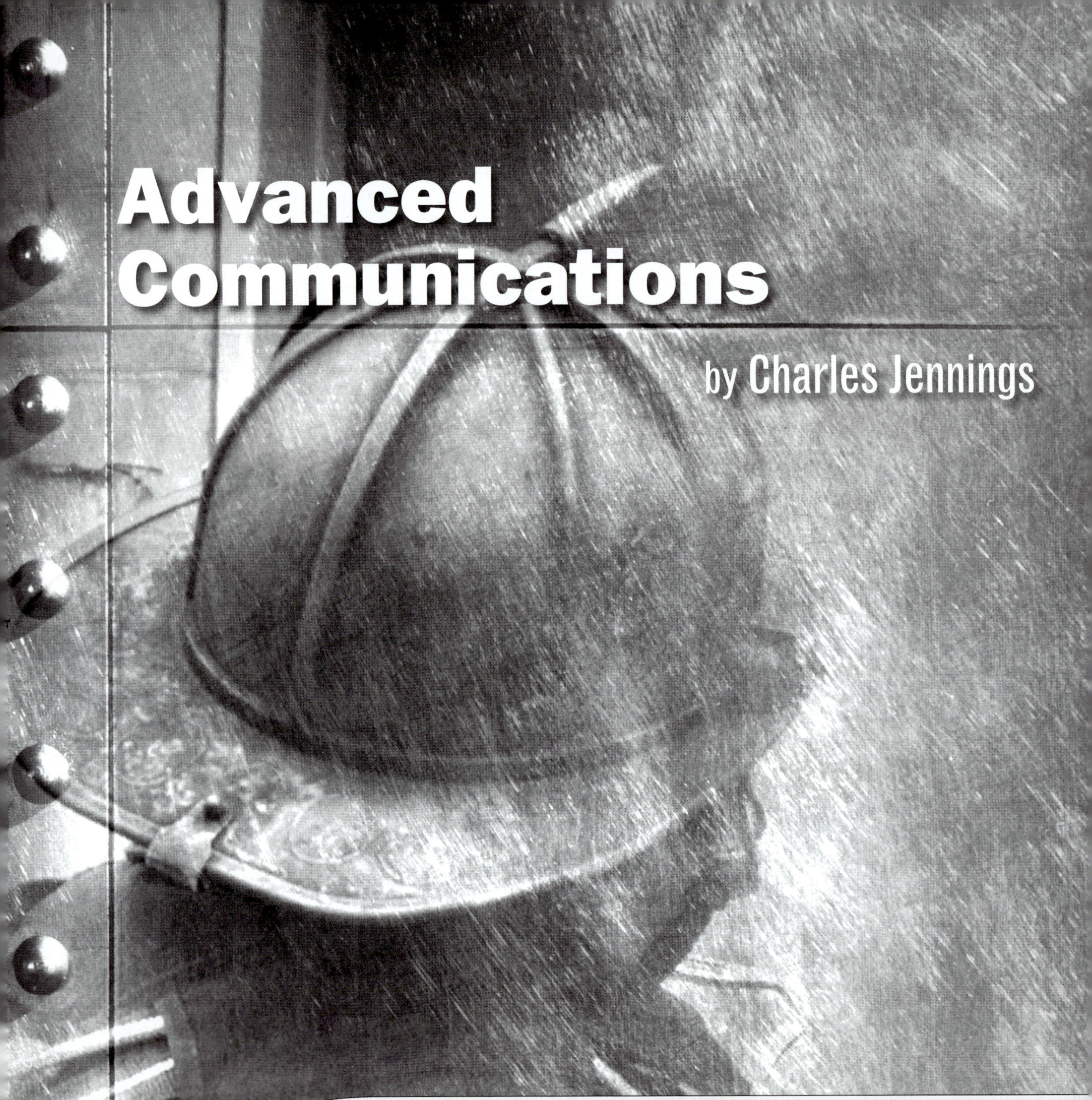

Advanced Communications

by Charles Jennings

This chapter provides required knowledge items for the following NFPA Standard 1001 Job Performance Requirements:

FFII 5.1.1

FFII 5.2.2

This chapter contains Skill Drills. When you see this icon, refer to your Skill Drill book for step-by-step instructions. SKILL DRILL

OBJECTIVES

Upon completion of this chapter, you should be able to do the following:

- Describe how fire departments are dispatched to emergency incidents
- Describe the term interoperability as it applies to incident management communications
- Describe the National Fire Incident Reporting System
- Identify standard forms and reports used by local fire departments
- Know how to communicate the need for assistance on the fireground or incident scene
- Identify emerging technologies available to the fire department to improve emergency scene communications

UNDERSTANDING COMMUNICATIONS

Communications is an important but often overlooked aspect of fire services. It is so embedded in our routine activities that we often take it for granted. On a daily basis, proper understanding and use of communication equipment, policy, and procedures can ensure smooth operations. When errors or lapses occur, the results can be major, exposing the public and fire service personnel to risk and liability and can lead to unnecessary loss of life and property. Studies of major incidents have shown that well-managed communications are essential to achieving positive outcomes. This chapter builds on information in chapter 4, Fire Department Communications.

Shannon and Weaver are credited with developing the leading model of communications using technology (fig. 24–1). Their model consists of several components. First, an **information source** produces a message. This message is sent to a **transmitter**, which encodes the message. The transmitter then sends the message via a **channel** (**radio frequency**, telephone line, etc.). During this stage, noise is introduced into the transmission. This noise can be caused by static, interference, and reflects limitations of the technology being used. The message then reaches a **receiver**, which decodes the message, where it then is received at a **destination**.[1]

This model of communications can be very useful in understanding the limitations of communications via common methods used in the fire service—telephone, radio, even face-to-face. Poor connections, background noise, and related problems can

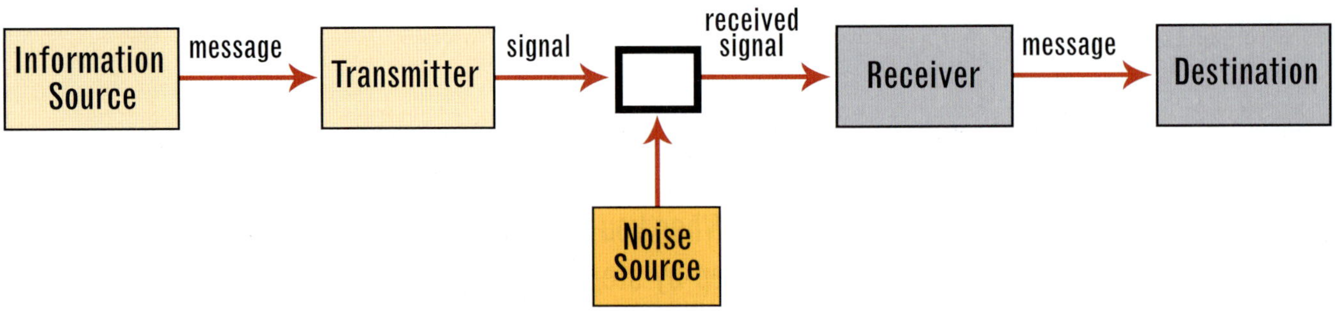

Fig. 24–1. Shannon and Weaver's model of communications

impair the ability to communicate and cause introduction of errors.

What is important to remember is that users of communications systems must be aware of the limitations of those systems, and allow for these limitations when communicating to ensure that their messages are properly transmitted and understood by the receiver at the other end of that system.

FIRE REPORTS

SKILL DRILL

FFII 5.1.1 Fire departments are required to maintain all records related to their daily operations. This includes the safeguarding of all fire and medical emergency reports along with daily activity logs. The use of computers in fire departments has made the data collection, storage, and retrieval process and the transfer of information much faster and more accurate. At a time when multiple copies of the same incident information must be filed with fire department headquarters, the station house, and other involved agencies, computer programs and applications have made compliance easier. The fire department not only retains records and information concerning emergency incidents for collection later, if needed, but also has the ability to store information concerning other business interests such has fire-prevention information and fire-investigation reports. The fire department also retains department personnel records, inventory information, and infrastructure requirements as needed by the fire agency.

Documenting the activity of your fire company is an important component of communications. A basic incident report establishes such milestones as the time of alarm and time your company went en route, arrived on the scene, and went back into service. The location of the alarm, nature of alarm, and damage or casualties are also recorded. Units equipped with MCTs may have the capability of completing reports from the field.

After an incident, it is important to be able to accurately record all of the important information about the incident.

The incident report can be handwritten or computerized. Record all of the important information such as the following:

- Time of the call
- Response time of the arriving apparatus
- Incident address
- Property owner information
- Incident outcome
- Any other information deemed important according to your observations or department procedures

To standardize reporting, incident report codes are often used to indicate information such as incident type, actions taken, area of fire origin, and property use. Written guides provide codes and explanation of those codes. Many incident report computer programs have code guides built into the reporting system. It is important to know where your department incident codes and regulation manuals are located, and whether they are in book form or incorporated into the computer system.

Many departments use an incident report based on the National Fire Incident Reporting System (NFIRS). Through state fire marshals' offices or their equivalents, state-specific versions of the NFIRS incident report are available for use throughout the United States. NFIRS-compliant reporting schemes are supported by numerous incident reporting and records-management software packages available for the fire services. The NFIRS Reference Guide can be found online at FEMA's NFIRS web page. Paper forms are also available for manual completion of these reports (fig. 24–2).[2]

Complete this side for all fires

A FDID ☆ | State ☆ | Incident Date (MM DD YYYY) ☆ | Station | Incident Number ☆ | Exposure ☆ | ☐ Delete ☐ Change | **NFIRS - 2 Fire**

B Property Details

B1 Estimated number of residential living units in building of origin *whether or not all units became involved* ☐ Not Residential

B2 Number of buildings involved ☐ Buildings not involved

B3 Acres burned (outside fires) ☐ None ☐ Less than one acre

C On-Site Materials or Products ☐ None

Complete if there were any significant amounts of commercial, industrial, energy or agricultural products or materials on the property, *whether or not they became involved*

Enter up to three codes. Check one box for each code entered.

On-site material (1)
1 ☐ Bulk storage or warehousing
2 ☐ Processing or manufacturing
3 ☐ Packaged goods for sale
4 ☐ Repair or service

On-site material (2)
1 ☐ Bulk storage or warehousing
2 ☐ Processing or manufacturing
3 ☐ Packaged goods for sale
4 ☐ Repair or service

On-site material (3)
1 ☐ Bulk storage or warehousing
2 ☐ Processing or manufacturing
3 ☐ Packaged goods for sale
4 ☐ Repair or service

D Ignition

D1 Area of fire origin ☆

D2 Heat source ☆

D3 Item first ignited ☆ 1 ☐ Check box if fire spread was confined to object of origin

D4 Type of material first ignited — Required only if item first ignited code is 00 or <70

E1 Cause of Ignition ☆

☐ Check box if this is an exposure report. ⇨ Skip to Section G

1 ☐ Intentional
2 ☐ Unintentional
3 ☐ Failure of equipment or heat source
4 ☐ Act of nature
5 ☐ Cause under investigation
U ☐ Cause undetermined after investigation

E2 Factors Contributing To Ignition ☐ None

Factor contributing to ignition (1)

Factor contributing to ignition (2)

E3 Human Factors Contributing To Ignition

Check all applicable boxes ☐ None

1 ☐ Asleep
2 ☐ Possibly impaired by alcohol or drugs
3 ☐ Unattended person
4 ☐ Possibly mentally disabled
5 ☐ Physically disabled
6 ☐ Multiple persons involved
7 ☐ Age was a factor

Estimated age of person involved

1 ☐ Male 2 ☐ Female

F1 Equipment Involved In Ignition

☐ None ⇨ If equipment was not involved, skip to Section G

Equipment Involved

Brand

Model

Serial #

Year

F2 Equipment Power

Equipment Power Source

F3 Equipment Portability

1 ☐ Portable
2 ☐ Stationary

Portable equipment normally can be moved by one person, is designed to be used in multiple locations, and requires no tools to install.

G Fire Suppression Factors

Enter up to three codes. ☐ None

Fire suppression factor (1)

Fire suppression factor (2)

Fire suppression factor (3)

H1 Mobile Property Involved

☐ None

1 ☐ Not involved in ignition, but burned
2 ☐ Involved in ignition, but did not burn
3 ☐ Involved in ignition and burned

H2 Mobile Property Type & Make

Mobile property type

Mobile property make

Mobile property model | Year

License Plate Number | State | VIN Number

Structure fire? Please be sure to complete the other side of this form.

Local Use

☐ Pre-Fire Plan Available

Some of the information presented in this report may be based upon reports from other agencies:

☐ Arson report attached
☐ Police report attached
☐ Coroner report attached
☐ Other reports attached

NFIRS-2 Revision 01/19/99

Fig. 24–2. A sample NFIRS incident report

The use of NFIRS reporting allows departments to have consistent data for comparison with other organizations, and permits analysis of response times, staffing, and fire losses in one database. Fire protection systems and occupancy data are used by code writing bodies, and by the fire service for justification of code requirements. Some organizations elect not to participate in NFIRS, primarily out of a perception that there are no benefits or because completion of reports is thought to be too time-consuming. The adoption of computer software for reporting greatly reduces the effort involved, and speeds the collection of data, making information available on an almost "real-time" basis. Functionality of software such as "pull-down" menus and context-specific help greatly speed the reporting process. Progressive organizations report their information—failure to do so holds back the entire fire service.

NFIRS-based incident reports are required for recipients of federally funded FIRE Act grants. As a nationwide standard, they contain extensive information of value in establishing trends in causes of fires, identifying consumer products with potential safety issues, establishing performance, and gathering comparative data useful for management and in justifying resources. NFIRS data are used as the basis for the development of national statistics on the fire problem by the U.S. Fire Administration. The voluntary reporting system is administered by an organization called the National Fire Information Council, composed primarily of State and local representatives of fire agencies.

Perhaps most importantly to firefighters, NFIRS data is the basis for determining trends in firefighter casualties and can be used to target equipment, activities, or types of properties that are particularly hazardous or involved in accident scenarios. The absence of accurate and complete NFIRS reports is a major impediment to a better understanding of the fire problem and firefighter safety.

NFIRS reporting includes separate modules that can be used for various situations including civilian casualties, fire service casualties, hazardous materials, emergency medical services, wildland fires, and personnel, among others.

Other types of forms and reports

Besides NFIRS reports, members of fire companies often prepare other types of forms and reports. These include equipment maintenance forms, injury forms, supply requests, and drill schedules, among others. Most fire departments have specific forms to be filled out; with the introduction of the computer, many of them have been converted to an electronic format.

One of the more extensive reports prepared by the fire department itself is an after-action report. These reports are completed after a major incident which resulted in death(s) or major property damage. These reports usually contain extensive use of narratives: essentially detailed descriptions of what took place. For example, this could involve describing how and where hoselines were placed, the description of a flashover, or a description of how a wall collapsed. Firefighters may be asked to prepare such narratives since they actually experienced the things described in the report.

It is important for firefighters to ensure that all forms and reports are properly prepared. This responsibility includes using correct grammar and spelling. All documents must be proofread. Written documents are a reflection of the author's professionalism.

COMMUNICATING THE NEED FOR TEAM ASSISTANCE

When working on the fireground or incident scene, portable radios can be extremely useful in enhancing safety for personnel who may work apart from their crew. Often, when a member of a crew is assigned to a task away from the rest of their unit, they may encounter a situation or circumstances requiring them to notify their supervisor, or on larger incidents, the operations section or incident commander. The firefighter should formulate a message to notify others on the fireground of the information and request assistance, as appropriate when such a situation is encountered.

In one example, a firefighter is assigned to operate a standpipe valve while the balance of his/her crew is advancing a line in preparation for forcible entry and fire attack. They may receive information from a civilian evacuating the building about conditions on a higher floor indicating a smoke condition or people in need of assistance. While continuing to perform the assigned task, the firefighter should inform the highest ranking person on the channel: "Engine 23-C, to operations—be advised I have a report from a civilian of a person overcome by smoke in the hallway of the sixth floor." The firefighter should wait for the message to be acknowledged for further instructions.

One of the most important communication-related activities is using the communications system to transmit the need for team assistance. These situations will be described in two cases: escalation of incidents, and distress or Mayday calls.

Firefighters should be familiar with their department's procedures and terminology for summoning additional resources or escalating an incident. They should be familiar with operation of the radio equipment and any accessories or attachments, such as speaker/microphones. The most important concept is to keep supervisory personnel and others operating on the incident scene appraised of important information. The goal is to assist in maintaining situation awareness among all persons operating at or monitoring the incident. Situation awareness is defined by Endsley as "perception of the elements in the environment within a volume of space and time, the comprehension of their meaning, the projection of their status into the near future, and the prediction of how various actions will affect the fulfillment of one's goals."[3] In plain terms, you must be aware of all those factors that bear on your effectiveness and safety at an incident. For example, before opening a hydrant valve to charge a supply line, it is important to know if the line has been clamped, or is connected to an inlet on the engine company. Failure to do so can result in a charged hose bed, an out-of-control line, and a delay in establishing a water supply.

The fireground and emergency scenes can be complex. Often, no single person has a complete understanding of all the critical information available or necessary to manage the incident. An incident commander outside a building may see fire extending into an attic, while a crew operating below may believe that they have the fire knocked down. Failure to communicate between the two may result in excess loss, risk to crews, and delays in controlling the incident.

Any change in condition, or information not readily observable to all those on the fireground and critical to successful management of the incident should be shared with the company officer. If in doubt, they should be communicated to command so that all on the incident are aware. In the event that additional help is needed, a request should be made clearly, calmly, specifying what is needed, or what conditions have evolved that would let command know that additional support may be needed. It is important that calls for escalation of an incident be routed in accordance with the chain of command, and done using the jurisdiction's terminology so that there is no confusion.

Calling the Mayday is reserved for situations where the firefighter is unable to exit a hazardous or immediate danger to life and health (IDLH) environment. This is most often due to becoming disoriented or lost, or trapped or entangled. Maintaining good situation awareness and crew integrity are the best ways to avoid such situations. Each department should have a well-defined procedure for calling a Mayday, and any communications-related components must be followed absolutely. Some portable radios are equipped with "emergency" buttons for use in such situations. Additional information on calling a Mayday can be found in chapter 17, Firefighter Safety and Survival.

EMERGING TECHNOLOGIES

New technologies hold great promise for improving fire service communications. New reporting methods will continue to evolve. Satellite phones, text-messaging devices, and **telematics** are likely to have major consequences for 9-1-1 systems in the future. For example, telematics, which is defined as the combined use of information technology and electronics (usually in mobile settings), is going to be used to generate reports of lost, stolen, or damaged vehicles directly to 9-1-1 centers.

These new devices will require investments in new technology for 9-1-1 centers, as well as changes to dispatching protocols and procedures to ensure that emergencies reported through these methods receive the appropriate responses.

Improved communications and tracking capabilities for fire service operations are likely to be introduced in the near future. These technologies include individual fireground text-messaging systems and even systems to determine the location of firefighters operating inside a building.

Some of these technologies are considered experimental, but could have a major impact on safety of firefighters if they prove to be practical.

The **Internet of Things (IoT)** is a term used to describe physical devices such as sensors, remote controlled valves, or actuators linked via wired or (usually) wireless networks. These sensors can be anything from personal health monitors, to temperature or other environmental sensors, to motion detectors that activate lights. The IoT is a cornerstone of concepts such as the "smart city," in which outputs from such devices are used to provide the basis for computer programs and real-time monitoring systems to improve physical processes such as traffic flow, energy efficiency, or even public safety issues. The IoT promises to provide great opportunities for improving fire and emergency services, especially when such technologies are linked with dispatch and incident command systems. For more information on this rapidly evolving area, refer to organizations such as the National Institute of Standards and Technology.

FirstNet, also known as the First Responder Network Authority, is a nationwide public safety broadband communication system designed or public safety priority use. The system is designed to provide interoperable communication to public safety agencies. The system was authorized by federal legislation in 2012, and is now in early stages of construction. Each state must opt in to the federal system or undertake their own projects.[4]

The FirstNet network will enable use of technologies such as smart phones, tablet computers, and applications tailored to first responders. In the near term, existing land mobile radio systems will continue to be used as the cornerstone of tactical incident communications.

OPERATIONAL COMMUNICATIONS CONCERNS FOR FIREFIGHTER II

Use of communications requires a combination of procedures, policies, and training in the use of technologies. Many agencies neglect training in such basic areas as use of radios, techniques for interfacing with microphones, and use of alarm-reporting equipment.

Users should be aware of the technology they are using and be familiar with techniques for using the equipment. Practices such as holding the microphone at an appropriate distance from the mouth, speaking in a clear and forceful tone without shouting, and avoiding transmitting in areas that will contribute to feedback (standing too near other speakers) should be adhered to. If necessary, ask personnel standing nearby to turn down or cover their speaker/microphones. Appropriate language should be used at all times.

Physical positioning of the portable radio on the user's body and in relation to protective attire can also affect the ability to transmit and receive clearly. Placing a radio in a dedicated pocket on the chest can improve transmitting capabilities over having the radio under a turnout coat at the user's waist level.

Additionally, users should always think about being succinct, calm, and clear in the verbal communication over the radio. Verbose messages, unclear language, and slang or jargon should be avoided in favor of well-considered, pointed messages. Once a report or message is given, it should not be necessary to elaborate unless asked for more information by the receiver.

Before transmitting, users should listen carefully to be sure that they are not interrupting a message between two other parties on the radio. This alone can make a great contribution to improving overall effectiveness of communications at an incident. If overloading of an operating channel is commonplace, then it is a sign of poor discipline or the possible need for additional frequencies.

Procedures for recognizing and establishing priority for urgent or emergency messages should be established in advance, and agreed upon by all agencies who will potentially be working on the same frequency at an incident. At any incident, radio channels will become overloaded, and the need to allow users to gain the attention of incident command will be critical.

UNIT AND PERSONNEL DESIGNATIONS

The question of unit and personnel designations seems trivial, but it can be very important to successful operations. Control of unit and personnel designations is usually maintained at the county or dispatch center level. Duplication of unit designations can lead to confusion, especially when those units may work together at an incident. Proper procedure for identifying units and personnel must be enforced.

For example, unit designations may be tied to a department, station, or unit type. Personnel assigned to a unit can be designated by their riding position. Usually, the designations of personnel are with letters, so that there is no confusion between personnel and equipment. That is, the officer can be designated as Engine 121-A, the driver as Engine 121-B, and other personnel as C, D, and so on. Some departments assign duties to personnel based on their riding position, which reinforces this system.

Personnel not assigned to a specific unit should be provided with a radio identifier, such as a badge number or rank and number (example: Lieutenant 7) to avoid the use of personal names on the radio. Once a person arriving at an incident is assigned, the designation should reflect the unit or function the person is assigned to.

INCIDENT MANAGEMENT

The National Incident Management System (NIMS) recognizes the importance of communications and provides guidance on how to establish communications at an incident (fig. 24–3). One of the key requirements of NIMS Incident Command is the establishment of common terminology and integration of planning for communications into the overall goals and objectives of the incident.

NIMS advocates the use of plain text in radio transmissions. Ten codes or other esoteric codes, especially when they are unique to a single agency, will compromise communications at an incident with multiple jurisdictions.

Large-scale incidents, particularly those that involve multiple agencies, typically require multiple frequencies, with command functions being on a separate channel from tactical operations. Many smaller organizations fail to establish an adequate number of frequencies at operations, and do not plan to assign appropriate resources to manage complex communications at an incident. The outcome is usually overloaded frequencies, difficulty in communicating without cross-talk and interruption by competing units. Failure to move incident scene communications off a main dispatch channel has been implicated in numerous cases where dispatch covers scene operations, resulting in impaired scene communications and "stepping on" units who may be requesting assistance.

RADIO FREQUENCY ASSIGNMENT WORKSHEET | 1. INCIDENT NAME | 2. DATE | 3. OPERATIONAL PERIOD (DATE/TIME)

4. INCIDENT ORGANIZATION

3. RADIO DATA

SOURCE	FUNCTION	CH#	FREQUENCY	BRANCH	DIVISION	DIVISION	DIVISION	BRANCH	DIVISION	DIVISION	DIVISION	BRANCH	DIVISION	DIVISION	DIVISION	INCIDENT COMMANDER	SAFETY OFFICER	OPERATIONS SECTION CHIEF	AIR OPERATIONS	AIR TACTICAL GROUP SUPERVISOR	PLANNING SECTION CHIEF	GROUND SUPPORT UNIT	BASE UNIT							COMM. CENTER	TOTAL BY REQ.
5. AGENCY ID.		CH#	FREQUENCY																												
6. TOTAL RADIOS REQUIRED																															

7. PREPARED BY (NAME/POSITION)

217 ICS (1/99) NFES 1340

Fig. 24–3. ICS Form 217 shows radio frequency assignments for all units working an incident.

INTEROPERABILITY

Interoperability is defined as the ability of emergency responders to work seamlessly with other systems or products without any special effort.[5] Wireless communications interoperability specifically refers to the ability of emergency response officials to share information via voice and data signals on demand, in real time, when needed, and as authorized.[6]

The local control of public safety radio systems often limits consideration of interoperability. Limited funding, aging equipment, and incompatible technologies already deployed in the field also hamper development of interoperability.

Solutions for communications interoperability can be as basic as providing radios for use by mutual aid partners, to programming of frequencies used by neighboring departments into your portable and mobile radios.

Gateways are devices that are used to link different frequencies together at a specific location. The gateway is often mobile and carried in a command or communications vehicle. Gateways can link wireless phones and public safety radios so that they can speak to each other. There are gateways that are hardware and others that are driven by software. The most desirable solution is to develop regional radio networks with shared channels so that communication between agencies becomes effortless and routine (fig. 24–4).

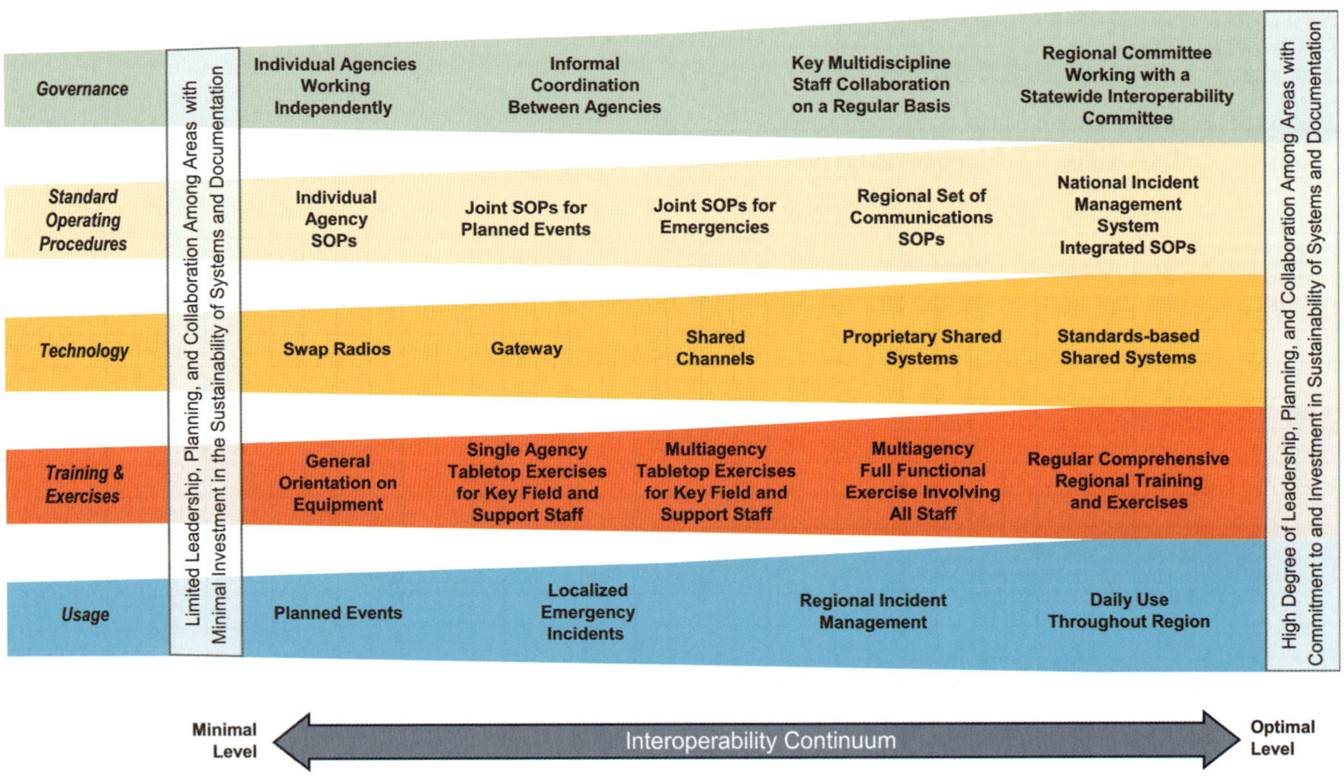

Fig. 24–4. SAFECOM's Interoperability Continuum provides a guide to assessing your organization's interoperability capabilities.

BIBLIOGRAPHY

Hawkins, Don. "Communications in the Incident Command System." *COPS Interoperable Communications Technology Program: Issue Brief Number 2,* May 2007.

Special report: Improving Firefighter Communications. USFA-TR-099. U.S. Fire Administration Technical report Series. Emmitsburg, Md.: U.S. Fire Administration, January 1999.

NOTES

1. Shannon, Claude E. and Warren Weaver. *A Mathematical Model of Communication.* Urbana, IL.: University of Illinois Press, 1949.
2. Federal Emergency Management Agency. NFIRS Reference page. https://nfirs.fema.gov/
3. Dow, Martha, Len Garis, and Larry Thomas. "Reframing Situational Awareness with the Fire Service Culture." Working paper. The Christian Regenhard Center for Emergency Response Studies, John Jay COllege of Criminal Justice of the City University of New York, October 2013. http://christianregenhardcenter.org/wp-content/uploads/2017/04/racerswp1301.pdf, 5.
4. FirstNet Facts. https://firstnet.gov/facts.
5. U.S. Department of Homeland Security. SAFECOM Continuum. https://www.dhs.gov/safecom/resources. http://www.safecomprogram.gov/SAFECOM/interoperability/default.htm.
6. Hawkins, Don. "Communications in the Incident Command System." *COPS Interoperable Communications Technology Program: Issue Brief Number 2,* May 2007.

QUESTIONS

1. Name the five steps in the communications model developed by Shannon & Weaver.
2. Why are unit and personnel designations important at an emergency operation?
3. Name three advantages of a MCT.
4. What is NFIRS, and why is it important to the fire service?
5. Explain the definition of situation awareness.
6. What is the Internet of Things? Why is it important to the future of fire service communications?

Pre-Incident Planning

by Jack J. Murphy

This chapter provides required knowledge items for the following NFPA Standard 1001 Job Performance Requirements:

FFII 5.5.3

This chapter contains Skill Drills. When you see this icon, refer to your Skill Drill book for step-by-step instructions. SKILL DRILL

OBJECTIVES

Upon completion of this chapter, you should be able to do the following:

- Define the term pre-incident plan
- Describe the 10 occupancy types identified in the International Building Code
- Identify the five types of building construction types
- Identify the benefits of pre-incident planning
- Identify the critical items to consider when conducting an exterior building reconnaissance survey
- Identify the critical items to consider when conducting an interior building reconnaissance survey
- Identify the types and components of fire protection systems found in a target hazard
- Describe the proper method for conducting a hydrant flow test
- Describe the building information card (BICard) point system
- Identify the proper symbols to use in a pre-incident planning sketch of a building

INTRODUCTION

The ability to successfully contend with a fire in a building is greatly enhanced by one critical factor: a **pre-incident plan**. Gathering detailed information about a specific building—known as a **pre-incident survey**—and formulating a plan of action *before an incident* gives a fireground commander *inside information* about the building and its contents, allowing the officer to anticipate problems and to best use his resources. A pre-incident plan that is implemented within the National Incident Management System (NIMS) structure will assist the incident commander with an incident action plan (IAP) to better reflect the overall fireground strategy, tactics, risk management, and firefighter safety.

Firefighters also benefit from gathering **building intelligence** in anticipation of preparing a pre-incident plan. By developing and improving their building intelligence-gathering skills, firefighters further enhance their tactical capabilities for actual specific buildings including anticipating fire behavior, water supply needs, search and rescue issues, forcible-entry challenges, proper hose and ladder placement, ventilation needs, and safety and survival concerns. Taking a closer look, we can see the particular benefits of this pre-incident information intelligence gathering and planning.

Pre-incident planning

Fire behavior. The basic principles of fire dynamics consider how the ignition, burning, and fire spread within a particular building relates to the characteristics of combustible materials present in the building. The need to understand building construction and the built-in fire protection features such as alarms, standpipe/sprinkler, fire-rated walls/partitions, fire doors, and so forth enable the firefighter to better understand the spread of fire and smoke within a structure.

Water supply. Water supply is a crucial component to controlling a fire. The best way to address a fire is by performing an exterior building reconnaissance to note the nearby water supply sources such as fire hydrants, the size of the water mains, and so forth. While performing an internal building reconnaissance, notice what built-in fire protection systems such as standpipe and sprinkler system can benefit to your response. We calculate the water flow requirements for a building and the actual water supply available later in the chapter.

Search and rescue. A vital component to performing a primary search for a quick investigation for victims is a good comprehension of the building floor layouts. Because the primary search is basically the first action taken on entry into a building, knowing a floor design ahead of time helps this specific phase of the search process.

Forcible entry. To carry out the fireground tasks to complete search, rescue, extinguishment, and so forth, firefighters must be able to access the structure. While performing a building reconnaissance, take a look at the various locking mechanisms, particularly new locking devices that offer more resistance to forcible entry. Take a minute and size up the situation for each; note on the reconnaissance from any locking devices that require more attention, especially multilock doors, metal gates, roll-up doors, window bars, interior security barriers, and so on.

Hose and ladder placement. Building intelligence obtained from conducting a pre-incident reconnaissance helps with proper hoseline selection, stretching, and placement within a particular building. It further assists in ensuring that the water discharge is of sufficient volume and that the line stretch can actually reach the seat of the fire.

Ventilation. When performing a building reconnaissance, look for easy paths of smoke travel to perform effective ventilation. Seek out ways to ventilate building smoke movement by vertical or horizontal means. Draw the fire away from the firefighters so they can access the building, perform a more effective search for victims, and suppress the fire.

Firefighter safety and survival. The firefighting tasks during the early fire growth stage have the highest degree of risk, and other fireground tactics that are being performed also present a level of risk. Hence, gathering building intelligence prior to an incident greatly helps in reducing firefighter death or injury.

Pre-incident planning involves evaluating the building fire protection systems, construction, contents, and operating procedures that could impact emergency operations. A pre-incident plan must address the structural integrity of walls, floors and the roof; live and dead loads; and obvious signs of deterioration or structural weakening.

The following are other conditions that could have an impact:

- Spread of the fire
- Ability of responding personnel to access the building's interior and to safely perform interior operations
- Potential for falling materials such as glass, curtain walls, exterior ornamentation, parapets, overhanging components and building exposures (fig. 25–1)

Fig. 25–1. Plenum areas above a drop ceiling may contain numerous multiple low-voltage cables that could entangle a firefighting crew in a collapse of the ceiling and its contents.

In deciding which building to target for a pre-incident plan, local fire departments typically target buildings in the community that present significant life safety concerns (for firefighters and occupants alike) or present a significant firefighting challenge. They are typically called **target hazards**.

The most important aspect of gathering building intelligence is to help responding firefighters effectively manage emergencies with available resources. This should not be confused with mandatory fire code inspections.

Accurate surveys and diagrams give firefighters the information needed to safely execute the most efficient operation. The pre-incident survey saves time during fire operations and helps firefighters prepare strategy and tactics to fight the fire.

Elements of the pre-incident survey

FFII 5.5.3 Department requirements for pre-incident surveys will vary. However, the common components typically found on the survey include the following:

- Date
- Address
- Owner/manager name
- Building name
- Floorplans
- Locations of
  - entrances/exits and internal accesses
  - utilities
  - water sources
    - hydrants
    - drafting locations
  - fire protection systems
    - fire alarm system
    - standpipe and sprinkler systems
    - fire department connections (FDCs)
- Response hazards

Accurate diagrams will provide critical information for fire personnel in the event of an incident. Diagrams can include a plot of the building's interior and accesses, as well as locations of fire alarm panel, standpipe, and sprinkler systems and water supply, interior and exterior building hazards, and any other factors that could affect firefighting operations at the location.

Gathering and distributing building intelligence

Building intelligence gathered during pre-incident survey of a structure can be accomplished by local fire companies, often accompanied by a chief officer. In the past, the plans that were subsequently developed often ended up in three-ring binders in the chief's car or on the dashboard of the fire engine. Although these **paper plans** continue to be used by many departments, electronic pre-incident plans offer quicker, more detailed information.

Today, local dispatch centers often have the ability to store and retrieve the building intelligence data and transmit the information to responding fire companies. This capability gives firefighters the necessary basic information concerning a building during a response and is followed by detailed diagrams/data and the actual pre-incident plan through mobile computer terminals in the apparatus. Distribution of building intelligence is advocated by the USFA to transform a fire department response with electronic building information card.

TYPICAL TARGET HAZARD PROPERTIES AND OCCUPANCY CONSIDERATIONS

To conduct a reconnaissance survey and identify the key elements for a building intelligence plan, it is first necessary to identify the occupancy of the building. The following occupancy types are based on the *International Building Code, 2018 Edition*:

- The various *assembly occupancies (Group A)* are divided into five subgroups for the gathering of persons from theaters, night clubs, exhibition halls, arenas, stadiums, and so forth.
- *Business (Group B) occupancies* are used for office, professional, or service transactions, including the storage of records and accounts, for businesses such as animal hospitals, clinic outpatient, laboratories, print shops, and telephone exchanges.
- *Educational (Group E) facilities* cover two basic occupancies from six or more persons at any one

time for educational purposes through grade 12 and a day care facility for more than five children older than 2½ years of age.

- *Factories (Group F)* are classified into two subgroups: industrial moderate and low-hazard occupancies. The use of the building or portion thereof is for assembling, fabricating, finishing, manufacturing, and the like. A moderate hazard occupancy may consists of aircraft, clothing, furniture, machinery, woodworking, and so on, whereas a low-hazard occupancy consists of beverages, foundries, gypsum, metal products, and so forth.
- *High-hazard (Group H) occupancies* are classified into three subgroups that include the manufacturing, processing, or storage of materials that constitute a physical or health hazard in quantities exceeding those allowed by the building and fire codes for occupancies, such as aerosol and compressed gas; combustible fibers and liquids; explosives, fireworks, flammable gas, and liquids; oxidizers; water-reactive materials; and so on.
- The various *Institution (Group I) occupancies* are classified into four subgroups in which people are cared for or live in a supervised environment, have physical limitations because of health or age, or are detained for penal or correctional purposes in assisted living facilities, halfway houses, alcohol and drug centers, hospitals, nursing homes, prisons, detention and correction centers, day-care facilities, and so forth (fig. 25–2).

Fig. 25–2. A fire in the area of the hospital where medical gases are stored may pose a threat to responding firefighters.

- *Mercantile (Group M) occupancies* are used for the display and sale of merchandise incidental to purposes that are accessible to the public for occupancy such as department stores, markets, and retail or wholesale facilities.
- The various *Residential (Group R) occupancies* are classified into four subgroups for sleeping purposes: boarding houses, hotels, and motels that are transient in nature; apartment buildings, dormitories, and hotels that are primarily permanent in nature; buildings with no more than two dwelling units and adult facilities for five or fewer candidates of any age for less than 24 hours; and residential occupancies arranged as residential care/assisted living facilities including more than five but not more than 16 occupants.
- *Storage (Group S) facilities* that are classified as not hazardous occupancies cover two basic groups: moderate hazard such as aerosols, books, furs, leather, and sugar; and low hazard for such occupancy as aircraft hangars, food products, glass, metal cabinets, and so forth.
- *Utility and miscellaneous (Group U)* are not classified in any specific occupancy but must conform to the building and fire codes as they pertain to fire and life hazard incidental to occupancies such as barns, carports, greenhouses, stables, tanks, and towers.[1]

CONDUCTING BUILDING RECONNAISSANCE SURVEYS

SKILL DRILL

Performing pre-incident **building reconnaissance** necessitates gathering information about both the exterior and interior of the structure. Having information about a building's features—such as floor layouts, exits, and construction—enhances your ability to more effectively perform a search and rescue as well ensure your own safety and survival.

Prior to conducting a pre-incident survey, it is recommended that an appointment be made with the building owner. Remember, this is not a fire inspection to enforce the fire code. Enforcement of the fire code must be done in a separate visit—usually unannounced—to the facility and identified as such. Fire inspections have **due process** protections for the building owner, including the ability to refuse entry to the inspector. Fire inspections and pre-incident surveys are two different things.

Exterior

Preceding an interior building reconnaissance, an exterior site survey should be conducted by assessing the following:

- Building address and cross streets
- Also known as (**AKA**)
- Primary and secondary entrances to the building
- Type of occupancy and construction
- Number of stories
- Primary/secondary hydrant locations as well as size of water mains; for rural area, nearest water supply
- Type of fire department connections, such as standpipe, sprinkler, combination sprinkler/standpipe; outside postindicator valves (PIV); outside stem and yoke (OS&Y)
- Obstructions to the building: trees, land depressions, overhead wires, roof setbacks, vehicle barriers, and so forth
- Exterior fire escape/fire stairs and exit discharge doors
- Window/door gates
- Key box location
- Basement exterior entrance and building exit discharge doors
- Fuel tanks, above and belowground locations
- Heating, ventilation, and air-conditioning (HVAC) air intake locations

The four sides to the building should be identified according to NIMS. The fireground sections are to be divided into four sides (A, B, C, D). The interior of a building is divided into divisions (floors, i.e., 7th division). For large buildings each division (floor) can be further reduced into subsector(s) (i.e., sectors 1, 2, 3, 4 and subsectors 1a, 1b, 1c, 1d). Identify the four sides of the building in figure 25–3, considering what is present on the various sides of the structure. Is the building attached or detached, a vacant lot or parking garage? Also note the topography of each side if different from side A, and so on.

Fig. 25–3. Building section identification in accordance with NIMS.

Design building descriptions

Side A: 236 Overlook Ave. Two-and-a-half story wood frame, two-family house

Side B: Three-story detached 20 ft. [6 m] away an ordinary construction tenement building

Side C: One-story detached (50 ft [15 m] away) mercantile stores (4)

Side D: Two-and-a-half-story wood frame detached 10 ft. [3 m] away from a two-family house.

Interior

Begin with a systematic building examination to evaluate the structure. For example, start from the roof level and work down to basement subgrade level(s).

Building statistics. Confirm the number of stories above and below grade, and identify the type of building construction (see chapter 7, Building Construction, for more details):

- A **fire-resistive** building is composed of structural members, walls, columns, beams, and divisions that are fire rated 2–4 hours with noncombustible materials.
- A **noncombustible** building is composed of metal frame, metal-clad, or concrete-block with unprotected or minimally protected metal bar joist (trusses).

- An building of **ordinary construction** is typically composed of masonry walls, while the interior framing system, including the roof and floors, are made of wood.
- A **heavy timber** building is composed of structural members that are large wooden columns and beams. They have masonry walls.
- A **wood frame** building may be composed of a balloon (vertical openings between exterior wall studs from basement level to attic) or a platform construction (floor joist and assembly are placed on top of the exterior bearing wall studs).
- A **hybrid building** is composed of two or more different construction classifications, e.g., Class-1 fire resistive, floor 1 and Class-5 wood frame, floors 2 to 5.

As shown in figure 25–4 (a, b, and c), from an ordinary construction foundation, a new addition constructed of pre-engineered cold-form galvanized steel is added to the upper levels. This will make it very difficult to identify where one section begins and the other ends. Adding a stucco exterior coating further obscures the different construction, making it more complicated for firefighters to determine how the building will react in a fire.

Fig. 25–4a. Top of building cold form steel addition

Fig. 25–4b. Lower portion of building/ordinary construction

Fig. 25–4c. Stucco exterior of the building

LIFE SAFETY FEATURES

Confirm the occupant load for the day (normal business hours), night, and weekend hours. For a residential occupancy, confirm the number of dwelling units on each floor. Identify the building staircase(s) as being open or enclosed as well as which stairs lead to the roof and basement. If access stairs (an open stair that is located between two or more common floors and separated from the main building staircase) are present, indicate what floors they serve. Identify the side(s) of the building where an exterior fire escape and/or fire stairs may be present. Identify if windows and doors are protected by bars or metal gates.

People with disabilities pose a challenge to firefighters. Knowing ahead of time the location of such a person within a building enables firefighters to help position themselves for a more effective evacuation. In many large new construction buildings, firefighters should note the **area of rescue assistance** that enables a physically challenged person to seek refuge and await instructions or assistance during an emergency evacuation. For existing buildings, local fire departments may create a safe haven area to implement an effective rescue for a person with special needs (fig. 25–5). Asking a person with special needs how best you can assist in their evacuation empowers them.

Fig. 25–5. Knowing where the areas of refuge assistance are located makes for a more effective evacuation for people with special needs.

The National Fire Incident Reporting System (NFIRS) identifies the property use as one of the following: assembly, educational, health care, detention and correction, residential, mercantile, business, industrial, utility, defense, agriculture, mining, manufacturing, processing, storage, outside or special property. If the building is a mixed-use occupancy, indicate the primary and secondary occupants.

WATER SUPPLY AND FIRE PROTECTION SYSTEMS

FFII 5.5.3 For your community water supply, if you are a suburban and urban area, identify the primary and secondary fire hydrants and the size of the water mains. Where private hydrants and private water tanks exist, identify the size of water main and tank capacity, as well as who is responsible to maintain the systems. For rural areas identify the main drafting water supply: lake, pond, pool, and river as well as the water shuttle travel time from the main water supply to portable water tanks. If your community has both types of water supplies, incorporate them into your plan. Conduct a water supply test on the hydrant system to establish the amount of water actually available. See the sidebar "Fire Flow Testing" on the following pages.

To access the water supply needed to control a fire, get an approximate structure measurement (length and width) of the building; this provides you with the needed water flows for a fire on a floor by multiplying the length by the width and dividing by 3 so as to get the total gallons per minute (gpm) per floor involvement. This also enables a fire officer and pump operator to effectively supply the water flow. If less of the total floor is involved in fire, reduce accordingly (i.e., if 25% of fire floor is involved, decrease the water supply needed; or if more than one floor is involved, increase the fire flow accordingly). For each exposure problem (building) add 50% of the 100% building involvement to the gpm figure.

When the building is equipped with a fire protection system, identify the types and particular system components that may be found.

FIRE FLOW TESTING

Fire flow testing is a relatively easy procedure. To conduct a fire **flow test**, the following equipment is needed:

- Two fire hydrants with at least one 2½-in. (65-mm) nozzle butt on each
- A map of the water supply piping serving the hydrants
- Pitot gauge
- 2½-in. (65-mm) cap gauge with bleed-off cock
- Hydrant wrench
- Ruler accurate to 1⁄16 in. (2 mm)
- Pad of paper and pencil

To conduct a flow test, follow the steps below in numerical order. (For safety, always stand *behind* an open hydrant.)

1. Select the two hydrants nearest the property in question; designate one as a *pressure hydrant* and the other as a *flow hydrant*.
2. Sketch the location of the hydrants and the mains supplying them.
3. View the area around the flow hydrant to ensure that precautions are taken to avoid property, pedestrian, and vehicular damage from discharging water.
4. Take one 2½-in. (65-mm) cap off the pressure hydrant and tighten its other caps.
5. Attach the cap gauge to pressure hydrant.
6. Open the pressure hydrant fully and bleed off trapped air.
7. Record the pressure (psi) shown on cap gauge.
8. Remove one 2½-i. (65-mm) cap from the flow hydrant and tighten its other caps.
9. Measure the orifice size (inside diameter) to the nearest 1⁄16 in. (2 mm).
10. Feel the interior of the nozzle butt where it is attached to the barrel and compare it with those pictured in figure 25–6 to determine its discharge coefficient.
11. Open flow hydrant fully.

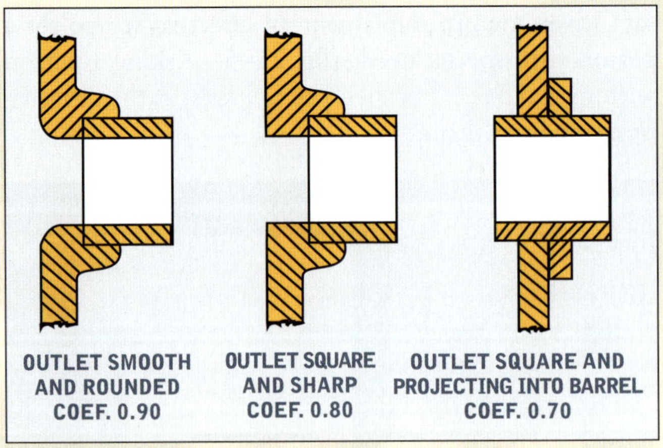

Fig. 25–6. Coefficients of discharge for hydrants

12. Wait for constant *clear* and consistent flow (no trapped air) and then place the pitot tube in the stream following the placement criteria. Place the orifice on the pitot blade in the center of the stream, and place the edge of the pitot blade one-half the diameter of the nozzle butt away from the nozzle butt. Take the velocity pressure reading (psi) and record.
13. Take and record a second pressure reading (residual pressure) on the pressure hydrant while taking the pitot reading on the flow hydrant.
    a. Additional step to increase accuracy (if necessary): If the pressure drop of the residual pressure on the pressure hydrant is less than 25% of the original static pressure, shut down the flow hydrant and open two 2½-in. (65-mm) nozzle butts and take a new set of pitot readings. Continue opening 2½-in. (65-mm) nozzle butts on additional flow hydrant(s) until the 25% figure is reached. (A pressure drop of at least 25% is a commonly accepted practice for these tests.) Remember that each open nozzle butt represents flow. You must record pitot readings for every nozzle flow to determine the total water flow.
14. Slowly close both hydrants (at the same time, open the bleed-off cock on the pressure hydrant to allow for proper drainage).
15. Verify that both hydrants have drained fully by making sure water has passed below the level of the *steamer* (pumper connection).
16. Replace caps and tighten.

After you have performed the flow test and have recorded the information, the next step is to calculate the actual flow and plot the date.

To calculate the actual flow for each hydrant butt, first convert the velocity pressure reading(s) on the pitot gauge to gallons per minute by using the pitot gauge conversion chart in figure 25–7. Find the hydrant's pitot pressure reading on the left side. Next, move along the hydrant butt orifice diameter columns until the correct hydrant butt size is reached. Record the flow specified. Finally, multiply this flow by the nozzle butt discharge coefficient (determined earlier) to find the final fire flow for that nozzle butt.

Pitot Gauge Conversion Chart

	Hydrant Butt Orifice Diameters (in inches)							
Pitot Pressure (psi)	$2^{3/8}$	$2^{7/16}$	$2^{1/2}$	$2^{9/16}$	$2^{5/8}$	$2^{3/4}$	4	$4^{1/2}$
6	412	434	457	480	503	553	1169	1480
8	476	501	527	554	581	638	1350	1709
10	532	560	590	619	650	713	1509	1910
12	583	614	646	679	712	781	1653	2093
14	630	663	698	733	769	844	1786	2260
16	673	709	746	784	822	902	1909	2416
18	714	752	791	831	872	957	2025	2562
20	752	793	834	876	919	1009	2134	2701
22	789	831	874	919	964	1058	2239	2833
24	824	868	913	960	1007	1105	2338	2959
26	858	904	951	999	1048	1150	2434	3080
28	890	938	987	1036	1088	1194	2526	3196
30	922	971	1021	1073	1126	1236	2614	3309
32	952	1003	1055	1108	1163	1276	2700	3417
34	981	1033	1087	1142	1199	1315	2783	3522
36	1010	1063	1119	1175	1233	1354	2864	3624
38	1037	1093	1149	1207	1267	1391	2942	3724
40	1064	1121	1179	1239	1300	1427	3019	3820
42	1090	1149	1209	1269	1332	1462	3093	3915
44	1116	1176	1237	1299	1363	1496	3166	4007
46	1141	1202	1264	1328	1394	1530	3237	4097
48	1166	1228	1292	1357	1424	1563	3307	4185
50	1190	1253	1318	1385	1453	1595	3375	4271
52	1213	1278	1344	1412	1482	1627	3442	4356
54	1236	1302	1370	1439	1510	1658	3507	4439
56	1259	1326	1395	1466	1538	1688	3572	4520
58	1281	1350	1420	1492	1565	1718	3635	4600
60	1303	1373	1444	1517	1592	1747	3697	4679
62	1325	1396	1468	1542	1618	1776	3758	4756
64	1346	1418	1491	1567	1644	1805	3818	4832
66	1367	1440	1515	1591	1670	1833	3877	4907
68	1388	1461	1537	1615	1695	1860	3936	4981
70	1408	1483	1560	1639	1720	1887	3993	5054
72	1428	1504	1582	1662	1744	1914	4050	5126
74	1447	1525	1604	1685	1768	1941	4106	5196
76	1467	1545	1625	1708	1792	1967	4161	5266
78	1486	1565	1647	1730	1815	1992	4215	5335
80	1505	1585	1668	1752	1838	2018	4269	5403

Fig. 25–7. Pitot gauge conversion chart

If you opened more than one hydrant butt, add all the *converted* flows together for each residual pressure reading. This is expressed in gallons per minute as some specific residual pressure.

For example, let us say fire flow testing is conducted at a paper processing plant, and two nearby hydrants were selected. The static pressure in the *pressure hydrant* is found to be 80 psi (560 kPa). A pitot gauge reading from the one open 2½-in. (65-mm) hydrant butt (with a 0.8 discharge coefficient) on the *flowing hydrant* is 45 psi (310 kPa) and the residual pressure is 60 psi (420 kPa).

With the pitot reading pressure, the flow rate from a 2½-in. (65-mm) hydrant butt is found; for the pitot reading of 45 psi (315 kPa) the flow is roughly 1,250 gpm (4,731 L/min). This is multiplied by 0.80 to find the actual flow of 1,000 gpm (3,784 L/min).

Next, plot the static and residual pressures and corresponding flows on Q1.85, semiexponential paper (see fig. 25–8). Remember, a static pressure reading means a 0.0 gpm (0.0 L/min) flow.

Finally, draw a straight line through the points and extend the line to 20 psi (140 kPa). This is the available water supply line (curve).

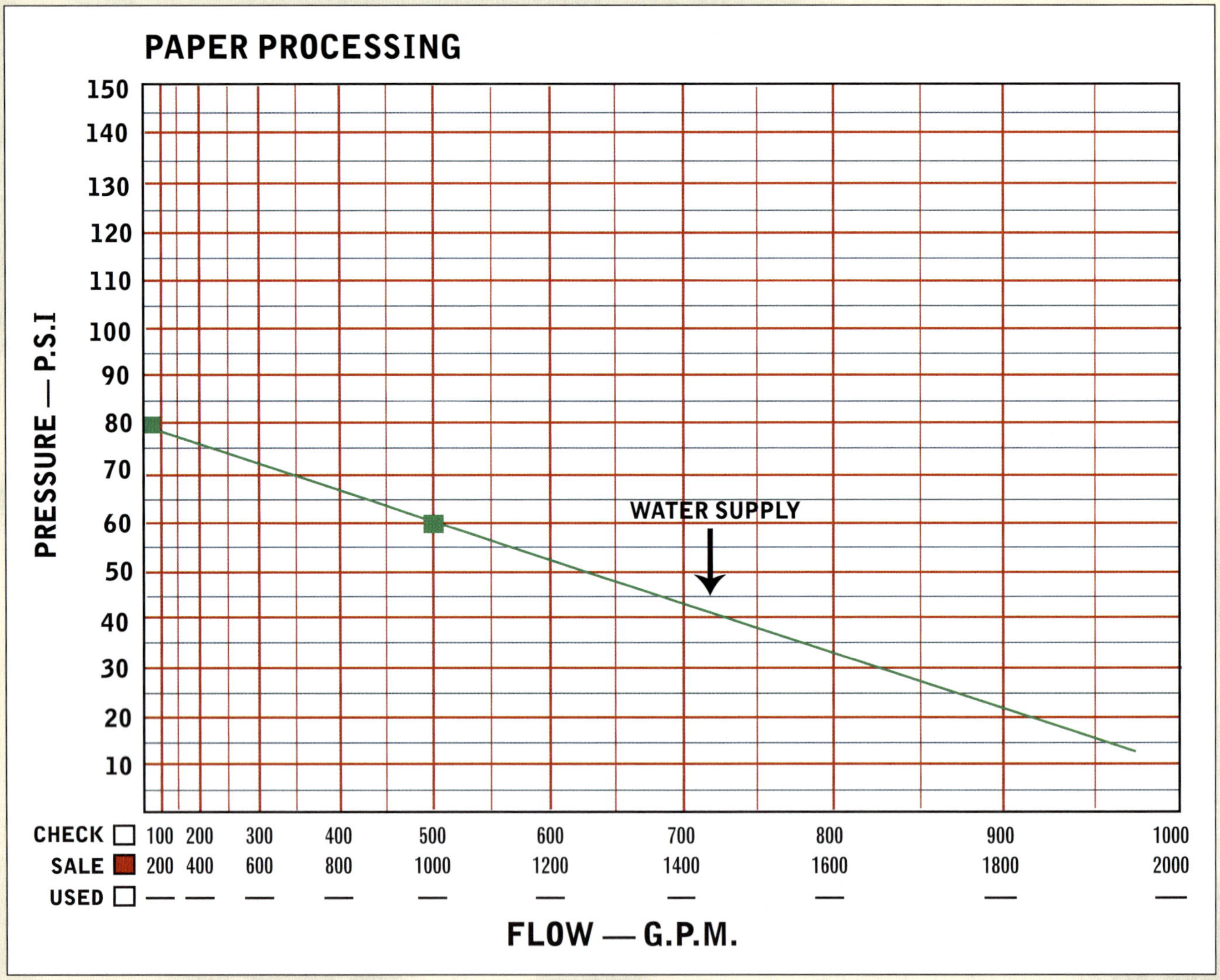

Fig. 25–8. Plot of statistical and residual pressures

Fire protection systems

Fire detection and alarm system (FAS). A building or local fire alarm system provides an audible and/or visual signal; it may range from single-station smoke detectors found in residential occupancies to a single-panel and multipanel hardwired fire alarm system. For the larger FAS systems, access the main building panel and see how the fire detection devices report to the panel so as to further assist the firefighters in locating the device that has been activated.

Sprinkler system. A building sprinkler piping system is a series of pipes that discharge water immediately from a sprinkler head once it is opened by heat from a fire. A sprinkler system may be a wet, dry, deluge, or a preaction arrangement. The system is fed from a water utility main. Identify what type of sprinkler system the building may have: full or partial. If partial, identify what area(s) it covers. A smaller **limited-area sprinkler system** consisting of 20 sprinkler heads or fewer may be tapped off the domestic water line.

Fire pump. A building fire pump is designed to increase water pressure. Where it is required, identify the fire pump's capacity in gallons per minute and the location of the pump that augments the building sprinkler and standpipe system (fig. 25–9a). For large buildings jockey pumps and pressure-reducing valves may be added to the system.

Standpipe and hose system. Standpipe and hose system are usually located within a building staircase. In a larger building, some auxiliary standpipe connections may be found within the floor area (fig. 25–9b). In either case, identify the size of the hose thread connection so firefighters can deploy hoselines.

Fire department connections (FDCs). Other than a limited sprinkler system, when a building is equipped with FDCs, the fire department augments this interior fire protection system by supplying more water off one or more supply hoselines connected to an engine company pumper. This extra water pressure supplies one of three types of systems: sprinkler, standpipe or combination sprinkler/standpipe.

Fire extinguishing systems. Fixed fire (nonwater) extinguishing systems are designed to control flammable and combustible liquids hazards. Extinguishing agents may be dry or wet chemical, carbon dioxide, clean agent, and halon (no new halon systems are being installed; they are found in older, existing systems). They protect areas such as kitchen cooking hood system, computer data centers, dip tanks, and electrical and electronic

Fig. 25–9. Building intelligence provides the firefighter with the fire pump gallons per minute and identifies the staircase where a standpipe connection can be found. (a) Fire pump (b) Standpipe connection

equipment. Identify the type of system, the location within the building, and what hazard the fire extinguishing system is protecting.

Fire wall and fire partition. A fire wall restricts the spread of fire and extends continuously from the foundation to the under side of the roof or through the roof. Fire walls essentially separate one building into two; it is the intent of a fire wall to allow complete burnout and possibly collapse on one side, but not allow fire to affect the other side of the wall. From the exterior at the roof level, fire walls will protrude no less than 18 in. (46 cm) above the roofline. However, a fire partition is a fire-rated assembly designed to restrict the spread of fire from one area of the building to an adjacent area. Openings in fire walls and partitions have openings (doors, etc.) that are protected as well as the following building components and features.

Water supply system. For urban and suburban areas, locate the nearest locations of fire hydrants to the building and the size of water main. For rural areas, determine locations of drafting areas and distance from a building to support a shuttle water system.

HAZARDOUS MATERIALS

The use, dispensing, and handling of hazardous materials within a building should be kept on the premises at an approved fire department location. Note the location and amount of hazardous material in the building. For example, a building maintenance shop may contain flammable/combustible liquids, a fuel oil tank, and so forth.

Seek support from the fire department hazmat unit to assist in assessing the hazmat risk. Each hazardous material should have a material safety data sheet (MSDS) and other hazardous material documents as required by the authority having jurisdiction.

Hazardous materials management plans (HMMP) and **hazardous materials inventory statements (HMIS)**, required under many fire codes, are particularly useful documents for pre-incident planning. They will tell you how the materials are stored/used and the quantities of the hazardous materials in the facility.

Temporary considerations

It is vital for fire department response to know when an existing building is undergoing construction alteration projects or temporary fire protection system impairments. Using an electronic building information system will provide first-due fire units with a daily building status.

Building emergency contact information

Emergency contract information for the building owner or management staff should be readily available to the fire department, particularly during the off-hour times.

Vertical risers

A building riser diagram shows firefighters the floors served by stairs, standpipe system, elevators, and HVAC zones.

Building footprint map

A grade-level building footprint map provides building intelligence as to locations of interior stairs from the main and secondary entrances to access the roof and below-grade areas, elevator cars, stairways with standpipe hose connections, the fire alarm panel/fire command center, and vertical void shafts. For the exterior it provides locations of FDC connections, fire hydrants, and building exposures.

BUILDING INFORMATION CARD (BIC) SYSTEM

Create a **BICard** to list the unique characteristic(s) within the building that are a potential threat to firefighters and/or occupants. **Safety and precautions** independently rank them according to their threat potential (e.g., truss roof: lightweight metal; 1,500 gal. (5,678 L) aboveground oil tank in basement: side D, sector 4; an open vertical roof shaft: side B, sector 2, etc.).

Building information

Identify if the building is equipped with a fire department **key box** system for off-hour access or if a **master key** system is readily available as well as the following building components and features:

- Address
- Also known as (aka: the Pentagon)
- Building occupancy
- Building population: daytime/night/weekends

Building statistics

The building's structural members should be identified according to the type: material supporting the horizontal load of the roof and floor (e.g., truss: metal/wood [light/heavyweight], cold-formed galvanized steel, poured concrete/re-bar, precast concrete, solid wood rafter/joist, laminated wood I-beams, and steel I-beams).

The floor decking may consist of precast or poured concrete, metal decking, wood such as tongue and groove planking, plywood, and oriented strand board (OSB) as well as the following building components and features:

- Number of stories
- Building height
- Building length and width
- Construction classification
- Horizontal connections (e.g., sky bridge, floor 2)
- Vertical void (e.g., basement to floor 6)

Transportation modes

Building transportation modes may consist of interior public stairways, private access/convenience stairs between the same occupancy floors, exterior fire escapes, escalators, and elevator banks.

Stairways. How is the basement access achieved? Check for an interior staircase(s) or an exterior access door. Access could be though both means. Note if the basement or other subbasement levels run the full length and width of the building or is partial. Also note any crawl space and entry locations.

How is the roof access achieved? Check the interior staircase or hatch. An exterior access may also be found via a fire escape gooseneck ladder. In some buildings, if a staircase does not enter into an attic space, look for an access port at the top floor level for entry into an attic or cockloft area (void space between the roof and the top floor apartment ceiling).

Elevator banks. When a building is equipped with an elevator or a group of elevators, note the location of the elevator mechanical room (EMR), the pit room (usually found below the elevator), and on what level the firefighter phase II key access control (described in chapter 31, Advanced Fire Attack) is located. For tall buildings, note if sky lobbies and blind shafts are present.

Ventilation systems

A large building may have mechanical equipment room (MER) for the various motorized equipment such as HVAC, air compressors, chillers, and so on. When checking the roof area, look for automatic or manual roof vents, exhaust systems, the staircase penthouse door, and elevator mechanical rooms that may assist with truck company ventilation. Note any heavy equipment on the roof such as refrigeration, HVAC units, and so forth.

Many buildings may be cooled or heated with either individual air-conditioning package units (IACPs) or an HVAC system. The IACP units usually cover a limited area or a floor. The HVAC units may cover either one floor or several floors via a duct system with fire and smoke damper to limit air movement. Some fire alarm systems may be equipped with a smoke management system to purge a smoke condition from the fire floor and pressurize the HVAC zones around the fire area. This can be achieved once the incident commander has received reports from the fire companies on the conditions in the affected areas. Other buildings may have a building management system (BMS) that automatically controls the warm and cool temperatures within a structure. The BMS system may assist the incident commander by shutting down the affected area from a remote location within the building. Understanding how the building IAPC/HVAC systems operate further assists firefighters in their efforts to control a fire.

Utilities

Utility systems may consist of building heating systems powered by electric, natural gas, liquefied petroleum/gas, oil, high-pressure utility steam lines, or a combination gas/oil supply that gives the building owner an option to select the power supply source. Other utility systems include water and steam as well as electric that may consist of alternative power sources (e.g., solar panel energy with high direct current voltage, and so on). Locating the building's main utility emergency shut-off valves and fuel tank locations helps the firefighter isolate the energy power supply source.

COLLECTING DATA AND PREPARING PRE-INCIDENT PLAN

FFII 5.5.3 To begin collecting the data for the pre-incident plan, a local fire department should develop a standard recon form to gather the building information in the field, which enables the firefighters to collect the pertinent data for a fire department response. Collecting the data should include making quick sketches of the building plot plan, access into the structure for both primary and secondary, fire department connections, stair and elevator locations, taking photos of the four sides of the building, listing the unique characteristic(s) within the building that are a potential threat to firefighters and taking notes, etc.

For additional fire protection symbols representing detection and suppression systems, see Table 30-3 in chapter 30, Fire Protection Systems. Also refer to NFPA 170, *Fire Safety and Emergency Symbols* to make the pre-incident survey document easier for users, in particular chapter 7, Symbols for Use in Pre-Incident Planning Sketches (fig. 25–10).

FIRE PROTECTION SYMBOLS

SYMBOL	DESCRIPTION
⊗	Riser
	Public fire hydrant: two hose outlet & pumper connection
(AS)	Fully sprinklered space
AS	Partially sprinklered space
NS	Non-sprinklered space
FACP	Fire alarm control panel
HD	Heat detector
SD	Smoke detector

Fig. 25–10. Some NFPA 170 symbols for use in pre-incident planning

Complete building intelligence data (Section examples)

Once the pre-incident plan recon form is complete, the building data should be entered into a building intelligence format showing only the relative information that is pertinent to this building address (fig. 25–11). should be divided into three manageable levels: (1) initial fireground operations (2) supplementary building information for the vertical risers (fig. 25–12) and building footprint map (fig. 25–13) and (3) comprehensive more detailed building information.

Depending on the fire department's data record capability (paper document (BIC) or computer data base (eBIC), the pre-incident planning information should be available for a fire department response.

BUILDING INFORMATION

Address: 1212 Hudson Street / Hoboken, New Jersey
Also Known As: (AKA) Maxwell House Building
Office Floors: 16
Public Assembly: Lobby to Mezzanine (498 OCC.)
Floor-17 Outdoor Roof Patio (120 OCC.)

Building Population: Day: 200 Night: 10 Weekend: 3

BUILDING STATISTICS

Stories: 18 Height: 235 Ft.
Length: 135 Feet X Width: 40 Feet
Construction Class: Class-1 Fire-Resistive

Truss Construction: Roof/Floors None
Horizontal Connections: Sky-bridge Floor-2 to 346 New St.
Vertical Voids: Basement to Floor-16 near Elevator Bank-A

Roof Setbacks: Floor-7 and Floor-12 (Side-C)

STAIRWAY

Designation	Floors Served	Pressurized	Standpipe
A	SB to Floor-18	No	Yes
B	B to Floor-17	No	No

Re-Entry Floors: 4, 8, 12, 16
Access/Convenience Stairs: Basement to Lobby
Floor-3 to Floor-4

Roof Access: Stair-A
Below Grade: Stair-B

ELEVATOR BANKS

Bank Designation:	Car Nos.	Floors Served:
A	1 to 2	1 to 16
B	3 to 4	1 to 17
C	5	Basement to 17

Freight Elevators: Elevator Bank-C

VENTILATION SYSTEMS

HVAC Floor Zones: 3 / 20-Ton units on Roof
Building Management System: None
Off-Site Emergency Number: None
Smoke Control System: Fire Command Center

FIRE PROTECTION SYSTEMS

Standpipe Location: Stair-A
Standpipe Isolation Valves: Stair-A
(Floor16, 8 and Sub-Basement)
Building Full Sprinkler System: Yes
Sprinkler PRV Valves: Stair-A on each floor
Fire Pump: 750 gpm off Stair-B / Room B04
Water Supply System: 24 inch fire hydrant on 2nd St

UTILITY SYSTEMS

Electric: Floor-17 Room 1701 off Stair-A
Basement Room B-03 off Stair-B
Emergency Generator: Sub-basement Room-003
Natural Gas: None
Stream: None
Water: Sub–basement Room-005

HAZARDOUS MATERIALS & LOCATION

UPS Battery Room.
Lead-acid battery rack system
Floor-14 off Stair-B MER Room 1402

Diesel Fuel
5000 Gals. Basement off Stair-A Room B-003

TEMPORARY CONSIDERATIONS BULDING

FIRE PROTECTION SYSTEM

Sprinkler system OOS / Floor-7 / Date: 1-06 to 1-30-2017

CONSTRUCTION PROJECT

Alteration project / Floor-7 / Date: 1-06 to 4/03-2017

OOS Out-of-service

EMERGENCY CONTACT INFORMATION

		Telephone
Fire/Life Safety Director	Jim Ellson	201-746-1929
Building Engineer	Eli Verasco	201-777-2212
Property Manager	Steve Auch	646-427-2224

Fig. 25–11. Sample of record of building intelligence card

Sample Building Information Card / Vertical Risers

816 Washington Street / Also Known As: **Ellson Building** **Date:**

Intermediate Level / Vertical Risers

Floors	Stairway Risers: Standpipe (A)	Stairway Risers (B)	Access Stair	Elevators Banks 1	Elevators Banks 2	Elevators Banks 3	HVAC Zones	Floors Occupancy Load	♿	Floor Plans	Floors
				EMR	EMR	EMR					
17 / Roof	SP							PA/110			17 / Roof
16 Re-entry	SP							Office/5			16 Re-entry
15	SP							Office/10	2		15
14/MER	SP						4	Office/10			14/MER
13	SP							Office/6			13
12 Re-entry	SP							Office/12			12 Re-entry
11	SP							Office/10			11
10	SP							Office/5			10
9	SP							Office/21			9
8 Re-entry	SP							Office/18			8 Re-entry
7	SP						3	Office/5	1		7
6	SP							Office/9			6
5	SP							Office/7			5
4 Re-entry	SP		4					Office/10			4 Re-entry
3	SP		3					Office/6			3
Mezzanine	SP						2	PA/498			Mezzanine
Lobby	SP		L								Lobby
Basement/MER	SP		C				1	PA/84			Basement/MER
Sub-basement	SP		Sub-C								Sub-Basement
	A	B	Access	1	2	3					
Floors	Standpipe		Stair	Elevators Banks			HVAC Zones	Floors Occupancy Load	♿	Floor Plans	Floors
	Stairway Risers										

grade level

LEGEND		
		Blind Shaft
EMR		Elevator Machine Rm.
HVAC	HVAC	Zones
MER		Mechanical Equipment Rm.
PA		Public Assembly

♿	Person With Disability
Re-entry	Stairwell Doors
SP	Standpipe outlet

Fig. 25–12. Building vertical risers

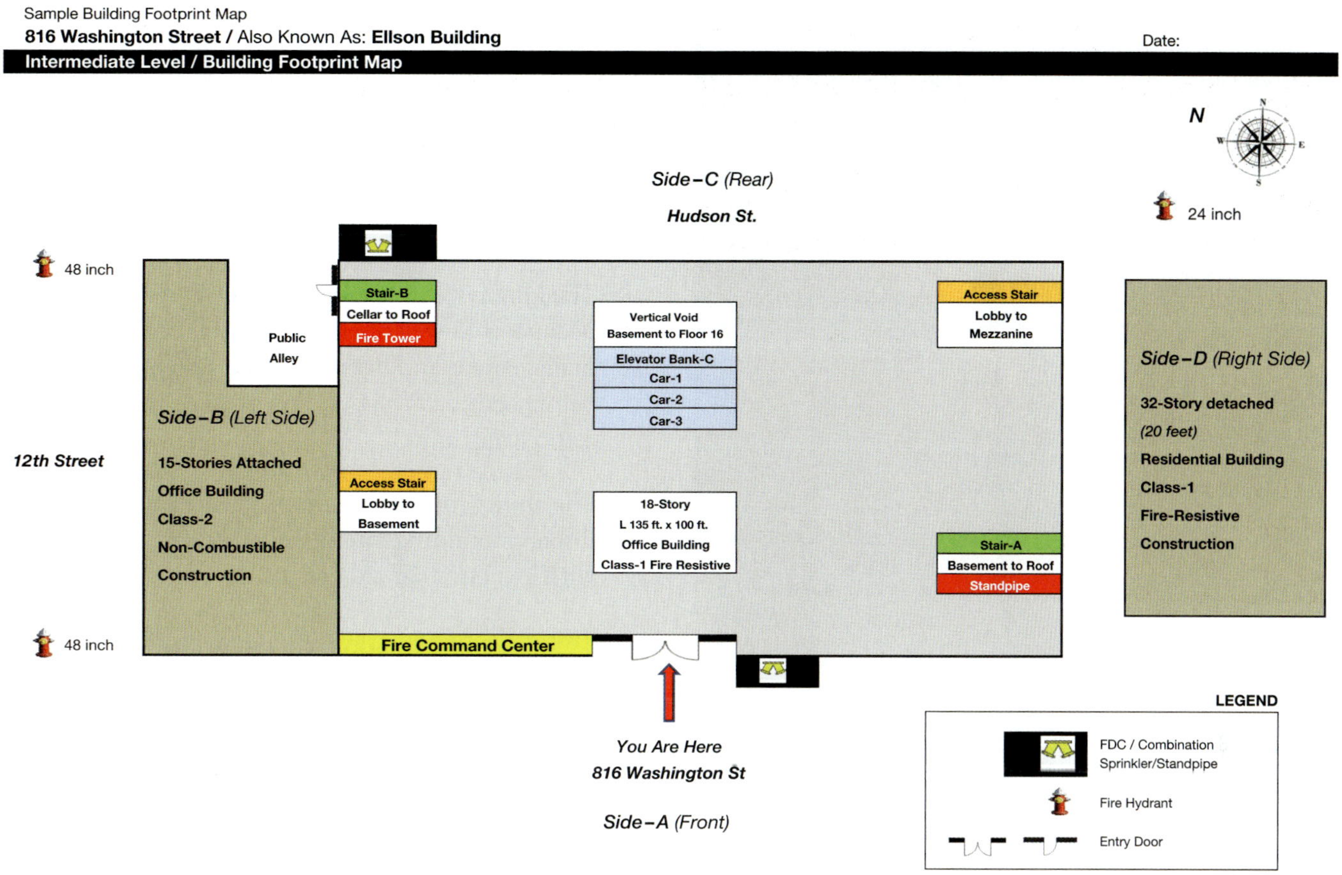

Fig. 25–13. Building footprint map

OTHER BUILDING CONSIDERATIONS

Vacant buildings

While a vacant building may not contain threats to an occupant, it does present a high risk to responding firefighters. When permitted, perform an interior building reconnaissance, and if the risk is too great, conduct an exterior ground level recon as well as a tower ladder observation from above. On completion, mark the building accordingly.

Temporary conditions

Buildings under construction, renovations, and demolitions create a distinctive challenge to firefighters (fig. 25–14). The conditions are suitable for rapid fire spread both vertically upward and downward. During a building reconnaissance, firefighters should note vehicle access; the location and accessibility of the fire department connection; that the standpipe riser is within two floors of the highest level of work; that the top of the standpipe riser is not left open without a shutoff valve; that floor outlet valves are closed, especially at the below-grade level; the storage of flammable and combustible liquids/gases; explosive materials; temporary heating equipment; cutting and welding operations; and the floor level on which the staircase ends before encountering makeshift ladders to access upper floors.

The danger of structural problems is heightened when the building is missing fire protection features such as spray-on fireproofing, fire walls and partitions, temporary erected wooden mold forms, steel bracing cables, and so forth.

Buildings undergoing floor renovations within an existing structure pose an even greater life safety threat to all occupants. Means of egress paths may also be temporarily altered but must be maintained during this phase of construction.

Fig. 25–14. A building under construction (Jersey City, NJ) creates distinctive challenges to firefighters. Conditions rapidly spread fire both upward and downward. Wind conditions may create the need for ember patrols. (Courtesy of Ron Jeffers)

BUILDING INTELLIGENCE PLANNING FOR A FIRE DEPARTMENT RESPONSE

When a complete building intelligence planning program is incorporated into the fire department management system, the local dispatch center can provide more decision-based knowledge about the building while the fire companies are responding. This "know before you go" tactic for critical building information enhances the fire officer's thinking-in-time process with the initial size-up and also empowers firefighter crews with their assigned tasks within a specific structure.

Because the fire service is a supportive organization by default, a department mission preparedness statement should include pre-incident building intelligence to further safeguard firefighter as well as occupant wellness.

PRE-INCIDENT PLANNING OF THE FUTURE

For future emergency responses, the fire service must wholeheartedly embrace a broad base-data entry and retrieval computer system that includes the following:

- Tracking systems not only for firefighters, but occupants as well operational status of building management and fire protection systems, even before arrival at the building (transmitted to responding units).
- The ability for fire communication centers to dispatch building intelligence that is also shared with local mutual-aid groups.
- This collective emergency database will enable the fire department to provide real-time decision-based information to all first responders while responding to any incident and will continue to support fireground operations throughout the event.

This collective emergency database will enable the fire department to provide real-time decision-based information to the fire and emergency services as well as the IC while responding to any incident and will continue to support fireground operations throughout the event.

NOTES

1. National Fire Protection Association. NFPA 1001, *Standard for Firefighter Professional Qualifications.*
2. International Building Code. 2018. *Use and Occupancy Classification and Building Information Card.*
3. International Fire Code (2018 edition) Building Information Card.
4. NFPA 170, *Standard for Fire Safety and Emergency Symbols*, 2006 edition.
5. NFPA 1620, *Standard for Pre-Incident Planning.*
6. United States Fire Administration (USFA), "Transforming Your Department's Response with Electronic Pre-Incident Planning," No. FM–2012-1. April 2, 2012. www.usfa.fema.gov/nfa/coffee-break.
7. Insurance Services Office (ISO), Fire Suppression Rating System, 2013 edition.

QUESTIONS

1. What is pre-incident planning? Why is it important to the incident commander?
2. What are some topics that should be included in a pre-incident plan?
3. Explain the difference between a "building reconnaissance" and a "building inspection."
4. When performing a building reconnaissance, you should look at both the ____________________ and the ________________________ of the building.
5. Name six items that should be identified while looking at the outside of the building.
6. When looking at the interior of the building, use a ______________________ approach to ensure that all areas are seen.
7. What are some unique concerns when pre-planning a building undergoing alteration/renovations construction or fire protection system impairments?
8. A quick equation to determine the need fire flow is __________ times __________ divided by _________.
9. Describe the steps when fire flow testing a hydrant system.
10. Once a building pre-incident plan is complete, what agency should receive a copy to assist the fire department when responding to the building?

Fire Protection Systems

by Bob Till

This chapter provides required knowledge items for the following NFPA Standard 1001 Job Performance Requirements:

FFII 5.5.3

OBJECTIVES

Upon completion of this chapter, you should be able to do the following:

- Identify the value of protective systems in protecting life and property and the reasons why firefighters should understand them
- Identify the common symbols used on drawings to diagram them
- Describe the basic functions and components of a fire alarm system
- Describe the basic types of fire alarm-initiation devices and how they operate
- Describe the basic types of fire alarm-notification appliances
- Explain the ways fire alarms can be transmitted to a fire department
- Identify types of sprinkler heads and how they operate
- Identify types of sprinkler systems and their main components
- Identify styles of indicating valves
- Identify standpipe classes and types of systems
- Describe problems firefighters encounter with standpipe valves in high-rise buildings
- Identify special hazard-extinguishing systems and how they work
- Identify specific problems specialized systems can introduce to responding firefighters

INTRODUCTION

Detection and suppression systems assist building occupants and fire departments. Detection systems warn occupants to flee burning buildings and can notify fire departments so fires can be safely controlled. Upon activation by a detection system, an automatic suppression system can limit fire spread and prevent unnecessary danger to building occupants by suppressing or controlling a fire before it gets large, thereby limiting smoke. These systems protect lives and property, and the public expects firefighters to know how they work.

Firefighters must know the names of detection and suppression system components. Not understanding them can result in communication breakdown between system designers and the fire service, and also within the fire service itself. These breakdowns can occur on the fireground or years before a fire when the system is initially designed. A firefighter can't speak with authority about a component or understand another's description of it if he or she doesn't know its name. Part of identifying components is knowing what they look like, but it's also important to know the different names they may be called. As with other areas of firefighting, communication breakdowns can cause injuries and death to firefighters and victims. Once you know something's name, you can get to how it works, and when it doesn't work properly, you can communicate that correctly.

FIRE ALARM SYSTEMS

FFII 5.5.3 Buildings protected by fire alarm systems are called **protected premises**. Without fire-detection systems, occupants might not discover a fire until they are overcome by smoke or heat, and fire departments might not arrive in time to save lives and property. Firefighters must know how systems work, the language to identify their components, and how to keep them reliably in service.

Fire alarm systems send and receive signals. They can do this electrically, visibly, audibly, wirelessly, or by other means. An **alarm signal** indicates that an emergency condition needs immediate attention. The two basic kinds of alarm signals are the following:

1. Local alarm signals
2. Remotely transmitted alarm signals

A **local alarm signal** notifies a building's occupants of fire. A **remotely transmitted alarm signal** notifies the fire department directly or indirectly of fire (for example, indirectly through a separate monitoring facility such as a central station, described below). Fire alarm systems also generate trouble, supervisory, and annunciation signals. A **trouble signal** identifies system problems that can influence the system's reliability. A **supervisory signal** provides a way of monitoring an alarm system status and indicating abnormal conditions.

Fire alarm system components

Key components of fire alarm systems are described using a set of terms defined by NFPA 72, the *National Fire Alarm Code*.[1] These include:

Alarm-initiating devices: automatic or manual mechanisms that cause a system to indicate an alarm condition.

Alarm-notification appliances: mechanisms that make building occupants aware of an alarm condition.

Fire alarm control panel (FACP) or **fire alarm control unit (FACU)**: the brains of these systems that connect and supervise all of this hardware, or the main control unit for a fire alarm system.

The following three sections discuss how alarm-initiating devices work and activate the alarm-notification appliances through the FACP (fig. 26–1).

Fig. 26–1. Fire alarm control panel

Alarm-initiating devices

An **alarm-initiation device** is a hardware component used to determine if there is a fire. There are many kinds of alarm-initiating devices and technologies. People activate some, such as manual-pull stations, and others automatically activate when they detect smoke or heat. The method used to connect and control these devices helps determine what a fire department knows about the fire when firefighters arrive.

Some initiating devices use similar ways to detect fires but work in different areas. **Spot-type detectors** detect fire at a particular point or location. Typically, these devices are heat, smoke, or rate-of-rise detectors. A **linear detector** is a device in which detection is continuous along a path, essentially a cable that may run through a large portion of a building or loop around the ceiling of a room.

Manual-pull stations. **Manual-pull stations** are really just switches. Most of the time they are off, but if there is a fire, people flip them on to sound an alarm. NFPA 72 refers to them as **manual fire alarm boxes**, or equipment that activates a fire alarm by opening or closing a circuit. Even system designers, however, call them pull stations. There are two kinds.

Single-action pull stations (fig. 26–2) are manual fire alarm boxes that require only one step to activate, usually pulling down a switch or handle.

Double-action pull stations (fig. 26–3) are manual fire alarm boxes that require two steps to activate. Generally one step, such as lifting a cover or breaking a glass, exposes another switch that must be activated. The double action helps prevent **false alarms**, the activation of fire alarm systems in the absence of an emergency or fire condition.

One might be tempted to call a manual-pull station a form of spot detector. It can be difficult to pinpoint fire in a building based on the location of the activated pull station. Someone far from the fire might smell smoke and pull the station as he or she leaves the building.

Fig. 26–2. Alarm-initiating devices—single-action pull station

Fig. 26–3. Alarm-initiating devices—double-action pull station

Heat detectors. A **heat detector** warns of a fire condition by sensing a high temperature, a high rate of rise in temperature, or both. A **fixed-temperature detector** activates when a predetermined high temperature is reached. Fixed-temperature detectors can be spot-type detectors or linear detectors (fig. 26–4). (The most common form of heat detector is the sprinkler head, which we will discuss later.)

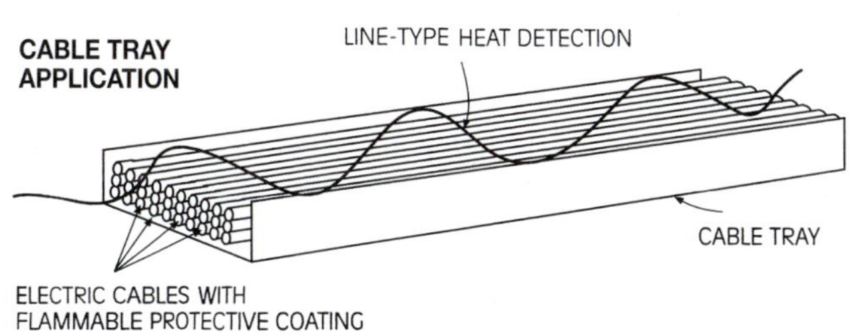

Fig. 26–4. Alarm-initiating devices—linear heat detector

Another type of heat detector, a **rate-of-rise** detector (fig. 26–5), activates when the temperature surrounding it rises faster than a predetermined value. Rate-of-rise detectors are almost always spot-type detectors, although some cutting-edge, fiber-optic, linear detectors can be put into a rate-of-rise mode.

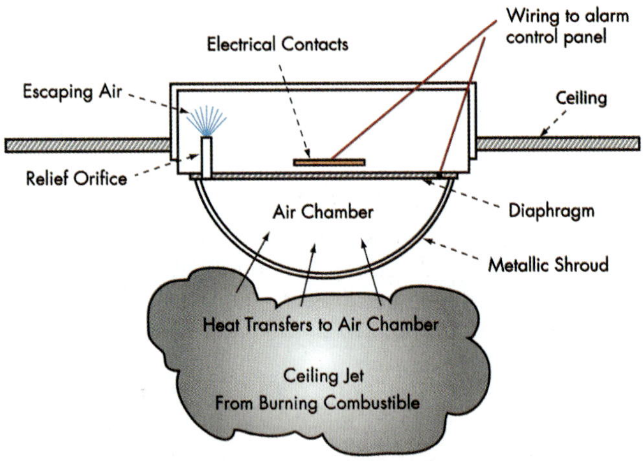

Fig. 26–5. Alarm-initiating devices—rate-of-rise detector. (Courtesy of John Morrison)

Another type of heat detector called a **rate compensation detector** (fig. 26–6) accurately senses surrounding air temperature regardless of temperature change in the air around it.

Fig. 26–6. Alarm-initiating devices—rate compensation detector

This kind of detector can be particularly useful in spaces where large temperature changes or cooling air movements occur.

Smoke detectors. A **smoke detector** (fig. 26–7) activates when it detects products of combustion, commonly called smoke. The most familiar is a **single-station smoke alarm device**, a combination smoke detector and alarm available in stores and used typically in existing private homes without detection-system code requirements. These units run on batteries and have built-in alarms. When they are activated, a local alarm sounds. They are commonly used in existing dwellings.

Fig. 26–7. A typical spot-type smoke detector

NFPA 72 has a number of requirements for new residences, including the following requirements for device locations:

- At least one device located on each level
- One device in each bedroom
- One device in the corridor outside each sleeping area

Newer residential systems are required to have all devices interconnected so that they can sound when one is activated. Newer systems also require a connection to household current.

Smoke detectors used in most modern buildings do not have built-in alarms. When they activate, they send a signal to an FACP, which then activates notification appliances throughout the building.

There are two basic types of smoke detector; ionization and photoelectric.

Ionization smoke detectors (fig. 26–8) use a small amount of radioactive material, usually americium, to charge the air passing through two oppositely charged electrodes. They are always spot detectors because the electrodes must be very close to each another. When the detector is powered, conductance between the electrodes is constantly measured.

Without smoke, this charge allows a constant current to flow between the electrodes. If smoke presents between the electrodes, this current drops. Circuits in the detector sense this, and the detector activates.

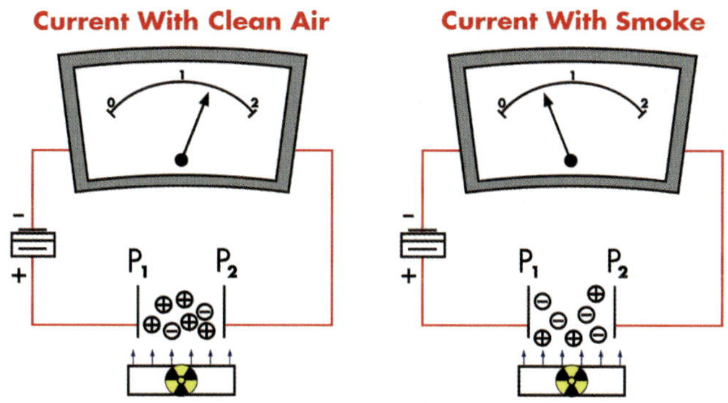

Fig. 26–8. Operation of an ionization smoke detector. (Courtesy of John Morrison)

Photoelectric smoke detectors use light beams to determine if smoke is present. There are two types of photoelectric detectors.

The spot-type generally uses **light scattering** (fig. 26–9). Smoke's deflection of the light beam sends the detector into alarm. Light normally projects inside the detector. Only when smoke is present does it reflect onto a sensor that activates the detector.

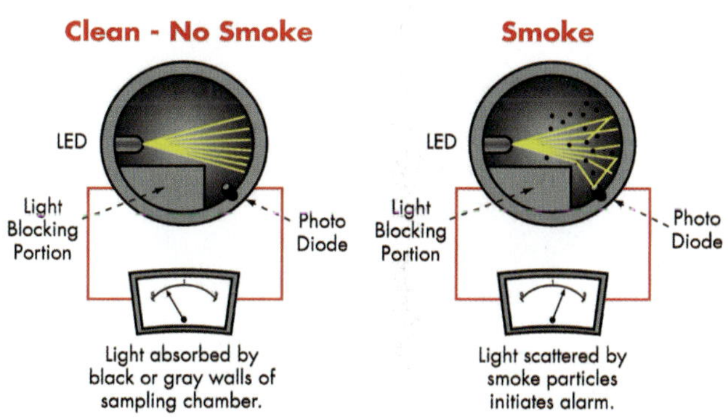

Fig. 26–9. Operation of a light-scattering smoke detector. (Courtesy of John Morrison)

The other type of photoelectric smoke detector is the **photoelectric light-obscuration smoke detector**, which measures how much of a light beam is blocked by smoke. This method is commonly used in a **projected-beam detector** (fig. 26–10), a type of linear photoelectric light-obscuration smoke detector in which a light beam is projected from a sending unit to a receiving unit. These detectors are not spot detectors. They are used to cover large areas such as an office building's atrium or a warehouse; the sending unit is mounted on one wall, and the receiving unit is mounted directly across on the opposite wall. When smoke blocks the beam, the receiver detects a reduction in transmitted light and actuates the alarm.

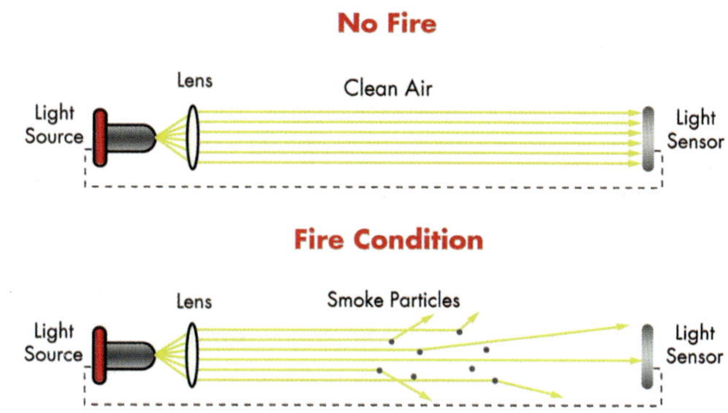

Fig. 26–10. Operation of a projected-beam smoke detector. (Courtesy of John Morrison)

An **air-sampling detector** (fig. 26–11), sometimes called a VESDA® system, uses suction tubes to draw air from different locations within a protected space. Air drawn into a smoke detector or gas-analysis device determines if fire is present (fig. 26–12). VESDA was one of the first innovators in this area, and the name sometimes describes the technology, although there are other vendors. Such systems are very sensitive, and often the tubes are mounted inside electrical cabinets and other spaces where rapid detection is necessary to prevent damage.

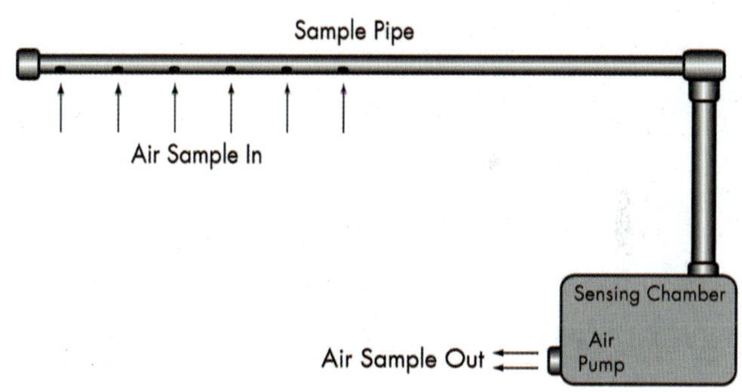

Fig. 26–11. Operation of an air-sampling detector. (Courtesy of John Morrison)

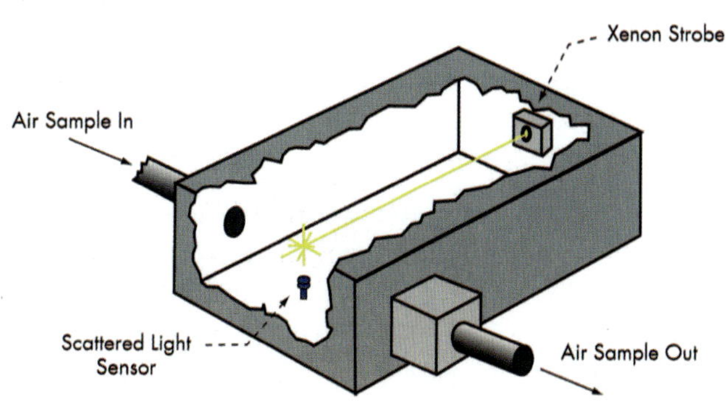

Fig. 26–12. Operation of an air-sampling detector sensing chamber. (Courtesy of John Morrison)

Gas detectors. **Gas detectors** detect dangerous chemical components a fire produces. **Single-station carbon monoxide (CO) detectors** are common in residential occupancies and detect the incomplete burning of solid, liquid, and gaseous fuels. They can detect CO exposure from cars running in connected garages and improperly vented appliances, such as furnaces and water heaters, as well as from fires. Vendors are introducing units to connect to building fire alarm systems as well.

Flame detectors. A **flame detector** activates an alarm when it detects radiant energy. There are two types: infrared, often called IR, and ultraviolet, often called UV (fig. 26–13). Sometimes, light from other sources causes unwanted alarms. Sunlight can accidentally activate an IR detector. Lightning, welding, and other electrical arcing can activate a UV detector. Combination UV/IR detectors ensure alarms activate only when they detect light from actual fires. The different kinds of flame detectors can look similar. Check a detector's information plate to determine its type.

Fig. 26–13. Alarm-initiating devices—UV detector

Alarm verification. If allowed by local authorities, some fire detection and alarm systems can reduce false alarms by allowing confirmation of smoke alarm operation. Sometimes alarms are confirmed by resetting smoke detectors that have activated and waiting for reactivation. If reactivation occurs within a given time, the system notification devices are activated. Another way alarms are confirmed is to require smoke detectors to be activated for a minimum time before accepting them as valid alarm-initiation signals and activating notification devices. Alarm verification is typically used on systems for commercial structures like office buildings. It can prevent frequent false activation of systems that can result in occupants ignoring alarms during an actual fire (the system crying wolf). However, they can also delay fire department response. The use of alarm verification should be carefully considered.

Some systems use a **cross-zoned alarm system** in which two detection devices must activate before initiating a fire alarm condition. This is commonly used where a suppression system will be activated to avoid false activation of the system.

Notification appliances

When a manual- or automatic-detection device activates, there must be a way to notify the building's occupants of a fire and evacuate them. **Alarm-notification appliances** accomplish this. There are two types: Audible appliances that are heard, and visible appliances that are seen.

Audible appliances include bells, horns, sirens, and speakers (figs. 26–14, 26–15, and 26–16). A **temporal 3 pattern** is a standard audible alarm signal to alert a building's occupants to evacuate.

Fig. 26–14. Alarm-notification appliances—alarm bell

Fig. 26–15. Alarm-notification appliances—horn

Fig. 26–16. Alarm-notification appliances—siren

Visible-notification devices commonly called strobes (fig. 26–17) warn occupants visually of an emergency condition.

Fig. 26–17. Alarm-notification appliances—strobe

Types of fire alarm systems

NFPA 72 defines a fire alarm system as: "A system or portion of a combination system that consists of components and circuits arranged to monitor and annunciate the status of fire alarm or supervisory signal-initiating devices and to initiate the appropriate response to those signals."[2]

We just discussed "the components and circuits arranged to monitor and annunciate the status of a fire alarm." They are the alarm-initiating devices, alarm-notification devices, and FACP. We even discussed supervisory signals. We haven't discussed how everything ties together to "initiate the appropriate response to those signals."

1. How is the fire department notified of a fire? What happens to the signal when an FACP receives word that a detector has activated?
2. What happens if the FACP indicates a trouble condition? Who, if anyone, is directly notified, and where are they located?

Fire alarm systems can be configured a couple of ways to ensure these things happen. There are trade-offs in cost and reliability, and different protected areas have different requirements. The following are the basic types of fire alarm systems:

- Local fire alarm systems
- Auxiliary municipal fire alarm systems
- Remote supervising-station fire alarm systems
- Central-station fire alarm systems
- Proprietary supervising-station systems

Local fire alarm systems. A **local fire alarm system** only warns occupants of a building that there is a fire. It sounds an alarm within the building where it is activated but does not send a signal out of the building. Occupants learn of the fire when a detector activates. A local alarm system does not call a fire department. Building occupants must contact the fire department after they exit the building. Signs often located near or on fire alarm bells remind occupants to call the fire department.

Auxiliary fire alarm system. When a system transmits a fire alarm signal through to local fire service communication center using the same equipment and methods as alarms transmitted from street level alarm boxes, it is called an **auxiliary fire alarm system** (as described in chapter 28, Advanced Fire Department Communications). A system like this is usually created by connecting the set of fire detection devices inside the protected building to a "master box" located on the outside of the building. The master box acts in a similar manner to a manually activated municipal fire alarm system pull box. However, instead of being manually activated, it is activated by the building detection system. When activated it transmits a unique code through the municipal fire alarm system wiring to fire department headquarters when a fire detection device within the protected building is activated.

Remote-supervising station. When a system transmits alarm, supervisory, and trouble signals from a protected location to a remote location where suitable action is taken, it is called a **remote supervising-station alarm system.** It allows trouble and supervisory signals to go elsewhere than the alarm signals. With these systems, the supervising and trouble signals can be transmitted to a maintenance facility while the alarm signals go to an alarm-dispatch facility. These requirements are less strict than those required of a central-station system, described below.

Central station. A **central-station alarm system** is one in which the operations of the circuits and devices are:

1. Transmitted automatically to a central station
2. Recorded in a central station
3. Maintained by a central station
4. Supervised from a central station

The station is staffed by operators who receive a signal and take action as required by the National Fire Alarm Code, such as notifying the fire department. All the devices and their supervision and trouble signals going to one place make these systems different from remote supervising stations.

Proprietary. A system that transmits a signal from a protected facility to a monitoring station operated by the facility's owner is a **proprietary system**. These systems are common on college and office campuses, where the same group manages and is responsible for protecting many buildings.

Fire alarm annunciation systems

Fire alarm annunciation systems notify the fire department and others concerned of a fire's location (table 26–1). Some systems provide no information at the FACP about a fire's location within the protected building. These systems are called **non-coded alarms**. Other systems annunciate a fire's **zone**, the area within a protected facility in which a signal such as a fire alarm usually can be received or a form of control can be maintained, such as activating notification appliances. For example, a zone might be the third floor, or it might be more specific depending on the design of the system. A **zoned coded alarm**, according to NFPA 72, is "an audible or visible signal that conveys several discrete bits or units of information. Notification signal examples are numbered strokes of an impact-type appliance and numbered flashes of a visible appliance." In other words, a coded device creates a signal based on pulses of light or sound that indicate which zone is involved. A zoned non-coded alarm system tells a firefighter what zone it is in at the FACP or **remote annunciation panel**, a secondary FACP generally more conveniently located than the main panel such as at the main entrance to the building. A zoned coded system indicates what zone the fire is in on the FACP and through a coded audio or visual signal.

Table 26–1. Types of alarm coding

Category	Description
Non-coded alarm	No information is given on the location of the activated device.
Zoned non-coded alarm	The fire alarm control panel shows the zone where the activated device is located, and it may indicate the specific device that was activated.
Zoned coded alarm	The fire alarm system indicates over an audible device the zone that has been activated. This type of system is often used in special occupancies like hospitals and prisons, where special evacuation strategies are used, and some occupants will not be evacuated.
Master coded alarm	The system is both zoned and coded, but the audible devices are used for a number of purposes, such as a change of class periods in a school. These systems are not installed in new buildings, and are being replaced.

Some newer systems have addressable devices. Again, using the NFPA definition, an **addressable device** is "a fire alarm system component with discrete identification that can have its status individually identified or that is used to individually control other functions." Think of each device, whether it's a detector, horn, strobe, some switch attached to the sprinkler system, or something else, as a simple computer on a large network. The computer isn't that smart, but it can continuously talk to the main computer on the network, which is the FACP. If it is a detection device, it can send a message that it has been activated to the FACP. If it is an initiation appliance, it may receive a message from the FACP to sound or light up. An addressable system has the following advantages:

1. Firefighters can quickly identify the specific detection devices that have activated when they arrive at a building.
2. The devices and appliances can continuously tell the panel they are working properly. In other words, the panel can supervise them.
3. Because each device or appliance has its own network address, the exact location of every activated device can be determined by looking at the FACP.

4. People servicing the system know which detectors need to be serviced because the detectors either report their own failures to the FACP or do not communicate with the panel.
5. If the designers want only certain appliances such as horns or strobes in specific areas of large buildings activated when the panel is in alarm, it is easily achieved because only certain network addresses can be activated.

Emergency voice alarm and fire department communication systems. **Emergency voice alarm** service is used when verbal instructions can be helpful to speed up evacuation and provide direction to building occupants. Voice alarm systems provide a recorded message that gives instructions to building occupants after receiving a detection signal. **Fire department communications systems**—often taking the form of a fixed, dedicated "firefighters telephone system" permit two way communications between firefighters in different locations, including stairwells, elevator cabs, and elevator lobbies in high-rise buildings for example.

Firefighting considerations for fire alarm systems

There are some things that should be thought about when working with fire alarm systems. These include:

- Making sure you don't reset the system until the device of origin has been determined. Modern smoke detectors usually have a solid red light on the individual detection device that indicates the device has been activated.
- Use the silence switch on the panel only if you are sure that the activation is a false alarm.
- Modern panels can usually store a history of alarm activations. You can usually scroll through them to see the history of activations.
- Systems with addressable devices will be able to tell you the exact device that has been activated at the FACP.
- Older zoned systems will only be able to tell you what zone has been activated. With zoned non-coded fire alarm systems you will only be able to find out what zone has been activated at the FACP. With coded fire alarm systems you can still tell what zone has been activated at the FACP, but you can also use the audio or visual signal provided to tell you what zone has been activated. You have to know how to interpret the code though—yet another reason to make sure you know the buildings in your jurisdiction.
- More recent versions of the fire alarm code allow for the use of wireless fire alarms. Wireless devices sometimes can easily be added to older alarm systems. Battery life of devices may now be a concern in situations where they are deployed.

FIRE SUPPRESSION SYSTEMS

FFII 5.5.3 **Fire-extinguishing agents** are liquid, solid, or gaseous materials applied to extinguish or control fires. All suppression systems in this chapter do nothing but ensure that a fire-extinguishing agent is put on a fire. Detection systems that work with suppression systems ensure an agent is applied as soon as a fire's existence is established.

Fire suppression systems can extinguish or control fires without firefighter intervention, and therefore, save lives and property. They can be an excellent first defense against fire. The public expects that firefighters understand how these systems operate. The most common automatic suppression systems are automatic sprinkler systems.

Water as an extinguishing agent

Water is the most common suppression agent applied by fire departments and activated sprinkler systems. Water puts out fires by removing heat from a fire, lowering the fire's temperature and thereby collapsing side one of the tetrahedron. Most of the temperature change is caused by water changing to steam; however the water temperature increases too, and that also removes heat from the fire.

Sprinkler systems

An **automatic sprinkler system**[3] is a system of pipes and sprinkler heads that, when activated, discharge water onto a fire without involving people. Most systems use **automatic sprinkler heads** that open when heated to a certain temperature. Deluge systems use open heads and rely on valves to open when nearby detection systems activate. Sprinkler systems are designed and installed under NFPA 13, *Standard for the Installation of Sprinkler Systems*. All water-based suppression systems are inspected according to **NFPA 25, *Standard for the***

Inspection, Testing, and Maintenance of Water-Based Fire Protection Systems.[4]

Sprinklers and life safety. Sprinkler systems improve life safety by preventing flashover and fire spread outside of the room of origin.[5] This suppresses smoke, the number-one killer of people in fires. Even when a sprinkler system is not able to extinguish, the fire is usually confined to the room in which it started. This makes firefighters' jobs much easier.

Sprinkler heads. An **automatic sprinkler head** is essentially a heat detector that activates upon reaching a predetermined temperature. Unlike the heat detectors we have discussed, automatic sprinkler heads connect to piping. When the heads open, the pipes release water in them. Because sprinkler heads around a fire heat first, water flows to them and quickly controls and possibly extinguishes the fire.

Sprinkler operating elements. A sprinkler head that opens when the solder holding two pieces of metal together fuses or melts is called a **fusible link sprinkler head** (fig. 26–18). This fusing releases two levers that prevent water flow from the sprinkler head. Table 26–2 contains a list of sprinkler operating temperatures. A sprinkler head with a glass bulb works similarly. The glass bulb that prevents water from flowing breaks when it reaches the activation temperature. This type of sprinkler is called a **frangible bulb sprinkler head**.

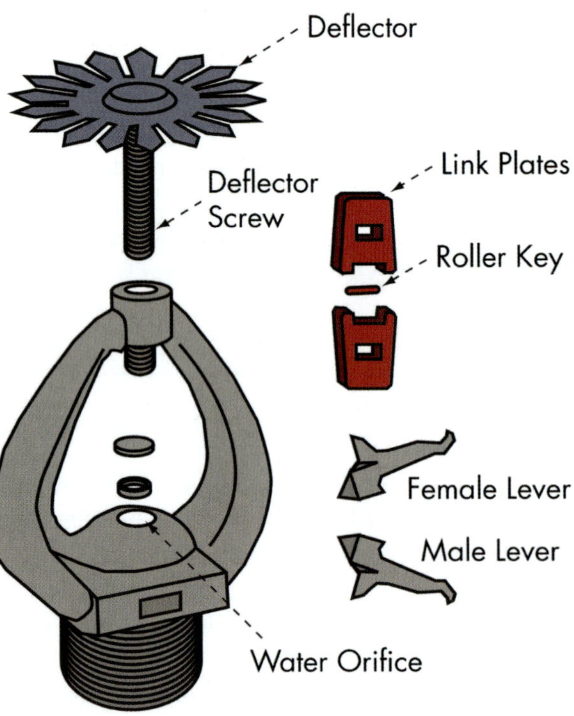

Fig. 26–18. Operation of a fusible-link sprinkler. (Courtesy of John Morrison)

Table 26–2. Typical sprinkler head activation temperatures

Temperature rating °F	Temperature rating °C	Temperature classification	Color coding	Glass bulb colors
135–170	57–77	Ordinary	Uncolored or black	Orange or red
175–225	79–107	Intermediate	White	Yellow or green
250–300	121–149	High	Blue	Blue
325–375	163–191	Extra high	Red	Purple
400–475	204–246	Very extra high	Green	Black
500–575	260–302	Ultra high	Orange	Black
625	343	Ultra high	Orange	Black

Sprinkler deflectors. All sprinklers have **deflectors** to break the water stream into droplets and push them to cover an area larger than a straight stream of water would.

Sprinkler heads are not distinguished only by operating elements. They are also identified by mounting position, either upright or pendent. An **upright sprinkler head** (fig. 26–19) is mounted above sprinkler piping, so a large deflector must divert the upward-flowing water stream to the floor. Upright sprinklers have the gravitational advantage that any dirt, rust, or debris in the pipes probably will not accumulate in the head to block the flow of water.

Fig. 26–19. Upright frangible-bulb sprinkler head. (Courtesy of Viking)

A **pendent sprinkler head** (figs. 26–20 and 26–21) is mounted on the underside of sprinkler piping. As with all modern sprinkler heads, deflectors direct the water stream down. Deflectors tend to be much smaller than in upright sprinklers because they only break water into droplets to increase coverage and do not change flow direction. Pendent sprinklers are common. They mount to unseen pipes via ceiling holes often covered with caps. These caps, when heated, drop to expose the sprinkler heads.

Fig. 26–20. Pendent frangible-bulb sprinkler head. (Courtesy of Viking)

Fig. 26–21. Pendent fusible-link sprinkler head

Sidewall sprinklers are designed for installation on one wall of a rectangular room. The sidewall sprinkler has a distinct deflector that distributes water sideways along the length of a room. These types of heads are often found in hotel guest rooms.

In buildings where sprinklers were installed before 1953, you may encounter **old-style sprinkler heads**. This type of head directs some water toward the ceiling. These are different from modern sprinklers that direct the entire flow down. Old-style heads discharge about half the water to the ceiling and half on the fire. Newer sprinklers flow almost all water to the floor, where it can control fires.

Special sprinkler heads. **Dry sprinkler heads** are designed for installation in areas subject to freezing. An extension has a seal in the inlet end that prevents water from entering the extension until the sprinkler operates. A **dry pendent sprinkler** consists of a standard pendent sprinkler head attached to a section of pipe known as a drop nipple. A shaft inside the top of the drop nipple holds a cap that prevents water from entering. As a result, the inside of the drop nipple is normally dry, so it can't freeze. When the sprinkler head opens, the shaft holding the cap releases and allows water to flow through the nipple to the sprinkler head and the refrigerated room. Because water flows only through the drop nipples where the sprinkler head fused, only the activated sprinkler/drop nipple assemblies must be replaced.

Early-suppression, fast-response (ESFR) sprinkler heads quickly discharge a density of water droplet sizes that provide fast suppression. They are usually used in warehouse rack storage and other high-challenge areas. ESFR sprinkler heads are installed to specifically avoid the problems presented by sprinklers built into storage racks. In-rack sprinklers can damaged when stock is moved on and off the racks. In-rack sprinkler heads also make rearrangement of the racks extremely difficult, since rearrangement of the racks would require rearrangement of the sprinkler pipes as well.

A **deluge sprinkler head** has no release mechanism. This type of head is always open. They are used in special systems where all the heads flow at the same time. Deluge systems are uncommon because they require a lot of water, but they are used in special situations—for example when sprinkler systems are used to apply foam onto a hazard.

A **residential sprinkler** activates quickly to enhance occupant survivability in the room of origin. As residential spacing of heads is usually limited, they are designed to control fires with as few heads activating as possible.

Types of sprinkler systems

There are a few types of sprinkler systems. Each has its own purpose and uses a discussed sprinkler head:

- A **wet-pipe system** is a system of pipes filled with water that flow water onto a fire when an automatic sprinkler head is fused.
- A **dry-pipe system** is similar to a wet-pipe system except its piping system is filled with air or nitrogen instead of water. Water still flows from fused heads, but the pipes must fill with water first. These systems are commonly used when freezing is a consideration.

- A **deluge system** consists of pipes with open sprinkler heads. All heads flow when a deluge system is activated. Such systems are used in high-hazard areas where much water is necessary to extinguish a fire.
- A **preaction system** uses elements of deluge and dry-pipe systems and minimizes water damage. It resembles a dry-pipe system except it requires another detection system and a sprinkler head's fusing for water to flow.
- A **residential system** prevents flashover in rooms of origin, improves chances for occupant escape, and prevents fire spread to adjacent rooms.

Wet-pipe systems. The most common and simple sprinkler is the **wet-pipe sprinkler system** (fig. 26–22). A series of automatic sprinkler heads connects to a system of pipes. If the sprinkler head opens because it is near a fire, it flows water and suppresses the fire. An alarm system activates in two basic ways when a wet-pipe system starts to flow water.

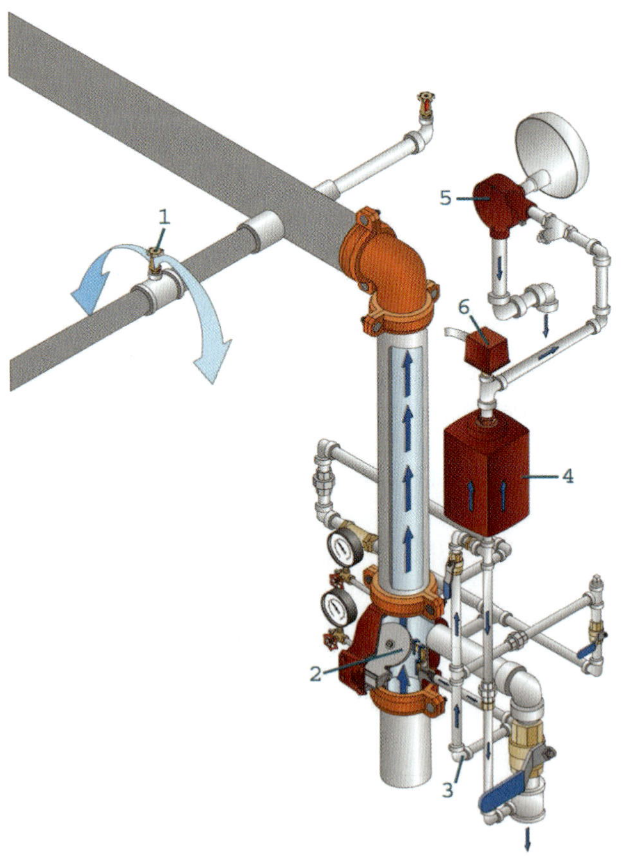

Fig. 26–22. Operation of a wet-pipe sprinkler system. When a sprinkler (1) opens, the discharging water lifts the alarm valve clapper (2) and flows through the alarm port (3) to the retard chamber (4). When the retard chamber is filled, water flows to the water motor alarm (5) and/or the optional pressure switch (6) which signals an electric alarm bell. (Courtesy of Viking)

The first method is from a water-flow indicator or **flow switch** that changes position when fluid such as water flows past it (fig. 26–23). A flow switch uses a paddle or vane to detect water flow in sprinkler pipes. These water-flow indicators normally contain a retard feature that prevents the water-flow indicator from sending a signal as a result of pressure surges in the system. It is distinct from a flowing sprinkler head.

The second method is from a **sprinkler alarm valve** that indicates water flow. This is essentially a **check valve** that permits flow in only one direction and detects that flow into the full set of sprinkler system pipes. It can activate an electric alarm device or operate a **water motor gong**, a type of local **water-flow alarm** powered by water moving through a sprinkler system. A water-flow detector or alarm-check valve activates the water-flow alarm. Sprinkler alarm valves are also subject to pressure surges. A **retard chamber** between the sprinkler valve and water-flow alarm prevents false alarms by allowing pressure surges to dissipate before setting off the water-flow alarm.

Fig. 26–23. Flow switch

Dry-pipe systems. A dry-pipe sprinkler system (fig. 26–24) is similar to a wet-pipe system and is used generally where freezing is an issue. Without water, the pipes can't freeze. Instead of all the pipes in the system being filled with water, they are filled with compressed air or nitrogen. When a sprinkler head fuses, the gas leaks out and water flows into the pipe.

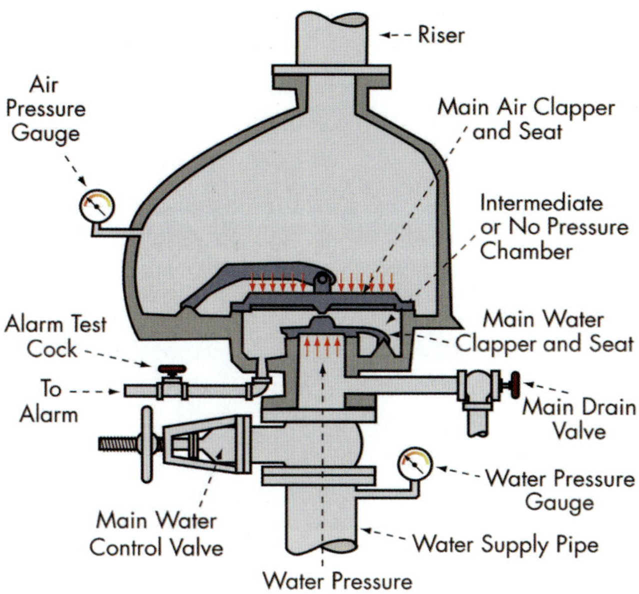

Fig. 26–24. Dry-pipe valve. (Courtesy of John Morrison)

A dry-pipe sprinkler system uses a **dry-pipe valve.** Balanced forces of air and water on opposing sides of the valve prevent it from opening under normal circumstances. Usually the water pressure is much higher, but pressure is a force exerted on an area, and a larger area of the valve face is exposed on the gas side. The forces balance, and the valve opens only when the air pressure drops.

When a sprinkler head opens, the pressure on the gas side decreases because all the air in the system leaks out the sprinkler head. Because the force from the gas side decreases, the forces are no longer balanced and the dry-pipe valve opens. **Accelerators** and **exhausters** speed opening of the valve and expulsion of air from the pipes so water can flow. They activate when the system is activated.

Dry-pipe systems are supervised by measuring the gas pressure within the sprinkler pipes. Under normal circumstances a small amount of gas will leak from the pipes over time. A pressure maintenance device (typically an air compressor) is used to add pressure back into the pipes when it drops below a certain level. This air is added so that the system will not discharge accidentally due to low pressure in the piping system. The device can add only a small amount of air and will not influence normal system activation.

Deluge systems. A **deluge sprinkler system** (fig. 26–25) uses open sprinkler heads called deluge sprinkler heads. The deluge system connects to a local detection system. The type of initiating device is based on the hazard being protected and so it might be a smoke, heat, or flame detector (the system can also be activated manually). When the initiating device is activated, it signals the fire alarm control panel, which in turn opens the deluge valve, allowing water to flow through the piping to the open sprinkler heads. Because all heads flow simultaneously, a large volume of water may be required to function properly, and fire pumps are often necessary to maintain pressure.

Preaction systems. Preaction systems (fig. 26–26) use elements of deluge and dry-pipe systems. A preaction system resembles a dry-pipe system, except it requires another detection system (typically using smoke detectors as initiating devices) for water to flow in addition to a sprinkler head's fusing. Water flows onto a fire only if the detection system opens the preaction valve and a sprinkler head fuses. In other words, two methods of detection, typically a smoke detector and a sprinkler head must, activate for water to flow, which minimizes the possibility of water damage.

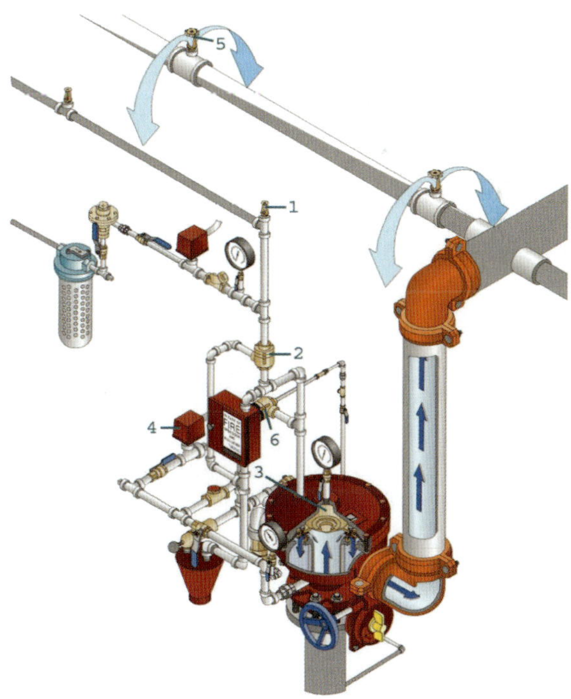

Fig. 26–25. Operation of a deluge system. (Courtesy of Viking)

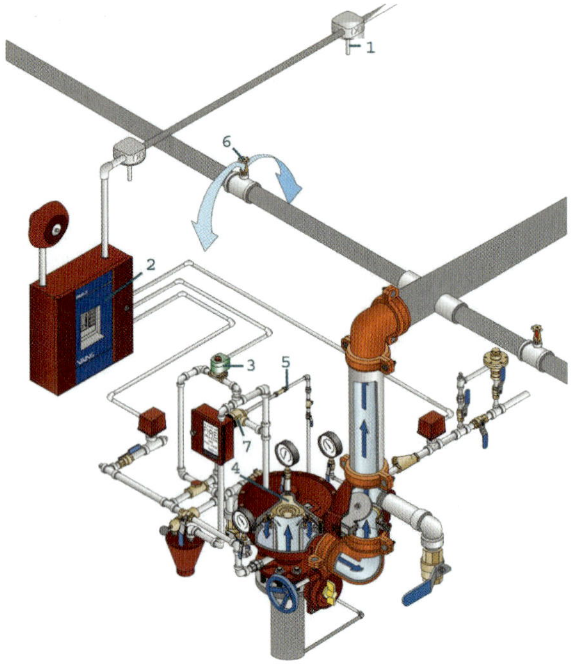

Fig. 26–26. Operation of a preaction system—single interlock electric release. When the detector (1) is activated by fire, a signal is sent to the release control panel (2). The panel sends appropriate alarm and trouble signals and, at the same time, signals the release of the solenoid valve (3). The deluge valve priming chamber (4) is then vented faster than water is supplied through the restricted orifice(5), allowing the deluge valve to open. The water enters the system piping, but until a sprinkler (6) activates, no water is discharged. When the deluge valve operates, pressure closes the PSOV (7) cutting off the supply of water to the priming chamber, latching the deluge valve in the open position. (Courtesy of Viking)

Residential systems. A residential system prevents flashover in rooms of origin, improves chances for occupant escape, and prevents fire spread to adjacent rooms in residential occupancies such as single- or two-family homes, as well as low-rise apartment buildings and hotels. **Residential sprinkler systems** (fig. 26–27) in homes and other occupancies are becoming more common. Their main objectives are to prevent flashover in rooms of origin, improve chances for occupant escape, and prevent fire spread to adjacent rooms. These systems have fewer heads compared with conventional sprinkler systems and an entire system can easily be seen in one figure. In one- and two-family homes, these systems often share a common water supply with the home's domestic water system (sinks, toilets, etc.). Sprinkler heads are omitted in small closets and bathrooms, as well as unused attics in residential systems.

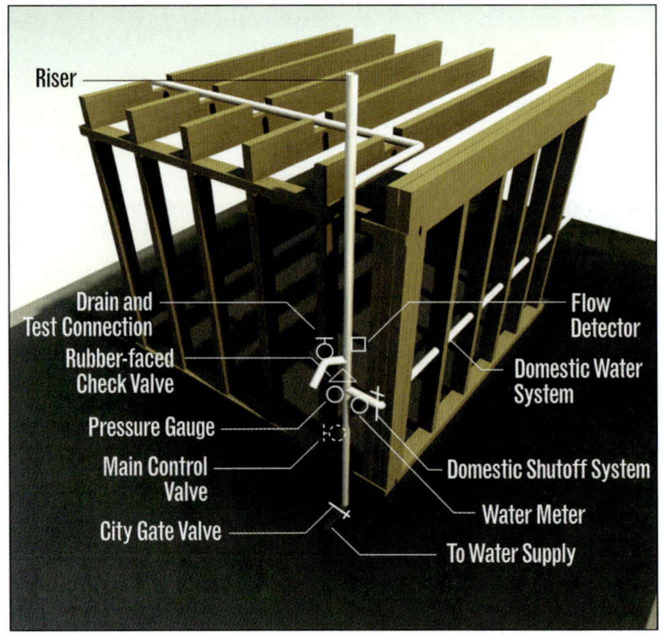

Fig. 26–27. Typical residential sprinkler system

Sprinkler system connections, piping, and control devices. Sprinkler systems need reliable water supplies. To ensure a supply is available, a continuous series of pipes from the source to the flowing head must be built and maintained. The main water supply may come from public or private companies, tanks on an owner's property, and local ponds, streams, or other bodies of water.

Fig. 26–28. Electric fire pump. (Courtesy of Paul Dansbach)

A fire pump might be necessary to maintain pressure within the activated system (fig. 26–28). This type of pump is permanently installed and connected to the sprinkler system. They are activated when a pressure drop caused by water flow is detected upstream of the pump. This type of fire pump is usually located in the basement of the building being protected. A much smaller capacity high-pressure pump called a jockey pump is sometimes installed next to the main pump. Its job is to maintain the pressure upstream of the pump so that small pressure variations don't cause main fire pump activation.

A fire department can supplement the water supply using a **fire department connection**, also called a **Siamese connection** (fig. 26–29). If the fire department supplements the water supply used by the sprinkler system, it is best to use a supply separate from the one the sprinkler system uses.

Fig. 26–29. FD connection

Every sprinkler system must have at least one main water supply control valve. In the United States, sprinkler systems fail about 5% of the time (one time out of every 20). The major reason for failure is that the main water supply control valve has been left shut accidentally. This valve must be an indicating type, meaning a firefighter can tell if the valve is open or closed just by looking at it. The two most common examples are the **outside stem and yoke (OS&Y) valve** that exposes a stem to indicate it is open (figs. 26–30 and 26–31) and the **post-indicator valve** that reads either open or shut depending on its position (fig. 26–32). Similarly, **wall post indicator valves** also provide view panels with signs to indicate "open" or "shut" (fig. 26–33).

Fig. 26–30. OS&Y valve with tamper switch

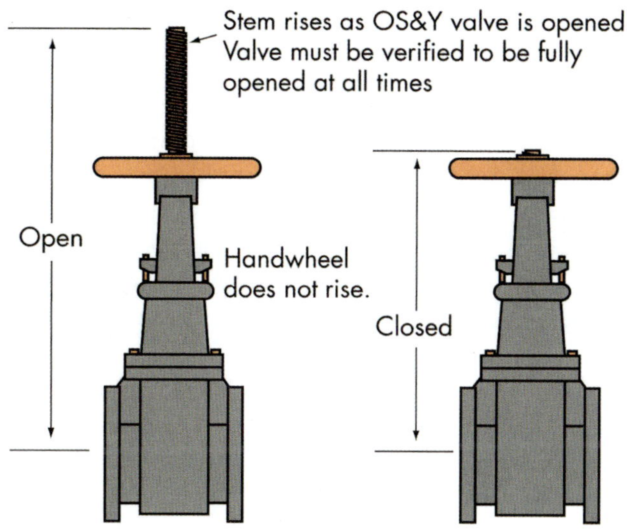

Fig. 26–31. Operation of an OS&Y valve. (Courtesy of John Morrison)

Fig. 26–32. Post-indicator valve

Fig. 26–33. Wall post indicator valve. (Courtesy of Paul Dansbach)

Supply-control valves are also supervised using **tamper switches** on valves that notify the FACP of a trouble condition if the valve is in the wrong position.

The main sprinkler system valve is found on all of the previously mentioned systems: residential, wet, dry, deluge, or preaction. Main sprinkler valves come in three types: alarm valve, dry-pipe, or deluge. The valve is attached to a riser that connects to the **sprinkler piping**, the network of pipes in a sprinkler system that delivers water to sprinkler heads. The sprinkler system has pipes called crossmains and branchlines if the system is a tree system (fig. 26–34); crossmains and branchlines if it is a grid system (fig. 26–35); and crossmains and loopmains if it is a looped system (fig. 26–36). Gridded systems and looped systems are used in many instances because they are more efficient than tree systems; they allow water flow to an outlet from more than one direction. All pipe systems except grid systems can be dry because the volume inside these grid systems takes too long to empty the air (or nitrogen in some cases) from the system.

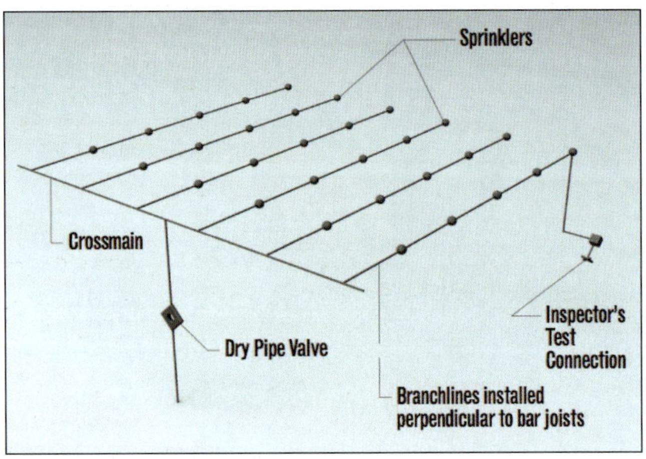

Fig. 26–34. Tree system

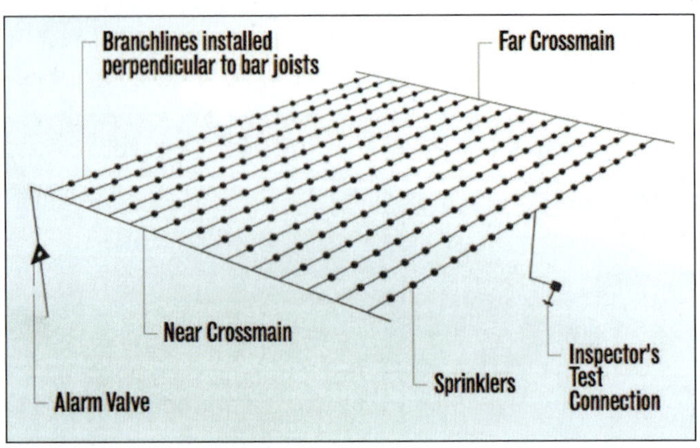

Fig. 26–35. Grid system

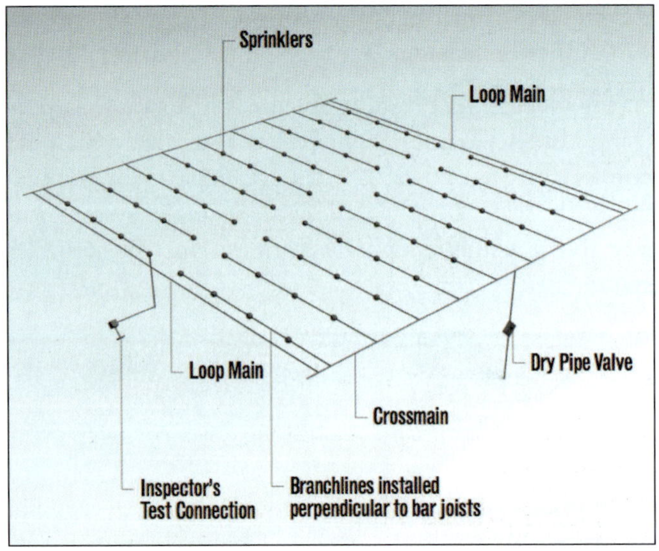

Fig. 26–36. Looped system

Firefighting considerations for operating sprinkler systems

There are some things that should be thought about when working with sprinkler systems. These include the following:

- Make sure that the fire department connection (FDC) is supplied.
- Make sure that the main riser control is open. This is usually an OS&Y valve.
- Make sure that, if there is a fire pump, it is working. Read the gauges on the pump and manually start it if you have to.
- Be sure to note the pressure on the main riser gauges on both side of the alarm valve, dry pipe valve or deluge valve. The gauges on the both the supply and discharge sides should provide appropriate readings. You shouldn't be seeing zero pressure for example—an indication that a valve may be closed somewhere and water isn't flowing.
- Make sure you don't close the riser control valve too soon! Only close the valve upon the order of the incident commander (the fire has been controlled) Leave a firefighter at the valve with a radio to quickly reopen the valve if it is necessary.
- Let a licensed contractor handle replacing fused sprinkler heads or resetting dry pipe valves. Post a fire watch if necessary until the system is fully restored.

Standpipes and standpipe classifications

Standpipe systems are systems of pipes and hose outlet valves used to deliver water to different building locations. These systems are designed and installed in accordance with NFPA 14, *Standard for the Installation of Standpipes and Hose Systems*. The following are types or classes of standpipes:

- Class I: designed exclusively for fire department use
- Class II: designed for occupant use
- Class III: designed for both fire department and occupant use

Most standpipe systems are supplied with a permanent water supply as well as a fire department connection; however, some dry systems will rely completely on the fire department connection for water (fig. 26–37). Some standpipes in tall buildings will have multiple fire department connections that serve different zones (groups of floors). It is important to supply the proper fire department connection.

Fig. 26–37. Fire department connections

Class I standpipes. Each **Class I standpipe** outlet has a 2½-in. (65-mm) male coupling and a valve to control water flow. They are usually designed so water accessible to fire department apparatus can be pumped to areas that need water within a building. Hose outlet valves connect to attack lines, and then personnel within a burning structure can deliver water.

Floor-to-floor openings are probably the most readily available fire paths for fire spread in tall buildings. Stairs are one of the primary reasons these openings are needed. It is clear that any stair in a high-rise building will need a fire-rated enclosure, often called a "fire-rated stair tower" to prevent this flame spread from occurring in the event of a fire. For high-rise fires, these standpipes are usually located within fire-rated stair towers so departments can have protected access to them. A hoseline can be connected in the protected area of the stair tower and then stretch the line from the stair tower to the fire floor.

Some taller high-rise systems have **pressure-control valves**, sometimes called **pressure-regulating valves**, that control pressure at the outlet of a standpipe to prevent excessive nozzle pressure in hoselines. Because **fire pumps** are often used to pump water to higher floors, a PRV keeps water pressure on lower floors manageable. The following are the three basic types of PRV:

- Pressure restricting
- Pressure reducing
- Pressure control

Pressure restricting devices restrict pressures in flowing conditions only so they don't control pressure in static conditions. They also don't compensate for changes in input pressure to maintain a constant discharge pressure. For these two reasons they are not always desirable. A commonly used pressure restricting device is an orifice plate. If they are used they should be installed at hose couplings so they can be removed. Other devices for regulating pressures are pressure control and pressure-reducing valves. A typical pressure reducing valve is shown in figure 3–38. These valves can regulate outlet pressures in both flowing and non-flowing conditions. They are required for outlets of systems where the static pressure can exceed 175 psi (1,225 kPa).

PRV operating pressure should reflect whether fog nozzles will be used because they often require higher pressure than straight bore nozzles. PRVs have caused difficulty in many major high-rise fires, usually because they restrict water flow when not adjusted properly. In newer systems they are designed so that the device can be frequently tested, and standpipe systems that use these devices have dedicated drainage systems with connections at every other floor. If buildings within your jurisdiction use PRVs, be sure to understand operation of their manual settings. If pressure restricting devices are used, be sure to know if they can be removed and how to remove them.

Class II standpipe. Although a **Class II standpipe** is designed for use by building occupant and firefighters, firefighters should not use them. They typically do not have an FDC and operate at pressures and flows that are usually too low to be of use to firefighters. Each hose outlet is usually preconnected with a 1.5-in. (38-mm) hose and nozzle. Because Class II standpipes require water to flow before a fire department arrives, they often require additional fire pumps and water supplies to maintain volume and pressure. Preconnected hoselines are often not maintained, and their presence and serviceability should not be assumed. See an example of a typical cabinet used to store hoses in figure 26–39.

Class III Standpipe. A **Class III standpipe** combines the services of Class I and Class II standpipes. It has outlets for fire department use and a preconnected hose for building occupants.

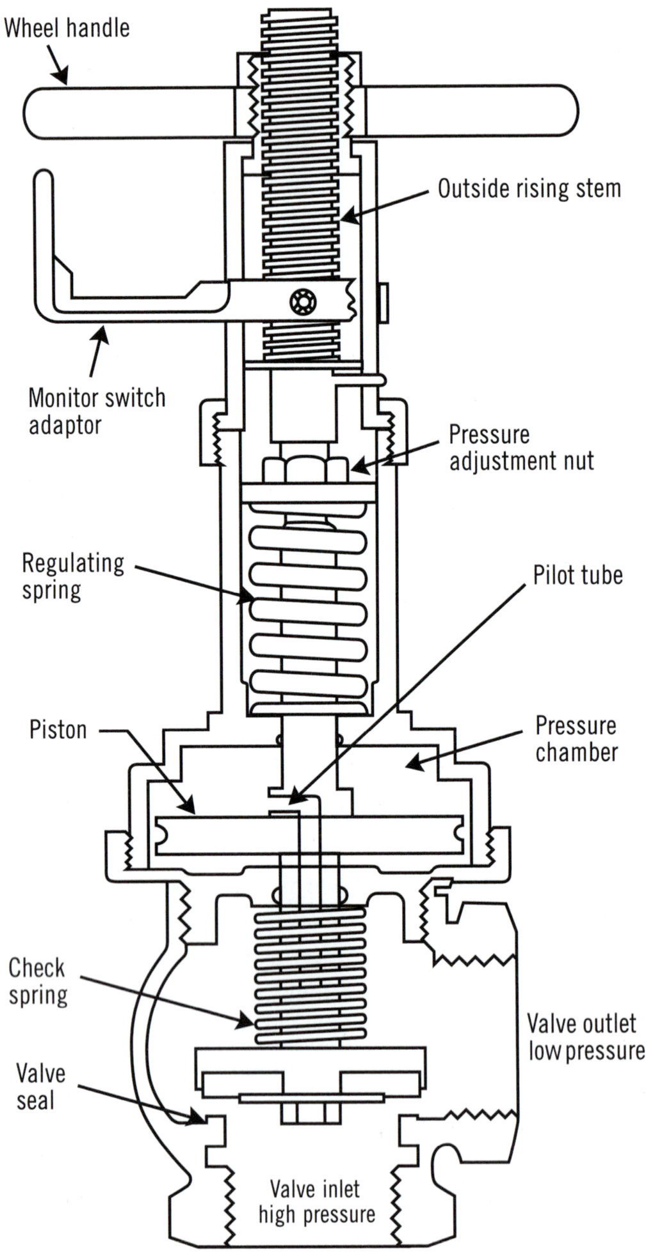

Fig. 26–38. Pressure control valve

Fig. 26–39. Hose cabinet

Firefighting considerations for operating standpipe systems

The following are some things that should be thought about when working with standpipe systems:

- Make sure that the FDC is supplied. If it is blocked or clogged, you may be able to pump back through a hose valve in a stairwell as a backup, although this won't work if the hose valve is a pressure-reducing valve.
- Make sure that the main control valve and riser isolation valve are open.
- Make sure that if there is a fire pump it is working. Read the gauges on the pump and manually start it if you have to.
- Be sure to note the pressure on the main riser gauge at the ground or basement level to make sure the system is working properly. There should be a pressure gauge at the top of the riser as well.
- Be wary of dry standpipe systems. They will take longer to fill, they may have been vandalized, or, depending on where they are, just plain neglected. Be wary of hose valves that have been opened by vandals before the fire.

Fixed-foam fire-protection systems. Although not found in all jurisdictions, it is possible to encounter fixed-foam fire-suppression systems. These locations include tank farms, bulk terminals, and flammable liquid storage warehouses.

At tank farms/bulk terminals, foam systems may be used to protect the storage tanks. These include systems that use **foam chambers** (also known as type 2 devices) that apply foam to the surface of the liquid (fig. 26–40). Some tanks apply foam at the base of the tank using a **subsurface technique**. The tanks may also be protected with fixed foam nozzles on the ground, surrounding the tanks.

Loading racks are where delivery trucks are loaded with the flammable and combustible liquids (fig. 26–41). They are often protected with foam sprinkler systems. Flammable liquid warehouses use foam sprinkler systems as well (fig. 26–42).

Fig. 26–40. The use of foam chamber on a tank can help to rapidly and effectively introduce foam onto a working tank fire.

Fig. 26–41. Loading racks are often protected with foam suppression systems.

Fig. 26–42. Many standard sprinkler systems can be retrofitted to supply foam, such as this one. This type of system would be effective in a flammable liquid warehouse. (Courtesy of Buckeye Fire Equipment Company)

OTHER FIRE PROTECTION SYSTEMS

Other fire protection systems are generally applied locally or as total-flooding systems. A **total-flooding system** completely fills a room with suppression agent. A **local-application system** is a suppression system designed to protect only the area directly around a hazard.

Total-flooding gaseous systems

Halon. Halon is a fire-extinguishing gas that chemically interrupts the combustion reaction. In other words, it destroys the sustained chemical-reaction side of the fire tetrahedron. It is used in total-flooding systems, some local-application systems, and fire extinguishers.

Halon systems consist of a series of pipes and nozzles that connect to pressurized halon storage tanks. It is considered safe for humans.

New halon production is banned in most countries because it destroys the ozone layer. Recycled halon from systems taken out of commission is still legally available from halon banks, but it is expensive. The U.S. military still uses halon extensively. It has been used to protect large computer systems, telephone switches, and other electronic gear, as well as art and historical collections.

Clean agents. Clean-agent gases are designed to replace halon as total-flooding suppression agents without harming the ozone layer. These gases use a system of tanks, pipes, and nozzles to deliver the gas, just like halon. Because they have different properties than halon, however, the chemical tanks cannot be simply swapped for halon in existing halon systems. If a halon system is being replaced, a whole new set of pipes, tanks, and nozzles must be installed.

Halocarbons. One type of clean agent is **halocarbons**, which work by cooling the fire and chemically reacting with the fire. Halocarbon gases have a NOAEL concentration. The NOAEL acronym stands for "no observed adverse effect level." This is the concentration below which testing authorities consider safe. These systems should be designed so this concentration is never exceeded when the gas is discharged.

All halocarbons contain the element fluorine. When exposed to flame, they usually produce hydrofluoric acid, or HF, which is not safe for humans to breathe. While HF is also produced by halon systems, the volumes produced by newer halocarbon agents can be much higher. Newer halocarbon systems should use sensitive detection systems such as air sampling to activate quickly for small fires. This way, only a small amount of agent will be needed to extinguish a fire, and HF production will be kept as low as possible.

Inert gases. An **inert gas** doesn't react with most chemicals. Inert gas mixtures used as fire suppressants include helium, neon, argon, nitrogen, and carbon dioxide. They extinguish fires by lowering a room's oxygen concentration from the normal level of 21% to below 15%. The number of tanks to store enough inert gas to extinguish fires is usually much higher than the number of halocarbon tanks. In building designs where space is a consideration, this can be a concern.

Carbon dioxide. **Carbon dioxide (CO_2)** can be used in total-flooding and local-application systems. The colorless, odorless, nonreactive gas works well in extinguishing fires by reducing the oxygen concentration in a room to the point a fire can no longer burn. To do this, total-flooding carbon dioxide systems require agent application at concentrations above 30%. At this concentration, carbon dioxide kills very quickly by interfering with central nervous system functions.[6] Accidentally activated total-flooding carbon dioxide systems that have discharged into occupied areas have been responsible for many deaths.

The local-application systems work in a similar way to carbon dioxide fire extinguishers. It should be noted that while the CO_2 concentrations produced near activated local application systems may not be as high as those for total flooding systems, they might produce localized concentrations that are hazardous to humans as well.

Total-flooding water-mist systems

Water-mist systems use low quantities of water at very high pressure to suppress fires. Generally they are installed in total-flooding systems similar to deluge sprinkler systems, but the quantity of water applied is much lower. Piping diameter is also small, and the piping used is flexible, making for easy installation in existing buildings.

These systems are commonly installed on cruise ships in cabins and engine rooms because they can fight fires without flooding a ship. They have also been installed in other special applications, where avoiding interference with existing building features is an issue.

Local-application systems

Dry-chemical systems. **Dry-chemical extinguishing systems** use the same types of powdered agents as dry-chemical fire extinguishers. The powder is kept in pressurized tanks or systems in which nitrogen or carbon dioxide pressurizes the system when activated. Typical systems are used around gas pumps (fig. 26–43). These systems are essentially large, dry-chemical fire extinguishers attached to piping systems. They must be specially designed to ensure correct flow from the pipes.

Fig. 26–43. Dry-chemical suppression system

Wet-chemical extinguishing systems.

Wet-chemical extinguishing systems use a proprietary liquid agent, often a potassium solution. (The chemical symbol for potassium is the letter K, so sometimes these systems have "K" in their names.) They are usually local-application systems and are often used in restaurant hood systems to extinguish fires in deep fryers and other cooking equipment. When the wet-chemical agent is applied on top of a pool of grease, it chemically breaks down the grease by a process called saponification. Saponification refers to the formation of soap like layer of foam on top of the grease that smothers the fire and stops combustion. Restaurants like to use wet-chemical systems because they are much easier to clean up after than dry-chemical systems (fig. 26–44).

Fig. 26–44. Wet chemical system. (Courtesy of Joseph LoScalzo)

Smoke management systems

Smoke is responsible for the majority of fire deaths. In building fire situations, smoke often flows to places far away from the fire. A **smoke control system** uses air pressure differences to keep smoke away from building occupants so they can be evacuated safely. Systems that use fans are called mechanical smoke control systems. Other systems just rely on the "chimney effect" to move smoke out of an area. These are called passive smoke control systems. Hotel atriums, casinos, warehouses and other large areas often have built in systems for smoke extraction and smoke control.

A specific application of smoke control is the pressurization of stairwells. This is often used to prevent smoke infiltration of stairwells so that the public can evacuate safely. The concept of the "pressure sandwich," which is venting the fire floor and pressurizing the surrounding floors to prevent smoke movement from the fire floor, is yet another way that smoke control is frequently used in high-rise buildings. Smoke control systems are also used to remove smoke that may accumulate in the large atriums that are now frequently found in hotels and office buildings. Often such systems can be controlled locally from a panel such as the one shown in figure 26–45.

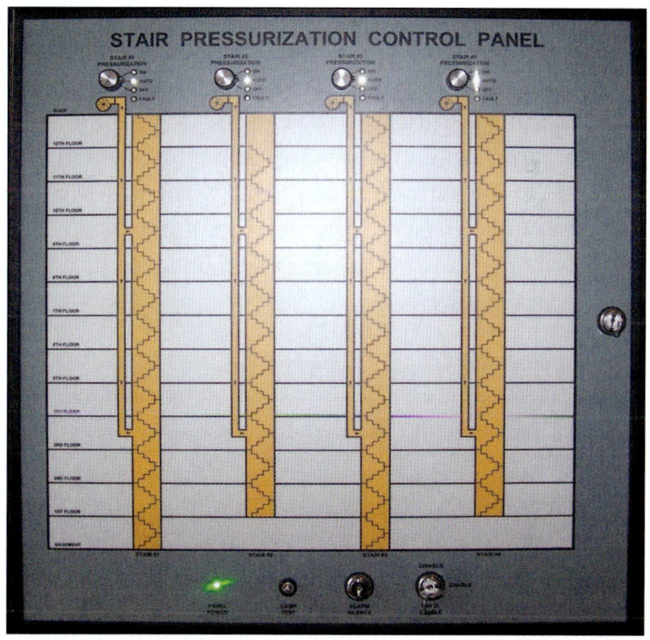

Fig. 26–45. Smoke control system control panel. (Courtesy of David Phelan)

It is important to know how a smoke control system for a building is activated. In the case of hotel atriums or casinos, they are usually connected directly to the fire alarm control panel and are activated automatically in the event that a detector in the protected area is activated. Generally in the case of warehouses the systems are activated manually.

Very large systems are used to clear the smoke out of train stations and rail or road tunnels. Often these systems are not under the direct control of the fire department. They are usually under the control of the operators of the rail or road system from a central control area. This occurs for a number of reasons. Like sprinkler systems, they often need to be activated before fire services personnel arrive. It may take the fire service a very long time to respond to a train stuck in a tunnel that goes under a river, for example. Passengers on the train may perish if they are not provided with clean air before the fire services personnel arrive.

Most of these systems also have more than one purpose. For example, in road tunnels the fans may be run every day to remove car exhaust from the tunnel—also a safety critical use. In rail tunnels they may also be run frequently to remove heat from the tunnels from train air conditioners and brakes. In all these cases the systems are designed to meet the requirements of both uses. The fact that the systems are used every day usually means that they are more reliable, since this means that they are tested every day; and if they break down, they must be fixed right away.

Clearly in these situations communications between the fire service and the system operators are very important. Joint training is critical so that, in the event there is an incident, the response will be properly coordinated. Fire department personnel should train with such systems so that they understand the design criteria (what it is supposed to achieve), how the system operates, and how to control the system.

Ammonia diffusion systems

Some cold-storage buildings storing food products use anhydrous ammonia as a refrigerant. Some hockey rinks also use ammonia. While many know that anhydrous ammonia is toxic, many do not know that it can be explosive under the right conditions (its explosive range is approximately 16%–25%). A 1984 anhydrous ammonia explosion in a Shreveport, Louisiana, cold-storage facility led to the death of a firefighter.

Ammonia refrigeration systems are required to be provided with ammonia diffusion systems under most fire codes. Basically, these systems allow firefighters to remove ammonia from the system, mix/dilute it by pumping water into a special fire department connection (specifically for this purpose), and finally discharge it down a sanitary sewer. While these diffusion systems perform the same function, they can differ in design. It is important to determine exactly how to operate the systems in your jurisdiction (fig. 26–46). When operating the system, it is imperative to notify the sewage treatment facility that will be receiving the ammonia mix.

Fig. 26–46. An ammonia diffusion system. You must understand how to operate this type of system before an incident occurs. (Courtesy of Ronald Spadafora)

Firefighter air replenishment systems (FARS)

While the challenge of filling empty self-contained breathing apparatus cylinders during high-rise fires has existed for some time, it was the 9/11 attacks on the World Trade Center towers that focused attention on the issue. The result of this attention was the creation of a firefighter air replenishment system (FARS) (fig. 26–47). In essence, the system is an "air standpipe," a vertical stainless steel pipe that incorporates fill stations on selected floors of a high-rise.

Fig. 26–47. Two firefighters use a FARS to recharge air bottles without uncovering their SCBA face pieces. Two air bottles can be filled in two minutes or less at every bottle-filling station throughout the building.

In addition to the vertical air supply riser and filling stations, the system often includes an on-site compressed air storage system, an exterior mobile air connection to allow the use of a mobile fire department air compressor vehicle to "pump" into the system, as well as an air monitoring system to check for proper pressures, moisture, and carbon monoxide levels of the air in the system.

While not yet common, FARS have been installed in some existing high-rises as well as new high-rise buildings in specific cities across the United States. The requirement for installing such systems is typically enacted by local city governments.

COMPONENT AND MATERIAL QUALITY IN DETECTION AND SUPPRESSION SYSTEMS

We have simplified descriptions of some components of modern detection and suppression systems. For example, a manual-pull station was compared to a switch, but not any switch will do. All components connected to a detection or suppression system must be **approved**, meaning they must meet the requirements of the **authority having jurisdiction (AHJ)**. An AHJ is required to enforce local law concerning the construction or life-safety issues of a particular building or structure, including adopted building codes and standards.

The NFPA does not test devices itself but requires components of a detection or suppression system to be listed by an approved testing laboratory, such as Underwriters Laboratory or Factory Mutual. These labs deem equipment, materials, or services **listed**, or acceptable to the AHJ, in a document published by the laboratory stating they were tested. A **listed system** ensures that a system works as designed.

Common diagramming symbols

FFII 5.5.3 All elements presented in the previous sections of this chapter are represented in construction drawings by symbols. An extensive list of typical symbols is shown in table 26–3.

Table 26–3. Typical symbols used to represent detection and suppression systems in construction drawings.[7]

FIRE PROTECTION SYMBOLS

SYMBOL | DESCRIPTION

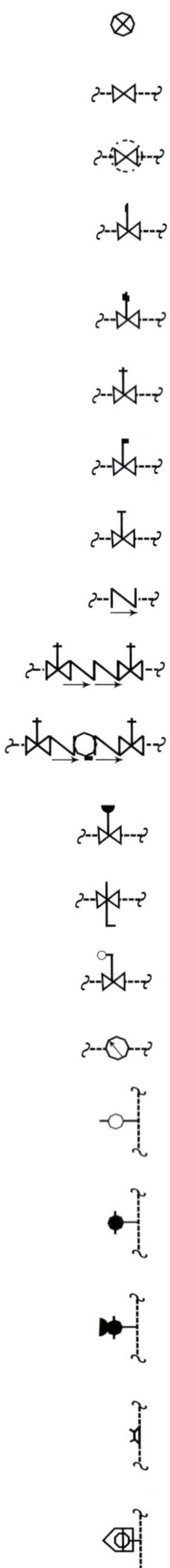

- RISER
- VALVES (GENERAL)
- VALVE IN PIT
- POST-INDICATOR VALVE
- KEY OPERATED VALVES
- OS & Y VALVE (OUTSIDE SCREW AND YOKE (RISING STEM)
- INDICATING BUTTERFLY VALVE
- NONINDICATING VALVE (NONRISING-STEM VALVE)
- CHECK VALVE
- BACKFLOW PREVENTER - DOUBLE CHECK TYPE
- BACKFLOW PREVENTER - REDUCED PRESSURE ZONE (RPZ) TYPE
- PRESSURE REGULATING VALVE
- PRESSURE RELIEF VALVE
- FLOAT VALVE
- METER
- PRIVATE HYDRANT, ONE HOSE OUTLET
- PUBLIC FIRE HYDRANT - TWO HOSE OUTLET
- PUBLIC FIRE HYDRANT - TWO HOSE OUTLET & PUMPER CONNECTION
- WALL HYDRANT - TWO HOSE OUTLET
- PRIVATE HOUSED HYDRANT- TWO HOSED OUTLET

SYMBOL | DESCRIPTION

- FIRE DEPARTMENT- TWO HOSE OUTLET, FREE STANDING
- FIRE DEPARTMENT-TWO HOSE OUTLET
- FIRE DEPARTMENT-ONE HOSE OUTLET
- FIRE PUMP WITH DRIVES
- FREESTANDING TEST HEADER
- WALL-MOUNTED TEST HEADER
- SCREEN/STRAINER
- XXX — **CONTROL PANEL (ATTRIBUTED)**
  - FCP FIRE ALARM CONTROL PANEL
  - FSA FIRE SYSTEM ANNUNCIATOR
  - FTR FIRE ALARM TRANSPONDER OR TRANSMITTER
  - ESR ELEVATOR STAUS/RECALL
  - FAC FIRE ALARM COMMUNICATOR
  - HCP HALON CONTROL PANEL
  - HVA CONTROL PANEL FOR HEATING, VENTILATION, AIR-CONDITIONING, EXHAUST STAIRWELL PRESSURIZATION, OR SIMILAR EQUIPMENT.
- XX — **MANUAL STATION**
  - HL - HALON
  - DC - DRY CHEMICAL
  - FO - FOAM
  - WC - WET CHEMICAL
  - P - PULL STATION
  - CA - CLEAN AGENT
  - WM - WATER MIST
  - DL - DELUGE SPRINKLER
- CO_2 — CARBON DIOXIDE
- REDUCER
- ALARM GONG
- WATER MOTOR ALARM GONG
- TANK, ABOVE GROUND, HORIZ.

SYMBOL | DESCRIPTION

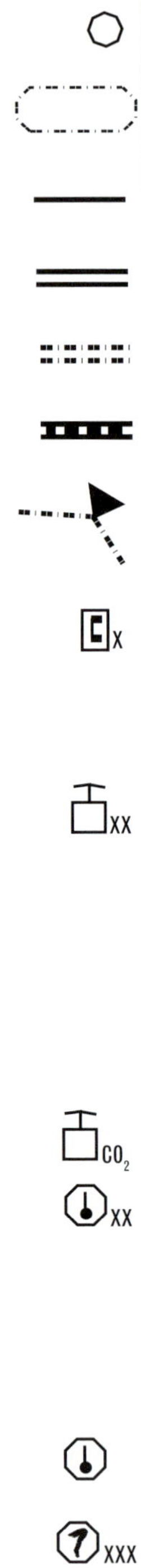

- TANK, ABOVE GROUND, VERT.
- TANK, BELOW GROUND, HORIZ.
- PUBLIC WATER MAIN
- PRIVATE WATER MAIN
- UNDER WATER MAIN BUILDING
- SUCTION PIPE
- THRUST BLOCK
- X — **FIRE SERVICE OR EMERGENCY TELEPHONE STATION**
  - A – ACCESSABLE
  - J – JACK
  - H – HANDSET
- XX — **ABORT SWITCH**
  - HL – HALON
  - DC – DRY CHEMICAL
  - FO – FOAM
  - WC – WET CHEMICAL
  - CA – CLEAN AGENT
  - WM – WATER MIST
  - DL – DELUGE SPRINKLER
- CO_2 — ABORT SWITCH (CO_2)
- XX — **HEAT DETECTOR (TERMINAL DETECTOR)**
  - R/F – COMBINATION - RATE AND FIXED TEMPERATURE
  - R/C – RATE COMPENSATION
  - F – FIXED TEMPERATURE
  - R – RATE OF RISE ONLY
- LINE OF TYPE DETECTOR (HEAT-SENSITIVE CABLE)
- XXX — **SMOKE DETECTOR**
  - P – PHOTOELECTRIC PRODUCTS OF COMBUSTION DETECTOR
  - I – IONIZATION PRODUCTS OF COMBUSTION DETECTOR
  - BT – BEAM TRANSMITTER
  - BR – BEAM RECEIVER
  - ASD – A/T SAMPLING

FIRE PROTECTION SYMBOLS, CONTINUED

SYMBOL	DESCRIPTION
	SMOKE DETECTOR FOR DUCT
	GAS DETECTOR
	FLAME DETECTOR
	FLOW DETECTOR/SWITCH
	PRESSURE DETECTOR/SWITCH
	LEVEL DETECTOR SWITCH
	TAMPER DETECTOR
	VALVE WITH TAMPER DETECTOR/SWITCH
	DOOR HOLDER
	ALARM, MANUAL CONTROL
	SPEAKER/HORN (ELECTRIC HORN)
M	ALARM, MINI HORN
	ALARM, HORN/LIGHT SEPARATE ASSEMBLY
	ALARM, HORN/LIGHT, ONE ASSEMBLY
	ALARM, LAMP/LIGHT, SIGNAL LIGHT STROBE
	WET CHARGED SYSTEM AUTOMATICALLY ACTUATED
	WET CHARGED SYSTEM MANUALLY ACTUATED
	DRY CHARGED SYSTEM AUTOMATICALLY ACTUATED
	DRY CHARGED SYSTEM MANUALLY ACTUATED
	FOAM SYSTEM AUTOMATICALLY ACTUATED
	FOAM SYSTEM MANUALLY ACTUATED
	WATER MIST EXTINGUISHING SYSTEM AUTOMATICALLY ACTUATED
	WATER MIST EXTINGUISHING SYSTEM MANUALLY ACTUATED
	DRY CHEMICAL SYSTEMS FOR LIQUID, GAS AND ELEC. FIRES AUTOMATICALLY ACTUATED
	DRY CHEMICAL SYSTEMS FOR LIQUID, GAS AND ELEC. FIRES MANUALLY ACTUATED
	FOR FIRES OF ALL TYPES (EXCEPT METAL) AUTOMATICALLY ACTUATED
	FOR FIRES OF ALL TYPES (EXCEPT METAL) MANUALLY ACTUATED
	SYSTEMS UTILIZING A GASIOUS MEDIUM AUTOMATICALLY ACTUATED
	SYSTEMS UTILIZING A GASIOUS MEDIUM MANUALLY ACTUATED
	HALON SYSTEM OR CLEAN AGENT EXTINGUISHING SYSTEM AUTOMATICALLY ACTUATED
	HALON SYSTEM OR CLEAN AGENT EXTINGUISHING SYSTEM MANUALLY ACTUATED
AS	FULLY SPRINKLERED SPACE
AS	PARTIALLY SPRINKLERED SPACE
NS	NONSPRINKLERED SPACE
WS	WATER SPRAY SYSTEM
	UPRIGHT SPRINKLER
	PENDENT SPRINKLER
	UPRIGHT SPRINKLER; NIPPLE UP
	PENDENT SPRINKLER; ON DROP NIPPLE
	SPRINKLER WITH GUARD
	SIDEWALL SPRINKLER
	OUTSIDE SPRINKLER
	PIPE HANGER
	ANGLE VALVE (ANGLE HOSE VALVE)
	ALARM CHECK VALVE
	DRY PIPE VALVE
	DRY PIPE VALVE WITH QUICK OPENING DEVICE (ACCELERATOR OR EXHAUSTER)
	DELUGE VALVE
	PREACTION VALVE
	CO_2 REEL STATION
	DRY CHEMICAL REEL STATION
	FOAM REEL STATION
	HOSE STATION, DRY STANDPIPE
	HOSE STATION, CHARGED STANDPIPE
	MONITOR HOSE, DRY
	MONITOR HOSE, CHARGED
	AGENT STORAGE CONTAINER FO - FOAM HL - HALON CL - CLEAN AGENT DC - DRY CHEMICAL WM - WATER MIST
CO_2	AGENT STORAGE CONTAINER CARBON DIOXIDE
	FUSIBLE LINK
ETL	FUSIBLE LINK W/ ELECTROTHERMAL FEATURE
SOV	SOLENOID VALVE
	UTILITY SHUTOFFS E - ELECTRIC SHUTOFF W - DOMESTIC WATER SHUTOFF G - GAS SHUTOFF LPG - LP-GAS SHUTOFF NG - NATURAL GAS SHUTOFF CNG - COMPRESSED NATURAL GAS SHUTOFF
XXX	**DETECTION/EXTINGUISHING EQUIPMENT** DD - DUCT DETECTOR HD - HEAT DETETOR SD - SMOKE DETECTOR FS - FLOW SWITCH (WATER) PS - MANUAL PULL STATION TS - TAMPER SWITCH HL - HALON SYSTEM DC - DRY CHEMICAL WC - WET CHEMICAL SYSTEM FO - FOAM SYSTEM CA - CLEAN AGENT SYSTEM BSD - BEAM SMOKE DETECTOR
CO_2	DETECTION/EXTINGUISHING EQUIPMENT CARBON DIOXIDE SYSTEM

NOTES

1. NFPA. 2007. NFPA 72, *The National Fire Alarm-*Code. Quincy, Mass.: National Fire Protection Association.

2. NFPA. 2007. NFPA 72, *The National Fire Alarm-Code*. Quincy, Mass.: National Fire Protection Association.

3. NFPA. 2007. NFPA 13, *Standard for the Installation of Sprinkler Systems*. Quincy, Mass.: National Fire Protection Association.

4. NFPA. 2008. NFPA 25, *Standard for the Inspection, Testing, and Maintenance of Water-Based Fire Protection Systems*. Quincy, Mass.: National Fire Protection Association.

5. *Sprinkler Hydraulics*, Harold S. Wass, Jr.

6. "Review of the Use of Carbon Dioxide Total Flooding Extinguishing Systems," Robert T. Wickham, P.E., Downloaded from http://www.epa.gov/ozone/ snap/fire/co2/co2report2.pdf on 6/29/2009.

7. NFPA. 2006. NFPA 170. *Standard for Fire Safety and Emergency Symbols*. Quincy, Mass.: National Fire Protection Association.

QUESTIONS

1. ______________________________ buildings are protected by fire alarm systems.
2. A ______________________________ alarm system notifies only the building occupants of fire.
3. A _____________________ signal identifies system problems that can influence the system's reliability.
4. Automatic or manual mechanisms that cause a system to indicate an alarm condition are known as _____________________________________.
5. What types of manual pull stations exist? Which one helps prevent false alarms?
6. Which NFPA standard deals with fire alarm systems?
7. A ___________________________ warns of a fire condition by sensing a high temperature, high rate of rise in temperature, or both.
8. List the types of detectors and how they react to different heat conditions.
9. What are photoelectric detectors and how do they work?
10. A flame detector activates when it detects ___________________________.
11. What is the most common gas detector?
12. List the three basic types of fire alarm systems.
13. Discuss the components of a central-station alarm system.
14. A sprinkler head that opens when the solder holding two pieces of metal together fuses or melts is called a ______________________________.

14. Discuss the difference between a pendant sprinkler head and an upright sprinkler head.

15. ________________________________ sprinkler heads are designed to be in areas below freezing.
16. Which of the sprinkler systems use low quantities of water at very high pressure to suppress fires?

Advanced Fire Attack

by Christopher Flatley with Jerry Kna

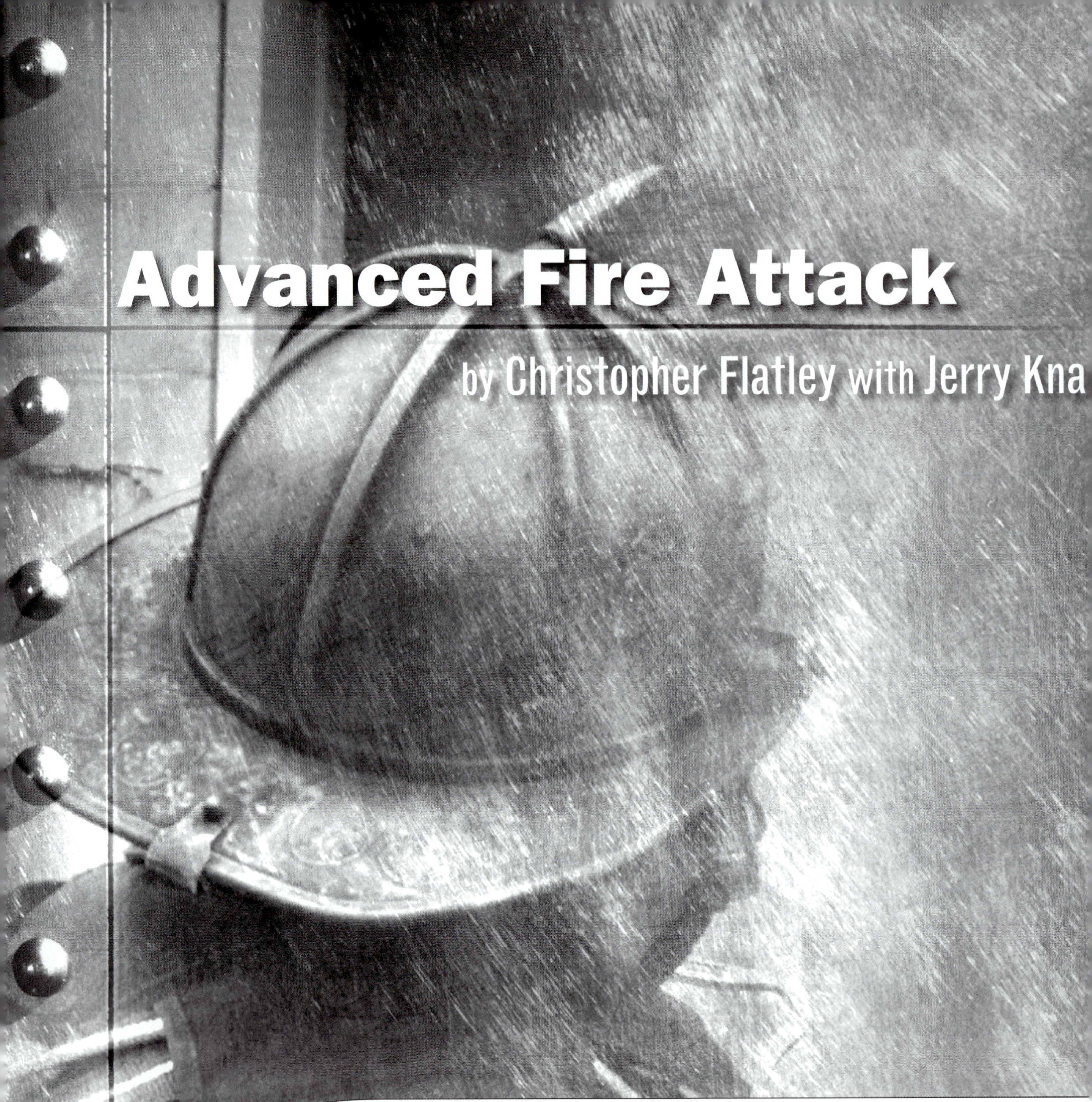

This chapter provides required knowledge items for the following NFPA Standard 1001 Job Performance Requirements:

FFII 5.3.1

FFII 5.3.2

FFII 5.3.3

27

This chapter contains Skill Drills. When you see this icon, refer to your Skill Drill book for step-by-step instructions. SKILL DRILL

OBJECTIVES

Upon completion of this chapter, you should be able to do the following:

- Identify the role of Firefighter II in carrying out assignments as team leader
- Describe offensive and defensive strategies used at fires involving flammable gases and liquids
- Define the five foam types as they apply to fire service operations
- Describe how firefighting foam is produced and applied to the surface of burning liquids
- List the different types of foam concentrate available to the fire service
- List the types of equipment used to develop foam that are available to the fire service
- List and describe the techniques for applying foam to a burning liquid
- List safety considerations when applying firefighting foam
- Describe the proper methods to determine a proper application rate of foam on a burning liquid
- Describe the two types of foam systems

INTRODUCTION

Now that you have received some training and gained experience as a firefighter, it is important to reinforce your basic firefighting skills. If you read fatal fire reports, (and you should) you will see that many injuries and deaths result from difficulties in fire company operations. Remember, other firefighters work in areas adjoining and above the fire, and their safety—as well as the safety of you and your crew—depends on your ability to execute your assignment. As you become a more seasoned firefighter, you will be tasked with greater responsibility. With that responsibility comes accountability. You will no longer be able to say "I didn't know" because at this point, you should. Firefighters *have to make it happen, not make excuses.*

This chapter builds on lessons learned from your training in Firefighter I. The new concepts introduced here guide you toward becoming a more experienced firefighter and advancing into the officer rank. You must thoroughly understand the concepts in this chapter before you complete the course. Once you complete the course, this manual will become an important reference. You will refer to it often during your career.

In Firefighter I you were taught how to perform specific firefighting techniques. Now in Firefighter II you learn how to decide which technique to employ as well as *where* and *when*. After a superior officer assigns tasks such as ventilation, fire attack, search, and so forth, you must be able to carry them out without immediate supervision. Although you do not yet have an officer rank, you are responsible for carrying out a variety of assigned duties as a team leader.

In many cases, your department's standard operating procedures (SOPs) will dictate the decision. For example, some departmental SOPs will dictate the size of a hoseline to be used for certain types of fires or their location, for example, commercial buildings. In other cases, your SOPs will help you make your decision, but it may be a judgment call, for example, selecting a larger line based on the amount of fire. These are decisions you may be empowered to make as a Firefighter II. This text is not intended to replace your department's SOPs. Any issues raised that deviate from departmental SOPs should be guided by the SOPs.

TYPES OF DUTIES

As a Firefighter II, you will be assigned tasks, and your responsibility is to complete them in a safe, efficient, and effective manner. Examine the tasks listed in the following text and consider the responsibilities and types of decisions you must make while completing each one:

- Coordinate an interior fire attack team given the appropriate amount of equipment and properly protected personnel.
- Identify and communicate hazards to your team.
- Decide the point of entry and the appropriate method to gain entry.
- Determine the method of fire attack based on its type, size, and location (basement, grade level, and upper levels of a building), as well as deciding on nozzle and hoseline selection and line placement based on conditions.
- Maintain proper communications between your team members, other teams operating in the immediate fire area, teams operating in adjoining areas that may be effected, as well as the incident commander. (Promptly notify the incident commander of any changing conditions.)
- Maintain team coordination and ensure proper team integrity through communications and accountability.
- Constantly evaluate changes in fire growth and behavior, and anticipate fire spread and the influence of flow paths.
- Decide the **priority areas** for search; and, should a victim be found, determine the most appropriate removal method.
- Decide where, how, and when to ventilate the building. Determine the need for proper ventilation and coordinate it with the fire attack.
- Monitor building structural conditions, paying particular attention to signs of structural distress or impending collapse.
- Scrutinize changes in fire conditions, including increase in fire intensity, impending flashover or backdraft, and other dangerous conditions.

SIZE-UP

As discussed in *Firefighter I*, size-up is a critical, basic function for incident commanders. Size-ups allow firefighters to make the best possible decisions on how to deal with the situation. However, it shouldn't be limited to officers. Firefighters, particularly those coordinating and carrying out tactical objectives, should also conduct their own size-up. With your knowledge and understanding of fire behavior and its effects on firefighting operations, you must carefully consider the seven specific issues in performing your size up:

- **Location of fire and type of construction.** Where is the fire now, where is it spreading, and where will it ultimately extend? How does construction type confine the fire or allow it to extend?
- **Life hazard.** What are the rescue priorities? Who might occupy the building given the time of day? Where are people located? Who is in the most danger and needs rescue first?
- **Fire control.** What is the fuel load in the building, what are the extinguishing capabilities of the attack line(s), and is there fire protection equipment in place to control a fire?

- **Forcing entry.** What is the best way to gain access to the building, support the rescue, and advance the hoseline?
- **Ventilation.** What type of ventilation will have to be performed (horizontal or vertical, mechanical positive pressure ventilation, etc.) and at what location? How will the fire's flow paths be affected?
- **Stretching considerations.** Where should the first and subsequent hoselines be placed?
- **Water supply.** Where is your hoseline's water supply coming from? Are you operating solely with water from your engine's booster tank or is the water supply coming from a hydrant, drafting source, or tanker shuttle?

SINGLE-FAMILY HOME FIRES

SKILL DRILL

Basement

FFII 5.3.2 As mentioned previously, the first step in developing any fire attack plan is an accurate size-up. The attack strategy may be based on the first arriving officer's size-up, but everyone on the fireground should conduct a size-up based on his or her assignment. For example, the nozzle firefighter may be concerned with stretching the shortest distance to the fire, while other firefighters may be concerned about whether there is enough hoseline to reach the fire. Estimating the stretch is a skill that all engine company members must be able to accomplish, even if hose loads are preconnected. You must be able to identify areas to "flake out" the extra length to avoid kinks and tell if the stretch will come up short and be able to correct the problem.

Location of fire and type of construction. Some general points of basement fire size-up may include confirming the fire is actually in the basement. If fire is suspected in the basement, break a small pane of glass in one of the basement windows. This small opening either vents smoke or allows a clear look into the basement. It also restricts the amount of air that can enter, reducing the possibility of causing a backdraft. If the fire has extended to upper floors and there is a possibility of extension to the attic through voids in balloon construction, then more personnel are required to stretch additional hand lines to cut off the fire spread. Is this home built of lightweight construction? Does that affect how long firefighters can commit to interior operations before being withdrawn over concern of possible collapse? Time is the only thing we don't have control over on the fireground. We don't know how long the fire has been burning. We can't tell with any certainty how quickly the structure is being destroyed. The one thing we do know is that every minute the fire burns, more of the building is being compromised.

Also consider the contents of the basement. If it is packed with a family's lifetime accumulation of valuables, does that increase the fire load and reduce your ability to reach the seat of the fire? Could the fire have been started by defective heating equipment or damaged utilities in the basement? If the fire involves an oil burner, can it be remotely shut off? How does that impact on the fire attack? If the fire has affected the utilities, can they be controlled from the outside? Is the gas meter outside and can it be shut down? Can a utility crew be called to cut the electric service to the house? For firefighters in the basement, what clues would indicate a gas feed fire or a fire where electricity may be involved?

Life hazard. As in every fire or emergency, life safety is the most important consideration. Is the house occupied or vacant (figs. 27–1 and 27–2)? **Occupied** often refers to a home that someone lives in. If no one is home the house is **unoccupied**. A vacant building is an abandoned building that has fallen into disrepair, and the owner no longer maintains the structure in a livable condition. Members should be aware of these important differences in the definitions. This does not preclude the possibility of vagrants or squatters occupying a vacant building. All structures must be searched within the limits of safety. Determining if the structure is occupied or vacant is more art than science. In the long run, it comes down to your observations and experience. That is why every structure must be searched within the limits of safety. Vacant building fires are discussed later in the chapter.

Fig. 27–1. Abandoned structures can still be inhabited by vagrants or juveniles.

Fig. 27–2. Determining whether a structure is vacant can be a difficult task.

For our discussion at this time, answer the following questions: Is someone home right now? What time of day are persons sleeping, unaware of the conditions in the basement? Is the house a single-family dwelling, and where are sleeping areas? Is there a possibility of a basement apartment? What clues indicate a separate apartment? Generally, features like separate mailboxes or gas and electric meters may indicate a separate living unit or apartment in the home, but may not direct you to the basement. The lack of separate mailboxes or utilities should not rule out an apartment being present.

Even if you are unfamiliar with the room layout of the house, what clues can be gleaned from just looking at the house? What does the size and configuration of the windows tell you? Bathrooms generally have smaller windows; they may have frosted or obscured glass for privacy. Bathrooms may adjoin bedrooms. Larger windows next to these smaller ones may be the location of bedrooms. Air conditioning units in windows in warm climates or during the summer season can also indicate sleeping areas. Windows that are off height from others on the same side of the house may indicate the location of the half landing on stairs. Although this window would not be advantageous for entrance from the outside when conducting a primary search, it may be beneficial as a route to get an additional hoseline above the fire.

What is the risk to firefighters? Has the fire developed to the point where it is unsafe to allow firefighters to enter? Has the fire escalated to the point where it is unsafe for members to remain within the structure? When firefighters occupy the structure, they must be considered as part of the life hazard within the building.

Fire control. Controlling and extinguishing basement fires are some of the most difficult tasks you may face. Improper application of water creates large amounts of steam from which there is no relief. Yet, the inability to apply water to the burning material allows the fire to intensify and expose more of the structure to damage. The cramped conditions that often exist in basements require operations with a minimum number of personnel. This is physically demanding and leads to the need for frequent and timely relief of members.

Remember, a basement fire has the opportunity to vertically extend to all portions of the house: up the interior stairs; up pipe chases; and in the case of balloon frame structures, up the stud channels of the exterior walls. Knocking down the fire in the basement only controls the fire in that location; it is imperative to check for extension of the fire into the upper portions of the building.

Recent research indicates that applying water through a basement exterior window as other attack hoselines are being stretched offers the potential benefit of quick reduction in the size of the fire. In addition, the increasing use of unprotected lightweight wood construction (with no sheathing, such as gypsum board) has led to quick collapse of the first floor into the basement. The use of such a technique, when ordered by your officer, may delay collapse of such floors as well as provide an easier final attack with interior handlines.

Forcing entry. Generally, gaining entry into private dwellings is easier than commercial structures. But in all cases, the fire will not go out unless firefighters can get in to extinguish it. Gaining entry is also the way we enter the house to search for victims.

When confronted with a basement fire, what points of access do we have to the basement? Is our only access through the front door and down the interior stairs? These stairs are generally lightweight and narrow. The fire may have weakened these stairs, so do not bunch up and move quickly to the base of the stairs. Descending these stairs requires a determined crew of firefighters to *make the push* to the bottom of the stairs where conditions will be better.

If the line can't advance into the basement, another option is to consider any outside entrances. Does the house have an exterior entrance? Basements and cellars may have either a walk-out door or a set of stairs through a metal Bilco® door and down to the basement door below grade. These entries may provide another avenue for fire attack.

A word of caution: deploying lines from outside entrances may drive fire and smoke up through the interior, exposing interior crews to more heat and fire. A line must be stretched to the top of the interior stairs to prevent extension to the floors above. If the exterior line can knock down the fire, the interior line should advance to the basement for extinguishment. Before the interior line advances into the basement, be sure no fire has passed you; if fire is extending above, call for a third line. If a third line is required, consider bringing it in from either another door (rear or side) or through a window near the point of attack. Having three hoselines stretched through the front door or up the stairs clutters the stairs and complicates entry and exit.

If fire conditions are so severe that the nozzle team cannot make the push into the basement based on the stability of the first floor, a small hole can be cut in the floor over the fire or as close as possible. A cellar nozzle or distributor is placed in it to knock down some fire so hand lines can advance into the basement. This is a dangerous operation as well, but it may be necessary to gain control of the fire. After the fire has been knocked down and hand lines advanced into the basement, the hole can be expanded for ventilation, if necessary.

Ventilation. Basement fires are arguably one of the most difficult fires you will fight. Ventilation in these fires is limited. Small basement or cellar windows do not relieve the large amounts of heat generated by these fires, and normally low basement ceilings add to the amount of heat banking down on firefighters. A tactic that is sometimes employed is to cut a hole in the floor on the first floor near a window to vent the smoke and heat. A hole cut in the living room near the picture window could relieve the basement of heat. Compare it to cutting a hole in the roof when horizontal venting of windows is not enough to relieve the fire floor of smoke and heat.

As previously discussed, basement fires in buildings of balloon frame construction offer challenges as well (fig. 27–3). Fires that expose the base of the outside walls can allow heat and fire to travel up the stud bays to the attic (fig. 27–4). Any fire in the basement of a balloon frame building requires early and frequent inspection of the top floor and attic for extension. The heat and smoke enter the attic between the outside wall siding and the interior finished wall.

Remember, any time you create an opening, you may be creating a flow path as well.

Stretching considerations. When stretching hoselines, the goals are as follows:

- Protect life
- Protect property
- Confine the fire
- Extinguish the fire

Fig. 27–3. In balloon frame construction, fire can travel vertically through the stud channel as seen above a basement window.

It has been said that the best life-saving measure is the prompt attack of the fire; put the fire out and many other problems go away. It has also been said that more lives are saved by the proper position and operation of a hoseline than any other life-saving technique. If we continue with that logic, the engine officer has the greatest influence on the outcome of the fire than any other firefighter on the fireground. Even with the incident commander's best plan, it will fail without firefighters skilled and trained to execute it.

Fig. 27–4. The attic of a balloon-framed house must be examined during every fire.

On the fireground all effort should be directed to getting the first line into position before any other line is stretched. Communication between engine company officers as to the difficulties and need for assistance will ensure this happens.

The most important considerations when stretching are to select the correct size hoseline and to make sure you have enough hose to reach the fire before the line is charged. Adding length to a charged line, although not impossible, requires close cooperation between all members of the engine company. One of my best pieces of advice when it comes to stretching is as follows: *It is better to be a length over, than a foot short* (fig. 27–5). If considering a private dwelling fire, how much area is available to us on the front lawn to lose some hose even if we *overstretch*? If we can *lose* the extra hose on the front lawn this is not *overstretching*. If we have more hose in the street than we can manage, that is overstretching. If the pump pressure required for the stretch exceeds the limit of the line, that is overstretching.

Fig. 27–5. Keeping extra attack hose near the point of ingress aids in fire attack.

For a basement fire, the line must be able to get to the base of the stairs quickly, so enough hose must be nearby when the advance is made. Hose may have to be flaked out inside the house to be available for rapid deployment to the basement. Discuss with your company officers where the **drop point** (the point where the lead length of hose is dropped before it is charged to facilitate the advance) is located. Always refer to your own department's SOPs for clarification.

Ensuring there is enough hose to reach the fire involves several steps. Correct line placement is crucial. As a general rule, lines should be placed between the fire and those endangered by it. That can be either civilians or firefighters. That said, the line must protect the means of egress. In the absence of any life hazard, the line must protect exposures—in this case property. When protecting property consider its value. Protect the most valuable property first, even if it is not where the fire will likely spread. For example, you may need to give up the vehicle to save the house. Judgment is key when deciding where to deploy available resources.

Water supply. At this point in your firefighting career, you may think that water supply is someone else's problem. Maybe it is, but this is where your grasp of where you fit into the puzzle is the key. You may be the firefighter spraying the water, but shouldn't you know where that water is coming from and how much you have? Is the pump operator supplying you with the water from a limited booster tank or using a good hydrant with an unlimited supply of water? Is your water coming from a tanker shuttle? You must know this before you get face to face with the fire. To use a military analogy, is all the ammunition you have available to shoot merely the ammunition you brought with you? Do you shoot and then have to wait for resupply or is supply of ammunition unlimited? How engaged are you going to be in the firefight given those parameters?

Pump operators are the unsung heroes of the fire service. They are often left alone with the simplest instruction, "get me water," which is usually the most difficult thing to produce and produce *on demand*. One story from the New York City Fire Department (FDNY) tells of a resourceful pump operator who stretched a supply line from a high-rise building's standpipe system to supply his engine with water for a fire in the building next door because the hydrant in the street was frozen.

Kitchen

Considering our seven-point size-up, how would our attack plan need to be adjusted for a fire in the kitchen? Many of our considerations are similar, whereas some present different challenges. A fire in the kitchen most often begins on the stove. Defective stoves can cause fires. If the stove is gas or electric, it presents its own unique problems. Hot oils and such can be dangerous and may require specific extinguishing methods. If the fire has extended to the cabinets and affected the structure, the fire is more serious. Kitchens and bathrooms commonly have many void spaces for plumbing and lighting. If the fire has entered the void space, it must be completely examined. If unchecked, these voids can cause fire to extend to remote areas. A thermal imaging camera can be useful in reducing the damage, but it is no substitute for a thorough examination.

Living room

If the fire begins in some other location on the first floor, such as the living room, does the fire meet you at the front door? Do you have to push your way into the house? Is there a firefighter available to complete horizontal ventilation opposite the line and drive the heat and smoke out of the building rather than up the interior stairs? Can you get past the fire to the stairs to get above the fire and search the bedrooms? Will the hoseline be able to control the fire until you complete our search?

Bedrooms

A bedroom fire in a single-family dwelling is common. Fires in the bedroom can be caused by space heaters, electric blankets, and hot lamps contacting fabric surfaces, even candles. Victims are closer to the main body of this type of fire than any other. Rescue requires a rapid entry to the fire area, many times ahead of the hoseline. In spite of our best efforts, the results are not always successful. A large fire in a bedroom can extend into the attic, either from the failure of the gypsum board ceiling or through the eaves when the fire has vented out the windows (fig. 27–6). Once fire is in the attic, roof ventilation is required and ceilings must be pulled to allow a hose stream to apply water to the fire.

Fig. 27–6. Fire lapping through a window can extend into the attic and soffit space in wood-frame residential dwellings. (Note the location of the soffit vent.)

GARDEN APARTMENTS OF WOOD FRAME CONSTRUCTION

The discussion of a fire in a garden apartment- or townhouse-type building follows the same parameters as the private dwelling. From our perspective, garden apartments are typical modern apartment buildings (built since the 1960s), usually two to three stories high, set back from the street, and constructed of wood frame.

The size-up a garden apartment begins much the same way as a private dwelling: determining the type of construction and the location of the fire. The wood frame construction typically used in garden apartments plays a big role in the fire attack because void spaces are often large and run the length of the structure. A small fire that enters a void can quickly involve the whole structure, which leads to the question, "Where's the fire?" The fire must be located quickly, and the route of fire extension must be cut off. This often requires identifying the locations of fire walls and opening up in adjoining units or apartments to get ahead of the fire. On the roof, the hole must also be small enough so it does not draw the fire toward the hole, which would increase the fire problem by spreading the fire throughout the attic space.

On the top floor, inspection holes must be made in the top floor ceiling created ahead of the fire and be large enough to determine conditions yet small enough to avoid introducing the oxygen that will intensify the fire. Examination holes should be made with a charged hoseline in place. If you are looking for fire and your examination reveals fire extension, the line must be able to drive the fire back and prevent the fire from passing your position. The hole may need to be expanded to give the stream an opening large enough to operate and drive the fire back toward the burned side. Additional firefighters with tools to pull the ceiling may be required, so be sure you have adequate resources before attempting this operation. Remember, from this position the goal of the line is to contain the fire and not let the fire pass you.

Knowledge of the apartment layout helps determine avenues of fire travel. Are the apartments duplexes, or is there a different apartment above? If there is a separate apartment above, the layout is probably the same as the one below. This means the voids for plumbing and drains are above one another. If the unit is a duplex, the upper level of the unit consists of bedrooms and bathrooms, reducing voids that continue vertically into the attic. In the duplex style, firewalls may limit horizontal extension to that unit.

Also consider whether the building has a public hall or if there are entrances to individual units from the outside (fig. 27–7). A public hall in the center of the building can act as a firebreak on floors below the attic (fig. 27–8). This hall may seal voids caused by the construction of stairs and other features and prevent extension across the hall.

Fig. 27–7. The use of party walls and private entrances can help eliminate concealed spaces in apartments.

Fig. 27–8. A common centralized entrance in this apartment complex can act as a fire break between individual apartment units.

Fig. 27–9. Firefighters must be prepared to extend attack lines to reach buildings with long setbacks from the roadway.

When assessing the life hazard in garden apartments, consider the six sides of the fire: the four sides on the same level as the fire and above and below. Fire can drop down in voids as well. Smoke and carbon monoxide (CO) can permeate a large area and effect many units; even low levels of CO require the area to be searched.

Gaining entry into multiple apartments can be time consuming. For a forcible entry, it is best to open any doors that may be necessary before being assigned other tasks. This ensures that doors are open if entry is needed quickly for an expanding fire situation or as an area of refuge.

When fighting fires in garden apartments, expect to find more people than a private dwelling. How do the escaping occupants affect your ability to stretch hoselines (fig. 27–9)? Your stretch must not interfere with the escaping occupants. Stretch the lines as close to the walls as possible to leave the area in the center of the hall clear. Where possible, avoid leaving hoselines on stairs. The line should be hauled up alongside stairs and secured to newel posts and railings with hose straps.

The exact location of the fire must be known before a line is committed to the apartment. Just because there is smoke in the apartment does not mean the fire is in that apartment. When the fire apartment is found, engine crews must have enough hose to cover the entire fire area. Some apartments are large, and, as with a duplex, with entrances on the lower level; you may need to advance to the upper level via an interior stair. This may require more than one length of hose (fig. 27–10).

Fig. 27–10. Even a small fire in this building could necessitate several hoselines due to the number of concealed voids and exposures.

Stretches in garden apartments are often longer than those in private dwellings. Extending hoselines by adding rolled or folded hose as well as operating from a manifold or water thief may be necessary. Rope stretches and other alternative stretching methods may be employed to reduce the number of lengths in the stretch. Firefighters with garden apartments in their area must know their department's procedure for completing these long stretches. As the stretches become longer, the pump pressure must increase to overcome friction loss. The effects of kinks are even more dramatic. All firefighters must be aware of kinks and remove any and all kinks when they develop.

With these long stretches, there is a greater possibility of the line getting caught on obstacles. Turning corners and going up stairs with a charged hoseline can be difficult. The rate of advance into the fire area is directly related to the nozzle crews' ability to move the hoseline. Locating the drop point is critical. Having sufficient hose near the point of attack makes the difference in a successful fire attack. Having hose laid in the public hall or in an adjoining apartment on fire floor speeds the advance of the nozzle team. When every foot of hose that enters the fire area has to be dragged in from the sidewalk, the advance is slow and arduous.

Water supply in these complexes can be a challenge. Are there hydrants on site? Where are they located? Are they blocked by cars or shrubbery (fig. 27–11)? Are they privately owned and maintained, making their operability questionable? Do they provide the required fire flow if the whole structure became involved? Any delay in getting a reliable water supply established compounds the fire problem.

Fig. 27–11. Local codes must be strictly enforced to prohibit landscaping from obscuring fire hydrants.

LARGER APARTMENT BUILDINGS OF ORDINARY CONSTRUCTION

Larger apartment buildings of ordinary construction, using brick load–bearing walls and wood joists, are often found in more urban areas than suburban or rural parts of the country. Two common types of these buildings are tenements and H-types.

Tenements

Tenements are smaller than H-types (fig. 27–12). They are usually three to five stories high and have two or four apartments per floor. There may be a store located on the first floor (fig. 3–13). Older tenements may have three or four windows and no front fire escape. This indicates a **railroad flat** room layout. The rooms are arranged one behind the other front to rear, and the rear fire escape is the secondary means of egress. Larger tenements have four windows and a fire escape on the front and the rear. These generally have four apartments per floor and the fire escapes serve front or rear apartments. An exception to this rule may be corner buildings.

Fig. 27–12. A typical tenement building

Fig. 27–13. It may be difficult for firefighters to determine the layout and compartmentation of mercantile aspects in tenement buildings.

First-floor store fires. When a store occupies the first floor of a tenement, a door may lead from the public hall into the store. When these buildings were built, the store owner was generally a tenant of the building or the building owner who would come down stairs and enter the store from this door. These doors may be sealed. From a firefighting perspective, this door provides a path for fire travel from the store in the occupancy above. When the store is heavily involved in fire, monitor conditions on the first-floor hall to ensure fire does not expose the public hall.

Many first-floor stores have storage in the basement accessible from a trap door in the floor for a bar-type occupancy. A narrow stair may have been installed in the rear of the store for access to the basement (fig. 27–14). You may find that a one-story addition has been built in the rear of these buildings. If the store is a restaurant, this is where the kitchen is located. The refrigeration and pantry are in the basement. Many times, food preparation areas are located in the basement, so don't overlook the possibility of a life hazard in the basement.

Fig. 27–14. Elderly or infirm persons may have limited means to exit the rear yard.

If the one-story addition creates a setback, people using the original rear fire escape may come down on this setback and have no egress to the ground level. A firefighter should always be sent to the rear and relay observations to the incident commander.

H-Types. **H-type** buildings get their name from the fact that they are predominantly in the shape of an H (fig. 27–15). The apartments are in the **wings** of the H and are connected by a **throat** to form the H shape. They have brick masonry walls and wood joist floors and roofs. The larger (newer) H-types usually have incorporated some steel in the construction to increase the size of the building and provide a more open floor plan.

Fig. 27–15. A typical H-type building has pipe recesses and voids that can run from the basement through the roof.

These buildings range in height from 4 to 7 stories. Pipe recesses and voids can run from the basement through the roof. Although these buildings are larger and the voids extensive, all the rules for examining voids still apply and may be even more critical because of the number of apartments that adjoin the void.

Three different types of stairways exist in these buildings. The **transverse stair** allows occupants to transverse from one wing to another. This is the best from the fire department's point of view, because occupants can be directed away from heat and smoke and members can stretch hoselines while avoiding escaping occupants.

Wing stairs are typically a problem for firefighters in that the stairs only serve the wing. This reduces the number of occupants that may be on the stairs, but it does not allow members to cross over on each floor.

Isolated stairs only provide access to the individual apartments from a small landing. This landing can become crowded with firefighters trying to access the apartment. When these stairs are identified, the information must be relayed to all members on the fireground, and a hoseline should not be committed until the exact fire apartment is known.

Fire escapes are the most common form of secondary means of egress from these buildings. Some large buildings may have been built with **fire-rated doors** in the throat, making the stairs in the opposite wing the secondary means of egress.

The apartment layouts of H-types differ than tenements. In H-types, the layout is based on their position in the wing and proximity to the throat. Larger H-types can have an apartment in the throat with entrance doors on either side of the fire-rated doors.

Gaining entry into affected apartments is dependent on the number of apartments involved. It may require more than two teams to gain entry, search, and examine for extension. It is common to have to enter every apartment in the vertical line of apartments to examine for extension.

Success at fires in multistory multiple dwellings—both tenements and H-types—is often a result of proper ventilation. Vertical venting and roof operations are a key to that success. The interior stairs act like a chimney. Heat and smoke are carried up the stairs and start to bank down on the top floor. Venting skylights and bulkheads at roof level reduce the smoke and heat, making the stairway more tenable for escaping occupants and firefighters. In these buildings, vertical ventilation must be coordinated with the hoseline advance, particularly when considering flow path issues.

For fires on floors below the top floor, good horizontal ventilation is normally sufficient to prevent extension. Often this is accomplished by a member with a hook or a Halligan tied to a piece of utility rope from the floor above.

For fire on the top floor, the combination of horizontal ventilation and a large vent hole in the roof reduces the chance of fire entering the cockloft. Extension to the cockloft is a major concern. The cockloft is the attic space above all the top floor apartments. Because of the size of the building and the need to achieve pitch on the roof for drainage, the height of the cockloft can vary from several inches to several feet. Once fire has entered the cockloft, it can race through the space fueled by the enormous amount of timber necessary for the roof support. Firestopping may be limited and ineffective. A trench cut may be used as a defensive measure (if the primary vent hole does not stop the spread) to prevent the fire from completely taking over the cockloft. A trench cut is a cut in the roof several feet wide that becomes a fire break between firewalls. If a trench is considered, it should be cut early in the operation and not pulled until it is evident that fire has entered the cockloft (fig. 27–16). A hoseline should be in position on the safe side of the roof and all members removed before the trench is pulled. The trench is not a second vent hole. A hoseline can be operated into the trench. By sweeping the stream side to side, you can prevent fire from crossing the trench. Operating the hoseline in a trench cut works like operating a hoseline into the attic space from the top floor of a garden apartment—by preventing the fire from crossing a void space and affecting additional units.

Fig. 27–16. A trench cut is a cut in the roof several feet wide that becomes a fire break between firewalls.

Stretches in these buildings can require two or more engine companies to complete. Long stretches are common. Few were constructed with standpipe systems. Adaptation of SOPs may be necessary. Supply lines with water thief fittings or manifolds deployed in similar fashion to those used for garden apartments may be practical to reduce the number of lengths in the stretch, thereby reducing friction loss. To get the line into position quickly, take full advantage of opportunities to use **well hole**, rope (fig. 27–17), and fire escape stretches (fig. 27–18).

Fig. 27–17. A rope was deployed from the front apartment to haul the line to the fourth floor for the fire in the rear.

Fig. 27–18. A line was brought up the rear to the fourth-floor balcony and advanced to the fifth-floor apartment.

RETAIL COMMERCIAL BUILDINGS

Taxpayers and strip malls

The term **taxpayer** is used to describe an older, inexpensively constructed commercial building put up on a piece of land to generate income for the property owner in order to offset the cost of taxes. That is why they are referred to as "taxpayers" (fig. 27–19). Strip malls are the modern descendent of the taxpayer. Although similar in outward appearance, the differences between them are a function of when they were constructed.

Original taxpayers were typically constructed of brick load-bearing walls and wood joist floors and roofs. Although some were built before 1920, most were built between 1920 and 1960. Some of the new taxpayers were two stories, with apartments or office spaces on the second floor. Many had basements with access through a trap door in the floor inside or sidewalk doors outside. The size of building and number of stores varied and depended on the size of the building lot. They are built directly adjacent to the street with no setback, unlike their modern descendent.

Fig. 27–19. Single-story taxpayers typically have 5–10 individual tenants.

Strip malls have been and continue to be built to accommodate the needs of people as they move from the city into the suburbs. These buildings are much larger than urban taxpayers because of the availability of land, with setbacks and contiguous parking lots. They are most commonly built of ordinary construction: exterior masonry walls of concrete block with wood joists/ lightweight wood trusses. They may also be constructed of concrete block walls or tilt walls with steel bar-joist roof trusses and a built-up metal deck roof. The roof structures are not typically fire-rated. As tenants come and go, alteration may have left the tenant separation walls in disrepair; most separations are non–fire-rated (fig. 27–20). (Commonly, the only fire-rated separations are between different types of occupancies that pose a life safety hazard such as places of public assembly—restaurants and night clubs—and retail stores that deal in hazardous materials, such as swimming pool supply stores.) They are generally single story with relatively high ceilings. Heating, ventilation, and air-conditioning (HVAC) equipment is roof mounted and ductwork is run between the trusses and above the drop ceiling. Service areas and loading docks are in the rear, and customer parking lots are in front. They may have fire sprinklers, but the water supply is often from a supply pipe that also supplies private hydrants and may be a dead-end main. Carefully preplan to avoid disaster.

Fig. 27–20. Walls and partitions between occupancies in strip malls must not be trusted for their fire integrity without proper reconnaissance.

Big box retail stores

Big box retail stores are as the name implies: a big box, square, or rectangular shape for maximum efficiency of floor space. They include the national home centers and the warehouse clubs. In many ways the construction of big box stores is similar to strip malls, concrete walls, and steel truss roofs. They may have 100,000 sq ft (9,290 sq m) of retail space indoors and outdoor space of 20,000 sq ft (1,858 sq m). Typically, there is no fire-resistive rating to the steel bar joists; they are bare steel. A moderate or large fire impinging on the trusses will affect their structural integrity. The height of these stores is larger than strip malls, so expect inside ceiling heights of 30 ft (9 m) (fig. 27–21). HVAC equipment may be located inside the store. High-rack storage of a wide variety of products is the norm.

Fig. 27–21. High ceiling heights with truss components for roof construction are common in big box retailers.

Now that the types of retail stores have been identified, we begin with the size-up. How big is the store or stores? What is the occupancy? Is the store a high-hazard occupancy full of chemicals, flammable liquids, or paints? Do not be fooled; even a seemingly benign storefront may be full of hazards. Remember, these are commercial buildings built and operated for the sole purpose of making money. Life safety may not be a top priority, including the safety of firefighters.

So what *should* be considered in a life hazard assessment? Is the fire during the day? Is the store occupied? Does a retail store have a true and accurate accountability of the number of persons who may be in the store? They should know how many employees are working, but can they account for them all? This requires a thorough search of all building areas. Do not overlook bathrooms and stockrooms. Make sure all areas near exits are checked, even rear exits that may only be used by employees. Be aware that many big box stores have required exits along the sides of the store that are often illegally locked. Make sure people are not trapped behind these doors.

Firefighters entering commercial buildings should use a search line or hoseline when entering the structure. Conditions can change rapidly and the rope or the hoseline may be the only guide to the exit. Thermal imaging cameras are a critical life safety tool. High ceilings and large open areas may make detection of actual fire conditions difficult.

Can you determine what is actually on fire? Fires in commercial building often start in the rear storage areas and in service areas where utilities are located. There may be a maze of halls and corridors to reach these areas, so remember the search rope! A search rope also allows firefighters advancing a hoseline to follow your path. Would you be able to tell them how many right or left turns you made? Or is it easier to tell them to follow the search rope? The search rope also helps others find you if you become trapped or entangled in the stock or contents and can't free yourself.

One advantage in the newer, larger commercial buildings is that they are required to have a fire sprinkler system. Sprinkler systems can be valuable in controlling a fire at an incipient stage. They can also help you locate the fire. Water flowing out of an area or the sound of the sprinklers flowing can direct the search to the fire area.

If the fire is at night, entry into the store may be difficult. Store owners have a great deal of money invested in stock, and to protect it they lock the doors. Depending on their insurance rates, they may also have roll-down gates or other security devices to protect their investment. Store owners only want people in their store when they can sell them something. Some commercial buildings have electronic fire detection systems—smoke and heat detectors—but even with these systems gaining entry into the store delays fire extinguishment. Be aware that some big box retail stores have cleaning and stock people in the store late at night.

An important point to emphasize is that when a store is locked and limits our way in, it also affects our way out. Rear doors may be bolted or barred. In the case of big box stores, the side exit doors may also be locked illegally. Firefighters assigned to forcible entry in commercial buildings may require additional tools to complete their tasks. When operating at commercial buildings, open all doors and advise all members of their location; if a firefighter must make an emergency exit, he or she can choose the closest exit. Also, rear doors often perform the critical function of serving as point for horizontal ventilation as a hoseline is advanced from the front of the store.

Ventilation in such commercial buildings is limited at best. Roof operations are restricted to removal of skylights and scuttles, *unless fire has heated the steel bar joist roof truss*. If fire has compromised the metal joists or is large enough to do so, conduct no roof operations, because the moment of failure cannot be predicted. These trusses warp, twist, and lose strength early on in a fire when exposed to heat. Unprotected steel lacks fire resistance. Switch to a defensive, **exterior attack**.

Tilt slab walls depend on the roof system to hold them upright. Any roof system warping or twisting that would lead to roof failure can allow the walls to lean. The entire structure could collapse.

If the structure is built with a corrugated metal roof deck, many fire departments avoid sending firefighters to cut the roof. The metal panels that make up the roof system do not always end on a roof beam or truss. When they are cut, sections can hinge inward and drop a firefighter into the fire. Review your department's SOPs to determine its policies for ventilation of metal deck roofs.

Other roof systems of gypsum plank and gypsum concrete should not be cut as well. These systems are prone to moisture and can deteriorate leading to failure. They should be identified during the planning and approval process, but they can be identified during operations when the saw cut reveals a white powdery residue. In this case, notify the incident commander and prepare to evacuate the roof.

You may ask, "So how are we going to know this?" Buildings are built over a time of weeks and months. You need to be aware when new construction is going on in your area and get out and look at how these buildings are being built, when they are being built. Go to the job site, speak to the people who are putting it together, and find out how it is done. If you know how it goes together, you can understand how it will fall apart! Pre-incident planning is critical.

So how do you ventilate commercial building? It is usually best to remove the available windows. If the building is sprinklered and they are flowing, sprinklers cool the smoke, preventing it from lifting vertically. A fan may assist in horizontal ventilation once the fire has been knocked down. Positive pressure ventilation is popular in many areas of the country and requires close coordination to be effective.

In the case of sprinklered buildings, incident commanders do not order the sprinkler system shutdown until the fire has been knocked down and hoselines are in place. In all cases, a firefighter with a radio is positioned at the valve to reopen it if the fire rekindles.

When operating hand lines in commercial buildings, consider the fire load of the store. With all the material in the store, the fire quickly progresses past the extinguishing capability of a 1¾-in. (45-mm) line. Also, judge the reach of the stream; for all reasons just stated, an incident commander may not allow firefighters to enter the building. A 2½-in. (65-mm) hand line is usually required. In the case of a large fire, a rapid attack monitor may be required. These devices may have the capability to reach the rear of the store without placing firefighters in the collapse zone.

If the decision is made to enter, the exact location of the fire must be known and enough line stretched to reach the fire, as with all fires. Select the entrance point closest to the fire, which puts the shortest amount of hose inside the building. The firefighter following the hoseline to the exit is out of the building in the shortest time. Do not let firefighters come off the line if they may be unable to return to it. In commercial buildings with aisles, displays, and cubicles, landmarks and reference points may be difficult to establish. In commercial buildings the hoseline is a *lifeline*.

If the multistory, older department store has a standpipe system, what size is the riser supply pipe? Is the hose valve in a protected location such as an enclosed stairwell? Can the standpipe supply a large hand line? Is the system actually a class II occupant use–only standpipe system (fig. 27–22)? Are there 2½-in. (65-mm) hose valves in the stairwell? If the decision is made to stretch lines from the standpipe, carefully consider their use in a large floor area store. How much hose is needed to reach the fire area given the layout of the sales area?

Fig. 27–22. 1½-in. (38-mm) piping is inadequate to supply a large hand line; this hoseline is intended for occupant use.

In the case of large big box stores, 1½-in. hose stations may be placed in the storage rack areas; these feed directly off of the sprinkler system, intended as mop-up lines. They are not considered a standpipe system. In large buildings with multiple sprinkler risers, each hose connection is typically fed with piping from an adjacent sprinkler riser (allowing shutdown of the sprinkler riser in the fire area while still maintaining a water supply to the hose connection from another *uninvolved* sprinkler system). These small hoselines are also designed to operate at the same pressure as the sprinkler system—typically insufficient for normal firefighting use. If 1½-in. (38-mm) connections are to be used by firefighters, the fire department connection must be supplied to provide adequate pressure.

When a commercial building has a sprinkler system or if the standpipe will be used for firefighting, a line must always stretch to supply the system through the fire department connection (Siamese). This should be one of the first lines stretched, ensuring that as additional heads fuse or the line is charged, enough water contains the fire and prevents the fire from over-running the sprinklers; however, it still could happen if the store owner has loaded the store with more stock than the system was designed for.

If the water supply for the sprinklers or standpipe comes from the same main as the hydrant to which the fire department pumper is connected, could you affect the others' water supply? Communications with interior crews on the performance of the sprinklers determine if it is so. Consult water district maps or public works employees if a problem exists.

COMMERCIAL OFFICE BUILDINGS

Low- and mid-rise commercial office buildings (structures up to seven stories) are being built throughout the country (figs. 27–23 and 27–24). They have all the same conditions as the taller urban predecessors, beginning with HVAC systems, firefighting operations from a standpipe system, and large-area searches.

Fig. 27–23. Low-rise commercial office buildings are common throughout the country.

Fig. 27–24. Mid-rise offices pop up as a result of suburban sprawl.

Let's examine the construction of these buildings using our seven-point methodology. The construction of these buildings is generally a steel frame with poured concrete floors over a steel Q-Deck (fig. 27–25). The exterior walls can be prefabricated **curtain walls**, where the preassembled panels interlock with each other and are attached to the floor slabs (fig. 27–26). This installation method can produce a space between the outside wall and edge of the floor slab. If this gap is not properly fire stopped, it can allow extension to the floor above the fire. Walls can also be constructed of concrete block with a brick veneer or made completely of glass to meet the desires of the architect.

Fig. 27–25. A steel frame supports the Q-deck.

Fig. 27–26. Before the concrete is poured into the Q-deck, the slab is reinforced with wire mesh and studs are welded to the deck to anchor the concrete to the decking.

The floor layout often uses a **center core** design, where all mechanical/electrical equipment and stairwells are located. Elevator and stair shafts as well as bathrooms and electrical and communications closets are configured in the center of the floor plan, leaving maximum area around the perimeter for the tenant space. The tenant space generally has a perimeter of walled offices for executives, and the space between the offices and core contains cubicles for assistants and support persons.

Large office buildings, including all high-rises, have a fire command center near the main lobby of the building. The main fire alarm control panel, elevator status panel, emergency generator controls, public address system, and other firefighting equipment are located here.

Hundreds of people can be on each floor of an office building. Thus, accounting for missing persons during emergencies can present a challenge. Security-conscious firms may have some sort of electronic tracking system for employees that may provide some accountability; use it as a guide and search the fire area within the limits of safety.

A large-area search plan is crucial to maintain crew integrity. Reliable aids include search ropes and tag lines for firefighters who need to expand the search area into cubicles. It is important for the search team leader to have a thermal imaging camera. It not only enables the leader to look for victims, but it will help maintain accountability for the crew. A leader with a clear picture in a camera can guide a firefighter toward a victim or back to safety. The camera is an important tool while searching to monitor the heat conditions. Viewing the thermal layer in conjunction with a fixed object indicates whether the heat is banking down. Occasionally monitoring conditions overhead is also important.

Fire extinguishment in these buildings requires a high-volume, far-reaching hoseline. The rate of heat release by combustibles in this type of occupancy is much higher than in a single-family dwelling. You wind up with a hotter fire and more of it. Fire, heat, and smoke can easily be pushed around the core if the engine crew is not careful. The area overhead must be monitored to ensure the fire does not pass the advancing crew by means of the plenum area above the drop ceiling.

Entry into these buildings is unique. Some owners have provided the fire department with keys to lock boxes outside that contain master keys to the building for access during off hours. When these are unavailable, many locking devices can be encountered, from a simple tubular lock in an aluminum frame door to electronic magnetic locking devices. Many office buildings, particularly multitenant high-rises, use locks on stairwell doors to prevent reentry back into the tenant space. Buildings equipped with such locks are required to be outfitted with release devices that unlock the doors when they receive a signal from the fire alarm system or a *manual* signal from the fire command center.

Building service areas, mechanical equipment rooms, fire pump rooms, and the like will probably be locked and require forcible entry. *Note*: Just as you were taught to try the knob first, gaining entry is often not about force. A little finesse goes along way!

Access to upper floors of these buildings requires the use of elevators. Members walking more than several floors with all the required equipment is debilitating. The firefighters arriving on the fire floor must be prepared to commence operation quickly.

During a fire emergency, elevators are recalled to the first-floor lobby (or designated level), known as a **phase I operation**. The operational phase of elevator use by firefighters is called **phase II operation** or **firefighter's service.**

Under phase II operation, the firefighter has control of all the functions of the elevator cab: floor selection, up, down, door open, and door close, and so forth. Refer to your department's SOP on when the phase II operation's use of elevators is permitted. A thorough understanding of the function of phase II operation is required for the member operating the elevator to perform it safely. He or she is as responsible for delivering the firefighters to the upper floors as the driver of the apparatus that drove them to the building.

When operating elevators, it is a good safety tip to look up the shaft before you get into the elevator. Shine a flashlight between the elevator car and the hoistway door to see if there is any smoke or water entering the shaft. If the shaft is clear and you are ready to leave the lobby, select a floor, and as the car begins to move, press the *call cancel* button. This clears the floor selection and the car should stop at the next available floor. This verifies that the phase II operation function works properly. When selecting a floor to get off the elevator, two floors below the reported fire floor can be appropriate. If the elevator does not behave properly, *do not use the elevator.*

Be aware that many tall buildings use multiple **banks** of elevators that serve different groups of floors or **zones** (floors 2–20, floors 21–40, etc.). In many cases elevators serving the upper floors pass through **blind shafts**, which have no doors on the lower floors.

Using elevators during a fire can be dangerous. The key to safe firefighting operations using elevators is understanding the features of elevators, the potential hazards, and your department's elevator SOPs.

Beware of **access stairs**, which are open (unenclosed) stairs within a tenant space that serve adjacent floors. A smoke detector could activate on an upper floor served by the access stairs from a fire below, which could have you exiting the elevator on the fire floor. If the occupancy has access stairs, take the elevator two floors below the lowest floor served by those stairs. How would you know if access stairs exist? Ask building personnel. If they don't know, look at the tenant directory. If the same tenant has offices on two floors, it is possible that they installed an access stair. Law firms often install access stairs.

Usually, these buildings cannot be ventilated by breaking windows. Broken glass from floors high above the street can sail a great distance, injuring innocent bystanders. Glass dropped from lower floors can damage firefighters' operating apparatus and cut hoselines, leading to water loss issues. As a general rule, before any glass is broken in a high-rise fire the incident commander should be notified so the street below can be secured. Some high-rises, however, are equipped with safer, tempered glass, or have window panels that are openable with keys designed specifically for firefighting ventilation. Pre-incident planning is critical to establish the presence of these windows. (*Note:* wind conditions at the level ventilation is required may be very different that at street level. Carefully consider any actions that can't be reversed.)

Depending on the sophistication of the air handling equipment, fans can be turned on and dampers turned to an exhaust mode to clear out smoke. This should only be ordered by the incident commander and performed by qualified building personnel. Smoke detectors required by your local building code may be located inside the HVAC ductwork and prevent you from using the system. If the HVAC system is used for such operations, all firefighting units should be made aware so they can report any adverse effects. In some cases, the building may actually have a smoke control system. Make sure you gain an understanding of how the system operates *before the fire*, while conducting pre-incident planning.

Hoselines usually are stretched from a standpipe in the stairwell. Some jurisdictions have the standpipe connection in a hose cabinet on the floor itself. Never hook up to these outlets on the fire floor. The floor below the fire floor provides a safe place to hook up the initial line. If additional lines are required, they must come from floors below. The line must be flaked out up the stairwell; and, if the attack on the fire begins at the door to the occupancy, the line must be charged. Never enter the fire area with a dry hoseline. It is safer for the fire attack to begin from the protection of the stairwell as this will provide an area of refuge if conditions change on the fire floor.

Before the attack begins, position the line on the stairs for a rapid advance: Flake the line up the stairs so the weight of the line assists the advance. Be sure to check the stairs for escaping occupants before opening the door to the fire area. You don't want to expose people to the heat and smoke when the fire attack begins from the **attack stairwell**. Any occupants in the attack stairwell must be brought in on floors above and transferred to the other stair *before the attack begins*. If possible, with the building's public address system, alert occupants to use another specific stair for egress, called an **evacuation stairwell**. The evacuation stair door should never be opened on the fire floor to prevent contamination, and all entries on to the fire floor must come from the attack stair.

Do not overlook the possibility of stretching lines from the apparatus for fires on lower floors. Remember, the goal of the line is to get water on the fire as quickly as possible. A good expression to remember is that "everything is a possibility until it has been eliminated."

RESTAURANTS

Restaurants can be found in almost any type of building. The construction characteristics of the building do not change because of the restaurant; they are only compounded by it. Most restaurant fires are a result of issues from the kitchen, but smoking materials carelessly discarded into garbage cans can easily start a fire. We look at the kitchen in depth.

Commercial kitchens are an obstacle course of hazards: high-voltage electricity, high-volume gas piping, stoves with open flames with pots full of boiling liquids, fryers full of superheated oils, ovens, microwaves, and slippery floors. And that is all *before* the fire starts!

Before operating a hoseline in a commercial kitchen, control the utilities. Turn off the gas before you put the fire out. Fires involving the fryers may require the use of a specific extinguisher. They are no longer *deep fat* fryers; because of health concerns, restaurants cook in a light oil. Simply cooling the oil is not enough to extinguish the fire and prevent it from reigniting. A type K extinguisher should be available on premise for this purpose. If the fire-suppression system has not extinguished the fire, you may need to employ the K extinguisher.

All cooking equipment that produces **grease-laden vapors** is required to be located under the protection of an exhaust hood with an approved fire-extinguishing system. There are **filters** in the hood that are designed to trap the grease and oily vapor (fig. 27–27). These filters must be cleaned on a regular basis as set forth by the jurisdiction. Although the filters may be cleaned, some grease and oil can enter the exhaust ductwork. Exhaust ducts are a heavy-gauge welded design to prevent fire from breaking out of the ductwork. If a fire is not extinguished by the fixed system or is hot enough to ignite the grease build-up in the ductwork, they become very messy to put out.

Fig. 27–27. The grease filters under the hood trap grease vapor, and the vents in the front are for the make-up air.

Some fire codes require that a diagram of the ductwork be posted in the kitchen. A firefighter can use this diagram to trace the duct to its termination outside the building (fig. 27–28). These codes also call for an access panel to be provided every place the duct makes a turn, providing a point for contractors to clean the duct and remove any buildup. However, it works more effectively as a point for the fire department to apply water.

Fig. 27–28. The exhaust duct exits the building into a shaft. Note the access panel on the turn and the location of the make-up air louver above

Using a thermal imaging camera to determine if the fire has entered the ductwork can be challenging because almost everything in the kitchen is hot or giving off some heat. The duct may be naturally hot because it is exhausting the air from over a hot stove. It takes a skilled firefighter to interpret the thermal image correctly. It also provides an excellent training opportunity when the fire is under control.

The range hood of a commercial kitchen has another set of ductwork for **make-up air**, which replaces the air being sucked out by the hood's exhaust. The fans in this ductwork provide more volume than exhaust fans to create a positive pressure in the kitchen and keep the grease vapor under the hood. When following the run of the duct, do not get confused and follow the wrong duct. The make-up air duct may not terminate in the same location as the exhaust. The grease duct is made up of heavy-gauge welded steel, while the make-up air duct more resembles air-conditioning ductwork with lightweight assembled in sections with noticeable seams.

Commercial kitchens floors are notoriously slippery. Even the smallest amount of water on greasy floor creates a dangerous condition. Salt applied to the floor effectively improves traction. Dump out a few shakers of salt in the area where you are standing; the effect will be noticeable. If more salt is needed, find the bulk items storage and use what you need. It is an easy way to prevent an injury.

VACANT BUILDINGS

Vacant building fires pose one of the most serious threats to firefighters. Fires in vacant building can occur in every community across America. With the ebb and flow of the economy, some areas see this more than others. The vacant building has been abandoned by the owner and fallen into disrepair. The extent of the damage is unknown. Floors, walls, and stairs are all of questionable stability.

So why do we fight these fires? Do we ever know if the building is truly vacant until it is searched? Does letting it burn create a hazard greater than the building's worth? Do we extinguish the fire so the property owner has more to haul away? The obligation to extinguish all fires must be balanced with the moral obligation to the safety of our firefighters.

Fires in vacant buildings should be considered suspicious: They happen for a reason. They may be set to cover up a crime, as an act of vandalism, or as arson for profit. A fire investigator determines arson after the fire is extinguished and all accidental causes have been ruled out. The reason the building is on fire is of less importance than the fact it is on fire. In a fire of this type, the intent is to do as much damage in the shortest period of time as possible. Even if the fire department extinguishes the blaze, the damage would have been maximized.

How does that affect how to operate? Conduct a good, comprehensive size-up. Looking closely at the house may indicate several key elements in fire attack plan. If the building has been abandoned by the owner, the electric service wires may be cut, the meter may be removed and the gas service may be locked out. This may eliminate many *accidental* causes of fires, leading to a greater suspicion of arson.

Maximum effort must be directed to extinguishing the fire quickly. All firefighters must work to get water on the fire as quickly as possible. In addition, a minimum commitment of firefighters is assigned to the hazard area.

This is one type of fire where we adjust our aggressive interior attack methods in favor of a more cautious approach. Water may be prudent if it can be applied from the outside without exposing firefighters to the unknown dangers of an interior attack.

In many cases, it is simply too dangerous to commit interior forces to fight a fire in a vacant or abandoned building because of the size of the fire, the existing poor conditions of the building, and the lack of fire protection features. No firefighter's life is worth that of any building, particularly vacant or abandoned ones.

Firefighting operations in vacant buildings must be conducted methodically with an emphasis on safety, concentrating on the known life hazard, the firefighters. All the voids that are present when a building is occupied are still present while it is vacant. Additional paths for fire spread may be present from deterioration from elements, either natural or man-made. Vagrants and squatters may have taken shelter in the building and caused damage. Others may have vandalized the building by removing fixtures and other items strictly for their salvage value.

Every department should have a policy for firefighting operations in vacant buildings. The first arriving officer generally determines the fire attack plan, based on fire conditions, age and condition of the structure, exposures, and especially life hazard. Firefighters must remember that firefighting operations are conducted *within the limits of safety* with a minimal commitment of firefighters.

An important consideration in vacant buildings is what the previous occupant left behind. Take time in your own size-up to note the areas of concern. Maintain situational awareness: is this building now vacant for a reason? For firefighters with portable radios, listen to all the transmissions and try to piece together the puzzle, as the incident commander is doing.

FLAMMABLE GAS FIRES SKILL DRILL

FFII 5.3.3 Responding to flammable gas fires and emergencies are a part of your responsibility as a firefighter. You may be faced with several different gases. To effectively respond to these alarms, you must first understand the properties of the gas. As you will recall from your Firefighter I training, some gases are heavier than air and some are lighter than air, based on vapor density (air = 1). Obviously, this affects where vapors collect when they are released. Gases heavier than air collect in low areas, and vapors lighter than air tend to disperse in the air. Flammable gases are stored in cylinders or delivered by utility gas systems.

We are not discussing specific gas systems that may be found in hospitals or specific manufacturing facilities that may use gases in their processes. If these industrial facilities are in your response area, discuss the operating procedures with your company officers.

We will discuss the most common gases and where they can be found. One of the most common gases is natural gas. It is delivered to homes and businesses for cooking and heating. It may also be found as a compressed gas stored in cylinders as an alternative motor vehicle fuel. These vehicles are placarded as CNG (compressed natural gas) vehicles.

Natural gas that is piped into your community is made up almost exclusively of methane, but it may have trace amounts of other gases. Natural gas is lighter than air, colorless, and odorless in its natural state. For consumer use, the chemical ethyl mercaptan is added to give the gas a distinctive odor. The odor of mercaptan can be detected by our senses at a concentration of 20% of the lower explosive limit, well below the explosive range of the gas. The explosive range of natural gas is between 5% and 15%, so the odor can be detected by a person with a normal sense of smell at around 2% gas in air. However, do not rely on your sense of smell to determine the percentage of gas in a compartment.

Natural gas ignites at 1,163°F (628°C). The gas is considered a simple asphyxiant. Inhaling the gas (in higher concentration) can displace the oxygen in your lungs and cause asphyxiation.

Natural gas is delivered to homes and businesses through underground piping. Often the gas piping is steel, but some newer installations are using plastic gas pipes. The gas is supplied from a main to individual houses or industrial services. There are several points of control on the distribution system; firefighters do not shut down any gas valves that are in the roadway. These shutoff valves often have control of large portions of the distribution system and affect many homes and businesses.

Low-pressure gas is delivered to homes and businesses, except for some commercial facilities. Controlling leaks can be relatively simple. In attempting to control a gas leak, the first point of control is at the appliance. Typically, each appliance has a quarter-turn gas valve to shut the off the flow of gas. When the handle of the valve is perpendicular to the piping, the valve is closed (fig. 27–29). When it is in line with the piping, the valve opens, allowing gas to flow. Inside a structure the gas usually is in steel threaded pipe (black iron). There may be a stainless steel flexible connection to the individual appliance from this black iron pipe. The shutoff valve is located behind the appliance or in an adjoining cabinet. Note the position of the valve indicating the gas is on (fig. 27–30). This appliance connects with a flexible line to the stove.

Fig. 27–29. Two gas valves: the one on the left is open, the valve is parallel to the pipe; the one on the right is off, the valve handle is perpendicular to the pipe.

Fig. 27–30. The gas supply to any appliance is required to have a shutoff.

Following the gas pipe back toward the street, the next point of control is a valve on the main line where the main enters the building (fig. 27–31). The size of the main and whether the building is residential or commercial dictates the gas valve style (fig. 27–32). The head of service valve is still a quarter-turn design. Generally, they are larger versions of the quarter-turn valve at the appliance.

All gas services have a gas meter that can also be shut off. The meter also often serves as a pressure regulator. The meter may be located inside or outside the building. A valve on the meter can be used to shut down the gas, or it may be on the riser where the gas piping comes out of the ground (outside the building) or through the exterior wall (inside the building), before it enters the meter. Commercial and residential meter piping may be different in style but are similar in function. Note the gas valve where the riser comes out of the ground and the other before the gas meter (figs. 27–33 and 27–34). Both valves have rings, so the meter can be locked out by the utility. Note the size of the gas line as it comes from the ground. The smaller pipe indicates that the service line to this building is higher pressure. Note the size of the pipe where it enters the building.

Fig. 27–31. A residential gas line is controlled by a quarter-turn valve.

Fig. 27–32. Commercial gas lines distribute a much larger volume of gas and thus have much larger valves and piping.

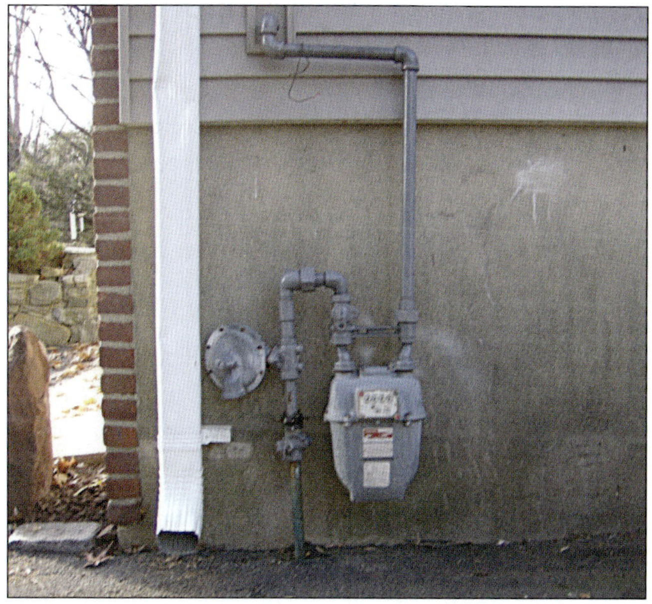

Fig. 27–33. Firefighters should become familiar with residential gas meters since they may be called upon to control gas during emergencies.

Fig. 27–34. Commercial gas meters are more complex than the residential variety.

Valves located outside may be difficult to turn because of exposure to the elements. When these valves need to be closed, a small pipe wrench or the fork of a Halligan tool is generally enough to get the valve to turn. If it can't be turned, don't force it. Breaking off the valve at this location severely compounds the problem. Summon the utility crew for assistance.

The last point of control for shutting off the gas is the **curb valve**, located between the gas main in the street and the building. This is the fire department's last way of stopping the flow of gas to a building if all other options are impossible—it is located outside of the building. It is important to note that some fire department SOPs do not allow firefighters to operate this valve—this is left to the utility company. The curb valve is located under a steel cover and in a pipe casing. The valve is at the bottom. A long-handled wrench is required to operate this valve. In newer installations the valve body may be plastic and break easily when forced. Contact the local utility company in your area for more specific information on their underground equipment.

It is important to remember that whenever the gas is shut down to an appliance or building, the fire department does not reopen the valves. Some gas-fueled equipment has specific relighting instructions, and relighting must be done by licensed individuals. Other areas require that the utility be notified whenever the gas is shut off to any appliance. Check with the gas utility for local requirements.

When called to a natural gas leak, *ensure that the gas utility has been called to respond*. You must remember the properties of the gas. The gas is lighter than air and mixes easily in a confined space. In a confined space, the gas air mixture is explosive. For a suspected gas leak, firefighters should not turn on any lights, ring doorbells, or cause any sparks within the building. Turning off lights can cause a spark as well. If lights are on, leave them on; if lights are off, leave them off. Firefighter handheld radios and flashlights may also be a source of ignition. Turn on flashlights before entering, and do not speak on the radio until the atmosphere is deemed safe.

Obviously, when attempting to locate the source of the leak, never check for it with an open flame. Use a combustible-gas meter or detection equipment (fig. 27–35). Decide how to handle the incident based on the concentration of gas in the building and its ability to be controlled. With low concentrations, it is possible to locate the source of the leak and stop it. Uncontrollable leaks or leaks resulting in moderate or high concentrations of gas are too dangerous to allow firefighters to remain inside or near the building; firefighting units must be evacuated along with people in surrounding buildings and areas. In such cases, gas utility crews must shut off the gas remote from the structure (fig. 27–36). Liquid soap mixed with water to make soap bubbles can be used to check for small gas leaks on piping. When the solution is applied to a leaking pipe, the bubbles expand, indicating a leak.

Whenever the presence of gas is detected, full protective equipment is required. Self-contained breathing apparatus (SCBA) use is mandatory when investigating gas leaks within a structure. When the source of the leak is located, it is best to turn off the gas at the control point closest to the point of the leak. This minimizes the disruption to the home or business owner. Sealants and putty only slow the leak until it can be shut off at a valve. Necessary repairs may not be made if the leak can be controlled without turning off the gas.

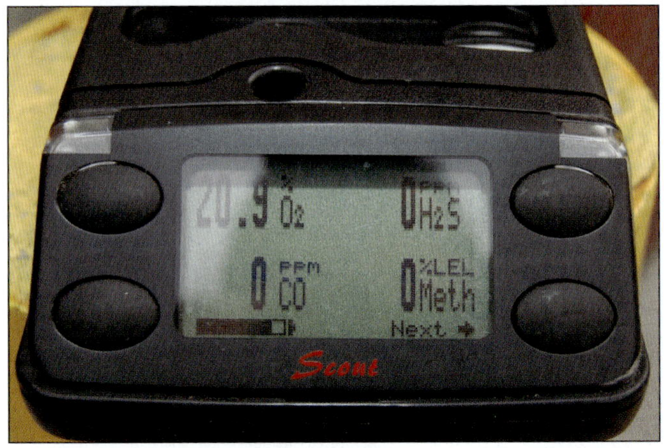

Fig. 27–35. Firefighters should learn how to operate all detection meters available in their department.

Fig. 27–36. Gas explosions can be extremely violent and destructive. (Courtesy of James O'Connell)

Once the leak has been controlled, the area should be well ventilated. Allow the area to vent naturally. Vent the upper floors first; this allows the gas to dissipate more readily, because the gas is lighter than air. Work your way down to the lowest level, allowing fresh air to enter. Do not use power vent fans unless they are intrinsically safe for explosive atmospheres. If it is necessary to use fans, they must be used in a positive-pressure mode to blow fresh air into the building and avoid a source of ignition.

Gas leaks located outdoors may involve the high-pressure side of the distribution system. There are important safety considerations when dealing with high-pressure gas. A relatively small pipe can dispel a large amount of gas, creating a large vapor cloud. When the gas leak is outside coming up through the ground, the odorant ethyl mercaptan does not pass as easily through the soil as the gas. The odorant is said to have been **scrubbed out**. In all cases, gas leaks must be verified by metering equipment.

Many times, outside gas leaks are the result of construction accidents, often from heavy equipment digging near underground utilities. When the gas line has been pulled out of the ground, resist the urge to plug it. The escaping gas causes static electricity to form on the piping. This small electrical charge may be enough to generate a spark and ignite. When the pipe was pulled from the ground, it may have created a void in the soil, which can cause gas to follow the pipe back into structures. Pulling the pipe may also have caused another break further down the line. Gas may be escaping into the ground undetected. If the gas leak runs under asphalt or concrete, the gas may travel great distances before being detected.

Utility crews have equipment to test for underground gas leaks. Firefighters must check all surrounding properties and areas below grade, because gas may permeate storm drains and septic systems. Eliminate all sources of ignition in the area of the vapor cloud. A fog stream may help disperse the vapor. If a hoseline is used near an excavation, use caution and prevent water runoff from entering the trench. This complicates the repair process.

If the outside gas leak has been caused by construction, evacuate as required. Stretch a precautionary line to protect exposures and possibly disperse the vapor. Remove all sources of ignition, staging apparatus upwind. As with all flammable gas, do not extinguish the fire until the flow of gas has been turned off.

Compressed natural gas can be used as a motor fuel in passenger vehicles, but identifying these vehicles may be difficult. Some states do not require these vehicles to be labeled, whereas others require a small CNG placard to be affixed to only the front and rear of the vehicle. If your response is to an alternative fuel vehicle powered by CNG, once you have confirmed that it is CNG-powered, the first task is to shut down the gas flow. There is a main shutoff located near the fuel cylinder. The cylinder may be located in the trunks of sedans and other cars or in the beds of pickup trucks. Another inline shutoff may be located under the **rocker panel** where the fuel lines run along the frame rails to the engine.

If the vehicle is involved in fire, let the fire burn off and protect exposures. The CNG cylinder is protected with a **burst disc** (for overpressures) and a **fusible plug** (which melts under fire attack, preventing cylinder overpressure because of heating from the fire). A hand line used from a safe distance prevents the disk from rupturing or the fusible plug from melting. If the vehicle is involved in an accident, control any fuel leaks and eliminate sources of ignition.

Propane is another common flammable gas. It is used for home barbecues, industrial forklifts, and other equipment. In rural areas propane may be used for home heating and cooking. It may be stored in large cylinders outside the house and piped to individual appliances. Controlling leaks is much the same as described for natural gas emergencies, except remembering that propane is heavier than air.

Propane is odorized much like natural gas. The cylinders for home barbecues are typically 20-lb (9-kg) cylinders, so there are 20 lb (9 kg) of propane by weight in the cylinder. A 20-lb (9 kg) cylinder contains approximately 4.75 gal (17 L) of liquid. One gallon of liquid produces 13,500 gal (51,103 L) of combustible vapor with a lower explosive limit of 2.15% and an upper limit of 9.60%. This means that it takes little propane to create an explosive mixture in air. One 20-lb (9-kg) cylinder can create an explosive mixture to completely fill a 1,100-sq-ft (102-sq-m) house. Thus the propane cylinders should never be stored indoors.

Newer propane cylinders are equipped with an **overfill protection device (OPD)**. Built into the valve is a float to shut off the valve when the cylinder contains the correct amount of liquid. The plastic tube is connected to the screw on the side of the valve. If the cylinder must be filled without a scale, the screw is removed and the cylinder is filled until liquid flows from the screw indicating the cylinder is full (fig. 27–37).

Fig. 27–37. Overfill protection devices (installed inside tanks) help ensure that propane cylinders are not overfilled.

Most liquefied petroleum gas cylinders are designed to accept 80% capacity as a liquid. This allows 20% of the gas to expand under normal temperature conditions. Propane gas expands approximately 1.5% for every 10°F (5 5/9°C) rise in temperature. The pressure relief device on a barbecue cylinder is located in the valve, pointing in a direction opposite the hose connection. When approaching the cylinder, avoid approaching in a direction that places you in the path of the pressure relief device.

Cylinders equipped with the OPD device can be identified by the triangle hand wheel and the large external threads for connecting to the appliance (fig. 27–38). The older-style valves have a star-shaped wheel and internal threads. Some older cylinders can't be refilled based on the date of manufacture, so check with your local propane dealer for details (fig. 27–39). Outdated cylinders must not be discarded in the trash; check with the propane dealer or local government offices. Some municipalities have hazardous waste collections to accept this type of material.

Fig. 27–38. The triangular-shaped handle on this propane cylinder is indicative of it having an overfill protection device.

Fig. 27–39. Older propane valves without the overfill protection device (OPD) will have a star shaped handle.

Forklifts use 40-lb (18-kg) motor fuel cylinders. These cylinders are designed to be used and stored in the horizontal position and flow liquid to the vaporizer in the engine (fig. 27–40). Their pressure relief valves work in much the same way, but there are important differences in the two types of cylinders. Because these cylinders are designed to contain liquid propane, a fire or emergency involving forklifts often involve a liquid leak (fig. 27–41). Escaping liquid is a more serious situation than escaping gas because of the vaporization rate.

Fig. 27–40. Forklifts use 40-lb motor fuel cylinders designed to deliver liquid propane to the vaporizer on the engine.

Fig. 27–41. Forklift cylinders have ports for liquid service as well as vapor. Note: The cylinder is equipped with a liquid level gauge.

It is important to note that, like all stored liquid cylinders, propane cylinders are susceptible to BLEVE (boiling liquid expanding vapor explosion). When heated, the liquid inside the cylinder increases the internal pressure (fig. 27–42). If the relief device cannot relieve the pressure, the cylinder could fail. Pressure relief valves in propane cylinders are spring-loaded, operate at 375 psi (2,625 kPa), and close when the pressure in the cylinder is reduced. If the cylinder is exposed to fire, the pressure relief valve could operate several times before the cylinder fails. Because the contents are flammable, a failed cylinder releases large quantities of an explosive mixture. That could result in an explosion and fire. Propane should always be stored and secured outdoors (fig. 27–43).

Fig. 27–42. A BLEVE is a very dangerous situation, and all too common with the widespread use of portable gas grills. These are the remnants of a gas grill cylinder.

Fig. 27–43. Storage of propane cylinders must be in a marked secured location in compliance with applicable codes and standards.

Let's look at some general fires and emergencies you may be called to. This discussion is not intended to replace any standard operating procedures of your department but to give insight into successfully handling these situations. Consult your department policies for clarification if necessary.

The first involves a propane gas leak inside a home. As previously mentioned propane cylinders have no place in the home. If a leak is suspected by the distinctive smell of the odorant, confirm with combustible gas detectors. How bad it smells is not an indication of the concentration of gas in the air or indicative of the explosive range.

For a leak with no fire, the procedure is as follows:

- Eliminate any sources of ignition
- Ventilate the house
- Find the leaking cylinder
- Close the valve
- Remove the cylinder from the house.

A leaking propane cylinder with *low concentrations of gas* in the house will also necessitate the stretching of a handline. Teams of two or more should enter the home and search for the cylinder. Remember all the sources of ignition. However, moderate or high concentrations of propane usually call for the evacuation of all firefighting personnel as well as surrounding areas, such as in natural gas leaks.

When venting the house, only a positive pressure method should be used. Air must be blown in so as not to pull the explosive mixture toward a fan that is not intrinsically safe. Remember, when venting a gas heavier than air in areas where the vapor collects, fans may be necessary to disperse the gas.

Note: Before approaching a flammable gas cylinder fire, firefighters should identify a safe haven or area of refuge that will provide protection from heat, concussion, or flying debris in event of a rupture or explosion. They should seek an area or shelter that can provide protection from heat, and far enough away from the failure areas of a cylinder to prevent or limit injury.

With all cylinders involved in fire, the best course of action is to use large mounts of water applied from a safe distance and allow the contents to burn off while protecting exposures. If the valve must be shut down or a rescue is required, approach under the protection of two properly flowing fog lines. Approach the cylinder to close the valve; otherwise stop the flow of the gas, or effect the rescue. Should the fire be extinguished before the valve is closed, the valve should still be closed completely as soon as it is reached.

Once the valve is closed and fire extinguished, the hose team can retreat by backing away. Firefighters should never turn their backs on a flammable gas cylinder or tank fire in the event the fire should suddenly reignite. Once the hose team has retreated, they should continue to cool the cylinder from a distance using a straight stream.

Fires are common in gas barbecue grills for several reasons. For example, spiders nesting in venturi tubes cause gas to back up in the tube and ignite when the grill is lit. The hose from the cylinder dry rots and leaks, and the escaping gas ignites. These types of fire are common at the beginning of the grilling season when the equipment has not be used for several months and been exposed to the elements.

If responding to a gas grill fire and the grill is 10 ft (3 m) from the house as recommended, you might allow the cylinder to burn off while protecting exposures. If the fire is exposing the house, this may develop into a structural fire and require two hand lines to control. Both hand lines should work in unison to advance firefighters close enough to shut down the flow of gas. Once the gas has been controlled, one hand line should continue to cool the cylinder while the other goes to work extinguishing the structural fire.

When attempting to shut down the valve, the hand wheel may be plastic or light white metal; they may have melted from the impinging fire. If attempting to shut down the valve, a pair of locking pliers may be necessary to turn the stem of the valve. The valves on these cylinders turn off as others turn them clockwise to shut them down. The fittings that attach to the valve's internal threads are left-hand thread. Fittings that attach to the large external threads of an OPD valve are right-hand thread.

Fires in forklifts can be handled in much the same way: two hand lines advancing to control the leak and extinguish the fire. Use caution not to extinguish the fire until the flow of gas has been shut off. If the fire goes out before the flow of gas is shut, use the hose stream to disperse the vapor. With forklifts a liquid release is more likely, but this liquid vaporizes quickly. Once the liquid vaporizes, use metering equipment to check for any concentrations of gas in surrounding properties.

Another common type of flammable gas is acetylene. Acetylene is often used by plumbers for soldering pipe and by ironworkers for cutting and welding operations. The acetylene gas is dissolved in acetone in the cylinder to make the gas more stable for consumer use.

Acetylene cylinders come in several sizes ranging from small *MC* tanks and larger *B* tanks used by plumbers to several larger size tanks for oxyacetylene welding or cutting torches (fig. 27–44). The important difference with these tanks is their relief devices. MC and B tanks have a fusible metal plug built into the valve assembly (fig. 27–45). Older cylinders for cutting and welding have two lead plugs that melt out if the cylinder is exposed to heat, one on top and the other on the bottom. Newer models have both plugs on top (figs. 27–46, 27–47, and 27–48).

Fig. 27–44. A collection of plumber's acetylene tanks. Note the larger B tanks on the left and the smaller MC tanks on the right.

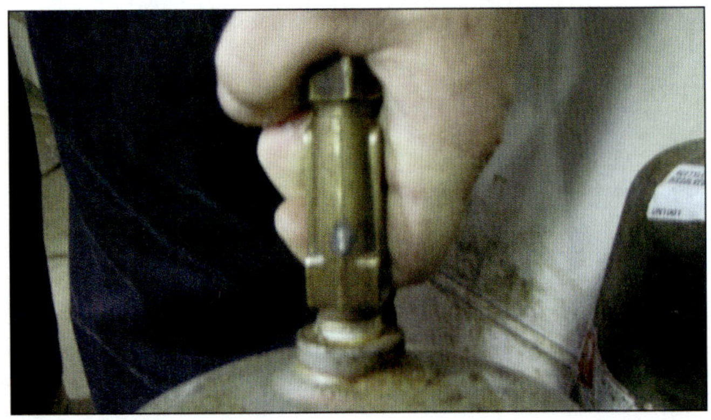

Fig. 27–45. On MC and B tanks, the relief device is a lead plug in the valve assembly. The small gray spot is the relief plug.

Fig. 27–46. Older torch cylinders have one plug on top and one on the bottom.

Fig. 27–47. The plug in the center is a lead plug, relief device.

Fig. 27–48. Newer cylinders for cutting and welding torches have two relief plugs on top of the cylinder.

Acetylene torches have the gas dissolved into liquid acetone. The relief device for these tanks is in the valve assembly in the form of a burst disk. The significance is that these relief devices don't reset or close; a thin metal wafer melts open (figs. 27–49 and 27–50). Once they open, the gas escapes until the cylinder is empty. Cooling the cylinder does not stop the flow of gas, but it may prevent an explosion. The oxygen used in conjunction with the acetylene in these torches is not itself flammable, but it supports combustion (it is an oxidizer) and can make for a very intense fire.

Fig. 27–49. The relief device may be built into the valve assembly. Here the relief device is on the right.

Fig. 27–50. The thin metal wafer is all that protects the cylinder from rupture. The relief cap on the right has the copper intact while the one on the left has melted open. (Courtesy of Lampton Welding Supply)

FIREFIGHTING FOAM

SKILL DRILL

FFII 5.3.1 Firefighters encounter class B fires—flammable or combustible liquids—on a fairly regular basis. We often see these fires where a product has been accidentally spilled, has ignited, and is burning vigorously. High radiant heat and large volumes of black smoke are common characteristics for hydrocarbon fuels such as gasoline and diesel fuel. Firefighting foams are used to control and extinguish class B fires. In addition, some special foams are used to fight class A fires.

NFPA 11 *Standard for Low-, Medium-, and High-Expansion Foam* covers the use of portable foam equipment, as well as fixed fire protection systems such as those that protect tank farms and **bulk terminals.**

A few important terms are common to all ignitable liquids:

Ignition temperature: The minimum temperature at which the fuel/air mixture must be heated to start self-sustained combustion

Lower flammable limit: The minimum concentration of vapor/air mixture below which propagation of a flame does not occur in the presence of an ignition source (measured as a percentage)

Upper flammable limit: Maximum concentration of a vapor/air mixture above which propagation of flame does not occur (measured as a percentage)

Flammable range: The range of vapor/air concentration between the lower flammable limit and upper flammable limit (measured as a percentage)

Flash point: The lowest temperature at which vapor pressure of the liquid is sufficient to produce a flammable mixture

Flammable liquid: A substance having a flash point below 100°F (38°C)

Combustible liquid: Any liquid having a flash point at or above 100°F (38°C)

Types of Class B fuels

Class B fuels are generally divided up into two types: hydrocarbon-based fuels and polar solvents. Each requires a unique type of foam for successful extinguishment. Hydrocarbons are oil based and include gasoline, motor oil, kerosene, jet fuel, and diesel fuel. Hydrocarbons do not mix with water and generally float on water. The old expression, "oil and water don't mix," holds true (fig. 27–51).

Fig. 27–51. Oil and water don't mix.

Polar solvents are the second major type of ignitable liquids we must understand. Polar solvents are distilled or synthetic products and will mix with water. Examples of polar solvents are alcohol, acetone, methanol, and ethanol. These polar solvent vapors are heavier than air and stay low to the ground and collect in low-lying areas. They burn with a pale blue flame and produce little smoke. Burning alcohol may not be easy to detect. Before approaching a spill of alcohol, look for heat waves or thermal currents rising from the spill. If a thermal image camera is unavailable, use a wooden pike pole or straw broom pushed out in front of you when approaching to confirm if the spill is on fire.

Alcohol-resistant foam must be used for fuels that contain polar solvents, since polar solvents are alcohol-based and will otherwise break down the foam. Full personal protective equipment, including SCBA, should be worn by all firefighters operating near an alcohol spill, even after the fire has been extinguished. Exposure to the vapors in sufficient concentration can produce dizziness, drowsiness, or apparent drunkenness. The effects of exposure may be delayed for many hours and could lead to unconsciousness or death.

Agents for extinguishing Class B fires

Several agents can be used to control and extinguish flammable and combustible liquid fires. We can use carbon dioxide (CO_2), dry chemical, or water on fuel with a high flash point such as diesel fuel. But each of these agents has major weaknesses: CO_2 simply blows away with the wind and is an asphyxiant in closed spaces; dry chemical has limited holding power; and water is unsuitable for low flash point liquids such as gasoline. Additionally, all are in limited supply (except water) on first-due apparatus. Specialty apparatus like those used at airports may have large quantities of specialty agents described earlier.

Class B foam is the agent of choice for flammable and combustible liquid spills and fires by engine companies. It is effective because it blankets the flammable liquid with an aerated, relatively stable foam blanket that extinguishes the fire, suppresses the vapors of the fuel, and most importantly provides a high degree of security to protect against reignition.

Class A foam lowers the surface tension of water molecules, which allows superior penetration into dense Class A fuel, but it is ineffective in fighting liquid fuel fires. In contrast to the penetrating qualities of Class A foam, Class B foam floats on top of the liquid hydrocarbon fuel, creating a barrier between the fuel and atmospheric oxygen, thus curtailing combustion. A Class B foam blanket must suppress fuel vapors for at least 15 minutes without reapplication.

Foam is an important tool in the fire service, and as such has specific uses (flammable liquids) and specific times when it should not be used.

Foam should not be used on electrical fires. Foam is 94% to 97% water, the remaining percentage is foam concentrate. Since water conducts electricity, so do foams. The chemicals in the foam can actually cause the foam to be more conductive than plain water. Electrical fires should be de-energized and treated as class A or B fires.

Foam is ineffective on **three-dimensional fires** (liquids that are spraying and burning at the same time) such as a broken pipeline or burning fuel leaking under pressure from an elevated tank. In a three-dimensional fire, dry chemical should be used to first extinguish the fire followed by foam which will cool and secure the situation.

Foam is ineffective on pressurized gas fires such as propane or butane. These fuels are normally vapor at room temperatures and are cooled and compressed for storage and transportation economies. Foam is not effective on products that are above their boiling point at normal temperatures and pressures.

Foam is also not effective on class D fires, combustible metals such as magnesium, titanium, sodium, or potassium because of the amount of water contained in the finished foam. A simple pneumonic summarizes this in general: "If the chemical ends in *ide* or *ium*, water is not their chum." It's silly but a great way to remember what *not* to use water or foam on.

Types of foam and foam terms. Foam is a specialized agent for a specific type of fire. There are several types of foam concentrate available. In addition, there are specific terms used when dealing with foams. It is critical to know these terms to avoid confusion on the fireground.

Chemical foam: In the 1920s chemical foams were developed and relied on mixing of two powders in hoppers that were attached to a hoseline. Although successful, systems were difficult to use during emergencies. Chemical foam systems are no longer made or used in the fire service.

Mechanical foams: Developed in the 1930s for the U.S. Navy and are the type used today. Mechanical foams require four major components: foam concentrate, water, air, and mechanical aeration.

Foam concentrate: The chemical component purchased in 5-gal (19-L) buckets, 55-gal (208-L) drums, or bulk amounts.

Foam solution: Foam concentrate and water, properly mixed by the foam eductor before it reaches the nozzle.

Finished foam: Produced after the air has been injected (usually at the nozzle) (fig. 27–52).

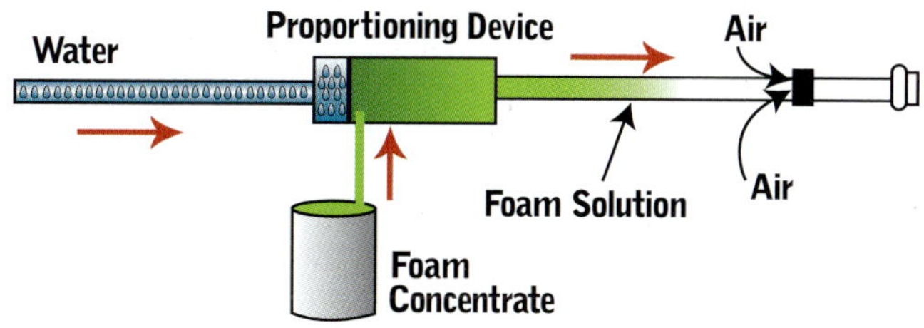

Fig. 27–52. The foam process introduces a set percentage of foam concentrate into the water supply, as well as utilized a foam nozzle to inject air into the foam solution.

Foam production. A common type of foam proportioning and delivery system is called an **inline eductor**. The eductor uses a venturi principle to pull foam concentrate from the foam container and mix it with the appropriate amount of water to make foam solution. Inside the eductor, there is a restriction in the metering orifice that creates a **venturi effect**, causing the water to flow faster and creating a lower pressure area in the line, resulting in suction. The foam pickup tube is attached to the venturi area of the eductor and goes in the foam bucket. A metering device allows the passage of the correct amount of foam concentrate to enter the hoseline. When water and foam concentrate are mixed it is called foam solution.

Many apparatus are being made with on-board foam systems. These systems can either incorporate an inductor in the piping of a particular discharge outlet or into the piping leading to the pump itself. Prepiped inductors have a disadvantage in that finished foam can only be discharged from that particular outlet. When foam concentrate is educted into the pump itself, foam can be delivered from any discharge outlet or apparatus-mounted deluge gun. All these systems require an increased level of cleaning and maintenance to ensure proper performance.

If an engine company needs to deliver finished foam and has no foam proportioning capability, foam concentrate in proper proportion (15 gal [57 L] foam to 500 gal [1,893 L] water) can be added to the apparatus booster tank to create foam solution. This is a crude method and depends completely on the mixing of the concentrate and water in the tank. The solution can be applied from an aerating nozzle to create finished foam. Foam solution flows to the foam nozzle where air is added to create finished foam.

Finished foam: How it works. It is important to understand several properties of finished foam. Like anything in life, these properties are a balance between several factors. In addition, there are several terms that you must know:

Knockdown: The foam blanket must spread across the fuel to control and ultimately extinguish the fire. The speed at which this occurs is called knockdown.

Heat resistance: How well the foam stands up to the heat of the fire. The blanket should be stable enough to withstand the extreme heat of the fire, hot metal, and debris that maybe in the spilled or burning area.

Fuel tolerance: The ability to shed any fuel that may be on top of the foam blanket as a result of the foam stream splashing the fuel. If the foam bubbles bring fuel to the surface, the fire can continue.

Vapor suppression: The ability of the foam blanket to keep fuel vapors from pushing up through the foam blanket where they can burn.

Burnback resistance: The foam's ability to withstand the heat of the fire without breaking down.

Drainage time: The time it takes for the water to separate from the foam blanket.

Types of foam concentrates. There are a variety of foam concentrates to choose from. Each has its advantages and disadvantages.

Protein foam was the first type of mechanical foam and has been used since World War II. It is primarily made from animal protein with other chemicals added to minimize corrosion and bacterial decomposition. It has excellent heat resistance and vapor suppression but is slow to spread over the fuel surface and it lacks fuel tolerance. It is inexpensive.

Fluoroprotein foam was developed by the addition of fluorochemical surfactants to protein foam. This foam flows better than protein foam, giving it a better knockdown speed and has better fuel tolerance. It is often used for **subsurface injection** at bulk tank fires. In this application, the foam is forced in under pressure below the liquid level of the tank using piping installed in the tank. The foam floats to the surface, creating a blanket and eventually extinguishing the fire.

Synthetic detergent foam is what we know as **high-expansion foam.** These foam concentrates are synthetic foaming chemicals and stabilizers. They are used in high-expansion systems for total flooding of basement fires and other large space fires that are difficult to enter to extinguish class A fires. High-expansion foams have limited use on class B fires because fuel vapors easily penetrate the lightweight foam, heat rapidly degrades the foam bubbles, and the reach of the streams is limited. High-expansion foams are considered unsuitable for class B fires.

Aqueous film-forming foam (AFFF) synthetic foaming agents replace the natural protein base and are designed for rapid knockdown (fig. 27–53). Developed for the U.S. Navy to rapidly extinguish shallow spill fires, AFFF sacrifices long-term heat resistance for rapid fire knockdown. This is one of the most common foams modern firefighters will use in day-to-day operation. The important additional capability of this concentrate is that it forms an aqueous film on the fuel surface. Rapidly spreading across the fuel surface, this thin layer of foam solution (that has drained out of the finished foam) results in rapid fire knockdown by vapor suppression. This is where the name *aqueous film-forming foam* is derived. Just as oil floats on water, AFFF floats on hydrocarbon fuels. AFFFs are effective on hydrocarbons such as diesel, kerosene, and jet fuels. The higher flash point of these fuels present less of a reignition hazard. Fuels such as gasoline and gasoline-alcohol blends require more frequent application of foam because of the lower flash point and the threat of reignition.

Fig. 27–53. Foam can be very effective when used properly.

Alcohol-resistant aqueous film-forming foam (AR-AFFF) was developed because fuels containing alcohol aggressively pull the water out of the foam, destroying it and reducing its fire-suppression effectiveness. These foams function much like a non-alcohol-resistant foam forming an aqueous film on the surface of the fuel. If alcohol is present in the fuel, the membrane is a bit more viscous and called a **polymeric membrane.** This type of foam is the most versatile in use today; and with the widespread use of oxygenated automobile fuels and alternate fuels with increasing amounts of alcohol, it is becoming the standard foam in use in first-response units.

Foam concentrates should never be mixed. Because of the differences in chemical composition of different brands of foam, mixing of concentrates can result in chemical reactions that congeal the concentrate. The congealed foam is semisolid and immediately clogs eductors and nozzles and stops your foam application operation.

Finished foams can be mixed on the scene without a negative effect. However, recall that each type of foam has strengths and weaknesses, which may result in confusion on the security and safety of the foam blanket.

Foam proportioning. Foam is a specific agent for a specific type of fire. As such it must be used properly. Proportioning or mixing the right amount of foam concentrate with the right amount of water and air at the right pressure results in good quality foam. Foam is not just a collection of bubbles; it is a specific mix of concentrate, water, and air.

It is also critical that there are sufficient quantities of foam concentrate to extinguish the fire and secure the spill. Running out of concentrate before the fire has been extinguished or has been secured allows the fire to quickly burn back to its original size and intensity—this is unlike a building fire. It is as if no attack had taken place at all.

Proper proportioning is important. If there is not enough concentrate in the finished foam, the overall quality and effectiveness will decrease. It may not extinguish the fire as expected, heat resistance decreases, and burnback resistance decreases.

If your finished foam has too much concentrate in it, foam will not flow around debris and obstacles. Additionally, limited foam concentrate will be depleted rapidly, not allowing enough for complete extinguishment and reapplication as necessary.

Proportioning may take place in one of several ways: induction (through eductors), injection, premixing, and batch-mixing. The most common method is induction using a portable inline eductor in conjunction with a hand line. Large, self-educting foam nozzles are also available with flows up over 2,000 gpm (7,571 L/min) (fig. 27–54). These larger flows must be supported with bulk (55 gal [208 L] or more) containers of foam concentrate. In addition, some foam is proportioned through eduction on apparatus using around the pump proportioners (fig. 27–55) and balanced-pressure proportioners (fig. 27–56).

Foam injection systems use specially designed foam pumps and are often found at fixed installations (fig. 27–57). Premixing is used in foam fire extinguishers where the foam concentrate is mixed, in specific measured amounts, with water inside the tank. Similarly, batch mixing is simply placing foam concentrate in a larger tank, an engine's tank, or a foam trailer (fig. 27–58). Although this type of proportioning is common for **compressed air foam systems (CAFS)**, described in the following text, take care when adding the foam concentrate, because it is corrosive to metal parts.

Fig. 27–54. Large self-educting nozzles can supply streams of up to 2,000 gpm (7,571 L/min).

Fig. 27–55. The around-the-pump proportioner is one way to supply the correct mixture of water to foam concentrate.

Fig. 27–56. This balanced-pressure proportioner is mounted inside the apparatus.

Fig. 27–57. A foam bladder tank

Fig. 27–58. Foam trailer

Inline eductors on hand lines. An inline eductor is the most common method for engine companies to proportion foam. As previously described, the eductor is placed in a hoseline. These generally have a flow of 65 or 95 gpm (246 or 360 L/min) for 1¾-in. (45-mm) lines. Two-and-a-half inch handlines will flow approximately 150 gpm (568 L/min).

Eductors are generally energy inefficient and consume approximately 65% of their inlet pressure to create the suction. Check your specific eductor, but generally the inlet pressure is designed to be approximately 200 psi (1,400 kPa). With about 70 psi (490 kPa) used to pick up concentrate, there is about 130 psi (910 kPa) available on the discharge side of the eductor. We know friction loss in 1¾-in. (44-mm) hose flowing 95 gpm (360 L/min) is 14 psi (98 kPa) per 100 ft (30 m), and we know we need 100 psi (700 kPa) at the nozzle to provide enough energy to aerate the foam and provide adequate stream reach; therefore 200 ft (60 m) is generally the maximum length of hose permitted after the eductor. When excessive back pressure occurs at the eductor, it causes incorrect proportioning and a less effective finished foam.

If a longer hose stretch is required, use 2½-in. (65-mm) hose to supply the eductor to ensure that proper inlet pressure is maintained. The position of the eductor in relation to the nozzle is another consideration. If the nozzle operates above the elevation of the eductor, the difference in height can slow the flow of foam solution and affect the proper aeration of finished foam. It is best to have the nozzle on the same level as or below the eductor.

Churning water from the pumps can affect the eductor's ability to draw foam concentrate. Churning water from the pumps may create turbulence that can affect the eductors ability to draw foam concentrate. To avoid this place the eductor one length from the apparatus. Firefighters encounter class B fires on a fairly regular basis. They are often the result of an accidental spill and when ignited can burn vigorously. There are some important points to consider when using an in-line eductor. If your in-line eductor setup is designed with the following critical factors in mind, you are likely be successful in proportioning and delivering foam at your next alarm:

- The eductor gpm must match the nozzle's rated flow in gpm
- Generally, pickup tubes are less than 6 ft (2 m) long, so don't modify or extend the pickup tube
- Avoid improper pump pressures to the eductor.
- Kinked hoses on either side of the eductor cause excessive friction loss and improper flows and pressures
- Nozzles should be fully open
- Air leaks suck in air instead of foam concentrate
- Keep your hands away from the air intake of the nozzle to ensure an adequate supply of air to make finished foam
- Always run clear water through the eductor for an extended period of time to flush it out after use

A bypass balanced pressure **proportioner** (eductor) works similar to the inline eductor (fig. 27–59). The bypass proportioner allows for a full flow of water if foam is not desired. This type of eductor is built into the plumbing of the engine. Along with such built-in systems, there is an additional foam tank as well with 25- to 100-gal (95- to 3,79-L) foam concentrate capacity.

Fig. 27–59. The inline eductor is mounted after the first length of hose.

Causes for poor foam generation can include the following:

- Not matching flow ratings between educator and nozzle
- Proportioning equipment that has not been cleaned, resulting in clogging
- Air leaks in fittings
- Excess back pressure resulting from too long hose lay on discharge side of the eductor
- Mixing different foams together, which can make foam too thick to pass through eductor

Foam application. The nozzle is the last step in the foam making process. Here the foam solution is aspirated and finished foam is produced. The air is entrained in the foam by some type of mechanical means, usually directed at a baffle, or by foam solution streams directed at each other in the nozzle, then directed down the foam tube and out the end forming the foam stream (fig. 27–60).

Fig. 27–60. The nozzle is the last determining factor in making foam. It can assist in entraining the foam with air.

The best quality foam blanket is generated with a nozzle designed for foam making and distribution. During the 1990s it was common to apply AFFF through standard combination (fog) nozzles. Now this is the exception rather than the rule, because all foams (other than AFFF) must be aspirated to be effective. Nonaerating AFFF was acceptable when applying AFFF for many applications. Because the foam solution was so effective at extinguishing the fire by the film-forming component, the poor quality aspiration was acceptable. However with the advent of the addition of ethanol and the addition of alcohol to standard gasoline, NFPA 11 recommends aeration of all types of foams including AFFF and AR-AFFF.

It is the choice of the nozzle that determines the expansion ratio of the foam concentrate into finished foam (fig. 27–61). Standard fog nozzles produce 4:1 to 6:1, standard fog nozzle with low expansion foam attachment 10:1, foam nozzles with tubes 10:1, and mid-expansion foam nozzles 30:1 to 50:1. Remember, the foam nozzle must match the designed eductor flow and pressures. Additionally, the foam concentrate must be correct, matching the eductor and nozzle. Read the foam bucket label and manufacturer's instructions for specific details.

Fig. 27–61. This foam nozzle helps to introduce air into the stream, thus aspirating the foam.

Foam application techniques. It is the nozzle operator's responsibility to apply the foam as gently as possible to allow the finished foam to form a cohesive but flowing blanket on the fuel to smother the fire, allow the best development of the film-forming actions (stop vapor production), and maintain the security to prevent reignition. There are several techniques that make that possible:

- **Bank-down method:** Bounce the foam off an object or wall adjacent to the spill, allowing the foam to bank down gently onto the spill (fig. 27–62).

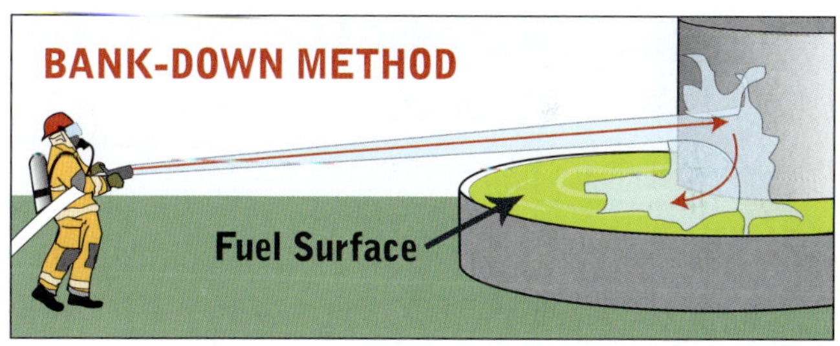

Fig. 27–62. The bank-down method allows the firefighter to attack the flame from a different direction.

- **Roll-on method:** Roll the foam in by piling up finished foam in front of you and pushing or sweeping the foam onto the burning fuel (fig. 27–63).

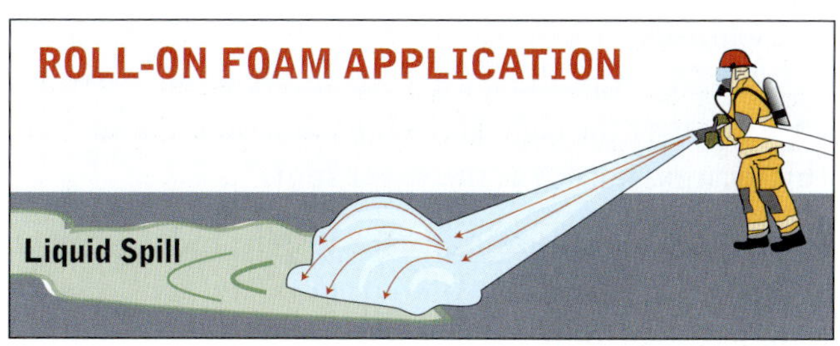

Fig. 27–63. The roll-on method is an effective way to push the foam onto a fuel spill.

- **Rain down (or snow storm) method:** The stream is directed almost vertically above the liquid's surface, allowing the finished foam to fall like rain or snow. An intense thermal column from the fire may prevent this method from being effective. If the foam stream is directed directly toward the fire, move rapidly back and forth horizontally to break up the stream and prevent it from plunging into the fuel (fig. 27–64).

Fig. 27–64. The snowstorm method may not be effective if there is an intense thermal column. Courtesy of Williams Fire and Hazard Control Corporation)

Foam is often used on tank fires at bulk terminals. The key to extinguishing tank fires is to use a technique known as **over the top**, developed several years ago by the Williams Fire and Hazard Control Corporation of Texas. First apply the foam at the location where air is being drawn into the fire (determine this location by conducting a 360-degree survey of the tank). A cooling steam of water (not foam) is then applied to the side of the tank, just below the edge of the tank, to cool the tank shell and allow the foam applied in the tank to establish a **beachhead**. Finally, the foam is laid down like snow (fig. 27–65). Aim the master stream foam nozzle to let it fall just over the edge of the tank and onto the surface. This location known as the **sweet spot**.

To apply foam to a burning, running spill (and to attempt to contain it as well), work ahead of the spill. Direct foam in the path of the liquid so the liquid runs under the foam blanket, extinguishing the fire. Then work the foam blanket toward the source of the leak until the liquid is covered. Bouncing the foam off the ground or **pushing** the foam across the liquid prevents the liquid from splashing and breaking the foam blanket.

Fig. 27–65. The "over the top" method was developed by Williams Fire and Hazard Control Corporation. It is an effective way to attack large petroleum tank fires.

In many cases, a flammable liquid fire is the result of an accident. There may be obstacles in the spill. This may require additional hand lines or monitors to apply foam to cover the spill and extinguish fires within the voids. A combination of application methods may be required in this situation. Coordination is required to ensure the foam blanket is properly applied.

All these techniques are designed to prevent the nozzle operator from plunging the foam into the fuel. This could cause flaming liquid to be spread around the scene, possibly onto other firefighters, with deadly results. Never direct a stream into the spill or fire.

The nozzle operator must also be aware of when discharging water versus foam. When the eductor system begins flowing, water comes out first. Direct the water away from the spill or fire, it only makes it worse or spreads the fire. It takes a minute or so for the eductor to pick up foam concentrate, so that is flows through the hose to reach the nozzle. Listen to the sound: Water flows with a crackling sound, foam is softer and quieter. It also looks different. The nozzle operator must be aware when foam begins and when the stream turns back to water. The firefighter assigned to the eductor must communicate with the nozzle operator before running out of foam. A water stream focused on the foam quickly washes away the foam just applied, washing away all the fire suppression just achieved.

When flowing 95 gpm (360 L/min) of foam solution, (for easy math, let's say we are flowing 100 gpm [379 L/min]), at 1% that means for every 99 gal (375 L) of water, we are flowing 1% or 1 gal (3.8 L) foam concentrate. Thus the foam concentrate lasts approximately 5 min (at 1%). An officer should be assigned to the foam eductor to ensure that there is an adequate supply of foam concentrate available. Consider that if the foam is used at 3%, for every 100 gal foam solution flowing, 3 gal (11 L) foam is consumed each minute. Thus a 5-gal (19-L) foam bucket must be replaced every 2 min. This is a significant logistics challenge. How many foam buckets does your engine carry? How long will it last? It is important to do it right the first time; you only have one chance!

The firefighter assigned to the eductor must monitor the foam concentrate in the buckets to ensure the pickup tube is placed in the next full bucket before the concentrate runs out. Buckets positioned for use should have all caps removed prior to operation. The foam operation should not be interrupted because you cannot remove a cap. Some caps are difficult to remove, if this is the case and wrenches are unavailable, use the point of a Halligan tool to punch a hole in the cap to insert the pickup tube. Do not punch the hole directly in the container; if the foam from that container is not needed, there will be no way to secure the container. If the cap is punctured, it is possible to find a replacement cap.

As the foam stream leaves the nozzle, the stream begins to break up. This limits the nozzle operator's ability to see where the foam is landing. It is always a good idea to have a spotter off to the side to direct the nozzle operator and ensure that the foam reaches the appropriate area.

Foam blankets break down over time and must be reapplied, often in as little as 20 min. Ensure that you have enough foam concentrate to be able to reapply foam as necessary until the spill is cleaned up.

Foam safety. Flammable liquid fires are dangerous and dynamic. A structure fire will not move unless the building collapses or exposures are rapidly lighting up; however, a flammable liquid can flow down street or into storm drains endangering firefighters, apparatus, and numerous exposures. Always be aware of where the spill or liquid is and where it is or may be going to avoid being trapped and burned. Because of the possibility of reignition, always retreat facing the hazard or fire. Also be aware that when you stop application of agent, the advantage returns to the fire and the threat of it reigniting or becoming a more significant hazard gains momentum. The threat of reignition requires that members operating near the spill or in the path of any vapor cloud be in full protective equipment, including SCBA (fig. 27–66).

Fig. 27–66. It is important to wear full PPE at all times when applying foam.

If possible, avoid walking through a foam blanket. Your feet open the blanket and flammable vapors can rise around you. Should they ignite you will be standing in a flammable liquid fire. If you must walk through a foam blanket, be sure foam is reapplied to seal up any openings you made in the blanket. Do not direct the foam stream on other firefighters, but be aware of the danger you or they are in.

Foam concentrates are not especially toxic, but they burn if they get in your eyes. Goggles or face shields *with* gloves are a good idea for those who may come in contact with foam.

Application rates. Just like structural fire attack, when water must be applied in sufficient quantity (gal/min) to overcome the heat produced by the fire, foam must be applied fast enough to extinguish the flammable liquid fire. An ignitable liquid must be applied at a rate per square foot of fire or spill. Using some simple calculations one can determine how much foam, water, and application rate is correct for any size flammable liquid fire.

When a foam hand line is required, it is important to understand the extinguishing capability of the foam operation. Before any foam operation is begun, an adequate volume of foam concentrate must be on the scene. If the foam blanket does not completely extinguish the fire or there is inadequate foam to reapply to overcome drainage and burnback, all of the foam applied will have been wasted.

Suppose there is a 40- by 50-ft hydrocarbon spill burning. NFPA 11 recommends an application rate of 0.10 gpm/ft^2:

- $40 \times 50 = 2{,}000$ ft^2
- Area × rate = gpm foam solution required
- $2{,}000 \times 0.10 = 200$ gpm solution flow

Note that this is the critical first step. This spill fire requires at least *two* 95-gpm eductors to meet this required flow, if you expect rapid extinguishment and lasting security from reignition.

Gallons of concentrate required:

- gpm × % concentrate = gpm concentrate
- × 0.03 = 6 gpm concentrate

To find rate of water needed (gpm):

- gpm × % water = gpm water
- 200 gpm × 0.97 = 194 gpm water

To find total volume (gal) of concentrate (*note:* per NFPA 11, standard recommended application time is 15 min):

- gpm concentrate × time = total gallons of concentrate
- 6 × 15 = 90 gal concentrate

To find total gallons of volume (gal) water:

- gpm × time
- 194 × 15 = 2,910 gal water

A good rule of thumb is that if you add a 0 to your eductor flow, it will give you an *estimate* of the maximum spill fire you can control with that one line in operation.

For example a 95-gpm eductor can handle a 950 ft^2 ignited hydrocarbon spill fire. Recall that 950 ft^2 is a fire less than 20 × 50 ft (1,000 ft^2). To estimate the size of a spill fire of polar solvent, add a 0 and divide by 2. For example, a 95-gpm eductor and nozzle control a polar solvent spill fire of approximately 475 ft^2: 950 ft^2 divided by 2 = 475 ft^2 or about 22 × 22 ft.

Controlled pit fires like the one shown here are excellent training for foam operations (fig. 27–67). It is important to note that this spill fire is not realistic. In a real scenario, such as a vehicle accident, there will be burning tires, hot metal, and other reignition sources. This training fire will be extinguished rapidly with minimum amounts of foam because the pit has no debris. Actual fires may take several times the amount of agent to control as training fires. The additional amounts of foam will be necessary to flow around debris and to replace what is being destroyed by the heat of the fire, hot metal, and so forth.

Fig. 27–67. Controlled pit fires like the one shown here are excellent training for foam operations.

Compressed air foam system (CAFS). CAFS are a relatively new class A firefighting capability. First developed for the **urban-wildland interface**, CAFS has found acceptance in structural firefighting nationwide. Essentially, CAFS uses a fire engine's centrifugal pump to provide a supply of water that is subsequently injected with foam concentrate (on the discharge side of the pump) and mixed with air through the use of an air compressor (fig. 27–68). The type of foam produced has the ability to stick to vertical surfaces (a great advantage in structural firefighting) and has been shown to have superior knockdown capability when compared with plain water (saving water in the process).

Fig. 27–68. CAFS has found acceptance in structural firefighting nationwide.

QUESTIONS

1. What is the first step in developing a fire attack plan?
2. What is the most important consideration at every fire or emergency?
3. What can happen if a basement fire is attacked from an exterior entrance into the basement?
4. When attacking a basement fire from an exterior entrance door, where should a second hoseline be placed to prevent extension of the fire to upper floors?
5. What piece of firefighting equipment can be used to knock down a fire if hoselines cannot enter into the basement to attack the fire?
6. What tactic can be used to ventilate a basement if basement windows and doors are not ventilating sufficiently?
7. What are the goals of stretching hoselines?
8. What is the most important consideration when stretching a hoseline?
9. As a general rule, where should hoselines be placed?
10. What hazards do the void spaces in townhomes present?
11. What tactic is used to locate the fire and route of extension in a townhome fire?
12. Generally, where do fires start in commercial buildings?
13. What are the benefits of having a search line when you enter a building?
14. What is an advantage commercial buildings have that is not normally found in residential buildings?
15. What are the limitations of roof operations when ventilating in commercial buildings?
16. What happens when steel bar joist roof trusses are exposed to heat?
17. What should firefighters do if they see a white powdery residue when cutting a hole in a roof?
18. When a commercial building has a sprinkler system or if the standpipe will be used for firefighting, what should be one of the first lines stretched to ensure there will be enough water to contain the fire?
19. What type of occupancy fire poses one of the most serious threats to firefighters?
20. What are the seven fireground priorities?

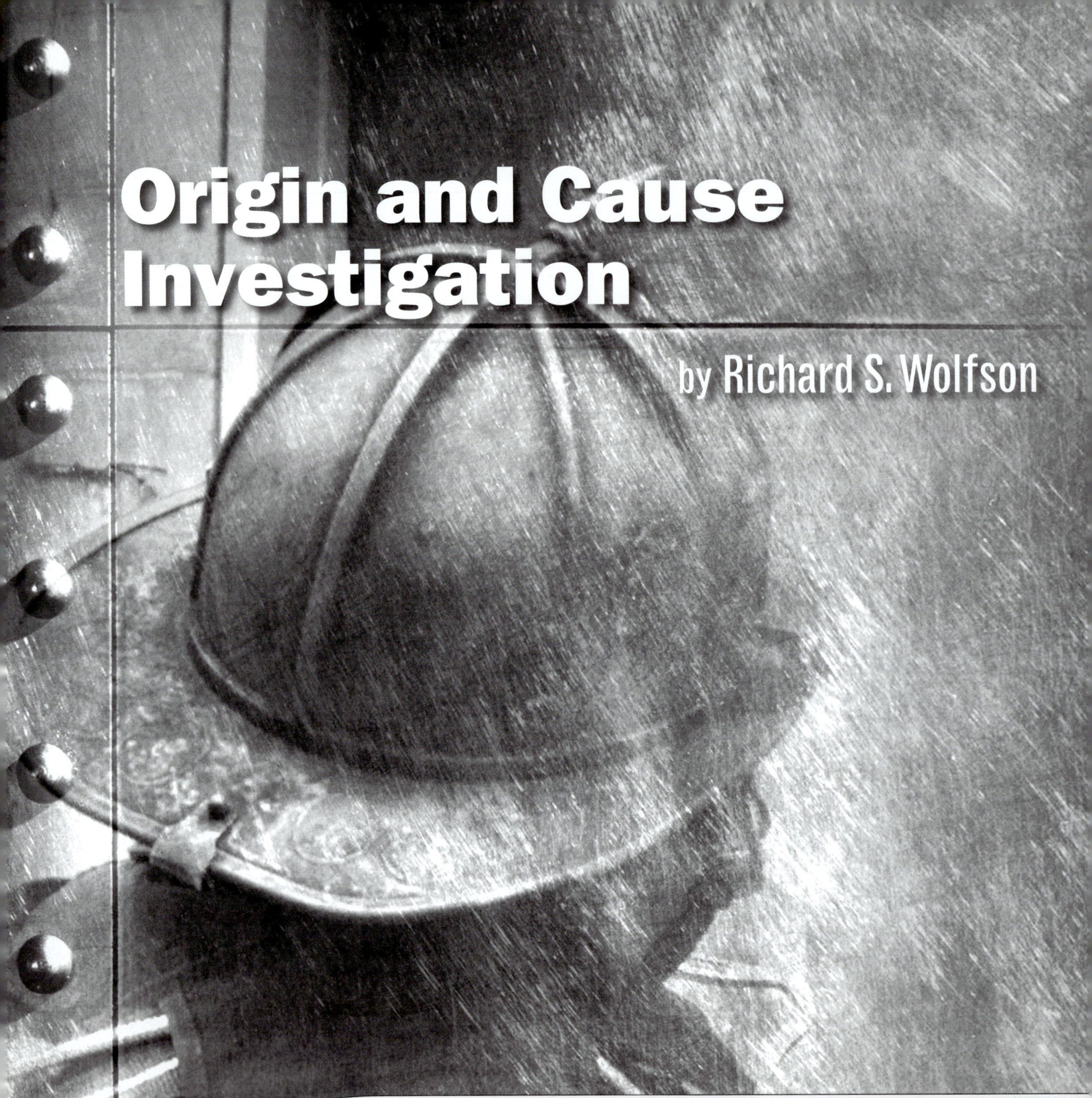

Origin and Cause Investigation

by Richard S. Wolfson

This chapter provides required knowledge items for the following NFPA Standard 1001 Job Performance Requirements:

FFII 5.3.4

This chapter contains Skill Drills. When you see this icon, refer to your Skill Drill book for step-by-step instructions. SKILL DRILL

OBJECTIVES

Upon completion of this chapter, you should be able to do the following:

- Describe the role of the firefighter in fire investigations
- Describe spoliation and its impact on evidence
- List common sources of ignition
- Identify the parts of the scientific method used to investigate fires
- List at least three type of fire patterns
- Identify common causes of electrical fires
- Identify common ignition sources found at motor vehicle fires
- List at least three types of fire setters
- List the three types of evidence
- Describe the importance of maintaining the "chain of evidence"
- Describe the importance of documenting the fire scene
- Describe the importance of securing the fire scene

WHY FIRES ARE INVESTIGATED

Fire is a powerful force of nature. As such, it is capable of killing and injuring people as well causing extensive damage to property. Fires may be accidental or intentional. Thus it is important to find out how and why fires are started. This is why fires are investigated.

The investigation of fires has many benefits, including the identification of defective products such as electrical appliances, which can malfunction and start fires. Fire investigations often lead to changes in **fire and building codes** when it is determined that a particular code does not properly address a certain issue such as the flame spread of a particular **interior finish**. Finally, fire investigations sometimes uncover the crime of arson and lead to the arrest of an arsonist, removing the arsonist from society.

ROLE OF THE FIREFIGHTER IN A FIRE INVESTIGATION

FFII 5.3.2 Firefighters who understand the procedures of a fire investigation are valuable tools to the fire investigator. The observations of the firefighter are an integral part of the fire investigation. Although the role of the first-due firefighter is rescue and fire suppression, firefighters can provide information about their response and fire scene observations that will greatly assist fire investigators, including insurance and criminal or arson investigators. Fire investigators utilize NFPA 921, *Guide to Fire and Explosion Investigations*, to carry out their investigations. It is considered the nationally recognized standard of practice that investigators adhere to while conducting investigations. Sections of this guide apply to firefighters as well, particularly when it comes to scene preservation and protection of evidence.

Insurance investigators are independent investigators hired by insurance companies to determine how the fire started and if the insurance claim is valid. Criminal investigators, who hold state certification with authority to arrest people, determine if the fire was caused intentionally. Arson investigators are criminal investigators who gather evidence to lead to the arrest and prosecution of a suspect. Firefighters should document their observations to provide information to investigators as needed.

Document the date and time of the alarm and how the alarm was received, the name and address of the person reporting the fire, and any information received regarding his or her initial observations. The firefighter's observations should include the fire scene conditions on arrival (fig. 28–1). First, what were the weather conditions, including the wind direction and speed? Were there any persons or vehicles leaving the fire scene at the time of the first unit's arrival? Where was the initial first line stretched and what was the location of the fire? Was there any unusual difficulty in attacking the fire? Did the fire seem to actually spread during the application of the first line? Were there any unusual odors detected by the first arriving units?

Multiple areas of fire should be noted and this information relayed to the chief officer, who will in turn notify the fire investigator. During fire suppression, document any unusual circumstances such as holes in the floor, difficulty in gaining access to the area of origin, and any evidence of vandalism or forced entry to the structure (fig. 28–2). The location of any victims should be fully documented.

If the occupants are out of the structure and awaiting the firefighters at the time of the fire department's arrival, firefighters should note the attire of the occupants. If the fire occurred in the early morning hours and the occupants are fully clothed, that is obviously out of place. This scenario should be documented and this information relayed to the chief officer and investigator.

Fig. 28–1. A gas can found in an arson fire

Fig. 28–2. Vandalism was the cause of this fire.

OVERHAULING AND THE FIRE INVESTIGATION

During the fire suppression, firefighters should be aware of excessive overhaul procedures and refrain from causing any additional damage to the area of origin. During the overhaul procedures, it is imperative that all evidence be preserved. Minimizing the overhaul will help the fire investigator determine the cause of the fire occurrence; although all remaining hot spots of the fire must be extinguished, take care to avoid excessive overhaul. If possible, overhaul of the scene should be delayed if there is any evidence of the fire being **incendiary** in nature. If this determination is made, the area of origin should be preserved and treated as a crime scene. Upon the investigator's arrival, the observations of the firefighters should be relayed to the investigator through the officer in charge.

According to NFPA 921, **spoliation** is the "loss, destruction, or material alteration of an object or document that is evidence or potential evidence in a legal proceeding by one who has the responsibility for its preservation." What this means to firefighters is that you should protect any evidence that you might encounter while performing your duties as well as notifying your company officer of the evidence that you have found. In some cases, senior officers may order the evidence removed from the area of origin when it is believed that the evidence might be damaged or destroyed due to weather or other conditions. A discussion of evidence collection techniques is provided later in the chapter.

The chief officer may request the electrical utility within the structure to be shut off (fig. 28–3). The firefighter performing this duty should identify the **electrical distribution panel** and be familiar if the panel has a **main disconnect**. When put in the *off* position, the main disconnect will shut down the entire panel, and moving the individual branch circuits into the *off* position will be unnecessary. If the distribution panel has no main disconnect, the individual branch circuit breakers should be put in the *off* position; if electrical panel is older and uses fuses, individual fuses should be removed. Any circuit breakers that are tripped should be left in that position; fuses that have blown can be left in place. Firefighters must take note of the individual circuit breakers and/or the main disconnect that they shut off. This is important information for the investigator.

Additional observations during the suppression of the fire may include open doors and windows and any unusual openings within the structure. The blockage of any exit doors and/or passages should be noted. Are the smoke alarms activated and operating at the time of the firefighter's arrival? Is the sprinkler system operative or disabled?

Fig. 28–3. An electrical panel after a fire

FUELS

To have a fire, obviously, a fuel must be present. Fuels take various forms but can be generally grouped into one of following categories: *Solid materials* include things like wood, plastics, and even many types of metals when finely divided into powder form. *Liquids* include such materials as gasoline, acetone, and motor oil. Natural gas and hydrogen are examples of flammable *gases/vapors*. It is important to note that some materials will not support combustion, such as nitrogen (gas), concrete (solid), and water (liquid).

It is important that a fire investigator research the combustion traits of different types of fuels. When studying a particular fire, the investigator must establish the characteristics of the material in question such as **ignition temperature** for solid objects, flammable limits for gases and vapors, and **flash point** for **flammable** and **combustible liquids**.

SOURCES OF IGNITION

Sources of ignition are the different types of heat energy that initiate a fire by causing a fuel to ignite. These sources must be of sufficient heat energy and temperature (recall that heat and temperature are two different characteristics, as discussed in chapter 5, Fire Behavior) to be considered competent, meaning they are capable of causing ignition. Sources of ignition can take various forms, but they can be grouped into a variety of different types (fig. 28–4). These include open flames such as a matches, lighters, stove burner gas jets, or a water heater burner flame; hot objects such as a hot plate used for cooking, an exhaust manifold on a car engine, a car's catalytic converter, or a hot halogen lamp bulb igniting bedding materials; hot molten metals such as **slag** from a welding operation or sparks from a metal grinding wheel; frictional sources such as *locked* truck brakes or rotating objects such as a clothes dryer drum rubbing against a piece of clothing jammed between the drum and the dryer frame; and electrical system malfunctions such as a short circuit, an arc, or even static electricity.

Fig. 28–4. An ignition source

A handful of materials are subject to **spontaneous ignition**. According to NFPA 921, *The Guide for Fire and Explosion Investigations*, spontaneous ignition is the "initiation of combustion of a material by an internal chemical or biological reaction that has produced sufficient heat to ignite the material." Therefore, spontaneous ignition is a chemical or biological process that allows certain materials to generate heat and ignite on their own, without the presence of an external heat source, such as a match. Examples of materials susceptible to spontaneous ignition are wet hay (where a biological reaction takes place inside the bale of hay over a period of time, generating heat, and sometimes igniting the hay as the heating process reaches the outside of the bale) and a pile of linseed-oil–soaked rags (where a chemical reaction takes place inside the pile of rags; the pile provides insulation for the heat, trapping it inside and not allowing it to escape).

BASIC FIRE INVESTIGATION PROCEDURES

An **origin and cause investigation** should be conducted using the scientifically established procedures set forth in NFPA 921. An origin and cause investigation is just that: finding out what started the fire and where it started. This is the fundamental duty of a fire investigator. Additionally, fire reports submitted to the National Fire Incident Reporting System database (NFIRS is covered in chapter 28, Advanced Communications) must include information such as the source of ignition and the first material ignited. However, it must be noted that such NFIRS reports are considered incident reports, not fire investigation reports. In terms of cause and origin, NFIRS reports are preliminary reports prepared by the fire department.

The origin and cause investigation should also be conducted using the **scientific method**. NFPA 921 identifies the following as the scientific method:

1. Recognize the need (identify the problem).
2. Define the problem.
3. Collect data.
4. Analyze the data (inductive reasoning).
5. Develop a hypothesis.
6. Test the hypothesis (deductive reasoning).
7. Select final hypothesis.

Because a fire has occurred, indicating that a problem exists, the cause should be determined. Defining the problem, an origin and cause investigation should be conducted by the examination of the fire scene and the collection of data, which includes interviewing witnesses and the results of scientific testing.

CREATING A PLAN

Once assigned to investigate a fire, the investigator must develop a plan. There are several considerations:

- Site safety: Is the building stable enough to enter and is the atmosphere free of smoke, carbon monoxide, and other toxins?
- The number of investigators and equipment needed: Is this a large fire scene or one that requires large amounts of resources?
- The type(s) of investigator(s) needed: Are specialists in electrical systems, fire protection systems, forensic anthropologists specializing in dealing with fatalities, and so forth needed?
- The type of scene recording that will be performed: video, still photos, and so on.

Once the appropriate resources have been put in place, most investigators will begin the investigation. Most investigators approach a fire scene using a the same, systematic approach every time. For example, many investigators will move from areas of less fire damage toward areas of more fire damage. Many will also search the outside area of a building looking for evidence (such as studying exterior building fire damage around windows or finding an accelerant container) before they go inside the building. What is important is that investigators use a systematic method so they don't miss anything.

FIRE PATTERNS

Fire patterns can be considered the **fingerprints** of a fire. NFPA 921 defines fire patterns as "the visible or measurable physical changes or identifiable shapes formed by a fire effect or group of fire effects. Fire effects are the observable or measurable changes in or on a material as a result of exposure to the fire." Fire patterns tell a fire investigator about the movement of a fire.

There are a variety of patterns, the most common being the charring of wood and other materials as well as soot stains on walls and other surfaces. A specific, common type of char pattern is called the **V pattern**, a pattern often burned into a wall surface that is shaped like the letter V (fig. 28–5a). The base of the V points toward an object on the floor that ignited at some point on the fire. Sometimes the object has been moved during firefighting; the V pattern tells the investigator that a burning object was at that location. In some cases, the V pattern may be pointing at the object that started the fire!

Fig. 28–5a. A V pattern on a wall

Other patterns include **crazed** (cracked) glass, deformed objects such as light bulbs, spalled (chipped away) concrete, melted metals like aluminum, distortion like a twisted metal beam, and many, many more. Some patterns are actually created by the absence of heat such as a **protected** floor area underneath a piece of furniture that is shielded from the heat of a fire (fig. 28–5b). Although the details of what each pattern means is beyond the scope of this book, it is important for firefighters to understand that it is vital to try to maintain the fire scene as closely as possible as to how it was when the fire started. Fire patterns are very important indicators that, when taken collectively, tell the story of a fire.

Fig. 28–5b. Protected floor area underneath a bed

ELECTRICAL FIRES

Electrical services and electrical equipment in buildings sometimes cause fires. Electrical wiring can be overloaded (meaning the wire is carrying more electricity than it is rated to carry, causing it to overheat). A **short circuit** is a situation when the insulation around *two conductors* in a wire is broken and the conductors come in contact with each other, creating sparks and heat (fig. 28–6). In arcing, electricity *jumps* from one conductor to another when the conductors are not in direct contact, creating sparks and heat.

Fig. 28–6. A short circuited wire

For fire investigators to properly determine that fire was started by a piece of electrical equipment, he or she must have physical evidence to support the hypothesis. In the past, fires were often classified as *electrical* in nature if all other sources of ignition were eliminated. This is inadequate today—the fire investigator must prove how the fire started, using actual evidence.

As noted earlier in the chapter, a fire investigator may ask if you switched off a circuit breaker in an electrical panel for safety reasons during a fire. Having made a mental note of which breakers you changed may help the investigator determine whether a particular circuit was involved in starting the fire.

MOTOR VEHICLE FIRES

You will likely respond to a large number of vehicle fires in your career. Although many vehicle fires start in the engine compartment, it is not unusual for them to start in the passenger compartment or other areas (fig. 28–7). In addition to the obvious use of flammable and combustible fuels, the increasing use of synthetic materials provides even larger quantities of combustibles for a fire. There are a variety of ignition sources: overheated wires, hot surfaces such as exhaust manifolds and catalytic converters, mechanical sparks (metal-metal or metal-road surface contact) such as a broken drive shaft or a wheel rim that has lost a tire and is in contact with the roadway, and smoking materials.

Be wary of automobile fires with the presence of gasoline odors in the passenger compartment or trunk; this *may* be an indication of arson. Notify your company officer if you think this is the case.

Fig. 28–7. Aftermath of an engine compartment fire

ARSON

Arson fires, also known as **incendiary fires**, are intentionally set fires that are a crime in virtually every jurisdiction in the United States. Different states classify arson in slightly different ways, so it is important to understand the arson laws of your state.

People who set fires are called **arsonists**. People set fires for different reasons: for the excitement, as an act of vandalism, as an act of revenge, to gain insurance monies through fraud, or for **vanity** (fire setter starts a fire and extinguishes it to receive praise). Unfortunately, even a small number of firefighters themselves start fires to fulfill a need to respond to fires or because of some other deviant, behavioral reason. *If you suspect a fellow firefighter (or any one else for that matter) is starting fires, it is important to notify your local police.*

The evidence of arson may be visible after a fire has been extinguished. In addition to previous arson indicators discussed in this chapter, also look for the presence of gasoline or other accelerants, located in a gasoline can or the unexpected odor of gasoline where it shouldn't be. Check for the presence of **trailers**, bed sheets or clothing tied into a long rope soaked in gasoline and laid in such a way as to spread the fire from one area of a building to another. (fig. 28–8). Check for multiple locations of separate fires with no logical connection, and fires in locations that typically lack an ignition source such as on stairwell treads and on top of beds.

If you suspect a fire may be intentionally set, you must notify your company officer who will, in turn, notify the incident commander. It is important not to disturb the suspected arson location once the fire has been extinguished.

Fig. 28–8. A trailer is used to spread fire from one area of a building to another

EVIDENCE

SKILL DRILL

FFII 5.3.2 There are three types of evidence: **demonstrative evidence** (tangible physical evidence such as a defective water heater), a fire pattern (described previously), or a photo of a fire scene. **Documentary evidence** is any evidence in written form such as an inventory list for a store or an insurance policy. **Testimonial evidence** includes such things as witness statements and expert witness testimony in a trial.

It is critical that the **chain of custody** be maintained for demonstrative evidence, meaning that an unbroken chain of possession is maintained. If a gasoline container is collected at a fire scene, for example, it is important that the container is never tampered with at any time. If the gasoline container must be transferred to a laboratory for testing, it must be *signed off* by the laboratory technician and so on as it moves from official to official. A chain of custody form accompanying the gasoline container will have the signatures of every person who has handled it. A piece of evidence must be protected from damage and alteration at all times.

When collecting evidence, proper precautions must be taken to protect the evidence from further damage and spoilation as described earlier. For example, gloves are worn when handling evidence which may contain fingerprints; precautions are taken to handle the evidence so that the fingerprints are not smudged. Delicate objects (like a water-soaked book of matches) must be handled appropriately. Evidence is then placed into labeled containers that are properly sized to protect the object (fig. 28–9). Containers should not allow evidence to "roll around" inside the container. In addition, the evidence must be compatible with the container—for example, chemicals which react with metal should be placed in non-reactive glass containers. All containers must be properly sealed to prevent contamination.

Fig. 28–9. Several evidence containers of varying shape, size, and material.

DETERMINING ORIGIN AND CAUSE

The collection of data is identified as the facts collected by observations, experiments, or other direct data-gathering means. This type of data is called **empirical data** because it is based on observation or experience and can be verified. When a fire investigator is investigating a fire scene, he is gathering a series of observable facts.

All the collected and observed information is analyzed by **inductive reasoning**, which should include only facts that can be clearly proven by observation or experiment. The development of a **hypothesis** is based on the data analysis. The investigator should now produce a hypothesis or group of hypotheses to explain the origin and cause of the fire. This hypothesis should be based solely on the empirical data that the investigator has collected.

The successful testing of the hypothesis (**deductive reasoning**) can only be achieved if it can withstand a careful and serious challenge. Test the hypothesis by the principal of deductive reasoning, in which the investigator compares the hypothesis to all known facts. If the hypothesis cannot withstand an examination by deductive reasoning, it should be discarded and a new hypothesis should be tested. If specific physical evidence from the fire scene contradicts the hypothesis that a fire investigator has developed, for example, the hypothesis must be rejected and a new hypothesis must be developed. In some cases, a hypothesis cannot be developed (because of lack of evidence, for example) and the fire is classified as **undetermined**. It is critical that an investigator never assign an origin and cause to a fire unless he can prove it using the scientific method.

DOCUMENTING THE FIRE SCENE

Every fire scene should be fully documented with photography (fig. 28–10). With today's digital media, the cost factor is negligible and should not be a concern to the investigator. When taking photographs, the investigator should use a **systematic approach**, identical to the systematic approach used to conduct the origin and cause investigation described earlier. First, the investigator will include overall views of the exterior of the structure, the structure's utilities including the electric service meter, distribution panel, heating systems, and any additional utilities identified within the structure. Fire protection systems—alarm systems and/or fire-suppression systems—should be photographed next. Finally, all rooms and areas involved in the fire should be documented (including the suspected room/area of origin) prior to the commencement of the investigation and the removal of debris. Photographs should be taken to document the actual removal of debris and reconstruction. The photographs will demonstrate to a jury how the investigation was conducted.

Fig. 28–10. A fire scene photographer

Fire investigators take large amounts of notes, recording all details of the investigation. This includes names and contact information of witnesses, observations about the building, details about fire damage, notes about photographs they take, and many other facts. Many investigators use a brand new newspaper-reporter-type note pad for each new investigation.

Investigators make sketches of the fire scene as well (fig. 28–11). These freehand drawings provide details on room layouts, furniture locations, and the like. Measurements are made of room dimensions and the distances between objects. Often, investigators are asked about details of a fire scene many years later while testifying in court about the fire. The investigator is permitted to use his notebook during the court proceedings to recall details of the fire scene for testimony.

Once the fire investigator has completed the investigation and has determined the origin and cause, he or she will prepare a final report. This report will be the *official* report of the fire, providing not only the origin and cause but many other details.

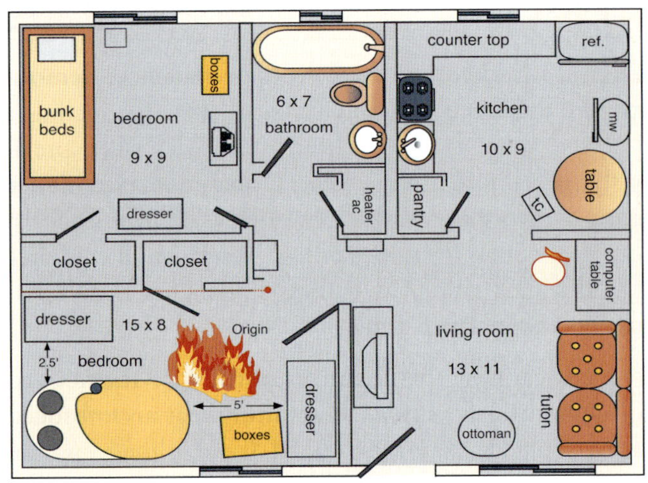

Fig. 28–11. Sketch of a fire scene

SECURING THE FIRE SCENE

It is imperative that a fire scene awaiting an investigation be properly secured, preventing the removal of evidence and alteration of the fire scene (fig. 28–12). Although building owners and occupants may want to enter the fire building, it is important that the investigators be allowed to do their work first.

Fig. 28–12. This fire scene has been secured following an incident.

Because investigators may not arrive while the fire is being fought, incident commanders may often leave a fire company to keep *control* of the building once the fire has been extinguished. Firefighters will be expected to secure the building and to keep an eye out for **rekindles** (smoldering fires which flare up back into a free-burning fire.)

An important 1978 United States Supreme Court decision, **Michigan v. Tyler (436 U.S. 499)**, affirmed a fire department's ability to maintain control of a fire scene, awaiting a fire investigator's arrival. This control means that a fire investigator does not need to get a **warrant** to conduct the investigation even though he or she arrives after the fire has been extinguished.

QUESTIONS

1. What is the most important benefit of investigating fires?
2. What is the role of the first-due firefighter in an investigation?
3. For what reason or reasons is it imperative to avoid excessive overhaul procedures as it pertains to fire investigations?
4. Why is it important for circuit breakers that are tripped to remain in the tripped position?
5. What could blockage of exit doors, and/or passages indicate?
6. What are some typical motives for arson?
7. What does the collection of data mean?
8. When documenting a fire scene using photography, what type of approach should be utilized?
9. Why is it important to use the scientific method to assist you in determining the origin and cause of a fire?
10. What does empirical data mean?

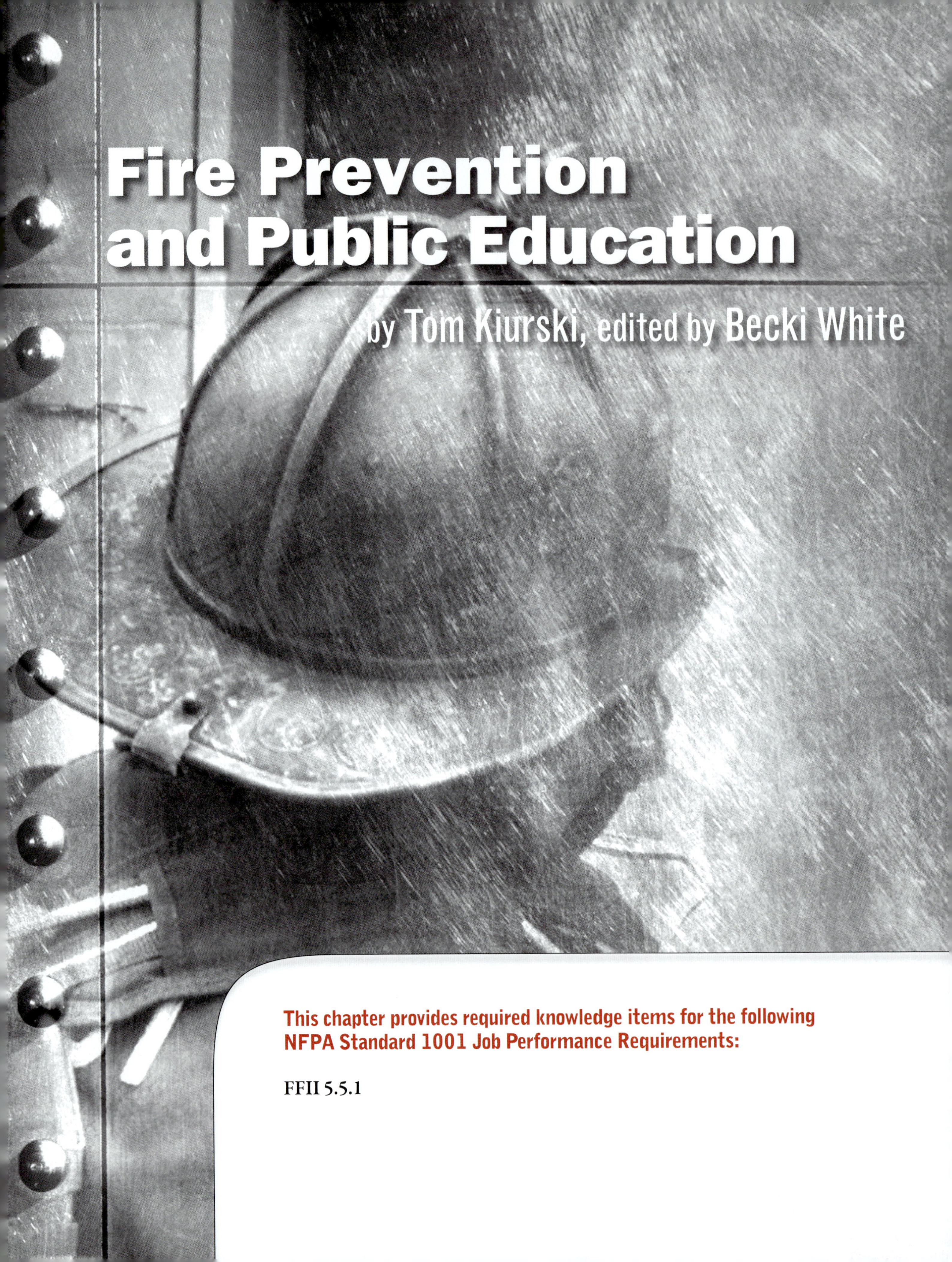

Fire Prevention and Public Education

by Tom Kiurski, edited by Becki White

This chapter provides required knowledge items for the following NFPA Standard 1001 Job Performance Requirements:

FFII 5.5.1

29

Fire Prevention and Fire-Safety Information

FIREFIGHTER II

This chapter contains Skill Drills. When you see this icon, refer to your Skill Drill book for step-by-step instructions.

SKILL DRILL

OBJECTIVES

Upon completion of this chapter, you should be able to do the following:

- Define the term fire prevention
- Describe the role of the fire department in fire prevention activities
- Define the terms building code and fire code
- Identify the major building and fire codes used in the United States today
- Identify the major features of a building code
- Identify the major feature of a fire code
- Describe the process for conducting a fire prevention inspection
- List organizations with fire safety interests
- Describe the procedure for conducting a Home Fire Safety Survey
- List common causes of fire found in a dwelling
- Describe the proper procedure for the placement of smoke alarms in a dwelling
- Describe methods of teaching fire safety to residents of the community
- Describe the proper procedures for conducting a tour of a fire station
- Describe the proper methods for presenting a talk on fire safety to community residents

INTRODUCTION

"Among the many measures that can be taken to reduce fire losses, perhaps none is more important than educating people about fire." This quote was taken from the landmark national report called ***America Burning***, which was first published in 1973 (fig. 29–1). The document was a reaction to the high number of fire deaths and injuries in the United States, while other countries had a significantly smaller fire problem than we did. Members of the National Commission on Fire Prevention and Control, including top fire officials, educators, burn injury specialists, and researchers, were charged with finding out what could be done to reduce our fire risk factor, and the finished product was *America Burning*. This document told us that we had to shift our focus from suppression to prevention, a drastic change from what had been going on in this country (fig. 29–2).

Fig. 29–1. *America Burning* was the definitive study of the American fire problem when published in 1973. Its publication led to the creation of the United States Fire Administration and the National Fire Academy. While some of its objectives were met, many others have yet to be fulfilled.

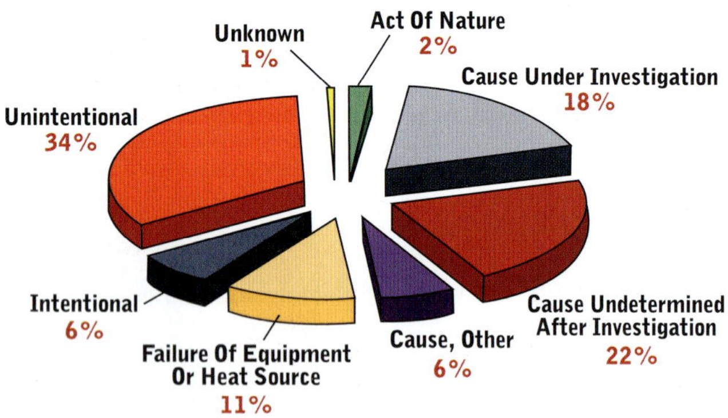

Fig. 29–2. Major causes of fires in structures

Although *America Burning* helped pave the way for fire-safety education in this country, this report also moved some of our thinking from the *reaction* phase of putting out fires to a more *proactive* phase that put more emphasis on fire prevention. Although there were fire departments back in the 1960s that did a great job at fire-prevention and fire-safety education, they were few and far between. *America Burning* suggested a U.S. Fire Administration (USFA; it exists today largely due to those recommendations) that seeks out great programs and makes them available to the fire service to learn from and improve on. It is interesting to note that this report also outlined the need for a **National Fire Academy** (also in existence based on the recommendations of the report).

The results of that study are contained in the pages of *America Burning*. Although the forward has been updated several times to show how our fire statistics have declined, the body of the report remains unchanged. It is still available from the USFA. There were 90 recommendations made throughout the report, and it provides a basis for how our fire-prevention and fire-safety education efforts are grounded.

Throughout the remainder of this chapter, we will look at the topics of fire-prevention and fire-safety education separately. While they are often linked, they can be broken down into smaller elements for study. Fire prevention will look at the code enforcement efforts that are designed to control the hazards within buildings found through an inspection, while fire-safety education will focus on educating the community about the dangers associated with fire and how to make minor changes to be more fire safe where they live, work, and play.

FIRE INSPECTIONS

Entry-level firefighters have some dealings with fire inspections. Although they may not be immediately transferred or promoted into a fire department's fire-inspections division, they will be involved in issues pertaining to fire code in buildings in their community. Some fire departments hire entry-level firefighters and provide them with a fire inspector course shortly after hire. Once that course is completed, they may take part in regular building inspections, under the direction of their fire-inspections division. It may take one inspector a good portion of the day to inspect a standard strip mall, whereas a fire company with trained inspectors may divide up and tackle the businesses individually, and get done in less than an hour.

FIRE AND BUILDING CODES

FFII 5.5.1 Model fire and building codes are designed to ensure the safety of buildings. The first building regulations in this country were in reaction to fires by early settlers. These codes prohibited the building of fireplaces with wood and also did not allow covering roofs with thatch, which is highly flammable. With settlements that had houses built close by each other, the chance of a single fire destroying multiple buildings was a possibility. With the size and complexity of today's buildings, it is not acceptable to lose a building, or several buildings, in an area. The concept of fire-resistive construction, fire areas within buildings (the area enclosed by rated fire walls), and fixed fire-suppression systems are all in use today to contain or control fires within buildings. Most of these codes are tied back to large, historical events that resulted in a loss of life.

Basically, building codes regulate how a building is to be constructed or renovated. Fire codes, on the other hand, regulate what activities take place inside the building in terms of fire safety. Building codes regulate how a fire wall should be built, while fire codes regulate such things as overcrowding in a nightclub or paint spray finishing in an auto body shop. Fire codes also ensure that the fire protection features built into a structure under the building code are maintained. There is overlap between these codes; they are a matched set of documents intended to be used together.

The **National Fire Protection Association (NFPA)** and the **International Code Council (ICC)** both prepare models of building and fire codes. The NFPA produces NFPA 5000: *Building Construction and Safety Code* as well as **NFPA 1:** ***Uniform Fire Code***. Similarly, the ICC produces the **International Building Code** and the **International Fire Code** (fig. 29–3). It is important for you to find out which of these codes is used in your jurisdiction—it is possible that your city or state does not have a building and/or a fire code at all, or uses a predecessor to these model codes. These codes are updated every three years by the NFPA and ICC.

Fig. 29–3. Publications of ICC and NFPA codes

These model codes are essentially ready for adoption and use; all that needs to be done is to fill in the name of the jurisdiction (city or state) that is adopting it. However, most jurisdictions make local amendments to the model code to customize it to their needs. When adopted at the local level by cities or counties, they are called ordinances; when adopted by a state, they are typically called statutes. Codes must be legally adopted by a jurisdiction to be enforceable.

Model **standards** are documents that are prepared by organizations like the NFPA to regulate the installation of individual building systems like automatic fire sprinkler systems (NFPA 13, *Standard for the Installation of Sprinkler Systems*) or standpipe systems (NFPA 14, *Standard for the Installation of Standpipes and Hose Systems*). See chapter 30, Fire Protection Systems, for a complete list of fire protection system standards.

Building code and fire code are two separate documents. They usually reference each other but cannot contradict each another. Correct use of the building code ensures that the building will be built in accordance to the regulations that the jurisdiction uses for new buildings. Factors such as the construction of walls, allowable distance between exits, and flame spread ratings of interior finishes are all covered in building codes. A business that wishes to construct a building within a jurisdiction starts off by giving a copy of its proposed building to the city for a review, usually in blueprint form. The city building code officials ascertain whether the specified building meets the applicable codes and ordinances. If a city has a fire marshal or fire inspector, that person may be the one reviewing the fire protection and notification systems. This original blueprint will get sent back to the company with an approval or with notes that outline the changes necessary in the building to meet codes or ordinances that may have been overlooked.

Corrections are made, and any necessary clarifications are exchanged until a final plan is submitted that is agreed upon by the jurisdiction. A building permit is issued at this point, allowing construction to begin. As the building is in progress, the jurisdiction can send building inspectors out to make sure the building is being constructed as agreed upon. Once the building is completed, the jurisdiction conducts the final inspection to determine whether the building has been constructed and completed as submitted. Once the building meets inspection, it will be ready to open for business, and the jurisdiction will issue a certificate of occupancy stating such.

Once the building opens for business, the owner is then expected to maintain a level of fire safety for the employees and patrons. This is where the fire codes come in, to implement adequate maintenance of buildings. Some items covered in these codes are fire extinguishers in the building that have been inspected and are acceptable; illumination of exit signs in dark, well-traveled areas; and exit doors that are kept unobstructed. The **fire prevention inspection** is a legal inspection that should be completed annually or more frequently for high-risk hazardous occupancies.

In general, the building plan review and subsequent inspections are handled by the jurisdiction's building department, while the fire codes that cover the building's use and maintenance are part of the fire department's fire-prevention division's responsibility.

The fire inspectors' task is to make sure that the buildings within their jurisdiction are maintaining the level of fire safety outlined in the fire-prevention code. Some cities use fire-prevention division personnel to perform this function, while others train on-duty firefighters to conduct regular inspections.

Any building that is not in compliance with the jurisdiction's fire-prevention code will need to be corrected. It is for this reason that a building representative should accompany fire inspectors during the inspection. Although a receptionist may offer to accompany the inspectors, firefighters should suggest a manager or supervisor be summoned. Ideally, it should be someone who can authorize changes in the building that may be needed to bring the business up to code compliance. In most companies, the receptionist is in an entry-level position and does not have the authority to spend money to correct building issues.

We begin the building inspection by entering the main entrance door and asking to speak to a business representative. Once introductions are complete and the nature of the visit is outlined, the inspection can begin. It is important to note that a fire inspection is an official government action; building owners can deny entry to fire inspectors under the "search and seizure" clause of the United States Constitution. In such cases, fire inspectors must obtain a warrant from an appropriate court to gain entry. It is best practice to notify the business prior to the inspection so they can arrange to have someone available to walk through on the inspection. It's also a good idea to provide guidance on what types of things you are going to be looking at, so they can potentially solve those problem ahead of your visit, which will save you both time and may alleviate the need for a return visit to ensure violations are corrected.

The inspection actually begins on the outside of the building. A walk around the building will let the inspector understand the layout of the building, how accessible the building is from the four sides, and any additional concerns, like dumpsters too close to the building, overgrown brush growing up the side of the building, and the like. The inspector can also note from the outside the building the access to the closest fire hydrant and the utilities coming into the building.

The course of the investigation is determined by the inspector. If the inspector chooses to move from the front of the business to the rear, or a bottom to top inspection on a multistory building, then stick to that plan. Doors that are not opened by the building representative must be noted at the time and inspected when the keys become available. Any confidential business that the inspector may see or hear is to be kept in confidence.

During the inspection process, inspectors should identify where the business is exceeding fire code requirements, as well as the deficiencies. Formally, these deficiencies are documented, and a reinspection is needed to verify that the building is meeting the fire-prevention code. If a simple fix is undertaken by the building representative, that also should be noted. Some business owners may feel they want things done a certain way, and will make changes during the inspection, and then change back to their original way once the inspectors leave the premises. A well-documented fire inspection report is vital to enforce fire-prevention code compliance (fig. 29–4).

Fig. 29–4. Fire inspectors arrive at a business, ready to conduct a fire prevention inspection.

If the building has been in operation, the annual inspection starts off with reviewing past fire prevention inspections, to note the history of the building and business. If a business was found on each inspection to have continually violated fire codes, this matter should be discussed with the head of the fire-prevention division, usually called the fire marshal.

If any deficient items necessitate a reinspection, the inspector will need to visit the property again to ascertain fire code compliance. This inspection is usually brief and includes checking only the area or items that did not meet code during the inspection. If the inspector is satisfied that the business is now in compliance with the fire codes, the inspection is complete. Interestingly, businesses tend to fail their inspections in very similar areas that are common to almost all buildings. Burned-out exit lights, failure to maintain fire extinguishers and fire-suppression systems, improperly maintained emergency exits, and use of extension cords as permanent wiring are violations that are found quite frequently.

The subject of exits comes up so often in discussing large loss-of-life fires that it warrants further study here. Many patrons of businesses are familiar with the main entrance as their primarily intended exit. If an emergency arises, additional exits must be clearly identified and maintained to be used successfully. The **exit system** is made up of three components: the **exit access** (includes aisles within rooms or spaces as well as corridors), including how it can be seen and followed in an unobstructed manner; the **exit** itself (a stairwell in multistory buildings), consisting of the proper egress door mounted in the proper frame in the proper direction of swing; and the **exit discharge** (the door to the exterior of a building), which must also be kept clear to allow occupants to continue their travel to a safe area away from the structure.

If a complaint comes in to the fire-prevention division, the inspectors will determine if a visit is necessary. Complaints must be taken seriously and investigated, so the inspector must listen to the complaint and determine the course of action.

In addition to the duties just described, the fire-prevention division is often the fire department's investigative section. If fire company personnel are called to a fire that they would be of interest to the fire-prevention division, such as a large fire or one that seems to be intentionally set, they should call the division to investigate the fire to determine the fire's origin and cause.

FIRE SAFETY ORGANIZATIONS WITH AN ALLIED INTEREST IN SAFETY

FFII 5.5.1 The concept of fire safety is addressed by a number of organizations, and each will in some way affect firefighters and their lessons of fire safety for the public they serve.

After numerous fire disasters, fire insurance companies have become involved in fire safety efforts for their insured clients. **Factory Mutual (FM Global)**, an insurance company that insures industrial properties, maintains a laboratory for the testing of fire protection equipment, which may bear the FM seal of approval if so recognized.

Likewise, **Underwriters Laboratories (UL)** is an independent organization that does testing, listing, and examining items as they relate to fire protection and safety. The UL listing is another recognizable symbol that consumers look for when purchasing products. There is also a division of UL called the Fire Safety Research Institute (UL-FSRI), which researches fire behavior. UL-FSRI releases material for fire department operations as well as public safety education based on their findings.

The **Insurance Services Office (ISO)** is familiar to many fire departments, as they have developed and utilize the Fire Suppression Rating Schedule that determines a community's capabilities for firefighting, which is used to determine fire insurance rates for residents of the community. The Fire Suppression Rating Schedule does not incorporate an individual fire department's fire code enforcement or public education efforts into its rating. ISO does have another program called the **Building Code Effectiveness Grading Schedule**, however, which rates building code enforcement in individual communities.

The standards relating to fire safety are written by a number of agencies. The **American Society for Testing and Materials (ASTM)** writes standards that pertain to the flammability characteristics of wall finishes (ASTM Test Standard E84-08a *Standard Test Method for Surface Burning Characteristics of Building Materials*) and the hourly fire rating of wall, floor, and roof assemblies (ASTM E119-08a *Standard Test Methods for Fire Tests of Building Construction and Materials*). The **American Society of Mechanical Engineers (ASME)** writes codes that help maintain safety in elevators and escalators across the country. The NFPA, as described earlier, writes a variety of codes and standards that apply to many facets of fire prevention and suppression used throughout the world.

FIRE SAFETY EDUCATION

FFII 5.5.1 It is the responsibility of the firefighters to educate residents in their district about fire safety at any opportunity that arises. This may be while visiting schools, at the fire station conducting a tour, or doing a **home fire-safety survey**.

A home fire-safety survey should not be confused with a fire prevention inspection. The fire prevention inspection is a legal inspection that demands compliance; whereas the home fire-safety survey is a voluntary inspection done by firefighters to help educate the occupants about the dangers of fire and how to be safer in their everyday lives.

The purpose of a home fire safety survey to identify common risk factors for fires and suggest safer practices. It is informational in nature; it does not imply that by following your suggestions the occupants will never have a fire in their home. In any inspection or survey, we are looking to educate and make a safer environment. Firefighters should not balk at conducting a home fire safety survey. The mission of the fire department is to protect the community from fire, and a home fire safety survey is a way to meet that mission.

Some departments may use nonfirefighting staff to conduct these surveys, like members of the Community Emergency Response Team (CERT). CERT members are community volunteers who assist the fire department during large scale events by being a liaison in their communities. Tapping into the CERT teams to conduct home safety surveys is a good way to engage CERT members and put homeowners at ease about the home fire safety survey having any violations or other consequences tied to it.

The purpose of the home fire safety survey is to educate. Individuals assigned to home survey details should be able to articulate safety information clearly and concisely and be neat in appearance. Conduct this survey only if the occupants have initiated contact with the fire department, or if your department does a door-to-door campaign and asks the owner for permission to conduct the survey beforehand. Let the occupants know that the survey is voluntary, and they can stop the survey if they wish. Inform them that there will be no follow-up inspections to make sure the suggestions you make are put into action, and the only paperwork the fire department will fill out is a simple form, a copy of which the occupants will receive at the completion of the inspection (fig. 29–5). The form should be in duplicate, so that one can stay on file with the fire department, for statistical purposes only. At the very least, the inspection should be noted in the fire department record management or computer aided dispatch (CAD) system.

The home fire-safety survey should address the common causes of home fires and personal injuries, like slips and falls.

Once you have made contact with the occupants and received permission to conduct the survey, start from outside the house. Begin by looking at the property itself. If the home is located in a wildland–urban interface, it is suggested that the occupants keep a "green space" of living grass or vegetation at least 30 feet in all directions from their home. This allows firefighters an area to set up to try and save the home if there is a wildfire.

Next, look at the fixtures and storage items on the property. Outbuildings and sheds are often places where flammable liquids are stored. Explain the importance of keeping these areas secured, and that flammable liquids should be stored high and out of reach of children in approved containers. If vandals frequent the area, locks should be installed.

Storage of flammable and combustible materials should be considered, and the amount of flammable liquids (ignitable liquids with a flash point below 100°F [38° C]) should be kept to a minimum. In the arson field, it is interesting to note that majority of fires are set by juveniles. These misguided minors usually are looking for excitement rather than revenge or a statement of protest. By having plenty of combustible materials on the property, the missing ingredient is a simple match or lighter. Getting rid of accumulated flammable liquids, or locking them in a shed, barn, or a garage will deter the risk. Since these children usually live in the area and do not want to be seen or recognized, installing motion-sensor lighting makes good sense, too.

The firefighters conducting the survey need to note the condition of the chimney, to see if it is in need of repair. Cracked or broken bricks are a sign that a chimney may need some repair. The condition of the roof should also be noted. Wooden shingles were once popular, but have fallen out of favor due to their vulnerability for fire spread. After years of use, they become thin pieces of wood that ignite easily, possibly from flying embers produced by an outdoor fire pit, if allowed.

Although some jurisdictions have a ban on any outdoor fires, others may not. Inform the residents in your area about your local laws for outdoor fires. If your area does allow outdoor cooking or pit fires, offer people information about how to do it safely. Proper use of camp fires, cooking fires, and the increased use of commercial fire pits are often the subject of questions that some residents may have. They are usually hesitant to bring this up, because they feel it may put them in a bad light.

You may be asked to look inside barns and garages. Inside these structures, you can address the same concerns of stored flammable liquids as you did with the sheds. You can also offer advice on accumulated storage items, if they are blocking access to and from the doors. If the garage is attached to the house, then the need for controlling storage and flammable liquids increases due to the close proximity to residents.

As you approach the entrance to the house, make sure that the address is clearly visible from the street. Night calls when people usually do not come outside to direct us can be difficult if the houses are not properly marked. The front porch light should be turned on if an emergency call has been made, and firefighters are expected to arrive.

The interior survey should include the main living areas of the home, as well as any additional areas the occupant may want inspected, such as a basement. If the survey does include the basement, housekeeping is a major issue. The basement should be clear of accumulated flammable combustible materials, and clear access should be maintained to the furnace, hot water heater and electrical panel, if located in the basement, as these are some of the main hazard areas common to basements.

If the basement is finished off into rooms, smoke alarms should be present throughout the basement to give early warning to occupants in case of fire (more on smoke alarms later in this chapter). If the basement is used for occupants to sleep within, certain safety considerations should be brought up to the occupants. First, if there are two clearly identified exits, they should be kept clear for easy access. This may include the basement door and stairs and large windows or a sliding door for backyard access. If the windows are smaller and high on the basement wall (unusable for egress), an appropriately sized escape window must also be installed near floor level as a required secondary means of egress (as required by the model building codes described earlier). Since you may not get homeowners to install this code-compliant secondary escape window (or stairwell to grade) in the near future, tell them of the dangers of using the basement as a sleeping area and how it is in violation of the building code.

If the basement is used often, such as a woodworking shop, laundry room, or a game room, installing a fire extinguisher is recommended. Regardless of whether the occupants choose an inexpensive model or a more expensive, rechargeable model, it should be an A-B-C rated fire extinguisher, and the adults should know how to use it if necessary. A short explanation on how fire extinguishers work may be necessary.

As you move through the home, find time to talk about safety related to some of the main causes of fire, such as arson (discussed earlier when outside), cooking-related, electrical and electrical distribution, heating equipment, and smoking materials.

Moving to the main level of the home, you can talk about kitchen fire safety. Safety topics in the kitchen are numerous, and many are common sense that need to be reinforced. Keep the cooking areas clean from grease and other combustibles that can easily ignite. Suggest that the occupants cook on back burners of the stove whenever possible, to prevent small children from reaching up and pulling hot items down onto themselves. Keep pot handles turned to the side to avoid them being bumped and the pot contents being spilled. Children should not be allowed to play in or around the cooking areas when cooking equipment is being used, commonly referred to as a 3-foot "kid free" zone. Since the main cause of home fires is cooking-related incidents, having a fire extinguisher near the kitchen area also makes good sense. The fire extinguisher should be stored between the stove and the exit. Occupants will likely struggle to find extinguishers stored under the sink or in the back of a cabinet when there is a fire. Besides the fire extinguisher, let the occupants know that if they are cooking with grease, keeping the lid to the pot or pan they are cooking with handy is recommended. If a grease fire occurs, the occupant should use an oven mitt to place the lid over the flaming pot or pan. This takes away the oxygen, putting out the fire. Tell the occupant to keep the lid on and shut off the heat to the burner, then call the fire department to search for any fire spread that may have occurred. The pan should not be moved from the stove because the burning grease inside may

HOME FIRE SAFETY SURVEY

This form is designed to help you check your home for fire hazards. All the questions are worded so a "YES" answer means the area is safe. A "NO" answer means a fire hazard exists and needs to be corrected. If you have any questions about how to correct a hazard, call your local fire and rescue station for assistance.

HEATING HAZARDS	YES	NO
1. Combustibles like clothes, curtains, and furniture are away from stoves and heaters.	_____	_____
2. Heating appliances are tested every year.	_____	_____
3. There is a screen in front of the fireplace.	_____	_____
4. Ashes are emptied into covered metal containers.	_____	_____
5. All alternative heating devices used are U.L. listed.	_____	_____

ELECTRICAL HAZARDS	YES	NO
1. Electric cords are checked to be sure they are not worn, frayed, or damaged.	_____	_____
2. Extension cords are not run under rugs, over nails, through water, or across passageways.	_____	_____
3. Proper size fuses are used in all sockets.	_____	_____
4. Each circuit is properly marked.	_____	_____
5. Electrical outlets are not overloaded by multiple plug devices.	_____	_____
6. Electric appliances are not used near water.	_____	_____

GENERAL HOUSEKEEPING HAZARDS	YES	NO
1. Closets, basements, attics, and garage areas are clean of old paper, boxes, paint cans, and other trash.	_____	_____
2. Rubbish is thrown away regularly.	_____	_____
3. Everyone in the family knows NOT to use a flammable liquid, like gasoline, for cleaning.	_____	_____
4. Flammable liquids and oily rags are disposed of after use or stored in a tightly closed metal container.	_____	_____
5. All combustibles are stored a safe distance away from any possible heat source.	_____	_____

SMOKING HAZARDS	YES	NO
1. Matches and lighters are kept out of the reach of children.	_____	_____
2. Smokers use ashtrays that are large, deep, and designed for use with smoking materials.	_____	_____

KITCHEN HAZARDS	**YES**	**NO**
1. The kitchen is kept clean all the time.	_____	_____
2. The stove and oven are in good working condition.	_____	_____
3. The stove area is kept clean and free of grease.	_____	_____
4. Combustible materials (like towels and pot holders) are kept away from open flames and heat sources.	_____	_____
5. Someone is always in the kitchen when food is cooking.	_____	_____

GENERAL FIRE SAFETY	**YES**	**NO**
1. The house address is clearly posted in large numbers that are easy to read from the street.	_____	_____
2. There is a working smoke alarm near the sleeping areas.	_____	_____
3. There is a working smoke alarm on every level.		
4. The smoke alarm is cleaned regularly and tested once a month.	_____	_____
5. You know the emergency telephone number for fire and rescue. (9-1-1)	_____	_____
6. This number is posted on all telephones for immediate reference.	_____	_____
7. You have portable fire extinguishers suitable for use in your home and family members know how to use them properly.	_____	_____
8. You have a home escape plan and practice it regularly.	_____	_____

If you have any "NO" answers, sit down with your family and discuss the problem and how it can be corrected. Take this opportunity to create fire safety awareness in ALL family members and, at the same time, make your home a safer place to live.

Fig. 29–5. Home fire safety survey form

spill and cause severe burns. Remind homeowners never to put water on a grease fire, because oil and water don't mix. Adding water to a grease fire actually spreads the fire. Past practice suggested using baking soda or salt on pan fires, but this is no longer recommended because of the amount of product that would be needed is generally not sufficient and because people would mistakenly use flour, which would likely spread the fire.

Also on the main level, you may find a room that is used by the family to spend quite a bit of time relaxing in. This may be called the living room, family room, or go by a different name. This may be a good time to discuss heating equipment, such as keeping furnaces and fireplaces clean and inspected annually. The screen of the fireplace should always be pulled into place when a fire is burning to keep any sparks contained. Portable space heaters are commonly used for small area heating throughout the home (fig. 29–6). Although great for warming up a small area, like any open flame from cooking or a fireplace, they should never be left unattended, and there should be an adult supervising at all times. The safety of today's space heaters is a vast improvement over earlier models. Many of the newer models have a "tip over" switch that turns the unit off if it is tipped over. The safe distance that is recommended for combustibles to be kept away from any heating source (furnace, fireplace, space heater, etc.) is 3 ft (1 m).

Fig. 29–6. Homeowners need to be reminded of the rules of fire safety in their homes to build lasting changes in their behavior.

Electricity is a luxury that we have become accustomed to, and we enjoy its benefits every day. Whether it is listening to music, using our personal computers, or taking a cold drink out of the refrigerator, we all draw electricity into our homes. Although many safety measures are built into the electrical system in our homes and the products we purchase, we must be vigilant in our use of electricity at home. A good rule of thumb is one major appliance or power strip should be the maximum for any one outlet, and the power strips that are common today behind entertainment centers and computers should include a built-in circuit breaker that will shut off power to the appliances if a problem is encountered in the system. Most national codes today insist on outlets installed with ground fault circuit interrupters (GFCIs) when there is a chance of water being splashed with resultant electrocution, such as in kitchens and bathrooms.

If electrical extension cords are used in the house as part of a permanent wiring system designed by the homeowners, they should be informed about the danger of this practice. Extension cords are designed to bring electricity to a location where it is needed for a short time period, and then should be disconnected. An electrician may be needed to come out and install a few more electrical outlets in the home if they feel the need for more electrical outlets. Extension cords should never be 'daisy chained' or plugged from one to another. Power strips should also not be plugged into extension cords.

Ask the occupants if anyone smokes in the house. If so, some of the safety tips that we can tell them include having large, tip-proof ashtrays in the home that can hold the entire cigarette in case it is lit and forgotten, avoid emptying ashtrays with hot ashes into garbage containers, and never smoke in bed or when taking medicines that may cause drowsiness. If the homeowners smoke outside, talk to them about having a metal container to put their cigarette butts. They should never use planters to extinguish their cigarettes because many planters have peat moss which includes bits of wood that can catch fire and smolder for hours. This can lead to a deck or exterior fire hours after the smoker has left the area.

The items we use to light tobacco products, usually matches and lighters, are a problem themselves. Too often, children have easy access to these items. Children experimenting with matches and lighters account for far too many fires, some of which have injured or killed family members. Matches and lighters should be stored high up and out of the reach of children. Children should be taught that matches and lighters are tools used by adults, and are not toys.

Candles fires are among the leading causes of home fires across the nation. The development of gel candles and other decorative candles that provide aromatic pleasure has created additional risks in the home. Candles should be treated as the open flame they are, supervised by an adult, and with combustibles kept at least 1 foot away.

Tell the occupants to make sure that no blowing curtains or other objects can come into contact with the open flame. Homeowners should be encouraged to use flameless, or battery-operated candles in place of real candles. You can find a wide-variety of flameless candles with scents and visually realistic flames.

Bathrooms and bedrooms contain hazards that have been outlined, but if a special hazard is brought up by the occupants, listen and provide suggestions that address their worries. The main concern of parents is their children, so focus on this when discussing smoke alarms and fire safety behaviors with the family.

Smoke alarm technology has come a long way since the 1970s, when a single unit cost over $50. Smoke alarms have been introduced to give occupants time to react and leave a building that has smoke (and probably fire) in it. All alarm activations should be taken as real, and family members should be encouraged to leave any building that has an alarm sounding. This may be going against the actions of the crowd, but it is always safest to leave first, then reenter if the "all clear" is given.

Smoke alarms should be installed throughout the homes in your area, with an absolute minimum of one smoke alarm per floor of the home (including the basement), and one outside of sleeping areas should be maintained in existing homes. Newer fire code in most areas also requires smoke alarms in each bedroom. The best level of fire protection, of course, includes installing alarms throughout the home. By doing this, wherever a fire may start, a smoke alarm (or heat detector in kitchens and bathrooms) is close by to alert the occupants of the smoke inside the home (fig. 29–7).

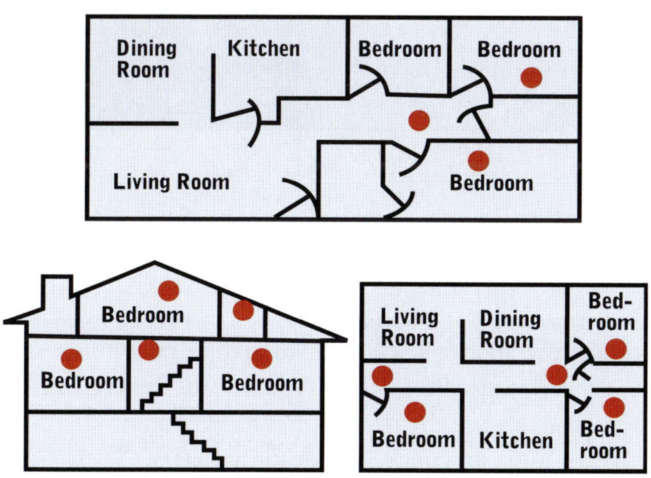

Fig. 29–7. Smoke alarm placement

The most common type of smoke alarm is the battery-operated single-station unit that is a sensor and alarm all in one unit. Because they are operated by batteries, they can be easily installed almost anywhere in the home. Consult manufacturers' recommendations for suggested locations, but in general, smoke alarms should be mounted on the ceiling or high on a wall if a decorative ceiling is in the room. The corners of rooms have a "dead air" space where the walls meet the ceiling; air movement is restricted, so keep ceiling-mounted alarms at least 4 in. (10 cm) away from the wall, and wall-mounted alarms between 4 in. (10 cm) and 12 in. (30 cm) below the ceiling.

Many of the building codes that have been updated since the widespread use of smoke alarms have included the mandate of smoke alarms in the homes that are newly built or renovated. Current building codes call for hard-wired (which uses a home's electrical supply with battery backups), interconnected smoke alarms in the following locations: on all levels, outside all sleeping areas, and in all sleeping rooms. Wiring the smoke alarm to the home's electricity solves the problem of dead batteries rendering smoke alarm inoperable. The problem comes in during power outages, when many open flames from candles and fireplaces are in the home that is now without smoke alarm protection. For homes built before the inter-connected alarm code was introduced, there are now wireless smoke alarms that can speak to each other using a wireless connection. These are slightly more expensive than the single-station alarms, but provide better notification to homeowners if there is a fire in a different part of the home.

Smoke alarms today come with many options, and different models should be considered carefully to see which ones are useful to the family. One option is a smoke alarm testing mechanism that allows a beam from a flashlight to test the alarm. This may be helpful to keep an elderly occupant from climbing on a chair to test the alarms. Others may have light beams that can be pointed at an exit, to help occupants find the way out of the house when smoke may obstruct their vision. Others have "silence switches" that will disarm the alarm mechanism for a short period, to allow time to investigate or clear out smoke that may have come from food cooking on the stove. Still other smoke alarms have 10-year batteries.

Another option should be considered by parents of children from the ages of about 5–12. This is a "vocal" smoke alarm that allows the parents to prerecord a message that will play after the alarm sounds. It has been found that children in this age group are so used to sleeping through unwanted sounds that they often sleep through smoke alarms going off. They are, however, more likely to awaken when they hear the familiar voice of a parent giving instructions. If this alarm makes sense for the family, it should be installed near the sleeping areas of the home where the children sleep.

Smoke alarm and smoke detector are terms often used interchangeably by those who don't understand the difference. Smoke alarms are generally single-station and act as a sensor and an alarm. Smoke detectors are part of a system and only detect the presence of smoke and send a signal to an alarm panel and annunciators or speakers which alarm. It may seem like an insignificant difference, but since fire safety is part of our business we should use the terms correctly. If you suggested a homeowner purchase a smoke detector rather than an alarm they would not be getting the right product.

Smoke alarms and smoke detectors can be part of a home security package and are monitored by an outside alarm company. There should be a way to alert the alarm company or temporarily disconnect the alarm from the company during testing periods. The smoke alarms as part of a home security system should be individualized to meet the family's needs, and the system should be developed by both the occupants and the alarm company.

Regardless of type, smoke alarms cannot do their job if they are not working. Each smoke alarm should be tested monthly by pressing and holding the test button until the alarm sounds. The whole family should take part in this activity to instill fire safety behaviors in the next generation. Pick an easy day to remember and mark the calendars in the house. Maybe the first day of each month, the first Tuesday of the month, or some other system that works with the family should be selected. If alarms are too high to reach, suggest the use of a broom or other long-handled item around the home.

The batteries in single-station units should be changed annually to ensure that fresh batteries will enable the unit to respond to smoke. The International Association of Fire Chiefs and the Eveready Battery Company instituted the "Change Your Clock, Change Your Battery" campaign years ago, which tells consumers to change the batteries in their smoke alarms when the time changes for daylight savings time each fall. Although the daylight savings time may not be used by all areas, it is advertised widely enough for firefighters to remember to get their campaign underway to alert their citizens. During the spring clock change, residents should be encouraged to check their alarms when they are changing their clocks. Since many clocks today change themselves, it's a good idea to change your batteries during annually events, like fire prevention week open house, New Year's Day or other significant annual community events. A 2004 telephone survey found that 96% of the households surveyed had at least one smoke alarm. According to the NFPA, in one out of every five homes equipped with at least one smoke alarm installed, not a single one was working. The main reason for smoke alarms failing to work is missing, disconnected, or dead batteries. The statistics arm us with good information to tell residents about the importance of having working smoke alarms in the home (fig. 29–8). Smoke alarms expire after 10 years. Encourage homeowners to check the dates on the back of the alarms when they are changing batteries or doing alarm tests.

Fig. 29–8. Firefighters should test the smoke alarms in the homes. This shows residents how to test them and what the alarm sounds like if they haven't checked them recently. Show them how to replace the battery on battery-operated units.

Many fire departments make it a standard operating procedure to check the smoke alarms in a home after every call, regardless of the type of call. This is a public service to ensure the proper operation of a home's smoke alarms and to check for dead batteries. These same departments carry extra batteries and smoke alarms on the fire apparatus, ready to be installed. A notation is made in the incident report that the battery or alarm was installed. This provides some legal protection, especially in the case of a call for a "smoke odor" that cannot be located by firefighters, followed hours later by a working fire in the same home. Although this may not provide immunity from lawsuits, it does show that firefighters took appropriate action for the residents' safety.

Regardless of the options purchased on smoke alarms, today's smoke alarms operate on one of two principles. The sensing chamber is either an ionization type or photoelectric type. Much debate has taken place on which is better, but the research shows us that the ionization smoke alarm responds faster to flaming fires, while the photoelectric units respond faster to smoldering fires.

Ionization smoke alarm sensing technology consists of a small radioactive chip located between two electrically-charged panels, which ionize the air and move current between the panels. Smoke entering the sensing chamber disrupts the flow of ions, reducing the flow of the current, and sets the unit into the alarm mode.

Photoelectric sensing technology has a sensing chamber with a light-emitting source and a light-receiving source. As long as the two do not meet, there is no alarm. When smoke enters this sensing chamber, the light beam is refracted into the light-receiver, and the alarm is set into the alarm mode.

Carbon monoxide (CO) alarms are another important component is the overall safety picture of today's families (fig. 29–9). Carbon monoxide is the by-product of combustion of natural gas, which is present in most homes, as the use of natural gas is common for many stoves, furnaces, and water heaters.

Fig. 29–9. Carbon monoxide alarm

Since carbon monoxide is similar in weight to air, some manufacturers suggest they be mounted on the ceiling or high on a wall, similar to smoke alarms. Others suggest a lower mounting, as evidenced by the plug-in type carbon monoxide alarms that are commonly available today. Either location works, and homeowners should be encouraged to follow the manufacturer's recommendations. Homeowners can also get alarms that are combination smoke and CO alarms.

The CO alarm will sound when elevated levels of CO are attained within the home; families must be told to evacuate when the alarm sounds. Most fire departments respond first when a call for a CO alarm activation is received. Responding firefighters will use metering equipment to establish the levels of CO in the home.

Home fire sprinkler systems are getting more popular today. In some jurisdictions, building code requires that homes have residential sprinkler systems, and where they are not required, some homeowners may choose to have them installed. It is important that firefighters have information to give homeowners on what these systems are and how they would go about installing them in their homes. This can also educate the firefighters if they find homes that have fire sprinkler systems that they were unaware existed in their response district. This information should be documented and disseminated to the other firefighters in the area. Fire sprinklers can go a long way if we want to reduce the number of fire deaths in this country, and firefighters should be prepared to educate homeowners if approached (fig. 29–10).

Fig. 29–10. Residential sprinklers are required in some communities, so be prepared to offer information about them if asked.

YOUTH FIRESETTER PREVENTION AND INTERVENTION PROGRAMS

During the early phase of the home survey, arson was mentioned. Arson is a leading cause of dollar losses due to fire every year. While arson is the deliberate act of setting a fire, the majority of arson is juvenile arson. Do not assume a natural curiosity about fire is an unhealthy "arsonist" attitude. Many children are curious about fire. It is important to teach children that using fire as a tool can be useful in the home and should not be played with or used for other reasons. This natural curiosity should not be used as an excuse for children who are involved in fire-play. All fire-plan incidents should be addressed and documented to ensure there isn't a pattern.

FFII 5.5.1 It is important for today's fire departments to be ready for the call or visit from an adult or guardian who is concerned about catching a child playing with matches or lighters. The firefighter must know what action is appropriate to take when this occurs. Today, many parts of the country have youth **firesetter programs**, following the lead of organizations such as the Rochester, New York, Fire Department, which created one of the first such program in the 1980s. These programs bring together fire safety, law enforcement, mental health, and child safety advocates to provide community-based treatment to address this problem.

If your fire prevention division is not equipped with the appropriate training to handle this situation, arrangements must be made. This may include a neighboring department having a youth firesetter program, a program through the state fire marshal's office, or you may suggest the family make arrangements to visit a mental health professional.

FIRE SAFETY EDUCATION

Fire safety behaviors can be taught at community events and in the schools in your jurisdiction. They should also be reinforced in the families that you visit during your home surveys. There are two different levels of fire safety education: fire prevention and fire safety and survival. Much education is referred to as fire prevention, when it is actually safety and survival messages.

Consider the common messages you teach when you are in front of audiences:

- Stop, drop, and roll
- Cooking with grease (having a lid or extinguisher nearby)
- Having working smoke alarms
- Having a meeting place outside
- Calling 9-1-1

All of these are very important messages, but they are all things people must do after a fire has happened, they are not teaching people how to prevent a fire from occurring in the first place. Fire safety education needs to focus more on fire prevention, but should also include fire survival messaging as an added message rather than the initial message.

There are many fire safety/survival and prevention messages that we want to pass along to audiences. Evaluate the risks that affect each audience.

Since most of our emergency calls are initiated by dispatch, one of the most important messages that should be taught is the proper use of the 9-1-1 system. Children should be taught about knowing how to identify the address they are calling from. This can be done easily by teaching them that every building has a name, the first name of the building is in numbers, since numbers are hard to pronounce and remember, the buildings wear nametags to help us out (the numbers that are on the outside of most buildings). The last name of the building is the street that it is on. So, if my house was 1234 Maple Street, the first name would be 1234 and the last name would be Maple Street. Students can practice identifying addresses on buildings around their community, their school, the fire station, their homes, etc. Children should be taught about when they should use this number and be reminded that it's not okay to call 9-1-1 unless they are having an emergency.

Many children around the country are taught to stop, drop, and roll (S-D-R). This is a good practice if they are commonly near open flame where their clothes might catch fire. However, a better message would be teaching children to stay at least three feet from bonfires, grills, fireplaces, and other areas with open flame, turning the safety/survival message to a prevention message. If you are speaking to an audience that is commonly around open flames, the scouts for instance, you can teach them the practice of S-D-R. When you explain the practice of S-D-R, explain fires need three things to keep going:

heat, fuel, and oxygen. The practice of S-D-R is taking away the oxygen. If clothing (the fuel) catches on fire, the person needs to stop (the natural instinct of running will add more oxygen making the fire increase), drop to the ground and smother the fire between themselves and ground by rolling or rubbing the fire into the ground. If your audience is not at risk of their clothes catching on fire, you can address this subject, but don't spend a large amount of your presentation time on it.

It is important to understand the cognitive level of children when you are presenting to them. Using messages like 'Crawl Low Under Smoke' makes no sense to children who have only seen smoke come from a firepit or fireplace. In every interaction they have with fire, it is always right above the fire and goes straight up. So when firefighters tell them to crawl under the smoke, they can get confused. Some fire personnel will have children crawl under a sheet to solidify the crawling low concept, however this doesn't clear up the confusion in their mind. You should explain smoke and fire behavior to children before instructing them to crawl under smoke. This illustration can be used to easily explain smoke and fire behavior. Smoke and fire works like a bathtub, when you turn on the water, it goes straight down and fills up the bottom of the tub or sink. Once the whole tub is filled up it starts getting deeper. Children have seen water filling a tub or sink and can easily picture this in their mind. Tell them that smoke acts the same way, except upside down. Smoke goes straight up until it hits the ceiling, then it fills the whole room and starts getting deeper (at this time motion that deeper actually brings the smoke closer to the floor) which causes the less smoky air to be closer to the floor. If children can see the exit, they should crawl under the smoke to reach the exit.

A very important message that has been added by fire safety educators is the 'Close Your Door' message. After explaining smoke behavior, as noted above, the close your door message can be added. Encouraging family members to sleep with their door closed at night will keep that smoke and heat away and give them more time to escape or survive in a home fire. If doors are open, when the smoke travels it will continue into their room as if it's part of the bathtub example above. With a closed door, the smoke and heat stay on the outside and their rooms will be clear of smoke and heat when the smoke alarms sound outside their room. With modern building materials and furnishings burning more rapidly, a closed door will increase the amount of survival time the occupants have.

All families should be encouraged to create a home fire escape plan, which is a document that has a plan for escape in case of fire, and it must be discussed with all family members present. The escape plan should include a map of the home, including the outside and inside walls and doors to the rooms, and it should show two separate ways out of each room in the home. Usually, this includes a door and a window. If a room does not have two ways out, such as a bathroom, it should not be used for extended or overnight periods. The windows should be inspected by the family to confirm that they open properly and to see if children can open them if an emergency arises. Small children and others living in the home may need someone to look after them in case of fire, and this should be discussed during the home fire escape plan. Children unable to open windows, infants in cribs, and elderly in need of assistance should all be included in the plan. Each plan should also include a meeting place outside the home, so that everyone escaping on their own goes to the same location.

Bedrooms that are up a flight of stairs can be handled in several ways. First, look out the window to see what is below the window. If a roof is there from a lower part of the home, such as a garage, it can be used as a secondary exit. If this is not the case, commercially made escape ladders can be kept under the bed in the bedroom. These take up very little room in their box, but deploy from the window down to the ground in case of a fire emergency. They are available in different lengths for use from second and third stories in the home.

The home escape plan should also include the locations of where the home's fire extinguishers, first aid kit, and family disaster kit are kept (fig. 29–11). This can help out during an emergency, or if a babysitter is watching the children. The location of smoke alarms should also be marked on the plan, to make certain that all smoke alarms are tested at least once a month. The escape plan must also include a family meeting place, where all members of the family know exactly where to go during a fire emergency, regardless of which exit they used to get out of the house (fig. 29–12). This allows for an accurate head count to be given to first-arriving fire companies by the occupants.

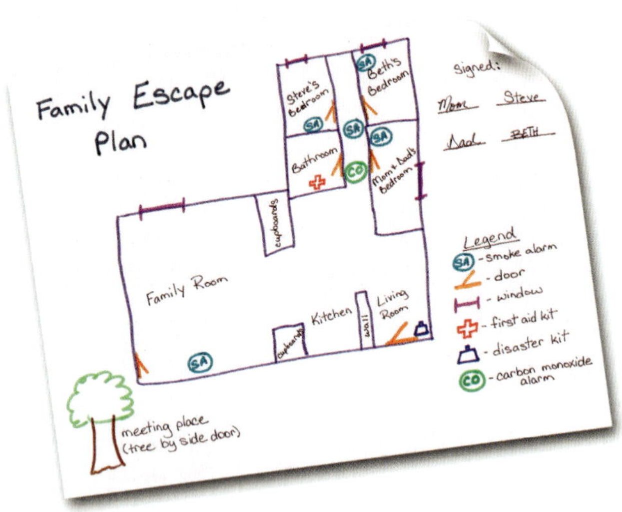

Fig. 29–11. The completed home fire escape plan should be prominently displayed in the home and practiced at least twice a year.

Fig. 29–12. All family members should know where their family meeting place is, and it should be easily accessible.

The most important part of the home fire escape plan is to practice the plan at least twice a year. There is nothing like some actual practice to make behaviors permanent, and this is a plan that needs to be remembered in an emergency. To make sure the correct behaviors are likely to take place, practice the plan at night every once in a while. This may wake children up, but it's the best way to find out if they know what to do when confronted by a fire emergency.

The comprehensive home fire escape plan may also be called E.D.I.T.H., which stands for Exit Drills in the Home. Remember, once the plan is made, the most important part is practicing the plan twice a year.

Although home fire escape plans have to do with the homes where people reside, make sure they can make them up when they go on vacation and stay at hotels and motels. Tell them to study the map of the building on the back of the hotel or motel room's door. A good meeting place can be the family car or under the hotel or motel sign, which is usually close to the main roadway.

Encourage families to discuss fire escape plans when they are planning a vacation or just going out for the evening, at popular places families frequent, such as restaurants and movie theatres. Although alarms may or may not be present, make sure everyone is encouraged to let one another know if unwanted smoke or flame is spotted. Many people are reluctant to alert others or take prompt action, often leading to devastating results.

COMMUNITY RISK REDUCTION

While the primary duty of a fire department is to respond to and extinguish fires, another important responsibility is to reduce the burden of fires in the community through fire prevention efforts such as code enforcement and public education. More broadly, these fire prevention activities are very often part of a larger endeavor called **community risk reduction**. The goal of community risk reduction is to reduce the number of lives lost and the amount of people injured (including firefighters) as well as the dollar loss of destroyed and damaged property. As a firefighter, you play an important role in community risk reduction, including implementing policies to directly address the risks that have been identified.

Community risk reduction efforts may be part of a fire department's strategic plan. Strategic plans are developed to set overall goals, with specific goals to be accomplished within a specific time range. Such plans cover many things in addition to community risks such as new training initiatives, improved equipment, and staffing levels. Not every fire department has a strategic plan but may have a community risk reduction program.

Community risk reduction entails taking a proactive approach to reducing the losses in a community to fires and other types of fire department responses. It requires conducting an assessment of the current risks within the community and developing strategies to combat them to lessen their negative impact. More specifically, this involves developing a list of the various risks, measuring them in terms of their size and magnitude, and coming up with targeted goals to lessen the risk.

Specific risks themselves can affect an entire city or particular groups within a community. For example, the risk of kitchen grease fires may be found throughout your jurisdiction, while playing with matches is typically limited to homes with children. It's important to understand that to reduce risk you first understand the parameters of each risk and which part of your community is affected.

A community risk reduction program is an integrated approach to the specific problems identified within the community. It incorporates five distinct strategies: emergency response, engineering, enforcement, public education, and economic incentives.

Reviewing each of the five strategies, particular examples include:

- **Emergency response.** Providing thermal imaging cameras for each firefighter to increase victim survivability by expediting search operations and to minimizing chances of firefighters getting lost in a structure, providing advanced training and equipment for collapse rescue in earthquake prone areas, adding a new fire station in a rapidly developing area to reduce response times, and providing power lift stretchers for ambulances to reduce back injuries to EMTs.
- **Engineering.** Installing residential sprinkler systems in new one- and two-family dwellings to reduce fire fatalities, improving building codes to provide increased structural fire resistance, installing new pumps to increase pressure for hydrants supplied by public water mains, and installing emergency generators at firehouses to allow them to operate during power outages.
- **Enforcement.** Creating a new inspection program for inspecting and monitoring old fire escapes, training new firefighters (like you) in the basics of the fire code, creating a program of mandated inspections for high-risk occupancies of schools and hospitals, deploying inspectors on weekend nights to inspect nightclubs to reduce chances of overcrowding.
- **Public education.** Providing neighborhood fire safety education in the aftermath of a multiple fatality fire, providing scheduled cardiopulmonary resuscitation training for local residents in the firehouse to increase heart attack survivability, deploying public safety educators to shopping malls during the Christmas holidays to provide fire safety literature and tips to homeowners about cut live trees and lighting, and providing children car seat education to new parents to increase a child's survivability in an accident.
- **Economic incentives.** Encouraging local insurance brokers to provide reductions in fire insurance homeowner policies for the installation of residential fire sprinkler systems; increasing fines for violations of the fire code as an incentive for compliance, increasing fines for false fire alarms emanating from commercial buildings to encourage fixing the problems with such systems, and allowing increases in fire hydrant spacing in new land developments to encourage the voluntary installation of fire sprinkler systems.

Let's look at an actual example from San Antonio, Texas, which is subjected to flash flooding from heavy downpours, often overwhelming drainage systems in the city and covering roadways. This includes "low water crossings" in roadways (roads that slope down over creek beds that are normally dry but fill with swift flowing water during heavy storms).

For many years, the San Antonio Fire Department was called out to rescue people trapped in or on their cars stuck in the swift flowing waters. The drivers drove into flood waters thinking they could pass through but instead ended up needing rescue. It was not only dangerous for the victims, but for the firefighters called to save them. Something had to be done with this specific risk.

The City of San Antonio studied the problem and took specific steps to address it that incorporate the five components identified above, including:

- Passage of an ordinance to charge a fee/fine of \$400 for each rescue that is conducted, paid by the individuals rescued (economic incentive to make people think twice about driving into flood waters)
- Created a public education campaign to identify the risks associated with driving in flood waters and the imposition of the \$400 fee/fine (education)
- Installed warning signage at low water crossings to highlight the problem (engineering)
- Collected the \$400 fee/fine (enforcement)
- Equipped/trained firefighters in swiftwater rescue response (emergency response).

When creating specific community risk reduction program goals, it's important to set aggressive but reasonable goals. For example, a goal might be to reduce the number of unattended candle-related fires from 350 per year to 200 such fires in five years through targeted public education efforts in neighborhoods where such fires typically occur. Another example might be a goal to provide smoke detectors to 400 low-income households annually.

It is important that you as a firefighter understand what community risk reduction programs may be in place in your community, the overall goals of the program(s), and your role in carrying out the program(s). Learn how these programs affect your duties and responsibilities, including any standard operating guidelines that may be affected by such programs.

FIRE STATION TOURS

SKILL DRILL

FFII 5.5.1 ours of the fire station should be a routine and expected part of any firefighter's daily activity. Preparation is the key to a successful program, and station tours are no exception. Knowing what time the tour is scheduled for, what group will be coming, the number of people in the group, and any specific information they have requested should all be noted on the tour form that the fire company should have in their possession prior to the tour.

Make sure the fire station has been properly cleaned and the trucks and tools are clean as well. Special attention should be given to the path the tour is scheduled to take, removing any standing water and other hazards from the area. Greet the tour group promptly and start by discussing the rules for the tour, including the procedures to follow on the tour and what they should do if there is an emergency call during their visit. Start this off by telling them what their role will be when the fire station alarm sounds, so they know exactly where to go to be out of the way, and when it is okay to move from that spot, which is usually when the last bay door goes down fully. Make sure the group stays together for the tour, and that children are not allowed to wander off on their own.

Remember to instill safety into your talk, by pointing out the fire station's working smoke alarms. This is a good time to discuss smoke alarms with the group, and all groups should hear our smoke alarm messages. While you are walking through or near the kitchen you can talk about how the firefighters practice safe and attentive cooking, using lids and staying nearby, turning pot handles, etc.

Mix up some fun with the educational components of the tour. While it is always a good idea to discuss and practice fire safety behaviors, show them the fire trucks and the tools on the trucks. Having a firefighter in full turnout gear at some point in the tour is a very personal way for children to recognize what firefighters look like when they go into a burning building. If children are allowed to touch the firefighter or put on gear, it is extremely important that the gear be clean and free from carcinogens that firefighter gear is regularly exposed to. Having a set of demo gear is best practice. If this is not feasible, explain that the gear has been in a fire and shouldn't be worn or touched because it could expose them to harmful chemicals.

If your fire department provides emergency medical services (EMS), then show them the EMS vehicle and some of the supplies that are kept on that vehicle as well. Explain to the group how your fire department is staffed and describe the normal schedule you may have for station duties and vehicle maintenance.

Handout materials for the group should be assembled prior to the beginning of the tour, and let the tour group representative know where they are in case you have to leave on an emergency call (fig. 29–13). When the tour is over, the main points can be summarized and handout materials passed out. If time allows, you can take several questions from the group and allow time during the tour for pictures to be taken by the parents, teachers, or caregivers in attendance. The fire station is a very interesting place for tour groups to visit. Be proud to show it off!

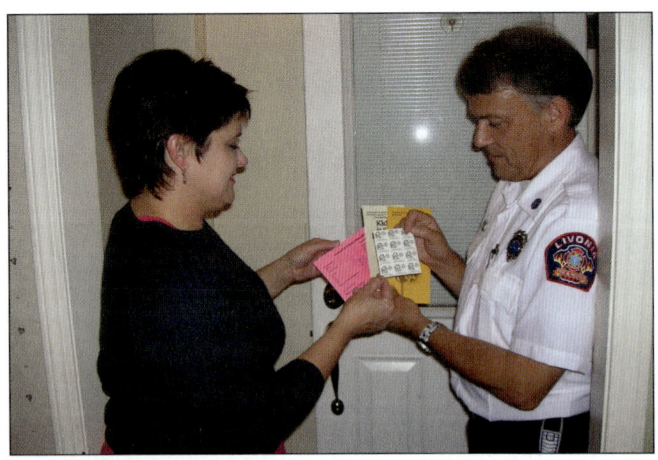

Fig. 29–13. Firefighters should have materials ready to hand out when conducting a residential survey.

FIRE SAFETY RESOURCES

FFII 5.5.1 Some very recognizable figures in the field of fire safety are **Smokey Bear** and **Sparky the Fire Dog**®. Smokey has been telling audiences his message of forest fire safety since 1944, while Sparky entered the fire-safety scene for residential homes in 1951. Children love both characters, and fire safety material that features them can be purchased and given as tour handouts to the children. You can also suggest that children visit their websites, smokeybear.com and sparky.org which are educational and very child friendly.

Some fire departments have educational and fun websites of their own. If that is the case, let your community know about it at every opportunity. Advertising your website address on fire department vehicles, on recorded phone messages, and in the media can help draw attention to your efforts.

Traditional media can be very helpful for getting the message of fire and life safety out to your community. Television, local access cable television, newspapers, and community newsletters have all been used effectively for educating people about fires. For example, one topic could be recent fires in the news in your area, or some general fire safety education tips. There are numerous resources for topics, including the USFA website. It has several topics to choose from, and you can add local flavor by including some local statistics, a background of a recognizable landmark, or your fire department uniform or patch prominently displayed. Some topics are general and can be used any time of the year, such as smoke alarms and cooking safety, while others may be timely, such as Halloween safety tips in October.

Social media has been growing in popularity for fire and other emergency service providers to educate the public. Make sure you understand the guidelines your jurisdiction has for social media use. You can partner with your jurisdiction's communications department if there is one. Otherwise, spend some time investigating with other agencies do to promote their messages on social media. Have a plan before you start your account. Make sure you have messages ready to post on a regular basis, so that people are drawn to your page as a resource for providing relevant, timely information. If you stop posting messages for long periods of time, less people will see your information when you do post it and the platform won't work to your advantage.

Many fire departments celebrate **Fire Prevention Week**, in recognition of the Great Chicago Fire on 1871. The NFPA is the official sponsor of Fire Prevention Week, and they have plenty of material for fire departments to use on their website, NFPA.org. During Fire Prevention Week, many fire departments hold a community open house event. This is a great way to showcase your fire department and firefighters, and teach fire safety in a fun way. Children love to put on a clean fire coat and operate a fire hose, but you should also make it fun for them to practice the fire safety behaviors discussed earlier in the chapter (fig. 29–14).

Fig. 29–14. The fire department open house should be entertaining *and* educational.

GIVING A FIRE SAFETY PRESENTATION

FFII 5.5.1 Many times, firefighters are called out to deliver a **fire safety presentation** to the public at a location other than the fire station. It may be at a school, mall, or business. Regardless, it is important to deliver appropriate safety information to the target group. Find out as much as possible about your audience prior to your presentation in order to customize the presentation to that specific group (age, particular fire safety issues, etc.). It would be good to know approximately how many people are expected to be in your audience (for handout material distribution), how long this talk is expected to last, any specific topics that should be covered, the availability of a room to present, and a system to meet any computer needs you may have. Topics can be recent newsworthy items such as fires in your area that have been covered in the news, or seasonal issues such as bicycle safety, swimming pool safety, or heating season safety.

The proper attire for this type of presentation is your fire department's approved uniform, so that you are easily recognized as the fire service professional to the group. While your firefighting gear is a must to show lower elementary school children, you may be able to skip the gear if you are addressing an adult audience.

Bring in a visual display or two to help you deliver your messages. A small display case containing a few objects destroyed in a house fire can create a visual message and be small enough to easily transport in a vehicle. When speaking to a group of adults that may have small children at home, such as a homeowners association meeting, a toolkit of common fire prevention and fire safety items is helpful. This can include a 3-foot ruler, a fire extinguisher, smoke and CO alarms, a flameless candle, some novelty lighters (so people can see that lighters are sometimes manufactured to look like children's toys), and a collection of fire safety and prevention brochures (fig. 29–15).

Fig. 29–15. Small displays, such as this novelty lighter awareness case, can provide a good visual message.

Remember to use simple descriptions here, and avoid our trade's jargon. Adult audiences can be quick to lose attention when they do not understand the message you are trying to get across.

Always schedule time to take questions. Questions make the person feel important enough that you take time to speak to their concerns directly. The best way to handle this for lower elementary school-aged children is by telling them that you will schedule a few minutes after your presentation for any questions. This will save time and effort during your presentation. Older audiences usually ask questions pertaining to your subject, so let them know at the outset to ask questions if and when they have them.

QUESTIONS

1. What is the name of the document that has helped pave the way for fire-safety education in the United States?
2. What is a fire inspection?
3. Does the fire company need to ask for the permission of the business to perform a fire inspection? Explain your answer.
4. Why it is important to sell the merits of the fire inspection to the business owner or manager?
5. What is the difference between home fire-safety and a fire-prevention inspection?
6. How can firefighters educate residents in their district about fire safety?
7. What items are important for firefighters to cover when conducting a fire station tour?
8. Name the main causes of home fires that firefighters can talk to families about when conducting home safety surveys.
9. How would you teach lower elementary school-aged children fire safety behaviors? How does that differ when you discuss the same behaviors with upper elementary school-aged children? Adults? Seniors?

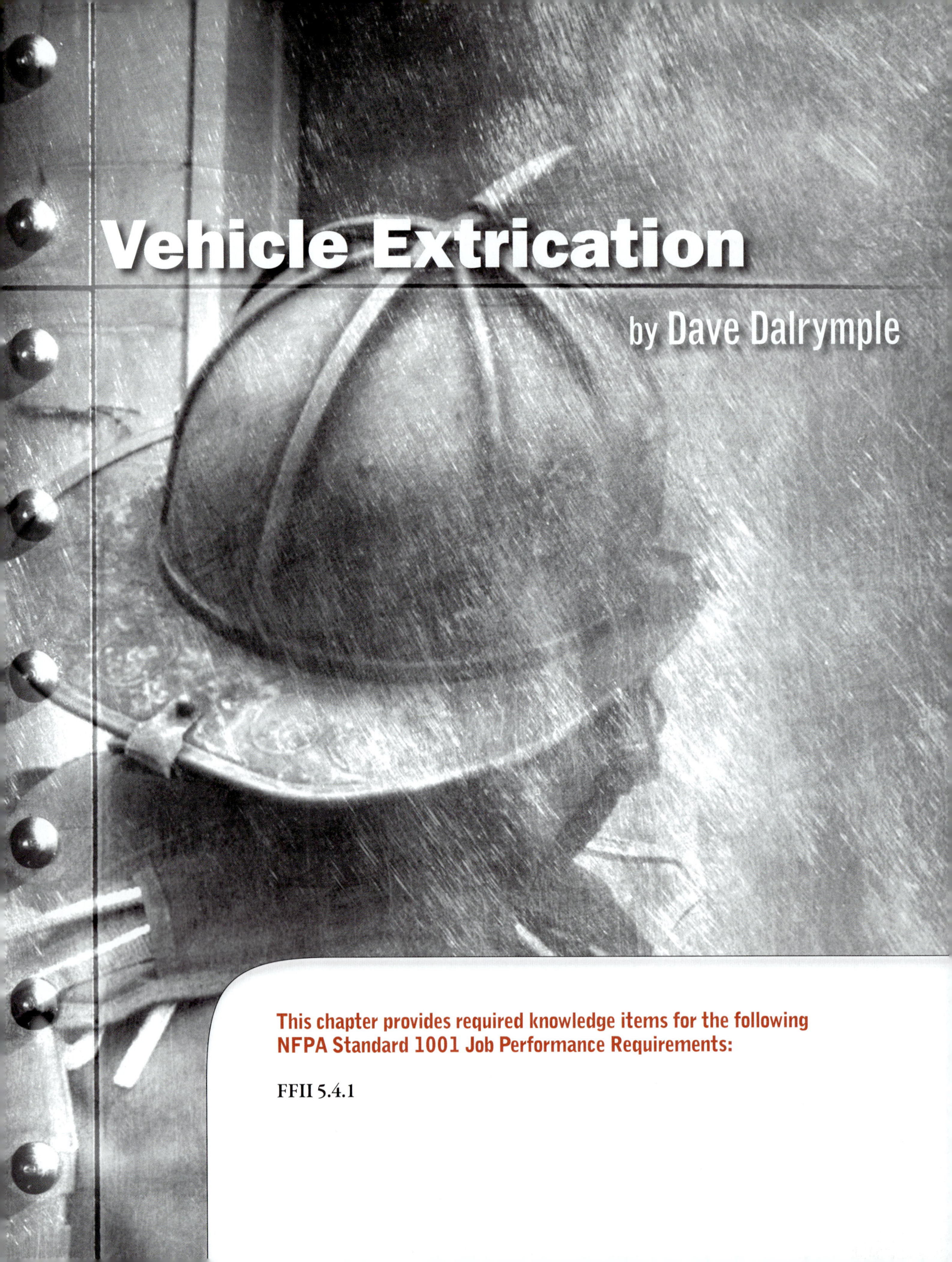

Vehicle Extrication

by Dave Dalrymple

This chapter provides required knowledge items for the following NFPA Standard 1001 Job Performance Requirements:

FFII 5.4.1

This chapter contains Skill Drills. When you see this icon, refer to your Skill Drill book for step-by-step instructions. SKILL DRILL

30

OBJECTIVES

Upon completion of this chapter, you should be able to do the following:

- Describe the importance of the "Golden Hour"
- List modern vehicle technologies that have positive and negative impacts on vehicle extrication
- List common hand tools used by the fire service for vehicle extrication
- List the power tools used by the fire service for vehicle extrication
- List the equipment used by the fire service for stabilizing a vehicle in order to perform a rescue
- Describe the "operational cycle" used at vehicle incidents by the fire service
- List the hazards that need to be identified during the initial size-up at a vehicle rescue incident
- Identify the proper protective clothing to be used by firefighters at a vehicle incident
- Describe the importance of patient management and care at a vehicle incident
- Identify the proper procedures to be used by firefighters for packaging and disentangling a victim at a vehicle incident
- Describe the proper procedures to be used to safely remove a victim from a vehicle
- Identify the proper procedures to be used by firefighters for performing preventive maintenance on vehicle tools

INTRODUCTION

After EMS calls, **motor vehicle crashes (MVCs)** are one of the most common types of incidents you will respond to. So it is a safe bet to call an MVC as a common, everyday emergency. However, is it a routine type of call? Every MVC is different, and yet the vehicle itself is the dynamic concern here. When vehicles crash, there are many systems in them to protect the occupants from harm, but these same systems can disrupt our efforts to disentangle the occupants and even injure us in the process.

FFII 5.3.4 Yet, we are tasked with rapidly managing the incident, mitigating the hazards, making patient contact, and managing that patient while displacing the vehicle components to create a proper pathway for the patient to be disentangled and transported to the appropriate medical facility. We clean up the scene, restock our apparatus, and return to service.

The golden hour

Responding to an MVC is a time-sensitive task. We need to remember that while dealing with occupants who have been involved in an impact; there is great potential for trauma. There are procedures that can only be performed by doctors in a controlled facility. Therefore, it is imperative that occupants involved in an MVC get medical care from a trauma team as soon as possible. As rescuers, we pay a vital role in the outcome.

A term used in the emergency services for many years is the **Golden Hour**. This term describes the time that elapses between the moment of the patient's injury and the patient's arrival at the hospital and preparation for emergency surgery. The clock starts ticking once the injury occurs, not when the responders arrive on scene. It is a race against time, as notification of the incident is beyond our control, as well as the overall response time needed to get a crew on scene (fig. 30–1).

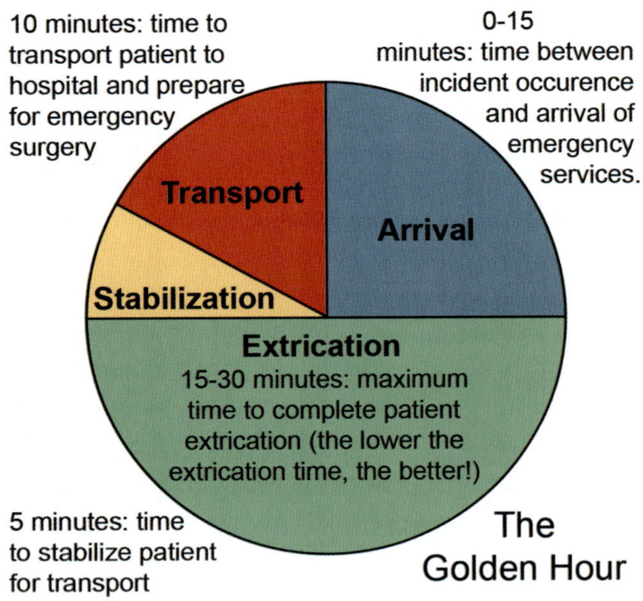

Fig. 30–1. The Golden Hour represents the 60-minute window of optimum response time for arrival, extrication, transport, and preparation of an injury victim for emergency surgery.

The only component of the Golden Hour within our control is the 15–30 minute window from the companies' first arrival on scene and through completion of the extrication operations. Keep in mind; this is a goal, not a definitive time frame. It may not always be possible to meet the components of the 60-minute window, as MVCs present many unpredictable challenges. The only way we can be most effective with extrication is through aggressive and consistent training.

In this chapter we present basic principles and methodology for vehicle extrication as a primer for the firefighter. First, let's look at some specific vehicle design issues and how they will impact our rescue efforts. We will cover tools and specific extrication techniques later in the chapter.

VEHICLE DESIGN AND TECHNOLOGIES

FFII 5.3.4 Today's motor vehicles challenge the emergency responder and pose many concerns (fig. 30–2). Each type of vehicle—car, pickup truck, sport-utility vehicle (SUV), or minivan—presents issues specific to the vehicle type, but they are also similar in some ways.

Fig. 30–2. New vehicle technology creates new challenges as well as tactics for extricating patients.

Although much of the focus on vehicles today centers on safety systems such as air bags, we need to be aware of many other components, materials, and systems. We need to take into account not just motor vehicle crashes but also vehicle fires and even the simple medical response to an individual in a vehicle. To keep pace with the ever-changing components and features of these vehicles, we must keep updating our foundational knowledge.

Construction

Most vehicles are now built with a **unibody** configuration. This type of construction is strong and lightweight, and it is built around the concept of structural integrity: All its components must be intact.

So, when we remove or displace the vehicle's components (doors, side, roof, etc.) or if they are displaced by the crash, the vehicle's integrity is weakened. This can help or hinder us. The key here is to consider this in our on-scene action plan because it can assist our space-making. The two other construction types found today are **full-frame construction,** which was typical of older vehicles, and **space-frame construction,** which is much like a bird or roll cage in which the cage itself is the vehicle's structural integrity (fig. 30–3).

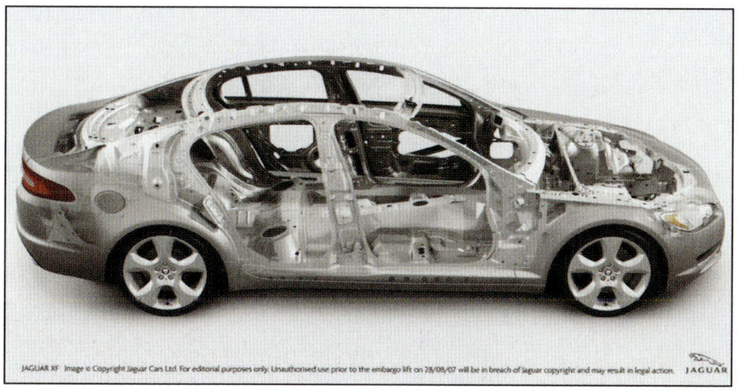

Fig. 30–3. This photo reveals the space-frame construction in a new vehicle.

Let's look at the **energy absorption areas (crumple zones)** in the vehicle. Throughout the vehicle are engineered areas designed to absorb energy (crush zones) during a crash; however, the largest area, and the one that has the greatest effect on us as responders, is the vehicle front. As the vehicle absorbs crash energy, the nose of the vehicle compresses, the engine/drivetrain drops downward, and often the front tires are forcibly deflated. This basically places the nose of the vehicle on the ground.

Now, add into this mix the significant reinforcement built into the dash area. This reinforcement has gained significant strength (high-strength steels/alloys) over the recent past. The reinforcement is *tied* to the vehicle's sides, as well as to the floor and forward to the firewall.

So coupling the issue of crush zones with dash reinforcements, how does this construction affect moving a dash (fig. 30–4)?

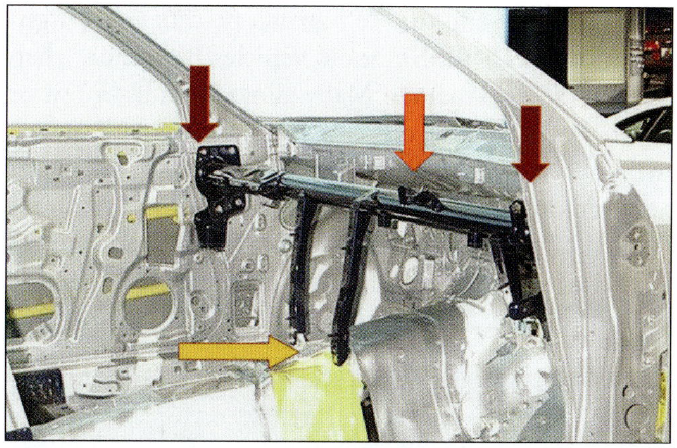

Fig. 30–4. Notice the new dash reinforcement in this vehicle.

Maybe we need to rethink how we perform **dash displacements.** Instead of performing a **dash roll-up,** perhaps a vertical lift of the dash would be more in order. Also, what if we place another relief cut to assist our dash displacement in a new location? We already make relief cuts into the lower dash—a post-area point for the dash—to move as well as pivot.

This **relief cut** might only look like a cut into the fender; the concept for the evolution is to *sever* the crush zone area behind the fender, thus allowing the dash area to move *independently* of the rest of the vehicle when displaced.

Therefore, even if the nose of the vehicle is on the ground, you can effectively displace the dash (fig. 30–5).

Fig. 30–5. Notice relief cut behind the fender on this 2007 Escape Hybrid.

Reinforcements

Because we've touched on reinforcements, let's delve into this area a little further. Although many vehicles are now made of lighter-weight materials than in the past, many more are reinforced with alternative materials and methodology. More points of reinforcement are found today, in addition to a greater percentage of **high-strength steels (HSS)**. Some vehicles have more than 63% of HSS and alloys. Many alloys, such as boron or microalloy steel, are difficult to cut, even with power hydraulic cutters. Even some of vehicle components and reinforcements are constructed of these alternative materials. Imagine what might occur if a door side impact bar were displaced forward or rearward. Most vehicle components are made of these materials (fig. 30–6).

Fig. 30–6. Notice the reinforcements revealed in the cutouts on a Mazda 6.

Another popular reinforcement in today's vehicles is **blown-in** polyester foam. Very similar to foam used to fill cracks and voids, this material adds to the structural integrity of the vehicle without adding significant weight.

However, think about how this material reacts when you are trying to sever it with a powered hydraulic cutter or a reciprocating saw! This material makes tool evolutions difficult because of the additional force needed to compress and then shear (with the powered hydraulic cutter), or the blade becomes dull and gummed up as the material adheres to it (with a reciprocating saw). Plastics are used significantly in vehicles today, even before you add in the large percentage of the vehicle's plastic-filled structural voids. Also, certain components, because of their location, have been given **extra duty** by changing their configuration and materials.

The seat belt attachment or bracket on the B-post is the best example of this type of reinforcement. Because most seat belt–attachment points on the B-post are adjustable, they are in the perfect location to reinforce the upper B-post area (fig. 30–7).

With all these construction concerns, what about our door displacement evolutions? Is just **popping** (or removing) the door enough? Is it any easier to pop the door these days? We used to accomplish the access we needed for patient care and removal with a door displacement, but now it is more difficult. Maybe we have to consider a side removal instead of a door pop. Using our knowledge of the vehicle's construction, we look for weaker areas and/or new pathways to displace, such as a B-post tear or rip evolution (fig. 30–8).

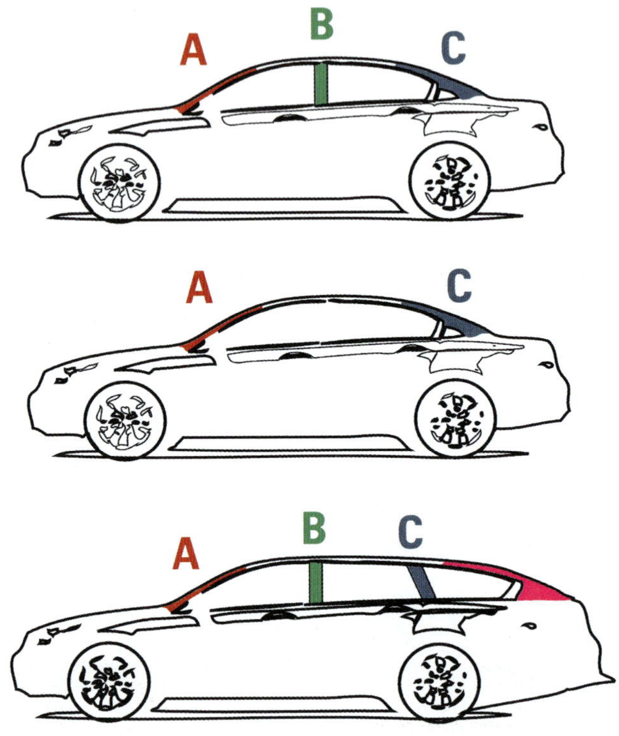

Fig. 30–7. The A, B, and C posts of a passenger vehicle

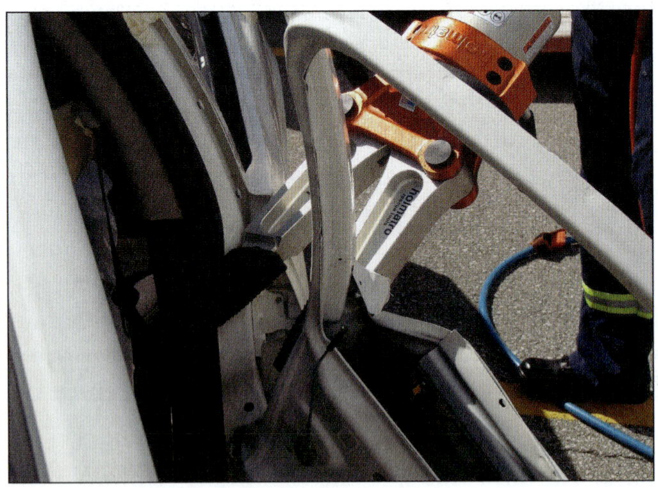

Fig. 30–8. Removing the door by tearing the door from the latch

Windows

The two types of glass to which we have become accustomed, **laminated and tempered safety glass,** are still the most widely used. Laminated glass is a "sandwich", a series of layers of glass & plastic laminated together while tempered glass (safety glass) is a type of glass that breaks into small granular fragments when shattered. However, there are a few new curves out there. Enhanced protective glass (EPG) is basically a form of laminated glass placed in the side and rear windows. Dual-paned glass and polycarbonate glazing are also used. Some of these materials require a change in our methodology and tools for removing such windows (fig. 30–9).

Also, add dust masks to your personal protective equipment (PPE) for glass management (fig. 30–10). And factor in the issues of rear glass hatches in SUVs and minivans.

These glass hatches have a nasty habit of flying apart when broken because of the tension placed on them by the multiple hatch struts and the energy absorbed by the vehicle during the crash.

Fig. 30–9. Notice how the laminated side glass (EPG) in this photo is shattered worse at the point of impact.

Fig. 30–10. Many departments now use reciprocating saws in their glass management.

Batteries

Although often helpful, by and large power in vehicles is a bane for rescuers. Today's vehicles have so much technology that the existing 12-volt systems cannot keep up with the demands. Experiments have been performed with different voltages (such as 24, 32, or 42) as well as multiple batteries. Remember, in approximately 42% of vehicles the battery is located outside the engine compartment (fig. 30–11). However, voltage is not the greatest concern; it is the difficulty in finding the battery. Moreover, it is possible that there are alternative power sources in the vehicle. You must disconnect both battery cables at the MVC, and remove the key from the vehicle. Keep the key inside your apparatus.

Fig. 30–11. As previously mentioned, batteries may be located in many locations such as under the rear seat found in the 2007 Cadillac CTS .

Motive power

The buzzword here is **gasoline-electric hybrid vehicle.** Their high-voltage drive systems pose concerns, but there are other alternatively fueled vehicles out there. Compressed natural gas (CNG), E85, bio-diesel, and even hydrogen fuels are available today and on the street. Each vehicle is referenced in the *ERG (U.S. Department of Transportation Emergency Response Guidebook)*, which can be downloaded from the manufacturer's web site. The ERG reviews a wide variety of issues and concerns and contains background information for emergency responders.

Take note of the following *hybrid 101* points:

- All hybrids start on electric power alone
- Some travel up to certain speeds on electric power alone
- High-voltage (HV) power is direct current (DC)
- Some hybrids also generate 110 alternating current (AC)
- The HV circuit is self-contained (does not ground out to the vehicle chassis)
- Drivetrain power is computer-controlled and managed
- A crash and/or change in voltage results in a shutdown of the HV system
- All hybrids have a HV drivetrain circuit as well as a normal 12-volt automotive electric system
- The gas/diesel side of drivetrain is load dependent
- Even when the hybrid is shut down and secured, there is *always* power in the HV battery
- As emergency responders, there is *no need* to interact with the HV battery. Leave it alone!
- Colors to remember:
  - Bright orange = HV power (p-IR16)
  - Bright blue = *intermediate* HV power (GM)

Safety systems

Frontal supplemental restraint system (SRS). We have become accustomed to **frontal SRS**; however, we should look at them more closely (fig. 30–12). Generally, in the past if an air bag was deployed, it was considered spent. Many vehicles today are equipped with dual-stage frontal SRSs. Basically, this means the system has two inflation modules: a module designed to deploy at a lower speed or impact and a second for a higher rate of speed and crash severity. Also, a high-speed impact usually deploys both modules. Now the issue is how to know which modules have deployed. We can't tell by observation.

These systems can be found across the board in many vehicles of various types and brands. Disconnecting the battery drains the energy storage system in the air bag computer, but other energy (an alternate power source or static electricity, for example) might interact with the system.

Many large vehicles are equipped with frontal SRSs. In fact, one manufacturer has provided some of its models with inflatable tubular systems (a head protection air bag) and even seat belt pretensioners. Think about disarming the battery system in these vehicles!

Side safety systems. These systems are becoming more prevalent. They have expanded from the Volvo seat-mounted **side impact protection system (SIPS)** in 1995, to a door-mounted system, and then to a rear-seat and door-mounted system. Now, we have head-thorax side systems and **roof rail/side curtain head protection systems**. Unlike frontal SRS, these systems and side curtain systems work on the principal of stored energy caused by their reaction time (fig. 30–13). Many side systems are multi-chambered to give additional protection to the occupant's side and head. Side-curtain systems protect the head and give added protection during a rollover, keeping the occupants inside the vehicle. Disconnecting the battery to drain the energy storage system helps ensure a safe working environment.

Fig. 30–12. Has this air bag been inflated once or twice?

Fig. 30–13. Notice how both the Taurus side curtain and side impact bag were deployed.

However, we must keep in mind the following:

- Some side systems are mechanically activated; disarming the battery does not make these systems safe.
- Certain mechanical systems have emergency guides for disarming them.
- When operating power hydraulic tools in close proximity to these side systems, be extremely cautious, especially when the systems are not deployed. When you perform door and side tool evolutions, do not put yourself in the path of undeployed side systems. Be careful about where you place your tools during these evolutions.

Side curtain/head protection systems. A few years ago, only a handful of vehicles had side curtain/head protection systems (fig. 30–14). Now, they can be found in more than 65% of vehicle models, across all types of vehicles.

Fig. 30–14. With the trim of the 2007 Fusion removed, the undeployed side curtain is revealed.

The bag part of these systems is benign; the inflation system is of concern. The nature of the system is that it must react and deploy one-third of the time as frontal SRS; the inflation module is a pressure vessel topped with a small pyrotechnic charge. The gas stored in the module is an inert gas, but it is stored at approximately 3,000 to 10,000 psi (21,000 to 70,000 kPa). What could happen if you cut through one of these undeployed devices? At a minimum, there would be a release of high-pressure gas in close proximity to the tool operator and possibly the interior rescuer and even the patients. The worse-case scenario would be the pyrotechnic charge deploying, coupled with the release of high-pressure gas. There is also the potential for creating debris and even shrapnel. You must avoid cutting these devices. Even disconnecting the battery and electrically disarming the system does not help if you cut through the device (fig. 30–15).

Fig. 30–15. An accidental cut of live side curtain cylinder could be extremely dangerous for both the rescuer(s) and the patient(s).

Where are these inflation devices located? A few years ago, the inflation device was in the rear roof post. In 2002 all that changed. They now can be found in any of the roof posts, the roof structure itself, or even in the vehicle body.

How do we find them? How do we know in which vehicles to look for them? One clue is if one of the side curtains is deployed. Another clue is if the vehicle is a recent and high-end model. These facts should at least alert us to use more caution when dealing with roof posts. However, the best and surest approach is to strip interior trim before cutting the roof posts (fig. 30–16).

Fig. 30–16. Stripping the trim back shows side curtain cylinder.

You need not remove the trim completely, just move the trim enough so you can see the roof post structure and make sure a side curtain module is not mounted there.

Remember, this does not apply to models with these devices in the roof structure or the body.

What do you do if you find such a device? Ensure that you have disconnected the vehicle's battery to isolate the electrical power. Try to sever the roof post below the module if at all possible. If you cannot, cut the post well above the top of the module to be sure you do not cut through the pyrotechnic charge. The bottom line is that these systems are a very real concern for emergency responders and should be treated with caution (fig. 30–17).

Seat belt pretensioners. **Seat belt pretensioners** are powered devices that remove slack from the seat belt, ensuring that the occupant is seated properly when the air bag systems deploy. They are usually powered by a pyrotechnic charge; but some are powered by a gas cylinder and even high-torque electric motors. These devices are used in all front and rear seating positions in most vehicles. These devices are fairly benign to the rescuer, but it is a good idea to be sure that you cut the seat belts when working close to the belt. Even if you disarm the pretensioner by disconnecting the battery (just as with air bag systems), there is the potential for accidental deployment. Avoid cutting through these devices during tool evolutions, especially in the area of the B-post (figs. 30–18 and 30–19).

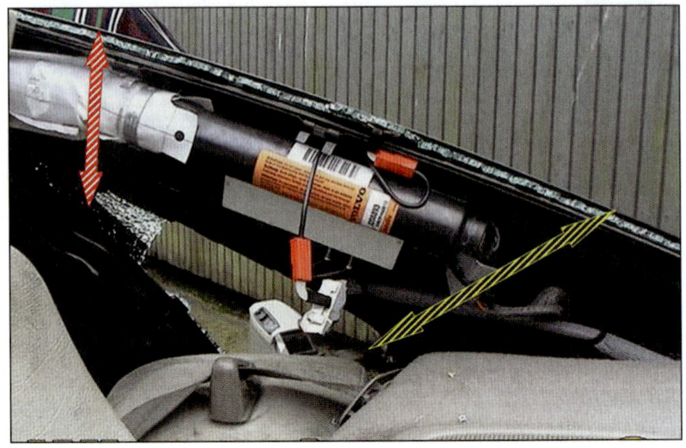

Fig. 30–17. Once the trim is stripped back, locate and cut around side curtain cylinder.

Fig. 30–18. The buckle side seat belt pretensioner is shown here.

Fig. 30–19. The spool side seat belt pretensioner is shown in this photo.

Other safety system locations. Last, let's take a look at the new locations we might find safety systems. We have had **knee air bag systems** since 1998 with Kia's Sportage, but we can now find such systems in a variety of vehicles.

These systems need careful attention when performing dash displacement evolutions. In the future look for new systems in the rear seat belts, head rests, and seat cushions, among other systems. Also, many convertibles have **pop-up rollover protective structures (ROPSs)** in conjunction with the other safety systems.

Construction, materials, and components affect our rescue efforts more and more as time progresses. We should review our methodology and tool evolutions more often and in conjunction with all vehicle concerns.

More background information is needed about vehicles and their components. Knowledge is power and key to our efforts when making changes and pushing the envelope is needed. Also remember that safety systems are becoming more widespread and comprehensive.

Safety systems already affect us and will continue to do so. Think about training issues and concerns. What do we get practice on today? Is it even fair to say the vehicles we encounter are much different from what we get to practice on? If we can't get the vehicles, can we simulate some of these hazards, issues, and concerns? The bottom line is this: Rescue teams must think and work smart when presented with issues and concerns of present and future vehicles (fig. 30–20).

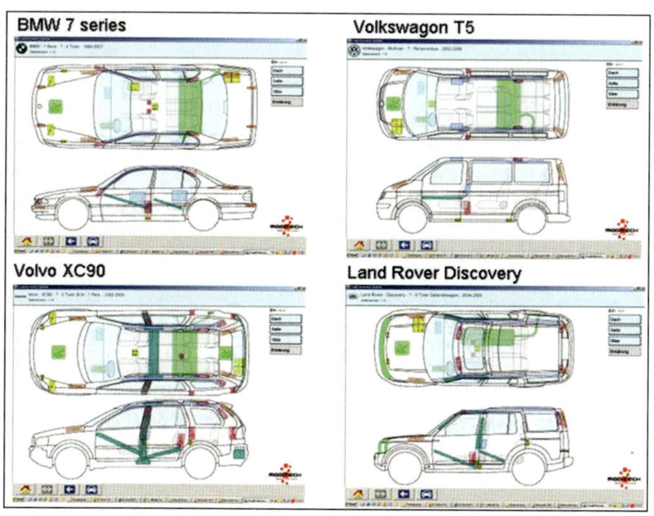

Fig. 30–20. It is important to review the crash recovery systems of as many vehicles as possible. These are a sampling of current crash recovery system information screens.

TOOLS AND EQUIPMENT

FFII 5.3.4 We use tools for a wide range of tasks at an MVC, whether extrication, hazard control, patient management, or fire suppression (fig. 30–21). Time and space limit the depth to which we can hit on all these areas and tools, so the focus is on the common power tools used for extrication operations. But before we do that, think about what tools actually do for us as rescuers:

- *Sever*—dividing into two or more parts by cutting, piercing, penetrating, splitting, breaking, sawing, and so forth.
- *Distort*—changing the shape of an object in a plastic flow or other permanent manner without severing or parting.
- *Displace*—moving an object or portion of an object from its original position. The motion may be in any plane and the effect of the force applied is generally controllable. Minor distortion may occur during displacement.
- *Disassemble*—reducing objects to basic component parts in the reverse manner from which originally assembled.

Fig. 30–21. Pictured above is a typical tool staging as used on a MVC. This allows tools to be quickly and easily identified and retrieved when needed.

Extrication evolutions can require either or both hand tools and power tools. Power tools have become the accepted norm for vehicle extrication. It is easy to forget about hand tools as we carry hydraulic tools capable of cutting and crushing metals meant to withstand impacts of up to 70 MPH. But power tools can and will fail; therefore, it is important that the rescuer has a working knowledge of the basic hand tools that can be utilized as a backup plan when all else fails. The following are just a few of the various hand and power tools that can be used during extrication (fig. 30–22).

Fig. 30–22. Common hand tools used in extrication

Hand tools

Hacksaw. A hacksaw is a fine-tooth saw blade held in a frame, designed to cut through various types of materials. A hacksaw can be a great primary tool for cutting through pillars or making a relief cut into the roof for a roof flap, especially while hydraulic tools are being set up. It is actually a good idea to have (4) hacksaws on scene. This can speed up the process while offering a backup saw.

Hand tool inventory. Here are a few of the hand tools that the rescue toolbox inventory should contain: tin snips, razor knives, hammers, chisels, crowbars, screwdrivers, and various types of wire cutters. It is also a good idea to carry a couple of rolls of duct tape to secure things such as tarps or cover sharp edges after components have been severed.

Center punch. This tool was used initially in the machine tool business. It has a pin that is operated by a spring. Once actuated the pin penetrates the material and leaves a dimple. The center punch is used in the rescue business to control glass of tempered windows. Once actuated, the glass will break into small pieces and access to the patient can be made.

Mechanic tools. Sometimes it is easier to "disassemble" the components of a vehicle than to cut or spread them. Socket sets, both metric and standard, ¼–½ in. (6–13 mm), as well as wrench sets are a great alternative for a door removal. As long as the hinges are exposed, you can remove the bolts that hold the hinges to the car. Once hinges are removed, you can roll the door off the "Nader Bolt."

Flathead axe & Halligan bar. As we know from our firefighting tools, these are two common tools carried on most apparatus today. The axe can be used as a striking tool in order to drive the Halligan into a tight crevice. The Halligan can be twisted or moved back and fourth to separate. These tools can be used to create **purchase points** or open a door completely by separating the door from the Nader Bolt.

Air impact wrench. This tool is also from the auto industry. It is operated by compressed air also and works great at disassembling vehicles bolt by bolt.

Bolt cutters. Bolt cutters are typically carried on most apparatus to cut locks. They can also be utilized to cut part of the steering wheel when the wheel is pining down the patient.

Come-along (hand winch). A come-along is designed to pull when it is used horizontally. It is designed to lift when used in a vertical situation. It may be used to displace a steering column when all other tools fail. It may also be used to hold tension on a vehicle on its side "pulling" the vehicle onto its crutches.

Power hydraulic tools

SKILL DRILL

Power hydraulic tools are a system consisting of a power unit, hydraulic fluid and hoses, and one or more tools. The power unit can use gasoline, diesel, electric, or compressed gas power driving a hydraulic pump. The power unit has a reservoir of hydraulic fluid, which it supplies under force through connected hydraulic hoses to a tool, providing that tool with force to cut, spread, push, pull, and so forth. Most manufacturers use a mineral oil base fluid but not all. And the system operating pressure can vary from 3,000–10,500 psi (21,000–73,500 kPa).

Cutter. Power hydraulic cutters are used to sever components and materials. The blades can have a range of size openings as well as blade shapes.

Spreader. Power hydraulic spreaders are used to spread or push components and materials apart (fig. 30–23). They also can be used to pull components and materials together.

Fig. 30–23. The photo displays a spreading tool and power unit.

Combination tool. Power hydraulic combination tools are literally a cutter and spreader put together (fig. 30–24). The cutting surface is on the inside of the spreader arms. This tool is good in certain applications because of its size and storage space on an apparatus. However, it is a compromise of both tools as far as use.

Fig. 30–24. A combination tool

Ram. Power hydraulic rams are tools used to push components or materials over a greater distance than a spreader (fig. 30–25). They also can be used to pull over a greater distance.

Fig. 30–25. A ram

Air tools

Air Chisel. An air-powered impact tool using a variety of bits to cut vehicle materials. This tool, usually the version used in an auto body or vehicle mechanic shop uses compressed air to move a piston forwards and sometimes backwards to hammer a bit that is inserted in the end of the tool.

Today we can see purpose-made rescue air chisels that use high pressures, upwards of 300 psi to impact with greater force on vehicle materials (figs. 30–26 and 30–27).

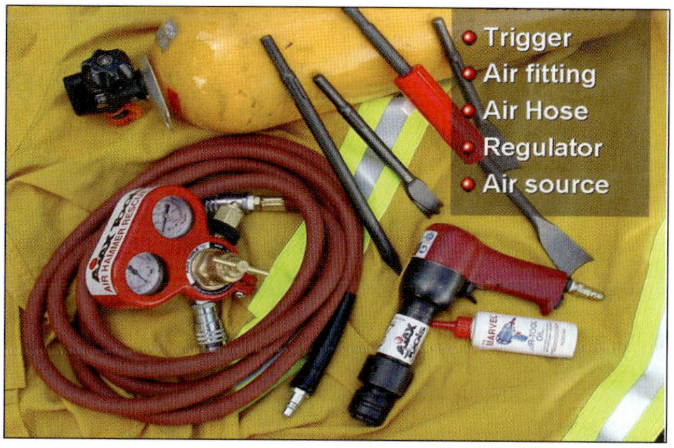

Fig. 30–26. The air tool displayed is an Ajax tool.

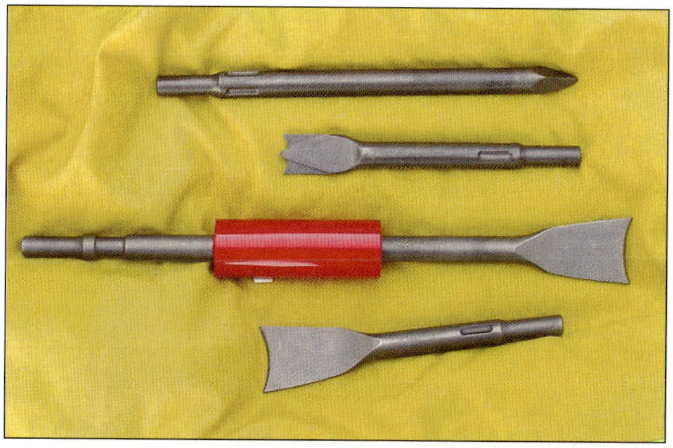

Fig. 30–27. A variety of Air chisel bits are good to have on hand to be used in many instances.

Impact wrench/gun. An air-powered wrench or gun that uses a socket to disassemble vehicle components (fig. 30–28).

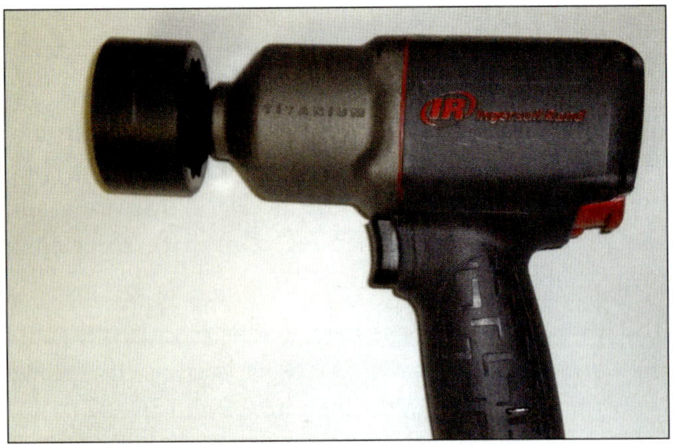

Fig. 30–28. An impact wrench

Reciprocating saw. A reciprocating saw is a powered saw that uses a blade in a forwards/backwards motion to sever vehicle materials.

The saw's motive power usually comes from an electrical cord but can be battery powered or air powered today as well. The saw itself has a trigger to control the blade speed, speed setting for high or low or even orbital and an adjustable shoe to allow the operator to gain blade teeth access. While the motive power on saw is not usually a major issue blade choice can be. Purpose made extrication or demolition blades are the key factor in your choice for reciprocating saw use. The average bi-metal reciprocating saw blade is of little use for the extrication scene as it is not up to dealing with the kinds of vehicle materials we will encounter nor the environment it will be used in.

Key note of reciprocating saw use is "the saw stays at 0 MPH" basically meaning the saw stays stationary against the component you are severing and allowing the blade itself to do the work (figs. 30–29 and 30–30).

Fig. 30–29. Many rescuers have a reciprocating saw on their apparatus for a wide variety of rescue situations.

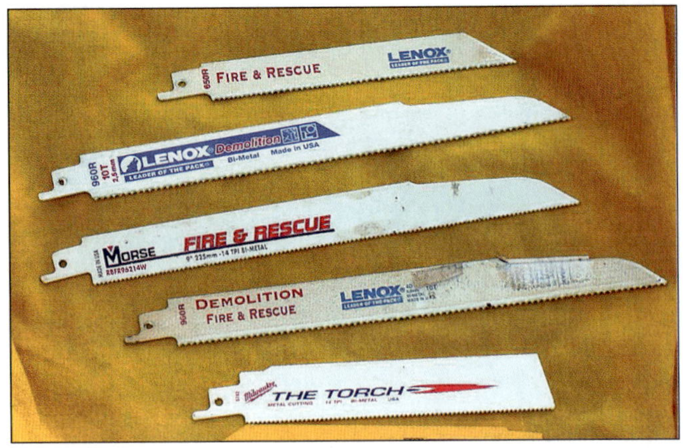

Fig. 30–30. There are many reciprocating saw blades used in today's fire/rescue service.

Stabilization equipment

SKILL DRILL

Cribbing. This is the basis of ALL of stabilization tasks. Cribbing is usually cut into a certain length and dimensions such as 4 × 4 × 18 in. (10 × 10 × 46 cm), however, can come in many shapes and sizes (fig. 30–31). Cribbing can be made of wood or plastic and can be made into specialty devices such as step chocks or quick chocks to better facilitate our stabilization needs.

Cribbing will be our first tool for stabilization as it will fill the gaps between the vehicle and the ground, thus building the foundation for our tool evolutions to work from. Also, when we lift the vehicle, cribbing will secure and stabilize the vehicle by "following" that lifting device up (i.e., "lift an inch, crib an inch").

Fig. 30–31. Cribbing of various sizes

Tension buttress system. This stabilization adjunct greatly assists us when we are faced with a vehicle in a position other than on its wheels such being on its side, overturned or some sort of other precarious position. The concept of **tension buttress stabilization** is simple—we are going to build a right triangle against the vehicle and extending the vehicle's contact with the ground (fig. 30–32). We place a "strut" or brace at a 45-degree angle against the vehicle. Then we take ratchet strap from the base of the brace and extend it to the vehicle staying as low and straight as possible. We then tighten the strap thus putting tension into the brace.

There are many different devices on the market that employ this type of stabilization, such as ground pads and vehicle rescue struts by various manufacturers.

Fig. 30–32. Tension buttress systems

Ratchet strap. These nylon straps or belts that can be "ratcheted" or tightened can greatly assist our stabilization tasks (fig. 30–33). They come in a variety of sizes and strengths and relatively inexpensive. We can use these to tie vehicles together, used in tension buttress systems, hold back vehicle components such as a roof flap or a door, etc.

Fig. 30–33. Ratchet straps

Lifting bag. Lifting bags are an enclosed cell made of rubber or neoprene compounds reinforced with Kevlar that we will fill with compressed air to lift or spread objects (fig. 30–34). **Lifting bags** are made in a variety of sizes and shapes and utilize a variety of inflation pressures. These are usually grouped into either high pressure lifting bags (100–124 psi [700–868 kPa]) or low pressure lifting bags (20–34 psi [140–238 kPa]). The lifting capability is directly related to contact area & inflation pressure. The lifting bag itself is just one part of the system, an air source, regulator, controller, and hoses complete the system.

Fig. 30–34. A lifting bag

Victim and rescuer protection equipment

Hard protection. **Hard protection** is a shatter-proof barrier empoyed when using tools in the occupant cell of the vehicle. These devices or cutting shields protect the patient and interior rescuer from flying debris and tool edges. In the past we used short boards and long spine boards for cutting shields but having a flexible shatter-proof barrier such as a Lexan™ or polycarbonate material shield is a better practice today. This is due to when severing high tensile strength reinforcements or even various vehicle components these can produce small fragments that have energy and can travel with force (fig. 30–35).

Fig. 30–35. There are various types of hard protection. This photo demonstrates how one type is used.

Soft protection. **Soft protection** is a barrier we will employ to protect the patient and interior rescuer when we remove glass or even as a barrier from the elements. This device usually takes the form of a tarp, blanket, or specialty extrication blanket (fig. 30–36).

Fig. 30–36. It is important to place sharps cover between the glass to be broken and all persons in the vehicle.

Edge protection. A crashed vehicle can have quite a few sharp, jagged edges. Our tool evolutions will produce even more. Many of the vehicle materials can slice through our PPE readily so we will need to employ some sort of **edge protection** as we work.

Edge protection will cover those sharp edges and give us a safer and cleaner workspace. Edge protection can run from simple everyday items such as duct tape and spent firehose sections to purpose-made devices such as Kevlar blankets with sewn-in magnets (figs. 30–37 and 30–38).

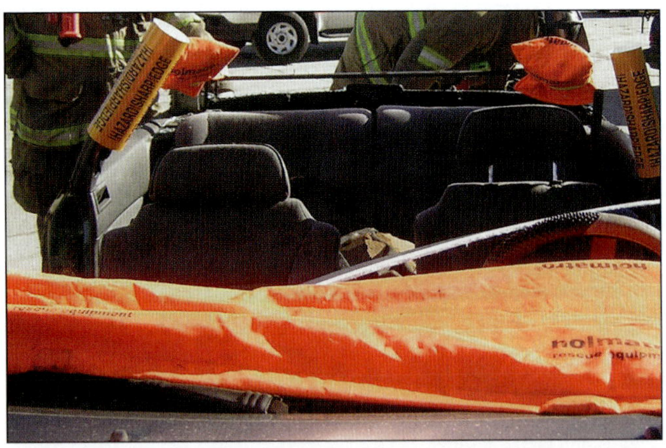

Fig. 30–37. It is very important to cover all sharp edges with some sharps covers.

Fig. 30–38. There are many ways to cover sharps from old hose sections to manufactured sharps covers.

MVC RESCUE OPERATIONAL CYCLE

Now let's take a look at our overall operational cycle at the MVC. Although many of the incidents to which we respond share many of the same components, MVCs have some unique issues all their own. The basic operational cycle includes the following steps:

- Size-up, initial actions, and scene management
- Hazard identification and control
- Vehicle stabilization and glass management
- Patient management
- Extrication evolutions
- Return to service

Now break down each part of the cycle. At every incident we provide for a continuous evaluation. Some steps may be necessary at a particular incident, and others may not. What we do or don't do—and what hazards are present—constantly change.

Our officers take charge, provide accountability, and develop sustained plans of action. Hazards are discovered and mitigated. Resources are requested to assist in mitigating the incident. We control the unwanted movement of the vehicles involved, create access to begin patient management, facilitate a path to remove our patient as we care for them, send the patient to the appropriate medical facility, and clear the highway before returning to service. The key here is to facilitate simultaneous operations on scene. That is what saves us precious time on scene.

Scene size-up, scene management, and initial actions

SKILL DRILL

Arriving to an MVC can be overwhelming to even the most experienced incident commander. Motor vehicle crashes can be chaotic to say the least. People, multiple vehicles, and debris can make any country road or city street look like a combat zone. To properly size up one of these scenes there must be methodical approach in order to not miss any critical details; power lines, fuel leaks or multiple victims. This should all be part of the first arriving companies size up. In order to conduct a size up you must be able to look at the overall picture and not become too task oriented.

While not your responsibility as a Firefighter II, it is important to understand the responsibilities of the incident commander. These duties include the following:

1. Assume command and establish an Incident Management System
2. Conduct a size up identifying the number of victims, and their condition as well as their possibility of survival (manage the scene)
3. Recognize and manage common hazards: power lines, fuels, hazardous materials, etc.
4. Request additional resources as needed
5. Appoint a safety officer

Scene Size-up. The incident commander's size-up will establish the following:

- Number of vehicles
- Number of injuries
- Number of victims trapped
- Rapid patient assessment
- Condition and type of vehicle heavy damage to a SUV or compact sedan
- Hazards such as power lines, fuel leaks, or hazardous materials

Scene Management. As we do with hazmat incidents, we will create three zones in order to create a safe and organized scene. Once the incident has become organized and operations can begin, a circle should be established around the vehicle or vehicles involved. It is important to manage the scene by creating three "work zones": the hot, warm, and cold zones.

The extrication area ("hot zone"). The hot zone should extend out approximately 10–20 ft (3–6 m) in every direction or as large that is necessary. The only people allowed in the "hot zone" are the people directly involved in the extrication operation. Here again, our safety is key, we must ensure that the action circle is as free of tools and debris as possible. A tool staging area should be established in order to keep the action circle safe and keep personnel operating as efficiently as possible.

Tool staging area ("warm zone"). Establishing a staging area outside the "hot zone" will increase the overall effectiveness of the extrication. This is the place where only the rescuers supporting the extrication and patient removal will work. The tool staging area should be set up close to the vehicle involved with the extrication operation in order to ensure a place where tools can be found when needed, while ensuring a work zone with no trip hazards.

Setting up an organized work zone in the beginning of the incident can make or break the overall operation. Seconds count while trying to keep the extrication time to a minimum in order to work within the "golden hour."

Command post ("cold zone"). The "cold zone" is where the command post is established and where the PIO can work with the press. The "cold zone" is also the buffer between onlookers and the overall rescue operation.

Staging area. Having a staging area during the incident is the most effective way of managing resources. It keeps from having the incident commander bombarded with multiple units arriving simultaneously. Staging keeps any uncommitted apparatus and personnel available to assist in the extrication operation. Having a designated staging area can make the job of an IC much easier. It eliminates freelancing of incoming units as they are held back from the scene until requested. Staging also sets the tone for the incident by requiring members to stop and await instruction from the IC. This keeps the officer, from having to feel like they have to make up their own assignment in order to look busy. Just keep one thing in mind; it is easier to turn the companies back after the incident is over than it is to call for them when things start going bad.

Initial actions. Over time, think of how our response and the hazards we encounter have changed (figs. 30–39 and 30–40). The vehicle is still the dynamic hazard in the equation. However, think of what we need to look for: **supplemental restraint systems (SRSs)**, batteries and their subsequent locations, motive power, and vehicle glazing. Additionally, we must do this size-up quickly and completely to ensure an effective and safe plan of action.

Fig 30–39. After securing the scene and hazards are mitigated, the victim is secured and a pathway is created to extricate the patient.

Fig. 30–40. Initial verbal patient contact must be made during the initial response size-up.

What initial actions will be taken? As we approach the vehicle, let's take in how the vehicle appears. What orientation is the vehicle: upright, on its side, or overturned? What kind of *stabilization* should we consider? How is the vehicle damaged? How much *crush* do you observe? This information should clue in potential *entrapment* and possible injuries. Do you see any deployed SRSs? Is there a fuel leak, presenting a fire hazard? We should make visual then verbal contact with the victims. Once we find the victims and establish contact, we should maintain it throughout the incident. We have assessed the scene and located and started to mitigate the hazards. When we are assured that the scene is safe, we must stabilize the vehicle to manage unwanted vehicle movement. This ensures a stable foundation for our space-making evolutions and minimizes movement to our patient.

Once we have the vehicle stabilized, we make access and begin hands-on victim management: airway, breathing, circulation (ABC); manual cervical spine (C-spine) management; cervical collar (C-collar); and oxygen (O_2) as indicated. Take a good look at the vehicle's interior. Where are any SRSs, deployed or undeployed? Is there damage to the interior? Is there any physical entrapment to our patient? Next we need to secure the vehicle's power.

Make sure that the vehicle is shut off, and remove the vehicle's keys. Remember, today's vehicles can have **proximity keys** that need to be more than 15 ft (4.5 m) from the vehicle, so it can't be accidentally started. Then disconnect the 12-volt battery. However, it may be hard to find the battery or even *batteries*, depending on the vehicle.

Remember, a substantial percentage of vehicles have the battery elsewhere than the engine compartment. But even there it can be hidden (fig. 30–41). Each vehicle has unique emergency procedures. However, if you do the following two actions they will disable the HV circuit and disarm or discharge the SRS system.

1. Turn the key to the OFF position, remove the key, and keep it AWAY from the vehicle.
2. Disconnect the 12V battery, both positive (+) and negative (–) cables.

Fig. 30–41. Batteries may be found in many locations such as under the rear seat, behind the front driver-side fender, under the floorboard, or, as in the BMW X3, in the trunk.

We need to make sure that the vehicle's power is secured for two important reasons. First, by shutting down the power, we start the process of draining the energy storage component of the SRS computer. This helps ensure our safety. Second, this also shuts down the high-voltage drive power in a **hybrid vehicle** (fig. 30–42). However, we also need to see what power accessories, such as windows and power seats, are in the vehicle before power is removed.

The officer in charge should document when power was shut down and document the items listed previously.

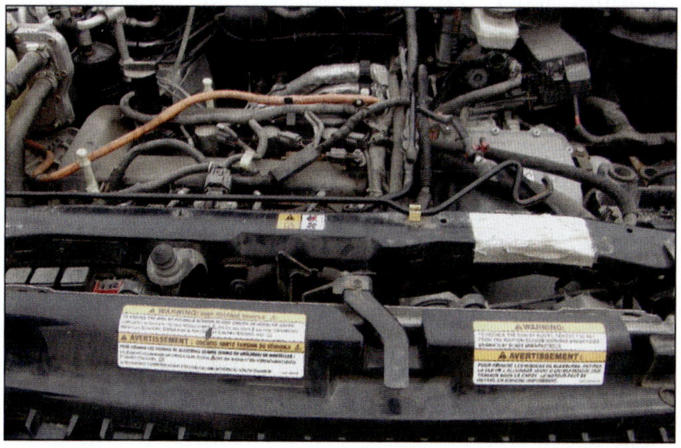

Fig. 30–42. This photo reveals the orange HV cable in the engine compartment of the 2007 Escape Hybrid.

How is our patient? Stable? What are the presented or potential injuries? And are they medically entrapped? Remember that we might need to make space to disentangle the patients, even if they are not physically pinned.

Do we need to extricate? Is the patient **physically entrapped** (encased in metal or pinned)? Consider the nature of the injuries and restricted space. Is the patient **medically entrapped**? Do we need to make space in order to prevent further injury by manipulating the patient during removal?

The officer in charge of the rescue effort must devise a tactical plan of action based on the information presented at the crash. The incident commander knows that the strategic goals at the MVC are life safety of our personnel and patient care of the injured. The tactical plan of action and various versions of it must account for many variables, much more than in the past, and must be performed much faster than ever before. We must protect ourselves and patients from traffic as well as evaluate scene hazards by following these steps:

- Scan the vehicle for damage, potential entrapment, and hazards.
- Find any patients and begin pre-hospital care for them.
- Stabilize the vehicle and then secure the vehicle's power.
- Evaluate the patient's injuries.
- Make space to disentangle the patient.
- Remove the patient to send to the appropriate medical facility.
- Check the equipment, service what is needed, and return to service.

So even if the MVC is a common emergency that we respond to regularly, how easy is it to manage today? Think about the present vehicle technology and various safety systems, as well as actual vehicle components, materials, and construction methodology. Let's take a close look at these concerns first.

Hazard identification and control

We need to create a safe working environment, so we must identify hazards. Some hazards we might face are as follows:

- Fire hazard (every MVC has fire potential)
- Sharp objects
- SRSs (deployed/undeployed)
- Utilities
- Unstable objects
- Loads/cargo
- Fuel leakage
- Weather
- Crowds
- Hazardous materials
- Batteries (both *normal* automotive and HV drivetrain)
- Adequate lighting
- Debris
- Materials cut/displaced
- Rescuers themselves

Take a look at the list. There could very well be others. However, the thing to remember is that the vehicle itself is the most dynamic hazard we face. Scene hazards can and will be different each time, but they are pretty much consistent in how we mitigate them. The potential for fire exists at every MVC today because of fuels, electrical issues, and even the vehicle's materials. Vehicle power is our enemy. Although it can assist us with moving power seats, windows, doors, and the like, it can harbor issues that are not always visible or even considered, at times.

While protecting the occupants, safety systems such as air bags in various locations, ROPS, and even high-strength glass can hinder our operations. Even displaced vehicle components can be a hazard.

Now besides hazards, we must read the vehicle for potential injuries to our patients, such as possible physical or medical entrapment. Make and maintain contact with patients. Command needs to take our operational guidelines, develop a sustained plan of action, and put it into motion.

How do we do this rapidly? Start by taking a good windshield *snapshot* and then take a good 360-degree sweep around the incident. One sweep should be out far enough to capture all concerns of the incident and another should be in close to the vehicles. The officer then should also do a 360-degree sweep to prepare a plan accordingly. As we find hazards they should be relayed to the incident commander and all personnel on scene.

Our apparatus should be positioned effectively to work as well as protect the patients and personnel alike from traffic concerns.

PPE. We know you are first. Your crew or partner is second, and the patient is third. MVC hazards are not what they used to be. We must use more caution than in the past; be more situationally aware than ever; and acquire more knowledge and, more importantly, up-to-date information on vehicles. Our tools generate hundreds of thousands of pounds of force and can injure us if we are not careful. Yet we need to work rapidly.

Many on-scene safety rules are common sense. Be aware of your surroundings. Do not get between tools and the vehicles. Always wear your PPE! (See fig. 30–43.) Use safety glasses or goggles for eye protection. Do not clear out windows with a gloved hand; use a tool. Verbalize your actions so the rest of your crew can hear what is going on.

Fig. 30–43. All rescuers must be wearing PPE to protect themselves from the many hazards associated with extrication incidents.

Always protect the patients with a barrier between them and whatever work is being done; the same goes for the interior rescuer. Cut wiring on a vehicle with a hand tool, not a power hydraulic cutter. Strip interior trim prior to cutting roof posts. Vehicle components we displace when removed are placed with the interior facing upwards because of possible undeployed safety systems.

Now what about fire protection? We must deploy a manned, charged hand line capable of at least 100 gpm (379 L/min) to provide adequate fire protection at an extrication scene.

Besides fuels, vehicle power and materials combined with vehicle fluids can make for a volatile mix. Rescuers must aggressively prepare for a potential fire at every MVC. Even with today's vehicles it is advantageous to have foam capability or a large quantity of water available because of these materials.

Vehicle Stabilization

Although neither glamorous nor high profile, vehicle stabilization is critical at an extrication scene and must be a consideration at all MVCs (fig. 30–44). Unwanted vehicle movement is a hazard at an MVC and needs to be addressed like any other scene hazard. Although most vehicles are upright and it is a very straightforward process to manage them, the vehicle in an unusual position is the situation that pushes the envelope and many times causes great stress. Additionally, as our tool evolutions remove and crash damage displaces vehicle components, the vehicle's structural integrity weakens.

Fig. 30–44. It is critical to stabilize any vehicle in which extrication is involved.

This weakening of the vehicle's structure can assist us, but it is vital to capture the vehicle and secure it on a foundation that is rock solid. Many tool evolutions require this foundation to work from. Think about it: How can you displace a dash if your vehicle is not sitting on a solid platform? So let's look at vehicle stabilization today in a variety of perspectives.

Vehicles. The vehicles of today and their construction affect every extrication effort, be it patient care, tool operations, or of course stabilization. The prevalent construction methodology of today is unibody construction. Basically the vehicle's framework is the occupant floor pan with subframes off the front and rear of the vehicle. The rest of the vehicle structure, including the glass, rounds out the structural integrity. In a crash, as vehicle components are damaged and displaced, the vehicle's strength weakens, as do the tool evolutions we perform to disentangle a patient. As we weaken the vehicle, the vehicle can twist, bend, and contort even if it is sitting on its wheels. This movement can trap doors that were previously open; add tension into glass, so when we break it, it *explodes* with magnified force; and even trap an occupants' feet and/or legs more by folding the footwell/firewall area. If we manage this movement by placing the vehicle on a *platform*, by securing it there we minimize the hazard, manage it better, and improve our tool evolutions.

OK, let's add another spin to the vehicles of today.

Consider ride height. Think about the ride height of today's vehicles or the distance between the underside of the vehicle and the ground. In the past most stabilization has been handled by step chocks (if the vehicle is upright). Step chocks are purpose-made cribbing blocks, built into a stair step fashion (fig. 30–45). But let's think about today's vehicle. Two of the most popular vehicle types today are SUVs and minivans. These vehicles tend to have a higher ground clearance (and higher center of gravity). Many times step chocks alone do not have enough height to properly capture and secure a vehicle alone; usually you must build a small box crib under the step chock for it to work well.

Fig. 30–45. A step chock

Conversely, many cars are much lower to the ground these days. In fact they are low enough that many times when we insert a step chock under the vehicle, we get only a few steps in and under the vehicle, leaving most of the step chock sticking outside the vehicle, becoming a trip hazard to us. Additionally, vehicles readily absorb energy when struck in a crash. This energy absorption principal can affect the vehicle by basically absorbing a certain amount of crash energy or **crush**, including the drivetrain and even tires; often the vehicle ends up sitting on the ground. How do we put a step chock under that? Well, you might ask why we bother, because it is sitting on the ground. However, we need to manage it and control ourselves to ensure it is mitigated properly. Last, add the addition of **run-flat tires** or tires with air monitoring/supplementation. How well can you deflate a run-flat tire today?

Tools and methodology. Cribbing, wedges, and step chocks make up our primary stabilization cache. We use cribbing of various sizes day in day out at a wide variety of incidents. Wedges fill up gaps and voids. Step chocks have been discussed a bit already. All these devices can be made of plastic or wood. However, I have mentioned the issue of both vehicle ride height and energy absorption principle, as well as the number of SUV and pickup trucks on the street. So I would like to take a look at some of the new and revisited concepts and tools for stabilization.

In the 1980s step chocks evolved from extrication challenges to assist with rapid stabilization of a vehicle on its wheels. Today's process has some additional options to help us. One of these is the rapid stair chock. This chock appears similar to a step chock but has a unique process to provide quick stabilization. It is a two-piece plastic device composed of a large chock but instead of stair steps it has an incline plane. The other part is a **traveling block** that has a mirrored incline plane to the large chock topped off with a large flat platform. Each incline plane has *teeth* that lock together when pressed. The rapid stair is placed much like a step chock, but the traveling block is placed near the base of the large stair. When the rapid stair is placed under the vehicle the traveling block is slid upward on the large stair, wedging the traveling block between the large stair and the underside of the vehicle.

This step maximizes contact between the vehicle underbody and the ground. This device makes for both rapid set up, stabilization, and securing of the vehicle.

Another type of chock is a **quick chock**. The quick chock is comprised of two pieces. The first piece is a platform of three or four 4 × 4–in. (10 × 10–cm) cribs, 18 in. (46 cm) long, screwed and glued together with a top plate of ½-in. (13-mm) plywood screwed on top of the platform. The second piece is a 6 × 6 or 4 × 6 in. (15 × 15 or 10 × 15 cm) wedge approximately 18 in. (46 cm) long. The quick chock is also placed under the vehicle like step chocks with the platform being the base and the wedge driven in between the platform and the vehicle underbody. This allows for rapid setup and a flexible approach. Depending on ride height and vehicle damage, you can add or subtract parts fill in space to stabilize.

Another idea that deserves a second visit is the lock blocks or Lego® block–type of cribbing. This crib made of plastic and manufactured by a variety of companies.

These cribs basically look like square or rectangular Lego-styled blocks of various heights that *nest* or lock together. These provide a large contact base, and because of various heights can be built and slid under the vehicle, placing a *column* of cribbing under the vehicle. Also, there are wedges that are also made to mesh with these blocks. All these adjuncts work well with the changes in vehicles; dynamics involved when these vehicles are crashed and interact with the tool operations as well as provide a more rapid and complete stabilization of vehicles upright on their wheels.

Also what about those vehicle tires? Think of the extremes we see out on the street today: from ultra-high performance tires that have almost no sidewall to the newest run-flat tires and the large sidewalls of SUV and pickup truck tires and then add aftermarket changes.

When we crib the vehicle, we capture the vehicle and its suspension, spreading that point of contact over a large area of the ground. However, depending on the vehicle's tires, we can still generate movement by *flexing* the tire sidewalls. Some sidewalls can be fairly large, so this movement can rock the vehicle off the platform we have created with our stabilization evolutions. This unwanted movement can be caused by accessing the vehicle or even tool evolutions. Remember, we need to minimize this unwanted movement in our patient care. So how do you deflate these tires? There are many options, but we want to strive for a controlled release of air from the tire. So stabbing the tire sidewall is not an advisable idea. Unscrewing the valve core from the stem, pulling or cutting valve stems, or using a purpose-made tire deflation tool are good choices.

Now let's look at stabilizing vehicles other than on their wheels. In the past, besides cribbing we relied on adjunct tools such as high-lift jacks, come-alongs, and even lifting bags to assist in our stabilization efforts. Although these items still give us options today, the trend today is the use of a technique called **tension buttress cribbing**, which resolves and mitigates stabilization of vehicles on their sides, overturned, or in an unusual situation. The concept is the placement of one or more right triangles against the vehicle. This shape ties a strut or brace with a ratchet strap to the vehicle, and then the strap is tensioned. This action basically extends the base of contact with the vehicle and the ground. There are many variations of this stabilization principal. From a simple homemade long crib with a ratchet strap, a section of cribbing with a manufactured base and strap combination, and crown or cap or a manufactured strut, base and strap combine all these to use the principal of tension buttress cribbing.

This principle allows us to *capture* a vehicle and hold it in place effectively and rapidly. This principal is also simple and straightforward to learn and apply in a variety of situations. The devices themselves also lend well to being stored on a variety of apparatus, and some have extensive accessories to enhance application options. Depending on vehicle orientation, this principal can be applied and vehicle stabilized in less than 2 minutes. Also depending on the device as the devices are adjusted they also can provide a source of controlled lift while stabilizing.

Some of the items mentioned chop away at that all-important scene time, making our extrications rapid and add in an enhanced margin of safety as well by making the vehicle we are working on rock solid.

We need to save time out there on the street so we can improve our patient's outcome. And this adds to our desire to always ensure a safe working environment. And remember, always recheck your cribbing and stabilization after every major action done on the vehicle (i.e., roof removal, side evolution, etc.).

Glass management. Primarily, we deal with two forms of glass or glazing in vehicles today. Laminated glass is found in windshields but is increasingly found in the side and rear windows as well. This type of glass is two layers of glass with a layer of plastic sandwiched in between the glass. This type of glass must be cut from the vehicle to be removed, which can be facilitated with an axe, a glass saw, or even a reciprocating saw by cutting the peripheral edge. This operation also produces glass dust, which is a respiratory hazard. Tempered glass usually found in side and rear windows when broken falls into small granular pieces. This breaking operation is facilitated by using an impact tool that imparts a large amount of force into a small concentrated area. We then clean out the window opening by pulling the glass onto the ground with a tool, not with our gloved hands.

Remember, today's vehicles can load the glass with energy from the crash, so when it's broken it can almost explode when that energy is released. We must protect our patient protective cover and even use hard protection to funnel glass pieces away from the patient. Make sure your own eye protection is in place, and you are wearing a dust mask to protect your airway from glass dust. Also, watch out for glass in hatches of vehicles. Not only might this glass become loaded with kinetic energy from the crash, but it might have up to four gas struts pushing on the glass and creating even more tension. When that glass is broken it might explode, and you might even have struts pushing out toward you.

Now let's think about managing all this broken and cut glass. Ideally, laminated glass we cut and remove should be folded slid in under the vehicle out of our way. If at all possible, tempered glass should be removed to the outside of the vehicle, away from our patient and the interior rescuer. However, in some vehicles, especially SUVs and minivans, we might encounter a lot of windows and/or a rather large window, which produce many glass particles.

A good practice is to sweep this under the vehicle, so we are not walking on it and our stabilization devices are on not on it either. Glass particles can be slippery, almost like marbles, when walked on.

Patient management

Vehicle rescue is a patient-care-driven skill. Although the focus of most education and training for mitigating issues at MVCs revolve around tool evolutions and vehicle concerns, let's consider patient care and considerations in the environment and our ability provide a better patient outcome. After all, the patient is the overall reason why rescue services are performed (fig. 30–46).

Fig. 30–46. Patient management begins as soon as the scene is safe enough to approach the vehicle and patient.

Good trauma care starts with good basic life support (BLS) care. Airway management, manual C-spine stabilization, application of C-collar, and oxygenation are basic BLS skills used in the initial care of a patient in a motor vehicle accident (MVA). Mass hemorrhage control comes next. It sounds simple and straightforward, but let's think about the many issues that affect our ability to provide the pre-hospital care necessary.

Think about how vehicles are configured today. Do you have enough room in the vehicle to perform basic patient care skills? Take a good look at how the dash and controls wrap around the driver and front passenger.

They form a cockpit-like configuration. How about seating? How many vehicles have bench seats, bucket seats, or racing styled seats? These make for tighter spaces, leaving less room and space for our equipment and ourselves (fig. 30–47).

Let's not forget about SRS items (fig. 30–48). We should observe space between those devices and ourselves, and our equipment as well. Visualize for a moment those spaces: frontal air bags, side impact air bags, and side curtain air bags. We even need to be aware of devices such as seat belt pretensioners and ROPS. Many times, the appropriate amount of space means displacing vehicle components. Consider packaging a patient in a **Kendrick extrication device (KED)** or similar device in today's cars, SUVs, and pickup trucks. Do you have the right amount of space to remove the patient with door access alone? Many mechanics of patient care and packaging taught today have been built around vehicles of the 1960s, 1970s, and even 1980s. Do they still apply in today's MVA environment? Recall that most education on access and patient removal revolves around either opening a door or displacing a roof. Although both are indeed viable, think of the additional variables that exist today. It is not as straightforward as it used to be.

Fig. 30–47. Smart cars, as shown here, are becoming more and more popular.

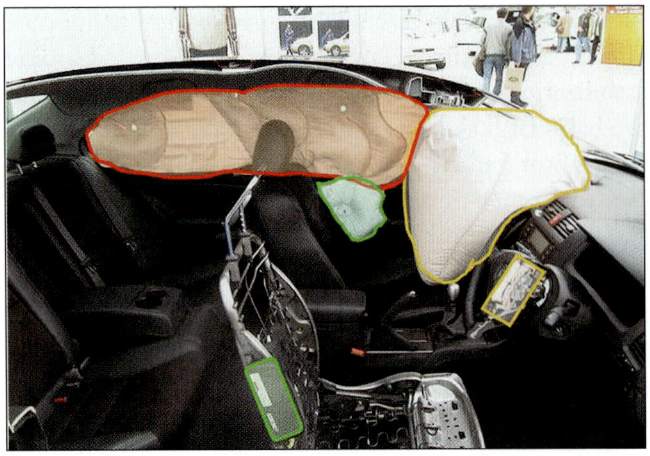

Fig. 30–48. This photo shows typical SRS locations deployed and undeployed.

Take a good look at patient injuries and focus on where they occur. Do you see the same head and chest trauma that you used to see decades ago? How about injuries to the lower extremities and the pelvis? Why is that? There are a few paths here: frontal SRS, driver, and passenger side and front air bags and knee air bag systems, although sometimes causing injury, can and do protect occupants from injury sustained from the crash, even when seat belts are not used. When seat belts are not used, however, frontal air bags have a tendency to push the occupant down and under the dash. Now connect this with how vehicles are constructed and configured today. Vehicles are designed to absorb energy from the crash and crumple inward. If the vehicle structure is crushing in and the frontal air bags push the unrestrained occupants down, is there any wonder why we see such injuries? Even when the front occupants are restrained, the vehicle's structure can become crushed so badly in a high-energy crash that responders may see some of the same injury patterns (fig. 30–49).

Fig. 30–49. The vehicle crumple zones are similar in most high-energy crashes.

Now how about side impact crashes? The same principle as with energy absorption still applies, but now the vehicle's structure is much stronger and more resistant to lateral forces than before; add to this side impact and side curtain air bags designed to protect vehicle occupants.

However, these days there is less space between the occupants and the impact. Although the various safety systems and the vehicle's construction and configuration do indeed effectively protect occupants from side impacts, think of how they hinder our work to provide care in the environment and package as well as extricate the patient from the wreck. We must strip interior trim prior to any sort of roof evolution today to avoiding cutting the gas generator for the side curtain air bags. We need to watch for side impact air bag systems when we force doors or perform side removal evolutions. Consider also vehicle construction material in these evolutions. We now end up pushing and prying against hardened materials for reinforcements and lightweight soft materials, such as enhanced and lightweight body skin alloys. Think of the necessary interactions to successfully mitigate the issues of both ends of the material spectrum.

To provide a better patient outcome, we must be aggressive when assessing MVCs and take a proactive approach in disentangling patient from the wreck and providing pre-hospital care. This requires us to perform simultaneous operations: patient care and space making, creating a pathway to remove that patient. EMS provides patient care in the same rescue real estate as the rescue team, which needs access to the exterior of the vehicle, especially the side between the wheels. EMS must be with the patient, comfortable working there, and keenly aware of the potential vehicle hazards as well. They must allow the rescue team to perform necessary tool evolutions to displace vehicle components to make appropriate space to remove the patient expediently.

The rescue team needs to be sharp. They must understand the issues and hazards involving the vehicle, their tools and evolutions, and how they will interact with the vehicle materials. Rescue personnel must also be aware of patient care considerations to make the best, safest, and most efficient pathway for that patient.

We need to provide prompt patient care simultaneously with our space-making efforts. Part of that requires us to get into the environment enhance patient care as well as create a rapid pathway for removal. A better patient outcome can be produced when these operations are fused together. Personnel operating tools and performing space-making evolutions must understand today's vehicle technology concerns, but so do EMS providers. As part of providing care, EMS providers operating in the environment interact with many vehicle hazards. They must have proper background knowledge of these concerns plus equip and use proper PPE for the environment in which they are working.

Patient care. We strive to perform proper pertinent patient care when we respond. We also focus on providing appropriate metal/material movement evolutions.

Yet how often do we combine these tasks, both in training and operationally? We need to practice these skills together, making the tempo of tasks at the MVC flow smoothly. The idea behind rescue real estate is that EMS must provide patient care in the environment (i.e., inside the vehicle) and the rescue team owns the vehicle's exterior, especially the vehicle's sides between the wheels. This concept allows both EMS and rescue team to operate together and enhance the speed of the rescue.

Next let's talk about patient care from our arrival on scene. Our visual survey approach should start our patient care process, even before we make contact. Reading the wreck, so to speak, gives us information about potential injuries, vehicle damage, and crush. Look at the windshield: Is it damaged or intact? Remember that frontal air bags cause damage to the windshield as part of their normal operating parameters. Check for interior damage: Are steering wheel and column intact? Are the air bags deployed? Don't forget side impact, side curtain, and knee bag systems. We should always try to approach our patients from the front of the vehicle. It is important to make visual contact first, keep their focus on us, and minimize movement prior to hands-on contact and C-spine management.

In addition to visual contact, we should attempt verbal contact. These contacts should be maintained throughout the rescue, ensuring communication to and from our patient as well as providing reassurance. This also provides a good presentation to our patient. Never leave the patient unless absolutely necessary. A good practice is having a rescuer maintain visual or verbal contact with our patient while another rescuer makes access to the vehicle to begin hands-on patient care.

After primary survey begins, ABCs are evaluated. Airway is a straightforward evaluation; with breathing remember to check quality as well as rate; for circulation, remember to look for and manage gross hemorrhage as needed.

C-spine management is next, both manual management and C-collar application. Oxygenation comes after that, managing proper fit and application of nonrebreather O_2 mask as warranted. The use of a nonrebreather O_2 mask on our patient also helps protect them from glass dust and other airborne particulates. As the rescue effort progresses, constantly reevaluate the patient for status changes, both good and bad. This information is communicated to the rescue team as well as the incident commander, facilitating appropriate decision making and operational planning. Communications are a key element no matter what part you play on scene, be it an EMS provider, tool operator, or incident commander.

This is just an overview of what basic care should be going on in the vehicle. However, depending on the skill level and ability of the caregivers on scene and what the patient needs, this care should be adjusted appropriately.

Space making/vehicle issues. Now let's take a look at the wreck so to speak. Even though we usually put tools to the vehicle to create a pathway for patient disentanglement, part of the concept of **space making** is to create space as required. That includes making space for us inside to work properly and safely. Re-proportion the inside of the vehicle as needed and performing strategic cutting operations. As vehicles become smaller, such operations are going to become the norm, not the exception.

There are many considerations. Displacing a roof off or away from a patient's head and/or face creates better airway control and C-spine management and has a positive psychological affect. How about cross-ramming the interior to make better access for us to work, to patients' lower extremities, or for that matter the patients themselves? The vehicles we have worked with in the past had plenty of space; with plenty of room for access, it was rarely necessary to consider such tool operations.

How about popping doors? Does such an operation have enough room in today's vehicles? Is a side removal more warranted these days? A side removal evolution is also many times easier than a door pop today as well. Why? With a door pop we are operating tools to force hardened components on both sides of the door. With a side removal we still need to force a latch. However, the balance we are working on usually has lesser strength materials. Remember that ultra-strength material reinforcements can now be found in the B-post areas as well (figs. 30–50 and 30–51).

Fig. 30–50. Rescuers are finding that the standard door pops are not enough space today.

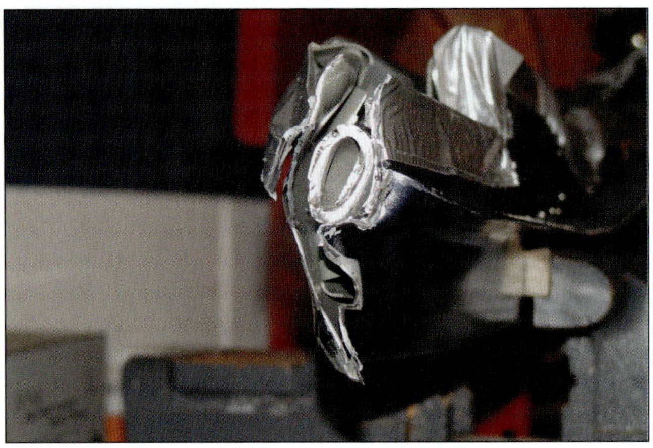

Fig. 30–51. Manufacturers are increasing the strength in vehicle supports such as B post with boron pipe reinforcement shown in this photo.

Also with this thread, let's add in strategic cutting. Many times today we are faced with dash areas that have dropped onto our patient's lower extremities as well as footwell areas that have folded up around these same extremities. Performing a fender evolution (strategically cutting the crumple or energy absorption zone) that severs the crumple or energy absorption area greatly helps the displacement of the dash and footwell. This same evolution also accesses the door hinges and the dash relief cut **cheat hole**. So by moving the fender as part of that evolution, we have helped with two other tool evolutions, thus saving time.

Remember, time saved helps produce better patient outcomes, especially if we can do these things on a consistent basis. If these operations must be provided often, we should go in with the notion to aggressively, proactively evaluate the wreck and set up for such evolutions, maybe even before we arrive on the scene or as part of our operational guidelines.

Note that you cannot forget that because we are cutting more often and materials are so much harder, it is increasingly important to provide a rigid barrier between our tool operation and the patient and interior rescuer. Also, the last roof post cut should be the one closest to the patient's head. Furthermore, don't forget to consider issues surrounding safety systems: air bags, seat belt pretensioner, or ROPSs. We should try to observe space between these systems and us and our patient. That also includes where we place our equipment.

Packaging/disentanglement. Next let's talk about removing our patient. Once we have created our pathway, cover up any sharp edges (created with tool operations and by the crash) we must work around. If we can facilitate this, this should be done as tool work progresses.

Take a look at what the patient is sitting in. How much space is found in today's vehicle interiors? Then think about this: Many techniques to package patients evolved from vehicles of the 1960s, 1970s, and even 80s. Take a good hard look at how interiors of those vehicles are configured versus the interior of today's vehicles. Although bucket seats were available then, think of how those seats have evolved into the prevalent seat style today.

Also consider the equipment we use to package those patients. Many devices were also developed during the same time frame. Seats have become more like motorsport- styled racing seats. They are designed to hold the body firmly in the seat as the operator drives, regardless (relatively) of the situation at hand. However, how difficult is it to package a patient in a KED or similar device in these seats? Additionally, how much space is needed just to apply these devices, let alone disentangle the patient once it has been applied? Maybe we need to look long and hard at these devices. What else is out there in the market and how well would it work? These are more reasons to make space and proactively look at this concept. From casual observation it appears that many patients are removed directly from the vehicle onto a long spineboard (that is, a rapid takedown or extrication derived from **pre-hospital trauma life support [PHTLS]**). However, think about the criteria for such a maneuver. Rapid takedown or extrication focuses on an unstable trauma patient, a small percentage in the grand scheme of the MVC. Otherwise the KED-type device is warranted. So are these all unstable patients or are we cutting corners because of space issues? Maybe we need to revisit the concept of space making, incorporating the idea of being medically entrapped and planning in the appropriate amount of space to properly disentangle the patient.

Training/PPE. How are you preparing for an MVC today? Many of us are out in the salvage yards, training centers, and such practicing tool evolutions. We are reading trade publications; attending programs and presentations; and surfing the Internet to keep pace with current trends, issues, and concerns with vehicle technology. But how many of you are out setting up scenarios and solving the problem? Think about such training situations. Besides the obvious tool operations, think about what other situations to practice:

- Command and control issues and concerns
- Power management (finding batteries in alternate locations let alone multiple batteries)
- Finding potential SRS locations by using adhesive marking simulations (i.e., decals)
- Hazard control as well as looking for and managing possible scene and vehicle hazards
- Patient care and patient protection. Many times, we use training manikins to substitute for patients. Or we use our fellow rescuers as patients.

Are these substitutes suited for simulating patients? Sometimes the scenarios we set up are too hazardous for a live person, so a training manikin is warranted. And depending on where you are, regulatory constraints may not allow us to use a live person. But let's think about it for a moment. It is important to know and understand the various facets of interaction with that patient. And how else are you going to become comfortable and functional working in that vehicle unless it becomes a trained and practiced ability? One concept to help resolve the patient care skills is a medically trained and oriented *patient* or *casualty*. This concept is a crossover from the extrication challenges held in North America and around the world and from the International Centre for Emergency Techniques (ICET) in the Netherlands. These training events regularly use simulators to interact with the rescue team and EMS providers, which provides reality to the scenario in a controlled environment. One of the items of note that goes along with patient care is the EMS providers themselves. Although not operating tools, the EMS provider is a critical component at the MVC. Obviously, we are there to provide for a better patient outcome. However, often training, education, and personal protective equipment in relation to the MVC is a neglected area.

Although the facets and skills of patient management are practiced, how often is access in a wrecked vehicle practiced? How about vehicle technology concerns? Because the EMS provider is in the vehicle working, aren't safety systems concerns just as important to them and the patient? Often, they are even closer to these systems than the tool operators. How many times do the EMS providers make access in the vehicle with minimal PPE? Will a helmet and maybe eye protection suffice today? Even if they have a bunker coat, does it fit or has it even been off the ambulance in weeks or months? Think of all the dynamic vehicle hazards of today and then add in working in an environment that doesn't always allow for quick exit. Shouldn't we be providing flash protection, proper head protection (helmet) that can stay on in the vehicle, proper gloves for the environment, and even respiratory protection from glass dust and other airborne particles? Take a good look at some the latest PPE that is on the market for EMS and technical rescue environments, and you will find some excellent solutions for these concerns.

Patient's rights. Patients have the right to a well-equipped, well-trained rescue effort. Think about that for a moment. It is not always about the best or newest tool out there. It is what you have and what you can and cannot do with it. Remember, vehicle extrication is a patient-care-driven skill. We are there to provide the best outcome we can in each and every response. This concept is carried through in all facets of our operations.

Plans are prepared with consideration for the vehicles we encounter today. We train as we operate: running rescue tools, providing patient care, operating a protective hoseline, or taking charge and mitigating the incident. Each part is important to providing a better patient outcome. Our equipment is always ready and prepared. We review our incidents to gain insight on the vehicles we work on, the hazards we face, and how we can enhance the patient's outcome.

Why do we extricate? Physical entrapment is what rescuers usually zero in on. The patient is physically trapped by a vehicle component, or we lack access to the patient (i.e., the doors are pinned). If our patients have relatively minor injuries and can move, they usually try and exit the vehicle prior to our arrival. In the past, most extrications revolved around such situations.

However, with the nature of vehicle construction and the vehicle's ability to absorb energy in a crash and crumple, think about another possibility that is actually more prevalent today: Vehicles are becoming smaller; add in the energy absorption principal; mix into this the way and position occupants actually sit today and the vehicle structure that surrounds them; finally, add into this injuries themselves. We are faced with a smaller vehicle that is now crushed closer to the occupants, giving us less space to work on the patient, not to mention create a proper pathway to disentangle them. We call this situation medically entrapped. We might have access to the patient but not the space to properly care for them, let alone package and remove them. So how do we figure out who is medically entrapped whether we need to create space? Ask yourself these questions:

- Do we have adequate room to provide patient care?
- Do we have adequate room to work?
- Do we have adequate room to properly remove the patient?

If we cannot answer *yes* to those questions, we need to create space. Always try the simple things first: lift the tilt wheel, push the door farther back, and move the seat. In today's vehicles, even these simple things do not always give us the space we need.

Extrication evolutions

SKILL DRILL

Roofs. Be they roof post cuts or relief cuts for a roof displacement, the big issue with all tool evolutions is the vehicle materials and construction. We have ultra-high strength reenforcements in the roof today. Some include boron steel, the same material as the door crash bar and beam.

We also can find side curtain air bag systems in the roof.

So before putting a cutting tool to the roof, displace the interior trim to look for side curtain air bag inflators and get a good look where the cut will be made (fig. 30–52).

Fig. 30–52. A reciprocating saw used to cut a C Post

The two primary tools we use for roof displacement is the power hydraulic cutter and the reciprocating saw or Sawzall®. These materials make the cutter twist and bend as it is weakened. Take a look at these cuts. The tool takes the path of least resistance, so anticipate where the cutter will go. Maybe even flip the tool over. Many times this methodology helps keep the cutter in line with the cut you wish to make, and it is even easier to cut. Make sure the blades go completely around the post and 90° to what is being cut. Don't forget to strip interior trim to locate the side curtain gas generators (this trim strip also allows us to look at the post and observe reinforcement locations and cut the post where it is weakest) (fig. 30–53).

Fig. 30–53. Always strip trim to find side curtain cylinder before cutting a roof or post.

Make sure there is a hard protection barrier between the tool and the patient and interior rescuer. This is even more important today; when hardened materials are cut, sometimes small hardened debris is created and sent flying about. When you make the cut, be sure the tool completely closes and makes the cut fully. Sometimes we have to wait for the second stage of the power unit to kick it to make the cut. Also, you must allow for foam reinforcement in the post to be compressed to make the cut. This means you need to visualize your cut, listen to the tool and its action as it cuts, and inspect the cut when the tool is removed to ensure it is complete. Always return the cutter blades to the tips touching position whenever a cut is complete for safety. With the reciprocating saw we must be sure the saw stays at 0 mph and the blade does the work. Don't try and saw with the saw itself. And like the cutter, make sure the blade is at a 90-degree angle to what's being cut. To minimize vibration make sure you are using a proper fire/rescue or demolition blade of 10 to 14 teeth per inch and that the shoe of the saw is tight against what is being cut. Remember to check the B-post for the adjustable seat belt bracket; we don't want to cut through it because it is also made of hardened material. Also, remember the vehicle's structure has crash energy stored in it at times. When we sever roof posts, sometimes they snap or pop as this energy is released.

Besides a roof removal, you can also flap or section a roof depending on the size and damage as well as the patient location. Also, on certain vehicles such as SUVs and minivans, it might be a better option to flap or section the roof because of the size of roof structure in such vehicles. One tip to remember always: The last roof post cut is the one closest to the patient's head!

Doors. The basic tool evolution to gain access to a patient is usually a door displacement or pop. This evolution is the forcing of either the hinges or the latches of the door with a power hydraulic spreader. To begin this we must make a purchase point for our spreader tips to make better access and bite. We can use hand tools such as a Halligan tool, the spreader itself using the tips to grab and pinch the edge and pull it away. Or we can use the existing space made as a result of the crash itself.

When we pop doors today more times than not the door materials shred, tear, and rip apart, leaving us to try and spread another way or try and cut the door off. Why does this happen? Well, couple the lightweight door materials with high-strength latches and hinges and those *coupled* by the door crash beam and the whole tied together tighter than before because of vehicle construction and design and its inherent ability to move crash energy throughout the vehicle structure (figs. 30–54, 30–55, and 30–56).

Today's cutters make short work of cutting hinges and latches. Why not just cut them in the first place? Think about it. Cutting these items to remove the door creates much less stress and strain on the vehicle, which eases stress on the patient and the tool operator. The evolution actually goes faster and smoother, which also makes it safer for all as well. The key for this cutting is the ability to *visualize what you seek* (i.e., observe the hinge and or the latch) (figs. 30–57, 30–58, and 30–59). Ensure the cutter blades get completely around the hinge or latch. We don't want to place the tips of the cutter on the hinge, which causes the tool to be tip loaded and damages the blades.

Fig. 30–54. Making a purchase point to remove the door at the hinge location

Fig. 30–55. Spreading the front of a door to remove the door by pulling the hinges off.

Fig. 30–56. Another angle of the purchase point.

Fig. 30–58. Cutting the door hinge

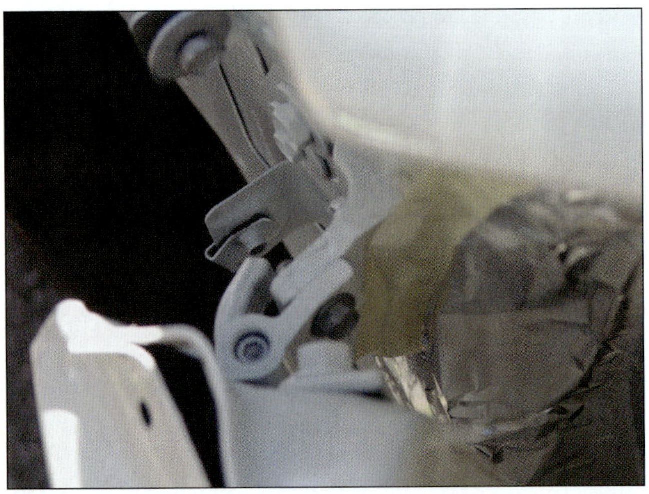

Fig. 30–57. Photo showing a top view of a door hinge

Fig. 30–59. Cutting the door hinge—close-up

To do this, we need to make space between the door and vehicle body so we can get the blades in and around the objects to be cut. Watch tool reaction here as well, because often we are cutting square or rectangular material. This causes the cutter to swing or torque quickly, so we need to watch closely. Also, you ensure hard protection is in place between the tool work and the patient and interior rescuer. Now, like with roof post cuts, as we find wires in our path, remember to cut these with a hand cutter like medic shears or battery cable cutters. Why? Well, with more vehicles with side impact air bags in doors and side curtains in roof edges, these devices are affected by static electricity as well as pressure and shock (fig. 30–60).

Fig. 30–60. Debris with live SRS

Devonian	Upper Devonian shale formations (Maple Mill)	Devonian confining units	shale, siltstone, dolomite	confining beds; aquitard	
	Lime Creek Fm.		shale, dolomite, limestone	confining beds; aquitard	
				local aquifers	north-central
	Cedar Valley Group	Devonian aquifer	dolomite, limestone, chert	regional aquifer	eastern
					north-central
	Wapsipinicon Group		limestone, dol., shale, gypsum	local aquitard	eastern
Silurian	Gower, Scotch Grove Hopkinton, Blanding, Tete des Morts, Mosalem formations	Silurian aquifer	dolomite, chert, limestone	regional aquifer	eastern
Ordovician	Maquoketa Fm.	Ordovician confining units	shale, dolomite, chert	confining beds; aquitard	
				local aquifer in northeast	northeast
	Galena Group		dolomite, limestone, chert	confining beds; aquitard	
				local aquifer in northeast	northeast
	Decorah, Platteville, Glenwood formations		shale, limestone, sandstone	confining beds; aquitard	
	St. Peter Sandstone	Cambrian - Ordovician aquifer ("Jordan aquifer")	sandstone	regional aquifer	statewide
	Prairie du Chien Group		dolomite, sandstone, chert	regional aquifer	
Cambrian	Jordan Sandstone		sandstone		
	St. Lawrence Fm.		dolomite		
	Lone Rock Fm.	Cambrian confining units	shale, siltstone, sandstone	confining beds; aquitard	
	Wonewoc Fm. Eau Claire Fm. Mt. Simon Sandstone	Dresbach aquifer	sandstone, shale, dolomite	regional aquifer	east-central northeast
Proterozoic	undifferentiated		igneous & metamorphic rocks; sandstone, shale	unknown	

Gary Hightshoe, Iowa State University

Floodplains are underlain by porous sand and gravel deposits that yield valuable groundwater supplies. These shallow alluvial aquifers are vulnerable to contamination from the land surface. (Iowa River meander loops, Iowa County)

SURFICIAL AQUIFERS (Quaternary age)
Alluvial aquifers

River valleys are one of Iowa's major landscape features. Along and beneath these valleys are shallow sand and gravel deposits containing *alluvial aquifers*. The alluvium itself was deposited during a river's past history. Such sediment may be as young as a sand bar from last year's flood or as old as a deep 10,000-year-old gravel deposit left by the last glacier's meltwater flow. Alluvial aquifers are important sources of moderate-to-large water supplies across the state, but their lateral extent is restricted to river valley corridors. The distribution of Iowa's alluvial aquifers is shown on the map to the right.

Alluvial materials can change abruptly in grain size both vertically and horizontally. This reflects rapid shifts in velocity and direction of a stream's flow during its history. Some alluvial deposits are dominated by fine-grained silts and clays, which reduce an aquifer's potential by restricting the ability of water to move through in response to pumping. In contrast, when alluvial materials are coarse, thick, and extensive, a productive and dependable source of groundwater can be expected. The advantages of alluvial aquifers are their abundant yields and shallow well depths, typically less than 100 feet deep. By the same token, their shallow depths and high

Alluvial Aquifers of Iowa

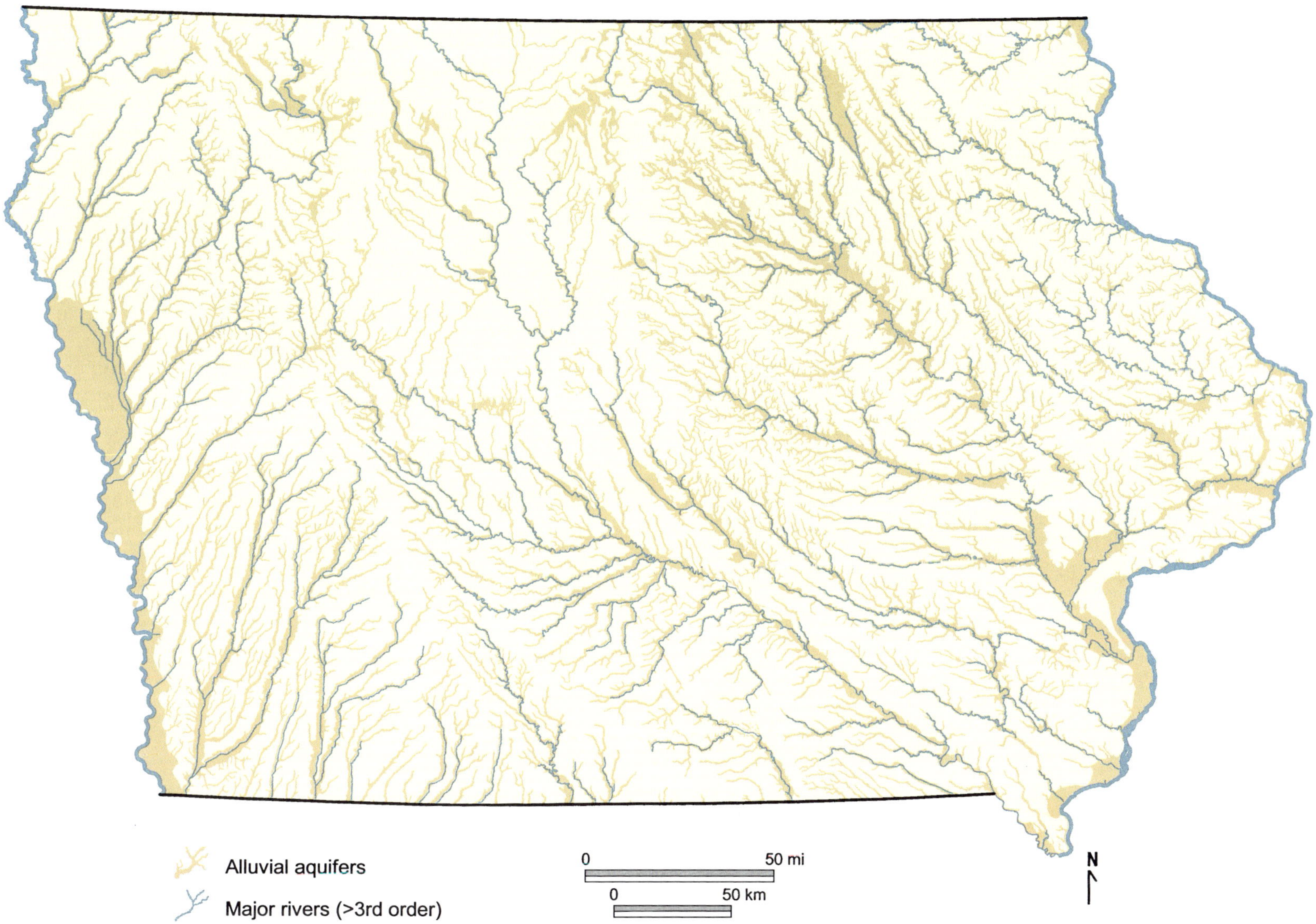

porosity make them particularly vulnerable to seasonal variations in precipitation, and thus subject to drought and fluctuating water table conditions. Their depth and porosity also leave these aquifers open to direct infiltration from the land surface, which increases their susceptibility to contamination problems.

Generally speaking, alluvial aquifers contain the least mineralized groundwater in Iowa (typically less than 500 milligrams per liter [mg/L] of dissolved solids) because the water has been in the ground for the shortest period. Quality can vary, however, since it also depends upon aquifer thickness, depth of wells, the character of an underlying aquifer or aquiclude, and the source of the groundwater (local precipitation, induced infiltration from a nearby river, or longer-term storage within the aquifer). Since sand and gravels usually are connected hydraulically with an adjoining river, these deposits also reflect the river's natural recharge and quality characteristics.

Significant alluvial aquifers occur along the broad Missouri and Big Sioux river valleys in western Iowa and along the Mississippi River corridor in eastern Iowa. These aquifers are also present along Iowa's larger interior streams (photo, p.16). The coarser, more permeable deposits in all these valleys result from sorting and depositing of outwash material transported during glacial meltwater flooding. Today, sand and gravel resources are frequently developed from these river valley deposits. Alluvial materials range from 50 to 160 feet thick along the Mississippi and Missouri rivers, and they store enormous quantities of groundwater. Thinner, narrower alluvial aquifers along major interior rivers may be from 30 to 70 feet thick. In isolated places, the alluvium may be thicker where modern river valleys coincide with older, deeper buried valleys cut into the underlying bedrock surface. Characteristically, alluvial deposits are thicker, coarser, and more productive in a downstream direction. Such deposits also tend to thin out near the valley sides and grade into finer grained, less productive sediments.

Groundwater and surface water interactions are closely tied in alluvial settings. As mentioned earlier, alluvial aquifers are unconfined; that is, they have no overlying impermeable layer to seal them under pressure, leaving the aquifer's water surface free to fluctuate up and down. The familiar backwater lakes seen along river *floodplains* are actually low places where the water table (groundwater surface) intersects the land surface. Lake Odessa at Mark Twain National Wildlife Area in Louisa County, Otter Creek Marsh State Wildlife Refuge in Tama County, Blue Lake at Lewis and Clark State Park in Monona County, and DeSoto Bend National Wildlife Refuge in Harrison County are good examples of these "windows" to the water table. These oxbow lakes fill low areas scoured by earlier cut-off meanders of the river channel.

A rising river level will cause a simultaneous rise in the water table of an adjacent alluvial aquifer, sometimes to visible heights in nearby fields. Similarly, a falling river stage will produce a falling water table, and shallow lakes may disappear. The subsurface gradient of the water table slopes toward the river during falling river levels and away during rising river levels. The continuous contribution of groundwater to streams, known as baseflow, keeps them flowing even after long periods without rainfall.

Most of the early Euro-American exploration of Iowa took place along its rivers, and many of Iowa's first settlements were nestled along river banks. Today many of Iowa's major metropolitan centers are along rivers. Consequently, Iowa has numerous municipal wells tapping the alluvial deposits beneath valley landscapes. These provide valuable, as well as vulnerable, water supplies for large numbers of urban-dwelling Iowans. Wells along the Mississippi and Missouri valleys are known to produce 1,000 to 2,000 gallons per minute (gpm) on a sustained basis for municipal, industrial, and irrigation uses. Alluvial aquifers along Iowa's interior streams commonly produce several hundred gallons per minute, with larger yields possible in some localities.

Buried-valley aquifers

A second type of surficial aquifer occurs along ancient river valleys once carved into the underlying bedrock surface across Iowa (map, right). Like alluvial aquifers, these deeper sources are composed of sand and gravel, but they are buried now by younger impermeable glacial tills, thus confining the aquifers and creating artesian pressures. These old valleys were partially filled with *outwash* sand and gravel prior to or between glacial episodes. The presence of these steep-sided valleys is easily recognized on topographic maps of the bedrock surface, but they usually show no visible expression on the modern landscape because they are blanketed by glacial drift. It is worth noting, however, that in places they do influence the routes taken by modern drainage. A number of Iowa rivers have segments that follow the course of one of these deeper buried valleys.

The glacial mantle above most *buried-valley aquifers* also keeps them from being recharged as quickly as the shallow alluvium along modern streams. If these buried valleys are filled with coarse alluvium, it is possible for wells to pump 100 gpm or more. Where the

Buried-Valley Aquifers of Iowa

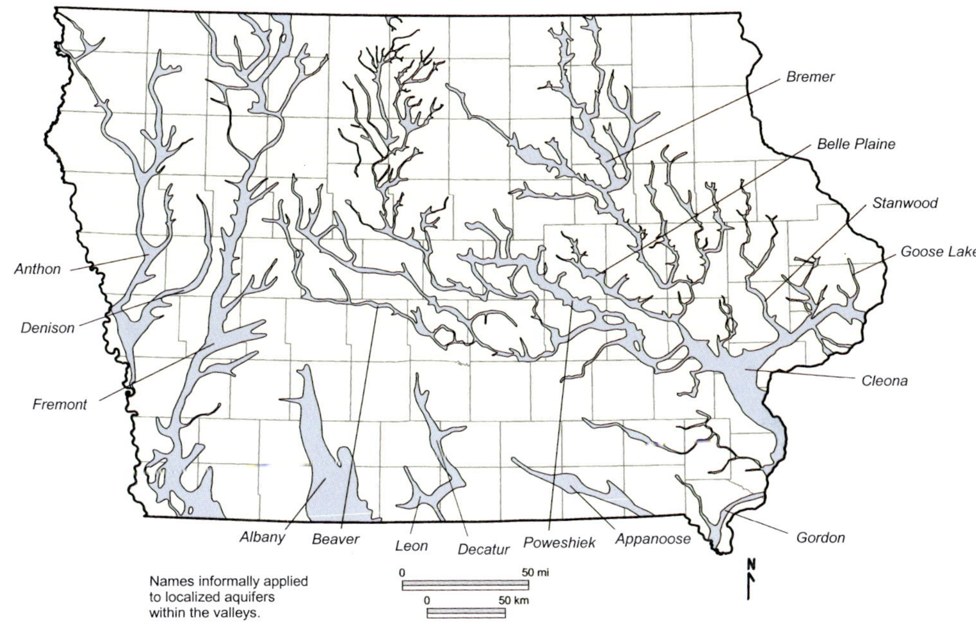

aquifers occasionally coincide with modern river valley deposits, yields of 500 to 1000 gpm are possible. The most widely used buried-valley aquifers occur in central and southeastern Iowa, while a few others are tapped in the western and southwestern part of the state. Iowa's buried valleys continue to be mapped in better detail as more data from new wells becomes available.

A buried-valley aquifer along the Belle Plaine Channel through Tama, Benton, and Iowa counties made history in 1886. The drilling of a new well for fire protection in Belle Plaine tapped buried sand

and gravel. Flows estimated at 30,000 to 50,000 gallons per minute gushed from the well, nicknamed "Jumbo," hurling chunks of fossil wood, two-pound stones, and tons of sand. Although the flow diminished to about 2,000 gpm after two weeks, efforts to stem the flow ultimately took over 13 months (photo, p.1).

Water quality in buried-valley aquifers is quite variable and is more highly mineralized than in shallow alluvial aquifers. This is because of the longer time that water has been in the ground and often because of its contact with an adjacent bedrock aquifer. Buried-valley aquifers tend to have high concentrations of total dissolved solids, with iron and ammonia being particular problems.

Glacial drift aquifers

A third type of surficial aquifer occurs widely beneath the broad upland landscapes that separate most of Iowa's river valleys. These are known as *glacial drift aquifers*. The term "drift" refers to the glacial deposits that mantle Iowa's deeper bedrock foundation. Though ranging from nearly zero to over 600 feet thick, glacial drift averages about 200 feet in thickness over the state. Drift includes not only pebbly clays known as *glacial till,* but also occasional sand and gravel deposits sandwiched within or at the bottom of these glacial clays. These porous sands and gravels, resulting from episodes of meltwater flow during glacial advances and retreats, tend to be thin, discontinuous, and quite unpredictable in their occurrence.

The suitability of drift aquifers as a source of groundwater varies considerably across the state. Some productive aquifers occur in north-central Iowa, where geologically youthful glacial materials lie at the land surface and contain abundant sand. Wells in this area report yields from less than 10 to over 90 gpm. In contrast, glacial drift is generally absent in the far northeastern Iowa counties. In eastern Iowa, the drift deposits are often bypassed in favor of deeper, more productive bedrock aquifers. In parts of southern and western Iowa, however, where alluvial aquifers may not be available and bedrock aquifers are deep, unproductive, or produce water of poor quality, a good glacial drift aquifer can be a life-saver. Such wells typically range from 50 to 200 feet deep but occasionally reach depths of 400 feet or more. They normally yield only a few gallons per minute, but a respectable 10 to 20 gpm may be obtained from some under favorable conditions. Though water production is low, their widespread potential in the abundant glacial deposits across the state make these aquifers valuable resources, especially for rural homeowners and especially in the western half of Iowa. In terms of natural water quality, shallow drift aquifers are consistently less mineralized than water from either deeper glacial drift aquifers or from the buried valley aquifers.

Falling in a special category are the interesting but enigmatic "salt and pepper" sand deposits known to occur across broad bedrock uplands of western Iowa. The term "salt and pepper" refers to contrasting grains of white quartz and dark volcanic glass, and these sands lie buried beneath 50 to 300 feet of glacial deposits. Although the sands' geological origins are uncertain, they appear to be derived from Rocky Mountain sources and transported to Iowa by eastward flowing ancestors of the Missouri River. The water-bearing potential of these sand deposits is not well known, but locally in northwestern Iowa they do generate moderate amounts of water.

A delicate balance of groundwater seepage from glacial drift

photos by Clay Smith

The permeability of soil and rock affects the movement of groundwater. Slow seepage to the land surface over long periods of geologic time produces fascinating ecological habitats. Hanging Bog State Preserve in Linn County is a forested fen noted for its abundance of heat-generating skunk cabbage (left).

aquifers is one of several hydrologic conditions that give rise to fens, a unique form of wetlands in Iowa (photos above). These spongy, groundwater-saturated peat deposits are only a few acres in size at most, and form low mounds on hill slopes and stream terraces, primarily across the northern portion of the state. Their water is highly mineralized compared to most wetlands, and as a result creates habitats for some of Iowa's rarest plants and insects. The flow of groundwater to the surface must be steady enough to prevent decay of plant material and enable peat accumulation, a process that takes several thousand years. If the flow is too strong, erosion prevails. Rowley Fen in Buchanan County and the Excelsior Fen complex in Dickinson County are fine examples of these groundwater-fed wetland habitats.

All of Iowa's wetlands, whether sustained by groundwater or surface water, are valuable hydrologic features of our landscape. They improve water quality by natural filtration; they reduce flood risks by storing water; they recharge groundwater supplies; and they provide ecological niches for diverse flora and fauna.

Loess-till contact aquifer

In many areas of southern and western Iowa, the surficial aquifers mentioned thus far are not available to residents, and often there are no bedrock aquifer options either. An alternate, but usually meager, water supply can be obtained from large-diameter seepage wells that tap the water table itself. Across this part of Iowa, loess often mantles the glacial till. Loess, being more porous, permits water to percolate easily downward until reaching the less permeable clayey till. Groundwater may be available to a seepage well near this geologic contact either when the water table has risen into the loess, or where groundwater moves easily along the contact's gradient. The availability of this shallow aquifer is affected by landscape position and by position of the seasonal water table.

A large-diameter well (typically 30 to 36 inches) intercepting groundwater near the loess-till contact yields about 1 to 3 gallons per minute. Such wells are most common in upland areas between major stream valleys. Recharge from precipitation and snowmelt occurs quickly to loess-till contact aquifers, and the natural water quality is quite good, given the brief period of contact with surrounding geologic materials. Obviously however, these wells are highly vulnerable to seasonal variations in the precipitation on which they depend, and also to contaminants infiltrating from the land surface.

The loess-till contact (photo, right) is often seen in roadcuts, especially across the southern half of Iowa. Look for places where a uniform-textured silt (loess) lies on top of more dense clayey material, usually with noticeable stones and cobbles mixed in (glacial till). Also, it is not unusual in this region to encounter hillside seeps

Ray Anderson

Infiltration of groundwater is affected by changes in geologic materials. A significant geologic contact in Iowa is shown here between loess (upper brown silt) and underlying glacial till (begins with terra cotta color). The clay-rich till includes numerous pebbles, is strongly rilled, has a paleosol at the top, and contains oxidized (yellow) and unoxidized (grey) zones. (Des Moines County)

These forested bluffs (left horizon) mark the leading edge of the Silurian bedrock aquifer across northeast Iowa. Recharge to this aquifer occurs across broad areas by infiltration through overlying geologic materials. The outcrop belt, known as the Silurian Escarpment, is a discharge zone for local flow systems within the aquifer and includes numerous large springs. (Clayton County)

Paul Horick

or perennial wet spots caused by groundwater movement along the loess-till geologic contact. Such wet, sticky areas are a familiar nuisance to local farmers driving heavy machinery across fields in the spring. Some of these seepage areas are textured with raised grassy tufts and unusual plants, particularly in uncropped fields.

In other places, especially uplands bordering streams and river valleys, the loess may be layered with blow-sand. Such fine-grained sand, carried by the wind, was supplied locally from nearby valleys. Being a highly permeable material in upland landscape positions, blow-sand may also contribute to the presence of hillside groundwater seeps and springs, especially in the southern half of Iowa. Hanging Bog in Linn County is a good example (photo, p.21).

BEDROCK AQUIFERS

Bedrock aquifers are another basic class of Iowa's groundwater resources. Bedrock consists of sedimentary rock layers, primarily limestone and dolomite (carbonate rocks), as well as shale, siltstone, and sandstone. These rocks originated as sediment deposited in ancient seas, rivers, and deltas that occupied Iowa between 75 million years ago (Cretaceous age) and 550 million years ago (Cambrian age). The total thickness of these formations ranges from 5,200 feet in southwest Iowa to about 800 feet in the northeast. This sequence of sedimentary rocks varies widely in its ability to store and transmit water.

Of the state's sedimentary rocks, sandstones and carbonates make the best aquifers. Sandstones often have abundant pore spaces between their grains, so groundwater moves easily following the interconnected openings. Problems with water yields from these

Brian Witzke

Massive dolomite of the Silurian aquifer forms scenic outcrops along the Cedar River at Palisades-Kepler State Park in Linn County.

Brian Witzke

Robert McKay

Brian Witzke

The Dakota Sandstone (Cretaceous aquifer) outcrops along the Middle Raccoon River in Guthrie County.

aquifers occur when there is an increase in mineral cements, which can clog the intergranular pore space. In contrast, carbonate rocks – limestone ($CaCO_3$) and dolomite ($CaMgCO_3$) – are finer grained, and groundwater actually follows vertical fractures, crevices, caves and other solutional openings as well as horizontal bedding planes (thin partings that separate sedimentary rock layers). Groundwater movement along these permeable pathways is especially noticeable at roadcuts in winter, when emerging groundwater forms frozen waterfalls (upper photo, p.9).

Solutional openings are peculiar to lime-rich carbonate rocks and result from the chemical dissolving action of circulating ground-

Middle left: Devonian limestone exposed in this Chickasaw County quarry is part of the Silurian-Devonian aquifer.

Near left: A major unit of the Cambrian-Ordovician aquifer is the colorful Jordan Sandstone, seen in bluffs at McGregor in Clayton County.

Near right: Blackhawk Spring issues from the Mississippian aquifer at Crapo Municipal Park near Burlington.

Far right: Bluffs of Prairie du Chien dolomite in Allamakee County are part of the Cambrian-Ordovician aquifer.

Ray Anderson

Brian Witzke

water. Such subterranean drainage accounts for sinkholes at the land surface, streams that disappear, caverns underground, and springs on hillsides – landscape features known as *karst* topography (photos, p.2, 3, above, 58). Other types of porosity in Iowa's carbonate rocks occur where fossiliferous zones have been dissolved out, or where the rock has been broken (*brecciated*) in the geologic past, or even between mineral crystals, particularly in the Silurian dolomites. Because these openings in limestone and dolomite are so irregular in occurrence, size, and orientation, yields to wells can be highly variable, even over very short distances.

The distribution of major bedrock aquifers and aquitards beneath Iowa is shown in three different cross-sectional views on pages 27-29. The general slope of these bedrock formations averages 15-18 feet per mile to the southwest, toward geologic basins centered in Kansas and Oklahoma. Note that these inclined strata are beveled by erosion across their upper leading edges, which affects their outcrop patterns across the state's bedrock surface (map, p.40). These conditions are the key to forecasting groundwater availability and quality throughout the range of each aquifer, and to explaining the artesian nature of most bedrock aquifers in Iowa.

For example, where a carbonate bedrock aquifer is the uppermost bedrock unit, natural water quality is at its best. This is where

Groundwater Zones of the Manson Crater

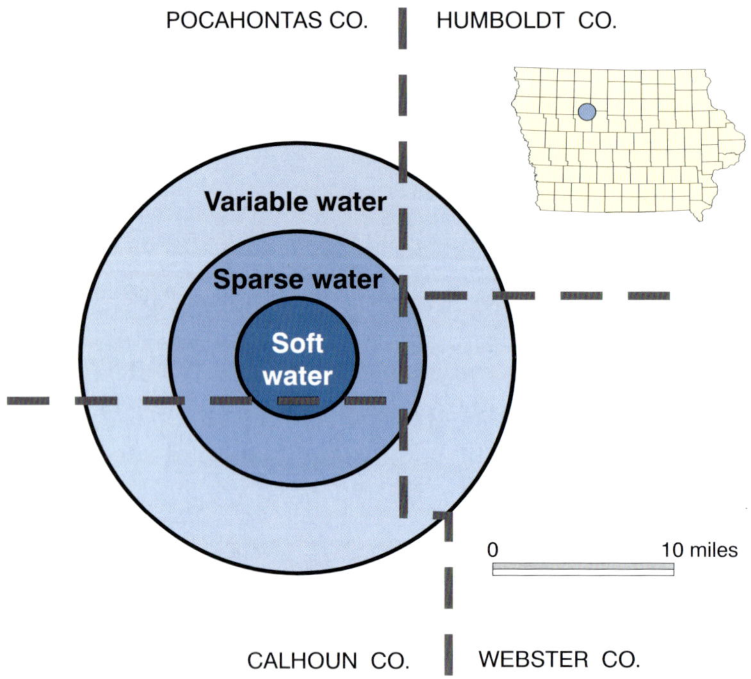

the aquifer received direct precipitation and weathering (prior to glaciation), as well as rapid groundwater circulation and extensive enlargement of rock openings by solutional activity. And now, even with a glacial mantle, the aquifer is still more readily recharged than where it is covered by other bedrock units. The result is much better naturally occurring water quality. The geographic region of greatest use of an aquifer coincides with this outcrop or *subcrop* area.

On the other hand, where a bedrock aquifer lies deeper and is overlain by younger bedrock units, natural water quality usually is poorer and use of the aquifer declines. These basic relationships will become more apparent as individual bedrock aquifers are discussed.

Though the deepest groundwater aquifers in the state are contained within this package of sedimentary rocks, the geological story does not end here. Thousands of feet of even more ancient, crystalline rock underlie all of Iowa's sedimentary strata. These so-called basement rocks are 1 to 2 billion-year-old igneous and metamorphic rocks of Precambrian age. In the extreme northwest corner of Iowa, the state's sedimentary rocks end abruptly against an ancient high ground known as the Sioux Ridge, which is composed of 1.6 billion-year-old Sioux Quartzite, a distinctly reddish silica-cemented sandstone. These rocks can be seen at Gitchie Manitou State Preserve in Lyon County.

Another unusual rise of ancient basement rocks in Iowa occurs in Pocahontas and Calhoun counties beneath the Manson Impact Structure (map, left). The sudden impact of a large meteorite 74 million years ago produced massive disruption of the Cretaceous and underlying sedimentary strata, and caused deep-lying igneous and metamorphic basement rocks to rebound to the land surface. The impressive 24-mile wide crater is now concealed from view by 100 to 200 feet of later-arriving glacial deposits. During the early 20th century, water well drillers reported finding unusually soft water from shallow fractured granite, which first brought this fascinating structural feature to geologists' attention. Although good quality water is obtained from the crater's 5-mile-wide granite core, the chaos of deformed strata surrounding it makes finding groundwater in the remainder of the area a "poke and hope" situation at best.

Bedrock Aquifer Systems across Iowa

Northwest to Southeast

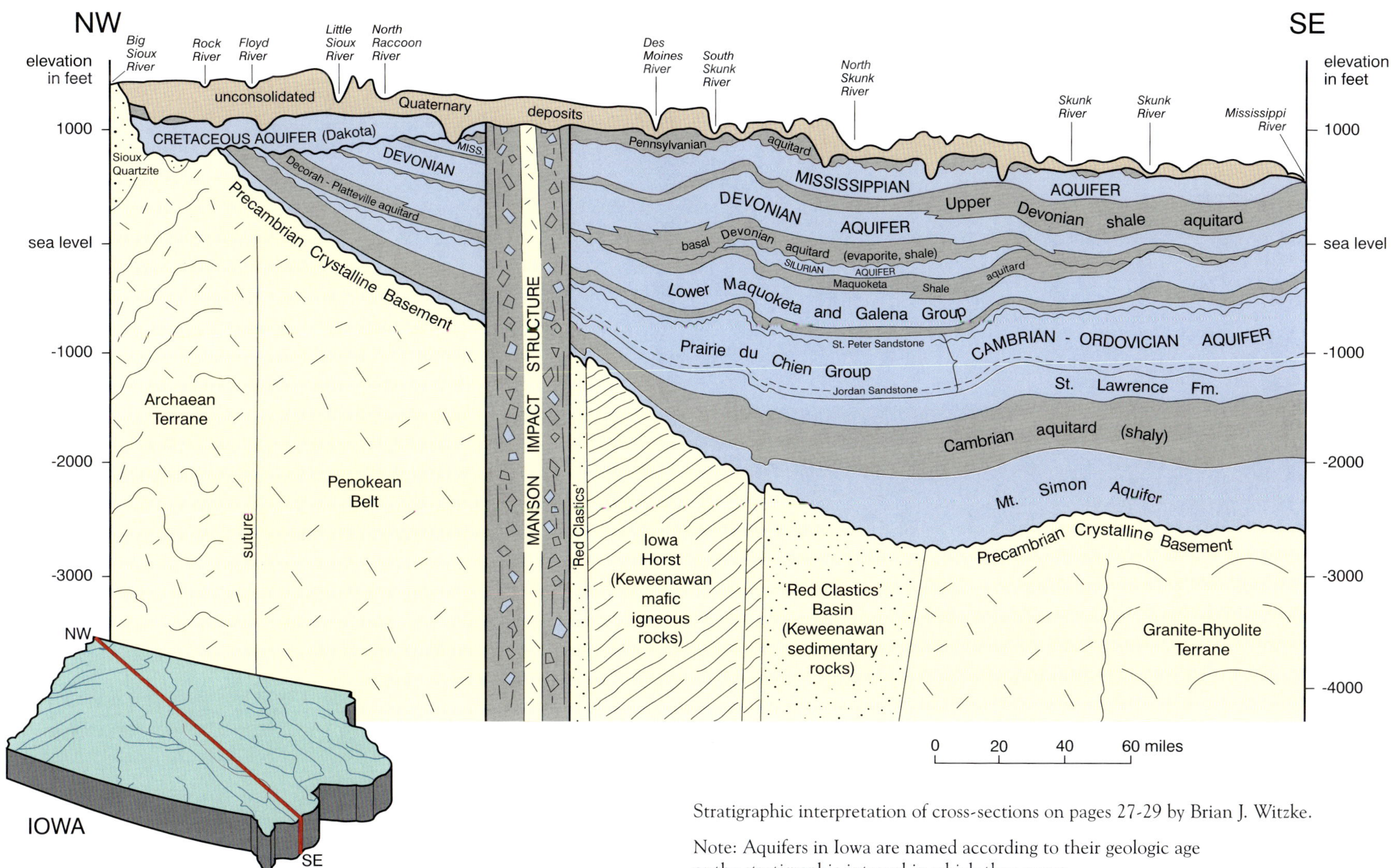

Stratigraphic interpretation of cross-sections on pages 27-29 by Brian J. Witzke.

Note: Aquifers in Iowa are named according to their geologic age or the stratigraphic interval in which they occur.

Bedrock Aquifer Systems across Iowa

Southwest to Northeast

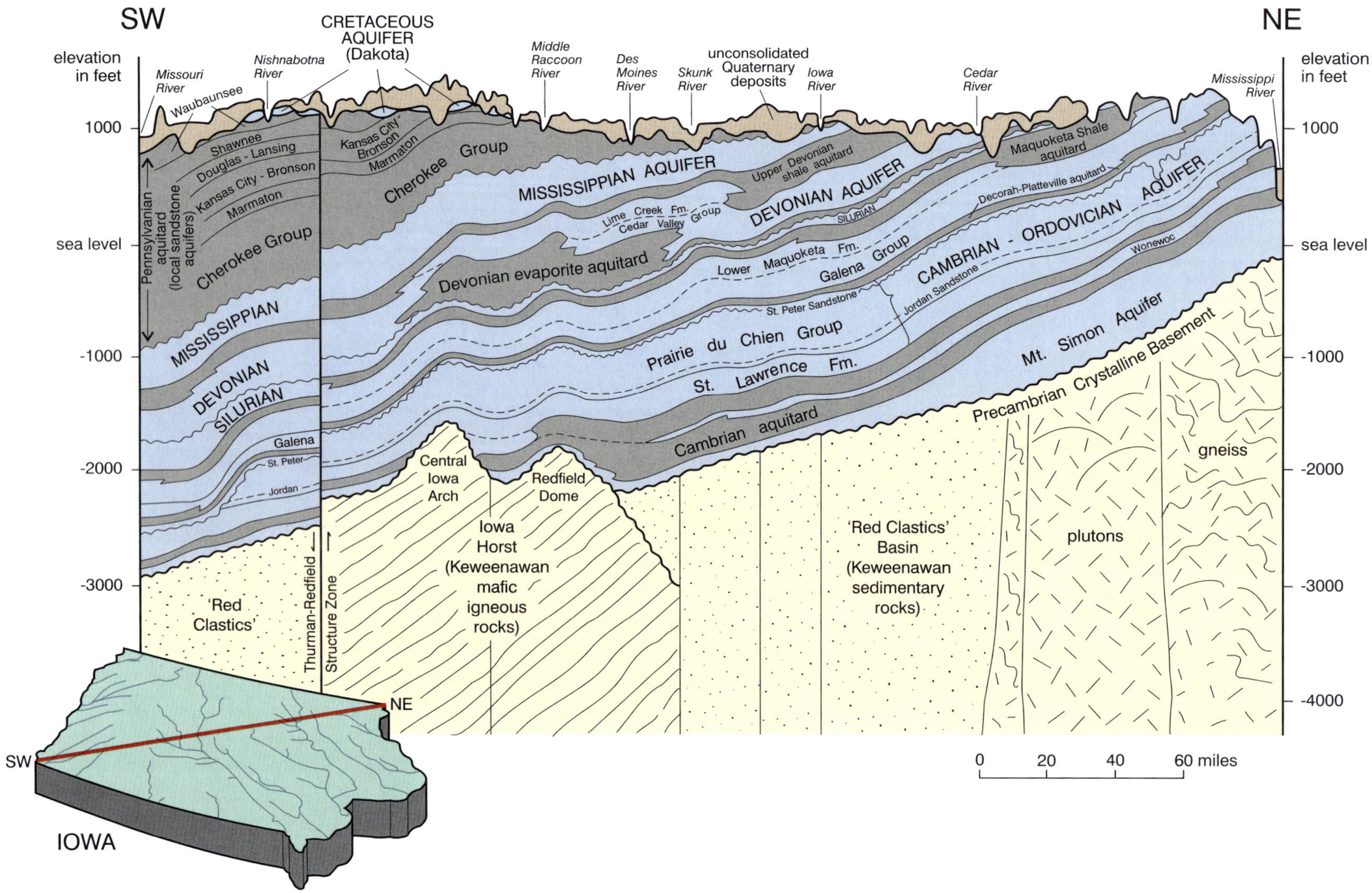

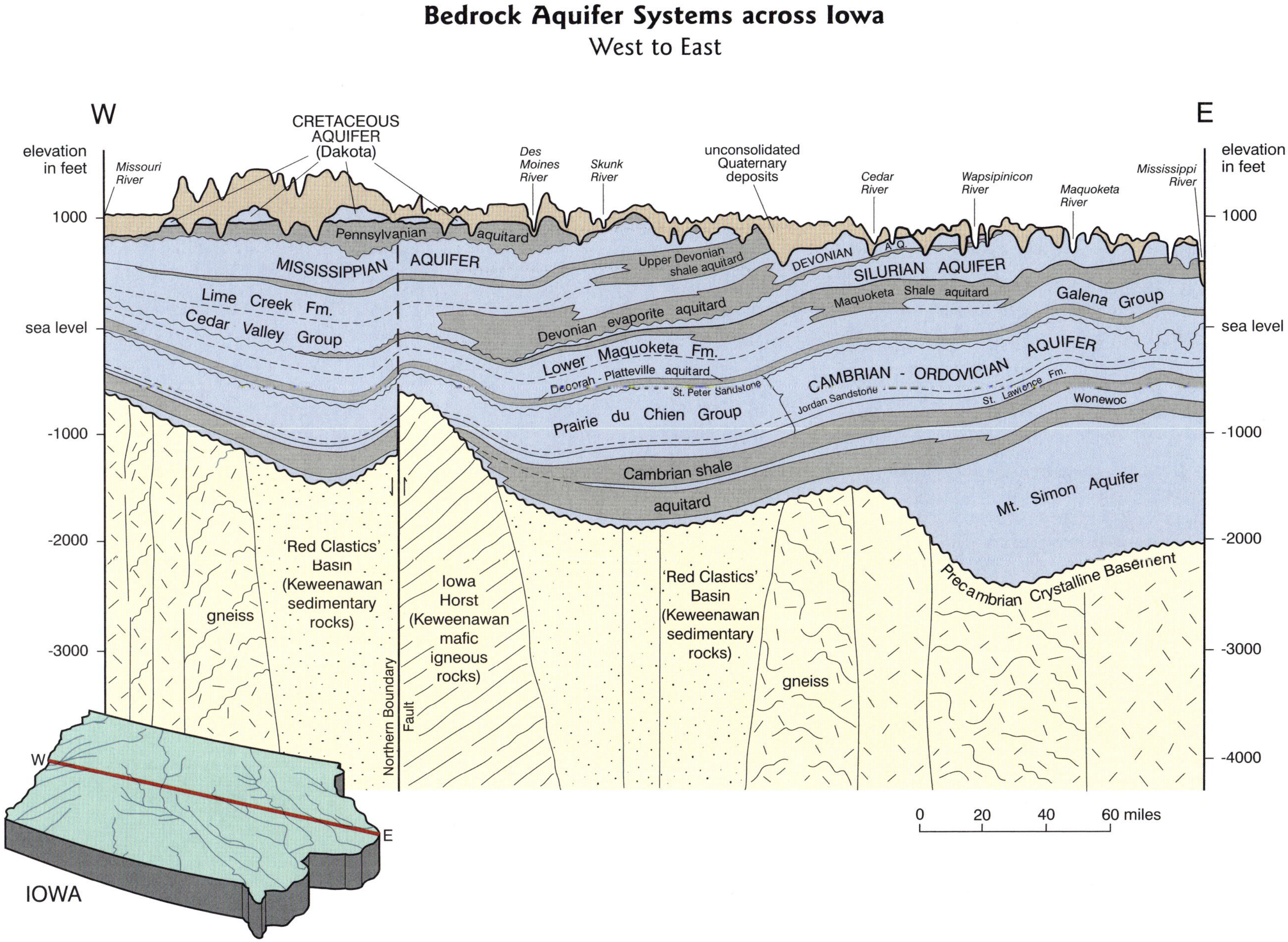
Bedrock Aquifer Systems across Iowa
West to East
W
E
elevation in feet
CRETACEOUS AQUIFER (Dakota)
Missouri River
Des Moines River
Skunk River
unconsolidated Quaternary deposits
Cedar River
Wapsipinicon River
Maquoketa River
Mississippi River
1000
sea level
-1000
-2000
-3000
-4000
Pennsylvanian aquitard
MISSISSIPPIAN AQUIFER
Upper Devonian shale aquitard
DEVONIAN
A. Q.
SILURIAN AQUIFER
Lime Creek Fm.
Cedar Valley Group
Devonian evaporite aquitard
Maquoketa Shale aquitard
Galena Group
Lower Maquoketa Fm.
Decorah - Platteville aquitard
St. Peter Sandstone
CAMBRIAN - ORDOVICIAN AQUIFER
Jordan Sandstone
St. Lawrence Fm.
Wonewoc
Prairie du Chien Group
Cambrian shale aquitard
Mt. Simon Aquifer
'Red Clastics' Basin (Keweenawan sedimentary rocks)
gneiss
Iowa Horst (Keweenawan mafic igneous rocks)
Northern Boundary Fault
Precambrian Crystalline Basement
0
20
40
60 miles
IOWA

Dakota Aquifer of Iowa

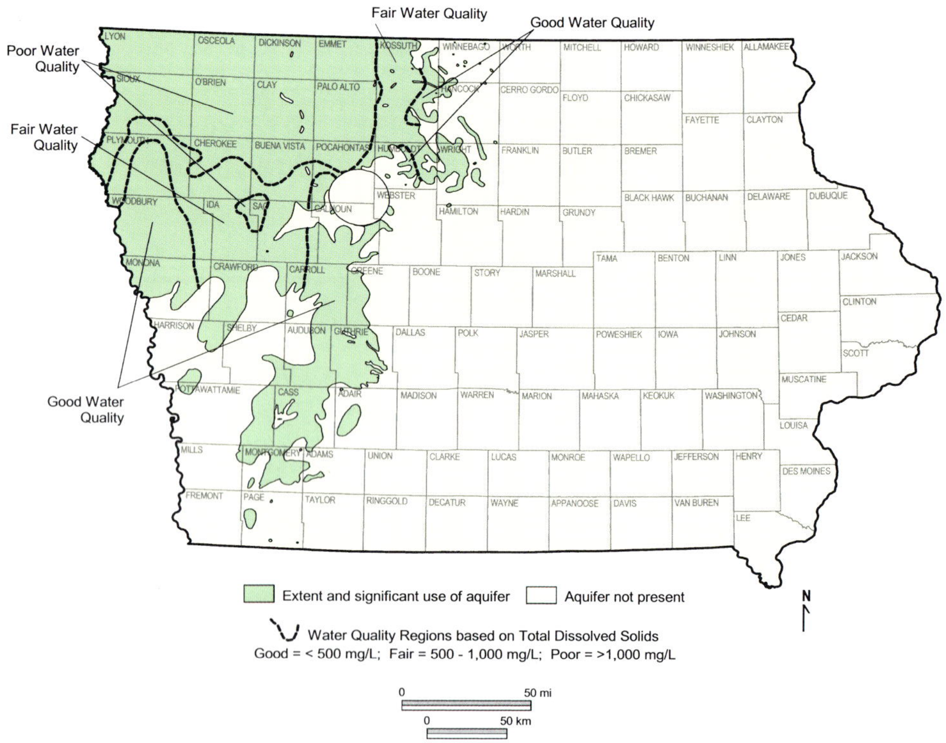

Dakota Aquifer (Cretaceous age)

The Dakota aquifer is composed of sandstone deposits 200 to 300 feet in thickness, and provides water for many rural and public water supplies in northwest and west-central Iowa (map, above). This aquifer is composed of the Woodbury and underlying Nishnabotna sandstones, which formed in riverine environments 100 million years ago. The sandstones range in grain-size from very coarse to fine, and they are poorly cemented, providing abundant pore space for groundwater storage. Over 80 percent of the Nishnabotna alone is composed of medium- to coarse-grained quartz sandstone. The sandstones are confined over most of their area by 200 to 400 feet of clay-rich glacial till as well as by thick shale, siltstone, thin chalky limestone, and lignite (low-grade coal).

Wells drilled into the Dakota range from 100 to 600 feet deep and yield 100 to 500 gpm. Some wells can produce as much as 800 to 1,500 gpm. Water quality tends to be fair to poor. The areas of poorer water quality result from high concentrations of dissolved solids (between 500 and 3,000 mg/L), particularly sulfate and calcium carbonate, which are common minerals picked up by groundwater in contact with the confining layers above. Areas of better water quality are found where confining layers are thin and some porous pathways allow for more rapid recharge by unmineralized surface waters. This occurs most notably in the southern area of the aquifer's extent.

The Dakota aquifer in Iowa is recharged by downward percolation through its confining units. Regional groundwater flow is from north to south, and natural discharge from the aquifer occurs into the lower reaches of major rivers in the region.

Views of the Dakota aquifer at the land surface are few and far between. Some of the best, however, are seen in Guthrie County along road cuts and outcroppings in the vicinity of Springbrook State Park and along the Middle Raccoon River valley (photo, p.24). The Dakota sandstone often displays a rust-colored coating formed by oxidation of iron-bearing minerals within the formation.

Mississippian Aquifer of Iowa

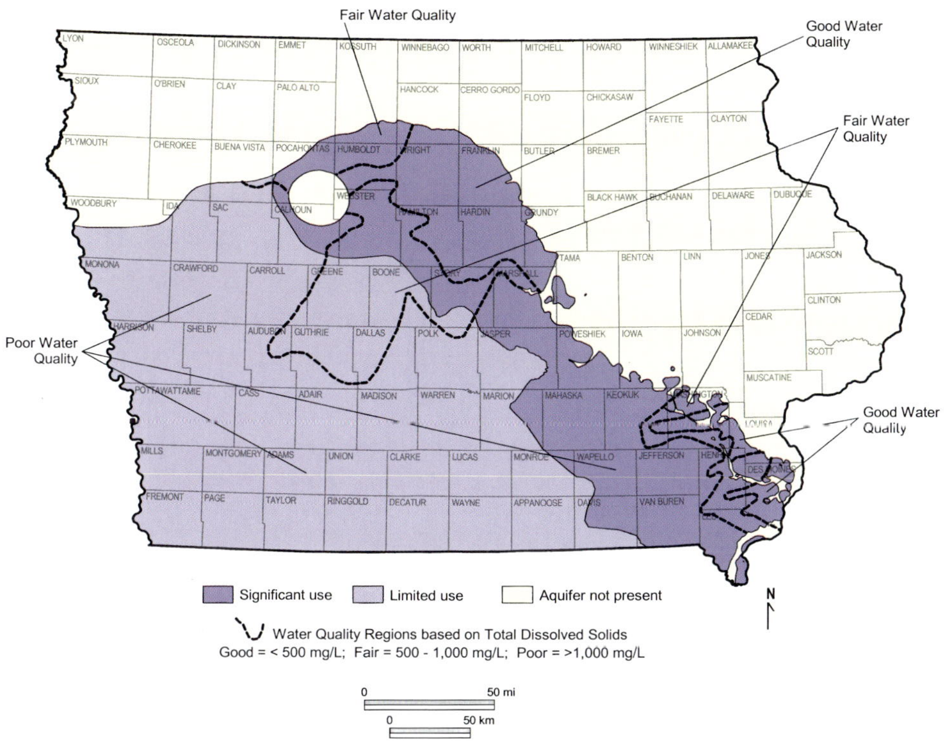

Mississippian Aquifer

Productive wells from the Mississippian aquifer supply private and public water supplies for much of the north-central part of the state (map above). The water quality is generally good. In contrast, the same aquifer produces much smaller yields of poorer quality water in central and southeastern Iowa. Throughout its range the aquifer consists of a thick sequence of limestone and dolomite, with thinner deposits of sandstone, shale, the silica-rich mineral chert (flint), and gypsum.

Along the outcrop belt of these rocks, the Mississippian aquifer is overlain by alluvium, loess, and glacial drift, while elsewhere the aquifer is first overlain by thick shale and lesser sandstone units (Pennsylvanian). Since recharge to the aquifer is primarily by infiltration from the land surface, this contrast in materials has a direct effect on the aquifer's water quality. For example, groundwater quality is very good in the north-central *subcrop* area, where dissolved solids concentrations are less than 500 mg/L. Less favorable conditions are present in the central and southeast parts of the area where Mississippian strata have fewer creviced openings and downward recharge is limited by the less permeable Warsaw shale. In these areas, domestic wells may produce only 5 to 10 gpm, and often less than that.

Like the bedrock aquifers beneath it, the entire Mississippian aquifer has a gradual overall slope to the southwest, so that hundreds of feet of younger Pennsylvanian strata and glacial clays overlie the aquifer in south-central and southwestern Iowa. Where these thick Pennsylvanian shales overlie the Mississippian aquifer, high concentrations of sodium, fluoride, and sulfate limit the aquifer's use as a drinking water source, even though yields of 10 to 50 gpm are sometimes possible. In western and southern Iowa the total dissolved solids content ranges from 1,500 to 5,000 mg/L, making the water unsuitable for human and livestock use.

The most productive wells from the Mississippian aquifer are concentrated in north-central Iowa where excellent yields of 500 to

900 gpm are known from municipal wells in Marshall, Story, Hardin, and Wright counties. These high yields coincide with an area of well-developed karst, where the limestone is highly dissolved. Typical well depths range from 100 to 300 feet and commonly yield from 50 to 100 gpm. In central and southeast Iowa, however, yields are as low as 2 to 3 gpm, owing to the scarcity of crevices in the rock formations. Well depths here extend about 100 to 400 feet below land surface.

Regional flow in the Mississippian aquifer is in a southerly direction, and it discharges into the Des Moines and Skunk rivers and their tributaries.

The Mississippian aquifer can be observed as picturesque limestone bluffs along the Mississippi River valley (its namesake), especially between Burlington and Keokuk in southeastern Iowa. In addition, the shallow carbonate aquifer here displays good examples of karst features, including sinkholes within the Burlington city limits, a well-developed solutional cavern at nearby Starr's Cave State Preserve, and Blackhawk Spring at Crapo Park (photo, p.25). Other natural exposures of Mississippian aquifer rocks are seen along river valleys and creek banks as well as roadsides and quarries from Lee, Des Moines, and Louisa counties at the southeast corner of the state to Humboldt and Hancock counties in north-central Iowa. Scenic Lacey-Keosauqua State Park in Van Buren County and Three Bridges County Park along the Iowa River in Marshall County, as well as limestone quarries near Le Grand (famous for their fossil crinoids) are in Mississippian carbonate rocks. Visitors can drink from a flowing artesian well at Benson Park, along Hwy. 3 just west of Clarion in Wright County. This well taps the Mississippian aquifer at a location of considerable artesian pressure. Mineralized groundwater from the Mississippian aquifer at Colfax in Jasper County was thought to have therapeutic effects, and so the ornate Hotel Colfax was built nearby in 1884 as a health spa (photo, p.1).

Silurian-Devonian Aquifer

This aquifer is composed of bedrock units from two major systems of geologic time, the Silurian and the Devonian. They are usually described as a single aquifer package because, regionally, they are composed of similar carbonate rocks that are hydraulically connected and yield similar water quality. Locally, however, there are differences in terms of rock types and water quality, and thus these two units are sometimes considered independently. For example, in eastern Iowa they are two discrete aquifers. In some places, there is no underlying Silurian, and in other places there is no overlying Devonian. And locally, shale and clayey dolomites act as aquicludes, further separating parts of the aquifer system.

This aquifer underlies the entire state except for the northwest and northeast corners (map, right). It is an important source of water primarily in eastern and northern Iowa, serving rural, public, and industrial users. The aquifer is composed mainly of porous dolomites of Silurian age, and limestone and thick shales of Devonian age. As is typical of carbonate rocks, porosity and permeability are dependent on such natural openings as fractures, *brecciated* zones, bedding planes, and solution caverns. These features vary in size, extent, and frequency of occurrence. This aquifer's porosity and permeability are best developed in a broad band about 65 to 70 miles wide and about 200 miles long across the northeastern part of the state (note area of "significant use," above right).

Silurian-Devonian Aquifer of Iowa

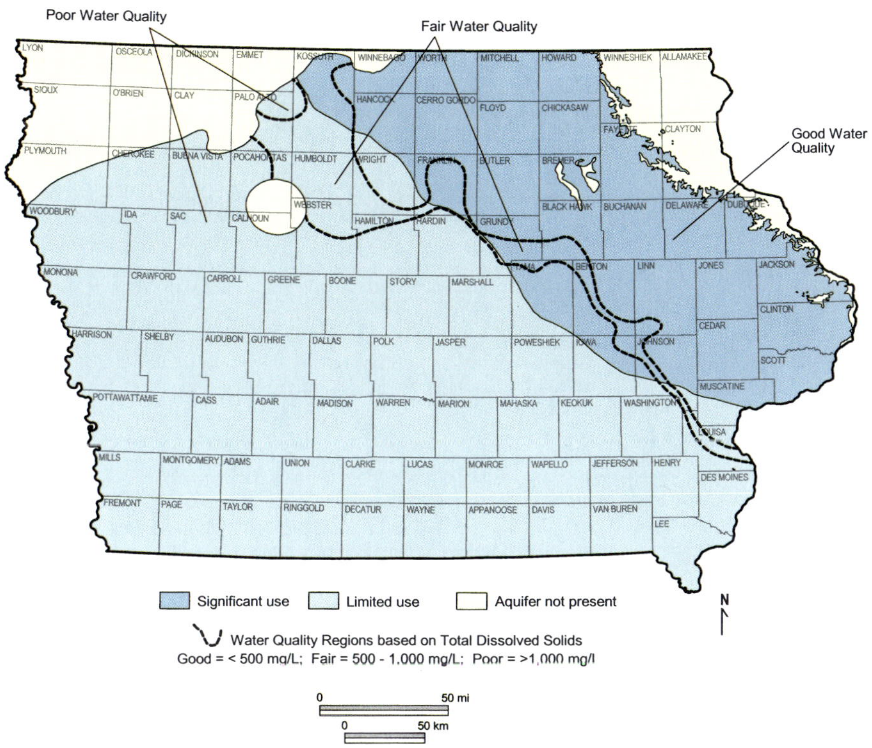

The typical thickness of the Silurian-Devonian aquifer in eastern and northern Iowa, its principal area of use, ranges from 200 to 400 feet (diagram, p.29). The aquifer reaches its greatest thickness, 500 to 700 feet, in the southwest part of the state. Most wells tapping the Silurian-Devonian aquifer are 100 to 700 feet deep, and many municipal and industrial wells yield 150 to 400 gpm. Domestic wells can deliver from 10 to 30 gpm, as long as the wells intercept a good fracture system. Interestingly, one of the most productive zones of any bedrock aquifer in Iowa occurs in a narrow zone along the Cedar River valley from Charles City to Waterloo, an area underlain by highly creviced and cavernous Devonian limestones. Yields of 2,000 to 4,000 gpm are known from wells at Cedar Falls, Waterloo, Waverly, and Charles City. The Silurian-Devonian aquifer in parts of this area receives induced recharge from the Cedar River, infiltrating first through a thin layer of alluvial sand and then through cavernous limestone.

Water quality from this aquifer is generally good in eastern and northern areas of the state, with a dissolved solids concentration between 300 and 500 mg/L. This good natural quality, however, decreases rapidly over short distances to the southwest, especially where covered by overlying shales. The aquifer is used much less in the central and southern parts of the state, where it includes thick deposits of gypsum and anhydrite in the Cedar Valley and Wapsipinicon rock units, as well as thick shales. Yields of 20 to 150 gpm can be obtained in these parts of Iowa, but high sulfate concentrations and total dissolved solids as high as 5,000 mg/L make the water unsuitable for human use. In western and southwestern Iowa, the aquifer is deeply buried beneath younger rocks, and the well-developed fracture systems so common in the eastern subcrop area are proportionately fewer and the yields smaller.

The regional flow of groundwater in the Silurian-Devonian aquifer is to the southeast. Natural discharge from the aquifer toward valleys contributes significantly to the amount of water flowing in the Iowa, Winnebago, Shell Rock, Cedar, and Maquoketa rivers. The aquifer thus serves as an important source of baseflow to these streams.

These river valleys are also good places to observe numerous scenic outcroppings of this aquifer (photo, p.24). Palisades-Kepler State Park in Linn County, Maquoketa Caves State Park in Jones County, and Backbone State Park in Delaware County are excellent and diverse examples of the Silurian aquifer. Fossiliferous Devonian rocks rim the margins of Coralville Lake near Iowa City, and are especially well exposed for visitors at Devonian Fossil Gorge in Johnson County. A portion of the Rockford Fossil and Prairie Center in Floyd County was the former site of the Rockford Brick and Tile quarry, which utilized the thick shale of the Upper Devonian aquitard.

Cambrian-Ordovician Aquifer

This is the only aquifer in Iowa that is widely used even where it is covered by younger bedrock units. The Cambrian-Ordovician aquifer is a widespread and dependable source of water for high-capacity wells, and it is used extensively by municipalities and industries in the eastern half of the state (map, right). The aquifer is composed of three separate formations – the Jordan Sandstone (Cambrian age) at the bottom, dolomite and sandstone of the Prairie du Chien Group in the middle (Ordovician age), and the St. Peter Sandstone at the top (diagrams, p.15, 27-29). All these bedrock units are water-bearing, so they usually are treated as a single aquifer. Wells that require large amounts of water typically tap the full range of the Prairie du Chien and Jordan units, while other wells may tap only the St. Peter or a combination of the St. Peter and upper Prairie du Chien. While often called the "Jordan aquifer," much groundwater also comes from the Prairie du Chien. Thus

Cambrian-Ordovician Aquifer of Iowa

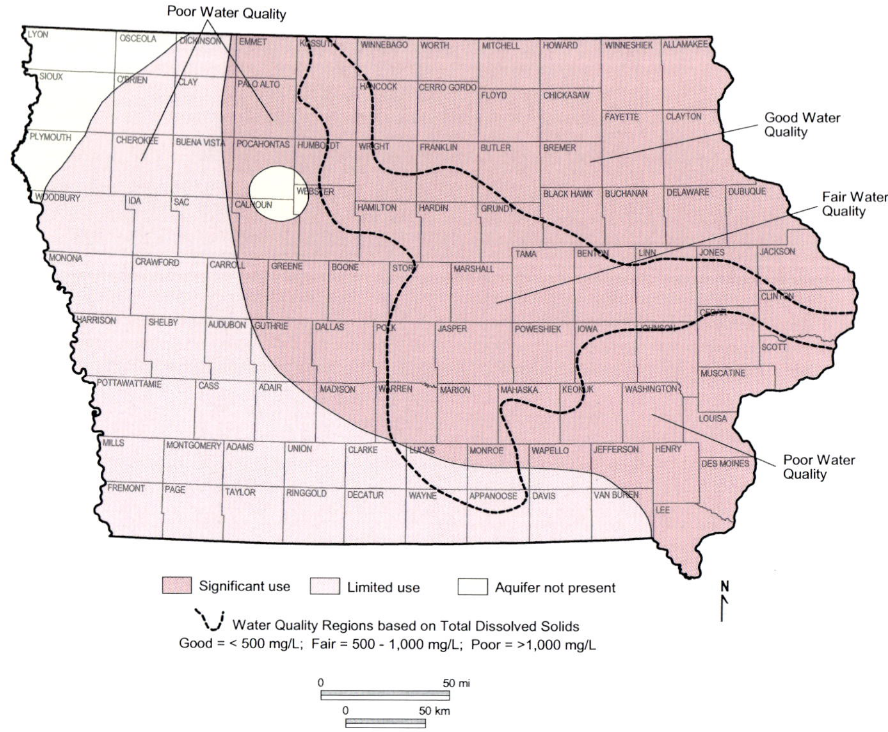

"Cambrian-Ordovician aquifer" is a more accurate term for this water-producing zone.

Typical well depths in this aquifer range from 300 to 2,000 feet, with some over 3,000 feet deep. As noted earlier, bedrock aquifers are inclined in a southwesterly direction toward Kansas and Oklahoma; thus, the deepest wells are in southwest Iowa. This depth range highlights another characteristic of water from this aquifer – its temperature. Since the natural temperature gradient within the earth increases with depth, water temperatures within the Cambrian-Ordovician aquifer range from 60 to 80 degrees F, and reach 85 degrees F in the 3,000-ft wells.

Wells that are properly developed in the Cambrian-Ordovician aquifer yield from several hundred to over 1,000 gpm. As with other sedimentary bedrock aquifers, yields are related to the amount of natural cements filling the pore space in sandstone, and to the presence or absence of fracture openings through denser limestone and dolomite. Water quality is generally good, with the best quality found in northeast Iowa, nearest the areas of outcrop and recharge. In western and southwestern areas, as the aquifer inclines deeper beneath the land surface, the dissolved mineral content increases to undesirable levels, with high sulfate, chloride, and fluoride in particular. In portions of southeast, central, and western Iowa, the aquifer sometimes contains excessive levels of radium and is treated before use in municipal water supplies.

The main area of aquifer recharge is in southern Minnesota and northern Iowa, via vertical leakage from overlying aquifers. Subsurface flow is then to the southeast, with discharge to the Mississippi Valley.

Historical use of the Cambrian-Ordovician aquifer by municipal and industrial pumping centers such as Ft. Dodge, Mason City, Des Moines, Washington, Fairfield, and Ottumwa caused the potentiometric surface of the aquifer to drop between 50 and 150 feet. Sharp recoveries occurred when these wells shut down, but they did not regain their original standing water levels. Despite this drop, the available water is not expected to be exhausted in the near future as the aquifer contains an estimated 80 trillion gallons of water. Local, smaller declines also are noted in the Silurian-Devonian aquifer.

Outcrops of Cambrian-Ordovician rocks produce some of the most spectacular scenery in Iowa, particularly from McGregor northward along the Mississippi Valley (photos, p.24, 25). The 480 to 510 million-year-old dolomites and sandstones that elsewhere comprise this aquifer dominate the landscape at such scenic points as Yellow River State Forest, Effigy Mounds National Monument, and Mt. Hosmer City Park at Lansing, all in Allamakee County.

OTHER GEOLOGIC UNITS

Galena Aquifer

The Galena Group dolomites are a much more restricted source of groundwater used locally in northeastern Iowa, especially in Winneshiek, Clayton, northeastern Fayette, southwestern Allamakee, and the eastern two-thirds of Dubuque counties (photo, p.12). In this area of primary use, the Galena is overlain by the younger Maquoketa Shale which serves to protect the shallow Galena aquifer from surface contamination problems. Most wells in this area yield between 10 and 30 gpm. Just to the east, in the area where the Galena strata are exposed, sinkholes and other karst features are abundant, and contamination from nitrate and bacteria is a significant problem. Locally, iron concentrations also can be a problem. Throughout the aquifer's geographic extent, groundwater flow is generally to the southeast, though local variations occur. Within its outcrop area, recharge occurs across the northeast Iowa uplands, and discharge occurs primarily to the Mississippi, Turkey, Upper Iowa, Volga, and Little Maquoketa rivers.

Brian Witzke

The fine-grained Maple Mill Shale forms part of the Upper Devonian aquitard in Iowa. These exposures occur along the English River in Washington County.

Dresbach Aquifer

Though the Dresbach (Mt. Simon) aquifer is a major water supply for parts of Minnesota, Wisconsin, and Illinois, this deep sandstone aquifer is important to only a few counties bordering the Mississippi River in northeast and east-central Iowa. For example, dependable supplies are tapped at New Albin, Lansing, Marquette, Dubuque, and Clinton. Because it lies beneath the more accessible Cambrian-Ordovician aquifer, it is not used as much, except for large industrial wells. Where tapped, however, it is a prolific source of water, especially near Clinton where pumping rates of 2,500 gpm are known. Quality problems within the aquifer's deep sandstone formations include salty water and high radium concentrations. Little is known about its water quality across the rest of Iowa.

Confining beds: aquitards

Bedrock units in Iowa not specifically highlighted as aquifers are referred to as aquitards and are noted in the tables on pages 14-15 and in the diagrams on pages 27-29. The importance of Iowa's aquitards, or confining beds, cannot be over-estimated. They are what define and separate aquifers from each other, and they provide the stratigraphic seals that place aquifers under artesian pressure. In general, these strata are thick, dense shale or siltstone units that inhibit vertical water movement, though they also may impart poorer water quality to adjacent aquifers. These aquitards include thick Cretaceous shales in western Iowa, Pennsylvanian shales across southern Iowa, Devonian shales in north-central Iowa, and Ordovician shales in northeastern Iowa. Within these broad aquitards, thin interbedded limestone and sandstone units occur and can sometimes be tapped as aquifers for local use. This is particularly true of Pennsylvanian deposits in central and southern Iowa, as well as Ordovician units in northeastern Iowa.

Because aquitards impede downward water movement, groundwater tends to flow laterally, along the paths of least resistance through more permeable aquifer materials. There still remains, however, a vertical component to this flow, and groundwater may very slowly discharge either upward or downward through aquitards, depending on the pressure gradients. This is how poorer natural water quality is imparted to some confined aquifers.

When shallower glacial drift aquifers were described earlier, we noted that their porous sands and gravels occur within more widespread fine-textured glacial till. It is this massive blanket of clayey glacial till that acts as an aquitard over much of Iowa's landscape, separating the land surface from underlying bedrock aquifers. These dense, low-permeability glacial deposits are the first line of defense in protecting Iowa's bedrock aquifers from sources of land surface contamination.

The glacial till aquitard varies considerably in thickness across the state. The thinner it is, the less protection it offers to underlying aquifers. Also, as impermeable as glacial till seems, it is known to be fractured in places. The open vertical "joints" are actually weathering features caused by dessication. These cracks, sometimes occurring in polygonal patterns similar to common mud cracks, formed as the post-glacial land surface was exposed to drying air. Fracture zones occur most often in the oxidized zone of glacial tills, and they can transmit water tens to hundreds of times faster than unfractured materials. In addition, these open pathways allow infiltrating water and contaminants to bypass the clayey matrix, to which some chemicals would otherwise adhere, and move easily through in an unfiltered and undiluted state to an aquifer beneath.

A final comment is warranted about well yields from the various aquifers described in this chapter. The figures given are from wells that are test-pumped in response to a specific, needed pumping rate. The true potential yields from many of these aquifers remain unknown.

IOWA'S GROUNDWATER PROVINCES

"[Groundwater's] importance does not need argument though it may need emphasis."

– Thomas C. Chamberlin
The Requisite and Qualifying Conditions of Artesian Wells, 1885

Brian Witzke

Each region of Iowa has a predictable pattern of available groundwater resources. The state's aquifers are mapped by examining outcrops like these and by studying samples and hydrologic data from drilled wells. (Silurian outcrops, Wapsipinicon River, Linn County)

Iowa's groundwater resources consist of five principal aquifers – four bedrock aquifers and an assortment of shallower sand and gravel deposits that overlie the bedrock and collectively are called surficial aquifers. Widespread confining beds retard movement of groundwater between the bedrock aquifers in particular.

The major bedrock aquifers extend over large geographic areas, well beyond Iowa's borders. These regional aquifers have been introduced as the Dakota (Cretaceous), the Mississippian, the Silurian-Devonian, and the Cambrian-Ordovician ("Jordan") aquifers. Each aquifer system includes several stratigraphic formations and water-bearing zones, as noted in the table on pages 14-15 and in the cross-sections on pages 27-29. Most of the groundwater flow in these aquifers can be described as deep, confined, and relatively unaffected by recharge from the surface. Because of the downward slope of the state's bedrock sequence toward the southwest, however, these aquifers do have areas of local flow where their upper edges lie closer to the land surface and where recharge occurs.

Each of these four bedrock aquifer systems serves as a primary groundwater source over some part of the state, or as a second option in other areas. As already noted, the smaller area of primary aquifer use coincides closely with its area of shallower depth and better natural water quality.

Now, with this overall framework of the bedrock and surficial aquifers in mind, it is possible to look at Iowa statewide and summarize its groundwater resource options as they occur in identifiable groundwater provinces. The general availability of groundwater (how accessible it is), plus its quantity and quality, form the basis for dividing Iowa into three districts: the Northeast, the Northwest, and the Southern Iowa groundwater provinces.

Northeast Iowa Groundwater Province

The Northeast Iowa Groundwater Province is rich in water resources and thus can be described as "good" (map, p.40). This region has several options for groundwater sources. Depending on location within the province, these options include the Mississippian aquifer, the Silurian-Devonian aquifer, and the Cambrian-Ordovician aquifer. Locally, the more restricted Galena and Dresbach (Mt. Simon) aquifers are also available.

Both glacial drift and river valley alluvium are often bypassed in favor of the relatively shallow and more productive underlying bedrock aquifers. Deposits of glacial drift are thin or absent in far northeast Iowa, but thicken to about 300 feet in the southern or western parts of the province. Alluvial aquifers are utilized along the larger rivers that cross the province in a northwest-to-southeast direction, especially the Iowa, Cedar, and Wapsipinicon.

The Mississippian aquifer is used primarily along its narrow outcrop belt that stretches from north-central to southeast Iowa. Where the Mississippian is absent, the Silurian-Devonian aquifer provides abundant water. Water quality in both these regional bedrock aquifers is generally good, but is typically hard because of dissolved minerals from the carbonate rocks and is occasionally prone to taste and odor problems. These naturally occurring quality problems increase in a westerly direction across the province.

In extreme northeast Iowa, both the Mississippian and Silurian-Devonian rocks are absent, having been removed by erosion in the geologic past. Here, the Cambrian-Ordovician aquifer is the most reliable groundwater source. A few counties in northeastern Iowa also use the Galena aquifer, a shallower Ordovician water-bearing formation. Use of the deep Dresbach aquifer is confined to a few Mississippi River counties, specifically eastern Allamakee, Clayton, Winneshiek, Dubuque, Jackson, Clinton, and Scott.

In parts of northeastern Iowa, the bedrock aquifers are highly susceptible to contamination, especially where they lie at or near the land surface and where the karst conditions described earlier are present. Newer wells frequently are drilled to greater depths to avoid these known contamination problems. Aquifers in the central and east-central portions of the province have a thicker cover of glacial drift, making them less susceptible to contamination problems from surface sources. Major aquitards in this region include the clay-rich glacial drift and the Maquoketa Shale.

Northwest Iowa Groundwater Province

Northwest Iowa can be characterized as "fair" in terms of groundwater availability, quantity, and quality. This area has fewer options for groundwater than northeast Iowa. The Dakota Sandstone, or Cretaceous aquifer, is the primary groundwater source because of the relatively shallow depth, generally good yields, and fair water quality of its sandstone units (except for local areas of exceptionally high sulfate). Regional bedrock aquifers lying below the Dakota Sandstone are very high in total dissolved solids and are unsuited for human or livestock use. Alluvial aquifers, on the other hand, are extensively used along major valleys, especially those of

Groundwater Provinces of Iowa

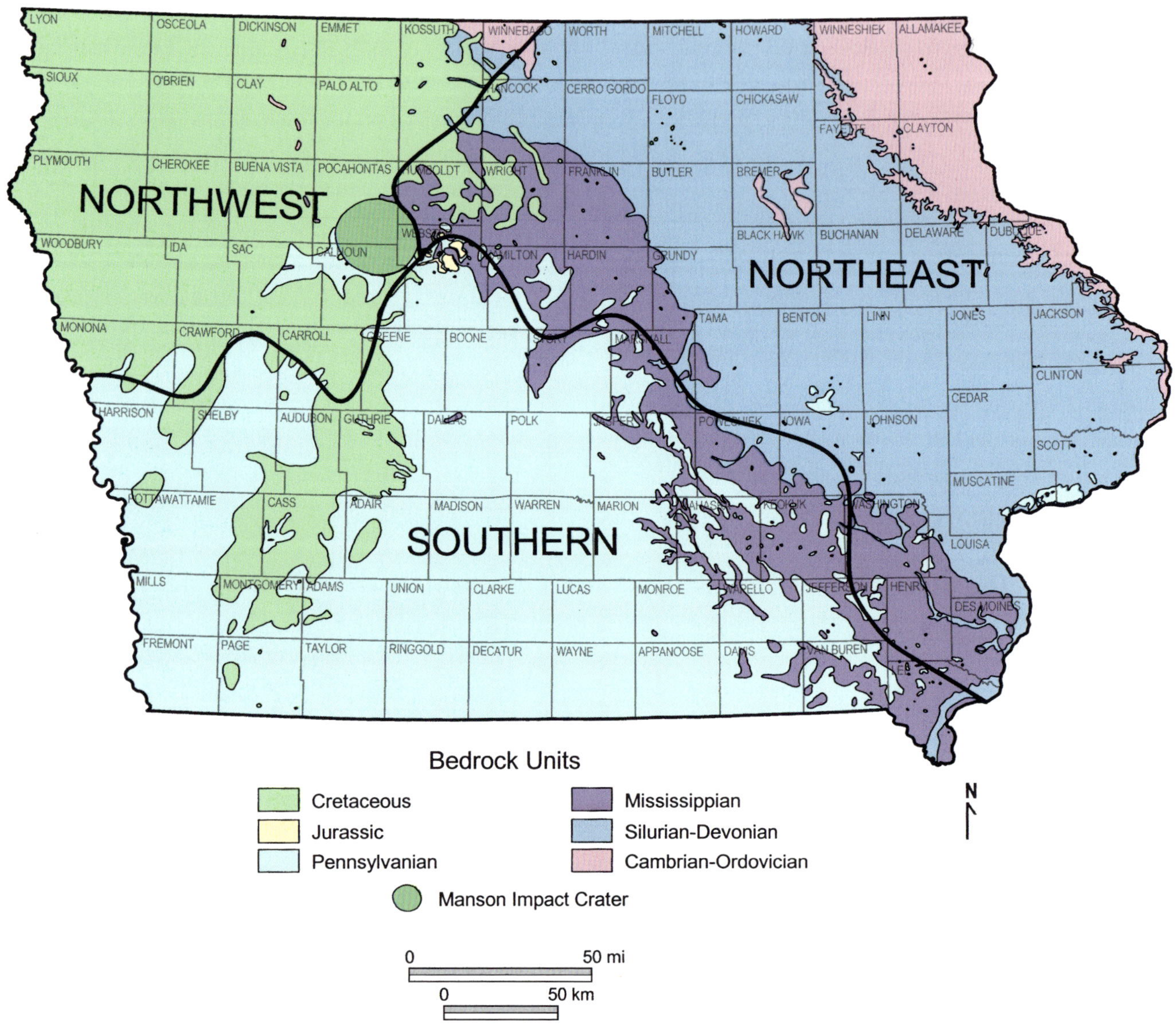

the Floyd, Rock, Little Sioux, Big Sioux, and Missouri rivers.

Buried valley aquifers are quite productive in certain locales of northwest Iowa (map, p.19). Glacial drift aquifers, while not as productive, are sufficient for many private and small public water supplies. The "salt and pepper" sands also occur within or at the base of the glacial tills in western Iowa, and locally these sand deposits produce moderate amounts of water.

The circular Manson Impact Structure in parts of Calhoun and Pocahontas counties features a massive disruption of the normal bedrock sequence, and finding groundwater within its boundaries can be difficult (map, p.26). Test holes are needed to search for water, and once found, the supply may not last long because the rocks often are no longer connected to the aquifers surrounding the meteor impact structure.

Northwest Iowa has substantial amounts of loess and glacial drift (an aquitard) covering the bedrock aquifers, thus protecting them from surface contamination. The major contamination issue in this province is the vulnerability of the widely used alluvial aquifers.

Southern Iowa Groundwater Province

The search for groundwater in southern Iowa can be challenging. Consequently this province is generally classified as a "poor" or difficult area in terms of finding water of sufficient quantity and quality. Pennsylvanian strata are the uppermost bedrock across large expanses of southern Iowa. These rock units, generally regarded as an aquitard, can sometimes be pressed into service if local occurrences of limestone and sandstone are favorable.

Three major bedrock aquifers are used in southern Iowa, the Mississippian carbonate as well as the Dakota and Cambrian-Ordovician sandstone aquifers. The Mississippian carbonate aquifer is used in areas where the Pennsylvanian bedrock is relatively thin or absent. The Dakota sandstone aquifer extends from northwest Iowa into several southwestern Iowa counties. The Cambrian-Ordovician aquifer has been relied on for decades in central and southeast Iowa as the only source of an adequate and reliable volume of water for towns and industries in particular. The introduction of Rural Water Districts has reduced or eliminated that reliance. Elsewhere, water in the deep regional bedrock aquifers is generally unusable because of very high levels of total dissolved solids.

For these reasons, both glacial drift aquifers and buried valley aquifers assume a more important role in providing water in southern Iowa. Alluvial aquifers also are valuable here, especially along the Skunk, Des Moines, and Nishnabotna rivers. These shallow sources are vulnerable to contamination problems from the land surface and present the most serious contamination issues in the province. The steeply rolling, well drained topography across much of southern Iowa is well suited to the installation of ponds and reservoirs, many of which are used to augment water supplies for both people and livestock.

Most areas of southern Iowa have a thick covering of glacial materials that keep surface contaminants from entering bedrock aquifers. The main exception is an outcrop area of the Mississippian aquifer in far southeastern Iowa, especially Des Moines County, where limestone is exposed near the land surface and local karst conditions are present.

WELLS
GETTING WATER from the GROUND

"The strata penetrated, the depth of water horizons, almost all the facts of use in artesian investigations must be gathered while the boring of any well is in progress, or not at all."

– William Harmon Norton
Artesian Wells of Iowa, 1897

Clay Smith

The internal pressure that causes artesian wells to flow comes from the arrangement of rock layers beneath the ground. (Black Hawk Bluff, Allamakee County)

Water flows from an aquifer into your drinking glass by means of a well. Techniques for well drilling and construction vary with aquifer depth, aquifer rock type, other strata encountered above the aquifer, and the amount of water needed. In Iowa, three major types of wells are used to produce groundwater: sandpoints, bored wells, and drilled wells. Common to most of them is a cylindrical hole excavated into the ground, a column of pipe or *casing* to keep the hole from collapsing, grout to seal the space between the hole and the casing, and a pump to lift the water to the surface.

The oldest method of obtaining groundwater is to dig a hole by hand to some depth below the water table. Historically such dug wells were lined with stones, bricks, or tiles and covered with a cap of wood, stone, or concrete. Modern versions of these dug wells are large-diameter seepage wells that are mechanically bored with power equipment. These wells usually range from 30 to 36 inches in diameter, and they are intended to expose a large surface area to slowly infiltrating groundwater. Such bored wells are useful in tapping shallow water-table conditions, often in fine-grained, slowly permeable materials such as loess (silt) or glacial till (clay).

Sandpoint wells, or "driven wells" as they are sometimes called, are small-diameter pipes (about two inches) pushed into the ground to tap shallow water-bearing sand or gravel. These pipes are usually fitted with a pointed screen at the bottom. Like the dug and bored wells, these shallow wells are typically less than 50 feet deep, and they go dry when the water table drops below the bottom of the well. They are also highly vulnerable to contamination sources originating at the land surface (diagram, p.45).

Most deeper wells today are constructed by truck-mounted, rotary drill rigs (photos, p.46). After a drill hole is completed, permanent steel or plastic casing is inserted. Typically the casing extends from just above ground level through the relatively soft glacial materials and into bedrock, usually several feet into the major water-bearing bedrock aquifer. The drilled hole below the casing is left exposed to provide water to the well (diagram, p.44). Wells drilled into shallower sand and gravel require casing plus an attached well screen to prevent the hole from collapsing and to keep fine sediment from entering the well and damaging the pump or clogging the pipe. When an aquifer consists of very fine sand, a gravel pack is added around the screen to prevent sand from entering the well. This technique allows use of larger slot openings in the screen, thus enabling increased water yields. Wells drilled into loosely cemented sandstone bedrock sometime need screens too. When the aquifer is in carbonate rocks, screens are unnecessary because groundwater movement is along fractures and bedding planes, and usually there is little sand or silt present.

Once the well casing is in place, grout is used to seal the *annular space* between the borehole and well casing. *Grout* is a slurry mixture of cement or *bentonite* and water that sets up to form an effective seal. This prevents surface water from draining down the outside of the casing and into the well and also prevents the movement of water between aquifers. In addition, grout seals off openings to undesirable formations and helps to prevent corrosion of the well casing (diagram, p. 44).

Modern wells are constructed in accordance with rules found in the *Iowa Administrative Code*. These rules fit the various well construction methods used in Iowa – drilled wells in unconsolidated (glacial) materials, drilled wells in bedrock materials, and bored and augered wells in unconsolidated materials (diagram p.45).

Most wells drilled in Iowa are for groundwater supplies. The deepest water well record on file at the Iowa Geological Survey in Iowa City is the 3,467-foot Greenfield Municipal Well #1 in Adair County drilled in 1929. Other wells are drilled to test for water-bearing conditions, to monitor aquifer conditions, and to explore for minerals, oil and gas (or its underground storage), and, historically, to drain wet ground (*ag-drainage wells*). For many years, the deepest well drilled in Iowa for any purpose was a 5,305-foot oil test in Page County completed in 1930. Then in 1987, an exploratory oil test hole in Carroll County near Halbur was drilled to a depth of 17,851 feet. This remains the deepest well drilled in Iowa to date.

Often wells are "developed" by drillers to increase the amount of water produced in a new well or to improve the performance of an older well. *Well development* may be as simple as the vigorous pumping of a new well to remove excess drilling mud and other fine sediments. At other times a solution of dilute hydrochloric acid may be pumped into a new well to enlarge openings in carbonate rock formations (called "acidizing") or into an older well to remove encrusting deposits of lime. The addition of polyphosphates, followed by vigorous surging of a well, is used to disperse silts and clays and remove oxides of iron and manganese.

Components of a Water Well

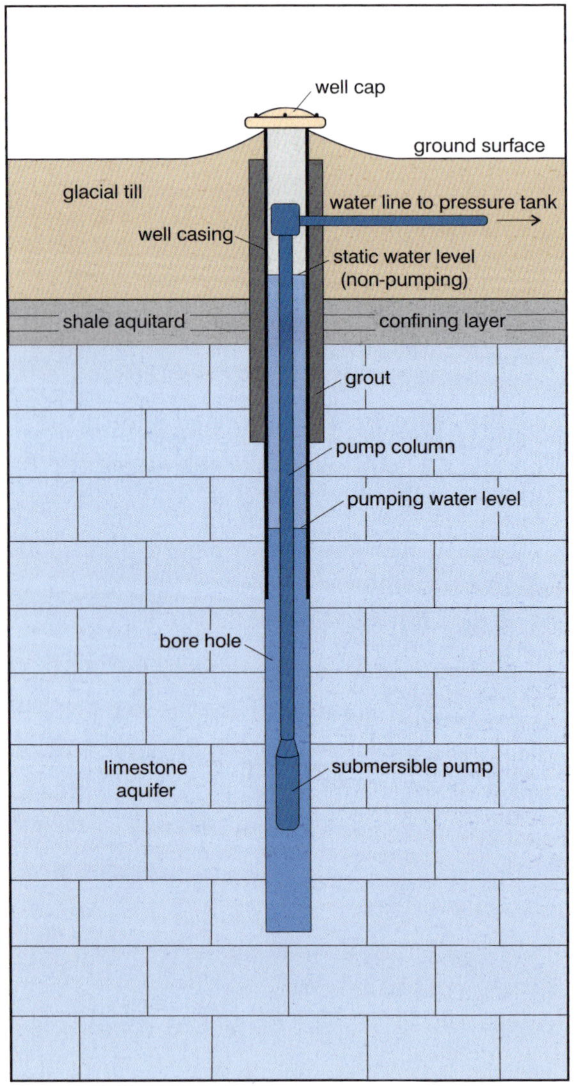

Yields from wells depend largely on the ability of an aquifer to store and transmit water. When water is pumped, water levels in the aquifer decline. The maximum decline occurs at the pumping well itself, and the decline, or *drawdown*, radiates from there, decreasing with increasing distance from the pumped well. In three dimensions, the drawdown is generally cone-shaped and therefore referred to as the well's *drawdown cone*, or *cone of depression* (diagram, right). In an unconfined aquifer, the drawdown cone reflects lowering of the water table. In a confined aquifer, it reflects lowering of the potentiometric (artesian pressure) surface around the well.

When a well no longer produces water or functions as a standby supply in good repair, then it should be properly plugged. Deteriorating wells become direct conduits for contaminants from the land surface to reach an aquifer and for cross-contamination to occur between different aquifers. Since 1987, about 44,000 abandoned wells in rural Iowa have been plugged, assisted by a Grants-to-Counties program funded by the Iowa Department of Natural Resources. The key aspects of proper well closure are to isolate the aquifers penetrated by a well from each other and from the land surface using impermeable filling material such as bentonite.

Since a well owner can't examine the inside of a well to troubleshoot problems, it is important to maintain basic information about its construction, performance, and water quality. Starting with construction records provided by the well drilling contractor, keep all records about a well's performance, routine pump maintenance, and needed repairs or remediation measures. Careful documentation of a well's history can be invaluable to diagnosing future well problems, suggesting appropriate treatment, and predicting the useful life of the well. Basic information about a well should include

Wells and Aquifers in Iowa

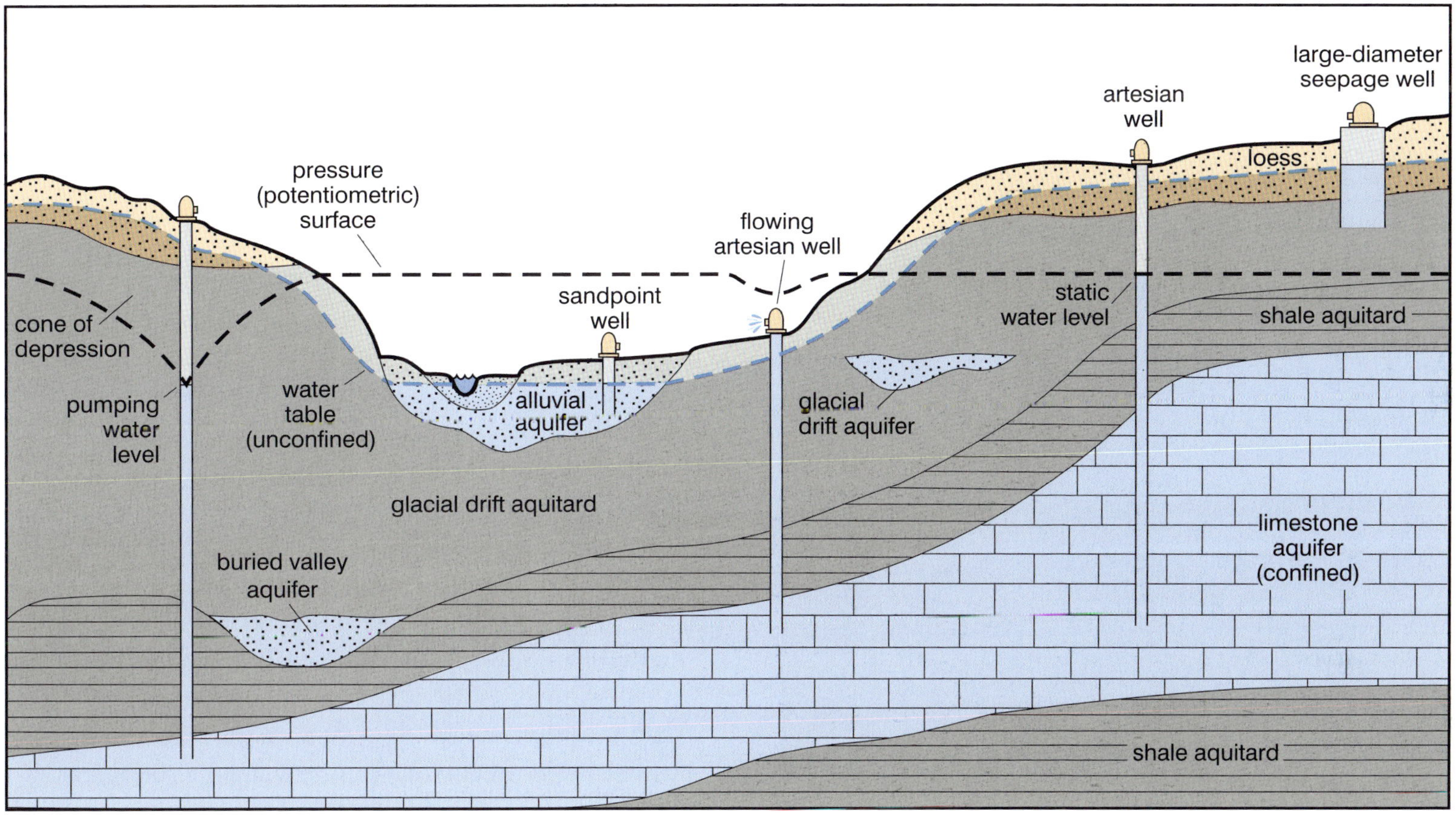

its precise location on a map (¼, ¼, ¼, ¼, Section, Township, and Range), well depth, *static water level*, pumping water level, estimated yield, and basic water quality.

A drilled well in Iowa can yield much more than water. Since the late 1800s, the Iowa Geological Survey has routinely archived well samples from cooperating drillers around the state. This

Wells were drilled at Briggs Woods Park in Hamilton County to test aquifer characteristics in the surficial and Mississippian aquifers. These nested wells enable water levels and quality of the groundwater supplies to be monitored over the long term.

photos by Clay Smith

library of 36,000 sets of drill chips and some 900 cores (solid cylinders of rock), totaling over 400,000 feet, account for over 90 percent of the information used to interpret the state's subsurface environments and their groundwater resources. Logs prepared from these samples are invaluable as baseline records of well conditions at the time of drilling (photo, right). They are solid pieces of geologic and hydrologic data to be used many times over as information to forecast conditions at prospective drilling sites or any site having engineering or contamination problems. When combined with sample sets from other wells, they become part of a database that is essential for both immediate and long-range use in solving the underground puzzles of Iowa's geologic strata and their associated groundwater resources.

Wells also reveal much about the health of an aquifer. By sampling well water periodically, changes can be noted in its quality and in the presence or absence of annoying or harmful contaminants. Also, keeping records of water levels can identify long-term trends in whether groundwater is being used faster than it is being naturally replenished. Such monitoring activities are valuable in identifying potential problems, tracking the geographic extent and depth of contaminant plumes, and supplying basic information about individual aquifers and how they differ from one another (photos, left).

One of the most interesting effects on water levels in Iowa wells occurred on March 27, 1964, following the Great Alaskan Earthquake, some 3,000 miles away. This earthquake, one of the largest

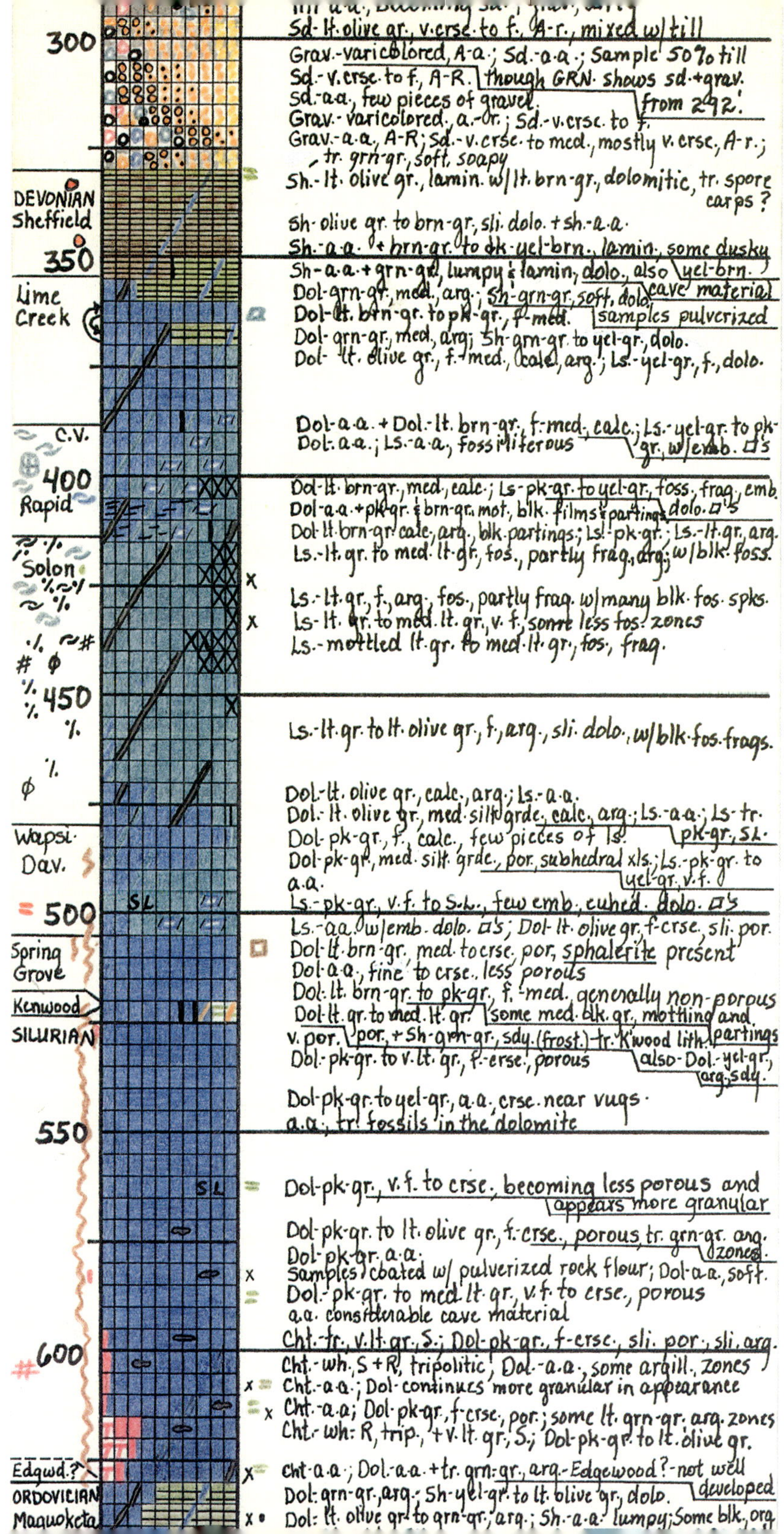

Logs of wells are prepared from microscopic study of samples collected during drilling. Details of soil and rock materials are keys to understanding the occurrence of groundwater resources in Iowa and to predicting conditions in untested areas.

ever recorded on the North American continent, disturbed water levels in all of Iowa's major aquifers. The effect in some wells was noted as sharp fluctuations in water levels on recorder charts. In other wells and in some springs, the water became turbid for a short time from disturbed silt and clay in the aquifer. Still other wells showed a permanent rise or fall in the water level of the aquifer, likely because the earthquake caused either compression or expansion of the porosity within the aquifer.

Then on the afternoon of November 3, 2002, another strong earthquake in central Alaska had noticeable repercussions in Iowa wells. The 7.9 Richter Scale quake along the Denali Fault struck at 4:12 p.m. Iowa time, and by 6:00 p.m. reports of discolored water in northeast Iowa started coming into local and state offices. Over the next two days, owners of limestone wells throughout northeastern Iowa reported water that ranged from lime-green to almost black (photo, p.51). Shock waves from the quake caused groundwater in Iowa's aquifers to slosh. This agitation stirred up fine sediment lying along rock fractures and in the bottoms of wells, and loosened the buildup of iron and manganese on submerged well pipes, casing, and pumps, contributing to the temporary discoloration of the water.

Water quality is a major factor in the development of a water

Significance of Commonly Occurring Constituents in Drinking Water

Constituent	Maximum contaminant level (MCL) for public water supplies	Recommended maximum level	Naturally occurring	Comments
Microorganisms				
Total coliform bacteria	No bacteria in 95% of samples collected		✓	Indicates pathway for potentially harmful microorganisms; inadequate well condition
Inorganic chemicals				
Arsenic (As)	0.01 mg/L		✓	Adverse health effects; carcinogenic
Chloride (Cl)		250 mg/L	✓	Salty taste when sodium present; corrosion of pipes
Fluoride (F)	4.0 mg/L	2.0 mg/L	✓	Affects dental health
Hardness: calcium (Ca) and magnesium (Mg) ($CaCO_3$)			✓	Limits lathering ability of soap; causes scale buildup
Iron (Fe)		0.3 mg/L	✓	Objectionable taste; stains laundry and porcelain
Manganese (Mn)		0.05 mg/L	✓	Objectionable taste; stains laundry and porcelain
Nitrate: as N as NO_3	10 mg/L 45 mg/L		✓	Land-applied fertilizer; leaching from septic tanks and sewage; adverse health effects; causes "blue baby syndrome" in infants
Sodium (Na) and potassium (K)			✓	Salty taste when combined with chloride

Sulfate (SO_4)		250 mg/L	✓	Objectionable taste; laxative effects; forms scale when combined with calcium
Total dissolved solids (TDS)		500 mg/L	✓	Refers to all material in solution; limits lathering of soap; objectionable taste
Suspended sediment			✓	Gives water a muddy or turbid appearance; pump wear
Dissolved gases				
Hydrogen sulfide (H_2S)			✓	Odor; corrosion of pipes, casing, and pumps
Methane (CH_4)			✓	Explosive
Organic chemicals	Atrazine 0.003 mg/L			
Pesticides	Alachlor 0.002 mg/L			Land-applied herbicides; adverse health effects
Benzene	0.005 mg/L			Leaking gasoline storage tanks; adverse health effects
Trichloroethylene (TCE)	0.005 mg/L			Industrial solvent and degreaser; adverse health effects
Radionuclides				
Gross alpha particles	15 pCi/L		✓	Formed by decay of radioactive elements; adverse health effects
Beta particles	4 millirems per year		✓	Formed by decay of radioactive elements; adverse health effects
Radium 226 & 228 (Ra)	5 pCi/L (combined)		✓	Formed by decay of radioactive elements; adverse health effects
Radon 222 (Rn)	4000 pCi/L (in review)		✓	Colorless gas formed by decay of radium; adverse health effects

mg/L = milligrams per liter; equivalent to parts per million (ppm)   pCi/L = picocuries per liter

Lowell Washburn

Abandoned wells are direct pathways for contaminants from the land surface to reach underground aquifers. Proper closure of these wells is essential to maintaining groundwater quality.

supply and problems can arise from a multitude of sources. As already noted, groundwater moving through soil and rock dissolves minerals, and these naturally occurring constituents can present health concerns and cause economic problems. In addition to mineral content, bacterial and chemical contamination can also affect water quality. Contaminants affecting health include bacteria, nitrate, pesticides, radionuclides, organic chemicals, arsenic, and lead. Contaminants that do not affect health, at least in small quantities, include sulfate, total dissolved solids, calcium and magnesium, hydrogen sulfide, iron, manganese, and iron bacteria.

The table on page 48-49 summarizes commonly occurring constituents that can cause problems in Iowa drinking water supplies, as well as those that are regulated and nonregulated ("recommended") for public water supplies.

Total *coliform bacteria*, which includes both fecal and non-fecal bacteria, is the most commonly reported health-related water quality problem in Iowa. Water containing coliform bacteria should not be consumed unless properly disinfected by boiling or chemical treatment.

Nitrate occurs naturally in soil, but concentrations exceeding 10 mg/L as NO_3-N often indicate pollution from sources of nitrogen such as fertilized cropland or human or animal waste-disposal sites.

Agricultural pesticides have been detected in Iowa groundwater since 1964. The U.S. Environmental Protection Agency (EPA) has established maximum contaminant levels (MCLs) for many commonly used pesticides. Two of those used in Iowa, atrazine and alachlor, have MCLs of 0.003 mg/L and 0.002 mg/L respectively.

Total dissolved solids are all the materials in water that are in solution. Water with less than 500 mg/L is considered excellent; 500 to 1,000 mg/L is good; 1,000 to 1,500 mg/L is fair; and greater than 1,500 mg/L is considered poor.

Calcium and magnesium are the primary cause of "hardness" and scale-forming properties in plumbing, and they also reduce the lathering ability of soap. Hardness is reported in terms of an equivalent amount of calcium carbonate ($CaCO_3$). Water is considered "soft" when the hardness is below 100 mg/L. Commercial water softening companies often use "grains per gallon" to report hardness. One "grain" equals 17.1 mg/L (or parts per million). "Very hard water" is reported as anything over 10.5 grains per gallon.

Adapted from photo by Michael Wade

Water discolored by suspended sediment appeared in northeast Iowa wells just hours after a major earthquake in central Alaska. Water in left jar from a 600 ft. well; water in right jar from a 200 ft. well.

Hydrogen sulfide (H_2S) is a dissolved gas that imparts a "rotten egg" odor to water and accelerates the corrosion of steel pipes, casing, and pumps. It is commonly produced by sulfate-reducing bacteria that live in aquifers or water distribution systems. Elimination of hydrogen sulfide odors in wells can be accomplished through aeration or chemical oxidation.

Following exposure to oxygen or chemical oxidizing agents, ferrous iron (soluble) is converted to ferric iron (insoluble), which imparts a yellow or reddish color to water. Iron concentrations exceeding 0.3 mg/L often will stain laundry or plumbing fixtures and can affect the taste of water. The accumulation of insoluble iron often causes deterioration of softening capacity and plugging of water softeners. Manganese concentrations exceeding 0.05 mg/L also cause brown stains on laundry and plumbing fixtures.

A common problem in Iowa is the clogging of well screens or pump components with iron bacteria. This nuisance aquatic bacteria, in the presence of dissolved oxygen, will precipitate an insoluble iron-oxide slime that can plug water distribution systems and impart an unpleasant odor to drinking water. Shock-chlorination can be effective in loosening such bacterial growths, though this procedure may have to be repeated on a regular basis. To distinguish whether the rusty color in water is caused by iron bacteria or simply the oxidation of dissolved iron, microscopic examination of the water is necessary.

Radioactivity is another natural contaminant in groundwater, derived from contact with rocks where the water is stored. While generally not excessive, it is highest in deep bedrock aquifers, especially in central and southeast Iowa, where it often exceeds drinking water standards.

After constructing a well, the water should be tested. In most areas the county health or sanitation department will work with well owners to test their water.

Good sources of information about well construction, maintenance, remediation, and abandonment include local well-drilling contractors, the Iowa Department of Natural Resources (IDNR) Water Supply Section, and county sanitarians. Publications about these topics are available from the U.S. Environmental Protection Agency and the U.S. Geological Survey. A local library can be another good source of water well information. The Iowa Geological Survey will provide individual well forecasts to landowners to help them estimate drilling depths, yields, and quality of water-bearing units beneath their property.

GROUNDWATER USE in IOWA

"Liquid Assets: A History of New York City's Water System."

– Diane Galusha, Communication Director for the Catskill Watershed Corporation, Walton, NY, 2002.

The book title above captures the value of water resources to any community today – they are indeed, "liquid assets." It is interesting to note that in 1870, not a single community water works existed in the state of Iowa. By the end of 1880, 16 community water systems were in operation, and that number had increased to 40 by the end of 1885. The *Manual of American Water Works* for 1890 and 1891 listed 85 Iowa towns as having a community water supply. Currently about 1,200 community supplies exist in Iowa, of which over 800 are municipal water supplies, 40 are rural water associations, and about 400 are unincorporated supplies, such as housing developments and mobile home parks. To understand something about the demand placed on Iowa's available water resources, it is helpful to take a look at how the resources are used.

Water use in Iowa includes both groundwater and surface water resources for a variety of purposes, including agriculture, commercial, domestic, industrial, irrigation, mining, power generation, and public water supplies. In these so-called "off-stream" uses, water is actually withdrawn or diverted from a groundwater or surface water source. This does not include such water uses as barge transportation, recreation, or fish hatcheries. The upper table on page 54 shows how groundwater and surface water withdrawals in Iowa compare for these various uses.

Information about water use usually is separated into water withdrawn and water consumed. Consumptive use is that part of water withdrawal that is removed from the environment by people

Paul VanDorpe

Groundwater pumped from aquifers supplies drinking water to many Iowa communities.

or livestock, by incorporation into manufactured products, by evaporation, or transpiration by crops and other vegetation. This water is no longer available for future use. Estimates indicate, for example, that less than 1 percent of the water used for power plant cooling is lost through evaporation, while nearly 100 percent of water used by agriculture and irrigation is consumed. Overall, only about 13 percent of the water withdrawn for use in Iowa is considered to be consumed. Note that groundwater accounts for about two-thirds of the water consumed in Iowa (lower table, p.54).

The largest amount of water used in Iowa is for power generation. A little over two-thirds of the state's total water withdrawn is for production of steam and for cooling of equipment in thermoelectric power plants. Almost 99 percent of this cooling water comes from surface sources such as the Mississippi and Missouri rivers, as well as larger interior streams such as the Des Moines and Cedar rivers.

Public water supplies account for the second largest amount of water withdrawn in Iowa. These include cities, towns, mobile home parks, housing developments, and rural water associations. On average, Iowans use about 1,100 gallons of water every day if indirect sources such as manufactured products, food production, and energy generation are included. Individually, however, we actually use about 100 gallons per day for drinking, food preparation, laundry, washing, and waste water. Nearly 70 percent of the water used by public water supplies in Iowa comes from groundwater. When the rural population on privately owned wells is added in, nearly 80 percent of Iowans depend on groundwater for their drinking water supplies.

Other major uses of groundwater in Iowa include irrigation,

Tim Kemmis

Deposits of sand and gravel often occur below the water table and are excavated by dredge and dragline operations. Materials are washed and sorted into different sizes for various uses, with little water being consumed. (Worth County)

with over 90 percent of needed water coming from groundwater sources, and livestock and poultry production, with 75 percent of their total withdrawals being from groundwater.

Sand and gravel production and limestone quarrying dominate mining activities in Iowa, and are one of the smaller categories of water use with less than 2 percent of the total. Most of the water withdrawn is for dewatering quarries and for washing aggregate products. The water consumed by these processes is negligible.

Which aquifers in Iowa supply the groundwater for each of the

Water Withdrawn in Iowa

Purpose	Groundwater withdrawals (Mgal/d)	Surface water withdrawals (Mgal/d)	Total withdrawals (Mgal/d)
Agriculture (livestock)	82	27	110
Commercial	18	25	43
Domestic	45	0	45
Industrial	74	184	258
Irrigation	35	4	39
Mining	1	42	43
Power generation	15	2110	2120
Public water supplies	257	116	373
Total water withdrawn	528	2510	3030

Mgal/d = million gallons per day

Source: *Estimated Use of Water in the United States in 1995.* U.S. Geological Survey Circular 1200, 1998.

Water Consumed in Iowa

Purpose	Groundwater consumed (Mgal/d)	Surface water consumed (Mgal/d)	Total consumed (Mgal/d)
Agriculture (livestock)	82	27	110
Commercial	6	8	14
Domestic	73	0	73
Industrial	12	41	53
Irrigation	35	4	39
Mining	<1	<1	<1
Power generation	<1	10	10
Public water supplies	52	36	89
Total consumptive use	261	127	388

Mgal/d = million gallons per day

Source: U.S. Geological Survey estimated water use in Iowa, 1995, unpublished data.

Figures above may not add to totals because of independent rounding.

various uses? The U.S. Geological Survey compiled figures for 1985 estimating the contribution for total groundwater withdrawals from each of the state's major aquifers. The left pie chart on page 55 totals these estimates, and we see that Iowa's relatively shallow surficial aquifers led the way in supplying 60 percent of the groundwater withdrawals in Iowa.

Another question often asked is, "Where does our drinking water come from?" In town or country, Iowans obtain their drinking water from ten major hydrogeologic sources as seen in the right pie chart on page 55. Taken together, wells drilled into bedrock aquifers supply about 40 percent of Iowa's drinking water, with wells completed in the shallower surficial aquifers accounting for another 40 percent, and surface water sources making up the remaining 20 percent.

Keep in mind that the numbers reported here for various aquifers and use categories change from year to year, and precise water use data are difficult to get a handle on over the entire state for any single year. While these collected data may reflect a span of several years, they still are useful in providing a perspective on the sources and use of water in Iowa.

In addition to the groundwater that society uses or consumes, there is another important aspect of this resource that is often overlooked. Throughout the Iowa landscape groundwater is constantly replenishing rivers, lakes, springs, and wetlands. In fact, this natural groundwater input, known as baseflow, accounts for about 50 percent of the total flow of Iowa's rivers and streams. The amount of baseflow varies across the state from 30 to 70 percent depending on whether the rivers and streams are cut into geologic materials having high or low permeability, as well as on the amount

Groundwater Withdrawals by Aquifer Source

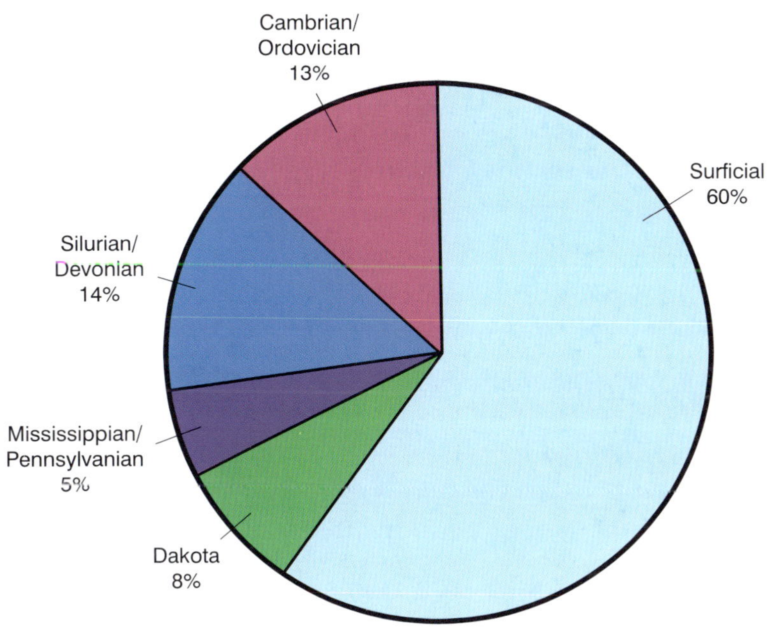

Source: U.S. Geological Survey, 1988. *Estimated Water Use in Iowa, USGS* Open File Report 87-704, p.11.

Drinking Water Supplies by Source

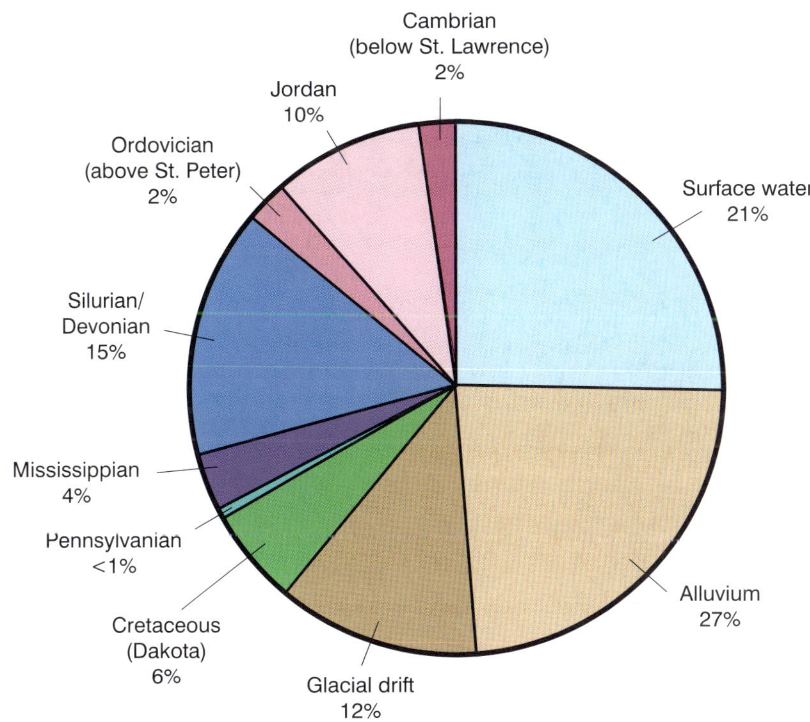

Source: Iowa Geological Survey (Unpublished data)

Clay Smith

Groundwater is used for irrigation of general farm crops and for orchards, nurseries, and golf courses. (Monona County)

J.R. Olson

A stream continues to flow during extended dry periods because of groundwater seepage into its channel. Known as baseflow, this condition affects water temperature, aquatic life, and delivery of pollutants to streams (obvious livestock sources notwithstanding).

of recharge to the groundwater system. During extended dry periods, groundwater may be the only contributor to a stream's flow (photo, bottom left).

A look at long-term trends in surface-water discharge reveals that baseflow contributes a greater percentage of water to rivers in the northeast and eastern parts of Iowa, compared to rivers in western and southern Iowa. Approximately 11.3 billion gallons are discharged every day to Iowa's surface waters from groundwater; that is about 3.5 times what we use and about 29 times what we consume. Without baseflow to sustain our streams, Iowa would not always be "the land between two rivers."

Studies at the Iowa Geological Survey show significant increases in baseflow to major Iowa rivers between 1940 and 2000. More of the precipitation falling across the Iowa landscape is infiltrating and being routed into streams as baseflow than is being shed as overland stormflow. This baseflow gain in the state's rivers are attributed to several causes: the effects of increased row-crop production and tile drainage, improved land management and conservation practices that increase absorption of surface water runoff, and the widespread down-cutting of streams below the water table during the post-settlement period in Iowa. A strong correlation exists between the significant increases in baseflow and the increasing nitrate concentrations in Iowa streams.

Another factor affecting the baseflow gains in Iowa streams is an increase in precipitation rates. The Environmental Protection Agency reports a 5 to 10 percent increase in precipitation over Iowa and the rest of the United States during the past century. Iowa's baseflow gain in the past 60 years, however, is greater than that accounted for by precipitation alone.

Electric generating plants are among the largest users of water in Iowa. Most of the water needed for steam production and cooling equipment comes from surface water sources along the state's major rivers, and most of this use is non-consumptive. Plant operations may be supple-mented by groundwater from alluvial wells. (Allamakee County)

Paul Horick

This interaction between underground aquifers and surface water is a dynamic part of the hydrologic cycle. Groundwater aquifers should not be viewed as end points in the hydrologic cycle. Rather, they are connected and contribute in a major way to water bodies on the land surface and to the water quality of those waters as well. It is increasingly important to assess water above and below ground as a single resource.

GROUNDWATER ISSUES
and PROTECTION STRATEGIES

"By the Way, You'll Need Water"

– Donivan Gordon
Iowa Geology 1981

Funnel shaped sinkholes at the land surface indicate shallow limestone and subterranean groundwater flow. Shallow aquifers are highly vulnerable to contamination problems in this geologic setting. (Clayton County)

Groundwater is usually out of sight, and for many people this keeps it "out of mind" as well. In many places, this attitude is understandable as good quality groundwater is in plentiful supply. At other locales, however, this is not the case. Numerous factors can affect the utility of groundwater as a water supply. Most important are the quality of water and the quantity that can be withdrawn, and these characteristics vary widely across the state depending on geological conditions.

In addition, different uses of water have different requirements for the quantity needed and the quality desired. For example, water for crop or turf-grass irrigation typically demands high yields, often in excess of several hundred gallons per minute. Water quality usually is not an issue. In contrast, a well producing just a few gallons per minute may be adequate for a household supply. Because the water is for human consumption, the home user has concerns about quality that the irrigator does not have. The water supply for a town or city must be adequate in terms of both quality and quantity. Water for some industrial and agricultural uses will have still other quality constraints. While many aspects of groundwater availability and quality are geologically controlled, society's needs and activities also have an impact.

Yields from wells depend largely upon the ability of the source aquifer to store and transmit water. These hydrogeological factors are fairly predictable for many aquifers in Iowa. The amount of groundwater any well can produce, however, may be affected by the presence of other nearby water wells. When two or more pumped wells are sufficiently close, their drawdown cones may intersect, causing further lowering of water levels – a condition commonly termed *well interference*. When interference is great, the pumping from one well can effectively render another well unusable. Therefore, the presence of other wells and the amount of drawdown they cause are factors that impact the quantity of water a new well can produce. How closely wells should be spaced to minimize this interference is a site-specific question because of the variable nature of aquifers.

Quarries and mines are often dewatered so that earth materials can be extracted more easily. Large volumes of water are pumped during these dewatering operations. A pumped quarry acts like a very large well, and therefore causes a drawdown of the water table in the surrounding area. Impacts on neighboring wells need to be evaluated before dewatering proceeds.

Typically, water level declines caused by pumping wells (or quarries) are relatively localized occurrences. Where numerous wells withdraw large volumes of groundwater over time, however, regional declines in water levels may occur. In Iowa, the most widespread of these declines has occurred in the extensive Cambrian-Ordovician aquifer. Regional water levels have dropped about 100 feet in this aquifer since use began in the late 1800s, with the greatest lowering of water-levels near major pumping centers. Other declines of a more local nature have occurred, such as those in the Cedar Rapids and Iowa City areas where the Silurian aquifer is heavily used.

Extended drought conditions can also impact groundwater levels and well yields. This effect is most significant in shallow water-table aquifers that recharge and discharge rapidly, and that supply similarly shallow wells. In Iowa, drought effects often hit hardest in the shallow alluvial aquifers of western Iowa, which are valuable sources of groundwater for many community supplies. The pinch is also felt in the shallow seepage wells that tap the water

Gary Hightshoe, Iowa State University

Contour farming, strip cropping, and grassed waterways are land management strategies to slow runoff from sloping land and to allow greater infiltration of water. (Near Iowa-Poweshiek County line)

table in western and southern Iowa. Such wells supply many rural residents. Water levels in the shallow, fractured bedrock aquifers of northeast Iowa also can fall significantly during extended dry periods. These very permeable aquifers are deeply cut by river valleys that drain the fractured rocks and drop water levels during drought conditions.

The quality of groundwater also affects its value for different water needs. The natural water quality in an aquifer results from interaction with rocks and minerals comprising the aquifer. Water slowly dissolves soluble minerals adding to the water's concentration

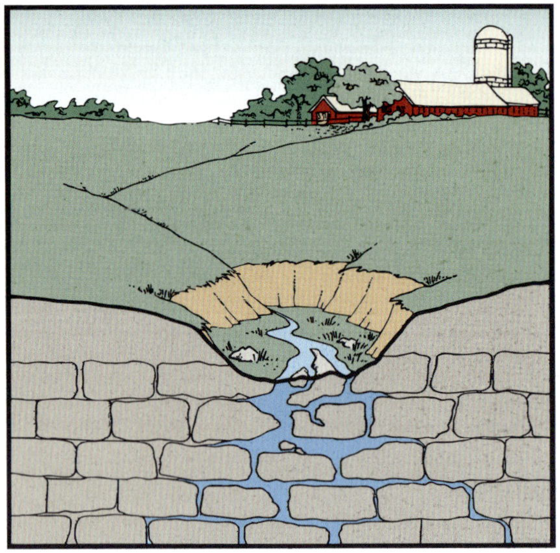

Sinkholes provide direct paths for silt, bacteria, and land-applied chemicals to reach shallow bedrock aquifers. Groundwater flow is rapid and undiluted in such karst areas. These natural circular pits should be cleaned of waste materials and buffered from cultivated land and surface runoff by permanent vegetation. Over 12,700 sinkholes have been mapped in northeast Iowa.

Underground tanks that store gasoline or diesel fuel are a risk to groundwater supplies. They lie near the water table; their metal corrodes with age; and leaks are hard to detect. Registration of tanks, new construction standards, early detection of leaks, and site cleanup are underway. Monitoring wells are used to determine product loss and establish flow direction and dimensions of a contaminant plume.

Old wells in disrepair or no longer in service provide direct pathways for contaminants to reach groundwater aquifers. If more than one aquifer is tapped, unwanted mixing of poor quality water with good water can occur. Finding and properly plugging and sealing the state's thousands of abandoned wells are important to maintaining groundwater quality.

of total dissolved solids, which is used as a general indicator of natural water quality. High concentrations impart a salty taste to water, and are usually associated with significant amounts of other constituents, such as sulfate and hardness, which may affect utility of the groundwater for some uses. High dissolved solids occur in aquifers that contain significant amounts of soluble minerals and have sluggish rates of groundwater movement. The table on pages 48-49 lists some of the more common troublesome constituents found in groundwater and the maximum concentrations allowable in public water supplies.

Widespread application of nitrogen fertilizer and pesticides to farm fields and their subsequent movement through soil to groundwater is an economic loss to farmers as well as an environmental loss to groundwater quality. More efficient use of these chemicals and best management practices on the land can help reduce these losses. Urban use of these materials is also a concern.

Agricultural drainage wells were intended to funnel excess surface water and tile-drainage deeper into the ground in poorly drained areas of north-central Iowa. Consequently they deliver sediment and land-applied chemicals directly into aquifers used for drinking water supplies. Some of Iowa's estimated 400 ag-drainage wells are being closed and sealed, with documented improvement in water quality.

Many kinds of waste have been buried in Iowa's land in the past, including toxic industrial wastes. Leachate carrying pathogens, heavy metals, and organic compounds still threaten groundwater supplies. Modern landfills must be sited in less vulnerable geologic settings, and monitoring of groundwater is required at all permitted sites to check for seepage. Recycling and composting can help reduce the volume of landfill waste.

Other natural constituents, such as hydrogen-sulfide, iron, manganese, ammonia, arsenic, and radioactive substances may also impair use. While some of these constituents tend to occur in groundwaters with high total dissolved-solids concentrations, they may also occur in less mineralized groundwater.

Groundwater quality can also be affected by society's day-to-day activities, many of which have the potential to release contaminants into the environment (illustrations, left and above). These include: nutrients, particularly nitrate, from fertilizers as well as animal and human wastes; pesticides, from agricultural, urban, and other uses;

photos by Clay Smith

Volunteers across Iowa are being trained to test streams and lakes for chemicals, aquatic organisms, and suspended sediment. The IOWATER monitoring program enables Iowa citizens to take an active role in protecting and restoring the state's waters.

road salts; diesel and gasoline-related petroleum compounds from leaking storage tanks and spills; and other organic chemicals and metals from industrial activities, waste storage, septic systems, and landfills.

The potential for a contaminant to affect groundwater at any given location depends, in part, on properties of the contaminant itself. These include: the chemical's mobility, or how likely the chemical is to attach to soil particles; the rate at which the chemical degrades to less harmful compounds; and the amount of the compound involved. As an example, consider nitrate nitrogen. It has virtually no tendency to attach to soil. In many important groundwater environments, it does not degrade. Large quantities are widely applied to row-crop lands in the form of fertilizers and manure. These factors, acting together, allow nitrate to be the most common contaminant in Iowa's groundwater. In contrast, many organic chemicals adhere tightly to soils and/or degrade quickly in the soil environment. The movement of significant amounts of such chemicals to groundwater is not likely unless large-scale, concentrated leaks or spills occur.

The sanitary condition of water from wells is another important consideration for drinking water supplies. Bacteria and other potential disease-causing agents may be present in shallow groundwater, but often are adsorbed and filtered out of the water before it travels far in the subsurface. Bacteria typically enter wells through defects in the casing or grouting of a well, rather than travelling with the groundwater itself. Wells are often tested for the presence of coliform bacteria. These bacteria are not necessarily a health concern; however, they show that bacteria are able to move from the land surface into a well. This is an indication that more harmful bacteria or other agents, if present, have the opportunity to do the same. Wells that produce water with coliform bacteria are sometimes tested for the presence of fecal coliforms, an indicator that water once in contact with human or animal waste has reached a well.

Along with contaminants present at the land surface, the potential for groundwater contamination is a function of the geologic setting of a given locale. The degree of vulnerability will always be a function of the enclosing earth materials and their hydrologic conditions. Much of Iowa's land surface is covered with a variable thickness of fine-textured glacial deposits. These deposits typically are slowly permeable, forming an aquitard. Where sufficiently thick, and if unfractured, they act as a protective blanket for underlying aquifers, limiting the downward movement of groundwater and contaminants. The diagram on page 64 summarizes common geological settings across Iowa and the relative protection provided to various groundwater aquifers.

Numerous studies in Iowa and adjacent states indicate that where an aquifer is covered by 50 feet or more of glacial deposits, the groundwater in the aquifer is usually free of contaminants from the surface. In many settings, the groundwater in these protected aquifers fell as rain and interacted with the land surface decades to

Paul VanDorpe

Long-term protection of a community's water supply includes the inventory of potential contaminant sites within the areas affecting their aquifer source.

hundreds of years ago, when contamination sources were considerably more scarce than today. The slow travel time of water through the glacial materials also allows for significant degradation and adsorption of some contaminants. In these areas, artificial pathways such as improperly constructed or abandoned wells, agricultural drainage wells, and underground mines may allow water and surficial contaminants to bypass the natural cover and reach otherwise protected aquifers.

Geological Conditions and Groundwater Vulnerability

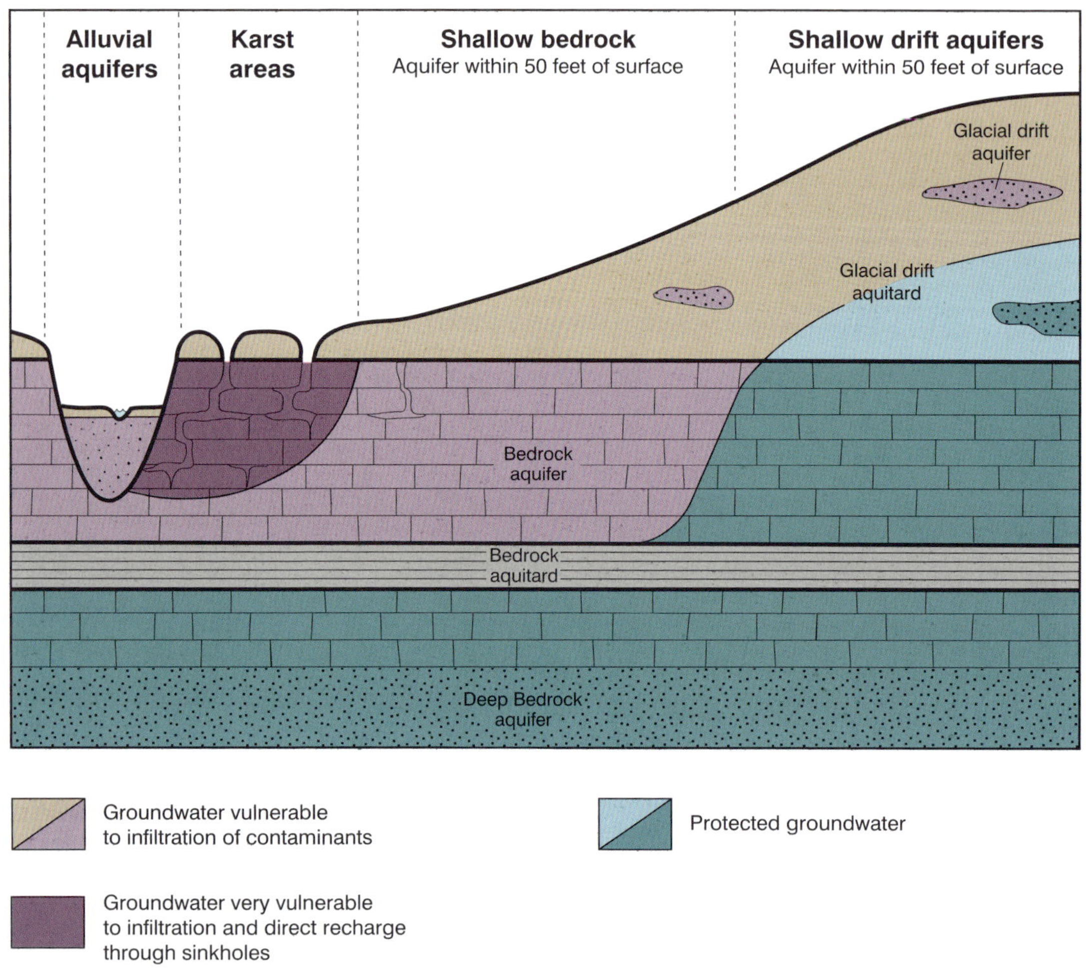

In contrast, aquifers with a thin protective cover are readily recharged by more recent water and contaminants. In particular, precipitation may infiltrate the land surface, mobilize contaminants, and enter alluvial aquifers, bedrock aquifers with less than 25 feet of glacial cover, and karst areas with sinkholes and fractured limestone on time-frames of hours to months, rather than many years. In addition, very shallow rock aquifers, very coarse-grained alluvium, or the presence of artificial pathways increase the potential for groundwater to carry bacteria and other biological contaminants into the subsurface. Extra precautions, planning, and management of contaminants are needed in these vulnerable areas.

With an improved understanding of groundwater vulnerability conditions, protection strategies today are aimed at finding and eliminating older existing sources of contamination. For example, *abandoned wells* need to be located and plugged with impermeable clay. Sinkholes need to be cleaned out and buffered from cultivated land with permanent vegetation. Other protection strategies aim to decrease and manage the contaminant sources that continue to enter our hydrogeologic environment. This means keeping a guardian eye on resource characteristics, gathering information, and watching for changes through long-term monitoring of all the aquifers that Iowans depend upon.

The value of monitoring wells installed in aquifers lies in the building of a long-term record whose value grows with the passage of time. If the usual patterns of water levels and water quality are known, then detection of something new provides a first alert. This protection strategy works just as effectively for an individual private well on a rural acreage as for a large operating farm well, a suburban housing development, or a major municipal or industrial well. It is

Sourcewater Delineation Map for Moville, Iowa

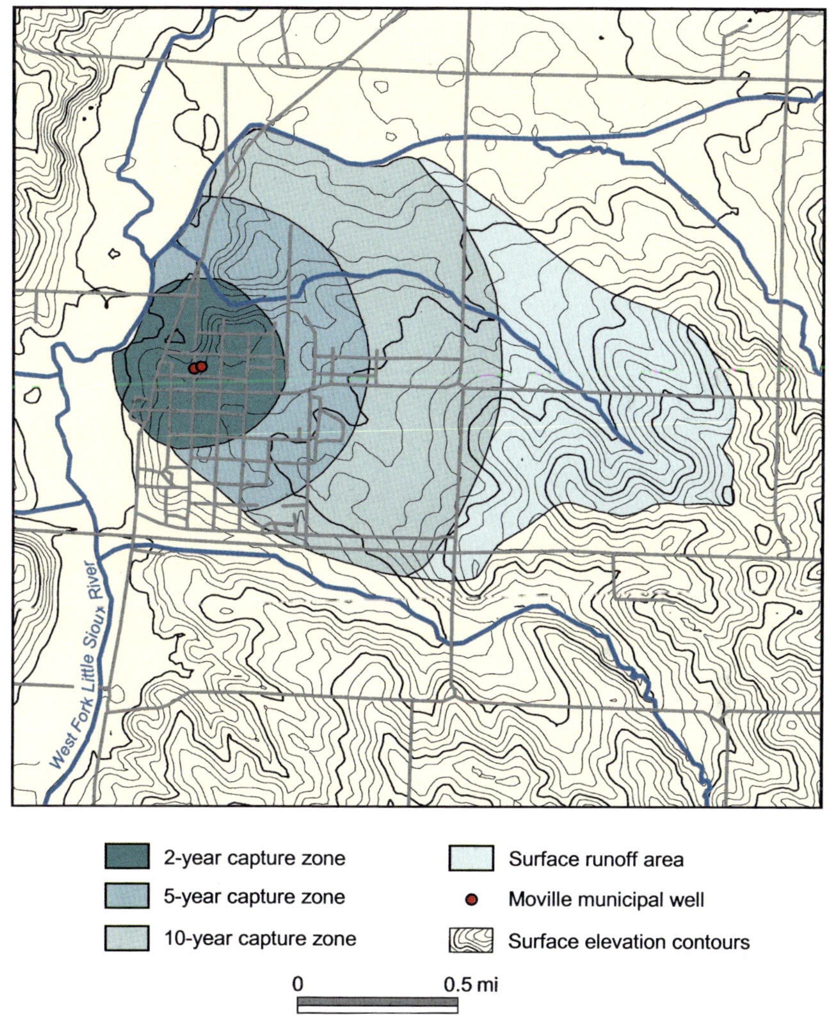

just a matter of scale; and whether small scale or large scale, monitoring data can build a valuable record and prompt action when necessary.

Contaminant sources in Iowa often stem from the long-term, widespread, commonly accepted practices of farmers, businesses, and homeowners. And once beneath the ground, the true extent of a problem is difficult to identify; it is often too late to apply corrective measures; and it takes a long time (if ever) for contaminants to be flushed from an aquifer. Thus, education and prevention are effective strategies to deal with both *point source* and *non-point source contamination* problems. Iowa's landmark Groundwater Protection Act of 1987 was significant for addressing growing evidence of groundwater degradation and public concern, and for focusing on solutions. It provided initiatives not only to deal with specific sources of contamination, but set out goals for long-term water monitoring, relevant research, and public education and action.

Federal government funding of the *Source Water Assessment and Protection* program in Iowa is prompting state action to delineate protection areas around community wells that tap various aquifers in Iowa. The idea is to preempt contamination problems and costly remediation procedures. This plan for long-term water supply protection includes identifying aquifer sources, examining their vulnerability to contamination, predicting the time-of-travel for contaminants moving in the groundwater flow system, and instructing the communities themselves on how to inventory contamination threats within areas needing protection (photo, p. 63). The map for the Woodbury County town of Moville on page 65 defines protection areas (capture zones) around the town's municipal wells based on the time-of-travel of water and direction of groundwater flow

Groundwater Vulnerability Map of the Cedar Rapids Area

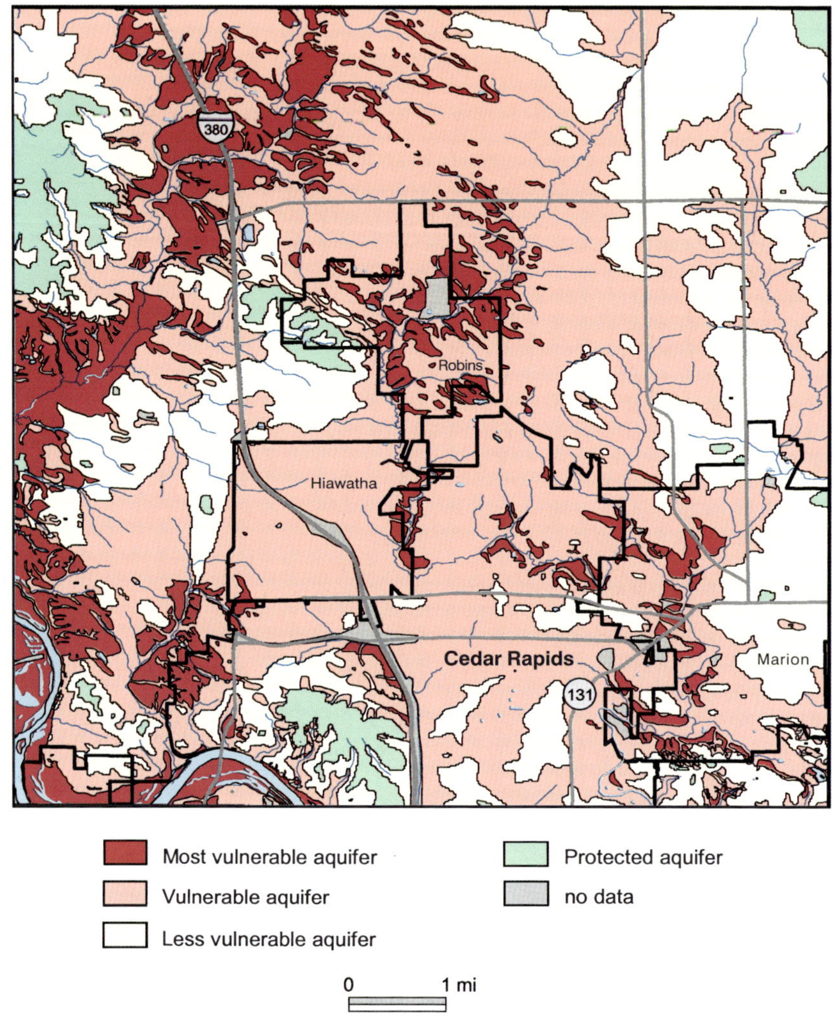

within the source aquifer. Defining these capture zones around a public water supply helps the community focus future source water protection efforts.

Many of the groundwater issues prompting these government actions will continue to be with us far into the future. They all involve use of Iowa's land. Our approach to safeguarding groundwater resources must always take into account their degree of vulnerability, which will always remain a function of the surrounding earth materials and the dynamic hydrologic conditions. Thus it is important to continue to investigate Iowa's geologic realms, to bring the most accurate and up-to-date information to bear on these issues as they arise at different times in different parts of the state.

In recognition of the key role of earth materials and their characteristics in groundwater contamination issues, the Iowa Geological Survey published a map depicting groundwater vulnerability regions of Iowa in 1991. It provides a regional summary of geologic conditions that affect aquifer recharge and contaminant transport, and it synthesized a long period of geologic research and record-keeping.

More detailed geologic mapping of shallow subsurface materials is underway in areas of the state experiencing rapid urban expansion and associated land use changes. Maps derived from this basic geologic mapping are available for use by local community leaders as well as the private sector to improve decisions on economic development, natural resource management, and environmental protection. An example of the varying susceptibility of aquifers to contamination from near-surface sources in part of Linn County is shown on page 66. The most vulnerable areas (red) are underlain by sandy alluvial aquifers, areas where bedrock aquifers lie near the land surface, and areas characterized by karst features such as sinkholes. In contrast, aquifers in protected areas (green) are overlain by 100 feet or more of slowly permeable glacial deposits, and are largely protected from surface-related contamination. This mapping effort is a direct spin-off from the U.S. Geological Survey funded STATEMAP program in Iowa.

To be effective, groundwater protection needs not only on-going monitoring and research efforts, but needs to include public action. Education is one of the most valuable tools to improve groundwater protection. Just as human activity on the land surface can be detrimental to the quality of underground waters, so it is possible for an informed public to intervene, both through individual efforts and teamed with government action, as seen with voluntary best management practices and with water monitor training programs, such as IOWATER (photos, p.62). Society must make the effort to control and manage potential sources of contamination.

Large quantities of groundwater are needed to support our modern society, as we have seen. At the same time, human activity can contaminate both surface water and groundwater. Agriculture, industry, urban development, and recreation can degrade groundwater quality. Unmanaged groundwater withdrawals can alter aquifers, damage wells, destroy wetlands, cause land subsidence, and change natural hydrologic flow conditions. Proper management of groundwater resources is essential to ensure that no one is without a useable supply of water. As long as Iowans rely on water from the ground, the discussion of protection issues and strategies must be underlain by an informed understanding of the hydrogeological dynamics of our groundwater resources.

GROUNDWATER
LEGAL and REGULATORY ASPECTS

The Missouri is a river " . . . that goes traveling sidewise, that interferes in politics, rearranges geography, and dabbles in real estate. It cuts corners, runs around at night, lunches on levees, and swallows islands and small villages for dessert."

– G. Fitch, 1907, as quoted in *Missouri River News,* January 2001, article by Ryan Roenfeld

The importance of water resources in Iowa has led to passage of laws to protect and regulate its use. The Iowa Code [section 455B.171] defines "water of the state" to mean: " . . . any stream, lake, pond, marsh, watercourse, waterway, well, spring, reservoir, aquifer, irrigation system, drainage system, and any other body or accumulation of water, surface or underground, natural or artificial, public or private, which are contained within, flow through or border upon the state or any portion thereof." Jurisdiction to prevent, abate, or control water pollution and to conduct the public water supply program is given to the Iowa Department of Natural Resources. The IDNR carries out responsibilities of the state related to public and private water supplies and sewage disposal systems for the protection of the environment and the health and safety of Iowa citizens. County boards of health adopt standards at least as stringent as those of the state and regulate both private water supplies and sewage disposal systems.

Clay Smith

The IDNR protects our water supplies through a set of environmental rules, the *Iowa Administrative Code* (IAC), Part 567. Some rules address drinking water quality concerns, including construction standards for private wells (567 IAC 38, 49), well plugging requirements (567 IAC 39), public water-supply design and operation (567 IAC 40, 41, 42, 43), the drinking-water revolving fund (567 IAC 44), whereby low-interest loans are made available to certain public drinking-water supplies for facility improvements, and the grants-to-counties program (567 IAC 47), whereby funds are made available for testing, rehabilitation, and closing of private wells. Other rules address water quantity concerns, including water withdrawals, water diversion, water storage, and water rights (567 IAC 50, 51, 52, 53, 54, 55). Protecting our water supplies also extends to how we handle wastewater effluent, wastewater plant construction and operation, septic tanks, wastewater treatment limitations and prohibitions, manure, land application of sewage sludge, and the application of aquatic pesticides to waters (567 IAC

Title IV). Rules are in place for certification of public water-supply operators, wastewater treatment plant operators, well contractors, and water-testing laboratories (567 IAC Title VI). Solid waste, land application of sludge and solid waste, spills and hazardous conditions, and hazardous waste are also regulated (567 IAC Title VIII, Title IX, Title X, Title XI).

Other rules enable the IDNR to assure that Iowa's water resources are put to beneficial use, that the resource is not used indiscriminately (567 IAC 50, 51, 52), and that conservation and protection are maintained. For example, users of large volumes of water (greater than 25,000 gallons per day) need a water withdrawal permit, regardless of the intended use and regardless of whether it is surface water or groundwater. Construction standards for private water wells insure that individual water supplies are safe from contamination and that aquifers are protected from contamination, thereby protecting public health. Certain criteria also apply to groundwater withdrawals. For example, wells in alluvial aquifers adjacent to streams with drainage areas of less than 50 square miles are not allowed to pump in excess of 200 gallons per minute (gpm). Wells close to such streams may not pump when stream flow falls below a protected flow threshold.

Additionally, two of the state's major sandstone aquifers are protected from overuse (567 IAC 52). Withdrawals from the Cambrian-Ordovician aquifer are limited to 200 gpm for new irrigation, recreational, and aesthetic-use permits. New permits for industrial and power generation are limited to less than 2,000 gpm. Also, the maximum allowable decline in the potentiometric level of the Cambrian-Ordovician aquifer is 200 feet from a 1977 baseline map. The Dakota aquifer is likewise protected from excessive use. For withdrawal rates in excess of 200 gpm, permit holders must monitor water levels within the aquifer using monitoring wells. If water level declines are a serious threat, then certain types of groundwater use have priority, for example public water supplies.

Though seldom used, a defined geographic area may be designated as " a protected source area" (567 IAC 53). Such areas are designated on a case-by-case basis to preserve water availability for beneficial use, to minimize migration of a contaminant plume in a groundwater aquifer, to maintain surface-water quality, or to preserve surface-water flow. Withdrawal of groundwater from a protected source area by non-regulated wells (e.g., private wells) can be restricted or denied, if necessary.

Private well construction permits are issued by individual counties (567 IAC 38), and well construction standards are set forth in 567 IAC 49. Construction standards notwithstanding, occasionally conflicts arise from well interference problems caused by simultaneous pumping of two or more nearby wells. 567 IAC 54 is used to resolve situations where a permitted use causes interference in a non-regulated well. Of course, this assumes that there was an adequate water supply for the non-regulated user prior to the well interference claim. Problems are resolved by modifying the affected well or by changing the conditions of permitted use.

The rules outlined here cover a wide range of activities related to specific uses of the state's groundwater, with other regulations designed for protection of the resource. One of the best ways that individual well owners can help protect the state's groundwater resources is by making sure that all abandoned wells on their property are properly plugged, thereby permanently sealing off contamination to individual aquifers (567 IAC 39).

EPILOGUE

" . . . blues would become the underground aquifer that would feed all the streams of American music, including jazz."

– Ken Burns, Iowa Public Television documentary film on *"Jazz"* (2001)

Silver Lake, Dickinson County. Photo by Clay Smith.

Groundwater surfaces to feed our lives in many ways, including its apt use in the metaphor above. To be fully appreciated, groundwater must be viewed as a vital natural resource, as a dynamic geologic process, and as a fundamental element of the natural environment. Groundwater influences the quality of our lakes and streams, the supply of our aquifers, the depth and purity of our well water, the appearance of our landscape, the sites of our cities and towns, and the ability of wetland habitats to sustain native forms of life. It has value both when extracted from the ground and when left in place.

The need for water resources drawn from wells has long been the essential motivation for groundwater studies in Iowa. The fact that nearly 80 percent of Iowans depend on groundwater for their drinking water supplies emphasizes this point. Many Iowa communities initially put down their roots along river valleys because of easy access to abundant shallow groundwater. Today, our ability to produce reliable forecasts of the availability, quantity, and quality of this resource for new wells requires increasingly accurate and detailed views of the geological conditions beneath Iowa's land. Similarly, as demand increases for larger capacity wells, information is needed to minimize the impact of such pumping on existing wells and to resolve conflicts among water users. If groundwater is to continue its essential role in Iowa's water resources picture, then it also needs protection from threats of contamination originating in both the past and present, from both point and non-point sources. The ability to forecast extends to identifying vulnerable groundwater aquifers and the geologic conditions that put them at risk from human activity.

With today's improved awareness and recognition of problems, and with more effective regulations in place, we are moving into an era that emphasizes contaminant prevention, in addition to after-the-fact cleanup and remediation. We must continue to investigate the diversity of geologic settings across Iowa and the complexities of groundwater travel through three-dimensional configurations of soil and rock in order to meet the demand for more detailed information about potential contaminant risk to groundwater resources.

In addition, groundwater is a slow but effective geologic agent currently shaping Iowa's landscape. The natural erosion of hillsides and stream banks is closely tied to subtle but persistent fluctuations of groundwater flow through time. The shifting intersection of water table with land slope contributes to slumps, collapse, and larger landslides. In fact, our own awareness of groundwater tends to fluctuate with the water table. Only when it rises well above its normal range, and floodwaters spread into our lives, do we take active notice of its presence. During the Flood of 1993 for example, many people found the water table rising into their basements to be as serious a problem as overland flooding was to others. Groundwater can be notorious during another extreme condition – when it's absent. During periods of drought, the water table falls and water disappears from creeks, river beds, and the reach of shallow wells. Still other expressions of groundwater as a geologic agent are the presence of solutional cave and karst features below ground, and springs and sinkholes at the land surface.

The movement of groundwater is a major contributor to natural hazards in Iowa. Such events occur when normal geologic processes are speeded up. For example, in the shallow limestone terrain of northeastern Iowa, underground flow systems persisting through time can result in the sudden collapse of land into a sinkhole. In western Iowa, groundwater saturation of the steep loess bluffs near the Missouri River creates slope instability that leads to landslides and other serious engineering problems.

Finally, groundwater is an important element of Iowa's natural environment. It is part of the dynamic hydrologic cycle, one of Earth's most fundamental and vital natural systems. Understanding the role of groundwater in this cycle is essential to interdisciplinary, regional, watershed-based, and site-specific assessments of biological habitats as well as environmental issues. Water-based habitats such as rivers, lakes, sloughs, springs, hillside seeps, and fens in Iowa are dependent in varying degrees on groundwater flow and replenishment. Distinct plant and animal communities reside in these ecological niches, and their long-term protection is dependent on maintenance of both the physical flow and chemical quality of the groundwater feeding them. The role of groundwater needs to be recognized and integrated into efforts to maintain and restore these habitats.

Watersheds are a keystone of natural resource management today. To improve the quality of a single body of water requires a combined, long-term effort over its entire watershed, a large, sometimes diverse drainage area where numerous individual decisions affect how land is used. Keep in mind that watersheds are more than three-dimensional features of the land surface. They extend below the ground and include the movement of groundwater through time. These added groundwater and time dimensions of watersheds in Iowa are essential to understanding the movement of contaminants below ground and their eventual delivery back to our surface streams and lakes.

Understanding the fundamentals of groundwater in its geologic realm provides valuable information and insight for the use, protection, and sustainable development of Iowa's land and waters.

GLOSSARY

abandoned well – a water well that is no longer in use or is in such a state of disrepair that continued use is unsafe or impractical.

ag(ricultural) drainage well – a drilled shaft that drains excess water into underlying bedrock; upper parts are often cistern-like structures that form a discharge point for tile-drainage lines.

alluvial aquifer – sand and gravel deposits that fill valleys along rivers and streams; important for public, industrial, and agricultural water supplies along large rivers.

alluvium – gravel, sand, silt, or clay deposited by flowing water.

annular space – the ring-shaped space between a borehole and the casing set within it.

aquifer – a body of earth materials that yields groundwater to wells or springs; a water-bearing formation.

aquitard (or aquiclude) – a body of earth materials capable of absorbing water but not transmitting it in sufficient quantities to supply a well; functions as the upper or lower boundary of an aquifer.

artesian – describes groundwater under pressure from the weight of water at higher elevations in a *confined aquifer*.

baseflow – that portion of stream flow originating from groundwater discharge into the stream channel.

bedrock – solid rock that underlies soil or other unconsolidated surficial materials.

bentonite – naturally occurring clays that swell greatly in volume as water is absorbed; used as an impermeable seal in well construction and abandonment.

brecciated – said of rock composed of broken, angular fragments.

buried-valley ("channel") aquifers – sand and gravel deposits along ancient river valleys, often carved into bedrock, and buried beneath other sediments.

capillary fringe – a zone in which water is drawn upward immediately above the *water table*, held by surface tension within tiny pores.

casing – the tubular steel or plastic lining of a well, installed to support the well opening and keep fluids and earth materials out.

coliform bacteria – a group of bacteria whose presence in well water indicates a direct path for potential contaminants from the land surface.

cone of depression (drawdown cone) – a conical depression in the *water table* or the *potentiometric surface* that forms in response to pumping groundwater from a well.

confined aquifer – an aquifer bounded above and below by impermeable strata and under *artesian* pressure.

discharge – the outflow of water from a stream or groundwater aquifer; opposite of *recharge;* also, outflow from a pumping well.

drawdown – the difference between water levels in a well before pumping and during pumping.

evapotranspiration – process by which water moves into the atmosphere by evaporation from land and water and by transpiration from plants.

fen – a special type of wetland sustained by mineralized groundwater flow and including saturated peat deposits, often in mounded positions on hill slopes.

floodplain – the relatively level land that lies adjacent to a river channel and periodically is covered with flood water.

geothermal gradient – the rate of increase in temperature with depth within the Earth.

glacial drift aquifers – pockets of water-bearing sand and gravel within pebbly clay material left by glacial and associated meltwater activity; their configurations are irregular and locations are often unpredictable.

glacial till – an unsorted and unstratified mixture of clay, silt, sand, gravel, and boulders deposited directly by a glacier without subsequent reworking by meltwater.

groundwater – subsurface water that occupies pores, fractures, or other openings within earth materials.

grout – slurry mixture of cement or bentonite and water that can be pumped through a pipe and placed as a protective fill or seal during well construction.

hydraulic conductivity – the capacity of porous earth materials to transmit groundwater; the rate is determined by size and shape of the pore spaces and their degree of interconnection.

hydraulic head – the pressure of groundwater at a given point caused by the height of groundwater higher in the aquifer; represented by the elevation to which water will rise in a well.

hydrogeology – the study of groundwater and its relationship to the geologic environment.

hydrologic cycle (water cycle) – the continuous circulation of water from the atmosphere to earth and into the ground by precipitation and infiltration, and its eventual return to the atmosphere from land and water surfaces by evaporation and *transpiration.*

impermeable – describes earth materials that will not readily transmit water.

infiltration – the downward movement of water into the ground; percolation.

karst – describes topography formed by the dissolving action of groundwater on underlying carbonate bedrock and characterized by sinkholes, caves, underground drainage, and springs.

leachate – a solution formed by groundwater percolating through shallow materials containing soluble minerals or chemicals.

loess – wind-blown silt, usually porous and friable, deposited as a by-product of glacial and meltwater activity.

mafic – describes igneous rocks composed mainly of dark-colored minerals (ferromagnesian).

nonpoint-source contamination – pollution stemming from diffuse, widespread sources such as runoff and percolation from agricultural or urban areas.

outwash – sand and gravel carried by glacial meltwater and deposited beyond the ice margin.

paleosol – a soil formed in the geologic past and buried by younger materials.

perched water table – the mounding of water above a low permeability material located above the water table.

permeable/permeability – the capacity of earth materials to allow free passage of water; a measure of the interconnectedness of porous openings in rock, sediment, or soil.

pluton – an intrusion of igneous rock formed at great depth.

point-source contamination – pollution stemming from a single, identifiable source, such as a drainpipe, sewer, ditch, or underground container.

porosity – a measure of the open spaces in a deposit of earth materials; capacity to store water.

potentiometric (or pressure) surface – an imaginary surface defined by the level to which water will rise in a well.

recharge – the replenishment of groundwater supplies by infiltrating precipitation and surface water.

recharge area – a portion of the land surface over which infiltrating water eventually reaches an aquifer.

runoff – precipitation that flows over land until it reaches surface water, infiltrates into the ground, or evaporates.

saturated zone (phreatic zone) – that portion of earth materials in which all subsurface openings are filled with water; the upper surface of this zone is the *water table*.

Source Water Assessment and Protection – a program to determine susceptibility of public water supplies to contamination; key components are defining the hydrogeologic area (source) that contributes to a water supply, and then assessing land surface contaminant sources in order to minimize risk.

spring – natural discharge of groundwater to the land surface or into a lake, stream, or wetland.

static water level – the standing water level in a well that is not being pumped.

subcrop – (in Iowa) an occurrence of bedrock lying directly beneath the much younger mantle of glacial deposits; a "subsurface outcrop."

surficial aquifer – occurs in earth materials lying above bedrock; typically includes alluvial, glacial drift, and buried valley groundwater sources.

total dissolved solids – the combination of all dissolved mineral constituents in groundwater.

transpiration – the process by which plants evaporate water into the atmosphere.

unconfined aquifer – an aquifer having a freely fluctuating water table, open to atmospheric pressure; not confined beneath *impermeable* strata.

unsaturated zone (zone of aeration; vadose zone) – an underground area where pore or fracture openings contain both air and water; lies between the land surface and the *water table*.

watershed – the land area drained by a stream; a drainage basin.

water table – the top (upper boundary) of the saturated zone in an *unconfined aquifer*.

well – a cylindrical hole drilled or excavated into a water-bearing zone so that water can be pumped or will flow to the land surface.

well development – a process to increase water yield by enlarging voids and flushing sediment from a well.

wellhead protection – a program that encourages communities to determine groundwater sources for and contamination threats to their public water wells; protection of well sources including the surface and subsurface area through which contaminants could move toward a well or well field.

well interference – when the *cone-of-depression* of one well overlaps with that of another well pumping from the same aquifer.

wetlands – lands where water-saturation during all or part of the year determines soil types and plant and animal communities.

REFERENCES and INFORMATION SOURCES

Cited References

Calvin, Samuel. 1895. *Geology of Allamakee County*. Iowa Geological Survey Annual Report, v. 4, p. 51.

Chamberlin, Thomas C. 1885. *The Requisite and Qualifying Conditions of Artesian Wells*. U.S. Geological Survey Annual Report 5, p. 131.

Clark, M.L. and J.N. Thamke. 1988. *Estimated Water Use in Iowa, 1985*. U.S. Geological Survey Open-File Report 87-704, 28 p.

Coble, Ronald W. 1965. *The Effects of the Alaskan Earthquake of March 27, 1964, on Ground Water in Iowa*. Iowa Academy of Science, v. 72, p. 323-332.

Galusha, Diane. 2002. *Liquid Assets: A History of New York City's Water System*. Purple Mountain Press, 303 p.

Manual of American Water Works, for 1890-1891, p.6. New York, 1892. As reported in Norton, William Harmon,1897. *Artesian Wells of Iowa*. Iowa Geological Survey Annual Report v. 6, p. 118-119.

Neilson, W.A., Knott, T.A., and Carhart, P.W. (eds). 1943. *Webster's New International Dictionary of the English Language, Second Edition, Unabridged*. G. & C. Merriam Company, Springfield, Massachusetts, 3210 p.

Norton, William Harmon. 1897. *Artesian Wells of Iowa*. Iowa Geological Survey Annual Report v. 6, p. 123.

Roenfeld, Ryan. 2001. "The Missouri River: A Council Bluffs Perspective," in *Missouri River News*, v. 2, no. 1. Jan. 2001. Contains 1907 quotation by G. Fitch.

Schilling, Keith. 2003. *Relationship of Increasing Baseflow, Changing Land Use and Nitrate Concentrations in Iowa's Streams*. Agricultural Hydrology and Water Quality. AWRA 2003 Spring Specialty Conference Proceedings, American Water Resources Association, Middleburg, Virginia.

Scott, Sir Walter. 1921. *St. Ronan's Well*. J.M. Dent & Sons Ltd. London, E.P. Dutton & Co. New York. *Everyman's Library* #143, Ernest Rhys Editor, p. 28. (First published in 1824.)

Solley, W.B., Pierce, R.R., and Perlman, H.A. 1998. *Estimated Use of Water in the United States in 1995*. U.S. Geological Survey Circular 1200, 71 p.

U.S. Environmental Protection Agency Website Fact Sheet. 5/20/2003. http://yosemite.epa.gov/OAR/globalwarming.nsf/content/ClimateTrendsPrecipitation.html.

County Groundwater Reports

Availability of Ground Water in Decatur County, Iowa, 1967, J.W. Cagle and W.L. Steinhilber, Iowa Geological Survey Water Atlas No. 2, 28 p.

Availability of Ground Water in Wayne County, Iowa, 1969, J.W. Cagle, Iowa Geological Survey Water Atlas No. 3, 33 p.

Geohydrology of Muscatine Island, Muscatine County, Iowa, 1977, R.E. Hansen and W.L. Steinhilber, Iowa Geological Survey Water-Supply Bulletin No. 11, 60 p.

Geology and Ground-Water Resources of Cerro Gordo County, Iowa, 1970, H.G. Hershey, K.D. Wahl, and W.L. Steinhilber, Iowa Geological Survey Water-Supply Bulletin No. 9, 75 p.

Geology and Ground-Water Resources of Clayton County, Iowa, 1961, W.L. Steinhilber, O.J. Van Eck, and A.J. Feulner, Iowa Geological Survey Water-Supply Bulletin No. 7, 142 p.

Geology and Ground-Water Resources of Linn County, Iowa, 1970, R.E. Hansen, Iowa Geological Survey Water-Supply Bulletin No. 10, 66 p.

Geology and Ground-Water Resources of Webster County, Iowa, 1955, W.E. Hale, Iowa Geological Survey Water-Supply Bulletin No. 4, 257 p.

I. Groundwater Quality and Hydrogeology of Devonian-Carbonate Aquifers in Floyd and Mitchell Counties, Iowa; II. Devonian Stratigraphy of North-Central Iowa, 1984, Iowa Geological Survey Open-File Report 84-2, 149 p.

I. Hydrogeologic Observations from Multiple Core Holes and Piezometers in the Devonian-Carbonate Aquifers in Floyd and Mitchell Counties, Iowa; II. Stratigraphic Framework for the Devonian Aquifers in Floyd-Mitchell Counties, Iowa, 1985, Iowa Geological Survey Open-File Report 85-2, 32 p.

Hydrogeology, Water Quality, and Land Management in the Big Spring Basin, Clayton County, Iowa, 1983, G.R. Hallberg, B.E. Hoyer, E.A. Bettis III, and R.D. Libra, Iowa Geological Survey Open File Report 83-3, 191 p., 6 appendices.

Hydrology of Carbonate Aquifers in Southwestern Linn County and Adjacent Parts of Benton, Iowa, and Johnson Counties, Iowa, 1986, K.D. Wahl and B.J. Bunker, Iowa Geological Survey Water-Supply Bulletin No. 15, 56 p.

Open-File County Groundwater Reports are available from the Iowa Geological Survey for the following counties: Appanoose, Boone, Dallas, Davis, Des Moines, Grundy, Hamilton, Hardin, Henry, Iowa, Jefferson, Keokuk, Lee, Louisa, Lucas, Madison, Mahaska, Marion, Monroe, Polk, Story, Van Buren, Wapello, Warren, and Washington.

Regional Water Atlases and Aquifer Reports

Availability and Quality of Water from the Alluvial, Glacial-Drift, and Dakota Aquifers and Water Use in Southwest Iowa, 1992, R.E. Hansen, C.A. Thompson, and P.E. VanDorpe, U.S. Geological Survey Water-Resources Investigations Report 91-4156, 187 p.

Availability and Quality of Water from the Dakota Aquifer, Northwest Iowa, 1984, M.R. Burkart, U.S. Geological Survey Water-Supply Paper 2215, 65 p., 8 plates.

Hydrogeology and Stratigraphy of the Dakota Formation in Northwest Iowa, 1983, J.A. Munter, G.A. Ludvigson, and B.J. Bunker, Iowa Geological Survey Water Supply Bulletin No. 13, 55 p.

Hydrology of the Alluvial, Buried-Channel, Basal Pleistocene, and Dakota Aquifers in West-Central Iowa, 1985, D.L. Runkle, U.S. Geological Survey Water-Resources Investigations Report 85-4239, 111 p.

Water Resources of East-Central Iowa, 1978, K.D. Wahl, G.A. Ludvigson, G.L. Ryan, and W.C. Steinkampf, Iowa Geological Survey Water Atlas No. 6, 91 p.

Water Resources of North-Central Iowa, 1985, R. Buchmiller, G. Gaillot, and P.J. Soenksen, Iowa Geological Survey Water Atlas No. 7, 93 p.

Water Resources of Northeast Iowa, 1989, P.J. Horick, Iowa Department of Natural Resources Water Atlas No. 8, 133 p.

Water Resources of South-Central Iowa, 1978, J.W. Cagle and A.J. Heinitz, Iowa Geological Survey Water Atlas No. 5, 97 p.

Water Resources of Southeast Iowa, 1971, R.W. Coble and J.V. Roberts, Iowa Geological Survey Water Atlas No. 4, 101 p.

The Water Story in Central Iowa, 1965, F.R. Twenter and R.W. Coble, Iowa Water Atlas No. 1, 89 p.

Statewide Groundwater Aquifer Reports

Deep Wells of Iowa (A Supplementary Report), with a chapter on Well Water Recessions in Iowa by James H. Lees, and a table of Iowa towns giving Municipal Water Supplies, 1928, W.H. Norton, Iowa Geological Survey Annual Report v. 33, 447 p.

Ground Water Atlas of the United States Segment 9 (Iowa, Michigan, Minnesota, Wisconsin), 1992, P.J. Olcott, U.S. Geological Survey Hydrologic Investigations Atlas 730-J, 31 map sheets.

Hydrology of Area 38, Western Region, Interior Coal Province, Iowa and Missouri, 1983, U.S. Geological Survey Water-Resources Investigations Open-File Report 82-1014, 85 p.

Iowa's Principal Aquifers: A Review of Iowa Geology and Hydrologic Units, 1990, Iowa Groundwater Association, 26 p.

Jordan Aquifer of Iowa, 1978, P.J. Horick and W.L. Steinhilber, Iowa Geological Survey Miscellaneous Map Series 6, 3 map sheets.

The Maquoketa Formation (Upper Ordovician) in Iowa, 1971, M.C. Parker, Iowa Geological Survey Misc. Map Series 1, 6 map sheets.

Mississippian Aquifer of Iowa, 1973, P.J. Horick, Iowa Geological Survey Miscellaneous Map Series 3, 3 map sheets.

Silurian-Devonian Aquifer of Iowa, 1984, P.J. Horick, Iowa Geological Survey Miscellaneous Map Series 10, 4 map sheets.

Water Quality in Iowa during 1998 and 1999, Part IV: Groundwater Assessment, p. 4-1 to 4-19. Iowa Department of Natural Resources, Environmental Protection Division, 2001.

Alluvial Aquifers in Iowa

Hydrogeology of the Surficial Aquifer in the Floyd River Basin, Iowa, 1982, K.D. Wahl, M.J. Meyer, and R.A. Karsten, Iowa Geological Survey Water Supply Bulletin No. 12, 53 p., 5 plates.

Hydrogeology and Water Quality of the Upper Des Moines River Alluvial Aquifer, 1984, C.A. Thompson, Iowa Geological Survey Open-File Report 84-5, 170 p., 2 plates.

Nitrate and Pesticide Distribution in the West Fork Des Moines River Alluvial Aquifer, 1990, C.A. Thompson, Iowa Department of Natural Resources, Geological Survey Bureau Technical Information Series 18, 43 p.

An Overview of Groundwater Quality in the Skunk River Basin, 1987, D.R. Bruner and G.R. Hallberg, Iowa Department of Natural Resources, Geological Survey Bureau Open-File Report 87-3, 36 p.

Water Resources of the Ocheyedan-Little Sioux Alluvial Aquifer, 1986, C.A. Thompson, Iowa Department of Natural Resources, Geological Survey Bureau Open-File Report 86-3, 90 p.

Water Resources of the Rock River Alluvial Aquifer, 1987, C.A. Thompson, Iowa Department of Natural Resources, Geological Survey Bureau Open-File Report 87-1, 109 p.

General Geology and Aquifer Information in Iowa

Agricultural Chemicals and Groundwater Quality in Iowa, 1984, G. Hallberg, Iowa State University Cooperative Extension Service CE-2081j, 6-page brochure.

Agricultural Drainage Wells in Iowa: Hydrogeologic Settings and Water-Quality Implications, 1993, R.D. Libra and G.R. Hallberg, Iowa Department of Natural Resources, Geological Survey Bureau Technical Information Series 24, 39 p.

Artesian Wells in Iowa, R. Ellsworth Call, 1887-1893. Proceedings of the Iowa Academy of Sciences for 1887-1893, v. 1, parts 1-4, p. 57-63.

Artesian Wells of Iowa: W. H. Norton, 1897, Iowa Geological Survey Annual Report v. 6, p. 113-428.

Bedrock Geologic Map of Iowa, 1998, Iowa Department of Natural Resources, Geological Survey Bureau, 1 map sheet.

Groundwater Primer for Iowa Issues, 1989, Iowa Department of Natural Resources, 24 p.

Groundwater Vulnerability Regions of Iowa, 1991, B.E. Hoyer and G.R. Hallberg, Iowa Department of Natural Resources, Geological Survey Bureau, Special Map Series II, 1 map sheet.

Hydrologic Unit Map – 1981 State of Iowa, 1982, U.S. Geological Survey, 1 map sheet.

Iowa Groundwater Education Strategy, 1988, Gail George, Iowa Department of Natural Resources, 33 p.

Iowa Groundwater Protection Strategy, 1987, B.E. Hoyer, J.E. Combs, R.D. Kelley, C. Cousins-Leatherman, and J.H. Seyb, Environmental Protection Commission, Iowa Department of Natural Resources, 106 p.

Iowa Ground-Water Quality, 1986, R.C. Buchmiller and P.J. Squillace, U.S. Geological Survey Water-Supply Paper 2325, p. 251-258.

Iowa Ground-Water Quality, 1987, R.C. Buchmiller, P.J. Squillace, and R.D. Drustrup, U.S. Geological Survey Open-File Report 87-0725, 9 p.

Iowa Ground-Water Resources, 1985, R.A. Karsten and M.R. Burkart, U.S. Geological Survey Water-Supply Paper 2275, p. 211-216.

The Iowa State-Wide Rural Well-Water Survey, Water-Quality Data: Initial Analysis, 1990, B.C. Kross, G. R. Hallberg, and others, Iowa Department of Natural Resources, Geological Survey Bureau Technical Information Series 19, 142 p.

The Iowa State-Wide Rural Well-Water Survey: Site and Well Characteristics and Water Quality, 1992, G.R. Hallberg, K. Woida, and others, Iowa Department of Natural Resources, Geological Survey Bureau Technical Information Series 23, 43 p.

Iowa Surface-Water Resources, 1984, O.G. Lara and P.J. Soenksen, U.S. Geological Survey Water-Supply Paper 2300, p. 229-236.

Iowa Water Plan '78 Framework Study Main Report, 1978, Iowa Natural Resources Council, Des Moines, 207 p.

Iowa's Geologic Past: Three Billion Years of Change, 1998, W.I. Anderson, University of Iowa Press, Iowa City, 424 p.

Landforms of Iowa, 1991, J.C. Prior, University of Iowa Press, Iowa City, 153 p.

Occurrence and Distribution of Ammonium in Iowa Groundwater, 2002, K.E. Schilling, Water Environment Research, v. 74, no. 2, p. 177-186.

Regional Water-Level Changes for the Cambrian-Ordovician Aquifer in Iowa, 1975-1997, 1999, M.J. Turco, U.S. Geological Survey Water-Resources Investigations Report 99-4134, 11 p.

The Role of Land Use and Geology in The Water Cycle, 1989, J. Littke and K. Bentley, Iowa State University Extension, 1-page map sheet.

Stratigraphic Column of Iowa, 1998, Iowa Department of Natural Resources, Geological Survey Bureau, 1 p.

Temporal Trends of Selected Agricultural Chemicals in Iowa's Groundwater, 1982-1995 – Are Things Getting Better?, 1997, D.W. Kolpin, D. Sneck-Fahrer, G.R. Hallberg, and R.D. Libra, Journal of Environmental Quality, v. 26, p. 1007-1017.

Water Resources of Iowa: A Symposium Sponsored by the Iowa Academy of Science, 1969, ed. P.J. Horick, University Printing Service, Iowa City,175 p.

General Groundwater Information

Basic Ground-Water Hydrology, 1983, U.S. Geological Survey Water-Supply Paper 2220, 84 p.

Groundwater, 1979, R.A. Freeze and J.A. Cherry, Prentice Hall, Inc., Englewood Cliffs, New Jersey, 604 p.

Groundwater, A Primer, 1995, J.E. Moore, A. Zaporozec, and J.W. Mercer, American Geological Institute Environmental Awareness Series: 1, 53 p.

Ground-Water Hydraulics, 1979, S.W. Lohman, U.S. Geological Survey Professional Paper 708, 69 p.

Ground Water Manual – A Water Resources Technical Publication, 1977, Bureau of Reclamation, First Edition, U.S. Government Printing Office, 480 p.

The Hidden Sea: Ground Water, Springs, and Wells, 1997, F.H. Chapelle, Geoscience Press Inc., Tuscon, Arizona, 238 p.

Manual of Individual Water Supply Systems, 1982, U.S. Environmental Protection Agency, 155 p.

Study and Interpretation of the Chemical Characteristics of Natural Water, 1989, J.D. Hein, U.S. Geological Survey Water-Supply Paper 2254, 263 p.

Sustainability of Ground-Water Resources, 1999, W.M. Alley, T.E. Reilly, and O.L. Franke, U.S. Geological Survey Circular 1186, 79 p.

Water Well Handbook, 1975, K.E. Anderson, Missouri Water Well & Pump Contractors Assn., Inc., 281 p.

Water Well Information

Assessing Your Water Well Condition and Maintenance Practices, 2000, Farm*A*Syst, 12-page brochure.

Coping with Contaminated Wells, 1993, T. Glanville, Iowa State University Cooperative Extension Service PM-1329, 6-page brochure.

Ground Water and the Rural Homeowner, 1991, R.M. Waller, U.S. Geological Survey, 36 p.

Guide to Conducting a Wellhead Contaminant Source Inventory, 2001, E.N. Nealson, C.A. Thompson, and P.E. VanDorpe, Iowa Department of Natural Resources, Geological Survey Bureau, 21 p., plus forms and worksheets.

Guidelines for Plugging Abandoned Water Wells, 1988, D.L. Gordon, Iowa Department of Natural Resources, Geological Survey Bureau Technical Information Series 15, 46 p.

Home Treatment Systems and Drinking Water Quality, 1990, University of Iowa Hygienic Laboratory, Public Service Information, 19 p.

Private Well Construction, 2000, Iowa Department of Natural Resources, Geological Survey Bureau, 1-page brochure.

Residential Water Treatment Systems, 1990, K.C. Choquette, Iowa Department of Public Health Consumer Information Pamphlet, 19 p.

Shock-Chlorinating Small Water Systems, 1988, T. Glanville, Iowa State University Cooperative Extension Service PM-899, 4-page brochure.

Source Water Assessment and Protection Program and Implementation Strategy for the State of Iowa, 2000, Iowa Department of Natural Resources, Des Moines, 221 p., plus appendices.

Successfully Plugging Your Abandoned Well, 1992, T. Glanville, Iowa State University Cooperative Extension Service PM 1328, 4-page brochure.

Water Well Construction, A Consumer Information Booklet, 1993, Iowa Department of Natural Resources, 31 p.

Well Forecasting and Aquifers in Iowa, 1999, Iowa Department of Natural Resources, Geological Survey Bureau, 1-page brochure.

Well Water Quality and Home Treatment Systems, 1993, University of Iowa Hygienic Laboratory, Public Service Information, 22 p.

Wellhead Protection in Iowa, Iowa Department of Natural Resources, 1-page folded brochure.

When Citizens in Iowa Need a Water Well, 1997, American Ground Water Trust and Iowa Water Well Association, 14 p.

Water Use in Iowa and U.S.

Estimated Use of Water in the United States in 1995, W.B. Solley, R.R. Pierce, and H.A. Perlman, U.S. Geological Survey Circular 1200, 71 p., 1998.

Estimated Water Use in Iowa, 1980, R.C. Buchmiller and R.A. Karsten, Iowa Geological Survey Miscellaneous Map Series 9, State of Iowa, 1 map sheet, 1983.

Estimated Water Use in Iowa, 1985, M.L. Clark and J.N. Thamke, U.S. Geological Survey Open-File Report 87-704, 28 p., 1988.

Permitted Water Use in Iowa, 1985, D.L. Runkle, J.L. Newman, and E.M. Shields, U.S. Geological Survey Open-File Report 86-302, 24 p., 1985.

Sources of Additional Information

Certified Well Contractors – The Iowa Department of Natural Resources will provide a list of certified well contractors in your area. The Iowa Water Well Association (515) 243-1558 will also provide information on well contractors in your area.

Iowa Department of Natural Resources – For information about state regulations, well permits, and a listing of certified well contractors call (515) 725-0275 or *www.iowadnr.com/*. For forecasts on well depth, water availability and quality, contact the Iowa Geological Survey at (319) 335-1575 *or www.igsb.uiowa.edu*.

Iowa Department of Public Health – For information about health effects of contaminants found in drinking water and regulations regarding the sale of water treatment systems in Iowa call (515) 281-7726 or *www.idph.state.ia.us*.

Local health departments (county sanitarians) – Information about local regulations for water well construction, water testing, well plugging, and well inspection services.

Local Iowa State University Extension offices – Publications about water testing and well construction.

U.S. Geological Survey – Information on surface water resources in Iowa and water level records from groundwater monitoring networks; District Office in Iowa City at (319) 337-4191 or *http://ia.usgs.gov*.

University of Iowa Hygienic Laboratory – For information about testing your water, the interpretation of test results, and water treatment systems, call (319) 335-4500 or *www.uhl.uiowa.edu*.